The CQ 1996 Amateur Radio Almanac

Third Edition

Edited By Doug Grant, K1DG

CQ Communications, Inc.

Library of Congress Catalog Card Number 93-074224
ISBN 0-943016-13-4

Editor: Doug Grant, K1DG
Managing Editor: Edith Lennon, N2ZRW
Assistant Editor: Charlie Morrison, WZ1R
Editorial Assistants: Deena Marie Amato, Nancy Barry,
                      Bernadette Schimmel, Kirstie Wickham

Published by CQ Communications, Inc.
76 North Broadway
Hicksville, NY 11801 USA

Printed in the United States of America

About the Editors...

Doug Grant, K1DG, was originally licensed in 1967 as WN1ICD. He holds the Amateur Extra Class license and has been active in nearly every facet of amateur radio at one time or another. He is best-known for his accomplishments as a contest operator and has held several world and national records in both single-operator and multi-operator categories. He holds DXCC on six bands, has worked all current DXCC countries, and has held reciprocal callsigns in a dozen countries. Doug maintains a competitive contest station in Windham, New Hampshire, where he lives with his wife, Karen, and their two children. He is a Life Member of ARRL, past member and Chairman of the ARRL Contest Advisory Committee, and a member of the CQ Contest Committee.

Edith Lennon, N2ZRW, has been a writer and editor for a variety of technical and other publications for over 10 years. She is the Editor of CQ's *Equipment Buyer's Guide* and the managing editor of *CQ VHF* and is involved in several special projects at CQ as well.

Charlie Morrison, WZ1R, was first licensed in 1977 as WB2RNT and in 1986 became KD2SX. Charlie discovered ham radio contesting as an SWL in 1976, and has since participated in contests under 35 different callsigns, achieving several W1 area and section single-operator and multi-operator contest records. Charlie is a member of the Yankee Clipper Contest Club, and has 300 countries confirmed, as well as WAC, WAS, and WAZ. He is very interested in radio wave propagation and antennas, and enjoys DXing—he met his wife while on a DXpedition in 1986. He currently lives in Pawtucket, Rhode Island with his wife, Yvonne, and their two children.

CONTENTS

PREFACE TO THE THIRD EDITION

Welcome to the Third Edition of the *CQ Amateur Radio Almanac*. The response to the first and second editions has continued to be very positive. Numerous organizations have referred to data found in the *Almanac* in articles and reports. We are pleased that the information we have published has been useful, and that the *Almanac* can serve as a convenient reference source. Nearly every chapter has been cited as a reference in the last year in one place or another. Several radio clubs have used it as the basis for ham radio trivia quizzes at club meetings.

The changes we made last year (larger type for text-intensive sections, census by city and country instead of ZIP code, etc.) were definite improvements. This year, we've continued to update databases (contest records, VHF/UHF distance records, propagation forecasts, etc.), as well as improve the page layouts for easier readability.

We've updated the continuing history series of 25, 50, and 75 years ago in amateur radio, as well as the extensive Year-in-Review Section. We've expanded the club listings, and added coverage of amateur radio in other countries. We've also added lots of information on where the hams are lurking on the Internet.

There were lots of changes to FCC Part, the rules of the Amateur Radio Service in the U.S.; you'll find a completely up-to-date copy in the 1996 Almanac. Frequency allocations for the U.S. amateur service were also changed (in the 219, 902, and 2400 MHz bands) the up-to-date tables are included here. You can also find a complete copy of PRB-1, which is useful in tower-permit applications, with the legal reference.

A lot of people contributed to the 1996 edition. Among them are K1AR, K1BV, W1FM, NX1G, KY1H, W1JR, K1MEM, K1TR, W2JGR, W3ASK, K3EST, W5YI, K5FUV, K5ZC, KD6GGD, AB6WM, NV6Z, WS7I, NØSS, NUØX, VP2ML, G3KMA, HB9DX, Ed Erwin of NGDC, Norbert Schroeder of NTIA, and Buckmaster Publishing. Additionally, Mark Wilson, AA2Z, and Maty Weinberg of ARRL graciously provided cooperation and support in the preparation of this volume. Numerous others offered corrections and suggestions for improvement.

The all-time contest records were compiled by Charlie "Wiz" Morrison, WZ1R, the Assistant Editor. With the assistance of WØUN, we believe the contest records to be the most accurate ever compiled in any one place. Wiz also managed to track down some truly obscure information sources again this year.

The staff at CQ Headquarters deserves all the credit for final editing and production of the *Almanac*. Managing Editor Edith Lennon, N2ZRW, kept the whole project on track this year, both prodding the Editor and Assistant Editor to meet deadlines, but also translating our often cryptic instructions into something the production staff could understand. Edith also knew when to ask and when to order us to get things done.

The actual production staff, headed by Dorothy Kehrweider, Production Manager, and Elizabeth Ryan, Art Director, succeeded in getting this book done in a year when CQ added several more books, two new magazines, and numerous other projects to their workload. They truly are miracle workers. Without their support and dedication, this edition would not have made it to the printer.

As in the previous editions, we are interested in your comments, criticisms, and suggestions. Use the comment card, or contact us by letter or e-mail.

We hope this edition will be a useful reference tool for you, and that you will enjoy owning it.

73,

Doug

Doug Grant, K1DG
Editor

THE YEAR IN REVIEW
September 1994 to August 1995

Ham Radio Gets 219-220 MHz

The FCC Commissioners released an Order on March 17, 1995 allocating a one megahertz slice of the 1.25-meter band to amateur point-to-point digital message forwarding systems including intercity packet backbone networks.

The FCC said," This allocation will alleviate congestion that amateurs are experiencing in certain areas of the country in the 222-225 MHz band and will facilitate establishment of regional and nationwide networks for amateur digital packet communications."

The lengthy Report and Order authorizing the spectrum allocation to the Amateur Service contained a rather complicated regulatory plan to ensure that amateurs do not interfere with other users of the band.

In 1988, the 220-225 MHz shared ham band was divided into two exclusive use segments. The lower two megahertz was allocated to ...spectrally efficient... narrow-band-business use with the Amateur Service obtaining 222-225 MHz on a primary basis. Amateurs were ordered to vacate the 220-222 MHz segment by August 27, 1991. More than 550 Petitions for Reconsideration were rejected by the Commission.

The FCC did say, however, that ...it would entertain a request for replacement spectrum ...acknowledging that in certain areas of the country, some relief was justified. [Any request] ...would need to provide support why an allocation is needed and show how amateur operators could use this band without causing interference to existing users...

After a series of tests, the American Radio Relay League petitioned the Commission seeking access to the 216-220 MHz band on a non-interference secondary basis it said it needed to relieve crowding in the 1.25 meter band and to establish digital packet backbone networks. The petition was accepted and assigned RM-7747.

The League said the loss of the 220-222 MHz band ...left the Amateur Service without a reasonable substitute for high speed links and the development of a truly unique nationwide communications system with unparalleled emergency preparedness and national defense capabilities.

ARRL said its analysis proved that amateurs would be able to peacefully co-exist with other users of the band. To prove their point, the League conducted laboratory tests involving TV Channel 13 reception and commissioned a private research study which looked into spectrum sharing possibilities at 216 to 220 MHz. The ARRL did agree, however, that amateur activity would have to be controlled. It would not be advisable for amateurs to be able to access the band without prior coordination.... The League said they were willing to assume this function. A power level of fifty watts PEP was proposed. The petition had been under consideration for nearly four years!

The 216-218 and 219-220 MHz band are currently allocated to the Maritime Mobile Service for Automated Maritime Telecommunications Systems (AMTS.) Coast and ship stations use these bands for inland waterway communications under Part 80 of the Commission's Rules.

In a February 26, 1993, Notice of Proposed Rulemaking, the FCC agreed that there is a need for additional spectrum for amateur wideband digital packet networks. The FCC concluded that amateurs could indeed share spectrum with the primary AMTS (inland waterway) traffic and adopted rules that preclude amateur 219-220 MHz operations within 80 kilometers (about 50 miles) of AMTS stations without the AMTS licensee's approval. The FCC also will require that AMTS stations be notified of all amateur operations within 640 km. (about 400 miles.)

The Commission agreed with the ARRL that only digital communications should be permitted in the 219-220 MHz band. It also went along with their 50 watt power limit and recommendation that only licensees holding Technician or higher class licenses be permitted to use the 219-220 MHz band, ...we find that 100 kilohertz channels are appropriate for amateur fixed point-to- point digital communications at 219-220 MHz.

The FCC felt that there should be a single national contact point to a database of operations which could be provided to anyone investigating interference. Since the ARRL volunteered, ...we will require that all amateurs notify the ARRL of operations in the 219-220 MHz band thirty days prior to initiation of operations, the FCC said.

Automatic Control of HF Packet Approved

The FCC Commissioners on April 17, 1995 amended the rules to permit automatic control of HF digital communications. The ruling came nearly a year after the FCC released the Notice of Proposed Rulemaking.

Automatic control is defined in the rules as the control of an amateur station without the operator being present at a control point. The HF bands allow for long distance communications. The variables affecting communications in the HF bands are highly complex. Ordinarily to avoid causing interference to other amateur stations, the control operator must constantly monitor the activity on the frequency being used.

The proceeding got its start when two petitions were filed by the American Radio Relay League (RM-8218) and the American Digital Radio Society (RM-8280). Even though the amateur service message forwarding rules were revised in June of 1994, automatic control of stations transmitting on the HF bands was not authorized.

Automatic control of amateur stations transmitting digital communications on the VHF and higher frequency bands was first authorized in 1986. This allowed amateur operators to use high-speed computer-based packet radio technology to quickly and accurately exchange messages. The FCC also said at the time they were interested in authorizing automatic control of stations transmitting digital communications in the HF bands.

An HF packet feasibility study was carried out under a Special Temporary Authority granted by the FCC. Some fifty amateur stations operated as an automated digital communications network on the 20 meter HF band. This STA continued in effect for some five years. The big concern was that HF packet" robot" stations could interfere with ongoing HF ham band operation. The matter became very controversial among HF phone operators. A compromise plan to permit semi-automatic control was criticized as unworkable and unacceptable by HF packeteers.

The ultimate answer came when an IARU (International Amateur Radio Union) Region 2 General Assembly meeting produced an HF band plan which provided for automatically controlled data communications subbands. These segments included all bands between 80 and 10 meters and provided for all digital modes including new CLOVER and PACTOR systems. Any HF data operation outside these segments had to be under local control. This subband plan was accepted by the ARRL who agreed that segregating HF packet to specific frequencies within the eight HF bands would alert other HF stations that they may receive interference from automatically controlled stations if they operated in these subbands.

The American Digital Radio Society agreed that communications between automatically controlled stations should be confined to subbands. ADRS also concurred with a provision in the IARU HF band plan which permitted HF packet communications on any frequency authorized for data and RTTY communications as long as one of the stations has a control operator who could terminate the transmissions from all the stations. The ARRL supported this position.

The new 97.221 rules provide that," A station may be automatically controlled while transmitting RTTY or data emissions on the 6 m or shorter wavelength bands, and on the 28.120- 28.189 MHz, 24.925-24.930 MHz, 21.090-21.100 MHz, 18.105-18.110 MHz, 14.0950-14.0995 MHz, 14.1005- 14.112 MHz, 10.140-10.150 MHz, 7.100-7.105 MHz, or 3.620-3.635 MHz segments. A station may also be automatically controlled while transmitting a RTTY or data emission on any other frequency authorized for such emission types provided that the station is responding to interrogation by a station under local or remote control: and no transmission from the automatically controlled station occupies a bandwidth of more than 500 Hz."

Revised Allocating In 2400 and 902 MHz Bands

The FCC has reallocated 50 MHz of spectrum, including 2390 to 2400 and 2402 to 2417 MHz, from government use. The Commission made these segments available for unlicensed, low-power devices including wireless LANs. At the same time, the FCC elevated Amateur Radio's status in those segments from secondary to primary. This means that amateurs will not have to protect any other user of those bands, and amateur stations are entitled to protection against interference. The Commission made available 2390 to 2400 MHz for use by unlicensed Personal Communications Services (PCS), and provided for continued use of 2402 to 2417 MHz by traditional unlicensed "Part 15" electronic devices.

The unlicensed PCS devices, which include wireless networking and data transfer devices, will be governed by the same rules that apply to PCS devices operating in the 1910 to 1920 MHz band.

The FCC said it believes these allocations will provide for the continued development and implementation of a new generation of advanced communication devices and services, including a new "on ramp" to the information superhighway.

The remainder of the 50 MHz of allocated spectrum is at 4660 to 4685 MHz, to fixed and mobile services, an allocation not directly affecting amateurs.

The FCC has voted to allocate spectrum from 2310 to 2360 MHz for satellite digital audio radio services (DARS).

This band is part of spectrum from 2300 to 2450 MHz, all of which is currently allocated to the Amateur and Amateur Satellite Services in International Telecommunication Union Region 2, on a secondary basis, with the exception of 2310 to 2390 MHz, which is not allocated to amateurs.

The FCC said the new DARS would allow direct satellite-to-ground radio to areas of the country and to minority and ethnic groups that are inadequately served by traditional, "terrestrial" radio.

The Commission said that service and licensing rules for the new allocation would be addressed in a later rule making. This was action in FCC General Docket 90-357.

The FCC also adopted rules for the future licensing and continued development of a number of services, including Amateur Radio, in the 902 to 928 MHz band. The new rules set standards for what had previously been called automatic vehicle monitoring (AVM) systems but which the Commission now refers to as the Location and Monitoring Service (LMS).

Amateurs will continue to have access to 902 to 928 MHz, on a secondary basis, to the new LMS systems; to industrial, scientific, and medical (ISM) systems; and to government users. Unlicensed low power "Part 15" users are on a secondary basis to all the above, including amateurs.

The FCC said it would adopt a plan to afford both amateurs and Part 15 users "a greater degree of protection to their operations" (from interference from other services). It also said it would clarify what constitutes harmful interference to LMS licensees by Part 15 devices and by amateurs. "Operational restrictions should be imposed to maintain the coexistence of the many varied users of the band," the Commission said.

The FCC also said it would use a "negative definition" to clearly establish the parameters under which amateurs and Part 15 users may operate without risk of being considered sources of interference to services with a higher allocation status.

In a petition in January 1994 the ARRL requested primary allocations for amateurs at 902 to 904 and 912 to 918 MHz. The FCC accepted this petition as comments in its AVM proceeding but denied it, saying that insufficient "quantitative support" for the petition had been shown.

Vanity Call Signs

It is a long story as to how specific amateur call signs chosen by the licensee got included as a regulatory fee in the Clinton budget. The process began in June 1986 when Dave Sumner, K1ZZ, of the ARRL, wrote the FCC Secretary expressing the ARRL's interest in working with the FCC to provide a means by which amateurs could request and subsequently be issued specific unassigned callsigns of their choice. The FCC was not in a position to offer this service, due to the administrative costs, and the lack of a license-fee mechanism to offset those costs. In 1987, the FCC issued a public notice (FCC 87-35), seeking input on the privatization of a special-callsign system. The FCC specifically mentioned a preference for a single organization to become the Special Call Sign Coordinator (SCSC), to minimize the number of points of contact between the FCC and the process. The ARRL was interested in becoming that organization. However, a total of 13 persons and organizations were interested, and the FCC withdrew the program from consideration, citing the administrative workload the program would involve as noted by the commenters.

In 1989, both the House and the Senate Budget

Reconciliation bills included amateur radio user fees ($30 in the House version, $35 in the Senate). An intensive effort, spearheaded by Perry WIlliams, W1UED, resulted in the fees being dropped from the final version.

Jim Wills, N5HCT of Tyler, Texas, filed a Petition for Rule Making in June 1990 requesting that amateurs be allowed to specify three call sign choices in order of preference in exchange for paying a $30.00 fee to the FCC. He said" The Federal Budget and the amateur community all gain from this proposal." President Wilson specifically wanted a way to have his previous WA4EHQ call sign reissued to him.

That petition was denied because of the statutory exemption of amateur service applications from fees. But it started the ball rolling. Wills later contacted his Congressman (Ralph Hall, D-TX) - and with the help of Telecommunications Subcommittee Chairman Edward J. Markey who wrote FCC Chairman Al Sikes, got the Vanity call sign proposal included in the Clinton Deficit Reduction Plan. ARRL President George Wilson, W4OYI, wrote to FCC Chairman Sikes endorsing the concept. President Clinton signed the measure into law on August 10, 1993.

It was an unbelievable accomplishment to pull off! The initial annual fee proposed in the Clinton budget for Vanity call signs was $7.00...or $70 for a ten year license term. It now appears certain that Jim Wills' original request for a $30.00 fee for an amateur call sign of choice will be the cost you will pay for the call sign you want! On June 14, 1995, the FCC adopted a revised schedule of Regulatory Fees and Vanity amateur call signs were reduced to $30.00.

The basic concept of regulatory fees is to charge those who benefit from FCC services rather than all taxpayers. The authority to impose and collect regulatory fees is contained in the Omnibus Budget Reconciliation Act of 1993. As of Sept. 1, 1995, the FCC Form 610-V "Vanity" call sign application form still has not yet been released. And the FCC can't release the form until five Petitions for Reconsideration of the vanity call sign rules are disposed of.

At press time, here are the vanity call sign rules that have been nailed down so far:

(1.) Any licensed amateur is eligible to choose a special call sign. Trustees of club stations are eligible, but RACES (Radio Amateur Civil Emergency Service) and military recreation stations are not.

(2.) The FCC will notify the amateur community when they may apply for a special call sign. Applications will be accepted in four phases.

Gate One: Applications for previously held call signs and for the call sign of a close deceased relative;

Gate Two: Extra Class amateurs may apply;

Gate Three: Advanced Class amateurs, and finally;

Gate Four: Would open the vanity call sign system to all others including club stations.

(3.) Amateurs may select up to 25 call signs in order of preference. These call signs must be currently unassigned and from the call sign group appropriate for your class of license. (See exception below.) The private sector will supply lists of available call signs.

(4.) Expired, canceled, voided, revoked, set-aside and call signs of deceased amateurs are available for reassignment after two years. (See exception below.)

(5.) Club stations with written permission of close relatives...and close relatives of deceased amateurs may immediately apply for their call sign. Club stations and close relatives do not have to conform to the appropriate call sign group requirement.

(6.) The first assignable callsign from the applicant's list will be shown on the license grant. The Form 610-V application and $30 payment must be sent to the FCC's bank. That address is: FCC, Amateur Vanity; P.O. Box 358924, Pittsburgh, PA 15251-5924. The Mellon Bank will process the payment and forward the application to the FCC in Gettysburg, PA, for vanity call sign assignment.

The final details of the Vanity Call Sign System will be available once the FCC Form 610-V is released, the Petitions for Reconsideration have been resolved and the Commission is ready begin accepting applications.

Special Event Call Signs

The FCC said when they adopted rules to implement Vanity Call Signs that they would be setting aside the one-by-one call sign block for special event call signs.

A one-by-one call sign consists of a single prefix letter (K, N or W), the region number (to 9) and a single suffix letter (A to Z). There are 780 such call signs.

In its comments concerning the vanity call sign system, the American Radio Relay League had requested that one-by-one call signs be reserved for assignment to stations operating in conjunction with short-term events of national significance.

The FCC wants stations wishing to obtain a special event call sign to indicate the nature of the event at least 120 days in advance and certify that it is of special significance to the amateur service community. In addition, the licensee would submit a list of one-by-one format call signs, in the order of preference.

This list could be included in a letter, on a form prepared by the applicant or supplied by an outside source. Unlike ten year term vanity call signs which are scheduled to cost $30, special event vanity call signs are proposed to be free. The first assignable call sign on the list would be stamped granted and a copy of the list would be returned to the person making the request.

The special event vanity call sign could be used for a period not to exceed that of the special event, or for 15 days, whichever is less. At press time, the FCC is still considering comments from the amateur community on this matter.

Self Assigned Indicators

The FCC will now allow amateurs to use any self-assigned indicator before, after, or before and after their station call sign.

The FCC said it received several informal requests for clarification of the station identification rules which provides that. An indicator may be included with the call sign. It must be separated from the call sign by the slant mark or by any suitable word that denotes the slant mark. If the indicator is self-assigned, it must be included after the call sign and must not conflict with any other indicator specified by the FCC Rules or with any prefix assigned to another country.

The FCC said it received requests to include a self-assigned indicator before rather than after the assigned call sign as provided in the current rule. For example, the licensee of amateur station W1AA in Boston, Massachusetts, decides to operate the station while vacationing in the Virgin Islands.

In order to direct more attention to the station, the licensee may include a self-assigned indicator such as /KP2, in the station identification announcement. (Stations located in the Virgin Islands are normally assigned a call sign with the prefix KP2, NP2 or WP2) The call sign given in the station announcement therefore, would be W1AA/KP2.

The FCC said "We propose to permit also the station announcement KP2/W1AA and KP2/W1AA/KP2. We believe that allowing indicators to be included before, after or both before and after, the assigned call sign will provide the amateur service community better flexibility when making the station identification announcement." After no objections were raised, the FCC adopted this new rule.

Privatizing Interference Handling

The FCC is looking into letting the private sector handle radio frequency interference complaints. The Commission would not get involved until an authorized service agent had determined that the problem could not be resolved at the local level.

The Federal Communications Commission alone receives approximately 30,000 complaints a year of radio frequency interference to home electronics equipment. And this may only represent 10% or 20% of the problem. Due to the FCC's limited resources, it is not possible for the Commission to resolve these individual RFI problems and it is now Commission policy not to further investigate them. The FCC also does not offer any protection from interference.

According to the Commission, interference to home electronic equipment is a major problem in the United States that they must deal with in order to ensure communications excellence for the American public.

The FCC is now looking into the possibility of having the private sector become involved in resolving these interference problems. The Tampa Office of the FCC's Compliance and Information Bureau is undertaking a pilot project to determine the feasibility of such a program. Canada and Great Britain al- ready have privatized RFI handling.

To get the project underway, a fact-finding meeting was held in Tampa on July 19, 1995. The meeting was moderated by FCC Engineer-in-Charge Ralph Barlow of the Tampa field office with Robert McKinney, EIC of the Vero Beach, Florida assisting. The Commission is now in the process of determining how to proceed with this program.

ARRL President Resigns for Health Reasons

On July 1, 1995, American Radio Relay League President George S. Wilson III, W4OYI, submitted his resignation from the office he has held since January 1992. Mr. Wilson suffered a stroke on February 11.

Mr. Wilson said that while he has made progress in rehabilitation, his medical condition prevents him from traveling and from devoting the energy required to perform the duties of the office for the remainder of his term, which expires at the January 1996 meeting of the ARRL Board of Directors. He expressed his appreciation to the members of the Board for the opportunity to serve.

In accordance with the ARRL By-Laws, First Vice President Rodney J. Stafford, KB6ZV, has been performing the duties of president since Mr. Wilson became incapacitated in February. With Mr. Wilson's resignation, Mr. Stafford becomes president for the remainder of the term. The next ARRL vice president in the order of succession, Jay A. Holladay, W6EJJ, becomes first vice president.

At the 1995 Second Meeting of the ARRL Board of Directors, the Board conferred upon Wilson the title of ARRL President Emeritus.

Amateur Satellites Lost

Two amateur satellites were lost when their launch vehicle exploded. The Israeli-built Gurwin-1 TechSAT and the Mexican UNAMSAT were part of the payload of a Russian SS-25 rocket originally built to carry ballistic missiles and recently converted to launch satellites. The Reuters news agency said the rocket, which was launched from Russia's Plesetsk cosmodrome, came down in the Russian Far East, in the Sea of Okhotsk, on March 28.

Both satellites were designed for packet radio repeater use. Gurwin-1 TechSAT was built at the Technion-Israel Institute of Technology in Haifa as a 9600-bit/s packet store-and-forward satellite. UNAMSAT was assembled by students at the Universidad Nacional Autonoma de Mexico (UNAM) in Mexico City. In addition to packet store-and-forward operation at 1200 bits per second, UNAMSAT carried a unique" meteor radar" experiment. Two Russian satellites also were lost in the failed launch.

In November 1993 the first test launch of a converted SS-25 was successful, but with a lighter payload. Radio amateurs, particularly in Europe, have hoped that the Russian rockets will provide an inexpensive way to launch Amateur Radio satellites.

International Morse Code Requirement

Proficiency in the radiotelegraphy is required in the Amateur Service due to Radio Regulation 2735 of Article 32 of the International Radio Regulations. A move is underway to eliminate this requirement.

The country of New Zealand is leading an effort to abolish RR-2735 in a view that the following regulation, RR-2736 contains "...ample scope to require competency in Morse code or not as deemed appropriate" . RR-2736 provides that "Administrations shall take such measures as they judge necessary to verify the qualifications of any person wishing to operate the apparatus of an amateur station."

Eliminating the Morse code requirement from the Amateur Service is opposed by the American Radio Relay League and the International Amateur Radio Union, an organization made up of national amateur radio societies, including the New Zealand Association of Radio Transmitters (NZART).

The New Zealand government's position seems to have evolved from a position promoted by the New Zealand-based ORACLE (Organization Requesting Alternatives by Code-Less Examinations), formed in 1994 by former NZART Council member ZL2CA. The suppression of RR-2735 will be brought up at the upcoming World Administration Radio Conference to be held beginning Oct. 23, 1995 in Geneva.

New Sunspot Cycle Begins

Astronomers at the California Institute of Technology identified the first new sunspot in the next sunspot cycle.

Scientists at Caltech's Big Bear Solar Observatory in Big Bear City, California, photographed the spot on August 12, 1995.

"This makes us happy," said Hal Zirin, professor of astrophysics at Caltech and director of the Big Bear facility. "The sun is a lot more interesting to study when things are going on."

Early in the 11-year sunspot cycle, sunspots appear rarely and at relatively high solar latitudes around 30 to 35 degrees, then increase in frequency and appear at lower latitudes until they reach sunspot maximum, Caltech said. After this peak in activity, the number of sunspots slowly declines, and they appear ever closer to the sun's equator until they reach a relatively quiet phase · called sunspot minimum.

The sun has been in a quiet period through much of 1994 and this year, with a few spots showing up near the equator. The new sunspot found on August 12 appeared at a solar latitude of 21 degrees, and its magnetic polarity is opposite to that seen over the last decade, a key to identifying it as the manifestation of the start of a new cycle, Caltech said.

Scientists at Caltech said they expected an early beginning to Cycle 23, but not this early. Sunspots in the new cycle should rapidly become more common and reach a high level of activity in 1998 or 1999, Caltech said.

Electronic Application Filing and Renewals

The FCC proposed to speed up application handling by implementing electronic filing of amateur radio operator applications and implementing a new renewal system. Electronic filing of amateur radio operator Form 610 applications has now been implemented by all VE Coordinators. Basically it works like this. Successful applications are received by the VECs from the volunteer examiners. There are sixteen VEC organizations, but W5YI and ARRL account for nearly 90% of all examinations administered. This transfer is either made by mail or via modem.

Once received, the application information is keyed into a computer program and then transmitted over the telephone lines directly into the FCC's computer database which is maintained in Washington, DC. The FCC's computer immediately authorizes a license grant and call sign and the examinee may begin operating on the ham bands as soon as this call sign is determined.

The FCC posts every license grant to the Internet and VECs download this information daily. Information may be retrieved using the following procedure:

Access: anonymous ftp.fcc.gov

Directory: pub/XFS-AlphaTest/amateur

Documentation: readme.txt

The FCC's Consumer Staff in Gettysburg, Pennsylvania, can answer questions at 800-322-1117 or 717-337-1212.

New call signs and license grants are also posted daily to the World Wide Web page of the University of Arkansas at Little Rock. Their web server address is: http://www.ualr.edu/doc/hamualr/callsign.html They download the new FCC database data at night and have it ready for access by the public the following morning.

Under the new policy, new amateur operators no longer have to wait until they have their license in their possession before getting on the air. You are considered licensed once the FCC's database has been updated...and that is immediately after transmission by the VEC into the FCC's computer.

Amateur radio operator license grants are being handled quicker than ever! Depending upon how fast the VE team handles their paperwork, you could even have your new license grant and call sign within a couple of days of testing. The average waiting time, however, is about a week. In any event, certainly a lot faster than the two to three months that new amateurs used have to wait!

At press time, a new amateur operator license renewal procedure was also scheduled to begin. A new Form 610-R return card renewal will be sent by the FCC to all licensees whose license expires in December and afterward. You need only sign and return the card to the FCC to renew your ticket.

International Amateur Radio Operation

The American Radio Relay League filed a 30-page petition for rule making on July 19, 1995 seeking to implement the Inter-American Convention on an International Amateur Radio Permit (IARP). It appears that international roaming is about to become a reality as various nations in our hemisphere become parties to the Treaty. The United States has already become a signatory.

The 25th General Assembly of the Organization of American States (OAS) held June 5, 1995 at Montrouis, Haiti, adopted the IARP. This permit allows radio amateurs who are citizens of and licensed in countries which are a party to the convention to operate temporarily in other OAS nations without further licensing. Included are most countries of North, Central and South America.

U.S. and foreign radio amateurs currently are able to operate their stations during temporary visits to other countries based upon certain bilateral agreements commonly referred to as reciprocal licensing. With the exception of Canadian citizens, the amateur radio operator must submit licensing paperwork to the FCC in Gettysburg or the licensing authority of a foreign nation. This procedure is time consuming and burdensome. Canadian and U.S. citizens may operate their equipment in the neighboring country without further licensing due to the existence of a 1952 Treaty.

The International Amateur Radio Union (IARU) has been trying for years to bring about a simpler method of permitting amateur radio operation across international borders. The IARU and the ARRL developed the concept of the IARP which is modeled after the successful International Driving Permit that has been in place for nearly 4 decades. The concept was presented to the Inter-American Telecommunication Commission (CITEL) in December of 1994 in Montevideo, Uruguay who recommended to the Organization of American States that it be approved. The OAS did so on June 8, 1995. The International Amateur Radio Permit is now a reality.

The ARRL has now asked the Commission to implement the IARP in the United States through either an immediate Report and Order or a Notice of Proposed Rule Making. The League believes that it is not necessary to go through the full notice-and-comment rule making process since the rules constitute a relief of restrictions and a reduction in unnecessary paperwork.

No fee or tax may be levied on holders of IARPs by the visited country. It will be issued for a one year term (or expiration of the national license whichever comes first) in the standard languages of the Americas: English,

French, Portuguese and Spanish...and other official languages if not one of these four. Permitees must carry both their home country license and the IARP and are expected to be familiar with the operating rules of the visited country. Enforcement will be by IARP cancellation or modification.

The operating conditions for IARP holders shall be specified by the administration of the visited country. The call sign to be used will be the amateur call sign prefix used in the home country, followed by the slant bar sign ("/") (or in telephony, the word "stroke") followed by the call sign of the amateur ope- rator's home license.

There are only two operating license classes: a Class 1 International Amateur Radio Permit authorizes all amateur privileges in the country visited and shall be issued only to those who have proven Morse code proficiency. The Class 2 IARP allows utilization of all amateur frequency bands above 30 MHz. The United States is also considering joining the CEPT (European community) agreement that permits temporary reciprocal amateur radio operation in certain Eastern European nations (including the United Kingdom and Scandinavia.) CEPT is an acronym for the European Conference of Postal and Telecommunications Administrations.

The CEPT nations have agreed to permit visiting amateurs of non-CEPT nations to also participate in the in the CEPT radio licensing agreements for short periods of time. Like the International Amateur Radio Permit planned for the Americas, there are only two license classes. The CEPT Class "A" license requires Morse code, the Class "B" VHF/UHF license does not.

Family Radio Service

On July 20, 1994, the Radio Shack division of the Tandy Corporation petitioned the FCC to create an unlicensed UHF-FM low power 2-way voice Family Radio Service (FRS.) The new service would share unused and little used General Mobile Radio Service (GMRS) spectrum at 462 and 467 MHz. Tandy told the FCC that FRS would help meet the growing public demand for an affordable and convenient way of direct communication among individuals. The FCC accepted the proposal as having merit and assigned it file No. RM-8499.

After a preliminary round of comments, the FCC proposed new rules on June 22, 1995, seeking to implement the Family Radio Service as requested by Radio Shack. The Commission said it will encourage rapid deployment and growth of inexpensive low power communications equipment for use by groups and families in which members need to communicate over very short distances.

The new service will share seven existing GMRS channels that are not used for repeater operation, as well as utilize seven unassigned channels that are located between certain GMRS channels. All fourteen channels are called interstitial frequencies since they are sandwiched in-between the existing eight GMRS duplex channel pairs.

Tandy says the current non-repeater channels are underutilized, a claim which is strongly disputed by the Personal Radio Steering Group, a GMRS user association. PRSG is vehemently opposed to mixing licensed and unlicensed operators on the same frequencies and they believe that FRS will disrupt and impair both current and future GMRS operations...

REACT International, basically a CB association -

supports the concept of the Family Radio Service, but opposes the use of GMRS spectrum for it. Motorola, on the other hand, supported the Tandy/ Radio Shack proposal ...the public interest is better served by the creation of a new unlicensed personal radio service that offers consumers improved communications options in a cost-effective manner.

The Telecommunications Industry Association foresees a potential strong market for FRS with applications varying from ...parents keeping in contact with children, local watch patrols monitoring neighborhood activities, small businesses improving efficiency through radio, and outdoor recreationists enhancing the enjoyment of their activity while increasing their safety as well.

FRS will be a Part 95 Personal Radio Service, the same as CB radio. The proposed direct frequencies are:

1. 462.5625 MHz 8. 467.5625 MHz
2. 462.5875 MHz 9. 467.5875 MHz
3. 462.6125 MHz 10. 467.6125 MHz
4. 462.6375 MHz 11. 467.6375 MHz
5. 462.6625 MHz 12. 467.6625 MHz
6. 462.6875 MHz 13. 467.6875 MHz
7. 462.7125 MHz 14. 467.7125 MHz

Family Radio Service users will use palm-size one-half watt output radio units to communicate while on outings, such as visiting shopping malls and amusement parks, attending sporting events, camping, and taking part in other recreational activities. Tandy stated that...many persons could benefit from such a service, particularly for personal security, due to the low cost of the units and the communication capability.

Tandy has already tested the feasibility of FRS under a five channel Special Temporary Authorization (or STA) conducted at the Disney World theme park in Orlando, Florida where visitors were provided with Radio Shack's own Model PRS-100 handheld which operates only on the interstitial frequencies.

The FCC also wanted to know whether interconnection to the public telephone network should be approved. The Commission agreed that selective calling would enhance the appeal of the Family Radio Service by allowing users to answer calls addressed only to them without having to also monitor all other communications on the channel. The new proposed rules allow equipment suppliers to incorporate this option.

Comments closed on the FRS proposal on October 1, 1995 and at press time, the FCC was reviewing them prior to issuing a Report and Order adopting the Family Radio Service.

Lifetime Operator License

The American Radio Relay League proposed on January 6, 1994, that all amateur operator (but not station) licenses should be issued for the lifetime of the operator. The League's petition (assigned RM-8418) also asked that the lifetime license be made retroactive to any previously licensed amateur whose license had already expired. The objective of this provision was to permit persons with new found interest in amateur radio to return to the service without the necessity for relicensing.

The ARRL also believed their proposal would also reduce the burden on the Volunteer Examiner program since previously licensed amateurs would not have to be re-examined. The FCC issued a Notice of Proposed Rulemaking on April 25, 1995. The comment period closed on July 14th. The Commission's version of a life-time license was far different from that of the League, however. Rather than to create a lifetime operator license, the FCC proposed to give examination credit for the fewest examination elements necessary for the license class previously held. The result was essentially the same, however. Under the Commission's plan, the former licensee will not have to retake the examination elements.

In their comments, the ARRL said that the VEC System has no authority to process renewal applications which is what the League believes that FCC's proposal amounts to. The ARRL requested that the Commission not adopt their proposal. Instead, the Commission should adopt the rule changes proposed in the League's petition, RM-8418, which provides for a lifetime operator license.

The National Conference of VECs, a corporation made up of representatives of all VEC organizations, also asked that the FCC not adopt their proposal, but for a different reason. NCVEC believes that amateurs with long term expired licenses - longer than the current two year grace period - should be re-examined. The NCVEC believes there is a fundamental difference between an individual who has let his/her license lapse 'years ago' and an amateur who has kept renewing. The difference is that currently licensed amateurs are more up-to-date on FCC rules and technology. Amateurs who have been away from the hobby for long periods of time will find that the amateur service and its regulations have changed drastically. They should undergo some sort of training or refresher course. The examination syllabus provides the needed curriculum. The FCC will now have to decide whether or not to implement any version of a lifetime license.

Club Definition

As of this past Spring, the FCC has resumed issuing club station licenses. Once the vanity call sign system is implemented, clubs will be able to trade this call sign in for a specific call sign chosen by the trustee.

The ARRL petitioned the FCC to change the rules to increase the minimum number of members required to constitute a club from two to four persons. Assigned RM-8462, the League points out that it is important for the FCC to determine that applicants for a club station license are legitimate clubs and not just persons pursuing an additional call sign.

The FCC agrees that there is merit to ARRL suggestion and has issued a Notice of Proposed Rulemaking seeking to increase the eligibility requirement to four persons for a club station license. The comment period on the NPRM has already closed.

While ARRL believes that this will discourage "...two individuals who simply wish to obtain a distinctive alternate call sign but who do not function as ...a normal amateur radio club", the W5YI Group pointed out that increasing the club eligibility from 2 to 4 club members would not in itself prevent abuse and that at least one family of four (the Tucker family of La Mirada, CA: Eric Tucker - AA6ET, Kathryn Tucker - AA6TK, Roy Tucker - N6TK, and Kent Tucker - AA6KT) has already obtained 23 different club call signs.

New RFI Book From the FCC

The Federal Communications Commission has released a new Interference Handbook for consumers.

The 24-page, full color book will be stocked by FCC field offices around the country to provide people experiencing interference to home electronic equipment with information and solutions to interference problems.

The book deals not only with interference to televisions from radio transmitters, but also illustrates and describes interference caused by poor antennas (weak signals, ghosting); electrical interference from home devices such as hair dryers; electrical interference from power lines; and interference from home computers and low power radio devices such as garage door openers.

In addition to interference to televisions, the handbook describes solutions to interference to hi-fi systems, telephones, and video cassette recorders. Techniques for solving problems include the use of ferrite cores, improving receiving antenna systems, checking cabling, and isolating interconnected units to find the one that is at fault.

The book lists addresses and phone numbers for sources of high pass filters, common mode filters, band reject filters, ferrites and beads, ac line filters, telephone filters, and interference resistant telephones, as well as an extensive list of manufacturers of home electronic equipment.

Page one of the new FCC Interference Handbook says, "Many interference problems are the direct result of poor equipment installation. Cost-cutting manufacturing techniques, such as insufficient shielding or inadequate filtering, may also cause your equipment to react to a nearby radio transmitter. This is not the fault of the transmitter and little can be done to the transmitter to correct the problem. If a correction cannot be made at the transmitter, actions must be taken to stop your equipment from reacting to the transmitter."

New Callsign Blocks for Hawaii, Alaska, and Puerto Rico

According to a new FCC fact sheet, an ARRL request for call sign relief for amateurs in Hawaii, Alaska, and Puerto Rico has been honored.

The "Amateur Station Sequential Call Sign System" Fact Sheet dated February 1995 replaces one dated June 1991 that is referenced in the FCC's recent vanity call sign report and order.

Systematically assigned call signs in Groups A (Amateur Extra Class) and C (General, Technician, and Technician Plus class) for Alaska, Hawaii, and Puerto Rico using the traditional numerals (7, 6, and 4, respectively) have been exhausted.

According to the new fact sheet, call signs in Alaska (Region 11), AL, KL, NL, and WL are no longer limited to the numeral 7. Any numeral, 0 through 9, is available. The call signs KL9KAA through KL9KHZ are reserved for assignment to US personnel stationed in Korea.

In the Caribbean (Region 12), KP3, NP3, and WP3 (or 4) call signs will indicate the Commonwealth of Puerto Rico (except Desecheo Island).

In Hawaii and the Pacific (Region 13), AH7, KH7, NH7, and WH7 (or 6) call signs will indicate Hawaii except that the letter K following the numeral 7 will indicate Kure Island.

Petition Would Prohibit One-Way Transmissions

A petition for rule making before the FCC would eliminate all one-way transmissions such as code practice and information bulletins on the amateur bands below 30 MHz, including those from W1AW.

The FCC has assigned file number RM-8626 to the petition, which was filed by Frederick O. Maia, W5YI, publisher of the commercial newsletter the W5YI Report. The petition seeks to eliminate the rules that permit one-way information bulletins and Morse code practice in the amateur bands below 30 MHz. The effect of the petition would be to silence W1AW bulletin and code practice transmissions, among others.

Maia calls the FCC rule that permits certain one-way transmissions on the amateur bands, "a very permissive category and taken in its broadest context, permits just about anything to be transmitted that is even remotely associated with the Amateur Service." The W5YI Report has carried numerous stories on the subject of the International Amateur Radio Network's bulletin transmissions from station K1MAN over the past few years.

Maia says that code practice is now available on computer software, and information bulletins about Amateur Radio can be had on various computer on-line services.

Possible New Spectrum for Hams?

A US government study concludes that amateurs could use 2180 additional kilohertz of spectrum in the future. The National Telecommunications and Information Administration (NTIA) reported this in a 10-year projection of spectrum requirements for all licensed radio services.

Here are the 2,180 kilohertz cited in the NTIA report:
30 to 50 MHz, 5 slots of 50 kHz each (250 kHz total);
160 to 190 kHz (30 kHz); Near 5 MHz (50 kHz);
Expansion of 10 MHz band (200 kHz);
Expansion of 14 MHz band (50 kHz);
Expansion of 18 MHz band (150 kHz);
Expansion of 24 MHz band (150 kHz);
219 to 220 MHz (1000 kHz, already in process);
Satellite downlink, 29.7 to 30.0 MHz (300 kHz).

Here are excerpts from the NTIA study:

"The amateur community commenters have suggested significant changes to the allocation table to accommodate expanded amateur operations. Many of the suggested allocation revisions are reasonable...."

"The amateur-satellite service will soon have a new generation of amateur satellites in orbit that will use all frequency bands allocated to the amateur-satellite service from 29 MHz through 24 GHz. For this reason, the retention of current allocations, additional amateur-satellite allocations, and the upgrading of certain current allocations is desirable."

"Amateur requests for international reallocations would be appropriate issues for ... future World Radiocommunication Conferences (WRCs). Additional allocations at 160-190 kHz, and near 5 MHz will require technical studies to determine the availability of these bands to support amateur use."

"The expansion and upgrading of amateur allocations in the 10, 14, 18, and 24 MHz bands are acceptable, but will depend on future decrease of requirements for the aeronautical mobile or the fixed services internationally. The alignment of the amateur 3.5 and 7 MHz bands worldwide will require the inclusion of these issues in U.S. preparations for future WRCs."

Finally, the study said that "any sharing of military radiolocation spectrum (e.g., 430-440 MHz) with the amateur services on a co-primary basis in current Federal radiolocation bands is not feasible because of the potential loss of operational flexibility for military radar systems. Further, the expansion of use in the 902-928 MHz band by federal and non-federal users, including the operation of wind profiling radars, may make this band untenable for amateur operations in the future."

FCC Denies Code Exemption for Seniors

The FCC denied a petition that proposed exemptions from Morse code exams for people age 65 and older.

Guy A. Matzinger, KB7PNQ, a 67-year-old Technician class licensee, argued in his petition that old age results in diminished faculties, and equated those people 65 and older with the severely disabled in terms of passing a CW exam. He proposed the exemption for speeds above 5 WPM.

The FCC said that similar petitions had been denied in the past, because the current Amateur Radio license structure was based on "the desires of the amateur service community," based on thousands of comments on previous petitions. The Commission also noted that its rules already provide for Morse code exemptions for certain recognized disabilities.

Although the petition got the usual public notice, the FCC said it received no comments on it.

VE Session Manager

In a Petition for Rulemaking filed July 15, 1993, the National Conference of VECs asked the FCC for a rule change that would recognize in the rules the existence of a volunteer examiner (VE) on-site manager at license examination sessions. The FCC issued a NPRM on April 26, 1995 looking toward adopting the NCVEC petition. In their formal comments, the NCVEC said that all VECs utilize the services of a VE who is considered to be in charge and accountable for the proper conduct of the test session. While VE's are organized into teams of three or more persons, it is almost always one examiner who organizes and supervises the activities of the other VEs.

This lead examiner usually has custody of the examination materials, submits the test results to the VEC and maintains the session records. The lead examiner organizes the test session, supervises the VEs and is responsible for the integrity of the test session. It thus follows that the VE who manages the examination session should be more accountable for its conduct and reliability than the other VEs who essentially assist. NCVEC noted that three VEs would still be required to conduct all examinations.

Holding the three certifying VEs equally accountable ...can make enforcement action difficult should an examiner team be found to be knowingly and willfully disregarding or circumventing proper examination practices. Having three examiners certify examinations often divides the responsibility to the point where no one can be held responsible.

The NCVEC also pointed out that VE teams often utilize a production line system of more than three examiners at large test sessions and that "...the three VEs who certify the FCC Form 610 application are frequently not the same ones who administered or observed all of the examinations to the examinee." The ARRL opposed the concept of the VE Session Manager and believes to permit a single person to bear the responsibility for the proper conduct of an examination session makes it far easier for an examination session to be compromised, without detection. "...The issue of a VE Session Manager should be a matter subject to the discretion of the VE teams, rather than a regulatory requirement."

The League said "...the system is not broken, doesn't need fixing, and the Commission's proposal is extreme overregulation." The comment period has closed on this item and at press time, the amateur community was awaiting a final FCC ruling on RM-8301.

AMATEUR RADIO EMERGENCY COMMUNICATIONS ACTIVITIES

Texas Floods

In October 1994, heavy rains caused flooding in Texas. The southeastern part of the state was hit especially hard. The Southwest Traffic Net, Central Gulf Coast Hurricane Net, Texas RACES Net 7290 Traffic Net, NTS Region 5 Net, and Texas Traffic Net were active in handling emergency and priority traffic related to the emergency.

The Federal Communications Commission, Houston office, designated the use of frequencies between 3967 and 3978 kHz and between 7245 and 7251 kHz for the handling of Emergency Traffic as part of a Voluntary Communications Emergency for several days.

Kobe Earthquake

More than 200 amateurs provided emergency communications in the Kobe, Japan area following an earthquake on January 16, 1995 that killed more than 5000 people and left tens of thousands homeless.

Amateur Radio was used to connect relief centers and to exchange information on road conditions and traffic; the health, welfare, and whereabouts of residents; and the availability of water and food, according to the Japan Amateur Radio League.

The JARL and the Japan Amateur Radio Equipment Industry Association (JAIA) conducted the effort, at the request of the Ministry of Posts and Telecommunications. 200 hand-held transceivers for 430 and 1260 MHz, as well as three repeater stations, were supplied by JAIA member companies for the operation. Each portable station was assigned a special call sign - 8J3AAA, 8J3AAB and so on.

Oklahoma City Bombing

Within minutes of the deadly explosion at the Oklahoma City federal office building, Amateur Radio operators set up an emergency coordination network. During the first few hours after the blast, telephone circuits were jammed and often inoperative. Amateur Radio provided vital emergency communications to rescue and relief organizations until regular telephone service was restored. Located at the Salvation Army Emergency Coordination Center, a net control station coordinated the efforts of more than 20 Amateur Radio stations in downtown Oklahoma City.

Volunteer operators were assigned to the five Salvation Army canteens, the Salvation Army Area Headquarters, the Red Cross Command Post, and the primary search and rescue command post. Using hand-held and mobile radio equipment, hams provided relief workers with reliable, mobile emergency communication. Volunteer hams also drove vital supplies to locations in the disaster area.

During early relief efforts, technical skills of volunteer hams were put to the test when it was determined the buildings in the downtown area were blocking radio signals. A mobile repeater station was established at a Salvation Army canteen, allowing for communications to be sent and received easily.

Amateurs maintained a network continuously from 9:15 AM on April 19 through 4 PM on May 2, with more than 300 operators from Oklahoma, Kansas and Texas participating.

ARRL Public Information Coordinator Tom Webb, WA9AFM, said that those involved in the post-bombing communication effort credited numerous disaster drills, established Amateur Radio networks, emphasis on proper on-air procedures, and experience in weather spotting nets for the success of this operation.

The Engineer in Charge of the FCC Field Office in Dallas declared a voluntary communications emergency, and assigned the frequencies between 3897 and 3903 kHz and between 3922 and 3928 kHz and between 7243.5 and 7249.5 kHz for the handling of emergency traffic. Health and Welfare traffic nets were established on 7273 and 7290 kHz (daytime) and 3873 and 3935 kHz (evening).

FCC ENFORCEMENT ACTIONS OF INTEREST TO AMATEURS

Three Volunteer Examiners' Licenses Suspended Over Exam Fraud

The FCC ordered three California amateurs to show cause why their station licenses should not be revoked, and ordered their operator licenses suspended.

The three (James B. Williams, AA6TC, of Wilmington, California; and Rose Marie Flores, N6WPR, and Robert L. Flores, N6WPQ, both of Santa Monica, California), all Extra Class volunteer examiners, are accused by the Commission of signing paperwork indicating that several persons had passed amateur exam elements when, the FCC says, they had not. The exams, the FCC said, were not administered at all.

The FCC also said that, in response to its inquiries, all three VEs replied with answers that the FCC called "misrepresentations of material fact."

The session was conducted under the coordination of the ARRL/VEC. The ARRL/VEC suspended the accreditations of the three volunteer examiners when questions arose in late 1993.

The three were given 30 days to request a hearing before an administrative law judge. Robert Flores and Rose Flores surrendered their amateur station and operator licenses. Williams agreed to a one-year suspension of his operator license and a "voluntary contribution" of $500 to the US Treasury. Williams also agreed to cooperate fully with the FCC's Private Radio Bureau's investigation of possible additional fraud in amateur license exams.

Four Ordered to be Retested

Four amateur licensees have been ordered by the Federal Communications Commission to be retested on suspect examinations.

The four took volunteer examination elements in Quapaw, Oklahoma, at two sessions in February and March, 1994. The ARRL/Volunteer Examiner Coordinator subsequently uncovered, in routine checking, apparent irregularities at the two sessions and reported them to the FCC.

On December 14, 1994, the ARRL/VEC suspended the accreditation of three volunteer examiners who were present at the two sessions.

The FCC ordered the four to retake, within 60 days, all exam elements that they passed at the sessions. As is standard procedure, they must retake the exams at sessions coordinated by a volunteer examiner coordinator other than the ARRL/VEC, and no volunteer examiner present at the earlier sessions may administer the reexaminations.

The license classes and call signs of the four amateurs will be adjusted according to the results of their reexaminations.

Fines Adjusted for Four

Four amateurs cited by the Federal Communications Commission for malicious interference on the 2-meter band have had the amounts of their monetary forfeitures adjusted.

The four, all in the New Orleans area, were cited for transmissions made in the spring of 1993. In September 1993 all four received Notices of Apparent Liability from the FCC's New Orleans office. The FCC, after reviewing responses from all four, then issued Notices of Forfeiture in November 1993, all in the amount of $2000. All four appealed the NOFs.

While these appeals were pending, the Court of Appeals for the D.C. Circuit vacated the forfeiture guidelines that the New Orleans Field Office had followed in the forfeitures, and on reconsideration, the FCC's Compliance and Information Bureau reduced them.

On July 10, 1995, the FCC issued Orders in all four cases, noting that the cases were based on information provided by the Amateur Auxiliary, and saying that "use of amateur volunteers for the purposes of monitoring violations in the amateur service is permitted by the (Communications) Act. In fact, the amateur radio community has distinguished itself for its self policing operations."

The FCC reduced the fine for Joseph F. Richard III, N5JNX, from $2000 to $1000, rejecting his contention that tape recordings used by the FCC in the case are "inaccurate or misleading."

Vernon Paroli, KA5OWW, had his forfeiture amount reduced to $700, based on his claim of being financially unable to pay the $2,000 forfeiture originally assessed.

The forfeiture amount for Will T. Blanton, N5ROC, was reduced from $2,000 to $1,000. That of John B. Genovese, WB5LOC, was lowered to $500, because of the shorter duration of his transmissions.

OTHER LEGISLATIVE ACTIONS OF INTEREST TO AMATEURS

Joint Resolution on Amateur Radio Becomes Law

The ARRL authored joint resolution supporting Amateur Radio was signed into law by President Bill Clinton on October 22, 1994.

The resolution passed both houses of Congress on October 7, 1994. The new law, PL 103-408, asks for "reasonable accommodation" in the operation of Amateur Radio in homes, automobiles and public places.

Amateur Elected to Congress

David Funderburk, K4TPJ, of Buies Creek, North Carolina has been elected to the U.S. House of Representatives. Funderburk, a Republican, was picked for North Carolina's 2nd District, to fill the seat of a retiring Democrat.

Funderburk, 50, holds a Ph.D. and taught history at several North Carolina colleges before being appointed by President Reagan as U.S. Ambassador to Romania in 1986. His expertise is in East European and Russian studies. He was a Fulbright Scholar.

K4TPJ described himself to *Roll Call*, a newspaper serving Capitol Hill, as an "avid ham radio operator." He holds a General class license.

NH Antenna Bill

New Hampshire Governor Stephen Merrill signed into law a bill protecting amateurs' rights to antennas.

The bill, HB 379, earlier passed both the state's House and Senate unanimously and was signed by the governor on June 5. It was originally written to include slightly stronger language regarding antennas than that contained in PRB-1; the language was then fine-tuned at the suggestion of members of the legislature who are "friendly to Amateur interests," according to ARRL New Hampshire Section Manager Al Shuman, N1FIK.

The bill was filed on the coattails of New Hampshire's HB 1380, passed in 1994, which provides protection from taxing Amateur Radio towers as real property. HB 379 reads:

" No city, town, or county in which there are located unincorporated towns or unorganized places shall adopt or amend a zoning ordinance or regulation with respect to antennas used exclusively in the amateur radio services that fail to conform to the limited federal preemption entitled Amateur Radio Preemption, 101 FCC 2nd 952 (1985) issued by the Federal Communications Commission."

NK2T (Hempstead, NY) Antenna Case

Mark Nadel, NK2T's request for a variance for his 55ft amateur radio tower was approved in April 1995. After more than 2 years, a NY State Supreme Court decision, 2 hearings before the Hempstead, NY Zoning Board of Appeals and a cost of more than $26,000, the case finally came to a positive conclusion. Over 150 hams attended the final hearing on the case. The decision set the groundwork for a new amateur radio ordinance which will be worked on by both the town and members of R.A.D.I.O., Inc. (the Radio Amateurs Defense & Information Organization). The previous ordinance contained a 30-foot restriction.

In the course of this process, the FCC wrote to inform the Zoning Board of Appeals that it was not within their authority to deny the variance based on interference concerns. That authority rests with the Federal Government, and PRB-1 was cited.

W1YG (Lyme, CT) Antenna Case

In October 1994, Parker Heinemann, W1YG, of Lyme, CT, won his PRB-1 based suit against the town's Planning and Zoning Commission after a three-year battle. The U. S. District Court in Bridgeport had ordered that a special exemption permit be issued to allow the installation of his 78 foot tower. Heinemann had been denied a special exception permit by the town planning and zoning commission. He then filed a civil suit against the town. In late October, US District Judge Gerard L. Goettel said that the town had acted in good faith based on the information it had at the time of the application (in March 1991) but ordered the tower approved because, the judge said, "'I think federal law is on the side of ham radio operators in this case."

(Compiled from contributions from W5YI, ARRL, and several other sources)

COMPLETE TEXT OF FCC REPORT AND ORDER REGARDING VANITY CALL SIGNS

Before the Federal Communications Commission Washington, D.C. 20554
PR Docket No. 93-305. In the Matter of Amendment of the Amateur Service Rules to Implement a Vanity Call Sign System.

REPORT AND ORDER
Adopted: December 23, 1994; Released: February 1, 1995

By the Commission:

I. INTRODUCTION

1. On December 13, 1993, we adopted a Notice of Proposed Rule Making (Notice)[1] in the above-captioned proceeding. In the Notice, we proposed to amend our rules to provide a system for the assignment of vanity call signs to amateur stations. This item adopts final rules implementing a vanity call sign system.

2. Each new amateur station licensed by the Commission is assigned a unique call sign.[2] An automated process selects the call sign according to our sequential call sign system.[3] Until recently, we have been unable to accommodate the many thousands of requests that we receive for call signs of the licensee's choice. One of our many steps in reinventing Government is to implement new licensing processing capabilities that make it practicable to grant such requests. To this end, we proposed a vanity call sign system, and asked for public comment on our proposal. We further proposed to use our increased capabilities to resume issuing new club and military recreation station licenses. We received one hundred and five timely comments and four timely reply comments. All of the comments have been carefully considered.

3. The comments confirm the ardent desire of many amateur operators to select the call signs for their stations and their willingness to pay a fee for this service. There were, moreover, several excellent improvements to the proposed system suggested. We hereby adopt rules for a vanity call sign system, incorporating several suggestions from the commenters as discussed below.

II. DISCUSSION

Fairness

4. A major concern of the amateur service community is that the system adopted for allocation of vanity call signs be fair and equitable. Specifically, many commenters suggest using a method of priority with respect to filing applications for vanity call signs.[4] The American Radio Relay League (ARRL) states that, in the interest of fairness and efficiency, the timing and priority in the filing of applications should be important facets of the system that we adopt.[5]

The ARRL favors giving the first priority in applying for a call sign to the former holder or, where the holder is deceased, to a close relative.[6] Several commenters favor giving high priority to those who hold the higher classes of operator license.[7] Other commenters favor giving priority to those who have held their licenses the longest.[8] 5. The ARRL's suggested method is to open the system gradually through four "starting gates."[9] Gate

One would allow a previous holder to apply for that call sign[10] or, where the holder is deceased, a close relative could apply. Gate Two would allow the 66,000 Amateur Extra Class operators, who have passed the most difficult license examinations, to apply. Gate Three would allow the 112,000 Advanced Class operators, who have passed the second most difficult license examinations, to apply. Gate Four would open the system to any licensee.[11] A club station license trustee would also be allowed to apply for the call sign of a deceased former holder.[12]

6. The suggestions regarding filing priority and fairness are persuasive. Given the strong interest in vanity call signs shown in the comments, it is obvious that the number of applications filed initially could be very large. We agree that a filing priority schedule would be helpful in maintaining fairness and efficiency during the initial implementation of the system, as well as ease administrative burdens on the Commission. The suggestions concerning giving the highest filing priority to former holders and close relatives of deceased holders appear to be perceived as fair by the amateur service community generally, as does the giving of high priority to those who hold the higher classes of operator license.[13] Information on the class of operator license held by each amateur operator, moreover, resides in our licensee data base and lends itself to an automated process. Information on the length of time a person has been an amateur operator is not readily available, thus making that criterion impracticable to use as the basis of a filing priority schedule. Thus, after all amateur operators have been given an opportunity to obtain call signs that they, or deceased close relatives, formerly held, we will use operator license class as the basis for the filing priority schedule. In this regard, we are adopting the ARRL's suggested starting gates.

7. We will announce the opening of each gate by a Public Notice. The first gate will open as soon as our new FCC Form 610-V is available and our licensing facility is prepared to begin processing the applications. Gate One will open the system to the smallest group, i.e., a few thousand prior holders and close relatives[14] of deceased former holders. This phase will provide validation of our system procedures and alert us to any adjustments needed. We will then open the subsequent gates at such times as it is clear that the system is ready to accommodate more applications. We will also continue our sequential call sign system for new licensees and for those who do not want vanity call signs.

Assignable call signs

8. In the Notice, the system we proposed would require applicants to file a form, together with the required fee with our fee collection contractor.[15] The applicants would request that their station licenses be modified to show vanity call signs. We further proposed that the applicant would list on the form a maximum of ten call signs in order of personal preference. After receiving the forms from our contractor, we would use an automated process to compare each applicant's list of preferred call signs with the list of call signs that are assignable at that time. The forms would be processed in the order they are received at the processor's work station. The first assignable call sign from the applicant's list would then he assigned to the station.

9. We requested comments on how the call signs that are already assigned could be made known to applicants so as to allow them to make prudent requests and thereby increase the probability that their requests can be granted. The ARRL states that private sector entities can easily provide applicants with lists of assignable call signs, but only after the initial surge of applications is completely processed. Until such time, even with starting gates, it foresees a very heavy demand for certain specific call signs so as to make it difficult for the applicant to determine which call signs are assignable. The ARRL suggests, therefore, that an applicant be permitted to submit a preferential list of twenty-five call signs, thus increasing the chances of requesting an assignable call sign.[16] We agree with ARRL that increasing the number of call signs that may be requested will reduce the number of unsuccessful applicants. We will allow, therefore, applicants to list up to twenty-five call signs in order of preference.

10. The ARRL prefers that an applicant be permitted to request only those call signs that are assignable to stations in the call sign region where the licensee resides.[17] We have carefully consid-

ered this suggestion. We have decided, however, not to impose that limitation. Otherwise, the applicant's choice of vanity call signs would be reduced to ten percent or less of the call signs that would otherwise be assignable to the station. A limitation based upon the person's place of residence, moreover, could easily be circumvented by using a mailing address in another call sign region.

11. We proposed that a call sign vacated by a licensee be made assignable immediately under the vanity call sign system. Several commenters, however, believe that a two-year period is necessary before a call sign again becomes assignable in order to avoid confusion in over-the-air station identification, to maintain accuracy in the licensee data base, and to accommodate QSL bureaus.[18] Further, they believe that it would preclude "trafficking in licenses" where a licensee, in exchange for some type of consideration, vacates a desirable call sign so that another licensee could immediately apply for it before its assignability becomes known generally. A two-year interval would, moreover, make the assignability of vacated call signs consistent with the assignability of a deceased person's station call sign, or a licensee's expired station call sign.[19] The comments are persuasive on this point. Therefore, the rules will reflect that a vacated call sign will not be assignable for a two-year waiting period.

Club stations

12. There was support in the comments for resumption of the issuance of new club station licenses.[20] In the Notice, we proposed that an applicant for a vanity call sign must be a current holder of a station license. The Hill Country Radio Club (Hill) suggests that applicants for new club station licenses be able to request a vanity call sign immediately, rather than having to wait and apply after they receive licenses. Hill considers a two-step procedure ponderous and unfair to new clubs that have been precluded from obtaining club licenses for many years.[21] The two-step process, however, is an administrative necessity because of the fee required for a vanity call sign. The application for a vanity call sign is the only amateur service application that must be filed with our fee collection contractor.[22] Persons not already holding a club station license, therefore, must first apply for and receive a license before they can file an application with the fee collection contractor requesting that the license be modified to show a vanity call sign. However, we will begin accepting applications[23] for club and military recreation station licenses on the date this Report and Order becomes effective. In many cases, therefore, the license trustee will be able to obtain a license document and thus will be eligible to apply for a vanity call sign for the club station when the starting gate for his or her class of operator license opens. The final gate will also allow a club station licensee trustee to apply for the call sign of a deceased former holder. The license trustee must obtain a written consent from a close relative of the deceased.

Special event stations

13. The ARRL requests that specific call signs in a unique call sign block be made assignable only to certain special event stations,[24] and suggests the one-by-one call sign block for this purpose.[25] The ARRL states that such specific call signs should be reserved for assignment to stations operating in conjunction with short term special events of national significance.[26] A special event vanity call sign system may meet the needs of amateur operators for temporary operation of their stations during events that are of special significance to the amateur service community. We will, therefore, set aside the one-by-one call sign block until the matter can be addressed in a separate proceeding.

Filing procedures

14. In our Notice, we asked for comments concerning alternative ways, such as magnetic computer disks, that applicants could use to apply directly to the Commission for a vanity call sign. Several ways were suggested.[27] One commenter suggests a procedure where applicants would file the application form by facsimile and provide a credit card number. His second suggestion is an electronic on-line filing procedure where the applicant answers a series of questions to search the Commission's data base for an assignable call sign. If the call sign is assignable, the applicant would file an application form after paying the fee by credit card.

The Commission's printed acknowledgment of the transaction would constitute a temporary license.[28] These suggestions were helpful and we will investigate these ideas for possible future use.

15. Some commenters believe that the fee charged for a vanity call sign should be charged on a one-time basis only, and that no fee should be required when the license is renewed.[29] The ARRL believes that a one-time fee is more appropriate because the Commission's additional workload occurs at the time of the initial processing of the vanity call sign.[30] Section 9(g) of the Communications Act of 1934, as amended, currently provides for the payment of an annual fee of $7.00 for an amateur station vanity call sign. Because the normal term of an amateur station license is ten years, a fee of $70.00 will have to be paid when requesting a new or renewed vanity call sign. At this time, under the Communications Act, we cannot provide a one-time fee for processing vanity call sign applications.

III. CONCLUSION

16. We have decided to offer a vanity call sign system to the amateur service community, in recognition of the strong sense of identity among amateur operators that is grounded in the call signs of their stations. We have also decided to resume issuing new club and military recreation station licenses. We see these actions as fundamental to our commitment to put the needs of people first in providing the services that they want. We are pleased to be able make this new system available to the amateur community. Therefore, we amend the amateur service rules to implement a vanity call sign system as set forth in the attached Appendix.

IV. ORDERING CLAUSES

17. Accordingly, IT IS ORDERED that effective March 24, 1995, Part 97 of the Commission's Rules, 47 C.F.R. Part 97, IS AMENDED as set forth in the Appendix hereto. Authority for this action is found in Section 4(i) and 303(r) of the Communications Act of 1934, as amended, 47 U.S.C. 154(i) and 303(r).

18. IT IS FURTHER ORDERED that this proceeding IS TERMINATED.

19. For further information, contact Maurice J. DePont, Wireless Telecommunications Bureau, 202-418-0690.

FEDERAL COMMUNICATIONS COMMISSION

William F. Caton Acting Secretary

APPENDIX
Part 97 of Chapter I of Title 47 of the Code of Federal Regulations is amended as follows:
Part 97 - Amateur Radio Service

1. The authority citation for Part 97 continues to read as follows:
Authority citation: 48 Stat. 1066, 1082, as amended; 47 U.S.C. 154, 303. Interpret or apply 48 Stat. 1064-1068, 1081-1105, as amended; 47 U.S.C. 151-155, 301-609, unless otherwise noted.

2. Section 97.3 is amended by redesignating paragraphs (a)(11) through (a)(45) as paragraphs (a)(12) through (a)(46) and adding new paragraph (a)(11) to read as follows:
97.3 Definitions
(a) ***
(11) Call sign system. The method used to select a call sign for amateur station over-the-air identification purposes. The call sign systems are:
(i) Sequential call sign system. The call sign is selected by the FCC from an alphabetized list corresponding to the geographic region of the licensee's mailing address and operator class. The call sign is shown on the license. The FCC will issue public announcements detailing the procedures of the sequential call sign system.
(ii) Vanity call sign system. The call sign is selected by the FCC from a list of call signs requested by the licensee. The call sign is shown on the license. The FCC will issue public announcements detailing the procedures of the vanity call sign system.

* * * * *

3. Section 97.17(f) is amended by revising paragraph (f) and adding paragraph (h) to read as follows:

97.17 Application for new license or reciprocal permit for alien amateur licensee.

* * * * *

(f) One unique call sign will be shown on the license of each new primary, club, and military recreation station. The call sign will be selected by the sequential call sign system.

* * * * *

(h) Each application for a new club or military recreation station license must be submitted to the FCC, 1270 Fairfield Road, Gettysburg, PA 17325-7245. No new license for a RACES station will be issued.

4. Section 97.19 is added to read as follows:

97.19 Application for a vanity call sign.

(a) A person who has been granted an operator/primary station license or a license trustee who has been granted a club station license is eligible to make application for modification of the license, or the renewal thereof, to show a call sign selected by the vanity call sign system. RACES and military recreation stations are not eligible for a vanity call sign.

(b) Each application for a modification of an operator/primary or club station license, or the renewal thereof, to show a call sign selected by the vanity call sign system must be made on FCC Form 610-V. The form must be submitted with the proper fee to the address specified in the Private Radio Services Fee Filing Guide.

(c) Only unassigned call signs that are available to the sequential call sign system are available to the vanity call sign system with the following exceptions:

(1) A call sign shown on an expired license is not available to the vanity call sign system for 2 years following the expiration of the license.

(2) A call sign shown on a surrendered, revoked, set aside, cancelled, or voided license is not available to the vanity call sign system for 2 years following the date such action is taken.

(3) Except for an applicant who is the spouse, child, grandchild, stepchild, parent, grandparent, stepparent, brother, sister, stepbrother, stepsister, aunt, uncle, niece, nephew, or in-law, and except for an applicant who is a club station license trustee acting with the written consent of at least one relative, as listed above, of a person now deceased, the call sign shown on the license of a person now deceased is not available to the vanity call sign-system for 2 years following the person's death, or for 2 years following the expiration of the license, whichever is sooner.

(d) Except for an applicant who is the spouse, child, grandchild, stepchild, parent, grandparent, stepparent, brother, sister, stepbrother, stepsister, aunt, uncle, niece, nephew, or in-law, and except for an applicant who is a club station license trustee acting with the written consent of at least one relative, as listed above, of a person now deceased who had been granted the license showing the call sign requested, the vanity call sign requested by an applicant must be selected from the groups of call signs designated under the sequential call sign system for the class of operator license held by the applicant or for a lower class.

(1) The applicant must request that the call sign shown on the current license be vacated and provide a list of up to 25 call signs in order of preference.

(2) The first assignable call sign from the applicant's list will be shown on the license grant. When none of those call signs are assignable, the call sign vacated by the applicant will be shown on the license grant.

(3) Vanity call signs will be selected from those call signs assignable at the time the application is processed by the FCC.

5. Section 97.21(a)(3) is revised to read as follows:

97.21 Application for a modified or renewed license.

(a) ***

(3) May apply for renewal of the license for another term. (The FCC may mail to the licensee an FCC Form 610-R that may be used for this purpose.)

(i) When the license does not show a call sign selected by the vanity call sign system, the application may be made on FCC Form 610-R if it is received from the FCC. If the Form 610-R is not received from the FCC within 30 days of the expiration date of the license for an operator/primary station license, the application may be made on FCC Form 610. For a club, military recreation, or RACES station license, the application may be made on FCC Form 610-B. The application may be submitted no more than 90 days before its expiration to: FCC, 1270 Fairfield Road, Gettysburg, PA 17325-7245. When the application for renewal of the license has been received by the FCC at 1270 Fairfield Road, Gettysburg, PA 17325-7245 prior to the license expiration date, the license operating authority is continued until the final disposition of the application.

(ii) When the license shows a call sign selected by the vanity call sign system, the application must be filed as specified in Section 97.19(b).

* * * * *

1 9 FCC Rcd 105 (1993)

2 Section 97.119(a), 17 C.F.R. 97.119(a), requires an amateur station to transmit its assigned call sign on its transmitting channel periodically for the purpose of making known, clearly, the source of the transmissions from the station to those receiving the transmissions.

3 Our sequential call sign system is described in the FACT SHEET PR-5000 #206 Amateur Station Call Sign Assignment System dated June, 1991. A new call sign is sequentially selected from alphabetized regional-group listings for the licensee's operator class and mailing address. Each call sign has a one or two letter prefix and a one, two, or three letter suffix separated by a number indicating the geographic region. Some examples are: W1AA, N3AAA, AA5A, AB7AA, and KA9AAA. There are almost 15 million possible combinations of letters and numbers for amateur station call signs.

4 For example. see comment of John Ward at 1, Dale Jones at 1 and 2, and Richard Bean at 1.

5 Comments of ARRL at 4.

6 Comments of ARRL at 7.

7 There are six classes of amateur operator license. In order of examination difficulty they are, lowest to highest Novice, Technician, Technician Plus, General, Advanced, and Amateur Extra. Comments of Michael C. and Nancy E. Bartlett at 1, Jim Monahan at 1, Ed Worst at 1, and Willard W. Wehe at 1.

8 Dale Jones at 1, Michael Dinkelman at 1, and R. W. Le Massena at 1.

9 Comments of ARRL at 6.

10 Previous holder priority was also recommended by other commenters. Comments of Dale Jones at 1, Robert Philbrook at 1, Kirby Brown at 1, and Donald Murray at 1.

11 As of May 31, 1994, there were 631,399 amateur stations licensed by the Commission.

12 Reply comments of ARRL at 5.

13 Comments of ARRL at 7, Michael C. and Nancy E. Bartlett at 1, Jim Monahan at 1, Ed Worst at 1, and Willard W. Wehe at 1.

14 We define close relatives as the spouse, child, grandchild, stepchild, parent, grandparent, stepparent, brother, sister, stepbrother, stepsister, aunt, uncle, niece, nephew, or in-law of the deceased.

15 The current contractor is the Mellon Bank. It accepts applications in Pittsburgh, Pennsylvania.

16 Comments of ARRL at 3 and 4.

17 Comments of ARRL at 10.

18 A QSL bureau is an organization that facilitates the exchange of confirmation cards between amateur operators whose stations have communicated with each other.

19 Comments of Thomas Johnston at 1, ARRL at 9, Stephan Sacco, Jr. at 2, and James Price at 2.

20 Comments of Metropolitan Amateur Radio Club at 1 and Portland Amateur Radio Club at 3.

21 Comments of The Hill Country Amateur Radio Club at 1 and 2.

22 Applications for licenses involving examinations are filed with the local volunteer examiners who forward them through a

coordinator to the Commission. All other amateur service license applications are filed directly with the Commission.

23 FCC Form 610-B must be used when applying for a club or military recreation station license.

24 Comments of ARRL at 11 and 12.

25 A one-by-one call sign consists of a single prefix letter (K, N, or W), the region number (0 to 9), and a single suffix letter (A to Z). There are 780 such call signs. They are not assigned under the sequential call sign system.

26 Comment of ARRL at 12.

27 See, for example, the comments of Ed Worst at 1.

28 Comments of Steven R. Kelly at 1 and 2.

29 Comments of Francis Vangeli at 1, John Chandler at 1, Ross Patterson at 1, and ARRL at 11.

30 Comments of ARRL at 11.

JOINT RESOLUTION IN SUPPORT OF AMATEUR RADIO

The following Joint Resolution passed both Houses of Congress in 1994, and was signed into law by President Bill Clinton as Public Law (P.L.) 103-408 on October 22, 1994.

Joint Resolution S.J.Res. 90

To recognize achievements of radio amateurs, and to establish support for such amateurs as National Policy

WHEREAS, Congress has expressed its determination in section 1 of the Communications Act of 1934 [47 U.S.C. 151] to promote safety of life and property through the use of radio communication;

WHEREAS, Congress, in section 7 of the Communications Act [47 U.S.C. 157], has established a policy to encourage the provision of new technologies and services;

WHEREAS, Congress, in section 3 of the Communications Act of 1934, defined radio stations to include amateur stations operated by persons interested in radio technique without pecuniary interest;

WHEREAS, the Federal Communications Commission has created an effective regulatory framework through which the amateur radio service has been able to achieve the goals of the service;

WHEREAS, these regulations, set forth in part 97, of title 47 of the Code of Federal Regulations, clarify and extend the purposes of the Amateur Radio Service as a —

(1) voluntary noncommercial communication service particularly with respect to providing emergency communications;

(2) as a contributing service to the advancement of the telecommunications infrastructure;

(3) as a service which encourages improvement of an individual's technical and operating skills;

(4) as a service providing a national reservoir of trained operators, technicians and electronics experts; and

(5) as a service enhancing international good will;

WHEREAS, Congress finds that members of the amateur radio service community has provided invaluable emergency communications services following such disasters as Hurricanes Hugo, Andrew and Iniki, the Mt. St. Helens eruption, the Loma Prieta earthquake, tornadoes, floods, wild fires and industrial accidents in great number and variety across the Nation; and

WHEREAS, Congress finds that the amateur radio service has made a contribution to our nation's communications by its crafting, in 1961, of the first earth satellite licensed by the Federal Communications Commission, by its proof-of- concept for search and rescue satellites, by its continued exploration of the low earth orbit in particular pointing the way to commercial use thereof in the 1990s, by its pioneering of communications using reflections from meteor trails, a technique now used for certain government and commercial communications, and by its leading role in development of low-cost, practical data transmission by radio which increasingly is being put to extensive use in, for instance, the land mobile service;

Now, therefore, be it Resolved by the Senate and House of Representatives of the United States of America in Congress assembled,

SECTION 1. FINDINGS AND DECLARATIONS OF CONGRESS.

Congress finds and declares that —

(1) radio amateurs are hereby commended for their contributions to technical progress in electronics, and for their emergency radio communications in times of disaster;

(2) the Federal Communications Commission is urged to continue and enhance the development of the amateur radio service as a public benefit by adopting rules and regulations which encourage the use of new technologies within the amateur radio service; and

(3) reasonable accommodation should be made for the effective operation of amateur radio from residences, private vehicles and public areas, and that regulation at all levels of government facilitate and encourage amateur radio operation as a public benefit.

TELEPHONE INTERFERENCE SURVEY
May 2, 1994

Prepared by Field Operations Bureau
Federal Communications Commission

Telephone Interference Survey

Background

The Federal Communications Commission (FCC) receives 25,000 complaints per year from individuals who are unable to use their telephones because nearby radio stations interfere with the proper operation of the telephones. Whenever the radio stations are on the air, the telephones pick up the stations' transmissions which then override any ongoing telephone conversation.

For about three years field offices have submitted narrative reports indicating that interference to telephones is an increasing problem among consumers. Because of those reports, FOB began a detailed sampling of calls and letters. The sampling is an actual count of specific complaints and inquiries, conducted one week each month by all offices. This sample is the basis of our count of 25,000 telephone interference complaints per year and is the most accurate measure of compliant categories now available.

The FCC's Field Operations Bureau (FOB) has done an informal survey to obtain information about telephone interference such as:

-which telephones are affected

-what type transmitting stations are involved

-the power levels at which the stations were transmitting

-whether commonly available filters would be effective in eliminating the interference

-whether specially designed telephones would be effective in eliminating the interference

In setting forth the results of its informal survey, FOB emphasizes that, because this survey is based on a random sample, it cannot be claimed that identical results would be derived under scientific surveying and testing, nor should the results be construed as FCC endorsement or criticism of any particular manufacturer's product. Rather, FOB believes these results to be a good "first look" at the problem.

Procedure

Thirty-five FCC field offices across the country participated in the survey. Each office was to choose three recent complaints of telephone interference on a random basis, and then to investigate the complaints.

At the transmitting station FOB staff would determine the type of station (i.e., amateur, citizens band, broadcast, etc.), measure the station power, and obtain information on antenna height, antenna gain, and distance from the complainant.

At the complainant's location, FOB would disconnect all telephones, take them to a chosen telephone jack, and plug them in one at a time while the station was transmitting. FOB would then record which telephones received interference and which did not.

FOB also tested the effectiveness of several commercially-available telephone filters, connecting the filters to telephones which were receiving interference while the radio station was transmitting.

Finally, FOB connected "interference free" telephones to the telephone jacks and observed whether these telephones received interference. "Interference free" telephones are those which the manufacturer claims to be immune from interference.

Findings

Types of Transmitting Stations

FOB inspected 108 transmitting stations which were involved in the telephone interference complaints, as follows:

Citizens Band	47
Amateur	27
AM Broadcast	23
FM Broadcast	10
International Broadcast	1

Power Levels

The power levels of the radio transmitting stations varied from two watts to half a million watts. One-third of the transmitting stations operated with less than ten watts, and one-third of the interfering stations were broadcast stations using between 3000 and half a million watts. Attachment 1 lists the power levels of the transmitting stations.

Types of Telephones

FOB tested 241 telephones found in the complainants' residences. Of the 241 telephones 68 percent received interference. Attachment 2 lists the telephones tested. FOB did not observe interference on 32 percent of the telephones it tested. These telephones are listed in Attachment 3.

Types of Filters

FOB tested the effectiveness of the AT&T Z100B1 filter on 138 telephones receiving interference. After connecting the filter to the telephones, 62 percent of the telephones continued to receive interference. The filter did eliminate interference on 38 percent of the telephones. A number of other filters were also tested on 82 telephones receiving interference. As a group these filters eliminated interference on 29 percent of the telephones. They did not eliminate interference on 71 percent of the telephones. Attachment 4 lists the filters which eliminated the interference.

"Interference Free" Telephones

FOB tested "interference free" telephones at 52 locations where the individuals were receiving interference to their telephones. The "interference free" telephones eliminated interference at 96 percent of the locations.

Conclusions

The transmitting stations most likely to cause telephone interference are citizens band, amateur, and broadcast stations. Citizens band stations accounted for half the telephone interference cases. Amateur stations and broadcast stations accounted for the other half.

The power levels used by the radio transmitting station did not appear to be a significant factor in causing telephone interference. Power levels of 10 watts or less caused telephone interference in a third of the cases.

A large portion of the residential telephones appeared to be susceptible to interference from nearby radio transmitting stations. Although some telephones did not receive interference, the limited nature of the testing performed in conjunction with this survey would not support the conclusion that they would always reject interference.

Filters cannot be relied upon to eliminate telephone interference: in two out of three cases in the test sample, they did not work. Manufacturers can, however, design telephones to be interference-free. The "interference free" telephones were immune from interference virtually all of the time.

A Final Note

Notwithstanding the 25,000 reports of telephone interference the FCC receives annually, it is FOB's experience that, as large as this number is, it probably represents only a fraction of the actual instances in which this interference occurs. given the enormous numbers of instances in which this type of interference is experienced by consumers, it is our hope that this survey, notwithstanding its informality, will serve as a catalyst for affected parties to productively address and resolve this problem. As always, FOB remains ready to assist in that effort.

Attachment 1

Field Operations Bureau
Telephone Interference Survey
Power Levels Of Interference
Sources

UNKNOWN	2 sources	75 watts	1 source	800 watts	1 source
2 watts	4 sources	80 watts	1 source	1,000 watts	3 sources
3 watts	7 sources	100 watts	5 sources	1,500 watts	2 sources
4 watts	14 sources	101 watts	1 source	3,000 watts	2 sources
5 watts	7 sources	110 watts	2 sources	5,000 watts	5 sources
6 watts	1 source	114 watts	1 source	8,000 watts	1 source
7 watts	4 sources	131 watts	1 source	10,000 watts	4 sources
12 watts	2 sources	150 watts	1 source	16,000 watts	1 source
26 watts	1 source	160 watts	1 source	25,000 watts	1 source
30 watts	1 source	200 watts	1 source	50,000 watts	11 sources
35 watts	1 source	250 watts	1 source	51,000 watts	4 sources
50 watts	1 source	438 watts	1 source	54,000 watts	1 source
53 watts	1 source	462 watts	1 source	100,000 watts	1 source
54 watts	1 source	500 watts	3 sources	500,000 watts	1 source
57 watts	1 source	600 watts	1 source	**Total sources:**	**108**

Attachment 2

Field Operations Bureau
Telephone Interference Survey
Telephones Receiving Interference

fo	fil	ix	power	att	oth	fcc	make	model
DL	N	AMA	800	N	N	Y	{UNKNOWN)	(UNKNOWN)
AL	N	CB	4	N	N	Y	(UNKNOWN)	(UNKNOWN)
PL	N	CB	2	N	N	Y	AT&T	(NONE)
DT	N	CB	5	N	N	Y	AT&T	(FCC AS593M-70230-TE-T)
LR	N	AM	1000	Y			AT&T	100
KI	N	FM	3000	N			AT&T	1306
KI	N	AMA	100	N			AT&T	1504
DV	N	AM	5000	Y	Y	Y	AT&T	150E
NY	N	CB	3	Y	Y		AT&T	1510
PO	N	AMA	35	Y	N		AT&T	1521
BS	N	AM	10000	Y	N		AT&T	1532
GI	N	AMA	53	N	N		AT&T	1611
DL	Y	CB	5	N		Y	AT&T	1618
KI	N	AMA	100	N			AT&T	210
KI	N	AMA	100	N			AT&T	210
DS	N	AM	5000			Y	AT&T	210
DS	Y	AMA	500	N	Y		AT&T	210
NF	Y	CB	3	Y	N	Y	AT&T	210
PL	N	AMA	1000	Y	Y	Y	AT&T	530
DL	N	AMA	800	Y	Y	Y	AT&T	5455
GI	N	AMA	101	N			AT&T	610
BF	N	AMA	100	N		Y	AT&T	710
KI	N	CB	5	N			AT&T	720
CG	N	AMA	1000	Y		Y	AT&T	720
KC	Y	CB	4	N		Y	AT&T	725
GI	N	AMA	101	N			AT&T	725
PA	N	CB	3	N		Y	AT&T	732
HL	N	AMA	462	N	N		AT&T	732
MA	Y	AMA	600	N			AT&T	AOM9RN 10451-TE-E
LR	Y	CB	4	N			AT&T	AS 550Z 71597
MA	N	AMA	50	N			AT&T	ASSS0Z-71597-TE-E
DT	N	AM	438		N		AT&T	MODEL 1506
NY	N	AM	25000	N	N		AT&T	PRINCESS
BM	N	CB	7	N	N	Y	AT&T	PRINCESS AS 550 Z-71322
PA	N	CB	3	N	Y		AT&T	REMOTE & ANS MACHINE COMB
PO	N	AM	50000	Y	Y		AT&T	SW204
BM	N	FM	16000	Y	N	Y	AT&T	TRIMLINE 210
BF	Y	AM	50000	N		Y	AT&T	TRIMLINE 210
DL	N	CB	5	Y		Y	AT&T	TRIMLINE 220
NF	N	AMA	1500	Y	Y	Y	AT&T	TRIMLINE AS550Z-71597-TE-E
ST	N	AM	10000	Y	N	Y	AUDIOVOX	AT-11
PO	N	AM	50000	Y	N		BELL	FAVORITE

							Manufacturer	Model
PL	N	AM	5000			Y	BELL TT	51490
KI	N	CB	5	N			BELL PHONES	100-5800
AT	Y	CB	160	N	N	Y	BELL SOUTH	2000
BS	N	AM	50000	Y	N		BELL SOUTH	227V
DT	N	CB	3				BELL SOUTH	473
KI	N	CB	5	N			BELL SOUTH	473
DS	N	AM	5000			Y	BELL SOUTH	473
MA	N	AMA	600				BELL SOUTH	CWIMLA-61669-ANN
OR	N	AMA	100	N	Y		BELL SOUTH	DPK-62W70687-TE-T
AT	Y	IB	500000	N		Y	BELL SOUTH	HAC 572
PS	Y	AMA	150	N	N	Y	BELL SOUTH	TP-201 L-10
SF	N	AM	50000	Y			CANON	FAX-350
PO	Y	AMA	500	Y			CODE-APHONE	1610
LA	Y	AM	50000	N	N		COMDIAL	25270-IY
SF	N	AM	50000				COMDIAL	VOICE EXPRESS
ST	N	CB	75	N	N	Y	CONAIR	PR 1000
AT	Y	CB				Y	CONAIR	PR 5001
LV	N	CB	57	N	N		CONAIR	SW 2502
TP	N	CB	5				CONAIR	SW204
BE	N	AMA	80		N		CONAIR	SW204A
PO	N	AM	50000	Y			CONAIR	TR1001
AN	N	FM	51000	Y	Y		CONAIR-PHONE	PR1006
PO	Y	AMA	500	Y			CONAIR-PHONE	PRIMA SERIES
DT	N	CB	3	N	N	Y	CONAIR-PHONE	SW-102
HL	N	AM	10000				DELTA	HEK-EB-200DSIHF-EXT
SD	N	CB	12	N			FORTEL	RECORD A CALL #2140
SF	N	CB	2	N	N		GENERAL ELECTRIC	12-MEMORY 2-9240A
PO	N	AM	50000	Y	N		GE	2-9051LRA
MA	N	CB	6	N			GE	2-9166B
CG	N	AM	50000	Y		Y	GE	2-9166B
TP	Y	CB	4	N		Y	GE	2-9167A
KC	N	CB	2	Y		Y	GE	2-9212BLB
LV	Y	CB	4	N	N		GE	2-9243
DT	N	AM	438				GE	2-9243C
KC	N	FM	8000	N			GE	2-9355A
LV	Y	CB	5	N	N		GE	2-9356
VB	N	AMA	100	Y			GE	2-9400C
HL	N	FM	54000				GE	2-94050A
HL	N	FM	54000				GE	2-94050A
HL	N	FM	54000				GE	2-94050A
HL	N	FM	54000				GE	2-94050A
MA	N	CB	6	N			GE	2-9420A
HU	N	FM	100000	N			GE	2-9675A
PS	Y	CB	5	Y	Y	Y	GE	2-9892A
DV	N	CB	2	Y	Y	Y	GE	2-9895A
LV	N	CB	4	N	N		GE	29270 A
DV	N	AM	5000	Y			GE	7-4700A
LA	N	AM	5000	Y	Y		GE	CRYSTAL CLEAR PLUS
BF	N	CB	5	N		Y	GE	MEMORY MODEL 21
MA	N	CB	6	Y			GE	NO 29420A
SD	N	CB	54	N	N		GE	SPEAKER PHONE2-9375F
CG	N	AMA	1500				GE	TL-5/TPS
PO	N	AMA	35	Y	N		GTE	58821
ST	Y	FM	51000	N	N	Y	GTE	7400
GI	N	AMA	101	Y			GTE	AUTO. ELEC.-WALL PHONE
GI	N	AMA	101	Y			GTE	AUTOMATIC ELECTRIC
PL	N	AM	5000		N		IMA	(NONE)
AL	Y	AM	5000	N	N	Y	ITT (CORINTH)	3480 FEATURE TELEPHONE
DS	N	FM	3000	Y	N	Y	LENOX SOUND	P4316
PL	N	AM	5000				LLOYDS	T510
HU	N	CB	110	N			LONE STAR	912
PA	N	CB	4			Y	LONE STAR	TRIMLINE
BS	N	CB	5	N	N		LOTEL	5020 DUEL LINE
BS	N	CB	5	N	N		LOTEL	5020 DUEL LINE
FE	N	AMA	131	N			N. TELECOM	FCC REG #AB6982-68817-TE-T
AN	N	FM	51000	Y	Y		N. TELECOM	NT3L00AA
DT	N	CB	5	Y	N	Y	N. TELECOM	SYMPHONY 3000
PS	N	AMA	200	N	N	Y	N.W. BELL	"EASY TOUCH"
SD	Y	CB	12	N			N.W. BELL	BASIC TGT 51450
LV	N	CB	4	N	N		PAC BELL	TR 2203 CHAC
PO	Y	AM	50000	N	N		PACTEL	FE5700
DS	N	AMA	500	N	Y		PANASONIC	EASA-PHONE VA-8205
NY	N	AMA	100	Y	Y		PANASONIC	EASA-PHONE-AUTO LOGIC
BS	N	AM	10000	N	N		PANASONIC	EASE PHONE KX-TX175-B

fo	fil	ix		att	oth	fcc	Manufacturer	Model
TP	Y	CB	3	N	N	Y	PANASONIC	KX-T-2420
SF	Y	AMA	250	Y	N		PANASONIC	KX-T-3155
DV	N	AMA	114	N	Y	Y	PANASONIC	KX-T2315
DL	N	CB	4	Y	Y	Y	PANASONIC	KX-T2315
PO	Y	AM	50000	Y	Y		PANASONIC	KX-T2355
CG	N	AMA	1500	N		Y	PANASONIC	KX-T2355
ST	Y	AM	10000	N	N	Y	PANASONIC	KX-T2622 EASAPHONE
HL	N	AMA	462				PANASONIC	KX-T3145
OR	N	AM	50000	N		Y	PANASONIC	KXT 2388
KI	N	CB	5	N			PHONE MATE	3900
NF	N	CB	7	Y	N	Y	PHONEMATE	3950
GI	Y	AMA	101	N			PHONEMATE	4650
BS	N	AM	10000	N	N		RADIO SHACK	209 DUOPHONE 43-617A
DS	N	FM	3000			Y	RADIO SHACK	43-384
PS	Y	CB	5				RADIO SHACK	43-384
DT	N	CB	3			Y	RADIO SHACK	43-625 DUOFONE 202
TP	Y	CB	4	N		Y	RADIO SHACK	43-430
PO	Y	AM	50000	Y	N		RADIO SHACK	DUOPHONE TAD-250
OR	N	CB	4	Y			RADIO SHACK	ET 200
LR	N	CB		N			SANYO	TA S 855
AT	N	CB	160	N		Y	SEARS	3293 4761550 SR 3000
KI	N	FM	3000	N			SEARS	34403
DL	N	AMA	800	Y	Y	Y	SEARS	34505
LV	N	CB	4	N	N		SEARS	SR-2000
KI	N	FM	3000	N			SOUND DESIGN	7255 AR
TP	Y	CB	3	N	N	Y	SOUND DESIGN	7339 IVY
NY	Y	AMA	100	Y	Y		S. W. BELL	FCC 2555
BE	N	CB	4	N			S. W.	FREEDOM PHONE
ST	Y	CB	75	N	N	Y	S. W.	FREEDOM PHONE FC 2570
BS	N	AM	10000	Y	N		S. W.	FT-325
SF	N	AM	50000				STROMBERG	IMAGE 1 AND IMAGE 2 DTMF
ST	Y	FM	51000	N	N	Y	TECHNICO	TT-2201
SJ	N	CB	26	N		Y	TELEFONICA HISPANO AMERICAN	TXE-1
SJ	N	CB	7	N		Y	TELEFONICA HISPANO AMERICAN	TXE-1
SJ	N	CB	7	N		Y	TELEFONICA HISPANO AMERICAN	TXE-1
TP	N	CB	3	N	N	Y	TELKO	221
ST	Y	CB	4	N	N	Y	THOMAS	PP9D NOSTALGIA PHONE
FE	N	AM	50000	Y			TIME LIFE	(NONE)
DS	N	AM	5000	Y	Y	Y	TISONIC	PT-91
LV	Y	CB	5	N	N		ULTRSONIC	019
KC	N	FM	8000	N			UNIDEN	TRIMLINE
FE	N	AM	50000	Y			UNISONIC	(NONE)
AL	Y	AM	5000	N	N	N	UNISONIC	(NONE)
VB	Y	CB	30	Y			UNISONIC	6434
BM	N	CB	4	N	N	Y	UNISONIC	9370
PL	Y	AM	5000	Y	Y	Y	WESTERN ELECTRIC	(NONE)
DT	N	AM	438	Y		Y	WESTERN ELECTRIC	(NONE) ROTARY DIAL PHONE
DT	N	AM	438				WESTERN ELECTRIC	(NONE) ROTARY DIAL PHONE
NF	N	AMA	1500	Y	Y	Y	WESTERN ELECTRIC	AS543M-62587-TE-T
PA	N	AMA	110	N		Y	WESTERN ELECTRIC	CS 2500 DMG84121

fo = office; **fil** = filter already attached?; **ix** = ix source (AM/FM/CB/AMAteur/International Broadcast); **att** = AT&T Z 100 B1 filter eliminate ix?; **oth** = other filters eliminate ix?; ; **fcc** = FCC "bulletproof" telephones free from ix?

Attachment 3

Field Operations Bureau
Telephone Interference Survey
Telephones Not Receiving Interference*

off	filter make	model
NF Y	AT&T	1316C
DL N	AT&T	1321
SF N	AT&T	1323
NY N	AT&T	5200
SF N	AT&T	5200
BF N	AT&T	5400
ST N	AT&T	5600
PA N	AT&T	60921
CG N	AT&T	700
ST N	AT&T	730
PA N	AT&T	ANS MACHINE
PA N	AT&T	AS5 THA-60921-MT-E
DL N	AT&T	CS8702A
DL Y	AT&T	NE-229A
OR N	AT&T	OLD DIAL STYLE
BF Y	AT&T	TRIMLINE
BM N	BELL SOUTH	665 CORDLESS PHONE
HU N	BELL SOUTH	HAE-701
OR N	BELL SYSTEM	TRIMLINE
AL N	COBRA	AN-8519
DT N	COBRA	ST-410
SD N	CODE A PHONE	8100
AL N	CONAIR CORP	XS2400
LV N	CONAIR	CORPORATE AMERICA
BS N	DAK INDUSTRIES	DAL-4000 AX
NF N	DAK INDUSTRIES	DK 4000 AX
CG N	GANDALF	COMDIAL
AL N	GENERAL ELECTRIC	2-9051LRB
KC N	GENERAL ELECTRIC	2-9800B
BM N	GTE	(UNKNOWN)
ST N	GTE	980 SLIMLINE
PS N	ITT	183499-236
FE N	ITT	500D
AN N	ITT	RJ11C

off	filter make	model
AL N	LENNOX SOUND	PH-319
HL N	UNKNOWN	
(CHINA)	FCC #: GEZ 364-16521 TE-E	
HL N	NORTHERN TELECOM	SYMPHONY 1000
DV N	NW BELL	FAVORITE PLUS DHT6CJ-14584-TE
BS N	PANASONIC	EASE PHONE-ONE LINE
HU N	PANASONIC	GASA-PHONE
DV N	PANASONIC	KX-T-3120
HL N	PANASONIC	KXT-2355
KC N	PHONEMATE	4300
ST N	RADIO SHACK	43-368A
FE N	RADIO SHACK	43-500
FE N	RADIO SHACK	43-5018
DV N	RADIO SHACK	43-621
OR N	SEARS	329 34451750
DT N	SEARS	34413
GI N	SEARS	HAC 3466
ST N	SEARS	SR 3000 SERIES
SF N	SHARP	UX-172
SF N	SOUND DESIGN	AM/FM CLOCK RADIO PHONE
AL N	SPECTRA PHONE	TL-4
DT N	STC TELECORP	SOUNDDESIGN 7255
ST N	TELECONCEPTS	(NONE)
NF N	WESTERNELECTRIC	(UNKNOWN)
BM N	WESTERN ELECTRIC	(UNKNOWN)
DL N	WESTERN ELECTRIC	(UNKNOWN)
KC Y	WESTERN ELECTRIC	(UNKNOWN)
LA N	WESTERN ELECTRIC	"LA OLYMPIC PHONE"
PA N	WESTERN ELECTRIC	2554 BMPG-82103
BS N	WESTERN ELECTRIC	RB 1311 (ROTARY)
NF Y	WESTERN ELECTRIC	TRIMLINE #CS 2224A
NF N	WESTERN ELECTRIC	TRIMLINE PHONE

*This list does not include 11 telephones which received interference in one case, but did not receive it in another case.

Attachment 4

Field Operations Bureau
Telephone Interference Survey
Other Filters Tested

K-COM RF-1	1
K-COM RF1	3
RADIO SHACK SNAPON	1
TCE	1
TCE 536	2
TCE TP12	4
TEC RF1400	2
TII 931-W1	2

The filters listed here deserve further study. During the project the FCC found that these filters eliminated interference when tried in individual cases. However, the FCC did not test any of these filters in a sufficient number of cases, and it cannot conclude whether any of these filters would be effective in eliminating interference most of the time.

Attachment 5

Specially Designed Interference Free Telephones

Pro Distributors
2811 B 74th Street
Lubbock, Texas 79423
(800) 658-2027
Model Tested: Western Electric/ATT Desk Model with
   touch tone
Cost: $79.95

TCE Laboratories, Inc.
2365 Waterfront Park Dr.
Canyon Lake, Texas 78133
(210) 899-4575
Model Tested: TPXL-D Desk Model,
Cost: $59.95

75 YEARS AGO IN AMATEUR RADIO—1921

It must be recalled that in the early days of wireless communications, there was no such thing as broadcasting. Radio, amateur or otherwise, was not a recreational activity, except for the experimenters who enjoyed building equipment and a very small few who used the medium for what we now call ragchewing. Most amateur activity was involved in the relaying of message traffic, somewhat in competition with the wired telegraph services and commercial radio services. All of this changed dramatically in 1921.

Since the amateurs were, well, amateurs, their ability to relay messages using their volunteer resources and home-built low-power equipment was somewhat at a disadvantage relative to paid professionals who were using high-power stations and much bigger antennas. Nonetheless, the amateurs were honing their message-relaying skills.

After months of planning, a transcontinental message-relay test was held in the evenings of January 14 through 18. On the final night, January 17 (actually January 18 in the wee hours of the morning), a record was set. The third of the five two-way messages of the night was originated at 4:13:45 a.m. by ARRL President Hiram Percy Maxim, owner of station "1AW" in Hartford, Connecticut. ARRL Traffic Manager F. H. Schnell, operating 1AW, sent it to R. H. G. Mathews, "9ZN," in Chicago, who relayed it to Louis Falconi, "5ZA," of Roswell, New Mexico, who passed it to Vernice Bitz, "6JD," of Los Angeles, California. The message asked "WHAT TIME DID YOU START MESSAGE?" The reply, following the same route back, stated "STARTED UR MSG AT 1:10 a.m.," and was received at 4:20:15 a.m. EST. This six-minute, 30-second round trip was almost matched by the final message of the test, which made the round-trip in 7 3/4 minutes, again between 1AW and 6JD via 9ZN and 5ZA. The success was almost certainly due to the publicity and organization that surrounded the event. This resulted in the cooperation of most of the amateurs in the country, who stood by patiently so as to keep the frequencies clear of interference. Recall that in those days, broadband spark transmission was used, and receiver selectivity was extremely poor.

Nearly all transmissions were made using Morse code and broadband spark-gap transmitters in those days. There were a few experiments in sending voice and music, but these were irregular and infrequent up until 1920. On November 2, 1920, Dr. Frank Conrad, 8XK, of Pittsburgh, Pennsylvania, began to broadcast music on a regular basis. While music broadcasts were made before, these were different. They were not simply intended to be heard by other hams. Conrad's employer, the Westinghouse Corporation, actually manufactured a number of receivers and sold them to the general public. Soon, thousands of non-hams in the Pittsburgh area had bought receivers to hear 8XK's music transmissions. Westinghouse recognized the commercial potential of transmitting to thousands of people simultaneously, and quickly established stations in Chicago, Illinois (KYW, on 360 meters) and Springfield, Massachusetts (WBZ, on 375 meters), and Newark, New Jersey (WJZ, on 360 meters). Westinghouse sold radios to the general public in these cities as well. Station 8XK became commercial

station KDKA, which is still on the air today. This was the beginning of true broadcasting.

It was also the beginning of broadcast interference problems for radio amateurs.

By late 1921, there were thousands of "broadcasting" stations on the air. Some were established by department stores, newspapers, and radio manufacturers, with the intent of making money from their transmissions (such as using their stations to drive demand for radio receivers). Others were hams just transmitting music for fun. The problem was that they were all using pretty much the same frequencies as the ragchewing and traffic-handling hams, with disastrous results for people (especially the non-technical public with poor receivers) wanting to tune in a single station. As the equipment was relatively expensive, the owners tended to be the more affluent and influential citizens of a community. These people took a very dim view of interference to their entertainment by amateurs.

In October 1921, letters to the editor of *QST* were received from people on both sides of the broadcast interference debate. The first letter complained of the interference caused by the hams to the fine concerts being broadcast in the Cleveland area. The second, from Don Mix, 1TS, complained about the interference to message relaying caused by the concert broadcasts.

Amateurs in some areas took some steps to alleviate the problem. Many adopted the "Chicago Plan," which restricted amateurs from using high power during prime radio-listening hours, and only calling CQ during the late hours. It was in this plan that the phrase "amateurs should use the minimum amount of power necessary" first appeared, as a voluntary way to reduce interference to other hams and to broadcasts.

In March 1921 *QST*, it was suggested that there was altogether too much CQing going on, and a full-page tongue-in-cheek advertisement announced the ARRL's "CQ Party." This event was ostensibly to allow hams to send CQ for the last time, since the ARRL had "declared CQ obsolete." Three-minute periods were assigned to each call area to call continuous CQs on the night of April 1, culminating in all call areas blasting away for the final three minutes. However, the idea proved irresistible, and hundreds of stations were heard on that night, all calling CQ. Some stations were amazed at the DX they were able to hear.

By 1922, the Department of Commerce took steps to control the situation by requiring broadcasters to procure "commercial" licenses and forbidding amateurs from "broadcasting." Separate frequencies were assigned for the exclusive use of broadcasters. At the end of 1921, Commerce Secretary (and later U.S. President) Herbert Hoover, offered an award (the "Hoover Cup") to the "best home-made amateur radio station of 1921." Hoover was to prove himself a champion of the amateur community many times over the years, including a 1922 confrontation between amateurs and corporate-backed broadcasting interests at the First National Radio Conference. Without his patronage, amateur radio in the U.S. would likely not have survived the 1920s.

The Hoover Cup winner for 1921 was Louis Falconi, 5ZA, who had been a participant in the record-setting January transcontinental relay test. His station consisted of:

1. A 200-watt CW/phone transmitter, using four 50-watt tubes;

2. A 1-kW rotary spark transmitter;
3. A variometer regenerative receiver with two-stage audio amplifier;
4. An antenna system of a four-wire flat-top, 90 feet long and 67 feet high, with a fan-style downlead and an extensive ground and counterpoise system.

During 1921, the primary mode used by amateurs switched from spark to CW, and this transition proceeded with blinding speed. Many hams returning from the European battlefront (where they had served as soldiers in World War I) had seen some of the early "true CW" transmitters in action. They soon figured out that a 5-watt CW transmitter could cover the same distance as a 500-watt spark rig. This was due to the bandwidth of the signal. An amateur spark station operating on 200 meters (1.500 MHz) actually spewed out a broadband mess covering from 150 to 250 meters (2.000 MHz to 1.25 MHz), while a CW signal concentrated all the transmitter's power into a tiny fraction of that bandwidth. Most hams couldn't operate CW, since one key ingredient of a CW transmitter was a vacuum tube (such as a DeForest audion), and most hams simply couldn't afford them. However, hams are well-known for finding ways to procure the components they need, and soon a few amateur CW signals were turning up on the bands. Hams then discovered that tuning a receiver to a CW signal was also beyond the capabilities of most of the crude receivers of the day, and new receiver techniques were developed to deal with these narrow, drifting signals.

An editorial in QST in early 1921 declared that while CW seemed to be technically superior, it was spark that was the most reliable for message relaying. In fact, a contest was held in early 1921 to select the best spark station, and it was won by the ARRL's Central Division Manager, R.H.G. Mathews, 9ZN.

In 1921, RCA made a "high-power" tube (the UV-202, good for about 5 watts of CW) available to hams for $8. Later in the year, the UV-203, good for 50 watts or so, was offered for $30. The migration to CW was aided by numerous transmitter construction articles in QST, as well as by numerous commercial receivers of better quality being manufactured in high volumes (mostly to serve the needs of the growing broadcast reception market). This was also reinforced because the interference produced by a 10-watt CW transmitter was quite a bit less than the interference generated by a 1000-watt broadband spark rig.

By the end of the year, the ARRL was actively encouraging the use of CW. This was inevitable, given the lower probability of interfering with broadcasters and the success of the new mode in long-distance communications.

Prior to 1921, there had been no confirmed two-way amateur contact across the Atlantic. In 1920, two amateurs in Aberdeen, Scotland, reported that they had copied radiophone signals from station 2QR in New Jersey. After many months, it was proven that the claimed reception had not occurred. Commercial traffic stations such as Telefunken's WSL in Sayville, New York, routinely passed traffic to POZ in Nauen, Germany, before World War I (on their 15,000-meter wavelength!), but amateurs, with their 1,000-watt limit, and 200-meter wavelength allocation had never successfully worked across the ocean. In 1919, M.B. Sleeper, editor of Everyday Engineering, had devised a plan to

organize transatlantic experiments. Unfortunately, the magazine ceased publication before the plan could be executed. Sleeper gave the idea to the ARRL in 1920, and plans were laid for the first amateur transatlantic tests to be held in 1921.

The transatlantic test conducted in February 1921 met with failure. The British receiving stations reported that they could hear only a few weak CW signals from the American side, and that none was strong enough to copy. Two conclusions were drawn from this test. First, only CW signals were heard. This was significant. Second, the quality of the receiving equipment on the British side was judged inferior for rejecting the stronger nearby broadcast signals. It was further complicated by the fact that many of the regenerative receivers were radiating enough signals to drown out the weak transatlantic signals!

In December 1921, the ARRL sent Paul Godley, 2ZE, one of the best receiver operators in the U.S., to a site at Ardrossan, Scotland, with state-of-the-art regenerative and the just-invented superheterodyne receivers to try to copy American amateur signals. Godley set up his receiving station in a tent, and installed a 1300-foot Beverage-type receiving antenna, along with several smaller antennas. He would stay up all night for the 10 days of the tests listening for American signals, and each morning would arrange for commercial wireless transmission of the night's results back to ARRL. The weather was cold, windy, and rainy for the whole period, and Godley was miserable. His account of the tests includes admissions that he arrived at the site late several times, and towards the end of the test found it hard to leave the comfort of the hotel and its warm fireplace to once again sit on a box in a drafty tent all night.

The ARRL held a series of qualifying tests to determine which stations had the best chance of getting a signal across the Atlantic. A final list of 27 stations was qualified. These stations were each assigned a specific time period each night to transmit to Godley and each was issued a five-letter code group, which would serve to confirm successful reception. An earlier time period each night was allocated for "free-for-all" operation, with each call district assigned a time slot in which any station was permitted to call Godley and hope for the best.

The first night of the tests, December 7–8, Godley copied only one station—a spark station on 270 meters signing the call 1AAW. No other stations were heard from the U.S. When Godley's report reached the ARRL, a team was dispatched to congratulate 1AAW on his achievement. However, it turned out that the original license-holder had moved to New Jersey, and held a "2" call. The 1AAW callsign had recently been re-issued. The new holder of the call was found, and discovered to have a very low-power station, quite incapable of crossing the Atlantic. Later investigation revealed that a station with a big signal had been heard numerous times signing the 1AAW callsign, apparently without the benefit of actually having a license. Thus it appears that the first U.S. amateur station successfully copied in Europe was illegal! The perpetrator was never identified.

The second night of the test was a failure. The third night, however, was a success, and with a known legitimate station. During the "free-for-all" period on December 9, Godley copied strong signals from station 1BCG in Greenwich, Connecticut. He copied 1BCG for over an hour, a remarkable accomplishment for both sending and receiving stations. The 1BCG station was owned by the Radio Club of America, and had been assembled primarily for this test. The station builders included Major Edwin H. Armstrong, who would go on to invent numerous techniques in radio transmission and reception. The station ran a full kilowatt on CW, with four 250-watt tubes in parallel, and an antenna system considered state-of-the-art. It was to become the standard by which other stations were measured.

Godley was so thrilled by the success that he decided it would be possible to copy entire messages from 1BCG. He sent a message directly to Armstrong that was supposed to read "signals wonderful send messages starting one Greenwich." Unfortunately, one of the commercial operators garbled the text, and Armstrong received a message that read "signals wonderful send mges starting one Greenwich." The next night, Armstrong and the crew at 1BCG dutifully began sending "MGES" at 0100 GMT, and continued to send the same thing over and over again all night long.

Conditions seemed good that night, and Godley heard about a dozen stations, using 1BCG as a beacon to help adjust the antenna and receiver. When the test was over, on December 17, Godley had copied signals from 26 stations—all but six of them on CW! Eight British amateurs had also logged American signals. A new era of ham radio had begun.

CW records were being broken weekly, as more stations became active and were amazed at the results they could obtain. On November 23, station 1ES in Brookline, Massachusetts, copied signals from the 50-watt transmitter of 6ALE in Reedley, California—the first time a transcontinental signal had been copied on CW. The big signal from 1BCG was heard in Scotland, England, France, Holland, Puerto Rico, and all over the U.S., including California.

The ARRL was the main amateur organization of the day. The stated goals of the ARRL were to provide an effective network for the purpose of relaying messages, for legislative protection, for orderly operating, and for practical improvement of shortwave radio communication. The ARRL had existed before World War I, but most of its members were called into service as military radio operators. The government had also shut down amateur radio for the duration of the War. When the War ended, a group of amateurs pooled their resources and started up the ARRL again. They published a mini-issue of QST in May 1919, and announced that they would sell ARRL bonds to raise the $7,500 needed to buy QST from its original owner, C.D. Tuska. In May 1921, the ARRL was on solid financial footing and announced they were calling the bonds and repaying the loans.

Other radio organizations sprang up. A competing CW network was being organized by the publishers of the magazine Radio (which had previously been known as Pacific Radio News). The western branch of the network was actually established, but the eastern branch never came together. The Radio Club of America was an organization devoted to the technical side of radio. And, as interest in radio spread beyond the amateur operators, the Radio Club of America likewise broadened its scope beyond the hams, focusing more on technical issues of interest to broadcasters. This left the ARRL as the primary organization serving the amateur radio enthusiast interested in both receiving and transmitting. An identity

crisis developed. Several leading amateurs coined the term "citizen wireless" or "citizen radio." Indeed, the QST covers of 1921 reflected this unstable period in the definition of the hobby, by claiming in various months to be a magazine devoted to "the wireless amateur," "citizen radio," "amateur wireless," "citizen wireless," and "the radio amateur." Circulation grew, since the public curiosity about radio was exploding.

The ARRL held its first National Convention in Chicago on August 30 through September 3. Over 1,200 amateurs attended, representing 80 affiliated clubs. Secretary of Commerce Hoover sent the head of the Radio Division" to learn where the Department can be of service." The convention included seminars on technical topics, operating procedures, and informal meetings lasting well into the night to discuss which stations were the best and to brag about working DX.

QST was up to well over 100 pages each month. Cover price was 20 cents, and membership in the ARRL was priced at $2.00. Articles covered topics including equipment construction, radio theory, traffic reports, station descriptions, new commercial equipment, and legislation. Advertisements ran the range from components for both CW and spark equipment to completely-assembled equipment to courses on radio. Most of the manufacturers' names are long since gone, such as Benwood (maker of spark gaps), Tuska, Paragon, Grebe, Murdock, and Moorhead. A few survive to this day—General Radio (now Genrad) and Magnavox, for example.

50 YEARS AGO IN AMATEUR RADIO—1946

In 1946, amateur radio was coming back to life after being shut down for the duration of World War II. In the U.S., the FCC had released the bands above 28 MHz for use, and additional bands were released throughout the year.

Amateur radio had been prohibited for the duration of the War. In contrast to the restrictions of World War I, which required hams to completely dismantle all transmitting and receiving equipment and antennas, some hams were allowed to operate in the War Emergency Radio Service (WERS). The WERS was organized in many cities in the 112–116 MHz (2 1/2 meter) ham band, for the purpose of providing an emergency communications system in the event of an attack on the mainland U.S. Prior to World War II, the VHF bands for hams were at 5 meters (56–60 MHz), 2 1/2 meters, and 1 1/4 meters (224–230 MHz). Actually, the unit of frequency of the day was not MHz, but "Mc," an abbreviation for Megacycles per second. When the end of the war seemed to be approaching, commercial interests were lobbying with the FCC and the Radio Technical Planning Board (RTPB), an industry group of which ARRL was a member, to set aside frequencies in the VHF spectrum for services like television broadcasting.

Meetings between the FCC and RTPB took place in late 1944, with proposals, counterproposals, and compromises. The RTPB was principally concerned with allocations in the VHF and UHF spectrum. When the FCC finally set down a plan for those bands, the RTPB decided that its work was over, and the FCC was left to decide what to do with the spectrum, and when to turn amateur radio back on. Most hams were anxious to get back on the HF bands.

The biggest disappointment to amateurs in the new VHF/UHF allocations was that the 56–60 Mc band was reduced to 58–60 Mc, the beloved 2 1/2-meter band from 112 to 116 Mc was moved to the 144–148 Mc band and 1 1/4 meters was moved from the 224–230 Mc band to 218–225 Mc

On August 21, 1945, the FCC issued Order No. 127, which dissolved the WERS and reinstated all amateur licenses which had been suspended since 1941, with permission to operate in the 112–115.5 MHz band. This was the first step in reopening the amateur service. This action took place just days after the announcement of Japanese acceptance of the surrender terms. It seems that the day after the announcement, the ARRL contacted both the FCC and the Bureau of War Communications to consider opening amateur radio again. Neither had any objection, and the hams were back on the air.

On November 9, 1945, the FCC issued Order No. 130, which reopened several more amateur bands, and officially canceled the 1941 and 1942 orders that prohibited all amateur operations. The bands which were opened, effective November 15, 1945, were,

- 28.0–29.7 Mc, using A1 emission
- 28.1–29.5 Mc, using A3 emission
- 28.95–29.7 Mc, for "special emission for frequency modulation (telephony)"
- 56.0–60.0 Mc, for A1, A2, A3, and A4 emissions
- 58.5–60.0 Mc, for "special emission for frequency modulation (telephony)"
- 144–148 Mc, for A1, A2, A3, and A4 emissions (except for 146.5–148 Mc, which would not be used by stations within 50 miles of Washington, DC and Seattle, Washington)
- 2300–2450 Mc, 5250–5650 Mc, 10,000–10,500 Mc, and 21,000–22,000 Mc, using A1, A2, A3, A4, and A5 emissions, and "special emission for frequency modulation (telephony and telegraphy)"

Discussions continued throughout 1945 on the final allocations. The area of most debate was the low-VHF region, with the hams clinging to the 56–60 Mc band, and commercial broadcasters eyeing the same region for a television channel. The FCC finally placed the amateur band at 50–54 Mc, effective March 1, 1946. One of the major issues of the spectrum allocations debates was where to put the new FM broadcasting service. The FCC wanted to put the FM band somewhere that was immune to the vagaries of potentially signal-quality-degrading propagation anomalies like Sporadic-E. The final decision was to put non-commercial FM at 88–92 Mc, and commercial FM between 92 and 106 Mc

Thus it was that the final amateur VHF allocations assigned by the FCC, and to be proposed for International approval in 1946, were,

- 28–29.7 Mc
- 50–54 Mc
- 144–148 Mc
- 220–225 Mc
- 420–450 Mc
- 1145–1245 Mc (later changed to 1215–1295 Mc)
- 2300–2450 Mc
- 5250–5650 Mc (later changed to two bands: 3300–3500 and 5650–5800)
- 10,000–10,500 Mc
- 21,000–22000 Mc

Over time, the remainder of the pre-War amateur bands were turned back to the hams, as other users related to the War activities vacated them. In mid-March, the 11-meter band (27.185–27.455 Mc) was opened for amateur use. Actually, this band was one of the first bands allocated for "operation of industrial, scientific, and medical apparatus," (along the lines of the 902–928 MHz band of today) and the FCC allowed hams to use it. The 235–240 Mc band was opened, with the understanding that it would be relocated to the 220–225 Mc range by 1949.

The 3625–4000 kc segment of the 80-meter band was opened on April 1, 1946, and the remaining 3500–3625 segment was opened on May 9. On July 1, half of the 40- and 20-meter bands were reopened (7150–7300 and 14150–14300). The remainder of the 40- and 20-meter bands (including 14300–14400) were opened in November. Thus, by the end of 1946, all the HF bands had been returned to amateur use. The 160-meter band was held back, however, due to its use for the new LORAN navigational service.

The first post-war amateur license was issued to Colonel Carl Hatch, W4IIT, on February 8, 1946. This was interesting since Virginia had previously been in the 3rd call area, and Hatch was the first "4" in Virginia. The call areas outlined in 1945 and 1946 are essentially those we know today, with only a few minor exceptions in some of the U.S. Possessions. The changes made are described in the table on the following page.

The first post-War contest was held February 22 through 25 and March 1 through 4, and was the "ARRL Band-Warming Party." It was held mostly on 10 meters, the only HF band available at the time. The top scorers were XE1A, with 829 QSOs and 112k points, followed by K6CGK (later KH6IJ) with 585 QSOs and 88k points. It was noted that XE1A (previously XE2N) set an all-time QSO rate record in this contest, with an average of 22.86 QSOs per hour for his 36 hours and 10 minutes of operation.

There Was also a Field Day in June 1946!

VHF operators made considerable progress in 1946. The first known successful moonbounce experiment was conducted on January 10, 1946, by a group at the Signal Corps Engineering Laboratories at Bradley Beach, New Jersey. A team headed by Lt. Col. John H. DeWitt, W4ERI, sent a 1/4-second pulse of 111.5 MHz energy at the moon and heard the echo. The transmitter ran 4 kW output to an array of 64 phased dipoles. W1NVL/2 and W9SAD/2, both employees of General Electric, successfully completed a two-way phone contact over an 800-foot path on 21,900 MHz—the highest frequency amateur band. Meteor-scatter QSOs were completed for the first time on (6 meters) during the Giacobini-Zimmer shower of October 9. On November 24, W1HDQ (on 6 meters) worked G6DH and G5BY (the Gs operating on 10 meters) for the first-ever transatlantic VHF QSO. Ironically, this was within a month of the 25th anniversary of the first successful reception of transatlantic amateur signals (on 200 meters, not 6!).

A 2-meter AM "Handie-Talkie" construction article by W2MYE appeared in January 1946 *CQ*. It was appar-

Area	States	Changes from pre-war status
1	New England	No Change
2	New York New Jersey	Transferred the W8 part of New York and the W3 part of New Jersey to W2
3	Delaware District of Columbia Maryland Pennsylvania	Transferred W8 part of Pennsylvania to W3, moved W3 part of NJ to W2, moved Virginia to W4
4	Alabama, Florida, Georgia, Kentucky, North & South Carolina, Tennessee, Virginia, U.S. Possessions in the Caribbean	Added Kentucky from W9 and Virginia from W3
5	Arkansas, Louisiana, Mississippi, New Mexico, Oklahoma, Texas	No changes
6	California, U.S. Possessions in the Pacific	Subtracted Arizona, Nevada and Utah and moved them to W7
7	Arizona, Idaho, Montana, Nevada, Oregon, Utah, Washington, Wyoming, Alaska	Added Arizona, Nevada, and Utah from W6
8	Michigan, Ohio, West Virginia	Added W9 part of Michigan, transferred W8 part of New York to W2, and W8 part of Pennsylvania to W3
9	Illinois, Indiana, Wisconsin	Transferred W9 part of Upper Michigan to W8, W9 part of Kentucky to W4. Moved Colorado, Iowa, Kansas, Minnesota, Missouri, Nebraska, North and South Dakota to form new Ø call area
Ø	New area with states removed from W9 area	

ently later marketed by Radio Transceiver Labs of Richmond Hill, New York, for $31.50, less tubes (1S4 and 6C4) and batteries (a 1.5V D-cell and three penlite cells for the filaments and bias, and a 67.5V "B" battery for the plate voltage). It was 10 3/4" by 2 3/4" by 2 3/4", and weighed four pounds with batteries.

New tubes were being introduced monthly, both for receiving and transmitting applications. Sylvania introduced (and several articles featured) the 1N34 and 1N35 germanium diodes, probably the first solid-state devices to appear in ham magazines.

QST was the primary amateur radio publication in the U.S. in 1945, having continued publication through the duration of the War. The fledgling *CQ* magazine, started in January 1945 by Radio Magazines, Inc., survived its first year, even though there was really no amateur radio

for most of the year. *CQ* filled the need for another amateur radio magazine, since the pre-War *Radio* had taken a turn towards the technical side of electronics and was no longer primarily focused on amateur radio.

CQ continued to grow through 1946, with regular columns such as a YL column and Herb Becker, W6QD's, "CQ DX" column beginning in the March issue. Becker had been the writer of the DX column (and "X-DX" when most countries including the U.S. suspended amateur radio during the war) for *Radio*, and his last column had been in March 1941. *CQ* also began a regular Propagation column by Oliver Perry Ferrell. Numerous articles appeared in both magazines on converting military surplus equipment for amateur use.

New amateur equipment of the day included the Hammarlund HQ-129X, RME 45, and National NC-46,

Hallicrafters SX-42, and Collins 75A receivers and the Collins 32V-1 transmitter. One interesting piece of equipment was named the "California Kilowatt"—a desk-mounted kilowatt transmitter, with clock, illuminated world map, speaker, and provision for installation of your choice of receiver. The product was manufactured by Kluge Electronics of Los Angeles. Some of the manufacturers, such as ElectroVoice, Astatic, Eimac, Vibroplex, and Radio Shack are still doing business, as are dealers such as Harrison Radio and Henry Radio. Others, like Echophone, Harvey, Heintz & Kaufman, RME, and Meissner are long since gone.

25 YEARS AGO IN AMATEUR RADIO—1971

The biggest change occurring in amateur radio in 1971 was unquestionably the explosive growth in 2-meter FM repeater operation. Most operators were using converted GE and Motorola commercial equipment, but the first dedicated amateur equipment was beginning to appear. The U.S. manufacturers of most 2-meter equipment (Clegg and Gonset) seemed to miss this trend and new manufacturers like Varitronics appeared. Varitronics, incidentally, was the U.S. marketing arm of Inoue Communications (now better known as Icom), and their FDFM-2 and IC-2F quickly gained popularity. One U.S. manufactured product was in the game early on—the Galaxy FM-210 was as popular as the Japanese rigs. Later in the year, Drake introduced the $329.95 "Marker Luxury" 2-meter FM rig, "manufactured for Drake in Japan." Other rigs from various manufacturers followed throughout the year.

The FCC was watching the growth of FM repeaters and issued a Notice of Proposed Rule making in 1970 to add regulations specific to FM repeaters. The proposed rules were widely criticized by the amateur community as being far too restrictive. For example, 2-meter repeaters would be restricted to outputs between 146.90 and 147.20 only. The numerous comments resulted in an extension to the comment deadline, and, even by the end of 1971, no action had been taken.

The FCC was continuing to review and revise other regulations relating to amateur radio. Incentive Licensing had taken effect in 1969, and many General class licensees were upset about the further loss of band space. Prior to 1968, General class licensees had the same frequency privileges as Advanced and Extra class licensees. The Incentive licensing program created exclusive subbands for the higher class license holders, implemented in two phases over a two-year period. The program was finalized in November 1969, and, with the exception of the changes to the Technician class license in the past few years, there have been few changes to the U.S. licensing structure.

In 1971, in response to 15 separate petitions, the FCC proposed to expand the phone subbands, reduce the Extra class CW segments, and create a 10-meter CW segment for Novices. The ARRL and numerous other commenters suggested modified subbands. By the end of 1971, the final rules had not yet been released. There were also numerous proposals for changes involving Technician class privileges. At the time, Techs did not have any HF privileges, and it was not permitted to simultaneously hold both Technician and Novice licenses. Furthermore, Techs did not have full 2-meter band privileges. This was all to change in the next few years.

There were proposals by various groups to re-allocate the 220-MHz band and even the 2-meter band to a CB-like service. The editorial in January 1971 *73* magazine accused the publishers of *CQ* magazine (who also published a CB magazine called *S9*) of being the driving force behind the 2-meter proposal. This was not the case, and a war of words ensued for several months between *73* and *CQ*

The FCC relaxed the restrictions on 160-meter operation, and for the first time all states (except Hawaii) were permitted operation in the 1800–1850 segment. Previously, some states were restricted to operation at the high end of the band only. Night-time power limits were lower to protect the LORAN system from interference.

The U.S. Senate had passed the "Goldwater Bill" and sent it to the House for action in late 1970; however, it did not make it through the lame duck Congress. This bill would allow immigrants to hold amateur licenses while awaiting U.S. citizenship. This was needed, since the previous rules allowed only U.S. citizens to hold U.S. licenses, and citizens of foreign countries could hold reciprocal licenses. Aspiring citizens were not permitted to hold U.S. licenses. Goldwater re-introduced the bill in 1971, and it was finally signed into law in August. Ironically, one of the first amateurs to propose the new rule, George Pataki, ex-YO2RO, completed the citizenship process while waiting for the bill to pass, and became WB2AQC. Mr. Pataki is the namesake and a relative of the present Governor of the State of New York.

The ARRL and several other groups filed Petitions for Reconsideration with the FCC to set aside the 1970 license fee increases (Five-year license fees were increased from $4 to $9, and special calls were $25). The FCC refused, and the new fees stood.

Everett Henry, W3BG, retired from his position as Chief of the Amateur and Citizens Radio Division of the FCC and was replaced by A. Prose Walker, W4BW. A letter from Walker to K4IIF, DX Editor of *CQ*, suggested that in the future, hams probably would "need more bands, say one at 10, 18, and 24 MHz." He hinted that an effort should be mounted at the proposed 1977 World Administrative Radio Conference (it was actually held in 1979) to secure these bands for hams. Bill Grenfell, W4GF, retired from his post as Chief, Rules and Legal Branch of the FCC, after more than 25 years of service. Ray Spence, W4QAW, and Merle Glunt, W3OKN, became FCC Chief and Assistant Chief Engineer, respectively.

In 1970, the FCC had taken an interest in the use of ham radio for the Lions-sponsored Eye Bank. Amateurs had formed a net to enable communications across the country to track and match eye donors to corneal transplant patients. The FCC felt that such communications, even though not conducted for money, were for the benefit of a non-amateur organization, and thus probably not within the intent of the rules for the amateur service. While no violation notices were ever issued for such operations, rumors spread through the amateur community that the FCC was going to crack down on ham operations for charitable organizations. In an address to the Washington, D.C. QCWA chapter in March 1971, FCC Chairman Dean Burch put the rumors to rest and encouraged the hams to propose suitably-worded rules through the petition process. The ARRL filed a lengthy proposal later in the year.

The ITU held a World Administrative Radio Conference (WARC) on Space Telecommunications in Geneva. There was considerable opposition–the continued use of certain bands by amateur satellites, since some frequencies were allocated on a shared basis in some countries. After lengthy negotiations, the "Amateur Satellite Service" was defined for the first time. The 40-, 20-, 15-, and 10-meter HF amateur bands were allocated–that service on a worldwide exclusive basis, as well as 144.0–146.0 MHz and 24.0–24.05 GHz. The 435–438-MHz segment was allocated on a shared basis.

An earthquake measuring 6.5 on the Richter scale struck the Los Angeles area on February 9, 1971, killing 62 persons and injuring hundreds. Amateur radio leaped into action, handling thousands of health and welfare requests and reports as well as helping with emergency communications in the stricken area.

A tower case was won in Ohio in 1971. W8ZCV, who wanted to install a 64-foot tower, was finally granted permission after an initial refusal by the town. This case served as precedent for many later zoning cases.

In 1971, there were four ham magazines in the U.S. QST and CQ were the biggest, with 73 and the more technical Ham Radio also doing well. The ARRL had over 75,000 members, and the recently-introduced life membership program grew to 1,000 life members. CQ began an FM column in January 1971, and announced the revised CQ DX Awards Program. Ken Sessions, K6MVH, left his position as Managing Editor of 73 magazine, and Wayne Green resumed the position, despite rumors earlier in the year that he was considering selling the magazine and retiring.

There were numerous operating achievements in 1971, especially on the VHF and UHF bands. A 432-MHz overland record was set on August 17 by K1PXE in Connecticut and WØDRL in Kansas—a 1205-mile path. The first two-way amateur laser QSO took place on February 25, 1971. WA8WEJ/Ø and W4UDS/Ø exchanged reports over a 950-foot path (including a mirror-assisted turn around a corner in a classroom building at the U.S. Air Force Academy) using 1 milliwatt of AM on a frequency of 475 teraHz (475 Mega-Mega Hertz) generated by a helium-neon laser. This claim was contested in a letter in QST from WB2MIC, who professed to have established a CW laser contact in 1969 with the St. Peter's College (Jersey City, New Jersey) club station W2GTF over a 1000-foot path.

At the other end of the amateur spectrum, KL7HEE in Alaska contacted VP8ME in Antarctica on July 16 for a new 160-meter distance record.

FG7XT Became the First Ham to Make DXCC on RTTY

The ARRL's 5BDXCC award had been claimed by 58 hams in its first two years, and the year-old 5BWAS had been achieved by 32. An elite group of four had completed both by early 1971: W1AX, W2PV, W4IC, and W8BT (now W8AH).

The ARRL added a DX Advisory Committee to the existing Contest and VHF Repeater Advisory Committees. Members of the first DXAC were Interim Chairman W4QCW (now W4DR), W1RAN, WA2FQG (now W2FG), W6RGG, W7LFA, W8BF, W9NN, WØELA, and VE3ACD.

There was also a lot of interesting DX activity in 1971. TI9 (Cocos Island) was activated by a group including TI2CF and K7CBZ (later K7ZZ/CT4AT, and now deceased). W9IGW and K9KNW operated from CE0Z. In March and again in September, ET3ZU/a activated Jabal at Tair and Abu Ail, which was added to the DXCC list later in the year. Intentional interference and other poor operating practices by U.S. hams during the July 3C1EG/3CØAN operation drew the attention of the FCC. Nonetheless, the 3CØ operation became the first from Annobon Island, which was then added to the DXCC list. OHØMA and OH2BHU/OHØ activated Market Reef, which had been added to the DXCC list in 1970. Perhaps the most unusual DXpedition was the KD2UMP operation of April 1, 1971, from Squaw Island, home of the garbage incinerator and sewage treatment plant for Buffalo, New York.

K2AGZ of 73 magazine wrote to the United Nations General Secretary, U Thant of Burma, suggesting that a ham station might be set up at the U.N. in New York. Mann suggested that it might also count for separate DXCC country status. The U.N. chose not to pursue setting up such a station, citing security concerns. Years later, of course, 4U1UN was established and the U.N was added to the DXCC list.

DX List operation on 14.220 (the International DX Association net) drew the same comments from hardcore DXers as it does today.

Equipment representing the state-of-the-art in 1971 included the Kenwood R-599 receiver ($298) and matching T-599 transmitter ($345), the Yaesu FT101 ($499.95), and the Galaxy GT-550. Drake introduced the "4B line": T-4XB transmitter for $495, and R-4B receiver for $475; and the TR-22 2-meter FM transceiver ($199.95). The National Radio NCX-1000 5-band kilowatt transceiver and Hallicrafters SR2000 "Hurricane" (2000-watt transceiver) marked a short-lived trend towards full-power desktop transceivers. Heathkit offered the SB-102 and HW-101 transceivers and the SB-303 receiver as improved versions of the successful SB-101, HW-100, and SB-301 models. At the high end, ETO added the air-cooled "Delta Seventy" to complement its vapor-phase-cooled Alpha 70V amplifier model. The ultimate transceiver remained the $2,195 Signal/One CX7A, whose ads proclaimed, "In a time when transceivers are a dime a dozen, this one costs an arm and a leg."

Mosley Electronics, makers of the TA-33 and "Classic" series of triband beams, celebrated its 25th anniversary. Kirk Electronics introduced a line of "Helicoidal" beams, whose elements consisted of metal tape wound helically on fiberglass cores attached to a fiberglass boom. Top of the line was a lightweight 3-element 40-meter beam. Hy-Gain introduced the Roto-Brake 400 antenna rotator.

The growing influx of low-priced equipment from Japan, marketed aggressively by importer/dealers, coupled with the shrinking number of amateur radio dealers, led Swan to begin selling its products direct from the factory to the amateur in 1971. This allowed Swan to keep prices competitive with the newer equipment. The company also allowed amateurs to purchase its equipment from dealers at a 10% markup, allowing hams to trade in their old equipment at the dealers as partial payment for new equipment.

COUNTRIES THAT SHARE THIRD-PARTY TRAFFIC AGREEMENTS WITH THE U.S.

V2	Antigua/Barbuda
LU	Argentina
VK	Australia
V3	Belize
CP	Bolivia
T9	Bosnia-Herzegovina
PY	Brazil
VE	Canada
CE	Chile
HK	Colombia
D6	Comoros
TI	Costa Rica
CO	Cuba
HI	Dominican Republic
J7	Dominica
HC	Ecuador
YS	El Salvador
V6	Federated States of Micronesia
C5	Gambia
9G	Ghana
J3	Grenada
TG	Guatemala
8R	Guyana
HH	Haiti
HR	Honduras
4X	Israel
6Y	Jamaica
JY	Jordan
EL	Liberia
V7	Marshall Islands
XE	Mexico
YN	Nicaragua
HP	Panama
ZP	Paraguay
OA	Peru
DU	Philippines
VR6	Pitcairn Island *
V4	St. Christopher/Nevis
J6	St. Lucia
J8	St. Vincent
9L	Sierra Leone
3DA	Swaziland
9Y	Trinidad/Tobago
GB	United Kingdom *
CX	Uruguay
YV	Venezuela
4U1ITU - ITU	Geneva
4U1VIC - VIC	Vienna

* Limited to special-event stations with callsign prefix GB (GB3 excluded) and informally to stations on Pitcairn Island (VR6).

COUNTRIES THAT SHARE RECIPROCAL OPERATING/LICENSING AGREEMENTS WITH THE U.S.

V2	Antigua/Barbuda
LU	Argentina
VK	Australia
OE	Austria
C6	Bahamas
8P	Barbados
ON	Belgium
V3	Belize
CP	Bolivia
A2	Botswana
PY	Brazil
VE	Canada
CE	Chile
HK	Colombia
TI	Costa Rica
5B	Cyprus
OZ	Denmark (incl. Greenland)
HI	Dominican Republic
J7	Dominica
HC	Ecuador
YS	El Salvador
V6	Federated States of Micronesia
3D	Fiji
OH	Finland
F	France *
DL	Germany
SV	Greece
J3	Grenada
TG	Guatemala
8R	Guyana
HH	Haiti
HR	Honduras
VS	Hong Kong
TF	Iceland
VU	India
YB	Indonesia
EI	Ireland
4X	Israel
I	Italy
6Y	Jamaica
JA	Japan
JY	Jordan
T3	Kiribati
9K	Kuwait
EL	Liberia
LX	Luxembourg
V7	Marshall Islands**
XE	Mexico
3A	Monaco
PA	Netherlands
PJ	Neth. Antilles
ZL	New Zealand
YN	Nicaragua
LA	Norway
HP	Panama
P2	Papua New Guinea
ZP	Paraguay
OA	Peru
DU	Philippines
CT	Portugal
J6	St. Lucia
J8	St. Vincent/Grenadines
S7	Seychelles
9L	Sierra Leone
H4	Solomon Island
ZS	South Africa
EA	Spain
PZ	Suriname
SM	Sweden
HB	Switzerland
HS	Thailand
9Y	Trinidad/Tobago
T2	Tuvalu
G	United Kingdom***
CX	Uruguay
YV	Venezuela
YU	Yugoslavia

Notes: An automatic reciporcal agreement exists between the U.S. and Canada, so there is no need to apply for a permit. Simply sign your U.S. call followed by a slantbar and the Canadian letter/number identifier.

* Including French Guiana, French Polynesia, Guadeloupe Amsterdam Island, Saint-Paul Island, Crozet Island, Kerguelen Island, Martinique, New Caledonia, Reunion, St. Pierre and Miquelon, and Wallis and Futuna Islands.
** The Marshall Islands are independent, but the FCC currently honors the previous reciprocal and third party agreements until formal agreements can be made.
*** Including Bermuda, British Virgin Islands, Cayman Islands, Channel Islands (including Guernsey and Jersey), Falkland Islands (including South Georgia Islands and South Sandwich Island), Gibraltar, Isle of Man, Montserrat, St. Helena, Gough Island, Tristan Da Cuhna Island, and the Turks and Caicos Islands.

AMATEUR STATION SEQUENTIAL CALLSIGN SYSTEM

Fact Sheet
Federal Communications Commission
1919 M Street, NW
Washington, DC 20554
PR-5000 #206
February 1995

Amateur Station Sequential Callsign System

A unique callsign is assigned to each amateur station during the processing of its license. The station is reassigned its same callsign upon renewal or modification of its license, unless the licensee applies for a change to a new callsign (FCC Form 610). Each new callsign is sequentially selected from the alphabetized regional-group list for the licensee's operator class and mailing address. The mailing address must be one where the licensee can receive mail delivery by the United States Postal Service.

Each callsign has a one letter prefix (K, N, or W) or a two letter prefix (AA-AL, KA-KZ, NA-NZ, WA-WZ) and a one, two, or three letter suffix separated by a numeral (1-10) indicating the geographic region. Certain combinations of letters are not used. When the callsigns in any regional-group list are exhausted, the selection is made from the next lower group. The groups are:

Group A. For Primary stations licensed to Amateur Extra Class operators.
Regions 1 through 10: prefix K, N or W, and two letter suffix; two letter prefix with first letter A, K, N or W, and one letter suffix; two letter prefix with first letter A, and two
letter suffix.
Region 11: prefix AL, KL, NL, or WL, and one letter suffix.
Region 12: prefix KP, NP, or WP, and one letter suffix.
Region 13: prefix AH, KH, NH, or WH, and one letter suffix.

Group B. For Primary stations licensed to Advanced Class operators.
Regions 1 through 10: two letter prefix with first letter K, N or W, and two letter suffix.
Region 11: prefix AL, and two letter suffix.
Region 12: prefix KP, and two letter suffix.
Region 13: prefix AH, and two letter suffix.

Group C. For Primary stations licensed to General, Technician, and Technician Plus Class operators.
Regions 1 through 10: prefix K, N, or W, and three letter suffix.
Region 11: prefix KL, NL, or WL, and two letter suffix.
Region 12: prefix NP or WP, and two letter suffix.
Region 13: prefix KH, NH, or WH, and two letter suffix.

Group D. For Primary stations licensed to Novice Class operators, and for club and military recreation stations.
Regions 1 through 10: two letter prefix with first letter K or W, and three letter suffix.
Region 11: prefix KL or WL, and three letter suffix.
Region 12: prefix KP or WP, and three letter suffix.
Region 13: prefix KH or WH, and three letter suffix.

The region and numerals are:

1. Connecticut, Maine, Massachusetts, New Hampshire, Rhode Island and Vermont. The numeral is 1.
2. New Jersey and New York. The numeral is 2.
3. Delaware, District of Columbia, Maryland and Pennsylvania. The numeral is 3.
4. Alabama, Florida, Georgia, Kentucky, North Carolina, South Carolina, Tennessee and Virginia. The numeral is 4.
5. Arkansas, Louisiana, Mississippi, New Mexico, Oklahoma and Texas. The numeral is 5.
6. California. The numeral is 6.
7. Arizona, Idaho, Montana, Nevada, Oregon, Utah, Washington and Wyoming. The numeral is 7.
8. Michigan, Ohio and West Virginia. The numeral is 8.
9. Illinois, Indiana and Wisconsin. The numeral is 9.
10. Colorado, Iowa, Kansas, Minnesota, Missouri, Nebraska, North Dakota and South Dakota. The numeral is Ø.
11. Alaska. The numeral is 1 through Ø. (KL9KAA-KL9KHZ is reserved for assignment to U.S. Personnel stationed in Korea.)
12. Caribbean Insular areas. The numeral 1 indicates Navassa Island; 2 indicates Virgin Islands; 3 or 4 indicates Commonwealth of Puerto Rico except Desecheo Island; and 5 indicates Desecheo Island.
13. Hawaii and Pacific Insular areas. The numeral 1 indicates Baker or Howland Island; 2 indicates Guam; 3 indicates Johnston Island; 4 indicates Midway Island; 5 indicates Palmyra or Jarvis Island; 5 followed by suffix letter K indicates Kingman Reef; 6 or 7 indicates Hawaii except Kure Island; 7 followed by the letter K indicates Kure Island; 8 indicates American Samoa; 9 indicates Wake, Wilkes or Peale Island; and Ø indicates the Commonwealth of Northern Mariana Islands.

Licensee information is retained in the licensee database for two years beyond expiration to provide a grace period during which persons who unintentionally fail to renew their licenses have additional time to do so. This Notice supersedes all previous Notices on this subject. FOR FURTHER INFORMATION, CONTACT THE FCC'S CONSUMER ASSISTANCE BRANCH, 1270 FAIRFIELD ROAD, GETTYSBURG, PA 17325-7245, (717) 337-1212.

QUESTION POOL ELEMENT 4A—ADVANCED CLASS
As released by Question Pool Committee, National Conference of
Volunteer Examiner Coordinators, December 1, 1994

For use in examinations beginning July 1, 1995. The answer to each question can be found in parenthesis following the question number. The Part 97 reference found within brackets in questions A1A01-A1F14 are the relevant FCC Rule citations.

A1–COMMISSION'S RULES [6 exam questions–6 groups]

A1A Advanced control operator frequency privileges; station identification; emissions standards

A1A01 (A) [97.301c]
What are the frequency limits for Advanced class operators in the 75/80-meter band (ITU Region 2)?
A. 3525–3750 kHz and 3775–4000 kHz
B. 3500–3525 kHz and 3800–4000 kHz
C. 3500–3525 kHz and 3800–3890 kHz
D. 3525–3775 kHz and 3800–4000 kHz

A1A02 (B) [97.301c]
What are the frequency limits for Advanced class operators in the 40- meter band (ITU Region 2)?
A. 7000–7300 kHz
B. 7025–7300 kHz
C. 7025–7350 kHz
D. 7000–7025 kHz

A1A03 (D) [97.301c]
What are the frequency limits for Advanced class operators in the 20- meter band?
A. 14000–14150 kHz and 14175–14350 kHz
B. 14025–14175 kHz and 14200–14350 kHz
C. 14000–14025 kHz and 14200–14350 kHz
D. 14025–14150 kHz and 14175–14350 kHz

A1A04 (C) [97.301c]
What are the frequency limits for Advanced class operators in the 15- meter band?
A. 21000–21200 kHz and 21250–21450 kHz
B. 21000–21200 kHz and 21300–21450 kHz
C. 21025–21200 kHz and 21225–21450 kHz
D. 21025–21250 kHz and 21270–21450 kHz

A1A05 (B) [97.119e3]
If you are a Technician Plus licensee with a Certificate of Successful Completion of Examination (CSCE) for Advanced privileges, how do you identify your station when transmitting on 14.185 MHz?
A. Give your call sign followed by the name of the VEC who coordinated the exam session where you obtained the CSCE
B. Give your call sign followed by the slant mark "/" followed by the identifier "AA"
C. You may not use your new frequency privileges until your license arrives from the FCC
D. Give your call sign followed by the word "Advanced"

A1A06 (B) [97.119a]
How must an Advanced class operator using Amateur Extra frequencies identify during a contest, assuming the contest control operator holds an Amateur Extra class license?
A. With his or her own call sign
B. With the control operator's call sign
C. With his or her own call sign followed by the identifier "AE"
D. With the control operator's call sign followed by his or her own call sign

A1A07 (D) [97.119d]
How must an Advanced class operator using Advanced frequencies identify from a Technician Plus class operator's station?
A. With either his or her own call sign followed by the identifier "KT", or the Technician Plus call sign followed by the identifier "AA"
B. With the Technician Plus call sign
C. The Advanced class operator cannot use Advanced frequencies while operating the Technician Plus station
D. With either his or her own call sign only, or the Technician Plus call sign followed by his or her own call sign

A1A08 (A) [97.307d]
What is the maximum mean power permitted for any spurious emission from a transmitter or external RF power amplifier transmitting on a frequency below 30 MHz?
A. 50 mW
B. 100 mW
C. 10 mW
D. 10 W

A1A09 (B) [97.307d]
How much below the mean power of the fundamental emission must any spurious emissions from a station transmitter or external RF power amplifier transmitting on a frequency below 30 MHz be attenuated?
A. At least 10 dB
B. At least 40 dB
C. At least 50 dB
D. At least 100 dB

A1A10 (C) [97.307e]
How much below the mean power of the fundamental emission must any spurious emissions from a transmitter or external RF power amplifier transmitting on a frequency between 30 and 225 MHz be attenuated?
A. At least 10 dB
B. At least 40 dB
C. At least 60 dB
D. At least 100 dB

A1A11 (D) [97.307e]
What is the maximum mean power permitted for any spurious emission from a transmitter having a mean power of 25 W or less on frequencies between 30 and 225 MHz?
A. 5 microwatts
B. 10 microwatts
C. 20 microwatts
D. 25 microwatts

A1B Definition and operation of remote control and automatic control; control link

A1B01 (D) [97.3a35]
What is meant by a remotely controlled station?
A. A station operated away from its regular home location
B. Control of a station from a point located other than at the station transmitter
C. A station operating under automatic control
D. A station controlled indirectly through a control link

A1B02 (D) [97.3a6]
What is the term for the control of a station that is transmitting without the control operator being present at the control point?
A. Simplex control
B. Manual control
C. Linear control
D. Automatic control

A1B03 (A) [97.201d,97.203d,97.205d]
Which kind of station operation may not be automatically controlled?
A. Control of a model craft
B. Beacon operation
C. Auxiliary operation
D. Repeater operation

A1B04 (B) [97.205d]
Which kind of station operation may be automatically controlled?
A. Stations without a control operator
B. Stations in repeater operation
C. Stations under remote control
D. Stations controlling model craft

A1B05 (A) [97.3a6]
What is meant by automatic control of a station?
A. The use of devices and procedures for control so that a control operator does not have to be present at a control point
B. A station operating with its output power controlled automatically
C. Remotely controlling a station such that a control operator does not have to be present at the control point at all times
D. The use of a control link between a control point and a locally controlled station

A1B06 (B) [97.3a6]
How do the control operator responsibilities of a station under automatic control differ from one under local control?
A. Under local control there is no control operator
B. Under automatic control a control operator is not required to be present at a control point
C. Under automatic control there is no control operator
D. Under local control a control operator is not required to be present at a control point

A1B07 (C) [97.205b, 97.301b,c,d]
What frequencies in the 10-meter band are available for repeater operation?
A. 28.0–28.7 MHz
B. 29.0–29.7 MHz
C. 29.5–29.7 MHz
D. 28.5–29.7 MHz

A1B08 (D) [97.205b, 97.301a]
What frequencies in the 6-meter band are available for repeater operation (ITU Region 2)?
A. 51.00–52.00 MHz
B. 50.25–52.00 MHz
C. 52.00–53.00 MHz
D. 51.00–54.00 MHz

A1B09 (A) [97.205b, 97.301a]
What frequencies in the 2-meter band are available for repeater operation (ITU Region 2)?
A. 144.5–145.5 and 146–148 MHz
B. 144.5–148 MHz
C. 144–145.5 and 146–148 MHz
D. 144–148 MHz

A1B10 (B) [97.205b, 97.301a]
What frequencies in the 1.25-meter band are available for repeater operation (ITU Region 2)?
A. 220.25–225.00 MHz
B. 222.15–225.00 MHz
C. 221.00–225.00 MHz
D. 223.00–225.00 MHz

A1B11 (A) [97.205b, 97.301a]
What frequencies in the 70-cm band are available for repeater operation (ITU Region 2)?
A. 420–431, 433–435 and 438–450 MHz
B. 420–440 and 445–450 MHz
C. 420–435 and 438–450 MHz
D. 420–431, 435–438 and 439–450 MHz

A1B12 (C) [97.301a]
What frequencies in the 23-cm band are available for repeater operation?
A. 1270–1300 MHz
B. 1270–1295 MHz
C. 1240–1300 MHz
D. Repeater operation is not permitted in the band

A1B13 (C) [97.213b]
If the control link of a station under remote control malfunctions, how long may the station continue to transmit?
A. 5 seconds
B. 10 minutes
C. 3 minutes
D. 5 minutes

A1B14 (C) [97.3a35, 97.3a36, 97.213a]
What is a control link?
A. A device that automatically controls an unattended station
B. An automatically operated link between two stations
C. The means of control between a control point and a remotely controlled station
D. A device that limits the time of a station's transmission

A1B15 (D) [97.3a35, 97.3a36, 97.213a]
What is the term for apparatus to effect remote control between a control point and a remotely controlled station?
A. A tone link
B. A wire control
C. A remote control
D. A control link

A1C Type acceptance of external RF power amplifiers and external RF power amplifier kits

A1C01 (D) [97.315a]
How many external RF amplifiers of a particular design capable of operation below 144 MHz may an unlicensed, non-amateur build or modify in one calendar year without obtaining a grant of FCC type acceptance?
A. 1
B. 5
C. 10
D. None

A1C02 (B) [97.315c]
If an RF amplifier manufacturer was granted FCC type acceptance for one of its amplifier models for amateur use, what would this allow the manufacturer to market?
A. All current models of their equipment
B. Only that particular amplifier model
C. Any future amplifier models
D. Both the current and any future amplifier models

A1C03 (A) [97.315b5]
Under what condition may an equipment dealer sell an external RF power amplifier capable of operation below 144 MHz if it has not been FCC type accepted?
A. If it was purchased in used condition from an amateur operator and is sold to another amateur operator for use at that operator's station
B. If it was assembled from a kit by the equipment dealer
C. If it was imported from a manufacturer in a country that does not require type acceptance of RF power amplifiers
D. If it was imported from a manufacturer in another country, and it was type accepted by that country's government

A1C04 (D) [97.317a1]
Which of the following is one of the standards that must be met by an external RF power amplifier if it is to qualify for a grant of FCC type acceptance?
A. It must produce full legal output when driven by not more than 5 watts of mean RF input power
B. It must be capable of external RF switching between its input and output networks
C. It must exhibit a gain of 0 dB or less over its full output range
D. It must satisfy the spurious emission standards when operated at its full output power

A1C05 (D) [97.317a2]
Which of the following is one of the standards that must be met by an external RF power amplifier if it is to qualify for a grant of FCC type acceptance?
A. It must produce full legal output when driven by not more than 5 watts of mean RF input power
B. It must be capable of external RF switching between its input and output networks
C. It must exhibit a gain of 0 dB or less over its full output range
D. It must satisfy the spurious emission standards when placed in the "standby" or "off" position, but is still connected to the transmitter

A1C06 (C) [97.317b]
Which of the following is one of the standards that must be met by an external RF power amplifier if it is to qualify for a grant of FCC type acceptance?

A. It must produce full legal output when driven by not more than 5 watts of mean RF input power
B. It must exhibit a gain of at least 20 dB for any input signal
C. It must not be capable of operation on any frequency between 24 MHz and 35 MHz
D. Any spurious emissions from the amplifier must be no more than 40 dB stronger than the desired output signal

A1C07 (B) [97.317a3]
Which of the following is one of the standards that must be met by an external RF power amplifier if it is to qualify for a grant of FCC type acceptance?
A. It must have a time-delay circuit to prevent it from operating continuously for more than ten minutes
B. It must satisfy the spurious emission standards when driven with at least 50 W mean RF power (unless a higher drive level is specified)
C. It must not be capable of modification by an amateur operator without voiding the warranty
D. It must exhibit no more than 6 dB of gain over its entire operating range

A1C08 (A) [97.317c1]
Which of the following would disqualify an external RF power amplifier from being granted FCC type acceptance?
A. Any accessible wiring which, when altered, would permit operation of the amplifier in a manner contrary to FCC Rules
B. Failure to include a schematic diagram and theory of operation manual that would permit an amateur to modify the amplifier
C. The capability of being switched by the operator to any amateur frequency below 24 MHz
D. Failure to produce 1500 watts of output power when driven by at least 50 watts of mean input power

A1C09 (C) [97.317c8]
Which of the following would disqualify an external RF power amplifier from being granted FCC type acceptance?
A. Failure to include controls or adjustments that would permit the amplifier to operate on any frequency below 24 MHz
B. Failure to produce 1500 watts of output power when driven by at least 50 watts of mean input power
C. Any features designed to facilitate operation in a telecommunication service other than the Amateur Service
D. The omission of a schematic diagram and theory of operation manual that would permit an amateur to modify the amplifier

A1C10 (D) [97.317c3]
Which of the following would disqualify an external RF power amplifier from being granted FCC type acceptance?
A. The omission of a safety switch in the high-voltage power supply to turn off the power if the cabinet is opened
B. Failure of the amplifier to exhibit more than 15 dB of gain over its entire operating range
C. The omission of a time-delay circuit to prevent the amplifier from operating continuously for more than ten minutes

D. The inclusion of instructions for operation or modification of the amplifier in a manner contrary to the FCC Rules

A1C11 (B) [97.317b2]
Which of the following would disqualify an external RF power amplifier from being granted FCC type acceptance?
A. Failure to include a safety switch in the high-voltage power supply to turn off the power if the cabinet is opened
B. The amplifier produces 3 dB of gain for input signals between 26 MHz and 28 MHz
C. The inclusion of a schematic diagram and theory of operation manual that would permit an amateur to modify the amplifier
D. The amplifier produces 1500 watts of output power when driven by at least 50 watts of mean input power

A1D Definition and operation of spread spectrum; auxiliary station operation

A1D01 (C) [97.3c8]
What is the name for emissions using bandwidth-expansion modulation?
A. RTTY
B. Image
C. Spread spectrum
D. Pulse

A1D02 (C) [97.311c]
What two spread spectrum techniques are permitted on the amateur bands?
A. Hybrid switching and direct frequency
B. Frequency switching and linear frequency
C. Frequency hopping and direct sequence
D. Logarithmic feedback and binary sequence

A1D03 (C) [97.311g]
What is the maximum transmitter power allowed for spread spectrum transmissions?
A. 5 watts
B. 10 watts
C. 100 watts
D. 1500 watts

A1D04 (D) [97.3a7]
What is meant by auxiliary station operation?
A. A station operated away from its home location
B. Remote control of model craft
C. A station controlled from a point located other than at the station transmitter
D. Communications sent point-to-point within a system of cooperating amateur stations

A1D05 (A) [97.3a6, 97.3a7, 97.3a35, 97.201, 97.205, 97.213a]
What is one use for a station in auxiliary operation?
A. Remote control of a station in repeater operation
B. Remote control of model craft
C. Passing of international third-party communications
D. The retransmission of NOAA weather broadcasts

A1D06 (B) [97.3a7]
Auxiliary stations communicate with which other kind of amateur stations?

A. Those registered with a civil defense organization
B. Those within a system of cooperating amateur stations
C. Those in space station operation
D. Any kind not under manual control

A1D07 (C) [97.201b]
On what amateur frequencies above 222.0 MHz (the 1.25-meter band) are auxiliary stations NOT allowed to operate?
A. 222.00–223.00 MHz, 432–433 MHz and 436–438 MHz
B. 222.10–223.91 MHz, 431–432 MHz and 435–437 MHz
C. 222.00–222.15 MHz, 431–433 MHz and 435–438 MHz
D. 222.00–222.10 MHz, 430–432 MHz and 434–437 MHz

A1D08 (B) [97.201a]
What class of amateur license must one hold to be the control operator of an auxiliary station?
A. Any class
B. Technician, Technician Plus, General, Advanced or Amateur Extra
C. General, Advanced or Amateur Extra
D. Advanced or Amateur Extra

A1D09 (C) [97.119b1]
When an auxiliary station is identified in Morse code using an automatic keying device used only for identification, what is the maximum code speed permitted?
A. 13 words per minute
B. 30 words per minute
C. 20 words per minute
D. There is no limitation

A1D10 (D) [97.119a]
How often must an auxiliary station be identified?
A. At least once during each transmission
B. Only at the end of a series of transmissions
C. At the beginning of a series of transmissions
D. At least once every ten minutes during and at the end of activity

A1D11 (A) [97.119b3]
When may an auxiliary station be identified using a digital code?
A. Any time the digital code is used for at least part of the communication
B. Any time
C. Identification by digital code is not allowed
D. No identification is needed for digital transmissions

A1E "Line A"; National Radio Quiet Zone; business communications; restricted operation; antenna structure limitations

A1E01 (A) [97.3a26]
Which of the following geographic descriptions approximately describes "Line A"?
A. A line roughly parallel to, and south of, the US-Canadian border
B. A line roughly parallel to, and west of, the US Atlantic coastline

C. A line roughly parallel to, and north of, the US-Mexican border and Gulf coastline
D. A line roughly parallel to, and east of, the US Pacific coastline

A1E02 (D) [97.303f1]
Amateur stations may not transmit in which frequency segment if they are located north of "Line A"?
A. 21.225-21.300 MHz
B. 53-54 MHz
C. 222-223 MHz
D. 420-430 MHz

A1E03 (C) [97.3a29]
What is the National Radio Quiet Zone?
A. An area in Puerto Rico surrounding the Aricebo Radio Telescope
B. An area in New Mexico surrounding the White Sands Test Area
C. An Area in Maryland, West Virginia and Virginia surrounding the National Radio Astronomy Observatory
D. An area in Florida surrounding Cape Canaveral

A1E04 (A) [97.203e,97.205f]
Which of the following agencies is protected from interference to its operations by the National Radio Quiet Zone?
A. The National Radio Astronomy Observatory at Green Bank, WV
B. NASA's Mission Control Center in Houston, TX
C. The White Sands Test Area in White Sands, NM
D. The space shuttle launch facilities in Cape Canaveral, FL

A1E05 (B) [97.113]
Which communication is NOT a prohibited transmission in the Amateur Service?
A. Sending messages for hire or material compensation
B. Calling a commercial tow truck service for a breakdown on the highway
C. Calling your employer to see if you have any customers to contact
D. Sending a false distress call as a "joke"

A1E06 (C) [97.113a3]
Under what conditions may you notify other amateurs of the availability of amateur station equipment for sale or trade over the airwaves?
A. You are never allowed to sell or trade equipment on the air
B. Only if this activity does not result in a profit for you
C. Only if this activity is not conducted on a regular basis
D. Only if the equipment is FCC type accepted and has a serial number

A1E07 (C) [97.113a2]
When may amateurs accept payment for using their own stations (other than a club station) to send messages?
A. When employed by the FCC
B. When passing emergency traffic
C. Under no circumstances
D. When passing international third-party communications

A1E08 (D) [97.113a2]
When may the control operator of a repeater accept payment for providing communication services to another party?
A. When the repeater is operating under portable power
B. When the repeater is operating under local control
C. During Red Cross or other emergency service drills
D. Under no circumstances

A1E09 (D) [97.113a3]
When may an amateur station send a message to a business?
A. When the total money involved does not exceed $25
B. When the control operator is employed by the FCC or another government agency
C. When transmitting international third-party communications
D. When neither the amateur nor his or her employer has a pecuniary interest in the communications

A1E10 (C) [97.15a]
What must an amateur obtain before installing an antenna structure more than 200 feet high?
A. An environmental assessment
B. A Special Temporary Authorization
C. Prior FCC approval
D. An effective radiated power statement

A1E11 (A) [97.15d]
From what government agencies must you obtain permission if you wish to install an antenna structure that exceeds 200 feet above ground level?
A. The Federal Aviation Administration (FAA) and the Federal Communications Commission (FCC)
B. The Environmental Protection Agency (EPA) and the Federal Communications Commission (FCC)
C. The Federal Aviation Administration (FAA) and the Environmental Protection Agency (EPA)
D. The Environmental Protection Agency (EPA) and National Aeronautics and Space Administration (NASA)

A1F Volunteer examinations: when examination is required; exam credit; examination grading; Volunteer Examiner requirements; Volunteer Examiner conduct

A1F01 (B) [97.505a]
What examination credit must be given to an applicant who holds an unexpired (or expired within the grace period) FCC-issued amateur operator license?
A. No credit
B. Credit for the least elements required for the license
C. Credit for only the telegraphy requirements of the license
D. Credit for only the written element requirements of the license

A1F02 (B) [97.503a1]
What ability with international Morse code must an applicant demonstrate when taking an Element 1(A) telegraphy examination?
A. To send and receive text at not less than 13 WPM
B. To send and receive text at not less than 5 WPM
C. To send and receive text at not less than 20 WPM
D. To send text at not less than 13 WPM

A1F03 (A) [97.503a]
Besides all the letters of the alphabet, numerals 0-9 and the period, comma and question mark, what additional characters are used in telegraphy examinations?
A. The slant mark and prosigns AR, BT and SK
B. The slant mark, open and closed parenthesis and prosigns AR, BT and SK
C. The slant mark, dollar sign and prosigns AR, BT and SK
D. No other characters

A1F04 (B) [97.507d]
In a telegraphy examination, how many letters of the alphabet are counted as one word?
A. 2
B. 5
C. 8
D. 10

A1F05 (C) [97.509b2]
What is the minimum age to be a Volunteer Examiner?
A. 16
B. 21
C. 18
D. 13

A1F06 (A) [97.509b4]
When may a person whose amateur operator or station license has ever been revoked or suspended be a Volunteer Examiner?
A. Under no circumstances
B. After 5 years have elapsed since the revocation or suspension
C. After 3 years have elapsed since the revocation or suspension
D. After review and subsequent approval by a VEC

A1F07 (B) [97.509b5]
When may an employee of a company engaged in the distribution of equipment used in connection with amateur station transmissions be a Volunteer Examiner?
A. When the employee is employed in the Amateur Radio sales part of the company
B. When the employee does not normally communicate with the manufacturing or distribution part of the company
C. When the employee serves as a Volunteer Examiner for his or her customers
D. When the employee does not normally communicate with the benefits and policies part of the company

A1F08 (A) [97.509a, b1, b2, b3i]
Who may administer an examination for a Novice license?
A. Three accredited Volunteer Examiners at least 18 years old and holding at least a General class license
B. Three amateur operators at least 18 years old and holding at least a General class license
C. Any accredited Volunteer Examiner at least 21 years old and holding at least a General class license
D. Two amateur operators at least 21 years old and holding at least a Technician class license

A1F09 (A) [97.509e]
When may Volunteer Examiners be compensated for their services? A. Under no circumstances

B. When out-of-pocket expenses exceed $25
C. When traveling over 25 miles to the test site
D. When there are more than 20 applicants attending an examination session

A1F10 (C) [97.509e]
What are the penalties that may result from fraudulently administering amateur examinations?
A. Suspension of amateur station license for a period not to exceed 3 months
B. A monetary fine not to exceed $500 for each day the offense was committed
C. Revocation of amateur station license and suspension of operator's license
D. Restriction to administering only Novice class license examinations

A1F11 (D) [97.509e]
What are the penalties that may result from administering examinations for money or other considerations?
A. Suspension of amateur station license for a period not to exceed 3 months
B. A monetary fine not to exceed $500 for each day the offense was committed
C. Restriction to administering only Novice class license examinations
D. Revocation of amateur station license and suspension of operator's license

A1F12 (A) [97.509h]
How soon must the administering Volunteer Examiners grade an applicant's completed examination element?
A. Immediately
B. Within 48 hours
C. Within 10 days
D. Within 24 hours

A1F13 (B) [97.509m]
After the successful administration of an examination, within how many days must the Volunteer Examiners submit the application to their coordinating VEC?
A. 7
B. 10
C. 5
D. 30

A1F14 (C) [97.509m]
After the successful administration of an examination, where must the Volunteer Examiners submit the application?
A. To the nearest FCC Field Office
B. To the FCC in Washington, DC
C. To the coordinating VEC
D. To the FCC in Gettysburg, PA

A2–OPERATING PROCEDURES [1 question–1 group]

A2A Facsimile communications; slow-scan TV transmissions; spread- spectrum transmissions; HF digital communications (i.e., PacTOR, CLOVER, HF packet); automatic HF Forwarding

A2A01 (D)
What is facsimile?

A. The transmission of characters by radioteletype that form a picture when printed
B. The transmission of still pictures by slow-scan television
C. The transmission of video by amateur television
D. The transmission of printed pictures for permanent display on paper

A2A02 (A)
What is the modern standard scan rate for a facsimile picture transmitted by an amateur station?
A. 240 lines per minute
B. 50 lines per minute
C. 150 lines per second
D. 60 lines per second

A2A03 (B)
What is the approximate transmission time per frame for a facsimile picture transmitted by an amateur station at 240 lpm?
A. 6 minutes
B. 3.3 minutes
C. 6 seconds
D. 1/60 second

A2A04 (B)
What is the term for the transmission of printed pictures by radio?
A. Television
B. Facsimile
C. Xerography
D. ACSSB

A2A05 (C)
In facsimile, what device converts variations in picture brightness and darkness into voltage variations?
A. An LED
B. A Hall-effect transistor
C. A photodetector
D. An optoisolator

A2A06 (D)
What information is sent by slow-scan television transmissions?
A. Baudot or ASCII characters that form a picture when printed
B. Pictures for permanent display on paper
C. Moving pictures
D. Still pictures

A2A07 (C)
How many lines are commonly used in each frame on an amateur slow-scan color television picture?
A. 30 or 60
B. 60 or 100
C. 128 or 256
D. 180 or 360

A2A08 (C)
What is the audio frequency for black in an amateur slow-scan television picture?
A. 2300 Hz
B. 2000 Hz
C. 1500 Hz
D. 120 Hz

A2A09 (D)
What is the audio frequency for white in an amateur slow-scan television picture?
A. 120 Hz
B. 1500 Hz
C. 2000 Hz
D. 2300 Hz

A2A10 (A)
Why are received spread-spectrum signals so resistant to interference?
A. Signals not using the spectrum-spreading algorithm are suppressed in the receiver
B. The high power used by a spread-spectrum transmitter keeps its signal from being easily overpowered
C. The receiver is always equipped with a special digital signal processor (DSP) interference filter
D. If interference is detected by the receiver it will signal the transmitter to change frequencies

A2A11 (D)
How does the spread-spectrum technique of frequency hopping (FH) work?
A. If interference is detected by the receiver it will signal the transmitter to change frequencies
B. If interference is detected by the receiver it will signal the transmitter to wait until the frequency is clear
C. A pseudo-random binary bit stream is used to shift the phase of an RF carrier very rapidly in a particular sequence
D. The frequency of an RF carrier is changed very rapidly according to a particular pseudo-random sequence

A2A12 (C)
What is the most common data rate used for HF packet communications?
A. 48 bauds
B. 110 bauds
C. 300 bauds
D. 1200 bauds

A3–RADIO-WAVE PROPAGATION [2 questions–2 groups]

A3A Sporadic-E; auroral propagation; ground-wave propagation (distances and coverage, and frequency vs. distance in each of these topics)

A3A01 (C)
What is a sporadic-E condition?
A. Variations in E-region height caused by sunspot variations
B. A brief decrease in VHF signal levels from meteor trails at E- region height
C. Patches of dense ionization at E-region height
D. Partial tropospheric ducting at E-region height

A3A02 (D)
What is the term for the propagation condition in which scattered patches of relatively dense ionization develop seasonally at E-region heights?
A. Auroral propagation
B. Ducting
C. Scatter
D. Sporadic-E

A3A03 (A)
In what region of the world is sporadic-E most prevalent?
A. The equatorial regions
B. The arctic regions
C. The northern hemisphere
D. The western hemisphere

A3A04 (B)
On which amateur frequency band is the extended-distance propagation effect of sporadic-E most often observed?
A. 2 meters
B. 6 meters
C. 20 meters
D. 160 meters

A3A05 (D)
What effect does auroral activity have upon radio communications?
A. The readability of SSB signals increases
B. FM communications are clearer
C. CW signals have a clearer tone
D. CW signals have a fluttery tone

A3A06 (C)
What is the cause of auroral activity?
A. A high sunspot level
B. A low sunspot level
C. The emission of charged particles from the sun
D. Meteor showers concentrated in the northern latitudes

A3A07 (B)
In the northern hemisphere, in which direction should a directional antenna be pointed to take maximum advantage of auroral propagation?
A. South
B. North
C. East
D. West

A3A08 (D)
Where in the ionosphere does auroral activity occur?
A. At F-region height
B. In the equatorial band
C. At D-region height
D. At E-region height

A3A09 (A)
Which emission modes are best for auroral propagation?
A. CW and SSB
B. SSB and FM
C. FM and CW
D. RTTY and AM

A3A10 (B)
As the frequency of a signal is increased, how does its ground-wave propagation distance change?
A. It increases
B. It decreases
C. It stays the same
D. Radio waves don't propagate along the Earth's surface

A3A11 (A)
What typical polarization does ground-wave propagation have?

A. Vertical
B. Horizontal
C. Circular
D. Elliptical

A3B Selective fading; radio-path horizon; take-off angle over flat or sloping terrain; earth effects on propagation

A3B01 (B)
What causes selective fading?
A. Small changes in beam heading at the receiving station
B. Phase differences between radio-wave components of the same transmission, as experienced at the receiving station
C. Large changes in the height of the ionosphere at the receiving station ordinarily occurring shortly after either sunrise or sunset
D. Time differences between the receiving and transmitting stations

A3B02 (C)
What is the propagation effect called that causes selective fading between received wave components of the same transmission?
A. Faraday rotation
B. Diversity reception
C. Phase differences
D. Phase shift

A3B03 (B)
Which emission modes suffer the most from selective fading?
A. CW and SSB
B. FM and double sideband AM
C. SSB and AMTOR
D. SSTV and CW

A3B04 (A)
How does the bandwidth of a transmitted signal affect selective fading?
A. It is more pronounced at wide bandwidths
B. It is more pronounced at narrow bandwidths
C. It is the same for both narrow and wide bandwidths
D. The receiver bandwidth determines the selective fading effect

A3B05 (D)
Why does the radio-path horizon distance exceed the geometric horizon?
A. E-region skip
B. D-region skip
C. Auroral skip
D. Radio waves may be bent

A3B06 (A)
How much farther does the VHF/UHF radio-path horizon distance exceed the geometric horizon?
A. By approximately 15% of the distance
B. By approximately twice the distance
C. By approximately one-half the distance
D. By approximately four times the distance

A3B07 (B)
For a 3-element Yagi antenna with horizontally mounted elements, how does the main lobe takeoff angle vary with

height above flat ground?
A. It increases with increasing height
B. It decreases with increasing height
C. It does not vary with height
D. It depends on E-region height, not antenna height

A3B08 (B)
For a 3-element Yagi antenna with horizontally mounted elements, how does the main lobe takeoff angle vary with a downward slope of the ground (moving away from the antenna)?
A. It increases as the slope gets steeper
B. It decreases as the slope gets steeper
C. It does not depend on the ground slope
D. It depends on F-region height, not ground slope

A3B09 (B)
What is the name of the high-angle wave in HF propagation that travels for some distance within the F2 region?
A. Oblique-angle ray
B. Pedersen ray
C. Ordinary ray
D. Heaviside ray

A3B10 (B)
Excluding enhanced propagation, what is the approximate range of normal VHF propagation?
A. 1000 miles
B. 500 miles
C. 1500 miles
D. 2000 miles

A3B11 (C)
What effect is usually responsible for propagating a VHF signal over 500 miles?
A. D-region absorption
B. Faraday rotation
C. Tropospheric ducting
D. Moonbounce

A3B12 (A)
What happens to an electromagnetic wave as it encounters air molecules and other particles?
A. The wave loses kinetic energy
B. The wave gains kinetic energy
C. An aurora is created
D. Nothing happens because the waves have no physical substance

A4—AMATEUR RADIO PRACTICE [4 questions—4 groups]

A4A Frequency measurement devices (i.e. frequency counter, oscilloscope Lissajous figures, dip meter); component mounting techniques (i.e. surface, dead bug raised, circuit board)

A4A01 (B)
What is a frequency standard?
A. A frequency chosen by a net control operator for net operations
B. A device used to produce a highly accurate reference frequency
C. A device for accurately measuring frequency to within 1 Hz
D. A device used to generate wide-band random frequencies

A4A02 (A)
What does a frequency counter do?
A. It makes frequency measurements
B. It produces a reference frequency
C. It measures FM transmitter deviation
D. It generates broad-band white noise

A4A03 (C)
If a 100 Hz signal is fed to the horizontal input of an oscilloscope and a 150 Hz signal is fed to the vertical input, what type of Lissajous figure should be displayed on the screen?
A. A looping pattern with 100 loops horizontally and 150 loops vertically
B. A rectangular pattern 100 mm wide and 150 mm high
C. A looping pattern with 3 loops horizontally and 2 loops vertically
D. An oval pattern 100 mm wide and 150 mm high

A4A04 (C)
What is a dip-meter?
A. A field-strength meter
B. An SWR meter
C. A variable LC oscillator with metered feedback current
D. A marker generator

A4A05 (D)
What does a dip-meter do?
A. It accurately indicates signal strength
B. It measures frequency accurately
C. It measures transmitter output power accurately
D. It gives an indication of the resonant frequency of a circuit

A4A06 (B)
How does a dip-meter function?
A. Reflected waves at a specific frequency desensitize a detector coil
B. Power coupled from an oscillator causes a decrease in metered current
C. Power from a transmitter cancels feedback current
D. Harmonics from an oscillator cause an increase in resonant circuit Q

A4A07 (D)
What two ways could a dip-meter be used in an amateur station?
A. To measure resonant frequency of antenna traps and to measure percentage of modulation
B. To measure antenna resonance and to measure percentage of modulation
C. To measure antenna resonance and to measure antenna impedance
D. To measure resonant frequency of antenna traps and to measure a tuned circuit resonant frequency

A4A08 (B)
What types of coupling occur between a dip-meter and a tuned circuit being checked?
A. Resistive and inductive
B. Inductive and capacitive
C. Resistive and capacitive
D. Strong field

A4A09 (A)
For best accuracy, how tightly should a dip-meter be coupled with a tuned circuit being checked?
A. As loosely as possible
B. As tightly as possible
C. First loosely, then tightly
D. With a jumper wire between the meter and the circuit to be checked

A4A10 (B)
What happens in a dip-meter when it is too tightly coupled with a tuned circuit being checked?
A. Harmonics are generated
B. A less accurate reading results
C. Cross modulation occurs
D. Intermodulation distortion occurs

A4A11 (D)
What circuit construction technique uses leadless components mounted between circuit board pads?
A. Raised mounting
B. Integrated circuit mounting
C. Hybrid device mounting
D. Surface mounting

A4B Meter performance limitations; oscilloscope performance limitations; frequency counter performance limitations

A4B01 (B)
What factors limit the accuracy, frequency response and stability of a D'Arsonval-type meter?
A. Calibration, coil impedance and meter size
B. Calibration, mechanical tolerance and coil impedance
C. Coil impedance, electromagnet voltage and movement mass
D. Calibration, series resistance and electromagnet current

A4B02 (A)
What factors limit the accuracy, frequency response and stability of an oscilloscope?
A. Accuracy and linearity of the time base and the linearity and bandwidth of the deflection amplifiers
B. Tube face voltage increments and deflection amplifier voltage
C. Accuracy and linearity of the time base and tube face voltage increments
D. Deflection amplifier output impedance and tube face frequency increments

A4B03 (D)
How can the frequency response of an oscilloscope be improved?
A. By using a triggered sweep and a crystal oscillator as the time base
B. By using a crystal oscillator as the time base and increasing the vertical sweep rate
C. By increasing the vertical sweep rate and the horizontal amplifier frequency response
D. By increasing the horizontal sweep rate and the vertical amplifier frequency response

A4B04 (B)
What factors limit the accuracy, frequency response and stability of a frequency counter?

A. Number of digits in the readout, speed of the logic and time base stability
B. Time base accuracy, speed of the logic and time base stability
C. Time base accuracy, temperature coefficient of the logic and time base stability
D. Number of digits in the readout, external frequency reference and temperature coefficient of the logic

A4B05 (C)
How can the accuracy of a frequency counter be improved?
A. By using slower digital logic
B. By improving the accuracy of the frequency response
C. By increasing the accuracy of the time base
D. By using faster digital logic

A4B06 (C)
If a frequency counter with a time base accuracy of +/- 1.0 ppm reads 146,520,000 Hz, what is the most the actual frequency being measured could differ from the reading?
A. 165.2 Hz
B. 14.652 kHz
C. 146.52 Hz
D. 1.4652 MHz

A4B07 (A)
If a frequency counter with a time base accuracy of +/- 0.1 ppm reads 146,520,000 Hz, what is the most the actual frequency being measured could differ from the reading?
A. 14.652 Hz
B. 0.1 MHz
C. 1.4652 Hz
D. 1.4652 kHz

A4B08 (D)
If a frequency counter with a time base accuracy of +/- 10 ppm reads 146,520,000 Hz, what is the most the actual frequency being measured could differ from the reading?
A. 146.52 Hz
B. 10 Hz
C. 146.52 kHz
D. 1465.20 Hz

A4B09 (D)
If a frequency counter with a time base accuracy of +/- 1.0 ppm reads 432,100,000 Hz, what is the most the actual frequency being measured could differ from the reading?
A. 43.21 MHz
B. 10 Hz
C. 1.0 MHz
D. 432.1 Hz

A4B10 (A)
If a frequency counter with a time base accuracy of +/- 0.1 ppm reads 432,100,000 Hz, what is the most the actual frequency being measured could differ from the reading?
A. 43.21 Hz
B. 0.1 MHz
C. 432.1 Hz
D. 0.2 MHz

A4B11 (C)
If a frequency counter with a time base accuracy of +/- 10 ppm reads 432,100,000 Hz, what is the most the actual

frequency being measured could differ from the reading?
A. 10 MHz
B. 10 Hz
C. 4321 Hz
D. 432.1 Hz

A4C Receiver performance characteristics (i.e., phase noise, desensitization, capture effect, intercept point, noise floor, dynamic range blocking and IMD, image rejection, MDS, signal- to-noise-ratio)

A4C01 (D)
What is the effect of excessive phase noise in a receiver local oscillator?
A. It limits the receiver ability to receive strong signals
B. It reduces the receiver sensitivity
C. It decreases the receiver third-order intermodulation distortion dynamic range
D. It allows strong signals on nearby frequencies to interfere with reception of weak signals

A4C02 (A)
What is the term for the reduction in receiver sensitivity caused by a strong signal near the received frequency?
A. Desensitization
B. Quieting
C. Cross-modulation interference
D. Squelch gain rollback

A4C03 (B)
What causes receiver desensitization?
A. Audio gain adjusted too low
B. Strong adjacent-channel signals
C. Squelch gain adjusted too high
D. Squelch gain adjusted too low

A4C04 (A)
What is one way receiver desensitization can be reduced?
A. Shield the receiver from the transmitter causing the problem
B. Increase the transmitter audio gain
C. Decrease the receiver squelch gain
D. Increase the receiver bandwidth

A4C05 (C)
What is the capture effect?
A. All signals on a frequency are demodulated by an FM receiver
B. All signals on a frequency are demodulated by an AM receiver
C. The strongest signal received is the only demodulated signal
D. The weakest signal received is the only demodulated signal

A4C06 (C)
What is the term for the blocking of one FM-phone signal by another stronger FM-phone signal?
A. Desensitization
B. Cross-modulation interference
C. Capture effect
D. Frequency discrimination

A4C07 (A)
With which emission type is capture effect most pronounced?

A. FM
B. SSB
C. AM
D. CW

A4C08 (D)
What is meant by the noise floor of a receiver?
A. The weakest signal that can be detected under noisy atmospheric conditions
B. The amount of phase noise generated by the receiver local oscillator
C. The minimum level of noise that will overload the receiver RF amplifier stage
D. The weakest signal that can be detected above the receiver internal noise

A4C09 (B)
What is the blocking dynamic range of a receiver that has an 8-dB noise figure and an IF bandwidth of 500 Hz if the blocking level (1-dB compression point) is -20 dBm?
A. -119 dBm
B. 119 dB
C. 146 dB
D. -146 dBm

A4C10 (B)
What part of a superheterodyne receiver determines the image rejection ratio of the receiver?
A. Product detector
B. RF amplifier
C. AGC loop
D. IF filter

A4C11 (B)
If you measured the MDS of a receiver, what would you be measuring?
A. The meter display sensitivity (MDS), or the responsiveness of the receiver S-meter to all signals
B. The minimum discernible signal (MDS), or the weakest signal that the receiver can detect
C. The minimum distorting signal (MDS), or the strongest signal the receiver can detect without overloading
D. The maximum detectable spectrum (MDS), or the lowest to highest frequency range of the receiver

A4D Intermodulation and cross-modulation interference

A4D01 (D)
If the signals of two transmitters mix together in one or both of their final amplifiers and unwanted signals at the sum and difference frequencies of the original signals are generated, what is this called?
A. Amplifier desensitization
B. Neutralization
C. Adjacent channel interference
D. Intermodulation interference

A4D02 (B)
How does intermodulation interference between two repeater transmitters usually occur?
A. When the signals from the transmitters are reflected out of phase from airplanes passing overhead
B. When they are in close proximity and the signals mix in one or both of their final amplifiers
C. When they are in close proximity and the signals

cause feedback in one or both of their final amplifiers
D. When the signals from the transmitters are reflected in phase from airplanes passing overhead

A4D03 (B)
How can intermodulation interference between two repeater transmitters in close proximity often be reduced or eliminated?
A. By using a Class C final amplifier with high driving power
B. By installing a terminated circulator or ferrite isolator in the feed line to the transmitter and duplexer
C. By installing a band-pass filter in the antenna feed line
D. By installing a low-pass filter in the antenna feed line

A4D04 (D)
What is cross-modulation interference?
A. Interference between two transmitters of different modulation type
B. Interference caused by audio rectification in the receiver preamp
C. Harmonic distortion of the transmitted signal
D. Modulation from an unwanted signal is heard in addition to the desired signal

A4D05 (B)
What is the term used to refer to the condition where the signals from a very strong station are superimposed on other signals being received?
A. Intermodulation distortion
B. Cross-modulation interference
C. Receiver quieting
D. Capture effect

A4D06 (A)
How can cross-modulation in a receiver be reduced?
A. By installing a filter at the receiver
B. By using a better antenna
C. By increasing the receiver RF gain while decreasing the AF gain
D. By adjusting the passband tuning

A4D07 (C)
What is the result of cross-modulation?
A. A decrease in modulation level of transmitted signals
B. Receiver quieting
C. The modulation of an unwanted signal is heard on the desired signal
D. Inverted sidebands in the final stage of the amplifier

A4D08 (C)
What causes intermodulation in an electronic circuit?
A. Too little gain
B. Lack of neutralization
C. Nonlinear circuits or devices
D. Positive feedback

A4D09 (A)
If a receiver tuned to 146.70 MHz receives an intermodulation-product signal whenever a nearby transmitter transmits on 146.52 MHz, what are the two most likely frequencies for the other interfering signal?
A. 146.34 MHz and 146.61 MHz
B. 146.88 MHz and 146.34 MHz
C. 146.10 MHz and 147.30 MHz
D. 73.35 MHz and 239.40 MHz

A4D10 (D)
If a television receiver suffers from cross modulation when a nearby amateur transmitter is operating at 14 MHz, which of the following cures might be effective?
A. A low-pass filter attached to the output of the amateur transmitter
B. A high-pass filter attached to the output of the amateur transmitter
C. A low-pass filter attached to the input of the television receiver
D. A high-pass filter attached to the input of the television receiver

A4D11 (B)
Which of the following is an example of intermodulation distortion? A. Receiver blocking
B. Splatter from an SSB transmitter
C. Overdeviation of an FM transmitter
D. Excessive 2nd-harmonic output from a transmitter

A5—ELECTRICAL PRINCIPLES [10 questions–10 groups]

A5A Characteristics of resonant circuits

A5A01 (A)
What can cause the voltage across reactances in series to be larger than the voltage applied to them?
A. Resonance
B. Capacitance
C. Conductance
D. Resistance

A5A02 (C)
What is resonance in an electrical circuit?
A. The highest frequency that will pass current
B. The lowest frequency that will pass current
C. The frequency at which capacitive reactance equals inductive reactance
D. The frequency at which power factor is at a minimum

A5A03 (B)
What are the conditions for resonance to occur in an electrical circuit?
A. The power factor is at a minimum
B. Inductive and capacitive reactances are equal
C. The square root of the sum of the capacitive and inductive reactance is equal to the resonant frequency
D. The square root of the product of the capacitive and inductive reactance is equal to the resonant frequency

A5A04 (D)
When the inductive reactance of an electrical circuit equals its capacitive reactance, what is this condition called?
A. Reactive quiescence
B. High Q
C. Reactive equilibrium
D. Resonance

A5A05 (D)
What is the magnitude of the impedance of a series R-L-C circuit at resonance?
A. High, as compared to the circuit resistance
B. Approximately equal to capacitive reactance
C. Approximately equal to inductive reactance
D. Approximately equal to circuit resistance

A5A06 (A)
What is the magnitude of the impedance of a circuit with a resistor, an inductor and a capacitor all in parallel, at resonance?
A. Approximately equal to circuit resistance
B. Approximately equal to inductive reactance
C. Low, as compared to the circuit resistance
D. Approximately equal to capacitive reactance

A5A07 (B)
What is the magnitude of the current at the input of a series R-L-C circuit at resonance?
A. It is at a minimum
B. It is at a maximum
C. It is DC
D. It is zero

A5A08 (B)
What is the magnitude of the circulating current within the components of a parallel L-C circuit at resonance?
A. It is at a minimum
B. It is at a maximum
C. It is DC
D. It is zero

A5A09 (A)
What is the magnitude of the current at the input of a parallel R-L-C circuit at resonance?
A. It is at a minimum
B. It is at a maximum
C. It is DC
D. It is zero

A5A10 (C)
What is the relationship between the current through a resonant circuit and the voltage across the circuit?
A. The voltage leads the current by 90 degrees
B. The current leads the voltage by 90 degrees
C. The voltage and current are in phase
D. The voltage and current are 180 degrees out of phase

A5A11 (C)
What is the relationship between the current into (or out of) a parallel resonant circuit and the voltage across the circuit?
A. The voltage leads the current by 90 degrees
B. The current leads the voltage by 90 degrees
C. The voltage and current are in phase
D. The voltage and current are 180 degrees out of phase

A5B Series resonance (capacitor and inductor to resonate at a specific frequency)

A5B01 (C)
What is the resonant frequency of a series R-L-C circuit if R is 47 ohms, L is 50 microhenrys and C is 40 picofarads?
A. 79.6 MHz
B. 1.78 MHz
C. 3.56 MHz
D. 7.96 MHz

A5B02 (B)
What is the resonant frequency of a series R-L-C circuit if R is 47 ohms, L is 40 microhenrys and C is 200 picofarads?
A. 1.99 kHz

B. 1.78 MHz
C. 1.99 MHz
D. 1.78 kHz

A5B03 (D)
What is the resonant frequency of a series R-L-C circuit if R is 47 ohms, L is 50 microhenrys and C is 10 picofarads?
A. 3.18 MHz
B. 3.18 kHz
C. 7.12 kHz
D. 7.12 MHz

A5B04 (A)
What is the resonant frequency of a series R-L-C circuit if R is 47 ohms, L is 25 microhenrys and C is 10 picofarads?
A. 10.1 MHz
B. 63.7 MHz
C. 10.1 kHz
D. 63.7 kHz

A5B05 (B)
What is the resonant frequency of a series R-L-C circuit if R is 47 ohms, L is 3 microhenrys and C is 40 picofarads?
A. 13.1 MHz
B. 14.5 MHz
C. 14.5 kHz
D. 13.1 kHz

A5B06 (D)
What is the resonant frequency of a series R-L-C circuit if R is 47 ohms, L is 4 microhenrys and C is 20 picofarads?
A. 19.9 kHz
B. 17.8 kHz
C. 19.9 MHz
D. 17.8 MHz

A5B07 (C)
What is the resonant frequency of a series R-L-C circuit if R is 47 ohms, L is 8 microhenrys and C is 7 picofarads?
A. 2.84 MHz
B. 28.4 MHz
C. 21.3 MHz
D. 2.13 MHz

A5B08 (A)
What is the resonant frequency of a series R-L-C circuit if R is 47 ohms, L is 3 microhenrys and C is 15 picofarads?
A. 23.7 MHz
B. 23.7 kHz
C. 35.4 kHz
D. 35.4 MHz

A5B09 (B)
What is the resonant frequency of a series R-L-C circuit if R is 47 ohms, L is 4 microhenrys and C is 8 picofarads?
A. 28.1 kHz
B. 28.1 MHz
C. 49.7 MHz
D. 49.7 kHz

A5B10 (D)
What is the resonant frequency of a series R-L-C circuit if R is 47 ohms, L is 1 microhenry and C is 9 picofarads?
A. 17.7 MHz
B. 17.7 kHz
C. 53.1 kHz
D. 53.1 MHz

A5B11 (C)
What is the value of capacitance (C) in a series R-L-C circuit if the circuit resonant frequency is 14.25 MHz and L is 2.84 microhenrys?
A. 2.2 microfarads
B. 254 microfarads
C. 44 picofarads
D. 3933 picofarads

A5C Parallel resonance (capacitor and inductor to resonate at a specific frequency)

A5C01 (A)
What is the resonant frequency of a parallel R-L-C circuit if R is 4.7 kilohms, L is 1 microhenry and C is 10 picofarads?
A. 50.3 MHz
B. 15.9 MHz
C. 15.9 kHz
D. 50.3 kHz

A5C02 (B)
What is the resonant frequency of a parallel R-L-C circuit if R is 4.7 kilohms, L is 2 microhenrys and C is 15 picofarads?
A. 29.1 kHz
B. 29.1 MHz
C. 5.31 MHz
D. 5.31 kHz

A5C03 (C)
What is the resonant frequency of a parallel R-L-C circuit if R is 4.7 kilohms, L is 5 microhenrys and C is 9 picofarads?
A. 23.7 kHz
B. 3.54 kHz
C. 23.7 MHz
D. 3.54 MHz

A5C04 (D)
What is the resonant frequency of a parallel R-L-C circuit if R is 4.7 kilohms, L is 2 microhenrys and C is 30 picofarads?
A. 2.65 kHz
B. 20.5 kHz
C. 2.65 MHz
D. 20.5 MHz

A5C05 (A)
What is the resonant frequency of a parallel R-L-C circuit if R is 4.7 kilohms, L is 15 microhenrys and C is 5 picofarads?
A. 18.4 MHz
B. 2.12 MHz
C. 18.4 kHz
D. 2.12 kHz

A5C06 (B)
What is the resonant frequency of a parallel R-L-C circuit if R is 4.7 kilohms, L is 3 microhenrys and C is 40 picofarads?
A. 1.33 kHz
B. 14.5 MHz
C. 1.33 MHz
D. 14.5 kHz

A5C07 (C)
What is the resonant frequency of a parallel R-L-C circuit if R is 4.7 kilohms, L is 40 microhenrys and C is 6 picofarads?
A. 6.63 MHz
B. 6.63 kHz
C. 10.3 MHz
D. 10.3 kHz

A5C08 (D)
What is the resonant frequency of a parallel R-L-C circuit if R is 4.7 kilohms, L is 10 microhenrys and C is 50 picofarads?
A. 3.18 MHz
B. 3.18 kHz
C. 7.12 kHz
D. 7.12 MHz

A5C09 (A)
What is the resonant frequency of a parallel R-L-C circuit if R is 4.7 kilohms, L is 200 microhenrys and C is 10 picofarads?
A. 3.56 MHz
B. 7.96 MHz
C. 3.56 kHz
D. 7.96 MHz

A5C10 (B)
What is the resonant frequency of a parallel R-L-C circuit if R is 4.7 kilohms, L is 90 microhenrys and C is 100 picofarads?
A. 1.77 MHz
B. 1.68 MHz
C. 1.77 kHz
D. 1.68 kHz

A5C11 (D)
What is the value of inductance (L) in a parallel R-L-C circuit if the circuit resonant frequency is 14.25 MHz and C is 44 picofarads?
A. 253.8 millihenrys
B. 3.9 millihenrys
C. 0.353 microhenrys
D. 2.8 microhenrys

A5D Skin effect; electrostatic and electromagnetic fields

A5D01 (A)
What is the result of skin effect?
A. As frequency increases, RF current flows in a thinner layer of the conductor, closer to the surface
B. As frequency decreases, RF current flows in a thinner layer of the conductor, closer to the surface
C. Thermal effects on the surface of the conductor increase the impedance
D. Thermal effects on the surface of the conductor decrease the impedance

A5D02 (C)
What effect causes most of an RF current to flow along the surface of a conductor?
A. Layer effect
B. Seeburg effect
C. Skin effect
D. Resonance effect

A5D03 (A)
Where does almost all RF current flow in a conductor?
A. Along the surface of the conductor
B. In the center of the conductor
C. In a magnetic field around the conductor
D. In a magnetic field in the center of the conductor

A5D04 (D)
Why does most of an RF current flow within a few thousandths of an inch of its conductor's surface?
A. Because a conductor has AC resistance due to self-inductance
B. Because the RF resistance of a conductor is much less than the DC resistance
C. Because of the heating of the conductor's interior
D. Because of skin effect

A5D05 (C)
Why is the resistance of a conductor different for RF currents than for direct currents?
A. Because the insulation conducts current at high frequencies
B. Because of the Heisenburg Effect
C. Because of skin effect
D. Because conductors are non-linear devices

A5D06 (C)
What device is used to store electrical energy in an electrostatic field?
A. A battery
B. A transformer
C. A capacitor
D. An inductor

A5D07 (B)
What unit measures electrical energy stored in an electrostatic field?
A. Coulomb
B. Joule
C. Watt
D. Volt

A5D08 (B)
What is a magnetic field?
A. Current through the space around a permanent magnet
B. The space around a conductor, through which a magnetic force acts
C. The space between the plates of a charged capacitor, through which a magnetic force acts
D. The force that drives current through a resistor

A5D09 (D)
In what direction is the magnetic field oriented about a conductor in relation to the direction of electron flow?
A. In the same direction as the current
B. In a direction opposite to the current
C. In all directions; omnidirectional
D. In a direction determined by the left-hand rule

A5D10 (D)
What determines the strength of a magnetic field around a conductor?
A. The resistance divided by the current
B. The ratio of the current to the resistance
C. The diameter of the conductor
D. The amount of current

A5D11 (B)
What is the term for energy that is stored in an electromagnetic or electrostatic field?
A. Amperes-joules
B. Potential energy
C. Joules-coulombs
D. Kinetic energy

A5E Half-power bandwidth

A5E01 (A)
What is the half-power bandwidth of a parallel resonant circuit that has a resonant frequency of 1.8 MHz and a Q of 95?
A. 18.9 kHz
B. 1.89 kHz
C. 189 Hz
D. 58.7 kHz

A5E02 (D)
What is the half-power bandwidth of a parallel resonant circuit that has a resonant frequency of 3.6 MHz and a Q of 218?
A. 58.7 kHz
B. 606 kHz
C. 47.3 kHz
D. 16.5 kHz

A5E03 (C)
What is the half-power bandwidth of a parallel resonant circuit that has a resonant frequency of 7.1 MHz and a Q of 150?
A. 211 kHz
B. 16.5 kHz
C. 47.3 kHz
D. 21.1 kHz

A5E04 (D)
What is the half-power bandwidth of a parallel resonant circuit that has a resonant frequency of 12.8 MHz and a Q of 218?
A. 21.1 kHz
B. 27.9 kHz
C. 17 kHz
D. 58.7 kHz

A5E05 (A)
What is the half-power bandwidth of a parallel resonant circuit that has a resonant frequency of 14.25 MHz and a Q of 150?
A. 95 kHz
B. 10.5 kHz
C. 10.5 MHz
D. 17 kHz

A5E06 (D)
What is the half-power bandwidth of a parallel resonant circuit that has a resonant frequency of 21.15 MHz and a Q of 95?
A. 4.49 kHz
B. 44.9 kHz
C. 22.3 kHz
D. 222.6 kHz

A5E07 (B)
What is the half-power bandwidth of a parallel resonant circuit that has a resonant frequency of 10.1 MHz and a Q of 225?
A. 4.49 kHz
B. 44.9 kHz
C. 22.3 kHz
D. 223 kHz

A5E08 (A)
What is the half-power bandwidth of a parallel resonant circuit that has a resonant frequency of 18.1 MHz and a Q of 195?
A. 92.8 kHz
B. 10.8 kHz
C. 22.3 kHz
D. 44.9 kHz

A5E09 (C)
What is the half-power bandwidth of a parallel resonant circuit that has a resonant frequency of 3.7 MHz and a Q of 118?
A. 22.3 kHz
B. 76.2 kHz
C. 31.4 kHz
D. 10.8 kHz

A5E10 (C)
What is the half-power bandwidth of a parallel resonant circuit that has a resonant frequency of 14.25 MHz and a Q of 187?
A. 22.3 kHz
B. 10.8 kHz
C. 76.2 kHz
D. 13.1 kHz

A5E11 (B)
What term describes the frequency range over which the circuit response is no more than 3 dB below the peak response?
A. Resonance
B. Half-power bandwidth
C. Circuit Q
D. 2:1 bandwidth

A5F Circuit Q

A5F01 (A)
What is the Q of a parallel R-L-C circuit if the resonant frequency is 14.128 MHz, L is 2.7 microhenrys and R is 18 kilohms?
A. 75.1
B. 7.51
C. 71.5
D. 0.013

A5F02 (B)
What is the Q of a parallel R-L-C circuit if the resonant frequency is 14.128 MHz, L is 4.7 microhenrys and R is 18 kilohms?
A. 4.31
B. 43.1
C. 13.3
D. 0.023

A5F03 (C)
What is the Q of a parallel R-L-C circuit if the resonant frequency is 4.468 MHz, L is 47 microhenrys and R is 180 ohms?
A. 0.00735
B. 7.35
C. 0.136
D. 13.3

A5F04 (D)
What is the Q of a parallel R-L-C circuit if the resonant frequency is 14.225 MHz, L is 3.5 microhenrys and R is 10 kilohms?
A. 7.35
B. 0.0319
C. 71.5
D. 31.9

A5F05 (D)
What is the Q of a parallel R-L-C circuit if the resonant frequency is 7.125 MHz, L is 8.2 microhenrys and R is 1 kilohm?
A. 36.8
B. 0.273
C. 0.368
D. 2.73

A5F06 (A)
What is the Q of a parallel R-L-C circuit if the resonant frequency is 7.125 MHz, L is 10.1 microhenrys and R is 100 ohms?
A. 0.221
B. 4.52
C. 0.00452
D. 22.1

A5F07 (B)
What is the Q of a parallel R-L-C circuit if the resonant frequency is 7.125 MHz, L is 12.6 microhenrys and R is 22 kilohms?
A. 22.1
B. 39
C. 25.6
D. 0.0256

A5F08 (B)
What is the Q of a parallel R-L-C circuit if the resonant frequency is 3.625 MHz, L is 3 microhenrys and R is 2.2 kilohms?
A. 0.031
B. 32.2
C. 31.1
D. 25.6

A5F09 (D)
What is the Q of a parallel R-L-C circuit if the resonant frequency is 3.625 MHz, L is 42 microhenrys and R is 220 ohms?
A. 23
B. 0.00435
C. 4.35
D. 0.23

A5F10 (A)
What is the Q of a parallel R-L-C circuit if the resonant frequency is 3.625 MHz, L is 43 microhenrys and R is 1.8 kilohms?

A. 1.84
B. 0.543
C. 54.3
D. 23

A5F11 (C)
Why is a resistor often included in a parallel resonant circuit?
A. To increase the Q and decrease the skin effect
B. To decrease the Q and increase the resonant frequency
C. To decrease the Q and increase the bandwidth
D. To increase the Q and decrease the bandwidth

A5G Phase angle between voltage and current

A5G01 (A)
What is the phase angle between the voltage across and the current through a series R-L-C circuit if XC is 25 ohms, R is 100 ohms, and XL is 100 ohms?
A. 36.9 degrees with the voltage leading the current
B. 53.1 degrees with the voltage lagging the current
C. 36.9 degrees with the voltage lagging the current
D. 53.1 degrees with the voltage leading the current

A5G02 (B)
What is the phase angle between the voltage across and the current through a series R-L-C circuit if XC is 25 ohms, R is 100 ohms, and XL is 50 ohms?
A. 14 degrees with the voltage lagging the current
B. 14 degrees with the voltage leading the current
C. 76 degrees with the voltage lagging the current
D. 76 degrees with the voltage leading the current

A5G03 (C)
What is the phase angle between the voltage across and the current through a series R-L-C circuit if XC is 500 ohms, R is 1 kilohm, and XL is 250 ohms?
A. 68.2 degrees with the voltage leading the current
B. 14.1 degrees with the voltage leading the current
C. 14.1 degrees with the voltage lagging the current
D. 68.2 degrees with the voltage lagging the current

A5G04 (B)
What is the phase angle between the voltage across and the current through a series R-L-C circuit if XC is 75 ohms, R is 100 ohms, and XL is 100 ohms?
A. 76 degrees with the voltage leading the current
B. 14 degrees with the voltage leading the current
C. 14 degrees with the voltage lagging the current
D. 76 degrees with the voltage lagging the current

A5G05 (D)
What is the phase angle between the voltage across and the current through a series R-L-C circuit if XC is 50 ohms, R is 100 ohms, and XL is 25 ohms?
A. 76 degrees with the voltage lagging the current
B. 14 degrees with the voltage leading the current
C. 76 degrees with the voltage leading the current
D. 14 degrees with the voltage lagging the current

A5G06 (C)
What is the phase angle between the voltage across and the current through a series R-L-C circuit if XC is 75 ohms, R is 100 ohms, and XL is 50 ohms?

A. 76 degrees with the voltage lagging the current
B. 14 degrees with the voltage leading the current
C. 14 degrees with the voltage lagging the current
D. 76 degrees with the voltage leading the current

A5G07 (A)
What is the phase angle between the voltage across and the current through a series R-L-C circuit if XC is 100 ohms, R is 100 ohms, and XL is 75 ohms?
A. 14 degrees with the voltage lagging the current
B. 14 degrees with the voltage leading the current
C. 76 degrees with the voltage leading the current
D. 76 degrees with the voltage lagging the current

A5G08 (D)
What is the phase angle between the voltage across and the current through a series R-L-C circuit if XC is 250 ohms, R is 1 kilohm, and XL is 500 ohms?
A. 81.47 degrees with the voltage lagging the current
B. 81.47 degrees with the voltage leading the current
C. 14.04 degrees with the voltage lagging the current
D. 14.04 degrees with the voltage leading the current

A5G09 (D)
What is the phase angle between the voltage across and the current through a series R-L-C circuit if XC is 50 ohms, R is 100 ohms, and XL is 75 ohms?
A. 76 degrees with the voltage leading the current
B. 76 degrees with the voltage lagging the current
C. 14 degrees with the voltage lagging the current
D. 14 degrees with the voltage leading the current

A5G10 (D)
What is the relationship between the current through and the voltage across a capacitor?
A. Voltage and current are in phase
B. Voltage and current are 180 degrees out of phase
C. Voltage leads current by 90 degrees
D. Current leads voltage by 90 degrees

A5G11 (A)
What is the relationship between the current through an inductor and the voltage across an inductor?
A. Voltage leads current by 90 degrees
B. Current leads voltage by 90 degrees
C. Voltage and current are 180 degrees out of phase
D. Voltage and current are in phase

A5H Reactive power; power factor

A5H01 (A)
What is reactive power?
A. Wattless, nonproductive power
B. Power consumed in wire resistance in an inductor
C. Power lost because of capacitor leakage
D. Power consumed in circuit Q

A5H02 (D)
What is the term for an out-of-phase, nonproductive power associated with inductors and capacitors?
A. Effective power
B. True power
C. Peak envelope power
D. Reactive power

A5H03 (B)
In a circuit that has both inductors and capacitors, what happens to reactive power?
A. It is dissipated as heat in the circuit
B. It goes back and forth between magnetic and electric fields, but is not dissipated
C. It is dissipated as kinetic energy in the circuit
D. It is dissipated in the formation of inductive and capacitive fields

A5H04 (A)
In a circuit where the AC voltage and current are out of phase, how can the true power be determined?
A. By multiplying the apparent power times the power factor
B. By subtracting the apparent power from the power factor
C. By dividing the apparent power by the power factor
D. By multiplying the RMS voltage times the RMS current

A5H05 (C)
What is the power factor of an R-L circuit having a 60 degree phase angle between the voltage and the current?
A. 1.414
B. 0.866
C. 0.5
D. 1.73

A5H06 (D)
What is the power factor of an R-L circuit having a 45 degree phase angle between the voltage and the current?
A. 0.866
B. 1.0
C. 0.5
D. 0.707

A5H07 (C)
What is the power factor of an R-L circuit having a 30 degree phase angle between the voltage and the current?
A. 1.73
B. 0.5
C. 0.866
D. 0.577

A5H08 (B)
How many watts are consumed in a circuit having a power factor of 0.2 if the input is 100-V AC at 4 amperes?
A. 400 watts
B. 80 watts
C. 2000 watts
D. 50 watts

A5H09 (D)
How many watts are consumed in a circuit having a power factor of 0.6 if the input is 200-V AC at 5 amperes?
A. 200 watts
B. 1000 watts
C. 1600 watts
D. 600 watts

A5H10 (B)
How many watts are consumed in a circuit having a power factor of 0.71 if the apparent power is 500 watts?

A. 704 W
B. 355 W
C. 252 W
D. 1.42 mW

A5H11 (A)
Why would the power used in a circuit be less than the product of the magnitudes of the AC voltage and current?
A. Because there is a phase angle greater than zero between the current and voltage
B. Because there are only resistances in the circuit
C. Because there are no reactances in the circuit
D. Because there is a phase angle equal to zero between the current and voltage

A5I Effective radiated power, system gains and losses

A5I01 (B)
What is the effective radiated power of a repeater station with 50 watts transmitter power output, 4-dB feed line loss, 2-dB duplexer loss, 1-dB circulator loss and 6-dBd antenna gain?
A. 199 watts
B. 39.7 watts
C. 45 watts
D. 62.9 watts

A5I02 (C)
What is the effective radiated power of a repeater station with 50 watts transmitter power output, 5-dB feed line loss, 3-dB duplexer loss, 1-dB circulator loss and 7-dBd antenna gain?
A. 79.2 watts
B. 315 watts
C. 31.5 watts
D. 40.5 watts

A5I03 (D)
What is the effective radiated power of a station with 75 watts transmitter power output, 4-dB feed line loss and 10-dBd antenna gain?
A. 600 watts
B. 75 watts
C. 150 watts
D. 299 watts

A5I04 (A)
What is the effective radiated power of a repeater station with 75 watts transmitter power output, 5-dB feed line loss, 3-dB duplexer loss, 1-dB circulator loss and 6-dBd antenna gain?
A. 37.6 watts
B. 237 watts
C. 150 watts
D. 23.7 watts

A5I05 (D)
What is the effective radiated power of a station with 100 watts transmitter power output, 1-dB feed line loss and 6-dBd antenna gain?
A. 350 watts
B. 500 watts
C. 20 watts
D. 316 watts

A5I06 (B)
What is the effective radiated power of a repeater station with 100 watts transmitter power output, 5-dB feed line loss, 3-dB duplexer loss, 1-dB circulator loss and 10-dBd antenna gain?
A. 794 watts
B. 126 watts
C. 79.4 watts
D. 1260 watts

A5I07 (C)
What is the effective radiated power of a repeater station with 120 watts transmitter power output, 5-dB feed line loss, 3-dB duplexer loss, 1-dB circulator loss and 6-dBd antenna gain?
A. 601 watts
B. 240 watts
C. 60 watts
D. 79 watts

A5I08 (D)
What is the effective radiated power of a repeater station with 150 watts transmitter power output, 2-dB feed line loss, 2.2-dB duplexer loss and 7-dBd antenna gain?
A. 1977 watts
B. 78.7 watts
C. 420 watts
D. 286 watts

A5I09 (A)
What is the effective radiated power of a repeater station with 200 watts transmitter power output, 4-dB feed line loss, 3.2-dB duplexer loss, 0.8-dB circulator loss and 10-dBd antenna gain?
A. 317 watts
B. 2000 watts
C. 126 watts
D. 300 watts

A5I10 (B)
What is the effective radiated power of a repeater station with 200 watts transmitter power output, 2-dB feed line loss, 2.8-dB duplexer loss, 1.2-dB circulator loss and 7-dBd antenna gain?
A. 159 watts
B. 252 watts
C. 632 watts
D. 63.2 watts

A5I11 (C)
What term describes station output (including the transmitter, antenna and everything in between), when considering transmitter power and system gains and losses?
A. Power factor
B. Half-power bandwidth
C. Effective radiated power
D. Apparent power

A5J Replacement of voltage source and resistive voltage divider with equivalent voltage source and one resistor (Thevenin's Theorem)

A5J01 (B)
In Figure A5-1, what values of V2 and R3 result in the same voltage and current as when V1 is 8 volts, R1 is 8

kilohms, and R2 is 8 kilohms?
A. R3 = 4 kilohms and V2 = 8 volts
B. R3 = 4 kilohms and V2 = 4 volts
C. R3 = 16 kilohms and V2 = 8 volts
D. R3 = 16 kilohms and V2 = 4 volts

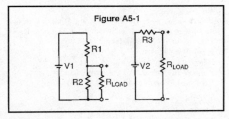
Figure A5-1

A5J02 (C)
In Figure A5-1, what values of V2 and R3 result in the same voltage and current as when V1 is 8 volts, R1 is 16 kilohms, and R2 is 8 kilohms?
A. R3 = 24 kilohms and V2 = 5.33 volts
B. R3 = 5.33 kilohms and V2 = 8 volts
C. R3 = 5.33 kilohms and V2 = 2.67 volts
D. R3 = 24 kilohms and V2 = 8 volts

A5J03 (A)
In Figure A5-1, what values of V2 and R3 result in the same voltage and current as when V1 is 8 volts, R1 is 8 kilohms, and R2 is 16 kilohms?
A. R3 = 5.33 kilohms and V2 = 5.33 volts
B. R3 = 8 kilohms and V2 = 4 volts
C. R3 = 24 kilohms and V2 = 8 volts
D. R3 = 5.33 kilohms and V2 = 8 volts

A5J04 (D)
In Figure A5-1, what values of V2 and R3 result in the same voltage and current as when V1 is 10 volts, R1 is 10 kilohms, and R2 is 10 kilohms?
A. R3 = 10 kilohms and V2 = 5 volts
B. R3 = 20 kilohms and V2 = 5 volts
C. R3 = 20 kilohms and V2 = 10 volts
D. R3 = 5 kilohms and V2 = 5 volts

A5J05 (C)
In Figure A5-1, what values of V2 and R3 result in the same voltage and current as when V1 is 10 volts, R1 is 20 kilohms, and R2 is 10 kilohms?
A. R3 = 30 kilohms and V2 = 10 volts
B. R3 = 6.67 kilohms and V2 = 10 volts
C. R3 = 6.67 kilohms and V2 = 3.33 volts
D. R3 = 30 kilohms and V2 = 3.33 volts

A5J06 (A)
In Figure A5-1, what values of V2 and R3 result in the same voltage and current as when V1 is 10 volts, R1 is 10 kilohms, and R2 is 20 kilohms?
A. R3 = 6.67 kilohms and V2 = 6.67 volts
B. R3 = 6.67 kilohms and V2 = 10 volts
C. R3 = 30 kilohms and V2 = 6.67 volts
D. R3 = 30 kilohms and V2 = 10 volts

A5J07 (B)
In Figure A5-1, what values of V2 and R3 result in the same voltage and current as when V1 is 12 volts, R1 is

10 kilohms, and R2 is 10 kilohms?
A. R3 = 20 kilohms and V2 = 12 volts
B. R3 = 5 kilohms and V2 = 6 volts
C. R3 = 5 kilohms and V2 = 12 volts
D. R3 = 30 kilohms and V2 = 6 volts

A5J08 (B)
In Figure A5-1, what values of V2 and R3 result in the same voltage and current as when V1 is 12 volts, R1 is 20 kilohms, and R2 is 10 kilohms?
A. R3 = 30 kilohms and V2 = 4 volts
B. R3 = 6.67 kilohms and V2 = 4 volts
C. R3 = 30 kilohms and V2 = 12 volts
D. R3 = 6.67 kilohms and V2 = 12 volts

A5J09 (C)
In Figure A5-1, what values of V2 and R3 result in the same voltage and current as when V1 is 12 volts, R1 is 10 kilohms, and R2 is 20 kilohms?
A. R3 = 6.67 kilohms and V2 = 12 volts
B. R3 = 30 kilohms and V2 = 12 volts
C. R3 = 6.67 kilohms and V2 = 8 volts
D. R3 = 30 kilohms and V2 = 8 volts

A5J10 (A)
In Figure A5-1, what values of V2 and R3 result in the same voltage and current as when V1 is 12 volts, R1 is 20 kilohms, and R2 is 20 kilohms?
A. R3 = 10 kilohms and V2 = 6 volts
B. R3 = 40 kilohms and V2 = 6 volts
C. R3 = 40 kilohms and V2 = 12 volts
D. R3 = 10 kilohms and V2 = 12 volts

A5J11 (D)
What circuit principle describes the replacement of any complex two- terminal network of voltage sources and resistances with a single voltage source and a single resistor?
A. Ohm's Law
B. Kirchhoff's Law
C. Laplace's Theorem
D. Thevenin's Theorem

A6–CIRCUIT COMPONENTS [6 questions–6 groups]

A6A Semiconductor material: Germanium, Silicon, P-type, N-type

A6A01 (B)
What two elements widely used in semiconductor devices exhibit both metallic and nonmetallic characteristics?
A. Silicon and gold
B. Silicon and germanium
C. Galena and germanium
D. Galena and bismuth

A6A02 (C)
In what application is gallium arsenide used as a semiconductor material in preference to germanium or silicon?
A. In bipolar transistors
B. In high-power circuits
C. At microwave frequencies
D. At very low frequencies

A6A03 (C)
What type of semiconductor material might be produced

by adding some antimony atoms to germanium crystals?
A. J-type
B. MOS-type
C. N-type
D. P-type

A6A04 (B)
What type of semiconductor material might be produced by adding some gallium atoms to silicon crystals?
A. N-type
B. P-type
C. MOS-type
D. J-type

A6A05 (A)
What type of semiconductor material contains more free electrons than pure germanium or silicon crystals?
A. N-type
B. P-type
C. Bipolar
D. Insulated gate

A6A06 (A)
What type of semiconductor material might be produced by adding some arsenic atoms to silicon crystals?
A. N-type
B. P-type
C. MOS-type
D. J-type

A6A07 (D)
What type of semiconductor material might be produced by adding some indium atoms to germanium crystals?
A. J-type
B. MOS-type
C. N-type
D. P-type

A6A08 (B)
What type of semiconductor material contains fewer free electrons than pure germanium or silicon crystals?
A. N-type
B. P-type
C. Superconductor-type
D. Bipolar-type

A6A09 (C)
What are the majority charge carriers in P-type semiconductor material?
A. Free neutrons
B. Free protons
C. Holes
D. Free electrons

A6A10 (B)
What are the majority charge carriers in N-type semiconductor material?
A. Holes
B. Free electrons
C. Free protons
D. Free neutrons

A6A11 (B)
What is the name given to an impurity atom that provides excess electrons to a semiconductor crystal structure?

A. Acceptor impurity
B. Donor impurity
C. P-type impurity
D. Conductor impurity

A6A12 (C)
What is the name given to an impurity atom that adds holes to a semiconductor crystal structure?
A. Insulator impurity
B. N-type impurity
C. Acceptor impurity
D. Donor impurity

A6B Diodes: Zener, Tunnel, Varactor, Hot-carrier, Junction, Point contact, PIN and Light-emitting

A6B01 (B)
What is the principal characteristic of a Zener diode?
A. A constant current under conditions of varying voltage
B. A constant voltage under conditions of varying current
C. A negative resistance region
D. An internal capacitance that varies with the applied voltage

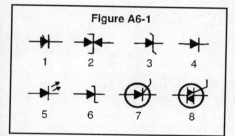

Figure A6-1

A6B02 (D)
In Figure A6-1, what is the schematic symbol for a Zener diode?
A. 7
B. 6
C. 4
D. 3

A6B03 (C)
What is the principal characteristic of a tunnel diode?
A. A high forward resistance
B. A very high PIV
C. A negative resistance region
D. A high forward current rating

A6B04 (C)
What special type of diode is capable of both amplification and oscillation?
A. Point contact
B. Zener
C. Tunnel
D. Junction

A6B05 (C)
In Figure A6-1, what is the schematic symbol for a tunnel diode?
A. 8

B. 6
C. 2
D. 1

A6B06 (A)
What type of semiconductor diode varies its internal capacitance as the voltage applied to its terminals varies?
A. Varactor
B. Tunnel
C. Silicon-controlled rectifier
D. Zener

A6B07 (D)
In Figure A6-1, what is the schematic symbol for a varactor diode?
A. 8
B. 6
C. 2
D. 1

A6B08 (D)
What is a common use of a hot-carrier diode?
A. As balanced mixers in FM generation
B. As a variable capacitance in an automatic frequency control circuit
C. As a constant voltage reference in a power supply
D. As VHF and UHF mixers and detectors

A6B09 (B)
What limits the maximum forward current in a junction diode?
A. Peak inverse voltage
B. Junction temperature
C. Forward voltage
D. Back EMF

A6B10 (D)
How are junction diodes rated?
A. Maximum forward current and capacitance
B. Maximum reverse current and PIV
C. Maximum reverse current and capacitance
D. Maximum forward current and PIV

A6B11 (A)
Structurally, what are the two main categories of semiconductor diodes?
A. Junction and point contact
B. Electrolytic and junction
C. Electrolytic and point contact
D. Vacuum and point contact

A6B12 (C)
What is a common use for point contact diodes?
A. As a constant current source
B. As a constant voltage source
C. As an RF detector
D. As a high voltage rectifier

A6B13 (D)
In Figure A6-1, what is the schematic symbol for a semiconductor diode/rectifier?
A. 1
B. 2
C. 3
D. 4

A6B14 (C)
What is one common use for PIN diodes?
A. As a constant current source
B. As a constant voltage source
C. As an RF switch
D. As a high voltage rectifier

A6B15 (B)
In Figure A6-1, what is the schematic symbol for a light-emitting diode?
A. 1
B. 5
C. 6
D. 7

A6B16 (B)
What type of bias is required for an LED to produce luminescence?
A. Reverse bias
B. Forward bias
C. Zero bias
D. Inductive bias

A6C Toroids: Permeability, core material, selecting, winding

A6C01 (D)
What material property determines the inductance of a toroidal inductor with a 10-turn winding?
A. Core load current
B. Core resistance
C. Core reactivity
D. Core permeability

A6C02 (B)
By careful selection of core material, over what frequency range can toroidal cores produce useful inductors?
A. From a few kHz to no more than several MHz
B. From DC to at least 1000 MHz
C. From DC to no more than 3000 kHz
D. From a few hundred MHz to at least 1000 GHz

A6C03 (A)
What materials are used to make ferromagnetic inductors and transformers?
A. Ferrite and powdered-iron toroids
B. Silicon-ferrite toroids and shellac
C. Powdered-ferrite and silicon toroids
D. Ferrite and silicon-epoxy toroids

A6C04 (B)
What is one important reason for using powdered-iron toroids rather than ferrite toroids in an inductor?
A. Powdered-iron toroids generally have greater initial permeabilities
B. Powdered-iron toroids generally have better temperature stability
C. Powdered-iron toroids generally require fewer turns to produce a given inductance value
D. Powdered-iron toroids are easier to use with surface-mount technology

A6C05 (C)
What is one important reason for using ferrite toroids rather than powdered-iron toroids in an inductor?

A. Ferrite toroids generally have lower initial permeabilities
B. Ferrite toroids generally have better temperature stability
C. Ferrite toroids generally require fewer turns to produce a given inductance value
D. Ferrite toroids are easier to use with surface-mount technology

A6C06 (B)
What would be a good choice of toroid core material to make a common-mode choke (such as winding telephone wires or stereo speaker leads on a core) to cure an HF RFI problem?
A. Type 61 mix ferrite (initial permeability of 125)
B. Type 43 mix ferrite (initial permeability of 850)
C. Type 6 mix powdered iron (initial permeability of 8)
D. Type 12 mix powdered iron (initial permeability of 3)

A6C07 (C)
What devices are commonly used as parasitic suppressors at the input and output terminals of VHF and UHF amplifiers?
A. Electrolytic capacitors
B. Butterworth filters
C. Ferrite beads
D. Steel-core toroids

A6C08 (A)
What is a primary advantage of using a toroidal core instead of a linear core in an inductor?
A. Toroidal cores contain most of the magnetic field within the core material
B. Toroidal cores make it easier to couple the magnetic energy into other components
C. Toroidal cores exhibit greater hysteresis
D. Toroidal cores have lower Q characteristics

A6C09 (D)
What is a bifilar-wound toroid?
A. An inductor that has two cores taped together to double the inductance value
B. An inductor wound on a core with two holes (binocular core)
C. A transformer designed to provide a 2-to-1 impedance transformation
D. An inductor that uses a pair of wires to place two windings on the core

A6C10 (C)
How many turns will be required to produce a 1-mH inductor using a ferrite toroidal core that has an inductance index (A sub L) value of 523?
A. 2 turns
B. 4 turns
C. 43 turns
D. 229 turns

A6C11 (A)
How many turns will be required to produce a 5-microhenry inductor using a powdered-iron toroidal core that has an inductance index (A sub L) value of 40?
A. 35 turns
B. 13 turns
C. 79 turns
D. 141 turns

A6D Transistor types: NPN, PNP, Junction, Unijunction

A6D01 (B)
What are the three terminals of a bipolar transistor?
A. Cathode, plate and grid
B. Base, collector and emitter
C. Gate, source and sink
D. Input, output and ground

A6D02 (C)
What is the alpha of a bipolar transistor?
A. The change of collector current with respect to base current
B. The change of base current with respect to collector current
C. The change of collector current with respect to emitter current
D. The change of collector current with respect to gate current

A6D03 (A)
What is the beta of a bipolar transistor?
A. The change of collector current with respect to base current
B. The change of base current with respect to emitter current
C. The change of collector current with respect to emitter current
D. The change of base current with respect to gate current

A6D04 (D)
What is the alpha cutoff frequency of a bipolar transistor?
A. The practical lower frequency limit of a transistor in common emitter configuration
B. The practical upper frequency limit of a transistor in common emitter configuration
C. The practical lower frequency limit of a transistor in common base configuration
D. The practical upper frequency limit of a transistor in common base configuration

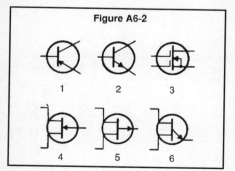

Figure A6-2

1   2   3

4   5   6

A6D05 (B)
In Figure A6-2, what is the schematic symbol for an NPN transistor?
A. 1
B. 2
C. 4
D. 5

A6D06 (A)
In Figure A6-2, what is the schematic symbol for a PNP transistor?
A. 1
B. 2
C. 4
D. 5

A6D07 (D)
What term indicates the frequency at which a transistor grounded base current gain has decreased to 0.7 of the gain obtainable at 1 kHz?
A. Corner frequency
B. Alpha rejection frequency
C. Beta cutoff frequency
D. Alpha cutoff frequency

A6D08 (B)
What does the beta cutoff of a bipolar transistor indicate?
A. The frequency at which the grounded base current gain has decreased to 0.7 of that obtainable at 1 kHz
B. The frequency at which the grounded emitter current gain has decreased to 0.7 of that obtainable at 1 kHz
C. The frequency at which the grounded collector current gain has decreased to 0.7 of that obtainable at 1 kHz
D. The frequency at which the grounded gate current gain has decreased to 0.7 of that obtainable at 1 kHz

A6D09 (A)
What is the transition region of a transistor?
A. An area of low charge density around the P-N junction
B. The area of maximum P-type charge
C. The area of maximum N-type charge
D. The point where wire leads are connected to the P- or N-type material

A6D10 (A)
What does it mean for a transistor to be fully saturated?
A. The collector current is at its maximum value
B. The collector current is at its minimum value
C. The transistor alpha is at its maximum value
D. The transistor beta is at its maximum value

A6D11 (C)
What does it mean for a transistor to be cut off?
A. There is no base current
B. The transistor is at its operating point
C. No current flows from emitter to collector
D. Maximum current flows from emitter to collector

A6D12 (D)
In Figure A6-2, what is the schematic symbol for a unijunction transistor?
A. 3
B. 4
C. 5
D. 6

A6D13 (C)
What are the elements of a unijunction transistor?
A. Gate, base 1 and base 2
B. Gate, cathode and anode
C. Base 1, base 2 and emitter
D. Gate, source and sink

A6E   Silicon controlled rectifier (SCR); Triac; neon lamp

A6E01 (B)
What are the three terminals of a silicon controlled rectifier (SCR)?
A. Gate, source and sink
B. Anode, cathode and gate
C. Base, collector and emitter
D. Gate, base 1 and base 2

A6E02 (A)
What are the two stable operating conditions of a silicon controlled rectifier (SCR)?
A. Conducting and nonconducting
B. Oscillating and quiescent
C. Forward conducting and reverse conducting
D. NPN conduction and PNP conduction

A6E03 (A)
When a silicon controlled rectifier (SCR) is triggered, to what other solid-state device are its electrical characteristics similar (as measured between its cathode and anode)?
A. The junction diode
B. The tunnel diode
C. The hot-carrier diode
D. The varactor diode

A6E04 (D)
Under what operating conditions does a silicon controlled rectifier (SCR) exhibit electrical characteristics similar to a forward-biased silicon rectifier?
A. During a switching transition
B. When it is used as a detector
C. When it is gated "off"
D. When it is gated "on"

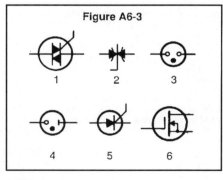

Figure A6-3

1   2   3

4   5   6

A6E05 (C)
In Figure A6-3, what is the schematic symbol for a silicon controlled rectifier (SCR)?
A. 1
B. 2
C. 5
D. 6

A6E06 (B)
What is the name of the device that is fabricated as two complementary silicon controlled rectifiers (SCRs) in parallel with a common gate terminal?
A. Bilateral SCR
B. TRIAC
C. Unijunction transistor
D. Field-effect transistor

A6E07 (B)
What are the three terminals of a TRIAC?
A. Emitter, base 1 and base 2
B. Gate, anode 1 and anode 2
C. Base, emitter and collector
D. Gate, source and sink

A6E08 (A)
In Figure A6-3, what is the schematic symbol for a TRIAC?
A. 1
B. 2
C. 3
D. 5

A6E09 (D)
What will happen to a neon lamp in the presence of RF?
A. It will glow only in the presence of very high frequency radio energy
B. It will change color
C. It will glow only in the presence of very low frequency radio energy
D. It will glow

A6E10 (C)
If an NE-2 neon bulb is to be used as a dial lamp with a 120 V AC line, what additional component must be connected to it?
A. A 150-pF capacitor in parallel with the bulb
B. A 10-mH inductor in series with the bulb
C. A 150-kilohm resistor in series with the bulb
D. A 10-kilohm resistor in parallel with the bulb

A6E11 (C)
In Figure A6-3, what is the schematic symbol for a neon lamp?
A. 1
B. 2
C. 3
D. 4

A6F   Quartz crystal (frequency determining properties as used in oscillators and filters); monolithic amplifiers (MMICs)

A6F01 (B)
For single-sideband phone emissions, what would be the bandwidth of a good crystal lattice band-pass filter?
A. 6 kHz at -6 dB
B. 2.1 kHz at -6 dB
C. 500 Hz at -6 dB
D. 15 kHz at -6 dB

A6F02 (C)
For double-sideband phone emissions, what would be the bandwidth of a good crystal lattice band-pass filter?
A. 1 kHz at -6 dB
B. 500 Hz at -6 dB
C. 6 kHz at -6 dB
D. 15 kHz at -6 dB

A6F03 (D)
What is a crystal lattice filter?
A. A power supply filter made with interlaced quartz crystals
B. An audio filter made with four quartz crystals that resonate at 1- kHz intervals
C. A filter with wide bandwidth and shallow skirts made using quartz crystals
D. A filter with narrow bandwidth and steep skirts made using quartz crystals

A6F04 (D)
What technique is used to construct low-cost, high-performance crystal filters?
A. Choose a center frequency that matches the available crystals
B. Choose a crystal with the desired bandwidth and operating frequency to match a desired center frequency
C. Measure crystal bandwidth to ensure at least 20% coupling
D. Measure crystal frequencies and carefully select units with less than 10% frequency difference

A6F05 (A)
Which factor helps determine the bandwidth and response shape of a crystal filter?
A. The relative frequencies of the individual crystals
B. The center frequency chosen for the filter
C. The gain of the RF stage preceding the filter
D. The amplitude of the signals passing through the filter

A6F06 (A)
What is the piezoelectric effect?
A. Physical deformation of a crystal by the application of a voltage
B. Mechanical deformation of a crystal by the application of a magnetic field
C. The generation of electrical energy by the application of light
D. Reversed conduction states when a P-N junction is exposed to light

A6F07 (C)
Which of the following devices would be most suitable for constructing a receive preamplifier for 1296 MHz?
A. A 2N2222 bipolar transistor
B. An MRF901 bipolar transistor
C. An MSA-0135 monolithic microwave integrated circuit (MMIC)
D. An MPF102 N-junction field-effect transistor (JFET)

A6F08 (A)
Which device might be used to simplify the design and construction of a 3456-MHz receiver?
A. An MSA-0735 monolithic microwave integrated circuit (MMIC).
B. An MRF901 bipolar transistor
C. An MGF1402 gallium arsenide field-effect transistor (GaAsFET)
D. An MPF102 N-junction field-effect transistor (JFET)

A6F09 (D)
What type of amplifier device consists of a small "pill sized" package with an input lead, an output lead and 2 ground leads?

A. A gallium arsenide field-effect transistor (GaAsFET)
B. An operational amplifier integrated circuit (OAIC)
C. An indium arsenide integrated circuit (IAIC)
D. A monolithic microwave integrated circuit (MMIC)

A6F10 (B)
What typical construction technique do amateurs use when building an amplifier containing a monolithic microwave integrated circuit (MMIC)?
A. Ground-plane "ugly" construction
B. Microstrip construction
C. Point-to-point construction
D. Wave-soldering construction

A6F11 (A)
How is the operating bias voltage supplied to a monolithic microwave integrated circuit (MMIC)?
A. Through a resistor and RF choke connected to the amplifier output lead
B. MMICs require no operating bias
C. Through a capacitor and RF choke connected to the amplifier input lead
D. Directly to the bias-voltage (VCC IN) lead

A7–PRACTICAL CIRCUITS [10 questions–10 groups]

A7A Amplifier circuits: Class A, Class AB, Class B, Class C, amplifier operating efficiency (i.e., DC input vs. PEP); transmitter final amplifiers

A7A01 (B)
For what portion of a signal cycle does a Class A amplifier operate?
A. Less than 180 degrees
B. The entire cycle
C. More than 180 degrees and less than 360 degrees
D. Exactly 180 degrees

A7A02 (A)
Which class of amplifier has the highest linearity and least distortion?
A. Class A
B. Class B
C. Class C
D. Class AB

A7A03 (A)
For what portion of a signal cycle does a Class AB amplifier operate?
A. More than 180 degrees but less than 360 degrees
B. Exactly 180 degrees
C. The entire cycle
D. Less than 180 degrees

A7A04 (D)
For what portion of a signal cycle does a Class B amplifier operate?
A. The entire cycle
B. Greater than 180 degrees and less than 360 degrees
C. Less than 180 degrees
D. 180 degrees

A7A05 (A)
For what portion of a signal cycle does a Class C amplifier operate?

A. Less than 180 degrees
B. Exactly 180 degrees
C. The entire cycle
D. More than 180 degrees but less than 360 degrees

A7A06 (C)
Which class of amplifier provides the highest efficiency?
A. Class A
B. Class B
C. Class C
D. Class AB

A7A07 (A)
Where on the load line should a solid-state power amplifier be operated for best efficiency and stability?
A. Just below the saturation point
B. Just above the saturation point
C. At the saturation point
D. At 1.414 times the saturation point

A7A08 (A)
What is the formula for the efficiency of a power amplifier?
A. Efficiency = (RF power out / DC power in) x 100%
B. Efficiency = (RF power in / RF power out) x 100%
C. Efficiency = (RF power in / DC power in) x 100%
D. Efficiency = (DC power in / RF power in) x 100%

A7A09 (C)
How can parasitic oscillations be eliminated from a power amplifier?
A. By tuning for maximum SWR
B. By tuning for maximum power output
C. By neutralization
D. By tuning the output

A7A10 (D)
What is the procedure for tuning a vacuum-tube power amplifier having an output pi-network?
A. Adjust the loading capacitor to maximum capacitance and then dip the plate current with the tuning capacitor
B. Alternately increase the plate current with the tuning capacitor and dip the plate current with the loading capacitor
C. Adjust the tuning capacitor to maximum capacitance and then dip the plate current with the loading capacitor
D. Alternately increase the plate current with the loading capacitor and dip the plate current with the tuning capacitor

A7A11 (B)
How can even-order harmonics be reduced or prevented in transmitter amplifiers?
A. By using a push-push amplifier
B. By using a push-pull amplifier
C. By operating Class C
D. By operating Class AB

A7A12 (D)
What can occur when a nonlinear amplifier is used with a single- sideband phone transmitter?
A. Reduced amplifier efficiency
B. Increased intelligibility
C. Sideband inversion
D. Distortion

A7B Amplifier circuits: tube, bipolar transistor, FET

A7B01 (C)
How can a vacuum-tube power amplifier be neutralized?
A. By increasing the grid drive
B. By feeding back an in-phase component of the output to the input
C. By feeding back an out-of-phase component of the output to the input
D. By feeding back an out-of-phase component of the input to the output

A7B02 (B)
What is the flywheel effect?
A. The continued motion of a radio wave through space when the transmitter is turned off
B. The back and forth oscillation of electrons in an LC circuit
C. The use of a capacitor in a power supply to filter rectified AC
D. The transmission of a radio signal to a distant station by several hops through the ionosphere

A7B03 (B)
What tank-circuit Q is required to reduce harmonics to an acceptable level?
A. Approximately 120
B. Approximately 12
C. Approximately 1200
D. Approximately 1.2

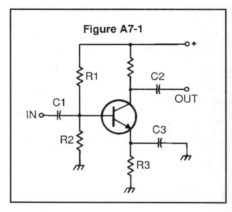

Figure A7-1

A7B04 (C)
What type of circuit is shown in Figure A7-1?
A. Switching voltage regulator
B. Linear voltage regulator
C. Common emitter amplifier
D. Emitter follower amplifier

A7B05 (B)
In Figure A7-1, what is the purpose of R1 and R2?
A. Load resistors
B. Fixed bias
C. Self bias
D. Feedback

A7B06 (D)
In Figure A7-1, what is the purpose of C1?
A. Decoupling
B. Output coupling
C. Self bias
D. Input coupling

A7B07 (D)
In Figure A7-1, what is the purpose of C3?
A. AC feedback
B. Input coupling
C. Power supply decoupling
D. Emitter bypass

A7B08 (D)
In Figure A7-1, what is the purpose of R3?
A. Fixed bias
B. Emitter bypass
C. Output load resistor
D. Self bias

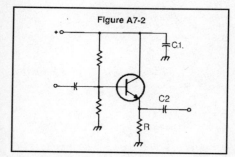

Figure A7-2

A7B09 (B)
What type of circuit is shown in Figure A7-2?
A. High-gain amplifier
B. Common-collector amplifier
C. Linear voltage regulator
D. Grounded-emitter amplifier

A7B10 (A)
In Figure A7-2, what is the purpose of R?
A. Emitter load
B. Fixed bias
C. Collector load
D. Voltage regulation

A7B11 (D)
In Figure A7-2, what is the purpose of C1?
A. Input coupling
B. Output coupling
C. Emitter bypass
D. Collector bypass

A7B12 (A)
In Figure A7-2, what is the purpose of C2?
A. Output coupling
B. Emitter bypass
C. Input coupling
D. Hum filtering

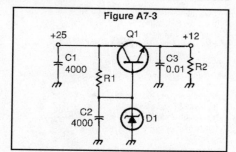

Figure A7-3

A7B13 (C)
What type of circuit is shown in Figure A7-3?
A. Switching voltage regulator
B. Grounded emitter amplifier
C. Linear voltage regulator
D. Emitter follower

A7B14 (B)
What is the purpose of D1 in the circuit shown in Figure A7-3?
A. Line voltage stabilization
B. Voltage reference
C. Peak clipping
D. Hum filtering

A7B15 (C)
What is the purpose of Q1 in the circuit shown in Figure A7-3?
A. It increases the output ripple
B. It provides a constant load for the voltage source
C. It increases the current-handling capability
D. It provides D1 with current

A7B16 (D)
What is the purpose of C1 in the circuit shown in Figure A7-3?
A. It resonates at the ripple frequency
B. It provides fixed bias for Q1
C. It decouples the output
D. It filters the supply voltage

A7B17 (A)
What is the purpose of C2 in the circuit shown in Figure A7-3?
A. It bypasses hum around D1
B. It is a brute force filter for the output
C. To self resonate at the hum frequency
D. To provide fixed DC bias for Q1

A7B18 (A)
What is the purpose of C3 in the circuit shown in Figure A7-3?
A. It prevents self-oscillation
B. It provides brute force filtering of the output
C. It provides fixed bias for Q1
D. It clips the peaks of the ripple

A7B19 (C)
What is the purpose of R1 in the circuit shown in Figure A7-3?

A. It provides a constant load to the voltage source
B. It couples hum to D1
C. It supplies current to D1
D. It bypasses hum around D1

A7B20 (D)
What is the purpose of R2 in the circuit shown in Figure A7-3?
A. It provides fixed bias for Q1
B. It provides fixed bias for D1
C. It decouples hum from D1
D. It provides a constant minimum load for Q1

A7C Impedance-matching networks: Pi, L, Pi-L

A7C01 (D)
What is a pi-network?
A. A network consisting entirely of four inductors or four capacitors
B. A Power Incidence network
C. An antenna matching network that is isolated from ground
D. A network consisting of one inductor and two capacitors or two
inductors and one capacitor

A7C02 (B)
Which type of network offers the greater transformation ratio?
A. L-network
B. Pi-network
C. Constant-K
D. Constant-M

A7C03 (D)
How are the capacitors and inductors of a pi-network arranged between the network's input and output?
A. Two inductors are in series between the input and output and a capacitor is connected between the two inductors and ground
B. Two capacitors are in series between the input and output and an inductor is connected between the two capacitors and ground
C. An inductor is in parallel with the input, another inductor is in parallel with the output, and a capacitor is in series between the two
D. A capacitor is in parallel with the input, another capacitor is in parallel with the output, and an inductor is in series between the two

A7C04 (B)
What is an L-network?
A. A network consisting entirely of four inductors
B. A network consisting of an inductor and a capacitor
C. A network used to generate a leading phase angle
D. A network used to generate a lagging phase angle

A7C05 (A)
Why is an L-network of limited utility in impedance matching?
A. It matches a small impedance range
B. It has limited power-handling capabilities
C. It is thermally unstable
D. It is prone to self resonance

A7C06 (B)
What is a pi-L-network?
A. A Phase Inverter Load network
B. A network consisting of two inductors and two capacitors
C. A network with only three discrete parts
D. A matching network in which all components are isolated from ground

A7C07 (C)
A T-network with series capacitors and a parallel (shunt) inductor has which of the following properties?
A. It transforms impedances and is a low-pass filter
B. It transforms reactances and is a low-pass filter
C. It transforms impedances and is a high-pass filter
D. It transforms reactances and is a high-pass filter

A7C08 (A)
What advantage does a pi-L-network have over a pi-network for impedance matching between the final amplifier of a vacuum-tube type transmitter and a multiband antenna?
A. Greater harmonic suppression
B. Higher efficiency
C. Lower losses
D. Greater transformation range

A7C09 (C)
Which type of network provides the greatest harmonic suppression?
A. L-network
B. Pi-network
C. Pi-L-network
D. Inverse-Pi network

A7C10 (C)
Which three types of networks are most commonly used to match an amplifying device and a transmission line?
A. M, pi and T
B. T, M and Q
C. L, pi and pi-L
D. L, M and C

A7C11 (C)
How does a network transform one impedance to another?
A. It introduces negative resistance to cancel the resistive part of an impedance
B. It introduces transconductance to cancel the reactive part of an impedance
C. It cancels the reactive part of an impedance and changes the resistive part
D. Network resistances substitute for load resistances

A7D Filter circuits: constant K, M-derived, band-stop, notch, crystal lattice, Pi-section, T-section, L-section, Butterworth, Chebyshev, elliptical

A7D01 (A)
What are the three general groupings of filters?
A. High-pass, low-pass and band-pass
B. Inductive, capacitive and resistive
C. Audio, radio and capacitive
D. Hartley, Colpitts and Pierce

A7D02 (B)
What value capacitor would be required to tune a 20-microhenry inductor to resonate in the 80-meter band?
A. 150 picofarads
B. 100 picofarads
C. 200 picofarads
D. 100 microfarads

A7D03 (D)
What value inductor would be required to tune a 100-picofarad capacitor to resonate in the 40-meter band?
A. 200 microhenrys
B. 150 microhenrys
C. 5 millihenrys
D. 5 microhenrys

A7D04 (A)
What value capacitor would be required to tune a 2-microhenry inductor to resonate in the 20-meter band?
A. 64 picofarads
B. 6 picofarads
C. 12 picofarads
D. 88 microfarads

A7D05 (C)
What value inductor would be required to tune a 15-picofarad capacitor to resonate in the 15-meter band?
A. 2 microhenrys
B. 30 microhenrys
C. 4 microhenrys
D. 15 microhenrys

A7D06 (A)
What value capacitor would be required to tune a 100-microhenry inductor to resonate in the 160-meter band?
A. 78 picofarads
B. 25 picofarads
C. 405 picofarads
D. 40.5 microfarads

A7D07 (C)
What are the distinguishing features of a Butterworth filter?
A. The product of its series- and shunt-element impedances is a constant for all frequencies
B. It only requires capacitors
C. It has a maximally flat response over its passband
D. It requires only inductors

A7D08 (B)
What are the distinguishing features of a Chebyshev filter?
A. It has a maximally flat response over its passband
B. It allows ripple in the passband
C. It only requires inductors
D. The product of its series- and shunt-element impedances is a constant for all frequencies

A7D09 (D)
Which filter type is described as having ripple in the passband and a sharp cutoff?
A. A Butterworth filter
B. An active LC filter
C. A passive op-amp filter
D. A Chebyshev filter

A7D10 (C)
What are the distinguishing features of an elliptical filter?
A. Gradual passband rolloff with minimal stop-band ripple
B. Extremely flat response over its passband, with gradually rounded stop-band corners
C. Extremely sharp cutoff, with one or more infinitely deep notches in the stop band
D. Gradual passband rolloff with extreme stop-band ripple

A7D11 (B)
Which filter type has an extremely sharp cutoff, with one or more infinitely deep notches in the stop band?
A. Chebyshev
B. Elliptical
C. Butterworth
D. Crystal lattice

A7E Voltage-regulator circuits: discrete, integrated and switched mode

A7E01 (D)
What is one characteristic of a linear electronic voltage regulator? A. It has a ramp voltage as its output
B. The pass transistor switches from the "off" state to the "on" state
C. The control device is switched on or off, with the duty cycle proportional to the line or load conditions
D. The conduction of a control element is varied in direct proportion
to the line voltage or load current

A7E02 (C)
What is one characteristic of a switching electronic voltage regulator?
A. The conduction of a control element is varied in direct proportion
to the line voltage or load current
B. It provides more than one output voltage
C. The control device is switched on or off, with the duty cycle proportional to the line or load conditions
D. It gives a ramp voltage at its output

A7E03 (A)
What device is typically used as a stable reference voltage in a linear voltage regulator?
A. A Zener diode
B. A tunnel diode
C. An SCR
D. A varactor diode

A7E04 (B)
What type of linear regulator is used in applications requiring efficient utilization of the primary power source?
A. A constant current source
B. A series regulator
C. A shunt regulator
D. A shunt current source

A7E05 (D)
What type of linear voltage regulator is used in applications requiring a constant load on the unregulated voltage source?
A. A constant current source
B. A series regulator

C. A shunt current source
D. A shunt regulator

A7E06 (C)
To obtain the best temperature stability, approximately what operating voltage should be used for the reference diode in a linear voltage regulator?
A. 2 volts
B. 3 volts
C. 6 volts
D. 10 volts

A7E07 (A)
How is remote sensing accomplished in a linear voltage regulator?
A. A feedback connection to an error amplifier is made directly to the load
B. By wireless inductive loops
C. A load connection is made outside the feedback loop
D. An error amplifier compares the input voltage to the reference voltage

A7E08 (D)
What is a three-terminal regulator?
A. A regulator that supplies three voltages with variable current
B. A regulator that supplies three voltages at a constant current
C. A regulator containing three error amplifiers and sensing transistors
D. A regulator containing a voltage reference, error amplifier, sensing resistors and transistors, and a pass element

A7E09 (B)
What are the important characteristics of a three-terminal regulator?
A. Maximum and minimum input voltage, minimum output current and voltage
B. Maximum and minimum input voltage, maximum output current and voltage
C. Maximum and minimum input voltage, minimum output current and maximum output voltage
D. Maximum and minimum input voltage, minimum output voltage and maximum output current

A7E10 (A)
What type of voltage regulator limits the voltage drop across its junction when a specified current passes through it in the reverse- breakdown direction?
A. A Zener diode
B. A three-terminal regulator
C. A bipolar regulator
D. A pass-transistor regulator

A7E11 (C)
What type of voltage regulator contains a voltage reference, error amplifier, sensing resistors and transistors, and a pass element in one package?
A. A switching regulator
B. A Zener regulator
C. A three-terminal regulator
D. An op-amp regulator

A7F Oscillators: types, applications, stability

A7F01 (D)
What are three major oscillator circuits often used in Amateur Radio equipment?
A. Taft, Pierce and negative feedback
B. Colpitts, Hartley and Taft
C. Taft, Hartley and Pierce
D. Colpitts, Hartley and Pierce

A7F02 (C)
What condition must exist for a circuit to oscillate?
A. It must have a gain of less than 1
B. It must be neutralized
C. It must have positive feedback sufficient to overcome losses
D. It must have negative feedback sufficient to cancel the input

A7F03 (A)
How is the positive feedback coupled to the input in a Hartley oscillator?
A. Through a tapped coil
B. Through a capacitive divider
C. Through link coupling
D. Through a neutralizing capacitor

A7F04 (C)
How is the positive feedback coupled to the input in a Colpitts oscillator?
A. Through a tapped coil
B. Through link coupling
C. Through a capacitive divider
D. Through a neutralizing capacitor

A7F05 (D)
How is the positive feedback coupled to the input in a Pierce oscillator?
A. Through a tapped coil
B. Through link coupling
C. Through a neutralizing capacitor
D. Through capacitive coupling

A7F06 (D)
Which of the three major oscillator circuits used in Amateur Radio equipment uses a quartz crystal?
A. Negative feedback
B. Hartley
C. Colpitts
D. Pierce

A7F07 (B)
What is the major advantage of a Pierce oscillator?
A. It is easy to neutralize
B. It doesn't require an LC tank circuit
C. It can be tuned over a wide range
D. It has a high output power

A7F08 (B)
Which type of oscillator circuits are commonly used in a VFO?
A. Pierce and Zener
B. Colpitts and Hartley
C. Armstrong and deForest
D. Negative feedback and Balanced feedback

A7F09 (C)
Why is the Colpitts oscillator circuit commonly used in a VFO?
A. The frequency is a linear function of the load impedance
B. It can be used with or without crystal lock-in
C. It is stable
D. It has high output power

A7F10 (A)
What component is often used to control an oscillator frequency by varying a control voltage?
A. A varactor diode
B. A piezoelectric crystal
C. A Zener diode
D. A Pierce crystal

A7F11 (B)
Why must a very stable reference oscillator be used as part of a phase-locked loop (PLL) frequency synthesizer?
A. Any amplitude variations in the reference oscillator signal will prevent the loop from locking to the desired signal
B. Any phase variations in the reference oscillator signal will produce phase noise in the synthesizer output
C. Any phase variations in the reference oscillator signal will produce harmonic distortion in the modulating signal
D. Any amplitude variations in the reference oscillator signal will prevent the loop from changing frequency

A7G Modulators: Reactance, Phase, Balanced

A7G01 (D)
What is meant by modulation?
A. The squelching of a signal until a critical signal-to-noise ratio is reached
B. Carrier rejection through phase nulling
C. A linear amplification mode
D. A mixing process whereby information is imposed upon a carrier

A7G02 (B)
How is an F3E FM-phone emission produced?
A. With a balanced modulator on the audio amplifier
B. With a reactance modulator on the oscillator
C. With a reactance modulator on the final amplifier
D. With a balanced modulator on the oscillator

A7G03 (C)
How does a reactance modulator work?
A. It acts as a variable resistance or capacitance to produce FM signals
B. It acts as a variable resistance or capacitance to produce AM signals
C. It acts as a variable inductance or capacitance to produce FM signals
D. It acts as a variable inductance or capacitance to produce AM signals

A7G04 (B)
What type of circuit varies the tuning of an oscillator circuit to produce FM signals?
A. A balanced modulator
B. A reactance modulator
C. A double balanced mixer
D. An audio modulator

A7G05 (C)
How does a phase modulator work?
A. It varies the tuning of a microphone preamplifier to produce FM signals
B. It varies the tuning of an amplifier tank circuit to produce AM signals
C. It varies the tuning of an amplifier tank circuit to produce FM signals
D. It varies the tuning of a microphone preamplifier to produce AM signals

A7G06 (C)
What type of circuit varies the tuning of an amplifier tank circuit to produce FM signals?
A. A balanced modulator
B. A double balanced mixer
C. A phase modulator
D. An audio modulator

A7G07 (B)
What type of signal does a balanced modulator produce?
A. FM with balanced deviation
B. Double sideband, suppressed carrier
C. Single sideband, suppressed carrier
D. Full carrier

A7G08 (A)
How can a single-sideband phone signal be generated?
A. By using a balanced modulator followed by a filter
B. By using a reactance modulator followed by a mixer
C. By using a loop modulator followed by a mixer
D. By driving a product detector with a DSB signal

A7G09 (D)
How can a double-sideband phone signal be generated?
A. By feeding a phase modulated signal into a low-pass filter
B. By using a balanced modulator followed by a filter
C. By detuning a Hartley oscillator
D. By modulating the plate voltage of a Class C amplifier

A7G10 (D)
What audio shaping network is added at a transmitter to proportionally attenuate the lower audio frequencies, giving an even spread to the energy in the audio band?
A. A de-emphasis network
B. A heterodyne suppressor
C. An audio prescaler
D. A pre-emphasis network

A7G11 (A)
What audio shaping network is added at a receiver to restore proportionally attenuated lower audio frequencies?
A. A de-emphasis network
B. A heterodyne suppressor
C. An audio prescaler
D. A pre-emphasis network

A7H Detectors; filter applications (audio, IF, Digital signal processing DSP)

A7H01 (B)
What is the process of detection?
A. The masking of the intelligence on a received carrier

B. The recovery of the intelligence from a modulated RF signal
C. The modulation of a carrier
D. The mixing of noise with a received signal

A7H02 (A)
What is the principle of detection in a diode detector?
A. Rectification and filtering of RF
B. Breakdown of the Zener voltage
C. Mixing with noise in the transition region of the diode
D. The change of reactance in the diode with respect to frequency

A7H03 (C)
What does a product detector do?
A. It provides local oscillations for input to a mixer
B. It amplifies and narrows band-pass frequencies
C. It mixes an incoming signal with a locally generated carrier
D. It detects cross-modulation products

A7H04 (B)
How are FM-phone signals detected?
A. With a balanced modulator
B. With a frequency discriminator
C. With a product detector
D. With a phase splitter

A7H05 (D)
What is a frequency discriminator?
A. An FM generator
B. A circuit for filtering two closely adjacent signals
C. An automatic band-switching circuit
D. A circuit for detecting FM signals

A7H06 (A)
Which of the following is NOT an advantage of using active filters rather than L-C filters at audio frequencies?
A. Active filters have higher signal-to-noise ratios
B. Active filters can provide gain as well as frequency selection
C. Active filters do not require the use of inductors
D. Active filters can use potentiometers for tuning

A7H07 (B)
What kind of audio filter would you use to attenuate an interfering carrier signal while receiving an SSB transmission?
A. A band-pass filter
B. A notch filter
C. A pi-network filter
D. An all-pass filter

A7H08 (D)
What characteristic do typical SSB receiver IF filters lack that is important to digital communications?
A. Steep amplitude-response skirts
B. Passband ripple
C. High input impedance
D. Linear phase response

A7H09 (A)
What kind of digital signal processing audio filter might be used to remove unwanted noise from a received SSB signal?

A. An adaptive filter
B. A notch filter
C. A Hilbert-transform filter
D. A phase-inverting filter

A7H10 (C)
What kind of digital signal processing filter might be used in generating an SSB signal?
A. An adaptive filter
B. A notch filter
C. A Hilbert-transform filter
D. An elliptical filter

A7H11 (B)
Which type of filter would be the best to use in a 2-meter repeater duplexer?
A. A crystal filter
B. A cavity filter
C. A DSP filter
D. An L-C filter

A7I Mixer stages; Frequency synthesizers

A7I01 (D)
What is the mixing process?
A. The elimination of noise in a wideband receiver by phase comparison
B. The elimination of noise in a wideband receiver by phase differentiation
C. The recovery of the intelligence from a modulated RF signal
D. The combination of two signals to produce sum and difference frequencies

A7I02 (C)
What are the principal frequencies that appear at the output of a mixer circuit?
A. Two and four times the original frequency
B. The sum, difference and square root of the input frequencies
C. The original frequencies and the sum and difference frequencies
D. 1.414 and 0.707 times the input frequency

A7I03 (B)
What are the advantages of the frequency-conversion process?
A. Automatic squelching and increased selectivity
B. Increased selectivity and optimal tuned-circuit design
C. Automatic soft limiting and automatic squelching
D. Automatic detection in the RF amplifier and increased selectivity

A7I04 (A)
What occurs in a receiver when an excessive amount of signal energy reaches the mixer circuit?
A. Spurious mixer products are generated
B. Mixer blanking occurs
C. Automatic limiting occurs
D. A beat frequency is generated

A7I05 (C)
What type of frequency synthesizer circuit uses a stable voltage- controlled oscillator, programmable divider, phase detector, loop filter and a reference frequency source?

A. A direct digital synthesizer
B. A hybrid synthesizer
C. A phase-locked loop synthesizer
D. A diode-switching matrix synthesizer

A7I06 (A)
What type of frequency synthesizer circuit uses a phase accumulator, lookup table, digital to analog converter and a low-pass antialias filter?
A. A direct digital synthesizer
B. A hybrid synthesizer
C. A phase-locked loop synthesizer
D. A diode-switching matrix synthesizer

A7I07 (B)
What are the main blocks of a phase-locked loop frequency synthesizer?
A. A variable-frequency crystal oscillator, programmable divider, digital to analog converter and a loop filter
B. A stable voltage-controlled oscillator, programmable divider, phase detector, loop filter and a reference frequency source
C. A phase accumulator, lookup table, digital to analog converter and a low-pass antialias filter
D. A variable-frequency oscillator, programmable divider, phase detector and a low-pass antialias filter

A7I08 (D)
What are the main blocks of a direct digital frequency synthesizer?
A. A variable-frequency crystal oscillator, phase accumulator, digital to analog converter and a loop filter
B. A stable voltage-controlled oscillator, programmable divider, phase detector, loop filter and a digital to analog converter
C. A variable-frequency oscillator, programmable divider, phase detector and a low-pass antialias filter
D. A phase accumulator, lookup table, digital to analog converter and a low-pass antialias filter

A7I09 (B)
What information is contained in the lookup table of a direct digital frequency synthesizer?
A. The phase relationship between a reference oscillator and the output waveform
B. The amplitude values that represent a sine-wave output
C. The phase relationship between a voltage-controlled oscillator and the output waveform
D. The synthesizer frequency limits and frequency values stored in the radio memories

A7I10 (C)
What are the major spectral impurity components of direct digital synthesizers?
A. Broadband noise
B. Digital conversion noise
C. Spurs at discrete frequencies
D. Nyquist limit noise

A7I11 (A)
What are the major spectral impurity components of phase-locked loop synthesizers?
A. Broadband noise
B. Digital conversion noise

C. Spurs at discrete frequencies
D. Nyquist limit noise

A7J Amplifier applications: AF, IF, RF

A7J01 (B)
For most amateur phone communications, what should be the upper frequency limit of an audio amplifier?
A. No more than 1000 Hz
B. About 3000 Hz
C. At least 10,000 Hz
D. More than 20,000 Hz

A7J02 (A)
What is the term for the ratio of the RMS voltage for all harmonics in an audio-amplifier output to the total RMS voltage of the output for a pure sine-wave input?
A. Total harmonic distortion
B. Maximum frequency deviation
C. Full quieting ratio
D. Harmonic signal ratio

A7J03 (D)
What are the advantages of a Darlington pair audio amplifier?
A. Mutual gain, low input impedance and low output impedance
B. Low output impedance, high mutual inductance and low output current
C. Mutual gain, high stability and low mutual inductance
D. High gain, high input impedance and low output impedance

A7J04 (B)
What is the purpose of a speech amplifier in an amateur phone transmitter?
A. To increase the dynamic range of the audio
B. To raise the microphone audio output to the level required by the modulator
C. To match the microphone impedance to the transmitter input impedance
D. To provide adequate AGC drive to the transmitter

A7J05 (A)
What is an IF amplifier stage?
A. A fixed-tuned pass-band amplifier
B. A receiver demodulator
C. A receiver filter
D. A buffer oscillator

A7J06 (C)
What factors should be considered when selecting an intermediate frequency?
A. Cross-modulation distortion and interference
B. Interference to other services
C. Image rejection and selectivity
D. Noise figure and distortion

A7J07 (D)
Which of the following is a purpose of the first IF amplifier stage in a receiver?
A. To improve noise figure performance
B. To tune out cross-modulation distortion
C. To increase the dynamic response
D. To provide selectivity

A7J08 (B)
Which of the following is an important reason for using a VHF intermediate frequency in an HF receiver?
A. To provide a greater tuning range
B. To move the image response far away from the filter passband
C. To tune out cross-modulation distortion
D. To prevent the generation of spurious mixer products

A7J09 (B)
How much gain should be used in the RF amplifier stage of a receiver?
A. As much gain as possible, short of self oscillation
B. Sufficient gain to allow weak signals to overcome noise generated in the first mixer stage
C. Sufficient gain to keep weak signals below the noise of the first mixer stage
D. It depends on the amplification factor of the first IF stage

A7J10 (C)
Why should the RF amplifier stage of a receiver have only sufficient gain to allow weak signals to overcome noise generated in the first mixer stage?
A. To prevent the sum and difference frequencies from being generated
B. To prevent bleed-through of the desired signal
C. To prevent the generation of spurious mixer products
D. To prevent bleed-through of the local oscillator

A7J11 (A)
What is the primary purpose of an RF amplifier in a receiver?
A. To improve the receiver noise figure
B. To vary the receiver image rejection by using the AGC
C. To provide most of the receiver gain
D. To develop the AGC voltage

A8–SIGNALS AND EMISSIONS
[6 questions–6 groups]

A8A FCC emission designators vs. emission types

A8A01 (A)
What is emission A3C?
A. Facsimile
B. RTTY
C. ATV
D. Slow Scan TV

A8A02 (B)
What type of emission is produced when an AM transmitter is modulated by a facsimile signal?
A. A3F
B. A3C
C. F3C
D. F3C

A8A03 (C)
What does a facsimile transmission produce?
A. Tone-modulated telegraphy
B. A pattern of printed characters designed to form a picture
C. Printed pictures by electrical means
D. Moving pictures by electrical means

A8A04 (D)
What is emission F3C?
A. Voice transmission
B. Slow Scan TV
C. RTTY
D. Facsimile

A8A05 (A)
What type of emission is produced when an FM transmitter is modulated by a facsimile signal?
A. F3C
B. A3C
C. F3F
D. A3F

A8A06 (B)
What is emission A3F?
A. RTTY
B. Television
C. SSB
D. Modulated CW

A8A07 (B)
What type of emission is produced when an AM transmitter is modulated by a television signal?
A. F3F
B. A3F
C. A3C
D. F3C

A8A08 (D)
What is emission F3F?
A. Modulated CW
B. Facsimile
C. RTTY
D. Television

A8A09 (C)
What type of emission is produced when an FM transmitter is modulated by a television signal?
A. A3F
B. A3C
C. F3C
D. F3C

A8A10 (D)
What type of emission is produced when an SSB transmitter is modulated by a slow-scan television signal?
A. J3A
B. F3F
C. A3F
D. J3F

A8A11 (A)
What emission is produced when an AM transmitter is modulated by a single-channel signal containing digital information without the use of a modulating subcarrier, resulting in telegraphy for aural reception?
A. CW
B. RTTY
C. Data
D. MCW

A8B Modulation symbols and transmission characteristics

A8B01 (A)
What International Telecommunication Union (ITU) system describes the characteristics and necessary bandwidth of any transmitted signal?
A. Emission Designators
B. Emission Zones
C. Band Plans
D. Modulation Indicators

A8B02 (C)
Which of the following describe the three most-used symbols of an ITU emission designator?
A. Type of modulation, transmitted bandwidth and modulation code designator
B. Bandwidth of the modulating signal, nature of the modulating signal and transmission rate of signals
C. Type of modulation, nature of the modulating signal and type of information to be transmitted
D. Power of signal being transmitted, nature of multiplexing and transmission speed

A8B03 (B)
If the first symbol of an ITU emission designator is J, representing a single-sideband, suppressed-carrier signal, what information about the emission is described?
A. The nature of any signal multiplexing
B. The type of modulation of the main carrier
C. The maximum permissible bandwidth
D. The maximum signal level, in decibels

A8B04 (D)
If the first symbol of an ITU emission designator is G, representing a phase-modulated signal, what information about the emission is described?
A. The nature of any signal multiplexing
B. The maximum permissible deviation
C. The nature of signals modulating the main carrier
D. The type of modulation of the main carrier

A8B05 (A)
If the first symbol of an ITU emission designator is P, representing a sequence of unmodulated pulses, what information about the emission is described?
A. The type of modulation of the main carrier
B. The maximum permissible pulse width
C. The nature of signals modulating the main carrier
D. The nature of any signal multiplexing

A8B06 (A)
If the second symbol of an ITU emission designator is 3, representing a single channel containing analog information, what information about the emission is described?
A. The nature of signals modulating the main carrier
B. The maximum permissible deviation
C. The maximum signal level, in decibels
D. The type of modulation of the main carrier

A8B07 (C)
If the second symbol of an ITU emission designator is 1, representing a single channel containing quantized, or digital information, what information about the emission is described?
A. The maximum transmission rate, in bauds

B. The maximum permissible deviation
C. The nature of signals modulating the main carrier
D. The type of information to be transmitted

A8B08 (D)
If the third symbol of an ITU emission designator is D, representing data transmission, telemetry or telecommand, what information about the emission is described?
A. The maximum transmission rate, in bauds
B. The maximum permissible deviation
C. The nature of signals modulating the main carrier
D. The type of information to be transmitted

A8B09 (B)
If the third symbol of an ITU emission designator is A, representing telegraphy for aural reception, what information about the emission is described?
A. The maximum transmission rate, in words per minute
B. The type of information to be transmitted
C. The nature of signals modulating the main carrier
D. The maximum number of different signal elements

A8B10 (B)
If the third symbol of an ITU emission designator is B, representing telegraphy for automatic reception, what information about the emission is described?
A. The maximum transmission rate, in bauds
B. The type of information to be transmitted
C. The type of modulation of the main carrier
D. The transmission code is Baudot

A8B11 (D)
If the third symbol of an ITU emission designator is F, representing television (video), what information about the emission is described?
A. The maximum frequency variation of the color-burst pulse
B. The picture scan rate is fast
C. The type of modulation of the main carrier
D. The type of information to be transmitted

A8C Modulation methods; modulation index; deviation ratio

A8C01 (C)
How can an FM-phone signal be produced?
A. By modulating the supply voltage to a Class-B amplifier
B. By modulating the supply voltage to a Class-C amplifier
C. By using a reactance modulator on an oscillator
D. By using a balanced modulator on an oscillator

A8C02 (A)
How can the unwanted sideband be removed from a double-sideband signal generated by a balanced modulator to produce a single-sideband phone signal?
A. By filtering
B. By heterodyning
C. By mixing
D. By neutralization

A8C03 (B)
What is meant by modulation index?
A. The processor index

B. The ratio between the deviation of a frequency modulated signal and the modulating frequency
C. The FM signal-to-noise ratio
D. The ratio of the maximum carrier frequency deviation to the highest audio modulating frequency

A8C04 (D)
In an FM-phone signal, what is the term for the ratio between the deviation of the frequency modulated signal and the modulating frequency?
A. FM compressibility
B. Quieting index
C. Percentage of modulation
D. Modulation index

A8C05 (D)
How does the modulation index of a phase-modulated emission vary with RF carrier frequency (the modulated frequency)?
A. It increases as the RF carrier frequency increases
B. It decreases as the RF carrier frequency increases
C. It varies with the square root of the RF carrier frequency
D. It does not depend on the RF carrier frequency

A8C06 (A)
In an FM-phone signal having a maximum frequency deviation of 3000 Hz either side of the carrier frequency, what is the modulation index when the modulating frequency is 1000 Hz?
A. 3
B. 0.3
C. 3000
D. 1000

A8C07 (B)
What is the modulation index of an FM-phone transmitter producing an instantaneous carrier deviation of 6 kHz when modulated with a 2-kHz modulating frequency?
A. 6000
B. 3
C. 2000
D. 1/3

A8C08 (B)
What is meant by deviation ratio?
A. The ratio of the audio modulating frequency to the center carrier frequency
B. The ratio of the maximum carrier frequency deviation to the highest audio modulating frequency
C. The ratio of the carrier center frequency to the audio modulating frequency
D. The ratio of the highest audio modulating frequency to the average audio modulating frequency

A8C09 (C)
In an FM-phone signal, what is the term for the maximum deviation from the carrier frequency divided by the maximum audio modulating frequency?
A. Deviation index
B. Modulation index
C. Deviation ratio
D. Modulation ratio

A8C10 (D)
What is the deviation ratio of an FM-phone signal having

a maximum frequency swing of plus or minus 5 kHz and accepting a maximum modulation rate of 3 kHz?
A. 60
B. 0.16
C. 0.6
D. 1.66

A8C11 (A)
What is the deviation ratio of an FM-phone signal having a maximum frequency swing of plus or minus 7.5 kHz and accepting a maximum modulation rate of 3.5 kHz?
A. 2.14
B. 0.214
C. 0.47
D. 47

A8D Electromagnetic radiation; wave polarization; signal-to-noise (S/N) ratio

A8D01 (C)
What are electromagnetic waves?
A. Alternating currents in the core of an electromagnet
B. A wave consisting of two electric fields at right angles to each other
C. A wave consisting of an electric field and a magnetic field at right angles to each other
D. A wave consisting of two magnetic fields at right angles to each other

A8D02 (A)
At approximately what speed do electromagnetic waves travel in free space?
A. 300 million meters per second
B. 468 million meters per second
C. 186,300 feet per second
D. 300 million miles per second

A8D03 (C)
Why don't electromagnetic waves penetrate a good conductor for more than a fraction of a wavelength?
A. Electromagnetic waves are reflected by the surface of a good conductor
B. Oxide on the conductor surface acts as a magnetic shield
C. The electromagnetic waves are dissipated as eddy currents in the conductor surface
D. The resistance of the conductor surface dissipates the electromagnetic waves

A8D04 (D)
Which of the following best describes electromagnetic waves traveling in free space?
A. Electric and magnetic fields become aligned as they travel
B. The energy propagates through a medium with a high refractive index
C. The waves are reflected by the ionosphere and return to their source
D. Changing electric and magnetic fields propagate the energy across a vacuum

A8D05 (A)
What is meant by horizontally polarized electromagnetic waves?
A. Waves with an electric field parallel to the Earth

B. Waves with a magnetic field parallel to the Earth
C. Waves with both electric and magnetic fields parallel to the Earth
D. Waves with both electric and magnetic fields perpendicular to the Earth

A8D06 (B)
What is meant by circularly polarized electromagnetic waves?
A. Waves with an electric field bent into a circular shape
B. Waves with a rotating electric field
C. Waves that circle the Earth
D. Waves produced by a loop antenna

A8D07 (C)
What is the polarization of an electromagnetic wave if its electric field is perpendicular to the surface of the Earth?
A. Circular
B. Horizontal
C. Vertical
D. Elliptical

A8D08 (D)
What is the polarization of an electromagnetic wave if its magnetic field is parallel to the surface of the Earth?
A. Circular
B. Horizontal
C. Elliptical
D. Vertical

A8D09 (A)
What is the polarization of an electromagnetic wave if its magnetic field is perpendicular to the surface of the Earth?
A. Horizontal
B. Circular
C. Elliptical
D. Vertical

A8D10 (B)
What is the polarization of an electromagnetic wave if its electric field is parallel to the surface of the Earth?
A. Vertical
B. Horizontal
C. Circular
D. Elliptical

A8D11 (D)
What is the primary source of noise that can be heard in an HF-band receiver with an antenna connected?
A. Detector noise
B. Man-made noise
C. Receiver front-end noise
D. Atmospheric noise

A8D12 (A)
What is the primary source of noise that can be heard in a VHF/UHF- band receiver with an antenna connected?
A. Receiver front-end noise
B. Man-made noise
C. Atmospheric noise
D. Detector noise

A8E   AC waveforms: Sine wave, square wave, sawtooth wave
A8E01 (B)
What is a sine wave?
A. A constant-voltage, varying-current wave
B. A wave whose amplitude at any given instant can be represented by a point on a wheel rotating at a uniform speed
C. A wave following the laws of the trigonometric tangent function
D. A wave whose polarity changes in a random manner

A8E02 (C)
Starting at a positive peak, how many times does a sine wave cross the zero axis in one complete cycle?
A. 180 times
B. 4 times
C. 2 times
D. 360 times

A8E03 (D)
How many degrees are there in one complete sine wave cycle?
A. 90 degrees
B. 270 degrees
C. 180 degrees
D. 360 degrees

A8E04 (A)
What is the period of a wave?
A. The time required to complete one cycle
B. The number of degrees in one cycle
C. The number of zero crossings in one cycle
D. The amplitude of the wave

A8E05 (B)
What is a square wave?
A. A wave with only 300 degrees in one cycle
B. A wave that abruptly changes back and forth between two voltage levels and remains an equal time at each level
C. A wave that makes four zero crossings per cycle
D. A wave in which the positive and negative excursions occupy unequal portions of the cycle time

A8E06 (C)
What is a wave called that abruptly changes back and forth between two voltage levels and remains an equal time at each level?
A. A sine wave
B. A cosine wave
C. A square wave
D. A sawtooth wave

A8E07 (D)
What sine waves added to a fundamental frequency make up a square wave?
A. A sine wave 0.707 times the fundamental frequency
B. All odd and even harmonics
C. All even harmonics
D. All odd harmonics

A8E08 (A)
What type of wave is made up of a sine wave of a fundamental frequency and all its odd harmonics?

A. A square wave
B. A sine wave
C. A cosine wave
D. A tangent wave

A8E09 (B)
What is a sawtooth wave?
A. A wave that alternates between two values and spends an equal time
at each level
B. A wave with a straight line rise time faster than the fall time (or vice versa)
C. A wave that produces a phase angle tangent to the unit circle
D. A wave whose amplitude at any given instant can be represented by a point on a wheel rotating at a uniform speed

A8E10 (C)
What type of wave has a rise time significantly faster than the fall time (or vice versa)?
A. A cosine wave
B. A square wave
C. A sawtooth wave
D. A sine wave

A8E11 (A)
What type of wave is made up of sine waves of a fundamental frequency and all harmonics?
A. A sawtooth wave
B. A square wave
C. A sine wave
D. A cosine wave

A8F   AC measurements: peak, peak-to-peak and root-mean-square (RMS) value; peak-envelope-power (PEP) relative to average

A8F01 (B)
What is the peak voltage at a common household electrical outlet?
A. 240 volts
B. 170 volts
C. 120 volts
D. 340 volts

A8F02 (C)
What is the peak-to-peak voltage at a common household electrical outlet?
A. 240 volts
B. 120 volts
C. 340 volts
D. 170 volts

A8F03 (A)
What is the RMS voltage at a common household electrical power outlet?
A. 120-V AC
B. 340-V AC
C. 85-V AC
D. 170-V AC

A8F04 (A)
What is the RMS value of a 340-volt peak-to-peak pure sine wave?

A. 120-V AC
B. 170-V AC
C. 240-V AC
D. 300-V AC

A8F05 (C)
What is the equivalent to the root-mean-square value of an AC voltage?
A. The AC voltage found by taking the square of the average value of the peak AC voltage
B. The DC voltage causing the same heating of a given resistor as the peak AC voltage
C. The AC voltage causing the same heating of a given resistor as a DC voltage of the same value
D. The AC voltage found by taking the square root of the average AC
value

A8F06 (D)
What would be the most accurate way of determining the RMS voltage of a complex waveform?
A. By using a grid dip meter
B. By measuring the voltage with a D'Arsonval meter
C. By using an absorption wavemeter
D. By measuring the heating effect in a known resistor

A8F07 (A)
For many types of voices, what is the approximate ratio of PEP to average power during a modulation peak in a single-sideband phone signal?
A. 2.5 to 1
B. 25 to 1
C. 1 to 1
D. 100 to 1

A8F08 (B)
In a single-sideband phone signal, what determines the PEP-to-average power ratio?
A. The frequency of the modulating signal
B. The speech characteristics
C. The degree of carrier suppression
D. The amplifier power

A8F09 (C)
What is the approximate DC input power to a Class B RF power amplifier stage in an FM-phone transmitter when the PEP output power is 1500 watts?
A. 900 watts
B. 1765 watts
C. 2500 watts
D. 3000 watts

A8F10 (B)
What is the approximate DC input power to a Class C RF power amplifier stage in a RTTY transmitter when the PEP output power is 1000 watts?
A. 850 watts
B. 1250 watts
C. 1667 watts
D. 2000 watts

A8F11 (D)
What is the approximate DC input power to a Class AB RF power amplifier stage in an unmodulated carrier transmitter when the PEP output power is 500 watts?

A. 250 watts
B. 600 watts
C. 800 watts
D. 1000 watts

A9 — ANTENNAS AND FEED LINES
[5 questions–5 groups]

A9A   Basic antenna parameters: radiation resistance and reactance (including wire dipole, folded dipole, gain, beamwidth, efficiency)

A9A01 (C)
What is meant by the radiation resistance of an antenna?
A. The combined losses of the antenna elements and feed line
B. The specific impedance of the antenna
C. The equivalent resistance that would dissipate the same amount of power as that radiated from an antenna
D. The resistance in the atmosphere that an antenna must overcome to be able to radiate a signal

A9A02–This question was inadvertently omitted. This number is vacant.

A9A03 (A)
Why would one need to know the radiation resistance of an antenna?
A. To match impedances for maximum power transfer
B. To measure the near-field radiation density from a transmitting antenna
C. To calculate the front-to-side ratio of the antenna
D. To calculate the front-to-back ratio of the antenna

A9A04 (B)
What factors determine the radiation resistance of an antenna?
A. Transmission-line length and antenna height
B. Antenna location with respect to nearby objects and the conductors' length/diameter ratio
C. It is a physical constant and is the same for all antennas
D. Sunspot activity and time of day

A9A05 (C)
What is the term for the ratio of the radiation resistance of an antenna to the total resistance of the system?
A. Effective radiated power
B. Radiation conversion loss
C. Antenna efficiency
D. Beamwidth

A9A06 (D)
What is included in the total resistance of an antenna system?
A. Radiation resistance plus space impedance
B. Radiation resistance plus transmission resistance
C. Transmission-line resistance plus radiation resistance
D. Radiation resistance plus ohmic resistance

A9A07 (C)
What is a folded dipole antenna?
A. A dipole one-quarter wavelength long
B. A type of ground-plane antenna
C. A dipole whose ends are connected by a one-half wavelength piece of wire

D. A hypothetical antenna used in theoretical discussions to replace the radiation resistance

A9A08 (D)
How does the bandwidth of a folded dipole antenna compare with that of a simple dipole antenna?
A. It is 0.707 times the bandwidth
B. It is essentially the same
C. It is less than 50%
D. It is greater

A9A09 (A)
What is meant by antenna gain?
A. The numerical ratio relating the radiated signal strength of an antenna to that of another antenna
B. The numerical ratio of the signal in the forward direction to the signal in the back direction
C. The numerical ratio of the amount of power radiated by an antenna compared to the transmitter output power
D. The final amplifier gain minus the transmission-line losses (including any phasing lines present)

A9A10 (B)
What is meant by antenna bandwidth?
A. Antenna length divided by the number of elements
B. The frequency range over which an antenna can be expected to perform well
C. The angle between the half-power radiation points
D. The angle formed between two imaginary lines drawn through the ends of the elements

A9A11 (A)
How can the approximate beamwidth of a beam antenna be determined?
A. Note the two points where the signal strength of the antenna is down 3 dB from the maximum signal point and compute the angular difference
B. Measure the ratio of the signal strengths of the radiated power lobes from the front and rear of the antenna
C. Draw two imaginary lines through the ends of the elements and measure the angle between the lines
D. Measure the ratio of the signal strengths of the radiated power lobes from the front and side of the antenna

A9A12 (B)
How is antenna efficiency calculated?
A. (radiation resistance / transmission resistance) x 100%
B. (radiation resistance / total resistance) x 100%
C. (total resistance / radiation resistance) x 100%
D. (effective radiated power / transmitter output) x 100%

A9A13 (A)
How can the efficiency of an HF grounded vertical antenna be made comparable to that of a half-wave dipole antenna?
A. By installing a good ground radial system
B. By isolating the coax shield from ground
C. By shortening the vertical
D. By lengthening the vertical

A9B   Free-space antenna patterns: E and H plane patterns, (i.e., azimuth and elevation in free-space); gain as a function of pattern; antenna design (computer modeling of antennas)

A9B01 (C)
What determines the free-space polarization of an antenna?
A. The orientation of its magnetic field (H Field)
B. The orientation of its free-space characteristic impedance
C. The orientation of its electric field (E Field)
D. Its elevation pattern

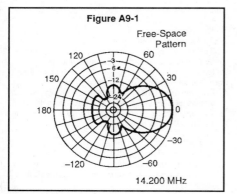

Figure A9-1

Free-Space
Pattern

14.200 MHz

A9B02 (B)
Which of the following describes the free-space radiation pattern shown in Figure A9-1?
A. Elevation pattern
B. Azimuth pattern
C. Bode pattern
D. Bandwidth pattern

A9B03 (B)
In the free-space H-Field radiation pattern shown in Figure A9-1, what is the 3-dB beamwidth?
A. 75 degrees
B. 50 degrees
C. 25 degrees
D. 30 degrees

A9B04 (B)
In the free-space H-Field pattern shown in Figure A9-1, what is the front-to-back ratio?
A. 36 dB
B. 18 dB
C. 24 dB
D. 14 dB

A9B05 (D)
What information is needed to accurately evaluate the gain of an antenna?
A. Radiation resistance
B. E-Field and H-Field patterns
C. Loss resistance
D. All of these choices

A9B06 (D)
Which is NOT an important reason to evaluate a gain antenna across the whole frequency band for which it was designed?

A. The gain may fall off rapidly over the whole frequency band
B. The feedpoint impedance may change radically with frequency
C. The rearward pattern lobes may vary excessively with frequency
D. The dielectric constant may vary significantly

A9B07 (B)
What usually occurs if a Yagi antenna is designed solely for maximum forward gain?
A. The front-to-back ratio increases
B. The feedpoint impedance becomes very low
C. The frequency response is widened over the whole frequency band
D. The SWR is reduced

A9B08 (A)
If the boom of a Yagi antenna is lengthened and the elements are properly retuned, what usually occurs?
A. The gain increases
B. The SWR decreases
C. The front-to-back ratio increases
D. The gain bandwidth decreases rapidly

A9B09 (B)
What type of computer program is commonly used for modeling antennas?
A. Graphical analysis
B. Method of Moments
C. Mutual impedance analysis
D. Calculus differentiation with respect to physical properties

A9B10 (A)
What is the principle of a "Method of Moments" analysis?
A. A wire is modeled as a series of segments, each having a distinct value of current
B. A wire is modeled as a single sine-wave current generator
C. A wire is modeled as a series of points, each having a distinct location in space
D. A wire is modeled as a series of segments, each having a distinct value of voltage across it

A9B11 (B)
In the free-space H-field pattern shown in Figure A9-1, what is the front-to-side ratio?
A. 12 dB
B. 14 dB
C. 18 dB
D. 24 dB

A9C Antenna patterns: elevation above real ground; ground effects as related to polarization; take-off angles as a function of height above ground.

A9C01 (A)
What type of antenna pattern over real ground is shown in Figure A9-2?
A. Elevation pattern
B. Azimuth pattern
C. E-Plane pattern
D. Polarization pattern

Figure A9-2

Over real ground

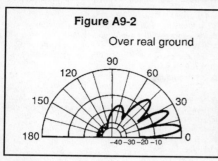

A9C02 (B)
How would the electric field be oriented for a Yagi with three elements mounted parallel to the ground?
A. Vertically
B. Horizontally
C. Right-hand elliptically
D. Left-hand elliptically

A9C03 (A)
What strongly affects the shape of the far-field, low-angle elevation pattern of a vertically polarized antenna?
A. The conductivity and dielectric constant of the soil
B. The radiation resistance of the antenna
C. The SWR on the transmission line
D. The transmitter output power

A9C04 (D)
The far-field, low-angle radiation pattern of a vertically polarized antenna can be significantly improved by what measures?
A. Watering the earth surrounding the base of the antenna
B. Lengthening the ground radials more than a quarter wavelength
C. Increasing the number of ground radials from 60 to 120
D. None of these choices

A9C05 (D)
How is the far-field elevation pattern of a vertically polarized antenna affected by being mounted over seawater versus rocky ground?
A. The low-angle radiation decreases
B. The high-angle radiation increases
C. Both the high- and low-angle radiation decrease
D. The low-angle radiation increases

A9C06 (B)
How is the far-field elevation pattern of a horizontally polarized antenna affected by being mounted one wavelength high over seawater versus rocky ground?
A. The low-angle radiation greatly increases
B. The effect on the radiation pattern is minor
C. The high-angle radiation increases greatly
D. The nulls in the elevation pattern are filled in

A9C07 (B)
Why are elevated-radial counterpoises popular with vertically polarized antennas?
A. They reduce the far-field ground losses

B. They reduce the near-field ground losses, compared to on-ground radial systems using more radials
C. They reduce the radiation angle
D. None of these choices

A9C08 (C)
If only a modest on-ground radial system can be used with an eighth- wavelength-high, inductively loaded vertical antenna, what would be the best compromise to minimize near-field losses?
A. 4 radial wires, 1 wavelength long
B. 8 radial wires, a half-wavelength long
C. A wire-mesh screen at the antenna base, an eighth-wavelength square
D. 4 radial wires, 2 wavelengths long

A9C09 (C)
In the antenna radiation pattern shown in Figure A9-2, what is the elevation angle of the peak response?
A. 45 degrees
B. 75 degrees
C. 7.5 degrees
D. 25 degrees

A9C10 (B)
In the antenna radiation pattern shown in Figure A9-2, what is the front-to-back ratio?
A. 15 dB
B. 28 dB
C. 3 dB
D. 24 dB

A9C11 (A)
In the antenna radiation pattern shown in Figure A9-2, how many elevation lobes appear in the forward direction?
A. 4
B. 3
C. 1
D. 7

A9D Losses in real antennas and matching: resistivity losses, losses in resonating elements (loading coils, matching networks, etc. i.e., mobile, trap); SWR bandwidth; efficiency

A9D01 (A)
What is the approximate input terminal impedance at the center of a folded dipole antenna?
A. 300 ohms
B. 72 ohms
C. 50 ohms
D. 450 ohms

A9D02 (A)
For a shortened vertical antenna, where should a loading coil be placed to minimize losses and produce the most effective performance?
A. Near the center of the vertical radiator
B. As low as possible on the vertical radiator
C. As close to the transmitter as possible
D. At a voltage node

A9D03 (C)
Why should an HF mobile antenna loading coil have a

high ratio of reactance to resistance?
A. To swamp out harmonics
B. To maximize losses
C. To minimize losses
D. To minimize the Q

A9D04 (D)
Why is a loading coil often used with an HF mobile antenna?
A. To improve reception
B. To lower the losses
C. To lower the Q
D. To tune out the capacitive reactance

A9D05 (A)
What is a disadvantage of using a trap antenna?
A. It will radiate harmonics
B. It can only be used for single-band operation
C. It is too sharply directional at lower frequencies
D. It must be neutralized

A9D06 (D)
What is an advantage of using a trap antenna?
A. It has high directivity in the higher-frequency bands
B. It has high gain
C. It minimizes harmonic radiation
D. It may be used for multiband operation

A9D07 (B)
What happens at the base feedpoint of a fixed length HF mobile antenna as the frequency of operation is lowered?
A. The resistance decreases and the capacitive reactance decreases
B. The resistance decreases and the capacitive reactance increases
C. The resistance increases and the capacitive reactance decreases
D. The resistance increases and the capacitive reactance increases

A9D08 (D)
What information is necessary to design an impedance matching system for an antenna?
A. Feedpoint radiation resistance and loss resistance
B. Feedpoint radiation reactance
C. Transmission-line characteristic impedance
D. All of these choices

A9D09 (A)
How must the driven element in a 3-element Yagi be tuned to use a "hairpin" matching system?
A. The driven element reactance is capacitive
B. The driven element reactance is inductive
C. The driven element resonance is higher than the operating frequency
D. The driven element radiation resistance is higher than the characteristic impedance of the transmission line

A9D10 (C)
What is the equivalent lumped-constant network for a "hairpin" matching system on a 3-element Yagi?
A. Pi network
B. Pi-L network
C. L network
D. Parallel-resonant tank

A9D11 (B)
What happens to the bandwidth of an antenna as it is shortened through the use of loading coils?
A. It is increased
B. It is decreased
C. No change occurs
D. It becomes flat

A9D12 (D)
What is an advantage of using top loading in a shortened HF vertical antenna?
A. Lower Q
B. Greater structural strength
C. Higher losses
D. Improved radiation efficiency

A9E Feed lines: coax vs. open-wire; velocity factor; electrical length; transformation characteristics of line terminated in impedance not equal to characteristic impedance.

A9E01 (D)
What is the velocity factor of a transmission line?
A. The ratio of the characteristic impedance of the line to the terminating impedance
B. The index of shielding for coaxial cable
C. The velocity of the wave on the transmission line multiplied by the velocity of light in a vacuum
D. The velocity of the wave on the transmission line divided by the velocity of light in a vacuum

A9E02 (A)
What is the term for the ratio of the actual velocity at which a signal travels through a transmission line to the speed of light in a vacuum?
A. Velocity factor
B. Characteristic impedance
C. Surge impedance
D. Standing wave ratio

A9E03 (B)
What is the typical velocity factor for a coaxial cable with polyethylene dielectric?
A. 2.70
B. 0.66
C. 0.30
D. 0.10

A9E04 (C)
What determines the velocity factor in a transmission line?
A. The termination impedance
B. The line length
C. Dielectrics in the line
D. The center conductor resistivity

A9E05 (D)
Why is the physical length of a coaxial cable transmission line shorter than its electrical length?
A. Skin effect is less pronounced in the coaxial cable
B. The characteristic impedance is higher in the parallel feed line
C. The surge impedance is higher in the parallel feed line
D. RF energy moves slower along the coaxial cable

A9E06 (C)
What would be the physical length of a typical coaxial transmission line that is electrically one-quarter wavelength long at 14.1 MHz? (Assume a velocity factor of 0.66.)
A. 20 meters
B. 2.33 meters
C. 3.51 meters
D. 0.25 meters

A9E07 (B)
What would be the physical length of a typical coaxial transmission line that is electrically one-quarter wavelength long at 7.2 MHz? (Assume a velocity factor of 0.66.)
A. 10.5 meters
B. 6.88 meters
C. 24 meters
D. 50 meters

A9E08 (C)
What is the physical length of a parallel conductor feed line that is electrically one-half wavelength long at 14.10 MHz? (Assume a velocity factor of 0.95.)
A. 15 meters
B. 20.2 meters
C. 10.1 meters
D. 70.8 meters

A9E09 (A)
What is the physical length of a twin lead transmission feed line at 3.65 MHz? (Assume a velocity factor of 0.8.)
A. Electrical length times 0.8
B. Electrical length divided by 0.8
C. 80 meters
D. 160 meters

A9E10 (B)
What parameter best describes the interactions at the load end of a mismatched transmission line?
A. Characteristic impedance
B. Reflection coefficient
C. Velocity factor
D. Dielectric Constant

A9E11 (D)
Which of the following measurements describes a mismatched transmission line?
A. An SWR less than 1:1
B. A reflection coefficient greater than 1
C. A dielectric constant greater than 1
D. An SWR greater than 1:1

A9E12 (A)
What characteristic will 450-ohm ladder line have at 50 MHz, as compared to 0.195-inch-diameter coaxial cable (such as RG-58)?
A. Lower loss in dB/100 feet
B. Higher SWR
C. Smaller reflection coefficient
D. Lower velocity factor

FCC FIELD OPERATIONS BUREAU AND MONITORING STATIONS

The lists of addresses and service areas that follows will probably change during 1996.

In an effort to reduce the FCC's budget, Chairman Reed Hundt has proposed actions including personnel reductions and facility closings.

Hundt said that although the FCC currently has fewer than its authorized personnel ceiling of 2,271, steps would be taken to reduce the number of FCC employees to about 2,050. In addition to retirements and buyouts, the closing of some regional and field offices would result in about 120 jobs lost, of which some 50 would be involuntary. Hundt said these 50 "reductions in force," or RIFs, would be the first in FCC history.

Regional offices would be closed in Atlanta, Boston, and Seattle, leaving their functions to regional offices in Chicago, Kansas City, and San Francisco.

The following field offices would be closed (leaving 16 still open): Buffalo, Miami, St Paul, Norfolk (Virginia), Portland (Oregon), Houston, San Juan, Anchorage and Honolulu.

All nine monitoring stations would be closed, as well as monitoring operations within four FCC field offices. "Fortunately," Hundt said, "technological advances will permit us to replace these monitoring stations with a national automated monitoring network by the summer of 1996."

These nine monitoring stations are at Vero Beach, Florida; Belfast, Maine; Allegan, Michigan; Douglas, Arizona; Livermore, California; Ferndale, Washington; Grand Island, Nebraska; Kingsville, Texas; and Powder Springs, Georgia.

Hundt said "No monitoring function will be impaired." One facility, in Laurel/Columbia, Maryland, would be the central site for "electronic monitoring."

If approved by the full Commission, the monitoring stations and field offices would close by July 1996. No timetable was given for closing of the four regional offices.

At presstime, the final approval of these actions had not been issued.

FCC FIELD OPERATIONS BUREAU (FOB) FIELD SERVICE BOUNDARIES (BY STATE)

State	County	Office
ALABAMA	All	ATLANTA
ALASKA	All	ANCHORAGE
AMERICAN SAMOA	All	HONOLULU
ARIZONA	All except La Paz and Yuma	DOUGLAS

ARIZONA	La Paz and Yuma	SAN DIEGO
ARKANSAS	All	NEW ORLEANS
CALIFORNIA	Alpine, Amador, Calaveras, Inyo, Mono, San Joaquin, Stanislaus and Tuolumne	LIVERMORE
	Kern, Los Angeles, Orange, San Bernardino, San Luis Obispo, Santa Barbara and Ventura	LOS ANGELES
	Imperial, Riverside, San Diego	SAN DIEGO
CALIFORNIA	Alameda, Butte, Colusa, Contra Costa, Del Norte, El Dorado, Fresno, Glenn, Humboldt, Kings, Lake, Lassen, Madera, Marin, Mariposa, Mendocino, Merced, Modoc, Monterey, Napa, Nevada, Placer, Plumas, Sacramento, San Benito, San Francisco, San Mateo, Santa Clara, Santa Cruz, Shasta, Sierra, Siskiyou, Solano, Sonoma, Sutter, Tehama, Trinity, Tulare, Yolo and Yuba	SAN FRANCISCO
COLORADO	All	DENVER
CONNECTICUT	All	BOSTON
DELAWARE	Kent, Sussex, New Castle (below C & D Canal)	LAUREL
	New Castle (above C & D Canal)	PHILADELPHIA
DISTRICT OF COLUMBIA	All	LAUREL
FLORIDA	Escambia and Santa Rosa	ATLANTA
	Broward, Collier, Dade, Hendry, Lee, Monroe, Palm Beach	MIAMI
	Brevard, Flagler, Indian River, Martin, Okeechobee, Orange, Osceola, St. Lucie, Seminole, Volusia	VERO BEACH
	Duval with assistance from Vero Beach plus all counties not covered by Miami or Vero Beach.	TAMPA
GEORGIA	All	ATLANTA
GUAM	All	HONOLULU
HAWAII	All	HONOLULU
IDAHO	Benewah, Bonner, Boundary, Clearwater, Idaho, Kootenai, Latah, Lewis, Nez Perce, Shosone.	SEATTLE
	All others	PORTLAND
ILLINOIS	All	CHICAGO
INDIANA	Allen, De Kalb, Elkhart, Fulton, Kosciusko, La Grange, Marshall, Noble, St. Joseph, Steuben, Whitley	ALLEGAN
	All others	CHICAGO
IOWA	All	KANSAS CITY
KANSAS	All	KANSAS CITY
KENTUCKY	Bath, Bell, Boone, Bourbon, Boyd, Bracken, Breathitt, Campbell, Carter, Clark, Clay, Elliott, Estill, Fayette, Fleming, Floyd, Franklin, Gallatin, Garrard, Grant, Greenup, Harlan, Harrison, Jackson, Jessamine, Johnson, Kenton, Knox, Knott, Larel, Lawrence, Lee, Leslie, Letcher, Lewis, Lincoln, Madison, Magoffin, Martin, Mason, McCreary, Menifee, Montgomery, Morgan, Nicholas, Owen, Ownsley, Pendleton, Perry, Pike, Powell, Pulaski, Robertson, Rockcastle, Rowan, Wayne, Whitley, Wolfe, Woodford, Scott.	DETROIT
	All others	CHICAGO

LOUISIANA	All	NEW ORLEANS
MAINE	York and Cumberland	BOSTON
	All others	BELFAST
MARIANA ISLANDS	All	HONOLULU
MARYLAND	All	LAUREL
MASSACHUSETTS	All	BOSTON
MICHIGAN	Allegan, Antrim, Barry, Benzie, Berrien, Branch, Calhoun, Cass, Charlevoix, Clare, Eaton, Grand Traverse, Ionia, Isabella, Kalamazoo, Kalkaska, Kent, Lake, Leelanau, Manistee, Mason, Mecosta, Missaukee, Montcalm, Muskegon, Newaygo, Oceana, Osceola, Ottawa, St. Joseph, Van Buren, Wexfor.	ALLEGAN
	Alger, Baraga, Delta, Dickinson, Gogebic, Houghton, Iron, Keweenaw, Marquette, Menominee, Ontonagon, Schoolcraft	ST. PAUL
	All others	DETROIT
MIDWAY ISLANDS	All	HONOLULU
MINNESOTA	All	ST. PAUL
MISSISSIPPI	All	NEW ORLEANS
MISSOURI	All	KANSAS CITY
MONTANA	All	SEATTLE
NEBRASKA	All	GRAND ISLAND
NEVADA	All	LIVERMORE
NEW HAMPSHIRE	Coos	BELFAST
	All others	BOSTON
NEW JERSEY	Bergen, Essex, Hudson, Hunterdon, Mercer, Middlesex, Monmouth, Morris, Passaic, Somerset, Sussex, Union, Warren.	NEW YORK
	Atlantic, Burlington, Camden, Cape May, Cumberland, Gloucester, Ocean, and Salem.	PHILADELPHIA
NEW MEXICO	All	DENVER
NEW YORK	Allegany, Broome, Cattaraugus, Cayuga, Chautauqua, Chemung, Chenango, Clinton, Cortland, Erie, Essex, Franklin, Fulton, Genesee, Hamilton, Herkimer, Jefferson, Lewis, Livingston, Madison, Monroe, Montgomery, Niagra, Oneida, Onondaga, Ontario, Orleans, Oswego, Otsego, St. Lawrence, Saratoga, Schoharie, Schuyler, Steuben, Tioga, Tomkins, Warren, Washington, Wayne, Wyoming, and Yates.	BUFFALO
	Albany, Bronx, Columbia, Delaware, Dutchess, Greene, Kings, Nassau, New York, Orange, Putnam, Queens, Rensselaer, Richmond, Rockland, Schenectady, Suffolk, Sullivan, Ulster, Westchester.	NEW YORK

NORTH CAROLINA	All	NORFOLK
NORTH DAKOTA	All	ST. PAUL
OKLAHOMA	All	DALLAS
OHIO	All	DETROIT
OREGON	All	PORTLAND
PACIFIC TRUST TERRITORIES & COMMONWEALTH	All	HONOLULU
PENNSYLVANIA	All	PHILADELPHIA
PUERTO RICO	All	SAN JUAN
RHODE ISLAND	All	BOSTON
SOUTH CAROLINA	All	ATLANTA
SOUTH DAKOTA	All	DENVER
SWAINS ISLAND	All	HONOLULU
TENNESSEE	All	ATLANTA
TEXAS	Aransas, Atascosa, Bandera, Bee, Brooks, Calhoun, Cameron, Dimmit, Duval, Edwards, Frio, Goliad, Hidalgo, Jim Hogg, Jim Wells, Karnes, Kenedy, Kinney, Kleberg, La Salle, Live Oak, Maverick, McMullen, Medina, Nueces, Real, Refugio, San Patricio, Starr, Uvalde, Val Verde, Webb, Willacy, Wilson, Zapata and Zavala.	KINGSVILLE
	Angelina, Austin, Bastrop, Bexar, Blanco, Brazoria, Brazos, Burleson, Caldwell, Chambers, Colorado, Comal, De Witt, Fayette, Fort Bend, Galveston, Gillespie, Gonzales, Grimes, Guadalupe, Hardin, Harris, Hays, Jackson, Jasper, Jefferson, Kendall, Kerr, Lavaca, Lee, Liberty, Madison, Matagorda, Montgomery, Nacogdoches, Newton, Orange, Polk, Sabine, San Augustine, San Jacinto, Travis, Trinity, Tyler, Victoria, Walker, Waller, Washington, Wharton, and Williamson.	HOUSTON
	All others	DALLAS
UTAH	Emery, Garfield, Grand, Kane, Piute, San Juan, Sevier and Wayne.	DOUGLAS
	All others	LIVERMORE
VERMONT	All	BOSTON
VIRGIN ISLANDS	All	SAN JUAN
VIRGINIA	Arlington, Fairfax, Loudoun, Prince William.	LAUREL
	All others	NORFOLK
WAKE ISLAND	All	HONOLULU
WASHINGTON	Clark, Cowlitz, Klickitat, Skamania, and Wahkiakum	PORTLAND
	Whatcom, San Juan, Skagit	FERNDALE
	All others	SEATTLE

WEST VIRGINIA	All	LAUREL
WISCONSIN	Brown, Calumet, Columbia, Crawford, Dane, Dodge, Door, Fond du'Lac, Grant, Green, Iowa, Jefferson, Kenosha, Kewaunee, LaFayette, Manitowoc, Milwaukee, Outagamie, Ozaukee, Racine, Richland, Rock, Sauk, Sheboygan, Walworth, Washington, Waukesha, Winnebago.	CHICAGO
	All others	ST. PAUL
WYOMING	All	DENVER

FCC OFFICE ADDRESSES

ALASKA, Anchorage Office
Federal Communications Commission
6721 West Raspberry Road
Anchorage, Alaska 99502
Phone: (907) 243-2153

ARIZONA, Douglas Office
Federal Communications Commission
P.O. Box 6
Douglas, Arizona 85608
Phone: (602) 364-8414

CALIFORNIA, San Diego Office
Federal Communications Commission
4542 Ruffner Street
Room 370
San Diego, California 92111-2216
Phone: (619) 557-5478

CALIFORNIA, Livermore Office
Federal Communications Commission
P.O. Box 311
Livermore, California 94551-0311
Phone: (415) 447-3614

CALIFORNIA, Los Angeles Office
Federal Communications Commission
Cerritos Corporate Tower
18000 Studebaker Road, Room 660
Cerritos, California 90701
Phone: (213) 809-2096

CALIFORNIA, San Francisco Office
Federal Communications Commission
424 Customhouse
555 Battery Street
San Francisco, California 94111
Phone: (415) 705-1101

COLORADO, Denver Office
Federal Communications Commission
12477 West Cedar Drive
Denver, Colorado 80228
Phone: (303) 236-8026

FLORIDA, Vero Beach Office
Federal Communications Commission
P.O. Box 1730
Vero Beach, Florida 32961-1730
Phone: (407) 778-3755

FLORIDA, Miami Office
Federal Communications Commission
Rochester Building, Room 310
8390 N.W. 53rd Street
Miami, Florida 33166
Phone: (305) 526-7420

FLORIDA, Tampa Office
Federal Communications Commission
Room 1215
2203 N. Lois Avenue
Tampa, Florida 33607-2356
Phone: (813) 228-2872

GEORGIA, Atlanta Office
Federal Communications Commission
Massell Building, Room 440
1365 Peachtree Street, N.E.
Atlanta, Georgia 30309
Phone: (404) 347-2631

GEORGIA, Powder Springs Office
Federal Communications Commission
P.O. Box 85
Powder Springs, Georgia 30073
Phone: (404) 943-5420

HAWAII, Honolulu Office
Federal Communications Commission
P.O. Box 1030
Waipahu, Hawaii 96797
Phone: (808) 677-3318

ILLINOIS, Chicago Office
Federal Communications Commission
Park Ridge Office Center, Rm 306
1550 Northwest Highway
Park Ridge, Illinois 60068
Phone: (312) 353-0195

LOUISIANA, New Orleans Office
Federal Communications Commission
800 West Commerce Rd,. Room 505
New Orleans, Louisiana 70123
Phone: (504) 589-2095

*MAINE, Belfast Office
Federal Communications Commission
P.O. Box 470
Belfast, Maine 04915
Phone: (207) 338-4088

MARYLAND, Laurel Office
Federal Communications Commission
P.O. Box 250
Columbia, Maryland 21045
Phone: (301) 725-3474

MASSACHUSETTS, Boston Office
Federal Communications Commission
NFPA Building
1 Batterymarch Park
Quincy, Massachusetts 02169
Phone: (617) 770-4023

MICHIGAN, Allegan Office
Federal Communications Commission
P.O. Box 89
Allegan, Michigan 49010
Phone: (616) 673-2063

MICHIGAN, Detroit Office
Federal Communications Commission
24897 Hathaway Street
Farmington Hills, Michigan 48331-4361
Phone: (313) 226-6078

MINNESOTA, Street Paul Office
Federal Communications Commission
693 Federal Building & US Courthouse
316 North Robert Street
St. Paul, Minnesota 55101
Phone: (612) 290-3819

MISSOURI, Kansas City Office
Federal Communications Commission
Brywood Office Tower, Room 320
8800 East 63rd Street
Kansas City, Missouri 64133
Phone: (816) 926-5111

NEBRASKA, Grand Island Office
Federal Communications Commission
P.O. Box 1588
Grand Island, Nebraska 68802
Phone: (308) 382-4296

NEW YORK, Buffalo Office
Federal Communications Commission
1307 Federal Building
111 West Huron Street
Buffalo, New York 14202
Phone: (716) 846-4511

NEW YORK, New York Office
Federal Communications Commission
201 Varick Street
New York, New York 10014-4870
Phone: (212) 620-3437

OREGON, Portland Office
Federal Communications Commission
1782 Federal Office Building
1220 S.W. 3rd Avenue
Portland, Oregon 97204
Phone: (503) 326-4114

PENNSYLVANIA, Philadelphia Office.
Federal Communications Commission
One Oxford Valley Office Building
2300 East Lincoln Highway
Room 404
Langhorne, Pennsylvania 19047
Phone: (215) 752-1324

PUERTO RICO, San Juan Office
Federal Communications Commission
747 Federal Building
Hato Rey, Puerto Rico 009118-2251
Phone: (809) 766-5567
TEXAS, Dallas Office
Federal Communications Commission
9330 LBJ Expressway, Room 1170
Dallas, Texas 75243
Phone: (214) 767-4827

TEXAS, Houston Office
Federal Communications Commission
1225 North Loop West, Rm 900
Houston Texas 77008
Phone: (713) 229-2748

TEXAS, Kingsville Office
Federal Communications Commission
P.O. Box 632
Kingsville, Texas 78363-0632
Phone: (512) 592-2531

VIRGINIA, Norfolk Office
Federal Communications Commission
1200 Communications Circle
Virginia Beach, Virginia 23455-3725
Phone: (804) 441-6472

WASHINGTON, Ferndale Office
Federal Communications Commission
1330 Loomis Trail Road
Custer, Washington 98240
Phone: (206) 354-4892

WASHINGTON, Seattle Office
Federal Communications Commission
One Newport, Room 414
3605 132nd Avenue, S.E.
Bellevue, Washington 98006
Phone: (206) 764-3324

U.S. VOLUNTEER-EXAMINER COORDINATORS IN THE AMATEUR SERVICE

Anchorage Amateur Radio Club, 2628 Turnagain Parkway, Anchorage, AK 99517; (907) 786-8121 (Day), (907) 243-2221 (Night), (907) 892-6365

ARRL/VEC*, 225 Main Street, Newington, CT 06111-1492; (203) 666-1541, (208) 665-7531 (FAX)

Central Alabama VEC, Inc., 1215 Dale Drive SE, Huntsville, AL 35801-2031; (205) 536-3904; (205) 534-5557 (FAX)

Golden Empire Amateur Radio Society, P.O. Box 508, Chico, CA 95927-0508; (916) 342-1180

Great Lakes Amateur Radio Club VEC, Inc., 3040 Harrison St., Glenview, IL 60025-0273

Greater Los Angeles Amateur Radio Group, 9737 Noble Avenue, North Hills, CA 91343-2403; (818) 892-2068, (818) 892-9855 (FAX)

Jefferson Amateur Radio Club, P.O. Box 24368, New Orleans, LA 70184-4368

Koolau Amateur Radio Club, 45-529 Nakuluai Street, Kaneohe, HI 96744-2224; (808) 235-4132

Laurel Amateur Radio Club, Inc., P.O. Box 3039, Laurel, MD 20709-0039; (301) 572-5124 (0900-2100); (301) 317-7819; (301) 588-3924 (1800-2100 hours)

The Milwaukee Radio Amateurs Club, Inc., 1737 N. 116th Street, Wauwatosa, WI 53226-3003; (414) 774-6999

Mountain Amateur Radio Club, P.O. Box 10, Burlington, WV 26710-0010; (304) 289-3576

PHD Amateur Radio Association, Inc., P.O. Box 11, Liberty, MO 64068-0011; (816) 781-7313

Sandarc-VEC, P.O. Box 2446, La Mesa, CA 91943-2446; (619) 465-3926

Sunnyvale VEC Amateur Radio Club, P.O. Box 60307, Sunnyvale, CA 94088-0307; (408) 255-9000 (24 hours)

Triad Emergency Amateur Radio Club, 3504 Stonehurst Place, High Point, NC 27265-2106; (919) 841-7576

Western Carolina Amateur Radio Society VEC, Inc., 5833 Clinton Hwy., Suite 203, Knoxville, TN 37912-2545; (615) 688-7771, (615) 689-7062 (FAX)

W5YI-VEC*, P.O. Box 565101, Dallas, TX 75356-5101; (817) 461-6443, (817) 548-9594 (FAX)

*This VEC regularly offers monthly exams in all parts of the country.

FCC PART 97

The rules governing all telecommunication services in the United States are contained in Title 47 of the Code of Federal Regulations. The entire text is published in five volumes, containing Parts 0-19, 20-39, 40-69, 70-79, and 80 to End. The published version of Title 47 is revised yearly, with regulations current as of October of that year. Any volume (or all five) can be purchased by contacting the Superintendent of Documents, U.S. Government Printing Office, Washington, DC, 20402, telephone (202) 783-3238. Each volume costs approximately $25, and credit cards are accepted.

From time to time, rules are changed. Since they often become effective before the next printing of the appropriate Title of the CFR, a monthly List of Sections Affected (LSA) is also published by the Government. Subscriptions are available to this publication for $21 per year, but be advised that it is only a list of changes to all 50 Titles of the U.S. Code (which documents all the rules and regulations administered by the Executive departments and agencies of the Federal Government). The LSA indicates which page of the Federal Register contains the complete text of any changed rule. The Federal Register is published daily (with a monthly index and Reader Aids section). Subscriptions are also available to the Federal Register, but be advised that it contains approximately 150 pages per day. Fortunately, most large libraries subscribe to both the LSA and the Federal Register, making it relatively easy to find the exact wording of a particular rule regarding amateur radio (or anything else regulated by the Government).

Part 97 (referred to as 47 CFR 97) contains all the rules relating to Amateur Radio in the U.S. The most current available version is presented here in its entirety.

RULES AND REGULATIONS

PART 97—AMATEUR RADIO SERVICE
(current as of July 1, 1995)

Subpart A—General Provisions

S 97.1 Basis and purpose

The rules and regulations in this Part are designed to provide an amateur radio service having a fundamental purpose as expressed in the following principles:

(a) Recognition and enhancement of the value of the amateur service to the public as a voluntary noncommercial communication service, particularly with respect to providing emergency communications.

(b) Continuation and extension of the amateur's proven ability to contribute to the advancement of the radio art.

(c) Encouragement and improvement of the amateur service through rules which provide for advancing skills in both the communications and technical phases of the art.

(d) Expansion of the existing reservoir within the amateur radio service of trained operators, technicians, and electronics experts.

(e) Continuation and extension of the amateur's unique ability to enhance international goodwill.

S 97.3 Definitions

(a) The definitions of terms used in Part 97 are:

(1) Amateur operator. A person holding a written authorization to be the control operator of an amateur station.

(2) Amateur radio services. The amateur service, the amateur-satellite service and the radio amateur civil emergency service.

(3) Amateur-satellite service. A radiocommunication service using stations on Earth satellites for the same purpose as those of the amateur service.

(4) Amateur service. A radiocommunication service for the purpose of self-training, intercommunication and technical investigations carried out by amateurs, that is, duly authorized persons interested in radio technique solely with a personal aim and without pecuniary interest.

(5) Amateur station. A station in an amateur radio service consisting of the apparatus necessary for carrying on radiocommunications.

(6) Automatic control. The use of devices and procedures for control of a station when it is transmitting so that compliance with the FCC Rules is achieved without the control operator being present at a control point.

(7) Auxiliary station. An amateur station, other than in a message forwarding system, that is transmitting communications point-to-point within a system of cooperating amateur stations.

(8) Bandwidth. The width of a frequency band outside of which the mean power of the transmitted signal is attenuated at least 26 dB below the mean power of the transmitted signal within the band.

(9) Beacon. An amateur station transmitting communications for the purposes of observation of propagation and reception or other related experimental activities.

(10) Broadcasting. Transmissions intended for reception by the general public, either direct or relayed.

(11) Call sign system. The method used to select a call sign for amateur station over-the-air identification purposes.

The call sign systems are:

(i) Sequential call sign system. The call sign is selected by the FCC from an alphabetized list corresponding to the geographic region of the licensee's mailing address and operator class. The call sign is shown on the license. The FCC will issue public announcements detailing the procedures of the sequential call sign system.

(ii) Vanity call sign system. The call sign is selected by the FCC from a list of call signs requested by the licensee. The call sign is shown on the license. The FCC will issue public announcements detailing the procedures of the vanity call sign system.

(12) Control operator. An amateur operator designated by the licensee of a station to be responsible for the transmissions from that station to assure compliance with the FCC Rules.

(13) Control point. The location at which the control operator function is performed.

(14) CSCE. Certificate of successful completion of an examination.

(15) Earth station An amateur station located on, or within 50 km of the Earth's surface intended for communications with space stations or with other Earth stations by means of one or more other objects in space.

(16) EIC. Engineer in Charge of an FCC Field Facility.

(17) External RF Power Amplifier. A device capable of increasing power output when used in conjunction with, but not an integral part of, a transmitter.

(18) External RF power amplifier kit. A number of electronic parts, which, when assembled, is an external RF power amplifier, even if additional parts are required to complete assembly.

(19) FAA. Federal Aviation Administration.

(20) FCC. Federal Communications Commission.

(21) Frequency coordinator. An entity, recognized in a local or regional area by amateur operators whose stations are eligible to be auxiliary or repeater stations, that recommends transmit/receive channels and associated operating and technical parameters for such stations in order to avoid or minimize potential interference.

(22) Harmful interference. Interference which endangers the functioning of a radionavigation service or of other safety services or seriously degrades, obstructs or repeatedly interrupts a radio-communication service operating in accordance with the Radio Regulations.

(23) Indicator. Words, letters or numerals appended to and separated from the call sign during the station identification.

(24) Information bulletin. A message directed only to amateur operators consisting solely of subject matter of direct interest to the amateur service.

(25) International Morse code. A dot-dash code as defined in International Telegraph and Telephone Consultative Committee (CCITT) Recommendation F.1 (1984), Division B, I. Morse Code.

(26) ITU. International Telecommunication Union.

(27) Line A. Begins at Aberdeen, WA, running by great circle arc to the intersection of 48 N, 120 W, thence along parallel 48 N, to the intersection of 95 W, thence by great circle arc through the southernmost point of Duluth, MN, thence by great circle arc to 45 N, 85 W, thence southward along meridian 85 W, to its intersection with parallel 41 N, thence along parallel 41 N, to its intersection with meridian 82 W, thence by great circle arc through the southernmost point of Bangor, ME, thence by great circle arc through the southernmost point of Searsport, ME, at which point it terminates.

(28) Local control. The use of a control operator who directly manipulates the operating adjustments in the station to achieve compliance with the FCC Rules.

(29) Message forwarding system. A group of amateur stations participating in a voluntary, cooperative, interactive arrangement where communications are sent from the control operator of an originating station to the control operator of one or more destination stations by one or more forwarding stations.

(30) National Radio Quiet Zone. The area in Maryland, Virginia and West Virginia bounded by 39 15' N on the north, 78 30' W on the east, 37 30' N on the south and 80 30' W on the west.

(31) Physician. For the purposes of this Part, a person who is licensed to practice in a place where the amateur service is regulated by the FCC, as either a Doctor of Medicine (MD) or a Doctor of Osteopathy (DO).

(32) Question pool. All current examination questions for a designated written examination element.

(33) Question set. A series of examination questions on a given examination selected from the question pool.

(34) Radio Regulations. The latest ITU Radio Regulations to which the United States is a party.

(35) RACES (radio amateur civil emergency service). A radio service using amateur stations for civil defense communications during periods of local, regional or national civil emergencies.

(36) Remote control. The use of a control operator who indirectly manipulates the operating adjustments in the station through a control link to achieve compliance with the FCC Rules.

(37) Repeater. An amateur station that simultaneously retransmits the transmission of another amateur station on a different channel or channels.

(38) Space station. An amateur station located more than 50 km above the Earth's surface.

(39) Space telemetry. A one-way transmission from a space station of measurements made from the measuring instruments in a spacecraft, including those relating to the functioning of the spacecraft.

(40) Spurious emission. An emission, on frequencies outside the necessary bandwidth of a transmission, the level of which may be reduced without affecting the information being transmitted.

(41) Telecommand. A one-way transmission to initiate, modify, or terminate functions of a device at a distance.

(42) Telecommand station. An amateur station that transmits communications to initiate, modify, or terminate functions of a space station.

(43) Telemetry. A one-way transmission of measurements at a distance from the measuring instrument.

(44) Third-party communications. A message from the control operator (first party) of an amateur station to another amateur station control operator (second party) on behalf of another person (third party).

(45) VE. Volunteer examiner.

(46) VEC. Volunteer-examiner coordinator.

(b) The definitions of technical symbols used in this Part are:

(1) EHF (extremely high frequency). The frequency range 30-300 GHz.

(2) HF (high frequency). The frequency range 3-30 MHz.

(3) Hz. Hertz.

(4) m. Meters.

(5) MF (medium frequency). The frequency range 300-3000 kHz.

(6) PEP (peak envelope power). The average power supplied to the antenna transmission line by a transmitter during one RF cycle at the crest of the modulation envelope taken under normal operating conditions.

(7) RF. Radio frequency.

(8) SHF (super-high frequency). The frequency range 3-30 GHz.

(9) UHF (ultra-high frequency). The frequency range 300-3000 MHz.

(10) VHF (very-high frequency). The frequency range 30-300 MHz.

(11) W. Watts.

(c) The following terms are used in this Part to indicate emission types. Refer to S 2.201 of the FCC Rules, Emission, modulation and transmission characteristics, for information on emission type designators.

(1) CW. International Morse code telegraphy emissions having designators with A, C, H, J or R as the first symbol; 1 as the second symbol; A or B as the third symbol; and emissions J2A and J2B.

(2) Data. Telemetry, telecommand and computer communications emissions having designators with A, C, D, F, G, H, J or R as the first symbol; 1 as the second symbol; D as the third symbol; and emission J2D. Only a digital code of a type specifically authorized in this Part may be transmitted.

(3) Image. Facsimile and television emissions having designators with A, C, D, F, G, H, J or R as the first symbol; 1, 2 or 3 as the second symbol; C or F as the third symbol; and emissions having B as the first symbol; 7, 8 or 9 as the second symbol; W as the third symbol.

(4) MCW. Tone-modulated international Morse code telegraphy emissions having designators with A, C, D, F, G, H or R as the first symbol; 2 as the second symbol; A or B as the third symbol.

(5) Phone. Speech and other sound emissions having designators with A, C, D, F, G, H, J or R as the first symbol; 1, 2 or 3 as the second symbol; E as the third symbol. Also speech emissions having B as the first symbol; 7, 8 or 9 as the second symbol; E as the third symbol. MCW for the purpose of performing the station identification procedure, or for providing telegraphy practice interspersed with speech. Incidental tones for the purpose of selective calling or alerting or to control the level of a demodulated signal may also be considered phone.

(6) Pulse. Emissions having designators with K, L, M, P, Q, V or W as the first symbol; 0, 1, 2, 3, 7, 8, 9 or X as the second symbol; A, B, C, D, E, F, N, W or X as the third symbol.

(7) RTTY. Narrow-band direct-printing telegraphy emissions having designators with A, C, D, F, G, H, J or R as the first symbol; 1 as the second symbol; B as the third symbol; and emission J2B. Only a digital code of a type specifically authorized in this Part may be transmitted.

(8) SS. Spread-spectrum emissions using bandwidth-expansion modulation emissions having designators with A, C, D, F, G, H, J or R as the first symbol; X as the second symbol; X as the third symbol. Only a SS emission of a type specifically authorized in this Part may be transmitted.

(9) Test. Emissions containing no information having the designators with N as the third symbol. Test does not include pulse emissions with no information or modulation unless pulse emissions are also authorized in the frequency band.

S 97.5 Station license required.

(a) The person having physical control of the station apparatus must have been granted a station license of the type listed in paragraph,

(b) or hold an unexpired document of the type listed in paragraph,

(c) before the station may transmit on any amateur service frequency from any place that is:

(1) Within 50 km of the Earth's surface and at a place where the amateur service is regulated by the FCC

(2) Within 50 km of the Earth's surface and aboard any vessel or craft that is documented or registered in the United States; or

(3) More than 50 km above the Earth's surface aboard any craft that is documented or registered in the United States.

(b) The types of station licenses are:

(1) An operator/primary station license. One, but only one, operator/primary station license is granted to each person who is qualified to be an amateur operator. The primary station license is granted together with the amateur operator license. Except for a representative of a foreign government, any person who qualifies by examination is eligible to apply for an operator/primary station license. The operator/primary station license document is printed on FCC Form 660.

(2) A club station license. A club station license is granted only to the person who is the license trustee designated by an officer of the club.The trustee must be a person who has been granted an Amateur Extra, Advanced, General, Technician Plus, or Technician operator license. The club must be composed of at least two persons and must have a name, a document of organization, management, and a primary purpose devoted to amateur service activities consistent with this Part. The club station license document is printed on FCC Form 660.

(3) A military recreation station license. A military recreation station license is granted only to the person who is the license custodian designated by the official in charge of the United States military recreational premises where the station is situated. The person must not be a representative of a foreign government. The person need not have been granted an amateur operator license. The military recreation station license document is printed on FCC Form 660.

(4) A RACES station license. A RACES station license is granted only to the person who is the license custodian designated by the official responsible for the governmental agency served by that civil defense organization. The custodian must be the civil defense official responsible for coordination of all civil defense activities in the area concerned. The custodian must not be a representative of a foreign government. The custodian need not have been granted an amateur operator license. The RACES station license document is printed on FCC Form 660.

(c) The types of documents are:

(1) A reciprocal permit for alien amateur licensee (FCC Form 610-AL) issued to the person by the FCC.

(2) An amateur service license issued to the person by the Government of Canada. The person must be a Canadian citizen.

(d) A person who has been granted a station license of the type listed in paragraph (b), or who holds an unexpired document of the type listed in paragraph (c), is authorized to use in accordance with the FCC Rules all transmitting apparatus under the physical control of the station licensee at points where the amateur service is regulated by the FCC.

S 97.7 Control operator required.

When transmitting, each amateur station must have a control operator.The control operator must be a person who has been granted an amateur operator/primary station license, or who holds an unexpired document of the following types:

(a) A reciprocal permit for alien amateur licensee (FCC Form 610-AL) issued to the person by the FCC, or

(b) An amateur service license issued to the person by the Government of Canada. The person must be a Canadian citizen.

S 97.9 Operator license.

(a) The classes of amateur operator licenses are: Novice, Technician, Technician Plus (until such licenses expire, a Technician Class license granted before February 14, 1991, is considered a Technician Plus Class license), General, Advanced, and Amateur Extra. A person who has been granted an operator license is authorized to be the control operator of an amateur station with the privileges of the operator class specified on the license.

(b) A person who has been granted an operator license of Novice, Technician, Technician Plus, General, or Advanced class and who has properly submitted to the administering VEs an application document, FCC Form 610, for an operator license of a higher class, and who holds a CSCE indicating that the person has completed the necessary examinations within the previous 365 days, is authorized to exercise the rights and privileges of the higher operator class until final disposition of the application or until 365 days following the passing of the examination, whichever comes first.

S 97.11 Stations aboard ships or aircraft.

(a) The installation and operation of an amateur station on a ship or aircraft must be approved by the master of the ship or pilot in command of the aircraft.

(b) The station must be separate from and independent of all other radio apparatus installed on the ship or aircraft, except a common antenna may be shared with a voluntary ship radio installation. The station's transmissions must not cause interference to any other apparatus installed on the ship or aircraft.

(c) The station must not constitute a hazard to the safety of life or property. For a station aboard an aircraft, the apparatus shall not be operated while the aircraft is operating under Instrument Flight Rules, as defined by the FAA, unless the station has been found to comply with all applicable FAA Rules.

S 97.13 Restrictions on station locations.

(a) Before placing an amateur station on land of environmental importance or that is significant in American history, architecture or culture, the licensee may be required to take certain actions prescribed by S 1.1301 - 1.1319 of the FCC Rules.

(b) A station within 1600 m (1 mile) of an FCC monitoring facility must protect that facility from harmful interference. Failure to do so could result in imposition of operating restrictions upon the amateur station by an EIC pursuant to S 97.121 of this Part. Geographical coordinates of the facilities that require protection are listed in S 0.121(c) of the FCC Rules.

S 97.15 Station antenna structures.

(a) Unless the amateur station licensee has received prior approval from the FCC, no antenna structure, including the radiating elements, tower, supports and all appurtenances, may be higher than 61 m (200 feet) above ground level at its site.

(b) Unless the amateur station licensee has received prior approval from the FCC, no antenna structure, at an airport or heliport that is available for public use and is listed in the Airport Directory of the current Airman's Information Manual or in either the Alaska or Pacific Airman's Guide and Chart Supplement; or at an airport or heliport under construction that is the subject of a notice or proposal on file with the FAA, and except for military airports, it is clearly indicated that the airport will be available for public use; or at an airport or heliport that is operated by the armed forces of the United States; or at a place near any of these airports or heliports, may be higher than:

(1) 1 m above the airport elevation for each 100 m from the nearest runway longer than 1 km within 6.1 km of the antenna structure.

(2) 2 m above the airport elevation for each 100 m from the nearest runway shorter than 1 km within 3.1 km of the antenna structure.

(3) 4 m above the airport elevation for each 100 m from the nearest landing pad within 1.5 km of the antenna structure.

(c) An amateur station antenna structure no higher than 6.1 m (20 feet) above ground level at its site or no higher than 6.1 m above any natural object or existing manmade structure, other than an antenna structure, is exempt from the requirements of paragraphs (a) and (b) of this Section.

(d) Further details as to whether an aeronautical study and/or obstruction marking and lighting may be required, and specifications for obstruction marking and lighting, are contained in Part 17

of the FCC Rules, Construction, Marking, and Lighting of Antenna Structures. To request approval to place an antenna structure higher than the limits specified in paragraphs (a), (b), and (c) of this Section, the licensee must notify the FAA on FAA Form 7460-1 and the FCC on FCC Form 854.

(e) Except as otherwise provided herein, a station antenna structure may be erected at heights and dimensions sufficient to accommodate amateur service communications. [State and local regulation of a station antenna structure must not preclude amateur service communications. Rather, it must reasonably accommodate such communications and must constitute the minimum practicable regulation to accomplish the state or local authority's legitimate purpose. See PRB-1, 101 FCC 2d 952 (1985) for details.]

S 97.17 Application for new license or reciprocal permit for alien amateur licensee.

(a) Any qualified person is eligible to apply for an amateur service license.

(b) Each application for a new amateur service license must be made on the proper document:

(1) FCC Form 610 for a new operator/primary station license;
(2) FCC Form 610-A for a reciprocal permit for alien amateur licensee; and
(3) FCC Form 610-B for a new amateur service club or military recreation station license.

(c) Each application for a new operator/primary station license must be submitted to the VEs administering the qualifying examination.

(d) Any eligible person may apply for a reciprocal permit for alien amateur licensee. The application document, FCC Form 610-A, must be submitted to the FCC, 1270 Fairfield Road, Gettysburg, PA 17325-7245.

(1) The person must be a citizen of a country with which the United States has arrangements to grant reciprocal operating permits to visiting alien amateur operators is eligible to apply for reciprocal permit for alien amateur licensee.

(2) The person must be a citizen of the same country that issued the amateur service license.

(3) No person who is a citizen of the United States, regardless of any other citizenship also held, is eligible for a reciprocal permit for alien amateur licensee.

(4) No person who has been granted an amateur operator license is eligible for a reciprocal permit for alien amateur licensee.

(e) No person shall obtain or attempt to obtain, or assist another person to obtain or attempt to obtain, an amateur service license or reciprocal permit for alien amateur licensee by fraudulent means.

(f) One unique call sign will be shown on the license of each new primary, club, and military recreation station. The call sign will be selected by the sequential call sign system.

(g) No new license for a club, military recreation, or RACES station will be granted.

(h) Each application for a new club or military recreation station license must be submitted to the FCC, 1270 Fairfield Road, Gettysburg, PA 17325-7245. No new license for a RACES station will be issued.

S 97.19 Application for a vanity call sign.

(a) A person who has been granted an operator/primary station license or a license trustee who has been granted a club station license is eligible to make application for modification of the license, or the renewal thereof, to show a call sign selected by the vanity call sign system. RACES and military recreation stations are not eligible for a vanity call sign.

(b) Each application for a modification of an operator/primary or club station license, or the renewal thereof, to show a call sign selected by the vanity call sign system must be made on FCC Form 610-V. The form must be submitted with the proper fee to the address specified in the Private Radio Services Fee Filing Guide.

(c) Only unassigned call signs that are available to the sequential call sign system are available to the vanity call sign system with the following exceptions:

(1) A call sign shown on an expired license is not available to the vanity call sign system for 2 years following the expiration of the license.

(2) A call sign shown on a surrendered, revoked, set aside, cancelled, or voided license is not available to the vanity call sign system for 2 years following the date such action is taken.

(3) Except for an applicant who is the spouse, child, grandchild, stepchild, parent, grandparent, stepparent, brother, sister, stepbrother, stepsister, aunt, uncle, niece, nephew, or in-law, and except for an applicant who is a club station license trustee acting with the written consent of at least one relative, as listed above, of a person now deceased, the call sign shown on the license of a person now deceased is not available to the vanity call sign system for 2 years following the person's death, or for 2 years following the expiration of the license, whichever is sooner.

(d) Except for an applicant who is the spouse, child, grandchild, stepchild, parent, grandparent, stepparent, brother, sister, stepbrother, stepsister, aunt, uncle, niece, nephew, or in-law, and except for an applicant who is a club station license trustee acting with the written consent of at least one relative, as listed above, of a person now deceased who had been granted the license showing the call sign requested, the vanity call sign requested by an applicant must be selected from the groups of call signs designated under the sequential call sign system for the class of operator license held by the applicant or for a lower class.

(1) The applicant must request that the call sign shown on the current license be vacated and provide a list of up to 25 call signs in order of preference.

(2) The first assignable call sign from the applicant's list will be shown on the license grant. When none of those call signs are assignable, the call sign vacated by the applicant will be shown on the license grant.

(3) Vanity call signs will be selected from those call signs assignable at the time the application is processed by the FCC.

S 97.21 Application for a modified or renewed license.

(a) A person who has been granted an amateur station license that has not expired:

(1) Must apply for a modification of the license as necessary to show the correct mailing address, licensee name, club name, license trustee name, or license custodian name. The application document must be submitted to: FCC, 1270 Fairfield Road, Gettysburg, PA 17325-7245. For an operator/primary station license, the application must be made on FCC Form 610. For a club, military recreation, or RACES station license, the application must be made on FCC Form 610-B.

(2) May apply for a modification of the license to show a higher operator class. The application must be made on FCC Form 610 and must be submitted to the VEs administering the qualifying examination.

(3) May apply for renewal of the license for another term. (The FCC may mail to the licensee an FCC Form 610-R that may be used for this purpose.)

(i) When the license does not show a call sign selected by the vanity call sign system, the application may be made on FCC Form 610-R if it is received from the FCC. If the Form 610-R is not received from the FCC within 30 days of the expiration date of the license for an operator/primary station license, the application may be made on FCC Form 610. For a club, military recreation, or RACES station license, the application may be made on FCC Form 610-B. The application may be submitted no more than 90 days before its expiration to: FCC, 1270 Fairfield Road, Gettysburg, PA 17325-7245. When the application for renewal of the license has been received by the FCC at 1270 Fairfield Road, Gettysburg, PA 17325-7245 prior to the license expiration date, the license operating authority is continued until the final disposition of the application.

(ii) When the license shows a call sign selected by the vanity call sign system, the application must be filed as specified in Section 97.19(b).

(4) May apply for a modification of the license to show a different call sign selected by the sequential call sign system. The application document must be submitted to: FCC, 1270 Fairfield Road, Gettysburg, PA 17325-7245. The application must be made on FCC Form 610. This modification is not available to club, military recreation, or RACES stations.

(b) A person who had been granted an amateur station license,

but the license has expired, may apply for renewal of the license for another term during a 2 year filing grace period. The application document must be received by the FCC at 1270 Fairfield Road, Gettysburg, PA 17325-7245 prior to the end of the grace period. For an operator/primary station license, the application must be made on FCC Form 610. For a club, military recreation, or RACES station license, the application must be made on FCC Form 610-B. Unless and until the license is renewed, no privileges in the Part are conferred.

(c) Each application for a modified or renewed amateur service license must be accompanied by a photocopy (or the original) of the license document unless an application for renewal using FCC Form 610-R is being made, or unless the original document has been lost, mutilated or destroyed.

(d) Unless the holder of a station license requests a change in call sign, the same call sign will be assigned to the station upon renewal or modification of a station license.

(e) A reciprocal permit for alien amateur licensee cannot be renewed. A new reciprocal permit for alien amateur licensee may be issued upon proper application.

S 97.23 Mailing address.

(a) Each application for a license and each application for a reciprocal permit for alien amateur licensee must show a mailing address in an area where the amateur service is regulated by the FCC and where the licensee or permittee can receive mail delivery by the United States Postal Service. Each application for a reciprocal permit for alien amateur licensee must also show the permittee's mailing address in the country of citizenship.

(b) When there is a change in the mailing address for a person who has been granted an amateur operator/primary station license, the person must file a timely application for a modification of the license. Revocation of the station license or suspension of the operator license may result when correspondence from the FCC is returned as undeliverable because the person failed to provide the correct mailing address.

(c) When a person who has been granted a reciprocal permit for alien amateur licensee changes the mailing address where he or she can receive mail delivery by the United States Postal Service, the person must file an application for a new permit. Cancellation of the reciprocal permit for alien amateur licensee may result when correspondence from the FCC is returned as undeliverable because the permittee failed to provide the correct mailing address.

S 97.25 License term.

(a) An amateur service license is normally granted for a 10-year term.

b) A reciprocal permit for alien amateur licensee is normally granted for a 1-year term.

S 97.27 FCC modification of station license.

(a) The FCC may modify a station license, either for a limited time or for the duration of the term thereof, if it determines:

(1) That such action will promote the public interest, convenience, and necessity; or

(2) That such action will promote fuller compliance with the provisions of the Communications Act of 1934, as amended, or of any treaty ratified by the United States.

(b) When the FCC makes such a determination, it will issue an order of modification. The order will not become final until the licensee is notified in writing of the proposed action and the grounds and reasons therefor. The licensee will be given reasonable opportunity of no less than 30 days to protest the modification; except that, where safety of life or property is involved, a shorter period of notice may be provided. Any protest by a licensee of an FCC order of modification will be handled in accordance with the provisions of 47 U.S.C. S 316.

S 97.29 Replacement license document.

Each person who has been granted an amateur station license or reciprocal permit for alien amateur licensee whose original license document or permit document is lost, mutilated or destroyed must request a replacement. The request must be made to: FCC, 1270 Fairfield Road, Gettysburg, PA 17325-7245. A statement of how the document was lost, mutilated, or destroyed must be attached to the request. A replacement document must bear the same expiration date as the document that it replaces.

Subpart B—Station Operation Standards

S 97.101 General standards.

(a) In all respects not specifically covered by FCC Rules each amateur station must be operated in accordance with good engineering and good amateur practice.

(b) Each station licensee and each control operator must cooperate in selecting transmitting channels and in making the most effective use of the amateur service frequencies. No frequency will be assigned for the exclusive use of any station.

(c) At all times and on all frequencies, each control operator must give priority to stations providing emergency communications, except to stations transmitting communications for training drills and tests in RACES.

(d) No amateur operator shall willfully or maliciously interfere with or cause interference to any radio communication or signal.

S 97.103 Station licensee responsibilities.

(a) The station licensee is responsible for the proper operation of the station in accordance with the FCC Rules. When the control operator is a different amateur operator than the station licensee, both persons are equally responsible for proper operation of the station.

(b) The station licensee must designate the station control operator. The FCC will presume that the station licensee is also the control operator, unless documentation to the contrary is in the station records.

(c) The station licensee must make the station and the station records available for inspection upon request by an FCC representative. When deemed necessary by an EIC to assure compliance with FCC Rules, the station licensee must maintain a record of station operations containing such items of information as the EIC may require in accord with S 0.314(x) of the FCC Rules.

S 97.105 Control operator duties.

(a) The control operator must ensure the immediate proper operation of the station, regardless of the type of control.

(b) A station may only be operated in the manner and to the extent permitted by the privileges authorized for the class of operator license held by the control operator.

S 97.107 Alien control operator privileges.

(a) The privileges available to a control operator holding an amateur service license issued by the Government of Canada are:

(1) The terms of the Convention Between the United States and Canada (TIAS no. 2508) Relating to the Operation by Citizens of Either Country of Certain Radio Equipment or Stations in the Other Country;

(2) The operating terms and conditions of the amateur service license issued by the Government of Canada; and,

(3) The applicable provisions of the FCC Rules, but not to exceed the control operator privileges of an FCC-issued Amateur Extra Class operator license.

(b) The privileges available to a control operator holding an FCC-issued reciprocal permit for alien amateur licensee are:

(1) The terms of the agreement between the alien's government and the United States;

(2) The operating terms and conditions of the amateur service license issued by the alien's government;

(3) The applicable provisions of the FCC Rules, but not to exceed the control operator privileges of an FCC-issued Amateur Extra Class operator license; and

(4) None, if the holder of the reciprocal permit has obtained an FCC- issued operator/primary station license.

(c) At any time the FCC may, in its discretion, modify, suspend, or cancel the amateur service privileges within or over any area where radio services are regulated by the FCC of any Canadian amateur service licensee or alien reciprocal permittee.

S 97.109 Station control.

(a) Each amateur station must have at least one control point.

(b) When a station is being locally controlled, the control operator must be at the control point. Any station may be locally controlled.

(c) When a station is being remotely controlled, the control operator must be at the control point. Any station may be remotely controlled.

(d) When a station is being automatically controlled, the control operator need not be at the control point. Only stations specifically designated elsewhere in this Part may be automatically controlled. Automatic control must cease upon notification by an EIC that the station is transmitting improperly or causing harmful interference to other stations. Automatic control must not be resumed without prior approval of the EIC.

(e) No station may be automatically controlled while transmitting third party communications, except a station transmitting a RTTY or data emission. All messages that are retransmitted must originate at a station that is being locally or remotely controlled.

S 97.111 Authorized transmissions.

(a) An amateur station may transmit the following types of two-way communications:

(1) Transmissions necessary to exchange messages with other stations in the amateur service, except those in any country whose administration has given notice that it objects to such communications. The FCC will issue public notices of current arrangements for international communications;

(2) Transmissions necessary to exchange messages with a station in another FCC-regulated service while providing emergency communications;

(3) Transmissions necessary to exchange messages with a United States government station, necessary to providing communications in RACES; and

(4) Transmissions necessary to exchange messages with a station in a service not regulated by the FCC, but authorized by the FCC to communicate with amateur stations. An amateur station may exchange messages with a participating United States military station during an Armed Forces Day Communications Test.

(b) In addition to one-way transmissions specifically authorized elsewhere in this Part, an amateur station may transmit the following types of one-way communications:

(1) Brief transmissions necessary to make adjustments to the station;

(2) Brief transmissions necessary to establishing two-way communications with other stations;

(3) Telecommand;

(4) Transmissions necessary to providing emergency communications;

(5) Transmissions necessary to assisting persons learning, or improving proficiency in, the international Morse code;

(6) Transmissions necessary to disseminate information bulletins;

(7) Transmissions of telemetry.

S 97.113 Prohibited transmissions.

(a) No amateur station shall transmit:

(1) Communications specifically prohibited elsewhere in this Part;

(2) Communications for hire or for material compensation, direct or indirect, paid or promised, except as otherwise provided in these rules;

(3) Communications in which the station licensee or control operator has a pecuniary interest, including communications on behalf of an employer. Amateur operators may, however, notify other amateur operators of the availability for sale or trade of apparatus normally used in an amateur station, provided that such activity is not conducted on a regular basis;

(4) Music using a phone emission except as specifically provided elsewhere in this Section; communications intended to facilitate a criminal act; messages in codes or ciphers intended to obscure the meaning thereof, except as otherwise provided herein; obscene or indecent words or language; or false or deceptive messages, signals or identification;

(5) Communications, on a regular basis, which could reasonably be furnished alternatively through other radio services.

(b) An amateur station shall not engage in any form of broadcasting, nor may an amateur station transmit one-way communications except as specifically provided in these rules; nor shall an amateur station engage in any activity related to program production or news gathering for broadcasting purposes, except that communications directly related to the immediate safety of human life or the protection of property may be provided by amateur stations to broadcasters for dissemination to the public where no other means of communication is reasonably available before or at the time of the event.

(c) A control operator may accept compensation as an incident of a teaching position during periods of time when an amateur station is used by that teacher as a part of classroom instruction at an educational institution.

(d) The control operator of a club station may accept compensation for the periods of time when the station is transmitting telegraphy practice or information bulletins, provided that the station transmits such telegraphy practice and bulletins for at least 40 hours per week; schedules operations on at least six amateur service MF and HF bands using reasonable measures to maximize coverage; where the schedule of normal operating times and frequencies is published at least 30 days in advance of the actual transmissions; and where the control operator does not accept any direct or indirect compensation for any other service as a control operator.

(e) No station shall retransmit programs or signals emanating from any type of radio station other than an amateur station, except propagation and weather forecast information intended for use by the general public and originated from United States Government stations and communications, including incidental music, originating on United States Government frequencies between a space shuttle and its associated Earth stations. Prior approval for shuttle retransmissions must be obtained from the National Aeronautics and Space Administration. Such retransmissions must be for the exclusive use of amateur operators. Propagation, weather forecasts, and shuttle retransmissions may not be conducted on a regular basis, but only occasionally, as an incident of normal amateur radio communications.

(f) No amateur station, except an auxiliary, repeater or space station, may automatically retransmit the radio signals of other amateur stations.

S 97.115 Third party communications.

(a) An amateur station may transmit messages for a third party to:

(1) Any station within the jurisdiction of the United States.

(2) Any station within the jurisdiction of any foreign government whose administration has made arrangements with the United States to allow amateur stations to be used for transmitting international communications on behalf of third parties. No station shall transmit messages for a third party to any station within the jurisdiction of any foreign government whose administration has not made such an arrangement. This prohibition does not apply to a message for any third party who is eligible to be a control operator of the station.

(b) The third party may participate in stating the message where:

(1) The control operator is present at the control point and is continuously monitoring and supervising the third party's participation; and

(2) The third party is not a prior amateur service licensee whose license was revoked; suspended for less than the balance of the license term and the suspension is still in effect; suspended for the balance of the license term and relicensing has not taken place; or surrendered for cancellation following notice of revocation, suspension or monetary forfeiture proceedings. The third party may not be the subject of a cease and desist order which relates to amateur service operation and which is still in effect.

(c) At the end of an exchange of international third party communications, the station must also transmit in the station identification procedure the call sign of the station with which a third party message was exchanged.

S 97.117 International communications.

Transmissions to a different country, where permitted, shall be made in plain language and shall be limited to messages of a technical nature relating to tests, and, to remarks of a personal character for which, by reason of their unimportance, recourse to the public telecommunications service is not justified.

S 97.119 Station identification.

(a) Each amateur station, except a space station or telecommand station, must transmit its assigned call sign on its transmitting channel at the end of each communication, and at least every ten minutes during a communication, for the purpose of clearly making the source of the transmissions from the station known to those receiving the transmissions. No station may transmit unidentified communications or signals, or transmit as the station call sign, any call sign not authorized to the station.

(b) The call sign must be transmitted with an emission authorized for the transmitting channel in one of the following ways:

(1) By a CW emission. When keyed by an automatic device used only for identification, the speed must not exceed 20 words per minute;

(2) By a phone emission in the English language. Use of a standard phonetic alphabet as an aid for correct station identification is encouraged;

(3) By a RTTY emission using a specified digital code when all or part of the communications are transmitted by a RTTY or data emission;

(4) By an image emission conforming to the applicable transmission standards, either color or monochrome, of S 73.682(a) of the FCC Rules when all or part of the communications are transmitted in the same image emission; or

(5) By a CW or phone emission during SS emission transmission on a narrow bandwidth frequency segment. Alternatively, by the changing of one or more parameters of the emission so that a conventional CW or phone emission receiver can be used to determine the station call sign.

(c) An indicator may be included with the call sign. It must be separated from the call sign by the slant mark or by any suitable word that denotes the slant mark. If the indicator is self-assigned it must be included after the call sign and must not conflict with any other indicator specified by the FCC Rules or with any prefix assigned to another country.

(d) When the operator license class held by the control operator exceeds that of the station licensee, an indicator consisting of the call sign assigned to the control operator's station must be included after the call sign.

(e) When the control operator who is exercising the rights and privileges authorized by S 97.9(b) of this Part, an indicator must be included after the call sign as follows:

(1) For a control operator who has requested a license modification from Novice to Technician Class: KT;

(2) For a control operator who has requested a license modification from Novice or Technician Class to General Class: AG;

(3) For a control operator who has requested a license modification from Novice, Technician, or General Class operator to Advanced Class: AA; or

(4) For a control operator who has requested a license modification from Novice, Technician, General, or Advanced Class operator to Amateur Extra Class: AE.

(f) When the station is transmitting under the authority of a reciprocal permit for alien amateur licensee, an indicator consisting of the appropriate letter-numeral designating the station location must be included before the call sign issued to the station by the licensing country. When the station is transmitting under the authority of an amateur service license issued by the Government of Canada, a station location indicator must be included after the call sign. At least once during each intercommunication, the identification announcement must include the geographical location as nearly as possible by city and state, commonwealth or possession.

S 97.121 Restricted operation.

(a) If the operation of an amateur station causes general interference to the reception of transmissions from stations operating in the domestic broadcast service when receivers of good engineering design, including adequate selectivity characteristics, are used to receive such transmissions, and this fact is made known to the amateur station licensee, the amateur station shall not be operated during the hours from 8 p.m. to 10:30 p.m., local time, and on Sunday for the additional period from 10:30 a.m. until 1 p.m., local time, upon the frequency or frequencies used when the interference is created.

(b) In general, such steps as may be necessary to minimize interference to stations operating in other services may be required after investigation by the FCC.

Subpart C—Special Operations

S 97.201 Auxiliary station.

(a) Any amateur station licensed to a holder of a Technician, General, Advanced or Amateur Extra Class operator license may be an auxiliary station. A holder of a Technician, General, Advanced or Amateur Extra Class operator license may be the control operator of an auxiliary station, subject to the privileges of the class of operator license held.

(b) An auxiliary station may transmit only on the 1.25 m and shorter wavelength bands, except the 219-220 MHz, 222.000-222.150 MHz, 431-433 MHz and 435-438 MHz segments.

(c) Where an auxiliary station causes harmful interference to another auxiliary station, the licensees are equally and fully responsible for resolving the interference unless one station's operation is recommended by a frequency coordinator and the other station's is not. In that case, the licensee of the non-coordinated auxiliary station has primary responsibility to resolve the interference.

(d) An auxiliary station may be automatically controlled.

(e) An auxiliary station may transmit one-way communications.

S 97.203 Beacon station.

(a) Any amateur station licensed to a holder of a Technician, General, Advanced or Amateur Extra Class operator license may be a beacon. A holder of a Technician, General, Advanced or Amateur Extra Class operator license may be the control operator of a beacon, subject to the privileges of the class of operator license held.

(b) A beacon must not concurrently transmit on more than 1 channel in the same amateur service frequency band, from the same station location.

(c) The transmitter power of a beacon must not exceed 100 W.

(d) A beacon may be automatically controlled while it is transmitting on the 28.20-28.30 MHz, 50.06-50.08 MHz, 144.275-144.300 MHz, 222.05-222.06 MHz, or 432.300-432.400 MHz segments, or on the 33 cm and shorter wavelength bands.

(e) Before establishing an automatically controlled beacon in the National Radio Quiet Zone or before changing the transmitting frequency, transmitter power, antenna height or directivity, the station licensee must give written notification thereof to the Interference Office, National Radio Astronomy Observatory, P.O. Box 2, Green Bank, WV 24944.

(1) The notification must include the geographical coordinates of the antenna, antenna ground elevation above mean sea level (AMSL), antenna center of radiation above ground level (AGL), antenna directivity, proposed frequency, type of emission, and transmitter power.

(2) If an objection to the proposed operation is received by the FCC from the National Radio Astronomy Observatory at Green Bank, Pocahontas County, WV, for itself or on behalf of the Naval Research Laboratory at Sugar Grove, Pendleton County, WV, within 20 days from the date of notification, the FCC will consider all aspects of the problem and take whatever action is deemed appropriate.

(f) A beacon must cease transmissions upon notification by an EIC that the station is operating improperly or causing undue interference to other operations. The beacon may not resume transmitting without prior approval of the EIC.

(g) A beacon may transmit one-way communications.

S 97.205 Repeater station.

(a) Any amateur station licensed to a holder of a Technician, General, Advanced or Amateur Extra Class operator license may be a repeater. A holder of a Technician, General, Advanced or Amateur Extra Class operator license may be the control operator of a repeater, subject to the privileges of the class of operator license held.

(b) A repeater may receive and retransmit only on the 10 m and shorter wavelength frequency bands except the 28.0-29.5 MHz, 50.0-51.0 MHz, 144.0-144.5 MHz, 145.5-146.0 MHz, 222.00-222.15 MHz, 431.0-433.0 MHz and 435.0-438.0 MHz segments.

(c) Where the transmissions of a repeater cause harmful interference to another repeater, the two station licensees are equally and fully responsible for resolving the interference unless the operation of one station is recommended by a frequency coordinator and the operation of the other station is not. In that case, the licensee of the noncoordinated repeater has primary responsibility to resolve the interference.

(d) A repeater may be automatically controlled.

(e) Ancillary functions of a repeater that are available to users on the input channel are not considered remotely controlled functions of the station. Limiting the use of a repeater to only certain user stations is permissible.

(f) Before establishing a repeater in the National Radio Quiet Zone or before changing the transmitting frequency, transmitter power, antenna height or directivity, or the location of an existing repeater, the station licensee must give written notification thereof to the Interference Office, National Radio Astronomy Observatory, P.O. Box 2, Green Bank, WV 24944.

(1) The notification must include the geographical coordinates of the station antenna, antenna ground elevation above mean sea level (AMSL), antenna center of radiation above ground level (AGL), antenna directivity, proposed frequency, type of emission, and transmitter power.

(2) If an objection to the proposed operation is received by the FCC from the National Radio Astronomy Observatory at Green Bank, Pocahontas County, WV, for itself or on behalf of the Naval Research Laboratory at Sugar Grove, Pendleton County, WV, within 20 days from the date of notification, the FCC will consider all aspects of the problem and take whatever action is deemed appropriate.

(g) The control operator of a repeater that retransmits inadvertently communications that violate the rules in this Part is not accountable for the violative communications.

S 97.207 Space station.

(a) Any amateur station may be a space station. A holder of any class operator license may be the control operator of a space station, subject to the privileges of the class of operator license held by the control operator.

(b) A space station must be capable of effecting a cessation of transmissions by telecommand whenever such cessation is ordered by the FCC.

(c) The following frequency bands and segments are authorized to space stations:

(1) The 17 m, 15 m, 12 m and 10 m bands, 6 mm, 4 mm, 2 mm and 1 mm bands; and

(2) The 7.0-7.1 MHz, 14.00-14.25 MHz, 144-146 MHz, 435-438 MHz, 1260-1270 MHz and 2400-2450 MHz, 3.40-3.41 GHz, 5.83-5.85 GHz, 10.45-10.50 GHz and 24.00-24.05 GHz segments.

(d) A space station may automatically retransmit the radio signals of Earth stations and other space stations.

(e) A space station may transmit one-way communications.

(f) Space telemetry transmissions may consist of specially coded messages intended to facilitate communications or related to the function of the spacecraft.

(g) The licensee of each space station must give two written, pre-space station notifications to the Private Radio Bureau, FCC, Washington, DC 20554. Each notification must be in accord with the provisions of Articles 11 and 13 of the Radio Regulations.

(1) The first notification is required no less than 27 months prior to initiating space station transmissions and must specify the information required by Appendix 4, and Resolution No. 642 of the Radio Regulations.

(2) The second notification is required no less than 5 months prior to initiating space station transmissions and must specify the information required by Appendix 3 and Resolution No. 642 of the Radio Regulations.

(h) The licensee of each space station must give a written, in-space station notification to the Private Radio Bureau, F, Washington, DC 20554, no later than 7 days following initiation of space station transmissions. The notification must update the information contained in the pre-space notification.

S 97.209 Earth station.

(a) Any amateur station may be an Earth station. A holder of any class operator license may be the control operator of an Earth station, subject to the privileges of the class of operator license held by the control operator.

(b) The following frequency bands and segments are authorized to Earth stations:

(1) The 17 m, 15 m, 12 m and 10 m bands, 6 mm, 4 mm, 2 mm and 1 mm bands; and

(2) The 7.0-7.1 MHz, 14.00-14.25 MHz, 144-146 MHz, 435-438 MHz, 1260-1270 MHz and 2400-2450 MHz, 3.40-3.41 GHz, 5.65-5.67 GHz, 10.45-10.50 GHz and 24.00-24.05 GHz segments.

S 97.211 Space Telecommand station.

(a) Any amateur station designated by the licensee of a space station is eligible to transmit as a telecommand station for that space station, subject to the privileges of the class of operator license held by the control operator.

(b) A telecommand station may transmit special codes intended to obscure the meaning of telecommand messages to the station in space operation.

(c) The following frequency bands and segments are authorized to telecommand stations:

(1) The 17 m, 15 m, 12 m and 10 m bands, 6 mm, 4 mm, 2 mm and 1 mm bands; and

(2) The 7.0-7.1 MHz, 14.00-14.25 MHz, 144-146 MHz, 435-438 MHz, 1260-1270 MHz and 2400-2450 MHz, 3.40-3.41 GHz, 5.65-5.67 GHz, 10.45-10.50 GHz and 24.00-24.05 GHz segments.

(d) A telecommand station may transmit one-way communications.

S 97.213 Telecommand of an amateur station.

An amateur station on or within 50 km of the Earth's surface may be under telecommand where:

(a) There is a radio or wireline control link between the control point and the station sufficient for the control operator to perform his/her duties. If radio, the control link must use an auxiliary station. A control link using a fiber optic cable or another telecommunication service is considered wireline.

(b) Provisions are incorporated to limit transmission by the station to a period of no more than 3 minutes in the event of malfunction in the control link.

(c) The station is protected against making, willfully or negligently, unauthorized transmissions.

(d) A photocopy of the station license and a label with the name, address, and telephone number of the station licensee and at least one designated control operator is posted in a conspicuous place at the station location.

S 97.215 Telecommand of model craft.

An amateur station transmitting signals to control a model craft may be operated as follows:

(a) The station identification procedure is not required for transmissions directed only to the model craft, provided that a label indicating the station call sign and the station licensee's name and address is affixed to the station transmitter.

(b) The control signals are not considered codes or ciphers intended to obscure the meaning of the communication.

(c) The transmitter power must not exceed 1 W.

S 97.217 Telemetry.

Telemetry transmitted by an amateur station on or within 50 km of the Earth's surface is not considered to be codes or ciphers intended to obscure the meaning of communications.

S 97.219 Message forwarding system.

(a) Any amateur station may participate in a message forwarding system, subject to the privileges of the class of operator license held.

(b) For stations participating in a message forwarding system, the control operator of the station originating a message is primarily accountable for any violation of the rules in this Part contained in the message.

(c) Except as noted in paragraph (d) of this section, for stations participating in a message forwarding system, the control operators of forwarding stations that retransmit inadvertently communications that violate the rules in this Part are not accountable for the violative communications. They are, however, responsible for discontinuing such communications once they become aware of their presence.

(d) For stations participating in a message forwarding system, the control operator of the first forwarding station must:

(1) Authenticate the identity of the station from which it accepts communication on behalf of the system; or

(2) Accept accountability for any violation of the rules in this Part contained in messages it retransmits to the system.

S 97.221 Automatically controlled digital station.

(a) This rule section does not apply to an auxiliary station, a beacon station, a repeater station, an earth station, a space station, or a space telecommand station.

(b) A station may be automatically controlled while transmitting a RTTY or data emission on the 6m or shorter wavelength bands, and on the 28.120-28.189 MHz, 24.925-24.930 MHz, 21.090-21.100 MHz, 18.105-18.110 MHz, 14.0950-14.0995 MHz, 14.1005-14.112 MHz, 10.140-10.150 MHz, 7.100-7.105 MHz, or 3.620-3.635 MHz segments.

(c) A station may be automatically controlled while transmitting a RTTY or data emission on any other frequency authorized for such emission types provided that:

(1) The station is responding to interrogation by a station under local or remote control; and

(2) No transmission from the automatically controlled station occupies a bandwidth of more than 500 Hz.

Subpart D—Technical Standards

S 97.301 Authorized frequency bands.

The following transmitting frequency bands are available to an amateur station located within 50 km of the Earth's surface, within the specified ITU Region, and outside any area where the amateur service is regulated by any authority other than the FCC.

(a) For a station having a control operator who has been granted an operator license of Technician, Technician Plus, General, Advanced, or Amateur Extra Class:

S 97.303 Frequency sharing requirements.

The following is a summary of the frequency sharing requirements that apply to amateur station transmissions on the frequency bands specified in S 97.301 of this Part. (For each ITU Region, each frequency band allocated to the amateur service is designated as either a secondary service or a primary service. A station in a secondary service must not cause harmful interference to, and must accept interference from, stations in a primary service. See SS 2.105 and 2.106 of the FCC Rules, United States Table of Frequency Allocations for complete requirements.)

(a) Where, in adjacent ITU Regions or Subregions, a band of frequencies is allocated to different services of the same category, the basic principle is the equality of right to operate. The stations of each service in one region must operate so as not to cause harmful interference to services in the other Regions or Subregions. (See ITU Radio Regulations, No. 346 (Geneva, 1979).)

(b) No amateur station transmitting in the 1900-2000 kHz segment, the 70 cm band, the 33 cm band, the 13 cm band, the 9 cm band, the 5 cm band, the 3 cm band, the 24.05-24.25 GHz segment, the 76-81 GHz segment, the 144-149 GHz segment and the 241-248 GHz segment shall cause harmful interference to, nor is protected from interference due to the operation of, the Government radiolocation service.

(c) No amateur station transmitting in the 1900-2000 kHz segment, the 3 cm band, the 76-81 GHz segment, the 144-149 GHz segment and the 241-248 GHz segment shall cause harmful interference to, nor is protected from interference due to the operation

Wavelength band	ITU Region 1	ITU Region 2	ITU Region 3	Sharing requirements See S 97.303, Paragraph:
VHF	MHz	MHz	MHz	
6 m	—	50-54	50-54	(a)
2 m	144-146	144-148	144-148	(a)
1.25 m	—	219-220	—	(a), (e)
-do-	—	222-225	—	(a)
UHF	MHz	MHz	MHz	
70 cm	430-440	420-450	420-450	(a), (b), (f)
33 cm	—	902-928	—	(a), (b), (g)
23 cm	1240-1300	1240-1300	1240-1300	(h), (i)
13 cm	2300-2310	2300-2310	2300-2310	(a), (b), (j)
-do-	2390-2450	2390-2450	2390-2450	(a), (b), (j)
SHF	GHz	GHz	GHz	
9 cm	—	3.3-3	-3.5	(a), (b), (k), (l)
5 cm	5.650-5.850	5.650-5.925	5.650-5.850	(a), (b), (m)
3 cm	10.00-10.50	10.00-10.50	10.00-10.50	b), (c), (i), (n)
1.2 cm	24.00-24.25	24.00-24.25	24.00-24.25	(a), (b), (h), (o)
EHF	GHz	GHz	GHz	
6 mm	47.0-47.2	47.0-47.2	47.0-47.2	
4 mm	75.5-81.0	75.5-81.0	75.5-81.0	(b), (c), (h)
2.5 mm	119.98-120.02	119.98-120.02	119.98-120.02	(k), (p)
2 mm	142-149	142-149	142-149	(b), (c), (h), (k)
1 mm	241-250	241-250	241-250	(b), (c), (h), (q)
—	above 300	above 300	above 300	(k)

(b) For a station having a control operator who has been granted an operator license of Amateur Extra Class:

Wavelength band	ITU Region 1	ITU Region 2	ITU Region 3	Sharing requirements See S 97.303, Paragraph:
MF	kHz	kHz	kHz	
160 m	1810-1850	1800-2000	1800-2000	(a), (b), (c)
HF	MHz	MHz	MHz	
80 m	3.50-3.75	3.50-3.75	3.50-3.75	(a)
75 m	3.75-3.80	3.75-4.00	3.75-3.90	(a)
40 m	7.0-7.1	7.0-7.3	7.0-7.1	(a)
30 m	10.10-10.15	10.10-10.15	10.10-10.15	(d)
20 m	14.00-14.35	14.00-14.35	14.00-14.35	
17 m	18.068-18.168	18.068-18.168	18.068-18.168	
15 m	21.00-21.45	21.00-21.45	21.00-21.45	
12 m	24.89-24.99	24.89-24.99	24.89-24.99	
10 m	28.0-29.7	28.0-29.7	28.0-29.7	

(c) For a station having a control operator who has been granted an operator license of Advanced Class:

Wavelength band	ITU Region 1	ITU Region 2	ITU Region 3	Sharing requirements See S 97.303, Paragraph:
MF	kHz	kHz	kHz	
160 m	1810-1850	1800-2000	1800-2000	(a), (b), (c)
HF	MHz	MHz	MHz	
80 m	3.525-3.750	3.525-3.750	3.525-3.750	(a)
75 m	3.775-3.800	3.775-4.000	3.775-3.900	a)
40 m	7.025-7.100	7.025-7.300	7.025-7.100	(a)
30 m	10.10-10.15	10.10-10.15	10.10-10.15	(d)
20 m	14.025-14.150	14.025-14.150	14.025-14.150	
-do-		14.175-14.350	14.175-14.350	14.175-14.350
17 m	18.068-18.168	18.068-18.168	18.068-18.168	
15 m	21.025-21.200	21.025-21.200	21.025-21.200	
-do-	21.225-21.450	21.225-21.450	21.225-21.450	
12 m	24.89-24.99	24.89-24.99	24.89-24.99	
10 m	28.0-29.7	28.0-29.7	28.0-29.7	

(d) For a station having a control operator who has been granted an operator license of General Class:

Wavelength band	ITU Region 1	ITU Region 2	ITU Region 3	Sharing requirements See S 97.303, Paragraph:
MF 160 m	kHz 1810-1850	kHz 1800-2000	kHz 1800-2000	(a), (b), (c)
HF 80 m	MHz 3.525-3.750	MHz 3.525-3.750	MHz 3.525-3.750	(a)
75 m	—	3.85-4.00	3.85-3.90	(a)
40 m	7.025-7.100	7.025-7.150	7.025-7.100	(a)
-do-	—	7.225-7.300 —	(a)	
30 m	10.10-10.15	10.10-10.15	10.10-10.15	(d)
20 m	14.025-14.150	14.025-14.150	14.025-14.150	
-do-	14.225-14.350	14.225-14.350	14.225-14.350	
17 m	18.068-18.168	18.068-18.168	18.068-18.168	
15 m	21.025-21.200	21.025-21.200	21.025-21.200	
-do-	21.30-21.45	21.30-21.45	21.30-21.45	
12 m	24.89-24.99	24.89-24.99	24.89-24.99	
10 m	28.0-29.7	28.0-29.7	28.0-29.7	

(e) For a station having a control operator who has been granted an operator license of Novice or Technician Plus Class:

Wavelength band	ITU Region 1	ITU Region 2	ITU Region 3	Sharing requirements See S 97.303, Paragraph:
HF 80 m	MHz 3.675-3.725	MHz 3.675-3.725	MHz 3.675-3.725	(a)
40 m	7.050-7.075	7.10-7.15	7.050-7.075	(a)
15 m	21.10-21.20	21.10-21.20	21.10-21.2o	
10 m	28.1-28.5	28.1-28.5	28.1-28.5	

(f) For a station having a control operator who has been granted an operator license of Novice Class:

Wavelength band	ITU Region 1	ITU Region 2	ITU Region 3	Sharing requirements See S 97.303, Paragraph:
VHF 1.25 m	MHz —	MHz 222-225	MHz —	(a)
UHF 23 cm	MHz 1270-1295	MHz 1270-1295	MHz 1270-1295	(h), (I)

of, stations in the non-Government radiolocation service.

(d) No amateur station transmitting in the 30 meter band shall cause harmful interference to stations authorized by other nations in the fixed service. The licensee of the amateur station must make all necessary adjustments, including termination of transmissions, if harmful interference is caused.

(e) In the 1.25 m band:

(1) Use of the 219-220 MHz segment is limited to amateur stations participating, as forwarding stations, in point-to-point fixed digital message forwarding systems, including intercity packet backbone networks. It is not available for other purposes.

(2) No amateur station transmitting in the 219-220 MHz segment shall cause harmful interference to, nor is protected from interference due to operation of Automated Maritime Telecommunications Systems (AMTS), television broadcasting on channels 11 and 13, Interactive Video and Data Service systems, Land Mobile Services systems, or any other service having a primary allocation in or adjacent to the band.

(3) No amateur station may transmit in the 219-220 MHz segment unless the licensee has given written notification of the station's specific geographic location for such transmissions in order to be incorporated into a data base that has been made available to the public. The notification must be given at least 30 days prior to making such transmissions. The notification must be given to: The American Radio Relay League, 225 Main Street, Newington, CT 06111-1494

(4) No amateur station may transmit in the 219-220 MHz segment from a location that is within 640 km of an AMTS Coast Station unless the amateur station licensee has given written notification of the station's specific geographic location for such transmissions to the AMTS licensee. The notification must be given at least 30 days prior to making such transmissions. AMTS Coast Station locations may be obtained either from: The American Radio Relay League, 225 Main Street, Newington, CT 06111-1494, or Interactive Systems, Inc., Suite 1103, 1601 North Kent Street, Arlington, VA 22209, Fax: (703) 812-8275, Phone: (703) 812-8270

(5) No amateur station may transmit in the 219-220 MHz segment from a location that is within 80 km of an AMTS Coast Station unless the amateur station licensee holds written approval from that AMTS licensee.

(f) In the 70 cm band:

(1) No amateur station shall transmit from north of Line A in the 420-430 MHz segment.

(2) The 420-430 MHz segment is allocated to the amateur service in the United States on a secondary basis, and is allocated in the fixed and mobile (except aeronautical mobile) services in the International Table of allocations on a primary basis. No amateur station transmitting in this band shall cause harmful interference to, nor is protected from interference due to the operation of, stations authorized by other nations in the fixed and mobile (except aeronautical mobile) services.

(3) The 430-440 MHz segment is allocated to the amateur service on a secondary basis in ITU Regions 2 and 3. No amateur station transmitting in this band in ITU Regions 2 and 3 shall cause harmful interference to, nor is protected from interference due to

the operation of, stations authorized by other nations in the radiolocation service. In ITU Region 1, the 430-440 MHz segment is allocated to the amateur service on a co-primary basis with the radiolocation service. As between these two services in this band in ITU Region 1, the basic principle that applies is the equality of right to operate. Amateur stations authorized by the United States and radiolocation stations authorized by other nations in ITU Region 1 shall operate so as not to cause harmful interference to each other.

(4) No amateur station transmitting in the 449.75-450.25 MHz segment shall cause interference to, nor is protected from interference due to the operation of stations in, the space operation service and the space research service or Government or non-Government stations for space telecommand.

(g) In the 33 cm band:

(1) No amateur station shall transmit from within the States of Colorado and Wyoming, bounded on the south by latitude 39 N, on the north by latitude 42 N, on the east by longitude 105 W, and on the west by longitude 108 W.1 This band is allocated on a secondary basis to the amateur service subject to not causing harmful interference to, and not receiving protection from any interference due to the operation of, industrial, scientific and medical devices, automatic vehicle monitoring systems or Government stations authorized in this band.[ARRL Note: 97.303(g)(1) was waived in part by the FCC on July 2, 1990 to permit amateurs in the restricted areas to transmit on the following segments: 902.0-902.4, 902.6-904.3, 904.7-925.3, 925.7-927.3, and 927.7-928.0 MHz.]

(2) No amateur station shall transmit from those portions of the States of Texas and New Mexico bounded on the south by latitude 31 41' N, on the north by latitude 34 30' N, on the east by longitude 104 11' W, and on the west by longitude 107 30' W.

(h) No amateur station transmitting in the 23 cm band, the 3 cm band, the 24.05-24.25 GHz segment, the 76-81 GHz segment, the 144-149 GHz segment and the 241-248 GHz segment shall cause harmful interference to, nor is protected from interference due to the operation of, stations authorized by other nations in the radiolocation service.

(i) In the 1240-1260 MHz segment, no amateur station shall cause harmful interference to, nor is protected from interference due to the operation of, stations in the radionavigation-satellite service, the aeronautical radionavigation service, or the radiolocation service.

(j) In the 13 cm band:

(1) The amateur service is allocated on a secondary basis in all ITU Regions. In ITU Region 1, no amateur station shall cause harmful interference to, and is not protected from interference due to the operation of, stations authorized by other nations in the fixed service. In ITU Regions 2 and 3, no station shall cause harmful interference to, and is not protected from interference due to the operation of, stations authorized by other nations in the fixed, mobile and radiolocation services.

(2) In the United States, the 2300-2310 MHz segment is allocated to the amateur service on a co-secondary basis with the Government fixed and mobile services. In this segment, the fixed and mobile services must not cause harmful interference to the amateur service. No amateur station transmitting in the 2400-2450 MHz segment is protected from interference due to the operation of industrial, scientific and medical devices on 2450 MHz.

(k) No amateur station transmitting in the 3.332-3.339 GHz and 3.3458-3525 GHz segments, the 2.5 mm band, the 144.68-144.98 GHz, 145.45-145.75 GHz and 146.82-147.12 GHz segments and the 343-348 GHz segment shall cause harmful interference to stations in the radio astronomy service. No amateur station transmitting in the 300-302 GHz, 324-326 GHz, 345-347 GHz, 363-365 GHz and 379-381 GHz segments shall cause harmful interference to stations in the space research service (passive) or Earth exploration-satellite service (passive).

(l) In the 9 cm band:

(1) In ITU Regions 2 and 3, the band is allocated to the amateur service on a secondary basis.

(2) In the United States, the band is allocated to the amateur service on a co-secondary basis with the non-Government radiolocation service.

(3) In the 3.3-3.4 GHz segment, no amateur station shall cause harmful interference to, nor is protected from interference due to

the operation of, stations authorized by other nations in the fixed and fixed-satellite service.

(4) In the 3.4-3.5 GHz segment, no amateur station shall cause harmful interference to, nor is protected from interference due to the operation of, stations authorized by other nations in the fixed and fixed-satellite service.

(m) In the 5 cm band:

(1) In the 5.650-5.725 GHz segment, the amateur service is allocated in all ITU Regions on a co-secondary basis with the space research (deep space) service.

(2) In the 5.725-5.850 GHz segment, the amateur service is allocated in all ITU Regions on a secondary basis. No amateur station shall cause harmful interference to, nor is protected from interference due to the operation of, stations authorized by other nations in the fixed-satellite service in ITU Region 1.

(3) No amateur station transmitting in the 5.725-5.875 GHz segment is protected from interference due to the operation of industrial, scientific and medical devices operating on 5.8 GHz.

(4) In the 5.650-5.850 GHz segment, no amateur station shall cause harmful interference to, nor is protected from interference due to the operation of, stations authorized by other nations in the radiolocation service.

(5) In the 5.850-5.925 GHz segment, the amateur service is allocated in ITU Region 2 on a co-secondary basis with the radiolocation service. In the United States, the segment is allocated to the amateur service on a secondary basis to the non-Government fixed-satellite service. No amateur station shall cause harmful interference to, nor is protected from interference due to the operation of, stations authorized by other nations in the fixed, fixed-satellite and mobile services. No amateur station shall cause harmful interference to, nor is protected from interference due to the operation of, stations in the non-Government fixed-satellite service.

(n) In the 3 cm band:

(1) In the United States, the 3 cm band is allocated to the amateur service on a co-secondary basis with the non-government radiolocation service.

(2) In the 10.00-10.45 GHz segment in ITU Regions 1 and 3, no amateur station shall cause interference to, nor is protected from interference due to the operation of, stations authorized by other nations in the fixed and mobile services.

(o) No amateur station transmitting in the 1.2 cm band is protected from interference due to the operation of industrial, scientific and medical devices on 24.125 GHz. In the United States, the 24.05-24.25 GHz segment is allocated to the amateur service on a co-secondary basis with the non-government radiolocation and Government and non-government Earth exploration-satellite (active) services.

(p) The 2.5 mm band is allocated to the amateur service on a secondary basis. No amateur station transmitting in this band shall cause harmful interference to, nor is protected from interference due to the operation of, stations in the fixed, inter-satellite and mobile services.

(q) No amateur station transmitting in the 244-246 GHz segment of the 1 mm band is protected from interference due to the operation of industrial, scientific and medical devices on 245 GHz.

S 97.305 Authorized emission types.

(a) An amateur station may transmit a CW emission on any frequency authorized to the control operator.

(b) A station may transmit a test emission on any frequency authorized to the control operator for brief periods for experimental purposes, except that no pulse modulation emission may be transmitted on any frequency where pulse is not specifically authorized.

(c) A station may transmit the following emission types on the frequencies indicated, as authorized to the control operator, subject to the standards specified in S 97.307(f) of this part:

S 97.307 Emission standards.

(a) No amateur station transmission shall occupy more bandwidth than necessary for the information rate and emission type being transmitted, in accordance with good amateur practice.

(b) Emissions resulting from modulation must be confined to the band or segment available to the control operator. Emissions outside the necessary bandwidth must not cause splatter or keyclick interference to operations on adjacent frequencies.

(c) All spurious emissions from a station transmitter must be reduced to the greatest extent practicable. If any spurious emission, including chassis or power line radiation, causes harmful interference to the reception of another radio station, the licensee of the interfering amateur station is required to take steps to eliminate the interference, in accordance with good engineering practice.

(d) The mean power of any spurious emission from a station transmitter or external RF power amplifier transmitting on a frequency below 30 MHz must not exceed 50 mW and must be at least 40 dB below the mean power of the fundamental emission. For a transmitter of mean power less than 5 W, the attenuation must be at least 30 dB. A transmitter built before April 15, 1977, or first marketed before January 1, 1978, is exempt from this requirement.

(e) The mean power of any spurious emission from a station transmitter or external RF power amplifier transmitting on a frequency between 30-225 MHz must be at least 60 dB below the mean power of the fundamental. For a transmitter having a mean power of 25 W or less, the mean power of any spurious emission supplied to the antenna transmission line must not exceed 25 uW and must be at least 40 dB below the mean power of the fundamental emission, but need not be reduced below the power of 10 uW. A transmitter built before April 15, 1977, or first marketed before January 1, 1978, is exempt from this requirement.

(f) The following standards and limitations apply to transmissions on the frequencies specified in S 97.305(c) of this Part.

(1) No angle-modulated emission may have a modulation index greater than 1 at the highest modulation frequency.

(2) No non-phone emission shall exceed the bandwidth of a communications quality phone emission of the same modulation type. The total bandwidth of an independent sideband emission

Wavelength band	Frequencies authorized	Emission types	Standards See §97.307(f), Paragraph:
MF:			
160 m	Entire band	RTTY, data	(3)
-do-	-do-	Phone, image	(1), (2)
HF:			
80 m	Entire band	RTTY, data	(3), (9)
75 m	Entire band	Phone, image	(1), (2)
40 m	7.000-7.100 MHz	RTTY, data	(3), (9)
-do-	7.075-7.100 MHz	Phone, image	(1), (2), (9), (11)
-do-	7.100-7.150 MHz	RTTY, data	(3), (9)
-do-	7.150-7.300 MHz	Phone, image	(1), (2)
30 m	Entire band	RTTY, data	(3)
20 m	14.00-14.15 MHz	RTTY, data	(3)
-do-	14.15-14.35 MHz	Phone, image	(1), (2)
17 m	18.068-18.110 MHz	RTTY, data	(3)
-do-	18.110-18.168 MHz	Phone, image	(1), (2)
15 m	21.0-21.2 MHz	RTTY, data	(3), (9)
-do-	21.20-21.45 MHz	Phone, image	(1), (2)
12 m	24.89-24.93 MHz	RTTY, data	(3)
-do-	24.93-24.99 MHz	Phone, image	(1), (2)
10 m	28.0-28.3 MHz	RTTY, data	(4)
-do-	28.3-28.5 MHz	Phone, image	(1), (2), (10)
-do-	28.5-29.0 MHz	Phone, image	(1), (2)
-do-	29.0-29.7 MHz	Phone, image	(2)
VHF:			
6 m	50.1-51.0 MHz	RTTY, data	(5)
-do-	-do-	MCW, phone, image	(2)
-do-	51.0-54.0 MHz	RTTY, data, test	(5), (8)
-do-	-do-	MCW, phone, image	(2)
2 m	144.1-148.0 MHz	RTTY, data, test	(5), (8)
-do-	-do-	MCW, phone, image	(2)
1.25 m	219-220 MHz	Data	(13)
-do-	222-225 MHz	MCW, phone, image, RTTY, data, test	(2),(6),(8)
UHF:			
70 cm	Entire band	MCW, phone, image, RTTY, data, SS, test	(6), (8)
33 cm	Entire band	MCW, phone, image, RTTY, data, SS, test, pulse	(7), (8), (12)
23 cm	Entire band	MCW, phone, image, RTTY, data, SS, test	(7), (8), (12)
13 cm	Entire band	MCW, phone, image, RTTY, data, SS, test, pulse	(7), (8), (12)
SHF:			
9 cm	Entire band	MCW, phone, image, RTTY, data, SS, test, pulse	(7), (8), (12)
5 cm	Entire band	MCW, phone, image, RTTY, data, SS, test, pulse	(7), (8), (12)
3 cm	Entire band	MCW, phone, image, RTTY, data, SS, test	(7), (8), (12)
1.2 cm	Entire band	MCW, phone, image, RTTY, data, SS, test, pulse	(7), (8), (12)
EHF:			
6 mm	Entire band	MCW, phone, image,RTTY, data, SS, test, pulse	(7), (8), (12)
4 mm	Entire band	MCW, phone, image, RTTY, data, SS, test, pulse	(7), (8), (12)
2.5 mm	Entire band	MCW, phone, image, RTTY, data, SS, test, pulse	(7), (8), (12)
2 mm	Entire band	MCW, phone, image, RTTY, data, SS, test, pulse	(7), (8), (12)
1 mm	Entire band	MCW, phone, image, RTTY, data, SS, test, pulse	(7), (8), (12)
—	Above 300 GHz	MCW, phone, image, RTTY, data, SS, test, pulse	(7), (8), (12)

(having B as the first symbol), or a multiplexed image and phone emission, shall not exceed that of a communications quality A3E emission.

(3) Only a RTTY or data emission using a specified digital code listed in S 97.309(a) of this Part may be transmitted. The symbol rate must not exceed 300 bauds, or for frequency-shift keying, the frequency shift between mark and space must not exceed 1 kHz.

(4) Only a RTTY or data emission using a specified digital code listed in S 97.309(a) of this Part may be transmitted. The symbol rate must not exceed 1200 bauds. For frequency-shift keying, the frequency shift between mark and space must not exceed 1 kHz.

(5) A RTTY, data or multiplexed emission using a specified digital code listed in S 97.309(a) of this Part may be transmitted. The symbol rate must not exceed 19.6 kilobauds. A RTTY, data or multiplexed emission using an unspecified digital code under the limitations listed in S 97.309(b) of this Part also may be transmitted. The authorized bandwidth is 20 kHz.

(6) A RTTY, data or multiplexed emission using a specified digital code listed in S 97.309(a) of this Part may be transmitted. The symbol rate must not exceed 56 kilobauds. A RTTY, data or multiplexed emission using an unspecified digital code under the limitations listed in S 97.309(b) of this Part also may be transmitted. The authorized bandwidth is 100 kHz.

(7) A RTTY, data or multiplexed emission using a specified digital code listed in S 97.309(a) of this Part or an unspecified digital code under the limitations listed in S 97.309(b) of this Part may be transmitted.

(8) A RTTY or data emission having designators with A, B, C, D, E, F, G, H, J or R as the first symbol; 1, 2, 7 or 9 as the second symbol; and D or W as the third symbol is also authorized.

(9) A station having a control operator holding a Novice or Technician Class operator license may only transmit a CW emission using the international Morse code.

(10) A station having a control operator holding a Novice or Technician Class operator license may only transmit a CW emission using the international Morse code or phone emissions J3E and R3E.

(11) Phone and image emissions may be transmitted only by stations located in ITU Regions 1 and 3, and by stations located within ITU Region 2 that are west of 130 West longitude or south of 20 North latitude.

(12) Emission F8E may be transmitted.

(13) A data emission using an unspecified digital code under the limitations listed in 97.309(b) of this Part also may be transmitted. The authorized bandwidth is 100 kHz.

S 97.309 RTTY and data emission codes.

(a) Where authorized by S 97.305(c) and 97.307(f) of this Part, an amateur station may transmit a RTTY or data emission using the following specified digital codes:

(1) The 5-unit, start-stop, International Telegraph Alphabet No. 2, code defined in International Telegraph and Telephone Consultative Committee Recommendation F.1, Division C (commonly known as Baudot).

(2) The 7-unit code, specified in International Radio Consultative Committee Recommendation CCIR 476-2 (1978), 476-3 (1982), 476-4 (1986) or 625 (1986) (commonly known as AMTOR).

(3) The 7-unit code defined in American National Standards Institute X3.4-1977 or International Alphabet No. 5 defined in International Telegraph and Telephone Consultative Committee Recommendation T.50 or in International Organization for Standardization, International Standard ISO 646 (1983), and extensions as provided for in CCITT Recommendation T.61 (Malaga-Torremolinos, 1984) (commonly known as ASCII).

(b) Where authorized by S S 97.305(c) and 97.307(f) of this Part, a station may transmit a RTTY or data emission using an unspecified digital code, except to a station in a country with which the United States does not have an agreement permitting the code to be used. RTTY and data emissions using unspecified digital codes must not be transmitted for the purpose of obscuring the meaning of any communication. When deemed necessary by an EIC to assure compliance with the FCC Rules, a station must:

(1) Cease the transmission using the unspecified digital code;

(2) Restrict transmissions of any digital code to the extent instructed;

(3) Maintain a record, convertible to the original information, of all digital communications transmitted.

S 97.311 SS emission types.

(a) SS emission transmissions by an amateur station are authorized only for communications between points within areas where the amateur service is regulated by the FCC. SS emission transmissions must not be used for the purpose of obscuring the meaning of any communication.

(b) Stations transmitting SS emission must not cause harmful interference to stations employing other authorized emissions, and must accept all interference caused by stations employing other authorized emissions. For the purposes of this paragraph, unintended triggering of carrier operated repeaters is not considered to be harmful interference.

(c) Only the following types of SS emission transmissions are authorized (hybrid SS emission transmissions involving both spreading techniques are prohibited):

(1) Frequency hopping where the carrier of the transmitted signal is modulated with unciphered information and changes frequency at fixed intervals under the direction of a high speed code sequence.

(2) Direct sequence where the information is modulo-2 added to a high speed code sequence. The combined information and code are then used to modulate the RF carrier. The high speed code sequence dominates the modulation function, and is the direct cause of the wide spreading of the transmitted signal.

(d) The only spreading sequences that are authorized are from the output of one binary linear feedback shift register (which may be implemented in hardware or software).

(1) Only the following sets of connections may be used:

Number of stages in shift register	Taps used in feedback
7	7, 1
13	13, 4, 3, and 1
19	19, 5, 2, and 1

(2) The shift register must not be reset other than by its feedback during an individual transmission. The shift register output sequence must be used without alteration.

(3) The output of the last stage of the binary linear feedback shift register must be used as follows:

(i) For frequency hopping transmissions using x frequencies, n consecutive bits from the shift register must be used to select the next frequency from a list of frequencies sorted in ascending order. Each consecutive frequency must be selected by a consecutive block of n bits. (Where n is the smallest integer greater than $\log 2X$.)

(ii) For direct sequence transmissions using m-ary modulation, consecutive blocks of $\log 2$ m bits from the shift register must be used to select the transmitted signal during each interval.

(e) The station records must document all SS emission transmissions and must be retained for a period of 1 year following the last entry. The station records must include sufficient information to enable the FCC, using the information contained therein, to demodulate all transmissions. The station records must contain at least the following:

(1) A technical description of the transmitted signal;

(2) Pertinent parameters describing the transmitted signal including the frequency or frequencies of operation and, where applicable, the chip rate, the code rate, the spreading function, the transmission protocol(s) including the method of achieving synchronization, and the modulation type;

(3) A general description of the type of information being conveyed (voice, text, memory dump, facsimile, television, etc.);

(4) The method and, if applicable, the frequency or frequencies used for station identification; and

(5) The date of beginning and the date of ending use of each type of transmitted signal.

(f) When deemed necessary by an EIC to assure compliance with this Part, a station licensee must:

(1) Cease SS emission transmissions;

(2) Restrict SS emission transmissions to the extent instructed; and

(3) Maintain a record, convertible to the original information (voice, text, image, etc.) of all spread spectrum communications transmitted.

(g) The transmitter power must not exceed 100 W.

S 97.313 Transmitter power standards.

(a) An amateur station must use the minimum transmitter power necessary to carry out the desired communications.

(b) No station may transmit with a transmitter power exceeding 1.5 kW PEP.

(c) No station may transmit with a transmitter power exceeding 200 W PEP on:

(1) The 3.675-3.725 MHz, 7.10-7.15 MHz, 10.10-10.15 MHz and 21.1-21.2 MHz segments;

(2) The 28.1-28.5 MHz segment when the control operator is a Novice or Technician operator; or

(3) The 7.050-7.075 MHz segment when the station is within ITU Regions 1 or 3.

(d) No station may transmit with a transmitter power exceeding 25 W PEP on the VHF 1.25 m band when the control operator is a Novice operator.

(e) No station may transmit with a transmitter power exceeding 5 W PEP on the UHF 23 cm band when the control operator is a Novice operator.

(f) No station may transmit with a transmitter power exceeding 50 W PEP on the UHF 70 cm band from an area specified in footnote US7 to S 2.106 of the FCC Rules, unless expressly authorized by the FCC after mutual agreement, on a case-by-case basis, between the EIC of the applicable field facility and the military area frequency coordinator at the applicable military base. An Earth station or telecommand station, however, may transmit on the 435-438 MHz segment with a maximum of 611 W effective radiated power (1 kW equivalent isotropically radiated power) without the authorization otherwise required. The transmitting antenna elevation angle between the lower half-power (3 dB relative to the peak or antenna bore sight) point and the horizon must always be greater than 10 .

(g) No station may transmit with a transmitter power exceeding 50 W PEP on the 33 cm band from within 241 km of the boundaries of the White Sands Missile Range. Its boundaries are those portions of Texas and New Mexico bounded on the south by latitude 31 41' North, on the east by longitude 104 11' West, on the north by latitude 34 30' North, and on the west by longitude 107 30' West.

(h) No station may transmit with a transmitter power exceeding 50 W PEP on the 219-220 MHz segment of the 1.25 m band.

S 97.315 Type acceptance of external RF power amplifiers.

(a) No more than 1 unit of 1 model of an external RF power amplifier capable of operation below 144 MHz may be constructed or modified during any calendar year by an amateur operator for use at a station without a grant of type acceptance. No amplifier capable of operation below 144 MHz may be constructed or modified by a non-amateur operator without a grant of type acceptance from the FCC.

(b) Any external RF power amplifier or external RF power amplifier kit (see S 2.815 of the FCC Rules), manufactured, imported or modified for use in a station or attached at any station must be type accepted for use in the amateur service in accordance with Subpart J of Part 2 of the FCC Rules. This requirement does not apply if one or more of the following conditions are met:

(1) The amplifier is not capable of operation on frequencies below 144 MHz. For the purpose of this part, an amplifier will be deemed to be incapable of operation below 144 MHz if it is not capable of being easily modified to increase its amplification characteristics below 120 MHz and either:

(i) The mean output power of the amplifier decreases, as frequency decreases from 144 MHz, to a point where 0 dB or less gain is exhibited at 120 MHz; or

(ii) The amplifier is not capable of amplifying signals below 120 MHz even for brief periods without sustaining permanent damage to its amplification circuitry.

(2) The amplifier was manufactured before April 28, 1978, and

has been issued a marketing waiver by the FCC, or the amplifier was purchased before April 28, 1978, by an amateur operator for use at that amateur operator's station.

(3) The amplifier was:

(i) Constructed by the licensee, not from an external RF power amplifier kit, for use at the licensee's station; or

(ii) Modified by the licensee for use at the licensee's station.

(4) The amplifier is sold by an amateur operator to another amateur operator or to a dealer.

(5) The amplifier is purchased in used condition by an equipment dealer from an amateur operator and the amplifier is further sold to another amateur operator for use at that operator's station.

(c) A list of type accepted equipment may be inspected at FCC headquarters in Washington, DC or at any FCC field location. Any external RF power amplifier appearing on this list as type accepted for use in the amateur service may be marketed for use in the amateur service.

S 97.317 Standards for type acceptance of external RF power amplifiers.

(a) To receive a grant of type acceptance, the amplifier must satisfy the spurious emission standards of S 97.307(d) or (e) of this Part, as applicable, when the amplifier is:

(1) Operated at its full output power;

(2) Placed in the "standby" or "off" positions, but still connected to the transmitter; and

(3) Driven with at least 50 W mean RF input power (unless higher drive level is specified).

(b) To receive a grant of type acceptance, the amplifier must not be capable of operation on any frequency or frequencies between 24 MHz and 35 MHz. The amplifier will be deemed incapable of such operation if it:

(1) Exhibits no more than 6 dB gain between 24 MHz and 26 MHz and between 28 MHz and 35 MHz. (This gain will be determined by the ratio of the input RF driving signal (mean power measurement) to the mean RF output power of the amplifier); and

(2) Exhibits no amplification (0 dB gain) between 26 MHz and 28 MHz.

(c) Type acceptance may be denied when denial would prevent the use of these amplifiers in services other than the amateur service. The following features will result in dismissal or denial of an application for the type acceptance:

(1) Any accessible wiring which, when altered, would permit operation of the amplifier in a manner contrary to the FCC Rules;

(2) Circuit boards or similar circuitry to facilitate the addition of components to change the amplifier's operating characteristics in a manner contrary to the FCC Rules;

(3) Instructions for operation or modification of the amplifier in a manner contrary to the FCC Rules;

(4) Any internal or external controls or adjustments to facilitate operation of the amplifier in a manner contrary to the FCC Rules;

(5) Any internal RF sensing circuitry or any external switch, the purpose of which is to place the amplifier in the transmit mode;

(6) The incorporation of more gain in the amplifier than is necessary to operate in the amateur service; for purposes of this paragraph, the amplifier must:

(i) Not be capable of achieving designed output power when driven with less than 40 W mean RF input power;

(ii) Not be capable of amplifying the input RF driving signal by more than 15 dB, unless the amplifier has a designed transmitter power of less than 1.5 kW (in such a case, gain must be reduced by the same number of dB as the transmitter power relationship to 1.5 kW; This gain limitation is determined by the ratio of the input RF driving signal to the RF output power of the amplifier where both signals are expressed in peak envelope power or mean power);

(iii) Not exhibit more gain than permitted by paragraph (c)(6)(ii) of this Section when driven by an RF input signal of less than 50 W mean power; and

(iv) Be capable of sustained operation at its designed power level.

(7) Any attenuation in the input of the amplifier which, when removed or modified, would permit the amplifier to function at its designed transmitter power when driven by an RF frequency input signal of less than 50 W mean power; or

(8) Any other features designed to facilitate operation in a telecommunication service other than the Amateur Radio Services, such as the Citizens Band (CB) Radio Service.

Subpart E-Providing Emergency Communications

S 97.401 Operation during a disaster.

(a) When normal communication systems are overloaded, damaged or disrupted because a disaster has occurred, or is likely to occur, in an area where the amateur service is regulated by the FCC, an amateur station may make transmissions necessary to meet essential communication needs and facilitate relief actions.

(b) When normal communication systems are overloaded, damaged or disrupted because a natural disaster has occurred, or is likely to occur, in an area where the amateur service is not regulated by the FCC, a station assisting in meeting essential communication needs and facilitating relief actions may do so only in accord with ITU Resolution No. 640 (Geneva, 1979). The 80 m, 75 m, 40 m, 30 m, 20 m, 17 m, 15 m, 12 m, and 2 m bands may be used for these purposes.

(c) When a disaster disrupts normal communication systems in a particular area, the FCC may declare a temporary state of communication emergency. The declaration will set forth any special conditions and special rules to be observed by stations during the communication emergency. A request for a declaration of a temporary state of emergency should be directed to the EIC in the area concerned.

(d) A station in, or within 92.6 km of, Alaska may transmit emissions J3E and R3E on the channel at 5.1675 MHz for emergency communications. The channel must be shared with stations licensed in the Alaska-private fixed service. The transmitter power must not exceed 150 W.

S 97.403 Safety of life and protection of property.

No provision of these rules prevents the use by an amateur station of any means of radiocommunication at its disposal to provide essential communication needs in connection with the immediate safety of human life and immediate protection of property when normal communication systems are not available.

S 97.405 Station in distress.

(a) No provision of these rules prevents the use by an amateur station in distress of any means at its disposal to attract attention, make known its condition and location, and obtain assistance.

(b) No provision of these rules prevents the use by a station, in the exceptional circumstances described in paragraph (a), of any means of radiocommunications at its disposal to assist a station in distress.

S 97.407 Radio amateur civil emergency service.

(a) No station may transmit in RACES unless it is an FCC-licensed primary, club, or military recreation station and it is certified by a civil defense organization as registered with that organization, or it is an FCC-licensed RACES station. No person may be the control operator of a RACES station, or may be the control operator of an amateur station transmitting in RACES unless that person holds a FCC-issued amateur operator license and is certified by a civil defense organization as enrolled in that organization.

(b) The frequency bands and segments and emissions authorized to the control operator are available to stations transmitting communications in RACES on a shared basis with the amateur ser-

vice. In the event of an emergency which necessitates the invoking of the President's War Emergency Powers under the provisions of S 706 of the Communications Act of 1934, as amended, 47 U.S.C. S 606, RACES stations and amateur stations participating in RACES may only transmit on the following frequencies:

(1) The 1800-1825 kHz, 1975-2000 kHz, 3.50-3.55 MHz, 3.93-3.98 MHz, 3.984-4.000 MHz, 7.079-7.125 MHz, 7.245-7.255 MHz, 10.10-10.15 MHz, 14.047-14.053 MHz, 14.22-14.23 MHz, 14.331-14.350 MHz, 21.047-21.053 MHz, 21.228-21.267 MHz, 28.55-28.75 MHz, 29.237-29.273 MHz, 29.45-29.65 MHz, 50.35-50.75 MHz, 52-54 MHz, 144.50-145.71 MHz, 146-148 MHz, 2390-2450 MHz segments;

(2) The 1.25 m, 70 cm and 23 cm bands; and

(3) The channels at 3.997 MHz and 53.30 MHz may be used in emergency areas when required to make initial contact with a military unit and for communications with military stations on matters requiring coordination.

(c) A RACES station may only communicate with:

(1) Another RACES station;

(2) An amateur station registered with a civil defense organization;

(3) A United States Government station authorized by the responsible agency to communicate with RACES stations;

(4) A station in a service regulated by the FCC whenever such communication is authorized by the FCC.

(d) An amateur station registered with a civil defense organization may only communicate with:

(1) A RACES station licensed to the civil defense organization with which the amateur station is registered;

(2) The following stations upon authorization of the responsible civil defense official for the organization with which the amateur station is registered:

(i) A RACES station licensed to another civil defense organization;

(ii) An amateur station registered with the same or another civil defense organization;

(iii) A United States Government station authorized by the responsible agency to communicate with RACES stations; and

(iv) A station in a service regulated by the FCC whenever such communication is authorized by the FCC.

(e) All communications transmitted in RACES must be specifically authorized by the civil defense organization for the area served. Only civil defense communications of the following types may be transmitted:

(1) Messages concerning impending or actual conditions jeopardizing the public safety, or affecting the national defense or security during periods of local, regional, or national civil emergencies;

(2) Messages directly concerning the immediate safety of life of individuals, the immediate protection of property, maintenance of law and order, alleviation of human suffering and need, and the combating of armed attack or sabotage;

(3) Messages directly concerning the accumulation and dissemination of public information or instructions to the civilian population essential to the activities of the civil defense organization or other authorized governmental or relief agencies; and

(4) Communications for RACES training drills and tests necessary to ensure the establishment and maintenance of orderly and efficient operation of the RACES as ordered by the responsible civil defense organizations served. Such drills and tests may not exceed a total time of 1 hour per week. With the approval of the chief officer for emergency planning the applicable State, Com-

Topics

	Element 2	3(A)	3(B)	4(A)	4(B)
(1) FCC Rules for the amateur radio services	10	5	4	6	8
(2) Amateur station operating procedures	2	3	3	1	4
(3) Radio wave propagation characteristics of amateur service frequency bands	1	3	3	2	2
(4) Amateur radio practices	4	4	5	4	4
(5) Electrical principles as applied to amateur station equipment	4	2	2	10	6
(6) Amateur station equipment circuit components	2	2	1	6	4
(7) Practical circuits employed in amateur station equipment	2	1	1	10	4
(8) Signals and emissions transmitted by amateur stations	2	2	2	6	4
(9) Amateur station antennas and feed lines	3	3	4	5	4

monwealth, District or territory, however, such tests and drills may be conducted for a period not to exceed 72 hours no more than twice in any calendar year.

Subpart F—Qualifying Examinations Systems

S 97.501 Qualifying for an amateur operator license.

Each applicant for the grant of a new amateur operator license or for the grant of a modified license to show a higher operator class, must pass or otherwise receive credit for the examination elements specified for the class of operator license sought:

(a) Amateur Extra Class operator: Elements 1(C), 2, 3(A), 3(B), 4(A) and 4(B)

(b) Advanced Class operator: Elements 1(B) or 1(C), 2, 3(A), 3(B) and 4(A)

(c) General Class operator: Elements 1(B) or 1(C), 2, 3(A) and 3(B)

(d) Technician Plus Class operator: Elements 1(A) or 1(B) or 1(C), 2, and 3(A)

(e) Technician Class operator: Elements 2 and 3(A)

(f) Novice class operator: Elements 1(A) or 1(B) or 1(C), and 2

S 97.503 Element standards.

(a) A telegraphy examination must be sufficient to prove that the examinee has the ability to send correctly by hand and to receive correctly by ear texts in the international Morse code at not less than the prescribed speed, using all the letters of the alphabet, numerals 0-9, period, comma, question mark, slant mark and prosigns AR, BT and SK.

(1) Element 1(A): 5 words per minute;

(2) Element 1(B): 13 words per minute;

(3) Element 1(C): 20 words per minute.

(b) A written examination must be such as to prove that the examinee possesses the operational and technical qualifications required to perform properly the duties of an amateur service licensee. Each written examination must be comprised of a question set as follows:

(1) Element 2: 30 questions concerning the privileges of a Novice Class operator license. The minimum passing score is 22 questions answered correctly.

(2) Element 3(A): 25 questions concerning the additional privileges of a Technician Class operator license. The minimum passing score is 19 questions answered correctly.

(3) Element 3(B): 25 questions concerning the additional privileges of a General Class operator license. The minimum passing score is 19 questions answered correctly.

(4) Element 4(A): 50 questions concerning the additional privileges of an Advanced Class operator license. The minimum passing score is 37 questions answered correctly.

(5) Element 4(B): 40 questions concerning the additional privileges of an Amateur Extra Class operator license. The minimum passing score is 30 questions answered correctly.

(c) The topics and number of questions required in each question set are listed below for the appropriate examination element:

S 97.505 Element credit.

(a) The administering VEs must give credit as specified below to an examinee holding any of the following documents:

(1) An unexpired (or expired but within the grace period for renewal) FCC-granted Advanced Class operator license document: Elements 1(B), 2, 3(A), 3(B), and 4(A).

(2) An unexpired (or expired but within the grace period for renewal) FCC-granted General Class operator license document: Elements 1(B), 2, 3(A), and 3(B).

(3) An unexpired (or expired but within the grace period for renewal) FCC-granted Technician Plus Class operator license document (including a Technician Class operator license granted before February 14, 1991) license document: Elements 1(A), 2, and 3(A).

(4) An unexpired (or expired but within the grace period for renewal) FCC-granted Technician Class operator license document: Elements 2 and 3(A).

(5) An unexpired (or expired but within the grace period for renewal) FCC-granted Novice Class operator license document: Elements 1(A) and 2.

(6) A CSCE: Each element the CSCE indicates the examinee passed within the previous 365 days.

(7) An unexpired (or expired for less than 5 years) FCC-issued commercial radiotelegraph operator license document or permit: Element 1(C).

(8) An expired or unexpired FCC-issued Technician Class operator license document granted before March 21 1987: Element 3(B).

(9) An expired or unexpired FCC-issued Technician Class license document granted before February 14, 1991: Element 1(A).

(10) An unexpired (or expired but within the grace period for renewal), FCC-granted Novice, Technician Plus (including a Technician Class operator license granted before February 14, 1991), General, or Advanced Class operator license document, and a FCC Form 610 containing:

(i) A physician's certification stating that because the person is an individual with a severe handicap, the duration of which will extend for more than 365 days beyond the date of the certification, the person is unable to pass a 13 or 20 words per minute telegraphy examination; and

(ii) A release signed by the person permitting the disclosure to the FCC of medical information pertaining to the person's handicap: Element 1(C).

(b) No examination credit, except as herein provided, shall be allowed on the basis of holding or having held any other license grant or document.

S 97.507 Preparing an examination.

(a) Each telegraphy message and each written question set administered to an examinee must be prepared by a VE who has been granted an Amateur Extra Class operator license. A telegraphy message or written question set, however, may also be prepared for the following elements by a VE who has been granted an FCC operator license of the class indicated:

(1) Element 3(B): Advanced Class operator.

(2) Elements 1(A) and 3(A): Advanced or General Class operator.

(3) Element 2: Advanced, General, Technician, or Technician Plus Class operator.

(b) Each question set administered to an examinee must utilize questions taken from the applicable question pool.

(c) Each telegraphy message and each written question set administered to an examinee for an amateur operator license must be prepared, or obtained from a supplier, by the administering VEs according to instructions from the coordinating VEC.

(d) A telegraphy examination must consist of a message sent in the international Morse code at no less than the prescribed speed for a minimum of 5 minutes. The message must contain each required telegraphy character at least once. No message known to the examinee may be administered in a telegraphy examination. Each 5 letters of the alphabet must be counted as 1 word. Each numeral, punctuation mark and prosign must be counted as 2 letters of the alphabet.

S 97.509 Administering VE requirements.

(a) Each examination for an amateur operator license must be administered by 3 administering VEs at an examination session coordinated by a VEC. Before the session, the administering VEs must make a public announcement stating the location and time of the session. The number of examinees at the session may be limited.

(b) Each administering VE must:

(1) Be accredited by the coordinating VEC;

(2) Be at least 18 years of age;

(3) Be a person who has been granted an FCC amateur operator license document of the class specified below:

(i) Amateur Extra, Advanced, or General Class in order to administer a Novice, Technician, or Technician Plus Class operator license examination;

(ii) Amateur Extra Class in order to administer a General, Advanced, or Amateur Extra Class operator license examination.

(4) Not be a person whose grant of an amateur station license or amateur operator license has ever been revoked or suspended.

(5) Not own a significant interest in, or be an employee of, any company or other entity that is engaged in the manufacture or dis-

tribution of equipment used in connection with amateur station transmissions, or in the preparation or distribution of any publication used in preparation for obtaining amateur operator licenses. (An employee who does not normally communicate with that part of an entity engaged in the manufacture or distribution of such equipment, or in the preparation or distribution of any publication used in preparation for obtaining amateur operator licenses, may be an administering VE.)

(c) Each administering VE must be present and observing the examinee throughout the entire examination. The administering VEs are responsible for the proper conduct and necessary supervision of each examination. The administering VEs must immediately terminate the examination upon failure of the examinee to comply with their instructions.

(d) No VE may administer an examination to his or her spouse, children, grandchildren, stepchildren, parents, grandparents, stepparents, brothers, sisters, stepbrothers, stepsisters, aunts, uncles, nieces, nephews, and in-laws.

(e) No VE may administer or certify any examination by fraudulent means or for monetary or other consideration including reimbursement in any amount in excess of that permitted. Violation of this provision may result in the revocation of the grant of the VE's amateur station license and the suspension of the grant of the VE's amateur operator license.

(f) No examination that has been compromised shall be administered to any examinee. Neither the same telegraphy message nor the same question set may be re-administered to the same examinee.

(g) Passing a telegraphy receiving examination is adequate proof of an examinee's ability to both send and receive telegraphy. The administering VEs, however, may also include a sending segment in a telegraphy examination.

(h) Upon completion of each examination element, the administering VEs must immediately grade the examinee's answers. The administering VEs are responsible for determining the correctness of the examinee's answers.

(i) When the examinee is credited for all examination elements required for the operator license sought, the administering VEs must certify on the examinee's application document that the applicant is qualified for the license.

(j) When the examinee does not score a passing grade on an examination element, the administering VEs must return the application document to the examinee and inform the examinee of the grade.

(k) The administering VEs must accommodate an examinee whose physical disabilities require a special examination procedure. The administering VEs may require a physician's certification indicating the nature of the disability before determining which, if any, special procedures must be used.

(l) The administering VEs must issue a CSCE to an examinee who scores a passing grade on an examination element.

(m) Within 10 days of the administration of a successful examination for an amateur operator license, the administering VEs must submit the application document to the coordinating VEC.

S 97.511 Examinee conduct.

Each examinee must comply with the instructions given by the administering VEs.

S 97.513 [Removed and Reserved]

S 97.515 [Reserved]

S 97.517 [Reserved]

S 97.519 Coordinating examination sessions.

(a) A VEC must coordinate the efforts of VEs in preparing and administering examinations.

(b) At the completion of each examination session, the coordinating VEC must collect the FCC Forms 610 documents and test results from the administering VEs. Within 10 days of collecting the FCC Forms 610 documents, the coordinating VEC must screen and, for qualified examinees, forward electronically or on diskette

the data contained on the FCC Forms 610 documents, or forward the FCC Form 610 documents to: FCC, 1270 Fairfield Road, Gettysburg, PA 17325-7245. When the data is forwarded electronically, the coordinating VEC must retain the FCC Forms 610 documents for at least fifteen months and make them available to the FCC upon request.

(c) Each VEC must make any examination records available to the FCC, upon request.

(d) The FCC may:

(1) Administer any examination element itself;

(2) Readminister any examination element previously administered by VEs, either itself or under the supervision of a VEC or VEs designated by the FCC; or

(3) Cancel the operator/primary station license of any licensee who fails to appear for readministration of an examination when directed by the FCC, or who does not successfully complete any required element that is readministered. In an instance of such cancellation, the person will be granted an operator/primary station license consistent with completed examination elements that have not been invalidated by not appearing for, or by failing, the examination upon readministration.

S 97.521 VEC qualifications.

No organization may serve as a VEC unless it has entered into a written agreement with the FCC. The VEC must abide by the terms of the agreement. In order to be eligible to be a VEC, the entity must:

(a) Be an organization that exists for the purpose of furthering the amateur service;

(b) Be capable of serving as a VEC in at least the VEC region (see Appendix 2) proposed;

(c) Agree to coordinate examinations for any class of amateur operator license

(d) Agree to assure that, for any examination, every examinee qualified under these rules is registered without regard to race, sex, religion, national origin or membership (or lack thereof) in any amateur service organization;

(e) Not be engaged in the manufacture or distribution of equipment used in connection with amateur station transmissions, or in the preparation or distribution of any publication used in preparation for obtaining amateur licenses, unless a persuasive showing is made to the FCC that preventive measures have been taken to preclude any possible conflict of interest.

S 97.523 Question pools.

All VECs must cooperate in maintaining one question pool for each written examination element. Each question pool must contain at least 10 times the number of questions required for a single examination. Each question pool must be published and made available to the public prior to its use for making a question set. Each question on each VEC question pool must be prepared by a VE holding the required FCC-issued operator license. See S 97.507(a) of this Part.

S 97.525 Accrediting VEs.

(a) No VEC may accredit a person as a VE if:

(1) The person does not meet minimum VE statutory qualifications or minimum qualifications as prescribed by this Part;

(2) The FCC does not accept the voluntary and uncompensated services of the person;

(3) The VEC determines that the person is not competent to perform the VE functions; or

4) The VEC determines that questions of the person's integrity or honesty could compromise the examinations.

(b) Each VEC must seek a broad representation of amateur operators to be VEs. No VEC may discriminate in accrediting VEs on the basis of race, sex, religion or national origin; nor on the basis of membership (or lack thereof) in an amateur service organization; nor on the basis of the person accepting or declining to accept reimbursement.

S 97.527 Reimbursement for expenses.

(a) VEs and VECs may be reimbursed by examinees for out-of-pocket expenses incurred in preparing, processing, administering, or coordinating an examination for an amateur operator license.

(b) The maximum amount of reimbursement from any one examinee for any one examination at a particular session regardless of the number of examination elements taken must not exceed that announced by the FCC in a Public Notice. (The basis for the maximum fee is $4.00 for 1984, adjusted annually each January 1 thereafter for changes in the Department of Labor Consumer Price Index.).

(c) Each VE and each VEC accepting reimbursement must maintain records of out-of-pocket expenses and reimbursements for each examination session. Written certifications must be filed with the FCC each year that all expenses for the period from January 1 to December 31 of the preceding year for which reimbursement was obtained were necessarily and prudently incurred.

(d) The expense and reimbursement records must be retained by each VE and each VEC for 3 years and be made available to the FCC upon request.

(e) Each VE must forward the certification by January 15 of each year to the coordinating VEC for the examinations for which reimbursement was received. Each VEC must forward all such certifications and its own certification to the FCC on or before January 31 of each year.

(f) Each VEC must disaccredit any VE failing to provide the certification. The VEC must advise the FCC on January 31 of each year of any VE that it has disaccredited for this reason.

Appendix 1:
Places Where the Amateur Service is Regulated by the FCC

In ITU Region 2, the amateur service is regulated by the FCC within the territorial limits of the 50 United States, District of Columbia, Caribbean Insular areas [Commonwealth of Puerto Rico, United States Virgin Islands (50 islets and cays) and Navassa Island], and Johnston Island (Islets East, Johnston, North and Sand) and Midway Island (Islets Eastern and Sand) in the Pacific Insular areas.

In ITU Region 3, the amateur service is regulated by the FCC within the Pacific Insular territorial limits of American Samoa (seven islands), Baker Island, Commonwealth of Northern Mariannas Islands, Guam Island, Howland Island, Jarvis Island, Kingman Reef, Kure Island, Palmyra Island (more than 50 islets) and Wake Island (Islets Peale, Wake and Wilkes).

Appendix 2:

VEC Regions
1. Connecticut, Maine, Massachusetts, New Hampshire Rhode Island and Vermont.
2. New Jersey and New York.
3. Delaware, District of Columbia, Maryland and Pennsylvania.
4. Alabama, Florida, Georgia, Kentucky, North Carolina, South Carolina, Tennessee and Virginia..
5. Arkansas, Louisiana, Mississippi, New Mexico Oklahoma and Texas,
6. California.
7. Arizona, Idaho, Montana, Nevada, Oregon, Utah, Washington and Wyoming.
8. Michigan, Ohio and West Virginia.
9. Illinois, Indiana and Wisconsin.
10. Colorado, Iowa, Kansas, Minnesota, Missouri, Nebraska, North Dakota and South Dakota.
11. Alaska.
12 Caribbean Insular areas.
13. Hawaii and Pacific Insular areas.

PRB-1

In 1985, the FCC issued a landmark Memorandum regarding the Federal preemption of State and Local regulations pertaining to Amateur Radio facilities, in particular, amateur radio antennas. It was issued in response to a growing number of communities attempting to impose restrictions on the height of ham antennas,

and the ensuing flurry of lawsuits that arose. It has been used successfully in many cases in local jurisdictions since it was issued.

The entire text is reproduced here.

Its full legal citation (which makes it easier for lawyers on both sides of an antenna dispute to locate) is:

FCC Order PRB-1, 101 FCC 2d 952, 50 Fed. Reg. 38813 (September 25, 1985).
Adopted as a Federal Regulation in 47 CFR Section 97.15(e)]

Before the
Federal Communications Commission
Washington, DC 20554

FCC 85-506
36149

In the Matter of

Federal preemption of state and   PRB-1
local regulations pertaining
to Amateur radio facilities.

MEMORANDUM OPINION AND ORDER

Adopted: September 16, 1985;
Released: September 19, 1985

By the Commission: Commissioner Rivera not participating.
Background

1. On July 16, 1984, the American Radio Relay League, Inc. (ARRL) filed a Request for Issuance of a Declaratory Ruling asking us to delineate the limitations of local zoning and other local and state regulatory authority over Federally-licensed radio facilities. Specifically, the ARRL wanted an explicit statement that would preempt all local ordinances which provably preclude or significantly inhibit effective, reliable amateur radio communications. The ARRL acknowledges that local authorities can regulate amateur installations to insure the safety and health of persons in the community, but believes that those regulations cannot be so restrictive that they preclude effective amateur communications.

2. Interested parties were advised that they could file comments in the matter[1]. With extension, comments were due on or before December 26, 1984[2], with reply comments due on or before January 25, 1985[3]. Over sixteen hundred comments were filed.

Local Ordinances

3. Conflicts between amateur operators regarding radio antennas and local authorities regarding restrictive ordinances are common. The amateur operator is governed by the regulations contained in Part 98 of our rules. Those rules do not limit the height of an amateur antenna but they require, for aviation safety reasons, that certain FAA notification and FCC approval procedures must be followed for antennas which exceed 200 feet in height above ground level or antennas which are to be erected near airports. Thus, under FCC rules some amateur antenna support structures require obstruction marking and lighting. On the other hand, local municipalities or governing bodies frequently enact regulations limiting antennas and their support structures in height and location, e.g. to side or rear yards, for health, safety or aesthetic considerations. These limiting regulations can result in conflict because the effectiveness of the communications that emanate from an amateur radio station are directly dependent upon the location and the height of the antenna. Amateur operators maintain that they are precluded from operating in certain bands allocated for their use if the height of their antennas is limited by a local ordinance.

4. Examples of restrictive local ordinances were submitted by several amateur operators in this proceeding. Stanley J. Cichy, San Diego, California, noted that in San Diego amateur radio antennas

come under a structures ruling which limits building heights to 30 feet. Thus, antennas there are also limited to 30 feet. Alexander Vrenios, Mundelein, Illinois wrote that an ordinance of the Village of Mundelein provides that an antenna must be a distance from the property line that is equal to one and one-half times its height. In his case, he is limited to an antenna tower for his amateur station just over 53 feet in height.

5. John C. Chapman, an amateur living in Bloomington, Minnesota, commented that he was not able to obtain a building permit to install an amateur radio antenna exceeding 35 feet in height because the Bloomington city ordinance restricted "structures" heights to 35 feet. Mr. Chapman said that the ordinance, when written, undoubtedly applied to buildings but was now being applied to antennas in the absence of a specific ordinance regulating them. There were two options open to him if he wanted to engage in amateur communications. He could request a variance to the ordinance by way of a hearing before the City Council, or he could obtain affidavits from his neighbors swearing that they had no objection to the proposed antenna installation. He got the building permit after obtaining the cooperation of his neighbors. His concern, however, is that he had to get permission from several people before he could effectively engage in radio communications for which he had a valid FCC amateur license.

6. In addition to height restrictions, other limits are enacted by local jurisdictions—anti-climb devices on towers or fences around them; minimum distances from high voltage power lines; minimum distances of towers from property lines; and regulations pertaining to the structural soundness of the antenna installation. By and large, amateurs do not find these safety precautions objectionable. What they do object to are the sometimes prohibitive, nonrefundable application filing fees to obtain a permit to erect an antenna installation and those provisions in ordinances which regulate antennas for purely aesthetic reasons. The amateurs contend, almost universally, that "beauty is in the eye of the beholder." They assert that an antenna installation is not more aesthetically displeasing than other objects that people keep on their property, e.g. motor homes, trailers, pick-up trucks, solar collectors and gardening equipment.

Restrictive Covenants

7. Amateur operators also oppose restrictions on their amateur operations which are contained in the deeds for their homes or in their apartment leases. Since these restrictive covenants are contractual agreements between private parties, they are not generally a matter of concern to the Commission. However, since some amateurs who commented in this proceeding provided us with examples of restrictive covenants, they are included for information. Mr. Eugene O. Thomas of Hollister, California included in his comments an extract of the Declaration of Covenants and Restrictions for Ridgemark Estates, County of San Benito, State of California. It provides:

> No antenna for transmission or reception of radio signals shall be erected outdoors for use by any dwelling unit except upon approval of the Directors. No radio or television signals or any other form of electromagnetic radiation shall be permitted to originate from any lot which may unreasonably interfere with the reception of television or radio signals upon any other lot.

Marshall Wilson, Jr. provided a copy of the restrictive covenant contained in deeds for the Bell Martin Addition #2, Irving, Texas. It is binding upon all of the owners or purchasers of the lots in the said addition, his or their heirs, executors, administrators or assigns. It reads:

> No antenna or tower shall be erected upon any lot for the purposes of radio operations.

William J. Hamilton resides in an apartment building in Gladstone, Missouri. He cites a clause in his lease prohibiting the erection of an antenna. He states that he has been forced to give up

operating amateur radio equipment except a hand-held 2 meter (144–148 MHz) radio transceiver. He maintains that he should not be penalized just because he lives in an apartment.

Other restrictive covenants are less global in scope than those cited above. For example, Robert Webb purchased a home in Houston, Texas. His deed restriction prohibited "transmitting or receiving antennas extending above the roof line."

8. Amateur operators generally oppose restrictive covenants for several reasons. They maintain that such restrictions limit the places that they can reside if they want to pursue their hobby of amateur radio. Some state that they impinge on First Amendment rights of free speech. Others believe that a constitutional right is being abridged because, in their view, everyone has a right to access the airwaves regardless of where they live.

9. The contrary belief held by housing subdivision communities and condominium or homeowner's associations is that amateur radio installations constitute safety hazards, cause interference to other electronic equipment which may be operated in the home (televisions, radio, stereos) or are eyesores that detract from the aesthetic and tasteful appearance of the housing development or apartment complex. To counteract these negative consequences, the subdivisions and associations include in their deeds, leases or by-laws restrictions and limitations on the location and height of antennas or, in some cases, prohibit them altogether. The restrictive covenants are contained in the contractual agreement entered into at the time of the sale or lease of the property. Purchasers or lessees are free to choose whether they wish to reside where such restrictions on amateur antennas are in effect or settle elsewhere.

Supporting Comments

10. The Department of Defense (DOD) supported the ARRL and emphasized in its comments that continued success of existing national security and emergency preparedness telecommunications plans involving amateur stations would be severely diminished if state and local ordinances were allowed to prohibit the construction and usage of effective amateur transmission facilities. DOD utilizes volunteers in the Military Affiliate Radio Service (MARS)[4], Civil Air Patrol (CAP) and the Radio Amateur Civil Emergency Service (RACES). It points out that these volunteer communicators are operating radio equipment installed in their homes and that undue restrictions on antennas by local authorities adversely affect their efforts. DOD states that the responsiveness of these volunteer systems would be impaired if local ordinances interfere with the effectiveness of these important national telecommunication resources. DOD favors the issuance of a ruling that would set limits for local and state regulatory bodies when they are dealing with amateur stations.

11. Various chapters of the American Red Cross also came forward to support the ARRL's request for a preemptive ruling. The Red Cross works closely with amateur radio volunteers. It believes that without amateurs' dedicated support, disaster relief operations would significantly suffer and that its ability to serve disaster victims would be hampered. It feels that antenna height limitations that might be imposed by local bodies will negatively affect the service now rendered by the volunteers.

12. Cities and counties from various parts of the United States filed comments in support of the ARRL's request for a Federal preemption ruling. The comments from the Director of Civil Defense, Port Arthur, Texas are representative:

> The Amateur Radio Service plays a vital role with our Civil Defense program here in Port Arthur and the design of these antennas and towers lends greatly to our ability to communicate during times of disaster.
>
> We do not believe there should be any restrictions on the antennas and towers except for reasonable safety precautions. Tropical storms, hurricanes and tornadoes are a way

of life here on the Texas Gulf Coast and good communications are absolutely essential when preparing for a hurricane and even more so, during recovery operations after the hurricane has past.

13. The Quarter Century Wireless Association took a strong stand in favor of the issuance of a declaratory ruling. It believes that Federal preemption is necessary so that there will be uniformity for all Amateur radio installations on private property throughout the United States.

14. In its comments, the ARRL argued that the Commission has the jurisdiction to preempt certain local land use regulations which frustrate or prohibit amateur radio communications. It said that the appropriate standard in preemption cases is not the extent of state and local interest in a given regulation, but rather the impact of that regulation on Federal goals. Its position is that Federal preemption is warranted whenever local governmental regulations relate adversely to the operational aspects of amateur communication. The ARRL maintains that localities routinely employ a variety of land use devices to preclude the installation of effective amateur antennas, including height restrictions, conditional use permits, building setbacks and dimensional limitations on antennas. It sees a declaratory ruling of Federal preemption as necessary to cause municipalities to accommodate amateur operator needs in land use planning efforts.

15. James C. O'Connell, an attorney who has represented several amateurs before local zoning authorities, said that requiring amateurs to seek variances or special use approval to erect reasonable antennas unduly restricts the operation of amateur stations. He suggested that the Commission preempt zoning ordinances which impose antenna height limits of less than 65 feet. He said that this height would represent a reasonable accommodation of the communication needs of most amateurs and the legitimate concerns of local zoning authorities.

Opposing Comments

16. The City of La Mesa, California has a zoning regulation which controls amateur antennas. Its comments reflected an attempt to reach a balanced view.

This regulation has neither the intent, nor the effect, of precluding or inhibiting effective and reliable communications. Such antennas may be built as long as their construction does not unreasonably block views or constitute eyesores. The reasonable assumption is that there are always alternatives at a given site for different placement, and/or methods for aesthetic treatment. Thus, both public objectives of controlling land use for the public health, safety, and convenience, and providing an effective communications network, can be satisfied.

A blanket ruling to completely set aside local control, or a ruling which recognizes control only for the purpose of safety of antenna construction, would be contrary to...legitimate local control.   .

17. Comments from the county of San Diego state:

While we are aware of the benefits provided by amateur operators, we oppose the issuance of a preemption ruling which would elevate 'antenna effectiveness' to a position above all other considerations. We must, however, argue that the local government must have the ability to place reasonable limitations upon the placement and configuration of amateur radio transmitting and receiving antennas. Such ability is necessary to assure that the local decision-makers have the authority to protect the public health, safety and welfare of all citizens.

In conclusion, I would like to emphasize an important difference between your regulatory powers and that of local governments. Your Commission's approval of the preemptive requests would establish a 'national policy.'

However, any regulation adopted by a local jurisdiction could be overturned by your Commission or a court if such regulation was determined to be unreasonable.

18. The City of Anderson, Indiana, summarized some of the problems that face local communities:

I am sympathetic to the concerns of these antenna owners and I understand that to gain the maximum reception from their devices, optimal location is necessary. However, the preservation of residential zoning districts as 'liveable' neighborhoods is jeopardized by placing these antennas in front yards of homes. Major problems of public safety have been encountered, particularly vision blockage for auto and pedestrian access. In addition, all communities are faced with various building lot sizes. Many building lots are so small that established setback requirements (in order to preserve adequate air and light) are vulnerable to the unregulated placement of these antennas.

...the exercise of preemptive authority by the FCC in granting this request would not be in the best interest of the general public.

19. The National Association of Counties (NACO), the American Planning Association (APA) and the National League of Cities (NLC) all opposed the issuance of an antenna preemption ruling. NACO emphasized that federal and state power must be viewed in harmony and warns that Federal intrusion into local concerns of health, safety and welfare could weaken the traditional police power exercised by the state and unduly interfere with the legitimate activities of the states. NLC believed that both Federal and local interests can be accommodated without preempting local authority to regulate the installation of amateur radio antennas. The APA said that the FCC should continue to leave the issue of regulating amateur antennas with the local government and with the state and Federal courts.

Discussion

20. When considering preemption, we must begin with two constitutional provisions. The tenth amendment provides that any powers which the constitution either does not delegate to the United States or does not prohibit the states from exercising are reserved to the states. These are the police powers of the states. The Supremacy Clause, however, provides that the constitution and the laws of the United States shall supersede any state law to the contrary. Article III, Section 2. Given these basic premises, state laws may be preempted in three ways: First, Congress may expressly preempt the state law. See Jones v. Rath Packing Co., 430 U.S. 519, 525 (1977). Or, Congress may indicate its intent to completely occupy a given field so that any state law encompassed within that field would implicitly be preempted. Such intent to preempt could be found in a congressional regulatory scheme that was so pervasive that it would be reasonable to assume that Congress did not intend to permit the states to supplement it. See Fidelity Federal Savings & Loan Ass'n v. de la Cuesta, 458 U.S. 141, 153 (1982). Finally, preemption may be warranted when state law conflicts with federal law. Such conflicts may occur when "compliance with both Federal and state regulations is a physical impossibility," Florida Lime & Avocado Growers, Inc. v. Paul, 373 U.S. 132, 142, 143 (1963), or when state law "stands as an obstacle to the accomplishment and execution of the full purposes and objectives of Congress," Hines v. Davidowitz, 312 U.S. 52, 67 (1941). Furthermore, federal regulations have the same preemptive effect as federal statutes. Fidelity Federal Savings & Loan Association v. de la Cuesta, supra.

21. The situation before us requires us to determine the extent to which state and local zoning regulations may conflict with federal policies concerning amateur radio operators.

22. Few matters coming before us present such a clear dichotomy of viewpoint as does the instant issue. The cities, counties,

local communities and housing associations see an obligation to all of their citizens and try to address their concerns. This is accomplished through regulations, ordinances or covenants oriented toward the health, safety and general welfare of those they regulate. At the opposite pole are the individual amateur operators and their support groups who are troubled by local regulations which may inhibit the use of amateur stations or, in some instances, totally preclude amateur communications. Aligned with the operators are such entities as the Department of Defense, the American Red Cross and local civil defense and emergency organizations who have found in Amateur Radio a pool of skilled radio operators and a readily available backup network. In this situation, we believe it is appropriate to strike a balance between the federal interest in promoting amateur operations and the legitimate interests of local governments in regulating local zoning matters. The cornerstone on which we will predicate our decision is that a reasonable accommodation may be made between the two sides.

23. Preemption is primarily a function of the extent of the conflict between federal and state and local regulation. Thus, in considering whether our regulations or policies can tolerate a state regulation, we may consider such factors as the severity of the conflict and the reasons underlying the state's regulations. In this regard, we have previously recognized the legitimate and important state interests reflected in local zoning regulations. For example, in Earth Satellite Communications, Inc., 95 FCC 2d 1223 (1983), we recognized that

> ...countervailing state interests inhere in the present situation...For example, we do not wish to preclude a state or locality from exercising jurisdiction over certain elements of an SMATV operation that properly may fall within its authority, such as zoning or public safety and health, provided the regulation in question is not undertaken as a pretext for the actual purpose of frustrating achievement of the preeminent federal objective and so long as the non-federal regulation is applied in a nondiscriminatory manner.

24. Similarly, we recognize here that there are certain general state and local interests which may, in their even-handed application, legitimately affect amateur radio facilities. Nonetheless, there is also a strong federal interest in promoting amateur communications. Evidence of this interest may be found in the comprehensive set of rules that the Commission has adopted to regulate the amateur service.[5] Those rules set forth procedures for the licensing of stations and operators, frequency allocations, technical standards which amateur radio equipment must meet and operating practices which amateur operators must follow. We recognize the Amateur radio service as a voluntary, noncommercial communication service, particularly with respect to providing emergency communications. Moreover, the amateur radio service provides a reservoir of trained operators, technicians and electronic experts who can be called on in times of national or local emergencies. By its nature, the Amateur Radio Service also provides the opportunity for individual operators to further international goodwill. Upon weighing these interests, we believe a limited preemption policy is warranted. State and local regulations that operate to preclude amateur communications in their communities are in direct conflict with federal objectives and must be preempted.

25. Because amateur station communications are only as effective as the antennas employed, antenna height restrictions directly affect the effectiveness of amateur communications. Some amateur antenna configurations require more substantial installations than others if they are to provide the amateur operator with the communications that he/she desires to engage in. For example, an antenna array for international amateur communications will differ from an antenna used to contact other amateur operators at shorter distances. We will not, however, specify any particular height limitation below which a local government may not regulate, nor will we suggest the precise language that must be contained in local ordinances, such as mechanisms for special exceptions, variances, or conditional use permits. Nevertheless, local regulations which involve placement, screening, or height of antennas based on health, safety, or aesthetic considerations must be crafted to accommodate reasonable amateur communications, and to represent the minimum practicable regulation to accomplish the local authority's legitimate purpose.[6]

26. Obviously, we do not have the staff or financial resources to review all state and local laws that affect amateur operations. We are confident, however, that state and local governments will endeavor to legislate in a manner that affords appropriate recognition to the important federal interest at stake here and thereby avoid unnecessary conflicts with federal policy, as well as time-consuming and expensive litigation in this area. Amateur operators who believe that local or state governments have been overreaching and thereby have precluded accomplishment of their legitimate communications goals, may, in addition, use this document to bring our policies to the attention of local tribunals and forums.

27. Accordingly, the Request for Declaratory Ruling filed July 16, 1984, by the American Radio Relay League, Inc., IS GRANTED to the extent indicated herein and, in all other respects, IS DENIED.

FEDERAL COMMUNICATIONS COMMISSION

William J. Tricarico
Secretary

Footnotes

1. Public Notice, August 30, 1984, Mimeo. No. 6299, 49 F.R. 36113, September 14, 1984.

2. Public Notice, December 19, 1984, Mimeo. No. 1498.
3. Order, November 8, 1984, Mimeo. No. 770.

4. MARS is solely under the auspices of the military which recruits volunteer amateur operators to render assistance to it. The Commission is not involved in the MARS program.

5. 47 CFR Part 97.

6. We reiterate that our ruling herein does not reach restrictive covenants in private contractual agreements. Such agreements are voluntarily entered into by the buyer or tenant when the agreement is executed and do not usually concern this Commission.

THE "CONSIDERATE OPERATOR'S FREQUENCY GUIDE"

Frequencies generally recognized for certain modes or activities
(all frequencies are in MHz):

1.800–1.830	CW, RTTY and other narrowband modes
1.830–1.840	CW, RTTY and other narrowband modes, intercontinental QSOs only
1.840–1.850	CW, SSB, SSTV and other wideband modes, intercontinental QSOs only
1.840–2.000	CW, phone, SSTV and other wideband modes
3.590	RTTY DX
3.580–3.620	RTTY
3.620–3.635	Packet
3.790–3.800	DX window
3.845	SSTV
3.885	AM calling freq.
7.040	RTTY DX
7.080–7.100	RTTY
7.171	SSTV
7.290	AM
10.130–10.140	RTTY
10.140–10.150	Packet
14.070–14.095	RTTY
14.095–14.0995	Packet

14.100	NCDXF beacons
14.1005–14.112	Packet
14.230	SSTV
14.286	AM calling freq.
18.100–18.105	RTTY
18.105–18.110	Packet
21.070–21.090	RTTY
21.090–21.100	Packet
21.340	SSTV
24.920–24.925	RTTY
24.925–24.930	Packet
28.070–28.120	RTTY
28.120–28.189	Packet
28.190–28.225	Beacons
28.680	SSTV
29.000–29.200	AM
29.300–29.510	Satellite downlinks
29.520–29.580	Repeater inputs
29.600	FM simplex
29.620–29.680	Repeater outputs

Courtesy of ARRL

U.S. Amateur Frequency Allocations by Class License

160 METERS

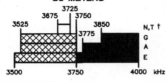

1800   1900   2000   kHz

Amateur stations operating at 1900–2000 kHz must not cause interference to the radiolocation service and are afforded no protection from radiolocation operations.

KEY

▨ = CW, RTTY and data

▨ = CW, RTTY, data, MCW, test, phone and image

■ = CW, phone and image

▨ = CW, RTTY, data, phone and image

▨ = CW and SSB

☐ = CW only

E = AMATEUR EXTRA
A = ADVANCED
G = GENERAL
T = TECHNICIAN
N = NOVICE

† Only Technician–class licensees who have passed a 5–WPM code test may use these frequencies.

✱✱ Geographical and power restrictions apply to these bands. See The FCC Rule Book for more information about your area.

---

80 METERS

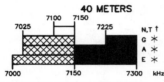

3675
3725
3750
3525   3850   N,T †
3775   G
          A
          E
3500   3750   4000   kHz

5,167.5 kHz (SSB only) Alaska emergency use only.

40 METERS

7100   7150
7025   7225   N,T †
          G ✱
          A ✱
          E ✱
7000   7150   7300   kHz

✱Phone operation is allowed on 7075–7100 kHz in Puerto Rico; US Virgin Islands and areas of the Caribbean south of 20 degrees north latitude; and in Hawaii and areas near ITU Region 3, including Alaska.

30 METERS

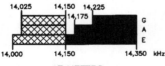

10,100   10,150   kHz   E,A,G

Maximum power on 30 meters is 200 watts PEP output. Amateurs must avoid interference to the fixed service outside the US.

20 METERS

14,025   14,150 14,225
          14,175
          G
          A
          E
14,000   14,150   14,350   kHz

17 METERS

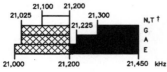

18,068   18,110   18,168   kHz   E,A,G

15 METERS

21,100   21,200
21,025   21,300   N,T †
21,225   G
          A
          E
21,000   21,200   21,450   kHz

12 METERS

24,890   24,930   24,990   kHz   E,A,G

10 METERS

28,100   28,500
          N,T †
          E,A,G
28,000 28,300   29,700   kHz

Novices and Technicians are limited to 200 watts PEP output on 10 meters.

6 METERS

50.1
          E,A,G,T
50.0   54.0   MHz

2 METERS

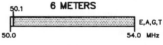

144.1
          E,A,G,T
144.0   148.0   MHz

1.25 METERS

222.1   223.91
          N
          E,A,G,T
222.0   225.0   MHz

Novices are limited to 25 watts PEP output from 222.1 to 223.91 MHz.

70 CENTIMETERS ✱✱

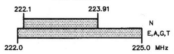

          E,A,G,T
420.0   450.0   MHz

33 CENTIMETERS ✱✱

          E,A,G,T
902.0   928.0   MHz

23 CENTIMETERS ✱✱

1270   1295
          N
          E,A,G,T
1240   1300   MHz

Novices are limited to 5 watts PEP output from 1270 to 1295 MHz.

TABLES OF FREQUENCY ALLOCATIONS

INTERNATIONAL			Band kHz	UNITED STATES		Remarks
Region 1 kHz	Region 2 kHz	Region 3 kHz		Government Allocation	Non-Government Allocation	
1800-1810 RADIOLOCATION 487 485 486	1800-1850 AMATEUR	1800-2000 AMATEUR FIXED MOBILE except aeronautical mobile RADIONAVIGATION Radiolocation 489	1800-1900		AMATEUR	
1810-1850 AMATEUR 490 491 492 493						
1850-2000 FIXED, MOBILE except aeronautical mobile 484 488 495	1850-2000 AMATEUR, FIXED MOBILE except aeronautical mobile RADIOLOCATION RADIONAVIGATION 494		1900-2000	RADIOLOCATION US290	RADIOLOCATION US290	
3500-3800 AMATEUR 510 FIXED MOBILE except aeronautical mobile 484	3500-3750 AMATEUR 510 509 511	3500-3900 AMATEUR 510 FIXED MOBILE	3500-4000	510	AMATEUR 510	
3800-3900 FIXED AERONAUTICAL MOBILE (OR) LAND MOBILE	3750-4000 AMATEUR 510 FIXED MOBILE except aeronautical mobile (R) 511 512 514 515					

| INTERNATIONAL | | | Band kHz | UNITED STATES | | |
Region 1 kHz	Region 2 kHz	Region 3 kHz		Government Allocation	Non-Government Allocation	Remarks
3900-3950 AERONAUTICAL MOBILE (OR) 513		3900-3950 AERONAUTICAL MOBILE BROADCASTING				
3950-4000 FIXED BROADCASTING		3950-4000 FIXED BROADCASTING 516				
7000-7100 AMATEUR 510, AMATEUR-SATELLITE 526 527			7000-7100		AMATEUR AMATEUR-SATELLITE 510	
7100-7300 BROADCASTING	7100-7300 AMATEUR 510 528	7100-7300 BROADCASTING	7100-7300	510 528	AMATEUR 510 528	
10100-10150 FIXED Amateur 510			10100-10150	US247 510	AMATEUR	
14000-14250 AMATEUR 510, AMATEUR-SATELLITE			14000-14250	510	AMATEUR AMATEUR-SATELLITE 510	
14250-14350 AMATEUR 510			14250-14350	510	AMATEUR 510	
18068-18168 AMATEUR 510, AMATEUR-SATELLITE 535			18068-18168	510	AMATEUR AMATEUR-SATELLITE 510	
21000-21450 AMATEUR 510, AMATEUR-SATELLITE 538			21000-21450	510	AMATEUR AMATEUR-SATELLITE 510	

INTERNATIONAL			Band kHz	UNITED STATES		Remarks
Region 1 kHz	Region 2 kHz	Region 3 kHz		Government Allocation	Non-Government Allocation	
24890-24990 AMATEUR 510, AMATEUR-SATELLITE 542	24890-24990 AMATEUR 510, AMATEUR-SATELLITE		24890-24990	510	AMATEUR AMATEUR-SATELLITE 510	
28-29.7 AMATEUR, AMATEUR-SATELLITE	28-29.7 AMATEUR, AMATEUR-SATELLITE		28-29.7		AMATEUR AMATEUR-SATELLITE	
47-68 BROADCASTING 553 554 555 559 561	50-54 AMATEUR 556 557 558 560		50-54		AMATEUR	
144-146 AMATEUR 510, AMATEUR-SATELLITE 605 606	144-146 AMATEUR 510, AMATEUR-SATELLITE		144-146	510	AMATEUR AMATEUR-SATELLITE 510	
146-148 FIXED MOBILE except aeronautical mobile (R) 607	146-148 AMATEUR 607	146-148 AMATEUR FIXED MOBILE 607	146-148		AMATEUR	
	220-225 AMATEUR FIXED, MOBILE Radiolocation 627		220-222	LAND MOBILE Radiolocation 627 G2	LAND MOBILE 627	The Channeling Plan for Land Mobile assignments in this band is shown in Section 4.3.15 of the NTIA Manual.
223-230 BROADCASTING Fixed Mobile 622 628 629 631 632 635	225-235 FIXED MOBILE	223-230 FIXED, MOBILE BROADCASTING AERONAUTICAL RADIONAVIGATION Radiolocation 636 637	222-225	Radiolocation 627 G2	AMATEUR 627	

| INTERNATIONAL | | | | UNITED STATES | | |
Region 1 kHz	Region 2 kHz	Region 3 kHz	Band kHz	Government Allocation	Non-Government Allocation	Remarks
420-430 FIXED, MOBILE except aeronautical mobile, Radiolocation — 651 652 653	*(spans Regions 1, 2, 3)*		420-450	RADIOLOCATION — US7 US87 US217 US228 US230 664 668 G2 G8	Amateur — US7 US87 US217 US228 US230 664 668 NG135	
430-440 AMATEUR RADIOLOCATION — 653 654 655 656 657 658 659 661 662 663 664 665	430-440 RADIOLOCATION Amateur — 653 658 659 660 660A 663 664					
	902-928 FIXED, Amateur Mobile except aeronautical mobile Radiolocation — 705 707 707A		902-928	RADIOLOCATION — US215 US218 US267 US275 707 G11 G59	US215 US218 US267 US275 707	ISM 915 ± 13 MHz
1240-1260 RADIOLOCATION, RADIONAVIGATION-SATELLITE (space-to-Earth) 710 Amateur — 711 712 712A 713 714			1240-1300	RADIOLOCATION — 664 713 714 G56	Amateur — 664 713 714	
1260-1300 RADIOLOCATION, Amateur — 664 711 712 712A 713 714						
2300-2450 FIXED, Amateur Mobile Radiolocation — 664 751A 752	2300-2450 FIXED, MOBILE, RADIOLOCATION Amateur — 664 750B 751 751B 752		2300-2310		Amateur	
			2310-2360	G123	US253	
			2360-2390	RADIOLOCATION MOBILE, Fixed — US276 G2 G120	MOBILE — US276	

Region 1 kHz	Region 2 kHz	Region 3 kHz	Band kHz	Government Allocation	Non-Government Allocation	Remarks
			2390-2400 2400-2402	RADIOLOCATION 664 752 G2	Amateur 664 752	ISM 2450 ± 50 MHz
3300-3400 RADIOLOCATION 778 779 780	3300-3400 RADIOLOCATION Amateur, Fixed, Mobile 778 780	3300-3400 RADIOLOCATION Amateur 778 779	3300-3500	RADIOLOCATION US108 664 778 G31	Amateur Radiolocation US108 664 778	
3400-3600 FIXED FIXED-SATELLITE (space-to-Earth) Mobile, Radiolocation 781 785	3400-3500 FIXED FIXED-SATELLITE (space-to-Earth) Amateur, Mobile Radiolocation 784 664 783	FIXED-SATELLITE (space-to-Earth)				
5650-5725 RADIOLOCATION, Amateur, Space Research (Deep Space) 664 801 803 804 805			5650-5850	RADIOLOCATION 664 806 808 G2	Amateur 664 806 808	ISM 5800 ± 75 MHz
5725-5850 FIXED-SATELLITE (Earth-to-space) RADIOLOCATION Amateur 801 803 805 806 807 808	5725-5850 RADIOLOCATION Amateur 803 805 806 808					
5850-5925 FIXED FIXED-SATELLITE (Earth-to-space) MOBILE 806	5850-5925 FIXED FIXED-SATELLITE (Earth-to-space) MOBILE, Amateur Radiolocation 806	5850-5925 FIXED FIXED-SATELLITE (Earth-to-space) MOBILE Radiolocation 806	5850-5925	RADIOLOCATION US245 806 G2	FIXED-SATELLITE (Earth-to-space) Amateur US245 806	
10-10.45 FIXED, MOBILE RADIOLOCATION Amateur 828	10-10.45 RADIOLOCATION Amateur 828 829	10-10.45 FIXED, MOBILE RADIOLOCATION Amateur 828	10-10.45	RADIOLOCATION US58 US108 828 G32	Amateur Radiolocation US58 US108 828 NG42	

INTERNATIONAL			Band	UNITED STATES		Remarks
Region 1 kHz	Region 2 kHz	Region 3 kHz	kHz	Government Allocation	Non-Government Allocation	
RADIOLOCATION Amateur, Amateur-Satellite 830			10.45-10.5	RADIOLOCATION US58 US108 G32	RADIOLOCATION Amateur Amateur-Satellite US58 US108 NG42 NG134	
AMATEUR, AMATEUR-SATELLITE 881			24-24.05	US211 881	AMATEUR AMATEUR-SATELLITE US211 881	
RADIOLOCATION Amateur, Earth Exploration-Satellite (Active) 881			24.05-24.25	RADIOLOCATION Earth Exploration-Satellite (Active) US110 881 G59	Amateur, Earth Exploration-Satellite (Active) Radiolocation US110 881	ISM 24.125 ± 125 MHz
AMATEUR, AMATEUR-SATELLITE			47-47.2		AMATEUR AMATEUR-SATELLITE	
AMATEUR, AMATEUR-SATELLITE, Space Research (space-to-Earth)			75.5-76		AMATEUR AMATEUR-SATELLITE	
RADIOLOCATION, Amateur, Amateur-Satellite, Space Research (space-to-Earth) 912			76-81	RADIOLOCATION 912	RADIOLOCATION Amateur Amateur-Satellite 912	
AMATEUR, AMATEUR-SATELLITE			142-144		AMATEUR AMATEUR-SATELLITE	
RADIOLOCATION Amateur, Amateur-Satellite 918			144-149	RADIOLOCATION 224 918	RADIOLOCATION Amateur Amateur-Satellite 918	
RADIOLOCATION Amateur, Amateur-Satellite 922			241-248	RADIOLOCATION 922	RADIOLOCATION Amateur Amateur-Satellite 922	ISM 245 ± 1 GHz
AMATEUR, AMATEUR-SATELLITE			248-250		AMATEUR AMATEUR-SATELLITE	

Footnotes:

G2—In the bands 216-225, 420-450 (except as provided by US217), 890-902, 928-942, 1300-1400, 2300-2450, 2700-2900, 5650-5925, and 9000-9200 MHz, the Government radiolocation is limited to the military services.

G11—Government fixed and mobile radio services including low power radio control operations, are permitted in the band 902-928 MHz on a secondary basis.

G31—In the bands 3300-3500 MHz, the Government radiolocation is limited to the military services, except as provided by footnote US108.

G32—Except for weather radars on meteorological-satellites in the band 9975-10025 MHz and for Government survey operations (see footnote US108), Government radiolocation in the band 10000-10500 MHz is limited to the military services.

G56—Government radiolocation in the bands 1215-1300, 2900-3100, 5350-5650 and 9300-9500 MHz is primarily for the military services; however, limited secondary use is permitted by other Government agencies in support of experimentation and research programs. In addition, limited secondary use is permitted for surveyoperations in the band 2900-3100 MHz.

G59—In the bands 902-928 MHz, 3100-3300 MHz, 3500-3700 MHz, 5250-5350 MHz, 8500-9000 MHz, 9200-9300 MHz, 13.4-14.0 GHz, 15.7-17.7 GHz and 24.05-24.25 GHz, all Government non-military radiolocation shall be secondary to military radiolocation, except in the subband 15.7-16.2 GHz airport surface detection equipment (ASDE) is permitted on a co-equal basis subject to coordination with the military departments.

US7—In the band 420-450 MHz and within thefollowing areas, the peak envelope power output of a transmitter employed in the amateur service shall not exceed 50 watts, unless expressly authorized by the Commission after mutual agreement, on a case-by-case basis, between the Federal Communications Commission Engineer in Charge at the applicable district office and the military area frequency coordinator at the applicable military base. For areas (e) through (j), the appropriate military coordinator is located at Peterson AFB, CO.

(a) Those portions of Texas and New Mexico bounded on the south by latitude 31°45' North, on the east by 104°00' West, on the north by latitude 34°30' North, and on the west by longitude 107°30' West;

(b) The entire State of Florida including the Key West area and the areas enclosed within a 322 kilometer (200-mile) radius of Patrick Air Force Base, Florida (latitude 28°21' North, longitude 80°43' West), and within a 322 kilometer (200-mile) radius of Eglin Air Force Base, Florida (latitude 30°30' North, longitude 86°30' West);

(c) The entire State of Arizona;

(d) Those portions of California and Nevada south of latitude 37°10' North, and the areas enclosed within a 322 kilometer (200-mile) radius of the Pacific Missile Test Center, Point Mugu, California (latitude 34°09' North, longitude 119°11' West).

(e) In the State of Massachusetts within a 160-kilometer (100 mile) radius around locations at Otis Air Force Base, Massachusetts (latitude 41°45' North, longitude 70°32' West).

(f) In the State of California within a 240-kilometer (150 mile) radius around locations at Beale Air Force Base, California (latitude 39°08' North, longitude 121°26' West).

(g) In the State of Alaska within a 160 kilometer (100 mile) radius of Clear, Alaska (latitude 64 degrees, 17' North, longitude 149 degrees 10' West).

(h) In the State of North Dakota within a 160 kilometer (100 mile) radius of Concrete, North Dakota (latitude 48 degrees 43' North, longitude97 degrees 54' West).

(i) In the States of Alabama, Florida, Georgia and South Carolina within a 200 kilometer (124 mile) radius of Warner Robins Air Force Base, Georgia (latitude 32° 38' North, longitude 83° 35' West).

(j) In the State of Texas within a 200-kilometer (124 mile) radius of Goodfellow Air Force Base, Texas (latitude 31° 25' North, longitude 100° 24' West).

US58—In the band 10000-10500 MHz, pulsed emissions are prohibited, except for weather radars on board meteorological-satellites in the band 10000-10025 MHz. The amateur service and the non-Government radiolocation service, which shall not cause harmful interference to the Government radiolocation service, are the only non-Government services permitted in this band. The non-Government radiolocation service is limited to survey operations as specified in footnote US108.

US87—The frequency 450 MHz, with maximum emission bandwidth of 500 kHz, may be used by Government and non-Government stations for space telecommand at specific locations, subject to such conditions as may be applied on a case-by-case basis.

US108—Within the bands 3300-3500 MHz and 10000-10500 MHz, survey operations, using transmitters with a peak power not to exceed five watts into the antenna, may be authorized for Government and non-Government use on a secondary basis to other Government radiolocation operations.

US110—In the frequency bands 3100-3300 MHz, 3500-3700 MHz, 5250-5350 MHz, 8500-9000 MHz, 9200-9300 MHz, 9500-10000 MHz, 13.4-14.0 GHz, 15.7-17.3 GHz, 24.05-24.25 GHz and 33.4-36 GHz, the non-Government radiolocation service shall be secondary to the Government radiolocation service and to airborne doppler radars at 8800 MHz, and shall provide protection to airport surface detection equipment (ASDE) operating between 15.7-16.2 GHz.

US211—In the bands 1670-1690, 5000-5250 MHz, and 10.7-11.7, 15.1365-15.35, 15.4-15.7, 22.5-22.55, 24-24.05, 31.0-31.3, 31.8-32, 40.5-42.5, 84-86, 102-105, 116-126, 151-164, 176.5-182, 185-190, 231-235, 252-265 GHz, applicants for airborne or space station assignments are urged to take all practicable steps to protect radio astronomy observations in the adjacent bands from harmful interference; however, US74 applies.

US215—Emissions from microwave ovens manufactured on and after January 1, 1980, for operation on the frequency 915 MHz must be confined within the band 902-928 MHz. Emissions from microwave ovens manufactured prior to January 1, 1980, for operation on the frequency 915 MHz must be confined within the band 902-940 MHz. Radiocommunications services operating within the band 928-940 MHz must accept any harmful interference that may be experienced from the operation of microwave ovens manufactured before January 1, 1980.

US217—Pulse-ranging radiolocation systems may be authorized for Government and non-Government use in the 420-450 MHz band along the shorelines of Alaska and the contiguous 48 States. Spread spectrum radiolocation systems may be authorized in the 420-435 MHz portion of the band for operation within the contiguous 48 States and Alaska. Authorizations will be granted on a case-by-case basis; however, operations proposed to be located within the zones set forth in US228 should not expect to be accommodated. All stations operating in accordance with this provision will be secondary to stations operating in accordance with the Table of Frequency Allocations.

US218—The band segments 902-912 MHz and918-928 MHz are available for Automatic Vehicle Monitoring (AVM) Systems subject to not causing harmful interference to the operation of Government stations authorized in these bands. These systems must tolerate any interference from the operation of industrial, scientific, and medical (ISM) devices and the operation of Government stations authorized in these bands.

US228—Applicants of operation in the band 420 to 450 MHz under the provisions of US217 should not expect to be accommodated if their area of service is within the following geographic areas:

(a) Those portions of Texas and New Mexico bounded on the south by latitude 31° 45' North, on the east by longitude 104° 00' West, on the north by latitude 34° 30' North, and on the West by longitude 107° 30' West.

(b) The entire State of Florida including the Key West area and the areas enclosed within a 322 kilometer (200-mile) radius of Patrick Air Force Base, Florida (latitude 28° 21' North, longitude 80° 43' West), and within a 322 kilometer (200-mile) radius of Eglin Air Force Base, Florida (Latitude 30° 30' North, Longitude 86° 30' West).

(c) The entire State of Arizona;

(d) Those portions of California and Nevada south of latitude 37° 10' North, and the areas enclosed within a 322 kilometer (200-mile) radius of the Pacific Missile Test Center, Point Mugu, California (latitude 34° 09' North, longitude 119° 11' West).

(e) In the State of Massachusetts within a 160-kilometer (100-mile) radius around locations at Otis Air Force Base, Massachusetts (latitude 41° 45' North, longitude 70° 32' West).

(f) In the State of California within a 240-kilometer (150-mile) radius around locations at Beale Air Force Base, California (latitude 39° 08' North, longitude 121° 26' West).

(g) In the State of Alaska within a 160 kilometer (100-mile) radius of Clear, Alaska (latitude 64 degrees, 17' North, longitude 149 degrees 10' West).

(h) In the State of North Dakota within a 160-kilometer (100-mile) radius of Concrete, North Dakota (latitude 48 degrees 43' North, longitude 97 degrees 54' West).

(i) In the States of Alabama, Florida, Georgia and South Carolina within a 200- kilometer (124-mile) radius of Warner Robins Air Force Base, Georgia (latitude 32° 38' North, longitude 83° 35' West).

(j) In the State of Texas within a 200-kilometer (124-mile) radius of Goodfellow Air Force Base, Texas (latitude 31° 25' North, longitude 100° 24' West).

US230—Non-government land mobile service is allocated on a primary basis in the bands 422.1875-425.4875 and 427.1875-429.9875 MHz within 80 kilometers (50 statute miles) of Detroit, MI, and Cleveland, OH, and in the bands 423.8125-425.4875 and 428.8125-429.9875 MHzwithin 80 kilometers (50 sta- tute miles) of Buffalo, NY.

US245—The Fixed-Satellite Service is limited to International inter-Continental systems and subject to case-by-case electromagnetic compatibility analysis.

US247—The band 10100-10150 kHz is allocated to the fixed service on a primary basis outside the United States and Possessions. Transmissions of stations in the amateur service shall not cause harmful interference to this fixed service use and stations in the amateur service shall make all necessary adjustments (including termination of transmission) if harmful interference is caused.

US253—In the band 2300-2310 MHz, the fixed and mobile services shall not cause harmful interference to the amateur service.

US267—In the band 902-928 MHz, amateur radio stations shall not operate within the States of Colorado and Wyoming, bounded by the area of: latitude 39° N to 42° N and longitude 103° W to 108° W.

US275—The band 902-928 MHz is allocated on a secondary basis to the amateur service subject to not causing harmful interference to the operations of Government stations authorized in this band or to the Automatic Vehicle Monitoring (AVM) systems. Stations in the amateur service must tolerate any interference from the operations of industrial, scientific and medical (ISM) devices, AVM systems, and the operations of Government stations authorized in this band. Further, the Amateur Service is prohibited in those portions of Texas and New Mexico bounded on the south by latitude 31° 41' North, on the east by longitude 104° 11' West, on the north by latitude 34° 30' North, and on the west by longitude 107° 30' West; in addition, outside this area but within 240 kilometers (150 miles) of these boundaries of White Sands Missile Range the service is restricted to a maximum transmitter peak envelope power output of 50 watts.

US276—Except as otherwise provided for herein, use of the band 2310-2390 MHz by the mobile service is limited to aeronautical telemetering and associated telecommand operations for flight testing of manned or unmanned aircraft, missiles or major components thereof. The following six frequencies are shared on a co-equal basis by Government and non-Government stations for telemetering and associated telecommand operations of expendable and re-usable launch vehicles whether or not such operations involve flight testing: 2312.5, 2332.5, 2352.5, 2364.5, 2370.5, and 2382.5 MHz. All other mobile telemetering uses shall be secondary to the above uses.

US290—In the band 1900-2000 kHz, amateur stations may continue to operate on a secondary basis to the Radiolocation Service, pending a decision as to their disposition through a future rule making proceeding in conjunction with implementation of the Standard Broadcasting Service in the 1625-1705 kHz band.

484—Some countries of Region 1 use radiodetermination systems in the bands 1606.5-1625 kHz,1635-1800 kHz, 1850-2160 kHz, 2194-2300 kHz, 2502-2850 kHz and 3500-3800 kHz. The establishment and operation of such systems are subject to agreement obtained under the procedures set forth in Article 14. The radiated mean power of these stations shall not exceed 50 W.

485—Additional allocation: in Angola, Bulgaria, Hungary, Mongolia, Nigeria, Poland, the German Democratic Republic, Chad, Czechoslovakia and the U.S.S.R., the bands 1625-1635 kHz, 1800-1810 kHz and 2160-2170 kHz are also allocated to the fixed and land mobile services on a primary basis subject to agreement obtained under the procedure set forth in Article 14.

486—In Region 1, in the bands 1625-1635 kHz, 1800-1810 kHz and 2160-2170 kHz (except in the countries listed in No. 485 and those listed in No. 499 in the band 2160-2170 kHz), existing stations in the fixed and mobile except aeronautical mobile, services (and stations of the aeronautical mobile (OR) service in the band 2160-2170 kHz) may continue to operate on a primary basis until satisfactory replacement assignments have been found and implemented in accordance with Resolution 38.

488—In the Federal Republic of Germany, Denmark, Finland, Hungary, Ireland, Israel, Jordan, Malta, Norway, Poland, the German Democratic Republic, the United Kingdom, Sweden, Czechoslovakia and the U.S.S.R., administrations may allocate up to 200 kHz to their amateur service in the bands 1715-1800 kHz and 1850-2000 kHz. However, when allocating the bands within this range to their amateur service, administrations shall, after prior consultations with administrations of neighboring countries, take such steps as may be necessary to prevent harmful interference from their amateur service to the fixed and mobile services of other countries. The mean-power of any amateur station shall not exceed 10 W.

489—In Region 3, the Loran system operates either on 1850 kHz or 1950 kHz, the bands occupied being 1825-1875 kHz and 1925-1975 kHz, respectively. Other services to which the band 1800-2000 kHz is allocated may use any frequency therein on condition that no harmful interference is caused to the Loran system operating on 1850 kHz or 1950 kHz.

490—Alternative allocation: in the Federal Republic of Germany, Angola, Austria, Belgium, Bulgaria, Cameroon, the Congo, Denmark, Egypt, Spain, Ethiopia, France, Greece, Italy, the Lebanon, Luxembourg, Malawi, the Netherlands, Portugal, Syria, the German Democratic Republic, Somalia, Tanzania, Tunisia, Turkey and the U.S.S.R., the band 1810-1830 kHz is allocated to the fixed and mobile, except aeronautical mobile, services on a primary basis.

491—Additional allocation: in Saudi Arabia, Iraq, Israel, Libya, Poland, Romania, Chad, Czechoslovakia, Togo and Yugoslavia, the band 1810-1830 kHz is also allocated to the fixed and mobile, except aeronautical mobile, services on a primary basis.

492—In Region 1, the use of the band 1810-1850 kHz by the amateur service is subject to the condition that satisfactory replacement assignments have been found and implemented in accordance with Resolution 38, for frequencies to all existing stations of the fixed and mobile, except aeronautical mobile, services operating in this band (except for the stations of the countries listed in Nos. 490, 491 and 493). On completion of satisfactory transfer, the authorization to use the band 1810-1830 kHz by the amateur service in countries situated totally or partially north of 40°N shall be given only after consultation with the countries mentioned in Nos. 490 and 491 to define the necessary steps to be taken to prevent harmful interference between amateur stations and stations of other services operating in accordance with Nos. 490 and 491.

493—Alternative allocation: in Burundi and Lesotho, the band 1810-1850 kHz is allocated to the fixed and mobile, except aeronautical mobile,services on a primary basis.

494—Alternative allocation: in Argentina, Bolivia, Chile, Mexico, Paraguay, Peru, Uruguay and Venezuela, the band 1850-2000 kHz is allocated to the fixed mobile, except aero-nautical mobile, radiolocation and radionavigation services on a primary basis.

495—In Region 1, in making assignments to stations in the fixed and mobile services in the bands 1850-2045 kHz, 2194-2498 kHz, 2502-2625 kHz and 2650-2850 kHz, administrations should bear in mind the special requirements of the maritime mobile service.

509—Additional allocation: in Honduras, Mexico, Peru and Venezuela, the band 3500-3750 kHz is also allocated to the fixed and mobile services on a primary basis.

510—For the use of the bands allocated to the amateur service at 3.5 MHz, 7.0 MHz, 10.1 MHz, 14.0 MHz, 18.068 MHz, 21.0 MHz, 24.89 MHz and 144 MHz in the event of natural disasters, see Resolution 640.

511—Additional allocation: in Brazil, the band 3700-4000 kHz is also allocated to the radiolocation service on a primary basis.

512—Alternative allocation: in Argentina, Bolivia, Chile, Ecuador, Paraguay, Peru and Uruguay, the band 3750-4000 kHz is allocated to the fixed andmobile, except aeronautical mobile, services on a primary basis.

513—Alternative allocation: in Botswana, Lesotho, Malawi, Mozambique, Namibia, South Africa, Swaziland, Zambia and Zimbabwe, the band 3900-3950 kHz is allocated to the broadcasting service on a primary basis. The use of this band by the broadcasting service is subject to agreement obtained under the procedure set forth in Article 14 with neighboring countries having services operating in accordance with the Table.

514—Additional allocation: in Canada, the band 3950-4000 kHz is also allocated to the broadcasting service on a primary basis. The power of broadcasting stations operating in this band shall not exceed that necessary for a national service within the frontier of this country and shall not cause harmful interference to other services operating in accordance with the Table.

515—Additional allocation: in Greenland, the band 3950-4000 kHz is also allocated to the broadcasting service on a primary basis. The power of the broadcasting stations operating in this band shall not exceed that necessary for a national service and shall in no case exceed 5 kW.

516—In Region 3, the stations of those services to which the band 3995-4005 kHz is allocated may transmit standard frequency and time signals.

526—Additional allocation: in Angola, Iraq, Kenya, Rwanda, Somalia and Togo, the band 7000-7050 kHz is also allocated to the fixed service on a primary basis.

527—Alternative allocation: in Egypt, Ethiopia, Guinea, Libya, Madagascar, Malawi and Tanzania, the band 7000-7050 kHz is allocated to the fixed service on a primary basis.

528—The use of the band 7 100-7 300 kHz in Region 2 by the amateur service shall not impose constraints on the broadcasting service intended for use within Region 1 and Region 3.

535—Additional allocation: in Afghanistan, China, the Ivory Coast, Iran and the U.S.S.R., the band 14 250-14 350 kHz is also allocated to the fixed service on a primary basis. Stations of the fixed service shall not use a radiated power exceeding 24 dBW.

538—Additional allocation: in the U.S.S.R., the band 18 068-18 168 kHz is also allocated to the fixed service on a primary basis for use within the boundary of the U.S.S.R., with a peak envelope power not exceeding 1 kW.

542—Additional allocation: in Kenya, the band 23 600-24 900 kHz is also allocated to the meteorological aids service (radiosondes) on a primary basis.

556—Alternative allocation: in New Zealand, the band 50-51 MHz is allocated to the fixed, mobile and broadcasting services on a primary basis, the band 53-54 MHz is allocated to the fixed and mobile services on a primary basis.

557—Alternative allocation: in Afghanistan, Bangladesh, Brunei, India, Indonesia, Iran, Malaysia, Pakistan, Singapore and Thai-land, the band 50-54 MHz is allocated to the fixed, mobile and broadcasting services on a primary basis.

558—Additional allocation: in Australia, China and the Democratic People's Republic of Korea, the band 50-54 MHz is also allocated to the broadcasting service on a primary basis.

560—Additional allocation: in New Zealand, the band 51-53 MHz is also allocated to the fixed and mobile services on a primary basis.

605—Additional allocation: in Singapore, the band 144-145 MHz is also allocated to the fixed and mobile services on a primary basis. Such use is limited to systems in operation on or before 1 January 1980, which in any case shall cease by 31 December 1995.

606—Additional allocation: In China, the band 144-146 MHz is also allocated to the aeronautical mobile (OR) service on a secondary basis.

607—Alternative allocation: in Afghanistan, Bangladesh, Cuba, Guyana and India, the band 146-148 MHz is allocated to the fixed and mobile services on a primary basis.

622—Different category of service: in the FederalRepublic of Germany, Austria, Belgium, Denmark, Spain, Finland, France, Israel, Italy, Liechtenstein, Luxembourg, Monaco, Norway, the Netherlands, Portugal, the United Kingdom, Sweden, Switzerland, and Yemen (P.D.R. of), the band 223-230 MHz is allocated to the land mobile service on a permitted basis (see No. 425). However, the stations of the land mobile service shall not cause harmful interference to, nor claim protection from, broadcasting stations, existing or planned, in countries other than those listed in this footnote.

627—In Region 2, the band 216-225 MHz is allocated to the radiolocation service on a primary basis until 1 January 1990. On and after 1 January 1990, no new stations in that service may be authorized. Stations authorized prior to 1 January 1990 may continue to operate on a secondary basis.

628—Additional allocation: in Somalia, the band 216-225 MHz is also allocated to the aero-nautical radionavigation service on a primary basis, subject to not causing harmful interference to existing or planned broad-casting services in other countries.

629—Additional allocation: in Oman, the United Kingdom and Turkey, the band 216-235 MHz is also allocated to the radiolocation service on a secondary basis.

631—Different category of service: in Spain and Portugal, the band 223-230 MHz is allocated to the fixed service on a permitted basis (see No. 425). Stations of this service shall not cause harmful interference to, or claim protection from, broadcasting stations of other countries, whether existing or planned, that operate in accordance with the Table.

632—Additional allocation: in Saudi Arabia, Bahrain, the United Arab Emirates, Israel, Jordan, Oman, Qatar and Syria, the band 223-235 MHz is also allocated to the aero-nautical radionavigation service on a permitted basis.

635—Alternative allocation: in Botswana, Lesotho, Namibia, South Africa, Swaziland and Zambia, the bands 223-238 MHz and 246-254 MHz are allocated to the broadcasting service on a primary basis subject to agreement obtained under the procedure set forth in Article 14.

636—Alternative allocation: in New Zealand, Western Samoa and the Niue and Cook Islands, the band 225-230 MHz is allocated to the fixed, mobile and aeronautical radio-navigation services on a primary basis.

637—Additional allocation: in China, the band225-235 MHz is also allocated to the radio astronomy service on a secondary basis.

651—Different category of service: in Australia, the United States, India, Japan and the United Kingdom, the allocation of the bands 420-

430 MHz and 440-450 MHz to the radiolocation service is on a primary basis (see No. 425).

652—Additional allocation: in Australia, the United States, Jamaica and the Philippines, the bands 420-430 MHz and 440-450 MHz are also allocated to the amateur service on a secondary basis.

653—Additional allocation: in China, India, the German Democratic Republic, the United Kingdom and the U.S.S.R., the band 420-460 MHz is also allocated to the aero-nautical radionavigation service (radio altimeters) on a secondary basis.

654—Different category of service: in France, the allocation of the band 430-434 MHz to the amateur service is on a secondary basis (see No. 424).

655—Different category of service: in Denmark, Libya, Norway and Sweden, the allocation of the bands 430-432 MHz and 438-440 MHz to the radiolocation service is on a secondary basis (see No 424).

656—Alternative allocation: in Denmark, Norway and Sweden, the bands 430-432 MHz and 438-440 MHz are allocated to the fixed and mobile, except aeronautical mobile, services on a primary basis.

657—Additional allocation: in Finland, Libya and Yugoslavia, the bands 430-432 MHz and 438-440 MHz are also allocated to the fixed and mobile except aeronautical mobile, services on a primary basis.

658—Additional allocation: in Afghanistan, Algeria, Saudi Arabia, Bahrain, Bangladesh, Brunei, Burundi, Egypt, the United Arab Emirates, Ecuador, Ethiopia, Greece, Guinea, India, Indonesia, Iran, Iraq, Israel, Italy, Jordan, Kenya, Kuwait, the Lebanon, Libya, Liechtenstein, Malaysia, Malta, Nigeria, Oman, Pakistan, the Philippines, Qatar, Syria, Singapore, Somalia, Switzerland, Tanzania, Thailand and Togo, the band 430-440 MHz is also allocated to the fixed service on a primary basis and the bands 430-435 MHz and 438-440 MHz are also allocated to the mobile, except aeronautical mobile, service on a primary basis.

659—Additional allocation: in Angola, Bulgaria, Cameroon, the Congo, Gabon, Hungary, Mali, Mongolia, Niger, Poland, the German Democratic Republic, Romania, Rwanda, Chad, Czechoslovakia and the U.S.S.R., the band 430-440 MHz is also allocated to the fixed service on a primary basis.

660—Different category of service: in Argentina, Colombia, Costa Rica, Cuba, Guyana, Honduras, Panama and Venezuela, the allocation of the band 430-440 MHz to the amateur service is on aprimary basis (see No. 425).

660A—Additional allocation: in Mexico, the bands 430-435 MHz and 438-440 MHz are also allocated on a primary basis to the land mobile service, subject to agreement obtained under the procedure set forth in Article 14.

661—In Region 1, except in the countries mentioned in No. 662, the band 433.05-434.79 MHz (center frequency 433.92 MHz) is designated for industrial, scientific and medical (ISM) applications. The use of this frequency band for ISM applications shall be subject to special authorization by the administration concerned, in agreement with other administrations whose radio-communication services might be affected. In applying this provision, administrations shall have due regard to the latest relevant CCIR Recommendations.

662—In the Federal Republic of Germany, Austria, Liechtenstein, Portugal, Switzerland and Yugoslavia, the band 433.05-434.79 MHz (center frequency 433.92 MHz) is designated for industrial, scientific and medical (ISM) applications. Radiocommunication services of these countries operating within this band must accept harmful interference which may be caused by these applications. ISM equipment operating in this band is subject to the provisions of No. 1815.

663—Additional allocation: in Brazil, France and the French Overseas Departments in Region 2, and India, the band 433.75-434.25 MHz is also allocated to the space operation service (Earthto-space) on a primary basis until 1 January 1990, subject to agreement obtained under the procedure set forth in Article 14. After 1 January 1990, the band 433.75-434.25 MHz will be allocated in the same countries to the same service on a secondary basis.

664—In the bands 435-438 MHz, 1260-1270 MHz, 2400-2450 MHz, 3400-3410 MHz (in Regions 2 and 3 only) and 5650-5670 MHz, the amateur-satellite service may operate subject to not causing harmful interference to other services operating in accordance with the Table (see No. 435). Administrations authorizing such use shall ensure that any harmful interference caused by emissions from a station in the amateur-satelliteservice is immediately eliminated in accordance with the provisions of No. 2741. The use of the bands 1260-1270 MHz and 5650-5670 MHz by the amateur-satellite service is limited to the Earth-to-space direction.

665—Additional allocation: in Austria, the band 438-440 MHz is also allocated to the fixed and mobile, except aeronautical mobile, services on a primary basis.

668—Subject to agreement obtained under the procedure set forth in Article 14, the band
449.75-450.25 MHz may be used for the space operation service (Earth-to-space) and the space research service (Earth-to-space).

705—Different category of service: in the UnitedStates, the allocation of the band 890-942 MHz to the radiolocation service is on a primary basis (see No. 425) and subject to agreement obtained under the procedure set forth in Article 14.

707—In Region 2, the band 902-928 MHz (center frequency 915 MHz) is designated for industrial, scientific and medical (ISM) applications. Radiocommunication services operating within this band must accept harmful interference which may be caused by these applications. ISM equipment operating in this band is subject to the provisions of No. 1815.

707A—Different category of service: in Chile, the band 903-905 MHz is allocated to the mobile, except aeronautical mobile, service on a primary basis and is subject to agreement obtained under the procedure set forth in Article 14.

710—Use of the radionavigation-satellite service in the band 1 215-1 260 MHz shall be subject to the condition that no harmful interference is caused to the radionavigation service authorized under No. 712.

711—Additional allocation: in Afghanistan, Angola, Saudi Arabia, Bahrain, Bangladesh, Cameroon, China, the United Arab Emirates, Ethiopia, Guinea, Guyana, India, Indonesia, Iran, Iraq, Israel, Japan, Jordan, Kuwait, the Lebanon, Libya, Malawi, Morocco, Mozambique, Nepal, Nigeria, Oman, Pakistan, the Philippines, Qatar, Syria, Somalia, Sudan, Sri Lanka, Chad, Thailand, Togo and Yemen (P.D.R. of), the band 1215-1300 MHz is also allocated to the fixed and mobile services on a primary basis.

712—Additional allocation: in Algeria, the Federal Republic of Germany, Austria, Bahrain, Belgium, Benin, Burundi, Cameroon, China, Denmark, the United Arab Emirates, France, Greece, India, Iran, Iraq, Kenya, Liechtenstein, Luxembourg, Mali, Mauritania, Norway, Oman, Pakistan, the Netherlands, Portugal, Qatar, Senegal, Somalia, Sudan, Sri Lanka, Sweden, Switzerland, Tanzania, Turkey and Yugoslavia, the band 1215-1300 MHz is allocated to the radionavigation service on a primary basis.

712A—Additional allocation: in Cuba, the band 1215-1300 MHz is also allocated to the radionavigation service on a primary basis subject to the agreement obtained under the procedure set forth in Article 14.

713—In the bands 1215-1300 MHz, 3100-3300 MHz, 5250-5350 MHz, 8550-8650 MHz, 9500-9800 MHz and 13.4-14.0 GHz, radiolocation stations installed on spacecraft may be employed for the earth exploration-satellite and space research services on a secondary basis.

714—Additional allocation: in Canada and the United States, the bands 1240-1300 MHz and 1350-1370 MHz are also allocated to the aeronautical radionavigation service on a primary basis.

750B WARC-92—Additional allocations: in the United States and India, the band 2310-2360 MHz is also allocated to the broadcasting-satellite service (sound) and complementary terrestrial sound broadcasting service on a primary basis. Such use is limited to digital audio broadcasting and is subject to the provisions of Resolution 528(WARC-92).**751**—In Australia, the United States and Papua New Guinea, the use of the band 2310-2390 MHz by the aeronautical mobile service for telemetry has priority over other uses by the mobile services.

751A WARC-92—In France, the use of the band 2310-2360 MHz by the aeronautical mobile service for telemetry has priority over other uses by the mobile service

751B WARC-92—Space stations of the broadcasting-satellite service in the band 2310-2360 MHz operating in accordance with No. 750B that may affect the services to which this band is allocated in other countries shall be coordinated and notified in accordance with Resolution 33 (WARC-79). Complementary terrestrial broadcasting stations shall be subject to bilateral coordination with neighboring countries prior to their bringing into use.

752—The band 2 400-2 500 MHz (center· frequency 2 450 MHz) is designated for industrial, scientific and medical (ISM) applications. Radio services operating within this band must accept harmful interference which may be caused by these applications. ISM equipment operating in this band is subject to the provisions of No. 1815.

778—In making assignments to stations of other services, administrations are urged to take all practicable steps to protect the spectral line observations of the radio astronomy service from harmful interference in the bands 3260-3267 MHz, 3332-3339 MHz, 3345.8-3352.5 MHz and 4825-4835 MHz. Emissions from space or airborne stations can be particularly serious sources of interference to the radio astronomy service (see Nos. 343 and 344 and Article 36).

779—Additional allocation: in Afghanistan, Saudi Arabia, Bahrain, Bangladesh, China, the Congo, the United Arab Emirates, India, Indonesia, Iran, Iraq, Israel, Japan, Kuwait, the Lebanon, Libya, Malaysia, Oman, Pakistan, Qatar, Syria, Singapore, Sri Lanka and Thailand, the band 3300-3400 MHz is also allocated to the fixed and mobile services on a primary basis. The countries bordering the Mediterranean shall not claim protection for their fixed and mobile services from the radiolocation service.

780—Additional allocation: in Bulgaria, Cuba, Hungary, Mongolia, Poland, the German Democratic Republic, Romania, Czechoslovakia and the U.S.S.R., the band 3300-3400 MHz is also allocated to the radio-navigation service on a primary basis.

781—Additional allocation: in the Federal Republic of Germany, Israel, Nigeria and the United Kingdom, the band 3 400-3 475 MHz is also allocated to the amateur service on a secondary basis.

783—Different category of service: in Indonesia, Japan, Pakistan and Thailand, the allocation of the band 3 400-3 500 MHz to the mobile, except aeronautical mobile, service is on a primary basis (see No. 425).

785—In Denmark, Norway and the United Kingdom, the fixed, radiolocation and fixed-satellite services operate on a basis of equality of rights in the band 3 400-3 600 MHz. However, these Administrations operating radiolocation systems in this band are urged to cease operations by 1985. After this date, these Administrations shall take all practicable steps to protect the fixed-satellite service and coordination requirements shall not be imposed on the fixed-satellite service.

801—Additional allocation: in the United Kingdom, the band 5 470-5 850 MHz is also allocated to the land mobile service on a secondary basis. The power limits specified in Nos. 2502, 2505, 2506 and 2507 shall apply in the band 5 725-5 850 MHz.

803—Additional allocation: in Afghanistan, Saudi Arabia, Bahrain, Bangladesh, Cameroon, the Central African Republic, China, the Congo, the Republic of Korea, Egypt, the United Arab Emirates, Gabon, Guinea, India, Indonesia, Iran, Iraq, Israel, Japan, Jordan, Kuwait, the Lebanon, Libya, Madagascar, Malaysia, Malawi, Malta, Niger, Nigeria, Pakistan, the Philippines, Qatar, Syria, Singapore, Sri Lanka, Tanzania, Chad, Thailand and Yemen (P.D.R. of), the band 5 650-5 850 MHz is also allocated to the fixed and mobile services on a primary basis.

804—Different category of service: in Bulgaria, Cuba, Hungary, Mongolia, Poland, the German Democratic Republic, Czechoslovakia and the U.S.S.R., the allocation of the band 5 670-5 725 MHz to the space research service is on a primary basis (see No. 425).

805—Additional allocation: in Bulgaria, Cuba, Hungary, Mongolia, Poland, the German Democratic Republic, Czechoslovakia and the U.S.S.R., the band 5 670-5 850 MHz is also allocated to the fixed service on a primary basis.

806—The band 5 725-5 875 MHz (center frequency 5 800 MHz) is designated for industrial, scientific and medical (ISM) applications. Radiocommunication services operating within this band must accept harmful interference which may be caused by these applications. ISM equipment operating in this band is subject to the provisions of No. 1815.

807—Additional allocation: in the Federal Republic of Germany and in Cameroon, the band 5 755-5 850 MHz is also allocated to the fixed service on a primary basis.

808—The band 5 830-5 850 MHz is also allocated in the amateur-satellite service (space-to-Earth) on a secondary basis.

811—Subject to agreement obtained under the procedure set forth in Article 14, the band 7 145-7 235 MHz may be used for Earth-to-space transmissions in the space research service. The use of the band 7 145-7 190 MHz is restricted to deep space; no emissions to deep space shall be effected in the band 7 190-7 235 MHz.

828—The band 9 975-10 025 MHz is also allocated to the meteorological-satellite service on a secondary basis for use by weather radars.

829—Additional allocation: in Costa Rica, Ecuador, Guatemala and Honduras, the band 10-10.45 GHz is also allocated to the fixed and mobile services on a primary basis.

830—Additional allocation: in the Federal Republic of Germany, Angola, China, Ecuador, Spain, Japan, Kenya, Morocco, Nigeria, Sweden, Tanzania and Thailand, the band 10.45-10.5 GHz is also allocated to the fixed and mobile services on a primary basis.

881—The band 24-24.25 GHz (center frequency 24.125 GHz) is designated for industrial, scientific and medical (ISM) applications. Radiocommunication services operating within this band must accept harmful interference which may be caused by these applications. ISM equipment operating in this band is subject to the provisions of No. 1815.

902—In the bands 43.5-47, 66-71, 95-100, 134-142, 190-200 and 252-265 GHz, stations in the land mobile service may be operated subject to not causing harmful interference to the space radiocommunication services to which these bands are allocated (see No. 435).

903—In the bands 43.5-47 GHz, 66-71 GHz, 95-100 GHz, 134-142 GHz, 190-200 GHz and 252-265 GHz, satellite links connecting land stations at specified fixed points are also authorized when used in conjunction with the mobile-satellite service or the radionavigation-satellite service.

912—In the band 78-79 GHz radars located on space stations may be operated on a primary basis in the Earth exploration-satellite service and in the space research service.

917—In the band 140.69-140.98 GHz all emissions from airborne stations, and from space stations in the space-to-Earth direction, are prohibited.

918—The bands 140.69-140.98 GHz, 144.68-144.98 GHz, 145.45-145.75 GHz and 146.82-147.12 GHz are also allocated to the radio astronomy service on a primary basis for spectral line observations. In making assignments to stations of other services to which the bands are allocated, administrations are urged to take all practicable steps to protect the radio astronomy service from harmful interference. Emissions from space or air-borne stations can be particularly serious sources of interference to the radio astronomy service (see Nos. 343 and 344 and Article 36).

922—The band 244-246 GHz (center frequency 245 GHz) is designated for industrial, scientific and medical (ISM) applications. The use of this frequency band for ISM applications shall be subject to special authorization by the administration concerned in agreement with other administrations whose radiocommunication services might be affected. In applying this provision administrations shall have due regard to the latest relevant CCIR Recommendations.

G123—The bands 2300-2310 and 2400-2402 MHz were identified for reallocation, effective August 10, 1995, for exclusive non-

Government use under Title VI of the Omnibus Budget Reconciliation Act of 1993. Effective August 10, 1995, any Government operations in these bands are on a non-interference basis to authorized non-Government operations and shall not hinder the implementation of any non-Government operations.

NG42—Non-Government stations in the radiolocation service shall not cause harmful interference to the amateur service.

NG134—In the band 10.45-10.5 GHz non-Government stations in the radiolocation service shall not cause harmful interference to the amateur and amateur-satellite services.

FCC EMISSION DESIGNATORS

Subpart C—Emissions

§ 2.201 Emission, modulation, and transmission characteristics.

The following system of designating emission, modulation, and transmission characters shall be employed.

(a) Emissions are designated according to their classification and their necessary bandwidth.

(b) A minimum of three symbols are used to describe the basic characteristics of radio waves. Emissions are classified and symbolized according to the following characteristics.

(1) First symbol—type of modulation of the main character;

(2) Second symbol—nature of signal(s) modulating the main carrier;

(3) Third symbol—type of information to be transmitted.

NOTE: A fourth and fifth symbol are provided for additional information and are shown in Appendix 6, part of the ITU Radio Regulations. Use of the fourth and fifth symbol is optional. Therefore, the symbols may be used as described in Appendix 6, but are not required by the Commission.

(c) First Symbol—types of modulation of the main carrier:

(1) Emission of an unmodulated carrier ....N

(2) Emission in which the main carrier is amplitude-modulated (including cases where sub-carriers are angle-modulated):

—Double-sidebanded ....A

—Single-sidebanded, full carrier...H

—Single-sidebanded, reduced or variable level carrier..R

—Single-banded, suppressed carrier..J

—Independent sidebands ....B

—Vestigal sideband..C

(3) Emission in which the main carrier is angle-modulated:

—Frequency modulation...F

—Phase modulation..G

NOTE: Whenever frequency modulation "F" is indicated, Phase modulation "G" is also acceptable.

(4) Emission in which the main carrier is amplitude and angle-modulated either simultaneously or in a pre-established sequence..........D

(5) Emission pulses:[1]

—Sequence of unmodulated pulses ....P

—A sequence of pulses:

—Modulated in amplitude..K

—Modulated in width/duration..L

—Modulated in position/phase ....M

—In which the carrier is angle-modulated during the period of the pulse ....Q

—Which is a combination of the foregoing or is produced by other means V

(6) Cases not covered above, in which an emission consists of the main carrier modulated, either simultaneously or in a pre-established sequence, in a combination of two or more of the following modes: amplitude, angle or pulse ....W

(7) Cases not otherwise covered X

[1] Emissions where the main carrier is directly modulated by a signal which has been coded into a quantized form (e.g. pulse code modulation) should be designated under (2) or (3).

(d) Second Symbol—nature of signal(s) modulating the main carrier:

(1) No modulating signal...0

(2) A single channel containing quantized or digital information without the use of a modulating sub-carrier, excluding time-division multiplex..1

(3) A single channel containing quantized or digital information with the use of a modulating sub-carrier, excluding time-division multiplex..2

(f) *Type B emission*: As an exception to the above principles, damped waves are symbolized in the Commission's rules and regulations as type B emission. The use of type B emissions is forbidden.

(g) Whenever the full designation of an emission is necessary, the symbol for that emission, as given above, shall be preceded by the necessary bandwidth of the emission as indicated in §2.202 (b)(1).

[49 FR 48697, Dec. 14, 1984]

[2] In this context the word "information" does not include information of a constant, unvarying nature such as is provided by standard frequency emissions, continuous wave and pulse radars, etc.

§ 2.202 Bandwidths.

(a) *Occupied bandwidth*. The frequency bandwidth such that, below its lower and above its upper frequency limits, the mean powers radiated are each equal to 0.5 percent of the total mean power radiated by a given emission. In some cases, for example multichannel frequency-division systems, the percentage of 0.5 percent may lead to certain difficulties in the practical application of the definitions of occupied and necessary bandwidth; in such cases a different percentage may prove useful.

(b) *Necessary bandwidth*. For a given class of emission, the minimum value of the occupied bandwidth sufficient to ensure the transmission of information at the rate and with the quality required for the system employed, under specific conditions. Emissions useful for the good functioning of the receiving equipment as, for example, the emission corresponding to the carrier of reduced carrier systems, shall be included in the necessary bandwidth.

(1) The necessary bandwidth shall be expressed by three numerals and one letter. The letter occupies the position of the decimal point and represents the unit of bandwidth. The first character shall be neither zero nor K, M, or G.

(2) Necessary bandwidths:

between 0.001 and 999 Hz shall be expressed in Hz (letter H);
between 1.00 and 999 kHz shall be expressed in kHz (letter K);
between 1.00 and 999 MHz shall be expressed in MHz (letter M);
between 1.00 and 999 GHz shall be expressed in GHz (letter G).

(3) Examples:
0.002 Hz—H002 ....180.5 kHz—181K
0.1 Hz—H100 ....180.7 kHz—181K
25.3 Hz—25H3 ....1.25 MHz—1M25
400 Hz—400H..2 MHz—2M00
2.4 kHz—2K40..10 MHz—10M0
6 kHz—6K00 ....202 MHz—202M
12.5 kHz—12K5 ....5.65 GHz—5G65
180.4 kHz—180K

(c) The necessary bandwidth may be determined by one of the following methods:

(1) Use of the formulas included in the table, in paragraph (g) of this section, which also gives examples of necessary bandwidths and designation of corresponding emission;

(2) For frequency modulated radio systems which have a substantially linear relationship between the value of input voltage to the modulator and the resulting frequency deviation of the carrier and which carry either single sideband suppressed carrier frequency division multiplex speech channels or television, computation in accordance with provisions of paragraph (f) of this section and formulas and methods indicated in the table, in paragraph (g) of this section;

(3) Computation in accordance with Recommendations of the International Radio Consultative Committee (C.C.I.R.);

(4) Measurement in cases not covered by paragraph (c) (1), (2), or (3) of this section.

(d) The value so determined should be used when the full designation of an emission is required. However, the necessary bandwidth so determined is not the only characteristic of an emission to be considered in evaluating the interference that may be caused that emission.

(e) In the formulation of the table in paragraph (g) of this section, the following terms are employed:

B_n = Necessary bandwidth in hertz
B = Modulation rate in bauds
N = Maximum possible number of black plus white elements to be transmitted per second, in facsimile
M = Maximum modulation frequency in hertz
C = Sub-carrier frequency in hertz

D = Peak frequency deviation, i.e., half the difference between the maximum and minimum values of the instantaneous frequency. The instantaneous frequency in hertz is the time rate of change in phase in radians divided by 2

t = Pulse duration in seconds at half-amplitude

t_r = Pulse rise time in seconds between 10% and 90% of maximum amplitude

K = An overall numeric factor which varies according to the emission and which depends upon the allowable signal distortion

N_c = Number of baseband telephone channels in radio systems employing multi-channel multiplexing

P = Continuity pilot sub-carrier frequency (Hz) (continuos signal utilized to verify performance of frequency-division multiplex systems).

(f) Determination of values of D and B_n for systems specified in paragraph (c) (2) of this section:

(1) Determination of D in systems for multi-channel telephony:

(i) The rms value of the per-channel deviation for the system shall be specified. (In the case of systems employing preemphasis or phase modulation, this value of per-channel deviation shall be specified at the characteristic baseband frequency.)

(ii) The value of D is then calculated by multiplying the rms value of the per-channel deviation by the appropriate factors, as follows:

Number of message circuits	Multiplying factors	Limits of X (P_{avg} (dBmO))
More than 3, but less than 12	4.47 x [a factor specified by the equipment manufacturer or station licensee, subject to Commission approval].	
At least 12, but less than 60	$\dfrac{3.76 \text{ antilog } (X+2 \log_{10} N_c)}{20}$	X: −2 to +2.6.
At least 60, but less than 240	$\dfrac{3.76 \text{ antilog } (X+4 \log_{10} N_c)}{20}$	X: −5.6 to −1.0.
240 or more	$\dfrac{3.76 \text{ antilog } (X+10 \log_{10} N_c)}{20}$	x: −19.6 to −15.0.

Where X represents the average power in a message circuit in dBmO; N_c is the number of circuits in the multiplexed message load; 3.76 corresponds to a peak load factor of 11.5 dB.

(2) The necessary bandwidth (B_n) normally is considered to be numerically equal to:

(i) $2M+2DK$, for systems having no continuity pilot subcarrier or having a continuity pilot subcarrier whose frequency is not the highest modulating the main carrier;

(ii) $2P+2DK$, for systems having a continuity pilot subcarrier whose frequency exceeds that of any other signal modulating the main carrier, unless the conditions set forth in paragraph (f)(3) of this section are met.

(3) As an exception to paragraph (f)(2)(ii) of this section, the necessary bandwidth (B_n) for such systems is numerically equal to $2P$ or $2M+2DK$, whichever is greater, provided the following conditions are met:

(i) The modulation index of the main carrier due to the continuity pilot subcarrier does not exceed 0.25, and

(ii) In a radio system of multichannel telephony, the rms frequency deviation of the main carrier due to the continuity pilot subcarrier does not exceed 70 percent of the rms value of the per-channel deviation, or , in a radio system for television, the rms deviation of the main carrier due to the pilot does not exceed 3.55 percent of the peak deviation of the main carrier.

(g) Table of necessary bandwidths:

Description of emission	Necessary bandwidth		Designation of emission
	Formula	Sample calculation	
I. NO MODULATING SIGNAL			
Continuous wave emission.		NON (zero)	

II. AMPLITUDE MODULATION

1. Signal With Quantized or Digital Information

Continuous wave telegraphy.	$B_n=BK$, k=5 for fading circuits, K=3 for non-fading circuits	25 words per minute; B=20, K=5, Bandwidth: 100 Hz	100HA1A
Telegraphy by on-off keying of a tone modulated carrier.	$B_n=BK+2M$, K=5 for fading circuits, K=3 for non-fad ing circuits	25 words per minute; B=20, M=1000, K=5, Bandwidth: 2100 Hz=2.1 kHz	2K10A2A
Selective calling signal, single-sideband full carrier.	$B_n=M$	Maximum code frequency is: 2110 Hz, M=2110, Bandwidth: 2110 Hz=2.11 kHz	2K11H2B
Direct-printing telegraphy using a frequency shifted modulating sub-carrier single-sideband suppressed carrier.	$B_n=2M+2DK$, M=B÷2	B=50, D=35 Hz (70 Hz shift), K=1.2, Bandwidth: 134 Hz.	134HJ2B
Telegraphy, single sideband reduced carrier.	B_n=central frequency+M+DK, M=B÷2	15 channels; highest central frequency is: 2805 Hz, B=100, D=42.5 Hz (85 Hz shift), K=0.7 Bandwidth: 2885 Hz=2.885 kHz	2K89R7B

2. Telephony (Commercial Quality)

Telephony double-sideband.	$B_n=2M$	M=3000, Bandwidth: 6000 Hz=6 kHz	6K00A3E
Telephony, single-sideband, full carrier.	$B_n=2M$	M=3000, Bandwidth: 3000 Hz=3 kHz	3K00H3E
Telephony, single-sideband suppressed carrier.	$B_n=M$–lowest modulation frequency	M=3000, lowest modulation frequency is 3000Hz, 2700 Hz Bandwidth: 2700 Hz=2.7 kHz	2K70J3E
Telephony with separate frequency modulated signal to control the level of demodulated speech signal, single-sideband, reduced carrier.	$B_n=M$	Maximum control frequency is 2990 Hz, M=2990, Bandwidth: 2990 Hz=2.99 kHz	2K99R3E
Telephony with privacy, single-sideband, suppressed carrier (two or more channels).	$B_n=N_cM$–lowest modulation frequency in the lowest channel	N_c=2, M=3000 lowest modulation frequency is 250 Hz, Bandwidth: 5750 Hz=5.75 kHz.	5K75J8E
Telephony, independent side-band (two or more channels).	B_n=sum of M for each sideband	2 channels, M=3000, Bandwidth: 6000 Hz=6 kHz	6K00B8E

3. Sound Broadcasting

Sound broadcasting, double sideband.	$B_n=2M$, M may vary between 4000 and 10000 depending on the quality desired	Speech and music, M=4000, Bandwidth: 8000 Hz=8 kHz	8K00A3E
Sound broadcasting, single-sideband reduced carrier (single channel).	$B_n=M$, M may vary between 4000 and 10000 depending on the quality desired	Speech and music, M=4000, Bandwidth: 4000 Hz=4 kHz	4K00R3E
Sound broadcasting, single-sideband, suppressed carrier.	$B_n=M$–lowest modulation frequency	Speech and music, M=4500, lowest modulation frequency= 50 Hz, Bandwidth: 4450 Hz=4.45 kHz	4K45J3E

4. Television

Television, vision and sound.	Refer to CCIR documents for the bandwidths of the commonly used television systems	Number of lines=525; Nominal video bandwidth, 4.2 MHz, Sound carrier relative to video carrier=4.5 MHz.	5M75C3F
		Total vision bandwidth: 5.75 MHz; FM aural bandwidth including guardbands: 250,000 Hz	250KF3E
		Total bandwidth: 6 MHz	6M25C3F

5. Facsimile

Analogue facsimile by sub-carrier frequency modulation of a single-sideband emission with reduced carrier.	$B_n=C-N\div2+DK$, K=1.1 (typically)	N=1100, corresponding to an index of cooperation of 352 and a cycler rotation speed of 60 rpm. Index of cooperation is the product of the drum diameter and number of lines per unit length C=1900, D=400 Hz, Bandwidth: 2.890 Hz+2.89 kHz	2K89R3C
Analogue facsimile; frequency modulation of an audio frequency sub-carrier which modulates the main carrier, single-sideband suppressed carrier.	$B_n=2M+2DK$, $M=N/_2$, K=1.1 (typically)	N=1100, D-400 Hz, Bandwidth: 1980 Hz=1.98 kHz	1K98J3C

6. Composite Emissions

Double-sideband, television relay.	$B_n=2C+2M+2D$	Video limited to 5 MHz, audio on 6.5 MHz frequency modulated sub-carrier deviation =50 kHz: $C=6.5\times10^6$ $D=50\times10^3$, M=15,000, Bandwidth: 13.13×10^6 Hz=13.13 MHz	13M2A8W
Double-sideband radio relay system.	$B_n=2M$	10 voice channels occupying baseband between 1 kHz and 164 kHz; M=164,000 bandwidth=328,000 Hz=328 kHz	328KA8E
Double-sideband emission of VOR with voice (VOR=VHF omnidirectional radio range).	$B_n=2C_{max}+2M+2DK$, K=1 (typically)	The main carrier is modulated by:—a 30 Hz sub-carrier—a carrier resulting from a 9960 Hz tone frequency modulated by a 30 Hz tone—a telephone channel—a 1020 Hz keyed tone for continual Morse identification. C_{max}=9960, M=30, D=480 Hz, Bandwidth: 20,940 Hz=20.94 MHz	20K9A9W
Independent sidebands; several telegraph channels together with several telephone channels.	B_n=sum of M for each sideband	Normally composite systems are operated in accordance with standardized channel arrangements (e.g. CCIR Rec. 348-2) 3 telephone channels require and 15 telegraphy channels require the bandwidth 12,000 Hz=12 kHz	12K0B9W

III-A. FREQUENCY MODULATION

1. Signal With Quantized or Digital Information

Telegraphy without error-correction (single-channel).	$B_n=2M+2DK$, $M=B \div 2$, $K=1.2$ (typically)	B=100, D=85 Hz (170 Hz shift), Bandwidth: 304 Hz	304HF1B
Four-frequency duplex telegraphy.	$B_n 2M+2DK$, B=Modulation rate in bands of the faster channel. If the channels are synchronized: $M=B \div 2$, otherwise $M=2B$, $K=1.1$ (typically)	Spacing between adjacent frequencies= 400 Hz; Synchronized channels; B=100, M=50, D=600 Hz, Bandwidth: 1420 Hz= 1.42 kHz	1K42F7B

2. Telephony (Commercial Quality)

Commercial telephony	$B_n=2M+2DK$, $K=1$ (typically, but under conditions a higher value may be necessary)	For an average case of commercial telephony, M=3000, Bandwidth: 16,000 Hz=16 kHz	16K0F3E

3. Sound Broadcasting

Sound broadcasting	$B_n=2M+2DK$, $K=1$ (typically)	Monaural, D=75,000 Hz, M=15,000, Bandwidth: 18,000 Hz=180 kHz	180KF3E

4. Facsimile

Facsimile by direct frequency modulation of the carrier; black and white.	$B_n=2M+2DK$, $M=N \div 2$, $K=1.1$ (typically)	N=1100 elements/sec; D=400 Hz, Bandwidth: 1980 Hz=1.98 kHz	1K98F1C
Analogue facsimile	$B_n=2M+2DK$, $M=N \div 2$, $K=1.1$ (typically)	N=1100 elements/sec; D=400 Hz, Bandwidth: 1980 Hz=1.98 kHz	1K98F3C

5. Composite Emissions (See Table III-B)

Radio-relay system, frequencydivision multiplex.	$B_n=2P+2DK$, $K=1$	Microwave radio relay system specifications: 60 telephone channels occupying baseband between 60 and 300 kHz; rms per-channel deviation 200 kHz; pilot at 331 kHz produces 200 kHz rms deviation of main carrier. Computation of B_n:$D=(200 \times 10^3 \ 3 \times 3.76 \times 1.19)$, Hz=$0.895 \times 10^6$, $P=0.331 \times 10^6$ Hz; Bandwidth: 2.452×10^6 Hz.	2M45F8E
Radio-relay system, frequency division multiplex.	$B_n+2M+2DK$, $K=1$	Microwave radio relay system specifications: 1200 telephone channels occupying base-band between 60 and 5564 kHz; rms per channel deviation 200 kHz; continuity pilot at 6199 kHz produces 140 kHz rms deviation of main carrier. Computation of B_n:$D=(20^0 \times 10^3 \times 3.76 \times 3.63)=2.73 \times 10^6$, $M=5.64 \times 10^6$ Hz; $P=6.2 \times 10^6$ Hz; $(2M+2DK<2P$; Bandwidth 16.59×10^6 Hz	16M6F8E
Radio-relay system, frequency division multiplex.	$B_n=2P$	Microwave radio relay system specifications: Multiplex 600 telephone channels occupying baseband between 60 and 2540 kHz; continuity pilot at 8500 kHz produces 140 kHz rms deviation of main carrier. Computation of B_n:$D=(200 \times 10^3 \times 3.76 \times 2.565) = 1.93 \times 10^6$ Hz; $M=2.54 \times 10^6$ Hz; $2DK) \leq 2P$ Bandwidth: 17×10^6 Hz	17M0F8E
Unmodulated pulse emission.	$B_n=2K \div t$, K depends upon the ratio of pulse rise time. Its value usually falls between 1 and 10 and in many cases it does not need to exceed 6	Primary Radar Range resolution: 150 m, K=1.5 (triangular pulse where $t \simeq t_r$, only components down to 27 dB from the strongest are considered). Then $t=2 \times$ range resolution $\div$ velocity of light=$2 \times 150 \div 3 \times 10^8=1 \times 10^{-6}$ seconds, Bandwidth: 3×10^6 Hz=3 MHz	3M00P0N

6. Composite Emissions

Radio-relay system	$B_n=2K \div t$, $K=1.6$	Pulse position modulated by 36 voice channel baseband; pulse width at half amplitude=0.4 μs, Bandwidth: 8×10^6 Hz=8 MHz (Bandwidth independent of the number of voice channels)	8M00M7E

[28 FR 12465, Nov. 22, 1963, as amended at 37 FR 8883, May 2, 1972; 37 FR 9996, May 18, 1972; 48 FR 16492, Apr. 18, 1983; 49 FR 48698, Dec. 14, 1984]

Subpart D—Call Signs and Other Forms of Identifying Radio Transmissions

AUTHORITY: Secs. 4, 5, 303, 48 Stat., as amended, 1066, 1068, 1082; 47 U.S.C. 154, 155, 303.

§2.301 Station identification requirement.

Each station using radio frequencies shall identify its transmissions according to the procedures prescribed by the rules governing the class of station to which it belongs with a view to the elimination of harmful interference and the general enforcement of applicable radio treaties, conventions, regulations, arrangements and agreements in force, and the enforcement of the Communications Act of 1934, as amended, and the Commission's rules.

[34 FR 5104, Mar. 12, 1969]

§ 2.302 Call Signs.

The table which follows indicates the composition and blocks of international call signs available for assignment when such call signs are required by the rules pertaining to particular classes of stations. When stations operating in two or more classes are authorized to the same licensee for the same location, the Commission may elect to assign a separate call sign to each station in a different class. (In addition to the U.S. call sign allocations listed below, call sign blocks AAA through AEZ and ALA through ALZ have been assigned to the Department of the Army; call sign block AFA through AKZ has been assigned to the Department of the Air Force; and call sign block NAA through NZZ has been assigned jointly to the Department of the Navy and the U.S. Coast Guard.

Class of station	Composition of call sign	Call sign blocks
Coast (Class I) except for coast telephone in Alaska.	3 letters	KAA through KZZ. WAA through WZZ.
Coast (Classes II and III) and maritime radio-determination.	3 letters, 3 digits	KAA2ØØ through KZZ999. WAA2ØØ through WZZ999.
Coast telephone in Alaska	3 letters, 2 digits 3 letters, 3 digits (for stations assigned frequencies above 30 MHz).	KAA2Ø through KZZ99. WAA2Ø through WZZ99. WZZ2ØØ through WZZ999
Fixed	3 letters, 2 digits 3 letters, 3 digits (for stations assigned frequencies above 30 MHz).	KAA2Ø through KZZ99. WAA2Ø through WZZ99. WAA2ØØ through WZZ999.
Marine receiver test	3 letters, 3 digits, (plus general geographic location when required).	KAA2ØØ through KZZ999. WAA2ØØ through WZZ999.
Ship telegraph	4 letters[1]	KAAA through KZZZ. WAAA through WZZZ.
Ship telephone	2 letters, 4 digits or 3 letters, 4 digits[1].	WA2ØØØ through WZ9999, through WZZ9999.
Ship telegraph plus telephone	4 letters	KAAA through KZZZ. WAAA through WZZZ.
Ship radar	Same as ship telephone and/or telegraph call sign, or, if ship has no telephone or telegraph: 2 letters, 4 digits, or 3 letters, 4 digits.	WA2ØØØ through WZ9999, through WZZ9999.
Ship survival craft	Call sign of the parent ship followed by 2 digits.	KAAA2Ø through KZZZ99. WAAA2Ø through WZZZ99.
Cable repair ship marker buoy	Call sign of the parent ship followed by the letters "BT" and the identifying number of the buoy.	
Marine utility	2 letters, 4 digits	KA2ØØØ through KZ9999.
Shipyard mobile	2 letters, 4 digits	KA2ØØØ through KZ9999.
Aircraft telegraph	5 letters	KAAAA through KZZZZ. WAAAA through WZZZZ.

Aircraft telegraph and telephone	5 letters[2].	KAAAA through KZZZZ. WAAA through WZZZZ.
Aircraft telephone	5 letters[2] (whenever a call sign is assigned).	KAAAA through KZZZZ. WAAA through WZZZZ.
Aircraft survival craft	Whenever a call sign[2] is assigned, call sign of the parent aircraft followed by a single digit other than 0 or 1.	
Aeronautical	3 letters, 1 digit[2]	KAA2 through KZZ9. WAA2 through WZZ9.
Land mobile (base)	3 letters, 3 digits	KAA2ØØ through KZZ999. WAA2ØØ through WZZ999.
Land mobile (mobile telegraph)	4 letters, 1 digit	KAAAA2 through KZZZ9. WAAA2 through WZZZ9
Land mobile (mobile telephone)	2 letters, 4 digits	KA2ØØØ through KZ9999. WA2ØØØ through WZ9999
Broadcasting (standard)	4 letters[3] (plus location of station)	KAAA through KZZZ. WAAA through WZZZ
Broadcasting (FM)	4 letters plus location of station)	KAAA through KZZZ. WAAA through WZZZ.
Broadcasting with suffix "FM"	6 letters[3] (plus location of station)	KAAA-FM through KZZZ-FM. WAAA-FM through WZZZ-FM.
Broadcasting (television)	4 letters (plus location of station)	KAAA through KZZZ. WAAA through WZZZ.
Broadcasting with suffix "TV"	6 letters[3] (plus location of station)	KAAA-TV through KZZZ-TV. WAAA-TV through WZZZ-TV.
Television broadcast translator	1 letter—output channel number—2 letters	KØ2AA through K83ZZ. WØ2AA through W83ZZ.
Disaster station, except U.S. Government	4 letters, 1 digit	KAAA2 through KZZZ9. WAAA2 through WZZZ9.
Experimental (letter "X" follows the digit)	2 letters, 1 digit, 3 letters	KA2XAA through KZ9XZZ. WA2XAA through WZ9XZZ.
Amateur (letter "X" may not follow digit)	1 letter, 1 digit, 1 letter[4]	K1A through KØZ. N1A through NØZ. W1A through WØZ.
Amateur	1 letter, 1 digit, 2 letters[4]	K1AA through KØZZ. N1AA through NØZZ. W1AA through WØZZ.
Amateur	1 letter, 1 digit, 3 letters[4]	K1AAA through KØZZZ. N1AAA through NØZZZ. W1AAA through WØZZZ.
Amateur	2 letters, 1 digit, 1 letter[4]	AA1A through ALØZ. KA1A through KZØZ. NA1A through NZØZ. WA1A through WZØZ.
Amateur	2 letters, 1 digit, 2 letters[4]	AA1AA through ALØZZ. KA1AA through KZØZZ. NA1AA through NZØZZ. WA1AA through WZØZZ.
Amateur (letter "X" may not follow digit)	2 letters, 1 digit, 3 letters[4]	AA1AAA through ALØZZZ. KA1AAA through KZØZZZ. NA1AAA through NZØZZZ. WA1AAA through WZØZZZ.
Standard frequency		WWV, WWVB through WWVI, WWVL, WWVS.

Personal radio	3 letters, 4 digits or 4 letters, 4 digits	KAAØØØ1 through KZZ9999, WAAØØØ1 through WPZ9999, KAAAØØØ1 through KZZZ9999.
Personal radio, temporary permit	3 letters, 5 digits	KAAØØØØØ through KZZ99999.
Personal radio in trust territories	1 letter, 4 digits	KØØØ1 through K9999.
Business radio temporary permit	2 letters, 7 digits	WT plus local telephone number.
Part 90 temporary permit	2 letters, 7 digits	WT plus local telephone number.
Part 90 conditional permit	2 letters, 7 digits	WT plus local telephone number.
General Mobile Radio Service, temporary permit	2 letters, 7 digits	WT plus business or residence telephone number.

NOTE: The symbol Ø indicates the digit zero.

[1] Ships with transmitter-equipped survival craft shall be assigned four letter call signs.

[2] See § 2.303.

[3] A letter call sign now authorized for and in continuous use by a licensee of a standard broadcasting station may continue to be used by that station. The same exception applies also to frequency modulation and television broadcasting stations using 5 letter call signs consisting of 3 letters with the suffix "FM" or "TV."

[4] Plus other identifying data as may be specified.

[34 FR 5104, Mar. 12, 1969; as amended at 54 50239, Dec. 5, 1989]

§ 2.303 Other forms of identification of stations.

a) The following table indicates forms of identification which may be used in lieu of call signs by the specified classes of stations. Such recognized means of identification may be one or more of the following: name of station, location of station, operating agency, official registration mark, flight identification number, selective call number or signal, selective call identification or signal, characteristic signal, characteristic of emission or other clearly distinguishing form of identification readily recognized internationally. Reference should be made to the appropriate part of the rules for complete information or identification procedures for each service.

Class of station	Identification, other than assigned call sign
Aircraft (U.S. registry) telephone	Registration number preceded by the type of the aircraft, or the radiotelephony designator of the aircraft operating agency followed by the flight identification number.
Aircraft (foreign registry) telephone	Foreign registry identification consisting of five characters. This may be preceded by the radiotelephony designator of the aircraft operating agency or it may be preceded by the type of aircraft.
Aeronautical	Name of the city, area, or airdrome served together with such additional identification as may be required.
Aircraft survival craft	Appropriate reference to parent aircraft, e.g., the air carrier parent aircraft flight number or identification, the aircraft registration number, the name of the aircraft manufacturer, the name of the aircraft owner, or any other pertinent information.
Ship telegraph	When an official call sign is not yet assigned: Complete name of the ship and name of licensee. On 156.65 MHz: Name of ship. Digital selective call.
Ship telegraph	Digital selective call.
Public coast (radiotelephone) and Limited Coast (radiotelephone).	The approximate geographic location in a format approved by the Commission. Coast station identification number.
Public coast (radiotelegraph)	Coast station identification number.
Fixed	Geographic location. When an approved method of superimposed identification is used, QTT DE (abbreviated name and company or station).
Fixed: Rural subscriber service	Assigned telephone number.
Land mobile: Public safety, forestry conservation, highway maintenance, local government, shipyard, land transportation, and aviation services.	Name of station licensee (in abbreviated form if practicable), or location of station, or name of city, area, or facility served. Individual stations may be identified by additional digits following the more general identification.

Land mobile:Industrial service	Mobile unit cochannel with its base station: Unit identifier on file in the base station records. Mobile unit not co-channel with its base station: Unit identifier on file in base station records and the assigned call sign of either the mobile or base station. Temporary base station: Unit designator in addition to base station identification.
Land mobile: Domestic public and rural radio	Special mode unit designation assigned by licensee or by assigned telephone number.
Land mobile: Railroad radio service	Name of railroad, train number, caboose number, engine number, or name of fixed wayside station or such other number or name as may be specified for use of railroad employees to identify a specific fixed point or mobile unit. A railroad's abbreviated name or initial letters may be used where such are in general usage. Unit designators may be used in addition to the station identification to identify an individual unit or transmitter of a base station.
Land mobile: Broadcasting (remote pickup)	Identification of an associated broadcasting station.
Broadcasting (Emergency Broadcast System)	State and operational area identification.
Broadcasting (aural STL and intercity relay)	Call sign of the TV broadcasting station with which it is associated.
Broadcasting (television auxiliary)	Call sign of the TV broadcasting station with which it is licensed as an auxiliary, or call sign of the TV broadcasting station whose signals are being relayed, or by network identification.
Broadcasting (television booster)	Retransmission of the call sign of the primary station.
Disaster station	By radiotelephony: Name, location, or other designation of station when same as that of an associated station in some other service. Two or more separate units of a station operated at different locations are separately identified by the addition of a unit name, number, or other designation at the end of its authorized means of identification.

(b) Digital selective calls will be authorized by the Commission and will be formed by groups of numbers (0 through 9), however, the first digit must be other than 0, as follows:

(1) Coast station identification number: 4 digits.

(2) Ship station selective call number: 5 digits.

(3) Predetermined group of ship stations: 5 digits.

(c) Ship stations operating under a temporary operating authority shall identify by a call sign consisting of the letter "K" followed by the vessel's Federal or State registration number, or a call sign consisting of the letters "KUS" followed by the vessel's documentation number. However, if the vessel has no registration number or documentation number, the call sign shall consist of the name of the vessel and the name of the licensee as they appear on the station application form.

[28 FR 12465, Nov. 22, 1963, as amended at 40 FR 57675, Dec. 11, 1975; 41 FR 44042, Oct. 6, 1976; 42 FR 31008, June 17, 1977; 44 FR 62284, Oct. 30, 1979]

INTERNATIONAL CALL SIGN PREFIX ALLOCATIONS
(Sorted by Prefix)

Prefix	Country		Prefix	Country
AAA–ALZ	United States of America		HTA–HTZ	Nicaragua
AMA–AOZ	Spain		HUA–HUZ	El Salvador
APA–ASZ	Pakistan		HVA–HVZ	Vatican
ATA–AWZ	India		HWA–HYZ	France
AXA–AXZ	Australia		HZA–HZZ	Saudi Arabia
AYA–AZZ	Argentina		H2A–H2Z	Cyprus
A2A–A2Z	Botswana		H3A–H3Z	Panama
A3A–A3Z	Tonga		H4A–H4Z	Solomon Islands
A4A–A4Z	Oman		H6A–H7Z	Nicaragua
A5A–A5Z	Bhutan		H8A–H9Z	Panama
A6A–A6Z	United Arab Emirates		IAA–IZZ	Italy
A7A–A7Z	Qatar		JAA–JSZ	Japan
A8A–A8Z	Liberia		JTA–JVZ	Mongolia
A9A–A9Z	Bahrain		JWA–JXZ	Norway
BAA–BZZ	China		JYA–JYZ	Jordan
CAA–CEZ	Chile		JZA–JZZ	Indonesia
CFA–CKZ	Canada		J2A–J2Z	Djibouti
CLA–CMZ	Cuba		J3A–J3Z	Grenada
CNA–CNZ	Morocco		J4A–J4Z	Greece
COA–COZ	Cuba		J5A–J5Z	Guinea-Bissau
CPA–CPZ	Bolivia		J6A–J6Z	St. Lucia
CQA–CUZ	Portugal		J7A–J7Z	Dominica
CVA–CXZ	Uruguay		J8A–J8Z	St. Vincent and the Grenadines
CYA–CZZ	Canada		KAA–KZZ	United States of America
C2A–C2Z	Nauru		LAA–LNZ	Norway
C3A–C3Z	Andorra		LOA–LWZ	Argentina
C4A–C4Z	Cyprus		LXA–LXZ	Luxembourg
C5A–C5Z	Gambia		LYA–LYZ	Lithuania
C6A–C6Z	Bahamas		LZA–LZZ	Bulgaria
C7A–C7Z	World Meteorological Organization		L2A–L9Z	Argentina
C8A–C9Z	Mozambique		MAA–MZZ	Great Britain and N.Ireland
DAA–DRZ	Germany		NAA–NZZ	United States of America
DSA–DTZ	Korea, South		OAA–OCZ	Peru
DUA–DZZ	Philippines		ODA–ODZ	Lebanon
D2A–D3Z	Angola		OEA–OEZ	Austria
D4A–D4Z	Cape Verde		OFA–OJZ	Finland
D5A–D5Z	Liberia		OKA–OKZ	Czech Republic
D6A–D6Z	Comoros		OLA–OMZ	Slovakia
D7A–D9Z	Korea		ONA–OTZ	Belgium
EAA–EHZ	Spain		OUA–OZZ	Denmark
EIA–EJZ	Irelsnd		PAA–PIZ	Netherlands
EKA–EKZ	Armenia (was UG)		PJA–PJZ	Netherlands Antilles
ELA–ELZ	Liberia		PKA–POZ	Indonesia
EMA–EOZ	Ukraine (was UB)		PPA–PYZ	Brazil
EPA–EQZ	Iran		PZA–PZZ	Suriname
ERA–ERZ	Moldova (was UO)		P2A–P2Z	Papua New Guinea
ESA–ESZ	Estonia		P3A–P3Z	Cyprus
ETA–ETZ	Ethiopia		P4A–P4Z	Aruba
EUA–EWZ	Belarus (was UC)		P5A–P9Z	Korea, North
EXA–EXZ	Kirghiz (was UM)		RAA–RZZ	Russian Federation
EYA–EYZ	Tadzhik (was UJ)		SAA–SMZ	Sweden
EZA–EZZ	Turkoman (was UH)		SNA–SRZ	Poland
E2A–E2Z	Thailand		SSA–SSM	Egypt
E3A–E3Z	Eritrea		SSN–SSZ	Sudan
FAA–FZZ	France		STA–STZ	Sudan
GAA–GZZ	Great Britain and N.Ireland		SUA–SUZ	Egypt
HAA–HAZ	Hungary		SVA–SZZ	Greece
HBA–HBZ	Switzerland		S2A–S3Z	Bangladesh
HCA–HDZ	Ecaudor		S5A–S5Z	Slovenia
HEA–HEZ	Switzerland		S6A–S6Z	Singapore
HFA–HFZ	Poland		S7A–S7Z	Seychelles
HGA–HGZ	Hungary		S9A–S9Z	Sao Tome and Principe
HHA–HHZ	Haiti		TAA–TCZ	Turkey
HIA–HIZ	Dominican Republic		TDA–TDZ	Guatemala
HJA–HKZ	Colombia		TEA–TEZ	Costa Rica
HLA–HLZ	Korea, South		TFA–TFZ	Iceland
HMA–HMZ	Korea, North		TGA–TGZ	Guatemala
HNA–HNZ	Iraq		THA–THZ	France
HOA–HPZ	Panama		TIA–TIZ	Costa Rica
HQA–HRZ	Honduras		TJA–TJZ	Cameroon
HSA–HSZ	Thailand		TKA–TKZ	Corsica

TLA–TLZ	Central African Republic
TMA–TMZ	France
TNA–TNZ	Congo
TOA–TQZ	France
TRA–TRZ	Gabon
TSA–TSZ	Tunisia
TTA–TTZ	Chad
TUA–TUZ	Ivory Coast
TVA–TXZ	France
TYA–TYZ	Benin
TZA–TZZ	Mali
T2A–T2Z	Tuvalu
T3A–T3Z	Kiribati
T4A–T4Z	Cuba
T5A–T5Z	Somalia
T6A–T6Z	Afghanistan
T7A–T7Z	San Marino
T9A–T9Z	Bosnia and Herzegovina
UAA–UIZ	Russian Federation
UJA–UMZ	Uzbek (was UI)
UNA–UQZ	Kazakh (was UL)
URA–UZZ	Ukraine (was UB)
VAA–VGZ	Canada
VHA–VNZ	Australia
VOA–VOZ	Canada
VPA–VSZ	Great Britain and Territories
VTA–VWZ	India
VXA–VYZ	Canada
VZA–VZZ	Australia
V2A–V2Z	Antigua and Barbuda
V3A–V3Z	Belize
V4A–V4Z	St. Kitts and Nevis
V5A–V5Z	Namibia
V6A–V6Z	Federated States of Micronesia
V7A–V7Z	Marshall Islands
V8A–V8Z	Brunei
WAA–WZZ	Untied States of America
XAA–XIZ	Mexico
XJA–XOZ	Canada
XPA–XPZ	Denmark
XQA–XRZ	Chile
XSA–XSZ	China
XTA–XTZ	Burkino Faso
XUA–XUZ	Cambodia
XVA–XVZ	Vietnam
XWA–XWZ	Laos
XXA–XXZ	Macau
XYA–XZZ	Myanmar
YAA–YAZ	Afghanistan
YBA–YHZ	Indonesia
YIA–YIZ	Iraq
YJA–YJZ	Vanuatu
YKA–YKZ	Syria
YLA–YLZ	Latvia
YMA–YMZ	Turkey
YNA–YNZ	Nicaragua
YOA–YRZ	Romania
YSA–YSZ	El Salvador
YTA–YUZ	Yugoslovia
YVA–YYZ	Venezuela
YZA–YZZ	Yugoslovia
ZAA–ZAZ	Albania
ZBA–ZJZ	Great Britain and Territories
ZKA–ZMZ	New Zealand
ZNA–ZOZ	Great Britain and Territories
ZPA–ZPZ	Paraguay
ZQA–ZQZ	Great Britain and Territories
ZRA–ZUZ	South Africa
ZVA–ZZZ	Brazil
Z2A–Z2Z	Zimbabwe
Z3A–Z3Z	Macedonia
2AA–2ZZ	Great Britan and N.Ireland
3AA–3AZ	Monaco
3BA–3BZ	Mauritius
3CA–3CZ	Equitorial Guinea
3DA–3DM	Swaziland
3DN–3DZ	Fiji Islands
3EA–3FZ	Panama
3GA–3GZ	Chile
3HA–3UZ	China
3VA–3VZ	Tunisia
3WA–3WZ	Vietnam
3XA–3XZ	Guinea
3YA–3YZ	Norway
3ZA–3ZZ	Poland
4AA–4CZ	Mexico
4DA–4IZ	Philippines
4JA–4KZ	Azerbaijan (was UD)
4LA–4LZ	Georgia (was UF)
4MA–4MZ	Venezuela
4NA–4OZ	Yugoslovia
4PA–4SZ	Sri Lanka
4TA–4TZ	Peru
4UA–4UZ	United Nations
4VA–4VZ	Haiti
4XA–4XZ	Israel
4YA–4YZ	International Civil Aviation Organization
4ZA–4ZZ	Israel
5AA–5AZ	Libya
5BA–5BZ	Cyprus
5CA–5GZ	Morocco
5HA–5IZ	Tanzania
5JA–5KZ	Colombia
5LA–5MZ	Liberia
5NA–5OZ	Nigeria
5PA–5QZ	Denmark
5RA–5SZ	Madagascar
5TA–5TZ	Mauritania
5UA–5UZ	Niger
5VA–5VZ	Togo
5WA–5WZ	Western Samoa
5XA–5XZ	Uganda
5YA–5ZZ	Kenya
6AA–6BZ	Egypt
6CA–6CZ	Syria
6DA–6JZ	Mexico
6KA–6NZ	Korea, South
6OA–6OZ	Somalia
6PA–6SZ	Pakistan
6TA–6UZ	Sudan
6VA–6WZ	Senegal
6XA–6XZ	Madagascar
6YA–6YZ	Jamaica
6ZA–6ZZ	Liberia
7AA–7IZ	Indonesia
7JA–7NZ	Japan
7OA–7OZ	Yemen
7PA–7PZ	Lesotho
7QA–7QZ	Malawi
7RA–7RZ	Algeria
7SA–7SZ	Sweden
7TA–7YZ	Algeria
7ZA–7ZZ	Saudi Arabia
8AA–8IZ	Indonesia
8JA–8NZ	Japan
8OA–8OZ	Botswana
8PA–8PZ	Barbados
8QA–8QZ	Maldives
8RA–8RZ	Guyana
8SA–8SZ	Sweden
8TA–8YZ	India
8ZA–8ZZ	Saudi Arabia
9AA–9AZ	Croatia
9BA–9DZ	Iran
9EA–9FZ	Ethiopia
9GA–9GZ	Ghana
9HA–9HZ	Malta
9IA–9JZ	Zambia

9KA–9KZ	Kuwait
9LA–9LZ	Sierra Leone
9MA–9MZ	Malaysia
9NA–9NZ	Nepal
9OA–9TZ	Zaire

9UA–9UZ	Burundi
9VA–9VZ	Singapore
9WA–9WZ	Malaysia
9XA–9XZ	Rwanda
9YA–9ZZ	Trinidad and Tobago

WORLD CALL SIGN PREFIX ALLOCATIONS
(Sorted by Country Name)

Afghanistan	T6A–T6Z, YAA–YAZ
Albania	ZAA–ZAZ
Algeria	7RA–7RZ, 7TA–7YZ
Andorra	C3A–C3Z
Angola	D2A–D3Z
Antigua and Barbuda	V2A–V2Z
Argentina	AYA–AZZ, LOA–LWZ, L2A–L9Z
Armenia (was UG)	EKA–EKZ
Aruba	P4A–P4Z
Australia	AXA–AXZ, VHA–VNZ, VZA–VZZ
Austria	OEA–OEZ
Azerbaijan (was UD)	4JA–4KZ
Bahamas	C6A–C6Z
Bahrain	A9A–A9Z
Bangladesh	S2A–S3Z
Barbados	8PA–8PZ
Belarus (was UC)	EUA–EWZ
Belgium	ONA–OTZ
Belize	V3A–V3Z
Benin	TYA–TYZ
Bhutan	A5A–A5Z
Bolivia	CPA–CPZ
Bosnia and Herzegovina	T9A–T9Z
Botswana	A2A–A2Z, 8OA–8OZ
Brazil	PPA–PYZ, ZVA–ZZZ
Brunei	V8A–V8Z
Bulgaria	LZA–LZZ
Burkino Faso	XTA–XTZ
Burundi	9UA–9UZ
Cambodia	XUA–XUZ
Cameroon	TJA–TJZ
Canada	CFA–CKZ, CYA–CZZ, VAA–VGZ, VOA–VOZ, VXA–VYZ, XJA–XOZ
Cape Verde	D4A–D4Z
Central African Republic	TLA–TLZ
Chad	TTA–TTZ
Chile	CAA–CEZ, XQA–XQZ, 3GA–3GZ
China	BAA–BZZ, XSA–XSZ, 3HA–3UZ
Colombia	HJA–HKZ, 5JA–5KZ
Comoros	D6A–D6Z
Congo	TNA–TNZ
Corsica	TKA–TKZ
Costa Rica	TEA–TEZ, TIA–TIZ
Croatia	9AA–9AZ
Cuba	CLA–CMZ, COA–COZ, T4A–T4Z
Cyprus	C4A–C4Z, H2A–H2Z, P3A–P3Z, 5BA–5BZ
Czech Republic	OKA–OKZ
Denmark	OUA–OUZ, XPA–XPZ, 5PA–5QZ
Djibouti	J2A–J2Z
Dominica	J7A–J7Z
Dominican Republic	HIA–HIZ
Ecuador	HCA–HDZ
Egypt	SSA–SSM, SUA–SUZ, 6AA–6AZ
El Salvador	HUA–HUZ, YSA–YSZ
Equatorial Guinea	3CA–3CZ
Eritrea	E3A–E3Z
Estonia	ESA–ESZ
Ethiopia	ETA–ETZ, 9EA–9FZ
Federated States of Micronesia	V6A–V6Z
Fiji Islands	3DN–3DZ
Finland	OFA–OJZ
France	FAA–FZZ, HWA–HYZ, THA–THZ, TMA–TMZ, TVA–TXZ, TOA–TQZ
Gabon	TRA–TRZ

Gambia	C5A–C5Z
Georgia (was UF)	4LA–4LZ
Germany	DAA–DRZ
Ghana	9GA–9GZ
Great Britain and N. Ireland	GAA–GZZ, MAA–MZZ, 2AA–2ZZ
Great Britain and Territories	VPA–VSZ, ZBA–ZJZ, ZNA–ZOZ, ZQA–ZQZ
Greece	J4A–J4Z, SVA–SZZ
Grenada	J3A–J3Z
Guatemala	TDA–TDZ, TGA–TGZ
Guinea	3XA–3XZ
Guinea-Bissau	J5A–J5Z
Guyana	8RA–8RZ
Haiti	HHA–HHZ, 4VA–4VZ
Honduras	HQA–HRZ
Hungary	HAA–HAZ, HGA–HGZ
Iceland	TFA–TFZ,
India	ATA–AWZ, VTA–VWZ, 8TA–8YZ
Indonesia	JZA–JZZ, PKA–POZ, YBA–YHZ, 7AA–7IZ, 8AA–8IZ
International Civil Aviation Organization	4YA–4YZ
Iran	EPA–EQZ, 9BA–9DZ,
Iraq	HNA–HNZ, YIA–YIZ
Ireland	EIA–EJZ
Israel	4XA–4XZ, 4ZA–4ZZ
Italy	IAA–IZZ
Ivory Coast	TUA–TUZ
Jamaica	6YA–6YZ
Japan	JAA–JSZ, 7JA–7NZ, 8JA–8NZ
Jordan	JYA–JYZ
Kazakh (was UL)	UNA–UQZ
Kenya	5YA–5ZZ
Kirghiz (was UM)	EXA–EXZ
Kiribati	T3A–T3Z
Korea, South	DSA–DTZ, D7A–D9Z, HLA–HLZ, 6KA–6NZ
Korea, North	HMA–HMZ, P5A–P9Z
Kuwait	9KA–9KZ
Laos	XWA–XWZ
Latvia	YLA–YLZ
Lebanon	ODA–ODZ
Lesotho	7PA–7PZ
Liberia	A8A–A8Z, D5A–D5Z, ELA–ELZ, 5LA–5MZ, 6ZA–6ZZ
Libya	5AA–5AZ
Lithuania	LYA–LYZ
Luxembourg	LXA–LXZ
Macau	XXA–XXZ
Macedonia	Z3A–Z3Z
Madagascar	5RA–5SZ, 6XA–6XZ
Malawi	7QA–7QZ
Malaysia	9MA–9MZ, 9WA–9WZ
Maldives	8QA–8QZ
Mali	TZA–TZZ
Malta	9HA–9HZ
Marshall Islands	V7A–V7Z
Mauritania	5TA–5TZ
Mauritius	3BA–3BZ
Mexico	XAA–XIZ, 4AA–4CZ, 6DA–6JZ
Moldova (was UO)	ERA–ERZ
Monaco	3AA–3AZ

Country	Prefix
Mongolia	JTA–JVZ
Morocco	CNA–CNZ, 5CA–5GZ
Mozambique	C8A–C9Z
Myanmar	XYA–XZZ
Namibia	V5A–V5Z
Nauru	C2A–C2Z
Nepal	9NA–9NZ
Netherlands	PAA–PIZ
Netherlands Antilles	PJA–PJZ
New Zealand	ZKA–ZMZ
Nicaragua	HTA–HTZ, H6A–H7Z, YNA–YNZ
Niger	5UA–5UZ
Nigeria	5NA–5OZ
Norway	JWA–JXZ, LAA–LNZ, 3YA–3YZ
Oman	A4A–A4Z
Pakistan	APA–ASZ, 6PA–6SZ
Panama	HOA–HPZ, H3A–H3Z, H8A–H9Z, 3EA–3FZ
Papua New Guinea	P2A–P2Z
Paraguay	ZPA–ZPZ
Peru	OAA–OCZ, 4TA–4TZ
Philippines	DUA–DZZ, 4DA–4IZ
Poland	HFA–HFZ, SNA–SRZ, 3ZA–3ZZ
Portugal	CQA–CUZ
Qatar	A7A–A7Z
Romania	YOA–YRZ
Russian Federation	RAA–RZZ, UAA–UIZ
Rwanda	9XA–9XZ
San Marino	T7A–T7Z
Sao Tome and Principe	S9A–S9Z
Saudi Arabia	HZA–HZZ, 7ZA–7ZZ, 8ZA–8ZZ
Senegal	6VA–6WZ
Seychelles	S7A–S7Z
Sierra Leone	9LA–9LZ
Singapore	S6A–S6Z, 9VA–9VZ
Slovakia	OLA–OMZ
Slovenia	S5A–S5Z
Solomon Islands	H4A–H4Z
Somalia	T5A–T5Z, 6OA–6OZ
South Africa	ZRA–ZUZ
Spain	AMA–AOZ, EAA–EHZ
Sri Lanka	4PA–4SZ
St. Lucia	J6A–J6Z
St. Vincent and the Grenadines	J8A–J8Z
St. Kitts and Nevis	V4A–V4Z
Sudan	SSN–SSZ, STA–STZ, 6TA–6UZ
Suriname	PZA–PZZ
Swaziland	3DA–3DM
Sweden	SAA–SMZ, 7SA–7SZ, 8SA–8SZ
Switzerland	HBA–HBZ, HEA–HEZ
Syria	YKA–YKZ, 6CA–6CZ
Tadzhik (was UJ)	EYA–EYZ
Tanzania	5HA–5IZ
Thailand	E2A–E2Z, HSA–HSZ
Togo	5VA–5VZ
Tonga	A3A–A3Z
Trinidad and Tobago	9YA–9ZZ
Tunisia	TSA–TSZ, 3VA–3VZ
Turkey	TAA–TCZ, YMA–YM
Turkoman (was UH)	EZA–EZZ
Tuvalu	T2A–T2Z
Uganda	5XA–5XZ
Ukraine (was UB)	EMA–EOZ, URA–UZZ
United Arab Emirates	A6A–A6Z
United Nations	4UA–4UZ
United States of America	AAA–ALZ, KAA–KZZ, NAA–NZZ, WAA–WZZ
Uruguay	CVA–CXZ
Uzbek (was UI)	UJA–UMZ
Vanuatu	YJA–YJZ
Vatican	HVA–HVZ
Venezuela	YVA–YYZ, 4MA–4MZ
Vietnam	XVA–XVZ, 3WA–3WZ
Western Samoa	5WA–5WZ
World Meteorological Organization	C7A–C7Z
Yemen	7OA–7OZ
Yugoslavia	YTA–YUZ, YZA–YZZ, 4NA–4OZ
Zaire	9OA–9TZ
Zambia	9IA–9JZ
Zimbabwe	Z2A–Z2Z

AMATEUR CALL SIGN DISTRICTS BY COUNTRY

Prefix	Area
AA-AG, AI-AK, K, KA-KG, KI-KK, KM-KO, KQ-KZ, N, NA-NG, NI-NK, NM-NO, NQ-NZ, W, WA-WG, WI-WK, WM-WO, WQ-WZ	United States Contiguous Areas
AH1, KH1, NH1, WH1	Baker & Howland Islands
AH2, KH2, NH2, WH2	Guam
AH3, KH3, NH3, WH3	Johnston Island
AH4, KH4, NH4, WH4	Midway Island
AH5, KH5, NH5, WH5	Palmyra & Jarvis Islands
AH5K, KH5K, NH5K, WH5K	Kingman Reef
AH8, KH8, NH8, WH8	American Samoa
AH9, KH9, NH9, WH9	Wake, WIlkes, and Peale Islands
AHØ, KHØ, NHØ, WHØ	Commonwealth of Northern Mariana Islands
AH6, AH7, KH6, KH7, NH6, NH7, WH6, WH7	Hawaii
AH7K, KH7K, NH7K, WH7K	Kure Island
AL, KL, NL,. WL	Alaska
KP1, NP1, WP1	Navassa Island
KP2, NP2, WP2	U.S. Virgin Islands
KP3, KP4, NP3, NP4, WP3, WP4	Puerto Rico
KP5, NP5, WP5	Desecheo Island

Notes: 1. On two-by-three format U.S. callsigns, the first letter after the number is not allowed to be an "X"
2. Some KV4 callsigns are in use in the U.S. Virgin Islands
3. Some KG4 callsigns are in use at the U.S. Naval base at Guantanamo Bay, Cuba
4. Some KA-prefix callsigns are used by U.S. military personnel in Japan

District	State
1	Maine (ME), New Hampshire (NH), Vermont (VT), Massachusetts (MA), Rhode Island (RI), Connecticut (CT)
2	New York (NY), New Jersey (NJ)
3	Pennsylvania (PA), Delaware (DE), Maryland (MD), District of Columbia (DC)
4	North Carolina (NC), South Carolina (SC), Georgia (GA), Tennessee (TN), Alabama (AL), Florida (FL), Kentucky (KY), Virginia (VA)
5	Mississippi (MS), Louisiana (LA), Arkansas (AR), Texas (TX), Oklahoma (OK), New Mexico (NM)
6	California (CA)
7	Nevada (NV), Utah (UT), Arizona (AZ), Oregon (OR), Washington (WA), Idaho (ID), Montana (MT), Wyoming (WY)
8	Michigan (MI), Ohio (OH), West Virginia (WV)
9	Illinois (IL), Indiana (IN), Wisconsin (WI)
Ø	Colorado (CO), Iowa (IA), Kansas (KS), Minnesota (MN), Missouri (MO), Nebraska (NE), North Dakota (ND), South Dakota (SD)

AZ   See LU, LW, AZ

BA, BD, BG, BR, BT, BW, BY, BZ   China (People's Republic)
     BA: Class 1 individual home stations
     BD: Class 2 individual home stations
     BG: Class 3 individual home stations
     BR: Repeaters

	BT: Special Calls		ØX	San Felix Island
	BW: Reciprocal Licenses		ØX	San Ambrosio Island
	BY: Club Stations		ØZ	Juan-Fernandez Island
	BZ: Individual Callsigns for operators of Club Stations		CL, CM,	
1	Beijing		CO	Cuba
2A-I	Heilongjiang		1	Pinar del Rio
2J-Q	Jilin		2	Havana City
2R-Z	Liaoning		3	Havana
3A-F	Tianjin		4	Isla de la Juventud (Isle of Youth)
3G-L	Nei Mongol (Inner Mongolia) (Autonomous Region)		5	Matanzas
3M-S	Hebei		6	Cienfuegos, Sancti Spiritus, Villa Clara
3T-Z	Shanxi		7	Camaguey, Ciego de Avila
4A-I	Shanghai		8	Granma, Guantanamo, Holguin, Las Tunas,
4J-Q	Shandong			Santiago de Cuba
4R-Z	Jiangsu			
5A-I	Zhejiang			
5J-Q	Jiangxi		CP	Bolivia
5R-Z	Fujian		1	La Paz
6A-I	Henan		2	Chuquisaca
6J-Q	Anhui		3	Oruro
6R-Z	Hubei		4	Potosi
7A-I	Hunan		5	Cochabamba
7J-Q	Guangxi (Autonomous Region), Hainan		6	Santa Cruz
7R-Z	Guangdong		7	Tarija
8A-I	Sichuan		8	Beni
8J-Q	Guizhou		9	Pando
8R-Z	Yunnan		Ø	Special Calls
9A-F	Ningxia (Autonomous Region)		CU	Azores (Açores)
9G-L	Qinghai		1	Santa Maria
9M-S	Shaanxi		2	Sao Miguel
9T-Z	Gansu		3	Terceira
ØA-M	Xinjiang (Sinkiang) (Autonomous Region)		4	Graciosa
ØN-Z	Xizang (Tibet) (Autonomous Region)		5	Sao Jorge
			6	Pico
BO, BS, BV	Taiwan		7	Faial
	BO: Jin Men (Quemoy) Area of Taiwan; (Taiwanese Islands) BO2, BOØK: Kinmen Island, BOØM: Matsu Island		8	Flores
			9	Corvo
	BSØH: Huang Yan Do (Scarborough Reef)		Ø	Repeaters
1	Ilan, Keelung Shi		CW-CX	
2	Taipei, Taipei Shi		1–9	Uruguay
3	Hsinchu (Xinzhu), Hsinchu Shi, Miaoli, Taoyuan			The First Letter After the Number Designates:
4	Nantou, T'aichung, T'aichung Shi		A, B, C	Montevideo
5	Changhua, Chiai, Chiai Shi, Yun Ling		D	Canelones
6	T'ainan, T'ainan Shi		E	San Jose
7	Kaohsiung (Gaoxiong), Kaohsiung Shi, Pingtung		F	Colonia
8	Hualien (Hualian), Taitung		G	Soriano
9	Taiwanese Islands		H	Rio Negro
	9A: Penhu		I	Paysandu
	9C: Chi-Lung Yu		J	Salto
	9G: Lu-Dao		K	Artigas
	9H: Hua-Ping Yu		L	Florida
	9K: Kuei-Shan Dao		M	Flores
	9L: Liu-Chiu Yu		N	Durazno
	9M: Mien-Hua Yu		O	Tacuarembo
	9O: Lan-Yu		P	Rivera
	9P: Pratas		R	Maldonado
	9S: Spratly		S	Lavalleja
	9U: U-Chiu Dao		T	Rocha
	9W: Peng-Chia Yu		U	Treinta-y-Tres
Ø	Reciprocal Licenses		V	Cerro Largo
			X	Special Calls
BW, BY, BZ	See BA, BD, BG, BT, BW, BY, BZ		Z	Mobile Stations
CE, XQ	Chile		DA–DD, DG–DH,	
1	Antofagasta, Atacama, Tarapaca		DJ–DL, DP	Germany
2	Aconcagua, Coquimbo		DA1, 2, 4	Foreign Military personnel
3	Metropolitan		DBØ	Club and relay (repeater) stations
4	Maule, Liberatador		DB, DC,	
5	Bio-Bio		DD, DG	VHF License
6	Araucania		DCØFA-	
7	Aisen, Los Lagos		DCØJZ	Reciprocal VHF license
8	Magallanes		DC7AA-	
ØY	Easter Island		DC7ZZ	West Berlin VHF license

DD6AAA-	
DD6ZZZ	West Berlin VHF license
DH	Class A license VHF + limited HF
DH7AAA	
-DH7ZZZ	Class A license, West Berlin
DF, DJ,	
DK, DL	Class B license
DL7AAA-	
DL7ZZZ	West Berlin Class B license
DJØAAA-	
DJØZZZ	Reciprocal licenses
DFØ, DLØ,	
DKØ	Club Stations
DP	Stations outside Germany

Examples: DPØGVN Antarctica—Georg Von Neumeier Station; DP1MIR/DPØSL—Spacelab operations

Former East German licensees (Y2, Y3, etc. prefixes) are now being issued callsigns in the following blocks: DL1RAA-RZZ, DL2RAA-RZZ, DL6RAA-RZZ, DL6UAA-UZZ, DL6VAA-VZZ, DL7UAA-UZZ, DL7VAA-VZZ, and DF1YA-DF6YZ for Berlin.

EA, EB, EC, ED, EE,	
EF, EH	Spain (EB VHF Licenses, EC Novice Class, ED, EE, EF, EG, EH)
1	Asturias, Avila, Burgos, Cantabria, La Coruna, Leon, Lugo, Orense, Palencia, Pontevedra, La Rioja, Salamanca, Segovia, Soria, Valladolid, Zamora
2	Alava, Guipuzcoa, Huesca, Navarra, Teruel, Vizcaya, Zaragoza
3	Barcelona, Gerona, Lerida, Tarragona
4	Badajoz, Caceres, Ciudad Real, Cuenca, Guadalajara, Madrid, Toledo
5	Albacete, Alicante, Castellon, Murcia, Valencia
6	Baleares (Balearic Islands)
7	Almeria, Cadiz, Cordoba, Granada, Huelva, Jaen, Malaga, Sevilla
8	Canary Islands
9	Ceuta and Melilla
Ø	–

EL	Liberia
1	Grand Bassa, River Cess
2	Bomi, Federal District, Margibi, Montserrado
3	Grand Kru, Sinoe
4	Maryland
5	Loffa
6	Grand Gedeh
7	Bong
8	Nimba
9	Grand Cape Mount
Ø	Maritime Mobile Stations

ES	Estonia
1	Tallinn, Northern Islands
2	Harju Rajon
3	Haapsalu, Rapla, Paide
4	Kohtla-Jarve, Rakvere, Sillamae, Narva
5	Jogeva, Polca, Tartu
6	Valga, Voru
7	Viljandi
8	Parnu
9	Special Calls
Ø	Western Islands

EX	Kyrghyzstan (former UM8)
1	Special callsigns
2	All individual callsigns
3	Special callsigns
4	Special callsigns
5	All individual callsigns
6	All individual callsigns
7	All individual callsigns
8	All individual callsigns
9	Club stations

Ø	All individual callsigns

EY	Tajikistan
1	TARL stations
2	Reserved for special stations
3	Reserved for special stations
4	Gornii Badakhshan, ex-UJ8R
5	Khatlonskaya oblast Kulab region, ex-UJ8K
6	Khatlonskaya oblast Kurganz Tube region, ex-UJ8X
7	Leninabadskaya oblast, ex-UJ8S
8	Dushanbe city, ex-UJ8J
9	Dushanbe regions, ex-UJ8J
Ø	Reserved for special stations

HA, HG	Hungary
1	Gyor, Sopron, Vas, Zala
2	Komarom, Esztergom, Veszprem
3	Baranya, Somogy, Tolna
4	Fejer
5	Budapest (city)
6	Heves, Nograd
7	Pest, Jasz-Nagykun-Szolnok
8	Bacs-Kiskun, Bekes, Csongrad
9	Borsod-Abauj-Zemplen
Ø	Hajdu-Bihar, Szabolcs-Szatmar-Bereg

HC	Ecuador
1	Carchi, Imbabura, Pichincha
2	Guayaquil, Los Rios
3	El Oro, Loja
4	Esmeraldas, Manabi
5	Azuay, Canar, Chimborazo
6	Bolivar, Cotopaxi, Tungurahua
7	Morona-Santiago, Napo, Pastaza, Zamora
8	Galapagos Islands
Ø	Contest and special event stations

HI	Dominican Repulblic
1	Alto Velo Island, Beata Island
2	Catalina Island, Catalinita Island, Saona-Island
3	Duarte, Espaillat, La Vega, Puerto Plata, Salcedo, Ramirez, Santiago
4	Dajabon, Monte Cristi, Santiago Rodriguez, Valverde
5	Bahoruco, Barahona, Independencia, Pedernales
6	Azua, La Estrelleta, San Juan
7	El Seybo, Hato Mayor, La Altagracia, La Romana, San Pedro de Macoris
8	Districto Nacional, Peravia, San Cristobal
9	Maria Trinidad Sanchez, Samana

HJ, HK	Colombia (HJ Novice Class)
1	Atlantico, Bolivar, Cordoba, Sucre
2	Cesar, Guajira, Magdalena, Norte de Santander
3	Cundinamarca, Meta, Vichada
4	Antioquia, Choco
5	Cauca, Valle del Cauca
6	Caldas, Huila, Quindio, Risaralda, Tolima
7	Arauca, Boyaca, Casanare, Santander
8	Caqueta, Narino, Putumayo
9	Amazonas, Guainia, Guaviare, Vaupes
Ø	Bajo Nuevo, San Andres Island, Providencia Archipelago, Malpelo Island, Cayo de Rancador, Serrana Bank

HL	South Korea
1	Seoul
2	Kyonggi, Kangwon, Inchon
3	North Chungchong, South Chungchong
4	North Cholla, South Cholla, Cheju Island
5	North Kyongsang, South Kyongsang, Pusan, Taegu
8	Mobile Stations
9	Reciprocal Licenses

Ø	Club Stations
HP	Panama
1	Panama
2	Colon, San Blas
3	Chiriqui
4	Bocas del Toro, Cocle
5	Herrera, Los Santos
6	Veraguas
7	Darien
8	Perlas Archipelago
9	Maritime Mobile Stations
Ø	Coiba Island
HR	Honduras
1	Comayagua, Distrito Central, Francisco Morazan
2	Cortes, La Paz, Yoro
3	Atlantida, Colon, El Paraiso, Gracias a Dios, Olancho
4	Choluteca, Valle
5	Copan, Intibuca, Lempira, Ocotepeque, Santa Barbara
6	Bahia Island
HS, E2	Thailand
1	Bangkok and its neighboring provinces
2	The East of Thailand
3	The lower Northeast of Thailand
4	The upper Northeast of Thailand
5	The upper North of Thailand
6	The lower North of Thailand
7	The West of Thailand
8	The upper South of Thailand
9	The lower South of Thailand
Ø	Bangkok and its neighboring provinces
I, IA-IZ	Italy
1	LIGUREE
	Genova (GE), Imperia (IM), La Spezia (SP), Savona (SV).
	PIEMONT
	Alessandria (AL), Asti (AT), Cuneo (CN), Novara (NO), Torino (TO), Vercelli (VC).
2	LOMBARDI
	Bergamo (BG), Brescia (BS), Como (CO), Cremona (CR), Milano (MI), Mantova (MN), PAVIA (PV), Sondrio (SO), Varese (VA).
3	VENIZIA
	Belluno (BL), Padova (PD), Rovigo (RO), Treviso (TV), Venezia (VE), Verona (VR), Vicenza (VI).
4	EMILIA-ROMAGNA
	Bologna (BO), Ferrara (FE), Forli (FO), Modena (MO), Parma (PR), Piacenza(PC), Ravenna (RA), Reggio Emilia (RE).
5	TOSCANA
	Arezzo (AR), Firenze (FI), Grosseto (GR), Livorno (LI), Lucca (LU), Massa VCarrara (MS), Pisa (PI), Pistoia (PT), Siena (SI).
6	ABRUZZO
	Chieti (CH), L'Aquila (AQ), Pescara(PE), Teramo (TE).
	MARCHE Ancona (AN), Ascoli Piceno (AP), Macerata (MC), Pesaro (PS).
7	PUGLIA
	Bari (BA), Brindisi (BR), Foggia (FG), Lecce (LE), Taranto (TA).
	BASILICATA Matera (MT).
8	BASILICATA
	Potenza (PZ).
	CALABRIA Catanzaro (CZ), Cosenza (CS), Reggio Calabria (RC). CAMPANIA Avellino

	(AV), Benevento (BN), Caserta (CE), Napoli (NA), Salerno (SA). MOLISE Campobasso (CB), Isernia (IS).
Ø	LATINA
	Frosinone (FR), Latina (LT), Rieti (RI), Roma (ROMA), Viterbo (VT).
	UMBRIA Perugia (PG), Terni (TR).
IA2	Islands of Laco Maggiore.
IA5	Tuscan Archepeligo.
	Argentarlo, Capraia, Cerboli, Corallo, Corbella, Corbelli, Di liscoli, Di Vada, Elba, Falconcino, Formica di Burano, Formiche, Formiche di Grosseto, Formiche di Montecristo, Ogliera, Ortano, Palmarola, Pianosa, Porcelli, Remaiolo, Santa Lucia, Scoglietto, Scolglio d'Africa, Sedia di Paolina, Sparviero, Topi, Triglia.
IBØ	Latina Islands
	Botte, Calzone Muto, Calzone Parroco, Cappello, Evangelista, Faraglioni, Faraglioni Madonna, Forcina, Fucile, Gavi, La Nave, Le Formiche, La Galere, Mezzogiorno, Monaco, Pallante, Palmarola, Piatti, Ponza, Ravia, Rosso, San Silverio, Santo Stefano, Scogliatelle, Scuncillo, Spaccapolpi, Suvace, Torre Astura, Venotene, Zannone.
IC8	Neapolitan Islands.
	A Penna, Camerota, Capri, Faraglioni, Gaiola, Isca, Ischia, Li Galli, Licosa, Monacone, Nisida, Procida, Revigliano, Santo Ianni, Scrupolo, Vervece, Vetara, Vivara.
ID8	Calabrian Islands.
	Cirella, Dino, Formicola, Ianni, Le Castelle, Licosa, Pietra Grande, San Leonardo, Scoglio Galera, Scoglio Iscra.
ID9	Eolie Islands.
	Alicudi, Basiluzzo, Bottaro, Canna, Dattilo, Faraglione, Filicudi, Formiche, Galera, Imeratr, Lipari, Lisca bianca, Lisca Nera, Montenassari, Nave Panarea, Panarea, Panarelli, Pietra del Bagno, Pietra Lunga, Pietra Menalda, Quaedri, Salina, Santa Palomba, Spinazzola, Stromboli, Strombolicchio, Vulcano.
IE9	Ustican Islands.
	Banco Apollo, Colombara, Medico, Ustica.
IF9	Egadi Islands.
	Asinelli, Cammello, Correnti, Faraglione, Favignana, Formica, Galeotta, Galera, Levanzo, Maraone, Marettimo, Porcelli, Preveto, Scherchi.
IG9	Pelagie Islands.
	Conigli, Lampedusa, Lampione, Linosa.
IH9	Pantelleria Island.
IJ7	Cheradi Islands.
	San Paolo, San Pietro.
IJ9	Sicilian Islands.
	Bottazza, Ciclopi, Colombaia, Dei Porri, Delle Correnti, Delle Femmine, Dello Stagnone, Di Capo Passero, Di Patti, Di Pietra Patella, Dietro Isola, Due Fratelli, Formica, Grande di Marzameni, Iannuzzo, Isola Bella, La Scuola, Lachea, Le Pietre, Nere, Lunga, Mal Consiglio, Ognina, Ortigia, Piccola, San Biagio, San Pantaleo, Santa Maria, Scialandro, Scoglio di Brono, Tracino, Vendicari.
IL3	Grado Islands.
	Anfora, Ara Storta, Dei Belli, Dei Brusiari, Del Lovo, Gorgo, Grado, Gran Chiusa, Manzi, Marina di Macia, Martignano, Mezzano, Montaron, Morgo, Orbi, Pampiola, Panera, Ravaiarina, San Andrea, San Giuliano, San Pietro d'Oro, Santa Maria Barbana, Sian, Taglio Nuovo, Valerian, Villa Nova, Volpera, Volperossa.
IL3	Venetian Islands.
	Buel del Lovo, Burano, Campalto, Campana, Carbonera, Crevan, Falconera, Le Certosa, La Cura, La Giudecca e Sacca Fisola, La Grazia, Lazzaretto Nuovo, Lazzaretto Vecchio, Le Vignole, lido, Lio

Mazor, Lio Piccolo, Madonna del Monte, Mazzorbo, Monte del Oro, Motta del Cunigi, Motta San Lorenzo, Murano, Ottagono Alberoni, Pellestrina, Poveglia, Sacca Sessola, Salina, San Angelo di Polvere, San Ariano, SAn Biagio, San Clemente, San Cristina, San Cristoforo, San Erasmo, San Francesco del Desreto, San Giacomo in Paluo, San Giorgio in Alba, San Giorgio Maggiore, San Giuliano, San Lazzaro di Armeni, San Michele, San Secondo, San Servolo, Sant'Elena, Santo Spirito, Sette Soleri, Spignon, Tessera, Torcello, Torson, Trezze, Tronchetto, Val di Pozzo, Volpego.

IL4	Adriatic Islands.

Albarella, Barone, Chioggia, Polesine, Scanno, Piallazza.

IL7	Puglia Islands.

Cacio, Cavallo, Campi, Campo, Chianca, Del Capezzone, Della Malva, Due Sorelle, Fanciulla, Gallipoli, Giurlita, Grande, I Guaceto, I Pagliai, La Vecchia, Malva, Palombaro, Pazzi, Pedacine, Portonouvo, Sant'Andrea, Sant'Eufemia, Santo Emiliano, Scoglio Portonouvo, Tondo.

IL7	Tremiti Islands.

Caprara, Cretaccio, Pianosa, San Domino, San Nicola.

IMØ	Cagliari Islands.

Campianna, Cavoli, Corno, Di Cala Vinagra. Il Catalano, Il Toro, La Vacca, Mal di Ventre, Meli, Pgliastra, Padiglioni, Pan di Zucchero, Piana di San Pietro, Piscadeddus, Proci, Quirra, Ratti, Rossa di Teulada, San Macario, San Pietro, Sant'Antioco, Sant'Elmo, Serpantara, Su Cardulinu, Su Giudeu, Tuaredda, Variglioni, Vitello.

IMØ	Sassarian Islands.

Asinara, Barca Sconcia, Bianca, Cana, Cappuccini, Coluccia, Corona Niedda, Dei Poeri, Della Bocca, Della Pelosa, Delli Nibani, Di Zui Paulo, Figarolo, Foradada, Gabbia, Giardinelli, Isolotto Rosso, Le Camere Le Soffi, Lepre, Maddalena Alghero, Manna, Marmorata, Molara, Molarotto, Monica, Mortorio, Mortoriotto, Pagliosa, Patron Fiaso, Pecora, Pedrami, Piana di Alghero, Piana di Paolo, Porri, Porritula, Portisco, Poveri, Proratola, Reulino, Rossa, Rossa di Alghero, Ruia, Scoglio Businco, Scoglio Forani, Scoglio Paganetto, Tavolara, Verde.

IMØ	Maddelena Islands.

Abbatoggia, Barattinelli di Fuore, Barettinelli, Barrenttini, Bisce, Budelli, Camize, Camize, Cana, Capicciolu, Caprera, Carpa, Cavalli, Chiesa, Corcelli, Degli Italiani, Giardinelli, La Presa, Maddalena, Monaci, Paduleddi, Piana, Porco, Porraggi, Ratino, Razzoli, Roma, Santa Maria, Santo Stefano, Spargi, Spargiottello, Spargiotto, Stramanari.

IMØ	Sardia Islands.

Brecconi, Corona Niedda, Dei Mucchi Bianchi, Dei Pedrami, Della Olgiastra, Delle Bisce, La Ghinghetta, Mangiabarche, Ottiolo, Pan di Zucchero, Peloso, Piana.

IN3	TRIENT

Bolzano (BZ), Trento (TN).

IP1	Ligurian Islands.

Bergeggi, Fercale, Gallinara, Palmaria, Tino, Tinetto.

ISØ	SARDINIA

Cagliari (CA), Nuoro (NU), Oristano (OR), Sassari (SS).

IT9	SICILY

Agrigento (AG), Caltanisetta (CL), Catania (CT), Enna (EN), Messina (ME), Palermo (PA), Ragusa (RG), Siracusa (SR), Trapani (TP).

IV3	FRUILI-VENEZIA GIULIA

Gorizia (GO), Pordenone (PN), Trieste (TS), Udine (UD).

IW	VHF License
IX1	AOSTA

Aosta (AO).

JA, JE, JF, JG, JH, JI, JJ, JK, JL, JM, JN, JO, JP, JQ, JR, JS, 7J, 7K, 7L, 7M, 7N, 8J	Japan (7J Reciprocal Licenses, 8J Special Calls)
1	Chiba, Gumma, Ibaraki, Kanagawa, Saitama, Tochigi, Tokyo, Yamanashi
2	Aichi, Gifu, Mie, Shizuoka
3	Hyogo, Kyoto, Nara, Osaka, Shiga, Wakayama
4	Hiroshima, Okayama, Shimane, Tottori, Yamaguchi
5	Ehime, Kagawa, Kochi, Tokushima
6	Fukuoka, Kagoshima, Kumamoto, Miyazaki, Nagasaki, Oita, Okinawa, Saga
7	Akita, Aomori, Fukushima, Iwate, Miyagi, Yamagata
8	Hokkaido
9	Fukui, Ishikawa, Toyama
Ø	Nagano, Niigata
JT	Mongolia
1A-G	Ulan Bator (Ulaan Baator)
1H-Z	Nalajch
2A-M	Dornod
2N-Z	Suhbaatar
3A-M	Hentiy
3N-Z	Dornogobi
4A-M	Omnogobi
4N-Z	Dungobi
5A-M	Selenge
5N-Z	Tow
6A-M	Archangay
6N-Z	Oworchangay
7A-M	Hovsgol
7N-Z	Bulgan
8A-M	Govialtay
8N-Z	Bayanhongor
9A-M	Uvs
9N-Z	Dzavhan
ØA-M	Bayan-Olgiy
ØN-Z	Hovd
LU, LW, AZ	Argentina (LW Novice Class, AZ Special Calls) The First Letter After The Number Indicates The Region:
A, B, C	Buenos Aires (City)
D, E	Buenos Aires (Province)
F	Santa Fe
GA-GO	Chaco
GP-GZ	Formosa
H	Cordoba
I	Misiones
J	Entre Rios
K	Tucuman
L	Corrientes
M	Mendoza
N	Santiago del Estero
O	Salta
P	San Juan
Q	San Luis
R	Catamarca
S	La Rioja
T	Jujuy
U	La Pampa
V	Rio Negro
W	Chubut
XA-XO	Santa Cruz

XP-XZ	Tierra del Fuego
Y	Neuquen
LUØ	Mobile Stations
OA	Peru
1	Lambayeque, Piura, Tumbes
2	Cajamarca, La Libertad
3	Ancash, Huanuco
4	Callao, Junin, Lima, Pasco
5	Apurimac, Ayacucho, Huancavelica, Ica
6	Arequipa, Moquegua, Tacna
7	Cuzco, Madre de Dios, Puno
8	Loreto, Ucayali
9	Amazonas, San Martin
Ø	Special Calls
OE	Austria
1	Vienna
2	Salzburg
3	Lower Austria
4	Burgenland
5	Upper Austria
6	Styria
7	Tyrol
8	Carinthia
9	Vorarlberg
Ø	Special Calls
	First Letter After The Number:
X	Club Station, Repeater, Beacon
Z	Reciprocal Licenses
OG, OH, OI	Finland
1	Turku, Pori
2	Uusimaa
3	Hame
4	Mikkeli
5	Kymi
6	Keski-Suomi, Vaasa
7	Kuopio, Pohjois, Karjala
8	Oulu
9	Lappi (Lappland)
Ø	Aland Island
OJØ	Market Reef
	First Letter After The Number:
Z	Reciprocal Licenses
OK	Czech Republic
1	Bohemia
2	Moravia
3	Not used now
4	Usually Maritime Mobile stations
5	Special events
6	Contests
7	Not used now
8.	Foreigners
9	Experimentals
Ø	Repeaters, beacons, special events
OL	Special events, contests
PJ	Netherlands Antilles
1	Special Calls
2	Curacao
4	Bonaire
5	St. Eustatius
6	Saba
7	St. Maarten
8	St. Eustatius, St. Maarten, Saba (Reciprocal Licenses & Special Calls)
9	Bonaire, Curacao (Reciprocal Licenses & Special Calls)
Ø	Special Calls
PP, PQ, PR, PS, PT, PU, PV, PW,	

PY, ZV, ZW, ZX, ZY, ZZ	Brazil (PU Novice Class, ZV-ZZ Special Calls)
PP1, ZZ1	Espirito Santo
PP2, ZZ2	Goias
PP5, ZZ5	Santa Catarina
PP6, ZZ6	Sergipe
PP7, ZZ7	Alagoas
PP8, ZZ8A-G	Amazonas
PQ2, ZX2	Tocantins
PZ8, ZV8	Amapa
PR7, ZX7	Paraiba
PR8, ZX8	Maranhao
PS7, ZW7	Rio Grande do Norte
PS8, ZW8	Piaui
PT2, ZV2	Federal District
PT7, ZV7	Ceara
PT8, ZZ8H-M	Acre
PT9, ZV9	Mato Grosso do Sul
PU1A-I	Espirito Santo
PU1J-Z	Rio de Janeiro
PU2A-E	Federal District
PU2F-H	Goias
PU2I-J	Tocantins
PU2K-Z	Sao Paulo
PU3	Rio Grande do Sul
PU4	Minas Gerais
PU5A-L	Santa Catarina
PU5M-Z	Parana
PU6A-I	Sergipe
PU6J-Z	Bahia
PU7A-D	Alagoas
PU7E-H	Paraiba
PU7I-L	Rio Grande do Norte
PU7M-P	Ceara
PU7R-Z	Pernambuco
PU8A-C	Amazonas
PU8D-F	Rondonia
PU8G-I	Amapa
PU8J-L	Acre
PU8M-O	Maranhao
PU8P-S	Piaui
PU8T-V	Roraima
PU8W-Z	Para
PU9A-N	Mato Grosso do Sul
PU9O-Z	Mato Grosso
PUØ	(See PYØ)
PV8, ZZ8T-Z	Roraima
PW8, ZZ8N-S	Rondonia
PY1, ZY1	Rio de Janeiro
PY2, ZY2	Sao Paulo
PY3, ZY3	Rio Grande do Sol
PY4, ZY4	Minas Gerais
PY5, ZY5	Parana
PY6, ZY6	Bahia
PY7, ZY7	Pernambuco
PY8, ZY8	Para
PY9, ZY9	Mato Grosso
PYØ, ZWØ	Brazilian Islands
PYØF, PYØZF, ZYØ	Fernando de Noronha
PYØM, PYØZM, PYØ	Martin Vaz Island
PYØR, PYØZR, ZYØ	Atol das Rocas
PYØS, PYØZS, ZYØ	St. Peter and St. Paul Rocks
PYØT, PYØZT, ZYØ	Trindade-Island
	First Letter After The Number:
Z	Reciprocal Licenses
PZ	Suriname

1	Paramaribo, Sipaliwini	6	Amasya, Cankiri, Corum, Kastamonu, Kirsehir, Samsun, Sinop, Tokat, Yozgat
2	Nickerie		
3	Totness (Coronie)	7	Erzincan, Savastepe (Giresun), Gumushane, Kayseri, Ordu, Sivas, Trabzon, Tunceli
4	Saramacca		
5	Reciprocal Licenses	8	Adiyaman, Bingol, Diyarbakir, Elazig, Gaziantep, Kahranmaras, Malatya, Margin, Sanliurfa
6	Para		
7	Brokopondo	9	Agri, Artvin, Bitlis, Erzurum, Hakkari, Kars, Mus, Rize, Siirt, Van
8	Commewijne		
9	Marowijne	ØA-D	Mediterranean Islands
Ø	Special Calls	ØE-L	Aegean Islands

SI, SJ, SK,
SL, SM   Sweden (SK Club & Repeater Stations,
                SL Military Stations)

		ØM-X	Marmara Sea Islands
		ØY-Z	Black Sea Islands
SI8	Market Reef		
SJ9	Morokulien	TF	Iceland
SM1	Gotland	1	Arnessysla, Rangarvallasysla, West-Skaftafellssysla
SM2	Norrbotten, Vasterbotten	2	Borgarfjordur, Myrskyla, Snaefells and Hnappadalssysla, Dalasysla
SM3	Gavleborg, Jamtlands Sikas, Vasternorrland		
SM4	Kopparberg, Orebro, Varmland	3	Kjosarsysla, Reykjavik, Kopavogur, Gardabaer, Kaupstadur, Hafnarfjordur, Seltjarnarnesskaupsta-dur
SM5	Ostergotland, Sodermanland, Stockholm Lan, Uppsala, Vastmanland		
		4	East and West-Bardastrandarsyslur, West and North Isafjordur, Starandasysla
SM6	Alvsborg, Goteborg/Bohus, Hallands Vadero, Skaraborg		
		5	Eyafjordur, South and North Thingeyri
SM7	Blekinge, Jonkoping, Kalmar, Kristianstad, Kronoberg, Malmohus	6	North & South Mulasyslur, East Skaftafellssysla
		7	Vestmannaeyjar Islands
		8	Gullbringusysla
SMØ	Stockholm City	9	West and East-Hunavatnssyslur, Skagafjordur
		Ø	Uninhabited Areas in Central Iceland

SO, SP, SR	Poland (SO Reciprocal Licenses, SR Repeaters)		
1	Koszalin, Slupsk, Szczecin	TG	Guatemala
2	Bydgoszcz, Elblag, Gdansk, Torun, Wloclawek	4	Chimaltenango, Escuintla, Santa Rosa, Solola, Suchitepequez
3	Gorzow Wiekopolski, Kalisz, Konin, Leszno, Pila, Poznan, Zielona Gora		
		5	Huehuetenango, El Quiche,Tototnicapan
4	Bialystok, Lomza, Olsztyn, Suwalki	6	Chiquimula, Jutiapa, Zacapa
5	Ciechanow, Ostroleka, Plock, Siedlce, Warszawa	7	Alta Verapaz, Izabal, Peten
6	Jelenia Gora, Legnica, Opole, Walbrzych, Wroclaw	8	Quetzaltenango, Retalhuleu, San Marcos
7	Kielce, Lodz, Piotrkow Trybunalski, Radom, Sieradz, Skierniewice, Tarnobrzeg	9	Baja Verapaz, El Progreso, Guatemala, Jalapa, Sacatepequez
		Ø	Special Calls
8	Biala Podlaska, Chelm, Krosno, Lublin, Przemysl, Rzeszow, Zamosc		
		TI	Costa Rica
9	Bielsko Biala, Czestochowa, Katowice, Krakow, Nowy Sacz, Tarnow	1	Special Calls
		2	San Jose
Ø	Special Calls	3	Cartago
		4	Heredia
SV	Greece	5	Alajuela
1	Central Greece, Euboea, Athens	6	Limon
2	Macedonia	7	Guanacaste
3	Peloponnesus	8	Puntarenas
4	Thessaly	9	Cocos Island
5	Dodecanese (Alimia, Arkoi, Astipalaia, Khalki, Giali, Kalimnos, Karpathos, Kasos, Kos, Leros, Lipsos, Nisiros, Patmos, Pserimos, Rodhos (Rhodes), Saria, Simi, Tilos)	Ø	Special Calls
		VE, VO, VY	Canada
		VE1	Nova Scotia (NS), New Brunswick (NB)
6	Epirus	VE2	Quebec (PQ)
7	East Macedonia, Thrace	VE3	Ontario (ON)
8	Islands in Ionian Sea (Kefallinia, Corfu, Levkas, Zakinthos), Islands in Aegean Sea except Crete & Dodecancese (Khios (Chios), Kykladen, Lesvos (Lesbos), Samos, Vorioi Sporadhes (Northern Sporades), Thasos)	VE4	Manitoba (MB)
		VE5	Saskatchewan (SK)
		VE6	Alberta (AB)
		VE7	British Columbia (BC)
		VE8	Northwest Territories (NW)
		VEØ MAA- VEØ NZZ	Maritime Mobile Stations
9	Crete	VO1	Newfoundland (NF)
Ø	Reciprocal Licenses	VO2	Newfoundland (Labrador) (LB)
		VY1	Yukon Territory (YK)
TA, TB, TC	Turkey	VY2	Prince Edward Island (PE)
1	Edirne, Istanbul (European Sector), Kirklareli, Tekirdag	VY9	Department of Communications Club Stations
2	Ankara, Bilecik, Bolu, Eskisehir, Istanbul (Asian Sector), Izmit (Kocaeli), Sakarya, Zonguldak	VK	Australia
		1	Capital Territory
3	Balikesir, Bursa, Canakkale, Izmir, Manisa	2	New South Wales
4	Afyonkarahisar, Antalya, Aydin, Burdur, Denizli, Isparta, Kutahya, Mugla, Usak	3	Victoria
5	Adana, Hatay, Mercin (Icel), Konya, Nevsehir, Nigde	4	Queensland

5	South Australia		YS	El Salvador
6	Western Australia		1	San Salvador
7	Tasmania		2	Santa Ana
8	Northern Territory		3	San Miguel
9C, 9NC	Cocos (Keeling) Island		4	La Libertad
9L, 9NL	Lord Howe Island		5	Chalatenango, Cuscatlan, La Paz
9M, 9NM	Mellish Reef		6	Cabanas, San Vicente, Usulutan
9N, 9NN	Norfolk Island		7	Sonsonate
9W, 9NW	Willis Islets		8	Ahuachapan
9X, 9NX	Christmas Island		9	La Union, Morazan
Ø	Macquarie, Heard Island			

First letter after the number, if N or V, for 3 letter suffixes denotes
novice class license

			YV	Venezuela
			1	Falcon, Trujillo, Zulia
VO, VY	SEE VE, VO, VY		2	Barinas, Merida, Tachira
			3	Lara, Portuguesa, Yaracuy
V31, 2	Belize		4	Aragua, Carabobo, Cojedes
	The first letter after the number indicates the district		5	Federal District , Guarico, Miranda
A, B	Corozal		6	Anzoategui, Bolivar
C, D	Orange Walk		7	Nueva Esparta, Sucre
E-K	Belize		8	Delta Amacuro, Monagas
L, M	Stann Creek		9	Amazonas, Apure
N, O	Cayo		Ø	Islands (Coche, Cubagua, Islas de Aves, La
P, Q	Toledo			Blanquilla, Isla Orchila, La Tortuga, Los
				Hermanos, Los Monjes, Los Roques, Los Testigos)
XE, XF	Mexico			
XE1	Colima, Federal District , Guanajuato, Hidalgo,		ZA	Albania
	Jalisco, Mexico, Michoacan, Morelos, Nayarit,		1	Durres, Elbasan, Librazhd, Lushnje, Tirane
	Puebla, Queretaro, Tlaxcala, Veracruz		2	Berat, Fier, Gramsh, Korce, Pogradec, Skrapar
			3	Gjirokaster, Kolonje, Permet, Sarande, Tepelene,
XE2	Aguascalientes, Baja California, Baja California			Vlore
	Sur, Chihuahua, Coahuila, Durango, Nuevo Leon,		4	Dibre, Kruje, Lezhe, Mat, Mirdite
	San Luis Potosi, Sinaloa, Sonora, Tamaulipas,		5	Kukes, Puke, Shkoder, Tropoje
	Zacatecas			
XE3	Campeche, Chiapas, Guerrero, Oaxaca, Quintana		ZF	Cayman Islands
	Roo, Tabasco, Yucatan		1	Grand Cayman
XEØ	Special Callsigns		2	Guest Licenses
XF1	Pacific Islands North of 20° Latitude		8	Little Cayman Island
XF2	Islands in the Gulf of Mexico East of 90° Longitude		9	Cayman Brac Island
XF3	Caribbean Islands			
XF4	(Revillagigedo-Inseln) Pacific Islands South of 20°		ZP	Paraguay
	Latitude, including Revillagigedo		1	Boqueron, Chaco, Nueva Asuncion
			2	Alto Paraguay, Presidente Hayes
			3	Amambay, Concepcion
XQ	See CE, XQ		4	Canindeyu, San Pedro
			5	Asuncion
YB, YC,			6	Central, Cordillera, Paraguari
YD, YE,			7	Caaguazu, Caazapa, Guaira
YF, YG, YH	Indonesia		8	Misiones, Neembucu
1	West Java		9	Alto Parana, Itapua
2	Middle Java		Ø	Special Calls
3	East Java			
4	South Sumatra		ZR, ZS, ZU	Republic of South Africa (ZR VHF Licenses, ZU
5	West Sumatra			Novice Class)
6	North Sumatra		1	Cape Province (Western Sector)
7	Kalimantan (Borneo)		2	Cape Province (Eastern Sector)
8	Sulawesi (Celebes),Moluccas		3	Cape Province (North of 30° Latitude)
9	Bali, Irian Jaya, Nusa Tenggara Barat, Nusa		4	Orange Free State
	Tenggara Timur, Timor Timur		5	Natal
Ø	Jakarta		6	Transvaal
			7	Antarctic
YO	Romania		8	Prince Edward Island, Marion Island
2	Arad, Caras-Serverin, Hunedoara, Timis		Ø	Packet Bulletin Boards
3	Bucharest			
4	Braila, Constanta, Galati, Tulcea, Vrancea		ZV, ZW,	
5	Alba, Bihor, Bistrita-Nasaud, Cluj, Maramures,		ZX, ZY,	
	Salaj, Satu Mare		ZZ	See PP, PY
6	Brasov, Covasna, Harghita, Mures, Sibiu7 Arges,			
	Dolj, Gorj, Mehedinti, Olt, Vilcea		5N	Nigeria
8	Bacau, Botosani, Iasi, Neamt, Suceava, Vaslui		1	Ogun, Oyo
9	Buzau, Calarasi, Dimbovita, Giurgiu, Ialomita,		2	Kwara, Niger
	Prahova, Teleorman		3	Bendel, Ondo
Ø	Special Callsigns		4	Anambra, Rivers
			5	Akwa Ibom, Cross Rivers, Imo
			6	Benue, Plateau

7	Bauchi, Gongola		9H	Malta
8	Bornu, Kano		1	Malta
9	Kaduna, Katsina, Sokoto		3	Guest Licenses, Special Calls
Ø	Abuja, Lagos		4	Gozo
			5	VHF Licenses
6W	Senegal		8	Comino
1	Dakar			
2	Ziguinchor		9L	Sierra Leone
3	Diourbel		1	Western Region
4	Saint Louis		2	Northern Province
5	Tambacounda		3	Southern Province
6	Kaolack		4	Eastern Province
7	Thies		7	Novice Class
8	Louga		8	VHF License
9	Fatick			
Ø	Kolda			

7J, 7K,
7L, 7M,
7N, 8J   See JA

(Compiled from several sources, including Jahrbuch fur den Funk-Amateur, HB9DX, Editor)

THE NEW RUSSIAN OBLAST LIST

In the past, the Central Radio Club of the USSR issued various operating awards, including the R-100-O award for confirming QSOs with 100 "Oblasts," and endorsable through the ultimate achievement of confirming all 173 oblasts.

With the extreme changes in the political situation of the former USSR in the last few years, it is not surprising that many changes have occurred in Amateur Radio. One change is the formation of a new Russian amateur radio organization, the "SRR" (Russian-language abbreviation for the "Union of Radioamateurs of Russia").

The old list of numbered oblasts appears to now be obsolete. The new list of Russian-only Oblasts (with call sign blocks and official 2-letter abbreviation, replacing the old number system) is as follows:

Prefixes: RA, UA, UI, RV, RW, RZ, RN, RU, RX

Blocks	Oblast (Republic)	Abbr.
European Part of Russia		
1A, 1B	St.Petersburg	SP
1C, 1D	Leningradskaja obl.	LO
1N	Karelija Rep.	KL
1O	Archangelskaja obl.	AR
1P	Nenetskij AO	NO
1Q	Vologodskaja obl.	VO
1T	Novgorodskaja obl.	NV
1W	Pskovskaja obl.	PS
1Z	Murmanskaja obl.	MU
2F	Kaliningradskaja obl.	KA
3A, 3B	Moscow	MA
3D, 3F	Moscowskaja obl.	MO
3E	Orlovskaja obl.	OR
3G	Lipetskaja obl.	LP
3I	Tverskaja obl.	TV
3L	Smolenskaja obl.	SM
3M	Yaroslavskaja obl.	JA
3N	Kostpomskaja obl.	KS
3P	Tulskaja obl.	TL
3Q	Voronevzskaja obl.	VR
3R	Tambovskaja obl.	TB
3S	Rjazanskaja obl.	RA
3T	Nivzegorodskaja obl.	NN
3U	Ivanovskaja obl.	IV
3V	Vladimirskaja obl.	VL
3W	Kurskaja obl.	KU
3X	Kaluvzskaja obl.	KG

3Y	Brjanskaja obl.	BR
3Z	Belgorodskaja obl.	BO
4A	Volgogradskaja obl.	VG
4C	Saratovskaja obl.	SA
4F	Penzenskaja obl.	PE
4H	Samarskaja obl.	SR
4L	Uljanovskaja obl.	UL
4N	Kirovskaja obl.	KI
4P	Tatarstan Rep.	TA
4S	Mariy-El Rep.	MR
4U	Mordovskaja Rep.	MD
4W	Udmurtija Rep.	UD
4Y	Chuvashija Rep.	CU
6A, 6B	Krasnodar	KR
6E	Karacharovo-Cherkessija Rep.	KC
6H, 6F	Stavropol	ST
6I	Kalmyk Rep.	KM
6J	Severnaja Osetija Rep.	SO
6L, 6M	Rostovskaja obl.	RO
6P	Chechen Rep.	CN
6U	Astrachanskaja obl.	AO
6W	Dagestan Rep.	DA
6X	Kabardino-Balkarija Rep.	KB
6Y	Adygeja Rep.	AD
Asian part of Russia		
8T	Ust-Ordynskiy AO	UO
8V	Aginsky Burjatsky AO	AB
9A, 9B	Cheljabinskaja obl.	CB
9C, 9D	Sverdlovskaja obl.	SV
9F	Permskaja obl.	PM
9G	Komi-Permjatskiy AO	KP
9H	Tomskaja obl.	TO
9J	Chanty-Mansijsky AO	HM
9K	Yamalo-Nenetskiy AO	JN
9L, 9E	Tumenskaja obl.	TN
9M	Omskaja obl.	OM
9O	Novosibirskaja obl.	NS
9Q, 9R	Kurganskaja obl.	KN
9S	Orenburgskaja obl.	OB
9U	Kemerovskaja obl.	KE
9W	Bashkorstan Rep.	BA
9X	Komi Rep.	KO
9Y	Altaj	AL
9Z	Gorno-Altajskaja AO	GA
ØA	Krasnojarskiy Kr.	KK

ØB	Taymyrskiy AO	TM
ØC	Chabarovskiy Kr.	HK
ØD	Evrejskaj AO	EA
ØF	Sachalinskaja obl.	SL
ØH	Ewenkijskij AO	EW
ØI	Magadanskaja obl.	MG
ØJ	Amurskaja obl.	AM
ØK	Chukotka	CK
ØL	Primorskiy Kr.	PK
ØO	Burjatija Rep.	BU
ØQ	Sacha Rep.	YA
ØS	Irkutskaja obl.	IR
ØU	Chitinskaja obl.	CT
ØW	Chakassija Rep.	HA
ØX	Karjakskiy AO	KJ
ØY	Tuva Rep.	TU
ØZ	Kamchatka	KT

Total: 88

Source: UA3DPX

AMATEUR RADIO AROUND THE WORLD

Every country of the world has its own government regulations concerning amateur radio, including licensing requirements, restrictions on operation, etc. In addition, many countries' national organizations publish magazines or newsletters. This section of the *CQ Amateur Radio Almanac* describes amateur radio as it exists in several different countries.

Australia

The Wireless Institute of Australia (WIA) is the National Radio Society, and was founded in 1910, making it the world's oldest national amateur radio society. WIA provides a worldwide QSL service, weekly news broadcasts, license classes, sponsors awards, contests, and trophies, and publishes a monthly magazine called *Amateur Radio*. There are seven autonomous WIA divisions, each of which establishes its own membership fee (ranging from about $A60 to $A70). As of 1994, there were approximately 18,500 licensed amateurs in Australia.

Licenses are issued by the Spectrum Management Agency (SMA), and examinations are conducted by WIA-accredited amateurs. There are three license classes. The Novice license ("Novice Amateur Operator's Limited Certificate of Proficiency", or NAOLCP) requires a written exam and a Morse code test at 5 words per minute. It carries privileges in the 80, 15, 10, and 2 meter bands, both CW and voice modes at 10W mean power (30 W PEP) maximum. Novice callsigns use three letter suffixes beginning with L, M, N, P, or V. The next class of license, the Amateur Operator's Limited Certificate of Proficiency, or AOLCP, requires a more detailed technical exam but no Morse proficiency. It is a VHF-only license, and callsigns have three-letter suffixes beginning with T, U, X, Y, and Z. The full license, known as the Amateur Operator's Certificate of Proficiency, or AOCP, requires a 10 words-per-minute code test, and the same written exam as the Limited license. It carries full privileges up to 400 watts PEP on all amateur bands. The license fee is $A36 per year.

Foreign amateurs may obtain permission to operate quite simply. Visitors from countries with reciprocal agreements visiting for less than a year are issued temporary permits. Visitors intending to become residents are issued Australian Amateur permits. Visitors from countries without reciprocal agreements are issued an temporary (one-year, non-renewable) equivalent class license. Visitors from countries with no agreements and no equivalent license class will be issued a temporary (one-year, non-renewable) permit allowing operation on FM in the 146—148 MHz band with 10 W mean power.

VK amateurs have some restrictions on the use of several VHF/UHF bands, since they are shared with other services. For example,. 50—52 MHz is shared with television channel 0, and amateur radio is a secondary service. The 52—54 MHz segment is assigned Amateur Primary. There is no allocation at 222 MHz. However, they have the use of a band for Amateur Television (ATV) at 578—585 MHz (50 cm). This band was actually withdrawn from amateur use in 1989, however existing ATV repeaters are allowed continued use of the band until television channel 35 is required for local use. Approximately 11 such ATV stations are still operating.

In addition to *Amateur Radio*, another magazine, *Amateur Radio Action*, is published monthly, and several national electronics magazines also carry amateur radio columns.

(Courtesy WIA Federal Council - Donna Reilly, Manager)

Austria

The national society in Austria is the OEVSV (Oesterreichischer Versuchsenderverband). There are presently two license classes. The CEPT 1 Class permits all bands and modes and requires 12 wpm Morse. The CEPT 2 license is restricted to bands above 30 MHz and requires no Morse exam. There are four power classes (25, 100, and 400 W individual licenses and a 1000 W club class), and the annual fee is related to the power permitted, ranging from about $20 per year to $100 per year. Licenses are granted for life but must be renewed annually. Call signs all carry the OE prefix. The number after the prefix indicates the geographic district (1-9), and the suffix is one, two, or three letters. The OEØ prefix is used for Austrian radio amateurs operating in international areas (such as ships, or from Antarctica); only two or three of these call signs have ever been used. The one-letter suffixes are reserved for club stations and are usually used only for special events and contests. Suffixes beginning with "X" are also reserved for club stations, and suffixes beginning with "Z" are reserved for guest licenses (for foreigners staying a year or more). The CEPT license is honored for short visits. There is also a reciprocal agreement with the United States. Other visitors may apply for guest licenses, which are issued by four separate agencies depending on the area of the country of the proposed visit. Applications must include copy of home licenses, dates of visit, and main hotel of visit. The OEVSV (Theresiengasse 11, A-1180 Vienna, tel: 0222-408-55-35, fax: 0222-403-18-30) can provide complete details.

A third license class became effective in 1995. This is a "Newcomer-class," restricted to 70-cm operation only. There are presently approximately 5300 licensed amateurs in Austria in the present two classes. Third-party traffic is not permitted except in natural disasters and/or by special government approval. Contacts are not permitted with countries that do not permit amateur radio (according to the annually-issued ITU list).

The OEVSV operates a QSL bureau. Membership in the club is required. A monthly magazine, *QSP*, is published. The July 1994 issue was 62 pages in length. A national convention ("Amateurfunktage LAA") is held in early May in the small city of LAA, on the Czech border. Typical attendance is approximately 7000.

In 1995, new regulations are expected to come into effect—these will be the first major changes since 1955. In addition to the "Newcomer-class" license, the amateurs are expected to become part of the examination process (along the lines of the U.S. Volunteer Examiner program). Also, in 1996, Austria will celebrate its 1000-year anniversary. Austrian amateurs will celebrate with the use of the "OEM" call sign block ("M" for 1000). Austrian amateurs will optionally convert their call signs for 1996 according to the example that OE3REB will sign OEM3REB. Special awards will be issued for contacting OEM stations. Previously, the OEM prefix block was assigned to the Austrian weather service.

(Courtesy Ronald Eisenwagner, OE3REB, President - OEVSV, and Director of the Weather Service, Ministry of Defense.)

Czech Republic

In the Czech Republic, Amateur Radio is regulated under Decree 390 of the Federal Ministry of Communications, dated 23rd June 1992. The minimum age for an individual license is 15 years old. In addition to individuals, other "legal entities" with headquarters in the territory of the Czech and Slovak Federal Republic may receive amateur licenses. The legal entity is obliged to designate the chief operator who is responsible for the operation of the amateur station. The chief operator must have a license for an amateur station. Licenses are valid for five years and are renewable. The prefixes are as follows:

OK1 - Bohemia
OK2 - Moravia
OK3 - not used now
OK4 - usually Maritime Mobile stations
OK5 - special events
OK6 - contests
OK7 - not used now
OK8 - foreigners
OK9 - experimentals
OKØ - repeaters, beacons, special events
OL - special events, contests

Amateur radio examinations take place at the request of the applicant which must be connected with an application for an amateur station license. To the application is attached a document on the prescribed practice and proof of payment of examination fees. The applicant must pay to the Federal Ministry of Communications the costs of the examination, set at the overall sum of 100 Kfs. The examination takes place within a period of six months from the sending in of the application, usually in the headquarters of the Examination Commission. The applicant will be informed as to the time and place of the examination at least one week in advance. Before the start of the examination the applicant must prove his identity and present the document on the prescribed practical experience. The examination is not public. Apart from the members of the Examination Commission there may be present at the examination only persons whose presence the Chairman of the Examination Commission permits in substantiated cases. The Chairman of the

Examination Commission directs examinations and follows the testing in individual fields, during which he may ask applicants supplementary questions. A candidate who has successfully passed the examination must take an oath before he receives his certificate that he will observe telecommunications secrecy. The promise to observe telecommunications secrecy goes as follows: "I promise that, as an operator, I shall always observe telecommunications secrecy according to valid legal provisions."

There are four classes of licenses designated by the letters D, C, B, and A:

(1) Class D - Age at least 15 years, completed basic education and the showing of basic knowledge in the following examination subjects:

a) Legal regulations on the establishment, operation and keeping of amateur stations (basic provisions) and the licensing conditions for amateur stations.

b) Basic principles of electrotechnology and radiotechnology, basic types of antennas, their use and safety regulations.

c) Radio amateur operating rules.

(2) Class C - Age at least 15 years, completed basic education and the showing of basic knowledge in the following examination subjects:

a) Legal regulations on the establishment, operation and keeping of amateur stations and the licensing conditions for amateur stations.

b) Basic principles of electrotechnology and radiotechnology, basic types of antennas, their use and safety regulations.

c) Radio amateur operating rules.

d) The telegraphic alphabet (rate of at least 40 symbols a minute with three minutes transmission and three minutes receiving).

(3) Class B - At least one year of practical experience as an operator in Class C or D and the making of at least 500 radio amateur connections (apart from repeaters and competitions) and also the showing of detailed knowledge in the following examination subjects:

a) Legal regulations on the establishment, operation and keeping of amateur stations and the licensing conditions of amateur stations.

b) Radio amateur operating rules.

c) The telegraphic alphabet (rate of at least 80 symbols a minute with three minutes transmission and three minutes receiving).

(4) Class A - Two years practice in Class B and the making of at least 3000 radio amateur connections (apart from convertor and competitions).

Operators of Class A may operate a transmitter with an output up to 750 W. Operation with an output over 300 W must be announced to the licensing body. The provision also applies for the stations of legal entities. Operators of Class B may operate a transmitter with output up to 300 W. Class A and B operators may operate on all HF and VHF/UHF amateur bands. Operators of Class C may operate a transmitter with output up to 100 W, and are restricted to segments of the 160, 80, 30, 15, and 10-meter HF bands and all VHF/UHF bands. Operators of Class D may operate a transmitter with output up to 100 W on VHF/UHF bands only. Foreigners holding amateur radio licenses may be allowed to apply for a Czech amateur radio license. Send an application to:

Czech Telecommunication office, Amateur Radio Licenses, Klimentska 27, 125 02 PRAHA 1, Czech

Republic, Phone: +42 2 24911605. It is also possible to operate in Czech Republic according to CEPT Recommendation T/R 61-01.

Number of licensed amateurs at June 1st, 1994 in Czech Republic:

Class	A	B	C	D	Clubs
OK1	416	865	844	946	259
OK2	197	459	493	468	158
Total	613	1324	1337	1414	417

Total number of individual licenses 4687.

The Czech Radio Club operates a QSL Bureau, free of charge, for the members of the Czech Radio Club and also open to non-members. The magazine of The Czech Radio Club is *AMA - Magazine*. Exact subscription information is available from the publisher who is OK2FD. There is also a magazine called *Amaterske radio*. One year subscription is $38 by surface mail or $55 by air mail. The address of the publisher is: MAGNET - PRESS, OZO. 312, Vladislavova 26, 113 66 PRAHA 1, Czech Republic.

There is an International Hamvention in Holice (located 100 km east of Praha) every year on the second weekend in September.

(Courtesy OK1MP, President, Czech Radio Club.)

Ecuador

There are two license classes in Ecuador: Novice and General. The Novice has to pass an exam that covers basic knowledge of the various technical and operating subjects and a basic knowledge of Morse code. The Novice license is valid for one year. To ascend to General class, the operator has to pass an exam of both technical and operating subjects. The Novice ready to ascend must present 25 QSLs confirming contacts within the country and 25 QSLs confirming contacts outside the country, all made on 40 and 80 meters within two years of his initial licensing date. The General license is valid for three years. The license fee is the equivalent of $2.50 per year.

The licensing authority is the Superintendencia de Telecommunicationes, 9 de Octubre y Berlin, Quito (Phone: 5932 560700). Reciprocal agreements exist with the U.S. and several other countries, but the procedure effectively precludes casual operation by tourists. The licensing procedure requires sending copies of your original license and passport pages with personal information, three photos (1.5 x 1.5 cm, preferably color), and a certificate from the President or Secretary of the IARU member society of your country three months in advance of any planned operation. Approval must be issued by both the licensing authority and the military authorities, which takes anywhere from a week to three months. A personal visit to the authority to pay the fee and receive the license is also required.

The CEPT license is not yet honored in Ecuador, but the administration and the national IARU Society are working on incorporating Ecuadorian amateurs to the CEPT agreement. There are 2,250 licensed amateurs in Ecuador. Third party traffic is allowed as long as there is no intentional competition with the State telephone company. There are no restricted countries.

Call districts each encompass several provinces as follows:

HC1 - Carchi, Imbabura, and Pichincha
HC2 - Guayas and Los Rics

HC3 - El Oro and Loja
HC4 - Esmeraldas and Manabi
HC5 - Azuay, Canar, and Chimborazo
HC6 - Bolivar, Cotopaxi, and Tungurahua
HC7 - Morona Santiago, Napo, Pastaza, Sucumbios, and Zamora
HC8 - Galapagos
HC9 - formerly assigned to mobiles, but this practice was discontinued
HCØ - assigned to special call signs for contests and celebrations

The HD prefix is sometimes used for special and temporary stations, such as contests and special emissions.

The national IARU Member-Society is the Guayaquil Radio Club, which also operates the QSL bureau for all amateurs, whether members or not. No national magazines are published. In 1993, the GRC celebrated its 70th anniversary. Two of the founding members were able to attend the festivities. The National Government gave the National Order of Merit to the GRC's flag. The Director of Civil Defense also gave the Flag a decoration.

Prominent Ecuadorians holding amateur licenses include the Minister for Agriculture, Mariano Gonzales, HC2XX; Ambassador to Belgium Luis Orrantia, HC2LO; Ambassador to France Santiago Maspons, HC2VO; and well-known musician/composer Alfredo Solines HC2SL, who is also one of the country's best CW operators.

(Courtesy HC2RB, President - GRC, and HC2EE, Secretary - GRC.)

Fiji

The entire Fiji island group, including Rotuma and Conway Reef, is assigned the 3D2 prefix. Suffixes are the invention of the license holder. There are two license classes, a General Class and a non-Morse Technician license for bands above 30 MHz. Third-party traffic and phone patches are not permitted. The licensing authority is the Ministry of Information, Broadcasting and Telecommunications, P.O. Box 2225, Suva, Fiji. The fee is 13.75 Fijian dollars (approximately US $10) per year. There are no reciprocal licensing agreements, but licenses are routinely granted to visiting amateurs who produce a valid home license, passport, and a list of equipment with serial numbers. There are approximately 15 resident hams, six of whom are presently active. They can usually be found on the Pacific DX net on 14.222 MHz. Some are active on 6 meters when it is open, and some are active on 2 meters. A few use HF packet (3D2CM even experiments on 10 GHz). There is also a local 80 meter net once a week.

Visitors to Fiji have frequently stayed at the Turtle Island hotel, which has no objections to Amateur Radio. The QSL bureau accepts incoming cards only; however, most residents prefer to receive cards direct with SAE and return postage (a U.S. dollar is sufficient). QSLs sent to the bureau for DXpeditions will not receive answers.

(Courtesy of Dick Northcott, 3D2CM.)

Finland

There are 5777 licensed radio amateurs in Finland. There are four classes of license, which break down as follows:

Telecommunication class (Modules K and T1) - 432 MHz to 438 MHz. Low power.

Novice class (Modules K, T1 and T2) - Limited HF, 2 meter and 70 cm. Low power.

Technical class (Modules K, T2) - 6 meter and up. High power.

General class (Modules K, T2 and CW2) - Highest possible. All bands. High power.

The modules break down as follows:

K - module includes questions of regulations, radio law and amateur radio traffic including emergency traffic.

T1 - module includes questions of basic electronics and radio theory.

T2 - module includes specific questions of designing radio circuits and electronic circuits, knowledge of interference and how to deal with it.

CW1 - module includes both receiving and transmitting of Morse code at 8 wpm.

CW2 - module includes both receiving and transmitting of Morse code at 12 wpm. Before entering the CW2-module test you have to hold 300 certified CW-QSOs.

Licenses are issued by Telehallintokeskus, Radioamatooriluvat, P.O.Box. 53, FIN-00211 Helsinki, Finland. Tel: +358-0-69 661

The CEPT-license is honored and holders are allowed to operate in Finland by following the CEPT and national regulations for a period of three months. All non CEPT-licensees have to apply for a short time visitors license which will be effective for a maximum of three months. This costs FIM 80, payable in Finland (no checks in advance). The processing time for a temporary license takes about 60 days. For more information or an application form call or write to the league's office.

Callsign areas are as follows:

OH1 - district of Turku and Pori
OH2 - district of Uusimaa
OH3 - district of Hame
OH4 - district of Mikkeli
OH5 - district of Kymi
OH6 - districts of Vaasa and Central Finland
OH7 - districts of Kuopio and North Karelia
OH8 - district of Oulu
OH9 - district of Lappi
OHØ - The Aland Islands
OJØ and OHØM - Market Reef
OJ9 - Antarctica

The QSL-bureau address is: SRAL QSL-service, FIN 11311 Riihimaki, Finland or SRAL-QSL buro, P.O.Box 30, FIN-00381 Helsinki. The service is free for members. Non members may buy the service. The SRAL magazine is called RADIOAMATOORI. 12 issues per year. Subscriptions via the league office. A Winter Ham Vention is held annually.*(Courtesy of Norbert Kelzenberg, OH2AUM, Assistant Secretary - SRAL.)*

Greece

There are three license classes in Greece. All have the same frequency privileges, and differ only in power permitted. The "A" license allows 300 W, the "B" is 150 W, and the "C" license allows 50 W. Power is measured as an average at the output connector of the transmitter. The fee is approximately $25, and licenses are valid for three years. Minimum age is 18 years, and license examinations are conducted twice per year. All three license classes require knowledge of the Morse code. No third-party traffic is permitted. There are no restricted countries. Fixed, portable, mobile, and maritime-mobile operation are permitted, but not aeronautical-mobile.

The licensing authority is the Ministry of Transport and Communications, Directorate of Communication Technique, Department of Electronic Applications, 49 Syngrou Ave., GR-11780 Athens, Greece. Foreign amateurs holding licenses in compliance with CEPT Recommendation T/R 61-01 may operate for up to three months without applying for a license. There are currently approximately 2600 licensed amateurs in Greece, including aliens.

The usual prefix for radio amateurs in Greece is SV. The additional prefixes SW, SX, SY, SZ, and J4 are given occasionally to special event stations. Call areas are numbered 1 though 9, as follows:

SV1 - Central Greece
SV2 - Makedonia
SV3 - Peloponisos
SV4 - Thessalia
SV5 - Dodecanisos
SV6 - Ipiros
SV7 - East Makedonia, Thraki
SV8 - Greek Islands
SV9 - Kriti (Crete)

Call signs are issued for life, and an amateur who is initially licensed in one area and later moves to another must sign his original call followed by the new call area (e.g. SV1RL/5), or else apply for a change to the number of the call sign. No other change to a call sign is permitted, and the call sign is not given to any person after the death of the initial holder. Suffixes are one, two, or three letters. The one-letter suffixes are given for repeater stations only. Only three-letter suffixes have been issued since 1987, since all two-letter suffixes were exhausted. The call signs are not related to the license classes.

Operation of a radio amateur station within the district of Athos (Mount Athos), in addition to the normal rules and regulations concerning operation of foreign radio amateurs in Greece, is subject to the official written permission of the local administration of this district.

The SVØ prefix is reserved for alien operators who fulfill the conditions to possess a Greek amateur radio license. Again, the call sign is given for life, no change is permitted in any case, and the call sign cannot be reissued to any person after the death or departure of the first holder.

The Radio Amateur Association of Greece ("RAAG" in English, "EEP" in Greek) is the national society. It operates a QSL bureau, open to both members and nonmembers. RAAG members receive a free subscription to the bimonthly magazine, *SV-NEA* (SV-News), in Greek. The 40-page, 4-color glossy May-June 1994 issue featured an article on the Athens Hamfest 94, held May 15.
(Courtesy of SV1RL, Secretary, RAAG.)

Israel

The National Society in Israel is the Israel Amateur Radio Club (I.A.R.C.). It was founded in 1948. The address is I.A.R.C. Headquarters, P.O. Box 17600, Tel-Aviv, 61176.

At present there are four license classes:

A: Advanced - Extra, output power 1500 Watt
B: General, output power 150 Watt

C: Novice, restricted on HF band, requires 6 wpm Morse code

D: Technical, bands above 30 MHz.

The prefix indicates the license class as follows:

4X1	+ 2 letter suffix	Advanced - Extra
4Z1	+ 2 letter suffix	Advanced - Extra
4X4	+ 2 letter suffix	General
4X6	+ 2 letter suffix	General
4Z4	+ 2 letter suffix	General
4Z5	+ 2 letter suffix	General
4Z7	+ 3 letter suffix	Technical
4Z9	+ 3 letter suffix	Novice

Israel has implemented the CEPT Recommendation T/R. 61-01 as a Non-European country. CEPT class 1 permits all bands and modes and requires 12 wpm Morse code. CEPT class 2 is restricted to bands above 30 MHz.

Visitors from countries which have not implemented the CEPT Recommendation T/R. 61-01 can apply for a guest license. Allow sufficient time for application and processing before arriving. More information or application forms can be obtained by writing to the I.A.R.C. Headquarters.

The I.A.R.C. issues several Awards, the most precious is the Holyland Award. The I.A.R.C. sponsors the Annual Holyland DX Contest every April around Easter.

(Courtesy Joseph Obstfeld, 4X6KJ, President - IARC)

Kyrghyzstan

Kyrghyzstan is one of the former Republics of the USSR in central Asia. The ARUK (Amateur Radio Union of Kyrghyzstan) was founded May 17,1994. The President is Ivan Udovin EX2A (ex-UM8MBO) and Vice-President is George Galkin, EX8MAT (ex-UM8MAT). ARUK operates a QSL bureau at Box 1100, Bishkek, 720020, Kyrghyzstan.

There are four classes of license. The first letter of the suffix of Class 1, 2, and 3 licenses indicates the oblast, and is P, Q, N, V, T, or M. Extra Class call signs end in a single letter which does not always indicate the oblast.

Class	Suffix	Number	Example
3	3 Letters	8	EX8MAO
2	3 Letters	8	EX8MAT
1	2 Letters	6, 7, 8	EX7MM
Extra	1 Letters	2, 8, Ø	EXØV

Club stations use the EX9 prefix and three letter suffixes. EX9HQ is the ARUK Central club station.

There are 190 hams in Kyrghystan, as follows:

EX8P - 0
EX8Q - 10
EX8N, V - 25
EX8M - 154
EX8T - 1

Foreign stations are assigned call signs of EX/HOME-CALL without number.

The Class 3 license allows operation on 160 meters with 5 W power on CW, AM and SSB; 80 meter CW only with 10 W, and 10 meter CW, SSB, and AM with 10 W. The Class 2 license allows all modes on 160, 80, 40, 15, and 10 meters with 50 W maximum power (5 W on 160), and full VHF privileges. The Class 1 license increases the power on most bands to 250 W (except 10 W on 160, and 100 on 30 meters), and includes privileges

for all HF and VHF bands. The Extra class license increases power to 1 kW on all bands (except 10 W on 160 and 250 W on 30 meters).

The Class 3 license can be obtained by a short interview. The Class 2 license requires 50 letters per minute Morse and a technical exam. The Class 1 license requires 80 letters per minute Morse and 150 DXCC countries confirmed. The Extra class license requires 250 DXCC countries confirmed and "high activity in contests." Licenses are issued by the ARUK.

(Courtesy Yuri Minin, EXØA.)

Lebanon

There is a single license class in Lebanon. CW privileges are available to those who pass a CW exam (which is not compulsory). The licensing procedure begins with an application to an examination, including a standard application form, identity card (passport for foreigners), two passport-size photos, proof of stay in Lebanon for foreigners, and a fee of 3000 Lebanese pounds (approximately $2). Examinations are held twice every year, depending on the number of applicants. Both Lebanese nationals and foreigners must complete the exam. Those who pass the exam then apply for a station license and call sign, which usually takes six months. The license is issued for life, and an annual fee of 53,000 Lebanese pounds (about $32) must be paid.

Licenses are issued by the Directorate of Telecommunications at the Ministry of Posts and Telecommunications, with the approval of the Directorate of the Interior and Defense ministries. The CEPT license scheme has not been adopted. Third party traffic is not permitted, and contacts with stations in Israel are not permitted. Since 1952, 352 licenses have been issued. The total number of amateurs in Lebanon on June 1 of each of the last few years has been:

1992 - 40
1993 - 87
1994 - 147

The QSL bureau is free to all amateurs and handles both incoming and outgoing cards. There are no national magazines—a single attempt was made in 1992 with no success. In November 1993, Lebanese amateurs signed the prefix OD5Ø, followed by their normal suffix to celebrate the 50th Anniversary of Lebanon's independence. A certificate will be issued to those who worked five or 10 OD5Ø stations.

(Courtesy Aref N. Mansour, President, RAL.)

Luxembourg

The Grand Duchy of Luxembourg offers a single class of amateur radio license. The regulations were established for private LX radio stations on May 22, 1950. The present license is equivalent to the CEPT Class 1, and permits use of all amateur frequency bands. Minimum age is 18 years old, and no previous experience is required. Maximum power permitted for the first year is 35 W, then 100 W by request. The fee is 210 FLux for the 35 W license, and 420 FLux for the 100 W class. There is only one call area (LX), and the number after the prefix is as follows:

LX1 - Luxemburgers
LX2 - Foreigners living in Luxembourg
LX9 - Club Stations
LXØ - Official and Amateur Stations

Licenses are issued by the Direction des P&T, Service des Radiocommunications, L-2020, Luxembourg. CEPT licenses are honored. A letter of application is required to participate in an examination. As of June 1994, there were 464 licensed amateurs in Luxembourg: 386 LX1 and 78 LX2. Third party traffic is not allowed, and there are no restricted countries. The national society is the Reseau Luxembourgeois des Amateurs d'Ondes Courtes. A quarterly *Bulletin du R. L.* is published for members of the Reseau Luxembourgeois.

(Courtesy LX1KJ, Secretary, R.L.)

Oman

There is a single license class in the Sultanate of Oman. It requires a written examination and Morse code test at 12 wpm. Minimum age is 16, and privileges include 150 W power on all amateur bands except 10 and 50 MHz. Licenses are valid for one year, and the fee is 35 Omani Rials (approximately $90). Licenses are issued by the Royal Omani Amateur Radio Society (R.O.A.R.S.) and the overall controlling authority is the Ministry of Posts, Telegraphs, and Telephones. There are no reciprocal agreements with any country and no third party traffic agreements. As of June 1, 1994, there were 92 licensed amateurs.

Prefix allocations are as follows:

A41AA-A41ZZ - Local Omani Amateur Radio Stations

A42AA-A42ZZ - Reserved

A43AA-A43ZZ - Special Event Amateur Radio Stations

A45AA-A45ZZ - Expatriates and Visitors Amateur Radio Stations

A47AA-A47ZZ - Club Stations

There are several prominent Omani government officials who hold amateur licenses. These include:

HM Sultan Qaboos Bin Said, A41AA, Sultan of Oman

HH Sayyid Thuwainy Bin Shihab Al-Said, A41AB (Personal Representative to HM The Sultan)

HH Sayyid Thuwainy Bin Saif Al-Said, A41AC (Royal Family)

HE Noor Mohammed Abdulrahman, A41BB (Chief Executive of Telecommunications)

HE Abdulaziz Saud Hareb Al-Busaidi, A41BC (Undersecretary)

HE Bader Saud Hareb Al-Busaidi, A41BD (Minister)

HE Abdulla Hamed Ahmed Al-Busaidi, A41BE (Undersecretary)

HE Ahmed Suwaidan Al-Balushi, A41FK (Minister of P, T, &T; President - Royal Omani ARS)

HE Sayyid Khalid Bin Hilal Saud Al-Busaidi, A41KM (Undersecretary)

HE Sayyid Qahtan Bin Yaarub Al-Busaidi, A41KZ (Undersecretary)

(Courtesy A41KB, Acting Secretary General, R.O.A.R.S.)

Mauritius

Licenses in Mauritius are issued by the Chairman, Mauritius Telecommunications Authority, 6th Floor, Blendax House, Port Louis, Mauritius. There are apparently no reciprocal license agreements, but in 1994, the Authority has once again begun to issue licenses to visitors. Examinations are administered by the City and Guilds of London Institute. The Morse test at 12 wpm is given by the Mauritius Amateur Radio Society. Applicants for amateur licenses must be at least 18 years of age, and the license fee is 250 Mauritian Rupees (about $15) per year. Power is 400 W PEP maximum, and all amateur bands except 6 meters are permitted. Call signs begin with the prefix 3B8. The prefixes 3B6, 3B7, and 3B9 are used for operation on Agalega, St. Brandon, and Rodrigues Islands, respectively. The CEPT license is nominally accepted, but not automatic. A prospective visitor with a CEPT license must apply well in advance to allow the local secret service to perform security checks.

As of July 1994, there were approximately 40 licensed amateurs in Mauritius. The most recent call signs issued were 3B8 GD, GE, and GF. The Mauritius Amateur Radio Society operates a QSL bureau, but most active 3B8 amateurs prefer to receive their cards direct via the Callbook address

Prior to World War II, Amateur Radio existed in Mauritius and the Dependencies (including at the time the Chagos archipelago). At one point, call signs which should have belonged to the amateur service (VQ8AA and VQ8AB) were issued to two lighthouses. Protests from the international amateur radio community convinced the authorities to change the call signs.

As a consequence of the international events taking place at the time, war was officially declared in Mauritius on September 3, 1939. The authorities seized all transmitters on September 1 and 2. Some of the 17 amateurs were prosecuted for ownership of amateur equipment during wartime. However, it was noted that war had not been declared officially until after the confiscations, so the amateurs were merely fined 5 rupees each. A few weeks later, the authorities were asking the amateurs for help in providing communications services for the government since there was a shortage of trained operators capable of sending and receiving CW messages. One amateur, VQ8AD, was asked to sail to Diego Garcia to establish a station from which he could report any unusual shipping activities in the Chagos back to Mauritius. Through several links, messages were successfully transmitted regularly to the Police Station in Beau Bassin.

(Courtesy "Jacky" Mandary, 3B8CF.)

Monaco

Monaco is the smallest country in Europe, with 27,000 inhabitants in 1.97 km^2 (about 5 square miles). There are 55 hams in the country. Licenses are issued by the Direction des Telecommunications, 35 Bvd de Suisse, Monaco cedex. The CEPT license is recognized for foreigners. For non-CEPT licensees, reciprocal licenses are available without examination by supplying a copy of the passport and home-country's license.

There are two license classes. The technical exam is identical for the two classes, and is identical to the French class D & E. The 3A1 prefix is used for the VHF-only license, and a 10 wpm code test is required for the 3A2 prefix, which includes all IARU Region 1 bands (except 50 MHz). Licensees must be at least 16 years old, and the annual fee is 73 French Francs (about $15). Third-party traffic is not permitted, but there are no restricted countries. There is one peculiar restriction regarding mobile operation under CEPT rules—Monaco is surrounded by the French town Beausoleil, and only inhabitants know the exact location of the border. It is necessary to check in with the Direction des Telecom to determine the exact boundary to prevent inadvertent operation in the wrong country!

The national society is the Association des Radio-Amateurs de Monaco ("A.R.M."). The A.R.M. was founded in 1953, and celebrated its 40th anniversary in 1993 by inaugurating a new club house. Members are permitted free use of the QSL bureau, but cards for non-members are returned to the sender. Approximately 35 kilograms (77 pounds) of cards are sent each year. The A.R.M. is concerned about the behavior of some DXpeditions to the country, and encourages visiting hams to behave in "true ham spirit."

(Courtesy of 3A2AH.)

Netherlands

In the Netherlands there are no regional call sign assignments. The prefix indicates the license class or type of special licensed station as follows:

PAØ with 2 or 3 letters: In general A-license. Letters own choice. But till 1975 also possible for B & C licenses. Issued till 1975

PA1 with 2 or 3 letters: A-licenses held by officials, like PTT-employees. Letters own choice No longer being issued.

PA2 with 3 letters: A-license. Letters own choice. Only for people upgrading from PE -calls, issued between 1975 and 1977. (See also PEØ.)

PA3 with 3 letters: A-license since 1977. Letters issued in alphabetical order.

PA6 with 2 or 3 letters: Special call signs, like contest station PA6DX, PA6WW, etc. Letters own choice.

PBØ with 3 letters: B-license since 1981. Letters issued in alphabetical order.

PDØ with 3 Letters: D-license since 1976. Letters issued in alphabetical order.

PEØ with 3 Letters: C-license, issued between 1975 and 1977. Letters own choice. Upgrading direct to A-license one receives a PA2 call sign with the same suffix. Upgrading to B-license one is issued a call sign in alphabetical order. Your own suffix is lost when upgrading to or via B-license.

PE1 with 3 Letters: C-license since 1977. Letters issued in alphabetical order.

Special Prefixes:

PI1 - Till 1990 Licenses for educational institutions, like universities, etc. No longer issued. Since 1990 Packet node stations.

PI2 - FM-repeaters 70 centimeter band.

PI3 - FM-repeaters 2 meter band.

PI4 - A-license. Club stations.

PI5 - A- or C-license. Issued to educational institutions, schools, universities, etc.

PI6 - FM-repeaters in 23 centimeter band. Crossband repeaters. ATV-repeaters.

PI7 - Unattended beacons.

PI8 - Unattended mailboxes, digipeaters, gateway stations.

There are four license classes.

Class A - All bands, All modes, max. power: 100 watt or 400 watt PEP output. Annual fee is Hfl. 92.

Examination requirements include theory; regulations and Morse Code, sending by hand and receiving by ear, at a speed of 12 wpm.

Class B - HF: part of 3.5 & 21 MHz CW, 28 MHz All modes; power as A-license. VHF: All bands above 30 MHz; Max. power 30 watt or 120 watt PEP. output. Annual fee is Hfl. 82.

Examination requirements include theory; regulations and Morse Code, sending by hand and receiving by ear, at a speed of 8 wpm.

Class C - All bands above 30 MHz, All modes, max. power 30 watt or 120 watt PEP. output.

Examination requirements include theory and regulations only. No Morse code requirement. Annual fee is Hfl. 77.

Note: The theory and regulations examination for the Classes A, B & C is of the same technical standard.

Class D - Part of 144 MHz band, FM-only, max. power 15 watt output.

Examination requirements include theory and regulations, but at a lower technical standard than the previous mentioned classes; No Morse code requirement. Annual fee is Hfl. 57.

There is no minimum age limit in the Netherlands. Licenses have no time limit. However, one has to pay every year to receive a valid registration document. This document shows proof that one is the holder of a valid radio amateur license. If you do not pay you lose your license. Annual fee for club stations is Hfl. 103, and for educational institutes Hfl. 103. Third party traffic is not allowed under any circumstances.

Licenses are issued by the Ministry of Transport, Public Works and Water Management, Telecommunications and Post Department, P.O.Box 450, NL-9700 AL, Groningen, the Netherlands. The Dutch Administration has implemented CEPT rec. T/R 61-01 and T/R 61-02. The same address must be used to apply for information concerning the examinations. As of June 1994, the number of licensed amateurs was:

Class A - 5797
Class B - 103
Class C - 6100
Class D - 2815
Total - 14815

Incoming cards for members received by the Dutch QSL Bureau are distributed via regional or local QSL-managers. Incoming cards for non-members are obtainable via SASE from the DQB. Outgoing QSL-service for members only.

The national society for radio amateurs in the Netherlands is the Vereniging voor Experimenteel Radio Onderzoek in Nederland (VERON), Postbus 1166, NL-6801 BD Arnhem, Netherlands. Tel.: ++31 85 426760. VERON issues a monthly magazine named *Electron.* This is sent to every full member. For "specialists," VERON issues a weekly bulletin *DXPRESS/VHF-Bulletin.* Subscriptions are open to every amateur. Information via the Centraal Bureau VERON, P.O.Box 1166, NL-6801 BD Arnhem, the Netherlands.

VERON sponsors the annual PACC Contest on HF-bands on the second full weekend of February. The PA-Bekercontest (Beker = Cup) is held in November. This is a national contest for Dutch stations only, organized on 7 and 3.5 MHz, and is comparable with ARRL Sweepstakes.

(Courtesy of A. Jaap Dijkshoorn, PAØTO, IARU Liaison Officer - VERON.)

Panama

In Panama, there are three license classes. The Class C license requires a theory examination only (no Morse code) and permits operation on 80, 40, and 2 meters with 100 W maximum power. The Class C license is valid for one year, non-renewable. The Class B license requires a theory examination and Morse code sending and receiving ability. This license is valid for four years and is renewable. It allows operation on all bands with 500 W. The Class A license requires at least four years of experience with the Class B license and a theory examination. It allows power up to 1000 W on all bands, is valid for four years and is renewable.

The government agency responsible for issuing licenses is Lic. Carlos Raul Trujillo, Vice Ministro de Gobierno y Justicia y Presidente de la Junta nacional de Radio Aficionados, Direccion Nacional de Communicacion Social, P.O. Box 1628, Panama 1, Panama.

The total number of licensed amateurs in Panama in 1993 was 1,655; in 1994 it grew to 1,883. The breakdown by call area is as follows:

HP1 - 1,161
HP2 - 171
HP3 - 114
HP4 - 39
HP5 - 99
HP6 - 101
HP7 - 82
HP8 - 34
HP9 - 52

Prominent Panamanians holding amateur licenses include Archbishop Jose Dimas Cedeno, HP6JDC, and Monsignor Romulo Emiliani, HP7DHK.

The national society is the Liga Panamena de Radio Aficionados, founded in 1978. A QSL bureau is operated, and national contests are held on HF and VHF. A newsletter, *El Corto Circuito* (The Short Circuit) is published. The LPRA also holds club call signs in each call area, HP1LR through HP9LR. Many of these licenses are used for repeaters on 2-meter FM. There are approximately 40 repeaters in the country.

(Courtesy HP1BUM, Presidente, LPRA.)

San Marino

San Marino is a small country—only 61 square kilometers (about 24 square miles), with a population of 23,000. There are two classes of amateur radio license: Class A (all bands, HF and VHF/UHF) and Class B (VHF and above only). Both licenses are allowed 1 KW of power. A personal license is granted for 6 meters only with 40 W of power. Licensees are required to pass an examination following the guidelines of the HAREC (Harmonized Amateur Radio Examination Certificate). There is no minimum age requirement for either license. Class A licensees are issued call signs with the prefix T77; Class B holders are issued the T72 prefix. Before April 20, 1983, the unofficial prefix "M1" was used. License fee is approximately $3.25 per year, and the licensing authority is the Direzione Generale Poste e Telecommunicazioni, Contrada Omerelli, 6, 47031 San Marino (fax: +378 992760; tel: +378 882555). The CEPT license is not yet implemented, and non-residents are not allowed to take the license.

As of June 1, 1994, there were 75 licenses issued: 24 Class A, 49 Class B, 1 Club, and 1 Silent Key.

The national society is the Amateure Radio of the Republic of San Marino (ARRSM). It operates a QSL bureau open to any San Marino amateurs.

(Courtesy T77J, IARU Liaison Officer, ARRSM.)

South Africa

There are three license classes in South Africa. The Novice Class requires an examination of elementary radio and electricity, Q-signals, Radio Regulations, and operating procedure. In addition, a 5 wpm Morse code test is required. Holders of the Novice class license are issued call signs with the ZU1 prefix, independent of geographic location within the country. Minimum age is 12, and the license permits a maximum power of 5 W input to the final stage.

The Restricted Class license provides privileges from 50 MHz upwards. The examination requirements are identical to the CEPT licenses, with no Morse code test required. Minimum age is 16, and maximum power permitted is 150 W dc or 400 W PEP. Prefixes are ZR1 through ZR6, depending on station location.

The Unrestricted Class License requires the same examination as the Restricted Class, but with the addition of a 12 wpm Morse code test. Prefix for this class is ZS.

A total of 5000 amateurs are licensed in South Africa. The call areas are:

ZR1/ZS1 - Western Cape province
ZR2/ZS2 - Eastern Cape province
ZR3/ZS3 - Northern Cape province
ZR4/ZS4 - Orange Free State
ZR5/ZS5 - Natal
ZR6/ZS6 - Transvaal

ZS7 is used for stations in the Antarctic, and ZS8 is used on Marion Island. The ZS9 prefix was previously used in Walvis Bay, but that territory has been transferred to Namibia. Unattended Packet Radio Bulletin Boards are licensed as repeaters and are issued ZSØ-prefixed call signs. Voice repeaters are issued ZU7 and ZU8 call signs.

Licenses are renewable annually. The fee is adjusted annually, and was 24 rands in 1994. Examinations are held in May and November every year, and the South African Radio League (SARL) has taken over the duties of administering licenses, effective November 1994. The Amateur Radio Examination Certificate is prepared by the SARL, and certified by the Department of Posts and Telecommunications. Applications for all licenses, including repeaters, digipeaters, special licenses, and guest licenses must be routed through the SARL. Guest licenses are available to all licensed amateurs, regardless of whether a reciprocal licensing agreement exists between the guest's home country and South Africa. The application must include a copy of the amateur's current home license, copy of passport, and intended address while visiting South Africa. The fee is $25, and a license equivalent to the home license will be granted.

Third-party traffic is not permitted. The only situation where unlicensed persons may operate a station is the case of a club station, where an unlicensed person may operate under supervision for short periods and for instruction purposes. Contacts with all countries of the world are permitted.

The SARL operates a QSL bureau which accepts incoming cards for members only. A monthly magazine, *RADIO ZS*, is also published and distributed to members only.

(Courtesy Reno Faber, ZS6OF, General Manager - SARL.)

Sweden

There are approximately 11,500 amateurs in Sweden. As of October 1, 1994, Sweden adopted the CEPT licensing structure with two classes of license. License term is 10 years, and the license fee is 300 kroner (about $40) per year. Frequency allocations are per the ITU Region 1 allocations and amateurs are permitted 1000 watts output on all bands except 10 MHz where the power limit is 150 watts. The licensing authority is Post och Telestyrelsen, P.O. Box 5398, S-102, 49 Stockholm, Sweden.

There are eight call districts in Sweden, numbered Ø through 7. Their boundaries are as noted elsewhere in this chapter. In addition, several permanent special call-signs have been issued. SI8MI on Market Reef is located in the Ø call area. SJ9WL at the Morokulien facility on the Norwegian border is located in call area number 4 (and also is assigned Norwegian special callsign LG5LG).

The national society is the Sveriges Sandareamatorer (SSA). It operates a QSL bureau at Ostmarksgatan 43, S-123 Farsta, Sweden. Membership in the SSA is required to use the bureau. The SSA's magazine, *QTC*, has been issued regularly since 1928.

In September 1994, the SSA hosted the 7th Amateur Radio Direction Finding World Championships, with 290 entrants from 27 countries competing. A special event station, 8SØRDF, was activated.

There are two antenna manufacturers in Sweden: Vargarda Radio AB in Vargarda and Cue Dee Produkter AB in Robertsfors.

(Courtesy Rune Wande, SMØCOP/KB1Q, President, SSA)

Switzerland

Two classes of license are issued in Switzerland. The Class A license is indicated by call signs HB9AAA though HB9LZZ and all single- and two-letter suffixes. This is the all-band license, with an examination according to the Harmonized Amateur Radio Examination Certificate of the CEPT, including 12 wpm Morse code. The Class B license requires the same written examination, but no code test, and offers all privileges above 30 MHz. Call signs in the block HB9MAA though HB9ZZZ are issued for Class B licensees. All amateur bands for IARU Region 1 are permitted, but special permission is required for operation on 6 meters and parts of the 13 cm band. Power limit is 1000 W PEP. The license term is unlimited, and the annual fee is 120 Swiss Francs. There is no minimum age limit. Third-party traffic is prohibited. There are no restricted countries.

Army club stations are issued call signs from the HB4FA through HB4FZ block.

Reciprocal licenses are automatic for stays up to three months for visiting amateurs whose countries have implemented the CEPT Recommendation T/R 61-01. Call signs in the format HB9/G3ABC/P or /M should be used for portable or mobile operation. The "/P" designator is to be used even if operation is from a fixed location such as a hotel or camping site.

Licensees whose countries have not implemented T/R 61-01, but with which the Swiss Administration has concluded a reciprocal licensing agreement, may apply for a license valid for three months to Generaldirektion PTT, Direktion Radiocom, Sektion Konzession und Bewilligungen, Speichergasse 6, 3030 Berne. The fee is

50 Swiss Francs, and the application is available from Radiocom. The call sign will be assigned in a format similar to that for CEPT licensees, except that the /M and /P appendices are not mandatory.

Licensees whose countries have a reciprocal agreement with the Swiss Administration may apply for a permanent license without examination. If a reciprocal agreement is not in place, amateurs from other countries may apply to take the examination. There are 17 PTT Directorate offices located throughout the country.

The number of amateurs in Switzerland at year's end has been:

    1984 - 3784
    1985 - 4093
    1986 - 4249
    1987 - 4304
    1988 - 4442
    1989 - 4464
    1990 - 4542
    1991 - 4643
    1992 - 4794
    1993 - 4576

The national society is the Union Schweizer Kurzwellen-Amateurs (USKA), or, in English, the Union of Swiss Shortwave Amateurs. A QSL-bureau is operated for members only. The monthly magazine, *Old Man*, has been published for 62 years. The March 1994 issue was 56 pages long. Non-members may subscribe for 45 Swiss Francs per year. The USKA offers several awards, including the Helvetia Award (for contacts in all 26 Swiss cantons). Several annual contests are held, including Field Day, the annual H-26 contest, and numerous VHF/UHF contests where portable operation from Swiss mountaintops is quite popular. A national ranking system is maintained to determine the top operator based on scores in USKA-sponsored contests. HB9BXE was the 1993 champion, winning all four events he entered.

(Courtesy Etienne Heritier, HB9DX, IARU Relations Officer, USKA.)

Syria

The Technical Institute of Radio is the National Organization for Radio Amateurs in Syria. There is a single class of license, issued by the Syrian Telecommunications Establishment, requiring both a Morse code test and an examination in radio basics. The license is valid for life. No licenses are issued to foreigners except on very limited special occasions, and no third-party traffic is allowed. As of June 1, 1994, the number of licensed Syrian amateurs was 9.

There are four call areas in Syria:

    YK1 - Southern Syria
    YK2 - Northern Syria
    YK3 - Eastern Syria
    YK4 - Western Syria

The YKØ prefix is used for club stations, and the prefix 6C for special events.

There is no functioning QSL bureau at present because of the limited number of amateurs and lack of resources.

(Courtesy Dr. Omar Shabsigh, YK1AO, President - T. I. R.)

Tajikistan

Tajikistan is one of the former USSR republics in Central Asia. The Tajik Amateur Radio League (TARL)

was founded January 30, 1994. The President is Masud Tursoon-Zadeh, EY8AA. Vice-Presidents are Alex Rubtsov, EY8CQ, Nodir Tursoon-Zadeh, EY8MM, and V. Kazansky, EY7AT. TARL has applied for membership in the IARU as the Official member-society of Tajikistan. TARL operates a QSL bureau at Box 303, Dushanbe, 734025, Glavpochtamt, Tajikistan.

There four classes of license as follows:

Class 3 - 50 W power limit, limited operation is allowed in most bands. No privileges on 40, 30, and 20 meters.

Class 2 - 100 W power limit, operation on all amateur bands.

Class 1 - 200 W power limit, all bands.

Extra Class - 200 W (500 W in contests) and expanded frequencies on 160 meters (starting from 1810 kHz) and 80 meters (up to 3800 kHz).

There were 61 licenses issued as of mid-1994.

Class 1 - 28
Class 2 - 14
Class 3 - 11
Extra Class - 8

All calls have two letters in the suffix. Three letter call signs will be used after the two letter suffixes are depleted. Club station suffixes start with Z (i.e., EY9ZA). Call areas are as follows:

EY4 - Gornii Badakhshan, ex-UJ8R
EY5 - Khatlonskaya oblast Kulab region, ex-UJ8K
EY6 - Khatlonskaya oblast KurganzTube region, ex-UJ8X
EY7 - Leninabadskaya oblast, ex-UJ8S
EY8 - Dushanbe city, ex-UJ8J
EY9 - Dushanbe regions, ex-UJ8J
EY2, 3, Ø - Reserved for special stations
EY1 - TARL stations.

Portable or special calls are available to foreigners, "Portable EY" and corresponding region of Tajikistan (i.e., EY7/LY1DS etc.).

Tajikistan was the site of several expeditions which were organized to rare oblasts of the USSR, such as UJ8K and UJ8R. A number of photos, slides and films about these trips, featuring spectacular scenery of mountains on ice and rocks, is in the NCDXF library.

(Courtesy Nodir Tursoon-Zadeh, EY8MM.)

Thailand

Major Milestones in Thai Amateur Radio History

1955: Amateur radio was first defined in Thailand's Radio Communications Act, however the activity was not covered by any ministerial regulations in the Act, and thus the Post and Telegraph Department could not issue rules or licenses.

1964: A group of people interested in amateur radio asked that regulations be issued to permit amateur radio through the offices of the Post and Telegraph Department. The Director General of the Post and Telegraph Department sought government approval, which was not granted because the National Security Council objected, giving the reason that there was still a serious national security problem in the country. The same group of individuals then sought permission from the Research and Development Agency of the Army, characterizing the activity as military assistance from the United States, seeking to use HF amateur radio frequencies to help develop the communications equipment and

capabilities of the Thai military forces. The Defence Ministry proposed this to the Government and the Post and Telegraph Department gave its approval. The Radio Amateur Society of Thailand (RAST) was registered in 1964. Amateur radio activities were interrupted after the US military forces withdrew from Thailand following the end of the war in Vietnam

1981: The Post and Telegraph Department established a pilot project, using the 2-meter amateur band, by calling amateur radio operators "radio volunteers" under regulations issued by the Post and Telegraph Department, which termed the activity "voluntary radio," with the establishment of a Voluntary Radio Association. Operators were required to pass an examination of technical knowledge, communications laws and operating techniques. Call signs with the prefix "VR" and a serial number were then issued. At the beginning of this activity, there were some 312 operators.

1987: Regulations governing amateur radio activities were announced on National Communications Day (August 4). These regulations, drawn up by Thailand's National Frequency Allocation and Management Committee, were put into effect on January 1, 1988, and applied only in Bangkok and its neighboring provinces. The regulations put the Post and Telegraph Department in charge of amateur radio activities, and changed the status of Voluntary Radio (VR) to Amateur Radio. The "VR" call signs were withdrawn, and the internationally-recognized "HS" call sign series was issued. At that time, in Bangkok and in five neighboring provinces there were only 2,953 amateur radio operators.

1989: The Post and Telegraph Department expanded the area that the amateur radio regulations covered to all of the Kingdom of Thailand, leading to a rapid increase in the number of amateur radio operators. As of May 16, 1994, the number of those who have passed the novice amateur radio examination exceeds 200,000 persons, and those having received an operating license number 92,000.

There are three license classes: novice, intermediate, and advanced. Novice class operators, who do not have to take a Morse code test, are allocated the 144-146 MHz frequency band for transmissions in the FM mode with a maximum power of 10 W. Intermediate class operators who have to take a Morse code exam are allowed to use the two-meter band frequency for FM mode with a maximum transmission power of 10 W and to use four HF frequency amateur bands (10, 15, 20, and 40 meters) using CW and SSB modes with a maximum power of 200 W. Advanced class operators, who have to take and pass a Thai and English Morse code exam, may use the same frequencies as the intermediate class operators but are allowed a maximum transmission power of 500 W.

Currently, Thailand has signed bilateral reciprocal amateur radio operating agreements with the United States of America and Switzerland, and is now in the process of negotiating agreements with the United Kingdom and Sweden. Foreign nationals of countries not covered by such agreements may receive permission to be an amateur radio operator in Thailand with a temporary license that is granted on a case-by-case basis. Currently, there are some 50 foreign nationals who have received such licenses. Licenses are issued by the Post and Telegraph Department, Soi Sailom, Phaholyothin Road, Bangkok. Third-party traffic is not permitted. There are no countries with which communications are not permitted.

Call signs are issued in accordance with Thailand's two prefixes assigned by the ITU, which are HS and E2. The call areas are as follows:

Ø, 1 - Bangkok and its neighboring provinces
2 - The East of Thailand
3 - The lower Northeast of Thailand.
4 - The upper Northeast of Thailand.
5 - The upper North of Thailand.
6 - The lower North of Thailand.
7 - The West of Thailand.
8 - The upper South of Thailand.
9 - The lower South of Thailand.

The issuance of call signs for foreign amateur radio operators in Thailand is divided into two types: foreign operators whose country has signed an amateur radio agreement with Thailand are assigned a call sign in a series from HSØAA to HSØZZZ. Foreign operators whose country has not signed such an agreement are assigned a call sign HSØ/HOMECALL.

Casual operation in Thailand by visitors is difficult, with the exception that permission may be granted to individuals to operate from a RAST club station who apply for life membership of RAST beforehand, or who are life members, who notify the society two months before their arrival in Thailand. Such individuals may be allowed to operate from the RAST club station, depending on the availability of authorized RAST officials to allow access to the premises. Operating permission is for up to seven days only, using the callsign HSØ/HOME-CALL or the club station's call.

There are three different categories of license for every amateur radio operator in Thailand. These are:

1 - the operator's license which expires five years after having been issued;
2 - a license to both possess and to use communications equipment (the license to operate radio communication equipment) which is valid for the duration of possession; and,
3 - a license for setting up a station which is valid for the duration of the station's operation.

Applicants for the last type of license must have passed an examination conducted by the Post and Telegraph office and have been approved following an investigation into the applicant's background which is conducted by the National Intelligence Agency. Operators who want to operate a radio transmitter with a transmission power in excess of 5 W or who want to install a separate antenna for the transmitter must apply for a station license (otherwise no station license is required). Currently, about 20 percent of the licensed amateur radio operators have requested a license to set up a station.

As of May 16, 1994, there were 92,000 amateur radio operators in Thailand divided into these categories:

Novice class - 91,540
Intermediate class - 460
Foreign intermediate class - 46
Foreign novice class - 3

The Post and Telegraph Department has not conducted any examinations for the advanced class license, and thus there are no advanced class amateur radio operators in Thailand except for His Majesty, the King, who graciously accepted an advanced class license from the National Committee on Frequency Allocation and Management. His Majesty the King holds the call sign HS1A.

There were 53 intermediate class amateur radio stations using HF frequencies established in Thailand as of May 16, 1994. These include two club stations in Bangkok and one in the provinces managed and operated by the RAST, 12 stations operated by foreign amateur radio operators and 38 stations operated by Thai nationals.

The RAST is the national society. Membership in the RAST is required for use of the QSL Bureau (address: P.O. Box 2008, Bangkok 10501, Thailand). Lifetime membership is 2,100 baht or approximately US $84. RAST holds its monthly meeting at the Singha BierHaus on Asoke Road from 11 a.m. on the first Sunday of each month (includes a buffet luncheon). Singha BierHaus is well-known to any taxi driver and a short distance from any centrally-located hotel.

In the provinces, amateur radio is organized under the auspices of 55 associations approved by the Post and Telegraph Department. Another organization is the Voluntary Radio Association (VRA), a national organization that handles the novice operators' domestic activities, including offering assistance to society and to the various public agencies. The Voluntary Radio Association operates an internal novice radio network and played an important role in providing emergency communications, together with the government, in the case of the typhoon that struck Chumporn Province in southern Thailand in 1989, helping the injured and the homeless.

Amateur radio magazines include, *100 W*, *Ham Radio Magazine*, *CQ Amateur Radio* (Thailand)—not associated with CQ Communications' magazine of the same name. All are in Thai language. RAST members receive a monthly newsletter (also in Thai).

Various problems regarding amateur radio in Thailand include the fact that imported radio transceivers usually have specifications not in accordance with the Thai regulations, which causes delays in such equipment receiving permission for radio amateurs to use. HF transceivers pose a particular problem in that many models will transmit outside the frequency bands authorized for amateur radio, while some will even transmit on all HF frequencies. This inhibits or delays the growth of the number of intermediate class operators station licenses.

The RAST has been given the use of special callsign HS5ØA from 10 May 95 to 10 May 96 to commemorate the 50th anniversary of HS9A's accession to the throne.

(Much of the foregoing was excerpted from a paper by HSØ/G4UAV, presented at the Seminar on Amateur Radio Communication, held in Tokyo Japan, June 14-17, 1994.)

Turkey

There are 10 call areas in Turkey (TAØ through TA9). The TAØ prefix is used for all islands in Turkish territorial waters. The call sign prefix block YMØ through YM9 is used on national holidays (April 23, May 19, August 30, and October 29) and for special events on request. The TB and TC prefixes have not yet been used for amateur radio.

Licenses are issued by the T.G.M. (National Directorate for Telecommunications), 91 Sokak, Ulastirma Bakanligi Blokari, 06500 Emek, Ankara. There are also branch offices in Mersin, Samsun, Izmir, and Istanbul. There are three license classes in Turkey. Each license applicant must take an examination in three areas:

Technical subjects
Operating subjects
National laws and regulations.

The Class C license requires a 50% grade on the three exams and no Morse code. This license is valid for five years, non-renewable, and allows operation with 25 W maximum power on the bands above 144 MHz. Call signs are issued in a three-letter-suffix format, and the first letter after the number is always "C." The Class B license requires a 60% grade on the examination, and 6 wpm Morse code ability. In addition to the Class C privileges, permission to operate on the full 10-meter band is included (at 100 W maximum power). Class A licensees must achieve a grade of 75% on the exam, and send and receive 12 wpm Morse code. Privileges for all amateur bands are included, with a 400 W limit on 40-10 meters and some VHF/UHF bands, 150 W on 80 meters, 30 W on 160 meters, and 25 W on 2 meters. Operation on 6 meters is not allowed. Call signs for Class A and B licenses have one- or two-letter suffixes, and are renewable every three years. The fee is 50,000 TL. Temporary operation in a different call area than the licensed location requires a portable designator (e.g. TA3BC/6). All YL operators have call sign suffixes beginning with a "Y" (e.g., TA1YB). Foreigners staying in Turkey for longer than three months are issued call signs with suffixes beginning with "Z" (e.g., TA4ZA, who is a foreigner with a valid Turkish license). Prominent Turkish license holders include the Minister of Energy and Natural Sources, Mr. Veysel Atasoy, TA2BR.

The CEPT license is recognized in Turkey for visits of up to three months, and holders of CEPT license need not complete any sort of registration process. Call signs such as TA9/DL4ABM are used. As of June 1, 1994, there were 1250 licensed amateurs in Turkey. The QSL bureau is open to members only.

The 1992 earthquake in Erzincan (TA7) was reported by a Turkish amateur operator, and national and international communication during rescue operations was enabled by hams.

(Courtesy of TA1DF.)

RECIPROCAL LICENSING, CEPT, AND HAREC

Any discussion of international amateur regulations will eventually gravitate to the topic of reciprocal licensing. Reciprocal agreements between two countries allow operators from one country to visit the other and be granted permission to operate without the requirement of passing an examination. Like most international agreements, reciprocal agreements involve negotiations and bureaucracy, often become cumbersome for both governments to administer.

In 1985, the member countries of the CEPT (a French acronym for the European Conference of Postal and Telecommunications Administrations) drafted Recommendation T/R 61-01, which enabled amateurs from one member country to visit another and be granted an automatic reciprocal license for up to three months. Call signs take the format of F/G4CLF, for example, when G4CLF visits France. No paperwork is required.

The CEPT license has been a huge success with few abuses and a greatly-reduced workload among the licensing authorities in Europe. The program has worked out almost exactly according to the plans laid out in 1985. The CEPT Recommendation outlines two classes of license: the Class A license permits HF operation; the Class B license, requiring no Morse code examination, is valid for operation above 30 MHz. Each country determines the equivalent grade of local license which qualifies for each of the CEPT classes, and similarly determines the local privileges to be extended to holders of the two CEPT classes who may visit from other countries. A detailed chart of equivalencies is available from the ITU.

T/R 61-01 was amended in 1992 to allow non-CEPT member countries to participate in the CEPT two-class license system. Peru, New Zealand, Turkey, and Israel have already adopted these provisions, and other countries are expected to follow.

More recently, a new Recommendation (T/R 61-02E) has been approved, defining a "Harmonized Amateur Radio Examination Certificate," or "HAREC," that license authorities may issue and that will be mutually recognized by all governments adopting the Recommendation. The HAREC will be accepted as proof of satisfactory completion of an amateur license examination and will allow the holder an automatic reciprocal license for long-term stays, including permanent relocations. No further examination will be required in the new country. The Recommendation includes a detailed syllabus that should be followed by any governments planning to issue HARECs. The examination is to include four sections:

1. Technical matters
2. National and International Operating Rules and Procedures
3. National and International Radio Amateur and Radio Amateur Satellite Regulations
4. Sending and receiving Morse code at 12 wpm.

Successful completion of an examination on the first three sections will qualify the candidate for a CEPT Class B license, valid for operation above 30 MHz. Adding the Morse code examination will qualify the candidate for a CEPT Class A license, valid for use on all amateur frequencies.

The HAREC Certificate is printed in at least one language of the issuing country, as well as English, French, and German. It includes a statement that the holder has passed an examination meeting the requirements of the CEPT level A or B license, the holder's name and birth date, the equivalent local national license class, date of issue, and the details of the issuing authority.

HF DIGITAL RADIO

If you are new to the world of high frequency digital communications, you have a great new aspect of the hobby waiting for you. Teleprinting via radio (radiotele-type, or "RTTY") on the amateur bands was first authorized by the FCC in the early 1950s. Some stations still use the old mechanical machines with dripping oil and noisy gears, but the advent of the ubiquitous personal computer has completely revolutionized the mode. The development of inexpensive solid state modems (sometimes called TNCs, Terminal Units, or Interface Units) has put this aspect of the hobby within the reach of most amateurs. These interface devices enable communication between your radio and your keyboard and CRT display. More recently, the TNC function has been incorporated directly into the computer, thereby eliminating an extra piece of station hardware.

One of the unique aspects of digital radio is the convenience of printing out hard copy of your QSOs. This is great for your own personal scrapbook, for traffic handling, or other appropriate record. You can share your printout of a particularly interesting QSO with your pals at the local radio club or hamfest.

While competition for rare DX, and during contests, is as fierce on RTTY as on CW and SSB, there is a whole new dimension of friendliness that pervades digital radio. For years RTTY has been known as the "gentlemen's mode," and, to a great extent, it still is.

Even if you are not the world's greatest typist, that is no detriment to enjoying this fascinating aspect of our hobby.

RTTY

RTTY (radioteletype) is one of the more common communications protocols used in the HF bands. It is a half-duplex, non error-correcting mode that can be used by any number of stations on a frequency in a round-table fashion. Since there is no error-correction, it suffers from noisy environments, and a good signal-to-noise ratio is needed for reasonable copy. Note that transmit/receive switching is manual, so all parties in a roundtable need to agree on who transmits next. It is primarily used for single keyboard-to-keyboard contacts. RTTY uses Baudot character encoding (also known as ITA2), which is a 5-bit code. Note that 5 bits is only 32 possible combinations, which is not enough for a full alphanumeric set, much less mixed-case alphabetics. Baudot (and RTTY) gets around this limitation by defining two "shifts" which switch between a "letters case," and a "figures case." On older RTTY setups (ones which actually use a teletype, for instance), you have to worry about the letters/figures shift. For example, the RST signal report "599" is often sent both as "599" and "TOO" (which is the letter-case equivalent) to ensure copy if the receiving station has inadvertently missed the "figures-shift" character. However, most TNCs and multimode digital controllers now take care of those shift characters automatically, sending them as necessary for the data being transmitted. Lower case is not used on Baudot RTTY. ASCII can be used in RTTY as well, but it is very uncommon.

If you are migrating from CW or SSB to RTTY, you will find that contacts are made in very much the same manner with which you are accustomed. Conversational QSOs are established by calling CQ, answering a CQ, or a QRZ?. Selective area CQs (e.g., CQ Pacific) are common.

As with any new mode, the best way to learn the prevailing operating practices is to listen before transmitting.

Most RTTY QSOs are carried on at 60 wpm (45.45 baud), although you may sometimes find a foreign station using 67 wpm (50 baud). This will be evident if you are getting less than perfect copy even with a 599 QRM-free signal. In that case the sender is probably using old equipment and cannot change. Adjust your baud rate to match his. Another problem often arises when an operator inadvertently uses Upper Sideband—by convention, all RTTY and AMTOR operation is done on Lower Sideband, regardless of frequency band. If you cannot get meaningful copy on a loud station, the sender is probably on the wrong sideband. Try calling on the same sideband that he is using, and ask him to switch to the correct one.

Very little RTTY DXing is done on 160 and 80 meters, although 40 and 80 meters can be quite active during RTTY contests. Most favored bands are 15 and 20 meters. In favorable parts of the sunspot cycle, 10 meters can be active. There is practically no RTTY activity on 12, 17, and 30 meters, without having a prior agreed-upon schedule.

AMTOR

During the 1970s, the Dutch government developed a system of digital communication called SITOR. It was designed to provide reliable RTTY communication under adverse conditions, while still maintaining an extremely low error-rate. The SITOR protocol was meant primarily for maritime use. It was so reliable a system that it was soon in use worldwide. In the early 1980s, Peter Martinez, G3PLX, made several minor changes to the SITOR protocol and called it AMTOR. The FCC authorized its use by U. S. amateurs beginning in January of 1983. AMTOR is different from any digital mode you may have used, in that it is the only synchronous type of RTTY authorized at the present time. "Synchronous" means that the two stations in an AMTOR ARQ contact are synchronized (linked) with each other. Most AMTOR activity is found on 20 meters. Look for it around 14.075. You will also find activity around 3.650, 7.050, and 21.075.

There are three modes in AMTOR. The first is ARQ, an acronym for Automatic Repeat Request, sometimes called Mode A. It makes the familiar "chirp chirp" sound you have often heard. The second mode is FEC, an acronym for Forward Error Correcting. It is sometimes called Mode B. The sound of an FEC transmission is similar to that of Baudot RTTY. The third mode is called Mode L, the Listen mode. It allows an operator to decode an ARQ signal even though he is not one of the two stations that are linked.

This description of the AMTOR ARQ mode refers to the "information sending station" and the "information receiving station." This is because when two stations are linked in Mode A, *both* stations are transmitting at intervals, thus it would be incorrect to refer to one as the "transmitting station" and the other as the "receiving station." In an ARQ link, the timing is set by the station that initiated the contact. It is called the "master" station, with the other station being the "slave." In AMTOR, the "ARQ Cycle" is 450 milliseconds long. It starts with the information-sending station transmitting a data-burst consisting of three characters. Since AMTOR uses a seven-bit code, and the data-transfer rate is fixed at 100 baud, this data-burst takes 210 milliseconds. Next there is an 85 millisecond period of silence, during which the receiving station checks the three characters for validity. If the data-burst passes the check, the receiving station prints it, and sends back a single character control code. This takes 70 milliseconds. The control code (an ACK) says, "OK, I got it, send the next burst." If the data-burst does not pass the check, the receiving station prints nothing and returns a control code (a NAK) that says, "Hey, I didn't get it. Send it again." The sending station will continue to repeat the data-burst until it gets an ACK. The control code is followed by another 85 milliseconds of silence, and the cycle begins again with either the next data-burst if the information sending station received an ACK, or a repeat of the last data-burst if it got a NAK.

The AMTOR ARQ mode has often been referred to as error-free. That is not quite true. Errors are possible, but the error rate is so extremely low that it is *virtually* error free. The method of error detection used is quite simple. Five of the bits in each character of the AMTOR code are information-bearing bits. The other two are used for error detection. The five information-bearing bits in each character are identical to the five information-bearing bits for that character in the Baudot code. Each of the two error detection bits can be either a mark or a space. The code is arranged so that every AMTOR character contains four marks and three spaces. A three character data-burst then, contains a total of twelve marks and nine spaces. All the information receiving station must do, therefore, is count the total number of marks and spaces in the data-burst it received to detect whether or not there has been an error. While data is not being transferred, the information-sending station transmits idle signals to maintain the link.

The second AMTOR mode is FEC. This mode is not synchronous, and the stations involved are not linked, but they do operate in phase with each other. In order for them to stay in phase, each FEC transmission is started with several sets of "phasing pairs." These are repeated at regular intervals during the course of the transmission, so that the two stations can stay in phase. While no data is being transferred, idle signals are transmitted to keep the two stations in phase. In Mode B, each character is transmitted twice, 350 milliseconds apart. The receiving station prints a character the first time it is received if the mark/space count is correct. If it were received correctly the first time it was sent, it will be printed and ignored the second time it is received. If it were incorrect the first time it was received, it will be ignored, and will be printed the second time it is sent if it is received correctly. This method of error detection is much less effective than that used in ARQ, and the error-rate is considerably higher than it is in that mode. Although higher than in ARQ, the error-rate is still far lower than it is in other forms of RTTY.

The third AMTOR mode is called Mode L, or the "Listen" mode. It allows an operator to print ARQ data-bursts even though he is not linked to the information sending station. There is no error detection at all in the Listen mode, and if the station being monitored is asked for repeats (RQs), they will be printed.

To make a contact on AMTOR, you call CQ very much the same as on RTTY or CW, except that you must use the FEC (Forward Error Correction) mode, sometimes called Mode B. An FEC transmission sounds very much like fast RTTY. Most newcomers get frustrated when starting because they do not seem to be able to get answers to their CQs. The trick is to allow three to six

seconds of idles at the beginning of your transmission. Without sufficient idles, listening stations will not be able to "sync" with you. When you call CQ, you include your "Selcal" in addition to your regular callsign. Selcal is an acronym for Selective Call. The convention is to derive your selcal from the first and the three trailing letters of your callsign. In the case of a one-by-two callsign such as W1AW, the first letter is repeated, followed by the two trailing letters, so its selcal would be WWAW. If the callsign is a two-by-one, the first letter is repeated twice, followed by the trailing letter. The selcal of WZ1R, for example, would be WWZR. A two-by-two callsign such as UW9AR, results in a selcal of UWAR. For a two- by-three callsign like WN4KKN, the selcal would be WKKN. While it is not necessary to follow this convention, since any four letters can be a selcal it is a good idea to do so. If you do, then any operator who knows your callsign can deduce your selcal.

Following your CQ, return to AMTOR Standby and wait for a reply. If you receive a response to your CQ, it could be in FEC, although it is much more likely that it will be in ARQ. If the response is in ARQ, your station will begin the ARQ cycle, and you will see that idle signals are being received. At this point, the link is established and your station is ready to receive text. Now the other operator need only open his transmit buffer, type the text, and it will be sent to you. If text stops appearing on your screen, just stand by. Your station is probably requesting repeats because it received a data-burst incorrectly. You will soon receive the block correctly, and traffic will begin to flow again. When the other station is ready to receive information from you, the operator will turn the link over to you, and it will be your turn to send text to him. He will do this by sending you a turnover sequence. That is the two characters "Plus-Question mark" (+?). You will know when the turnover has been made by the change in rhythm of your station's transmissions. Before the turnover, your station was sending single character data-bursts 70 milliseconds in length. When the turnover occurs, it will begin transmitting three character data- bursts 210 milliseconds in length. When the turnover has been made, simply open your transmit buffer and begin typing text for transmission. When you are transmitting text in ARQ, you will see it pop onto your screen in three character blocks, as the blocks are acknowledged by the receiving station. If this text stops appearing and nothing seems to be happening, it just means that the receiving station is requesting repeats because it received the last data-burst your station sent incorrectly. Stand by. The incorrect block will soon be acknowledged, and traffic will flow again. When you are ready to receive text again, reverse the link by sending the turnover sequence (+?).

The second way to initiate a contact is to answer another station's CQ. When a station calls CQ in FEC, its selcal should be included. If it is not, you can deduce what it is. To answer the CQ, just start an ARQ transmission. When you do this, your program will ask for the selcal of the station you wish to call. When you enter it, the ARQ cycle will start, with your station transmitting the other station's selcal. You will soon hear the other station responding with control codes, and you will see an indication that your station is sending idle signals. Now the link is established. Just open your transmit buffer and type in text for transmission. When you are ready for the other station's reply, send the turnover (+?) and reverse the link.

The third way to initiate a contact is to call a specific station in ARQ. Usually, this will happen when you have been watching an ARQ contact in the Listen mode, and you want to call one of the stations when they finish. In this case, since you have been "eavesdropping," you already know the selcal of the station you want to call. You also know it is on frequency, since the contact has just been terminated. Call the station by starting an ARQ transmission just as though you were replying to a CQ call.

There are two important control codes used in AMTOR. The first is "Control-C." It is used to force a changeover. If the other station in an ARQ link is the information-sending station, and you want to reverse the link without waiting for it to send the changeover sequence, you can do so by sending it a Control-C. The second code is "Control-D." It is used to break an ARQ link. An ARQ contact should *always* be ended with a Control D.

The T/R switching time of your rig should be on the order of 25 milliseconds for best ARQ operation. All newer rigs and most older ones easily meet this standard. Some rigs designed before AMTOR came into use, however, may require simple modifications. Since the ARQ cycle requires your rig to return to full receive sensitivity quickly, always set your AGC for fast release, or turn it off entirely.

(This section was written from material supplied by W2JGR, KD6GGD, and a file on the ARRL BBS written by N9ANL.)

OPERATING ON THE VHF FREQUENCIES AND ABOVE

Some of the most interesting and exciting Ham Radio operating takes place on the VHF, UHF and higher frequency bands. The sheer volume of spectrum, accounting for greater than 99% of the U.S. allocations, allows for the peaceful coexistence of a nearly endless variety of operating modes and styles. The 16 bands above 30 MHz span frequencies between 50 and 250,000 MHz and total over 20,000 MHz of Amateur band space. This large range of frequencies, and the desire to make the best use of them, has drawn many Amateurs into exploring and enjoying the bands above 10 meters.

When most Hams think of the frequencies of VHF and above they might consider FM repeater or packet operations. Others may think of satellites, ATV or SSB/CW work. It all depends on what you have discovered that interests you. The more common applications of the VHF and UHF spectrum, such as FM repeater, packet, satellite, and ATV operations, are aptly described elsewhere in this Almanac. The principal subject of this section is to describe so called weak signal operation on the bands above 50 MHz.

Weak signal work is the art of contacting distant stations by making the best use of equipment, antennas, band conditions, and operating skills. Although there a number of enhanced propagation modes that allow for contacts with stations several hundred, if not thousands, miles distant, the normal mode of propagation on the VHF and above bands is ground wave. For stations that are well equipped and have well placed directive antennas, reliable contacts can be routine with similarly

equipped stations 300-400 miles away. Some of the key elements to successful and enjoyable weak signal work is to know where the activity is on the band and in what direction to point your antenna.

Calling Frequencies

The best place to start looking for VHF or UHF QSOs is on the calling frequencies. The following Table indicates the calling frequencies for the lower seven VHF/UHF bands allocated in North America.

Band (MHz)	Calling Frequency
50	50.110 for DX
	50.125 for Domestic QSOs
144	144.200
222	222.100
432	432.100
902	903.100
1296	1296.100
2304	2304.100

It is common practice to CQ on the calling frequency if it is not already in use. Amateurs active on the VHF and UHF bands often leave a receiver or two on the calling frequencies when they are in the shack so that they can keep track of band conditions and catch other VHFers without having to constantly tune the band.

Antennas

One of the advantages of VHF/UHF and Microwave operation is that an antenna comparable in size to, say, an average 10/15/20 meter triband Yagi has considerably more gain and directivity at VHF/UHF than the tribander does at HF. This is attributed to the inverse relationship between frequency and wavelength: the higher in frequency you go, the better the antenna performance at a given size.

For some very practical reasons, FM operation, which is primarily mobile, uses vertically polarized antennas. The use of vertically polarized antennas for FM mobile work has also been adopted for base station FM voice and packet operations. SSB and CW stations generally operate with horizontally polarized antennas, even when mobile. There is no clear advantage to one or the other in terms of performance for weak signal work. The important thing is that both stations use the same polarization. The most common antenna type is the long Yagi on the bands 50-432 MHz and the loop Yagi on 903 and 1296. Reflector antennas such as the common "dish" antenna are generally employed on the bands above 1296 Mhz.

Rigs for VHF and Above

The most popular and straightforward radios for 50-1296 MHz are the multimode (usually FM/SSB/CW) rigs. They come in basically three flavors: single-band mobile rigs, single-band base rigs, and multi-band base rigs. Most of these radios are quite similar to HF rigs in their features and operation, but generally transmit lower power levels in the 25 watt range. To boost output power levels, most amateurs couple their 25 watt radios to solid state "brick" amplifiers that are powered from 13.8 volts and amplify to typically about 150 watts. There are several commercial sources of these amplifiers, providing various output power levels and amplifier gains.

The easiest avenue to SSB/CW operation on 903, 1296,

2304 and 3456 is the "no-tune" transverter connected to a 144 MHz radio. These transverters take advantage of microwave monolithic integrated circuit technology in printed circuit boards employing etched tuned circuits. This design approach allows transverters to be constructed and tuned up with a minimum of adjustment as compared with earlier designs. Amateurs are getting on these higher bands like never before using these transverters, which are available in kit or fully constructed form.

Equipment for operation on 5760 and above is generally homebrew, but often employs surplus or used commercial pieces to keep costs down. The 10 GHz band is the most popular microwave band largely due to the availability of wide-band Gunn diode-based systems. There is even a commercially available 10 GHz FM transceiver with a built-in horn antenna for the amateur.

Grid Squares

While large antenna gain can help result in the ability to work over long distances, with it comes a commensurate increase in antenna directivity. Of course, you need to know where to point such directive antennas if they are to be useful. To make it easier to quickly know in what direction to turn your antenna, the earth has been mapped into 32,400 rectangles, called grid squares. Each square encompasses 1 degree of latitude and 2 degrees of longitude and has its own unique four-character designator. Almost all QSOs first begin with an exchange of grid squares, allowing both ends to quickly align their antennas.

Propagation

What follows describes a few common propagation modes capable of extending your range beyond what can be expected under average conditions.

E-Skip

This propagation mode, which is also called Sporadic-E, is common on the 6-meter band during the period of late May through July. E-skip usually produces contacts with stations that are 1100-1400 miles distant. During a Sporadic-E opening, radio waves propagate by refracting off of ionized regions in the E-layer, which is about 50 miles above the earth.

Tropo

Unlike E-skip, tropo openings have their greatest effect on the frequencies above 144 MHz. While there is a variety of types of tropo, they are all generally characterized by the refraction of radio waves at the temperature boundaries of lower atmospheres. Tropo openings can be local, providing enhancement within 100-200 miles, or long distance, allowing contacts with stations more than 1,000 miles distant. Unlike most of the propagation modes at HF, tropo is often developed and affected by the earth's weather patterns.

Aurora

The same aurora borealis that is such a treat to the eye can provide some exciting DX to the wary VHF operator on the bands 50-432. When charged particles thrown off by the sun during a solar storm interact with the earth's upper atmosphere over the north or south poles, there is the potential for the creation of large ionized polar regions. These regions can support extended VHF propagation for many hundreds of miles. If an aurora is suspected (if the WWV solar K index is above 4, for example), North American Hams should turn their antennas north and listen for **Au**. Unique to auroral propagation is the characteristic "buzz" or raspy sound to the signals. The signal distortion doesn't impact the effectiveness of CW, but it can make SSB contacts difficult.

Meteor-Scatter

Meteors strike and vaporize in the earth's upper atmosphere everyday. Meteors leave an ionized trail about 50-75 miles above the earth, supporting communications out to about 1400 miles. These ionized trails are such that signals on 10 and 6 meters are refracted particularly well and, to a lesser extent, signals on 2 meters and above. QSOs via meteor-scatter can be particularly challenging, as the ionized trail from a meteor may last anywhere from milliseconds to a minute or more. Stations often refer to meteor trail-induced receptions of distant stations as meteor bursts or pings if they are of very short duration. While skilled, persistent operators can work meteor-scatter almost any morning on 6 meters, the best chance to work meteor-scatter is during a meteor shower. The following table lists the meteor showers that occur every year during which amateurs can look for extended propagation.

Shower	Date(s)
Quadrantids	Jan 3-5
Lyrids	Apr 19-23
Eta Aquarids	May 1-6
Arietids	June 2-14
Delta Aquarids	July 26-31
Perseids	July 7-August 14
Orionids	Oct 18-23
Taurids	Oct 26-Nov 16
Leonids	Nov 14-18
Geminids	Dec 10-14
Ursids	Dec 22

Typically, the two most productive meteor showers for the amateur are the yearly Perseids and the Geminids.

In addition to the propagation modes described above, there are several others, such as double-hop E-skip, F-layer, and transequatorial, that can greatly extend your VHF horizon. Part of the fun of operating VHF and above is working distant stations by rare and exotic propagation.

Hilltopping and Contesting

One popular way to dramatically improve your QSO range on VHF and above is to operate from a hilltop, mountaintop, or any other elevated spot that provides a clear shot in the direction in which you want to operate. The improvement in range becomes more significant as your frequency of operation increases. Microwave contacts that are impossible when operating from a valley become easy if you move to a hilltop that provides a line-of-sight or nearly line-of-sight path.

Many stations enjoy hilltopping during VHF/UHF contests. Hilltopping contest stations can pull in many hundreds of QSOs without an elaborate setup. Small antennas, often short Yagis, can be very effective and easily mounted on a short mast that can be hand-rotated. Radios are usually multi-mode solid-state rigs plus solid-state "brick" amplifiers in the 50-150 watt class. Dozens of stations travel to hilltops in exotic grid squares during the contests to pass out rare multipliers and to feel the rush of being on the other end of a pile-up.

VHF/UHF Contests that Occur Every Year

ARRL VHF Sweepstakes	January
ARRL June VHF QSO Party	June
CQ VHF WPX Contest	July
ARRL UHF Contest	August
ARRL September VHF QSO Party	September
ARRL EME Contest	Oct-Nov

Courtesy of Ed Parsons, K1TR

VHF/UHF/MICROWAVE/LIGHTWAVE DX RECORDS

History

Amateur radio operators were quick to exploit the HF bands but before WW II the frequencies above 30 MHz were treated more as a curiosity or experimental area rather than a useful frequency spectrum. Furthermore, there weren't many high power tubes available, especially above 150 MHz.

After WW II when VHF/UHF equipment did become available, it was mostly obsolete RADAR gear that had poor frequency stability and low output power. Hence it was only usable on FM and later on AM. In the 1950's crystal controlled receivers and transmitters finally reached the 2-meter band and narrower band operation such as CW and later SSB became available.

In North America in the early years following WW II the primary DX was conducted on 6 and 2 meters in the quest for the WAS award. 6-meters of course was not too difficult since sporadic E and, at sunspot peaks, F2 propagation was available. However, 2-meter DXers had to primarily rely on tropospheric and meteors scatter propagation with no chance of completing a WAS award.

In the 1960's the Europeans invented the so called "Locator" system to pinpoint ones latitude and longitude. Later this evolved into a competition to see how many different locators you could contact. However, due to the way locators were originally designated, this system was not usable outside Europe.

Meanwhile, new propagation modes, better equipment, especially with the arrival of the transistor and FET improved operating techniques, higher gain antenna systems, and EME communications opened up new horizons but WAS and WAC were still the big awards here in North America. Some DX records were made but these were often overshadowed by "special" regions of the world that exhibited unusual propagation such as the tropo paths from California to Hawaii or over the Mediterranean Sea and the Great Australian Bight.

In 1983 the ARRL introduced the VUCC (VHF/UHF Century Club) award using the "Maidenhead" worldwide grid locator system. This is now well entrenched and has improved activity as well as DX chasing., However, there was still no real incentive to "improve the state of the art" in equipment and DXing since many of the long DX records were not easily exceeded in North America. Figure 1 shows the DX table as it appeared in 1984. Note that the only North American record were on 2 meters and the 135-cm band, the later not available outside of region 2!

Hence, in 1985 I introduced a new system whereby DX records in North America were treated as a separate enti-

ty. Furthermore, a separate record was established for each type of radio propagation on each VHF band above 6 meters. The 6-meter band was not listed since it frequently has mixed modes and even "long path" DX where the distances are greater than half-way around the world. Finally, in February 1989 I added the OL (overland) and OW (over-water) designations to further enhance the challenges. To qualify for an OW DX record, the path must be at least 75% over water and must exceed the equivalent OL path distance.

Why are VHF and above DX records important?

1. They provide a challenge as described by the old saw: "because it's there."
2. They improve the state of the art (SOA) in radio propagation.
3. They set the pace for the design of SOA antennas, receivers, & transmitters.
4. They help determine the best transmission mode (CW, SSB, FM, etc.).
5. They require improved operating techniques to make new DX records.

What are some of the improvements brought forth primarily because DX records were being pursued?

1. New propagation modes such as TE, EME, FAI, Aurora, etc.
2. High stability narrowband receivers and transmitters have been developed that permit CW and SSB operation as high as 48 GHz!
3. New almost noiseless receivers have been developed that operate beyond the microwave region (above 10 GHz).
4. High gain low side lobe antenna systems are now common place.
5. High power vacuum tube and solid state transmitters are now available.
6. Improved operating techniques using the most optimum communications mode now allow almost routine contacts using meteor scatter and EME.
7. You tell me!

Current DX record tables (See following information)

DX records serve a very useful purpose
1. They are sort of a "yard stick" that can be used to measure our accomplishments against those of our peers.
2. They give us a good idea of what can be accomplished on each of the bands above 50 MHz as well as the propagation modes available.
3. They tell us which Amateurs are in the forefront of technology thus providing a person or persons that can be contacted to help us along without having to "reinvent the wheel."

References

1. Joe Reisert, W1JR, "The VHF/UHF primer: an introduction to propagation," *Ham Radio Magazine*, July 1984, pg. 14.
2. Joe Reisert, W1JR, "Propagation update," *Ham Radio Magazine*, July 1985, pg. 86.
3. Joe Reisert, W1JR, "Propagation update—part 3," *Ham Radio Magazine*, July 1988, pg. 38.
4. Joe Reisert, W1JR, "DX records on 50 MHz and above," *Ham Radio Magazine*, Jan. 1989, pg. 48.
5. Joe Reisert, W1JR, "DX records on 50 MHz and above: part 2," *Ham Radio Magazine*, Feb. 1989, pg. 42.

Courtesy of Joe Reisert, W1JR (evhfc91), 16 May 1991

NORTH AMERICAN VHF/UHF RECORD HOLDERS

Note: The information and format used in this table is copyrighted in 1989 and registered at U.S. Patent Office by Joseph H. Reisert, W1JR, and may not be reprinted or used in any form without the written permission of the author.

Joseph. H. Reisert, W1JR (navhfuhf)
Revised 93-09-11

North American VHF and Above Claimed DX Records (notes 1, 2 & 3)

Frequency	Record Holders	Date	Mode	DX Miles (km)
50 MHz	(Note 4)			
EME	WA4NJP (EM84DG)-ZL2BGJ (RF7ØDX)	1988-09-08	cw	8258 (13288)
144 MHz				
Aurora	KA1ZE (FN31TU)-WAØTKJ/WBØDRL (EM18CT)	1986-02-08	cw	1347 (2167)
EME	VE1UT (FN63XV)-VK5MC (QFØ2EJ)	1984-04-07	cw	10,985 (17676)
FAI	KXØO (DM78PU)-WA4CHA (EL88QA)	1993-06-19	ssb	1472 (2370)
MS	K5UR (EM35WA)-KP4EKG (FK68VG)	1985-12-13	ssb	1960 (3153)
Spor. E.	WA4CQG (EM72FO)-W7YOZ (CN87VR)	1988-06-06	ssb	2172 (3495)
TE	KP4EOR (FK78AJ)-LU5DJZ (GF11LU)	1978-02-12	ssb	3933 (6328)
Tropo OL	WB4MJE (EL94HQ)-VE1KG (FN84CM)	1994-11-05	ssb	1685 (2714)
Tropo OW	KH6HME (BK29GO)-W7FI (CN87)	1995-07-01	ssb	2691 (4333)
220 MHz				
Aurora	WC2K (FM29PT)-WB5LUA (EM13QC)	1989-03-13	cw	1298 (2089)
EME	K1WHS (FM43MK)-KH6BFZ (BL11CJ)	1983-11-17	cw	5058 (8139)
MS	W1JR (FN42HN)-KØALL (EN16OU)	1988-08-13	ssb	1274 (2057)
Spor. E	K5UGM (EM12MS)-W5HUQ/4 (EM9ØGC)	1987-06-14	cw/ssb	932 (1499)
TE	KP4EOR (FK78AJ)-LU7DJZ (GFØ5RJ)	1983-03-09	cw/ssb	3670 (5906)
Tropo OL	K5SW (EM25HR)-W2SZ/1 (FN32KP)	1992-09-13	cw	1275 (2050)
Tropo OW	KH6HME (BK29GO)-XE2GXQ (DL29CX)	1989-07-15	NBFM	2575 (4143)
432 MHz				
Aurora	W3IP (FM19PD)-WB5LUA (EM13QC)	1986-02-08	cw	1182 (1901)
EME	K2UYH (FN2ØQG)-VK6ZT (OF78VB)	1983-01-29	cw	11,567 (18612)
MS	KD5RO (FN13FB)-WB5LUA (EM13QC)	1988-12-13	cw	1239 (1994)
Tropo OL	KM1H (FN42HR)-WB4MJE (EL94HQ)	1992-12-16	ssb	1370 (2204)
Tropo OW	KH6HME (BK29GO)-XE2/N6XQ (DL29CX)	1989-07-15	cw	2575 (4143)
902 MHz				
Aurora	K3HZO (FM18QP)-WA3NZL (FM19JG)	1991-11-08	cw	54 (87)
EME	AF1T (FN43ED)-K5JL (EM15DQ)	1990-05-27	cw	1476 (2376)
Tropo OL	WB5LUA (EM13QC)-W4WSR (EL96WW)	1989-03-14	cw	1071 (1725)
Tropo OW	KH6HME (BK29GO)-N6XQ (DM12JR)	1994-07-14	cw	2522 (4061)
1296 MHz				
EME	K2UYH (FN2ØQG)-VK5MC (QFØ2EJ)	1981-12-06	cw	10,562 (16995)
Tropo OL	WB3CZG (FN21AX)-KD5RO (EM13PA)	1986-11-29	cw	1287 (2070)
Tropo OW	KH6HME (BK29GO)-XE2GXQ (DL29CX)	1989-07-15	ssb	2575 (4143)
2304 MHz				
EME	W3IWI/8 (FMØ8CK)-ZL2AQE (RE78JS)	1987-10-18	cw	8658 (13931)
Tropo OL	WB5LUA (EM13QA)-WA8WZG (EN81OM)	1993-09-17	cw	952 (1533)
Tropo OW	KH6HME (BK29GO)-N6CA/6 (DMO3TR)	1994-07-11	cw	2469 (3973)
3456 MHz				
EME	W7CNK/5 (EM15FI)-KØKE/Ø (DM79NO)	1987-04-12	cw	498 (802)
Tropo OL	WB5LUA (EM13QC)-WAØBWE (XXXX)	1995-07-12	ssb	841 (1354)
Tropo OW	KH6HME (BK29GO)-N6CA/6 (DMØ3TR)	1991-07-28	cw	2469 (3973)
5760 MHz				
EME	WA5TNY (EM12KV)-W7CNK/5 (EM15FI)	1987-04-24	cw	174 (279)
Tropo OL	WB5LUA (EM13QC)-W9ZIH (EN51NV)	1994-11-12	cw	738 (1187))
Tropo OW	KH6HME (BK29GO)-N6CA/6 (DMØ3TR)	1991-07-29	cw	2469 (3973)
10.368 GHz				
EME	I6ZAU (JN63RJ) - WA7CJO (DM33XL)	1991-05-04	cw	6125 (9855)
Tropo OL	W6HCC/7 (CN82PB) - WA6EXV/6 (DMO4KT)	1993-07-18	CW	537 (865)
Tropo OW	WB6CWN (CM96QI) - XE2/N6XQ (DMØ3TR)	1994-08-25	MCW	698 (1124)
24.192 GHz				
LOS	KK6TG/6 (CM98MR)-WB7ABP/6 (CN8ØON)	1992-09-12	cw	159 (256)
	WA3RMX/7 (CN93IQ)-WB7UNU/7 (CN95DH)	1986-08-23	ssb	116 (186)

47.040 GHz
LOS WA3RMX/7 (CN82VW)-K7AUO/7 (CN82PB) 1988-08-06 cw/ssb 65.3 (105)

75-120 GHz None reported

120 GHz
LOS WA1MBA (FN32RJ)-WB2BYW (FN32RJ) 1994-07-11 cw 0.71 (1.15)

145 GHz WA1MBA/1 (FN32RI)-WB2BYW (FN32RJ) 1993-05-02 cw 2.4 (3.75)
300 GHz
and above (note 5)

474 THz
LOS K6MEP/6 (DMØ4MS)-WA6EJO/6 (DMØ5XA) 1991-06-09 Laser 57 (92)
678 THz
LOS KY7B/7 (DM42OK)-WA7LYI/7 (DM34TF) 1991-06-08 Laser 154 (248)

Notes:
1. The records are listed alphabetically by mode. Tropo OL is over land. Tropo OW is over water where at least 75% of the path is over water. Tropo OW records are only listed when they exceed the tropo OL DX record.
2. The information within the brackets ( ) following the call sign is the grid square locator.
3. Distances have been calculated assuming a spherical earth model using the actual latitude and longitude rather than grid square centers which is less accurate.
4. 6-meters records, excepting EME, were left off since the primary propagation mode is often hard to distinguish. Also long-path QSO's have been reported during solar cycles 19, 21, and 22 which exceed approximately 12430 miles (20000 km).
5. There have been very few reports of contacts in the wide open frequency allocation above 300 GHz. Therefore, at least for the time being, we will list those records that show considerable distance at widely different frequencies.
Reprinted with permission.

WORLD-WIDE VHF/UHF RECORD HOLDERS

Note: The information and format used in this table is copyrighted in 1989 and registered at the U. S. Patent Office by Joseph H. Reisert, W1JR, and may not be reprinted or used in any form without the written permission of the author.

Joseph H. Reisert, W1JR (wwvhfuhf),
Revised 94-09-23

World-Wide Claimed VHF/UHF/SHF Terrestrial DX Records (notes 1 & 2)

Frequency	Record Holders	Date	Mode	Miles (km)
50 MHz	Note 3			
70 MHz	GW4ASR/P(IO82JG)-5B4CY (KM64MR)	1981-06-07	Es	2153 (3465)
144 MHz	I4EAT (JN54VG)-ZS3B (JG73OI)	1979-03-30	TE	4884 (7860)
220 MHz	KP4EOR (FK68XM)-LU7DJZ (GFØ5RJ)	1983-03-09	TE	3670 (5906)
432 MHz	KH6HME (BK29GO)-XE2GXQ/P (DL29CX)	1989-07-15	Duct	2575 (4143)
902 MHz	W4WSR (EL96WW) - WB5LUA (EM13QC)	1989-03-14	Tropo	1072 (1726)
1296 MHz	KH6HME (BK29GO)-XE2GXQ/P (DL29CX)	1989-07-15	Duct	2575 (4143)
2304 MHz	KH6HME (BK29GO) - N6CA/6 (DMO3TR)	1994-07-11	Duct	2469 (3973)
3456 MHz	KH6HME (BK29GO)-N6CA/6 (DMØ3TR)	1991-07-28	Duct	2469 (3973)
5760 MHz	KH6HME (BK29GO)-N6CA/6 (DMØ3TR)	1991-07-29	Duct	2469 (3973)
10.3 GHz	VK5NY (XXXX) - VK6KZ (XXXX)	1994-12-30	Duct	1187 (1911)
24 GHz	IØSNY/IC8 (JN6ØWR)-I8YZO/8 (JM78WE)	1984-08-11	LOS	206 (331)
47 GHz	HB9MIN (JN47GF)-HB9MIO (JN37MD)	1992-09-15	LOS	103 (166)
76 GHz	OE9PMJ/9 (JN48—)-OE9FKI/9 (JN48—)	1989-09-18	LOS	1.3 (2.1)
120 GHz	WA1MBA (FN32RJ) - WB2BYW (FN32RJ)	1994-07-11	LOS	0.71 (1.15)
145 GHz	WA1MBA (FN32RI)-WB2BYW (FN32RJ)	1993-05-02	LOS	2.4 (3.75)
474 THz	K6MEP/6 (DMØ4MS)-WA6EJO/6 (DMØ5XA)	1991-06-09	LOS	57 (92)
678 THz	KY7B/7 (DM42MS)-WA7LYI/7 (DM34TF)	1991-06-08	LOS	154 (248)

Notes:
1. The information within the brackets ( ) after the call sign is the grid square locator.
2. Distances have been calculated assuming a spherical earth model. The actual latitude and longitude are used rather than grid square centers which are less accurate.
3. 6 meters has been left blank on this listing because long-path QSO's (those exceeding approximately 12430 miles or 20000 km) have been reported during solar cycles 19, 21, and 22.

Reprinted with permission.

WORLD-WIDE EARTH-MOON-EARTH RECORDS

Note: The information and format used in this table is copyrighted in 1989 and registered at U.S. Patent Office by Joseph H. Reisert, W1JR, and may not be reprinted or used in any form without the written permission of the author.

Joseph H. Reisert, W1JR (wwemerec)
Revised 93-09-08

World-Wide Claimed VHF/UHF/SHF EME DX Records (notes 1 & 2)

Frequency	Record holders	Date	Mode	Miles (km)
50 MHz	WA4NJP (EM84DG)-ZL2BGJ (RF7ØDX)	1988-09-08	cw	8258 (13288)
144 MHz	K6MYC/KH6 (BK29AO)-ZS6ALE (KG43RC)	1983-02-18	cw	12091 (19455)
220 MHz	K1WHS (FN43MK)-KH6BFZ (BL11CJ)	1983-11-17	cw	5058 (8139)
432 MHz	G3SEK (IO91IP)-ZL3AAD (RE66GR)	1989-03-12	cw	11724 (18864)
902 MHz	AF1T (FN43ED)-K5JL (EM15DQ)	1990-05-27	cw	1476 (2376)
1296 MHz	PAØSSB (JO11WI)-ZL3AAD (RE66GR)	1983-06-13	cw	11595 (18657)
2304 MHz	W3IWI/8 (FMØ8CK)- ZL2AQE (RE78JS)	1987-10-18	cw	8658 (13931)
3456 MHz	W7CNK/5 (EM15FI)-KØKE (DM79NO)	1987-04-06	cw	498 (802)
5760 MHz	WA5TNY (EM12KV)-W7CNK/5 (EM15FI)	1987-04-24	cw	174 (279)
10,368 MHz	I6ZAU (JN63RJ) - WA7CJO (DM33XL)	1991-05-04	cw	6125 (9855)
24,000 MHz and above:	None reported			

Notes:
 1. The information within the brackets ( ) following the call signs is the grid square locator.
 2. The distances shown have been calculated assuming a spherical earth model. The actual latitudes and longitude are use rather than grid square centers which are less accurate.

Reprinted with permission.

CONTESTS

Contest Operating—A Tutorial

Ham radio is a competitive hobby. Active amateurs often find themselves engaged in competition by working a rare country, satisfying the requirements for a new award, or setting a long distance record on a UHF band. However, even when considering these endeavors in our hobby, there is nothing that is more competitive than the sport of contesting.

Contesting's beginnings took place over 60 years ago. The concept was originally developed as an attempt to improve the operating ability of amateurs around the world and advance the state of the art. Those goals remain in effect in today's contests.

When considering contesting, the most fundamental question one can ask is: "What is a Contest?" The basic definition is simply a scheduled operating event designed to encourage amateurs to contact as many other stations as possible over a fixed period of time and frequency spectrum. These events range from domestic affairs to worldwide phenomena such as the CQ World Wide. With dozens of contests being held annually around the world, contests have become a fundamental part of amateur radio.

Contest operating ability and knowledge is an acquired skill. While many operators have "instinctive proficiency" (e.g., capability to copy fast CW, good hearing, ability to process information quickly, etc.), the fastest way to improve your skill set is to meet other contesters and dive right into the fray.

Contests can be intimidating to the first-timer. On CW, there are seemingly hundreds of stations sending at least 30 WPM. When listening to SSB, you find world class experts working each other at rates of 350 QSOs per hour. The new competitor says: "that's not an environment for me!" The key is to remember that each and every station you hear in a contest was once a newcomer! The champions of the 90s were the newcomers of the 70s.

Here are a few basic points to keep in mind while operating:

1) Always sign your entire callsign when calling another station.

2) Accuracy is more important than speed. However, try to say/send the minimum amount of information necessary to complete a valid QSO as defined by the rules. Inaccurate logging will be found by the contest sponsor, and disqualification can result!

3) Practice, practice, practice. Not every contest is a "free for all." Many competitions are low key and offer a more subdued environment to hone your operating skills.

4) Practice your operating skills outside of contests. There is no substitute for improved CW skills and good ears on SSB.

5) Never give up or feel dismayed. Success in contesting is always a relative term. Most competing stations will never win. Those having the most fun compete against their previous accomplishments or a variety of other personal goals.

Frequently-Asked Questions about Contest Operating

Are contests only for big stations?

NO! If contests were only for big stations, there wouldn't be contests; there just aren't enough big stations to go around. A good operator at a modest station can beat a mediocre operator at a big station. Skill counts more than size.

What kind of radio/options should I be using?

From a contesting perspective, there are some *basic* features you should demand when selecting the "right radio." These include:

* Solid state finals—Critical for fast band changing.

* Digital displays—Reduces eye strain while operating and critical when packet spotting or meeting schedules.

* Selectable filtering options—The ability to copy through heavy QRM is never more needed than during a contest.

* RIT/XIT—Many stations call you off frequency during QSO runs. The RIT/XIT functions solve this common problem.

* Dual VFOs—Absolutely mandatory for 80/40 meter SSB. Also, more DX contesters are using split operation on other bands.

* Selectable attenuation—This is a necessary function on 40 meter to eliminate intermodulation problems from overseas broadcast stations.

* Programmable memories—Useful to program pileup frequencies or packet radio spots.

* Speech compression—A clear differentiator for phone contest operating. Audio punch is absolutely critical, especially when running low power.

* Noise blanking capability—Helpful for copying weak stations during high noise periods.

Some of the more *advanced* features to consider include:

* Computer interface (RS-232)—Absolutely critical if you plan to interface your computer logging program or packet radio spotting system to the transceiver.

* Independent tuning dials for main/sub VFOs—Extremely practical feature for separately tuning each VFO to allow VFO-independent band scanning and multiplier chasing.

* Antenna tuner—Useful for odd high-SWR antenna scenarios or when the input SWR to your amplifier is too high on certain bands.

* Electronic Keyer—Although most stations have a separate keyer accessory, backup devices such as these are always useful.

* Front panel control access—Helps the ease of adjustment for commonly changed settings (e.g., VOX settings, keyer speed).

* Full break-in—Nice CW feature that enhances your operating style in pileups by allowing you to time your calls.

Many of these features define a quality transceiver for any use within amateur radio. However, their absence in the competitive world of contest operating can dramatically impact your score. Not unlike most consumer purchases, it is easy to get caught up in the feature/function war between manufacturers. Don't forget that there is nothing better than good old fashioned word of mouth, especially if you can get it from an experienced contest operator.

A second consideration in equipment has arisen in recent years—the computer. While operating from a big contest station, you can sometimes find yourself looking for the radio in the maze of 386s, 486s, TNCs, keyboards, displays, cables, and other "amateur" support equipment. The guidelines for computers in amateur radio should parallel the same requirements for any home system. In today's world, a 286 system is barely adequate, while a 386 (or 386SX) is the model of choice. Nowadays, a con-

test station without a computer is only slightly less inept than one without antennas!

The final area to contemplate is the myriad of accessories (mostly homebrew) that can help differentiate a contest station. Consider issues such as audio and antenna switching, remote access to station functions, etc.

What about antennas?

Antenna selection decisions are usually determined by several factors such as money, time, and/or real estate. The guidelines for choosing antennas should be no different for a contest station than any other amateur setup. After all, whether you are interested in contesting, DXing, or casual ragchewing, you want to have the best signal possible.

For years, contesters have led the field in antenna innovation. Although most amateurs have limited resources to play in the mega-station field, a simple setup with a 60-foot tower, tribander for 10/15/20, two-element short yagi for 40, and a few strategically placed dipoles can do amazing things. Another alternative is to focus on a single band and place all your efforts in that direction.

And, don't forget that remarkable accomplishments are possible with only wire antennas. A quick tour through the ARRL antenna handbook (and other publications) will not only educate you, but offer low cost alternatives for the newcomer.

How do I choose my operating category?

Choosing an operating category begins by reading the rules and understanding your options. Multioperator competition is an excellent entry point for the "novice" contester, by teaming up with an experienced contester who will allow you to "help out" at his station.

In recent years, the growth of packet radio (and resultant "Single-operator-assisted" categories) has led many to believe that operating with packet spotting assistance is probably the most fun for a new contester from a small station. You have the combined advantage of making lots of interesting QSOs while honing your natural skills. It also reduces your loneliness during slow-rate periods. Simply put, it is a lot of fun!

How can I maximize my score from a small station?

Most of the strategic skill in contest operating is fundamentally based on common sense. Unfortunately, not everyone has the opportunity to operate from a contest "superstation." The vast majority of competitors use tribanders and dipoles.

For most people, contesting is a sport that allows us to operate and "see what we can do." The issue of maximizing your score begins with an honest assessment of your station's strengths and weaknesses. If you are using a dipole on 40 meters, it is going to be difficult to compete against the crowd of stations with large beams at the low end of the band. However, tuning up and down the band and calling stations can be very productive. Secondly, selection of operating times is key. If you have limited operating time, try to choose a schedule that matches the times when conditions are optimum for peak rates (e.g., 15 meters in the morning for Europe in DX contests).

Operating from smaller stations actually forces you to be a better operator. It requires that you be more clever in signing your callsign in pileups (brute force just doesn't work). A strategic callsign placement during a lull in a pile often pays off! More important, the small station can still be very effective during the peak times of activity. Use your VFO to tune, tune, and tune again. And when you feel there is opportunity to make QSOs faster by calling CQ, stay high in the band and avoid the "big guns."

Where can I get more information about contests?

CQ Magazine and the ARRL Contest Branch can help you with additional information. The *National Contest Journal* is an excellent magazine that is exclusively focused on contesting. It includes interesting articles and features, and can help identify specific contesters who would be willing to answer questions and provide direction to the new operator.

Also, depending on your geographic location, there are a number of active contest clubs around the world that are interested in gaining new members. The following, while not a complete list of these clubs, can help you get started.

CLUBS DEVOTED TO CONTEST OPERATING

Club	Geography	Contact
USA Clubs		
Cascade Contest Club	WA/OR	N6TR
Carolina DX Association	NC/SC	WD4R, K2SD
Central Texas DX/Contest Club	Central TX	N5DDT
Cops Contest Club	San Francisco	KG6LF
Colorado Contest Conspiracy	CO	WØUN
Florida CW Contest Group	Florida	K1ZX, WC4E, AC1O
Florida Contest Club	Florida	WB4MAI, WB4EYX
Fort Wayne DX Assn.	Indiana	K9LA, NJØU
Frankford Radio Club	Philadelphia	KY3N
High Plains Contest Club	CO/WY	N2IC
High Plains Contest Group	Montana	AA7PD, AA7BG
Hoosier Contesters	Indiana	N9NS, WB9CIF
Hudson Valley Contesters	NY	W2XL
Kentucky Contest Group	KY/TN/So. IN	ND4Y, N4TG, N4TY, AB4RX
Licking County Contest Group	OH	AA8SM
MileHi DX Association	Colorado	WA8SWM, N3SL, WØPSY
Miss. Valley DX/Contest Club	MO/IL	WW9Q, KØOD
New Mexico Big River Contesters	Rio Grande Vly	AA5B, K5TA
North East Weak Signal Group	CT	WZ1V
North Coast Contesters	OH/PA	K3LR
North Texas Contest Club	Dallas	K5RX, KI5JC, NA5Q
Northern Calif. Contest Club	No. Cal./NV	WM2C, K2MM, AB6YL, W6ISQ
Oregon Nocturnal Chordal Corps	OR/WA	N7AVK, KC7EM
Potomac Valley Radio Club	MD/VA	KE3Q, KF3P
Redwood Empire DX Association	No. Cal.	WA8LLY
River City Contesters	Sacramento, CA	AA6WJ, KN6OX
Rochester DX Assoc.	Rochester, NY	WA2LCC
Salt City DX Assoc. Contest Group	Central NY	KE2VB
Society of Midwest Contesters	Upper Midwest	NØBSH, WB9TIY
Southeastern DX Club	Atlanta	KR4DL, KE4LDJ
Southern Calif. Contest Club	Los Angeles	WA6OTU
Tennessee Contest Group	TN	KØEJ
Texas DX Society	Houston	N5RP, KG
Thoroughbred DX Group	Central KY	K2YJL
Yankee Clipper Contest	New England	AA2DU, KB2R
DX Clubs		
Cuba Libre Contest Club	Belize	KI6WM, N6YRU
Chiltern DX Club	U.K.	GØORH
EDR Roskilde	Denmark	OZ1FTE
FPM Contest Group	England	GØRDI, GØJIM
Grupo DX Panamericano	South TX/Mex.	K5TSQ
Pretoria Contest Club	Pretoria	ZS6EZ
Red Dragons Contest Group	Wales	GW3NWS, GW3KYA
Top of Europe Contesters	Sweden	SM3SGP, SM3OJR
TuPY DX Group	Brazil	PY2NY

CONTEST SCHEDULE

Contest	Wkend./Mo.	Hours	Sponsor Address
ARRL Straight Key Night	1/JAN	varies	ARRL, 225 Main St., Newington, CT 06111
ARRL RTTY Roundup	1/JAN	24/30	ARRL, 225 Main St., Newington, CT 06111
AGCW-DL QRP CW Winter Contest	1/JAN	15	DJ7ST, Dr.H.Weber, Schesierweg 13,W-3320, Salzgitter, Germany
Michigan QRP Club CW Contest	1/JAN	36	N8CQA, 654 Georgia, Marysville, MI 48040
Hunting Lions in the air CW Contest	1/JAN	36	LIONS, P.O. Box 106, 64223 Flen, Sweden
NCJ N.A.QSO Party-CW	2/JAN	10/12	Bob Schbrede, W9NQ, 6200 Natoma Ave., Mojave, CA 93501
JA Int'l CW Contest (160,80,40M)	2/JAN	48	59 Magazine Contest, P.O. Box 59, Kamata, Tokyo 144, Japan
Hunting Lions in the air SSB Contest	2/JAN	36	LIONS, P.O. Box 106, 64223 Flen, Sweden
HA DX CW Contest	2/JAN	24	Budapest Hungary Radio Amateur Society, Box 86, Budapest, Hungary H-1581
NCJ N.A. QSO Party-SSB	3/JAN	10/12	Bob Schbrede, W9NQ, 6200 Natoma Ave., Mojave, CA 93501
ARRL VHF Sweepstakes	3/JAN	33	ARRL, 225 Main St., Newington, CT 06111
YL-ISSB CW QSO Party	3/JAN	24	Rhonda Livingston, N4KNF, 2160 Ivy St., Pt. Charlotte, FL 33952
REF CW Contest	4/JAN	36	R.E.F., P.O.Box 2129, F-37021 Tours Cedex, France
CQ WW DX 160m CW Contest	LAST/JAN	42	CQ 160 CW Contest, 76 N. Broadway, Hicksville, NY 11801
U.B.A. SSB Contest	LAST/JAN	24	UBA-HF, ON6JG Oude Gengarmeriestraat 62, B-2220 Heist Op Den Berg, Belgium
NCJ N.A. Sprint-CW	1/FEB	4	N6TR, Larry "Tree" Tyree, 15125 SE Bartell Rd., Boring, OR 97009
New England QSO Party	1/FEB	18	GEARS, P.O.Box 1076, Claremont, NH 03743-1076
Classic Radio Exchange	1/FEB	48	W8KGI, P.O.Box 581, Sandia Park, NM 87047
QCWA QSO Party CW	2/FEB	25	Bob Reed, WB2DIN, 597 Brewers Bridge Rd., Jackson, NJ 08527
NCJ's N.A. Sprint-SSB	2/FEB	4	K7GM, Rick Niswander, P.O.Box 3778, Greenville, NC 27836-1778
Utah 160m Challenge	2/FEB	48	WE7H, 68 South 300 West, Brigham City, UT 84302
Dutch "PACC" Contest	2/FEB	24	PACC, PAØINA, P.O. Box 499, 4600 AL Bergen op Zoom, The Netherlands
YLRL YL-OM SSB Contest	2/FEB	24	WO6X, 473 Palo Verde Dr., Sunnyvale, CA 94086
ARRL DX CW Contest	3/FEB	48	ARRL, 225 Main St., Newington, CT 06111
YL-ISSB SSB QSO Party	3/FEB	48	Rhonda Livingston, N4KNF, 2160 Ivy St., Pt. Charlotte, FL 33952
YLRL YL-OM CW Contest	4/FEB	24	WO6X, 473 Palo Verde Dr., Sunnyvale, CA 94086
CQ WW DX 160m SSB Contest	4/FEB	42	CQ 160 PH Contest, 76 N. Broadway, Hicksville, NY 11801
REF SSB Contest	4/FEB	36	R.E.F., P.O.Box 2129, F-37021 Tours Cedex, France
RSGB 7 MHz CW Contest	4/FEB	18	G3UFY, HF Contests, 77 Bensham Manor Rd, Thornton Heath, Surrey CR7 7AF, UK
U.B.A. CW Contest	4/FEB	24	UBA-HF, ON6JG, Oude Gengarmeriestraat 62,B-2220 Heist Op Den Berg, Belgium
ARRL DX SSB Contest	1/MAR	48	ARRL, 225 Main St., Newington, CT 06111
QCWA QSO Party-SSB	2/MAR	25	QCWA, W1EES, 75 Chestnut Circle, West Suffield, CT 06093
Wisconsin State QSO Party	2/MAR	7	WARAC, P.O. Box 1072, Milwaukee, WI 53201
Bermuda Contest	3/MAR	48	RSB, P.O.Box HM275, Hamilton HM AX, Bermuda
Virginia State QSO Party	3/MAR	32	W3FTG, 3627 Great Laurel Ln., Fairfax, VA 22033-1212
BARTG Spring RTTY Contest	3/MAR	30	G4SKA, 32 Wellbrook St., Tiverton, Devon, EX16 5JW, UK
CQ WW WPX SSB Contest	4/MAR	48	CQ WPX SSB Contest, 76 N. Broadway, Hicksville, NY 11801
Poisson d'Avril Contest	1/APR	42	Doug Grant, K1DG, 144 Kendall Pond Rd., Windham, NH 03087
ARCI QRP CW Spring QSO Party	1/APR	24	K5VOL, ARCI Contest Mgr, 835 Surryse Rd., Lake Zurich, IL 60047

Contest	Date	Hours	Address
SP DX Contest	1/APR	36	PZK, P.O.Box 98, 59-220, Legnica, 2, Poland
ARRL VHF/UHF Spring Sprint-144MHz	1/APR	4	ARRL 225 Main St., Newington, CT 06111
EA RTTY Contest	1/APR	24	EA1MV, EA RTTY, P.O.Box 240, 09400-Aranda,
Holyland DX Contest SSB,CW	1/APR	24	IARC, 4Z4UT, Box 3003, Beer-Shera, 84130,
JA Int'l CW Contest (20,15,10M)	2/APR	48	'59 Magazine Contest, P.O.Box 59, Kamata, Tokyo 144, Japan
MARAC SSB County Hunters Contest	2/APR	48	Bill Nash, WØ0WY, 13212 N. 37th Ave., Phoenix, AZ, 85029
ARRL VHF/UHF Spring Sprint - 222 MHz	2/APR	4	ARRL, 225 Main St., Newington, CT 06111
SARTG AMTOR Contest	3/APR	24	SM4CMG, Skulsta 1258, S-710 41 Fellingsbro, Sweden.
QST QSO Award Party	3/APR	24	VE3IAE, 19 Honeysuckle Crese, London, Ontario N5Y 4P3, Canada
Spring NWQRP Sprint-CW	3/APR	4	WU7F, 6822 131 Ave. SE, Bellevue, WA 98006
ARRL VHF/UHF Spring Sprint-432 MHz	3/APR	4	ARRL, 225 Main St., Newington, CT 06111
Swiss HELVETIA Contest SSB,CW	4/APR	24	HB9DDZ, Postfach 651, CH-4147 Aesch, Switzerland
Friendship Contest	4/APR	24	SRR, Box 59, Moscow, Russia
ARI Int'l DX Contest SSB, CW, RTTY	1/MAY	24	I2UIY, ARI Contest Mgr, P.O. Box 14, 27043 Broni (PV), Italy
MARAC CW County Hunters Contest	1/MAY	48	W3DYA, 3320 McMillan Dr., Tyler, TX 75701-8239
ARRL UHF Spring Sprint-902/1296/2304 MHz	1/MAY	4	ARRL, 225 Main St., Newington, CT 06111
10-X Int'l Spring CW QSO Party	1/MAY	48	N8FU, 4441 Andreas Ave., Cincinnati, OH 45211
CQ-M Contest SSB,CW	2/MAY	24	KCRC, P.O.Box 88, Moscow, Russia
ARI A.VOLTA RTTY Contest	2/MAY	24	P.O. Box 55, 22063 Cantu, Italy
Massachusetts QSO Party	2/MAY	30	FARA, PO Box 3005, Framingham, MA 01701
Georgia QSO Party	2/MAY	28	Sandy Walker III, WB4EVH, 411 Wilson Dr., Centerville, GA 31028
Nevada QSO Party	2/MAY	48	NW7O, 4120 Oakhill Ave., Las Vegas, NV 89121
Danish SSTV Contest	2/MAY	48	Carl Emkjer, Soborghus Park 8, DK-2860 Soborg, Denmark
Michigan QSO Party	3/MAY	24	K8ED, 27600 Franklin Rd., Apt 816, Southfield, MI 48034
ARRL Spring Sprint-50 MHz	3/MAY	4	ARRL, 225 Main St., Newington, CT 06111
CQ WW WPX CW Contest	4/MAY	48	CQ WPX CW Contest, 76 N. Broadway, Hicksville, NY 11801
RSGB National Field Day	1/JUN	24	G3UFY, HF Contests, 77 Bensham Manor Rd, Thornton Heath, Surrey, CR7 7AF, UK
PORTUGAL Day Contest	2/JUN	24	REP, P.O. Box 2483, 1112 Lisboa Codex, Portugal
ARRL June VHF Contest	2/JUN	33	ARRL, 225 Main St., Newington, CT 06111
ANARTS WW RTTY Contest	2/JUN	48	ANARTS, P.O. Box 93, Toongabbie, NSW 2146
ALL ASIAN CW Contest	3/JUN	48	J.A.R.L., P.O. Box 377, Tokyo Central, Japan
SMIRK 50 MHz QSO Party	3/JUN	48	W5OZI, P.O. Box 393, Junction, TX 78649
ARRL Field Day	4/JUN	27	ARRL, 225 Main St., Newington, CT 06111
RSGB 1.8 MHz Contest	4/JUN	48	
SP-QRP Int'l Contest	4/JUN	48	
Connecticut QSO Party	LAST/JULY	28	CARA, P.O. Box 3441, Danbury, CT 06813-3441
R.A.C. CANADA Day Contest	1/JUL	24	R.A.C., 614 Noris Ct., Unit 6, Kingston, ON K7P 2R9 Canada.
Venezuela Independence Day SSB Contest	1/JUL	48	RCV, P.O. Box 2285, Caracas 1010-A,Venezuela
IARU HF Championship SSB-CW	2/JUL	24	IARU HQ, Box 310905, Newington, CT 06131-0905
CQ WW WPX VHF Contest	2/JUL	27	N6CL, Joe Lynch, P.O. Box 73, Oklahoma City, OK 73101
ARCI QRP Summer Homebrew Sprint-CW	2/JUL	4	K5VOL, ARCI Contest Mgr, 835 Surryse Rd., Lake Zurich, IL 60047
Colombian Independence Day Contest	3/JUL	24	Columbian Independance Day Contest, Apartado 584, Santafe de Bogota, Columbia
AGCW-DL QRP CW Summer Contest	3/JUL	24	DJ7ST, Dr.H.Weber, Schesierweg 13, W-3320, Salzgitter, Germany
SEANET CW Contest	4/JUL	48	SEANET '94, Eshee Razak, 9M2FK, P.O. Box 13, 10700 Penang, Malaysia
Venezuela Independence Day CW Contest	4/JUL	48	RCV, P.O. Box 2285, Caracas 1010-A,Venezuela
RSGB IOTA HF Contest	LAST/JUL	24	G3UFY, HF Contests, 77 Bensham Manor Rd, Thornton Heath, Surrey CR7 7AF, UK

NCJ N.A. QSO Party-CW	1/AUG	10/12	Bob Schbrede, W9NQ, 6200 Natoma Ave., Mojave, CA 93501
QRP ARCI Summer SSB QSO Party	1/AUG	4	K5VOL, ARCI Contest Mgr, 835 Surryse Rd., Lake Zurich, IL 60047
10-X Summer Phone QSO Party	1/AUG	48	K0PVI, P.O. Box 112, Watkins, CO 80137
ARRL UHF Contest	1/AUG	24	ARRL, 225 Main St., Newington, Ct., 06111
YO DX HF Contest	1/AUG	20	RARF, P.O. Box 05-50, R-76100 Bucharest, Romania
WAE CW Contest	2/AUG	36	WAEDC, P.O.Box 1126, D-74370, Serhiem, Germany
Maryland-D.C. QSO Party	2/AUG	19	Antietam Radio Assn., P.O. Box 52, Hagerstown, MD 21741
SARTG WW RTTY Contest	3/AUG	10/12	SM4CMG Skulsta 1258, S-710 41 Fellingsbro, Sweden
NCJ N.A. QSO Party-SSB	3/AUG	10/12	Bob Schbrede, W9NQ, 6200 Natoma Ave., Mojave, CA 93501
SEANET SSB Contest	3/AUG	48	SEANET '94, Eshee Razak, 9M2FK, P.O. Box 13, 10700 Penang, Malaysia
New Jersey State QSO Party	3/AUG	17	EARA, P.O. Box 528, Englewood, NJ 07631-0528
ARRL 10Ghz Cumulative Contest-Part 1	3/AUG	24	ARRL, 225 Main St., Newington, CT 06111
West Virginia QSO Party	4/AUG	48	
TOEC Field CW Contest	4/AUG	24	TOEC, Box 2063, S-83102, Ostersund, Sweden
NCJ N.A. Sprint-CW	1/SEPT	4	K7GM, Rick Niswander, P.O. Box 3778, Greenville, NC 27836-1778
R.A.C. VHF/UHF Sprint-902/1296/2304 MHz	1/SEPT	4	R.A.C., P.O. Box 356, Kingston, Ontario, K7L 4W2, Canada
LZ DX-Contest	1/SEPT	48	BFRA, P.O. Box 830, Sofia 1000, Bulgaria
All-Asian SSB Contest	1/SEPT	48	J.A.R.L., P.O. Box 377, Tokyo Central, Japan
Panama Aniv.Contest	1/SEPT	24	RCP, P.O. Box 10745, Panama, 4, Panama
WAE DARC SSB Contest	2/SEPT	36	WAEDC, P.O. Box 1126, D-74370, Serhiem, Germany
ARRL VHF QSO Party	2/SEPT	33	ARRL 225 Main St., Newington, CT 06111
NCJ's N.A. Sprint-SSB	2/SEPT	4	N6TR, Larry "Tree" Tyree, 15125 SE Bartell Rd., Boring, OR 97009
R.A.C. VHF/UHF Sprint-432 MHz	2/SEPT	4	R.A.C., P.O. Box 356, Kingston, Ontario, K7L 4W2, Canada
YLRL Howdy Days Contest	2/SEPT	36	Carla Watson, WO6X, 473 Palo Verde Dr., Sunnyvale, CA 94086
ARRL 10 GHz Cumulative Contest-Part 2	3/SEPT	24	ARRL, 225 Main St., Newington, CT 06111
R.A.C. VHF/UHF Sprint-220MHz	3/SEPT	4	R.A.C., P.O.Box 356, Kingston, Ontario, K7L 4W2, Canada
Scandinavian CW Contest	3/SEPT	27	SRAL, Harri Mantila, OH6YF, P.O. Box 30, SF-64701, Teuva, Finland
CQ WW RTTY Contest	4/SEPT	48	KT1N, Roy Gould, CQ RTTY Contest Mgr, P.O. Box DX, Stow, MA 01775
R.A.C. VHF/UHF Sprint-144MHz	4/SEPT	4	R.A.C., 614 Noris Ct., Unit 6, Kingston, ON K7P 2R9 Canada.
Scandinavian SSB Contest	LAST/SEPT	27	SRAL, Harri Mantila, OH6YF, P.O. Box 30, SF-64701, Teuva, Finland
Washington State Salmon Run	4/SEPT	31	WWDXClub, W7FR, P.O. Box 224, Mercer Island, WA 98040
Classic Radio Exchange	4/SEPT	48	W8KGI, P.O.Box 581, Sandia Park, NM 87047
VK/ZL SSB DX Contest	1/OCT	24	NZART, ZL1AAS, 146 Sandpit Rd., Howick, 1705, New Zealand
California QSO Party (CQP)	1/OCT	30	NCCC, K6PU, P.O. Box 853, Pine Grove, CA 95665
F9AA Cup Contest	1/OCT	24	Coupe Fernand Raoult, 11 Rue de Bordeaux, 94700 Maisons Alfort, France
R.A.C. VHF/UHF Sprint-50MHz	1/OCT	48	R.A.C., 614 Noris Ct., Unit 6, Kingston, ON K7P 2R9 Canada.
Tennessee QSO Party	1/OCT	31	Douglas Smith, 1385 Old Carlesville Pike, Pleasant View, TN 34146-8098
RSGB 21/28 MHz SSB Contest	1/OCT	14	G3UFY, HF Contests, 77 Bensham Manor Rd, Thornton Heath, Surrey CR7 7AF, UK
XVII Iberoamencano Contest	1/OCT	48	Concurso Iberoamericano, c/Concepcion Arenal 5, 08027, Barcelona, Spain

Pennsylvania State QSO Party	2/OCT	22	Douglas Maddox, W3HDH, NARC, Box 614, State College, PA 16804-0614
VK/ZL CW DX Contest	2/OCT	24	NZART, ZL1AAS, 146 Sandpit Rd., Howick, 1705, New Zealand
YLRL Anniversary CW Party	2/OCT	48	Carla Watson, W06X, 473 Palo Verde Dr., Sunnyvale, CA 94086
Illinois State QSO Party	3/OCT	8	RAMS, KB9II, 7079 West Ave., Hanover Park, IL 60103
Texas State QSO Party	3/OCT	48	TDXS, P.O. Box 540291, Houston, TX 77254-0291
RSGB 21 MHz CW Contest	3/OCT	14	G3UFY, HF Contests, 77 Bensham Manor Rd, Thornton Heath, Surrey CR7 7AF, UK
ARCI QRP CW Contest	3/OCT	36	K5VOL, ARCI Contest Mgr, 835 Surryse Rd., Lake Zurich, IL 60047
W.A.G. Worked All Germany	3/OCT	24	DL1DTL, P.O.Box 427, 0-8072 Dresden, Germany
JARTS WW RTTY Contest	3/OCT	48	JARTS, Hiroshi, Aihera, JH1BIH, 1-29 Honcho 4 Shiki, Saitana, 353 Japan
CQ WW DX PHONE Contest	LAST/OCT	48	CQ WW PH Contest, 76 N. Broadway, Hicksville, NY 11801
YLRL Anniversary SSB Party	LAST/OCT	48	Carla Watson, 473 Palo Verde, Sunnyvale Dr., CA 94086
ARRL Sweepstakes CW	1/NOV	24/30	ARRL, 225 Main St., Newington, CT 06111
JA Int'l DX SSB Contest	1/NOV	48	59 Magazine Contest, P.O. Box 59, Kamata, Tokyo 144, Japan
OK-OM DX Contest	2/NOV	24	OK2FD, Gen. Srobody 636, 67401, Trebic, Czech Republic
WAE DARC RTTY Contest	2/NOV	36	WAEDC, P.O.Box 1328, D-8950 Kaufbeuren, Germany
ARRL E-M-E Contest	2/NOV	48	ARRL, 225 Main St., Newington, CT 06111
ARRL Sweepstakes SSB	3/NOV	24/30	ARRL, 225 Main St., Newington, CT 06111
CQ WW DX CW Contest	LAST/NOV	48	CQ WW CW Contest, 76 N. Broadway, Hicksville, NY 11801
ARRL 160m DX Contest	1/DEC	42	ARRL 225 Main St., Newington, CT 06111
ARRL 10m DX Contest	2/DEC	36/48	ARRL 225 Main St., Newington, CT 06111
RAC Winter Contest	LAST/DEC	24	R.A.C., 614 Noris Ct., Unit 6, Kingston, ON K7P 2R9 Canada.

HIGHEST REPORTED RATES IN CONTEST OPERATIONS (QSO'S PER HOUR)

DX Contests, DXpeditions—DX Side

Phone

457	KRØY at P4ØL	1993	CQWW	15M 00Z
387	CT1BOH at KP2A	1993	CQWW	19Z
385	K5ZD at NP4A	1980	CQWW	
380	CT1BOH at KP2A	1993	CQWW	First hour
379	K1DG at VP2E	1994	CQWW	10M 18Z D1
375	W6QHS at P4ØV	1994	ARRL DX	10M 1537Z
371	K1AR at TI1J	1989	CQWW	15M 00Z
370	K1DG at KP2A	1988	CQWW	15M 00Z
367	N6KT	1988		20M
365	K3NA at PJ9JR	1979	CQWW	
362	CT1BOH at KP2A	1991	CQWW	15Z
351	K1DG at VP2E	1994	CQWW	15M 19Z D2
350	K5RX at VP2EC	1987	CQWW	
347	CT1BOH at KP2A	1992	CQWW	20Z
343	N6KT at HC8A	1994	CQWW	10M 2202Z D2
342	CT1BOH at KP2A	1991	CQWW	15Z
341	KØGU at VP2KBU	1984	ARRL DX	10M
341	K1DG at VP2E	1994	CQWW	15M 12Z D1
340	WX9E at J6DX	1994	CQWW	20M 00Z
337	WB2CHO at VP2ML		ARRL DX	
337	K1MM at LU2MM	1983	ARRL DX	15M 00Z
335	CT1BOH at KP2A	1991	CQWW	Second hour
334	CT1BOH at KP2A	1992	CQWW	20M 01Z
334	K5TSQ at 6D2X	1993	ARRL DX	15M First hour
334	KA9FOX at PJ8Z	1994	CQWW	15M 15Z D1
325	K1MM at LU2MM	1983	ARRL DX	15M 01Z
324	CT1BOH at P4ØE	1994	CQWW	02Z

321	FS/KC1F	1989	CQWW	15M First hour
320	V31DX	1992	CQWW	10M
317	WB5VZL at HR6A	1987	ARRL DX	
315	AA5B at ZF2HM	1984	ARRL DX	15/20M
313	W2GD at P4ØW	1993	CQWW	20/15/10M 00Z
310	CT1BOH at KP2A	1992	CQWW	20M 00Z
310	AA5B at ZF2HM	1984	ARRL DX	
308	N6ZZ at KP2A	1980	CQWW	15M 03Z
308	K1DG at VP2E	1994	CQWW	20M 22Z
299	PJ1B	1992	CQWW	20M 00Z
290	W6QHS at EA9UK	1993	CQWW	10M 1618Z
289	WB5VZL at 6D2X	1993	ARRL DX	10M First hour

CW

249	W2GD at P4ØW	1993	CQWW	20M 00Z
248	N6TR at TI1C	1993	CQWW	20M 01Z
234	WN4KKN at HC8N	1992	CQWW	15M 21Z
231	KC1F at NP4A	1982	CQWW	
230	W2GD at P4ØW	1993	CQWW	20M 01Z
230	WB9JKI at V31A	1987	CQWW	
227	NL7GP at KL7Y	1995	ARRL DX	00Z
228	CT1BOH at PYØF	1993	CQWW	18Z
224	WD8IXE at J6DX	1993	CQWW	15M 15Z
223	LU8DQ at ZP5Y	1990	CQWW	10M 18Z
222	W2GD at P4ØW	1993	CQWW	10M 15Z
221	K3UA at NP4A	1982	CQWW	15M 00Z
219	NL7GP at KL7Y	1995	ARRL DX	01Z
217	CT1BOH at PYØF	1993	CQWW	13Z
215	W2GD at P4ØGD	1987	CQWW	
215	K3UA at NP4A	1980	CQWW	
212	WN4KKN at HC8N	1992	CQWW	10M 19Z
212	WN4KKN at TI1C	1993	ARRL DX	20M First hour
212	CT1BOH at 4M2BYT	1992	CQWW	14Z
211	W6QHS at P49V	1993	ARRL DX	15M
211	N6ZZ at XE2MX	1984	CQWW	40M 03Z
209	WN4KKN at TI1C	1993	ARRL DX	15M 20Z
208	WN4KKN at ZPØY	1990 and 1991	CQWW	both 10M 18Z
207	WN4KKN at TI1C	1993	ARRL DX	40M 03Z
206	WN4KKN at TI1C	1993	ARRL DX	10M 15Z
208	W6QHS at P49V	1993	ARRL DX	
206	N6ZZ at PZ5JR	1994	CQWW	15/10M 14Z
205	K7SS & NØAX at NH6T	1992	ARRL DX	
204	WN4KKN at TI1C	1993	ARRL DX	15M 21Z
201	WN4KKN at HC8N	1992	CQWW	10M 18Z
200	K7SS at K7SS/KH6		ARRL DX	
200	K6NA at KP6BD	1977	Kingman Reef DXpedition	
200	CT1BOH at HC5M	1991	CQWW	03Z
199	CT1BOH at PYØFF	1994	CQWW	13Z
198	K8MR at VP2E	1982	ARRL DX	
193	K1MM at 1S1DX	April 1979		15M 13Z
188	WØCP at AH1A	Jan. 1993	DXpedition	17M 19Z
185	NØAX at KH6RS	1995	ARRLDX	15M
180	K1MM at 1S1DX	April 1979		10M 04Z

DX Contests—U.S. Side

Phone

221	KC1F at K1EA	1988	CQWW	10M 12Z
220	K1AR at K1EA	1992	CQWW	10M 12Z
211	AD1C at KC1XX	1993	CQWW	10M 13Z D2
200	AA6TT	1992	CQWW	15M 00Z M/M
192	AD1C at KC1XX	1993	CQWW	10M 14Z D2
189	K5ZD	1994	ARRLDX	15M 1229Z
177	N2RM	1992	CQWW	15M 12Z M/M
175	K1DG	1992	CQWW	15M/10M 17Z
174	K3LR at K3TUP	1988	CQWW	15M/10M 12Z
173	K1AR at K1EA	1992	CQWW	10M/15M 13Z
172	K1AR at K1EA	1992	CQWW	20M 19Z
172	K1DG	1993	ARRL DX	
169	KC1XX/AD1C/KM3T at KC1XX	1994	ARRL DX	15M 12Z D1
169	K5ZD	1993	CQ WW	15M 1419Z
166	N6BV/1	1993	ARRL DX	10M 14Z
161	KI3V at NZ7E	1992	ARRL DX	
160	K1KI	1986	ARRL DX	20M 13Z
153	K5ZD	1994	CQ WPX	20M 1631Z
139	K3WW	1992	CQ WW	2209Z
131	K3WW	1993	ARRL	1626Z
118	K3WW	1992	ARRL	0100Z

CW

173	KM9P at N4RJ	1995	ARRL DX	15M 13Z
170	AD1C at KC1X	1994	CQWW	15M 13Z D2
168	K3WW	1992	ARRL	1257Z
160	K1EA using KC1XX at K1EA	1993	WAE	15M
153	K3WW	1993	CQ WW	1725
152	K1DG (K1DG, WZ1R oprs.)	1992	CQWW	15M/10M 12Z
147	K5ZD	1993	CQWW	15M 1301z
147	K5ZD	1995	ARRL DX	15M 1206Z
143	K3WW	1991	ARRL	1243
142	K1TO at K1KI	1992	CQWW	15/10M 12Z
142	K5ZD	1994	CQWW	15M 1420Z
136	K3WW	1992	CQ WW	1844
136	W2UP	1995	ARRL	
131	K3WW	1994	ARRL	40M 0059Z D1
128	K1KI	1985	CQM	20M
125	K1KI	1987	ARRL DX	20M 11Z
125	KC1XX using AD1C at KC1XX	1994	ARRL DX	15M 13Z D1
122	K3WW	1994	CQ WW	1735
115	K3WW	1991	CQ WW	1402Z

ARRL 10M Contest

273	K6LL	1994	ALL SSB	SUNDAY 1940-2039Z
244	AA5B	1990		
235	AA5B	1988		
181	K1KI	1980	SSB	16Z
152	K1KI	1992	CW	13Z

ARRL SS

Phone

189	N8RR/5	1994	First hour 15M
186	KI3V	1992	10M First hour
178	WB5VZL at W5KFT	1994	2105 - 2205 15M First day
175	WB5VZL at N5AU	1986	20/15M First hour
174	N2IC at KP2A	1988	10M First hour
173	K5ZD at N5AU	1980	First hour
170	AA5B (twice)		
153	N6BV at N6RO	1987	First hour
142	KØKR	1993	First Hour
140	K1KI	1985	40M 01Z

CW

116	K6LL	1993	21Z
108	N6TR	1994	21Z
108	N5RZ	1994	22Z
107	KM9P at N4RJ	1994	21Z
105	N6TR	1981	First hour
105	KM9P at N4RJ	1993	00Z
103	N2IC/Ø	1994	21Z
103	N5RZ	1994	21Z
100	KI3V	1992	15M First hour
100	K5ZD at N5AU (several times 1980-84)		First hour
93	K1KI	1990	20M 12Z

Courtesy of National Contest Journal Mar/Apr 1988, K1AR column in CQ Magazine, CQ-CONTEST@TGV.COM, and various reporters.

OPERATING ON THE AMATEUR HF BANDS

Amateur Radio as an avocation for experimenters and communicators began at the start of this century with the use of frequencies below the present High Frequency bands (1.8 to 30 MHz) and now includes allocations reaching to 250 GHz. Until the 1970s, the HF bands were the medium used for all international and coast-to-coast radio communications by hams as well as commercial services. These frequencies still carry a great deal of commercial worldwide traffic, as well as including all phases of ham interests from across town to cross-country ragchewing, DXing, contesting, public service work and experimentation. In the past, most hams were first exposed to HF; quite often they later began to utilize the VHF bands also for satellites, local repeaters, and easy antenna experimentation.

More recently, many hams have reversed this trend; they are first exposed to 2 meter repeater operation and then become curious about the worldwide capabilities of the HF bands. This section will provide an introduction to each of the Amateur bands from 10 to 160 meters.

10 Meters

One band that older and newer hams have in common is the 10 meter. It has long been appreciated by newcomers for the ease of worldwide communications with low power and small gain-type antennas. There is plenty of spectrum to accommodate modes from CW and SSB to FM, AM and satellites; and the band is available to all U.S. amateur license classes except the codeless Technician. Those who have been active through a few sunspot cycles remember how easy it was to work DX on this band when the MUF (Maximum Usable Frequency) was high and how dead the receiver can sound six years later. This phenomenon spurred the creation of a group, the 10-X International Net, whose purpose is to promote activity. Because 10 meters shows some of the propagation characteristics of the higher (VHF) bands, it is well known that listening is not proof that the band is closed. Short but excellent openings are always possible, waiting to be discovered by those who call CQ.

The most popular station setup for 10 meters is probably a barefoot (100 watt) transceiver and a small tribander (three-element 10,15 and 20 meter) or monoband 10 meter beam at rooftop level. Linear amplifiers and higher antennas are not really needed, but do put the antenna as high as you can. Small TV rotators will easily handle these antennas. Because of the large amount of spectrum available, it is important to tune slowly but completely. You may find an easily workable but weak station in a state you need 300 miles away or a new friend on an island in the South Pacific. You will be weak at his location also but, because of the lack of static and the avoidance of big pileups, a good contact can result.

15 Meters

As we tune lower in frequency (the WARC bands of 17, 12 and 30 meters are covered later), band conditions become more predictable and more congested. Fifteen meter long-haul DX is more common during low sunspot times than on 10 meters, and contacts across the U.S. are possible almost every day. There are more nets found on 15 than on 10 meters, including weather emergency, Red Cross, U.S. State Department hams overseas and DX nets.

There is no Novice/Technician SSB allocation and *all* license classes are limited to 200 watts in the Novice sub-band, unlike 10 meters. This doesn't mean that DXCC is impossible for Novice/Technicians because of the CW limitation. In 1957, KN4RID (now WØZV) became the first Novice to gain the DXCC award. He was limited by the regulations at that time to 75 watts, crystal control and a one-year maximum license term. Novice Worked All Zones has also been earned on this band.

More hams will be using higher power here because of the increased competition for less band space. The three-element tribander is still the most popular antenna, but it won't take long chasing DX or contest points to wish for a little more gain and height. As with 10 meters it is important to begin checking this band to the north and east from the time of your sunrise for the first DX openings. A good guide is to "follow the path of the sun" with your directional antenna; as your daylight begins to fade signals from the west and north will peak.

20 Meters

If any HF band could be labeled "workhorse" or the "band for all seasons" this is the one. Whether it is talking from Michigan to your brother in Florida every day, or reaching the DXCC Honor Roll using one band only, this is the place. There are good days and bad ones, and sunspots do take their effect, but for overall predictability no band is better than 20. Once again these bonuses attract more occupants, more power and bigger antennas. There are more DX nets, hurricane nets, maritime mobile nets and old friend nets on 20 meter SSB than elsewhere. Add the popular Slow Scan TV area around 14.230 MHz, HF packet links around 14.105 MHz, and the popularity of all of the digital modes below 14.1 MHz to good propagation characteristics and the possible congestion is obvious.

While plenty of contacts of all types can be made with 100 watts and a low tribander, you may find your frustration level raised often. The average three-band antenna is a compromise on 20; gain and front to back (F/B) both suffer in exchange for the small size. A six-element triband beam is one answer; a three-element or more 20 meter beam at 50 to 60 feet will definitely prove worthwhile. Front to back ratio takes on more importance on 20 meters and the lower bands because short skip propagation is common. This is most apparent on 20 during mid-day, when stations 300 to 500 miles away will sound like they are down the street. It becomes difficult to find a clear frequency to talk to your brother unless you can lessen interference in unwanted directions. Another advantage of improved gain and F/B ratio is the ability to more easily work stations via the long path. While long path contacts happen on 15 and 10 meters, especially from the west coast, they are more common on 20 meters. It is often easier to contact Asia and the Indian Ocean area from the U.S. in the morning by beaming the opposite direction from your beam heading chart; in the afternoon the same is true for the Pacific and Australia.

40 Meters

Here is the band that is most obviously shared with other services. It won't take very long to discover 500 KW international broadcasters. They make forty especially challenging for Novice and SSB operations, but they also indicate the distance that can be covered at this portion of the spectrum. During the day, 40 meters is a dependable band for contacts up to 1500 miles. That holds true at night, but it is also possible to contact any point on earth, and you don't need 500 KW to do it. Some hams have all the countries on the DXCC list on 40 meters, and it is a very important band for all types of contesting, domestic as well as DX. During the day there are regional nets on CW, SSB and digital modes. At night there are SSB nets to help overcome the obstacles of working DX stations with split frequencies and broadcast interference to contend with. Because almost all amateur stations outside ITU Region 2 are limited to 7.000 to 7.100 MHz, and the U.S. SSB allocation starts at 7.150 MHz, the majority of U.S. to DX contacts on SSB require the use of split frequency operation. CW and digital modes follow the practices of the higher bands, but they do tend to be compressed lower in frequency because of the presence of non-U.S. SSB hams.

There is no question that one of the shortened two-element beams is a big help in dealing with the competition of all kinds on 40. Other antennas do very well; dipoles and open-wire fed wires can be very useful, especially if some tall trees are handy. Single or multiple verticals also work well for some; make sure you understand their use on the lower bands before expecting good DX results. An amplifier is helpful at night, as it is on 80 and 160 meters, but is not necessary. Note that the majority of articles on building low power portable rigs focus on 40 meters. Look around 7.040 MHz for QRP operators.

80 Meters

The things that most hams think of when the 80 and 160 meter bands are discussed are the very large antennas that are needed and the annoying natural and man-made noise levels. Keep in mind that the first hams were restricted to 200 meters and down, and that these bands were very popular years before our sophisticated equipment and computer-aided antenna designs were available. Prior to the immense growth of two-meter repeater systems in the 1970s, most local ragchewing, traffic handling, and emergency communications took place on 80 (and 75 meters, as the phone portion is called). There are many ways to cope with the obstacles that appear to preclude enjoying "the low bands."

For regional communications, narrow bandwidth antennas like short loaded dipoles, verticals and other designs combined with an antenna tuner perform well. Even very low fully sized dipoles or inverted vee's bent to fit into the available space will put out very good signals for a few hundred miles, as well as providing less capture area for noise. For consistently greater distance coverage, as well as better bandwidth, it helps to have a full size dipole up higher; 50 feet or better can often be achieved by making use of trees, ropes and ingenuity. Those who hope to reach the 300 country mark on 80 usually use some variety of phased array: slopers, high wires, or verticals. Separate lower noise receiving antennas are also popular.

Eighty meters is usually the hardest band for those chasing the very popular Five Band Worked All Zones Award. Over 360 have completed all 40 zones on all five major HF bands; chasing the last few zones on 80 meters is an experience they all share. In common "the last one" was a zone with very low human population, few if any hams and on the other side of the world. Contacting that zone requires knowledge of propagation, operating times or plans of DXpeditions, good conditions at both ends of the contact, persistence on the part of both operators, very good ears and often lack of sleep. Once you have trained on this band, move down one to 160.

160 Meters

There seems to be two kinds of hams who operate on 160. Many find it a good band for nightly ragchews with much less congestion than 75 meters. Antenna experimentation and comparisons are a popular topic. The other operators are the DXers. During the day tuning across the band will usually reveal no signals at all; you may discover a strange noise every 15 kHz. That's a harmonic from a local TV set. In the summer you can hear the static from every thunderstorm within hundreds of miles. At night the band will seem to come alive, especially in the fall through spring seasons. You may hear many strong signals, mostly on CW, but not the station they are calling and working. If you tune above 1850 kHz you will find loud SSB stations ragchewing. To join in on local contacts the same compromise type antennas used on 80 will work. Not many suburbanites have room for anything else; even a dipole is 270 feet long on this band.

Working DX on 160 meters involves some discovery. You need to find those signals that you missed when you first tuned by; they are weak compared to most of the DX you are used to hearing on the higher bands. You also need to learn how to use your receiver to improve the signal-to-noise ratio and pull those weak signals out of the higher background noise level. Do some reading on Beverage or steering-wave longwire antenna types and on loop receiving antennas. Often, switching to a 40-meter or other antenna to listen will help in copying a weak signal. Towers used to support HF beams are often used successfully on 160 by shunt feeding them or as supports for inverted-L type antennas. The author has confirmed over 220 countries with a full-sized inverted vee, supported at the 70-foot level on a tower. Keep in mind that contrary to first impression, 160 conditions can be quite different from those on 80 meters. The noise, QSB and propagation may be the exact opposite of those found on the same night on 80 meters.

The "WARC" Bands—12, 17, and 30 Meters

The World Administrative Radio Conference held in 1979 included a major Amateur radio presence by the IARU. The Amateur bands were reconfirmed at an international level, and new allocations were added for the Amateur satellite service as well as three new HF bands. Access in the U.S. to those HF bands was fully gained in 1989.

As you would expect, the bands of 12, 17 and 30 meters share many of the characteristics of those bands that are their close neighbors, such as propagation and antenna size. Where they differ greatly is in their relatively small frequency allocations and lack of congestion. Each band at times has propagation advantages; 12 is often open when 10 is not. Seventeen meters can be the best long-haul DX and ragchew band when 15 is not really open and 20 is too crowded. Thirty meters, with a 200-watt power limit and a shared band outside of the U.S., still provides almost around the clock DX possibilities and supports a large amount of digital activity including HF packet links. Many hams are using unusual antennas on the WARC bands; 80-meter antennas on 30 meters, 40-meter antennas on 17 and 12 meters are common. There are many three-band antennas available, from simple wires to three-band yagis. A newly popular solution is the log periodic array, for coverage of 20, 17, 15, 12 and 10 meters, and sometimes 30 meters as well.

(Courtesy of Jim Dionne, K1MEM)

NET FREQUENCY LIST (160—10 METERS)

The abbreviation in parentheses after the net name refers to the state or province that the net is registered in. If the net name clearly defines the location, no abbreviation is given. Common abbreviations are used for states and provinces. WC stands for Wide Coverage Net. MM stands for Maritime Mobile Net. NTS stands for Area or Regional National Traffic System net. Further information on the net may be found in the appropriate part of the directory.

Listed in kHz:

1860	The Gateway 160 Meter Net (WC)
1877	Firebird Amateur Radio Club (WC)
1880	OM International Sideband Society (WC)
1936	1936 GCARA Net (OH)
1938	Fleet Radio Unit Pacific (WC)
1945	San Diego Section ARES Net (CA)
1984	Virgin Islands Net
1995	160 Meter Weather Net (NE)
3530	Eighth Region Net, Cycle 4 (NTS)
3530	New York State RACES
3538	Fleet Radio Unit Pacific (WC)
3539	Vermont-New Hampshire Traffic Net
3560	Disciples AR Fellowship (Australia) (WC)
3560	GLN (QPR ARCI) (WC)
3560	Iowa Tall Corn Net
3560	Seventh Region Net, Cycle 4 (NTS)
3560	WSN-80 (QRP ARCI) (WC)
3567	Fourth Region Net, Cycle 4 (NTS)
3567	West Virginia Early Net
3567	West Virginia Late Net
3570	Twelfth Region Net, Cycle 4 (NTS)
3573	Carolinas Net (NC)
3575	Alabama Section Net (CW)
3577	Buckeye Net (OH)
3577	Ohio Sunrise Slow Net
3585	Missouri Traffic Net
3585	Western Pennsylvania CW Traffic Net
3587	Oregon Section Net
3590	Empire Slow Speed Net (NY, WC)
3590	Tenth Region Net, Cycle 4 (NTS)
3590	Third Region Net, Cycles 3 & 4 (NTS)
3590	Washington State Net
3592	Arkansas CW Traffic Net
3596	Pine Tree Net (ME)
3598	Southern California Net 1, Cycle 4
3600	Kentucky CW Net
3602	First Region Net, Cycles 3 & 4 (NTS)
3605	Bukeye Net RTTY (OH)
3605	Minnesota Section Net
3610	Eastern Pennsylvania CW Net
3610	Kansas Section CW Traffic Net
3610	Pennsylvania Traffic Training Net
3620	Disciples AR Fellowship -New Zealand (WC)
3625	Oklahoma AMTOR Net (OK)
3630	Northern California Net
3635	Tennessee CW Net
3640	Connecticut Net
3640	Ninth Region Net, Cycle 4 (NTS)

3643	Maryland-Delaware- DC Net
3643	Texas CW Net
3645	Ontario Beaver Net
3645	Wisconsin Slow Speed Net
3647	IMN (Idaho Montana Net)
3650	Adventist Amateur Radio Association (WC)
3650	Fifth Region Net, Cycles 3 & 4 (NTS)
3650	South Dakota CW Net
3651	All Florida CW Traffic Net- QFN
3651	Florida Medium Speed Net (CW)
3651	Gator CW Traffic Net (FL)
3651	Pacific Area Net, Cycle 4 (NTS)
3652	British Columbia Emergency Net
3652	Eastern Canada Net, Cycle 4 (NTS)
3655	Sixth Region Net, Cycle 4 (NTS)
3656	Indiana Section CW Net
3658	East Mass/Rhode Island CW Net
3662	Disciples AR Fellowship CW (WC)
3662	Wisconsin Intrastate Net
3663	Michigan Net (QMN)
3665	Mississippi Traffic Net
3665	Illinois Section Net
3667	Ontario Quebec Net
3667	Quebec Section Net
3670	Eastern Area Net, Cycle 3 (NTS)
3670	Eastern Area Net, Cycle 4 (NTS)
3670	Central Area Net, Cycle 4 (NTS)
3677	New York State CW Net
3680	Illinois Training Net
3685	Maine Slow Speed Net
3680	Mississippi Slow Net
3680	Virginia Emergency Net, Charlie
3680	Virginia Late Net
3680	Virginia Net
3680	Virginia Slow Net
3685	Alberta Traffic Net
3690	Second Region Net, Cycle 4 (NTS)
3693	Oklahoma Training Net (CW)
3695	New Jersey Morning (CW) Net
3695	New Jersey Net
3700	South Dakota Novice Net
3701	Midwest RTTY Net (WC)
3705	Delaware Traffic Net
3705	Indiana Code Net
3705	Northern California Net/2- Slow Speed
3705	Southern California Net/2
3708	Ohio Slow Net
3710	Kansas Slow Speed Net
3710	Minnesota Slow Speed Net
3710	Puerto Rico Net
3712	Alabama Training Net- CW
3715	All Florida Slow CW Training Net
3715	Carolinas Slow Net (NC/SC)
3715	Colorado-Wyoming Net
3715	Eastern Mass/RI Slow Speed Net
3715	New Jersey Slow Net
3717	Maryland Slow Net
3719	Oklahoma/Texas Slow Net
3720	Adventist Amateur Radio Association (WC)
3720	San Diego Co ARES Novice Net (CA)
3720	San Diego Section ARES CW Net (CA)
3722	Professional Loafers Net CW (Quebec)
3723	Wisconsin Novice Net
3729	BC Group (SSB) (WC)
3729	British Columbia Public Service
3740	Alberta Public Service Net (WC)
3740	Walworth Co Net (SD)
3742	Ontario Phone Net
3750	Alberta ARES
3750	LARC Swap Net (Ontario)
3750	London ARC Net (Ontario)
3765	Happy Gang Joyeux Copin (Quebec)
3775	Quebec Radio Net
3780	VE2AQC Net (Quebec, Maritimes, Ontario)

3787	Professional Loafers Net (Phone) (Quebec)
3805	Adventist Amateur Radio Association (WC)
3815	Arkansas DX Association
3818	Central States VHF Society (WC)
3832	QCWA Chapter 49 (GA)
3840	AMSAT (WC)
3845	Sooner Traffic Net (OK)
3850	Random Net (WC)
3853	Disciples AR Fellowship (Midwest) (WC)
3855	New England QRP (SSB) (WC)
3855	Northern California Emergency Net
3855	QCWA Chapter #63 (OK)
3855	SKYWARN Net (CA)
3856	Mission Trail Net (AZ, CA, WC)
3860	Minnesota Section Phone Net
3860	Sheboygan Co ARES (WI)
3862	Coast Guard Auxiliary Amateur Radio Net (WC)
3862	North Conway District Emergency Net (NH)
3862.5	Magnolia Section Net
3862.5	Mississippi Section Phone Net
3865	Drake, Tube and Antique Radio Net (WC)
3865	Navy Amateur Radio Club (WC)
3865	West Virginia ARES/RACES Net
3865	West Virginia Fone Net
3866	Promise Keepers Ham Net (WC)
3870	South Dakota NJQ Noon Net
3873	North Texas ARES
3873	Texas ARES Net
3873	Texas Traffic Net
3875	Ohio Section ARES Net
3885	Daytime Oregon Section Net
3890	Northern Lights Chapter QCWA (AK)
3890	South Dakota QCWA Net
3900	Arkansas Phone Net
3900	Eastern Idaho Net
3900	Montana Traffic Net
3900	Oklahoma Phone Emergency Net
3900	Oklahoma Traffic & Weather Net
3900	Salvation Army (OK)
3902	Montana Section Net
3902	The "Used to" Net (WC)
3905	Delaware Emergency Phone Net
3905	Illinois Sideband Net
3905	The Old Buzzards Radio Club (WC)
3905	Pacific ARES Net (HI)
3905	San Diego Section ARES 75 Meter Net (CA)
3905	South Carolina Noontime Net (alternate)
3905	Washington Region Public Operations Net (MD)
3907	Virginia Traffic Net
3908	Beaver State Net (OR)
3908	Florida Public Operations Net (CA)
3909	Midwest United Church Amateur Net (WC)
3910	Indiana Traffic Net
3910	Indiana Wet Net
3910	Louisiana Traffic Net
3910	Missoula Area Emergency Net (MT)
3910	Virginia Emergency Net, Alpha
3912	North Central Phone Net (WC)
3913	New York Public Operations Net
3913	North Central Phone Net (WC)
3913	North Star Trade (WI)
3913	Third Region Net, Cycle 2 (NTS)
3915	East Mass/Rhode Island Phone Net
3915	Mass/Rhode Island Emergency Net
3915	South Carolina Single Sideband Net
3917	EPA Emergency Phone/Traffic Net (PA)
3917	River Rats (WI)
3918	Dog Biscuit Net (MI)
3918	San Joaquin Net (CA, WC)
3920	Alaska Snipers Net
3920	ARRL Technical Net (OR)
3920	Central States Traffic Net (KS)
3920	Kansas AM Weather Net
3920	Kansas Chapter 110 QCWA Net
3920	Kansas Phone Net
3920	Kansas Sideband Net
3920	Kansas Weather Net
3920	Maryland Emergency Phone Net
3920	Salvationist AR Operators Fellowship (WC)
3920	Van Wert Area Emergency Net (OH)
3921	Upper Peninsula Net (MI, WC)
3922	Big Bend Emergency Net (TX)
3923	North Carolina Evening Net
3923	Outhouse (WC)
3923	Pioneer Chapter #183 QCWA Net (MA)
3923	Tar Heel Emergency Net
3923	Twelfth Region Net, Cycle 2 (NTS)
3923	Wyoming ARES/RACES
3923	Wyoming Cowboy Net
3923	Wyoming Pony Express
3923	Wyoming Weather Net
3925	Central Vermont ARC Phone Net
3925	Clearing House Net (WC)
3925	Louisiana Emergency Net (alternate)
3925	New York Phone
3925	New York Public Operations Net (alternate)
3925	New York State Phone Traffic & Emergency Net
3925	Pico Net (MN)
3927	Great Lakes Evening Talk Net
3927	North Carolina Morning Net
3928	Colorado ARES HF
3928	Mockingbird Net (AR)
3930	Panhellenic Amateur Radio Federation Net (WC)
3930	Second Region Net, Cycles 2 & 3 (WC)
3932	Great Lakes Emergency & Traffic Net (WC)
3932	Michigan State ARPSC Net
3932	Wormy Net (WC)
3933	Arizona YL Cactus Keys (Coffee Cup)
3933	Green Mountain Net (VT)
3933	Motley Group (AK)
3934	Vermont Phone Net
3935	Carrier Net (WC)
3935	Louisiana Emergency Net
3935	Southwest Traffic Net (WC)
3935	Wolverine Net (MI)
3936	Barefoot Net (WC)
3937	Friendly Amateur Radio Missions Net (ID)
3937	W Mass Emergency ARES Net (MA)
3938	North Carolina SSB Net
3939	New Mexico Breakfast Club
3939	New Mexico Roadrunner Traffic Net
3940	ARRL Information Net (FL)
3940	Disciples AR Fellowship (WC)
3940	Eighth Region Net, Region 2 (NTS)
3940	Florida Amateur Sideband Traffic (FAST)
3940	Illinois Emergency Net
3940	Illinois Phone Net
3940	Maine Public Service Net
3940	Northern Maine Skidder Group
3940	OM International Sideband Society, Inc (WC)
3940	Sea Gull (ME)
3940	South Florida ARES
3940	Tropical Phone Traffic Net (FL)
3943	Auto State Young Ladies (MI)
3943	Granite State Phone Net (NH)
3943	Montana RACES Net
3945	Colorado Amateur Radio Weather Net (WC)
3945	Jackson Co ARES/RACES (MS)
3945	New England Phone Net (WC)
3945	Northwest Single Sideband Net (WA, WC)
3946	Riverside Co ARES/RACES (CA)
3947	Virginia Emergency Net, Bravo
3947	Virginia Fone Net
3947	Virginia Sideband Net
3948	First Region Net, Cycle 2 (NTS)
3950	Auto State Young Ladies (MI)
3950	Michigan Thumb Net
3950	New Jersey Phone Net

3950	North Florida ARES Net
3950	Northern Florida Phone Net
3952	Western Public Service System (WC)
3953	Michigan Amateur Communications System
3955	Adventist Amateur Radio Association (WC)
3955	Western New York Section Coordination
3958	Pennsylvania Phone Net (PA)
3960	BVARC Rag Chew Public Service (WC)
3960	Columbia Basin Net (OR)
3960	Good Morning Kentucky Phone Net
3960	Kentucky Rebel Net
3960	Kentucky Traffic Net
3960	Riverside Co Red Cross Emergency (CA)
3960	Northeast Coast Hurricane Net (WC)
3960	South Dakota Morning Weather Net
3960	South Dakota Sunday Emergency Net
3960	Southern Humboldt Co ARES/RACES (CA)
3962	Adventist Amateur Radio Association (WC)
3963	Fleet Radio Unit Pacific (WC)
3963	Missouri Emergency Operations & Weather
3963	Missouri Single Sideband Net
3963	Weather Emergency & Service Teams (NY)
3965	After Breakfast Club (WI)
3965	Alabama Emergency Net
3965	Alabama Traffic Net Mike
3965	Area 1 ARES/RACES HF Net (CT)
3965	Area 1 North ARES Net (CT)
3965	Connecticut Phone Net
3965	Connecticut Statewide ARES/RACES Net
3965	MANCORAD (WI)
3970	Iowa Traffic & Emergency Net
3970	Iowa 75 Meter Net
3970	Noontime Net (WA)
3970	Southern Section Country Cousins (WC)
3970	Washington Amateur Radio Traffic System
3970	Western Country Cousins (WC)
3972	Adventist Amateur Radio Association (WC)
3972.5	Midwest Country Cousins (WC)
3972.5	Ohio Single Sideband Net
3973	City of Mesa Emergency Group (AZ)
3975	Adventist Amateur Radio Association (WC)
3975	Amateur Radio Emergency Service (GA)
3975	Georgia Single Sideband Net
3975	Golden Bear Amateur Radio Net (WC)
3976	Adventist Amateur Radio Association (WC)
3976	Vermont Phone Emergency Net
3977	Firebird Amateur Radio Club Net (WC)
3980	Adventist Amateur Radio Association (WC)
3980	Nebraska Cornhusker Net
3980	Oregon Emergency Net
3980	Southeast Virginia ARES Net
3980	Tennessee Early Morning Phone Net
3980	Tennessee Evening Phone Net
3980	Tennessee Morning Phone Net
3982	Nebraska Storm Net
3983	Western Pennsylvania Emergency Net (alternate)
3983	WPA Phone & Traffic Net (PA)
3983	WPA Administration Net (PA)
3983.5	Nebraska ARES Net
3985	Badger Emergency Net (WI)
3985	Badger Weather Net (WI)
3985	Interstate Sideband Net (WC)
3985	Mercury Northern California Net
3985	NEWDXA (WI)
3985	QCWA Chapter 55 (WI)
3985	WD8OAA Memorial (WC)
3985	Wisconsin Side Band
3985	75 Meter Interstate Sideband Net (WC)
3987	D.B. and Friends (MI)
3987	Great Lakes Evening Talk Net (MI)
3987	Washington State Emergency Net
3987.5	Arkansas Emergency Communications Net
3987.5	Arkansas Section Net
3987.5	Emergency Coordinator Net (AR)

3987.5	Razorback Phone Net (AR)
3988	Panhellenic Amateur Radio Federation Net (Net)
3989	Colorado Columbine Net
3990	HF Mobile Experimenters Net (WC)
3990	Idaho Civil Defense Net
3990	Sunbelt Service Net (WC)
3990.5	Iowa RACES Net
3992	Arizona Traffic and Emergency Net
3993	New York State RACES
3993	Oklahoma Traffic Training Net)
3993.5	ARES/RACES Emergency Net (SC)
3993.5	Oregon ARES Traffic Net
3993.5	Oregon Emergency Management Net
3993.5	Oregon Section EC Rountable
3994	RACES (WI)
3995	Traffic Handlers Net (OR)
3996.5	Nevada State RACES Net
7030	SEN (QRP ARCI) (WC0
7033	Fleet Radio Unit Pacific US (WC)
7035	A.F.A.R. (CW) (WI)
7040	Eighth Region Net, Cycles 2 & 4 (NTS)
7040	New York State CW Net, Cycle 1 (alternate)
7040	Ontario Quebec Net
7040	WSN-40 (QRP-ARCI) (WC)
7048	Seventh Region Net, Cycle 4 (NTS)
7050	Eastern Area Net, Cycle 2 (NTS)
7052	Central Area Net, Cycle 4 (NTS)
7052	Fifth Region Nets, Cycles 3 & 4 (NTS)
7052	Pacific Area Net, Cycle 4 (NTS)
7055	Aurora Net (Manitoba)
7055	Eastern Area Net, Cycle 2 (NTS)
7055	Panhellenic Amateur Radio Federation Net (WC)
7055	Trans-Provincial Net (Ontario, WC)
7060	OK QRP Group
7061	All Florida CW Traffic Net- QFN (alternate)
7061	Florida Medium Speed Net (CW) (summer)
7061	Gator CW Traffic Net (FL)
7063	Twelfth Region Net, Cycle 4 (NTS)
7070	Third Region Net, Cycle 3 (NTS)
7087	Alaska Bush Net
7087	Midwest RTTY Net (WC)
7090	HAAM Radio (Friends of Bill W.) (WC)
7102	New York State RACES (alternate)
7105	Firebird Amateur Radio Club (WC)
7105	Texas High Noon CW Net
7114	Hit and Bounce Slow Traffic Net (WC)
7140	Triple States Slow CW Net (OH)
7140.5	Sloppy Code Net (IN)
7215	ARmenian Amateur Radio & Traffic Net (WC)
7225	Coastide ARC (CA0
7227.5	LY Net (WC)
7228	Daytime Oregon Section Net (alternate)
7228	Mercury Inter Mountain Net (CA)
7228	Western USA & Mexico Chaverim (WC)
7230	Bay Area Hospitals Disaster Preparedness Net (CA)
7230	New York Phone (alternate)
7233	Twelfth Region Net, Cycle 2 (NTS)
7233.5	Hams for Christ (WC)
7234	Fleet Radio Unit (WC)
7235	Colorado ARES HF (alternate)
7235	Louisiana Emergency Net
7235	West Virginia Mid Day Net
7237	Caribbean Maritime Mobile Net (VI, Maritime)
7237	MUFON Amateur Radio Net (WC)
7238	Fleet Radio Unit (WC)
7238	National Amateur Family Tracers Net (WC)
7238	Oldtime Radio Collectors and Traders Society (WC)
7238	Seventh Region Net, Cycle 2 (NTS)
7240	American Red Cross Disaster Comm Net (WC)
7240	Colorado High Noon Net
7240	Eighth Region Net, Cycle 2 (NTS)
7240	North Carolina Alligator Group (WC)
7240	Triple States North Carolina Chapter

7240	Triple States 40 M Net (WC)	14247.5	Displaced Peoria (IL) Group (WC)
7243	Alabama Day Net	14250	Clamdiggers (WC)
7243	Eastern Area Net, Cycle 2 (NTS)	14255	Flying Boat Amateur Radio Society (WC)
7243	Fourth Region Net, Cycle 2 (NTS)	14255	Tin Can Sailors Net (WC)
7243	South Carolina Noontime Net	14257	Old China Hands (WC)
7245	Jackson Co ARES/RACES Net	14260	ARmenian Amateur Radio and Traffic Net (WC)
7245	New York State RACES (alternate)	14262.5	Diamond State Net (WC)
7247.5	Florida Mid-Day Traffic Net	14263	Family Motor Coach Assoc Amateur Radio (WC)
7248	Friendly Triple "N" (WC)	14263	Family Motor Coach Assoc YL Net (WC)
7248	Texas RACES	14265	International Nude Net (WC)
7253.5	Central States Traffic Net (KS, WC)	14265	Salvation Army Team Emergency Radio Net (WC)
7255	Salvation Army Net (alternative) (OK)	14272.5	AYN Rand Admirers Net Discussion (WC)
7258	Tin Can Sailors Net (WC)	14277	Firebird Amateur Radio Club (WC)
7260	ARmenian Amateur Radio & Traffic Net (WC)	14280	International Assoc of Airlines Hams (WC)
7260	Flying Boat Net (WC)	14280	International Mission Radio Association (WC)
7260	Tin Can Sailors Net (WC)	14282	FAA/CAA Retirees Net (WC)
7260	Triple States 40 M Net (WC)	14287	Disciples Amateur Radio Fellowship/
7260	Virginia Traffic Net (alternative)	14287	United Church Amateur Net (WC)
7260	Wyoming Jackalope Net	14287	V.A. Amateur Radio Service Net (WC)
7262	Mississippi Baptist Ham Net	14290	Boy Scouts of America Net, International (WC)
7262	Professional Loafers Club (WC)	14290	Earthquake/Tsunami/ARES Net (AK)
7262.5	OM International Sideband Society, Inc (WC)	14290	HAAM Radio (Friends of Bill W.) (WC)
7265	Salvation AR Operators Fellowship (WC)	14290	Wisconsin/Florida
7265	Salvation Army Team Emergency Radio Net (WC)	14290	OM International Sideband Society, Inc (WC)
7265	Yonkers Amateur Radio Club Net (WC)	14300	Maritime Mobile Service Net (MM)
7268	Waterway Radio & Cruising Club Net (MM)	14303	Atlantic Region Net/ IATN (NTS/ WC)
7268.5	Noontime Net (WA, WC)	14303	CQ All Schools Net (WC)
7270	Adventist Amateur Radio Network (WC)	14303	International Assistance & Traffic Net (WC)
7270	Disciples AR Fellowship (WC)	14305	Adventist Amateur Radio Association Net (WC)
7272	Central Garden Net (WC)	14305	California/Hawaii Cocktail Net (WC)
7272	Western Pennsylvania Emergency Net	14306	Latvians World Wide (WC)
7273	City of Mesa Emergency Group (AZ)	14317	ICOM Users Net (WC)
7273	Daytime Texas Traffic Net	14325	Hurricane Watch Net (WC)
7273	North Texas ARES (alternate)	14326	Chaverim International (WC)
7275	Sixth Region Net, Cycle 2 (NTS)	14326	Myer Delnick Mishpulcha Net (WC)
7277	Firebird Amateur Radio Club (WC)	14326	Western USA & Mexico Chaverim (WC)
7277	RACES (WI)	14327	International Brotherhood of Electrical Workers'
7277.5	Tenth Region Net, Cycle 2 (NTS)		Network (WC)
7278	Rock River Radio Club (WI)	14328	Adventist Amateur Radio Association (WC)
7278	Sixth Region Net, Cycle 2 (NTS)	14329.5	Vermont Snowbird Net (WC)
7280	Fifth Region Net, Cycle 2 (NTS)	14336	20 Meter Mobile Emergency and Co Hunters Net (WC)
7280	Hambutchers (MO)	14340	California-Hawaii Net (WC)
7280	Louisiana Emergency Net (alternate)	14340	HAAM Radio (Friends of Bill W.) (WC)
7280	Riverside Co Red Cross Emergency (CA)	14342	Disciples AR Fellowship (USA) (WC)
7281	Nebraska 40 Meter Net	14345	Central Area Net, Cycle 2 (NTS)
7282	Ninth Region Net, Cycle 2 (NTS)	14345	Pacific Area Net, Cycles 1 & 2 (NTS)
7284	Experimental Aircraft (WI)	14345	Two-Meter EME Net (WC)
7284	Good Sam RV Radio Network (WC)	14345	VHF/UHF Coordination & EME Scheduling
7285	Arkansas Emergency Comm Net (alternate)	14345	432 and Above EME Net (WC)
7285	Emergency Coordinator Net (alternate) (AR)		
7288	Panhellenic Amateur Radio Federation Net (WC)	18140	OM International Sideband Society, Inc (WC)
7290	Hawaii Afternoon Net (HI Emergency Net)		
7290	Pacific ARES Net (HI)	21130	Yonkers ARC CW Net (NY)
7290	7290 Traffic Net	21175	Adventist Amateur Radio Association Net (WC)
7295	Rainbow Connection (FL)	21303	CQ All Schools Net (WC)
		21312	Kettle Moraine RA (WI)
10120	Fleet Radio Unit Pacific (WC)	21310	Pacific/Alaska/Taiwan (WC)
10123	Northwest QRP Club (WC)	21355	Flying Boat Amateur Radio Society (WC)
		21360	OM International Sideband Society (WC)
14052	Old China Hands (WC)	21375	Adventist Amateur Radio Association Net (WC)
14055	FAA/CAA Retirees Net (WC)	21375	Fleet Radio Unit Pacific (WC)
14058	Fleet Radio Unit Pacific (WC)	21377	Firebird and Winnebego Combined Net (WC)
14060	CW/SSB TCN (QRP ARCI) (WC)	21377	Panhellenic Amateur Radio Federation
14060	VE QRP (WC)	21380	International Association of Airline Hams (WC)
14087	Fleet Radio Unit Pacific (WC)	21405	Adventist Amateur Radio Association Net (WC)
14092	Family Motor Coach Assoc AMTOR Net (WC)	21410	International Police Association Radio Club (WC)
14192	Dayton Net (WC)		
14234	US Submarine Veterans (WC)	28165	Radoops of El Jebel Shrine (CO)
14235.5	Retired General Electric Employee Net (WC)	28195	Southern Patuxent RTTY Net (MD)
14238	City of Mesa Emergency Group (AZ)	28303	CQ All Schools Net (WC)
14240	Good Sam RV Radio Network (WC)	28309	Novice-Technician (+) Training Net (GA)
14240	International Police Assoc Radio Club (WC)	28310	Midwest United Church Amateur Net (WC)
14240	West Valley Vacation Net (WC)	28310	River Valley Net (MN)
14243	Fleet Radio Unit Pacific (WC)	28310	Webster Wireless Roundtable (MA)

28320	Minnetonka MN ARC 10 M Training Net (MN)
28325	GBM & K Chat (WI)
28325	Mission Trail Net (WC)
28325	Mobile ARC Net (AL)
28325	Morris Code Club (CA)
28330	Kitsap Co ARES/RACES Net (WA)
28332	TCSN [QRP ARCI-SSB] (WC)
28333	Bennington Co ARES 10 M Net (VT)
28333	Orlando Chapter "333" Mosquito Net (FL)
28333	Titusville "333" Mosquito Net (FL)
28340	Washington Co 10 Meter Net (PA)
28343	Firebird Amateur Radio Club (WC)
28345	Cradle of Confederacy (AL)
28350	The Broken Heart Net 10-10 (AL)
28350	Four Lakes ARC 10 Meter Net (WI)
28350	Lake Ozark ARC (MO)
28350	North Carolina Alligator Group
28350	Triple States North Carolina Chapter
28355	Flying Boat Amateur Radio Society (WC)
28355	International Police Association Radio Club (WC)
28360	OM International Sideband Society (WC)
28365	10-10 Milwaukee (WI)
28370	Milledgeville-Baldwin Co Area ARES Net (GA)
28375	Brattleboro ARC 10 M Net (VT)
28375	Fleet Radio Unit Pacific (WC)
28375	Northlake Radio Association (LA)
28375	Oneonta Roundtable Net (NY)
28377	Firebird Amateur Radio Club (WC)
28380	Ten Meter Maritime Mobile Net (MM)
28385	Lincoln Co Information & Emergency Net (ME)
28385	Santa Clara Co ARA (CA)
28400	All American Cities Net 10-10 (AL)
28400	Green Bay Area (WI)
28400	Milwaukee RAC (WI)
28400	Northeast Nebraska 10 M Ragchew Net
28400	NYC ARES Tactical Communications Net
28400	Saratoga ARES/RACES (CA)
28400	Taylor Co ARES Net (WV)
28400	Wilderness Road ARC 10 Meter Net (KY)
28410	Southern Patuxent 10 Meter Net (MD)
28416	Foreign Service Net (WC)
28425	Sailfish Net (MM)
28425	Verticle Nut Club (TN)
28429	South Carolina Golden Corner 10 M Net
28430	Grant Co 10-Meter Net (IN)
28450	Arrowhead Radio Amateur Club (MN)
28450	Central Kansas ARC Training Net
28450	Manatee Co ARES RACES 10 M Net (FL)
28450	New York/Long Island 10 Meter Traffic
28455	Flying Boat Amateur Radio Society (WC)
28456	Yonkers ARC (NY)
28460	City of Mesa Emergency Group (AZ)
28470	Bauxite Chapter 10-10 Net (AR)
28480	Triple States ARES CW Net (WV)
28480	Triple States 10 M Phone Net (OH)
28480	TSARC 40 M Net (WV)
28485	Los Gatos ARES/RACES (CA)
28488	Space Houston on Ten (10-10) Net (TX)
28490	Faulkner Co AR 10 M Net (AR)
28493	Fleet Radio Unit Pacific (WC)
28495	Fresno-Madera Amateur Net (CA)
28535	Wiregrass Peanut Pickers 10-10 (AL)
28577	Panhellenic Amateur Radio Federation Net (WC)
28700	Smoky Mountain 10 Meter Net (TN)
28747	Flying Boat Amateur Radio Society (WC)

(Source: ARRL)

QRP OPERATING

Before discussing the joy and frustration of operating at QRP power levels, let us first define a few terms.

QRP is a Q-signal that when used as an interrogatory

means "Shall I reduce power?" and when used as a statement means "reduce power."

QRP levels are defined by The QRP Amateur Radio Club, International, as a maximum of 5W output for CW, 10W PEP for SSB, and 5W maximum output for all other modes. In order for a station to sign that they are QRP they should be operating at power levels at or below these levels. A station usually indicates that they are operating QRP with the callsign followed by /QRP on CW and on voice communications their callsign followed by the letters QRP.

Another term that the radio amateur will see is QRPp meaning power levels at or less than 1W output. This type of operation is also called "milliwatting" to indicate that the power levels are measured in milliwatts.

Note that the above power levels are all that are needed in order for a radio amateur to operate QRP. This may be done with homebrew or commercial equipment capable of these power levels. Accurate means of determining the power levels should be used in order to accurately determine the actual output power while in operation. There are number of lower power wattmeters available either commercially or that can be built by the radio amateur to determine the exact power levels in operation.

There are many reasons for individual radio amateurs to be or become interested in QRP operations and the following list is by no means meant to be exhaustive, just to give you some ideas of why radio amateurs are interested in this area of communications.

1. By using the lowest power level possible for communication the radio amateur is pushing the limit of the equipment, antennas, propagation, and most importantly the operators on both ends of the information exchange loop. It is a challenge in the same way that athletes strive for the best they can do.

2. A lot of QRP equipment may be built by the amateur radio operators themselves, thus adding pride to the accomplishment of communicating with low power levels and furthering their knowledge of the internals of the equipment being used, propagation, effective antenna systems, etc.

3. A lot of QRP equipment is small and portable and may be used taken along and used with camping and other outdoor activities. The equipment may be small enough to throw into a suitcase for travel to other parts of the country or world (with proper licensing issues taken care of) in order that an individual may combine business/pleasure with amateur radio activities.

When you ask your fellow radio amateurs about QRP activities you will immediately see that everyone has a different view of how effectively you can communicate at QRP levels. Most likely their views will be dependent upon what they have heard from others or how they actually succeeded or failed at their attempts at QRP operating. If you are interested in QRP operating then there are some important plans that you must make in order to be successful.

You must first determine what goal(s) you want to accomplish. Do you just want to try it and see what it's all about? Is this your first rig on HF? Are you bored with high power communications and want to try something different? Do you already have equipment that is capable of operating at QRP levels? There are a number of good books available on QRP operating and give greater detail than given here.

QRP operating involves patience and a greater understanding of propagation and band conditions than operating at higher power levels. Your QRP level signal will be weaker than those at higher power levels. It may mean that if you call another station that you hear calling CQ they may not answer you because they do not hear you or you are too weak for them to bother with. Remember you are operating under conditions much more demanding than operating at higher power levels. Just be patient. When propagation conditions are excellent you can take pride in having a station come back to your low power signal and say that they find it rather remarkable that you are running such low power. During sunspot peaks it is not unusual to talk to all continents with less than 5W on 10 meters.

Operating at QRP power levels does not mean operating with inferior equipment. Use or build the best that your budget and time will allow. It is important to have a very good receiver with stability, selectivity, and sensitivity for communicating with other low power stations. The transmitter must also have a clean signal with excellent keying or voice characteristics in order to have others want to communicate with you. There are a number of QRP kits available for almost any budget that will allow almost anyone with interest to build their own transceiver at budget levels under $100, $200, and higher if you have an interest in building. Check with others or read as much as possible about what is available to make a choice that you will not be disappointed with and within your budget.

Needless to say, the antenna plays an important part in QRP operating and should not be neglected in any way. It is the antenna that transfers the power the power generated by the transmitter to the rest of the universe. An efficient antenna with low losses (and if possible, directivity to give you additional signal gain) will do a great deal to aid in your successful operation at QRP power levels. QRP operators are successful and have been successful with different types of antenna arrangements. From dipoles, random wires, and sophisticated yagis and long wire arrays, the QRP operator can successfully communicate with other radio amateurs around the world. It is also another area for experimentation for the QRPer and any radio amateur. It is just another component in the communication system.

After getting set up to operate QRP and experimenting to see how the combination of rig(s), antennas, etc. work, you might want to consider setting small goals to further your operating skills and performance of you station. You may want to see how many states you can work on a single band over a period of time. This requires furthering your knowledge of propagation conditions, sunspot activity, and other outside factors to achieve a given goal. You may even get ambitious and attempt working as many stations as possible during some of the contests that occur throughout the year. There even some QRP-only contests sponsored by various clubs throughout the year. There are a number of awards that the ARRL and others issue that you can chase after at QRP levels and even obtain special endorsements for these accomplishments.

You can work QRP anywhere in the ham bands, but you'll find a large number of active QRPers start around these frequencies in order to easily locate others running low power.

Band	CW	SSB
160	1.810	1.910
		1.843 (Europe)
80	3.560	3.985
	3.710 (Novice)	3.690 (SSB EU)
40	7.040	7.285
	7.030 (Europe)	7.090 (SSB EU)
	7.060 (Europe)	
	7.110 (Novice)	
30	10.106	
20	14.060	14.285
17	18.096	
15	21.060	21.385
	21.110 (Novice)	21.285 (SSB EU)
12	24.906	
10	28.060	28.885
	28.110 (Novice)	28.385 (Novice)
		28.360 (SSB EU)
6	50.060	50.885
		50.285 (SSB EU)
2	144.060	144.285
		144.585 (FM)

Another thing that an individual considering QRP operating should consider is to find other radio amateurs locally that have already been doing QRP operations. Attend local radio club meetings and ask around. There is usually one or more active QRPer in any given radio club. If you are not successful in finding anyone then consider starting a QRP club in your area. There are other clubs around the United States that will be glad to give you information on how to get started and may even have names of people close to you that can help. You'll find that the QRP community is both helpful and friendly. An excellent starting point for the beginner and seasoned QRP operator for information is the QRP Amateur Radio Club, International. Contact Michael Bryce, WB8VGE, Publicity Officer and Membership Chairman, 2225 Mayflower, N.W., Massilon, OH 44647.

QRP Amateur Radio Club International Operating Awards Program

The objective of the QRP ARCI Operating Awards Program is to demonstrate that "power is no substitute for skill." It encourages full enjoyment of Ham Radio while running the minimum power necessary to complete a QSO and thereby reducing QRM on our crowded bands. QRP is defined by the club as 5 watts output CW and 10 watts PEP output SSB. The following awards are available to any Amateur. Requirements are set forth below.

QRP-25 This award is issued to any Amateur for working 25 members of QRP ARCI while those members were running QRP. Endorsements are offered for 50, 100 and every 100 thereafter.

WAC-QRP: This award is issued to any Amateur for confirming QSOs with stations in all six continents while running QRP.

WAS-QRP: This award is issued to any Amateur for confirming QSOs with stations in 20 or more of the 50 states of the USA while running QRP. Endorsement seals are issued at 30, 40 and 50 states confirmed.

DXCC-QRP: This award is issued to any Amateur for

confirmed QSOs with 100 ARRL countries while running QRP.

1000-MILE-PER-WATT (KM/W): This award is issued to any Amateur transmitting from, or receiving the transmission of, a QRP station such that the Great Circle Bearing distance between the two stations, divided by the QRP stations power output equals or exceeds 1000 Miles-per-Watt. Additional certificates can be earned with different modes and bands.

QRP-NET (QNI-25): This award is issued to those members completing 25 check-ins into any individual QRP ARCI net. Subsequent 25 QNIs in another net will earn an endorsement seal. Net managers send a list of those qualifying to the Nets Manager at the end of the month. Awards are issued FREE to those qualifying by the Awards Chairman as information is received from the Nets Manager.

NOTES:

1. The fee for all awards, except QRP NET (QNI-25), is $2.00 US or 10 IRCs. Subsequent Endorsement Seals are $1.00 or 5 IRCs. Make checks or money orders (preferred) payable to QRP ARCI. Cash accepted, but not recommended through the mail.

2. GCR List (General Certificate Rule): QRP ARCI will accept as satisfactory proof of confirmed QSOs and that the QSLs are on hand as claimed by the applicant if the list is signed by: (a) a radio club official, OR (b) two amateur radio operators, general class or higher, OR (c) notary public, OR (d) CPA. If you must send QSLs, please include postage for their return. Neither QRP ARCI or the Awards Chairperson are responsible for lost or damaged QSLs. It is recommended that photocopies be sent first. If additional supporting information is required, the awards chairperson will request that it be sent.

3. QRP ARCI member numbers are not published. The Awards Program will accept as satisfactory proof for any of the club awards a QSO with a club member giving their membership number and power output in the log data. If the QRP number and power are not given a QSL is required for confirmation. See Note 2 above.

4. Endorsement seals are available for a) One Band, b) One Mode, c) Natural Power, d) Novice and e) Two-way QRP if log data so indicates.

Send Applications to:
QRP ARCI AWARDS CHAIRMAN, Chuck Adams, K5FO, PO Box 181150, Dallas,TX 75218-8150

QRP Miles-Per-Watt Distance Records

Band	Holder	Miles/Watt
1.8MHz	GW4AEC	13,300
3.5MHz	AA2U	851,339
7.0MHz	AA4XX	1,909,502
10.1MHz	NWØO	20,727
14.0MHz	OK1DKW	87,800,000
18.0MHz	K4TWJ	59,380
21.0MHz	WB6UNH	19,250,000
24.0MHz	JL1FXW	2,445
28.0MHz	K7IRK	218,333,333
50.0MHz	JO1XWH	134,200,000

(Contributed by Chuck Adams, K5FO, QRP ARCI Awards Chairman.)

FOREIGN LANGUAGES

Since HF operation often involves working stations in other countries, it is often helpful to be able to exchange radio-related information in several languages. The following translation guides should be helpful.

PHONETIC ALPHABET/PHRASES

	English	German	French	Italian	Spanish	Portuguese	Russian (Phonetic)	Russian (Phonetic)
A	Alpha	Anton	Alfa	Alfa	Alfa	Antena	Anatoli	Anna
B	Bravo	Berta	Bravo	Bravo	Brasil	Bateria	Baris	Borja
C	Charlie	Caser	Charlie	Canada	Canada	Condensador	Tsentralnii	Tsentr
D	Delta	Dora	Delta	Delta	Delta	Detector	Dimitri	Dima
E	Echo	Emil	Echo	Europa	Espana	Estatico	Jelena	Jelena
F	Foxtrot	Friedrich	Foxtrot	Firenze	Francia	Filamento	Fjodor	Fedja
G	Golf	Gustav	Golf	Guatemala	Guatemala	Grade	Grigorii	Galja
H	Hotel	Heinrich	Hotel	Hotel	Hotel	Hotel	Hariton	Hariton
I	India	Ida	India	Italia	Italia	Intensidade	Ivan	Ivan
J	Juliet	Julius	Juliett	Juventus	Japon	Juliete	Ivan krakii	Ivan krakii
K	Kilo	Konrad	Kilo	Kilometro	Kilo	Kilo	Kanstantin	Kilavat
L	Lima	Ludwig	Lima	Lima	Lima	Lampada	Leanit	Leanit
M	Mike	Martha	Mike	Messico	Mejico	Manipulador	Mihail	Misa
N	November	Nordpol	November	Novembre	Noviembre	Negativo	Nikalai	Nina
O	Oscar	Otto	Oscar	Otranto	Oscar	Onda	Olga	Olja
P	Papa	Paula	Papa	Palermo	Papa	Placa	Pavel	Pavel
Q	Quebec	Quelle	Quebec	Quebec	Quito	Quadro	Ssuka	Ssuka
R	Romeo	Richard	Romeo	Romeo	Radio	Radio	Raman	Roma
S	Sierra	Siegfried	Sierra	Santiago	Santiago	Sintonia	Sergei	Sergei
T	Tango	Theodur	Tango	Tango	Tango	Terra	Tamara	Tanja
U	Uniform	Ulrich	Uniform	Universita	Universidad	Unidade	Uljana	Uljana
V	Victor	Viktor	Victor	Venezia	Victor	Valvula	Zenja	Zenja
W	Whisky	Wilhelm	Whisky	Whisky	Whisky	Watt	Vasilii	Viktar
X	X-ray	Xanthippe	X-ray	Xilofono	Xilofono	Xilofono	Mjahkii znak	Znak
Y	Yankee	Ypsilon	Yankee	Yokohama	Yucatan	Yucatan	Igrek	Igrek
Z	Zulu	Zeppelin	Zulu	Zelanda	Zulu	Zulu	Zinaida	Zoja

From The Radio Amateur's Conversation Guide, *by Jukka Heikinheimo, OH1BR and Miika Heikinheimo, OH2BAD. Reprinted with permission.*

NUMBERS—CARDINALS AND ORDINALS

	English card.	ord.	German card.	ord.	French card.	ord.	Italian card.	ord.
0	zero		null		zero		zero	
1	one	first	eins	erste	un	premier	uno	primo
2	two	second	zwei	zweite	deux	deuxieme	due	secondo
3	three	third	drei	dritte	trois	troisieme	tre	terzo
4	four	fourth	vier	vierte	quatre	quartrieme	quattro	quarto
5	five	fifth	funf	funfte	cinq	cinquieme	cinque	quinto
6	six	sixth	sechs	sechste	six	sixieme	sei	sesto
7	seven	seventh	sieben	siebte	sept	septieme	sette	settimo
8	eight	eighth	acht	achte	huit	huitieme	otto	ottavo
9	nine	ninth	neun	neunte	neuf	neuvieme	nove	nono
10	ten	tenth	zwehn	zehnte	dix	dixieme	dieci	decimo
11	eleven		elf		onze		undici	
12	twelve		zwolf		douze		dodici	
13	thirteen		dreizehn		treize		tredici	
14	fourteen		veirzehn		quatorze		quattordici	
15	fifteen		funfzehn		quinze		quindici	
16	sixteen		sechzehn		seize		sedici	
17	seventeen		siebzehn		dix-sept		diciassette	
18	eighteen		achtzehn		dix-huit		diciotto	
19	nineteen		neunzehn		dix-neuf		diciannove	
20	twenty		zwanzig		vingt		venti	
21	twenty-one		einundzwanzig		vingt et un		ventuno	
22	twenty-two		zweiundzwanzig		vingt deux		ventidue	
23	twenty-three		dreiundzwanzig		vingt trois		ventitre	
30	thirty		dreissig		trente		trenta	
40	forty		vierzig		quarante		quaranta	
50	fifty		funfzig		cinquante		cinquanta	
60	sixty		sechzig		soixante		sessanta	
70	seventy		siebzig		soixante dix		settanta	
80	eighty		achzig		quatre-vingt		ottanta	
90	ninety		neunzig		quatre vingt dix		novanta	
100	one hundred		hundert		cent		cento	
1000	one thousand		tausend		mille		mille	

	Spanish card.	ord.	Portuguese card.	ord.	Russian card.	ord. (Phonetic)	Japanese card.
0	cero		zero		nol		zero
1	uno	primero	um	primeiro	adin	pjervyi	ichi
2	dos	segundo	dois	segundo	dva, dvie	ftaroi	ni
3	tres	tercero	tres	terceiro	tri	tretii	san
4	cuarto	cuarto	quatro	quarto	cetire	cetvjortyi	shi, yon
5	cinco	quinto	cinco	quinto	pjat	pjatyi	go
6	seis	sexto	seis	sexto	sest	sestoi	roku
7	siete	septimo	sete	setimo	siem	sedmoi	hichi, nana
8	ocho	octavo	oito	oitavo	vosiem	vasmoi	hachi
9	nueve	noveno	nove	nono	djevit	divjatyi	kyu
10	diez	decimo	dez	decimo	djesit	disjatyi	ju
11	once		onze		adinnatsat		juichi
12	doce		doze		dvenatsat		juni
13	trece		treze		trinatsat		jusan
14	catorce		quatorze		cetirnatsat		jushi, juyon
15	quince		quinze		pitnatsat		jugo
16	diez y seis		dezesseis		sestnatsat		juroku
17	diez y siete		dezessete		semnatsat		juhichi
18	diez y ocho		dezoito		vosemnatsat		juhachi
19	diez y nueve		dezenove		djevitnatsat		jukyu, juku
20	veinte		vinte		dvatsat		niju
21	veintiuno		vinte e um		dvatsat adin		nijuichi
22	veintidos		vinte e dois		dvatsat dva		nijuni
23	veintitres		vinte e tres		dvatsat tri		nijusan
30	treinta		trinta		tritsat		sanju
40	cuarenta		quarenta		sorak		yonju, shiju
50	cincuenta		cinquenta		pidisjat		goju
60	sesenta		sessenta		sestdesjat		rokuju
70	setanta		setenta		siemdesjat		nanaju,hichiju
80	ochenta		oitenta		vosemdesjat		hachiju
90	noventa		noventa		divjanosta		kyuju
100	cien		cem		sto		hyaku
1000	mil		mil		tisaca		sen

From The Radio Amateur's Conversation Guide, *by Jukka Heikinheimo, OH1BR and Miika Heikinheimo, OH2BAD. Provided with permission by Jukka Heikinheimo, OH1BR*

OFTEN USED RADIO PHRASES

Phrases will be presented in the following sequence: English, German, French, Italian, Spanish, Portuguese, Russian (Phonetic), and Japanese.

Please give me another [a long] call.
Bitte geben Sie mir noch einmal einen [langen] Anruf.
S'il vous plait, donnez a nouveau [longuement] votre indicatif.
Per favore ripeti la chiamata [a lungo].
Por favore deme [una] otra llamada [larga].
Por favor, de-me uma outra [longa] chamada.
Pazalsta, pazavitje minja jisso ras. Pazalsta, daitje mnie dlinnyi vyizaf.
Moichido call kudasai. [long call de onegai shimasu.]

Please repeat your call sign slowly several times.
Bitte wiederholen SieIhr Rufzeichen langsam mehrere Male.
S'il vous plait, repetez lentement votre indicatif plusieurs fois.
Per favore ripeti lentamente il tuo nominativo parecchie volte.
Por favor repita despacio varias veces su indicativo.
Por favor repita seu indicativo vagarosamente varias vezes.
Pazalsta, paftaritje mjedlenna vas pazyivnoi njeskalka ras.
Yukkurito nankaika call-sign o kurikaeshitekudasai.

What is your call sign?
Was ist Ihr Rufzeichen?
Quel est votre indicatif?
Qual' e il tuo nominativo?
Cual es su indicativo?
Qual e seu indicativo?
Kakoi vas pazyivnoi?
Call-sign wa nandesuka?

I cannot copy you at the moment.
Ich kann Sie im Augenblick nicht lesen.
Je ne peux pas vous copier en ce moment.
Al momento non riesco piu' a copiarti.
No le peudo copiar por lel momento.
Eu nao posso lhe copiar no momento.
Ja vas sicas ni razbiraju.
Genzai anatao copy-dekimasen.

Thank you very much for your call.
Ich danke Ihnen vielmals fur Ihern Anruf.
Merci beaucoup pour votre appel.
Grazie molte per la tua chiamata.
Muchas gracias por su llamada.
Muito obrigado pela sua chamada.
Ocen blagadaren vam zavas vyizaf.
Oyobidashi domo arigato gozaimasu.

Good morning	Good afternoon	Good evening	Good night
Guten Morgen	Guten Tag	Guten Abend	Gute Nacht
Bonjour	Bon apres-midi	Bonsoir	Bonne nuit
Buon giorno	Buon pomeriggio	Buona sera	Buona notte
Buenos dias	Buenas tardes	Buenos tardes	Buenas noches
Bom dia	Boa tarde	Boa noite	Boa noite
Dobraje utra	Dobryi djen	Dobryi vjecer	Dobrai noci
Ohayogozaimasu	Konnichiwa	Konbanwa	Sayonara

It is very nice to meet you for the first time.
Es ist sehr nett Sie das erste Mal zu treffen.
C'est tres agreable de vous rencontrer pour la premiere fois.
Molto lieto di incontrarti per la prima volta.
Encantado de conocerle por primera vez.
E muito agradavel encontrar voce pela primeira vez.
Ocen prijatna fstretit vas fpervyie.
Hajimete omenikakarete totemo ureshikuomoimasu.

Your report is five and nine.
Der rapport fur Sie ist funf and neun.
Votre report est cinq neuf.

Il tuo rapporto e cinque nove.
Su reporte es cinco nueve.
Sua reportagem e cinco e nove.
Vas silisu pjat djevit.
Anatano report wa go kyu [gojukyu] desu.

What is my report?
Was ist mein Rapport?
Quel est mon report?
Qual' e il mio rapporto?
Cual es mi reporte?
Qual e a minha reportagem?
Kak vyi minja prinimajetje?
Watashino report wa nandesuka?

My name is Laurie.
Mein Name ist Bernd.
Mon nom est Bernard.
Il mio nome e Toni.
Mi nombre es Jorge.
Meu nome e Gerson.
Majo imja pabukvam.
Watashino namae wa Nao desu.

What is your name?
Wie ist Ihr Name?
Quel est votre nom?
Qual' e il tuo nome?
Cual es su nombre?
Qual e o seu nome?
Vase imja ? [Kak vazzavut ?]
Anatano namae wa nandesuka?

My antenna is ten meters high.
Meine Antenne ist in einer Hohe von zehn Metern.
Mon antenne est a dix metres de hauteur.
La mia antenna si trova a dieci metri di altezza.
Mi antena esta a una altura de diez metros.
Minha antena esta numa altura de dez metros.
Maja antjenna navisatje disjati mjetraf.
Watashi no antenna wa takasa ju meter desu.

I have recently put up a new antenna.
Neulich habe ich eine Antenne aufgestellt.
J'ai recemment monte une nouvelle antenne.
Ho montato recentemente un antenna nuova.
Recientemente he colocado una nova antena.
Eu levantei recentemente uma novwa antena
Nidavna ja ustanavil novuju antjennu.
Saikin atara shii antenna o age mashita.

I'd like to compare two antennas. Could you give me a comparative report?
Ich wurde gerne zwei Antennen miteeinander vergleichen.
Konnten Sie mir einen Vergliech-Rapport geben?
Je voudrais comparar deux antennes. Pourriez vous me donner un report comparatif?
Desidererei paragonare due antenne. Puoi darmi un rapporto comparativo?
Querria comparar dos antenas. Podria darme un reporte comparativo?
Eu gostaria de comparar duas antenas. Voce poderia me dar uma reportagem comparativa?
Ja hatjel byi sravnit dvje antjennyi. Nje dadite li vyi sravnitelnu-ju atsenku?
Nihon no antenna o kurabe tai to omoimasu. Hikakushita report o itadakemasuka?

The weather here is 1. very fine 2. clear 3. cloudy 4. rainy 5. windy 6. foggy 7. warm 8. cold. It is snowing.
Das Wetter hier ist 1. sehr 2. klar 3. bewolkt 4. regnerisch 5. windig 6. neblig 7. warm 8. kalt. Es schneit gerade.
Ici le temps est 1. tres beau 2. beau 3. nuageux 4. pluvieux 5. venteux 6. brumeux 7. chaud 8. froid. Il neige.

Il tempo qui e 1. molto bello 2. sereno 3. nuvuloso 4. piovoso 5. ventoso 6. nebbioso 7. caldo 8. freddo. Sta nevicando.

El tempo aqui es 1. muy bueno 2. despejado 3. nublado 4. illu-vioso 5. ventoso 6. nebuloso 7. caluroso 8. frio. Esta nevando.

O tempo aqui esta 1. muito bom 2. claro 3. nublado 4. chuvoso 5. tempestuoso 6. nevoado 7. calorento 8. frio. Esta nevando.

Pagoda zdjes 1. prikrasnaja 2. jasnaja 3. Zdjes oblacna 4. Idjot dozd 5. Zdjes silnyi vjeter 6. Unas tuman 7. zaraja 8. halodnaja. Idjot snjek.

Kochirano tenkiwa 1. kaiseidesu 2. haredesu 3. kumoridesu 4. amedesu 5. kazega fuiteimasu 6. kiriga deteimasu 7. atatakadesu 8. samuidesu. Yukiga futteimasu.

The temperature here is minus 10 [plus 25] degrees centigrade.
Die Temperatur hier betragt minus 10 [plus 25] Grad celsius.
Ici la temperature qui est de moins 10 [plus 25] degres centi-grades.
La temperatura qui e meno 10 [plus 25] gradi centigradi.
La temperaturr es de 10 grados cent.

Foreign Ham Radio Phrases from The Radio Amateur's Conversation Guide,
by Jukka Heikinheimo, OH1BR and Miike Heikinheimo, OH2BAD. Courtesy of Jukka Heikinheimo, OH1BR.

REPEATER OPERATING PRACTICES

The following suggestions will assist you in operating a repeater like you've been doing it for years.

1) Monitor the repeater to become familiar with any peculiarities in its operation.

2) To initiate a contact simply indicate that you are on frequency. Various geographical areas have different practices on making yourself known, but, generally, "This is NUØX monitoring" will suffice. Please don't "ker-chunk" the repeater "just to see if it's working."

3) Identify legally; you must identify at the end of a transmission or series of transmissions and at least once each 10 minutes during the communication.

4) Pause between transmissions. This allows other hams to use the repeater (someone may have an emergency). On most repeaters a pause is necessary to reset the timer.

5) Keep transmissions short and thoughtful. Your "monologue" may prevent someone with an emergency from using the repeater. If you talk long enough, you may actually time out the repeater. Your transmissions are being heard by many listeners including non-hams with "public service band" monitors and scanners; don't give a bad impression of our service.

6) Use simplex whenever possible. If you can complete your QSO on a direct frequency, there is no need to tie up the repeater and prevent others from using it.

7) Use the minimum amount of power necessary to maintain communications. This FCC regulation [(97.313(a)] minimizes the possibility of accessing distant repeaters on the same frequency.

8) Don't break into a contact unless you have something to add. Interrupting is no more polite on the air than it is in person.

9) Repeaters are intended primarily to facilitate mobile operation. During the commuter rush hours, base stations should relinquish the repeater to mobile stations; some repeater groups have rules that specifically address this practice.

10) Many repeaters are equipped with autopatch facilities which, when properly accessed, connect the repeater to the telephone system to provide a public service. The FCC forbids using the autopatch for anything that could be construed as business communications. Nor shall an autopatch be used to avoid a toll call. Do not use an autopatch where regular telephone service is available. Autopatch privileges that are abused may be rescinded.

11) All repeaters are assembled and maintained at considerable expense and inconvenience. Usually an individual or group is responsible, and those who are regular users of a repeater should support the efforts of keeping the repeater on the air.

With an increase in the number of reports of repeater-to-repeater interference, the FCC is placing more emphasis on repeaters being coordinated. Repeater coordination is an example of voluntary self-regulation within the Amateur service. Non-coordinated repeater operation may imply non-conformance with locally recognized band plans (e.g., an unusual frequency split) or simply that the repeater trustee has not yet applied for, or received "official" recognition from the Frequency Coordinator.

(Courtesy of the ARRL)

ARRL AND ITU REGION 1 VHF-UHF BAND PLANS

Although the FCC rules set aside portions of some bands for specific modes, there's still a need to further organize our space among user groups by "gentlemen agreements." These agreements, or band plans, usually emerge by consensus of the band occupants, and are sanctioned by a national body like ARRL.

VHF-UHF Band Plans

When considering frequencies for use in conjunction with a repeater, be sure that both the input and output fall within subbands authorized for repeater use, and do not extend past the subband edges. FCC regulation 97.205(b) defines frequencies which are currently available for repeater use.

For example, a two-meter repeater on exactly 145.50 MHz would be "out-of-band," as the deviation will put the signal outside of the authorized band segment.

Packet-radio operations under automatic control should be guided by Section, 97.109(d) of the FCC rules.

Regional Frequency Coordination

The ARRL supports regional frequency coordination efforts by amateur groups. Band plans published in the ARRL Repeater Directory are recommendations based on a consensus as to good amateur operating practice on a nationwide basis. In some cases, however, local conditions may dictate a variation from the national band plan. In these cases, the written determination of the regional frequency coordinating body shall prevail and be considered good amateur operating practice in that region.

28.000-29.700 MHz
29.400–29.700 MHz
28.000–28.070 CW
28.070–28.150 RTTY
28.150–28.190 CW

28.190–28.200 New beacon subband (*)
28.200–28.300 Old beacon subband (*)
28.300–29.340 Phone
29.304–29.510 Satellites
29.510–29.590 Repeater inputs
29.600 FM simplex calling
29.610–29.700 Repeater,outputs

(*) User Note: The FCC states in 97.203(d) that automatically controlled beacons may only operate on 28.20–28.30 MHz.

50 –54 MHz

50.0–50.1 CW beacons
54.064–50.080 Beacon subband
50.1-50.3 SSB, CW
50.10-50.125 DX window
50.125 SSB calling
50.3-50.6 all modes
50.6-50.8 non-voice communications
50.62 digital (packet) calling
54.8-51.0 radio remote control (24-kHz channels)

Note: Activities above 51.14 MHz are set on 20-kHz-spaced "even channels"

51.4-51.1 Pacific DX window
51.12-51.48 repeater inputs (19 channels)
51.12-51.58 digital repeater inputs
51.5-51.6 simplex (6 channels)
51.62-51.98 repeater outputs (19 channels)
51.62-51.68 digital repeater outputs
52.0-52.48 repeater inputs (except as noted; 23 channels)
52.02, 52.04 FM simplex
52.2 TEST PAIR (input)
52.5-52.98 repeater output (except as noted; 23 channels)
52.525 primary FM simplex
52.54 secondary FM simplex
52.7 TEST PAIR (output)
53.0-53.48 repeater inputs (except as noted; 19 channels)
53.0 remote base FM simplex
53.02 simplex
53.1, 53.2 radio remote control
53.3, 53.4
53.5-53.98 repeater outputs (except as noted; 19 channels)
53.5, 53.6 radio remote control
53.7, 53.8
53.52-53.9 simplex

144–148 MHz

144.00-144.05 EME (CW)
144.05-144.10 General CW and weak signals
144.10-144.20 EME and weak-signal SSB
144.200 National calling frequency
144.20-144.275 General SSB operation
144.275-144.300 Propagation beacons
144.30-144.50 New OSCAR subband
144.50-144.60 Linear translator inputs
144.60-144.90 FM repeater inputs
144.90-145.10 Weak signal and fm simplex
(145.01,03,05,07,09 are widely used for packet radio)
145.10-145.20 Linear translator outputs
145.20-145.50 FM repeater outputs
145.50-145.80 Miscellaneous and experimental modes
145.80-146.00 OSCAR subband
146.01-147.37 Repeater inputs
146.40-146.58 Simplex
146.61-147.39 Repeater outputs
147.42-147.57 Simplex
147.60-147.99 Repeater inputs

Notes: The frequency 146.40 MHz is used in some areas as a repeater input.
1) Automatic/unattended operations should be conducted on 145.01, 145.03, 145.05, 145.07 and 145.09 MHz.
a) 145.01 should be reserved for inter-LAN use.

b) Use of the remaining frequencies should be determined by local user groups.
2) Additional frequencies within the two-meter band may be designated for packet radio use by local coordinators.

Simplex frequencies:
(*)146.415 (*)146.475 146.535 146.595 147.465 547.525
(*)546.430 (*)146.490 146.550 147.420 147.480 147.540
(*)146.445
(*)146.505 146.565 147.435 147.495 147.555 (*)146.460
(#)146.520 146.580 147.450 147.510 147.570 147.585

(*) May also be a repeater (input/output). See repeater pairs listing
(#) National Simplex Frequency

Several states have chosen to realign the 146-148 MHz band, using 20 kHz spacing between channels. This choice was made to gain additional repeater pairs.
The transition from 30 to 20 kHz spaking is taking place on a case-by-case basis as the need for additional pairs occurs.
Typically the repeater on an odd numbered pair, will shift to 14 kHz, up or down, creating a new set on an even numbered channel. For example, the pair of 146.13/73 would change to 146.12/72 or 146.14/74 while the pairs of 146.10/70 and 146.16/76 would be left unchanged.

222–225 MHz

222.04-222.15 Weak signal modes
222.00-222.425 UME
222.05-222.064 Propagation beacons
222.1 SSB & CW Calling
222.50-222.150 Weak signal CW & SSB
222.15-222.25 Local coordinator's option:
weak signal, ACSB, repeater inputs and control
222.25-223.38 FM repeater inputs only
223.40-223.52 FM simplex
223.54 Simplex calling
223.52-223.223.64 Digital packet
223.64-223.70 Links control
223.71-223.85 Local coordinator's option;
FM simplex packet repeater outputs
223.85-224.98 Repeater outputs only

Simplex frequencies:
223.42 223.52 223.62 223.72 223.82 223.44 223.54
223.64 223.74 223.84 223.46 223.56 223.66 223.76
223.48 223.58 223.68 223.78 (*)223.50 223.60 223.70
223.80

(*) National Simplex Frequency

420–450 MHz

420.00-426.00 ATV repeater or simplex with 421.25 MHz video carrier, control links and experimental
426.00-432.00 ATV simplex with 427.25 MHz video carrier frequency
432.00-432.07 EME (Earth-Moon-Earth)
432.07-432.10 Weak signal CW
432.10 Calling frequency
432.10-432.30 Mixed-mode and weak-signal work
432.30-432.40 Propagation beacons
432.40-433.00 Mixed-mode and weak signal work
433.00-435.00 Auxiliary/repeater links
435.00-438.00 Satellite only (internationally)
438.00-444.00 ATV repeater input with 439.250-MHz video carrier frequency and repeater links
442.00-445.00 Repeater inputs and outputs (local option)
445.00-447.00 Shared by auxiliary and control links, repeaters and simplex (local option)
446.00 National simplex frequency
447.00-450.00 Repeater inputs and outputs (local option)

902–928 MHz

902.0-903.0 Weak signal (902.1 calling frequency)
902.1 Calling frequency
903.0-906.0 Digital (903.1 alternate calling frequency)
906.0-909.0 FM repeater outputs
909.0-915.0 Amateur TV
915.0-918.0 Digital
918.0-921.0 FM repeater inputs
921.0-927.0 Amateur TV
927.0-928.0 FM simplex and links

Notes: Adopted by the ARRL Board of Directors in July, 1989. The following packet radio frequency recommendations were adopted by the ARRL Board of Directors in January, 1988 as interim guidance. Two 3-MHz-bandwidth channels are recommended for 1.5 Mbit/s links. They are 903-906 MHz and 914-917 MHz with 10.7 MHz spacing.

1) Extracts of FCC Rules & Regulations, 97.303(g)(1). No amateur station shall transmit from within the states of Colorado and Wyoming, bounded on the south by latitude 39 degrees North, on the north by latitude 42 degrees North, on the east by longitude 105 degrees West, and on the west by longitude 108 degrees West. This band is allocated on a secondary basis to the Amateur Service subject to not causing harmful interference to, and not receiving protection from any interference due to the operation of industrial, scientific and medical devices, automatic vehicle monitoring systems or government stations authorized in this band.

2) Coordinated frequency assignments are required.

3) ATV assignments should be made according to modulation type, e.g., VSB-ATV, SSB-ATV or combinations. Coordination of multiple users of a single channel in a local area can be achieved through isolation by means of cross-polarization and directional antennas.

4) Coordinated assignments at 100 kHz until allocations are filled, then assign 50 kHz until allocations are filled, before assigning 25 kHz channels.

5) Simplex services only; permanent users shall not be coordinated in this segment. High altitude repeaters or other unattended fixed operations are not permitted.

6) Voice and non-voice operation.

7) Spread-spectrum requires FCC authorization.

8) Consult FCC (97.307) for allowable data rates and bandwidths.

1240 - 1300 MHz

1244-1246 ATV #1
1246-1248 Narrow-bandwidth FM point-to-point links end digitel, duplex with 5258-1264.
1248-1252 Digital,Communications
1252-1258 ATV #2
1258-1260 Narrow-bandwidth FM point-to-point links and digital duplexed with 1246-1252
1264-1270 Satellite uplinks experimental simplex ATV
1270-1276 Repeater inputs, FM and linear, paired with 1282-1288. (239 pairs every 25 kHz, e.g., 1270.025, 450, etc.)
1271/1283 Non-coordinated test pair
1276-1282 ATV #3
1282-1288 Repeater outputs paired with 1270-1276
1288-1294 Wide Band experimental, simplex ATV
1294-1295 Narrow band FM simplex, 25-kHz channels
1294.50 National FM Simplex calling
1295-1297 Narrow band weak signal (no FM)
1295-1295.80 ATV, FAX, ACSSB experimental
1295-1296 Reserved for EME, CW expansion
1296-1296.05 EME exclusive
1296.07-1296.08 CW beacons
1296.1 CW/SSB calling frequency
1296.40-1296.60 Crossband linear translator output
1296.80-1297.00 Experimental beacons (exclusive)
1297-1300 Digital communication

2300 - 2310 and 2390 - 2450 MHz

2300-2302 High-rate data
2303-2303.5 Packet radio

2303.5-2303.8 RTTY packet
2303.8-2303.9 Packet, RTTY, CW, EME
2303.9-2304.1 CW, EME
2304.1 Calling frequency
2304.1-2304.2 CW, EME, SSB
2304.2-2304.3 SSB, SSTV, FAX, Packet, AM, AMTOR
2304.3-2304.32 Propagation beacon network
2304.32-2304.4 General propagation beacons
2304.4-2304.5 SSB, SSTV, ACSSB, FAX, Packet, AM, AMTOR experimental
2304.5-2304.7 Crossband linear translator input
2304.7-2304.9 Crossband linear translator output
2304.9-2305 Experimental beacons
2305-2305.2 FM simplex (25-kHz spacing)
2305.2 FM simplex calling frequency
2305.2-2306 FM simplex (25-kHz spacing)
2306-2309 FM repeaters (25-kHz) input
2309-2310 Control and auxiliary links
2390-2396 Fast-scan TV
2396-2399 High-rate data
2399-2399.5 Packet
2399.5-2400 Control and auxiliary links
2400-2403 Satellite
2403-2408 Satellite high-rate data
2408-2410 Satellite
2410-2413 FM repeaters (25-kHz spacing) output
2413-2418 High-rate data
2418-2430 Fast-scan TV
2430-2433 Satellite
2433-2438 Satellite high-rate data
2438-2450 Wideband FM, FSTV, FMTV, SS experimental

3300 - 3500 MHz

The following beacon subband was adopted by the ARRL Board of Directors in July, 1988.
3456.3-3456.4 Propagation beacons

5650 - 5925 MHz

The following beacon subband was adopted by the ARRL Board of Directors in July, 1988.
5760.3–5760.4 Propagation beacons

10.000 - 10.500 GHz

The following subband recommendation was adopted by the ARRL Board of Directors in January, 1987
10.368 GHz Narrow-band calling frequency

The following beacon subband was adopted by the ARRL Board of Directors in July, 1988.
10368.3-10368.4 Propagation beacons

ITU Region 1 (Europe and Africa) 144-146 MHz Band Plan

144-144.01 EME
144.05 CW calling, once contact has been established, move off frequency
144.10 CW random meteor scatter
144.15 Upper limit CW exclusive
144.20 SSB random meteor scatter
144.30 SSB calling, once contact has been established, move off frequency
144.40 SSB meteor scatter reference frequency
144.50 SSTV calling, upper limit CW/SSB exclusive
144.60 RTTY calling
144.675 Data transmission calling frequency
144.70 FMX calling
144.845-144.990 Beacon exclusive
145-145.175 Repeater inputs RØ/ to R7
145.2-145.575 Channelized FM simplex
145.5 FM calling frequency, once contact has been established, move off frequency
145.6-145.775 Repeater outputs RØ/ to R7
145.8-146 Space communications, CW/SSB uplink only

Repeater channels (input/output)

RØ/ 145.000/145.600
R1 145.025/145.625
R2 145.050/145.650
R3 145.075/146.675
R4 145.100/145.700
R5 145.125/145.725
R6 145.150/146.750
R7 145.175/145.775

Packet frequencies most widely used are:

144.650   144.675   432.675

Simplex Channels

S10 145.25
S11 145.275
S12 145.30
S13 145.325
S14 145.35
S15 145.375
S16 145.40
S17 145.45
S18 145.45
S19 145.475
S20 145.50–Most common simplex calling frequency, once contact has been established, move off frequency
S21 145.525
S22 145.55
S23 145.575

UK repeaters require a 1750-Hz (+/- 25-Hz) tone burst of about 200-ms duration to open the repeater. It is best to transmit this tone at the start of each transmission. Belgian repeaters require a 3-second tone burst. Swiss repeaters require an 1800-Hz tone burst.

Note: Several French repeaters are on nonstandard frequency pairs. Those pairs are:
144.725/146.325 144.825/145.425
144.750/145.350 44.850/145.450
144.800/145.400 144.875/145.475

(Courtesy of the ARRL)

CTCSS TONE FREQUENCIES

Frequency (Hz)	"P/L" Designator
67.0	XZ
69.3	WZ
71.9	XA
74.4	WA
77.0	XB
79.7	WB
82.5	YZ
85.4	YA
88.5	YB
91.5	ZZ
94.8	ZA
97.4	ZB
100.0	1Z
103.5	1A
107.2	1B
110.9	2Z
114.8	2A
118.8	2B
123.0	3Z
127.3	3A
131.8	3B
136.5	4Z
141.3	4A
146.2	4B
151.4	5Z
156.7	5A
162.2	5B
167.9	6Z
173.8	6A
179.9	6B
186.2	7Z
192.8	7A
203.5	M1
206.5	8Z
210.7	M2
218.1	M3
225.7	M4
229.1	9Z
233.6	M5
241.8	M6
250.3	M7
254.1	ØZ

REGIONAL FREQUENCY COORDINATION ORGANIZATIONS & STATE COORDINATORS:

MACC Officers:
President, Ken Enenbach, KCØWX, 110 E 51st St #14, Kansas City, MO 64112
Vice-President, Nels Harvey, WA9JOB, z104 W County Line Rd, Mequon, WI 53090
Treasurer, Denny Crabb, WBØGGI, 115 N 14th St, Denison, IA 51442-1452
Secretary, George Isely, WD9GIG, 736 Fellows St., St Charles, IL 60174

The following states are members of the Mid-America Coordination Council, Inc. (MACC):

ARKANSAS
Al Fisher, NI5A, 609 Honeysuckle Ln., Trumann, AR 72472
Dan Puckett, K5FXB, PO Box 2458 U-A, Fayetteville, AR 72701

COLORADO
Whitman E. Brown, WBØCJX, 14418 W. Ellsworth Place, Golden, CO 80401-5324

ILLINOIS
Jeremy D. Ruck, WM9C, PO Box 9274, Peoria, IL 61612-9274

INDIANA
Repeater Coordination, PO Box 615, Anderson, IN 46015
Chairman, Steve Riley, WA9CWE, RR 2, Box 225-A, Alexandria, IN 46001
52 MHz, Kevin Berlin, WB9QBR, 308 W 4th St., Clay City, IN 47841
29 MHz, Walter A. Breining, N9WB, RR 1, Spiceland, IN 47385
144 MHZ, Neil Rapp, WB9VPG, 1506 S Parker Dr., Evansville, IN 47714-5741
222 MHz, Martin Hensley, KA9PCT, 6426 Maidstone Rd. #206, Indianapolis, IN 46254
440 MHz, Jeff Tucker, KB9KIX, 1959 S. Brown Av., Terre Haute, IN 47803
902 & Above/Packet, Ron Pogue, KD9QB, 210 Westminister Dr., Noblesville, IN 46060

IOWA
Denny Crabb, WBØGGI, 115 N 14th St., Denison, IA 51442-1452

Al Groff, KØVM, 1446 Council St. NE, Cedar Rapids, IA 52402

KANSAS

State Frequency Coordinator, Wendell D. Wilson, WØTQ, 717 Second Av., Concordia, KS 66901

East Kansas Asst., Slim Cummings, WAØEDA, PO Box 298, Pittsburg, KS 66762-0298

MINNESOTA

Paul Emeott, KØLAV, 3960 Schuneman Rd., White Bear Lake, MN 55110

MISSOURI

Wayland N. "Mac" McKenzie, K4CHS, 8000 S. Barry Rd., Columbia, MO 65201

MONTANA

Kenneth G. Kopp, KØPP, Box 848, Anaconda, MT 59711, 406-797-3340

NEBRASKA

John Gebuhr, WBØCMC, 2340 N 64th St., Omaha, NE 68104

Billy McCollum, KEØXQ, 1314 Deer Park Bldvd., Omaha, NE 68108

NEVADA

(northern) CARCON ,PO Box 7523, Reno, NV 89510-7523

Jay Ranney, K7WYC (FC), 430 McClure Cir., Sparks, NV 89431-1242

NORTH DAKOTA

Stanley E. Kittelson, WDØDAJ, 261 10th St. E, Dickinson, ND 58601

OHIO

Ken Bird, WB8SMK, 244 N Parkway Rd., Delaware, OH 43015

OKLAHOMA

Vince Moore, N5RFW, 5613 South 66th West Av., Tulsa, OK 74107

OREGON

Frequency Coordination, PO Box 4402, Portland, OR 97208-4402

George Pell, KB7PSM, 27754 SW Strawberry Hill Dr., Hillsboro, OR 97123

SOUTH DAKOTA

Richard Neish, WØSIR, PO Box 100, Chester, SD 57016

UPPER PENINSULA MICHIGAN

UPARRA Coordinations, PO Box 9, McMillan, MI 49853

Noel Beardsley, WD8DON, W7021 CR 356, Stephenson, MI 49887

WESTERN WASHINGTON

Frequency Coordination - WWARA, PO Box 65492, Port Ludlow, WA 98365-0492

WISCONSIN

Nels Harvey, WA9JOB, Chairman WAR, 2104 W. County Line Rd., Mequon, WI 53090

Frequency Coordinator, Scot Thompson, WBØWOT, W7137 770th Av., Beldenville, WI 54003

WYOMING

Don Breazile, N7MYR, 4406 Greenhill Ct., Cheyenne, WY 82001

SERA Officers:

President, Dave Shiplett, AC4MU, 107 Mossy Lake Rd., Perry, GA 31069

Vice-President, Nita Wofford, N4DON, 2966 Cordell, Memphis, TN 38118

Treasurer, Raymond K. Adams, N4BAQ, 5833 Clinton Hwy, Suite 203, Knoxville, TN 37912

Secretary, H. Alex Hedrick, Jr., N8FWL, PO Box 417, Beckley, WV 25802-0417

The following states are members of the South Eastern Repeater Association (SERA):

GEORGIA

Director: Stu Sims, N4MXC, 112 Carol Dr., Cochran, GA 31014

Vice-Director: Mike Flammia, N4PLM, 5303 Driskell Dr., Winston, GA 30187

Asst. Director: Bert Coker, N4BZJ, 2102 Milican Ln., Dalton, GA 30721

Asst. Director: Scott Haner, KBØY, 1610 Marion St., Valdosta, GA 31602-3004

KENTUCKY

Director: Jeffrey Martin, N5KOL, 308 Marylan Av., Bowling Green, KY 42101

Asst. Director: Mark Smith, KM4IV, 153 Riverside Dr., Ivel, KY 41642

MISSISSIPPI

Director South: Steve Grantham, N5DWU, PO Box 127, Ellisville, MS 39437

Vice-Director North: Jim Akers, W5VZF,.21 Whispering Pines, Starkville, MS 39759

NORTH CAROLINA

Director: Danny Hampton, KM4OX, PO Box 19122, Raleigh, NC 27609-9797

Vice Director: Norman Harrill, N4NH, #7 Skylyn Ct., Asheville, NC 28806

Asst. Director: Pamela Glaub, KC4SWM, PO Box 19122, Raleigh, NC 27609-9122

SOUTH CAROLINA

Director: Bill Jones, N4MNH, 1609 Bur-Clare Rd., Charleston, SC 29412-8148

Vice Director: Gary Foster, WD8OXE, 670 Foster Rd., Inman, SC 29349

TENNESSEE

Director: Johnny Wofford, WA4ETE, 2966 Cordell, Memphis, TN 38118

Vice-Director: Tim Berry, WB4GBI, 214 Echodale Ln., Knoxville, TN 37920-5042

Freq. Coord (W-TN): Andy Masters, NU5O, 240 W White Rd., Collierville, TN 38017

Asst. Director: Brad Adams, N4PYI, 1706 Bender's Ferry Rd., Mt Juliet, TN 37122

VIRGINIA

Director: Wally Burkett, WA4KXV, 242 Raintree Rd., Virginia Beach, VA 23452

Vice-Director: Don Williams, WA4K, 412 Ridgeway Dr., Bluefield, VA 24605-1630

WEST VIRGINIA

Director: R.T. (Dick) Fowler, N8FMD, Route 3, Box 52, Clarksburg, WV 28301

Vice-Director: H. Alex Hedrick, Jr., N8FWL, South, PO Box 417, Beckley, WV 25802

Asst. Director: G. David Ramezan, KA8ZXP, PO Box 330, Glenville, WV 26351

The following are Directors/Zone Coordinators within the Texas VHF/FM Society (TVFS):

TEXAS

State Frequency Coordinator, Chairman, Coordinating Committee, Paul Z. Gilbert, KE5ZW, 2608 El Toro Dr., # 111-B, Huntsville, TX 77340

Assistant State Frequency Coordinator, Merle Taylor, WB5EPI, 910 Kingston Dr., Mansfield, TX 76063

Assistant Chairman, Coordinating Committee, Walt Wiederhold, W5OGZ, 2812 Pritchett, Irving, TX 75061

Link Frequency Coordinator, Jim Reese, WD5IYT, 2850 Wallingford Dr. #104, Houston, TX 77042

President, Louis Petit, WB5BMB, 1213 15th Av. N., Texas City, TX 77590

Secretary / Treasurer, Robert McWhorter, K5PFE, PO Box 461, Jasper, TX 75951

Director / Editor, John Johnson, N5NHH, 4124 Hollow Oak Ln., Dallas, TX 75287

Director at Large, Joe Makeever, W5EBJ, 8609 Tallwood Dr., Austin, TX 78759

Director at Large, Jay Maynard, K5ZC, 6027 Leaf Wood Dr., League City, TX 77573

Director at Large, Marty Plass, II, KA5LZG, 6721 Pintail, Corpus Christi, TX 78413

Director at Large, Mark Earl, WA2MCT, PO Box 3456, Corpus Christi, TX 78411

Director at Large, Dave Hammer, WJ5B, 25411 Morgan Dr., Tomball, TX 77375

Technical Coordinator, Greg Jurrens, WDØACD, 22819 Acorn Valley, Spring, TX 77389

Zone 1: 29-52-144-440 Coordinator, Send coordination requests to:, Paul Gilbert, KE5ZW, 2601 El Toro Dr. Apt 111B, Huntsville, TX 77340

Zone 1: 220 & 1.2 GHz Coordinator, Paul Baumgardner, KB5BFJ, PO Box 181912, Arlington, TX 76096-1912

Zone 2: 29-52-144-222 Coordinator, Howard Smith, KB5VAW, PO Box 2734, Bryan, TX 77805

Zone 2: 440-902-1.2 Coordinator, Joe Spagnoletti, N5HGL, 30603 Hummingbird, Conroe, TX 77385

Zone 3: Coordinator, Richard Norton, WB5FRO, 1011 Valley View Dr., Weslaco, TX 78596

Zone 4: Coordinator, Louis Bancook, WB5UUT, 2200 Logan Dr., Round Rock, TX 78664

Zone 5: Coordinator, Paul Gilbert, KE5ZW, 2608 El Toro Dr., Apt 111B, Huntsville, TX 77340

Mexican Liaison, Fernando J. Muguerza, XE2FL, Apartado 91, 66200 San Pedro Garza, NL MEXICO

The following are Canadian coordination groups presently known to be providing service:

ALBERTA
Ken Oelke, VE6AFO, 7136 Temple Dr. NE, Calgary, AB T1Y 4E7, CANADA
Don Moman, VE6JY, 61-52152 Range Rd 210, Sherwood Park, AB T8G 1A5, CANADA

BRITISH COLUMBIA
PRARFCA, Al Muir, VE7BEU, 871 Walfred Rd., Victoria, BC V9C 2P1, CANADA

MANITOBA
Thomas Blair, VE4TOM, 121 Miramar Rd. ,Winnipeg, MB R3R 1E4 ,CANADA

ONTARIO
(western New York and southern Ontario)
Western New York - Southern Ontario Repeater Council (WNYSORC) Frequency Coordinator, Paul Toth, VE3GRW, 4629 Queensway Gardens, Niagara Falls, ON L2E 6R2, CANADA

(eastern Ontario - St. Lawrence Seaway Valley) St. Lawrence Valley Repeater Association (SLVRA)
Peter de Wolfe, VE3YYY - VHF, RR 1, Braeside, ON K0A 1G0, CANADA
Luc Pernot, VE3LJC - UHF, 295 Tremblay Cres., Russell, ON K4R 1G3, CANADA

QUEBEC
RAQI , J. Leo Daigle, VE2LEO, 9450 Andre Grasset, Montreal, PQ H2M 2B5, CANADA

(Ottawa Valley-St. Lawrence , Seaway Valley) St. Lawrence Valley Repeater Association (SLVRA) , (SEE ABOVE)

SASKATCHEWAN
Ken Nyeste, VE5NR, 123 Holland Rd., Saskatoon, SK S7H 4Z5

OTHER AREAS
(areas not included above) In areas of Canada, where no frequency coordination exists, please notify the following of any new repeaters or changes.

Ken Oelke, VE6AFO, 7136 Temple Dr. NE, Calgary, AB T1Y 4E7, CANADA

Other Coordination Organizations — alphabetically by state:

ALABAMA
Alabama Repeater Council, Inc. Steve Flory, WA4OUE, PO Box 1305, Decatur, AL 35602
Frequency Coordinator, Dave Baughn, KX4I, 3926 Woodland Hills Dr., Tuscaloosa, AL 35405
Jack Flory, W4RXH (Sec), 1903 Woodmeade St., Decatur, AL 35601

ALASKA
Mel Bowns, KL7GG , HC 83, Box 1599 , Eagle River, AK 99577
(western interior and northern) Jerry Curry, KL7EDK , 940 Vide Way , Fairbanks, AK 99712
(southeast area) Edward Shilling, W6SJJ , PO Box 1087 , Petersburg, AK 99833

ARIZONA
Arizona Council of Amateur, Radio Clubs, Ralph Turk, W7HSG, Chairman , 5232 W. Calle Paint , Tucson, AZ 85704

CALIFORNIA NORTH
Northern Amateur Relay Council, of California, Inc (NARCC), PO Box 60531, Sunnyvale, CA 94088-0531

CALIFORNIA - SOUTH
Southern California Repeater , and Remote Base Association , SCRRBA, PO Box 5967 , Pasadena, CA 91117
Two Meter Area Spectrum , Management Association , TASMA, 358 S Main St. # 90, Orange, CA 92668
220 Spectrum Management , Association, 220SMA, 21704 Devonshire St. #220, Chatworth, CA 91311

CONNECTICUT
See New York City - Long Island

DELAWARE
See Maryland

DISTRICT OF COLUMBIA
See Maryland

FLORIDA
Florida Repeater Council, Inc. , FRC, Secretary, Charles Burkey, K4EVA, 810 E. Wisconsin Ave., DeLand, FL 32724
FRC President, Walt Maxwell, W2DU, 243 N. Cranor Ave., DeLand, FL 32720

HAWAII
Hawaii State Repeater , Advisory Council , Pat Corrigan, KH6DD, PO Box 67, Honolulu, HI 96810

IDAHO
(south and east) , Clark Lenz, N7QEO, 4475 E 1200 N, Ashton, ID 83420
Jim Sherman, N7VVG, 175-A N 4300 E, Rigby, ID 83442

Bud Dunn, KI7SJ, 770 S River Rd., St Anthony, ID 83445
Harold Short, WA7UHW, 923 10th St., Rupert, ID 83350
(north and west) Larry E. Smith, W7ZRQ , 8106 Bobran St ,
Boise, ID 83709

(panhandle and eastern Washington), Doug Rider, KC7JC , E
11516 Mission Ave , Spokane, WA 99206

LOUISIANA
Louisiana Council of Amateur , Radio Clubs , Tom Palko,
WB5ASD, PO Box 8762, Alexandria, LA 71306-1762
Dave Breeding, KF5JC, 17330 Sanders Rd., Franklinton, LA
70438

MAINE
See Massachusetts

MARYLAND
The Middle Atlantic FM and, Repeater Council, PO Box 1022,
Savage, MD 20763-1022
Owen Wormser, K6LEW, 406 N Pitt St., Alexandria, VA 22314

MASSACHUSETTS
New England Spectrum Management Council, (29 MHz), Roger
Perkins, W1OJ, Old Bay Rd., Bolton, MA 01740
(52 MHz), Alan D. Tasker, WA1NYR , 64 Dyer St, N Billerica,
MA 01862
(144 MHz), Bob Skinner, WA1YEG, 68 Governor Dinsmore Rd.,
Windham, NH 03087
(222 MHz), Tom Greenwood, N1JQB, 126 Haynes Rd., Sudbury,
MA 01776
(440 MHz), Ken Chilton, KA1TIH, 53 Alan Rd., Marlboro, MA
01752
(902 MHz and above), Lewis D. Collins, W1GXT, 10 Marshall
Terrace , Wayland, MA 01778

MICHIGAN - LOWER
Larry Tissue, N8QGE - President, 851 Wheaton Rd., Charlotte, MI
48813
(29 - 52 - 144 MHz), Larry French, NW8J, 12215 Myers Lake Av
NE, Cedar Springs, MI 49319
(222 - 440 MHz), Vince Vielhaber, KA8CSH, 790 Glaspie Rd.,
Oxford, MI 48371

NEVADA
(southern), R. Scott Fowler, WA7GIV, 2208 Jansen Ave , Las
Vegas, NV 89101
Joe Lambert, W8IXD, PO Box 61201, Boulder City, NV 89006

NEW HAMPSHIRE
See Massachusetts

NEW JERSEY
(northern), See New York (NYC-LI)
(southern), See Pennsylvania (eastern)

NEW MEXICO
New Mexico Frequency Coordinating Committee, Tom Ellis,
WD5JMA, 3232 San Mateo Blvd NE, # 54, Albuquerque, NM
87110

NEW YORK
(Connecticut, Northern New Jersey, New York City, Long Island,
Southern New York (914 Area Code))
Tri-State Amateur Repeater Council (TSARC), Frequency
Coordinations, PO Box 1022, Suffern, NY 10901-1022

TSARC - PRESIDENT, Thomas Raffaelli, WB2NHC, 544
Manhattan Ave. Thornwood, NY 10594
(upstate - except: [Clinton and N. Essex Counties—see Vermont]
[Franklin and St. Lawrence Counties—see St. Lawrence Seaway
Valley] Upper New York Repeater Council (UNYREPCO)
Jim Mozley, W2BCH, 126 Windcrest Dr., Amillus, NY 13031
Bill Reiter, WA2UKX, 3079 Ferguson Corners, Penn Yan, NY 14527

(western New York [west of Rochester] and southern Ontario)
Western New York - Southern Ontario Repeater Council
(WNYSORC)
Frequency Coordinator, Paul Toth, VE3GRW, 4629 Queensway
Gardens, Niagara Falls, ON L2E 6R2, CANADA

(eastern Ontario - St. Lawrence Seaway Valley) St. Lawrence
Valley Repeater Association (SLVRA)
(VHF)
· Peter de Wolfe, VE3YYY, RR 1, Braeside, ON K0A 1G0, CANA-
DA
(UHF)
Luc Pernot, VE3LJC, 295 Tremblay Cres., Russell, ON K4R 1G3,
CANADA

PENNSYLVANIA
(eastern Pennsylvania and southern New Jersey)
The Philadelphia Area Repeater Council (TPARC)
Jerry Smedley, WB3BLG, PO Box 36, Croydon, PA 19021,
VOICE-MAIL: 215-788-0759
(western)
Western Pennsylvania Repeater Council (WPRC)
WPRC Frequency Coordination Cmte., 10592 Perry Highway
#173, Wexford, PA 15090

Membership Information
WPRC Secretary, Dale Conrad, W3IXR, 10592 Perry Highway
#173, Wexford, PA 15090
Frequency Coordinator, Joseph A. McElhaney, KR3P, 319 Mt.
Vernon Dr., Apollo, PA 15613-8701
NOTE: All Coordination requests MUST be sent to the Wexford
address.

PUERTO RICO
PR/VI VFC, PO Box 191917, San Juan, PR 00919-1917
Guillermo B. Martinez, KP4BKY, PO Box 475, Mayaguez, PR
00681-0475

RHODE ISLAND
See Massachusetts

UTAH
Utah VHF Society John Lloyd, K7JL, 2078 Kramer Dr., Sandy,
UT 8409

VERMONT
(also Clinton and N. Essex Counties of New York)
Vermont Independent Repeater Coordinating Committee
VIRCC, PO Box 99, Essex, VT 05451

VIRGINIA
(northern)
See Maryland

WASHINGTON
IACC, (eastern, also Idaho panhandle)
Doug Rider, KC7JC , E 11516 Mission Ave , Spokane, WA 99206

WEST VIRGINIA
(eastern panhandle)
See Maryland

PACKET COORDINATION ORGANIZATIONS

ARKANSAS
Dan Puckett, K5FXB, PO Box 2458 U-A, Fayetteville, AR 72701

CALIFORNIA
Frequency Coordinator, PO Box 4425, Carson, CA 90749

CALIFORNIA - SOUTH
Karl Pagel, N6BVU, PO Box 6080, Anaheim, CA 92816-6080

FLORIDA
Florida Amateur Digital Communications Association (FADCA)
John P. Paxton, KB4RLL, 6333 NE 120th St., Okeechobee, FL
34972

ILLINOIS
Northern Illinois Packet, Radio Frequency Council, (NIPRFC),
Mark Thompson, WB9QZB, PO Box 357, Mount Prospect, IL
60056
(NIPRFC), Carl Bergstedt, K9VXW, 308 W. Osage Ln.,
Naperville, IL 60540-7821
(NIPRFC), William Davidson, KA9SWW, 3122 N Drake Ave.,
Naperville, IL 60618

INDIANA
Indiana Digital Experimenters Association (IDEA), John Gooldy,
WJ9U, 1712 Mulberry Cir., Noblesville, IN 46060

KANSAS
Kansas Digital Coordination Committee (KDCC), Karl Medcalf,
WK5M, 1544 N 1000 Dr., Lawrence, KS 66046-9610

KENTUCKY
Brian Walker, KC4FIE, 1428 Grapevine Rd., Madisonville, KY
42431

MAINE
James F. Ledger, N1PGH, PO Box 35, New Gloucester, ME 04260

MARYLAND
Digital Mid-Atlantic RadioCouncil (DMARC), Tom Abernethy,
WA3TAI, 1133 Apple Valley Rd., Accokeek, MD 20607

MICHIGAN
(MIPAC), Jay Nugent, WB8TKL, 3081 Braeburn Cir., Ann Arbor,
MI 48104

MICHIGAN
UPARRA, PO Box 9, McMillan, MI 49853

MINNESOTA
(MNPSD), Brian Klier, NØQVC, 8230 Cedar Lake Blvd.,
Faribault, MN 55021-7524

NEVADA
CARCON, PO Box 7523, Reno, NV 89510-7923

SOUTH CAROLINA
(SCARDS), Avery J. Wright, KD4GBA, 417 Lowrey Av. Box
1284, Shaw AFB, SC 29152-5060

TEXAS
Texas Packet Radio Society, (TPRS), Ronnie Franklin, WD5GIC,
RR 8 Box 141, Granbury, TX 76048
Jim Neely, WA5LHS, 505 East Huntland Dr. Suite 480, Austin,
TX 78752

WASHINGTON
Chuck Robertson, KG7WV, 13305 NE 171st St. #L375,
Woodinville, WA 98072

WISCONSIN
Wisconsin Amateur Packet, Radio (WAPR), Steve McDonough,
KE9LZ, 445 S. St Bernard Dr., Depere, WI 54115

MOST COMMON REPEATER OUTPUT FREQUENCIES IN NORTH AMERICA

29 MHz
(Inputs are normally 100 kHz below the outputs)

Freq.	Count
29.620	32
29.680	29
29.640	23
29.660	13
29.600	5
29.500	2
29.670	1
29.650	1
29.630	1
29.560	1

50-54 MHz
(Inputs are normally 500 kHz below the outputs)

Freq.	Count
53.010	25
53.030	20
53.050	17
53.250	16
53.170	16
53.130	16
53.070	15
53.110	14
53.210	13
53.150	12

144-148 MHz
(Inputs are normally 600 kHz below the outputs, except
above 147 MHz, where inputs are 600 kHz above the
outputs)

Freq.	Count
146.94	156
146.76	156
146.88	130
146.82	127
146.70	121
146.64	119
147.00	117
147.30	111
147.18	104
147.36	103
147.24	103
147.06	103

222-225 MHz
(Inputs are normally 1.6 MHz below the outputs)

Freq.	Count
224.940	59
224.100	47
223.940	47
224.820	42
224.880	41
224.900	40
224.760	40
224.660	39
224.960	38
224.860	38

420-450 MHz
(Inputs are normally 5 MHz below the outputs)

Freq.	Count
444.90	78
444.50	75
444.80	74
444.00	72
444.20	69
444.95	67
444.70	66
444.85	65
444.10	64
444.40	61

902 MHz
(Inputs are normally 12 MHz below the outputs)

Freq.	Count
920.000	7
921.200	5
919.025	4
927.700	3
927.600	3
921.500	3
921.100	3
921.000	3
920.600	3
919.500	3
919.200	3
919.100	3

1240 MHz
(Inputs are normally 12 MHz below the outputs)

Freq.	Count
1292.00	12
1285.00	9
1283.00	9
1292.10	8
1282.60	7
1282.10	7
1292.30	6
1282.30	6
1292.20	5
1292.40	4
1291.40	4
1288.00	4
1287.00	4
1285.50	4
1284.50	4
1284.05	4
1282.50	4
1282.40	4

STATES WITH THE MOST REPEATERS BY BAND

29 MHz

State	No. Repeaters
New York	12
California	9
Florida	8
Maryland	6
Pennsylvania	5
Ohio	5
Illinois	5
Wisconsin	4
North Carolina	4
Connecticut	4
New Jersey	3
Minnesota	3
Massachusetts	3
Georgia	3
Alabama	3
West Virginia	2
Washington	2
Virginia	2
Texas	2
Tennessee	2
Puerto Rico	2
Nebraska	2
Montana	2
Mississippi	2
Louisiana	2
Iowa	2
District Of Columbia	2
Colorado	2
Wyoming	1
Rhode Island	1
Oklahoma	1
Michigan	1
Kansas	1
Arkansas	1

50-54 MHz

State	No. Repeaters
California	114
New York	46
Pennsylvania	32
Florida	23
Massachusetts	22
Washington	21
New Jersey	21
Ohio	18
Indiana	18
Virginia	17
Wisconsin	16
Texas	16
Tennessee	15
Alabama	14
Georgia	13
Minnesota	12
North Carolina	11
Michigan	10
Maryland	10
District Of Columbia	9
Illinois	7
New Hampshire	6
Colorado	6
Arkansas	6

Missouri	5
Nevada	4
Louisiana	4
Kentucky	4
Iowa	4
Connecticut	4
Oregon	3
Nebraska	3
Arizona	3
West Virginia	2
Utah	2
South Carolina	2
Rhode Island	2
Puerto Rico	2
Oklahoma	2
Mississippi	2
Maine	2
Kansas	2
Hawaii	2
North Dakota	1
Idaho	1

144-148 MHz

State	No. Repeaters
California	574
Texas	493
Florida	364
New York	298
Ohio	270
Pennsylvania	266
Arizona	200
Tennessee	181
Illinois	174
North Carolina	165
Alabama	163
Georgia	162
Indiana	155
Michigan	151
Wisconsin	150
Missouri	148
Washington	145
Virginia	140
Minnesota	129
Colorado	127
Kansas	126
Oklahoma	120
Iowa	118
Arkansas	115
Oregon	114
New Jersey	111
Kentucky	109
Utah	98
Louisiana	97
Mississippi	92
West Virginia	90
Massachusetts	90
Nebraska	81
Nevada	77
South Carolina	76
New Mexico	76
Maryland	74
Hawaii	74
Puerto Rico	69
Montana	65
Connecticut	61

Maine	53
Idaho	53
North Dakota	44
Alaska	43
Wyoming	42
District Of Columbia	38
South Dakota	37
New Hampshire	33
Vermont	21
Rhode Island	14
Delaware	11
US Virgin Islands	8

222-225 MHz

State	No. Repeaters
California	208
Texas	104
New York	104
Ohio	90
Pennsylvania	84
Florida	81
Illinois	76
New Jersey	55
Michigan	51
Georgia	50
Indiana	49
Washington	45
Tennessee	45
Massachusetts	45
Virginia	42
North Carolina	40
Colorado	38
Hawaii	37
Connecticut	35
Maryland	34
Wisconsin	32
Puerto Rico	30
Missouri	27
Arizona	23
District Of Columbia	21
New Hampshire	20
Utah	19
Alabama	19
South Carolina	16
Rhode Island	16
Minnesota	16
Oregon	14
Kentucky	14
Oklahoma	12
Kansas	11
Nevada	10
Iowa	10
New Mexico	9
Delaware	9
Nebraska	8
Maine	8
Louisiana	8
Arkansas	8
Mississippi	7
West Virginia	5
Idaho	5
Alaska	3
Vermont	2
North Dakota	2
Montana	2

420-450 MHz

State	No. Repeaters
California	924
Texas	516
Florida	356
New York	335
Ohio	251
Washington	216
Arizona	213
Illinois	185
Michigan	161
Pennsylvania	150
Indiana	143
Colorado	137
New Jersey	136
Georgia	110
Tennessee	108
North Carolina	104
Utah	102
Wisconsin	100
Alabama	100
Virginia	99
Minnesota	88
Maryland	85
Oregon	78
Kansas	77
Missouri	76
Oklahoma	75
Massachusetts	75
Connecticut	74
Hawaii	67
Nevada	66
Arkansas	63
Iowa	52
Kentucky	51
District Of Columbia	50
Louisiana	46
Nebraska	42
New Mexico	38
West Virginia	34
Mississippi	34
Puerto Rico	33
New Hampshire	32
Idaho	26
South Carolina	24
Wyoming	21
Montana	19
North Dakota	18
Delaware	18
Rhode Island	16
Vermont	13
Maine	13
Alaska	10
South Dakota	6
US Virgin Islands	3

Virginia	5
Indiana	4
Illinois	4
Ohio	3
Maryland	3
Connecticut	3
Wisconsin	2
Pennsylvania	2
Georgia	2
District Of Columbia	2
West Virginia	1
Washington	1
Utah	1
Texas	1
Tennessee	1
Puerto Rico	1
Minnesota	1
Michigan	1
Hawaii	1
Delaware	1
Alabama	1

1240 MHz

State	No. Repeaters
California	204
New York	15
Michigan	14
Florida	14
Texas	13
Washington	12
New Jersey	8
Massachusetts	7
Illinois	7
Ohio	6
Connecticut	6
Virginia	5
Utah	4
Rhode Island	4
Kansas	4
Indiana	4
Wisconsin	3
Tennessee	3
Pennsylvania	3
Oregon	3
Oklahoma	3
North Carolina	3
District Of Columbia	3
Arkansas	3
New Hampshire	2
Nevada	2
Montana	2
Georgia	2
Colorado	2
West Virginia	1
Puerto Rico	1
Missouri	1
Maryland	1
Louisiana	1
Iowa	1
Delaware	1

902 MHz

State	No. Repeaters
California	20
New York	11
Florida	10
Rhode Island	9
New Jersey	8
Massachusetts	6

FREQUENTLY ASKED QUESTIONS FOR DIGITAL AMATEUR RADIO
(Well, it's still mostly just packet...)

This document is for unlimited distribution. Please send corrections and additions to:

digital-faq@wattres.sj.ca.us

Which will expand to a list of people who are familiar with most digital issues. I hope.

The Digital Radio FAQ list will be posted on a monthly basis to rec.radio.amateur.digital.misc, rec.radio.info, rec.answers, and news.answers. The current version of this document is available via anonymous FTP at ftp.cs.buffalo.edu.

Many FAQ's, including this one, are available on the archive site rtfm.mit.edu in the directory pub/usenet/news.answers. The name under which a FAQ is archived appears on the Archive-Name: line at the top of the article. This FAQ is archived as radio/amateur/digital-faq.

There is also a mail server on rtfm.mit.edu, which can be addressed as mail-server@rtfm.mit.edu. For details on how to operate this server, send a message to that address with the word "help" in the BODY of the message.

Table of contents:

1 Basic Packet Radio Information
1.1 What is packet radio?

Packet radio is one method of digital communications via amateur radio. Packet radio takes any digital data stream and sends that via radio to another amateur radio station. Packet radio is so named because it sends the data in small bursts, or packets.

1.2 What is amateur radio?

Amateur Radio (sometimes called Ham Radio) is individuals using specified radio frequencies for personal enjoyment, experimentation, and the continuation of the radio art. Amateur radio operators must be licensed by their government. In the United States, the Federal Communications Commission issues amateur radio licenses. Normally, a test on operating practices, radio theory, and in some cases Morse code proficiency test is administered. Amateur radio is not to be used for commercial purposes. Also, amateur radio operators are restricted from using profanity and using amateur radio for illegal purposes.

For more information on Amateur Radio in general, see the monthly frequently asked questions (FAQ) posting in rec.radio.amateur.misc. A copy of that FAQ is also available for FTP from ftp.cs.buffalo.edu and by mail from rtfm.mit.edu.

1.3 What can I do on packet radio?

Keyboard-to-Keyboard contacts: Like other digital communications modes, packet radio can be used to talk to other amateurs. For those who cannot use HF frequencies, two amateurs can talk to each other from long distances using the packet radio network.

Packet BBS operations: Many cities have one or more packet Bulletin Board System (BBS) available on the local packet network. Amateurs can check into the BBSes and read messages from other packet users on almost any topic. BBSes are networked together over the packet network to allow messages to reach a broader audience than just your local BBS users. Private messages may also be sent to other packet operators, either locally or who use other BBSes. BBSes have the latest ARRL, AMSAT, and propagation bulletins. Many BBSes have a file section containing various text files full of information on amateur radio in general.

DX Packet Cluster: A recent development is use of packet radio for DX spotting. HF operators connect to the local DX Packet Cluster for the latest reports on DX. Often a user will 'spot' some hot DX and distribute the DX report real time.

File Transfer: With special software, amateurs can pass any binary files to other amateurs. Currently, this is done with TCP/IP communications, YAPP, and other specialized protocols.

Satellite Communications: Many of the amateur radio satellites contain microcomputer systems that can provide special information to amateurs. Some satellites contain CCD cameras on board and you can download images of the earth and the stars. Others provide store and forward packet mailboxes to allow rapid message transfers over long distances. Some satellites use AX.25, some use special packet protocols developed for satellite communications. A few transmit AX.25 packets over FM transmitters, but most use SSB transmissions.

1.4 Why packet over other digital modes?

Packet has three great advantages over most of the other digital modes: transparency, error correction, and automatic control.

The operation of a packet station is transparent to the end user; connect to the other station, type in your message, and it is sent automatically. The Terminal Node Controller (TNC) automatically divides the message into packets, keys the transmitter and sends the packets. While receiving packets, the TNC automatically decodes, checks for errors, and displays the received messages. In addition, any packet TNC can be used as a packet relay station, sometimes called a digipeater. This allows for greater range by stringing several packet stations together.

Packet radio provides error free communications because of built in error detection schemes. If a packet is received, it is checked for errors and will be displayed only if it is correct.

With VHF/UHF packet, many countries allow packet operators to operate in automatic control mode. This means that you can leave your packet station on constantly. Other users can connect to you at any time they wish to see if you are home. Some TNC's even have Personal BBSes (sometimes called mailboxes) so other amateurs can leave you messages if you are not at home.

The most important advantage of packet over other modes is the ability for many users to be able to use the same frequency channel simultaneously. No other digital mode yet gives this ability.

1.5 What elements make up a packet station?

TNC (Terminal Node Controller): A TNC contains a modem, a CPU, and the associated circuitry required to convert between RS-232 and the packet radio protocol in use. It assembles a packet from some of the data on the serial line, computing an error check (CRC) for the packet, modulates it into audio frequencies, and puts out appropriate signals to transmit that packet over the connected radio. It also reverses the process, translating the audio that the connected radio receives into a byte stream on the RS-232 port.

Most TNC's currently use 1200 BPS (bits per second) for local VHF and UHF packet, and 300 BPS for longer distance, lower bandwidth HF communication. Higher speeds are available for use in the VHF, UHF, and especially microwave region, but they often require unusual hardware and drivers.

Computer or Terminal: This is the user interface. A computer running a terminal emulator program, a packet-specific program, or just a dumb terminal can be used. For computers, almost any phone modem communications program can be adapted for packet use, but there are also customized packet radio programs available.

A radio: For 1200 BPS UHF/VHF packet, commonly available narrow band FM voice radios are used. For HF packet, 300 BPS data is used over single side band modulation. For high speed packet (anything greater than 1200 BPS), special radios or modified FM radios must be used.

1.6 What do you mean we can all use the same channel?

Packet radio uses a protocol called AX.25. AX.25 specifies channel access (ability to transmit on the channel) to be handled by CSMA (Carrier Sense Multiple Access). If you need to transmit, your TNC monitors the channel to see if someone else is transmitting. If no one else is transmitting, then the TNC keys up the radio, and sends its packet. All the other stations hear the packet and do not transmit until you are done. Unfortunately, 2 stations could accidentally transmit at the same time. This is called a collision. If a collision occurs, neither TNC will receive a reply back from the last packet it sent. Each TNC will wait a random amount of time and then retransmit the packet.

In actuality, a more complex scheme is used to determine when the TNC transmits. See the "AX.25 Protocol Specification" for more information.

1.7 What is AX.25?

AX.25 (Amateur X.25) is the communications protocol used for packet radio. A protocol is a standard for how two computer systems are to communicate with each other, somewhat analogous to using business format when writing a business letter. AX.25 was developed in the 1970's and based of the wired network protocol X.25. Because of the difference in the transport medium (radios vs. wires) and because of different addressing schemes, X.25 was modified to suit amateur radio's needs. AX.25 includes a digipeater field to allow other stations to automatically repeat packets to extend the range of transmitters. One advantage of AX.25 is that every packet sent contains the sender's and recipient's amateur radio callsign, thus providing station identification with every transmission.

1.8 What is RTTY?

RTTY (Radio TeleTYpe) is one of the more common communications protocols used in the HF bands. It is a half-duplex, non error-correcting mode that can be used by any number of stations on a frequency in a round-table fashion. Note that transmit/receive switching is manual, so all parties in a roundtable need to agree on who transmits next. It is primarily used for single keyboard-to-keyboard contacts. RTTY uses Baudot character encoding (also known as ITA2), which is a 5 bit code. Those who can do advanced math will note that 5 bits is only 32 possible combinations, which is not enough for a full alphanumeric set, much less mixed-case alphabetics. Baudot (and RTTY) gets around this limitation by defining two "shifts" which switch between a "letters case," and a "figures case." On older RTTY setups (ones which actually use a teletype, for instance), you have to worry about the letters/figures shift. However, most TNCs and multimode digital controllers now do the "Right Thing(TM)" with respect to those shift characters, sending them as necessary for the data being transmitted. Lower case is not used on Baudot RTTY. ASCII can be used in RTTY as well, but it is very uncommon.

1.9 What is AMTOR?

AMTOR (AMateur Teleprinting Over Radio) is an error-correcting protocol used in the HF bands. It uses the same character set as Baudot (ITA2), but is encoded differently, so that each character has a constant mark to space ratio. This constant ratio is how errors are detected. Errors are corrected via either of two methods: ARQ (Automatic Retransmit reQuest), and FEC (Forward Error Correction).

In ARQ mode, exactly two stations connect to each other. The station with data to transmit (also known as the Information Sending Station, or ISS) transmits 3 characters, and then waits for the other station (called the Information Receiving Station, or IRS) to send back an acknowledgment that those 3 characters were correctly received. This back-and-forth activity makes for the characteristic "chirp-chirp-chirp" of AMTOR ARQ operation. This also means that each transceiver needs to be able to switch from sending to receiving mode fairly quickly. The first time that you operate AMTOR ARQ with a relay-switched rig, you will be convinced that the rig is going to self-destruct.

In FEC mode, one station can communicate with many others at once, since there is no back-and-forth acknowledging of data. FEC gets its error correction from time diversity, which is a fancy way of saying it sends each character twice. Actually, it interleaves the characters, so that the character is not repeated until 4 character times later. An example: The text string to send is "This is FEC." What comes out of the controller looks like "ThisThisis Fis FEC._EC._" The receiving controller looks for which of the characters have a proper mark to space ratio, and prints the one that does. If neither do, it prints an error symbol.

1.10 Definitions: Commonly used terms in Amateur Packet Radio.

44 net—The class A network designator for TCP/IP amateur packet radio. All numerical TCP/IP addresses on packet radio should be in the format 44.xxx.xxx.xxx.

AFSK—Audio Frequency Shift Keying. A method of representing digital information by using different audio frequencies modulated on a carrier.

AMPR—Amateur Packet Radio.

ampr.org—The high level domain recognized on Internet for amateur packet radio TCP/IP.

AMTOR—AMateur Teleprinting Over Radio. This protocol allows error-free point-to-point or multicast, single user per channel communications. Usually used on HF, but not VHF or above.

ARQ—Automatic Resend reQuest. This is the point-to-point error correcting mode for AMTOR. It works by sending bursts of 3 characters, and then the other station sends an OK/NotOK code for those 3 characters.

AX.25—Amateur X.25 protocol. The basis of most packet systems. See section 1.7 for more information.

CRC—Cyclic Redundancy Check. The error detection scheme included in each packet. Verify that the packet was received error free.

CSMA—Carrier Sense Multiple Access. A system allowing many stations to use the same radio frequency simultaneously for packet communications.

digi—Short name for a digipeater.

digipeater—A packet radio station used for repeating packets. See section 3.3.1 for more information.

FCC—Federal Communications Commission. Regulates and issues licenses for amateur radio in the United States.

FEC—Forward Error Correction. This is the multicast method of (almost) error-corrected communications on AMTOR. It works by sending each character twice.

FM—Frequency Modulation. The radio modulation scheme used for VHF and UHF packet communications.

FSK—Frequency Shift Keying. A method of representing digital information by shifting the radio carrier frequency different amounts to represent ones and zeros.

HDLC—(High-Level Data Link Control Procedures) A standard for high level link control. (ISO 3309)

KA9Q NOS—(KA9Q Network Operating System) A TCP/IP program originally developed by Phil Karn, KA9Q. Currently there are many different versions available. See section 3.2 for more information.

KA-Node—A simple networking scheme developed by TNC maker Kantronics. See section 3.3.2 for more info.

KISS—Keep It Simple Stupid. A simple interface developed for communications between TNCs and computers. This allows for most of the packet processing to be handled by the computer. Commonly used with packet TCP/IP software.

LAN—Local Area Network. A packet network developed for communications throughout a city or region. Often, the LAN uses separate frequencies from inter-city packet links.

modem—MODulator/DEModulator. Converts the analog signals into a binary data stream (a series of ones and zeros) for the TNC or a micro-computer. First step in decoding packets. It also converts binary data to analog, which is the last step in encoding packets.

NET/ROM—A scheme for packet radio networking. See section 3.3 for more information.

NODE—A network node. Often a network node running NET/ROM.

PPP—Point to Point Protocol. PPP is another protocol used for moving IP frames over a serial line. It supports host authentication, and non- transparent serial lines. It also has a standard way of negotiating header (and potentially data) compression over the line. See also SLIP.

protocol—A standard used for intercommunication between different computer systems.

RS-232 (RS-323C)—A (more or less) standard for interconnection of serial peripherals to small computer systems. In packet radio, RS-232 is the most common interface between TNC's and the Computer/Terminal.

RTTY—Radio TeleTYpe. This protocol allows point-to-point or multicast, single user per channel communications, without error correction.

SLIP—Serial Line Internet Protocol. A trivial protocol for putting IP frames over a serial line to do (potentially) cheap TCP/IP networking. Approximately the same as KISS, except over wireline networks. See also PPP.

SSB—Single Side Band. The radio modulation scheme used for HF packet and satellite packet communications.

TAPR—Tucson Amateur Packet Radio. Was the first group to create a packet radio TNC using AX.25. Soon a TAPR TNC became cloned by many others. TAPR continues development of packet radio equipment.

TCP/IP—Transmission Control Protocol/Internet Protocol. A set of utility programs used over AX.25. See sections 3.2 for more information.

TNC—Terminal Node Controller. See section 1.5 for more information.

1.12 Do's and Don'ts : Rules and Regulations

NOTE: These regulations apply only to amateurs regulated by the FCC (United States), but often are similar to regulations in other countries.

[Since I have no experience with amateur radio in other countries, I cannot make any comments. Please bring any notable exceptions to my attention. -ed]

Although there are no specific rules that apply to amateur packet radio, the general amateur radio rules force some restrictions on packet usage.

Can I set up a TNC at home and one at work so I can check my Electronic mail via packet?

This cannot be done without special restrictions. Amateur radio rules prohibit any business. Since you could have mail from your boss (or maybe even someone selling you something over Internet), that would constitute business activity and is specifically prohibited.

Profanity can also be a complication. Since you have no control over the language in E-mail, proper filtering is required. Since no filter scheme can catch every offense, it is best to say every message must be hand filtered.

I would like to set up a packet radio gateway between a land line computer network and the packet network. Is this possible?

Yes, and there are several such gateways in use, but they must be managed with caution. Electronic mail may be passed FROM the packet network INTO the land line network without intervention. However, mail passed TO packet radio is considered third party traffic (the sender is not an amateur) and these messages must be hand filtered to ensure that rules of message content are followed.

It's my license if I use packet radio illegally anyway, so what does anyone else care!?

Packet radio is one of the few NETWORKED systems in amateur radio. Many people have helped develop the network and there are many amateurs who own parts of the packet radio network. Sending packet BBS mail, digipeating, and sharing the channel involves the licenses of MANY people. Because of FCC rules stating that anything to come out of a transmitter (either in automatic mode or via your direct control) is the licensee's responsibility, one illegal message sent over the packet radio network could literally jeopardize the licenses of thousands of other amateurs. When in doubt, it is best to check with other amateurs about sending the message before it is sent.

I have some ideas on how to use packet radio in a new way, but I don't know if it is legal. Who could tell me if I can do it legally?

The worst thing you can do is talk to the FCC about such an issue. The FCC rules are written to be general enough to encompass but not restrict new radio activities. In the past, any non-thought-out requests sent to the FCC have meant a reduction of privilege for all amateur radio operators.

The best source for legal assistance is your national amateur radio association. In the United States, that is the American Radio Relay League (ARRL). Another good place for such conversations is over Usenet/packet mailing lists, or the amateur radio BBS network.

2 Computing Network Resources for Amateur Packet Radio.

This section summarizes the resources available on Internet for amateur packet radio operators.

2.1 What Newsgroups/mailing lists are available?

This is a list of all groups that regularly discuss amateur packet radio. For newsgroups, join the group through use of your news reader. For mailing lists, add a '-request' to the end of the list name to request subscriptions. For listserv groups, send mail to 'listserv' at the node which contains the list. The first line of the mail should be 'SUBSCRIBE groupname yourname'. Send the command 'help' for more information.

rec.radio.amateur.packet (Newsgroup): General discussions involving Packet Radio. This group was deleted on 21 September 93!

rec.radio.amateur.digital.misc (Newsgroup): General discussions about all aspects of digital transmissions over Amateur Radio. This group is a replacement for rec.radio.amateur.packet.

rec.radio.amateur.equipment (Newsgroup): May contain discussions about equipment related to digital amateur radio, specifically HF rigs that are good for HF digital modes, and the like.

rec.radio.amateur.homebrew (Newsgroup): Contains discussions on making your own gear, which includes packet, AMTOR, and RTTY equipment.

rec.radio.amateur.misc (Newsgroup): General amateur radio discussion. Usually does not contain any particular information about Digital Amateur Radio.

rec.radio.amateur.policy (Newsgroup): Discussion of regulation policies regarding every aspect of amateur radio. Occasionally deals with policies of packet coordination and legal issues of packet radio.

rec.radio.swap (Newsgroup): General For-Sale for any radio equipment. Occasionally will have packet equipment for sale. Recommended location for any amateur packet radio for-sale items.

info-hams@ucsd.edu (Listserv group): A digest redistribution of the rec.radio.amateur.misc Usenet discussion.

packet-radio@ucsd.edu (Listserv group): General discussions involving packet radio and packet-related issues.

ham-digital@ucsd.edu (Listserv group): A digest redistribution of the rec.radio.amateur.digital.misc Usenet discussion.

ham-policy@ucsd.edu (Listserv group): A digest redistribution of the rec.radio.amateur.policy Usenet discussion

hs-modem@wb3ffv.ampr.org (Mailing list): Discussion of high speed modems and radios available and future plans. Also includes discussion of networking using high speed modems. This list is not very active.

tcp-group@ucsd.edu (Mailing list): Group discussion of technical developments of TCP/IP over packet radio and use of the NOS TCP/IP programs.

gateways@uhm.ampr.org (Mailing list): Discussion

of current gateways and future plans for gateways. May deal with sensitive internetworking issues.

listserv@knuth.mtsu.edu has several interesting mailing lists available:

GRAPES-L: Discussions with GRAPES (Georgia Radio Amateur Packet Enthusiasts Society) on 56kb WAN's and the WA4DSY 56kb RF modem that they distribute.

TENNET-L: Tennessee's efforts at a coordinated high-speed RF packet network

GRACILIS-L: Discussions on Gracilis tcp/ip packet equipment. Includes some of the people from Gracilis.

KA9Q-UNIX: Discussions on porting and using various versions of KA9Q Unix/Xenix NET/NOS under any of a variety of Unix/Xenix variants.

TNV-HAMS: General discussions among email connected amateur radio operators in and surrounding Tennessee

Send a message with a body of "HELP" to get help from the list server. Also, Internet users may now INTERACTIVELY work with the Listserv there by:
telnet knuth.mtsu.edu 372
or
telnet 161.45.1.1 372

For all lists at ucsd.edu, archives may be found via anonymous FTP at ucsd.edu. Some listserv groups also have archives. Send the command 'help' to the group's listserv for more information. Digest mailings for the ucsd.edu discussions are also available. Send mail to listserv@ucsd.edu with the first line being 'longindex' for more information.

Terry Stader (KA8SCP) <tstader@aol.com> maintains a list of Mac packet-related software, and posts it periodically (somewhat less often than monthly) to rec.radio.amateur.digital.misc.

Carl Trommel <carl@codewks.nacjack.gen.nz> posts the weekly news bulletin of the ZL Data Group to rec.radio.amateur.digital.misc. This bulletin covers the current happenings in New Zeland's packet network.

There appears to be a mailing list about NOS on hydra.carleton.ca, but I don't have much more detail. Send a message to nos-bbs-request@hydra.carleton.ca. The list name is nos-bbs@hydra.carelton.ca.

2.2 What anonymous FTP sites and electronic mail servers are available for getting packet radio information and programs?

This is a sampling of FTP sites that carry amateur packet radio related files. Consult the Archie archive server for info on locating particular files. For more information on using Archie, send mail to archie@cs.mcgill.edu with the line 'help'.

ucsd.edu: Primary distribution site of KA9Q's derived TCP/IP packages. Also, general packet radio information in the /hamradio/packet subdirectory. UCSD is also the home of the Amateur Radio "Requests For Comments" directory. If you write something that you believe could be included in this directory (for example, a specification of AX.25, or maybe some other protocol that nobody's thought of before) contact Brian Kantor (brian@ucsd.edu) and let him know.

wuarchive.wustl.edu: Very large collection of amateur radio software. This stuff used to be on wsmr-simtel20.army.mil, but that system (and service) was discontinued as of 1 Oct 93 for budgetary reasons. WUArchive used to "just" mirror Simtel20, and still has all the files that were on Simtel20 when it shut down.

ftp.cs.buffalo.edu: Supplemental archive site for amateur radio information. Contains current copies of all rec.radio.amateur.* FAQ's.

akutaktak.andrew.cmu.edu: SoftKiss for the Mac, in /aw0g. Requires NET/Mac. SoftKiss is an init/cdev/driver that allows a MAC to do packet using a Poor Man's Packet modem.

sumex-aim.stanford.edu: NET/Mac is the port of KA9Q's NET program. It doesn't have nifty features like crolling or saving the windows, but the individual sessions can be recorded into a file.

As for mail servers, there is only one that I know of at the moment:

Ham-Server@GRAFex.Cupertino.CA.US

This mail server, which is run by Steve Harding (KA6ETB), has a wide variety of information on most aspects of amateur radio, not just digital modes. Send a message with the body HELP to get a fairly useful listing. Also note that Steve posts the index from the server approximately weekly to rec.radio.amateur.misc, rec.radio.amateur.digital.misc, and sbay.hams.

2.3 How do I contact the ARRL via electronic mail?

There are various addresses at ARRL HQ, but the most important one is probably info@arrl.org. That is where general information about the ARRL can be acquired. This address reaches a person, so you'll have to specify what/who you're looking for.

2.4 Are there any gateways for mail or news between Internet and Amateur Packet radio?

Internet / Packet Radio BBS Gateways: There are currently two comprehensive gateways between the Internet and the packet radio BBS system. One is run by Jim Durham, W2XO, in Pennsylvania and the other, which allows access to PBBS bulletins, as well as mail, is run by Bob Arasmith, NØARY, in California.

2.4.1 The W2XO Gateway.

To mail from Internet to Packet:
1. Mail to: "bbs@w2xo.pgh.pa.us"
2. Make the first line of the text a Packet BBS "send" command, i.e.: SP TOCALL @ BBSCALL.ROUTING-HINTS < FROMCALL
3. The "subject" line of the Internet mail becomes the "title" line of the Packet BBS mail.

NOTE: Because of FCC regulations, Jim must hand filter each message sent FROM Internet TO the Amateur Packet Radio BBS system. Messages should be of minimal length and appropriate content. Read Section 1.9 (Do's and Don'ts: Rules and Regulations) regarding appropriate usage of packet radio for more information. Always include the routing hints with the BBS callsign.

To mail from Packet to Internet:
1. The amateur radio operator must have his callsign registered in the gateway alias list. If you want to mail

from packet to a specific amateur on Internet, send mail to 'durham@w2xo.pgh.pa.us' (Internet) or 'W2XO @ W2XO.#WPA.PA.USA.NOAM' (Packet BBS mail) with his/her amateur callsign and their Internet address.

2. Once the above is accomplished, packet BBS mail should be sent to 'CALL @ W2XO.#WPA.PA. USA.NOAM'. The mail will automatically be forwarded to the Internet address of the amateur with the 'CALL' callsign.

Jim Durham's Internet address is 'durham @w2xo.pgh.pa.us'.

2.4.2 The NØARY Internet mail <-> full packet BBS gateway.

Bob Arasmith, NØARY, runs a gateway between the Internet mail system and the PBBS system. His gateway allows you to read, post, and respond to other postings on the PBBS system via email. For more information, send mail to gateway_info@arasmith.com.

2.4.3 The N6QMY Internet mail <-> full packet BBS gateway.

Patrick Mulrooney, N6QMY, also runs a gateway between the Internet mail system and the PBBS system. His gateway allows you to read, post, and respond to other postings on the PBBS system via email. For more information, send mail to gateway-info@lbc.com.

Users are required to register with the gateway to allow automatic forwarding of mail from the Internet to Packet. No registration is needed to send mail from Packet to the Internet. To register, send the following information to gateway-request@lbc.com:

CALL:
FIRST NAME:
LAST NAME:
CITY & ST:
COUNTRY
ZIP:
HOME BBS:

2.4.4 LAN Gateways (Packet wormholes via Internet).

Currently a group of amateurs are experimenting with connecting packet LANs together via Internet IP inside IP Encapsulation. Some of the gateways only accept TCP/IP packets, others AX.25 packets. These gateways uses the Internet as a transport medium, thus it is impossible to access the packet radio network from Internet. For more information, join the Gateways mailing list by sending mail to "gateways-request@uhm.ampr.org".

2.5 How do I contact TAPR?

The only route I currently know of is via US Mail and telephone.

US Mail:
TAPR (Tuscon Amateur Packet Radio)
8987-309 E. Tanque Verde Rd. #337
Tuscon, AZ 85749-9399

Telephone:
Voice: +1 817 383 0000
Fax: +1 817 566 2544

3 Networking and special packet protocols.

This is a sample of some of the more popular networking schemes available today. By far, there are more customized networking schemes used than listed. Consult your local packet network guru for specific network information.

3.1 Are there any other protocols in use other than AX.25?

AX.25 is considered the defacto standard protocol for amateur radio use and is even recognized by many countries as a legal operation mode. However, there are other standards. TCP/IP is used in some areas for amateur radio. Also, some networking protocols use other packet formats than AX.25.

Often, special packet radio protocols are encapsulated within AX.25 packet frames. This is done to insure compliance with regulations requiring packet radio transmissions to be in the form of AX.25. However, details of AX.25 encapsulation rules vary from country to country.

3.2 What is TCP/IP?

TCP/IP stands for Transmission Control Protocol/Internet Protocol. This is commonly used over the Internet wired computer network. The TCP/IP suite contains different transmission facilities such as FTP (File Transfer Protocol), SMTP (Simple Mail Transport Protocol), Telnet (Remote terminal protocol), and NNTP (Net News Transfer Protocol)

TCP/IP doesn't use all of the AX.25 protocol. Instead, it uses special AX.25 packets called Unnumbered Information (UI) packets and then puts its own special protocol (called IP) on top of AX.25. This is used to increase efficiency of its transmissions, since IP does not require packets to be "reliable", that is to say, guaranteed delivered error-free. TCP handles the retransmission of lost and garbled packets in its own way, at a higher level. Therefore the extra information in an AX.25 "VC" (virtual circuit) frame is not useful, and thus consuming needed bandwidth.

The KA9Q NOS program (also called NET) is the most commonly used version of TCP/IP in packet radio. NOS originally was written for the PC compatible. However, NOS has been ported to many different computers such as the Amiga, Macintosh, Unix, and others. Smaller computers like the Commodore 64 and the Timex-Sinclair do not currently have version of NOS available.

For more general information about IP (not necessarily over packet radio), try the newsgroup comp. protocols.tcp-ip, and any of the plethora of books on the subject. I have found the various books by Douglas Comer (the "Internetworking with TCP/IP series) to be excellent.

3.3 How do I get an IP address?

Brian Kantor (brian@ucsd.edu) is the IP address coordinator for the AMPR.ORG domain, also known as the "44 net." Brian (sensibly enough) delegates coordination for each state to sub-coordinators. This list(*) of coordinators is available via FTP from ftp.cs.buffalo.edu:/ pub/ham-radio/ampr_coordinators.

(*) The coordinator list doesn't have e-mail addresses. If you're willing to do the legwork required to gather them, please contact Brian Kantor.

3.4 Networking Schemes.

What are some of those other networking schemes?

During the early days of amateur packet radio, it became apparent that a packet network was needed. To this end, the following packet network schemes where created.

Digipeaters: The first networking scheme with packet radio was Digipeaters. Digipeaters would simply look at a packet, and if its call was in the digipeater field, would resend the packet. Digipeaters allow the extension of range of a transmitter by retransmitting any packets addressed to the digipeater.

This scheme worked well with only a few people on the radio channel. However, as packet became more popular, digipeaters soon were clogging up the airwaves with traffic being repeated over long distances. Also, if a packet got lost by one of the digipeaters, the originator station would have to retransmit the packet again, forcing every digipeater to transmit again and causing more congestion.

KA-Nodes: Kantronics improved on the digipeater slightly and created KA-Nodes. As with digipeaters, KA-Nodes simply repeat AX.25 frames. However, a KA-Node acknowledges every transmission each link instead of over the entire route. Therefore, instead of an end-to-end acknowledgment, KA-Nodes allow for more reliable connections with fewer timeouts, because acknowledgments are only carried on one link. KA-Nodes therefore are more reliable than digipeaters, but are not a true network. It is similar like having to wire your own telephone network to make a phone call.

NET/ROM: NET/ROM was one of the first networking schemes to try to address the problems with digipeaters. A user connects to a NET/ROM station as if connecting to any other packet station. From there, he can issue commands to instruct the station to connect to another user locally or connect to another NET/ROM station. This connect, then connect again, means that to a user's TNC, you are connected to a local station only and its transmissions do not have to be digipeated over the entire network and risk loosing packets. This local connection proved to be more reliable.

NET/ROM is a commercial firmware (software put on a chip) program that is used as a replacement ROM in TAPR type TNCs. Other programs are available to emulate NET/ROM. Among them are TheNet, G8BPQ node switch, MSYS, and some versions of NET.

NET/ROM nodes, at regular intervals, transmit to other nodes their current list of known nodes. This is good because as new nodes come on-line, they are automatically integrated in the network. However, if band conditions such as ducting occur, ordinarily unreachable nodes can be entered into node lists. This causes the NET/ROM routing software to choose routes to distant nodes that are impossible. This problem requires users to develop a route to a distant node manually defining each hop instead of using the automatic routing feature.

ROSE: ROSE is another networking protocol derived from X.25. Each ROSE node has a static list of the nodes it can reach. For a user to use a ROSE switch, he issues a connect with the destination station and in the digipeater field places the call of the local rose switch and the distant rose switch the destination station can hear. Other then that, the network is completely transparent to the user.

ROSE's use of static routing tables ensures that ROSE nodes don't attempt to route packets through links that aren't reliably reachable, as NET/ROM nodes often do. However, ROSE suffers from the inability to automatically update its routing tables as new nodes come online. The operators must manually update the routing tables, which is why ROSE networks require more maintenance.

3.5 BBS message transfer

Many of the BBS programs used in packet radio allow for mail and bulletins to be transferred over the packet radio network. The BBSes use a special forwarding protocol developed originally by Hank Oredsen, WØRLI.

Besides full service BBSes, many TNC makers have developed Personal BBS software to allow full service BBSes to forward mail directly to the amateur's TNC. This allows operators to receive packet mail at night and avoid tying up the network during busy hours.

(Courtest of Steven Watt, WD6GGD.)

CW OPERATING

The first radio operators, amateur and professional, used telegraphy for message transmission. Proficiency in the "International Morse Code" has long been a requirement for amateur licensing. CW has fallen somewhat in popularity and was determined to be an impediment to the growth of amateur radio, and several countries (including the United States) have developed classes of license that do not require Morse code proficiency.

Organizations Devoted to CW Operation

Several amateur radio organizations are dedicated to CW operation. The best-known is the prestigious "First-class CW Operators Club," more commonly known as the FOC. Membership in this organization is by invitation only, with a formal process of nomination and sponsorship. Selection is done on the basis of CW proficiency, activity, and operating skill. Maximum membership at any one time is limited, and there have been approximately 1,600 members in the club's history. FOC members can usually be found operating on or near the unofficial club frequencies, 25 kHz above the bottom of each CW band.

A somewhat less formal organization is called the "Chicken Fat Operators," or CFO. It is "dedicated to competent Morse code operations in the Amateur Service." Founded in 1979 by W9WBL and W9TO (designer of the "TO" keyer manufactured by Hallicrafters for many years), this group can be found between 7030 and 7035 most evenings. The informal membership qualification is comfortable copy by ear at 45 WPM. No formal submission or proof is required. There have been approximately 950 members of the CFO club in approximately 30 countries, and a newsletter called "Key Clucks" is published. Members occasionally end their QSOs with a distinctive signature which sounds like a chicken's cluck. WB9LTN is currently in charge of the CFO's affairs.

Perhaps the most elite club of CW operators was the "Five Stars" club, founded by Bill Eitel, W6UF, in the 1970s. This group was as informal as the CFO group, but required CW proficiency at 80 words per minute! Paper-copy was not required to qualify for membership—merely the ability to copy by ear and correctly answer questions sent by a carefully-calibrated keyboard. Eitel stopped issuing membership certificates after qualifying 10 members, not for lack of applicants, but from lack of time. The club has not been formally active since W6UF's death in 1989, but some of the members remain active, still casually conversing on the bands at 70-80 WPM. One member still active on the bands is Florence, W7QYA, who began her high-speed CW career as an operator for the Federal Aviation Administration, on radiotelegraph circuits in Alaska.

Q SIGNALS

Given below are a number of Q signals whose meanings most often need to be expressed with brevity and clarity in amateur work. (Q abbreviations take the form of questions only when each is sent followed by a question mark.)

QRA	What is the name of your station? The name of my station is ____.
QRB	How far approximately are you from my station? The approximate distance between our stations is____nautical miles (or kilometers).
QRD	Where are you bound and where are you coming from? I am bound for ____ from ____.
QRG	Will you tell me my exact frequency (or that of ____)? Your exact frequency (or that of ____) is ____ kHz.
QRH	Does my frequency vary? Your frequency varies.
QRI	How is the tone of my transmission? The tone of your transmission is ____ (1. Good; 2. Variable; 3. Bad).
QRJ	Are you receiving me badly? I cannot receive you. Your signals are too weak.
QRK	What is the intelligibility of my signals (or those of ____)? The intelligibility of your signals (or those of ____) is ____ (1. Bad; 2. Poor; 3. Fair; 4. Good; 5. Excellent).
QRL	Are you busy? I am busy (or I am busy with ____). Please do not interfere.
QRM	Is my transmission being interfered with? Your transmission is being interfered with ____ (1. Nil; 2. Slightly; 3. Moderately; 4. Severely; 5. Extremely.)
QRN	Are you troubled by static? I am trouble by static ____ (1–5 as under QRM).
QRO	Shall I increase power? Increase power.
QRP	Shall I decrease power? Decrease power.
QRQ	Shall I send faster? Send faster (____ WPM).
QRR	Are you ready for automatic operation? I am ready for automatic operation. Send at ____ words per minute.
QRS	Shall I send more slowly? Send more slowly (____ WPM).
QRT	Shall I stop sending? Stop sending.
QRU	Have you anything for me? I have nothing for you.
QRV	Are you ready? I am ready.
QRW	Shall I inform ____ that you are calling on ____ kHz? Please inform ____ that I am calling on ____ kHz.
QRX	When will you call me again? I will call you again at ____ hours (on ____ kHz).
QRY	What is my turn? Your turn is numbered ____.
QRZ	Who is calling me? You are being called by ____ (on ____ kHz).
QSA	What is the strength of my signals (or those of ____)? The strength of your signals (or those of ____) is ____ (1. Scarcely perceptible; 2. Weak; 3. Fairly good; 4. Good; 5. Very good).
QSB	Are my signals fading? Your signals are fading.
QSD	Is my keying defective? Your keying is defective.
QSG	Shall I send ____ messages at a time? Send ____ messages at a time.
QSJ	What is the charge to be collected per word to ____ including your internal telegraph charge? The charge to be collected per word to ____ including my internal telegraph charge is ____ francs (or other unit of currency).
QSK	Can you hear me between your signals and if so can I break in on your transmission? I can hear you between my signals; break in on my transmission.
QSL	Can you acknowledge receipt? I am acknowledging receipt.
QSM	Shall I repeat the last message which I sent you, or some previous message? Repeat the last message which you sent me (or message[s] number[s] ____).
QSN	Did you hear me (or ____) on ____ kHz? I did hear you (or ____) on ____ kHz.
QSO	Can you communicate with ____ direct or by relay? I can communicate with ____ direct (or by relay through ____).
QSP	Will you relay to ____? I will relay to ____.
QSQ	Have you a doctor on board (or is ____ [name of person] on board)? I have a doctor on board (or ____ [name of person] is on board).
QSU	Shall I send or reply on this frequency (or on ____ kHz)? Send or reply on this frequency (or____kHz).

QSV	Shall I send a series of Vs on this frequency (or ____ kHz/MHz)? Send a series of Vs on this frequency (or ____ kHz/MHz).	QUA	Have you news of ____ (call sign)? I have news of ____ (call sign).
QSW	Will you send on this frequency (or on ____ kHz)? I am going to send on this frequency (or on ____ kHz).	QUB	Can you give me, in the following order, information concerning: visibility, height of clouds, direction and velocity of ground wind at ____ (place of observation)? Here is the information you requested ____.
QSX	Will you listen to ____ on ____ kHz? I am listening to ____ on ____ kHz.	QUC	What is the number (or other indication) of the last message you received from me (or from ____ call sign)? The number (or other indication) of the last message I received from you (or from ____ call sign) is ____.
QSY	Shall I change to transmission on another frequency? Change to transmission on another frequency (or on ____ kHz).		
QSZ	Shall I send each word or group more than once? Send each word or group twice (or ____ times).		
QTA	Shall I cancel message number ____? Cancel message number ____.	QUD	Have you received the urgency signal sent by ____ (call sign of mobile station)? I have received the urgency signal sent by ____ (call sign of mobile station) at ____ hours.
QTB	Do you agree with my counting of words? I do not agree with your counting of words. I will repeat the first letter or digit of each word or group.	QUF	Have you received the distress signal sent by ____ (call sign of mobile station)? I have received the distress signal sent by ____ (call sign of mobile station) at ____ hours.
QTC	How many messages have you to send? I have ____ messages for you (or for ____).		
QTE	What is my true bearing from you? Your true bearing from me is ____ degrees (at ____ hours). OR What is my true bearing from ____ (call sign)? Your true bearing from ____ (call sign) was ____ degrees (at ____ hours). OR What is the true bearing of ____ (call sign) from ____ (call sign)? The true bearing of ____ (call sign) from ____ (call sign) was ____ degrees (at ____ hours).	QUG	Will you be forced to land? I am forced to land immediately OR I shall be forced to land at ____ (position or place).
		QUH	Will you give me the present barometric pressure at sea level? The present barometric pressure at sea level is ____ (units).
QTG	Will you send two dashes of 10 seconds each followed by your call sign (repeated ____ times) (on ____ kHz/MHz)? I am going to send two dashes of 10 seconds each followed by my call sign (repeated ____ times) (on ____ kHz/MHz). OR Will you request ____ (call sign) to send two dashes of 10 seconds each followed by his call sign (repeated ____ times) (on ____ kHz/MHz)? I have requested ____ (call sign) to send two dashes of 10 seconds each followed by his call sign (repeated ____ times) (on ____ kHz/MHz).	**Unofficial**	
		QLF	Are you sending CW with your left foot? Try sending CW with your left foot now.
		QST	General call preceding a message addressed to all amateurs and ARRL members. This is in effect "CQ ARRL."
		QTHR	Is your address listed in the RSGB Callbook? My address is listed in the RSGB Callbook. (British)
QTH	What is your location? My location is ____.		
QTI	What is your true track? My true track is ____ degrees.		

ARRL QN Signals for C.W. NET Use

These "Special QN Signals for Net Use" originated late in 1940, used by the Michigan QMN Net, and were first known to Headquarters through the then head traffic honcho W8FX. Ev Battey, W1UE, then ARRL assistant communications manager, thought enough of them to print them in QST and later to make them standard for ARRL nets, with a few modifications.

QTJ	What is your speed? My speed is ____ knots (or kilometers per hour).	QNA	Answer in prearranged order.
QTL	What is your true heading? My true heading is ____ degrees.	QNB	Relay between [____] and [____].
		QNC	All net stations copy.
QTN	At what time did you depart from ____ (place)? I departed from ____ (place) at ____ hours.	QND	Net is directed.
		QNE	Entire net stand by.
QTO	Have you left dock (or port)? I have left dock (or port). OR Are you airborne? I am airborne.	QNF	Net is free.
		QNG	Take over as net control.
QTP	Are you going to enter dock (or port)? I am going to enter dock (or port). OR Are you going to land? I am going to land.	QNH	Your Net frequency is high.
		QNI	Net stations report in.
		QNJ	Can you copy me?
QTQ	Can you communicate with my station by means of the International Code of Signals? I am going to communicate with your station by means of the International Code of Signals.	QNK	Send messages for [____] to [____].
		QNL	Your Net frequency is low.
		QNM	You are QRMing the net.
QTR	What is the correct time? The time is ____.	QNN	Net control station is [____].
QTS	Will you send your call sign for ____ minutes now (or at ____ hours) on ____ kHz/MHz so that your frequency may be measured? I will send my call sign for ____ minutes now (or at ____ hours) on ____ kHz/MHz so that my frequency may be measured.	QNO	Station is leaving the net.
		QNP	Unable to copy you.
		QNQ	Move frequency to [____] and wait for [____] to finish handling traffic. Then send him traffic for [____].
		QNR	Answer [____] and receive traffic.
QTU	What are the hours during which your station is open? My station is open from ____ to ____ hours.	QNS	Following stations are in the net. Request list of stations on the net.
QTV	Shall I stand guard for you on the frequency of ____ kHz/MHz (from ____ hours to ____ hours)? Stand guard for me on the frequency of ____ kHz/MHz (from ____ hours to ____ hours).	QNT	I request permission to leave the net.
		QNU	The net has traffic for you.
		QNV	Establish contact with [____] on this frequency. If successful, move to [____] and send him traffic for [____].
QTX	Will you keep your station open for further communication with me until further notice (or until ____ hours)? I will keep my station open for further communication with you until further notice (or until ____ hours).	QNW	How do I route messages for [____]?
		QNX	You are excused from the net.
		QNY	Shift to another frequency with [____].
		QNZ	Zero beat your frequency with mine.

Morse Code Character Set[1]

A	didah	▪▬	I	didit	▪▪
B	dahdididit	▬▪▪▪	J	didahdahdah	▪▬▬▬
C	dahdidahdit	▬▪▬▪	K	dahdidah	▬▪▬
D	dahdidit	▬▪▪	L	didahdidit	▪▬▪▪
E	dit	▪	M	dahdah	▬▬
F	dididahdit	▪▪▬▪	N	dahdit	▬▪
G	dahdahdit	▬▬▪	O	dahdahdah	▬▬▬
H	didididit	▪▪▪▪	P	didahdahdit	▪▬▬▪
			Q	dahdahdidah	▬▬▪▬
			R	didahdit	▪▬▪

S	dididit	▪▪▪	8	dahdahdahdidit	▬▬▬▪▪
T	dah	▬	9	dahdahdahdahdit	▬▬▬▬▪
U	dididah	▪▪▬	0	dahdahdahdahdah	▬▬▬▬▬
V	didididah	▪▪▪▬			
W	didahdah	▪▬▬			
X	dahdididah	▬▪▪▬			
Y	dahdidahdah	▬▪▬▬			
Z	dahdahdidit	▬▬▪▪			

1	didahdahdahdah	▪▬▬▬▬	4	didididah	▪▪▪▪▬
2	dididahdahdah	▪▪▬▬▬	5	didididit	▪▪▪▪▪
3	didididahdah	▪▪▪▬▬	6	dahdidididit	▬▪▪▪▪
			7	dahdahdididit	▬▬▪▪▪

Period [.]:	didahdidahdidah	▪▬▪▬▪▬
Comma [,]:	dahdahdididahdah	▬▬▪▪▬▬
Question mark or request for repetition [?]:	dididahdahdidit	▪▪▬▬▪▪
Error:	didididididididit	▪▪▪▪▪▪▪▪
Hyphen or dash [−]:	dahdididididah	▬▪▪▪▪▬
Double dash [=]:	dahdidididah	▬▪▪▪▬
Colon [:]:	dahdahdahdididit	▬▬▬▪▪▪
Semicolon [;]:	dahdidahdidahdit	▬▪▬▪▬▪
Left parenthesis [(]:	dahdidahdahdit	▬▪▬▬▪
Right parenthesis [)]:	dahdidahdahdidah	▬▪▬▬▪▬

Fraction bar [/]:	dahdidididahdit	▬▪▪▬▪
Quotation marks ["]:	didahdidididahdit	▪▬▪▪▬▪
Dollar sign [$]:	didididahdididah	▪▪▪▬▪▪▬
Apostrophe [']:	didahdahdahdahdit	▪▬▬▬▬▪
Paragraph [¶]:	didahdidahdidit	▪▬▪▬▪▪
Underline [_]:	didididahdahdidah	▪▪▬▬▪▬
Wait:	didahdidit	▪▬▪▪▪
End of message or cross [+]:	didahdidahdit	▪▬▪▬▪
Invitation to transmit [K]:	dahdidah	▬▪▬
End of work:	dididahdidah	▪▪▪▬▪▬
Understood:	dididahdidit	▪▪▬▪

Prosigns (right column overbar pairs):
DN, AF, SX, WG, AL, IQ, KA, AS, AR, K, SK, SN

Notes

1. Not all Morse characters shown are used in FCC code tests. License applicants are responsible for knowing, and may be tested on, the 26 letters, the numerals 0 to 9, the period, the comma, the question mark, AR, SK, BT and fraction bar (DN).

2. The following letters are used in certain European languages which use the Latin alphabet:

Ä, Á, Å, Â	didahdidah	▪▬▪▬
Ĉ, Ç, C̆	dahdidahdidit	▬▪▬▪▪
È, É, Ê, Ę	dididahdit	▪▪▬▪
Ĕ	dahdididah	▬▪▪▬

Ö, Ó, Ó	dahdahdah	▬▬▬
Ñ	dahdahdidahdah	▬▬▪▬▬
Ü, Ŭ	dididahdah	▪▪▬▬
Ż	dahdahdididah	▬▬▪▪▬
Ź	dahdidahdidah	▬▪▬▪▬
CH, Ŝ	dahdahdahdah	▬▬▬▬

3. Special Esperanto characters:

Ĉ	dahdidahdidah	▬▪▬▪▬
Ĝ	dididahdidit	▪▪▬▪
Ĥ	dahdahdahdahdit	▬▬▬▬▪
Ĵ	didahdahdahdit	▪▬▬▬▪
Ŝ	didahdah	▪▬▬

4. Signals used in other radio services:

Interrogatory	didahdidahdah	▪▬▪▬▬	INT
Emergency silence	didididahdahdah	▪▪▪▬▬	HM
Executive follows	dididahdidah	▪▪▬▪▬	IX
Break-in signal	dahdahdahdahdah	▬▬▬▬▬	TTTTT
Emergency signal	didididahdahdahdididit	▪▪▪▬▬▬▪▪▪	SOS
Relay of distress	dahdididahdididahdididit	▬▪▪▬▪▪▬▪▪	DDD

Morse Codes for Other Languages

Code	Japanese	Thai	Korean	Arabic	Hebrew	Russian	Greek
·	he	sara-a	a		vav	Е,Э	E epsilon
−	mu	tor-tow	ŏ	ta	tav	Т	T tau
··	nigori	sara-e	ya	ya	yod	И	I iota
·−	i	sara-r	o	alif	aleph	А	A alpha
−·	ta	nor-nu	yo	noon	nun	Н	N nu
−−	yo	mor-ma	m	meem	mem	М	M mu
···	ra	sor-sue	yŏ	seen	shin	С	Σ sigma
··−	na	mai-ek	yu	ta	tet	У	ΟΥ omicron ypsilon
·−·	ya	roe-rue	p(b)	ra	reish	Р	P rho
·−−	ho	vor-van	-ng	waw	tzadi	В	Ω omega
−··	wa	door-dek	s	dal	dalet	Д	Δ delta
−·−	ri	kor-kwai	p'	kaf	chaf	К	K kappa
−−·	re	go-kai	u	ghain	gimmel	Г	Γ gamma
−−−	nu	sara-o	r-(l)	kha	heh	О	O omicron
····	ku	hor-heep	n	ha	chet	Х	H eta
···−	tsi	mai-tho		dad		Ж	HΥ eta ypsilon
··−·	no	for-fun	k(g)	fa	feh	Ф	Φ phi
··−−	ka	sara-eu	ch(j)			Ю	AΥ alpha ypsilon
·−··	ro	law-ling	h	lam	lamed	Л	Λ lambda
·−·−	tu	sara-air	t(d)	ain		Я	AI alpha iota
·−−·	wo	por-pla	k'	jeem	peh	П	Π pi
·−−−	ha	yor-ying	ch'	ba	ayen	Й	ΥI ypsilon iota
−···	ma	bor-baimai	e	sad	bet	Б	B beta
−··−		chor-chang	t'	tha		ь,ъ mute	Ξ xi
−·−·	ni	kho-khai		za	samech	Ц	Θ theta
−·−−	ke	yor-yak				Ы	Υ ypsilon
−−··	hu	zor-zo		dhal	zain	З	Z zeta

ne
so
ko
to
mi
han-nigori
o
n
(w)i
te
(w)e
hyphen
se
me
mo
yu
ki
sa
ru
e
hi
si
a
su

pho-phueng
sara-u
choe-ching
sara-ie
sara-au
sara-aue
sara-ar
sara-i
mai-jatawa
row-rue
por-pan
mai-han agas
maitho han agas
or-ang
jor-jan
tor-tahan
tor-tung

ngor-ngoo

hor-nok hook

ae

qaf
zay
sheen
he

kof

SHCH
CH
SH

ET epsilon
psi
khi
ypsilon

lam·alif

THE FASTEST CW OPERATOR IN HISTORY

McElroy, Theodore R. (1901–1963)

Ted McElroy is well-known among hams as a maker of telegraph keys in the 1930s and as the "World's Champion Radio Telegrapher" who set the official record of 75.2 words per minute in 1939 at a hamfest in Asheville, NC. But behind those bare facts was one of the most colorful telegraphers the world has known.

"Teddy" McElroy, as he was called by his friends, was born in Somerville, MA, on September 15, 1901. He attended the local public schools and, by age 14, was working part-time as a messenger for the Western Union. While in the seventh grade, Teddy took a typing course, at which he excelled. After attending Somerville High School for a few months, he left at age 15 to become a full-time telegraph operator.

McElroy worked at various Western Union offices around Boston and southern New Hampshire, and quickly became one of the company's best telegraphers. After World War I, he took a job with RCA at its transatlantic station on Cape Cod. Because it was a radio station, he had to learn the Continental (International Morse) code. He did so easily. This experience would have far-reaching consequences for McElroy.

RCA moved its operation to New York. Unhappy in that city, McElroy became involved in union organizing on behalf of telegraphers. He soon found himself on the street, and returned to Boston and the Western Union.

In the spring of 1922, impressed with his speed and knowing of his experience with Continental code, McElroy's boss entered him in a code receiving competition at the Boston Radio Exposition. There, in his first official contest, McElroy beat the competition and set a new world's record of 51.5 words per minute. The same year, at contests in New York and Chicago, he successfully defended his title and increased the world's record to 56.5 words per minute.

Ted married Margaret Frances Coleman of South Boston in 1925. Their only child, John Charles, was born the following year. Now a husband and father, Ted left the field of telegraphy and took jobs as a sales representative for various companies. He had not heard the code for several years when the next important code contest was announced.

The World-Wide ARRL Convention was held during the Chicago World's Fair in August, 1933. The Western Union sponsored McElroy as their entry against the best telegraphers in the world. Even without recent professional experience, McElroy put up a good fight. But it wasn't quite good enough. Both Joseph W. Chaplin, representing Press Wireless, and McElroy achieved a speed of 54.1 words per minute—but Chaplin made only five errors, while McElroy made eight. Ted's 1922 record meant he was still the world's fastest telegrapher, but he was no longer its champion.

In 1934, "T.R. McElroy" started manufacturing telegraph keys at his home in Boston. Over the next seven years, he would come up with more than a dozen different variations or models of his semi-automatic "Mac-Key" and straight keys. Perhaps his two most famous and collectible keys are the Super Stream-speed bug, with its all-chrome teardrop shaped base, and the chrome-plated teardrop "Streamkey" straight key.

In 1935, Ted attended a meeting of the Cape Cod Radio Club to demonstrate his code receiving abilities. The awed members gave him a certificate attesting to his amazing run of 77 words per minute. Showing that this was no fluke, McElroy participated in the contest held six months later in Brockton, MA. This time, Joe Chaplin was the loser and McElroy set a new record of 69 words per minute.

The next official contest was held in Asheville, NC. It was there that McElroy again beat all challengers to his title with a new record of 75.2 words per minute. It was the last official competition ever held; McElroy's record still stands today. (The world record for sending with a straight key is 35 wpm, set on November 9, 1942 by Harry A. Turner, W9YZE, of the U.S. Army Signal Corps at Camp Crowder, MO.)

McElroy was licensed as W1JYN in 1936, but it is not known if he was active on the amateur bands. During WWII, McElroy manufactured code training equipment for the military. As an promotional device, he designed his famous "Chart of Codes and Signals," which he introduced at the 1943 annual meeting of the Veteran Wireless Operators of America. Thousands of the colorful charts were given to anyone who asked. Few of those charts survive today.

McElroy made millions of dollars from his military contracts during the war and, after the war, lost it as quickly. He sold his company in 1955 and became a sales representative for an electronics company in New York. Ted McElroy died at his home in Littleton, MA, on November 12, 1963.

(Contributed by Tom French, W1IMQ. Tom is the author of McElroy: World's Champion Radio Telegrapher, *an illustrated biography available for $19.95 plus $2.00 s/h from Artifax Books, Box 88, Maynard MA 01754.)*

CW ABBREVIATIONS

AA	All After
AB	All before
ABT	About
ADR	Address
AF	Africa
AGN	Again
AM	Amplitude Modulation
AMP	Unit of Current, Power Amplifier
ANI	Any
ANT	Antenna
AS	Asia
AU	Aurora
AWDH	Auf Wiederhoren (See you later) GERMAN
AWS	Auf Weidersehen (Good-bye) GERMAN
BCI	Broadcast Interference
BCL	Broadcast Listener
BCNU	Be seeing you
BCP	Beaucoup (Much) FRENCH
BD	Bad
BEV	Beverage Receive Antenna
BJR	Bonjour (Hello, Good Morning) FRENCH
BK	Break, Back
BLV	Believe
BN	All between, Been
BSR	Bonsoir (Good evening) FRENCH
BTR	Better
BUG	Semi-Automatic key
BURO	QSL Bureau
B4	Before
C	Yes (from Spanish "si"), Centigrade, Celsius
CFM, CFMD	Confirm, Confirmed

CK	Check	NIL	No, Nothing, Not in Log
CL	Closing my station	NR	Near, Number
CLD, CLG	Called, calling	NW	Now, Northwest
CNT	Cannot	OB	Old Boy
CONDX	Condition	OC	Old Chap
CONGRATS	Congratulations	OM	Old Man
CQ	General Call	OP, OPR, OPS	Operator(s)
CQDO	General Call for Japanese Morse contact	OT	Old timer
CUAGN	See you again	PA	Power/Final Amplifier
CUD	Could	PAC	Pacific
CUL	See you later	PHONE	Telephone, Voice transmission
CUM	Come	PKT	Packet
CW	Continuous Wave	PSE	Please
DE	From	PWR	Power
DF	Direction Finding	R	Received
DLD, DLVD	Delivered	RCD	Received
DR	Dear	RCVR, RX	Receiver
DS	Danke schon (Thanks) GERMAN	RFI	Radio Frequency Interference
DSW	Do svidaniya (Good-bye) RUSSIAN	RIG	Radio, Station equipment
DUPE	QSO B4, Duplicate	RPT	Repeat
DX	Long distance	RPRT	Report, Repeater
ENUF	Enough	RS	Radiosputnik
ES	And (originally American Morse "&")	RST	Readability, Strength, Tone
EU	Europe	RTTY	RadioTeletype
EX	Excellent, Formerly	SA	Say again, South America
FB	Fine Business	SAE	Self-Addressed-Envelope
FER	For	SAT	Saturday
FONE	Phone-Voice transmission	SASE	Self-Addressed-Stamped-Envelope
FM	Frequency Modulation	SSB	Single Side Band, Voice
FOT	Frequency of optimum traffic	SED	Said
FREQ	Frequency	SHACK	Radio equipment location
FT	Feet	SID	Sudden Ionospheric Disturbance
GA	Go ahead-or-Good afternoon	SIG, SIGS	Signal(s)
GB	Good-bye	SKED	Schedule
GD	Good day	SN	Soon, Serial Number
GE	Good evening	SO	Single-Operator
GG	Going	SP	Short path
GL	Good Luck	SRI	Sorry
GLD	Glad	STN	Station
GM	Good morning	SUM	Some
GN	Good night	SUN	Sunday
GND,GRND	Ground	SVP	S'il vous plait (Please) FRENCH
GP	Ground plane vertical antenna	SWL	Short Wave Listener
GUD	Good	TCVR, TRX	Transceiver
HEJ	Hey (Cheerio) SWEDISH, FINNISH	TEMP	Temperature
HI	C.W. Laughter	TEST	Contest
HLV	Hasta La vista (See you later) SPANISH	TFC	Traffic
HNY	Happy New Year	TKS, TNX	Thanks
HPE	Hope	TMW, TMRW	Tomorrow
HR	Here, Hear	TU	Thank you
HRD	Heard	TVI	Television Interference
HV	Have	TX	Transmitter
HVY	Heavy	TXT	Text
HW	How	U	You
INFO	Information	UR	You are, Your
KW	Kilowatt	USB	Upper side band
LID	Poor operator	UTC	Coordinated Universal Time (G.M.T.)
LIL	Little	VERT	Vertical Antenna
LOC	Location	VFO	Variable Frequency Oscillator
LP	Long path	VG	Very Good
LSB	Lower side band	VIA	QSL to (i.e. QSL VIA)
LSN	Listen	VY	Very
LTR	Letter	WID	With, Wide
LUF	Least usable frequency	WKD	Worked
M, MTRS	Meters	WKG	Working
MA, MILS	Milliamperes	WL	Will
MCI	Merci (Thank you) FRENCH	WPM	Words per minute
MIN	Minute	WUD	Would
MM	Multi-Operator Multi-Transmitter	WX	Weather
MNI	Many	XMAS	Christmas
MOD	Modulation	XMTR	Transmitter
MS	Meteor Scatter, Multi-operator Single-transmitter	XTAL	Crystal
		XYL	Wife
MSG	Message	YDAY	Yesterday
MUF	Maximum usable frequency	YL	Young Lady
N	No	55	Best Success—EASTERN EUROPE
NA	North America	73	Best Wishes
NCS	Net Control Station	88	Love and Kisses

W1AW schedule

Pacific	Mtn	Cent	East	Sun	Mon	Tue	Wed	Thu	Fri	Sat
6 am	7 am	8 am	9 am			Fast Code	Slow Code	Fast Code	Slow Code	
7 am	8 am	9 am	10 am	Code Bulletin						
8 am	9 am	10 am	11 am	Teleprinter Bulletin						
9 am	10 am	11 am	noon							
10 am	11 am	noon	1 pm	**Visiting Operator Time**						
11 am	noon	1 pm	2 pm							
noon	1 pm	2 pm	3 pm							
1 pm	2 pm	3 pm	4 pm	Slow Code	Fast Code	Slow Code	Fast Code	Slow Code	Fast Code	Slow Code
2 pm	3 pm	4 pm	5 pm	Code Bulletin						
3 pm	4 pm	5 pm	6 pm	Teleprinter Bulletin						
4 pm	5 pm	6 pm	7 pm	Fast Code	Slow Code	Fast Code	Slow Code	Fast Code	Slow Code	Fast Code
5 pm	6 pm	7 pm	8 pm	Code Bulletin						
6 pm	7 pm	8 pm	9 pm	Teleprinter Bulletin						
6^{45} pm	7^{45} pm	8^{45} pm	9^{45} pm	Voice Bulletin						
7 pm	8 pm	9 pm	10 pm	Slow Code	Fast Code	Slow Code	Fast Code	Slow Code	Fast Code	Slow Code
8 pm	9 pm	10 pm	11 pm	Code Bulletin						
9 pm	10 pm	11 pm	Mdnte	Teleprinter Bulletin						
9^{45} pm	10^{45} pm	11^{45} pm	12^{45} am	Voice Bulletin						

Note: W1AW's schedule is at the same local time throughout the year. The schedule according to your local time will change if your local time does not have seasonal adjustments that are made at the same time as North American time changes between standard time and daylight time. From the first Sunday in April to the last Sunday in October, UTC = Eastern Time + 4 hours. For the rest of the year, UTC = Eastern Time + 5 hours.

❏ **Morse code transmissions:**

Frequencies are 1.818, 3.5815, 7.0475, 14.0475, 18.0975, 21.0675, 28.0675 and 147.555 MHz.

Slow Code = practice sent at 5, 7¹/₂, 10, 13 and 15 wpm.

Fast Code = practice sent at 35, 30, 25, 20, 15, 13 and 10 wpm.

Code practice text is from the pages of QST. The source is given at the beginning of each practice session and alternate speeds within each session. For example, "Text is from July 1992 QST, pages 9 and 81," indicates that the plain text is from the article on page 9 and mixed number/letter groups are from page 81.

Code bulletins are sent at 18 wpm.

❑ *Teleprinter transmissions:*

Frequencies are 3.625, 7.095, 14.095, 18.1025, 21.095, 28.095 and 147.555 MHz.
Bulletins are sent at 45.45-baud Baudot and 100-baud AMTOR, FEC Mode B.
110-baud ASCII will be sent only as time allows.

On Tuesdays and Saturdays at 6:30 PM Eastern Time, Keplerian elements for
many amateur satellites are sent on the regular teleprinter frequencies.

❑ *Voice transmissions:*

Frequencies are 1.855, 3.99, 7.29, 14.29, 18.16, 21.39, 28.59 and 147.555 MHz.

❑ *Miscellanea:*

On Fridays, UTC, a DX bulletin replaces the regular bulletins.

W1AW is open to visitors during normal operating hours: from 1 PM until 1 AM on
Mondays, 9 AM until 1 AM Tuesday through Friday, from 1 PM to 1 AM on Saturdays, and from 3:30 PM to 1 AM on Sundays. FCC licensed amateurs may operate
the station from 1 to 4 PM Monday through Saturday. Be sure to bring your current
FCC amateur license or a photocopy.

In a communications emergency, monitor W1AW for special bulletins as follows:
voice on the hour, teleprinter at 15 minutes past the hour, and CW on the half hour.

Headquarters and W1AW are closed on New Year's Day, President's Day, Good
Friday, Memorial Day, Independence Day, Labor Day, Thanksgiving and the
following Friday, and Christmas Day. On the first Thursday of September, Headquarters and W1AW will be closed during the afternoon.

THE RST SYSTEM

Readability
1—Unreadable.
2—Barely readable, occasional words distinguishable.
3—Readable with considerable difficulty.
4—Readable with practically no difficulty.
5—Perfectly readable.

Signal Strength
1—Faint signals barely perceptible.
2—Very weak signals.
3—Weak signals.
4—Fair signals.
5—Fairly good signals.
6—Good signals.
7—Moderately good signals.
8—Strong signals.
9—Extremely strong signals.

Tone
1—Sixty-cycle AC or less, very rough and broad.
2—Very rough AC, very and broad.
3—Rough AC tone, rectified but not filtered.
4—Rough note, some trace of filtering.
5—Filtered rectified AC but strongly ripple-modulated.
6—Filtered tone, definite trace of ripple modulation.
7—Near pure tone, trace of ripple modulation.
8—Near perfect tone, slight trace of modulation.
9—Perfect tone, no trace of ripple or modulation of any kind.

The "tone" report refers only to the purity of the signal, and has
no connection with its stability or freedom from clicks or chirps. If
the signal has the characteristic steadiness of crystal control, add X
to the report (e.g., RST 469X). If it has a chirp or "tail" (either on
"make" or "break") add C (e.g., 469C). If it has clicks or notice-
able other keying transients, add K (e.g., 469K). Of course, a signal could have both chirps and clicks, in which case both C and K
could be used (e.g., RST 469CK).

ITU ZONE LIST
(Sorted by Country Prefix)

Prefix	Zone	Prefix	Zone	Prefix	Zone	Prefix	Zone
1A	28	CO	11	KG4	11	UA2	29
1S	50	CP	12, 14	KHØ	64	UA9, Ø	20–26, 30, 35, 75
3A	27	CT	37	KH1	61, 62	UK	30
3B6–9	53	CT3	36	KH2	64	UN–UQ	30
3C	47	CU	36	KH3	61	UR–UZ	29
3CØ	52	CX	14	KH4	61	V2	11
3D2–All	56	CY9, Ø	09	KH5	61	V3	11
3DAØ	57	D2	52	KH6	61	V4	11
3V	37	D4	46	KH7	61	V5	57
3W	49	D6	53	KH8	62	V6	65
3Y/B	67	DA–DL	28	KH9	65	V7	65
3Y/P	72	DU	50	KL7	01, 02	V8	54
4J, 4K	29	EA	37	KP1–5	11	VE	02, 03, 04, 09, 75
4L	29	EA6	37	LA	18	VK	55, 58, 59
4S	41	EA8	36	LU	14, 16	VKØ/H	68
4U1ITU	28	EA9	37	LX	27	VKØ/M	60
4U1UN	08	EI	27	LY	29	VK9/C, K	54
4U1VIC	28	EK	29	LZ	28	VK9/L	60
4X	39	EL	46	OA	12	VK9/M	56
5A	38	EP	40	OD	39	VK9/N	60
5B	39	ER	29	OE	28	VK9/W	55
5H	53	ES	29	OH	18	VK9/X	54
5N	46	ET	48	OHØ	18	VP2	11
5R	53	EU–EW	29	OJØ	18	VP5	11
5T	46	EX	30	OK	28	VP8	73
5U	46	EY	30	OM	28	VP8/F	16
5V	46	EZ	30	ON	27	VP9	11
5W	62	F	27	OX	05, 75	VQ9	41
5X	48	FG	11	OY	18	VR6	63
5Z	48	FH	53	OZ	18	VS6, VR2	44
6W	46	FJ, FS	11	P29	51	VU	41
6Y	11	FK	56	P4	11	VU/A	49
7O	39	FM	11	P5	44	VU/L	41
7P	57	FO	63	PA	27	XE	10
7Q	53	FO/C	10	PJ	11	XF4	10
7X	37	FP	09	PY	12, 13, 15	XT	46
8P	11	FR	53	PYØ/F	13	XU	49
8Q	41	FT5W	68	PYØ/P	13	XW	49
8R	12	FT5X	68	PYØ/T	15	XX	44
9A	28	FT5Z	68	PZ	12	XZ	49
9G	46	FW	62	R1F	75	YA	40
9H	28	FY	12	R1M	29	YB	51, 54
9J	53	G	27	SØ	37	YI	39
9K	39	H4	51	S2	41	YJ	56
9L	46	HA	28	S5	28	YK	39
9M2, 4	54	HB	28	S7	53	YL	29
9M6, 8	54	HC	12	S9	47	YN	11
9N	42	HH	11	SM	18	YO	28
9Q	52	HI	11	SP	28	YS	11
9U	52	HK	12	ST	47, 48	YU	28
9V	54	HKØ/A	11	SU	38	YV	12
9X	52	HKØ/M	12	SV, J4	28	YVØ	11
9Y	11	HL	44	T2	65	Z2	53
A2	57	HP	11	T3Ø	65	Z3	28
A3	62	HR	11	T31	62	ZA	28
A4	39	HS	49	T32	61, 63	ZB	37
A5	41	HV	28	T33	65	ZC	39
A6	39	HZ	39	T5	48	ZD7	66
A7	39	I	28	T7	28	ZD8	66
A9	39	J2	48	T9	28	ZD9	66
AP	41	J3	11	TA	39	ZF	11
BV	44	J5	46	TF	17	ZK1–3	62
BY	33, 42, 43, 44	J6	11	TG	11	ZL	60
		J7	11	TI	11	ZP	14
C2	65	J8	11	TJ	47	ZS	57
C3	27	JA	45	TK	28	ZS8	57
C5	46	JD/M	90	TL	47	ZL	60
C6	11	JD/O	45	TN	52	ZP	14
C9	53	JT	32, 33	TR	52	ZS	57
CE	14, 16	JW	18	TT	47	ZS8	57
CEØA	63	JX	18	TU	46		
CEØX, Z	14	JY	39	TZ	46		
CN	37	K	06, 07, 08	UA–UI, R	19, 20, 29, 30		
		KC6	64				

ITU ZONE LIST
(Sorted by Zones)

1	KL7	38	5A, SU
2	KL7, VE	39	4X, 5B, 7O, 9K, A4, A6, A7, A9, HZ, JY, OD, TA, YI,
3	VE		YK, ZC
4	VE	40	EP, YA
5	OX	41	4S, 8Q, A5, AP, S2, VQ9, VU, VU/L
6	K	42	9N, BY
7	K	43	BY
8	K, 4U1UN	44	BV, BY, HL, P5, VS6,VR2, XX
9	CY9, CYØ, FP, VE	45	JA, JD/O
10	FO/C, XE, XF4	46	5N, 5T, 5U, 5V, 6W, 9G, 9L, C5, D4, EL, J5, TU,
11	6Y, 8P, 9Y, C6, CO, FG, FJ, FS, FM, HH, HI, HKØ/A,		TZ, XT
	HP, HR, J3, J6, J7, J8, KG4, KP1-5, P4, PJ, TG, TI, V2,	47	3C, S9, ST, TJ, TL, TT
	V3, V4, VP2, VP5, VP9, YN, YS, YVØ, ZF	48	5X, 5Z, ET, J2, ST, T5
12	8R, CP, FY, HC, HK, HKØ/M, OA, PY, PZ, YV,	49	3W, HS, VU/A, XU, XW, XZ,
13	PY, PYØ/F, PYØ/P	50	1S, DU
14	CE, CEØX, Z, CP, CX, LU, ZP	51	H4, P29, YB
15	PY, PYØ/T	52	3CØ, 9Q, 9U, 9X, D2, TN, TR
16	CE, LU, VP8/F	53	3B6-9, 5H, 5R, 7Q, 9J, C9, D6, FH, FR, S7, Z2
17	TF	54	9M2, 4, 9M6, 8, 9V, V8, VK9/C, K, VK9/X, YB
18	JW, JX, LA, OH, OHØ, OJØ, OY, OZ, SM	55	VK, VK9W
19	UA-UI, R	56	3D2, FK, VK9M, YJ
20	UA-UI, R, UA9, Ø	57	3DAØ, 7P, A2, V5, ZS, ZS8
21	UA9, Ø	58	VK
22	UA9, Ø	58	VK
23	UA9, Ø	60	VKØ/M, VK9/L, VK9/N, ZL
24	UA9, Ø	61	KH1, KH3, KH4, KH5, KH6, KH7, T32
25	UA9, Ø	62	5W, A3, FW, KH1, KH8, T31, ZK1-3
26	UA9, Ø	63	CEØA, FO, T32, VR6
27	3A, C3, EI, F, G, LX, ON, PA	64	KC6, KHØ, KH2
28	1A, 4U1ITU, 4U1VIC, 9A, 9H, DA-DL, HA, HB, HV, I,	65	C2, KH9, T2, T3Ø, T33, V6, V7
	J4, LZ, OE, OK, OM, S5, SP, SV, T2, T7, T9, TK, YO,	66	ZD7, ZD8, ZD9
	YU, Z3, ZA	67	3Y/B
29	4J, 4K, 4L, EK, ER, ES, EU-EW, LY, R1M, UA-UI, R,	68	FT5W, FT5X, FT5Z, VKØH
	UA-UI,R, UR-UZ, YL	69	KC4 Antarctica
30	EX, EY, EZ, UA9, Ø, UK, UN-UQ	70	KC4 Antarctica
31	UN-UQ, EX	71	KC4 Antarctica
32	JT	72	KC4 Antarctica, 3Y/P
33	BY, JT	73	KC4 Antarctica, VP8
34	UA9, Ø	74	KC4 Antarctica
35	UA9, Ø	75	OX, R1F, UA9,Ø, VE
36	CT3, CU, EA8	90	JD/M
37	3V, 7X, CN, CT, EA, EA6, EA9, SØ, ZB		

LATITUDE AND LONGITUDE FOR SELECTED LOCATIONS

NOTES:
1. Table is in degrees and minutes.
2. South Latitudes and East Longitudes are shown as negative values.

PFX	City/State/Country	LAT	LONG
A4	MUSCAT, OMAN	23°23'	-58°30'
AP	KARACHI, PAKISTAN	24°59'	-68°56'
BV	TAIPEI, TAIWAN	25°02'	-121°38'
BY	CHUNKING, CHINA	29°38'	-107°30'
BY	BEIJING, CHINA	39°55'	-116°23'
C2	REP°OF NAURU	-0°30'	-167°00'
C3	ANDORRA	42°30'	-2°00'
C6	NASSAU, BAHAMAS	22°05'	77°20'
C9	MAPUTO, MOZAMBIQUE	-26°50'	-32°30'
CE	SANTIAGO, CHILE	-33°26'	70°40'
CEØA	EASTER ISLAND	-26°50'	109°00'
CN	CASABLANCA, MOROCCO	33°32'	7°41'
CP	LAPAZ, BOLIVIA	-16°31'	68°03'
CT	LISBON, PORTUGAL	38°42'	9°05'
CU	SAO MIGUEL IS°, AZORES	37°59'	26°38'
CX	MONTEVIDEO, URUGUAY	-34°50'	56°10'
D6	COMOROS	-12°30'	-42°45'
DJ	BONN, GERMANY	50°44'	-7°06'
DU	MANILA, PHILIPPINES	14°37'	-121°00'
EA6	BALEARIC ISLANDS	39°25'	-1°28'
EA8	CANARY ISLANDS	29°15'	16°30'
EI	DUBLIN, REP° IRELAND	53°20'	6°15'
ET	ADDIS ABABA, ETHIOPIA	9°00'	-38°44'
EZ	ASHKHABAD, TURKOMAN	39°45'	-58°13'
FT	KERGUELEN	-49°50'	-69°30'
FG	GUADELOUPE	16°40'	61°10'
FK	NOUMEA, NEW CALEDONIA	-22°18'	-166°48'
FO	CLIPPERTON ISL°	11°00'	110°00'
FO	PAPEETE, POLYNESIA	-17°30'	149°30'
FP	ST° PIERRE & MIQUELON	46°47'	56°11'
FR	REUNION	-21°06'	-55°36'
FY	CAYENNE, FR° GUIANA	4°56'	52°18'
G	LONDON, ENGLAND	51°30'	0°07'
HA	BUDAPEST, HUNGARY	47°30'	-19°05'
HC	QUITO, ECUADOR	-0°17'	78°32'
HC8	GALAPAGOS ISLS°	-0°17'	92°00'
HP	PANAMA CITY, PANAMA	8°35'	81°08'
HS	BANGKOK, THAILAND	13°50'	-100°29'
I	ROME, ITALY	41°52'	-12°37'
JA	TOKYO, JAPAN	35°41'	-139°44'
JD	MINAMI TORI-SHIMA	24°00'	-155°00'
JT	ULAN BATOR, MONGOLIA	47°56'	-107°00'
JW	SVALBARD	77°00'	-20°00'
KC4	McMURDO, ANTARCTICA	-78°00'	-167°00'
KC6	W. CAROLINE ISL.	9°00'	-138°20'
KH6	HILO, HAWAII	19°44'	155°01'
KL7	FAIRBANKS, ALASKA	64°50'	147°48'
KL7	JUNEAU, ALASKA	58°25'	134°30'
KP2	VIRGIN ISLANDS	18°15'	64°00'
LA	OSLO, NORWAY	59°56'	-10°41'
LZ	SOFIA, BULGARIA	42°43'	-23°20'
OH	HELSINKI, FINLAND	60°10'	-24°53'
OX	GODTHAAB, GREENLAND	64°10'	51°32'
P2	MADANG P.N.G.	-5°15'	-145°45'
PY	RIO DE JANEIRO, BRAZIL	-22°50'	43°20'
PY	NATAL, BRAZIL	-6°00'	35°13'
R1F	FRANZ JOSEF LAND	81°32'	-40°00'
SU	CAIRO, EGYPT	30°00'	-31°17'
T3	CENTRAL KIRIBATI	-4°00'	171°00'
T5	MOGADISHU, SOMALIA	2°08'	-45°22'
TA	ANKARA, TURKEY	39°55'	-32°50'
TF	REYKJAVIK, ICELAND	64°09'	21°39'
TJ	YAOUNDE, CAMEROON	3°52'	-11°31'
TT	FT.LAMY, CHAD	12°07'	-15°03'
UA3	MOSCOW, EUR. RUSSIA	55°45'	-37°37'
UA9	NOVOSIBIRSK, RUSSIA	55°09'	-82°58'
UA9	PERM, ASIATIC RUSSIA	58°00'	-56°15'
UAØ	KHABAROVSK, AS. RUSSIA	48°35'	-135°12'
UAØ	YAKUTSK, AS. RUSSIA	62°13'	-129°49'
UAØ	MAGADAN, AS. RUSSIA	59°28'	-143°32'
UAØ	USTOLENEK, AS. RUSSIA	72°52'	-120°15'
UN7	ALMA-ATA KAZAKH	43°19'	-77°08'
V3	BELIZE, BELIZE	17°31'	88°10'
V85	BRUNEI	5°00'	-114°59'
VE1	Halifax, N.S.	44°39'	63°36'
VE2	Quebec City, Que.	46°49'	71°13'
VE3	Ottawa, Ont.	46°05'	77°20'
VE4	Winnipeg, Mb.	49°53'	97°09'
VE5	Regina, Sk.	50°25'	104°39'
VE6	Edmonton, Al.	53°33'	113°28'
VE7	Prince Rupert, B.C.	54°19'	130°19'
VE8	Yellowknife, NWT	62°29'	114°38'
VO2	Goose Bay, Labrador	53°19'	60°33'
VY1	Whitehorse, Yukon	60°39'	135°01'
VK1	CANBERRA, AUSTRALIA	-35°21'	-149°10'
VK4	BRISBANE, AUSTRALIA	-27°30'	-153°10'
VK6	PERTH, AUSTRALIA	-31°50'	-116°10'
VK8	DARWIN, AUSTRALIA	-12°25'	-131°00'
VK9	COCOS (KEELING) IS.	-11°50'	-90°50'
VK9	NORFOLK ISLAND	-27°10'	-166°50'
VKØ	MACQUARIE ISLAND	-54°36'	-158°45'
VP8	FALKLAND ISLANDS	-50°45'	61°00'
VP8	SO. SANDWICH IS.	-58°00'	27°00'
VP8	SO. SHETLAND IS.	-62°00'	58°50'
VP9	BERMUDA	32°20'	64°45'
VQ9	DIEGO GARCIA CHAGOS	-7°30'	-72°40'
VR6	PITCAIRN IS.	-24°30'	133°00'
VU7	ANDAMAN IS.	11°38'	-92°17'
VU7	LACCADIVE IS.	11°00'	-73°02'
WØ	Denver, CO	39°44'	104°59'
WØ	DesMoines, IA	41°35'	93°37'
WØ	Topeka, KS	39°02'	95°41'
WØ	Minneapolis, MN	44°58'	93°15'
WØ	Jefferson City, MO	38°34'	92°10'
WØ	Lincoln, NE	40°49'	96°43'
WØ	Bismark, ND	46°48'	100°46'
WØ	Pierre, SD	44°22'	100°20'
W1	Augusta, ME	44°19'	69°42'
W1	Boston, MA	42°15'	71°07'
W1	Concord, NH	43°10'	71°30'
W1	Hartford, CT	41°45'	72°40'
W1	Montpelier, VT	44°20'	72°35'
W1	Providence, RI	41°50'	71°23'
W2	Albany, NY	42°40'	73°50'
W2	Trenton, NJ	40°13'	74°46'
W3	Dover, DE	39°10'	76°30'
W3	Harrisburg, PA	40°15'	76°50'
W3	Washington, D.C.	38°50'	77°00'
W4	Atlanta, GA	33°45'	84°23'
W4	Columbia, SC	34°00'	81°00'
W4	Frankfort, KY	38°10'	84°55'
W4	Montgomery, AL	32°23'	86°17'
W4	Nashville, TN	36°10'	86°48'
W4	Raleigh, NC	35°45'	78°39'
W4	Richmond, VA	37°35'	77°30'
W4	Tallahassee, FL	30°25'	84°17'
W5	Austin, TX	30°15'	97°42'
W5	Baton Rouge, LA	30°28'	91°10'
W5	Jackson, MS	32°17'	90°10'
W5	Little Rock, AR	34°42'	92°16'
W5	Oklahoma City, OK	35°27'	97°32'
W5	SantaFe, NM	35°10'	106°00'
W6	Los Angeles, CA	34°00'	118°15'
W6	San Francisco, CA	37°45'	122°26'
W7	Boise, ID	43°38'	116°12'
W7	Carson City, NV	39°10'	119°45'
W7	Cheyenne, WY	41°10'	104°49'
W7	Olympia, WA	47°02'	122°52'
W7	Phoenix, AZ	33°30'	112°00'

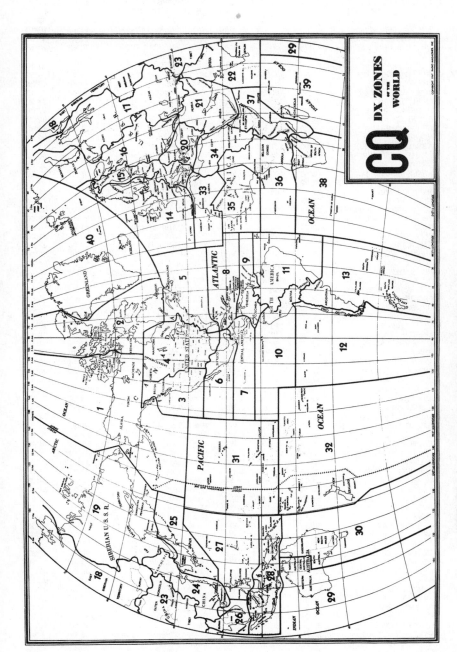

NOTE: This map is a reproduction of the official 1947 CQ WAZ Map, and is intended to be used for general reference only. Over the years numerous adjustments have been made to Zone boundaries for various reasons. Always consult the latest CQ WAZ Rules for the most up-to-date Zone delineations.

W7	Salem, OR	44°55'	123°03'
W7	Salt Lake City, UT	40°45'	111°52'
W8	Charleston, WV	38°20'	81°35'
W8	Columbus, OH	40°00'	83°00'
W8	Lansing, MI	42°45'	84°35'
W9	Indianapolis, IN	39°45'	86°08'
W9	Madison, WI	43°05'	89°23'
W9	Springfield, IL	39°45'	89°37'
XE1	MEXICO CITY, MEXICO	19°28'	99°09'
XW	VIENTIANE, LAOS	18°07'	-102°33'
YA	KABUL, AFGHANISTAN	34°39'	-69°14'
YB	JAKARTA, INDONESIA	-6°17'	-106°45'
ZD8	ASCENSION IS.	-8°00'	13°00'
ZD9	TRISTAN DA CUNHA	-35°30'	12°15'
ZF	CAYMAN IS.	19°30'	80°30'
ZL1	AUCKLAND, NEW ZEALAND	-36°53'	-174°45'
ZL4	DUNEDIN, NEW ZEALAND	-45°48'	-170°32'
ZP	ASUNCION, PARAGUAY	-25°25'	57°30'
ZS1	CAPETOWN, SO. AFRICA	-33°48'	-18°28'
ZS8	Pr. EDWARD & MARION	-46°36'	-37°57'
1S	SPRATLY IS.	8°38'	-111°54'
3V	TUNIS, TUNISIA	36°59'	-10°06'
3Y	BOUVET IS.	-54°26'	-3°24'
3Y	PETER 1ST IS.	-68°00'	91°00'
4J	BAKU, AZERBAIJAN	40°28'	-49°45'
4U	UNITED NATIONS HQ	40°40'	73°58'
4X	ISRAEL	32°40'	-34°00'
5R	TANANARIVE, MALAGASY	-18°51'	-47°40'
5W	WESTERN SAMOA	-14°30'	172°00'
5Z	NAIROBI, KENYA	-1°17'	-36°49'
6W	DAKAR, SENEGAL	14°40'	17°26'
7O	ADEN, YEMEN P. DEM.	12°48'	-45°00'
7P	LESOTHO	-29°45'	-28°07'
8Q	MALDIVE IS.	4°30'	-71°30'
9G	ACCRA, GHANA	5°33'	0°13'
9J	LUSAKA, ZAMBIA	-15°25'	-28°17'
9K	KUWAIT	29°00'	-48°45'
9N	KATMANDU, NEPAL	27°49'	-85°21'
9Q	KINSHASA, ZAIRE	-4°18'	-15°18'
9V	SINGAPORE	1°22'	-103°45'
9X	RWANDA	-2°10'	-29°37'
9Y	TRINIDAD	10°00'	61°00'

HOW TO DETERMINE YOUR GRID SQUARE WITHOUT A MAP

You will need your longitude and latitude rounded up to the nearest minute. Save the number obtained from each step, for it will be used in the next.

First use your longitude:

1. Convert the minutes portion of the longitude from minutes to decimal by dividing by 60.

2. For North America and locations of West longitude, subtract your longitude from 180 degrees. For location of East longitude, add 180 degrees.

3. Next divide this value by 20. The whole number result will be used to determine the first digit of your Grid, as follows:

0=A, 1=B, 2=C thru to 17=R .

4. For the the third digit, multiply this last number by 10. The digit immediately before the decimal point is the third digit of your Grid.

5. Use only the decimal portion of this last number and multiply by 24. The digit immediately before the decimal point is used to determine the fifth digit of your Grid.

0=A, B=1, C=2, thru to X=23.

Now use your Latitude:

1. If your Latitude is North, add 90. If your latitude is South, subtract your latitude from 90.

2. Divide this number by 10. The whole number result will be used to determine the second digit of your Grid, as follows:

0=A, B=1, C=2 thru to 17=R.

3. Now, multiply this number by 10. The digit immediately before the decimal point is the fourth digit of your Grid.

4. Use only the decimal portion of this last number and multiply by 24. The digit immediately before the decimal point is used to determine the sixth digit of your Gird, as follows:

0=A, 1=B, 2=C thru to X=24.

This completes your Six digit Grid Location.

Expressed in another form:

Longitude = (20A + 2C + E/12) –180 degrees East.
Latitude = (10B + 1D + F/24) –90 degrees North

Grid Locator of "ABCDEF," degrees where C and D are numbers and A, B, E and F are represented as letters.

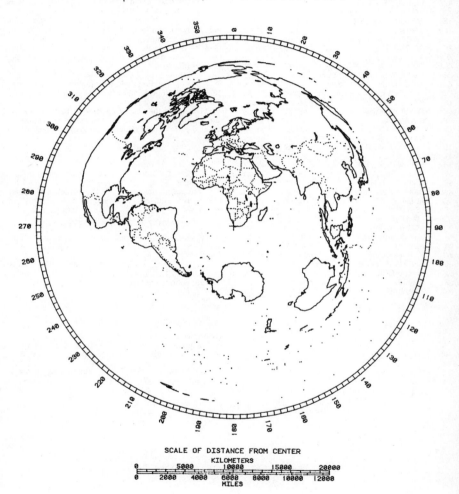

NSKR AZIMUTHAL EQUIDISTANT MAP CENTERED ON
Capetown, South Africa

SCALE OF DISTANCE FROM CENTER

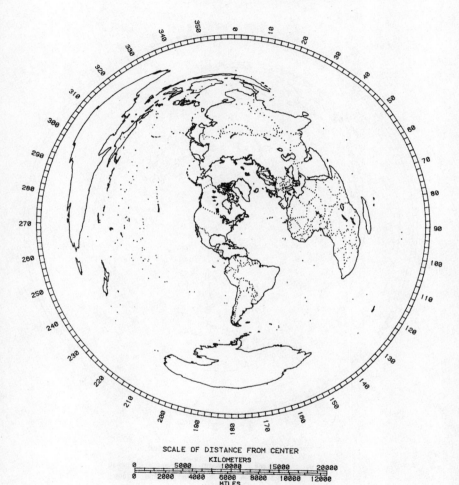

NSKR AZIMUTHAL EQUIDISTANT MAP CENTERED ON
- New York, New York -

SCALE OF DISTANCE FROM CENTER
KILOMETERS

N5KR AZIMUTHAL EQUIDISTANT MAP CENTERED ON
— Miami, Florida —

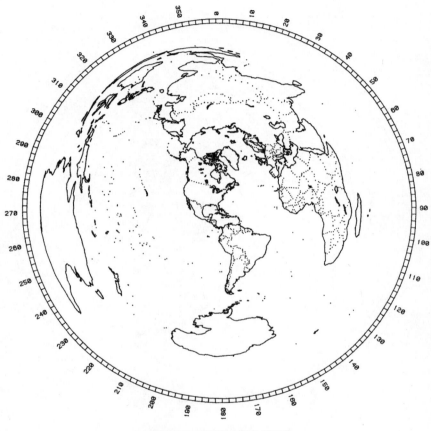

SCALE OF DISTANCE FROM CENTER

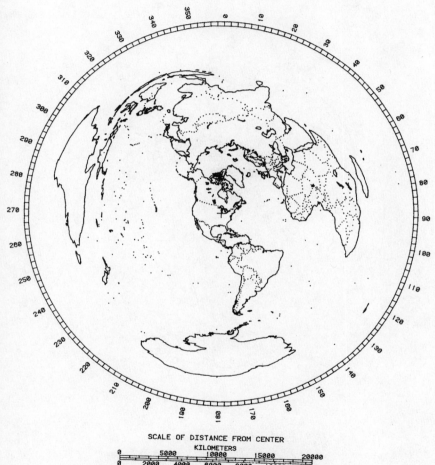

NSKR AZIMUTHAL EQUIDISTANT MAP CENTERED ON
— Chicago, Illinois —

SCALE OF DISTANCE FROM CENTER
KILOMETERS
MILES

N5KR AZIMUTHAL EQUIDISTANT MAP CENTERED ON
— Seattle, Washington —

SCALE OF DISTANCE FROM CENTER

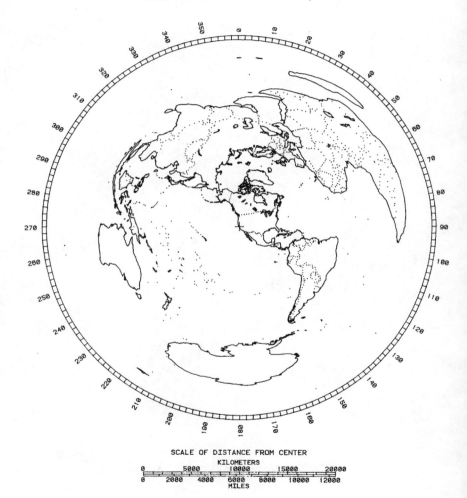

N5KR AZIMUTHAL EQUIDISTANT MAP CENTERED ON
— San Francisco, CA —

SCALE OF DISTANCE FROM CENTER
KILOMETERS

SCALE OF DISTANCE FROM CENTER

N5KR AZIMUTHAL EQUIDISTANT MAP CENTERED ON
— Denver, Colorado —

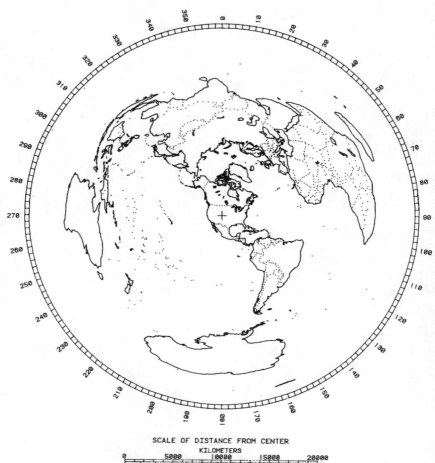

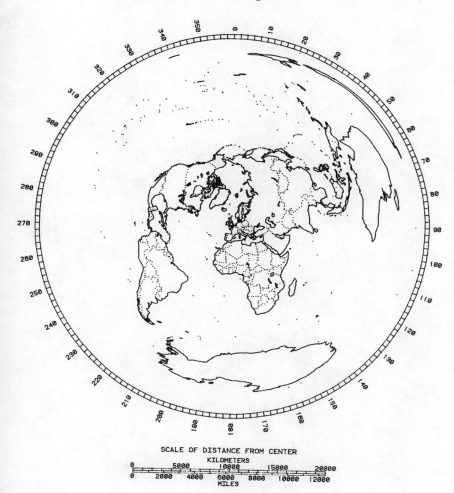

N5KR AZIMUTHAL EQUIDISTANT MAP CENTERED ON

– London, England (G) –

SCALE OF DISTANCE FROM CENTER

KILOMETERS

0   5000   10000   15000   20000

0   2000   4000   6000   8000   10000 12000

MILES

N5KR AZIMUTHAL EQUIDISTANT MAP CENTERED ON
Rome, Italy (I0)

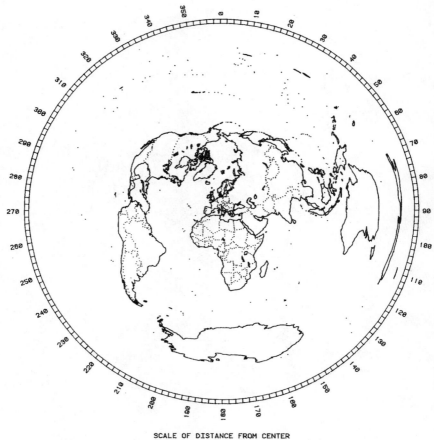

SCALE OF DISTANCE FROM CENTER

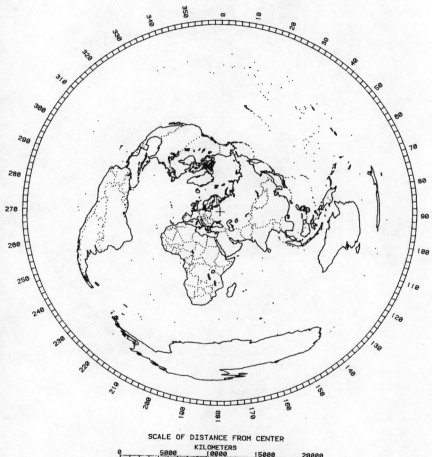

NSKR AZIMUTHAL EQUIDISTANT MAP CENTERED ON
- Moscow, Russia (UA3) -

SCALE OF DISTANCE FROM CENTER
KILOMETERS

N5KR AZIMUTHAL EQUIDISTANT MAP CENTERED ON
— Tokyo, Japan (JA1) —

SCALE OF DISTANCE FROM CENTER
KILOMETERS

N5KR AZIMUTHAL EQUIDISTANT MAP CENTERED ON
- San Juan, P.R. (KP4) -

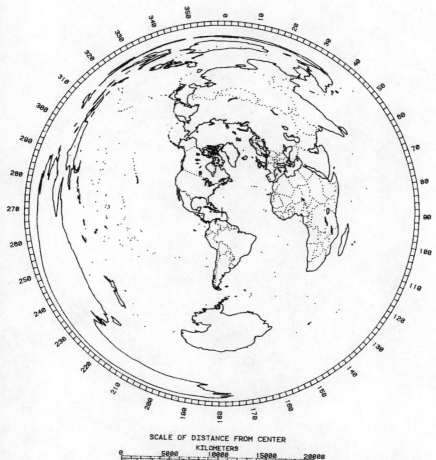

SCALE OF DISTANCE FROM CENTER

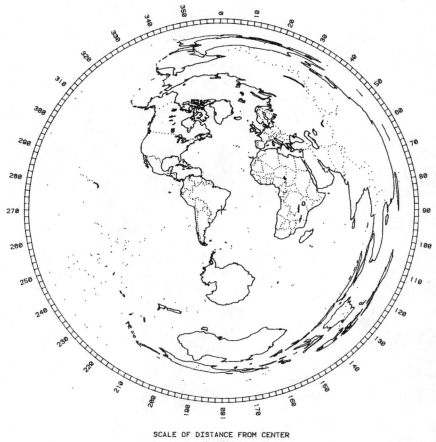

N5KR AZIMUTHAL EQUIDISTANT MAP CENTERED ON
— Rio de Janeiro (PY1) —

SCALE OF DISTANCE FROM CENTER
KILOMETERS

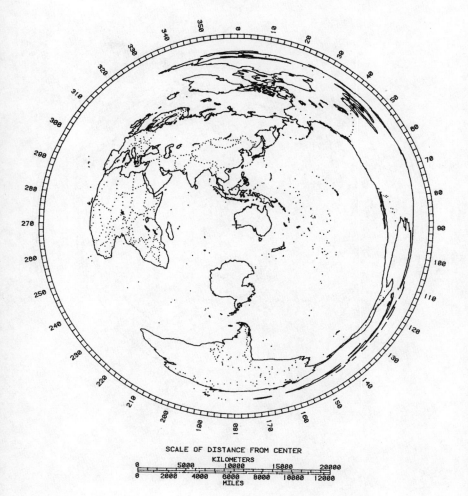

N5KR AZIMUTHAL EQUIDISTANT MAP CENTERED ON
Perth, Australia (VK6)

SCALE OF DISTANCE FROM CENTER
KILOMETERS
MILES

N5KR AZIMUTHAL EQUIDISTANT MAP CENTERED ON
Melbourne, Australia VK3

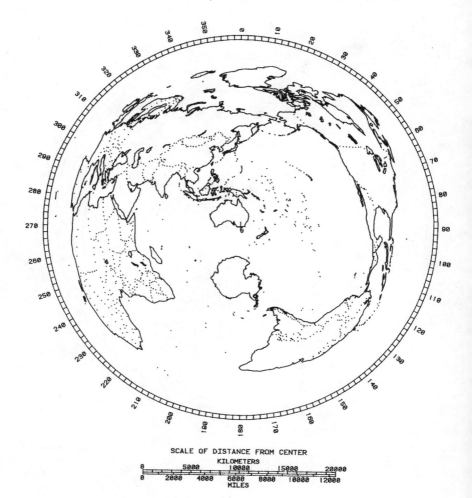

SCALE OF DISTANCE FROM CENTER

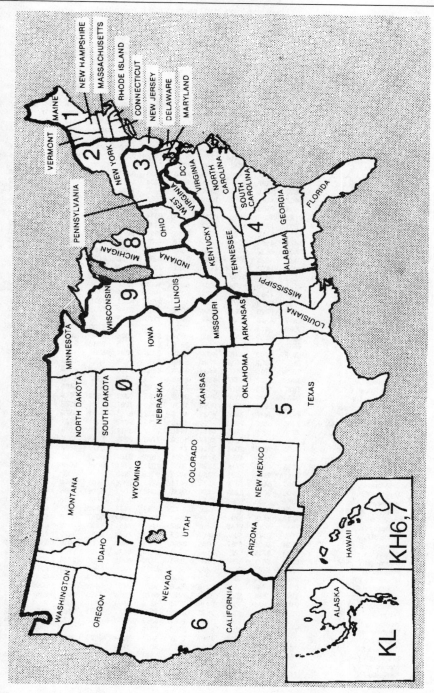

MAJOR CITY DISTANCE TABLE (In miles)

		A4	BY	CE	CT	CU	CX	DU	FO	FY	G	HA	HC
A4	MUSCAT OMAN	0	3522	9294	4044	4985	8475	4095	10594	7380	3627	2727	9137
BY	PEKING CHINA	3522	0	11834	6000	6562	11889	1768	7211	9249	5053	4554	9534
CE	SANTIAGO CHILE	9294	11834	0	6350	5682	832	10937	4908	2907	7246	7857	2344
CT	LISBON PORTUGAL	4043	6000	6350	0	950	5900	7537	9622	3566	983	1533	5128
CU	SAO MIGUEL IS AZORES	4984	6562	5682	950	0	5370	8234	8707	2794	1585	2371	4218
CX	MONTEVIDEO URUGUAY	8475	11889	832	5900	5370	0	11020	5695	2756	6855	7335	2786
DU	MANILA PHILIPPINES	4095	1768	10937	7537	8234	11020	0	6502	11002	6665	6012	1076
FO	PAPEETE POLYNESIA	10594	7211	4908	9622	8707	5695	6502	0	6767	9535	10251	4935
FY	CAYENNE FR. GUIANA	7380	9249	2907	3566	2794	2756	11002	6767	0	4378	5098	1844
G	LONDON ENGLAND	3627	5053	7246	983	1585	6855	6665	9535	4378	0	901	5732
HA	BUDAPEST HUNGARY	2726	4554	7857	1533	2371	7335	6012	10251	5098	901	0	6582
HC	QUITO ECUADOR	9136	9534	2344	5128	4218	2786	10767	4935	1844	5732	6582	0
HS	BANKOK THAILAND	2814	2042	10964	6626	7479	10374	1373	7814	10156	5913	5114	1144
I	ROME ITALY	2913	5040	7398	1160	2077	6849	6446	10431	4706	895	501	628
JA	TOYKO JAPAN	4819	1302	10699	6920	7267	11530	1859	5914	9517	5936	5617	896
JT	ULAN BATOR MONGOLIA	3134	722	11418	5278	5847	11176	2435	7737	8568	4331	3844	912
KH6	HILO HAWAII	8691	5278	6644	7830	7225	7467	5494	2600	6919	7280	7771	534
KL7	FAIRBANKS ALASKA	6176	3921	7926	4940	4656	8408	5337	5684	6066	4212	4639	563
KL7	JUNEAU ALASKA	6741	4527	7315	5034	4602	7834	5869	5296	5641	4427	4972	504
KP2	VIRGIN ISLANDS	7602	8411	3594	3559	2625	3700	10133	6285	1211	4117	4974	161
LA	OSLO NORWAY	3402	4361	7901	1699	2219	7554	6016	9330	5003	715	922	620
LZ	SOFIA BULGARIA	2406	4563	7912	1707	2610	7325	5911	10632	5255	1250	390	682
P2	MADANG P.N.G.	6183	3633	8851	9632	10117	9314	2179	4464	11185	8684	8145	935
PY	RIO DJANEIRO BRAZIL	7531	10754	1812	4790	4335	1134	11255	6704	2010	5758	6203	283
T5	MOGADISHU SOMALIA	1710	5119	7782	4247	5146	6950	5217	10978	6727	4299	3503	855
TF	REYKJAVIK ICELAND	4474	4888	7242	1830	1818	7087	6643	8395	4362	1167	1901	528
TJ	YAOUNDE CAMEROON	3407	6822	5911	2730	3362	5121	7440	10816	4391	3356	3046	621
UA3	MOSCOW EUR. RUSSIA	2473	3596	8772	2423	3138	8296	5127	9762	5928	1552	972	722
UA9	NOVOSIBIRSK RUSSIA	2524	1852	10464	4155	4782	10040	3453	8667	7560	3228	2705	849
UA9	PERM ASIATIC RUSSIA	2392	2879	9467	3130	3796	9014	4443	9344	6590	2221	1686	774
UAØ	YAKUTSK AS RUSSIA	4195	1638	10247	5095	5386	10519	3314	6996	7790	4114	3953	790
UAØ	OKHOTSK AS RUSSIA	4661	1789	10053	5471	5676	10508	3301	6498	7898	4502	4407	776
UAØ	UST OLENEK AS RUSSIA	4108	2278	9677	4359	4613	9798	4021	7378	7046	3388	3330	735
UL7	ALMA-ATA KAZAKH	1736	2020	10569	4294	5067	9891	3252	9202	7847	3482	2762	909
V3	BELIZE BELIZE	8817	8159	3701	4887	3940	4178	9450	4798	2572	5190	6091	139
V85	BRUNEI	3959	2411	10431	7759	8573	10291	779	6700	11314	6989	6233	114
VE1	Halifax N.S.	6482	6589	5408	2788	1953	5507	8325	6845	2826	2872	3769	323
VE2	Quebec City Que.	6657	6421	5539	3115	2319	5713	8119	6543	3103	3098	3979	328
VE3	Ottawa Ont.	6892	6430	5505	3409	2613	5739	8084	6255	3211	3365	4238	320
VE4	Winnipeg Mb.	7153	5902	5976	4164	3466	6361	7408	5625	4046	3912	4708	363
VE5	Regina Sk.	7229	5715	6148	4439	3773	6583	7158	5410	4349	4126	4887	382
VE6	Edmonton Al.	7093	5327	6527	4638	4046	6993	6734	5343	4779	4219	4918	421
VE7	Prince Rupert B.C.	7037	4824	7015	5124	4624	7556	6118	5081	5455	4580	5174	476
VE8	Yellowknife NWT	6486	4816	7054	4332	3869	7452	6327	5832	5049	3791	4419	471
VO2	Goose Bay Labrador	6007	5988	6019	2598	1921	6090	7737	7121	3372	2471	3341	384
VY1	Whitehorse Yukon	6574	4410	7435	4926	4526	7932	5791	5451	5675	4294	4822	515
VK1	CANBERRA AUSTRALIA	7165	5598	7026	11208	12131	7314	3910	3941	9921	10548	9761	849
VK4	BRISBANE AUSTRALIA	7210	5219	7335	11145	11703	7717	3617	3710	10141	10267	9612	850
VK6	PERTH AUSTRALIA	5385	4952	7899	9349	10292	7798	3222	5846	10425	8996	8102	100
VP8	SO. SHETLAND IS.	8499	10887	2044	7475	7128	1878	9155	5172	4632	8459	8674	
VP9	BERMUDA	7062	7439	4556	3111	2171	4670	9164	6540	2055	3446	4342	242
VQ9	DIEGO GARCIA CHAGOS	2336	4296	8718	6096	7036	7997	3644	9092	8647	5894	5002	103
WØ	Denver CO	7941	6343	5505	4850	4070	6013	7653	4868	4052	4682	5496	
WØ	DesMoines IA	7600	6495	5379	4275	3471	5783	7964	5401	3584	4183	5031	303
WØ	Topeka KS	7806	6613	5252	4456	3636	5688	8033	5208	3585	4385	5236	292
WØ	Minneapolis MN	7384	6286	5594	4152	3382	5976	7788	5555	3701	4008	4844	325
WØ	Jefferson City MO	7747	6717	5157	4308	3472	5564	8177	5342	3406	4275	5139	281
WØ	Lincoln NE	7718	6479	5387	4441	3641	5821	7904	5241	3693	4334	5175	305
WØ	Bismark ND	7411	6019	5846	4410	3687	6277	7461	5360	4073	4179	4977	351
WØ	Pierre SD	7562	6178	5684	4477	3725	6127	7599	5261	3973	4285	5098	336
W1	Augusta ME	6724	6600	5367	3086	2254	5529	8304	6552	2916	3134	4025	297
W1	Boston MA	6874	6735	5225	3194	2340	5404	8426	6436	2823	3275	4169	297
W1	Concord NH	6844	6671	5288	3194	2351	5470	8361	6440	2888	3253	4145	292
W1	Hartford CT	6955	6761	5192	3280	2423	5389	8439	6349	2834	3359	4252	292
W1	Montpelier VT	6825	6584	5369	3223	2396	5561	8269	6419	2984	3249	4136	310
W1	Providence RI	6904	6763	5196	3216	2358	5379	8451	6413	2805	3304	4198	294
W2	Albany NY	6947	6684	5263	3315	2472	5470	8356	6320	2924	3365	4254	296
W2	Trenton NJ	7107	6851	5090	3419	2551	5314	8509	6207	2806	3511	4405	280
W3	Dover DE	7223	6908	5025	3531	2658	5270	8547	6093	2805	3629	4522	272
W3	Harrisburg PA	7179	6830	5102	3520	2657	5347	8470	6109	2875	3594	4484	280
W3	Washington D.C.	7258	6925	5005	3565	2690	5256	8559	6060	2804	3664	4557	276
W4	Atlanta GA	7789	7172	4722	4087	3192	5074	8694	5535	2861	4203	5095	237
W4	Columbia SC	7656	7205	4703	3907	3006	5015	8772	5714	2721	4053	4949	237
W4	Frankfort KY	7561	6871	5026	3973	3112	5364	8416	5658	3073	4013	4895	276
W4	Montgomery AL	7931	7230	4655	4229	3331	5034	8716	5393	2896	4348	5239	231
W4	Nashville TN	7734	6972	4916	4129	3257	5282	8480	5499	3070	4183	5067	267

HS	I	JA	JT	KH6	KL7 (FBAK)	KL7 (JNAK)	KP2	LA	LZ	P2	PY	T5	TF	TJ	UA3	UA9 (WVSK)
2814	2914	4819	3134	8692	6177	6742	7603	3402	2407	6183	7531	1710	4474	3407	2474	2525
2042	5040	1302	722	5278	3921	4527	8411	4361	4563	3633	10754	5119	4888	6822	3596	1852
10964	7398	10699	11418	6644	7926	7315	3594	7901	7912	8851	1812	7782	7242	5911	8772	10464
6626	1160	6920	5278	7830	4940	5034	3559	1699	1707	9632	4790	4247	1830	2730	2423	4155
7479	2077	7267	5847	7225	4656	4602	2625	2219	2610	10117	4335	5146	1818	3362	3138	4782
10374	6849	11530	11176	7467	8408	7834	3700	7554	7325	9314	1134	6950	7087	5121	8296	10040
4373	6446	1859	2435	5494	5337	5869	10133	6016	5911	2179	11255	5217	6643	7440	5127	3453
7814	10431	5914	7737	2600	5684	5296	6285	9330	10632	4464	6704	10978	8395	10816	9762	8667
10156	4706	9517	8568	6919	6066	5641	1211	5003	5255	11185	2010	6727	4362	4391	5928	7560
5913	895	5936	4331	7280	4212	4427	4117	715	1250	8684	5758	4299	1167	3356	1552	3228
5114	501	5617	3844	7771	4639	4972	4974	922	390	8145	6203	3503	1901	3046	972	2705
1489	6285	6864	9120	5343	5633	5048	1615	6203	6826	9350	2830	8552	5280	6217	7221	8497
5471	0	6117	4334	8101	4984	5261	4703	1249	549	8603	5715	3410	2047	2625	1471	3196
2857	6117	0	1865	4064	3507	4015	8393	5220	5697	2852	11521	6372	5460	8114	4645	2959
2383	4334	1865	0	5565	3682	4309	7824	3641	3872	4356	10047	4817	4207	6268	2880	1138
6801	8101	4064	5565	0	3131	2853	5857	6868	8115	4373	8069	10381	6126	10559	7163	6245
5962	4984	3507	3682	3131	0	628	4895	3741	4987	5871	8056	7756	3118	7569	4097	3722
5569	5261	4015	4309	2853	628	0	4441	4052	5341	6152	7586	8258	3276	7747	4544	4326
9973	4703	8393	7824	5857	4895	4441	0	4599	5234	10230	3161	7431	3727	5180	5612	7001
5382	1249	5220	3641	6868	3741	4052	4599	0	1302	7982	6467	4394	1073	3871	1022	2562
4921	549	5697	3872	8115	4987	5341	5234	1302	0	8079	6192	3113	2290	2778	1102	2743
3369	8603	2852	4356	4373	5871	6152	10230	7982	8079	0	10394	6941	8312	9277	7213	5477
9980	5715	11521	10047	8069	8056	7586	3161	6467	6192	10394	0	6185	6117	4124	7167	8909
3844	3410	6372	4817	10381	7756	8258	7431	4394	3113	6941	6185	0	5402	2336	3727	4219
6253	2047	5460	4207	6126	3118	3276	3727	1073	2290	8312	6117	5402	0	4478	2045	3278
6081	2625	8114	6268	10559	7569	7747	5180	3871	2778	9277	4124	2336	4478	0	3862	5268
4381	1471	4645	2880	7163	4097	4544	5612	1022	1102	7213	7167	3727	2045	3862	0	1744
3003	3196	2959	1138	6245	3722	4326	7001	2562	2743	5477	8909	4219	3278	5268	1744	0
3807	2182	3940	2163	6769	3854	4379	6166	1595	1765	6502	7885	3902	2467	4443	717	1027
4637	4447	1881	1322	4538	2362	2988	6821	3404	4129	4733	9689	5905	3580	6853	3027	1708
4831	4895	1650	1662	4053	2133	2738	6842	3803	4601	4469	9876	6372	3858	7342	3499	2198
4152	3808	2655	1769	4807	2113	2726	6133	2690	3557	5507	8916	5768	2805	6332	2471	1611
2460	3195	3298	1467	6964	4578	5184	7515	2922	2667	5411	8788	3447	3819	4816	1933	857
10184	5944	7602	7813	4348	4245	3656	1587	5469	6421	8694	4113	8987	4426	6767	6471	7383
159	6612	2638	3001	6097	6112	6648	10820	6403	6063	2236	10526	4799	7161	7114	5438	3890
4268	3691	6692	6010	5333	3327	3033	1822	3119	4121	9181	4831	7029	2093	5296	4128	5265
209	3950	6417	5885	4943	2991	2662	2013	3259	4346	8815	5113	7331	2196	5675	4240	5229
288	4227	6330	5931	4661	2855	2476	2066	3489	4610	8614	5212	7618	2419	5966	4451	5344
7900	4807	5577	5523	3762	2072	1594	2847	3848	5097	7680	6001	8205	2807	6826	4685	5175
7738	5021	5313	5382	3452	1827	1303	3143	4002	5277	7350	6282	8390	2991	7125	4785	5126
7362	5106	4885	5032	3201	1415	872	3575	4008	5306	6947	6714	8394	3058	7359	4704	4864
6865	5435	4260	4621	2718	942	315	4246	4252	5550	6265	7365	8522	3417	7853	4798	4638
6829	4656	4516	4468	3514	1016	734	3881	3498	4801	6841	7041	7845	2625	7059	4128	4250
6657	3341	6164	5399	5308	2930	2744	2427	2608	3713	8792	5357	6744	1548	5264	3592	4656
6452	5119	3946	4171	2987	491	167	4484	3904	5188	6163	7640	8092	3150	7627	4381	4165
6651	10065	4941	6310	5271	7697	7801	10084	9926	9528	2089	8328	7074	10401	8962	8996	7364
6534	10004	4448	5940	4755	7106	7217	9988	9581	9454	1611	8783	7369	9886	9422	8744	7039
6318	8285	4904	5535	6855	8355	8742	11489	8613	7792	2649	8417	5174	9563	7212	7590	6318
6943	8176	10446	11304	7620	9859	9231	5548	9174	8516	7608	2797	6790	8913	5814	9616	10986
6094	4167	7471	6862	5510	4015	3625	973	3817	4656	9714	4062	7296	2857	5270	4840	6092
6408	5084	5264	4360	9130	8011	8637	9425	5736	4645	5012	7625	1994	6809	4284	4803	4365
6383	5572	5793	6065	3251	2421	1822	2849	4644	5884	7465	5864	8981	3597	7415	5479	5860
6497	5061	6117	6120	3857	2631	2103	2372	4224	5412	8017	5486	8466	3155	6809	5127	5745
6631	5260	6177	6264	3739	2719	2164	2375	4431	5616	7961	5446	8662	3362	6956	5333	5924
6278	4893	5952	5897	3892	2451	1951	2494	4021	5227	7961	5636	8304	2957	6738	4910	5512
6719	5142	6326	6341	3927	2847	2310	2195	4356	5515	8152	5286	8532	3284	6778	5280	5951
6497	5215	6050	6131	3693	2586	2035	2482	4357	5558	7877	5569	8623	3292	6978	5244	5802
6039	5073	5612	5677	3555	2132	1598	2864	4117	5366	7579	5989	8474	3077	7049	4947	5385
6202	5177	5746	5843	3542	2280	1730	2762	4248	5485	7639	5867	8584	3199	7084	5094	5554
6366	3969	6605	6055	5046	3174	2832	1829	3338	4384	8983	4926	7323	2288	5593	4333	5376
6518	4101	6705	6200	5004	3250	2881	1708	3492	4525	9019	4830	7438	2445	5657	4490	5532
6460	4086	6640	6137	4973	3187	2823	1774	3457	4504	8965	4895	7435	2405	5679	4450	5476
6565	4187	6702	6235	4932	3233	2849	1700	3570	4609	8976	4837	7525	2519	5736	4564	5585
6388	4092	6544	6057	4906	3090	2728	1867	3431	4499	8872	4992	7456	2372	5735	4417	5411
6549	4128	6726	6229	4996	3266	2892	1684	3524	4553	9026	4811	7461	2477	5670	4521	5563
6504	4201	6614	6166	4862	3143	2758	1786	3554	4615	8887	4926	7555	2497	5796	4542	5532
6683	4336	6748	6341	4843	3260	2850	1645	3721	4761	8946	4799	7666	2669	5845	4714	5713
6763	4425	6768	6410	4763	3270	2839	1626	3835	4878	8906	4786	7779	2780	5939	4825	5804
6689	4425	6693	6334	4734	3198	2772	1699	3786	4844	8852	4859	7765	2727	5954	4772	5734
6788	4488	6775	6432	4740	3274	2838	1620	3869	4914	8894	4782	7812	2814	5966	4859	5831
6126	5027	6852	6744	4380	3356	2841	1650	4393	5453	8669	4747	8333	3330	6402	5371	6243
6123	4865	6946	6749	4570	3440	2947	1509	4270	5301	8843	4639	8150	3214	6208	5259	6199
6820	4859	6586	6438	4321	3081	2587	1863	4159	5262	8531	4999	8216	3088	6396	5120	5945
6205	5172	6865	6822	4282	3383	2850	1692	4533	5598	8595	4749	8474	3467	6523	5507	6350
6943	5027	6637	6559	4227	3143	2625	1858	4331	5433	8483	4963	8375	3261	6517	5291	6094

		A4	BY	CE	CT	CU	CX	DU	FO	FY	G	HA	HC
W4	Raleigh NC	7477	7116	4804	3735	2839	5083	8720	5887	2707	3873	4769	2487
W4	Richmond VA	7340	7005	4922	3624	2740	5182	8629	5999	2756	3741	4636	2614
W4	Tallahassee FL	7969	7395	4496	4195	3281	4861	8894	5430	2722	4363	5260	2152
W5	Austin TX	8402	7113	4737	4878	3999	5242	8401	4750	3427	4911	5783	2455
W5	Baton Rouge LA	8201	7258	4609	4544	3647	5049	8652	5078	3077	4641	5528	2278
W5	Jackson MS	8063	7162	4710	4428	3539	5130	8590	5195	3085	4509	5394	2372
W5	Little Rock AR	7982	6962	4907	4446	3579	5335	8384	5186	3273	4471	5345	2572
W5	Oklahoma City OK	8073	6795	5062	4672	3828	5535	8150	4976	3561	4635	5493	2753
W5	Santa Fe NM	8259	6585	5257	5077	4263	5801	7816	4605	3996	4960	5789	3019
W6	Los Angeles CA	8459	6250	5583	5669	4901	6216	7291	4096	4651	5440	6211	3492
W6	San Francisco CA	8205	5902	5932	5665	4951	6563	6961	4191	4923	5352	6082	3830
W7	Boise ID	7788	5796	6043	5152	4460	6599	7031	4707	4686	4833	5572	3816
W7	Carson City NV	8107	5921	5913	5492	4776	6520	7039	4348	4796	5191	5931	3765
W7	Cheyenne WY	7844	6262	5588	4787	4019	6085	7593	4944	4082	4602	5410	3299
W7	Olympia WA	7564	5408	6429	5242	4620	6996	6633	4745	5054	4828	5510	4215
W7	Pheonix AZ	8450	6493	5342	5418	4617	5939	7612	4288	4293	5261	6068	3185
W7	Salem OR	7710	5511	6325	5348	4705	6909	6696	4611	5036	4954	5646	4135
W7	Salt Lake City UT	7956	6087	5753	5101	4361	6301	7327	4679	4418	4857	5634	3518
W8	Charleston WV	7443	6907	5003	3805	2937	5306	8491	5822	2947	3872	4760	2672
W8	Columbus OH	7398	6776	5130	3826	2976	5441	8354	5810	3082	3852	4733	2795
W8	Lansing MI	7292	6572	5332	3824	3006	5648	8147	5832	3272	3788	4657	2994
W9	Indianapolis IN	7508	6747	5148	3984	3140	5491	8288	5657	3194	3986	4861	2804
W9	Madison WI	7404	6476	5412	4036	3234	5771	8004	5637	3470	3955	4810	3068
W9	Springfield IL	7607	6688	5196	4148	3315	5572	8189	5501	3340	4120	4987	2851
XE1	MEXICO CITY MEXICO	9108	7733	4105	5387	4463	4689	8825	4244	3303	3543	6434	1950
YA	KABUL AFGANISTAN	1010	2585	10146	4208	5075	9370	3486	9789	7754	3540	2704	9273
ZD8	ASCENSION IS.	5282	8618	4056	3234	3293	3296	9285	8958	2849	4178	4303	4538
ZF	CAYMAN IS.	8378	8183	3711	4397	3448	4076	9656	5309	2146	4759	5657	1372
ZL1	AUCKLAND NEWZEALAND	8579	6464	6004	12183	11255	6495	4982	2564	8732	11387	10991	7120
ZP	ASUNCION PARAGUAY	8440	11365	964	5418	4811	655	11675	5802	2123	6343	6908	2230
ZS1	CAPE TOWN SO.AFRICA	4743	8036	4937	5306	5737	4149	7480	8793	5315	5996	5613	6603
4X	ISRAEL	1619	4447	8172	2429	3375	7461	5502	11362	5815	2147	1287	7523
5R	TANANARIVE MALAGASY	3006	5991	6996	5404	6201	6218	5493	9669	6973	5643	4911	8551
5W	WESTERN SAMOA	9089	5922	6296	10459	9765	7023	4996	1527	8282	9833	10061	6440
6W	DAKAR SENEGAL	4940	7633	4825	1735	1705	4256	8943	9228	2461	2718	3079	4294
9N	KATMANDU NEPAL	1695	1952	10953	5261	6110	10121	2457	8908	8814	4548	3744	10247
9Q	KINSHASA ZAIRE	3479	6990	5817	3351	3963	4996	7336	10587	4705	3958	3583	6475
9Y	TRINIDAD IS.	7739	8977	3064	3753	2873	3111	10722	6302	690	4436	5244	1397

		UA9 (PMAS)	UAØ (YKSK)	UAØ (OKSK)	UAØ (U.OK)	UL7	V3	V85	VE1	VE2	VE3	VE4	VE5
A4	MUSCAT OMAN	2392	4196	4662	4109	1737	8818	3959	6483	6657	6892	7153	722
BY	PEKING CHINA	2879	1638	1789	2278	2020	8159	2411	6589	6421	6430	5902	571
CE	SANTIAGO CHILE	9467	10247	10053	9677	10569	3701	10431	5408	5539	5505	5976	614
CT	LISBON PORTUGAL	3130	5095	5471	4359	4294	4887	7759	2788	3115	3409	4164	443
CU	SAO MIGUEL IS AZORES	3796	5386	5676	4613	5067	3940	8573	1953	2319	2613	3466	377
CX	MONTEVIDEO URUGUAY	9014	10519	10508	9798	9891	4178	10291	5507	5713	5739	6361	658
DU	MANILA PHILIPPINES	4443	3314	3301	4021	3252	9450	779	8325	8119	8084	7408	715
FO	PAPEETE POLYNESIA	9344	6996	6498	7378	9202	4798	6700	6845	6543	6255	5625	541
FY	CAYENNE FR. GUIANA	6590	7790	7898	7046	7847	2572	11314	2826	3103	3211	4046	434
G	LONDON ENGLAND	2221	4114	4502	3388	3482	5190	6989	2872	3098	3365	3912	412
HA	BUDAPEST HUNGARY	1686	3953	4407	3330	2762	6091	6233	3769	3979	4238	4708	484
HC	QUITO ECUADOR	7744	7905	7761	7351	9098	1391	11440	3234	3281	3201	3633	382
HS	BANKOK THAILAND	3807	3637	3831	4152	2460	10184	1159	8268	8209	8288	7900	773
I	ROME ITALY	2182	4447	4895	3808	3195	5944	6612	3691	3950	4227	4807	502
JA	TOYKO JAPAN	3940	1881	1650	2655	3298	7602	2638	6692	6417	6330	5577	531
JT	ULAN BATOR MONGOLIA	2163	1322	1662	1769	1467	7813	3001	6010	5885	5931	5523	538
KH6	HILO HAWAII	6769	4538	4053	4807	6964	4348	6097	5333	4943	4661	3762	345
KL7	FAIRBANKS ALASKA	3854	2362	2133	2113	4578	4245	6112	3327	2991	2855	2072	182
KL7	JUNEAU ALASKA	4379	2988	2738	2726	5184	3656	6648	3033	2662	2476	1594	130
KP2	VIRGIN ISLANDS	6166	6821	6842	6133	7515	1587	10820	1822	2013	2066	2847	314
LA	OSLO NORWAY	1595	3404	3803	2690	2922	5469	6403	3119	3259	3489	3848	400
LZ	SOFIA BULGARIA	1765	4129	4601	3557	2667	6421	6063	4121	4346	4610	5097	527
P2	MADANG P.N.G.	6502	4733	4469	5507	5411	8694	2236	9181	8815	8614	7680	735
PY	RIO DJANEIRO BRAZIL	7885	9689	9876	8916	8788	4113	10526	4831	5113	5212	6001	658
T5	MOGADISHU SOMALIA	3902	5905	6372	5768	3447	8987	4799	7029	7331	7618	8205	839
TF	REYKJAVIK ICELAND	2467	3580	3858	2805	3819	4426	7161	2093	2196	2419	2807	299
TJ	YAOUNDE CAMEROON	4443	6853	7342	6332	4816	6767	7114	5296	5675	5966	6826	712
UA3	MOSCOW EUR. RUSSIA	717	3027	3499	2471	1933	6471	5438	4128	4240	4451	4685	478
UA9	NOVOSIBIRSK RUSSIA	1027	1708	2198	1611	857	7383	3890	5265	5229	5344	5175	512
UA9	PERM ASIATIC RUSSIA	0	2409	2899	1967	1353	6832	4807	4549	4595	4767	4828	486
UAØ	YAKUTSK AS RUSSIA	2409	0	497	775	2461	6547	4022	5012	4811	4801	4271	410
UAØ	OKHOTSK AS RUSSIA	2899	497	0	1114	2925	6376	4051	5076	4832	4783	4150	394
UAØ	UST OLENEK AS RUSSIA	1967	775	1114	0	2465	6054	4690	4312	4146	4172	3766	365
UL7	ALMA-ATA KAZAKH	1353	2461	2925	2465	0	8125	3506	5891	5911	6060	5984	595
V3	BELIZE BELIZE	6832	6547	6376	6054	8125	0	10213	2351	2241	2067	2290	244
V85	BRUNEI	4807	4022	4051	4690	3506	10213	0	8998	8830	8820	8180	793

HS	I	JA	JT	KH6	KL7 (FBAK)	KL7 (JNAK)	KP2	LA	LZ	P2	PY	T5	TF	TJ	UA3	UA9 (NVSK)
9003	4686	6913	6639	4683	3406	2938	1503	4094	5121	8914	4657	7980	3042	6066	5087	6057
8875	4560	6836	6517	4726	3332	2884	1564	3955	4990	8912	4725	7871	2901	5997	4946	5921
9356	5170	7044	6973	4417	3559	3029	1523	4582	5611	8749	4569	8428	3525	6422	5570	6471
9152	5766	6545	6816	3622	3192	2595	2266	5010	6156	7974	5140	9124	3937	7211	5933	6531
9273	5475	6793	6899	4010	3360	2792	1897	4798	5891	8353	4856	8789	3728	6827	5756	6506
9167	5347	6733	6788	4057	3278	2722	1890	4660	5759	8380	4901	8675	3590	6748	5618	6377
8974	5325	6527	6599	3926	3073	2516	2071	4584	5717	8220	5104	8689	3510	6836	5522	6217
8826	5505	6291	6475	3628	2880	2299	2364	4695	5871	7918	5361	8897	3625	7115	5600	6172
8626	5846	5964	6340	3151	2672	2057	2814	4948	6174	7450	5729	9256	3894	7577	5795	6169
8258	6335	5473	6118	2452	2459	1834	3491	5324	6601	6759	6287	9715	4315	8245	6069	6145
7910	6245	5134	5778	2313	2134	1515	3742	5177	6471	6534	6610	9565	4203	8313	5865	5840
7835	5727	5175	5583	2775	1901	1276	3479	4673	5962	6865	6496	9067	3688	7821	5397	5515
7945	6085	5200	5766	2485	2098	1471	3607	5030	6321	6682	6520	9423	4047	8138	5740	5779
8301	5493	5733	5976	3278	2341	1749	2875	4554	5798	7459	5914	8900	3510	7372	5383	5762
7445	5712	4776	5219	2609	1540	912	3843	4594	5896	6544	6889	8958	3664	7955	5241	5214
8522	6151	5779	6310	2797	2628	2001	3131	5205	6456	7118	5956	9559	4167	7937	6010	6242
7543	5841	4843	5338	2519	1667	1040	3828	4732	6033	6527	6840	9102	3793	8052	5386	5353
8128	5752	5472	5859	2915	2182	1562	3217	4753	6024	7097	6205	9137	3735	7720	5528	5749
8823	4709	6677	6448	4499	3169	2698	1743	4044	5122	8691	4896	8051	2979	6215	5019	5909
8704	4700	6539	6325	4413	3031	2561	1879	3996	5100	8573	5033	8065	2926	6280	4959	5806
8507	4652	6353	6127	4320	2828	2368	2075	3892	5031	8418	5233	8041	2819	6340	4837	5628
8703	4842	6458	6320	4248	2953	2461	1985	4107	5232	8428	5125	8217	3034	6446	5057	5843
8448	4829	6174	6065	4077	2669	2179	2264	4017	5189	8186	5409	8232	2945	6576	4936	5630
8670	4985	6347	6288	4064	2851	2337	2128	4212	5361	8255	5249	8372	3138	6629	5144	5860
9769	6364	7016	7501	3616	3831	3212	2294	5705	6793	7966	4762	9597	4634	7458	6654	7274
2418	3053	3887	2135	7682	5274	5868	7657	3106	2503	5662	8335	2719	4114	4248	2085	1560
7912	3804	9856	7994	9763	7940	7801	3918	4866	4179	10701	2253	4078	5001	1876	5260	6901
10125	5479	7797	7748	4801	4336	3788	1080	5094	5972	9169	3850	8467	4082	6255	6110	7182
5947	11415	5489	7182	4376	7307	7188	8671	10687	10867	2849	7618	8421	10425	9912	10059	8315
10791	6441	11184	10659	7238	7800	7245	3045	7022	6950	9798	911	7078	6478	5033	7841	9569
6296	5237	9143	7836	11376	10208	10328	6493	6486	5291	8062	3774	3034	7105	2639	6284	7221
4344	1330	5700	3835	8761	5695	6115	5985	2161	904	7670	6367	2232	3186	2462	1603	2806
4241	4751	7075	5882	10949	9173	9704	7984	5820	4523	6622	5774	1456	6789	2914	5183	5512
6289	10517	4682	6547	2629	5622	5452	7681	9287	10277	2938	8107	9741	8720	11654	9175	7624
7752	2594	8644	6924	8694	6361	6278	3085	3432	3089	11113	3123	4360	3422	2104	4044	5785
1370	4110	3198	1809	7230	5430	6057	8644	4046	3561	4618	9226	3176	4978	5103	3031	1891
5962	3192	8292	6512	11178	8165	8564	5608	4441	3284	8947	4106	2120	5091	621	4335	5597
10349	4913	8996	8360	6238	5495	5029	603	4986	5460	10568	2563	7298	4195	4979	5973	7471

VE6	VE7	VE8	VO2	VY1	VK1	VK4	VK6	VP8	VP9	VQ9	WØ (CO)	WØ (IA)	WØ (KS)	WØ (MN)	WØ (MO)	WØ (NE)
7094	7037	6487	6008	7165	7210	5385	8498	7062		2336	7941	7600	7806	7384	7748	7718
5327	4824	4816	5988	4410	5598	5219	4952	10887	7439	4296	6343	6495	6613	6286	6717	6479
6527	7015	7054	6019	7435	7026	7335	7899	2044	4556	8718	5505	5379	5252	5594	5157	5387
4638	5124	4332	2598	4926	11208	11145	9349	7475	3111	6096	4850	4275	4456	4152	4308	4441
4046	4624	3869	1921	4526	12131	11703	10292	7128	2171	7036	4070	3471	3636	3382	3472	3641
6993	7556	7452	6090	7932	7314	7717	7798	1878	4670	7997	6013	5783	5688	5976	5564	5821
6734	6118	6327	7737	5791	3910	3617	3222	9155	9164	3644	7653	7964	8033	7788	8177	7904
5343	5081	5832	7121	5451	3941	3710	5846	5172	6540	9092	4868	5401	5208	5555	5342	5241
4779	5455	5049	3372	5675	9921	10141	10425	4632	2055	8647	4052	3584	3585	3701	3406	3693
4219	4580	3791	2471	4294	10548	10267	8996	8459	3446	5894	4682	4183	4385	4008	4275	4334
4918	5174	4419	3341	4822	9761	9612	8102	8674	4342	5002	5496	5031	5236	4844	5139	5175
4218	4769	4718	3841	5151	8499	8506	10011	4387	2424	10372	3226	3039	2923	3250	2817	3057
7362	6865	6829	7657	6452	4651	4534	3318	8943	9094	2408	8383	8497	8631	8278	8719	8497
5106	5435	4656	3341	5119	10065	10004	8285	8176	4167	5084	5572	5061	5260	4893	5142	5215
4885	4260	4516	6164	3946	4941	4448	4904	10446	7471	5264	5793	6117	6177	5952	6326	6050
5032	4621	4468	5399	4171	6310	5940	5535	11304	6862	4360	6065	6120	6264	5897	6341	6131
3201	2514	3514	5308	2987	5271	4755	6855	7620	5510	9130	3251	3857	3739	3892	3927	3693
1415	942	1016	2930	491	7697	7106	8355	9859	4015	8011	2421	2631	2719	2451	2847	2586
872	315	734	2744	167	7801	7217	8742	9231	3625	8637	1822	2103	2164	1951	2310	2035
3575	4246	3881	2427	4484	10084	9988	11489	5648	973	9425	2849	2372	2375	2494	2195	2482
4008	4252	3498	2608	3904	9926	9581	8613	9174	3817	5736	4644	4224	4431	4021	4356	4357
5306	5550	4801	3713	5188	9528	9454	7792	8516	4656	4645	5884	5412	5616	5227	5515	5558
6947	6265	6841	8792	6163	2089	1611	2649	7608	9714	5012	7465	8017	7961	7961	8152	7877
6714	7365	7041	5357	7640	8328	8783	8417	2797	4062	7625	5864	5486	5446	5636	5286	5569
8394	8522	7845	6744	8092	7074	7369	5174	6790	7296	1994	8981	8466	8662	8304	8532	8623
3058	3417	2625	1548	3150	10401	9886	9563	8913	2857	6809	3597	3155	3362	2957	3284	3292
7359	7853	7269	7627		8962	9422	7212	5814	5270	4284	7415	6809	6956	6738	6778	6978
4704	4798	4128	3592	4381	8996	8744	7590	9616	4840	4803	5479	5127	5333	4910	5280	5244
4864	4638	4250	4656	4165	7364	7039	6318	10986	6092	4365	5860	5745	5924	5512	5951	5802
4704	4664	4095	3969	4212	8342	8051	7097	10254	5234	4614	5593	5336	5534	5106	5515	5429
3731	3299	3192	4431	2852	6821	6327	6536	12158	5849	5676	4762	4867	4995	4654	5089	4860
3543	3037	3058	4527	2621	6553	6028	6494	11667	5887	6005	4554	4734	4840	4535	4954	4705
3335	3040	2742	3716	2568	7596	7098	7230	11676	5161	5927	4362	4362	4514	4136	4581	4383
5710	5496	5094	5301	5023	7065	6852	5743	10231	6677	3519	6692	6538	6724	6304	6735	6606
2833	3378	3355	2881	3761	8729	8493	10616	5724	1780	10963	1834	1692	1553	1917	1473	1687
7512	6898	7095	8393	6569	3559	3391	2543	8477	9849	3040	8430	8741	8812	8562	8955	8683

		UA9 (PMAS)	UAØ (YKSK)	UAØ (OKSK)	UAØ (U.OK)	UL7	V3	V85	VE1	VE2	VE3	VE4	VE5
VE1	Halifax N.S.	4549	5012	5076	4312	5891	2351	8998	0	395	672	1598	1928
VE2	Quebec City Que.	4595	4811	4832	4146	5911	2241	8830	395	0	295	1202	1533
VE3	Ottawa Ont.	4767	4801	4783	4172	6060	2067	8820	672	295	0	949	1283
VE4	Winnipeg Mb.	4828	4271	4150	3766	5984	2290	8180	1598	1202	949	0	333
VE5	Regina Sk.	4864	4101	3941	3651	5955	2447	7936	1928	1533	1283	333	0
VE6	Edmondton Al.	4704	3731	3543	3335	5710	2833	7512	2284	1894	1669	740	432
VE7	Prince Rupert B.C.	4664	3299	3037	3040	5496	3378	6898	2948	2563	2349	1425	1108
VE8	Yellowknife NWT	4095	3192	3058	2742	5094	3355	7095	2340	1988	1839	1091	913
VO2	Goose Bay Labrador	3969	4431	4527	3716	5301	2881	8393	613	650	896	1569	1860
VY1	Whitehorse Yukon	4212	2852	2621	2568	5023	3761	6569	3017	2656	2486	1638	1367
VK1	CANBERRA AUSTRALIA	8342	6821	6553	7596	7065	8729	3559	10594	10211	9923	9031	8713
VK4	BRISBANE AUSTRALIA	8051	6327	6028	7098	6852	8493	3391	10087	9694	9416	8496	8172
VK6	PERTH AUSTRALIA	7097	6536	6494	7230	5743	10616	2543	11542	11321	11209	10315	9990
VP8	SO. SHETLAND IS.	10254	12158	11667	11676	10231	5724	8477	7368	7545	7534	8011	8163
VP9	BERMUDA	5324	5849	5887	5161	6677	1780	9849	852	1056	1160	2050	2374
VQ9	DIEGO GARCIA CHAGOS	4614	5676	6005	5927	3519	10963	3040	8765	8977	9223	9442	9460
WØ	Denver CO	5593	4762	4554	4362	6692	1834	8430	2120	1751	1456	797	737
WØ	DesMoines IA	5336	4867	4734	4362	6538	1692	8741	1518	1161	866	597	805
WØ	Topeka KS.	5534	4995	4840	4514	6724	1553	8812	1683	1341	1049	752	898
WØ	Minneapolis MN	5106	4654	4535	4136	6304	1917	8562	1444	1063	772	385	648
WØ	Jefferson City MO	5515	5089	4954	4581	6735	1473	8955	1524	1200	915	819	1020
WØ	Lincoln NE	5429	4860	4705	4383	6606	1687	8683	1688	1329	1034	626	764
WØ	Bismark ND	5073	4401	4246	3936	6204	2145	8239	1780	1388	1111	269	305
WØ	Pierre SD	5233	4564	4402	4105	6372	1983	8378	1793	1408	1120	409	463
W1	Augusta ME	4715	4993	5018	4323	6042	2137	9010	301	187	390	1337	1671
W1	Boston MA	4874	5122	5133	4461	6200	1981	9184	411	314	405	1345	1678
W1	Concord NH	4826	5056	5067	4397	6150	2021	9077	406	252	350	1295	1629
W1	Hartford CT	4940	5141	5142	4490	6262	1907	9164	497	356	378	1297	1629
W1	Montpelier VT	4778	4967	4974	4312	6095	2060	8989	442	183	260	1210	1543
W1	Providence RI	4906	5148	5157	4489	6232	1949	9169	436	343	416	1349	1681
W2	Albany NY	4903	5062	5057	4416	6219	1934	9084	526	308	287	1209	1541
W2	Trenton NJ	5083	5225	5208	4587	6400	1759	9245	645	488	425	1272	1598
W3	Dover DE	5187	5276	5245	4651	6500	1650	9291	762	591	479	1250	1569
W3	Harrisburg PA	5125	5199	5169	4575	6434	1709	9214	738	532	403	1187	1509
W3	Washington D.C.	5218	5293	5258	4672	6530	1617	9305	797	623	500	1246	1562
W4	Atlanta GA	5695	5534	5436	4974	6980	1145	9468	1336	1134	928	1288	1539
W4	Columbia SC	5611	5567	5492	4980	6914	1221	9539	1180	1021	855	1368	1643
W4	Frankfort KY	5419	5233	5144	4668	6691	1439	9186	1186	915	669	1009	1283
W4	Montgomery AL	5821	5595	5480	5053	7097	1033	9493	1481	1274	1058	1330	1557
W4	Nashville TN	5584	5336	5229	4791	6848	1290	9255	1346	1087	842	1079	1324
W4	Raleigh NC	5449	5480	5426	4874	6760	1387	9478	1000	854	716	1345	1641
W4	Richmond VA	5308	5371	5330	4755	6620	1528	9380	870	713	586	1289	1599
W4	Tallahassee FL	5910	5758	5651	5203	7203	923	9671	1491	1329	1144	1501	1735
W5	Austin TX	6150	5523	5325	5088	7338	1064	9173	2095	1816	1545	1355	1437
W5	Baton Rouge LA	6035	5637	5483	5141	7280	913	9431	1785	1550	1309	1376	1543
W5	Jackson MS	5898	5536	5394	5027	7147	1027	9370	1659	1415	1171	1267	1453
W5	Little Rock AR	5772	5339	5191	4842	7000	1213	9163	1659	1374	1107	1076	1250
W5	Oklahoma City OK	5799	5190	5011	4737	6979	1365	8929	1886	1566	1280	996	1092
W5	Santa Fe NM	5913	5028	4796	4655	7006	1638	8587	2310	1962	1668	1109	1055
W6	Los Angeles CA	6067	4798	4493	4556	7002	2177	8040	2952	2582	2288	1531	1324
W6	San Francisco CA	5816	4459	4147	4241	6698	2496	7716	3026	2640	2353	1500	1235
W7	Boise ID	5397	4263	4010	3950	6368	2432	7806	2560	2168	1890	995	715
W7	Carson City NV	5715	4444	4152	4191	6636	2411	7802	2852	2466	2180	1327	1068
W7	Cheyenne WY	5494	4675	4474	4268	6594	1908	8371	2075	1700	1407	706	638
W7	Olympia WA	5175	3897	3626	3626	6071	2831	7409	2785	2390	2131	1187	859
W7	Pheonix AZ	6073	4989	4716	4684	7093	1842	8370	2663	2308	2013	1360	1225
W7	Salem OR	5322	4016	3734	3761	6211	2759	7468	2845	2450	2183	1251	932
W7	Salt Lake City UT	5575	4543	4300	4203	6596	2134	8101	2427	2045	1755	949	751
W8	Charleston WV	5344	5269	5202	4680	6635	1491	9253	1024	786	577	1105	1401
W8	Columbus OH	5264	5138	5067	4556	6543	1583	9117	1038	754	507	968	1267
W8	Lansing MI	5115	4933	4860	4357	6378	1755	8910	1052	714	421	773	1084
W9	Indianapolis IN	5337	5109	5017	4550	6598	1540	9058	1196	892	622	881	1159
W9	Madison WI	5174	4840	4738	4297	6404	1766	8775	1282	922	626	596	879
W9	Springfield IL	5396	5054	4939	4522	6632	1538	8964	1365	1041	757	789	1034
XE1	MEXICO CITY MEXICO	6896	6189	5948	5802	8087	731	9559	2675	2458	2216	2103	2157
YA	KABUL AFGANISTAN	1719	3193	3654	3169	732	8532	3561	6206	6290	6477	6532	6542
ZD8	ASCENSION IS.	5944	8243	8668	7556	6598	5417	8855	4801	5191	5428	6377	6711
ZF	CAYMAN IS.	6533	6546	6438	5979	7864	519	10435	1988	1957	1843	2288	2509
ZL1	AUCKLAND NEWZEALAND	9343	7268	6886	8000	8234	7314	4803	9395	9068	8774	8023	7752
ZP	ASUNCION PARAGUAY	8549	9864	9868	9150	9604	3609	10932	4852	5059	5089	5737	5972
ZS1	CAPE TOWN SO.AFRICA	6704	8925	9403	8669	6479	7815	6779	7457	7851	8099	9048	9381
4X	ISRAEL	2035	4421	4917	3995	2429	7274	5503	5008	5245	5512	5993	6161
5R	TANANARIVE MALAGASY	5328	7113	7533	7112	4680	9535	4860	8154	8509	8804	9541	9771
5W	WESTERN SAMOA	8536	6148	5676	6734	7940	6119	5173	7812	7446	7152	6326	6036
6W	DAKAR SENEGAL	4762	6830	7199	6090	5747	4669	8899	3391	3786	4055	4988	5316
9N	KATMANDU NEPAL	2511	3107	3472	3364	1163	9268	2501	7055	7071	7211	7061	6992
9Q	KINSHASA ZAIRE	4860	7250	7747	6799	5045	7188	6903	5881	6265	6553	7430	7734
9Y	TRINIDAD IS.	6576	7408	7442	6706	7904	1891	11355	2397	2611	2670	3439	3727

	VE6	VE7	VE8	VO2	VY1	VK1	VK4	VK6	VP8	VP9	VQ9	WØ (CO)	WØ (IA)	WØ (KS)	WØ (MN)	WØ (MO)	WØ (NE)
	2284	2948	2340	613	3017	10594	10087	11542	7368	852	8765	2120	1518	1683	1444	1524	1688
	1894	2563	1988	650	2656	10211	9694	11321	7545	1056	8977	1751	1161	1341	1063	1200	1329
	1669	2349	1839	896	2486	9923	9416	11209	7534	1160	9223	1456	866	1049	772	915	1034
	740	1425	1091	1569	1638	9031	8496	10315	8011	2050	9442	797	597	752	385	819	626
	432	1108	913	1860	1367	8713	8172	9990	8163	2374	9460	737	805	898	648	1020	764
	0	685	617	2125	943	8412	7852	9575	8517	2789	9229	1033	1232	1305	1080	1442	1173
	685	0	796	2725	472	7805	7231	8890	8917	3475	8953	1545	1882	1921	1753	2079	1796
	617	796	0	2014	683	8535	7951	9369	9076	3000	8613	1621	1679	1802	1476	1901	1668
	2125	2725	2014	0	2695	10530	9956	10935	7961	1463	8331	2263	1721	1919	1568	1804	1879
	943	472	683	2695	0	7864	7277	8725	9373	3643	8486	1932	2173	2248	2007	2385	2116
	8412	7805	8535	10530	7864	0	590	1904	5531	10481	5155	8475	9081	8928	9150	9100	8912
	7852	7231	7951	9956	7277	590	0	2229	6002	10146	5395	7995	8603	8467	8648	8648	8436
	9575	8890	9369	10935	8725	1904	2229	0	5942	12363	3256	10081	10658	10578	10610	10770	10509
	8517	8917	9076	7961	9373	5531	6002	5942	0	6521	6974	7484	7413	7276	7632	7192	7410
	2789	3475	3000	1463	3643	10481	10146	12363	6521	0	9207	2283	1705	1785	1753	1593	1854
	9229	8953	8613	8331	8486	5155	5395	3256	6974	9207	0	10197	9928	10133	9706	10083	10036
	1033	1545	1621	2263	1932	8475	7995	10081	7484	2283	10197	0	608	498	697	690	441
	1232	1882	1679	1721	2173	9081	8603	10658	7413	1705	9928	608	0	207	234	222	169
	1305	1921	1802	1919	2248	8928	8467	10578	7276	1785	10133	498	207	0	427	192	134
	1080	1753	1476	1568	2007	9150	8648	10610	7632	1753	9706	697	234	427	0	445	335
	1442	2079	1901	1804	2385	9100	8648	10770	7192	1593	10083	690	222	192	445	0	287
	1173	1796	1668	1879	2116	8912	8436	10509	7410	1854	10036	441	169	134	335	287	0
	727	1383	1210	1812	1669	8826	8306	10228	7866	2135	9686	532	504	594	382	716	459
	867	1497	1375	1882	1810	8803	8298	10286	7701	2077	9847	398	389	438	350	581	306
	2048	2723	2167	746	2831	10298	9802	11509	7366	869	9028	1822	1218	1382	1153	1223	1388
	2072	2754	2237	905	2889	10230	9756	11604	7231	768	9170	1761	1153	1301	1118	1132	1321
	2019	2700	2174	860	2829	10210	9728	11540	7296	832	9147	1737	1129	1285	1082	1121	1298
	2030	2714	2218	974	2863	10149	9681	11587	7206	782	9254	1686	1077	1221	1054	1049	1245
	1929	2608	2078	824	2733	10156	9662	11443	7382	929	9138	1680	1076	1241	1013	1085	1245
	2079	2761	2252	937	2902	10215	9746	11621	7204	750	9198	1751	1142	1287	1114	1115	1310
	1940	2624	2127	949	2772	10092	9614	11496	7282	873	9256	1621	1013	1165	976	1000	1181
	2012	2697	2243	1122	2873	10031	9580	11588	7114	777	9406	1591	986	1112	992	933	1149
	1990	2674	2255	1232	2870	9930	9490	11555	7056	808	9523	1512	913	1025	944	841	1072
	1927	2612	2182	1179	2801	9923	9471	11499	7133	865	9487	1483	878	1003	893	824	1041
	1986	2669	2260	1266	2871	9900	9463	11543	7037	820	9558	1490	894	1001	933	816	1051
	1969	2619	2378	1784	2906	9418	9025	11229	6766	1138	10097	1208	738	725	905	545	833
	2076	2740	2445	1666	3003	9608	9219	11422	6745	945	9945	1377	862	884	997	695	979
	1716	2383	2091	1558	2642	9472	9034	11161	7070	1202	9889	1080	517	583	636	393	654
	1981	2615	2421	1924	2924	9284	8900	11116	6697	1253	10242	1157	752	698	945	539	819
	1753	2404	2170	1730	2691	9336	8914	11082	6958	1282	10062	1018	523	524	694	337	623
	2072	2748	2399	1495	2983	9769	9364	11536	6841	828	9766	1458	899	959	994	767	1036
	2026	2707	2320	1354	2921	9858	9434	11557	6956	806	9635	1483	899	988	960	798	1049
	2160	2795	2592	1977	3102	9347	8988	11217	6539	1157	10253	1330	929	877	1116	718	999
	1791	2314	2350	2442	2704	8604	8211	10429	6731	1941	10737	772	814	616	1043	654	731
	1944	2534	2445	2196	2884	8968	8590	10814	6637	1557	10528	1007	779	644	1007	562	779
	1864	2473	2346	2060	2807	9069	8677	10890	6744	1479	10392	972	669	559	890	447	691
	1659	2268	2148	2004	2601	9013	8594	10780	6936	1587	10319	779	480	353	710	266	486
	1467	2034	2006	2165	2398	8751	8315	10484	7070	1882	10400	503	473	267	693	365	372
	1320	1762	1922	2516	2180	8303	7851	10007	7209	2359	10515	320	802	627	952	798	637
	1369	1519	1972	3073	1989	7645	7172	9307	7409	3058	10470	831	1436	1295	1522	1480	1267
	1171	1202	1739	3070	1675	7570	7068	9133	7744	3230	10135	948	1546	1445	1581	1636	1385
	696	972	1302	2563	1412	8046	7526	9505	7961	2857	9876	637	1152	1106	1133	1293	1012
	1036	1156	1624	2897	1624	7744	7240	9293	7766	3070	10109	787	1379	1285	1408	1478	1220
	944	1478	1526	2195	1854	8519	8029	10087	7573	2276	10098	99	580	503	638	693	422
	610	598	1113	2699	1061	7857	7313	9193	8319	3170	9548	1025	1483	1469	1418	1648	1361
	1386	1689	2004	2840	2142	7945	7490	9651	7231	2715	10607	579	1147	983	1271	1158	978
	735	724	1258	2789	1194	7782	7246	9173	8196	3190	9678	986	1489	1454	1448	1639	1356
	886	1263	1504	2503	1691	8167	7669	9719	7687	2644	10114	369	948	864	983	1055	791
	1832	2508	2165	1436	2744	9651	9215	11331	7045	1032	9761	1255	674	760	755	571	822
	1697	2375	2028	1395	2606	9599	9145	11220	7173	1143	9726	1161	564	678	623	499	722
	1509	2192	1822	1317	2409	9550	9070	11067	7376	1297	9628	1076	468	633	457	490	638
	1591	2261	1963	1519	2516	9433	8979	11069	7191	1294	9843	998	411	511	510	333	561
	1310	1985	1681	1488	2231	9314	8829	10835	7455	1528	9740	838	239	430	231	344	407
	1465	2122	1880	1649	2401	9254	8795	10890	7236	1473	9944	814	244	327	404	159	380
	2473	2910	3060	3105	3340	8180	7875	10082	6048	2300	11435	1441	1561	1366	1792	1383	1481
	6335	6176	5717	5652	5704	7057	6951	5508	9499	6924	2918	7278	7053	7249	6822	7234	7141
	7086	7747	7073	5059	7754	9216	9807	8095	4372	4395	5847	6670	6100	6172	6133	5980	6248
	2927	3537	3380	2564	3870	9230	9010	11089	5755	1316	10448	2014	1707	1623	1903	1489	1754
	7561	7088	7883	9588	7301	1426	1413	3305	4986	9086	6561	7334	7908	7726	8032	7875	7741
	6390	6980	6825	5437	7333	7885	8244	8453	2526	4014	8384	5436	5168	5085	5354	4952	5217
	9735	10358	9602	7634	10230	6689	7260	5422	3769	7068	3888	9324	8772	8830	8811	8639	8915
	6164	6350	5632	4616	5955	8779	8824	6957	8256	5496	3759	6783	6317	6521	6131	6419	6462
	9828	9976	9298	7997	9538	6082	6511	4300	5571	8183	1850	10254	9669	9837	9556	9670	9838
	5822	5342	6137	7894	5571	2799	2407	4635	6056	7740	7747	5695	6294	6133	6384	6302	6124
	5645	6268	5544	3543	6213	10774	11354	9282	5741	3204	6350	5443	4837	4956	4813	4769	5001
	6698	6373	6096	6465	5908	6021	5893	4588	9551	7834	2580	7717	7636	7812	7402	7842	7687
	7977	8474	7679	5874	8242	8377	8865	6689	5430	5795	3942	7992	7383	7521	7325	7339	7551
	4159	4824	4482	2990	5078	9848	9901	10909	4972	1560	9262	3383	2942	2927	3079	2754	3041

		WØ (ND)	WØ (SD)	W1 (ME)	W1 (MA)	W1 (NH)	W1 (CT)	W1 (VT)	W1 (RI)	W2 (NY)	W2 (NJ)	W3 (DE)	W3 (PA)
A4	MUSCAT OMAN	7412	7562	6725	6874	6845	6955	6825	6905	6947	7108	7224	7179
BY	PEKING CHINA	6019	6178	6600	6735	6671	6761	6584	6763	6684	6851	6908	6830
CE	SANTIAGO CHILE	5846	5684	5367	5225	5288	5192	5369	5196	5263	5090	5025	5102
CT	LISBON PORTUGAL	4410	4477	3086	3194	3194	3280	3223	3216	3315	3419	3531	3520
CU	SAO MIGUEL IS AZORES	3687	3725	2254	2340	2351	2423	2396	2358	2472	2551	2658	2657
CX	MONTEVIDEO URUGUAY	6277	6127	5529	5404	5470	5389	5561	5379	5470	5314	5270	5347
DU	MANILA PHILIPPINES	7461	7599	8304	8426	8361	8439	8269	8451	8356	8509	8547	8470
FO	PAPEETE POLYNESIA	5360	5261	6552	6436	6440	6349	6419	6413	6320	6207	6093	6109
FY	CAYENNE FR. GUIANA	4073	3973	2916	2823	2888	2834	2984	2805	2924	2806	2805	2875
G	LONDON ENGLAND	4179	4285	3134	3275	3253	3359	3249	3304	3365	3511	3629	3594
HA	BUDAPEST HUNGARY	4977	5098	4025	4169	4145	4252	4136	4198	4254	4405	4522	4484
HC	QUITO ECUADOR	3517	3360	3126	2972	3030	2924	3101	2941	2985	2805	2725	2800
HS	BANKOK THAILAND	8039	8202	8366	8518	8460	8565	8388	8549	8504	8683	8763	8689
I	ROME ITALY	5073	5177	3969	4101	4086	4187	4092	4128	4201	4336	4453	4425
JA	TOYKO JAPAN	5612	5746	6605	6705	6640	6702	6544	6726	6614	6748	6768	6693
JT	ULAN BATOR MONGOLIA	5677	5843	6055	6200	6137	6235	6057	6229	6166	6341	6410	6334
KH6	HILO HAWAII	3555	3542	5046	5004	4973	4932	4906	4996	4862	4843	4763	4734
KL7	FAIRBANKS ALASKA	2132	2280	3174	3250	3187	3233	3090	3266	3143	3260	3270	3198
KL7	JUNEAU ALASKA	1598	1730	2832	2881	2823	2849	2728	2892	2758	2850	2839	2772
KP2	VIRGIN ISLANDS	2864	2762	1829	1708	1774	1700	1867	1684	1786	1645	1626	1699
LA	OSLO NORWAY	4117	4248	3338	3492	3457	3570	3431	3524	3554	3721	3835	3786
LZ	SOFIA BULGARIA	5366	5485	4384	4525	4504	4609	4499	4553	4615	4761	4878	4844
P2	MADANG P.N.G.	7579	7639	8983	9019	8965	8976	8872	9026	8887	8946	8906	8852
PY	RIO DJANEIRO BRAZIL	5989	5867	4926	4830	4895	4837	4992	4811	4926	4799	4786	4859
T5	MOGADISHU SOMALIA	8474	8584	7323	7438	7435	7525	7456	7461	7555	7666	7779	7765
TF	REYKJAVIK ICELAND	3077	3199	2288	2445	2405	2519	2372	2477	2497	2669	2780	2727
TJ	YAOUNDE CAMEROON	7049	7084	5593	5657	5679	5736	5735	5670	5796	5845	5939	5954
UA3	MOSCOW EUR. RUSSIA	4947	5094	4333	4490	4450	4564	4417	4521	4542	4714	4825	4772
UA9	NOVOSIBIRSK RUSSIA	5385	5554	5376	5532	5476	5585	5411	5563	5532	5713	5804	5734
UA9	PERM ASIATIC RUSSIA	5073	5233	4715	4874	4826	4940	4778	4906	4903	5083	5187	5125
UAØ	YAKUTSK AS RUSSIA	4401	4564	4993	5122	5056	5141	4967	5148	5062	5225	5276	5199
UAØ	OKHOTSK AS RUSSIA	4246	4402	5018	5133	5067	5142	4974	5157	5057	5208	5245	5169
UAØ	UST OLENEK AS RUSSIA	3936	4105	4323	4461	4397	4490	4312	4489	4416	4587	4651	4575
UL7	ALMA-ATA KAZAKH	6204	6372	6042	6200	6150	6262	6095	6232	6219	6400	6500	6434
V3	BELIZE BELIZE	2145	1983	2137	1981	2021	1907	2060	1949	1934	1759	1650	1709
V85	BRUNEI	8239	8378	9010	9142	9077	9164	8989	9169	9084	9245	9291	9214
VE1	Halifax N.S.	1780	1793	301	411	406	497	442	436	526	645	762	738
VE2	Quebec City Que.	1388	1408	187	314	252	356	183	343	308	488	591	532
VE3	Ottawa Ont.	1111	1120	390	405	350	378	260	416	287	425	479	403
VE4	Winnipeg Mb.	269	409	1337	1345	1295	1297	1210	1349	1209	1272	1250	1187
VE5	Regina Sk.	305	463	1671	1678	1629	1629	1543	1681	1541	1598	1569	1509
VE6	Edmonton Al.	727	867	2048	2072	2019	2030	1929	2079	1940	2012	1990	1927
VE7	Prince Rupert B.C.	1383	1497	2723	2754	2700	2714	2608	2761	2624	2697	2674	2612
VE8	Yellowknife NWT	1210	1375	2167	2237	2174	2218	2078	2252	2127	2243	2255	2182
VO2	Goose Bay Labrador	1812	1882	746	905	860	974	824	937	949	1122	1232	1179
VY1	Whitehorse Yukon	1669	1810	2831	2889	2829	2863	2733	2902	2772	2873	2870	2801
VK1	CANBERRA AUSTRALIA	8826	8803	10298	10230	10210	10149	10156	10215	10092	10031	9930	9923
VK4	BRISBANE AUSTRALIA	8306	8298	9802	9756	9728	9681	9662	9746	9614	9580	9490	9471
VK6	PERTH AUSTRALIA	10228	10286	11509	11604	11540	11587	11443	11621	11496	11588	11555	11499
VP8	SO. SHETLAND IS.	7866	7701	7366	7231	7296	7206	7382	7204	7282	7114	7056	7133
VP9	BERMUDA	2135	2077	869	768	832	782	929	750	873	777	808	865
VQ9	DIEGO GARCIA CHAGOS	9686	9847	9028	9170	9147	9254	9138	9198	9256	9406	9523	9487
WØ	Denver CO	532	398	1822	1761	1737	1686	1680	1751	1621	1591	1512	1483
WØ	DesMoines IA	504	389	1218	1153	1129	1077	1076	1142	1013	986	913	878
WØ	Topeka KS	594	438	1382	1301	1285	1221	1241	1287	1165	1112	1025	1003
WØ	Minneapolis MN	382	350	1153	1118	1082	1054	1013	1114	976	992	944	893
WØ	Jefferson City MO	716	581	1223	1132	1121	1049	1085	1115	1000	933	841	824
WØ	Lincoln NE	459	306	1388	1321	1298	1245	1245	1310	1181	1149	1072	1041
WØ	Bismark ND	0	169	1501	1483	1442	1423	1365	1481	1342	1371	1327	1274
WØ	Pierre SD	169	0	1503	1467	1432	1400	1363	1462	1324	1331	1273	1228
W1	Augusta ME	1501	1503	0	158	119	231	142	191	233	383	498	459
W1	Boston MA	1483	1467	158	0	65	86	161	32	142	236	353	327
W1	Concord NH	1442	1432	119	65	0	113	97	92	121	264	379	340
W1	Hartford CT	1423	1400	231	86	113	0	178	66	91	152	269	240
W1	Montpelier VT	1365	1363	142	161	97	178	0	182	125	305	409	355
W1	Providence RI	1481	1462	191	32	92	66	182	0	140	208	325	303
W2	Albany NY	1342	1324	233	142	121	91	125	140	0	181	284	232
W2	Trenton NJ	1371	1331	383	236	264	152	305	208	181	0	117	109
W3	Dover DE	1327	1273	498	353	379	269	409	325	284	117	0	77
W3	Harrisburg PA	1274	1228	459	327	340	240	355	303	232	109	77	0
W3	Washington D.C.	1316	1258	533	388	413	304	442	360	317	152	35	98
W4	Atlanta GA	1242	1122	1069	928	950	843	964	900	843	692	574	611
W4	Columbia SC	1357	1252	932	781	812	701	842	751	717	549	434	488
W4	Frankfort KY	1000	905	894	779	781	693	767	758	661	560	458	455
W4	Montgomery AL	1254	1120	1213	1073	1094	988	1106	1045	986	837	719	755
W4	Nashville TN	1027	908	1059	936	943	850	935	913	826	709	599	609
W4	Raleigh NC	1370	1285	756	603	637	524	673	573	547	373	263	326

W3 (D.C.)	W4 (GA)	W4 (SC)	W4 (KY)	W4 (AL)	W4 (TN)	W4 (NC)	W4 (VA)	W4 (FL)	W5 (TX)	W5 (LA)	W5 (MS)	W5 (AR)	W5 (OK)	W5 (NM)	W6 (L.A.CA)	W6 (S.F.CA)
]7259	7790	7656	7562	7931	7734	7478	7341	7969	8402	8201	8063	7983	8073	8260	8460	8206
6925	7172	7205	6871	7230	6972	7116	7005	7395	7113	7258	7162	6962	6795	6585	6250	5902
5005	4722	4703	5026	4655	4916	4804	4922	4496	4737	4609	4710	4907	5062	5257	5583	5932
3565	4087	3907	3973	4229	4129	3735	3624	4195	4878	4544	4428	4446	4672	5077	5669	5665
2690	3192	3006	3112	3331	3257	2839	2740	3281	3997	3647	3539	3579	3828	4263	4901	4951
5256	5074	5015	5364	5034	5282	5083	5182	4861	5242	5049	5130	5335	5535	5801	6216	6563
8559	8694	8772	8416	8716	8480	8720	8629	8894	8401	8652	8590	8384	8150	7816	7291	6961
6060	5535	5714	5658	5393	5499	5887	5999	5430	4750	5078	5195	5186	4976	4605	4096	4191
2804	2861	2721	3073	2896	3070	2707	2756	2722	3427	3077	3085	3273	3561	3996	4651	4923
3664	4203	4053	4013	4348	4183	3873	3741	4363	4911	4641	4509	4471	4635	4960	5440	5352
4557	5095	4949	4895	5239	5067	4769	4636	5260	5783	5528	5394	5345	5493	5789	6211	6082
2701	2379	2372	2684	2310	2571	2487	2614	2152	2455	2278	2372	2572	2753	3019	3492	3830
8788	9126	9123	8820	9205	8943	9007	8875	9356	9152	9273	9167	8974	8826	8626	8258	7910
4488	5027	4865	4859	5172	5027	4686	4560	5170	5766	5475	5347	5325	5505	5846	6335	6245
6775	6852	6946	6586	6865	6637	6913	6836	7044	6545	6793	6723	6527	6291	5964	5473	5134
6432	6744	6749	6438	6822	6559	6639	6517	6973	6816	6899	6788	6599	6475	6340	6118	5778
3274	3356	3440	3081	3386	3143	3406	3332	3559	3192	3360	3278	3073	2880	2672	2459	2134
2838	2841	2947	2587	2850	2625	2938	2884	3029	2595	2792	2722	2516	2299	2057	1834	1515
1620	1650	1509	1863	1692	1858	1503	1564	1523	2266	1897	1890	2071	2364	2814	3491	3742
3869	4393	4270	4159	4533	4331	4094	3955	4582	5010	4798	4660	4584	4695	4948	5324	5177
4914	5453	5301	5262	5598	5433	5121	4990	5611	6156	5891	5759	5717	5871	6174	6601	6471
8894	8669	8843	8531	8595	8483	8914	8912	8749	7974	8353	8380	8220	7918	7450	6759	6534
4782	4747	4639	4999	4749	4963	4657	4725	4569	5140	4856	4901	5104	5361	5729	6287	6610
7812	8333	8150	8216	8474	8375	7980	7871	8428	9124	8789	8675	8689	8897	9256	9715	9565
2814	3330	3214	3088	3467	3261	3042	2901	3525	3937	3728	3590	3510	3625	3894	4315	4203
5966	6402	6208	6396	6523	6517	6066	5997	6422	7211	6827	6748	6836	7115	7577	8245	8313
4859	5371	5259	5120	5507	5291	5087	4946	5570	5933	5756	5618	5522	5600	5795	6069	5865
5831	6243	6199	5945	6350	6094	6057	5921	6471	6531	6506	6377	6217	6172	6169	6145	5840
5218	5695	5611	5419	5821	5584	5449	5308	5910	6150	6035	5898	5772	5799	5913	6067	6174
5293	5534	5567	5233	5595	5336	5480	5371	5758	5523	5637	5536	5339	5190	5028	4798	4459
5258	5436	5492	5144	5480	5229	5426	5330	5651	5325	5483	5394	5191	5011	4796	4493	4147
4672	4974	4940	4668	5053	4791	4874	4755	5203	5088	5141	5027	4842	4737	4655	4556	4241
6530	6980	6914	6691	7097	6848	6760	6620	7203	7338	7280	7147	7000	6979	7006	7002	6698
1617	1145	1221	1439	1033	1290	1387	1528	923	1064	913	1027	1213	1365	1638	2177	2496
9305	9468	9539	9186	9493	9255	9478	9380	9671	9173	9431	9370	9163	8929	8587	8040	7716
797	1336	1180	1186	1481	1346	1000	870	1491	2095	1785	1659	1659	1886	2310	2952	3026
623	1134	1021	915	1274	1087	854	713	1329	1816	1550	1415	1374	1566	1962	2582	2640
500	928	855	669	1058	842	716	586	1144	1545	1309	1171	1107	1280	1668	2288	2353
1246	1288	1368	1009	1330	1079	1345	1289	1501	1355	1376	1267	1076	996	1109	1531	1500
1562	1539	1643	1283	1557	1324	1641	1599	1735	1437	1543	1453	1250	1092	1055	1324	1235
1986	1969	2076	1716	1981	1753	2072	2026	2160	1791	1944	1864	1659	1467	1320	1369	1171
2669	2619	2740	2383	2615	2404	2748	2707	2795	2314	2534	2473	2268	2034	1762	1519	1202
2260	2378	2445	2091	2421	2170	2399	2320	2592	2350	2445	2346	2148	2006	1922	1972	1739
1266	1784	1666	1558	1924	1730	1495	1354	1977	2442	2196	2060	2004	2165	2516	3073	3070
2871	2906	3003	2642	2924	2691	2983	2921	3102	2704	2884	2807	2601	2398	2180	1989	1675
9900	9418	9608	9472	9284	9336	9769	9858	9347	8604	8968	9069	9013	8751	8303	7645	7570
9463	9025	9219	9034	8900	8914	9364	9434	8988	8211	8590	8677	8594	8315	7851	7172	7068
11543	11229	11422	11161	11116	11082	11536	11557	11217	10429	10814	10890	10780	10484	10007	9307	9133
7037	6766	6745	7070	6697	6958	6841	6956	6539	6731	6637	6744	6936	7070	7209	7409	7744
820	1138	945	1202	1253	1282	828	806	1157	1941	1557	1479	1587	1882	2359	3058	3230
9558	10097	9945	9889	10242	10062	9766	9635	10253	10737	10528	10392	10319	10400	10515	10470	10135
1490	1208	1377	1080	1157	1018	1458	1483	1330	772	1007	972	779	503	320	831	948
894	738	862	517	752	523	899	899	929	814	779	669	480	473	802	1436	1546
1001	725	884	583	698	524	959	988	877	616	644	559	353	267	627	1295	1445
933	995	907	636	945	694	994	960	1116	1043	1007	890	710	693	952	1522	1581
816	545	695	393	539	337	767	798	718	654	562	447	266	365	798	1480	1636
1051	833	979	654	819	623	1036	1049	999	731	779	691	486	372	637	1267	1385
1316	1242	1357	1000	1254	1027	1370	1342	1432	1154	1239	1147	944	801	848	1269	1265
1258	1122	1252	905	1120	908	1285	1273	1300	984	1081	997	792	632	702	1191	1232
533	1069	932	894	1213	1059	756	616	1245	1805	1508	1378	1366	1586	2009	2653	2735
388	928	781	779	1073	936	603	467	1094	1686	1374	1248	1251	1489	1927	2588	2690
413	950	812	781	1094	943	637	497	1126	1691	1390	1261	1253	1481	1912	2566	2660
304	843	701	693	988	850	524	385	1014	1599	1288	1161	1165	1404	1846	2511	2619
442	964	842	767	1106	935	673	531	1153	1676	1392	1259	1236	1449	1868	2511	2594
360	900	751	758	1045	913	573	437	1064	1663	1348	1222	1230	1470	1912	2576	2682
317	843	717	661	986	826	547	405	1028	1572	1276	1146	1134	1360	1792	2449	2548
152	692	549	560	837	709	373	233	862	1458	1139	1015	1029	1280	1732	2408	2533
35	574	434	458	719	599	263	121	747	1346	1023	899	921	1180	1639	2321	2458
98	611	488	455	755	609	326	187	797	1359	1050	922	926	1172	1623	2299	2425
0	539	400	429	684	567	231	90	712	1313	988	865	889	1151	1612	2296	2437
539	0	194	305	144	215	353	467	230	815	456	349	454	755	1232	1932	2135
400	194	0	360	324	359	179	315	313	1008	641	542	643	942	1419	2120	2313
429	305	360	0	406	172	383	405	536	910	639	502	472	721	1183	1872	2028
684	144	324	406	0	262	494	612	179	689	316	226	379	678	1145	1843	2063
567	215	359	172	262	0	456	522	422	749	466	329	323	602	1077	1775	1956
231	353	179	383	494	456	0	141	491	1164	809	700	770	1058	1533	2230	2403

		W0 (ND)	W0 (SD)	W1 (ME)	W1 (NH)	W1 (CT)	W1 (VT)	W1 (RI)	W2 (NY)	W2 (NJ)	W3 (DE)	W3 (PA)
W4	Richmond VA	1342	1273	616	467	497	385	531	437	405	233	121	187
W4	Tallahassee FL	1432	1300	1245	1094	1126	1014	1153	1064	1028	862	747	797
W5	Austin TX	1154	984	1805	1686	1691	1599	1676	1663	1572	1458	1346	1359
W5	Baton Rouge LA	1239	1081	1508	1374	1390	1288	1392	1348	1276	1139	1023	1050
W5	Jackson MS	1147	997	1378	1248	1261	1161	1259	1222	1146	1015	899	922
W5	Little Rock AR	944	792	1366	1251	1253	1165	1236	1230	1134	1029	921	926
W5	Oklahoma City OK	801	632	1586	1489	1481	1404	1449	1470	1360	1280	1180	1172
W5	Santa Fe NM	848	702	2009	1927	1912	1846	1868	1912	1792	1732	1639	1623
W6	Los Angeles CA	1269	1191	2653	2588	2566	2511	2511	2576	2449	2408	2321	2299
W6	San Francisco CA	1265	1232	2735	2690	2660	2619	2594	2682	2548	2533	2458	2425
W7	Boise ID	780	788	2279	2252	2215	2187	2141	2248	2110	2119	2059	2016
W7	Carson City NV	1090	1059	2562	2519	2488	2449	2421	2512	2377	2365	2292	2258
W7	Cheyenne WY	438	316	1779	1727	1699	1654	1637	1718	1585	1566	1493	1459
W7	Olympia WA	1038	1098	2521	2516	2472	2460	2390	2517	2376	2410	2363	2312
W7	Pheonix AZ	1090	975	2362	2285	2268	2204	2220	2270	2148	2092	1999	1983
W7	Salem OR	1076	1113	2573	2559	2517	2499	2439	2557	2418	2440	2386	2339
W7	Salt Lake City UT	692	636	2135	2089	2058	2018	1994	2081	1947	1933	1861	1827
W8	Charleston WV	1132	1054	740	613	623	526	623	590	508	386	279	286
W8	Columbus OH	1003	934	741	636	632	552	610	618	511	435	350	326
W8	Lansing MI	838	794	751	685	660	612	609	676	544	536	488	436
W9	Indianapolis IN	879	793	897	800	791	717	762	783	670	601	514	492
W9	Madison WI	611	553	983	927	899	855	841	919	785	778	721	675
W9	Springfield IL	740	633	1065	977	963	895	926	961	843	785	699	676
XE1	MEXICO CITY MEXICO	1889	1720	2408	2268	2289	2183	2297	2240	2179	2032	1915	1949
YA	KABUL AFGANISTAN	6769	6933	6398	6556	6513	6627	6471	6588	6597	6773	6881	6823
ZD8	ASCENSION IS.	6510	6468	5042	5034	5081	5088	5167	5033	5173	5140	5194	5241
ZF	CAYMAN IS.	2204	2060	1822	1663	1715	1603	1773	1631	1651	1469	1378	1448
ZL1	AUCKLAND NEWZEALAND	7776	7708	9097	8990	8989	8903	8960	8968	8868	8765	8653	8664
ZP	ASUNCION PARAGUAY	5668	5523	4874	4751	4816	4736	4908	4726	4819	4664	4623	4700
ZS1	CAPE TOWN SO.AFRICA	9189	9145	7710	7710	7756	7766	7839	7709	7850	7819	7871	7919
4X	ISRAEL	6262	6385	5278	5415	5396	5500	5395	5443	5509	5651	5768	5737
5R	TANANARIVE MALAGASY	9799	9879	8455	8538	8552	8620	8597	8554	8672	8739	8839	8847
5W	WESTERN SAMOA	6092	6044	7512	7434	7418	7352	7370	7418	7299	7232	7131	7125
6W	DAKAR SENEGAL	5167	5163	3666	3696	3730	3765	3803	3702	3838	3851	3931	3960
9N	KATMANDU NEPAL	7262	7431	7204	7363	7312	7424	7255	7395	7379	7560	7658	7591
9Q	KINSHASA ZAIRE	7647	7673	6174	6230	6255	6306	6316	6241	6371	6408	6497	6517
9Y	TRINIDAD IS.	3441	3330	2425	2309	2374	2303	2468	2286	2389	2249	2228	2302

		W7 (ID)	W7 (NV)	W7 (WY)	W7 (WA)	W7 (AZ)	W7 (OR)	W7 (UT)	W8 (WV)	W8 (OH)	W8 (MI)	W9 (IN)	W9 (WI)
A4	MUSCAT OMAN	7788	8108	7844	7565	7565	7711	7956	7444	7399	7293	7509	7405
BY	PEKING CHINA	5796	5921	6262	5408	5408	5511	6087	6907	6776	6572	6747	6476
CE	SANTIAGO CHILE	6043	5913	5588	6429	6429	6325	5753	5003	5130	5332	5148	5412
CT	LISBON PORTUGAL	5152	5492	4787	5242	5242	5348	5101	3805	3826	3824	3984	4036
CU	SAO MIGUEL IS AZORES	4460	4776	4019	4620	4620	4705	4361	2937	2976	3006	3140	3234
CX	MONTEVIDEO URUGUAY	6599	6520	6085	6996	6996	6909	6301	5306	5441	5648	5491	5771
DU	MANILA PHILIPPINES	7031	7039	7593	6633	6633	6696	7327	8491	8354	8147	8288	8004
FO	PAPEETE POLYNESIA	4707	4348	4944	4745	4745	4611	4679	5822	5810	5832	5657	5637
FY	CAYENNE FR. GUIANA	4686	4796	4082	5054	5054	5036	4418	2947	3082	3272	3194	3470
G	LONDON ENGLAND	4833	5191	4602	4828	4828	4954	4857	3872	3852	3788	3986	3955
HA	BUDAPEST HUNGARY	5572	5931	5410	5510	5510	5646	5634	4760	4733	4657	4861	4810
HC	QUITO ECUADOR	3816	3765	3299	4215	4215	4135	3518	2672	2795	2994	2804	3068
HS	BANKOK THAILAND	7835	7945	8301	7445	7445	7543	8128	8823	8704	8507	8703	8448
I	ROME ITALY	5727	6085	5493	5712	5712	5841	5752	4709	4700	4652	4842	4829
JA	TOYKO JAPAN	5175	5200	5733	4776	4776	4843	5472	6677	6539	6335	6458	6174
JT	ULAN BATOR MONGOLIA	5583	5766	5976	5219	5219	5338	5859	6448	6325	6147	6320	6065
KH6	HILO HAWAII	2775	2485	3278	2609	2609	2519	2915	4499	4413	4320	4248	4077
KL7	FAIRBANKS ALASKA	1901	2098	2341	1540	1540	1667	2182	3169	3031	2828	2953	2669
KL7	JUNEAU ALASKA	1276	1471	1749	912	912	1040	1562	2698	2561	2368	2461	2179
KP2	VIRGIN ISLANDS	3479	3607	2875	3843	3843	3828	3217	1743	1789	2075	1985	2264
LA	OSLO NORWAY	4673	5030	4554	4594	4594	4732	4753	4044	3996	3892	4107	4017
LZ	SOFIA BULGARIA	5962	6321	5798	5896	5896	6033	6024	5122	5100	5031	5232	5189
P2	MADANG P.N.G.	6865	6682	7459	6544	6544	6527	7097	8691	8573	8418	8428	8186
PY	RIO DJANEIRO BRAZIL	6496	6520	5914	6889	6889	6840	6205	4896	5033	5233	5125	5409
T5	MOGADISHU SOMALIA	9067	9423	8900	8958	8958	9102	9137	8051	8065	8041	8217	8232
TF	REYKJAVIK ICELAND	3688	4047	3510	3664	3664	3793	3735	2979	2926	2819	3034	2945
TJ	YAOUNDE CAMEROON	7821	8138	7372	7955	7955	8052	7720	6215	6280	6340	6446	6576
UA3	MOSCOW EUR. RUSSIA	5397	5740	5383	5241	5241	5386	5528	5019	4959	4837	5057	4936
UA9	NOVOSIBIRSK RUSSIA	5515	5779	5762	5214	5214	5353	5749	5909	5806	5628	5843	5630
UA9	PERM ASIATIC RUSSIA	5397	5715	5494	5175	5175	5322	5575	5344	5264	5115	5337	5174
UA0	YAKUTSK AS RUSSIA	4263	4444	4675	3897	3897	4016	4543	5269	5138	4933	5109	4840
UA0	OKHOTSK AS RUSSIA	4010	4152	4474	3626	3626	3734	4300	5202	5067	4860	5017	4738
UA0	UST OLENEK AS RUSSIA	3950	4191	4268	3626	3626	3761	4203	4680	4536	4357	4550	4297
UL7	ALMA-ATA KAZAKH	6368	6636	6594	6071	6071	6211	6596	6635	6543	6378	6598	6404
V3	BELIZE BELIZE	2432	2411	1908	2831	2831	2759	2134	1491	1583	1755	1540	1766
V85	BRUNEI	7806	7802	8371	7409	7409	7468	8101	9253	9117	8910	9058	8775
VE1	Halifax N.S.	2560	2852	2075	2785	2785	2845	2427	1024	1038	1052	1196	1282

W3 (D.C.)	W4 (GA)	W4 (SC)	W4 (KY)	W4 (AL)	W4 (TN)	W4 (NC)	W4 (VA)	W4 (FL)	W5 (TX)	W5 (LA)	W5 (MS)	W5 (AR)	W5 (OK)	W5 (NM)	W6 (L.A.CA)	W6 (S.F.CA)
90	467	315	405	612	522	141	0	628	1259	921	803	845	1118	1587	2277	2431
712	230	313	536	179	422	491	628	0	799	409	369	550	841	1299	1990	2223
1313	815	1008	910	689	749	1164	1259	799	0	389	466	441	359	589	1226	1501
988	456	641	639	316	466	809	921	409	389	0	138	299	504	919	1596	1847
865	349	542	502	226	329	700	803	369	466	138	0	206	475	929	1623	1853
889	454	643	472	379	323	770	845	550	441	299	206	0	302	777	1478	1686
1151	755	942	721	678	602	1058	1118	841	359	504	475	302	0	477	1177	1385
1612	1232	1419	1183	1145	1077	1533	1587	1299	589	919	929	777	477	0	700	928
2296	1932	2120	1872	1843	1775	2230	2277	1990	1226	1596	1623	1478	1177	700	0	348
2437	2135	2313	2028	2063	1956	2403	2431	2223	1501	1847	1853	1686	1385	928	348	0
2042	1831	1989	1665	1790	1630	2048	2051	1965	1370	1642	1609	1415	1139	797	673	520
2272	1986	2161	1867	1920	1802	2245	2268	2085	1389	1719	1716	1541	1243	804	365	174
1473	1227	1387	1075	1186	1028	1456	1472	1363	851	1060	1014	814	557	419	884	966
2350	2193	2340	2001	2163	1985	2382	2370	2341	1769	2032	1993	1794	1528	1195	931	641
1972	1583	1773	1544	1488	1435	1891	1947	1632	867	1236	1266	1129	833	360	360	654
2371	2179	2335	2005	2140	1976	2388	2385	2315	1705	1989	1959	1765	1487	1121	795	495
1841	1578	1746	1441	1524	1386	1822	1840	1695	1073	1356	1333	1145	858	500	582	601
249	352	300	180	488	323	240	228	567	1073	769	638	642	901	1364	2052	2203
329	438	428	163	556	335	377	339	601	1066	802	665	625	852	1297	1973	2107
479	621	634	317	721	470	577	516	852	1126	922	784	692	855	1259	1905	2006
491	425	487	127	508	250	492	488	653	924	701	562	484	689	1131	1807	1946
708	698	772	412	757	496	762	730	918	997	876	746	598	682	1041	1669	1763
676	505	619	275	542	291	658	669	711	799	647	517	378	525	950	1623	1762
1879	1340	1498	1547	1195	1374	1677	1804	1197	750	908	1045	1133	1108	1161	1542	1883
6914	7405	7311	7134	7534	7300	7144	7003	7616	7866	7754	7616	7492	7512	7597	7663	7373
5210	5475	5296	5594	5553	5654	5213	5201	5402	6178	5799	5770	5926	6228	6699	7393	7609
1350	1012	1001	1315	958	1212	1127	1261	789	1305	1007	1065	1271	1512	1891	2515	2805
8619	8099	8279	8210	7957	8055	8451	8561	7994	7306	7642	7757	7738	7511	7105	6516	6524
4610	4448	4381	4734	4416	4659	4442	4538	4241	4672	4453	4526	4732	4947	5245	5714	6055
7886	8116	7947	8261	8178	8306	7878	7874	8014	8743	8388	8383	8557	8855	9307	9965	10235
5804	6343	6186	6160	6488	6330	6007	5878	6494	7058	6786	6655	6618	6770	7076	7485	7335
8868	9316	9122	9297	9436	9426	8979	8904	9331	10125	9738	9662	9748	10019	10464	11069	10992
7102	6632	6825	6673	6502	6539	6978	7062	6584	5816	6190	6283	6216	5952	5506	4864	4828
3955	4342	4149	4377	4453	4479	4023	3971	4340	5137	4748	4680	4789	5081	5556	6249	6383
7687	8124	8067	7829	8235	7981	7917	7778	8350	8413	8397	8268	8108	8055	8017	7889	7557
6523	6933	6739	6950	7047	7062	6608	6547	6933	7731	7342	7273	7377	7664	8134	8816	8906
2222	2209	2084	2443	2231	2422	2094	2162	2054	2739	2396	2411	2604	2886	3312	3962	4239

W9 (IL)	XE1	YA	ZD8	ZF	ZL1	ZP	ZS1	4X	5R	5W	6W	9N	9Q	9Y
7608	9108	1011	5282	8379	8578	8440	4743	1619	3005	9089	4940	1696	3479	7739
6688	7733	2585	8618	8183	6464	11365	8036	4447	5991	5922	7633	1952	6990	8977
5196	4105	10146	4056	3711	6004	964	4937	8172	6996	6296	4825	10953	5817	3064
4148	5387	4208	3234	4397	12183	5418	5306	2429	5404	10459	1735	5261	3351	3753
3315	4463	5075	3293	3448	11255	4811	5737	3375	6201	9765	1705	6110	3963	2873
5572	4689	9370	3296	4076	6495	655	4149	7461	6218	7023	4256	10121	4996	3111
8189	8825	3486	9285	9656	4982	11675	7480	5502	5493	4996	8943	2457	7336	10722
5501	4244	9789	8958	5309	2564	5802	8793	11362	9669	1527	9228	8908	10587	6302
3340	3303	7754	2849	2146	8732	2123	5315	5815	6973	8282	2461	8814	4705	690
4120	5543	3540	4178	4759	11387	6343	5996	2147	5645	9833	2718	4548	3958	4436
4987	6434	2704	4303	5657	10991	6908	5613	1287	4911	10061	3079	3744	3583	5244
2851	1950	9273	4538	1372	7120	2230	6603	7523	8551	6440	4294	10247	6475	1397
8670	9769	2418	7912	10125	5947	10791	6296	4344	4241	6289	7752	1370	5962	10349
4985	6364	3053	3804	5479	11415	6441	5237	1330	4751	10517	2594	4110	3192	4913
6347	7016	3887	9856	7797	5489	11184	9143	5700	7075	4682	8644	3198	8292	8996
6288	7501	2135	7994	7748	7182	10659	7836	3835	5882	6547	6924	1809	6512	8360
4064	3616	7682	9763	4801	4376	7238	11376	8761	10949	2629	8694	7230	11178	6238
2851	3831	5274	7940	4336	7307	7800	10208	5695	9173	5622	6361	5430	8165	5495
2337	3212	5868	7801	3788	7188	7245	10328	6115	9704	5452	6278	6057	8364	5029
2128	2294	7657	3918	1080	8671	3045	6493	5985	7984	7681	3085	8644	5608	603
4212	5705	3106	4866	5094	10687	7022	6486	2161	5820	9287	3432	4046	4441	4986
5361	6793	2503	4179	5972	10867	6950	5291	904	4523	10277	3089	3561	3284	5460
8255	7966	5662	10701	9169	2849	9798	8062	7670	6622	2938	11113	4618	8947	10568
5249	4762	8335	2253	3850	7618	911	3774	6367	5774	8107	3123	9226	4106	2563
8372	9597	2719	4078	8467	8421	7078	3034	2232	1456	9741	4360	3176	2120	7298
3138	4634	4114	5001	4082	10425	6478	7105	3186	6789	8720	3422	4978	5091	4195
6629	7458	4248	1876	6255	9912	5033	2639	2462	2914	11654	2104	5103	621	4979
5144	6654	2085	5260	6110	10059	7841	6284	1603	5183	9175	4044	3031	4335	5973
5860	7274	1560	6901	7182	8315	9569	7221	2806	5512	7624	5785	1891	5597	7471
5396	6896	1719	5944	6533	9343	8549	6704	2035	5328	8536	4762	2511	4860	6576
5054	6189	3193	8243	6546	7282	9864	8925	4421	7113	6148	6830	3107	7250	7408
4939	5948	3654	8668	6438	6886	9868	9403	4917	7533	5676	7199	3472	7747	7442
4522	5802	3169	7556	5979	8000	9150	8669	3995	7112	6734	6090	3364	6799	6706
6632	8087	732	6598	7864	8234	9604	6479	2429	4680	7940	5747	1163	5045	7904
1538	731	8532	5417	519	7314	3609	7815	7274	9535	6119	4669	9268	7188	1891
8964	9559	3561	8855	10435	4803	10932	6779	5503	4860	5173	8899	2501	6903	11355
1365	2675	6206	4801	1988	9395	4852	7457	5008	8154	7812	3391	7055	5881	2397

		W7 (ID)	W7 (NV)	W7 (WY)	W7 (WA)	W7 (AZ)	W7 (OR)	W7 (UT)	W8 (WV)	W8 (OH)	W8 (MI)	W9 (IN)	W9 (WI)
VE2	Quebec City Que.	2168	2466	1700	2390	2390	2450	2045	786	754	711	892	922
VE3	Ottawa Ont.	1890	2180	1407	2131	2131	2183	1755	577	507	424	622	626
VE4	Winnipeg Mb.	995	1327	706	1187	1187	1251	949	1105	968	773	881	596
VE5	Regina Sk.	715	1068	638	859	859	932	751	1401	1267	1084	1159	879
VE6	Edmonton Al.	696	1036	944	610	610	735	886	1832	1697	1509	1591	1310
VE7	Prince Rupert B.C.	972	1156	1478	598	598	724	1263	2508	2375	2192	2261	1985
VE8	Yellowknife NWT	1302	1624	1526	1113	1113	1258	1504	2165	2028	1822	1963	1681
VO2	Goose Bay Labrador	2563	2897	2195	2699	2699	2789	2503	1436	1395	1317	1519	1488
VY1	Whitehorse Yukon	1412	1624	1854	1061	1061	1194	1691	2744	2606	2409	2516	2231
VK1	CANBERRA AUSTRALIA	8046	7744	8519	7857	7857	7782	8167	9651	9599	9550	9433	9314
VK4	BRISBANE AUSTRALIA	7526	7240	8029	7313	7313	7246	7669	9215	9145	9070	8979	8829
VK6	PERTH AUSTRALIA	9505	9293	10087	9193	9193	9173	9719	11331	11220	11067	11069	10835
VP8	SO. SHETLAND IS.	7961	7766	7573	8319	8319	8196	7687	7045	7173	7376	7191	7455
VP9	BERMUDA	2857	3070	2276	3170	3170	3190	2644	1032	1143	1297	1294	1528
VQ9	DIEGO GARCIA CHAGOS	9876	10109	10098	9548	9548	9678	10114	9761	9726	9628	9843	9740
WØ	Denver CO	637	787	99	1025	1025	986	369	1255	1161	1076	998	838
WØ	DesMoines IA	1152	1379	580	1483	1483	1489	948	674	564	468	411	239
WØ	Topeka KS	1106	1285	503	1469	1469	1454	864	760	678	633	511	430
WØ	Minneapolis MN	1133	1408	638	1418	1418	1448	983	755	623	457	510	231
WØ	Jefferson City MO	1293	1478	693	1648	1648	1639	1055	571	499	490	333	344
WØ	Lincoln NE	1012	1220	422	1361	1361	1356	791	822	722	638	561	407
WØ	Bismark ND	780	1090	438	1038	1038	1076	692	1132	1003	838	879	611
WØ	Pierre SD	788	1059	316	1098	1098	1113	636	1054	934	794	793	553
W1	Augusta ME	2279	2562	1779	2521	2521	2573	2135	740	741	751	897	983
W1	Boston MA	2252	2519	1727	2516	2516	2559	2089	613	636	685	800	927
W1	Concord NH	2215	2488	1699	2472	2472	2517	2058	623	632	660	791	899
W1	Hartford CT	2187	2449	1654	2460	2460	2499	2018	526	552	612	717	855
W1	Montpelier VT	2141	2421	1637	2390	2390	2439	1994	623	610	609	762	841
W1	Providence RI	2248	2512	1718	2517	2517	2557	2081	590	618	676	783	919
W2	Albany NY	2110	2377	1585	2376	2376	2418	1947	508	511	544	670	785
W2	Trenton NJ	2119	2365	1566	2410	2410	2440	1933	386	435	536	601	778
W3	Dover DE	2059	2292	1493	2363	2363	2386	1861	279	350	488	514	721
W3	Harrisburg PA	2016	2258	1459	2312	2312	2339	1827	286	326	436	492	675
W3	Washington D.C.	2042	2272	1473	2350	2350	2371	1841	249	329	479	491	708
W4	Atlanta GA	1831	1986	1227	2193	2193	2179	1578	352	438	621	425	698
W4	Columbia SC	1989	2161	1387	2340	2340	2335	1746	300	428	634	487	772
W4	Frankfort KY	1665	1867	1075	2001	2001	2005	1441	180	163	317	127	412
W4	Montgomery AL	1790	1920	1186	2163	2163	2140	1524	488	556	721	508	757
W4	Nashville TN	1630	1802	1028	1985	1985	1976	1386	323	335	470	250	496
W4	Raleigh NC	2048	2245	1456	2382	2382	2388	1822	240	377	577	492	762
W4	Richmond VA	2051	2268	1472	2370	2370	2385	1840	228	339	516	488	730
W4	Tallahassee FL	1965	2085	1363	2341	2341	2315	1695	567	665	852	653	918
W5	Austin TX	1370	1389	851	1769	1769	1705	1073	1073	1066	1126	924	997
W5	Baton Rouge LA	1642	1719	1060	2032	2032	1989	1356	769	802	922	701	876
W5	Jackson MS	1609	1716	1014	1993	1993	1959	1333	638	665	784	562	746
W5	Little Rock AR	1415	1541	814	1794	1794	1765	1145	642	625	692	484	598
W5	Oklahoma City OK	1139	1243	557	1528	1528	1487	858	901	852	855	689	682
W5	Santa Fe NM	797	804	419	1195	1195	1121	500	1364	1297	1259	1131	1041
W6	Los Angeles CA	673	365	884	931	931	795	582	2052	1973	1905	1807	1669
W6	San Francisco CA	520	174	966	641	641	495	601	2203	2107	2006	1946	1763
W7	Boise ID	0	359	604	399	399	349	297	1827	1714	1583	1564	1341
W7	Carson City NV	359	0	799	565	565	431	431	2040	1942	1836	1782	1593
W7	Cheyenne WY	604	799	0	979	979	953	368	1244	1143	1041	984	800
W7	Olympia WA	399	565	979	0	0	146	697	2152	2030	1876	1891	1642
W7	Pheonix AZ	734	581	659	1093	1093	982	500	1725	1657	1611	1490	1386
W7	Salem OR	349	431	953	146	146	0	634	2163	2046	1903	1900	1665
W7	Salt Lake City UT	297	431	368	697	697	634	0	1612	1511	1405	1352	1162
W8	Charleston WV	1827	2040	1244	2152	2152	2163	1612	0	138	343	262	523
W8	Columbus OH	1714	1942	1143	2030	2030	2046	1511	138	0	206	166	392
W8	Lansing MI	1583	1836	1041	1876	1876	1903	1405	343	206	0	222	243
W9	Indianapolis IN	1564	1782	984	1891	1891	1900	1352	262	166	222	0	284
W9	Madison WI	1341	1593	800	1642	1642	1665	1162	523	392	243	284	0
W9	Springfield IL	1390	1599	803	1727	1727	1730	1171	441	350	332	184	229
XE1	MEXICO CITY MEXICO	1938	1831	1534	2324	2324	2226	1650	1674	1710	1817	1599	1726
YA	KABUL AFGANISTAN	7012	7299	7179	6738	6738	6880	7221	7053	6978	6833	7055	6893
ZD8	ASCENSION IS.	7252	7457	6670	7548	7548	7580	7036	5427	5537	5676	5690	5915
ZF	CAYMAN IS.	2646	2694	2067	3039	3039	2991	2356	1301	1423	1622	1437	1707
ZL1	AUCKLAND NEWZEALAND	7038	6696	7398	6955	6955	6847	7085	8379	8357	8358	8199	8147
ZP	ASUNCION PARAGUAY	6041	5995	5503	6440	6440	6364	5743	4668	4804	5011	4861	5143
ZS1	CAPE TOWN SO.AFRICA	9925	10098	9333	10226	10226	10258	9693	8098	8212	8355	8361	8593
4X	ISRAEL	6835	7191	6696	6740	6740	6881	6911	6017	5999	5934	6133	6093
5R	TANANARIVE MALAGASY	10477	10836	10191	10409	10409	10550	10490	9117	9171	9206	9337	9430
5W	WESTERN SAMOA	5331	5002	5746	5220	5220	5117	5410	6852	6801	6761	6635	6531
6W	DAKAR SENEGAL	5945	6214	5417	6175	6175	6236	5783	4197	4281	4377	4446	4621
9N	KATMANDU NEPAL	7309	7529	7621	6969	6969	7097	7567	7783	7686	7513	7730	7521
9Q	KINSHASA ZAIRE	8423	8732	7954	8570	8570	8664	8308	6770	6843	6915	7009	7155
9Y	TRINIDAD IS.	4019	4115	3418	4395	4395	4369	3745	2333	2471	2670	2568	2850

W9 (IL)	XE1	YA	ZD8	ZF	ZL1	ZP	ZS1	4X	5R	5W	6W	9N	9Q	9Y
1041	2458	6290	5191	1957	9068	5059	7851	5245	8509	7446	3786	7071	6265	2611
757	2216	6477	5428	1843	8774	5089	8099	5512	8804	7152	4055	7211	6553	2670
789	2103	6532	6377	2288	8023	5737	9048	5993	9541	6326	4988	7061	7430	3439
1034	2157	6542	6711	2509	7752	5972	9381	6161	9771	6036	5316	6992	7734	3727
1465	2473	6335	7086	2927	7561	6390	9735	6164	9828	5822	5645	6698	7977	4159
2122	2910	6176	7747	3537	7088	6980	10358	6350	9976	5342	6268	6373	8474	4824
1880	3060	5717	7073	3380	7883	6825	9602	5632	9298	6137	5544	6096	7679	4482
1649	3105	5652	5059	2564	9588	5437	7634	4616	7997	7894	3543	6465	5874	2990
2401	3340	5704	7754	3870	7301	7333	10230	5955	9538	5571	6213	5908	8242	5078
9254	8180	7057	9216	9230	1426	7885	6689	8779	6082	2799	10774	6021	8377	9848
8795	7875	6951	9807	9010	1413	8244	7260	8824	6511	2407	11354	5893	8865	9901
10890	10082	5508	8095	11089	3305	8453	5422	6957	4300	4635	9282	4588	6689	10909
7236	6048	9499	4372	5755	4986	2526	3769	8256	5571	6056	5741	9551	5430	4972
1473	2300	6924	4395	1316	9086	4014	7068	5496	8183	7740	3204	7834	5795	1560
9944	11435	2918	10448	6561	8384	3888	3759	1850		7747	6350	2580	3942	9262
814	1441	7278	6670	2014	7334	5436	9324	6783	10254	5695	5443	7717	7992	3383
244	1561	7053	6100	1707	7908	5168	8772	6317	9669	6294	4837	7636	7383	2942
327	1366	7249	6172	1623	7726	5085	8830	6521	9837	6133	4956	7812	7521	2927
404	1792	6822	6133	1903	8032	5354	8811	6131	9556	6384	4813	7402	7325	3079
159	1383	7234	5980	1489	7875	4952	8639	6419	9670	6302	4769	7842	7339	2754
380	1481	7141	6248	1754	7741	5217	8915	6462	9838	6124	5001	7687	7551	3041
740	1889	6769	6510	2204	7776	5668	9189	6262	9799	6092	5167	7262	7647	3441
633	1720	6933	6468	2060	7708	5523	9145	6385	9879	6044	5163	7431	7673	3330
1065	2408	6398	5042	1822	9097	4874	7710	5278	8455	7512	3666	7204	6174	2425
977	2268	6556	5034	1663	8990	4751	7710	5415	8538	7434	3696	7363	6230	2309
963	2289	6513	5081	1715	8989	4816	7756	5396	8552	7418	3730	7312	6255	2374
895	2183	6627	5088	1603	8903	4736	7766	5500	8620	7352	3765	7424	6306	2303
926	2297	6471	5167	1773	8960	4908	7839	5395	8597	7370	3803	7255	6316	2468
961	2240	6588	5033	1631	8968	4726	7709	5443	8554	7418	3702	7395	6241	2286
843	2179	6597	5173	1651	8868	4819	7850	5509	8672	7299	3838	7379	6371	2389
785	2032	6773	5140	1469	8765	4664	7819	5651	8739	7232	3851	7560	6408	2249
699	1915	6881	5194	1378	8653	4623	7871	5768	8839	7131	3931	7658	6497	2228
676	1949	6823	5241	1448	8664	4700	7919	5737	8847	7125	3960	7591	6517	2302
676	1879	6914	5210	1350	8619	4610	7886	5804	8868	7102	3955	7687	6523	2222
505	1340	7405	5475	1012	8099	4448	8116	6343	9316	6632	4342	8124	6933	2209
619	1498	7311	5296	1001	8279	4381	7947	6186	9122	6825	4149	8067	6739	2084
275	1547	7134	5594	1315	8210	4734	8261	6160	9297	6673	4377	7829	6950	2443
542	1195	7534	5553	958	7957	4416	8178	6488	9436	6502	4453	8235	7047	2231
291	1374	7300	5654	1212	8055	4659	8306	6330	9426	6539	4479	7981	7062	2422
658	1677	7144	5213	1127	8451	4442	7878	6007	8979	6978	4023	7917	6608	2094
669	1804	7003	5201	1261	8561	4538	7874	5878	8904	7062	3971	7778	6547	2162
711	1197	7616	5402	789	7994	4241	8014	6494	9331	6584	4340	8350	6933	2054
799	750	7866	6178	1305	7306	4672	8743	7058	10125	5816	5137	8413	7731	2739
647	908	7754	5799	1007	7642	4453	8388	6736	9738	6190	4748	8397	7342	2366
517	1045	7616	5770	1065	7757	4526	8383	6655	9662	6283	4680	8268	7273	2411
378	1133	7492	5926	1271	7738	4732	8557	6618	9748	6216	4789	8108	7377	2604
525	1108	7512	6228	1512	7511	4947	8855	6776	10019	5952	5081	8055	7664	2886
950	1161	7597	6699	1891	7105	5245	9307	7076	10464	5506	5556	8017	8134	3312
1623	1542	7663	7393	2515	6516	5714	9965	7485	10969	4864	6249	7889	8816	3962
1762	1883	7373	7609	2805	6524	6055	10235	7335	10992	4828	6383	7557	8906	4239
1390	1938	7012	7252	2646	7038	6041	9925	6835	10477	5331	5945	7309	8423	4019
1599	1831	7299	7457	2694	6696	5995	10098	7191	10836	5002	6214	7529	8732	4115
803	1534	7179	6670	2067	7398	5503	9333	6696	10191	5746	5417	7621	7954	3418
1727	2324	6738	7548	3039	6955	6440	10226	6740	10409	5220	6175	6969	8570	4395
1309	1250	7721	7034	2160	6765	5414	9608	7353	10817	5150	5915	8044	8495	3605
1730	2226	6880	7580	2991	6847	6364	10258	6881	10550	5117	6236	7097	8664	4369
1171	1650	7221	7036	2356	7085	5743	9693	6911	10490	5410	5783	7567	8308	3745
441	1674	7053	5427	1301	8379	4668	8098	6017	9117	6852	4197	7783	6770	2333
350	1710	6978	5537	1423	8357	4804	8212	5999	9171	6801	4281	7686	6843	2471
332	1817	6833	5676	1622	8358	5011	8355	5934	9206	6761	4377	7513	6915	2670
184	1599	7055	5690	1437	8199	4861	8361	6133	9337	6635	4446	7730	7009	2568
229	1726	6893	5915	1707	8147	5143	8593	6093	9430	6531	4621	7521	7155	2850
0	1511	7115	5867	1499	8034	4951	8536	6265	9515	6457	4631	7751	7193	2701
1511	0	8614	6149	1213	6803	4172	8510	7683	10265	5474	5353	9149	7902	2623
7115	8614	0	6090	8193	8364	9216	5753	2019	3955	8403	5456	1060	4409	7947
5867	6149	6090	0	4963	9288	3159	2678	4172	4123	10310	1593	6978	1958	3525
1499	1213	8193	4963	0	7808	7463	7450	6805	9059	6639	4153	9026	6689	1455
8034	6803	8364	9288	7808	0	6948	7314	10198	7264	1746	10720	7305	9291	8505
4951	4172	9216	3159	3463	6948	0	4495	7219	6549	7234	3861	10136	5014	2456
8536	8510	5753	2678	7314	4495	4698	0	2069	3668	9022	4094	6099	2046	6003
6265	7683	2019	4172	6805	10198	7219	4698	0	3668	10369	3443	3051	2830	6128
9515	10265	3955	4123	9059	7264	6549	2069	3668	0	4992	4088	2401	7646	7646
6457	5474	8403	10310	6639	1746	7234	3881	10369	3881	0	10729	7436	11036	7777
4631	5353	5456	1593	4153	10720	3861	4094	3443	4092	10729	0	6495	2594	2951
7751	9149	1060	6978	9026	7305	10136	6099	3051	2401	7436	6495	0	5148	8982
7193	7902	4409	1958	6689	9291	5014	2046	2830	2401	11036	2594	5148	0	5337
2701	2623	7947	3525	1455	8505	2456	6003	6128	7646	7777	2951	8982	5337	0

VHF/UHF CENTURY CLUB
LOCATOR MAP
© 1983

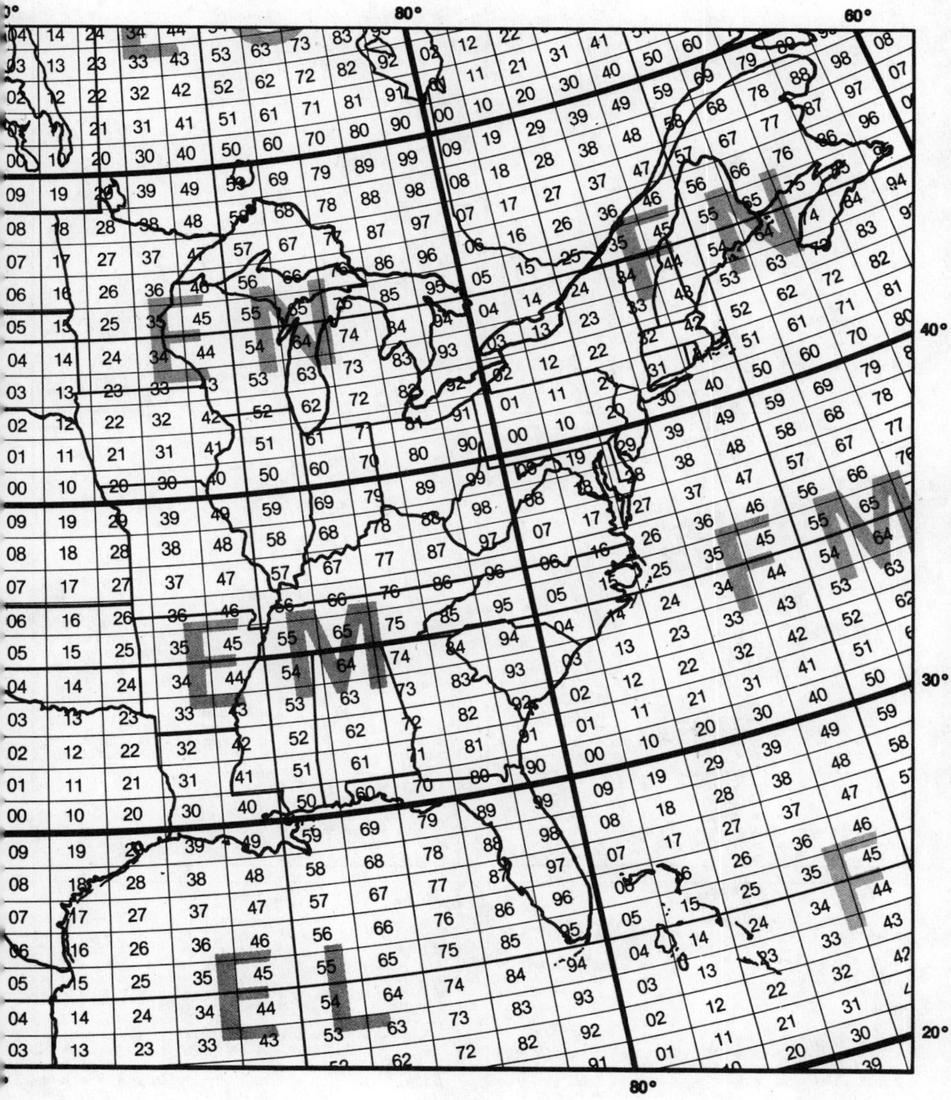
80°
60°
40°
30°
20°
80°
FN
EN
EM
FM
EL
F

1990

January							February							March						
S	M	T	W	T	F	S	S	M	T	W	T	F	S	S	M	T	W	T	F	S
	1	2	3	4	5	6					1	2	3					1	2	3
7	8	9	10	11	12	13	4	5	6	7	8	9	10	4	5	6	7	8	9	10
14	15	16	17	18	19	20	11	12	13	14	15	16	17	11	12	13	14	15	16	17
21	22	23	24	25	26	27	18	19	20	21	22	23	24	18	19	20	21	22	23	24
28	29	30	31				25	26	27	28				25	26	27	28	29	30	31

April							May							June						
S	M	T	W	T	F	S	S	M	T	W	T	F	S	S	M	T	W	T	F	S
1	2	3	4	5	6	7			1	2	3	4	5						1	2
8	9	10	11	12	13	14	6	7	8	9	10	11	12	3	4	5	6	7	8	9
15	16	17	18	19	20	21	13	14	15	16	17	18	19	10	11	12	13	14	15	16
22	23	24	25	26	27	28	20	21	22	23	24	25	26	17	18	19	20	21	22	23
29	30						27	28	29	30	31			24	25	26	27	28	29	30

July							August							September						
S	M	T	W	T	F	S	S	M	T	W	T	F	S	S	M	T	W	T	F	S
1	2	3	4	5	6	7				1	2	3	4							1
8	9	10	11	12	13	14	5	6	7	8	9	10	11	2	3	4	5	6	7	8
15	16	17	18	19	20	21	12	13	14	15	16	17	18	9	10	11	12	13	14	15
22	23	24	25	26	27	28	19	20	21	22	23	24	25	16	17	18	19	20	21	22
29	30	31					26	27	28	29	30	31		23	24	25	26	27	28	29
														30						

October							November							December						
S	M	T	W	T	F	S	S	M	T	W	T	F	S	S	M	T	W	T	F	S
	1	2	3	4	5	6					1	2	3							1
7	8	9	10	11	12	13	4	5	6	7	8	9	10	2	3	4	5	6	7	8
14	15	16	17	18	19	20	11	12	13	14	15	16	17	9	10	11	12	13	14	15
21	22	23	24	25	26	27	18	19	20	21	22	23	24	16	17	18	19	20	21	22
28	29	30	31				25	26	27	28	29	30		23	24	25	26	27	28	29
														30	31					

1991

January							February							March						
S	M	T	W	T	F	S	S	M	T	W	T	F	S	S	M	T	W	T	F	S
		1	2	3	4	5						1	2						1	2
6	7	8	9	10	11	12	3	4	5	6	7	8	9	3	4	5	6	7	8	9
13	14	15	16	17	18	19	10	11	12	13	14	15	16	10	11	12	13	14	15	16
20	21	22	23	24	25	26	17	18	19	20	21	22	23	17	18	19	20	21	22	23
27	28	29	30	31			24	25	26	27	28			24	25	26	27	28	29	30
														31						

April							May							June						
S	M	T	W	T	F	S	S	M	T	W	T	F	S	S	M	T	W	T	F	S
	1	2	3	4	5	6				1	2	3	4							1
7	8	9	10	11	12	13	5	6	7	8	9	10	11	2	3	4	5	6	7	8
14	15	16	17	18	19	20	12	13	14	15	16	17	18	9	10	11	12	13	14	15
21	22	23	24	25	26	27	19	20	21	22	23	24	25	16	17	18	19	20	21	22
28	29	30					26	27	28	29	30	31		23	24	25	26	27	28	29
														30						

July							August							September						
S	M	T	W	T	F	S	S	M	T	W	T	F	S	S	M	T	W	T	F	S
	1	2	3	4	5	6					1	2	3	1	2	3	4	5	6	7
7	8	9	10	11	12	13	4	5	6	7	8	9	10	8	9	10	11	12	13	14
14	15	16	17	18	19	20	11	12	13	14	15	16	17	15	16	17	18	19	20	21
21	22	23	24	25	26	27	18	19	20	21	22	23	24	22	23	24	25	26	27	28
28	29	30	31				25	26	27	28	29	30	31	29	30					

October							November							December						
S	M	T	W	T	F	S	S	M	T	W	T	F	S	S	M	T	W	T	F	S
		1	2	3	4	5						1	2	1	2	3	4	5	6	7
6	7	8	9	10	11	12	3	4	5	6	7	8	9	8	9	10	11	12	13	14
13	14	15	16	17	18	19	10	11	12	13	14	15	16	15	16	17	18	19	20	21
20	21	22	23	24	25	26	17	18	19	20	21	22	23	22	23	24	25	26	27	28
27	28	29	30	31			24	25	26	27	28	29	30	29	30	31				

1992

January							February							March						
S	M	T	W	T	F	S	S	M	T	W	T	F	S	S	M	T	W	T	F	S
			1	2	3	4							1	1	2	3	4	5	6	7
5	6	7	8	9	10	11	2	3	4	5	6	7	8	8	9	10	11	12	13	14
12	13	14	15	16	17	18	9	10	11	12	13	14	15	15	16	17	18	19	20	21
19	20	21	22	23	24	25	16	17	18	19	20	21	22	22	23	24	25	26	27	28
26	27	28	29	30	31		23	24	25	26	27	28	29	29	30	31				

April							May							June						
S	M	T	W	T	F	S	S	M	T	W	T	F	S	S	M	T	W	T	F	S
			1	2	3	4						1	2		1	2	3	4	5	6
5	6	7	8	9	10	11	3	4	5	6	7	8	9	7	8	9	10	11	12	13
12	13	14	15	16	17	18	10	11	12	13	14	15	16	14	15	16	17	18	19	20
19	20	21	22	23	24	25	17	18	19	20	21	22	23	21	22	23	24	25	26	27
26	27	28	29	30			24	25	26	27	28	29	30	28	29	30				
							31													

July							August							September						
S	M	T	W	T	F	S	S	M	T	W	T	F	S	S	M	T	W	T	F	S
			1	2	3	4							1			1	2	3	4	5
5	6	7	8	9	10	11	2	3	4	5	6	7	8	6	7	8	9	10	11	12
12	13	14	15	16	17	18	9	10	11	12	13	14	15	13	14	15	16	17	18	19
19	20	21	22	23	24	25	16	17	18	19	20	21	22	20	21	22	23	24	25	26
26	27	28	29	30	31		23	24	25	26	27	28	29	27	28	29	30			
							30	31												

October							November							December						
S	M	T	W	T	F	S	S	M	T	W	T	F	S	S	M	T	W	T	F	S
				1	2	3	1	2	3	4	5	6	7			1	2	3	4	5
4	5	6	7	8	9	10	8	9	10	11	12	13	14	6	7	8	9	10	11	12
11	12	13	14	15	16	17	15	16	17	18	19	20	21	13	14	15	16	17	18	19
18	19	20	21	22	23	24	22	23	24	25	26	27	28	20	21	22	23	24	25	26
25	26	27	28	29	30	31	29	30						27	28	29	30	31		

1993

January							February							March						
S	M	T	W	T	F	S	S	M	T	W	T	F	S	S	M	T	W	T	F	S
					1	2		1	2	3	4	5	6		1	2	3	4	5	6
3	4	5	6	7	8	9	7	8	9	10	11	12	13	7	8	9	10	11	12	13
10	11	12	13	14	15	16	14	15	16	17	18	19	20	14	15	16	17	18	19	20
17	18	19	20	21	22	23	21	22	23	24	25	26	27	21	22	23	24	25	26	27
24	25	26	27	28	29	30	28							28	29	30	31			
31																				

April							May							June						
S	M	T	W	T	F	S	S	M	T	W	T	F	S	S	M	T	W	T	F	S
				1	2	3							1			1	2	3	4	5
4	5	6	7	8	9	10	2	3	4	5	6	7	8	6	7	8	9	10	11	12
11	12	13	14	15	16	17	9	10	11	12	13	14	15	13	14	15	16	17	18	19
18	19	20	21	22	23	24	16	17	18	19	20	21	22	20	21	22	23	24	25	26
25	26	27	28	29	30		23	24	25	26	27	28	29	27	28	29	30			
							30	31												

July							August							September						
S	M	T	W	T	F	S	S	M	T	W	T	F	S	S	M	T	W	T	F	S
				1	2	3	1	2	3	4	5	6	7				1	2	3	4
4	5	6	7	8	9	10	8	9	10	11	12	13	14	5	6	7	8	9	10	11
11	12	13	14	15	16	17	15	16	17	18	19	20	21	12	13	14	15	16	17	18
18	19	20	21	22	23	24	22	23	24	25	26	27	28	19	20	21	22	23	24	25
25	26	27	28	29	30	31	29	30	31					26	27	28	29	30		

October							November							December						
S	M	T	W	T	F	S	S	M	T	W	T	F	S	S	M	T	W	T	F	S
					1	2		1	2	3	4	5	6				1	2	3	4
3	4	5	6	7	8	9	7	8	9	10	11	12	13	5	6	7	8	9	10	11
10	11	12	13	14	15	16	14	15	16	17	18	19	20	12	13	14	15	16	17	18
17	18	19	20	21	22	23	21	22	23	24	25	26	27	19	20	21	22	23	24	25
24	25	26	27	28	29	30	28	29	30					26	27	28	29	30	31	
31																				

1994

January							February							March						
S	M	T	W	T	F	S	S	M	T	W	T	F	S	S	M	T	W	T	F	S
						1			1	2	3	4	5			1	2	3	4	5
2	3	4	5	6	7	8	6	7	8	9	10	11	12	6	7	8	9	10	11	12
9	10	11	12	13	14	15	13	14	15	16	17	18	19	13	14	15	16	17	18	19
16	17	18	19	20	21	22	20	21	22	23	24	25	26	20	21	22	23	24	25	26
23	24	25	26	27	28	29	27	28						27	28	29	30	31		
30	31																			

April							May							June						
S	M	T	W	T	F	S	S	M	T	W	T	F	S	S	M	T	W	T	F	S
					1	2	1	2	3	4	5	6	7				1	2	3	4
3	4	5	6	7	8	9	8	9	10	11	12	13	14	5	6	7	8	9	10	11
10	11	12	13	14	15	16	15	16	17	18	19	20	21	12	13	14	15	16	17	18
17	18	19	20	21	22	23	22	23	24	25	26	27	28	19	20	21	22	23	24	25
24	25	26	27	28	29	30	29	30	31					26	27	28	29	30		

July							August							September						
S	M	T	W	T	F	S	S	M	T	W	T	F	S	S	M	T	W	T	F	S
					1	2		1	2	3	4	5	6					1	2	3
3	4	5	6	7	8	9	7	8	9	10	11	12	13	4	5	6	7	8	9	10
10	11	12	13	14	15	16	14	15	16	17	18	19	20	11	12	13	14	15	16	17
17	18	19	20	21	22	23	21	22	23	24	25	26	27	18	19	20	21	22	23	24
24	25	26	27	28	29	30	28	29	30	31				25	26	27	28	29	30	
31																				

October							November							December						
S	M	T	W	T	F	S	S	M	T	W	T	F	S	S	M	T	W	T	F	S
						1			1	2	3	4	5					1	2	3
2	3	4	5	6	7	8	6	7	8	9	10	11	12	4	5	6	7	8	9	10
9	10	11	12	13	14	15	13	14	15	16	17	18	19	11	12	13	14	15	16	17
16	17	18	19	20	21	22	20	21	22	23	24	25	26	18	19	20	21	22	23	24
23	24	25	26	27	28	29	27	28	29	30				25	26	27	28	29	30	31
30	31																			

1995

January							February							March						
S	M	T	W	T	F	S	S	M	T	W	T	F	S	S	M	T	W	T	F	S
1	2	3	4	5	6	7				1	2	3	4				1	2	3	4
8	9	10	11	12	13	14	5	6	7	8	9	10	11	5	6	7	8	9	10	11
15	16	17	18	19	20	21	12	13	14	15	16	17	18	12	13	14	15	16	17	18
22	23	24	25	26	27	28	19	20	21	22	23	24	25	19	20	21	22	23	24	25
29	30	31					26	27	28					26	27	28	29	30	31	

April							May							June						
S	M	T	W	T	F	S	S	M	T	W	T	F	S	S	M	T	W	T	F	S
						1		1	2	3	4	5	6					1	2	3
2	3	4	5	6	7	8	7	8	9	10	11	12	13	4	5	6	7	8	9	10
9	10	11	12	13	14	15	14	15	16	17	18	19	20	11	12	13	14	15	16	17
16	17	18	19	20	21	22	21	22	23	24	25	26	27	18	19	20	21	22	23	24
23	24	25	26	27	28	29	28	29	30	31				25	26	27	28	29	30	
30																				

July							August							September						
S	M	T	W	T	F	S	S	M	T	W	T	F	S	S	M	T	W	T	F	S
						1			1	2	3	4	5						1	2
2	3	4	5	6	7	8	6	7	8	9	10	11	12	3	4	5	6	7	8	9
9	10	11	12	13	14	15	13	14	15	16	17	18	19	10	11	12	13	14	15	16
16	17	18	19	20	21	22	20	21	22	23	24	25	26	17	18	19	20	21	22	23
23	24	25	26	27	28	29	27	28	29	30	31			24	25	26	27	28	29	30
30	31																			

October							November							December						
S	M	T	W	T	F	S	S	M	T	W	T	F	S	S	M	T	W	T	F	S
1	2	3	4	5	6	7				1	2	3	4						1	2
8	9	10	11	12	13	14	5	6	7	8	9	10	11	3	4	5	6	7	8	9
15	16	17	18	19	20	21	12	13	14	15	16	17	18	10	11	12	13	14	15	16
22	23	24	25	26	27	28	19	20	21	22	23	24	25	17	18	19	20	21	22	23
29	30	31					26	27	28	29	30			24	25	26	27	28	29	30
														31						

1996

```
     January                February                March
S  M  T  W  T  F  S    S  M  T  W  T  F  S    S  M  T  W  T  F  S
      1  2  3  4  5  6                1  2  3                   1  2
 7  8  9 10 11 12 13    4  5  6  7  8  9 10    3  4  5  6  7  8  9
14 15 16 17 18 19 20   11 12 13 14 15 16 17   10 11 12 13 14 15 16
21 22 23 24 25 26 27   18 19 20 21 22 23 24   17 18 19 20 21 22 23
28 29 30 31            25 26 27 28 29         24 25 26 27 28 29 30
                                              31

      April                   May                    June
S  M  T  W  T  F  S    S  M  T  W  T  F  S    S  M  T  W  T  F  S
    1  2  3  4  5  6             1  2  3  4                         1
 7  8  9 10 11 12 13    5  6  7  8  9 10 11    2  3  4  5  6  7  8
14 15 16 17 18 19 20   12 13 14 15 16 17 18    9 10 11 12 13 14 15
21 22 23 24 25 26 27   19 20 21 22 23 24 25   16 17 18 19 20 21 22
28 29 30               26 27 28 29 30 31      23 24 25 26 27 28 29
                                              30

      July                  August                September
S  M  T  W  T  F  S    S  M  T  W  T  F  S    S  M  T  W  T  F  S
    1  2  3  4  5  6                1  2  3    1  2  3  4  5  6  7
 7  8  9 10 11 12 13    4  5  6  7  8  9 10    8  9 10 11 12 13 14
14 15 16 17 18 19 20   11 12 13 14 15 16 17   15 16 17 18 19 20 21
21 22 23 24 25 26 27   18 19 20 21 22 23 24   22 23 24 25 26 27 28
28 29 30 31            25 26 27 28 29 30 31   29 30

     October               November               December
S  M  T  W  T  F  S    S  M  T  W  T  F  S    S  M  T  W  T  F  S
       1  2  3  4  5                   1  2    1  2  3  4  5  6  7
 6  7  8  9 10 11 12    3  4  5  6  7  8  9    8  9 10 11 12 13 14
13 14 15 16 17 18 19   10 11 12 13 14 15 16   15 16 17 18 19 20 21
20 21 22 23 24 25 26   17 18 19 20 21 22 23   22 23 24 25 26 27 28
27 28 29 30 31         24 25 26 27 28 29 30   29 30 31
```

1997

```
     January                February                March
S  M  T  W  T  F  S    S  M  T  W  T  F  S    S  M  T  W  T  F  S
          1  2  3  4                      1                      1
 5  6  7  8  9 10 11    2  3  4  5  6  7  8    2  3  4  5  6  7  8
12 13 14 15 16 17 18    9 10 11 12 13 14 15    9 10 11 12 13 14 15
19 20 21 22 23 24 25   16 17 18 19 20 21 22   16 17 18 19 20 21 22
26 27 28 29 30 31      23 24 25 26 27 28      23 24 25 26 27 28 29
                                              30 31

      April                   May                    June
S  M  T  W  T  F  S    S  M  T  W  T  F  S    S  M  T  W  T  F  S
       1  2  3  4  5             1  2  3    1  2  3  4  5  6  7
 6  7  8  9 10 11 12    4  5  6  7  8  9 10    8  9 10 11 12 13 14
13 14 15 16 17 18 19   11 12 13 14 15 16 17   15 16 17 18 19 20 21
20 21 22 23 24 25 26   18 19 20 21 22 23 24   22 23 24 25 26 27 28
27 28 29 30            25 26 27 28 29 30 31   29 30

      July                  August                September
S  M  T  W  T  F  S    S  M  T  W  T  F  S    S  M  T  W  T  F  S
       1  2  3  4  5                   1  2       1  2  3  4  5  6
 6  7  8  9 10 11 12    3  4  5  6  7  8  9    7  8  9 10 11 12 13
13 14 15 16 17 18 19   10 11 12 13 14 15 16   14 15 16 17 18 19 20
20 21 22 23 24 25 26   17 18 19 20 21 22 23   21 22 23 24 25 26 27
27 28 29 30 31         24 25 26 27 28 29 30   28 29 30
                       31

     October               November               December
S  M  T  W  T  F  S    S  M  T  W  T  F  S    S  M  T  W  T  F  S
          1  2  3  4                      1       1  2  3  4  5  6
 5  6  7  8  9 10 11    2  3  4  5  6  7  8    7  8  9 10 11 12 13
12 13 14 15 16 17 18    9 10 11 12 13 14 15   14 15 16 17 18 19 20
19 20 21 22 23 24 25   16 17 18 19 20 21 22   21 22 23 24 25 26 27
26 27 28 29 30 31      23 24 25 26 27 28 29   28 29 30 31
                       30
```

1998

```
     January                February                March
S  M  T  W  T  F  S    S  M  T  W  T  F  S    S  M  T  W  T  F  S
             1  2  3    1  2  3  4  5  6  7    1  2  3  4  5  6  7
 4  5  6  7  8  9 10    8  9 10 11 12 13 14    8  9 10 11 12 13 14
11 12 13 14 15 16 17   15 16 17 18 19 20 21   15 16 17 18 19 20 21
18 19 20 21 22 23 24   22 23 24 25 26 27 28   22 23 24 25 26 27 28
25 26 27 28 29 30 31                          29 30 31

      April                   May                    June
S  M  T  W  T  F  S    S  M  T  W  T  F  S    S  M  T  W  T  F  S
          1  2  3  4                1  2       1  2  3  4  5  6
 5  6  7  8  9 10 11    3  4  5  6  7  8  9    7  8  9 10 11 12 13
12 13 14 15 16 17 18   10 11 12 13 14 15 16   14 15 16 17 18 19 20
19 20 21 22 23 24 25   17 18 19 20 21 22 23   21 22 23 24 25 26 27
26 27 28 29 30         24 25 26 27 28 29 30   28 29 30
                       31

      July                  August                September
S  M  T  W  T  F  S    S  M  T  W  T  F  S    S  M  T  W  T  F  S
          1  2  3  4                      1          1  2  3  4  5
 5  6  7  8  9 10 11    2  3  4  5  6  7  8    6  7  8  9 10 11 12
12 13 14 15 16 17 18    9 10 11 12 13 14 15   13 14 15 16 17 18 19
19 20 21 22 23 24 25   16 17 18 19 20 21 22   20 21 22 23 24 25 26
26 27 28 29 30 31      23 24 25 26 27 28 29   27 28 29 30
                       30 31

     October               November               December
S  M  T  W  T  F  S    S  M  T  W  T  F  S    S  M  T  W  T  F  S
             1  2  3    1  2  3  4  5  6  7          1  2  3  4  5
 4  5  6  7  8  9 10    8  9 10 11 12 13 14    6  7  8  9 10 11 12
11 12 13 14 15 16 17   15 16 17 18 19 20 21   13 14 15 16 17 18 19
18 19 20 21 22 23 24   22 23 24 25 26 27 28   20 21 22 23 24 25 26
25 26 27 28 29 30 31   29 30                  27 28 29 30 31
```

1999

```
     January                February                March
S  M  T  W  T  F  S    S  M  T  W  T  F  S    S  M  T  W  T  F  S
                1  2       1  2  3  4  5  6       1  2  3  4  5  6
 3  4  5  6  7  8  9    7  8  9 10 11 12 13    7  8  9 10 11 12 13
10 11 12 13 14 15 16   14 15 16 17 18 19 20   14 15 16 17 18 19 20
17 18 19 20 21 22 23   21 22 23 24 25 26 27   21 22 23 24 25 26 27
24 25 26 27 28 29 30   28                     28 29 30 31
31

      April                   May                    June
S  M  T  W  T  F  S    S  M  T  W  T  F  S    S  M  T  W  T  F  S
             1  2  3                      1       1  2  3  4  5
 4  5  6  7  8  9 10    2  3  4  5  6  7  8    6  7  8  9 10 11 12
11 12 13 14 15 16 17    9 10 11 12 13 14 15   13 14 15 16 17 18 19
18 19 20 21 22 23 24   16 17 18 19 20 21 22   20 21 22 23 24 25 26
25 26 27 28 29 30      23 24 25 26 27 28 29   27 28 29 30
                       30 31

      July                  August                September
S  M  T  W  T  F  S    S  M  T  W  T  F  S    S  M  T  W  T  F  S
             1  2  3    1  2  3  4  5  6  7             1  2  3  4
 4  5  6  7  8  9 10    8  9 10 11 12 13 14    5  6  7  8  9 10 11
11 12 13 14 15 16 17   15 16 17 18 19 20 21   12 13 14 15 16 17 18
18 19 20 21 22 23 24   22 23 24 25 26 27 28   19 20 21 22 23 24 25
25 26 27 28 29 30 31   29 30 31               26 27 28 29 30

     October               November               December
S  M  T  W  T  F  S    S  M  T  W  T  F  S    S  M  T  W  T  F  S
                1  2       1  2  3  4  5  6             1  2  3  4
 3  4  5  6  7  8  9    7  8  9 10 11 12 13    5  6  7  8  9 10 11
10 11 12 13 14 15 16   14 15 16 17 18 19 20   12 13 14 15 16 17 18
17 18 19 20 21 22 23   21 22 23 24 25 26 27   19 20 21 22 23 24 25
24 25 26 27 28 29 30   28 29 30               26 27 28 29 30 31
31
```

2000

```
     January                February                March
S  M  T  W  T  F  S    S  M  T  W  T  F  S    S  M  T  W  T  F  S
                   1          1  2  3  4  5             1  2  3  4
 2  3  4  5  6  7  8    6  7  8  9 10 11 12    5  6  7  8  9 10 11
 9 10 11 12 13 14 15   13 14 15 16 17 18 19   12 13 14 15 16 17 18
16 17 18 19 20 21 22   20 21 22 23 24 25 26   19 20 21 22 23 24 25
23 24 25 26 27 28 29   27 28 29               26 27 28 29 30 31
30 31

      April                   May                    June
S  M  T  W  T  F  S    S  M  T  W  T  F  S    S  M  T  W  T  F  S
                   1       1  2  3  4  5  6             1  2  3
 2  3  4  5  6  7  8    7  8  9 10 11 12 13    4  5  6  7  8  9 10
 9 10 11 12 13 14 15   14 15 16 17 18 19 20   11 12 13 14 15 16 17
16 17 18 19 20 21 22   21 22 23 24 25 26 27   18 19 20 21 22 23 24
23 24 25 26 27 28 29   28 29 30 31            25 26 27 28 29 30
30

      July                  August                September
S  M  T  W  T  F  S    S  M  T  W  T  F  S    S  M  T  W  T  F  S
                   1          1  2  3  4  5                1  2
 2  3  4  5  6  7  8    6  7  8  9 10 11 12    3  4  5  6  7  8  9
 9 10 11 12 13 14 15   13 14 15 16 17 18 19   10 11 12 13 14 15 16
16 17 18 19 20 21 22   20 21 22 23 24 25 26   17 18 19 20 21 22 23
23 24 25 26 27 28 29   27 28 29 30 31         24 25 26 27 28 29 30
30 31

     October               November               December
S  M  T  W  T  F  S    S  M  T  W  T  F  S    S  M  T  W  T  F  S
 1  2  3  4  5  6  7             1  2  3  4                1  2
 8  9 10 11 12 13 14    5  6  7  8  9 10 11    3  4  5  6  7  8  9
15 16 17 18 19 20 21   12 13 14 15 16 17 18   10 11 12 13 14 15 16
22 23 24 25 26 27 28   19 20 21 22 23 24 25   17 18 19 20 21 22 23
29 30 31               26 27 28 29 30         24 25 26 27 28 29 30
                                              31
```

2001

```
     January                February                March
S  M  T  W  T  F  S    S  M  T  W  T  F  S    S  M  T  W  T  F  S
    1  2  3  4  5  6             1  2  3                1  2  3
 7  8  9 10 11 12 13    4  5  6  7  8  9 10    4  5  6  7  8  9 10
14 15 16 17 18 19 20   11 12 13 14 15 16 17   11 12 13 14 15 16 17
21 22 23 24 25 26 27   18 19 20 21 22 23 24   18 19 20 21 22 23 24
28 29 30 31            25 26 27 28            25 26 27 28 29 30 31

      April                   May                    June
S  M  T  W  T  F  S    S  M  T  W  T  F  S    S  M  T  W  T  F  S
 1  2  3  4  5  6  7          1  2  3  4  5                1  2
 8  9 10 11 12 13 14    6  7  8  9 10 11 12    3  4  5  6  7  8  9
15 16 17 18 19 20 21   13 14 15 16 17 18 19   10 11 12 13 14 15 16
22 23 24 25 26 27 28   20 21 22 23 24 25 26   17 18 19 20 21 22 23
29 30                  27 28 29 30 31         24 25 26 27 28 29 30

      July                  August                September
S  M  T  W  T  F  S    S  M  T  W  T  F  S    S  M  T  W  T  F  S
 1  2  3  4  5  6  7             1  2  3  4                      1
 8  9 10 11 12 13 14    5  6  7  8  9 10 11    2  3  4  5  6  7  8
15 16 17 18 19 20 21   12 13 14 15 16 17 18    9 10 11 12 13 14 15
22 23 24 25 26 27 28   19 20 21 22 23 24 25   16 17 18 19 20 21 22
29 30 31               26 27 28 29 30 31      23 24 25 26 27 28 29
                                              30

     October               November               December
S  M  T  W  T  F  S    S  M  T  W  T  F  S    S  M  T  W  T  F  S
    1  2  3  4  5  6             1  2  3                      1
 7  8  9 10 11 12 13    4  5  6  7  8  9 10    2  3  4  5  6  7  8
14 15 16 17 18 19 20   11 12 13 14 15 16 17    9 10 11 12 13 14 15
21 22 23 24 25 26 27   18 19 20 21 22 23 24   16 17 18 19 20 21 22
28 29 30 31            25 26 27 28 29 30      23 24 25 26 27 28 29
                                              30 31
```

HF SKYWAVE PROPAGATION BASICS

Skywave propagation in the high frequency (HF) band extending from 3 MHz to 30 MHz depends upon the apparent reflection of radio waves from various ionized layers of the earth's atmosphere. To begin to understand the basics of HF skywave radio wave propagation, it must be borne in mind that several interdependent factors come into play. These include time of day, season of the year, and solar activity level. These factors strongly influence such propagation due to its effect on the earth's ionosphere which plays a key role in determining the propagation path for a given HF circuit. In addition, station location at both ends of such a circuit is also important since the amount of daylight and darkness encountered along the signal path influences the propagation of the radio wave by affecting the choice of available frequencies which are capable of being supported by the path. The ionosphere is composed of several low-pressure regions or layers of the atmosphere where ionization consisting of free electrons and ions can be found before combining into neutral atoms. The primary cause of this ionization is radiation from the sun which encompasses the ultraviolet and x-ray regions of the spectrum. In addition, particle emissions from the sun, consisting of high-energy protons and alpha particles and low-energy protons and electrons also play a role in their influence on the ionosphere by providing varying effects on radio wave propagation. It is in such a rarefied atmosphere that radio waves can have their propagation directions altered, as it occurs when an electromagnetic wave passes through the junctions of mediums having different dielectric constants. As a radio wave passes through the ionosphere, the wave loses small but significant amounts of energy to electrons, resulting in the attenuation of the radio wave due to the collision process.

The ionosphere has historically been described by three distinct stratified ionized layers above the earth's surface using the letters of the alphabet D, E and F. Under certain conditions, these layers can also exhibit subdivisions, such as F1 and F2. The D, E and F layers, in order of increasing altitude, play an important role in beyond-line-of-sight propagation of radio waves at HF.

The D layer, ranging in altitude from 50-90 km, primarily acts as an absorber of signal energy, resulting in the attenuation of signals. The D layer also exhibits large daily or diurnal variations in ionization along with pronounced seasonal variations. The daily electron density in the D layer usually peaks shortly after local noon time and tends to exhibit small values at night; the seasonal peak occurs in the summer.

The E layer, ranging in altitude from 90-130 km and having a maximum ionization density at about 110 km, acts as a reflector of radio waves by bending these waves through the mechanism of refraction. The E layer also exhibits a strong solar zenith angle dependency whereby maximum electron density occurs near local noon and a seasonal peak occurs in the summer. Like the D layer, only a small amount of residual ionization remains at night for the E layer. Thus the E layer only maintains its ability to refract radio waves in sunlight and virtually disappears after sunset. There is also a dependence of the electron density in the E layer on sunspot activity, with maximum layer density occurring during the peak of the solar sunspot cycle. The normal or non-sporadic E layer ionization is considered to be important for propagation of daytime HF radio waves at distances under 2000 km. Sporadic E effects, produced by dense patches of ionization, can also occur and it results in enhancement of the E layer. Sporadic E sometimes prevents radio waves from penetrating the E layer and reaching into higher layers of the ionosphere. Sporadic E is also very strongly dependent on latitude whereby it is primarily a night-time occurrence at high latitudes and a daytime occurrence at low latitudes.

The F region of the ionosphere, depending on the season of the year, the time of day, and solar activity, varies in height and is not necessarily a single layer but may also split into two layers during daylight. The lower layer, existing at an altitude range of about 130-210 km, is known as the F1 layer. The higher layer is known as the F2 layer and ranges in height from 250 km to over 500 km. As such, the F2 region is the highest layer of the ionosphere and is the layer that is principally responsible for long distance HF communication. The F1 layer behaves more like the E layer and plays primarily a minor role in long distance communication and contributes additional absorption to radio waves. During the night the F1 layer, like the E layer, disappears by merging with the F2 layer while the F2 layer also tends to lower its altitude from its daytime value. The elevation angle of a transmitted radio wave is measured upward from the horizon and determines the hop distance between the transmitter and the corresponding receiver at the end of the path. If the elevation angle is small, the path of the transmitted wave and its hop distance measured along the surface of the earth will be long. If the elevation angle is large, the path of the radio wave will be short and the hop distance will be short. As the elevation angle becomes larger, a critical angle is reached beyond which the transmitted wave escapes by penetrating the ionospheric layers and will not be returned to the earth.

For single hop ranges of up to 2000 km, the maximum usable frequency (MUF) can be governed by the E, F1 or F2 layers. When the single hop range is increased to about 3000 km, the controlling ionospheric layers are F1 and F2. When the range is extended to about 4000 km, the single hop limit for a transmitted wave is reached with the F2 ionospheric layer. In general, the MUF encountered in each of the E and F propagating layers depends on how highly ionized each layer is.

To the radio amateur, solar sunspot activity is primarily indicated by the sunspot number because these numbers correlate closely with radio wave propagation. Sunspot cycles average approximately 10.7 years in length, with the peaks of the solar maximum and the lows of the solar minimum varying greatly in magnitude. The observations of solar sunspots are usually presented as averaged or statistically smoothed values for the period of six months before and six months after a particular month. International Sunspot Numbers (ISN) are always shown as six months behind the present date and are used as the basis for traditional HF propagation predictions.

(Courtesy of Jake Handwerker, W1FM)

PROPAGATION FORECASTS FOR 1996

The following pages are W3ASK's propagation predictions for Northern Hemisphere locations for 1996. While this data can be used for general guidelines, day-to-day propagation will vary considerably. To get a better idea of the propagation expected for any particular day, consult the monthly Propagation column in CQ and monitor WWV for unusual short-term events. The "NEW Shortwave Propagation Handbook," by W3ASK, N4XX, and K6GKU, published by CQ, is an excellent text on the subject, and is the source for the Propagation Charts included here.

HOW TO USE THE DX PROPAGATION CHARTS

1. Use chart appropriate to your transmitter location. The Eastern USA Chart can be used in the 1, 2, 3, 4, 8, KP4, KG4, and KV4 areas in the USA and adjacent call areas in Canada; the Central USA Chart in the 5, 9, and 0 areas; the Western USA Chart in the 6 and 7 areas; and with somewhat less accuracy in the KH6 and KL7 areas.

2. The predicted times of openings are found under the appropriate meter band column (10 through 80 meters) for a particular DX region, as shown in the left-hand column of the charts. An * indicates the best time to listen for 160 meter openings.

3. The propagation index is the number that appears in ( ) after the time of each predicted opening. The index indicates the number of days during the month on which the opening is expected to take place as follows:

(4) Opening should occur on more than 22 days
(3) Opening should occur between 14 and 22 days
(2) Opening should occur between 7 and 13 days
(1) Opening should occur on less than 7 days

4. Times shown in the charts are in the 24-hour system, where 00 is midnight; 12 is noon; 01 is 1 A.M.; 13 is 1 P.M., etc. Appropriate *daylight* time is used, not GMT. To convert to GMT, add to the times shown in the appropriate chart 7 hours in PDT Zone, 6 hours in MDT Zone, 5 hours in CDT Zone, and 4 hours in EDT Zone. For example, 14 hours in Washington, D.C. is 18 GMT. When it is 20 hours in Los Angeles, it is 03 GMT, etc.

5. The charts are based upon a transmitted power of 250 watts CW, or 1 kw, PEP on sideband, into a dipole antenna a quarter-wavelength above ground on 160 and 80 meters, and a half-wavelength above ground on 40 and 20 meters, and a wavelength above ground on 15 and 10 meters. For each 10 dB gain above these reference levels, the *propagation index* will increase by one level; for each 10 dB loss, it will lower by one level.

6. Propagation data contained in the charts has been prepared from basic data published by the Institute for Telecommunication Sciences of the U.S. Dept of Commerce, Boulder, Colorado 80302.

HOW TO USE THE SHORT-SKIP CHARTS

1. In the Short-Skip Chart, the predicted times of openings can be found under the appropriate distance column of a particular meter band (10 through 160 meters) as shown in the left-hand column of the chart. For the Alaska and Hawaii Charts the predicted times of openings are found under the appropriate meter band column (15 through 80 meters) for a particular geographical region of the continental USA as shown in the left-hand column of the charts. An * indicates the best time to listen for 80 meter openings.

2. The propagation index is the number that appears in ( ) after the time of each predicted opening. On the Short-Skip Chart, where two numerals are shown within a single set of parentheses, the first applies to the shorter distance for which the forecast is made, and the second to the greater distance. The index indicates the number of days during the month on which the opening is expected to take place, as follows:

(4) Opening should occur on more than 22 days
(3) Opening should occur between 14 and 22 days
(2) Opening should occur between 7 and 13 days
(1) Opening should occur on less than 7 days

3. Times shown in the charts are in the 24-hour system, where 00 is midnight; 12 is noon; 01 is 1 AM; 13 is 1 PM, etc. On the Short-Skip Chart appropriate daylight time is used at the path midpoint. For example on a circuit between Maine and Florida, the time shown would be EDT, on a circuit between New York and Texas, the time at the midpoint would be CDT, etc. Times shown in the Hawaii Chart are in HST. To convert to standard time in other USA time zones add 2 hours in the PDT zone; 3 hours in the MDT zone; 4 hours in the CDT zone; and 5 hours in the EDT zone. Add 10 hours to convert from HST to GMT. For example, when it is 12 noon in Honolulu, it is 14 or 2 PM in Los Angeles; 17 or 5 PM in Washington, D.C.; and 22 GMT. Time shown in the Alaska Chart is given in GMT. To convert to daylight time in other areas of the USA subtract 8 hours in the PDT zone; 7 hours in the MDT zone; 6 hours in the CDT zone; and 5 hours in the EDT zone. For example, at 20 GMT it is 15 or 3 PM in New York City.

4. The Short-Skip Chart is based upon a transmitted power of 75 watts CW or 300 watts PEP on sideband; the Alaska and Hawaii Charts are based upon a transmitter power of 250 watts CW or 1 kw PEP on sideband. A dipole antenna a quarter-wavelength above ground is assumed for 160 and 80 meters, a half-wave above ground on 40 and 20 meters, and a wavelength above ground on 15 and 10 meters. For each 10 dB gain above these reference levels, the propagation index will increase by one level; for each 10 dB loss, it will lower by one level.

5. Propagation data contained in the charts has been prepared from basic data published by the Institute for Telecommunication Sciences of the U.S. Dept. of Commerce, Boulder, Colorado 80302.

MASTER DX PROPAGATION CHART
SOLAR PHASE: LOW
SMOOTHED SUNSPOT RANGE: 0-30
SEASON: SUMMER
TIME ZONE: EST (24-Hour Time)
EASTERN USA TO:

Reception Area	10/15 Meters	20 Meters	40 Meters	80/160 Meters
Western & Central Europe & North Africa	15-18 (1)	05-06 (1) 06-09 (3) 09-13 (2) 13-15 (1) 15-17 (4) 17-19 (3) 19-20 (2) 20-21 (1)	19-22 (1) 22-00 (2) 00-02 (1)	21-23 (1) 23-00 (2) 00-01 (1) 23-01 (1)†
Northern & Eastern Europe	Nil	05-07 (1) 07-09 (2) 09-14 (1) 14-16 (2) 16-19 (1)	22-00 (1)	22-00 (1)
Eastern Mediterranean & East Africa	11-13 (1)	05-06 (1) 06-07 (2) 07-09 (1) 09-11 (2) 11-15 (1) 15-17 (2) 17-19 (1)	20-00 (1)	21-23 (1)
Western Africa	14-17 (1)	04-06 (1) 06-08 (2) 08-15 (1) 15-16 (2) 16-18 (3) 18-19 (2) 19-21 (1)	21-00 (1) 00-02 (2) 02-04 (1)	00-02 (1)
Central & South Africa	Nil	05-06 (1) 06-07 (2) 07-14 (1) 14-16 (2) 16-18 (2) 01-03 (1)	22-23 (1) 23-01 (2) 01-03 (1)	23-01 (1)
Central & South Asia	Nil	05-08 (1) 18-21 (1)	Nil	Nil
Southeast Asia	Nil	05-06 (1) 06-08 (2) 08-10 (1) 18-21 (1)	Nil	Nil
Far East	Nil	06-07 (1) 07-09 (2) 09-11 (1) 20-23 (1)	Nil	Nil
South Pacific & New Zealand	18-20 (1)	16-22 (1) 22-00 (2) 00-06 (1) 06-09 (2) 09-11 (1)	01-02 (1) 02-05 (2) 05-06 (1)	02-05 (1) 02-04 (1)†
Australasia	19-22 (1)	15-22 (1) 22-00 (2) 00-06 (1) 06-09 (2) 09-11 (1)	01-02 (1) 02-05 (2) 05-06 (1)	03-05 (1) 03-04 (1)†
Carribbean, Central America & Northern Countries of South America	13-15 (1)* 15-17 (2)* 17-18 (1)* 08-09 (1) 09-11 (3) 11-13 (2) 13-14 (3) 14-17 (4) 17-19 (3) 19-21 (1)	06-07 (3) 07-09 (4) 09-11 (3) 11-16 (2) 16-18 (3) 18-21 (4) 21-22 (3) 22-00 (2) 00-06 (1)	19-21 (1) 21-00 (2) 00-03 (3) 03-05 (2) 05-06 (1)	22-01 (1) 01-04 (2) 04-05 (1) 01-03 (1)†
Peru, Bolivia, Paraguay, Brazil, Chile, Argentina & Uruguay	14-17 (1)* 08-11 (1) 11-14 (2) 14-15 (3) 15-16 (4) 16-17 (3) 17-18 (2) 18-20 (1)	05-06 (1) 06-10 (2) 10-14 (1) 14-17 (2) 17-18 (3) 18-20 (4) 20-22 (3) 22-23 (2) 23-01 (1)	21-00 (1) 00-02 (2) 02-06 (1)	00-04 (1) 02-04 (1)†
McMurdo Sound, Antarctica	14-17 (1)	14-16 (1) 16-18 (2) 18-22 (1)	03-07 (1)	Nil

MASTER DX PROPAGATION CHART
SOLAR PHASE: LOW
SMOOTHED SUNSPOT RANGE: 0-30
SEASON: SUMMER
TIME ZONES: CST & MST (24-Hour Time)
CENTRAL USA TO:

Reception Area	10/15 Meters	20 Meters	40 Meters	80/160 Meters
Western & Central Europe & North Africa	15-17 (1)	05-06 (1) 06-08 (2) 08-12 (1) 12-14 (2) 14-17 (3) 17-19 (2) 19-20 (1)	20-22 (1) 22-00 (2) 00-01 (1)	21-23 (1)
Northern & Eastern Europe	Nil	05-07 (1) 07-09 (2) 09-13 (1) 13-15 (2) 15-18 (1)	21-23 (1)	Nil
Eastern Mediterranean & East Africa	Nil	05-06 (1) 06-07 (2) 07-14 (1) 14-16 (2) 16-18 (1)	20-23 (1)	Nil
Western Africa	Nil	05-06 (1) 06-09 (2) 09-14 (1) 14-16 (2) 16-17 (3) 17-18 (2) 18-20 (1)	21-00 (1) 00-01 (2) 01-03 (1)	00-01 (1)
Central & South Africa	Nil	05-06 (1) 06-07 (2) 07-14 (1) 14-16 (2) 16-18 (2) 00-02 (1)	22-23 (1) 23-00 (2) 00-02 (1)	22-00 (1)
Central & South Asia	Nil	05-09 (1) 18-21 (1)	Nil	Nil
Southeast Asia	Nil	05-06 (1) 06-09 (2) 09-11 (1) 18-20 (1) 20-22 (2) 22-23 (1)	Nil	Nil

Reception Area	10/15 Meters	20 Meters	30 Meters	80/160 Meters
Far East	21-23 (1)	06-07 (1)	04-06 (1)	Nil
		07-10 (2)		
		10-20 (1)		
		20-22 (2)		
		22-00 (1)		
South Pacific & New Zealand	14-18 (1)	02-06 (1)	00-02 (1)	01-06 (1)
	18-20 (2)	06-09 (2)	02-06 (2)	03-05 (1)†
	20-22 (1)	09-17 (1)	06-07 (1)	
		17-19 (2)		
		19-22 (3)		
		20-02 (2)		
Australasia	15-17 (1)	06-07 (1)	00-02 (1)	02-03 (1)
	17-19 (2)	07-09 (2)	02-05 (2)	03-05 (2)
	19-22 (1)	09-13 (1)	05-07 (1)	05-06 (1)
		13-15 (2)		03-05 (1)†
		15-18 (1)		
		18-20 (2)		
		20-23 (3)		
		23-01 (2)		
		01-03 (1)		
Caribbean, Central America & Northern Countries of South America	14-15 (1)*	06-07 (3)	19-21 (1)	21-23 (1)
	15-16 (2)*	07-09 (4)	21-23 (2)	23-02 (2)
	16-17 (1)*	09-11 (3)	23-02 (3)	02-04 (1)
	08-10 (1)	11-16 (2)	02-04 (2)	00-02 (1)†
	10-12 (2)	16-18 (3)	04-05 (1)	
	12-14 (3)	18-20 (4)		
	14-16 (4)	20-22 (3)		
	16-18 (3)	22-00 (2)		
	18-19 (2)	00-06 (1)		
	19-20 (1)			
Peru, Bolivia, Paraguay, Brazil, Chile, Argentina & Uruguay	13-16 (1)*	05-06 (1)	20-23 (1)	23-04 (1)
	08-11 (1)	06-09 (2)	23-01 (2)	01-03 (1)†
	11-13 (2)	09-14 (1)	01-05 (1)	
	13-14 (3)	14-16 (2)		
	14-16 (4)	16-17 (3)		
	16-17 (3)	17-19 (4)		
	17-18 (2)	19-21 (3)		
	18-19 (1)	21-22 (2)		
		22-00 (1)		
McMurdo Sound, Antarctica	13-15 (1)	12-16 (1)	03-07 (1)	Nil
		16-18 (2)		
		18-21 (1)		

MASTER DX PROPAGATION CHART
SOLAR PHASE: LOW
SMOOTHED SUNSPOT RANGE: 0-30
SEASON: SUMMER
TIME ZONE: PST (24-Hour Time)
WESTERN USA TO:

Reception Area	10/15 Meters	20 Meters	30 Meters	80/160 Meters
Western & Central Europe & North Africa	Nil	20-22 (1)	19-23 (1)	Nil
		05-06 (1)		
		06-08 (2)		
		08-13 (1)		
		13-16 (2)		
		16-17 (1)		
Northern & Eastern Europe	Nil	05-07 (1)	20-22 (1)	Nil
		07-09 (2)		
		09-16 (1)		
		20-22 (1)		
Eastern Mediterranean & East Africa	Nil	06-11 (1)	Nil	Nil
		11-14 (2)		
		14-16 (1)		
		20-22 (1)		
Western & Central Africa	09-11 (1)	21-23 (1)	20-23 (1)	Nil
		05-06 (1)		
		06-08 (2)		
		08-13 (1)		
		13-16 (3)		
		16-17 (2)		
		17-18 (1)		

Reception Area	10/15 Meters	20 Meters	30 Meters	80/160 Meters
Southern Africa	Nil	05-07 (1)	19-21 (1)	19-21 (1)
		07-08 (2)	20-21 (2)	
		08-13 (1)	21-22 (1)	
		21-23 (1)		
Central & South Asia	Nil	07-11 (1)	Nil	Nil
		17-18 (1)		
		18-20 (2)		
		20-21 (1)		
Southeast Asia	20-22 (1)	06-08 (1)	02-06 (1)	Nil
		08-09 (2)		
		09-13 (1)		
		19-21 (1)		
		21-23 (2)		
		23-00 (1)		
Far East	12-14 (1)	06-07 (1)	01-02 (1)	01-04 (1)
	20-22 (1)	07-09 (2)	02-05 (2)	
		09-18 (1)	05-07 (1)	
		18-20 (2)		
		20-22 (2)		
		22-00 (2)		
		00-02 (1)		
South Pacific & New Zealand	14-17 (1)	02-07 (1)	23-01 (1)	23-01 (1)
	17-20 (2)	07-09 (2)	01-04 (3)	01-04 (2)
	20-21 (1)	09-11 (1)	04-06 (2)	04-06 (1)
		11-17 (2)	06-07 (1)	02-04 (1)†
		17-18 (3)		
		18-22 (2)		
		22-00 (3)		
		00-02 (2)		
Australasia	14-17 (1)	01-07 (1)	23-01 (1)	00-02 (1)
	17-20 (2)	07-09 (2)	01-04 (2)	02-04 (2)
	20-22 (1)	09-12 (1)	04-07 (1)	04-06 (1)
		12-14 (2)		02-04 (1)†
		14-18 (1)		
		18-20 (2)		
		20-23 (3)		
		23-01 (2)		
Caribbean, Central America & Northern Countries of South America	14-17 (1)*	06-08 (3)	19-21 (1)	20-22 (1)
	08-10 (1)	08-10 (1)	21-23 (2)	22-00 (2)
	10-12 (2)	10-13 (1)	23-01 (3)	00-03 (1)
	12-14 (3)	13-15 (2)	01-03 (2)	00-02 (1)†
	14-16 (4)	15-17 (3)	03-04 (1)	
	16-18 (2)	17-20 (4)		
	18-19 (1)	20-22 (3)		
		22-23 (2)		
		23-04 (2)		
		04-06 (2)		
Peru, Bolivia, Paraguay, Brazil, Chile, Argentina & Uruguay	12-14 (1)*	05-06 (1)	20-22 (1)	22-04 (1)
	08-11 (1)	06-08 (2)	22-01 (2)	00-02 (1)†
	11-12 (2)	08-14 (1)	01-04 (1)	
	12-13 (3)	14-16 (2)		
	13-15 (4)	16-18 (4)		
	15-16 (2)	18-20 (2)		
	16-18 (1)	20-22 (1)		
McMurdo Sound, Antarctica	12-16 (1)	11-16 (1)	19-21 (1)	
		16-18 (2)	02-07 (1)	
		18-20 (1)		

MASTER DX PROPAGATION CHART
SOLAR PHASE: LOW
SMOOTHED SUNSPOT RANGE: 0-30
SEASON: WINTER
TIME ZONE: EST (24-Hour Time)
EASTERN USA TO:

Reception Area	10/15 Meters	20 Meters	40 Meters	80/160 Meters
Western & Central Europe & North Africa	09-11 (1)*	06-07 (1)	15-16 (1)	17-19 (1)
	08-09 (1)	07-08 (2)	16-17 (2)	19-20 (2)
	09-11 (2)	08-10 (3)	17-19 (3)	20-02 (3)
	11-13 (1)	10-12 (4)	19-01 (2)	02-03 (2)
		12-13 (3)	01-03 (3)	03-04 (1)
		13-14 (2)	03-04 (2)	20-00 (1)†
		14-16 (1)	04-05 (1)	00-02 (2)†
				02-03 (1)†

Reception Area	10/15 Meters	20 Meters	40 Meters	80/160 Meters
Northern & Eastern Europe	08-11 (1)	06-07 (1) 07-11 (2) 11-13 (1)	15-17 (1) 17-19 (2) 19-01 (1) 01-02 (2) 02-03 (1)	17-19 (1) 19-02 (2) 02-03 (1) 21-02 (1)†
Eastern Mediterranean & Middle East	09-10 (1)* 08-09 (1) 09-11 (2) 11-12 (1)	07-08 (1) 08-10 (2) 10-12 (3) 12-14 (2) 14-15 (1)	17-19 (1) 19-21 (2) 21-00 (1) 00-01 (2) 01-02 (1)	18-20 (1) 20-22 (1) 22-00 (1) 22-00 (1)†
Western Africa	10-12 (1)* 08-09 (1) 09-11 (2) 11-13 (3) 13-14 (2) 14-15 (1)	06-07 (1) 07-09 (2) 09-12 (1) 12-14 (2) 14-16 (3) 16-17 (2) 17-18 (1)	18-20 (1) 20-23 (2) 23-01 (1) 01-03 (2) 03-04 (1)	19-22 (1) 22-01 (2) 01-03 (1) 22-01 (1)†
Eastern & Central Africa	10-12 (1)* 08-11 (1) 11-13 (2) 13-14 (1)	07-13 (1) 13-16 (2) 16-18 (1)	18-20 (1) 20-23 (2) 23-01 (1)	19-00 (1)
Southern Africa	10-13 (1)* 08-09 (1) 09-11 (2) 11-13 (3) 13-14 (2) 14-15 (1)	07-09 (1) 12-14 (1) 14-15 (2) 15-16 (3) 16-17 (2) 17-19 (1)	18-20 (1) 20-22 (1) 22-00 (1)	19-22 (1)
Central & South Asia	16-18 (1)	07-10 (1) 19-21 (1)	06-08 (1) 18-22 (1)	06-07 (1) 18-20 (1)
Southeast Asia	16-18 (1)	07-10 (1) 17-20 (1)	06-08 (1) 18-21 (1)	06-07 (1) 18-20 (1)
Far East	16-18 (1)	06-07 (1) 07-09 (2) 09-11 (1) 15-17 (1) 17-19 (2) 19-21 (1)	05-08 (1) 17-18 (1)	05-08 (1) 17-18 (1)
South Pacific & New Zealand	13-15 (1)* 12-14 (1) 14-17 (2) 17-18 (1)	05-07 (1) 07-10 (2) 10-18 (1) 18-20 (2) 20-22 (1)	01-02 (1) 02-04 (2) 04-07 (3) 07-08 (2) 08-09 (1)	04-05 (1) 05-07 (2) 07-08 (1) 04-07 (1)†
Australasia	14-16 (1)* 12-15 (1) 15-17 (2) 17-18 (1)	06-07 (1) 07-10 (2) 10-12 (3) 15-16 (1) 16-19 (2) 19-21 (1)	03-05 (1) 05-08 (2) 08-09 (1) 17-19 (1)	05-06 (1) 06-07 (2) 07-08 (1) 17-18 (1) 05-07 (1)†
Caribbean, Central America & Northern Countries of South America	10-15 (1)* 08-09 (1) 09-12 (2) 12-16 (3) 16-17 (2) 17-18 (1)	05-07 (1) 07-08 (3) 08-09 (4) 09-11 (3) 11-15 (2) 15-17 (3) 17-18 (4) 18-19 (3) 19-20 (2) 20-02 (1)	17-18 (1) 18-19 (2) 19-21 (3) 21-03 (2) 03-06 (3) 06-07 (2) 07-08 (1)	18-20 (1) 20-21 (2) 21-04 (3) 04-06 (2) 06-07 (1) 21-03 (1)† 03-05 (2)† 05-06 (1)†
Peru, Bolivia, Paraguay, Brazil, Chile, Argentina & Uruguay	11-15 (1)* 08-09 (1) 09-11 (2) 11-13 (2) 13-14 (2) 14-16 (3) 16-17 (2) 17-18 (1)	06-07 (1) 07-09 (2) 09-10 (1) 12-14 (2) 14-15 (2) 15-16 (3) 16-18 (4) 18-19 (3) 19-20 (2) 20-22 (1) 22-00 (2) 00-02 (1)	19-21 (1) 21-02 (2) 02-05 (1) 05-06 (2) 06-07 (1)	21-03 (1) 03-05 (2) 05-06 (1) 03-05 (1)†
McMurdo Sound, Antarctica	15-17 (1)	07-09 (1) 17-18 (1) 18-20 (2) 20-22 (1) 22-00 (2) 00-02 (1)	22-00 (1) 00-02 (2) 02-06 (1)	Nil

MASTER DX PROPAGATION CHART
SOLAR PHASE: LOW
SMOOTHED SUNSPOT RANGE: 0-30
SEASON: WINTER
TIME ZONES: CST & MST (24-Hour Time)
CENTRAL USA TO:

Reception Area	10/15 Meters	20 Meters	40 Meters	80/160 Meters
Western Europe & North Africa	08-09 (1) 09-11 (2) 11-12 (1)	06-08 (1) 08-10 (2) 10-12 (3) 12-13 (2) 13-15 (1)	15-17 (1) 17-19 (2) 19-23 (3) 23-01 (2) 01-02 (1)	17-19 (1) 19-00 (2) 00-01 (1) 20-01 (1)†
Northern, Central, & Eastern Europe	08-11 (1)	07-08 (1) 08-11 (1) 11-12 (1)	16-18 (1) 18-19 (2) 19-22 (1) 22-00 (2) 00-01 (1)	18-00 (1) 20-00 (1)†
Eastern Mediterranean & Middle East	08-11 (1)	07-09 (1) 09-12 (2) 12-14 (1) 22-00 (1)	17-19 (1) 19-22 (2) 22-23 (1)	19-22 (1)
Western Africa	09-12 (1)* 08-09 (1) 09-11 (2) 11-13 (3) 13-14 (2) 14-15 (1)	06-07 (1) 07-09 (2) 09-11 (1) 11-13 (2) 13-15 (3) 15-16 (2) 16-18 (1) 22-00 (1)	17-20 (1) 20-23 (2) 23-01 (1)	19-22 (1) 22-23 (2) 23-00 (1) 21-23 (1)†
Eastern & Central Africa	10-12 (1)* 08-11 (1) 11-13 (2) 13-14 (1)	06-12 (1) 12-14 (2) 14-16 (3) 16-17 (1)	18-19 (1) 19-21 (2) 21-23 (1)	19-22 (1)
Southern Africa	10-12 (1)* 08-10 (1) 10-13 (2) 13-14 (1)	07-13 (1) 13-15 (2) 15-16 (3) 16-17 (2) 17-18 (1) 22-00 (1)	18-19 (1) 19-21 (2) 21-23 (1)	19-22 (1)
Central & South Asia	17-19 (1)	07-10 (1) 19-21 (1)	06-08 (1) 18-21 (1)	06-07 (1) 18-20 (1)
Southeast Asia	17-19 (1)	06-07 (1) 07-09 (2) 09-12 (1) 17-20 (1)	06-08 (1) 17-19 (1)	06-07 (1) 17-19 (1)
Far East	17-19 (1)	06-07 (1) 07-09 (2) 09-11 (1) 15-17 (1) 17-19 (2) 19-20 (1)	01-03 (1) 03-07 (2) 07-08 (1)	02-04 (1) 04-06 (2) 06-07 (1) 04-06 (1)†
South Pacific & New Zealand	12-16 (1)* 11-13 (1) 13-15 (2) 15-17 (3) 17-18 (2) 18-19 (1)	06-07 (1) 07-11 (2) 11-16 (1) 16-17 (2) 17-19 (3) 19-20 (2) 20-22 (1)	23-01 (1) 01-02 (2) 02-06 (3) 06-07 (2) 07-09 (1)	00-01 (1) 01-06 (2) 06-08 (1) 13-07 (1)†

Reception Area				
Australasia	14-17 (1)* 11-15 (1) 15-18 (2) 18-19 (1)	06-07 (1) 07-11 (2) 11-18 (1) 18-21 (2) 21-22 (1)	01-03 (1) 03-07 (3) 07-08 (2) 08-09 (1)	03-05 (1) 05-07 (2) 07-08 (1) 04-07 (1)†
Caribbean, Central America & Northern Countries of South America	10-15 (1)* 07-08 (1) 08-10 (2) 10-13 (3) 13-15 (4) 15-16 (3) 16-18 (1)	04-06 (1) 06-07 (2) 07-10 (3) 10-14 (2) 14-16 (1) 16-18 (4) 18-19 (3) 19-20 (2) 20-22 (1) 22-00 (1) 00-02 (1)	18-20 (1) 20-22 (2) 22-00 (3) 00-04 (2) 04-06 (3) 06-07 (1)	19-21 (1) 21-05 (2) 05-06 (1) 23-05 (1)†
Peru, Bolivia, Paraguay, Brazil, Chile, Argentina & Uruguay	11-15 (1)* 07-09 (1) 09-11 (2) 11-13 (1) 13-14 (2) 14-16 (3) 16-17 (2) 17-18 (1)	06-07 (1) 07-09 (2) 09-13 (1) 13-14 (2) 14-15 (3) 15-17 (4) 17-18 (3) 18-19 (2) 19-21 (1) 21-23 (2) 23-00 (1)	19-21 (1) 21-02 (2) 02-04 (1) 04-06 (2) 06-07 (1)	21-05 (1) 00-04 (1)†
McMurdo Sound, Antarctica	15-17 (1)	06-07 (1) 07-09 (2) 09-11 (1) 17-18 (1) 18-20 (1) 20-22 (1) 22-00 (1) 00-02 (1)	22-00 (1) 00-02 (2) 02-06 (1)	Nil

MASTER DX PROPAGATION CHART
SOLAR PHASE: LOW
SMOOTHED SUNSPOT RANGE: 0-30
SEASON: WINTER
TIME ZONE: PST (24-Hour Time)
WESTERN USA TO:

Reception Area	10/15 Meters	20 Meters	40 Meters	80/160 Meters
Western Europe & North Africa	08-10 (1)	06-07 (1) 07-11 (2) 11-13 (1) 23-01 (1)	17-21 (1) 21-23 (1) 23-01 (1)	18-20 (1) 20-22 (2) 22-23 (1) 19-22 (1)†
Northern, Central & Eastern Europe	08-10 (1)	06-07 (1) 07-10 (2) 10-12 (1) 23-01 (1)	17-00 (1)	19-22 (1) 19-21 (1)†
Eastern Mediterranean & Middle East	08-10 (1)	07-10 (1) 10-12 (2) 12-13 (1) 21-23 (1)	06-08 (1) 18-22 (1)	06-08 (1) 18-21 (1)
Western Africa	09-11 (1)* 08-09 (1) 09-12 (2) 12-13 (1)	07-10 (1) 10-13 (2) 13-16 (3) 16-17 (2) 17-18 (1)	18-23 (1)	19-22 (1)
Eastern & Central Africa	09-11 (1)	08-10 (1) 13-16 (1) 21-23 (1)	06-08 (1) 18-22 (1)	06-08 (1) 18-21 (1)
Southern Africa	08-10 (1) 10-12 (2) 12-14 (1)	09-13 (1) 13-16 (2) 16-18 (1) 23-01 (1)	18-21 (1)	18-20 (1)

Reception Area				
Central & South Asia	17-19 (1)	08-10 (1) 17-18 (1) 18-19 (2) 19-20 (1)	15-08 (1) 17-19 (1)	05-07 (1)
Southeast Asia	14-16 (1)* 14-15 (1) 15-17 (2) 17-18 (1)	08-09 (1) 09-11 (2) 11-16 (1) 16-19 (2) 19-20 (1)	01-04 (1) 04-07 (2) 07-09 (1)	04-07 (1)
Far East	14-15 (1) 15-17 (2) 17-19 (1)	08-10 (1) 13-14 (1) 14-15 (2) 15-17 (3) 17-18 (2) 18-19 (1)	22-00 (1) 00-02 (2) 02-06 (3) 06-08 (2) 08-10 (1)	23-01 (1) 01-06 (2) 06-08 (1) 01-06 (1)†
South Pacific & New Zealand	14-16 (1)* 11-13 (1) 13-14 (2) 14-16 (3) 16-18 (2) 18-19 (1)	07-08 (1) 08-13 (2) 13-15 (1) 15-16 (2) 16-18 (4) 18-19 (2) 19-21 (1)	20-22 (1) 22-00 (2) 00-07 (3) 07-08 (2) 08-09 (1)	00-03 (1) 03-06 (2) 06-08 (1) 03-06 (1)†
Australasia	14-16 (1)* 10-13 (1) 13-15 (2) 15-17 (3) 17-18 (1)	07-08 (1) 08-11 (2) 11-17 (1) 17-18 (2) 18-19 (3) 19-20 (2) 20-22 (1)	01-03 (1) 03-05 (2) 05-07 (3) 07-08 (2) 08-09 (1)	03-05 (1) 05-06 (2) 06-08 (1) 04-07 (1)†
Caribbean, Central America & Northern Countries of South America	10-14 (1)* 07-08 (1) 08-10 (2) 10-12 (3) 12-14 (4) 14-15 (3) 15-16 (2) 16-17 (1)	04-06 (1) 06-07 (2) 07-09 (3) 09-13 (2) 13-15 (3) 15-17 (4) 17-18 (3) 18-19 (2) 19-21 (1) 21-23 (2) 23-01 (1)	18-20 (1) 20-21 (2) 21-23 (3) 23-01 (2) 01-03 (3) 03-04 (2) 04-05 (1)	19-21 (1) 21-03 (2) 03-04 (1) 21-03 (1)†
Peru, Bolivia, Paraguay, Brazil, Chile, Argentina & Uruguay	11-14 (1)* 08-10 (1) 10-12 (2) 12-14 (3) 14-16 (2) 16-17 (1)	05-07 (1) 07-09 (2) 09-13 (1) 13-15 (2) 15-16 (3) 16-17 (4) 17-18 (3) 18-19 (2) 19-20 (1) 22-00 (1)	19-21 (1) 21-00 (2) 00-02 (1) 02-04 (2) 04-06 (1)	22-05 (1) 00-04 (1)†
McMurdo Sound, Antarctica	14-16 (1)	06-07 (1) 07-09 (2) 09-11 (1) 15-17 (1) 17-19 (2) 19-21 (1) 21-23 (2) 23-01 (1)	21-00 (1) 00-02 (2) 02-05 (1)	Nil

MASTER DX PROPAGATION CHART
SOLAR PHASE: LOW
SMOOTHED SUNSPOT RANGE: 0-30
SEASONS: SPRING & FALL
TIME ZONE: EST (24-Hour Time)
EASTERN USA TO:

Reception Area	10/15 Meters	20 Meters	40 Meters	80/160 Meters
Western & Central Europe & North Africa	08-10 (1) 10-12 (2) 12-14 (1)	06-07 (1) 07-08 (2) 08-12 (3) 12-13 (4)	16-18 (1) 18-19 (2) 19-23 (3) 23-02 (2)	18-20 (1) 20-23 (3) 23-01 (2) 01-03 (1)

Reception Area	10/15 Meters	20 Meters	40 Meters	80/160 Meters
	13-14 (3) 14-15 (2) 15-18 (1)	02-05 (1)		20-23 (1)† 23-01 (2)† 01-02 (1)†
Northern & Eastern Europe	09-12 (1)	06-07 (1) 07-10 (2) 10-14 (1)	18-02 (1)	20-00 (1) 21-23 (1)†
Eastern Mediterranean & East Africa	09-11 (1)	06-11 (1) 11-13 (2) 13-15 (1)	18-20 (1) 20-21 (2) 21-23 (1)	19-23 (1) 20-22 (1)†
Western Africa	08-10 (1) 10-12 (3) 12-13 (2) 13-16 (1)	06-07 (1) 07-09 (2) 09-12 (1) 12-14 (2) 14-16 (3) 16-17 (2) 17-19 (1)	18-19 (1) 19-22 (2) 22-01 (1)	19-21 (1) 21-22 (2) 22-00 (1) 20-22 (1)†
Central & South Africa	10-13 (1)* 07-10 (1) 10-12 (2) 12-14 (2) 14-15 (2) 15-18 (1)	07-14 (1) 14-15 (2) 15-17 (3) 17-18 (2) 18-21 (1)	18-20 (1) 20-22 (2) 22-00 (1)	19-22 (1) 19-21 (1)†
Central & South Asia	Nil	06-07 (1) 07-09 (2) 09-11 (1) 19-22 (1)	05-07 (1) 18-21 (1)	Nil
Southeast Asia	17-19 (1)	06-07 (1) 07-09 (2) 09-11 (1) 17-20 (1)	06-08 (1) 17-20 (1)	Nil
Far East	16-19 (1)	06-07 (1) 07-09 (2) 09-11 (1) 17-20 (1)	05-08 (1)	06-07 (1)
South Pacific & New Zealand Australasia	15-17 (1)* 12-16 (1) 16-18 (2) 18-20 (1) 12-16 (1) 16-18 (2) 18-20 (1)	07-09 (2) 09-20 (1) 20-22 (2) 22-07 (1) 06-07 (1) 07-09 (2) 09-15 (1) 20-23 (1)	00-02 (1) 02-06 (3) 06-07 (2) 07-08 (1) 03-05 (1) 05-07 (2) 07-09 (1)	02-03 (1) 03-05 (2) 05-07 (1) 02-06 (1)† 04-05 (1) 05-07 (2) 07-08 (1) 05-07 (1)†
Caribbean, Central America & Northern Countries of South America	12-16 (1)* 07-08 (1) 08-09 (2) 09-11 (4) 11-13 (2) 13-15 (4) 15-16 (3) 16-18 (2) 18-20 (1)	00-06 (1) 06-07 (2) 07-10 (3) 10-15 (2) 15-17 (3) 17-19 (4) 19-21 (3) 21-00 (2)	18-19 (1) 19-20 (2) 20-03 (3) 03-05 (2) 05-07 (1)	19-21 (1) 00-02 (2) 02-06 (1) 00-04 (1)†
Peru, Bolivia, Paraguay, Brazil, Chile, Argentina & Uruguay	12-15 (1)* 07-09 (1) 09-11 (2) 11-13 (1) 13-15 (2) 15-17 (3) 17-19 (1)	06-07 (1) 07-10 (2) 10-14 (1) 14-16 (2) 16-18 (3) 18-20 (4) 20-22 (2) 22-03 (1)	19-21 (1) 21-03 (2) 03-07 (1)	21-06 (1) 01-04 (1)†
McMurdo Sound, Antarctica	15-17 (1)	16-18 (1) 18-20 (2) 20-23 (1) 06-07 (1) 07-09 (2) 09-11 (1)	23-05 (1)	Nil

MASTER DX PROPAGATION CHART
SOLAR PHASE: LOW
SMOOTHED SUNSPOT RANGE: 0-30
SEASONS: SPRING & FALL
TIME ZONES: CST & MST (24-Hour Time)
CENTRAL USA TO:

Reception Area	10/15 Meters	20 Meters	40 Meters	80/160 Meters
Western & Central Europe & North Africa	06-08 (1) 08-09 (1) 09-12 (2) 12-14 (1)	06-08 (1) 08-12 (2) 12-14 (3) 14-15 (2) 15-17 (1)	16-19 (1) 19-22 (2) 22-02 (1)	18-20 (1) 20-22 (2) 22-00 (1) 20-00 (1)†
Northern & Eastern Europe	08-12 (1)	07-08 (1) 08-10 (2) 10-13 (1)	19-01 (1)	20-23 (1)
Eastern Mediterranean & East Africa	09-12 (1)	07-11 (1) 11-13 (2) 13-15 (1)	19-23 (1)	20-22 (1)
Western Africa	08-10 (1) 10-12 (2) 12-15 (1)	06-07 (1) 07-09 (2) 09-11 (2) 11-13 (2) 13-15 (3) 15-17 (2) 17-18 (1)	18-19 (1) 19-21 (2) 21-00 (1)	19-20 (1) 20-22 (2) 22-23 (1) 20-22 (1)†
Central & South Africa	11-13 (1)* 08-10 (1) 10-12 (2) 12-14 (3) 14-15 (2) 15-18 (1)	07-14 (1) 14-15 (2) 15-16 (3) 16-17 (2) 17-20 (1)	18-20 (1) 20-22 (2) 22-00 (1)	19-22 (1) 19-21 (1)†
Central & South Asia	Nil	06-07 (1) 07-09 (2) 09-11 (1) 19-21 (1)	06-08 (1) 19-21 (1)	Nil
Southeast Asia	10-14 (1) 17-20 (1)	06-07 (1) 07-09 (2) 09-12 (1) 19-21 (1)	06-08 (1) 17-19 (1)	Nil
Far East	16-19 (1)	06-07 (1) 07-09 (2) 09-11 (1) 16-18 (1) 18-20 (1) 20-22 (1)	02-09 (1)	05-07 (1)
South Pacific & New Zealand	14-17 (1)* 12-16 (1) 16-18 (2) 18-21 (1)	18-19 (2) 19-21 (3) 21-00 (2) 00-06 (1) 06-09 (2) 09-18 (1)	22-01 (1) 01-06 (3) 06-07 (2) 07-09 (1)	00-03 (1) 03-06 (2) 06-07 (1) 03-07 (1)†
Australasia	12-16 (1) 16-18 (2) 18-20 (1)	06-07 (1) 07-09 (2) 09-14 (1) 17-19 (1) 19-21 (2) 21-23 (1)	02-04 (1) 04-07 (2) 07-09 (1)	04-05 (1) 05-07 (2) 07-08 (1) 05-07 (1)†
Caribbean, Central America & Northern Countries of South America	11-15 (1)* 07-08 (1) 08-09 (2) 09-14 (2) 11-13 (2) 13-15 (4) 15-17 (2) 17-19 (1)	00-06 (1) 06-07 (2) 07-09 (3) 09-14 (2) 14-16 (3) 16-18 (4) 18-20 (3) 20-00 (2)	18-19 (1) 19-20 (2) 20-02 (3) 02-04 (2) 04-06 (1)	20-21 (1) 21-02 (2) 02-06 (1) 00-03 (1)†

Peru, Bolivia, Paraguay, Brazil, Chile, Argentina & Uruguay	12-14 (1)* 07-08 (1) 08-10 (2) 10-12 (1) 12-13 (2) 13-16 (3) 16-17 (2) 17-19 (1)	06-07 (1) 07-09 (2) 09-13 (1) 13-15 (2) 15-17 (3) 17-19 (4) 19-21 (2) 21-02 (1)	19-21 (1) 21-03 (2) 03-06 (1)	21-05 (1) 01-04 (1)†
McMurdo Sound, Antarctica	15-17 (1)	16-18 (1) 18-20 (2) 20-00 (1) 06-07 (1) 07-09 (2) 09-11 (1)	00-07 (1)	Nil

MASTER DX PROPAGATION CHART
SOLAR PHASE: LOW
SMOOTHED SUNSPOT RANGE 0-30
SEASONS: SPRING & FALL
TIME ZONE: PST (24-Hour Time)
WESTERN USA TO:

Reception Area	10/15 Meters	20 Meters	40 Meters	80/160 Meters
Western & Central Europe & North Africa	09-11 (1)	23-01 (1) 06-08 (1) 08-11 (2) 11-15 (1)	18-00 (1)	19-22 (1) 19-21 (1)†
Northern & Eastern Europe	07-10 (1)	23-01 (1) 06-07 (1) 07-09 (2) 09-12 (1)	18-23 (1)	20-23 (1)
Eastern Mediter- ranean & East Africa	Nil	07-12 (1) 19-21 (1)	18-21 (1)	Nil
West & Central Africa	07-08 (1) 08-10 (2) 10-13 (1)	06-10 (1) 10-13 (2) 13-15 (3) 15-16 (2) 16-18 (1)	18-22 (1)	19-21 (1) 19-21 (1)†
Southern Africa	09-11 (1)* 07-10 (1) 10-13 (2) 13-15 (1)	05-14 (1) 14-17 (2) 17-18 (1) 23-01 (1)	19-22 (1)	20-21 (1)
Central & South Asia	17-19 (1)	07-09 (1) 16-18 (1) 18-20 (2) 20-21 (1)	05-08 (1)	Nil
Southeast Asia	11-15 (1) 15-17 (2) 17-19 (1)	07-09 (1) 09-11 (2) 11-13 (1) 19-22 (1)	02-05 (1) 05-07 (2) 07-09 (1)	Nil
Far East	12-14 (1) 14-18 (2) 18-20 (1)	07-12 (1) 12-14 (2) 14-16 (1) 16-17 (2) 17-19 (3) 19-20 (2) 20-22 (1)	22-00 (1) 00-02 (2) 02-06 (3) 06-08 (2) 08-09 (1)	00-02 (1) 02-05 (2) 05-07 (1) 03-06 (1)†
South Pacific & New Zealand	15-17 (1)* 10-13 (1) 13-15 (2) 15-17 (4) 17-18 (2) 18-20 (1)	07-08 (1) 08-10 (2) 10-16 (1) 16-17 (2) 17-19 (4) 19-21 (2)	21-22 (1) 22-05 (2) 05-07 (2) 07-09 (1)	22-00 (1) 00-05 (2) 05-07 (1) 02-06 (1)†

Australasia	15-17 (1)* 13-17 (1) 17-19 (3) 19-20 (1)	07-08 (1) 08-10 (2) 10-17 (1) 17-18 (2) 18-20 (3) 20-21 (2) 21-22 (1)	00-03 (1) 03-05 (3) 05-07 (2) 07-08 (1)	02-03 (1) 03-05 (2) 05-07 (1) 04-06 (1)†
Caribbean, Central America & Northern Countries of South America	10-14 (1)* 06-08 (1) 08-13 (2) 13-15 (3) 15-16 (1) 16-18 (1)	00-05 (1) 05-06 (2) 06-08 (3) 08-14 (2) 14-16 (3) 16-18 (4) 18-20 (3) 20-00 (2)	18-20 (1) 20-00 (3) 00-03 (2) 03-05 (1)	20-21 (1) 21-01 (2) 01-04 (1) 23-02 (1)†
Peru, Bolivia, Paraguay, Brazil, Chile, Argentina & Uruguay	10-14 (1)* 06-08 (1) 08-10 (2) 10-12 (1) 12-13 (2) 13-15 (3) 15-16 (2) 16-18 (1)	05-07 (1) 07-09 (2) 09-13 (1) 13-15 (2) 15-17 (3) 17-19 (2) 19-00 (1)	19-21 (1) 21-02 (2) 02-05 (1)	21-04 (1) 00-03 (1)†
McMurdo Sound, Antarctica	08-10 (1) 14-16 (1)	15-17 (1) 17-19 (3) 19-00 (1) 05-06 (1) 06-08 (2) 08-11 (1)	00-06 (1)	Nil

MASTER SHORT-SKIP PROPAGATION CHART
SOLAR PHASE: LOW
SMOOTHED SUNSPOT RANGE: 0-30
SEASON: SUMMER
TIME ZONES: LOCAL STANDARD AT PATH MID-POINT
(24-Hour Time)

Band (Meters)	Distance Between Stations (Miles)			
	50-250	250-750	750-1300	1300-2300
10	Nil	07-09 (0-1)* 09-13 (0-3)* 13-17 (0-1)* 17-21 (0-2)* 21-23 (0-1)*	07-09 (1)* 09-13 (3)* 13-17 (1-2)* 17-21 (2-3)* 21-07 (1)*	07-09 (1-0)* 09-13 (3-0)* 13-17 (2-0)* 17-21 (2-0)* 21-07 (1-0)*
15	Nil	07-09 (0-2)* 09-13 (0-3)* 13-17 (0-2)* 17-19 (0-3)* 19-21 (0-2)* 21-07 (0-1)*	07-09 (2)* 09-13 (3)* 13-17 (2)* 17-19 (3)* 19-21 (2)* 21-23 (1-2)* 23-07 (1)*	07-09 (2-0) 09-13 (3-2) 13-17 (2-0) 17-19 (3-1) 19-20 (3-1) 20-23 (2-0) 23-07 (1-0)
20	09-00 (0-1)*	06-09 (0-2)* 09-15 (0-4)* 15-20 (0-2)* 20-00 (0-2)* 00-06 (0-1)*	06-09 (2-3)* 09-16 (4)* 16-21 (3-4)* 21-00 (2-3)* 00-06 (1-2)*	06-09 (2-0) 09-15 (3-2) 15-16 (4-2) 16-21 (4) 21-23 (3) 23-00 (3-2) 00-06 (2-1)
40	07-09 (1-2)* 09-15 (1-4)* 15-19 (2-4) 19-22 (1-2) 22-07 (0-1)*	07-09 (2)* 09-11 (4-2) 11-15 (4-1) 15-17 (4-3) 17-19 (4) 19-22 (2-4) 22-07 (1-3)*	07-09 (2-1) 09-11 (2-0) 11-15 (1-0) 15-17 (3-1) 17-20 (4-3) 20-22 (4) 22-05 (3-4) 05-07 (3)	07-09 (1-0) 09-15 (0) 15-17 (1-0) 17-20 (3-2) 20-05 (4) 05-07 (3-1)

Band (Meters)	50-250	250-750	750-1300	1300-2300
80	06-09 (3-4) 09-17 (4-3) 17-21 (4) 21-04 (3-4) 04-06 (3)	07-09 (4-1) 09-17 (3-0) 17-19 (4-0) 19-21 (4-2) 21-23 (4-3) 23-04 (4) 04-06 (3) 06-07 (4-2)	07-09 (1-0) 09-19 (0) 19-21 (2-1) 21-23 (3) 23-04 (4) 04-06 (3) 06-07 (2-1)	07-19 (0) 19-21 (1) 21-23 (3) 23-03 (4-3) 03-04 (4-2) 04-05 (3-2) 05-06 (3-1) 06-07 (1)
160	17-18 (1-0) 18-19 (1) 19-21 (3-1) 21-23 (4-2) 23-05 (4-3) 05-07 (3-2) 07-09 (1-0)	18-20 (1-0) 20-21 (1) 21-22 (2-1) 22-23 (2) 23-05 (3-0) 05-06 (2-1) 06-07 (2-0)	20-22 (1) 22-00 (2-1) 00-02 (2) 02-06 (2-1)	20-22 (1-0) 22-00 (1-0) 00-02 (2-1) 02-05 (1)

MASTER SHORT-SKIP PROPAGATION CHART
SOLAR PHASE: LOW
SMOOTHED SUNSPOT RANGE: 0-30
SEASON: WINTER
TIME ZONES: LOCAL STANDARD AT PATH MID-POINT
(24-Hour Time)

Band (Meters)	Distance Between Stations (Miles)			
	50-250	250-750	750-1300	1300-2300
10	Nil	Nil	11-16 (0-1)	11-16 (1-0)
15	Nil	10-16 (0-1)	09-10 (0-1) 10-12 (1) 12-16 (1-2) 16-17 (0-1)	09-10 (1) 10-12 (1-3) 12-14 (2-4) 14-15 (2-3) 15-16 (2) 16-17 (1) 17-18 (0-1)
20	Nil	09-11 (0-1) 11-16 (0-2) 16-19 (0-1)	08-09 (1-0) 09-11 (1-4) 11-16 (2-4) 16-17 (1-3) 17-18 (1-2) 18-19 (1) 19-21 (0-1)	07-08 (0-1) 08-09 (1-3) 09-11 (4) 11-15 (4-3) 15-16 (4) 16-17 (3) 17-18 (2-3) 18-19 (1-2) 19-20 (1)
40	07-09 (0-1) 09-10 (1-3) 10-15 (3-4) 15-16 (2-3) 16-18 (1-2) 18-20 (0-1)	07-09 (1-3) 09-10 (3) 10-15 (4-3) 15-16 (3-4) 16-18 (2-4) 18-20 (1-2) 20-00 (0-2) 00-07 (0-1)	07-09 (3) 09-14 (3-1) 14-15 (3-2) 15-16 (3) 16-18 (4) 18-20 (2-4) 20-22 (2-3) 22-00 (2) 00-04 (1-2) 04-07 (1-3)	07-08 (3-2) 08-09 (3-1) 09-14 (1-0) 14-15 (2-0) 15-16 (3-1) 16-17 (4-2) 17-18 (4-3) 18-20 (4) 20-22 (3-4) 22-00 (2-3) 02-04 (2-3) 04-06 (3)
80	08-16 (4) 16-18 (2-4) 18-20 (1-3) 20-06 (1-2) 06-08 (2-3)	08-09 (4-2) 09-16 (4-1) 16-18 (4-2) 18-20 (3-4) 20-06 (2-4) 06-07 (3-4) 07-08 (3)	08-09 (2-1) 09-16 (1-0) 16-18 (2-1) 18-20 (4-3) 20-06 (4) 06-07 (4-2) 07-08 (3-1)	08-09 (1-0) 09-16 (0) 16-18 (1-0) 18-20 (3-2) 20-04 (4-3) 04-06 (4-2) 06-07 (2-1) 07-08 (1)
160	07-09 (3-2) 09-11 (2-0) 11-17 (1-0) 17-19 (3-2) 19-07 (4)	07-09 (2-1) 09-17 (0) 17-19 (1-2) 19-04 (4) 04-05 (4-3) 05-07 (4-2)	06-07 (2-1) 07-09 (1-0) 17-19 (1-0) 19-20 (4-2) 20-21 (4-3) 21-04 (4) 04-06 (3-2)	06-07 (1-0) 07-19 (0) 19-20 (2-1) 20-21 (3-2) 21-04 (4-2) 04-06 (2-1)

MASTER SHORT-SKIP PROPAGATION CHART
SOLAR PHASE: LOW
SMOOTHED SUNSPOT RANGE: 0-30
SEASONS: SPRING & FALL
TIME ZONES: LOCAL STANDARD AT PATH MID-POINT
(24-Hour Time)

Band (Meters)	Distance Between Stations (Miles)			
	50-250	250-750	750-1300	1300-2300
10	Nil	09-13 (0-1) 13-21 (0-1)	07-09 (1) 09-13 (1-2)	07-09 (1-0) 09-11 (2-0) 11-13 (2-1) 13-17 (1) 17-21 (1-0)
15	Nil	07-09 (0-1) 09-13 (0-2) 13-21 (0-1)	07-09 (1) 09-13 (2) 13-17 (1-2) 17-21 (1) 21-07 (0-1)	07-09 (1) 09-15 (2) 15-17 (2-1) 17-19 (1) 19-07 (1-0)
20	Nil	07-09 (0-1) 09-11 (0-2) 11-14 (0-4) 14-16 (0-3) 16-18 (0-3) 18-07 (0-1)	07-09 (1-2) 09-11 (3-1) 11-14 (4) 14-16 (3-4) 16-18 (2-4) 18-20 (1-3) 20-22 (1-2) 22-07 (1)	07-09 (2) 09-13 (4-2) 13-15 (4-3) 15-18 (4) 18-20 (3) 20-22 (2) 22-00 (2) 00-05 (1-0) 05-07 (1)
40	07-09 (0-2) 09-11 (4-3) 11-15 (3-4) 15-17 (2-3) 17-19 (1-2) 19-21 (0-1)	07-09 (2-3) 09-11 (4-3) 11-15 (4-2) 15-17 (3) 17-19 (4-3) 19-21 (1-4) 21-23 (0-3) 23-02 (0-2) 02-05 (0-1) 05-07 (0-2)	07-09 (3-2) 09-11 (3-1) 11-15 (1-0) 15-17 (3-2) 17-19 (4-3) 19-21 (4) 21-23 (3-4) 23-02 (2-3) 02-05 (1-2) 05-07 (2-4)	07-09 (2-1) 09-15 (1-0) 15-17 (2-1) 17-19 (3-2) 19-23 (4) 23-02 (3-4) 02-05 (2-3) 05-07 (4-2)
80	06-08 (3-4) 08-21 (4) 21-03 (3-4) 03-06 (2-3)	06-08 (4-2) 08-16 (1-0) 16-18 (4-2) 18-21 (4-3) 21-03 (4) 03-05 (3-4) 05-06 (3)	06-08 (2-1) 08-16 (1-0) 16-18 (2-1) 18-21 (3-2) 21-03 (4) 03-05 (4-2) 05-06 (3-2)	06-08 (1) 08-16 (0) 16-18 (1) 18-21 (2) 21-03 (4-3) 03-06 (2)
160	16-18 (1-0) 18-20 (2-1) 20-05 (4) 05-07 (3-2) 07-09 (2-1) 09-11 (1-0)	17-19 (1-0) 19-20 (1) 20-02 (4-3) 02-05 (3-2) 05-07 (2-1) 07-09 (1-0)	19-20 (1-0) 20-22 (2-1) 22-02 (3) 02-05 (2-1) 05-07 (1)	20-22 (1-0) 22-02 (3-2) 02-05 (1) 05-07 (1-0)

FOOTNOTES

Worldwide Charts:
* Predicted times of 10 meter openings; all others in column are 15 meter openings.
† Predicted times of 160 meter openings; all others in column are 80 meter openings.

Short Skip Charts:
* Predominately sporadic-E openings.

SUNRISE/SUNSET FORMULAS

Each day is said to begin with the sunrise and end at sunset. For Amateur Radio DXers, these times can be the most exciting! Most other folks, though, they see these times on calendars or during the evening weather forecast, but otherwise take for granted that the sun will rise today and set tonight and return tomorrow.

There presently exist many methods of calculating sunrise and sunset times. With maps and slides, globes and lights, or maps that move and computer programs from crude to multi-colored, one can easily estimate a sunrise or sunset, but each has an inherent inaccuracy. Some are larger than others. Corrections may be needed for leap year, leap second, non-circular Earth orbit, and the varying speed of Earth's travel and rotation. For higher accuracy, and to determine the times for other locations, a computer program is the answer. Each program author uses a variation on a standard formula to increase accuracy, or correction factors must be added.

The declination of the sun, caused by the Earth's axis angle relative to the sun, is one of the important variables. The Earth axis is tilted 23.5 degrees relative to its solar orbit. As the Earth rotates around the sun, the North and South Poles can be the exact distance from the sun. This occurs on March 21st and September 21st, when the sun's declination is zero degrees. This marks the time on the Earth when the length of day (12 hours) equals the length of night (12 hours), giving us the term "Equinox": "Equi-" meaning equal and "-nox" meaning night.

When the Earth's North Pole is farther away from the sun than the South Pole, the declination of the sun is negative (up to -23.5 degrees on December 21st). When the South Pole is farther from the sun than the North pole, the declination is positive (up to +23.5 degrees on June 21st).

To calculate the sun's declination, you will need to know the date on the Julian calendar and have a trigonometric table or a calculator that will process trigonometric functions. Since the Julian calendar is the standard calendar we use, simple count the number of days since January 1 to your target day.

$$A = -23.5 \times \text{SIN} \, (\, 360 \times (d - 80/365))$$

Where A = Sun declination; d = Day of the Year

Next, this sun declination angle can be used to directly calculate the sunrise or sunset time of the target location on this target day. You'll also need to know the Latitude and Longitude of the target location.

$$\text{Sunrise} = \frac{1}{(\text{LONG}/15)} + \frac{\cos \, (\, \tan \, [A] \times \tan \, [\text{LAT}] \,)}{15}$$

Where A = Sun declination; LONG = Longitude in Degrees; LAT = Latitude in Degrees
Values for LAT and LONG must be positive for targets North and West.

$$\text{Sunset} = \frac{1}{(\text{LONG}/15)} - \frac{\cos \, (\, \tan \, [A] \times \tan \, [\text{LAT}] \,)}{15}$$

Where A = Sun declination; LONG = Target Longitude; LAT = Target Latitude

The answer will be a decimal representation of the time the terminator crosses the target location relative to the Zero meridian. (0, 0 - UTC). To convert the decimal portion of the calculated time, multiply by 60. (i. e., sunrise time was calculated as 10.88. Multiply .88 x 60 = 52.8 minutes. Therefore, the calculated sunrise would be 10:52.48 Z (UTC)).

Due to the previously mentioned variables, corrections are required to each sunrise and sunset formula result. Add or subtract the following number of minutes from your times for your day.

Jan01	+3	Apr01	+4	Jul01	+4	Oct01	-10
Jan16	+9	Apr16	0	Jul15	+6	Oct16	-14
Feb01	+14	May01	-3	Aug01	+6	Nov01	-16
Feb15	+14	May16	-4	Aug16	+4	Nov16	-15
Mar01	+13	Jun01	-2	Sep01	0	Dec01	-11
Mar15	+9	Jun16	0	Sep16	-5	Dec16	-5

The terminator is a line, perpendicular to the light rays from the sun, marking the transition on the Earth's surface between the nighttime portion and the daylight portion. This would mark the beginning of the twilight zone or Gray-Line area.

The sunrise and sunset periods are the most exciting times of radio propagation.

Other sunrise/sunset and Gray-Line reading includes *Low-Band DXing*, by John Devoldere, ON4UN, available through the ARRL and amateur bookstores; *Long-Path Propagation*, by Bob Brown, NM7M, available directly from the author; *An Introduction to Gray-Line DXing*, by Tom Russell, N4KG, published in *QST*, November 1992; *The ARRL Antenna Book*, edited by Gerald Hall, K1TD, published by the ARRL; *Ionospheric Radio*, by Kenneth Davies, published by Peter Peregrinus, Ltd. and available through the I.E.E.E.

SUNRISE/SUNSET CALCULATION SOFTWARE

```
9 CLS
10 CLEAR , 64000, 4096
15 REM ORIGINAL PROGRAM BY Van Heddegem ON4HW
17 D = 59
20 PI = 3.1415927#
30 SW = -.97599592#
40 CW = .21778881#
50 SE = .39777961#
60 CE = .9174811
61 K1 = -.014834754#
62 n$ = "b:\latlong"
63 OPEN n$ FOR INPUT ACCESS READ AS #1 LEN = 43
64 na$ = "a:\srssw_of.e28"
65 OPEN na$ FOR OUTPUT AS #2
67 IF EOF(1) THEN CLOSE : GOTO 500
69 FOR L = 1 TO 181
70 INPUT #1, L$
72 pf$ = LEFT$(L$, 4)
73 ct$ = MID$(L$, 6, 19)
74 at$ = MID$(L$, 30, 6)
75 on$ = RIGHT$(L$, 7)
79 REM PRINT #2, pf$;
80 REM PRINT #2, ct$;
88 RSET at$ = at$
89 LA = VAL(at$)
90 RSET on$ = on$
91 LO = VAL(on$)
120 LO = LO * PI / 180
130 LA = LA * PI / 180
160 M = (2 * PI * D + LO) / 365.24219# - .052708
170 L = M - 1.351248
180 C1 = 1 - .03343 * COS(M)
190 C2 = .99944 * SIN(M) / C1
200 C3 = (COS(M) - .03343) / C1
210 C4 = SW * C3 + CW * C2
220 C5 = CW * C3 - SW * C2
230 C6 = SE * C4: REM SINE OF SUN DECLINATION
240 B1 = K1 - C6 * SIN(LA)
250 B2 = (COS(LA)) ^ 2 * (1 - C6 ^ 2) - B1 ^ 2
260 IF B2 <= 0 THEN R$ = "NO.SR": S$ = "NO.SS": GOTO 345
270 B3 = ATN(B1 / SQR(B2)) - PI / 2
280 B4 = ATN((COS(L) * CE * C4 - SIN(L) * C5) / (SIN(L) * CE
    * C4 + COS(L) * C5))
290 GOSUB 370
300 R$ = STR$(B6)
310 B3 = -B3
320 GOSUB 370
330 S$ = STR$(B6)
345 WRITE #2, R$, S$
350 NEXT L: REM GOTO 440
370 B5 = B4 + B3 + LO + PI
380 IF B5 < 0 THEN B5 = B5 + 2 * PI
390 B5 = INT(B5 * 720 / PI + .5): REM MINUTES PAST 0000
    UTC
400 IF B5 > 1439 THEN B5 = B5 - 1440
410 B6 = .4 * INT(B5 / 60) + B5 / 100: REM TIME IN HH.MM
420 RETURN
440 REM IF D < 344 THEN LET D = D + 14: GOTO 160
450 REM IF D > 356 THEN GOTO 9
460 REM RETURN
500 END
```

Prefix City Country	Jan14 Sunrise	Jan 14 Sunset	Jan 28 Sunrise	Jan 28 Sunset	Feb 14 Sunrise	Feb 14 Sunset	Feb 28 Sunrise	Feb 28 Sunset	Mar 14 Sunrise	Mar 14 Sunset	Mar 28 Sunrise	Mar 28 Sunset
A4 MUSCAT OMAN	02:50	13:39	02:48	13:46	02:40	14:00	02:30	14:08	02:17	14:14	02:03	14:20
AP KARACHI PAKISTAN	02:11	12:55	02:09	13:05	02:00	13:17	01:49	13:25	01:36	13:32	01:21	13:38
BV TAIPEI TAIWAN	22:40	09:24	22:38	09:34	22:30	09:46	22:18	09:54	22:05	10:01	21:51	10:07
BY CHUNKING CHINA	23:46	10:11	23:42	10:23	23:31	10:37	23:18	10:47	23:03	10:57	22:46	11:05
BY BEIJING CHINA	23:35	09:11	23:27	09:27	23:10	09:47	22:51	10:03	22:30	10:18	22:07	10:33
C2 REP. OF NAURU	18:56	07:05	19:00	07:09	19:02	07:10	19:01	07:09	18:58	07:05	18:54	07:01
C3 ANDORRA	07:20	16:42	07:11	16:59	06:52	17:21	06:31	17:39	06:08	17:56	05:43	18:12
C6 NASSAU BAHAMAS	11:51	22:45	11:49	22:55	11:42	23:05	11:32	23:12	11:20	23:18	11:07	23:23
C9 MAPUTO MOZAMBIQUE	03:09	16:49	03:20	16:46	03:33	16:36	03:42	16:24	03:50	16:09	03:57	15:54
CE SANTIAGO CHILE	09:47	23:56	10:00	23:51	10:17	23:37	10:29	23:21	10:41	23:04	10:51	22:45
CE0 EASTER ISLAND	12:35	02:15	12:46	02:12	12:59	02:02	13:08	01:49	13:16	01:35	13:23	01:20
CN CASABLANCA MOROCCO	07:35	17:44	07:03	17:57	07:17	18:13	07:02	18:25	06:44	18:37	06:25	18:47
CP LA PAZ BOLIVIA	10:10	23:12	10:19	23:11	10:27	23:06	10:31	22:58	10:35	22:49	10:37	22:38
CT LISBON PORTUGAL	07:54	17:37	07:46	17:52	07:30	18:11	07:12	18:27	06:51	18:41	06:29	18:55
CU SAO MIGUEL IS.AZORES	09:02	18:49	08:55	19:04	08:39	19:23	08:21	19:38	08:01	19:51	07:39	20:05
CX MONTEVIDEO URUGUAY	08:45	23:02	08:59	22:56	09:17	22:41	09:30	22:25	09:42	22:06	09:53	21:47
D6 COMOROS	02:54	15:41	03:01	15:42	03:08	15:39	03:11	15:33	03:12	15:25	03:13	15:16
DJ BONN GERMANY	07:29	15:52	07:15	16:14	06:48	16:44	06:20	17:09	05:50	17:32	05:19	17:55
DU MANILA PHILIPPINES	22:24	09:45	22:25	09:52	22:21	10:00	22:14	10:04	22:05	10:06	21:55	10:08
EA6 BALEARIC ISLANDS	07:13	16:52	07:06	17:08	06:49	17:28	06:30	17:44	06:09	17:59	05:47	18:13
EA8 CANARY ISLANDS	08:01	18:28	07:58	18:40	07:47	18:54	07:34	19:04	07:18	19:13	07:02	19:21
EI DUBLIN REP IRELAND	08:34	16:33	08:18	16:58	07:47	17:31	07:17	17:58	06:44	18:25	06:11	18:50
ET ADDIS-ABABA ETHIOPIA	03:44	15:23	03:46	15:29	03:44	15:34	03:40	15:36	03:33	15:36	03:25	15:36
EZ ASHKHABAD TURKOMAN	03:27	13:04	03:20	13:14	03:02	13:40	02:44	13:56	02:22	14:11	02:00	14:26
FT8 KERGUELEN	23:33	15:28	23:55	15:14	00:25	14:48	00:49	14:21	01:12	13:51	01:34	13:21
FG GUADELOUPE	10:37	21:50	10:37	21:59	10:31	22:06	10:24	22:11	10:14	22:14	10:03	22:17
FK NOUMEA NEW CALEDON	18:20	07:43	18:30	07:41	18:41	07:33	18:48	07:23	18:54	07:11	18:59	06:58
FO CLIPPERTON ISL	13:43	01:15	13:44	01:22	13:41	01:27	13:36	01:30	13:28	01:31	13:20	01:31
FO PAPEETE POLYNESIA	15:33	04:38	15:42	04:37	15:50	04:31	15:55	04:23	15:59	04:13	16:02	04:02
FP ST PIERRE & MIQUELON	11:26	20:21	11:15	20:40	10:52	21:06	10:28	21:27	10:01	21:47	09:34	22:07
FR REUNION	01:47	15:05	01:57	15:04	02:07	14:57	02:14	14:47	02:19	14:35	02:23	14:23
FY CAYENNE FR.GUIANA	09:42	21:34	09:45	21:39	09:45	21:42	09:41	21:43	09:36	21:41	09:30	21:39
G LONDON ENGLAND	08:01	16:17	07:46	16:40	07:18	17:11	06:50	17:36	06:19	18:01	05:47	18:25
HA BUDAPEST HUNGARY	06:28	15:17	06:16	15:37	05:53	16:03	05:28	16:25	05:01	16:46	04:33	17:06
HC QUITO ECUADOR	11:19	23:27	11:23	23:31	11:25	23:32	11:23	23:30	11:20	23:27	11:16	23:23
HC8 GALAPAGOS ISLS	12:13	00:21	12:17	00:25	12:18	00:26	12:17	00:24	12:14	00:21	12:10	00:17
HP PANAMA CITY PANAMA	11:43	23:23	11:45	23:29	11:43	23:34	11:39	23:36	11:32	23:36	11:25	23:35
HS BANGKOK THAILAND	23:45	11:08	23:46	11:16	23:42	11:22	23:36	11:26	23:27	11:28	23:18	11:30
I ROME ITALY	06:36	16:01	06:27	16:18	06:08	16:39	05:48	16:57	05:25	17:13	05:01	17:29
JA TOYKO JAPAN	21:51	07:48	21:45	08:03	21:31	08:20	21:14	08:34	20:55	08:46	20:35	08:58
JD MINAMI TORI-SHIMA	20:25	07:12	20:23	07:22	20:15	07:34	20:04	07:41	19:51	07:48	19:37	07:54
JT ULAN-BATOR MONGOLIA	00:38	09:23	00:26	09:43	00:02	10:10	23:37	10:32	23:10	10:53	22:41	11:14
JW SVALBARD	NO:SR	NO:SS	NO:SR	NO:SS	09:50	11:59	07:11	14:34	05:25	16:14	03:45	17:46
KC4 McMurdo Antarctica	NO:SR	NO:SS	NO:SR	NO:SS	NO:SR	NO:SS	15:42	10:28	17:46	08:18	19:27	06:28

Prefix City Country	Apr 14 Sunrise	Apr 14 Sunset	Apr 28 Sunrise	Apr 28 Sunset	May 14 Sunrise	May 14 Sunset	May 28 Sunrise	May 28 Sunset	Jun 14 Sunrise	Jun 14 Sunset	Jun 28 Sunrise	Jun 28 Sunset
A4 MUSCAT OMAN	01:47	14:26	01:35	14:32	01:25	14:39	01:20	14:46	01:19	14:53	01:22	14:56
AP KARACHI PAKISTAN	01:04	13:46	00:52	13:52	00:41	14:00	00:35	14:07	00:34	14:14	00:37	14:18
BV TAIPEI TAIWAN	21:33	10:15	21:21	10:21	21:10	10:29	21:05	10:36	21:03	10:44	21:06	10:47
BY CHUNKING CHINA	22:26	11:15	22:12	11:24	21:59	11:34	21:52	11:42	21:49	11:51	21:52	11:54
BY BEIJING CHINA	21:40	10:50	21:20	11:04	21:02	11:20	20:51	11:32	20:45	11:44	20:48	11:47
C2 REP. OF NAURU	18:50	06:56	18:47	06:53	18:45	06:51	18:46	06:52	18:49	06:55	18:52	06:58
C3 ANDORRA	05:14	18:31	04:53	18:47	04:32	19:04	04:20	19:18	04:14	19:30	04:16	19:34
C6 NASSAU BAHAMAS	10:51	23:29	10:40	23:34	10:31	23:41	10:26	23:47	10:25	23:53	10:28	23:57
C9 MAPUTO MOZAMBIQUE	04:04	15:36	04:12	15:23	04:21	15:12	04:28	15:06	04:36	15:04	04:39	15:07
CE SANTIAGO CHILE	11:03	22:23	11:14	22:07	11:25	21:52	11:35	21:44	11:44	21:41	11:47	21:44
CE0 EASTER ISLAND	13:31	01:02	13:39	00:49	13:47	00:37	13:55	00:32	14:02	00:30	14:05	00:33
CN CASABLANCA MOROCCO	06:03	19:00	05:46	19:10	05:32	19:22	05:23	19:32	05:20	19:41	05:23	19:45
CP LA PAZ BOLIVIA	10:40	22:25	10:43	22:17	10:47	22:10	10:52	22:07	10:58	22:07	11:01	22:10
CT LISBON PORTUGAL	06:03	19:11	05:44	19:24	05:26	19:39	05:16	19:51	05:11	20:02	05:13	20:05
CU SAO MIGUEL IS.AZORES	07:14	20:20	06:55	20:33	06:38	20:48	06:28	20:59	06:23	21:10	06:26	21:13
CX MONTEVIDEO URUGUAY	10:07	21:24	10:18	21:07	10:30	20:52	10:40	20:43	10:49	20:40	10:53	20:43
D6 COMOROS	03:14	15:05	03:16	14:58	03:19	14:52	03:22	14:50	03:27	14:51	03:30	14:54
DJ BONN GERMANY	04:42	18:23	04:13	18:45	03:46	19:10	03:28	19:29	03:18	19:45	03:20	19:49
DU MANILA PHILIPPINES	21:44	10:10	21:35	10:12	21:29	10:16	21:26	10:20	21:26	10:25	21:29	10:29
EA6 BALEARIC ISLANDS	05:20	18:29	05:00	18:43	04:42	18:59	04:31	19:11	04:26	19:22	04:29	19:25
EA8 CANARY ISLANDS	06:42	19:31	06:28	19:39	06:15	19:49	06:08	19:58	06:06	20:06	06:09	20:09
EI DUBLIN REP IRELAND	05:30	19:21	04:59	19:46	04:28	20:14	04:08	20:36	03:57	20:53	03:59	20:57
ET ADDIS-ABABA ETHIOPIA	03:16	15:35	03:10	15:35	03:06	15:37	03:04	15:40	03:06	15:44	03:09	15:47
EZ ASHKHABAD TURKOMAN	01:33	14:43	01:13	14:57	00:55	15:12	00:44	15:25	00:38	15:36	00:41	15:39
FT8 KERGUELEN	02:00	12:45	02:22	12:18	02:45	11:52	03:03	11:36	03:17	11:26	03:21	11:29
FG GUADELOUPE	09:51	22:20	09:42	22:23	09:34	22:28	09:31	22:32	09:31	22:38	09:34	22:41
FK NOUMEA NEW CALEDON	19:05	06:43	19:10	06:31	19:16	06:22	19:23	06:17	19:29	06:16	19:32	06:19
FO CLIPPERTON ISL	13:10	01:31	13:03	01:32	12:58	01:35	12:56	01:38	12:57	01:43	13:00	01:46
FO PAPEETE POLYNESIA	16:05	03:49	16:09	03:40	16:13	03:33	16:18	03:29	16:24	03:30	16:27	03:32
FP ST PIERRE & MIQUELON	09:01	22:30	08:36	22:49	08:12	23:10	07:58	23:26	07:50	23:39	07:52	23:43
FR REUNION	02:29	14:08	02:33	13:57	02:39	13:48	02:45	13:44	02:52	13:43	02:55	13:46
FY CAYENNE FR.GUIANA	09:23	21:36	09:18	21:35	09:15	21:36	09:15	21:38	09:17	21:41	09:20	21:44
G LONDON ENGLAND	05:09	18:53	04:40	19:16	04:12	19:42	03:53	20:02	03:43	20:18	03:45	20:22
HA BUDAPEST HUNGARY	03:59	17:29	03:34	17:49	03:09	18:11	02:54	18:27	02:46	18:41	02:48	18:45
HC QUITO ECUADOR	11:11	23:18	11:09	23:15	11:07	23:14	11:08	23:14	11:11	23:17	11:14	23:20
HC8 GALAPAGOS ISLS	12:05	00:12	12:02	00:09	12:01	00:08	12:02	00:08	12:05	00:11	12:08	00:14
HP PANAMA CITY PANAMA	11:16	23:34	11:10	23:34	11:06	23:36	11:04	23:39	11:06	23:43	11:09	23:46
HS BANGKOK THAILAND	23:06	11:31	22:58	11:33	22:52	11:37	22:49	11:41	22:50	11:46	22:53	11:49
I ROME ITALY	04:33	17:48	04:11	18:03	03:51	18:20	03:40	18:34	03:34	18:45	03:36	18:49
JA TOYKO JAPAN	20:12	09:12	19:54	09:24	19:38	09:37	19:29	09:48	19:24	09:57	19:27	10:01
JD MINAMI TORI-SHIMA	19:21	08:00	19:09	08:07	18:59	08:14	18:53	08:21	18:52	08:28	18:54	08:31
JT ULAN-BATOR MONGOLIA	22:07	11:38	21:41	11:58	21:16	12:20	21:01	12:37	20:52	12:52	20:54	12:55
JW SVALBARD	01:24	19:57	NO:SR	NO:SS	NO:SR	NO:SS	NO:SR	NO:SS	NO:SS	NO:SS	NO:SR	NO:SS
KC4 McMurdo Antarctica	21:39	04:06	NO:SR	NO:SS	NO:SR	NO:SS	NO:SR	NO:SS	NO:SS	NO:SS	NO:SR	NO:SS

Prefix City Country	July 14 Sunrise	July 14 Sunset	July 28 Sunrise	July 28 Sunset	Aug 14 Sunrise	Aug 14 Sunset	Aug 28 Sunrise	Aug 28 Sunset	Sep 14 Sunrise	Sep 14 Sunset	Sept 28 Sunrise	Sept 28 Sunset
A4 MUSCAT OMAN	01:28	14:55	01:34	14:51	01:41	14:41	01:46	14:29	01:52	14:12	01:56	13:58
AP KARACHI PAKISTAN	00:43	14:17	00:49	14:12	00:57	14:01	01:03	13:49	01:09	13:31	01:15	13:16
BV TAIPEI TAIWAN	21:12	10:46	21:18	10:42	21:26	10:31	21:32	10:18	21:38	10:01	21:44	09:46
BY CHUNKING CHINA	21:59	11:53	22:06	11:47	22:16	11:33	22:24	11:19	22:33	10:59	22:41	10:42
BY BEIJING CHINA	20:57	11:43	21:08	11:34	21:24	11:15	21:37	10:55	21:53	10:28	22:06	10:05
C2 REP. OF NAURU	18:55	07:01	18:56	07:01	18:54	07:00	18:51	06:57	18:45	06:51	18:40	06:47
C3 ANDORRA	04:27	19:29	04:39	19:18	04:56	18:57	05:11	18:36	05:29	18:06	05:44	17:42
C6 NASSAU BAHAMAS	10:34	23:56	10:39	23:52	10:46	23:42	10:51	23:31	10:56	23:15	10:59	23:01
C9 MAPUTO MOZAMBIQUE	04:38	15:13	04:33	15:20	04:21	15:28	04:08	15:35	03:50	15:42	03:34	15:48
CE SANTIAGO CHILE	11:45	21:52	11:38	22:00	11:23	22:12	11:07	22:22	10:44	22:33	10:25	22:42
CE0 EASTER ISLAND	14:04	00:40	13:59	00:46	13:47	00:55	13:34	01:01	13:15	01:08	13:00	01:14
CN CASABLANCA MOROCCO	05:31	19:42	05:39	19:35	05:51	19:20	06:01	19:03	06:13	18:41	06:22	18:21
CP LA PAZ BOLIVIA	11:01	22:15	10:59	22:19	10:51	22:23	10:42	22:25	10:31	22:27	10:18	22:29
CT LISBON PORTUGAL	05:23	20:01	05:33	19:52	05:48	19:34	06:01	19:15	06:16	18:49	06:28	18:26
CU SAO MIGUEL IS.AZORES	06:35	21:10	06:45	21:01	07:00	20:42	07:12	20:24	07:26	19:59	07:38	19:37
CX MONTEVIDEO URUGUAY	10:50	20:51	10:43	21:00	10:27	21:12	10:10	21:22	09:45	21:34	09:27	21:45
D6 COMOROS	03:31	14:58	03:29	15:01	03:24	15:04	03:16	15:05	03:05	15:05	03:00	15:05
DJ BONN GERMANY	03:34	19:41	03:51	19:25	04:16	18:57	04:37	18:29	05:04	17:52	05:25	17:20
DU MANILA PHILIPPINES	21:34	10:29	21:38	10:27	21:41	10:20	21:43	10:12	21:45	10:00	21:45	09:49
EA6 BALEARIC ISLANDS	04:38	19:21	04:49	19:12	05:05	18:53	05:18	18:34	05:33	18:07	05:46	17:44
EA8 CANARY ISLANDS	06:16	20:08	06:23	20:02	06:33	19:49	06:41	19:34	06:50	19:14	06:57	18:57
EI DUBLIN REP IRELAND	04:14	20:47	04:33	20:30	05:01	19:58	05:25	19:28	05:55	18:47	06:19	18:13
ET ADDIS-ABABA ETHIOPIA	03:13	15:49	03:15	15:48	03:17	15:43	03:17	15:37	03:15	15:27	03:14	15:19
EZ ASHKHABAD TURKOMAN	00:50	15:35	01:01	15:26	01:17	15:07	01:30	14:47	01:46	14:20	01:59	13:57
FT8 KERGUELEN	03:14	11:41	03:00	11:57	02:34	12:20	02:07	12:40	01:31	13:05	01:00	13:26
FG GUADELOUPE	09:39	22:42	09:43	22:39	09:48	22:31	09:50	22:22	09:53	22:09	09:54	21:57
FK NOUMEA NEW CALEDON	19:32	06:25	19:28	06:30	19:19	06:37	19:08	06:41	18:52	06:46	18:38	06:50
FO CLIPPERTON ISL	13:04	01:47	13:07	01:46	13:10	01:40	13:10	01:33	13:10	01:22	13:09	01:13
FO PAPEETE POLYNESIA	16:27	03:38	16:24	03:42	16:17	03:46	16:07	03:49	15:54	03:52	15:42	03:53
FP ST PIERRE & MIQUELON	08:04	23:37	08:19	23:24	08:40	22:59	08:58	22:35	09:20	22:01	09:38	21:33
FR REUNION	02:55	13:52	02:51	13:57	02:42	14:03	02:31	14:07	02:16	14:11	02:03	14:15
FY CAYENNE FR.GUIANA	09:23	21:46	09:25	21:46	09:25	21:43	09:24	21:38	09:20	21:30	09:17	21:23
G LONDON ENGLAND	03:59	20:13	04:17	19:57	04:43	19:28	05:05	18:59	05:32	18:21	05:54	17:49
HA BUDAPEST HUNGARY	03:00	18:39	03:15	18:25	03:37	18:00	03:55	17:35	04:18	17:01	04:37	16:33
HC QUITO ECUADOR	11:17	23:23	11:17	23:24	11:16	23:22	11:12	23:19	11:07	23:13	11:02	23:09
HC8 GALAPAGOS ISLS	12:11	00:17	12:11	00:18	12:10	00:16	12:06	00:13	12:01	00:07	11:56	00:02
HP PANAMA CITY PANAMA	11:13	23:48	11:15	23:47	11:17	23:42	11:16	23:36	11:15	23:26	11:13	23:18
HS BANGKOK THAILAND	22:57	11:50	23:01	11:48	23:04	11:41	23:06	11:33	23:07	11:21	23:07	11:11
I ROME ITALY	03:46	18:44	03:58	18:34	04:15	18:13	04:30	17:52	04:47	17:24	05:02	16:59
JA TOKYO JAPAN	19:35	09:58	19:45	09:50	19:58	09:34	20:09	09:17	20:22	08:53	20:32	08:32
JD MINAMI TORI-SHIMA	19:01	08:31	19:07	08:26	19:14	08:16	19:19	08:04	19:25	07:47	19:30	07:32
JT ULAN-BATOR MONGOLIA	21:07	12:49	21:21	12:35	21:44	12:10	22:03	11:45	22:26	11:10	22:45	10:41
JW SVALBARD	NO:SR	NO:SS	NO:SR	NO:SS	NO:SR	NO:SS	00:55	20:28	03:13	17:59	04:44	16:18
KC4 McMurdo Antarctica	NO:SR	NO:SS	NO:SR	NO:SS	NO:SR	NO:SS	22:17	03:30	19:47	05:49	17:59	07:27

Prefix City Country	Oct 14 Sunrise	Oct 14 Sunset	Oct 28 Sunrise	Oct 28 Sunset	Nov 14 Sunrise	Nov 14 Sunset	Nov 28 Sunrise	Nov 28 Sunset	Dec 14 Sunrise	Dec 14 Sunset	Dec 28 Sunrise	Dec 28 Sunset
A4 MUSCAT OMAN	02:02	13:43	02:09	13:31	02:19	13:22	02:28	13:19	02:39	13:22	02:46	13:28
AP KARACHI PAKISTAN	01:21	13:00	01:29	12:48	01:39	12:38	01:49	12:34	02:00	12:37	02:07	12:43
BV TAIPEI TAIWAN	21:50	09:29	21:58	09:17	22:08	09:07	22:18	09:04	22:29	09:06	22:37	09:12
BY CHUNKING CHINA	22:50	10:23	22:59	10:09	23:12	09:56	23:24	09:51	23:36	09:52	23:43	09:58
BY BEIJING CHINA	22:22	09:39	22:37	09:20	22:57	09:01	23:12	08:52	23:27	08:50	23:35	08:56
C2 REP. OF NAURU	18:35	06:42	18:32	06:40	18:32	06:40	18:35	06:44	18:41	06:50	18:48	06:57
C3 ANDORRA	06:02	17:14	06:19	16:53	06:40	16:32	06:57	16:22	07:13	16:19	07:21	16:25
C6 NASSAU BAHAMAS	11:05	22:46	11:11	22:36	11:20	22:27	11:29	22:25	11:40	22:28	11:47	22:34
C9 MAPUTO MOZAMBIQUE	03:17	15:56	03:04	16:04	02:53	16:16	02:48	16:27	02:50	16:38	02:56	16:46
CE SANTIAGO CHILE	10:04	22:54	09:48	23:05	09:33	23:21	09:27	23:34	09:27	23:47	09:33	23:55
CE0 EASTER ISLAND	12:42	01:22	12:29	01:30	12:18	01:42	12:14	01:53	12:16	02:05	12:23	02:12
CN CASABLANCA MOROCCO	06:34	18:00	06:45	17:44	07:00	17:30	07:13	17:23	07:26	17:24	07:34	17:30
CP LA PAZ BOLIVIA	10:05	22:31	09:57	22:35	09:51	22:42	09:50	22:50	09:53	23:00	10:00	23:07
CT LISBON PORTUGAL	06:43	18:02	06:58	17:43	07:16	17:25	07:31	17:17	07:46	17:15	07:53	17:22
CU SAO MIGUEL IS. AZORES	07:53	19:13	08:07	18:54	08:25	18:37	08:40	18:29	08:54	18:28	09:01	18:34
CX MONTEVIDEO URUGUAY	09:05	21:57	08:48	22:09	08:32	22:26	08:25	22:39	08:25	22:53	08:31	23:00
D6 COMOROS	02:45	15:06	02:38	15:08	02:33	15:13	02:33	15:20	02:38	15:29	02:44	15:36
DJ BONN GERMANY	05:51	16:45	06:14	16:17	06:43	15:49	07:06	15:33	07:25	15:26	07:33	15:32
DU MANILA PHILIPPINES	21:47	09:38	21:50	09:30	21:56	09:25	22:03	09:24	22:12	09:28	22:19	09:35
EA6 BALEARIC ISLANDS	06:02	17:19	06:17	17:00	06:36	16:41	06:51	16:33	07:06	16:31	07:13	16:37
EA8 CANARY ISLANDS	07:06	18:39	07:15	18:25	07:28	18:13	07:39	18:08	07:51	18:09	07:59	18:16
EI DUBLIN REP IRELAND	06:48	17:35	07:14	17:04	07:46	16:33	08:11	16:15	08:32	16:06	08:40	16:12
ET ADDIS-ABABA ETHIOPIA	03:13	15:10	03:14	15:04	03:18	15:01	03:23	15:02	03:31	15:07	03:38	15:14
EZ ASHKHABAD TURKOMAN	02:15	13:32	02:30	13:12	02:49	12:54	03:05	12:45	03:19	12:43	03:27	12:49
FT8 KERGUELEN	00:26	13:51	23:58	14:14	23:30	14:42	23:14	15:05	23:08	15:24	23:13	15:33
FG GUADELOUPE	09:57	21:45	10:01	21:37	10:08	21:30	10:15	21:30	10:24	21:33	10:32	21:40
FK NOUMEA NEW CALEDON	18:23	06:56	18:12	07:02	18:03	07:11	18:00	07:21	18:02	07:31	18:09	07:38
FO CLIPPERTON ISL	13:09	01:04	13:11	00:57	13:15	00:53	13:21	00:54	13:30	00:59	13:37	01:06
FO PAPEETE POLYNESIA	15:29	03:57	15:20	04:01	15:14	04:09	15:13	04:16	15:16	04:26	15:23	04:33
FP ST PIERRE & MIQUELON	10:00	21:02	10:20	20:38	10:45	20:14	11:04	20:01	11:21	19:57	11:29	20:03
FR REUNION	01:48	14:20	01:38	14:25	01:29	14:34	01:27	14:43	01:30	14:54	01:36	15:01
FY CAYENNE FR.GUIANA	09:15	21:16	09:14	21:12	09:16	21:11	09:21	21:13	09:28	21:19	09:35	21:26
G LONDON ENGLAND	06:21	17:13	06:45	16:44	07:15	16:15	07:38	15:58	07:58	15:52	08:06	15:57
HA BUDAPEST HUNGARY	04:59	16:01	05:20	15:36	05:45	15:11	06:05	14:57	06:23	14:53	06:31	14:59
HC QUITO ECUADOR	10:57	23:04	10:54	23:02	10:55	23:02	10:58	23:06	11:04	23:13	11:11	23:19
HC8 GALAPAGOS ISLS	11:51	23:58	11:48	23:56	11:48	23:56	11:52	00:00	11:58	00:06	12:05	00:13
HP PANAMA CITY PANAMA	11:12	23:09	11:13	23:04	11:17	23:01	11:22	23:03	11:30	23:08	11:37	23:15
HS 3ANGKOK THAILAND	23:09	11:00	23:11	10:53	23:17	10:48	23:24	10:48	23:32	10:52	23:40	10:58
I ROME ITALY	05:19	16:32	05:36	16:11	05:56	15:51	06:13	15:41	06:28	15:39	06:36	15:45
JA TOYKO JAPAN	20:45	08:10	20:58	07:52	21:14	07:36	21:28	07:29	21:42	07:28	21:49	07:35
JD MINAMI TORI-SHIMA	19:36	07:16	19:43	07:05	19:53	06:55	20:03	06:52	20:14	06:54	20:21	07:01
JT ULAN-BATOR MONGOLIA	23:08	10:09	23:28	09:43	23:54	09:18	00:15	09:04	00:33	08:59	00:41	09:05
JW SVALBARD	06:33	14:20	08:49	11:59	NO:SR	NO:SS	NO:SR	NO:SS	NO:SR	NO:SS	NO:SR	NO:SS
KC4 McMurdo Antarctica	15:40	09:36	NO:SR	NO:SS	NO:SR	NO:SS	NO:SR	NO:SS	NO:SR	NO:SS	NO:SR	NO:SS

Prefix City Country	Jan 14 Sunrise	Jan 14 Sunset	Jan 28 Sunrise	Jan 28 Sunset	Feb 14 Sunrise	Feb 14 Sunset	Feb 28 Sunrise	Feb 28 Sunset	Mar 14 Sunrise	Mar 14 Sunset	Mar 28 Sunrise	Mar 28 Sunset
KC6 W CAROLINE ISL	21:06	08:45	21:08	08:51	21:06	08:56	21:01	08:58	20:55	08:58	20:47	08:57
KH6 HILO HAWAII	16:57	04:01	16:56	04:09	16:50	04:19	16:41	04:25	16:30	04:29	16:18	04:33
KL7 FAIRBANKS ALASKA	19:31	00:29	18:52	01:16	17:56	02:15	17:07	03:01	16:16	03:45	15:25	04:28
KL7 JUNEAU ALASKA	17:35	00:38	17:12	01:09	16:34	01:50	15:58	02:24	15:19	02:56	14:39	03:28
KP2 VIRGIN ISLANDS	10:51	21:59	10:50	22:07	10:44	22:16	10:36	22:21	10:26	22:25	10:14	22:29
LA OSLO NORWAY	08:07	14:45	07:42	15:18	07:01	16:02	06:22	16:39	05:40	17:13	04:58	17:48
LZ SOFIA BULGARIA	05:55	15:15	05:46	15:33	05:27	15:55	05:06	16:13	04:42	16:30	04:18	16:46
OH HELSINKI FINLAND	07:12	13:46	06:47	14:20	06:05	15:05	05:25	15:41	04:44	16:16	04:01	16:51
OX GODTHAAB GREENLAND	12:58	18:12	12:21	18:56	11:28	19:53	10:40	20:37	09:51	21:20	09:02	22:02
P2 MADANG P.N.G.	20:14	08:38	20:19	08:40	20:23	08:40	20:23	08:36	20:22	08:31	20:20	08:25
PY RIO DE JANEIRO BR.	08:20	21:44	08:30	21:42	08:41	21:34	08:48	21:24	08:54	21:11	08:59	20:58
PY NATAL BRAZIL	08:16	20:43	08:22	20:45	08:26	20:44	08:27	20:41	08:26	20:35	08:24	20:29
R1F FRANZ JOSEF LAND	NO:SR	NO:SS	NO:SR	NO:SS	NO:SR	NO:SS	07:42	11:24	04:26	14:33	01:51	17:00
SU CAIRO EGYPT	04:52	15:15	04:48	15:27	04:37	15:42	04:23	15:52	04:07	16:02	03:50	16:10
T3 CENTRAL KIRIBATI	17:23	05:43	17:28	05:46	17:31	05:45	17:31	05:42	17:29	05:38	17:27	05:32
T5 MOGADISHU SOMALIA	03:07	15:08	03:11	15:12	03:11	15:14	03:09	15:14	03:05	15:11	03:00	15:08
TA ANKARA TURKEY	05:09	14:46	05:01	15:01	04:44	15:22	04:25	15:38	04:04	15:53	03:41	16:07
TF REYKJAVIK ICELAND	10:58	16:12	10:22	16:57	09:29	17:53	08:41	18:38	07:52	19:20	07:02	20:02
TJ YAOUNDE CAMEROON	05:25	17:20	05:28	17:25	05:28	17:28	05:26	17:28	05:21	17:26	05:15	17:24
TT FT.LAMY CHAD	05:24	16:53	05:25	17:00	05:22	17:06	05:16	17:09	05:08	17:10	05:00	17:11
UA3 MOSCOW EUR.RUSSIA	05:52	13:25	05:33	13:52	04:59	14:29	04:26	14:59	03:50	15:28	03:14	15:56
UA9 NOVOSIBIRSK RUSSIA	02:47	10:27	02:29	10:53	01:56	11:29	01:24	11:58	00:49	12:27	00:13	12:54
UA9 PERM AS.RUSSIA	04:51	11:56	04:29	12:26	03:52	13:07	03:16	13:40	02:37	14:12	01:58	14:43
UA0 KHABAROVSK AS.RUSSIA	22:48	07:28	22:35	07:48	22:11	08:16	21:46	08:38	21:18	09:00	20:48	09:21
UA0 Ø YAKUTSK AS.RUSSIA	00:32	06:27	00:02	07:05	23:15	07:55	22:32	08:36	21:46	09:15	21:00	09:52
UA0 OKHOTSK AS.RUSSIA	23:13	05:56	22:49	06:28	22:08	07:12	21:30	07:47	20:50	08:21	20:08	08:55
UA0 UST-OLENEK AS.RUSSIA	NO:SR	NO:SS	NO:SR	NO:SS	01:19	07:08	23:54	08:30	22:37	09:41	21:20	10:49
UN7 ALMA-ATA KAZAKH	02:22	11:38	02:13	11:56	01:53	12:19	01:32	12:37	01:08	12:54	00:43	13:11
V3 BELIZE BELIZE	12:26	23:37	12:26	23:45	12:20	23:54	12:12	23:59	12:02	00:02	11:51	00:05
V85 BRUNEI	22:33	10:24	22:36	10:30	22:36	10:33	22:32	10:33	22:27	10:32	22:21	10:30
VE1 Halifax N.S.	11:49	20:57	11:39	21:16	11:18	21:40	10:55	21:59	10:30	22:17	10:04	22:35
VE2 Quebec City Que.	12:27	21:21	12:15	21:40	11:52	22:06	11:28	22:27	11:01	22:47	10:34	23:07
VE3 Ottawa Ont.	12:49	21:48	12:37	22:07	12:15	22:32	11:52	22:52	11:26	23:12	10:59	23:31
VE4 Winnipeg Mb.	14:22	22:53	14:08	23:15	13:42	23:44	13:15	00:07	12:46	00:30	12:16	00:52
VE5 Regina Sk.	14:54	23:21	14:40	23:43	14:13	00:12	13:46	00:37	13:16	01:00	12:45	01:23
VE6 Edmonton Al.	15:44	23:42	15:27	00:07	14:56	00:40	14:28	01:07	13:53	01:34	13:19	02:00
VE7 Prince Rupert B.C.	16:55	00:45	16:37	01:11	16:05	01:45	15:34	02:14	15:00	02:41	14:25	03:08
VE8 Yellowknife NWT	16:51	22:44	16:20	23:23	15:32	00:14	14:48	00:55	14:02	01:34	13:15	02:12
VO2 Goose Bay Labrador	12:11	20:23	11:54	20:36	11:24	21:09	10:54	21:36	10:21	22:02	09:47	22:28
VY1 Whitehorse Yukon	17:56	00:23	17:29	00:58	16:46	01:44	16:05	02:21	15:23	02:57	14:39	03:33
VK1 CANBERRA AUSTRALIA	19:02	09:30	19:16	09:16	19:34	09:01	19:48	08:45	20:00	08:26	20:12	08:06
VK4 BRISBANE AUSTRALIA	19:04	08:48	19:16	08:44	19:29	08:34	19:39	08:22	19:47	08:07	19:54	07:52
VK6 PERTH AUSTRALIA	21:23	11:25	21:36	11:20	21:51	11:08	22:03	10:53	22:13	10:37	22:23	10:19
VK8 DARWIN AUSTRALIA	21:01	09:48	21:08	09:49	21:15	09:46	21:18	09:40	21:19	09:32	21:20	09:23
VK9 COCOS (KEELING) IS	23:43	12:28	23:50	12:29	23:56	12:26	23:59	12:20	00:00	12:12	00:01	12:04

Prefix City Country	Apr 14 Sunrise	Apr 14 Sunset	Apr 28 Sunrise	Apr 28 Sunset	May 14 Sunrise	May 14 Sunset	May 28 Sunrise	May 28 Sunset	Jun 14 Sunrise	Jun 14 Sunset	Jun 28 Sunrise	Jun 28 Sunset
KC6 W CAROLINE ISL	20:38	08:57	20:32	08:57	20:27	08:59	20:26	09:02	20:27	09:06	20:30	09:09
KH6 HILO HAWAII	16:04	04:38	15:53	04:42	15:45	04:48	15:41	04:53	15:41	05:00	15:44	05:03
KL7 FAIRBANKS ALASKA	14:22	05:21	13:31	06:07	12:34	07:01	11:46	07:50	11:03	08:40	11:04	08:45
KL7 JUNEAU ALASKA	13:51	04:06	13:13	04:38	12:35	05:13	12:09	05:41	11:52	06:04	11:55	06:08
KP2 VIRGIN ISLANDS	10:01	22:32	09:51	22:36	09:43	22:41	09:40	22:47	09:40	22:52	09:42	22:56
LA OSLO NORWAY	04:07	18:29	03:26	19:03	02:45	19:43	02:15	20:14	01:55	20:40	01:56	20:44
LZ SOFIA BULGARIA	03:49	17:06	03:27	17:22	03:06	17:40	02:54	17:53	02:48	18:05	02:50	18:09
OH HELSINKI FINLAND	03:09	17:33	02:29	18:08	01:46	18:47	01:16	19:19	00:55	19:45	00:57	19:50
OX GODTHAAB GREENLAND	08:01	22:52	07:11	23:36	06:17	00:28	05:33	01:13	04:57	01:56	04:57	02:01
P2 MADANG P.N.G.	20:18	08:18	20:16	08:13	20:17	08:10	20:19	08:10	20:22	08:12	20:25	08:14
PY RIO DE JANEIRO BR.	09:05	20:42	09:11	20:31	09:18	20:21	09:24	20:16	09:31	20:16	09:34	20:18
PY NATAL BRAZIL	08:22	20:21	08:21	20:16	08:22	20:13	08:24	20:12	08:27	20:14	08:30	20:17
R1F FRANZ JOSEF LAND	NO:SR	NO:SS	NO:SR	NO:SS	NO:SR	NO:SS	NO:SR	NO:SS	NO:SR	NO:SS	NO:SR	NO:SS
SU CAIRO EGYPT	03:30	16:21	03:16	16:29	03:03	16:40	02:56	16:48	02:53	16:57	02:56	17:00
T3 CENTRAL KIRIBATI	17:24	05:25	17:22	05:21	17:22	05:19	17:24	05:19	17:27	05:21	17:30	05:24
T5 MOGADISHU SOMALIA	02:54	15:04	02:51	15:02	02:48	15:01	02:49	15:03	02:51	15:06	02:54	15:09
TA ANKARA TURKEY	03:14	16:24	02:54	16:39	02:36	16:54	02:25	17:07	02:19	17:18	02:22	17:21
TF REYKJAVIK ICELAND	06:02	20:53	05:12	21:36	04:18	22:28	03:34	23:13	02:57	23:56	02:58	00:01
TJ YAOUNDE CAMEROON	05:09	17:20	05:04	17:19	05:02	17:19	05:01	17:21	05:04	17:24	05:07	17:27
TT FT.LAMY CHAD	04:49	17:12	04:42	17:13	04:36	17:16	04:34	17:20	04:35	17:25	04:38	17:28
UA3 MOSCOW EUR.RUSSIA	02:30	16:56	01:56	16:59	01:22	17:30	00:59	17:54	00:45	18:14	00:47	18:18
UA9 NOVOSIBIRSK RUSSIA	23:30	13:27	22:57	13:55	22:24	14:25	22:02	14:49	21:48	15:08	21:50	15:13
UA9 PERM AS.RUSSIA	01:10	15:21	00:33	15:53	23:55	16:28	23:29	16:55	23:12	17:18	23:14	17:22
UA0 KHABAROVSK AS.RUSSIA	20:14	09:46	19:47	10:07	19:22	10:29	19:05	10:47	18:56	11:02	18:59	11:06
UA0 YAKUTSK AS.RUSSIA	20:04	10:39	19:19	11:18	18:31	12:03	17:56	12:40	17:29	13:12	17:29	13:18
UA0 OKHOTSK AS.RUSSIA	19:18	09:35	18:38	10:09	17:57	10:47	17:28	11:17	17:09	11:43	17:10	11:48
UA0 UST-OLENEK AS.RUSSIA	19:43	12:16	18:06	13:47	NO:SR	NO:SS	NO:SR	NO:SS	NO:SR	NO:SS	NO:SR	NO:SS
UN7 ALMA-ATA KAZAKH	00:13	13:31	23:51	13:48	23:30	14:06	23:17	14:20	23:10	14:32	23:13	14:36
V3 BELIZE BELIZE	11:38	00:08	11:29	00:12	11:21	00:17	11:18	00:22	11:18	00:28	11:20	00:31
V85 BRUNEI	22:14	10:27	22:09	10:26	22:06	10:27	22:06	10:29	22:08	10:32	22:11	10:35
VE1 Halifax N.S.	09:33	22:56	09:10	23:14	08:48	23:33	08:35	23:48	08:28	00:00	08:31	00:04
VE2 Quebec City Que.	10:01	23:30	09:36	23:49	09:12	00:10	08:58	00:26	08:50	00:40	08:52	00:43
VE3 Ottawa Ont.	10:26	23:53	10:02	00:12	09:39	00:32	09:25	00:48	09:17	01:01	09:20	01:05
VE4 Winnipeg Mb.	11:40	01:19	11:12	01:40	10:45	02:04	10:29	02:23	10:19	02:38	10:22	02:41
VE5 Regina Sk.	12:09	01:50	11:41	02:12	11:13	02:36	10:56	02:55	10:47	03:11	10:49	03:14
VE6 Edmondton Al.	12:38	02:31	12:07	02:56	11:36	03:25	11:16	03:46	11:04	04:04	11:06	04:07
VE7 Prince Rupert B.C.	13:44	03:40	13:11	04:06	12:40	04:35	12:19	04:58	12:06	05:16	12:09	05:20
VE8 Yellowknife NWT	12:19	02:59	11:34	03:39	10:45	04:25	10:09	05:03	09:42	05:35	09:43	05:40
VO2 Goose Bay Labrador	09:07	22:58	08:58	23:24	08:05	23:52	07:46	00:13	07:34	00:30	07:36	00:34
VY1 Whitehorse Yukon	13:47	04:15	13:05	04:51	12:22	05:32	11:51	06:05	11:30	06:32	11:32	06:36
VK1 CANBERRA AUSTRALIA	20:25	07:43	20:37	07:26	20:49	07:10	21:00	07:01	21:09	06:57	21:13	07:00
VK4 BRISBANE AUSTRALIA	20:03	07:33	20:10	07:20	20:19	07:08	20:27	07:02	20:34	07:00	20:38	07:03
VK6 PERTH AUSTRALIA	22:34	09:58	22:44	09:42	22:55	09:29	23:04	09:21	23:12	09:18	23:16	09:21
VK8 DARWIN AUSTRALIA	21:21	09:12	21:22	09:05	21:25	08:59	21:29	08:57	21:34	08:58	21:37	09:01
VK9 COCOS (KEELING) IS	00:01	11:53	00:03	11:46	00:05	11:41	00:09	11:39	00:13	11:40	00:17	11:43

Prefix City Country	July 14 Sunrise	July 14 Sunset	July 28 Sunrise	July 28 Sunset	Aug 14 Sunrise	Aug 14 Sunset	Aug 28 Sunrise	Aug 28 Sunset	Sep 14 Sunrise	Sep 14 Sunset	Sept 28 Sunrise	Sept 28 Sunset
KC6 W CAROLINE ISL	20:34	09:11	20:37	09:09	20:38	09:05	20:38	08:58	20:37	08:49	20:35	08:40
KH6 HILO HAWAII	15:49	05:03	15:54	04:59	16:00	04:50	16:03	04:40	16:07	04:25	16:10	04:12
KL7 FAIRBANKS ALASKA	11:49	08:05	12:35	07:21	13:32	06:20	14:17	05:29	15:08	04:26	15:49	03:35
KL7 JUNEAU ALASKA	12:15	05:53	12:40	05:29	13:16	04:49	13:47	04:12	14:24	03:24	14:54	02:44
KP2 VIRGIN ISLANDS	09:48	22:56	09:52	22:53	09:57	22:44	10:01	22:34	10:03	22:20	10:06	22:08
LA OSLO NORWAY	02:19	20:27	02:47	20:01	03:27	19:18	04:00	18:37	04:40	17:46	05:13	17:04
LZ SOFIA BULGARIA	03:01	18:04	03:13	17:53	03:31	17:32	03:46	17:11	04:04	16:41	04:19	16:16
OH HELSINKI FINLAND	01:20	19:33	01:48	19:06	02:29	18:22	03:03	17:42	03:43	16:50	04:16	16:07
OX GODTHAAB GREENLAND	05:35	01:29	06:17	00:48	07:11	23:51	07:54	23:02	08:43	22:01	09:23	21:11
P2 MADANG P.N.G.	20:27	08:18	20:27	08:20	20:24	08:20	20:19	08:18	20:11	08:15	20:04	08:12
PY RIO DE JANEIRO BR.	09:34	20:24	09:29	20:30	09:19	20:37	09:08	20:41	08:52	20:47	08:38	20:51
PY NATAL BRAZIL	08:32	20:21	08:32	20:23	08:28	20:23	08:23	20:22	08:15	20:19	08:08	20:16
RF1 FRANZ JOSEF LAND	NO:SR	NO:SS	NO:SR	NO:SS	NO:SR	NO:SS	NO:SR	NO:SS	01:03	17:29	03:31	14:51
SU CAIRO EGYPT	03:03	16:58	03:11	16:52	03:21	16:39	03:29	16:24	03:38	16:04	03:46	15:46
T3 CENTRAL KIRIBATI	17:32	05:27	17:32	05:27	17:29	05:28	17:25	05:26	17:17	05:22	17:11	05:19
T5 MOGADISHU SOMALIA	02:57	15:11	02:58	15:12	02:58	15:09	02:55	15:05	02:51	14:59	02:46	14:53
TA ANKARA TURKEY	02:31	17:17	02:42	17:08	02:58	16:49	03:11	16:29	03:27	16:02	03:41	15:39
TF REYKJAVIK ICELAND	03:36	23:29	04:17	22:49	05:11	21:52	05:54	21:02	06:43	20:02	07:24	19:12
TJ YAOUNDE CAMEROON	05:10	17:30	05:11	17:29	05:11	17:26	05:09	17:22	05:05	17:14	05:02	17:08
TT FT.LAMY CHAD	04:42	17:29	04:45	17:27	04:48	17:21	04:49	17:14	04:49	17:03	04:49	16:53
UA3 MOSCOW EUR.RUSSIA	01:04	18:07	01:25	17:47	01:57	17:12	02:24	16:38	02:57	15:54	03:24	15:17
UA9 NOVOSIBIRSK RUSSIA	22:06	15:02	22:27	14:42	22:58	14:08	23:24	13:36	23:56	12:52	00:22	12:16
UA9 PERM AS.RUSSIA	23:33	17:09	23:57	16:46	00:33	16:07	01:03	15:30	01:40	14:42	02:10	14:02
UA0 KHABAROVSK AS.RUSSIA	19:11	10:59	19:26	10:45	19:49	10:19	20:08	09:53	20:32	09:18	20:52	08:49
UA0 YAKUTSK AS.RUSSIA	17:57	12:56	18:30	12:24	19:17	11:34	19:55	10:50	20:40	09:54	21:16	09:08
UA0 OKHOTSK AS.RUSSIA	17:31	11:32	17:58	11:07	18:37	10:25	19:09	09:46	19:48	08:56	20:20	08:14
UA0 UST-OLENEK AS.RUSSIA	NO:SR	NO:SS	NO:SR	NO:SS	17:44	14:24	19:24	12:38	20:52	10:58	21:58	09:42
UN7 ALMA-ATA KAZAKH	23:23	14:31	23:36	14:20	23:54	13:59	00:09	13:37	00:28	13:07	00:44	12:41
V3 BELIZE BELIZE	11:26	00:31	11:30	00:28	11:35	00:20	11:38	00:10	11:40	23:57	11:42	23:45
V85 BRUNEI	22:14	10:37	22:16	10:37	22:16	10:34	22:15	10:29	22:11	10:21	22:08	10:14
VE1 Halifax N.S.	08:42	23:58	08:55	23:47	09:14	23:24	09:31	23:01	09:51	22:30	10:07	22:03
VE2 Quebec City Que.	09:04	00:37	09:19	00:24	09:40	23:59	09:58	23:35	10:20	23:02	10:38	22:33
VE3 Ottawa Ont.	09:32	00:59	09:46	00:46	10:06	00:22	10:24	23:58	10:45	23:26	11:03	22:58
VE4 Winnipeg Mb.	10:35	02:33	10:52	02:19	11:16	01:51	11:36	01:24	12:02	00:47	12:23	00:16
VE5 Regina Sk.	11:03	03:06	11:20	02:51	11:44	02:22	12:05	01:55	12:31	01:18	12:53	00:46
VE6 Edmonton Al.	11:22	03:57	11:41	03:39	12:10	03:07	12:34	02:36	13:04	01:55	13:29	01:21
VE7 Prince Rupert B.C.	12:25	05:09	12:45	04:50	13:15	04:17	13:40	03:45	14:11	03:03	14:36	02:28
VE8 Yellowknife NWT	10:13	05:16	10:47	04:43	11:35	03:52	12:13	03:07	12:59	02:10	13:36	01:23
VO2 Goose Bay Labrador	07:51	00:24	08:11	00:07	08:39	23:35	09:03	23:04	09:32	22:24	09:57	21:50
VY1 Whitehorse Yukon	11:56	06:17	12:25	05:49	13:07	05:03	13:42	04:22	14:23	03:29	14:57	02:46
VK1 CANBERRA AUSTRALIA	21:10	07:08	21:03	07:16	20:47	07:25	20:30	07:40	20:06	07:52	19:46	08:03
VK4 BRISBANE AUSTRALIA	20:37	07:09	20:32	07:11	20:20	07:25	20:07	07:31	19:48	07:39	19:32	07:45
VK6 PERTH AUSTRALIA	23:14	09:28	23:08	09:36	22:54	09:47	22:38	09:56	22:17	10:06	21:59	10:14
VK8 DARWIN AUSTRALIA	21:38	09:05	21:36	09:08	21:31	09:11	21:23	09:12	21:12	09:12	21:02	09:12
VK9 COCOS (KEELING) IS	00:18	11:47	00:16	11:50	00:11	11:53	00:03	11:53	23:52	11:53	23:43	11:53

Prefix City Country	Oct 14 Sunrise	Oct 14 Sunset	Oct 28 Sunrise	Oct 28 Sunset	Nov 14 Sunrise	Nov 14 Sunset	Nov 28 Sunrise	Nov 28 Sunset	Dec 14 Sunrise	Dec 14 Sunset	Dec 28 Sunrise	Dec 28 Sunset
KC6 W CAROLINE ISL	20:34	08:32	20:35	08:26	20:39	08:23	20:44	08:24	20:52	08:29	20:59	08:36
KH6 HILO HAWAII	16:14	03:58	16:19	03:49	16:28	03:41	16:36	03:40	16:46	03:43	16:53	03:50
KL7 FAIRBANKS ALASKA	16:38	02:37	17:23	01:47	18:20	00:51	19:07	00:11	19:49	23:42	19:58	23:46
KL7 JUNEAU ALASKA	15:30	01:59	16:02	01:22	16:42	00:43	17:12	00:19	17:38	00:07	17:46	00:12
KP2 VIRGIN ISLANDS	10:09	21:55	10:14	21:46	10:21	21:39	10:29	21:38	10:39	21:42	10:46	21:48
LA OSLO NORWAY	05:51	16:16	06:26	15:36	07:09	14:54	07:43	14:27	08:11	14:12	08:20	14:17
LZ SOFIA BULGARIA	04:37	15:49	04:54	15:27	05:16	15:06	05:33	14:56	05:48	14:53	05:56	14:59
OH HELSINKI FINLAND	04:55	15:19	05:30	14:39	06:14	13:56	06:48	13:28	07:16	13:13	07:25	13:17
OX GODTHAAB GREENLAND	10:10	20:14	10:53	19:27	11:48	18:33	12:33	17:55	13:12	17:29	13:21	17:33
P2 MADANG P.N.G.	19:57	08:10	19:53	08:09	19:51	08:12	19:53	08:16	19:58	08:24	20:05	08:31
PY RIO DE JANEIRO BR.	08:22	20:56	08:11	21:03	08:02	21:13	07:59	21:22	08:02	21:33	08:08	21:40
PY NATAL BRAZIL	08:00	20:14	07:56	20:14	07:54	20:17	07:56	20:21	08:01	20:29	08:08	20:36
RF1 FRANZ JOSEF LAND	06:43	11:29	NO:SR	NO:SS	NO:SR	NO:SS	NO:SR	NO:SS	NO:SR	NO:SS	NO:SR	NO:SS
SU CAIRO EGYPT	03:55	15:27	04:05	15:13	04:18	15:00	04:30	14:55	04:42	14:56	04:49	15:03
T3 CENTRAL KIRIBATI	17:05	05:16	17:01	05:15	17:00	05:17	17:02	05:22	17:08	05:29	17:15	05:36
T5 MOGADISHU SOMALIA	02:43	14:47	02:41	14:44	02:42	14:44	02:46	14:47	02:53	14:53	03:00	15:00
TA ANKARA TURKEY	03:57	15:13	04:12	14:54	04:31	14:35	04:47	14:26	05:01	14:24	05:09	14:30
TF REYKJAVIK ICELAND	08:11	18:15	08:54	17:27	09:48	16:33	10:33	15:56	11:12	15:30	11:22	15:34
TJ YAOUNDE CAMEROON	04:59	17:02	04:58	16:58	05:00	16:57	05:04	16:59	05:11	17:05	05:18	17:12
TT FT.LAMY CHAD	04:49	16:43	04:51	16:36	04:56	16:32	05:03	16:32	05:11	16:37	05:18	16:43
UA3 MOSCOW EUR.RUSSIA	03:56	14:36	04:25	14:02	05:00	13:27	05:28	13:07	05:51	12:56	05:59	13:02
UA9 NOVOSIBIRSK RUSSIA	00:53	11:36	01:21	11:03	01:56	10:29	02:23	10:09	02:46	09:59	02:54	10:04
UA9 PERM AS.RUSSIA	02:45	13:18	03:17	12:41	03:56	12:02	04:27	11:39	04:52	11:26	05:01	11:31
UA0 KHABAROVSK AS.RUSSIA	21:15	08:16	21:37	07:50	22:03	07:23	22:24	07:09	22:43	07:04	22:51	07:09
UA0 YAKUTSK AS.RUSSIA	21:59	08:16	22:38	07:32	23:26	06:44	00:05	06:12	00:38	05:51	00:48	05:55
UA0 OKHOTSK AS.RUSSIA	20:58	07:27	21:32	06:48	22:14	06:06	22:47	05:40	23:15	05:25	23:24	05:29
UA0 UST-OLENEK AS.RUSSIA	23:16	08:15	00:34	06:52	03:17	04:09	NO:SR	NO:SS	NO:SR	NO:SS	NO:SR	NO:SS
UN7 ALMA-ATA KAZAKH	01:02	12:13	01:20	11:51	01:42	11:30	01:59	11:19	02:15	11:16	02:23	11:22
V3 BELIZE BELIZE	11:45	23:32	11:49	23:24	11:57	23:17	12:05	23:16	12:14	23:20	12:21	23:26
V85 BRUNEI	22:06	10:07	22:05	10:03	22:07	10:01	22:12	10:04	22:19	10:09	22:26	10:16
VE1 Halifax N.S.	10:27	21:34	10:45	21:11	11:08	20:49	11:27	20:38	11:43	20:34	11:51	20:40
VE2 Quebec City Que.	11:00	22:02	11:20	21:38	11:45	21:14	12:04	21:01	12:22	20:57	12:29	21:03
VE3 Ottawa Ont.	11:24	22:28	11:43	22:03	12:07	21:40	12:26	21:28	12:43	21:24	12:51	21:30
VE4 Winnipeg Mb.	12:47	23:42	13:10	23:15	13:38	22:48	14:00	22:33	14:18	22:27	14:26	22:33
VE5 Regina Sk.	13:18	00:12	13:41	23:44	14:10	23:16	14:32	23:01	14:51	22:55	14:59	23:01
VE6 Edmonton Al.	13:58	00:43	14:24	00:12	14:56	23:40	15:21	23:22	15:42	23:14	15:50	23:20
VE7 Prince Rupert B.C.	15:06	01:49	15:33	01:17	16:07	00:44	16:32	00:26	16:54	00:17	17:02	00:23
VE8 Yellowknife NWT	14:19	00:31	14:58	23:46	15:48	22:58	16:27	22:25	17:00	22:05	17:09	22:10
VO2 Goose Bay Labrador	10:25	21:12	10:51	20:41	11:23	20:10	11:48	19:52	12:09	19:44	12:17	19:49
VY1 Whitehorse Yukon	15:37	01:57	16:13	01:16	16:58	00:32	17:33	00:04	18:02	23:48	18:10	23:53
VK1 CANBERRA AUSTRALIA	19:24	08:16	19:06	08:28	18:50	08:45	18:43	08:59	18:42	09:13	18:48	09:20
VK4 BRISBANE AUSTRALIA	19:14	07:54	19:01	08:02	18:49	08:14	18:44	08:25	18:46	08:37	18:52	08:44
VK6 PERTH AUSTRALIA	21:39	10:25	21:23	10:36	21:09	10:50	21:03	11:03	21:03	11:15	21:10	11:23
VK8 DARWIN AUSTRALIA	20:52	09:13	20:45	09:15	20:40	09:20	20:40	09:27	20:45	09:35	20:51	09:42
VK9 COCOS (KEELING) IS	23:33	11:53	23:26	11:55	23:22	12:00	23:22	12:06	23:26	12:15	23:33	12:22

Prefix City Country	Jan 14 Sunrise	Jan 14 Sunset	Jan 28 Sunrise	Jan 28 Sunset	Feb 14 Sunrise	Feb 14 Sunset	Feb 28 Sunrise	Feb 28 Sunset	Mar 14 Sunrise	Mar 14 Sunset	Mar 28 Sunrise	Mar 28 Sunset
VK9 NORFOLK ISLAND	18:10	07:52	18:22	07:49	18:35	07:39	18:44	07:27	18:52	07:13	18:59	06:57
VK0 MACQUARIE ISLAND	17:11	09:56	17:38	09:38	18:14	09:05	18:43	08:32	19:11	07:58	19:38	07:23
VK0 FALKLAND ISLANDS	08:11	00:14	08:34	23:55	09:05	23:31	09:30	23:03	09:54	22:33	10:17	22:02
VP8 SO.SANDWICH IS	05:12	22:42	05:44	22:18	06:26	21:38	07:00	21:01	07:33	20:22	08:04	19:43
VP8 SO.SHETLAND IS	06:40	01:25	07:22	00:52	08:14	00:02	08:56	23:18	09:35		10:12	21:46
VP9 BERMUDA	11:21	21:35	11:16	21:48	11:03	22:03	10:49	22:15	10:32	22:25	10:14	22:35
VQ9 DIEGO GARCIA	01:02	13:34	01:08	13:36	01:13	13:34	01:14	13:30	01:14	13:24	01:13	13:17
VR6 PITCAIRN IS	14:16	03:46	14:26	03:44	14:38	03:35	14:46	03:24	14:53	03:10	14:59	02:56
VU7 ANDAMAN IS	00:14	11:45	00:16	11:51	00:13	11:57	00:07	12:00	00:00	12:02	23:51	12:02
VU7 LACCADIVE IS	01:30	13:03	01:32	13:09	01:29	12:15	01:24	13:18	01:16	13:19	01:08	13:19
W0 Denver CO	14:20	23:58	14:12	00:14	13:55	00:34	13:36	00:50	13:14	01:05	12:52	01:19
W0 Des Moines IA	13:40	23:07	13:31	23:24	13:12	23:45	12:52	00:02	12:29	00:19	12:06	00:34
W0 Topeka KS	13:41	23:22	13:33	23:38	13:16	23:58	12:58	00:13	12:37	00:27	12:15	00:41
W0 Minneapolis MN	13:48	22:55	13:38	23:14	13:17	23:38	12:54	23:57	12:29	00:16	12:03	00:34
W0 Jefferson City MO	13:25	23:09	13:18	23:25	13:02	23:44	12:43	23:59	12:23	00:14	12:01	00:27
W0 Lincoln NE	13:50	23:22	13:41	23:38	13:23	23:59	13:04	00:15	12:42	00:31	12:18	00:46
W0 Bismark ND	14:25	23:19	14:13	23:39	13:50	00:04	13:26	00:25	12:59	00:46	12:32	01:05
W0 Pierre SD	14:15	23:25	14:05	23:44	13:44	00:07	13:22	00:26	12:57	00:45	12:31	01:02
W1 Augusta ME	12:12	21:23	12:02	21:41	11:41	22:05	11:19	22:24	10:55	22:42	10:29	22:59
W1 Boston MA	12:12	21:35	12:03	21:52	11:34	22:14	11:23	22:32	11:00	22:48	10:35	23:04
W1 Concord NH	12:16	21:34	12:06	21:51	11:46	22:14	11:25	22:32	11:01	22:50	10:37	23:06
W1 Hartford CT	12:16	21:43	12:07	21:59	11:49	22:21	11:28	22:38	11:06	22:55	10:42	23:10
W1 Montpelier VT	12:24	21:34	12:14	21:53	11:53	22:16	11:31	22:35	11:06	22:54	10:40	23:11
W1 Providence RI	12:11	21:37	12:03	21:54	11:44	22:24	11:23	22:42	11:01	22:49	10:37	23:05
W2 Albany NY	12:20	21:44	12:15	22:02	11:55	22:32	11:34	22:48	11:11	22:59	10:46	23:15
W2 Trenton NJ	12:24	21:55	12:12	22:11	11:55	22:41	11:35	22:56	11:14	23:03	10:51	23:18
W3 Dover DE	12:24	22:05	12:17	22:21	12:00	22:40	11:41	22:56	11:20	23:11	10:58	23:25
W3 Harrisburg PA	12:29	22:04	12:21	22:20	12:03	22:43	11:44	22:58	11:24	23:12	10:59	23:26
W3 Washington D.C.	12:26	22:08	12:18	22:24	12:01		11:43		11:22	23:13	11:23	00:26
W4 Atlanta GA	12:43	22:50	12:37	23:03	12:24	23:16	12:08	23:32	11:50	23:44	11:32	23:54
W4 Columbia SC	12:30	22:36	12:24	22:49	12:11	23:06	11:55	23:18	11:37	23:30	11:18	23:41
W4 Frankfort KY	12:55	22:41	12:48	22:57	12:32		12:14	23:31	11:54	23:45	11:32	23:58
W4 Montgomery AL	12:47	23:01	12:42	23:14	12:30	23:29	12:15	23:41	11:58	23:52	11:40	00:01
W4 Nashville TN	12:58	22:54	12:52	23:08	12:37	23:26	12:20	23:40	12:01	23:53	11:40	00:05
W4 Raleigh NC	12:24	22:22	12:18	22:37	12:04	22:54	11:47	23:08	11:28	23:20	11:08	23:32
W4 Richmond VA	12:24	22:13	12:17	22:28	12:02	22:47	11:44	23:02	11:24	23:15	11:03	23:28
W4 Tallahassee FL	12:35	22:57	12:31	23:09	12:19	23:24	12:05	23:34	11:49	23:44	11:32	23:53
W5 Austin TX	13:28	23:51	13:24	00:03	13:13	00:18	12:59	00:28	12:43	00:38	12:26	00:46
W5 Baton Rouge LA	13:02	23:24	12:58	23:37	12:47	23:51	12:33	00:02	12:17	00:11	12:00	00:20
W5 Jackson MS	13:02	23:16	12:58	23:29	12:45	23:45	12:30	23:56	12:13	00:07	11:55	00:17
W5 Little Rock AR	13:16	23:19	13:11	23:33	12:57	23:50	12:41	00:03	12:22	00:15	12:03	00:26
W5 Oklahoma City OK	13:39	23:39	13:33	23:53	13:19	00:10	13:02	00:23	12:43	00:36	12:24	00:47
W5 Santa Fe NM	14:12	00:13	14:06	00:27	13:52	00:44	13:36	00:58	13:17	01:10	12:57	01:21
W6 Los Angeles CA	14:59	01:05	14:53	01:19	14:39	01:35	14:24	01:48	14:06	01:59	13:47	02:10
W6 San Francisco CA	15:24	01:13	15:17	01:28	15:01	01:47	14:44	02:01	14:23	02:15	14:02	02:28
W7 Boise ID	15:16	00:31	15:06	00:49	14:46	01:12	14:24	01:31	14:00	01:48	13:35	02:05

Prefix City Country	Apr 14 Sunrise	Apr 14 Sunset	Apr 28 Sunrise	Apr 28 Sunset	May 14 Sunrise	May 14 Sunset	May 28 Sunrise	May 28 Sunset	Jun 14 Sunrise	Jun 14 Sunset	Jun 28 Sunrise	Jun 28 Sunset
VK9 NORFOLK ISLAND	19:08	06:39	19:15	06:26	19:24	06:14	19:32	06:08	19:39	06:06	19:42	06:09
VK0 MACQUARIE ISLAND	20:11	06:41	20:37	06:09	21:05	05:37	21:27	05:17	21:45	05:05	21:49	05:06
VP8 FALKLAND ISLANDS	10:44	21:25	11:06	20:57	11:30	20:30	11:49	20:13	12:04	20:04	12:07	20:06
VP8 SO.SANDWICH IS	08:41	18:56	09:12	18:20	09:45	17:43	10:11	17:19	10:32	17:04	10:36	17:06
VP8 SO.SHETLAND IS	10:57	20:52	11:34	20:09	12:16	19:25	12:49	18:53	13:16	18:32	13:21	18:33
VP9 BERMUDA	09:52	22:47	09:36	22:57	09:22	23:08	09:14	23:18	09:11	23:27	09:14	23:30
VQ9 DIEGO GARCIA	01:11	13:09	01:11	13:03	01:12	12:59	01:14	12:58	01:18	13:00	01:22	13:03
VR6 PITCAIRN IS	15:06	02:39	15:12	02:27	15:19	02:17	15:26	02:12	15:33	02:11	15:36	02:14
VU7 ANDAMAN IS	23:41	12:02	23:33	12:04	23:28	12:07	23:26	12:10	23:27	12:15	23:30	12:18
VU7 LACCADIVE IS	00:58	13:19	00:51	13:20	00:46	13:23	00:44	13:26	00:45	13:31	00:48	13:34
W0 Denver CO	12:25	01:36	12:05	01:50	11:47	02:06	11:36	02:18	11:31	02:29	11:34	02:32
W0 DesMoines IA	11:37	00:53	11:16	01:08	10:57	01:25	10:45	01:38	10:40	01:49	10:42	01:53
W0 Topeka KS	11:49	00:58	11:29	01:11	11:11	01:27	11:01	01:39	10:56	01:49	10:59	01:53
W0 Minneapolis MN	11:32	00:56	11:08	01:13	10:46	01:33	10:33	01:48	10:26	02:00	10:28	02:04
W0 Jefferson City MO	11:35	00:43	11:16	00:56	10:58	01:11	10:48	01:23	10:44	01:34	10:46	01:37
W0 Lincoln NE	11:51	01:04	11:30	01:19	11:11	01:35	11:00	01:48	10:55	01:59	10:57	02:03
W0 Bismark ND	11:59	01:28	11:34	01:47	11:10	02:08	10:56	02:24	10:48	02:38	10:51	02:41
W0 Pierre SD	12:01	01:23	11:38	01:40	11:16	01:59	11:03	02:14	10:56	02:26	10:59	02:30
W1 Augusta ME	09:58	23:20	09:35	23:37	09:14	23:56	09:01	00:11	08:54	00:24	08:57	00:27
W1 Boston MA	10:07	23:23	09:45	23:39	09:25	23:56	09:13	00:10	09:07	00:22	09:10	00:25
W1 Concord NH	10:07	23:26	09:45	23:42	09:24	00:00	09:12	00:14	09:06	00:26	09:08	00:30
W1 Hartford CT	10:13	23:29	09:52	23:44	09:33	00:01	09:21	00:15	09:15	00:26	09:18	00:30
W1 Montpelier VT	10:10	23:32	09:47	23:49	09:25	00:08	09:12	00:23	09:05	00:35	09:08	00:39
W1 Providence RI	10:08	23:24	09:47	23:39	09:27	23:56	09:16	00:10	09:10	00:21	09:12	00:25
W2 Albany NY	10:17	23:35	09:55	23:51	09:35	00:09	09:22	00:22	09:16	00:34	09:19	00:38
W2 Trenton NJ	10:24	23:35	10:04	23:50	09:45	00:06	09:34	00:18	09:29	00:29	09:31	00:33
W3 Dover DE	10:32	23:41	10:12	23:55	09:54	00:10	09:44	00:22	09:39	00:33	09:42	00:36
W3 Harrisburg PA	10:32	23:44	10:12	23:58	09:53	00:14	09:42	00:27	09:37	00:38	09:39	00:41
W3 Washington D.C.	10:34	23:43	10:15	23:56	09:57	00:11	09:47	00:23	09:42	00:34	09:45	00:37
W4 Atlanta GA	11:09	00:07	10:53	00:18	10:38	00:30	10:30	00:40	10:26	00:49	10:29	00:52
W4 Columbia SC	10:55	23:54	10:39	00:04	10:24	00:17	10:15	00:27	10:12	00:36	10:15	00:39
W4 Frankfort KY	11:07	00:14	10:48	00:27	10:30	00:41	10:20	00:53	10:16	01:03	10:18	01:07
W4 Montgomery AL	11:18	00:13	11:02	00:23	10:48	00:35	10:40	00:44	10:37	00:53	10:40	00:56
W4 Nashville TN	11:16	00:19	10:59	00:31	10:42	00:45	10:33	00:55	10:29	01:05	10:32	01:08
W4 Raleigh NC	10:44	23:46	10:27	23:58	10:11	00:11	10:02	00:22	09:58	00:31	10:00	00:35
W4 Richmond VA	10:38	23:43	10:19	23:56	10:02	00:11	09:52	00:22	09:48	00:32	09:51	00:35
W4 Tallahassee FL	11:12	00:03	10:57	00:12	10:44	00:23	10:37	00:32	10:34	00:40	10:37	00:43
W5 Austin TX	12:06	00:57	11:51	01:06	11:38	01:16	11:31	01:25	11:28	01:33	11:31	01:36
W5 Baton Rouge LA	11:39	00:31	11:25	00:40	11:11	00:50	11:04	00:59	11:02	01:08	11:04	01:11
W5 Jackson MS	11:34	00:29	11:18	00:39	11:04	00:50	10:56	00:59	10:53	01:08	10:56	01:11
W5 Little Rock AR	11:40	00:39	11:23	00:51	11:07	01:03	10:59	01:14	10:55	01:23	10:58	01:26
W5 Oklahoma City OK	12:00	01:01	11:43	01:13	11:27	01:26	11:18	01:37	11:14	01:46	11:17	01:49
W5 Santa Fe NM	12:34	01:35	12:17	01:46	12:01	01:59	11:53	02:10	11:49	02:19	11:52	02:22
W6 Los Angeles CA	13:24	02:23	13:08	02:33	12:53	02:46	12:44	02:56	12:41	03:05	12:44	03:08
W6 San Francisco CA	13:37	02:43	13:18	02:56	13:01	03:11	12:51	03:22	12:47	03:32	12:50	03:36
W7 Boise ID	13:05	02:26	12:43	02:42	12:22	03:01	12:09	03:15	12:03	03:27	12:05	03:30

Prefix City Country	July 14 Sunrise	July 14 Sunset	July 28 Sunrise	July 28 Sunset	Aug 14 Sunrise	Aug 14 Sunset	Aug 28 Sunrise	Aug 28 Sunset	Sep 14 Sunrise	Sep 14 Sunset	Sept 28 Sunrise	Sept 28 Sunset
VK9 NORFOLK ISLAND	19:42	06:15	19:36	06:22	19:25	06:30	19:12	06:37	18:53	06:44	18:37	06:51
VK0 MACQUARIE ISLAND	21:40	05:21	21:22	05:40	20:20	06:09	20:19	06:34	19:37	07:05	19:02	07:30
VP8 FALKLAND ISLANDS	12:00	20:20	11:45	20:36	11:17	21:00	10:50	21:21	10:13	21:47	09:41	22:09
VP8 SO.SANDWICH IS	10:23	17:24	10:02	17:47	09:24	18:21	08:48	18:51	08:02	19:26	07:22	19:56
VP8 SO.SHETLAND IS	13:02	18:57	12:34	19:27	11:48	20:10	11:05	20:46	10:11	21:29	09:25	22:05
VP9 BERMUDA	09:22	23:28	09:30	23:21	09:42	23:06	09:51	22:50	10:01	22:28	10:10	22:10
VQ9 DIEGO GARCIA	01:23	13:07	01:23	13:09	01:19	13:06	01:13	13:09	01:04	13:07	00:56	13:05
VR6 PITCAIRN IS	15:35	02:20	15:31	02:26	15:20	02:33	15:08	02:39	14:51	02:45	14:36	02:50
VU7 ANDAMAN IS	23:34	12:19	23:37	12:18	23:40	12:12	23:41	12:04	23:40	11:54	23:40	11:44
VU7 LACCADIVE IS	00:52	13:35	00:55	13:34	00:57	13:28	00:58	13:21	00:57	13:10	00:57	13:01
W0 Denver CO	11:43	02:28	11:54	02:18	12:10	01:59	12:23	01:39	12:39	01:12	12:52	00:49
W0 DesMoines IA	10:52	01:48	11:04	01:38	11:21	01:17	11:35	00:56	11:53	00:28	12:07	00:04
W0 Topeka KS	11:08	01:49	11:19	01:39	11:34	01:21	11:47	01:01	12:02	00:35	12:15	00:12
W0 Minneapolis MN	10:39	01:58	10:53	01:46	11:12	01:23	11:29	01:00	11:49	00:28	12:06	00:02
W0 Jefferson City MO	10:56	01:33	11:06	01:24	11:21	01:06	11:33	00:47	11:48	00:21	12:01	23:58
W0 Lincoln NE	11:07	01:58	11:19	01:48	11:35	01:28	11:49	01:08	12:05	00:40	12:19	00:16
W0 Bismark ND	11:03	02:35	11:17	02:22	11:38	01:57	11:56	01:33	12:18	01:00	12:37	00:31
W0 Pierre SD	11:10	02:24	11:23	02:12	11:42	01:50	11:58	01:27	12:18	00:56	12:34	00:30
W1 Augusta ME	09:08	00:22	09:21	00:10	09:40	23:48	09:56	23:25	10:15	22:54	10:32	22:28
W1 Boston MA	09:20	00:20	09:32	00:10	09:50	23:49	10:04	23:27	10:22	22:58	10:37	22:34
W1 Concord NH	09:19	00:25	09:31	00:14	09:49	23:52	10:05	23:30	10:23	23:00	10:39	22:35
W1 Hartford CT	09:28	00:25	09:40	00:14	09:57	23:54	10:11	23:33	10:27	23:04	10:43	22:40
W1 Montpelier VT	09:19	00:33	09:32	00:21	09:51	23:59	10:07	23:36	10:24	23:05	10:43	22:39
W1 Providence RI	09:23	00:20	09:34	00:10	09:52	23:49	10:06	23:39	10:24	22:59	10:38	22:35
W2 Albany NY	09:29	00:33	09:42	00:22	09:59	00:01	10:15	23:39	10:33	23:12	10:48	22:45
W2 Trenton NJ	09:41	00:29	09:52	00:19	10:08	23:59	10:22	23:39	10:45	23:10	10:51	22:49
W3 Dover DE	09:51	00:32	10:02	00:23	10:17	00:04	10:30	23:46	10:45	23:18	10:58	22:56
W3 Harrisburg PA	09:49	00:37	10:00	00:27	10:16	00:08	10:30	23:48	10:46	23:20	11:00	22:57
W3 Washington D.C.	09:54	00:33	10:05	00:24	10:20	00:06	10:32	23:46	10:48	23:20	11:00	22:58
W4 Atlanta GA	10:37	00:50	10:46	00:42	10:58	00:27	11:08	00:10	11:19	23:47	11:29	23:28
W4 Columbia SC	10:23	00:37	10:32	00:29	10:44	00:14	10:54	23:57	11:06	23:34	11:16	23:14
W4 Frankfort KY	10:28	01:03	10:38	00:54	10:53	00:36	11:05	00:17	11:20	23:51	11:32	23:29
W4 Montgomery AL	10:48	00:54	10:56	00:47	11:08	00:32	11:17	00:16	11:28	23:54	11:36	23:36
W4 Nashville TN	10:41	01:05	10:50	00:57	11:04	00:40	11:15	00:22	11:28	23:58	11:39	23:37
W4 Raleigh NC	10:09	00:32	10:19	00:24	10:32	00:07	10:43	23:49	10:56	23:22	11:06	23:05
W4 Richmond VA	10:00	00:32	10:10	00:23	10:24	00:06	10:36	23:47	10:50	23:25	11:02	23:00
W4 Tallahassee FL	10:44	00:41	10:52	00:35	11:03	00:21	11:11	00:06	11:20	23:46	11:28	23:28
W5 Austin TX	11:39	01:35	11:46	01:28	11:57	01:15	12:05	01:00	12:14	00:39	12:22	00:22
W5 Baton Rouge LA	11:12	01:09	11:20	01:02	11:30	00:49	11:38	00:34	11:48	00:13	11:56	23:55
W5 Jackson MS	11:04	01:09	11:12	01:02	11:23	00:47	11:32	00:32	11:43	00:10	11:52	23:51
W5 Little Rock AR	11:06	01:23	11:15	01:16	11:28	01:00	11:38	00:43	11:51	00:19	12:01	23:59
W5 Oklahoma City OK	11:25	01:46	11:35	01:38	11:48	01:22	11:59	01:05	12:11	00:41	12:22	00:20
W5 Santa Fe NM	12:00	02:19	12:09	02:12	12:22	01:55	12:33	01:38	12:45	01:14	12:56	00:54
W6 Los Angeles CA	12:52	03:06	13:01	02:58	13:13	02:42	13:23	02:26	13:35	02:03	13:45	01:43
W6 San Francisco CA	12:59	03:32	13:09	03:23	13:24	03:05	13:36	02:47	13:50	02:20	14:02	01:59
W7 Boise ID	12:16	03:25	12:29	03:14	12:47	02:52	13:03	02:30	13:22	01:59	13:38	01:34

Prefix City Country	Oct 14 Sunrise	Oct 14 Sunset	Oct 28 Sunrise	Oct 28 Sunset	Nov 14 Sunrise	Nov 14 Sunset	Nov 28 Sunrise	Nov 28 Sunset	Dec 14 Sunrise	Dec 14 Sunset	Dec 28 Sunrise	Dec 28 Sunset
VK9 NORFOLK ISLAND	18:19	06:59	18:06	07:07	17:55	07:19	17:50	07:30	17:52	07:42	17:58	07:49
VKØ MACQUARIE ISLAND	18:22	08:01	17:49	08:29	17:14	09:04	16:54	09:31	16:43	09:55	16:48	10:04
VP8 FALKLAND ISLANDS	09:06	22:35	08:37	22:59	08:08	23:28	07:52	23:51	07:45	00:11	07:51	00:19
VP8 SO.SANDWICH IS	06:37	20:32	05:59	21:04	05:19	21:46	04:54	22:18	04:40	22:45	04:45	22:53
VP8 SO.SHETLAND IS	08:33	22:48	07:48	23:28	06:58	00:19	06:23	01:00	06:00	01:15	06:04	01:46
VP9 BERMUDA	10:21	21:49	10:32	21:34	10:47	21:20	10:59	21:14	11:12	21:15	11:19	21:21
VQ9 DIEGO GARCIA	00:48	13:03	00:43	13:04	00:40	13:07	00:42	13:12	00:47	13:20	00:54	13:27
VR6 PITCAIRN IS	14:20	02:57	14:08	03:04	13:58	03:15	13:55	03:25	13:57	03:35	14:04	03:43
VU7 ANDAMAN IS	23:40	11:34	23:42	11:28	23:47	11:24	23:53	11:24	00:01	11:29	00:08	11:35
VU7 LACCADIVE IS	00:57	12:52	00:58	12:45	01:03	12:41	01:09	12:42	01:17	12:47	01:24	12:53
WØ Denver CO	13:08	00:24	13:23	00:05	13:42	23:46	13:58	23:37	14:12	23:36	14:20	23:42
WØ Des Moines IA	12:24	23:37	12:40	23:16	13:01	22:57	13:17	22:47	13:33	22:45	13:40	22:51
WØ Topeka KS	12:30	23:48	12:45	23:28	13:04	23:11	13:19	23:02	13:43	23:01	13:41	23:07
WØ Minneapolis MN	12:26	23:32	12:45	23:09	13:08	22:44	13:26	22:35	13:43	22:32	13:50	22:38
WØ Jefferson City MO	12:16	23:34	12:30	23:15	12:48	22:58	13:03	22:49	13:18	22:48	13:25	22:54
WØ Lincoln NE	12:36	23:50	12:52	23:30	13:12	23:11	13:28	23:01	13:43	23:00	13:50	23:06
WØ Bismark ND	12:58	00:00	13:18	23:36	13:43	23:17	14:02	23:05	14:20	22:55	14:27	23:01
WØ Pierre SD	12:54	00:01	13:12	23:38	13:35	23:17	13:53	23:05	14:09	23:02	14:17	23:08
W1 Augusta ME	10:51	21:59	11:09	21:36	11:32	21:14	11:50	21:03	12:06	21:00	12:14	21:06
W1 Boston MA	10:55	22:07	11:11	21:45	11:33	21:25	11:49	21:15	12:05	21:13	12:12	21:19
W1 Concord NH	10:57	22:07	11:14	21:45	11:36	21:24	11:54	21:14	12:09	21:11	12:17	21:17
W1 Hartford CT	11:01	22:13	11:17	21:52	11:38	21:32	11:54	21:23	12:09	21:20	12:17	21:27
W1 Montpelier VT	10:56	22:10	11:12	21:48	11:44	21:26	11:54	21:15	12:18	21:11	12:12	21:18
W1 Providence RI	11:03	22:08	11:23	21:47	11:33	21:27	11:49	21:17	12:04	21:22	12:25	21:21
W2 Albany NY	11:06	22:17	11:23	22:03	11:45	21:35	12:02	21:24	12:17	21:34	12:20	21:28
W2 Trenton NJ	11:08	22:23	11:28	22:12	11:47	21:44	11:58	21:35	12:13	21:44	12:24	21:40
W3 Dover DE	11:14	22:31	11:31	22:11	11:51	21:43	12:02	21:35	12:17	21:42	12:29	21:50
W3 Harrisburg PA	11:16	22:31	11:30	22:14	11:48	21:53	12:07	21:45	12:21	21:47	12:25	21:48
W3 Washington D.C.	11:15	22:33	11:30	22:13	11:49	21:50	12:03	21:43	12:18	21:45	12:24	21:53
W4 Atlanta GA	11:41	23:07	11:39	22:37	11:55	22:36	12:21	22:30	12:34	22:30	12:28	22:36
W4 Columbia SC	11:28	22:53	11:58	22:47	12:13	22:22	12:08	22:16	12:21	22:16	12:25	22:22
W4 Frankfort KY	11:46	23:05	12:00	23:00	12:22	22:29	12:33	22:21	12:48	22:20	12:55	22:27
W4 Montgomery AL	11:47	23:15	11:58	22:57	12:13	22:46	12:25	22:40	12:38	22:41	12:45	22:47
W4 Nashville TN	11:52	23:10	12:05	22:25	11:49	22:41	12:36	22:34	12:50	22:33	12:57	22:40
W4 Raleigh NC	11:19	22:42	11:32	22:18	11:48	22:09	12:02	22:02	12:16	22:02	12:23	22:08
W4 Richmond VA	11:16	22:36	11:30	22:54	12:01	22:09	12:02	21:53	12:16	21:52	12:24	21:59
W4 Tallahassee FL	11:38	23:09	11:48	22:51	11:48	22:42	12:13	22:37	12:25	22:38	12:32	22:44
W5 Austin TX	12:32	00:02	12:41	23:22	12:28	23:36	13:06	23:31	13:18	23:32	13:26	23:38
W5 Baton Rouge LA	12:06	23:36	12:15	23:15	12:19	23:06	12:41	23:04	12:53	22:57	13:00	23:12
W5 Jackson MS	12:03	23:31	12:14	23:10	12:25	23:06	12:41	22:56	13:00	22:50	13:00	23:03
W5 Little Rock AR	12:13	23:37	12:25	23:21	12:41	23:25	12:54	22:59	13:08	22:59	13:15	23:05
W5 Oklahoma City OK	12:35	23:58	12:47	23:41	13:04	00:00	13:17	23:18	13:31	23:18	13:38	23:25
W5 Santa Fe NM	13:08	00:32	13:21	00:15	13:37	00:05	13:50	23:53	14:04	23:53	14:11	23:59
W6 Los Angeles CA	13:57	01:22	14:08	01:06	14:24	00:51	14:37	00:45	14:50	00:45	14:57	00:51
W6 San Francisco CA	14:16	01:36	14:30	01:17	14:48	01:00	15:03	00:52	15:17	00:52	15:24	00:58
W7 Boise ID	13:57	01:05	14:14	00:43	14:36	00:22	14:54	00:11	15:10	00:08	15:17	00:15

Prefix City Country	Jan 14 Sunrise	Jan 14 Sunset	Jan 28 Sunrise	Jan 28 Sunset	Feb 14 Sunrise	Feb 14 Sunset	Feb 28 Sunrise	Feb 28 Sunset	Mar 14 Sunrise	Mar 14 Sunset	Mar 28 Sunrise	Mar 28 Sunset
W7 Carson City NV	15:17	00:58	15:10	01:14	14:53	01:34	14:34	01:49	14:13	02:04	13:51	02:18
W7 Cheyenne WY	14:23	23:53	14:15	00:10	13:56	00:31	13:36	00:48	13:14	01:03	12:51	01:19
W7 Olympia WA	15:54	00:47	15:42	01:06	15:19	01:33	14:55	01:54	14:28	02:14	14:00	02:34
W7 Phoenix AZ	14:32	00:55	14:27	00:54	14:14	01:11	13:58	01:23	13:41	01:34	13:22	01:45
W7 Salem OR	15:47		15:37	01:13	15:16	01:37	14:53	01:57	14:28	02:15	14:02	02:33
W7 Salt Lake City UT	14:50	00:22	14:42	00:39	14:24	01:00	14:04	01:16	13:42	01:32	13:19	01:47
W8 Charleston WV	12:43	22:28	12:35	22:43	12:19	23:02	12:01	23:17	11:40	23:31	11:19	23:45
W8 Columbus OH	12:53	22:29	12:45	22:45	12:27	23:05	12:08	23:17	11:47	23:36	11:24	23:51
W8 Lansing MI	13:07	22:47	12:58	22:45	12:38	23:07	12:17	23:25	11:54	23:42	11:29	23:58
W9 Indianapolis IN	13:04	22:42	12:57	22:58	12:39	23:18	12:20	23:34	11:59	23:49	11:36	00:03
W9 Madison WI	13:27	22:46	13:18	23:03	12:58	23:26	12:37	23:44	12:13	00:01	11:48	00:18
W9 Springfield IL	13:18	22:56	13:11	23:12	12:53	23:32	12:34	23:48	12:13	00:03	11:50	00:17
XE1 MEXICO CITY MEXICO	13:13	00:17	13:13	00:26	13:06	00:35	12:57	00:41	12:46	00:46	12:35	00:49
XW VIENTIANE LOAS	23:44	10:53	23:44	11:01	23:38	11:10	23:30	11:15	23:20	11:19	23:08	11:22
YA KABUL AFGANISTAN	02:30	12:33	02:25	12:47	02:11	13:04	01:55	13:17	01:37	13:29	01:17	13:40
YB JAKARTA INDONESIA	22:48	11:15	22:54	11:18	22:51	11:17	22:59	11:13	22:58	11:07	22:56	11:01
ZD8 ASCENSION IS.	06:44	19:17	06:50	19:19	06:55	19:17	06:57	19:13	06:56	19:07	06:55	18:59
ZD9 TRISTAN DA CUNHA	05:48	20:08	06:02	20:01	06:20	19:47	06:34	19:30	06:46	19:11	06:58	18:51
ZF CAYMAN IS.	11:59	23:03	11:58	23:12	11:52	23:21	11:43	23:27	11:32	23:31	11:20	23:35
ZL1 AUCKLAND NZ	17:16	07:43	17:31	07:37	17:49	07:21	18:04	07:04	18:17	06:44	18:30	06:24
ZL4 DUNEDIN NZ	17:05	08:28	17:24	08:17	17:50	07:54	18:10	07:31	18:30	07:05	18:49	06:38
ZP ASUNCION PARAGUAY	09:12	22:46	09:22	22:43	09:35	22:34	09:34	22:22	09:50	22:09	09:57	21:54
ZS1 CAPETOWN S.AFRICA	03:49	18:00	04:03	17:55	04:20	17:41	04:32	17:26	04:44	17:08	04:54	16:49
ZS8 Pr.EDWARD & MARION	01:53	17:21	02:13	17:09	02:39	16:46	03:00	16:22	03:20	15:55	03:40	15:28
1S SPRATLY IS.	22:51	10:31	22:53	10:37	22:51	10:42	22:47	10:44	22:40	10:44	22:33	10:43
3V TUNIS TUNISIA	06:32	16:24	06:26	16:39	06:11	16:57	05:53	17:11	05:34	17:25	05:13	17:37
3Y BOUVET IS.	03:34	20:16	04:01	19:57	04:37	19:25	05:06	18:52	05:34	18:18	06:01	17:43
3Y PETER 1ST IS.	NO:SR	NO:SS	08:18	04:16	09:44	02:53	10:43	01:51	11:36	00:51	12:27	23:52
4J BAKU AZERBAIJAN	04:03	13:36	03:55	13:52	03:37	14:13	03:18	14:30	02:56	14:45	02:33	15:00
4U UNITED NATIONS HQ	12:18	21:51	12:10	22:07	11:52	22:28	11:33	22:45	11:11	23:00	10:47	23:15
4X ISRAEL	04:47	14:59	04:42	15:12	04:29	15:27	04:14	15:39	03:57	15:50		16:00
5R TANANARIVE MALAGASY	02:23	15:33	02:32	16:04	02:41	15:26	02:47	15:17	02:51	15:07	02:55	14:55
5W WESTERN SAMOA	17:10	06:04	17:18	05:51	17:25	06:00	17:29	05:53	17:31	05:44	17:33	05:34
5Z NAIROBI KENYA	03:36	15:47	03:40		03:42	15:52	03:41	15:50	03:39	15:46	03:35	15:41
6W DAKAR SENEGAL	07:38	18:59	07:39	19:06	07:35	19:13	07:28	19:17	07:19	19:20	07:09	19:21
7O ADEN YEMEN P.DEM.	03:25	14:52	03:26	14:59	03:23	15:05	03:17	15:09	03:09	15:10	03:00	15:11
7P LESOTHO	03:20	17:13	03:32	17:09	03:47	16:57	03:57	16:44	04:06	16:28	04:15	16:11
8Q MALDIVE IS	01:26	13:19	01:29	13:24	01:29	13:27	01:26	13:28	01:21	13:26	01:15	13:24
9G ACCRA GHANA	06:15	18:04	06:17	18:10	06:17	18:13	06:13	18:14	06:08	18:13	06:02	18:11
9J LUSAKA ZAMBIA	03:47	16:44	03:55	16:44	04:02	16:40	04:07	16:33	04:10	16:23	04:12	16:13
9K KUWAIT	03:40	14:08	03:36	14:19	03:26	14:33	03:13	14:43	02:57	14:52	02:41	15:00
9N KATMANDU NEPAL	01:11	11:43	01:08	11:55	00:58	12:08	00:45	12:17	00:31	12:26	00:15	12:33
9Q KINSHASA ZAIRE	04:57	17:18	05:02	17:21	05:05	17:21	05:06	17:18	05:04	17:13	05:02	17:07
9V SINGAPORE	23:12	11:15	23:16	11:19	23:17	11:21	23:15	11:20	23:12	11:18	23:07	11:14
9X RWANDA	04:03	16:17	04:08	16:21	04:10	16:21	04:10	16:19	04:07	16:15	04:04	16:10
9Y TRINIDAD IS.	10:25	22:01	10:27	22:07	10:24	22:12	10:19	22:14	10:12	22:15	10:04	22:15

Prefix City Country	Apr 14 Sunrise	Apr 14 Sunset	Apr 28 Sunrise	Apr 28 Sunset	May 14 Sunrise	May 14 Sunset	May 28 Sunrise	May 28 Sunset	Jun 14 Sunrise	Jun 14 Sunset	Jun 28 Sunrise	Jun 28 Sunset
W7 Carson City NV	13:25	02:34	13:05	02:48	12:47	03:03	12:37	03:15	12:32	03:26	12:35	03:29
W7 Cheyenne WY	12:23	01:37	12:02	01:52	11:43	02:08	11:31	02:21	11:26	02:33	11:28	02:36
W7 Olympia WA	13:27	02:57	13:02	03:16	12:38	03:38	12:23	03:54	12:16	04:07	12:18	04:11
W7 Phoenix AZ	13:00	01:57	12:43	02:08	12:29	02:20	12:21	02:30	12:17	02:39	12:20	02:42
W7 Salem OR	13:31	02:55	13:07	03:12	12:45	03:32	12:32	03:47	12:25	03:59	12:28	04:03
W7 Salt Lake City UT	12:51	02:05	12:31	02:19	12:12	02:36	12:01	02:48	11:55	03:00	11:58	03:03
W8 Charleston WV	10:53	00:01	10:34	00:14	10:17	00:29	10:07	00:40	10:02	00:51	10:05	00:54
W8 Columbus OH	10:57	00:08	10:37	00:22	10:18	00:38	10:07	00:51	10:02	01:02	10:05	01:05
W8 Lansing MI	11:00	00:18	10:38	00:34	10:18	00:52	10:05	01:05	09:59	01:17	10:02	01:21
W9 Indianapolis IN	11:10	00:20	10:50	00:34	10:32	00:50	10:21	01:03	10:16	01:13	10:18	01:17
W9 Madison WI	11:19	00:37	10:57	00:54	10:36	01:12	10:24	01:26	10:17	01:38	10:20	01:41
W9 Springfield IL	11:24	00:34	11:04	00:48	10:45	01:04	10:35	01:16	10:30	01:27	10:32	01:31
XE1 MEXICO CITY MEXICO	12:20	00:54	12:10	00:58	12:02	01:04	11:58	01:09	11:58	01:15	12:01	01:19
XW VIENTIANE LOAS	22:55	11:26	22:45	11:30	22:37	11:35	22:34	11:40	22:34	11:46	22:36	11:49
YA KABUL AFGANISTAN	00:54	13:53	00:37	14:04	00:22	14:17	00:13	14:27	00:09	14:37	00:12	14:40
YB JAKARTA INDONESIA	22:54	10:53	22:53	10:48	22:54	10:45	22:56	10:44	23:00	10:46	23:03	10:49
ZD8 ASCENSION IS.	06:54	18:51	06:54	18:45	06:55	18:41	06:58	18:40	07:02	18:42	07:05	18:45
ZD9 TRISTAN DA CUNHA	07:11	18:28	07:23	18:10	07:36	17:55	07:46	17:46	07:55	17:42	07:59	17:45
ZF CAYMAN IS.	11:06	23:39	10:56	23:44	10:47	23:49	10:43	23:55	10:43	00:01	10:46	00:04
ZL1 AUCKLAND NZ	18:44	05:59	18:56	05:41	19:10	05:25	19:21	05:15	19:31	05:11	19:35	05:13
ZL4 DUNEDIN NZ	19:11	06:07	19:29	05:43	19:48	05:20	20:03	05:06	20:16	04:59	20:20	05:01
ZP ASUNCION PARAGUAY	10:04	21:37	10:11	21:25	10:19	21:14	10:26	21:08	10:33	21:07	10:36	21:10
ZS1 CAPETOWN S.AFRICA	05:07	16:26	05:18	16:10	05:29	15:55	05:39	15:47	05:48	15:44	05:52	15:47
ZS8 Pr.EDWARD & MARION	04:02	14:55	04:21	14:31	04:41	14:08	04:57	13:54	05:10	13:46	05:14	13:49
1S SPRATLY IS.	22:24	10:42	22:18	10:42	22:14	10:44	22:12	10:47	22:14	10:51	22:17	10:54
3V TUNIS TUNISIA	04:48	17:52	04:30	18:05	04:13	18:19	04:04	18:30	03:59	18:40	04:02	18:43
3Y BOUVET IS.	06:32	17:02	06:58	16:30	07:27	15:59	07:48	15:39	08:06	15:27	08:10	15:29
3Y PETER 1ST IS.	13:29	22:40	14:21	21:42	15:27	20:34	16:32	19:31	NO:SR	NO:SS	NO:SR	NO:SS
4J BAKU AZERBAIJAN	02:06	15:17	01:46	15:32	01:27	15:48	01:15	16:01	01:10	16:12	01:12	16:15
4U UNITED NATIONS HQ	10:20	23:33	10:00	23:47	09:41	00:04	09:29	00:16	09:24	00:28	09:27	00:31
4X ISRAEL	03:17	16:12	03:01	16:22	02:47	16:34	02:39	16:43	02:36	16:52	02:38	16:56
5R TANANARIVE MALAGASY	02:59	14:41	03:03	14:32	03:08	14:23	03:13	14:20	03:19	14:20	03:22	14:22
5W WESTERN SAMOA	17:35	05:22	17:37	05:14	17:40	05:08	17:45	05:06	17:50	05:06	17:53	05:09
5Z NAIROBI KENYA	03:31	15:36	03:28	15:33	03:27	15:31	03:28	15:31	03:31	15:34	03:34	15:37
6W DAKAR SENEGAL	06:57	19:23	06:49	19:26	06:42	19:30	06:40	19:34	06:40	19:39	06:43	19:42
7O ADEN YEMEN P.DEM.	02:49	15:12	02:41	15:14	02:35	15:17	02:33	15:21	02:34	15:26	02:37	15:29
7P LESOTHO	04:25	15:51	04:34	15:37	04:43	15:24	04:52	15:17	05:00	15:15	05:03	15:18
8Q MALDIVE IS	01:08	13:21	01:04	13:20	01:01	13:20	01:00	13:22	01:02	13:25	01:05	13:28
9G ACCRA GHANA	05:54	18:08	05:49	18:08	05:46	18:08	05:46	18:10	05:48	18:14	05:50	18:17
9J LUSAKA ZAMBIA	04:14	16:01	04:16	15:53	04:20	15:46	04:25	15:43	04:30	15:44	04:33	15:46
9K KUWAIT	02:21	15:10	02:07	15:18	01:55	15:28	01:48	15:36	01:46	15:44	01:48	15:48
9N KATMANDU NEPAL	23:56	12:42	23:43	12:50	23:31	12:59	23:24	13:07	23:22	13:15	23:25	13:18
9Q KINSHASA ZAIRE	04:59	17:00	04:57	16:56	04:57	16:53	04:59	16:53	05:02	16:55	05:05	16:58
9V SINGAPORE	23:01	11:10	22:58	11:08	22:56	11:07	22:56	11:08	22:59	11:11	23:02	11:14
9X RWANDA	04:00	16:04	03:58	16:01	03:57	15:59	03:58	15:59	04:02	16:01	04:05	16:04
9Y TRINIDAD IS.	09:55	22:15	09:48	22:15	09:43	22:17	09:42	22:21	09:43	22:25	09:46	22:28

Prefix City Country	July 14 Sunrise	July 14 Sunset	July 28 Sunrise	July 28 Sunset	Aug 14 Sunrise	Aug 14 Sunset	Aug 28 Sunrise	Aug 28 Sunset	Sep 14 Sunrise	Sep 14 Sunset	Sept 28 Sunrise	Sept 28 Sunset
W7 Carson City NV	12:44	03:25	12:55	03:16	13:10	02:57	13:23	02:38	13:39	02:11	13:51	01:49
W7 Cheyenne WY	11:38	02:32	11:50	02:21	12:07	02:01	12:21	01:41	12:38	01:12	12:52	00:49
W7 Olympia WA	12:30	04:04	12:45	03:51	13:06	03:26	13:24	03:01	13:47	02:28	14:05	02:00
W7 Phoenix AZ	12:28	02:39	12:37	02:32	12:49	02:17	12:59	02:00	13:10	01:38	13:20	01:18
W7 Salem OR	12:39	03:57	12:52	03:45	13:12	03:22	13:28	02:59	13:49	02:27	14:05	02:01
W7 Salt Lake City UT	12:08	02:59	12:19	02:48	12:36	02:29	12:50	02:08	13:06	01:40	13:20	01:17
W8 Charleston WV	10:14	00:50	10:24	00:41	10:39	00:23	10:55	00:04	11:06	23:38	11:18	23:16
W8 Columbus OH	10:15	01:01	10:26	00:51	10:42	00:32	10:58	00:12	11:11	23:53	11:24	23:22
W8 Lansing MI	10:12	01:13	10:25	01:05	10:42	00:44	10:58	00:22	11:16	23:57	11:31	23:27
W9 Indianapolis IN	10:28	01:16	10:39	01:03	10:55	00:44	11:08	00:24	11:24	00:12	11:37	23:34
W9 Madison WI	10:31	01:36	10:43	01:25	11:01	01:04	11:16	00:42	11:35	00:11	11:50	23:47
W9 Springfield IL	10:42	01:27	10:53	01:17	11:09	00:58	11:22	00:38	11:38	00:41	11:51	23:48
XE1 MEXICO CITY MEXICO	12:06	01:19	12:11	01:15	12:17	01:06	12:20	00:56	12:24	01:15	12:26	00:29
XW VIENTIANE LOAS	22:42	11:49	22:46	11:46	22:51	11:38	22:54	11:28	22:57	11:15	22:59	11:03
YA KABUL AFGANISTAN	00:20	14:37	00:29	14:30	00:42	14:14	00:52	13:57	01:04	13:34	01:14	13:14
YB JAKARTA INDONESIA	23:05	10:52	23:05	10:54	23:01	10:55	22:56	10:54	22:47	10:51	22:40	10:48
ZD8 ASCENSION IS.	07:07	18:49	07:06	18:51	07:02	18:52	06:56	18:51	06:47	18:49	06:39	18:47
ZD9 TRISTAN DA CUNHA	07:56	17:53	07:48	18:02	07:32	18:15	07:15	18:26	06:52	18:38	06:31	18:49
ZF CAYMAN IS.	10:51	00:04	10:56	00:01	11:02	23:52	11:06	23:41	11:09	23:27	11:12	23:14
ZL1 AUCKLAND NZ	19:32	05:22	19:24	05:31	19:07	05:45	18:49	05:56	18:25	06:09	18:04	06:21
ZL4 DUNEDIN NZ	20:15	05:12	20:03	05:25	19:41	05:45	19:17	06:02	18:45	06:23	18:18	06:40
ZP ASUNCION PARAGUAY	10:35	21:16	10:30	21:22	10:19	21:30	10:07	21:36	09:49	21:43	09:34	21:48
ZS1 CAPETOWN S.AFRICA	05:49	15:55	05:42	16:03	05:27	16:15	05:11	16:25	04:48	16:36	04:29	16:46
ZS8 Pr.EDWARD & MARION	05:08	14:00	04:56	14:14	04:32	14:34	04:08	14:51	03:35	15:13	03:08	15:31
1S SPRATLY IS.	22:20	10:56	22:23	10:55	22:25	10:50	22:24	10:44	22:23	10:34	22:21	10:26
3V TUNIS TUNISIA	04:11	18:40	04:21	18:32	04:34	18:14	04:46	17:56	05:00	17:31	05:11	17:10
3Y BOUVET IS.	08:00	15:44	07:42	16:03	07:11	16:32	06:39	16:57	05:58	17:27	05:22	17:52
3Y PETER 1ST IS.	16:51	19:28	15:50	20:31	14:37	21:40	13:38	22:33	12:27	23:32	11:29	00:21
4J BAKU AZERBAIJAN	01:22	16:11	01:33	16:02	01:49	15:42	02:03	15:22	02:19	14:55	02:33	14:31
4U UNITED NATIONS HQ	09:37	00:27	09:48	00:17	10:04	23:57	10:18	23:37	10:34	23:09	10:48	22:45
4X ISRAEL	02:46	16:53	02:54	16:47	03:06	16:32	03:15	16:16	03:26	15:54	03:35	15:35
5R TANANARIVE MALAGASY	03:22	14:28	03:19	14:32	03:11	14:37	03:01	14:41	02:47	14:44	02:35	14:46
5W WESTERN SAMOA	17:54	05:14	17:51	05:18	17:44	05:21	17:36	05:23	17:24	05:24	17:13	05:24
5Z NAIROBI KENYA	03:37	15:40	03:37	15:41	03:35	15:40	03:32	15:37	03:26	15:32	03:20	15:27
6W DAKAR SENEGAL	06:48	19:43	06:52	19:41	06:55	19:34	06:57	19:25	06:58	19:13	06:59	19:03
7O ADEN YEMAN P.DEM.	02:41	15:30	02:45	15:28	02:48	15:22	02:49	15:14	02:49	15:03	02:49	14:53
7P LESOTHO	05:02	15:25	04:56	15:32	04:43	15:42	04:28	15:50	04:08	15:59	03:51	16:06
8Q MALDIVE IS	01:09	13:31	01:11	13:30	01:11	13:27	01:09	13:22	01:05	13:15	01:02	13:08
9G ACCRA GHANA	05:54	18:19	05:56	18:19	05:56	18:15	05:55	18:10	05:52	18:02	05:49	17:55
9J LUSAKA ZAMBIA	04:34	15:51	04:32	15:55	04:25	15:59	04:16	16:01	04:04	16:02	03:53	16:03
9K KUWAIT	01:55	15:46	02:03	15:40	02:12	15:27	02:20	15:13	02:29	14:53	02:36	14:36
9N KATMANDU NEPAL	23:31	13:17	23:38	13:12	23:48	12:59	23:55	12:46	00:03	12:27	00:09	12:10
9Q KINSHASA ZAIRE	05:08	17:01	05:08	17:03	05:05	17:03	05:00	17:01	04:52	16:57	04:46	16:54
9V SINGAPORE	23:05	11:17	23:06	11:17	23:05	11:15	23:02	11:11	22:57	11:05	22:53	10:59
9X RWANDA	04:07	16:07	04:07	16:09	04:05	16:08	04:01	16:05	03:55	16:00	03:49	15:56
9Y TRINIDAD IS.	09:50	22:30	09:53	22:28	09:55	22:23	09:55	22:16	09:54	22:06	09:53	21:57

Prefix City Country	Oct 14 Sunrise	Oct 14 Sunset	Oct 28 Sunrise	Oct 28 Sunset	Nov 14 Sunrise	Nov 14 Sunset	Nov 28 Sunrise	Nov 28 Sunset	Dec 14 Sunrise	Dec 14 Sunset	Dec 28 Sunrise	Dec 28 Sunset
W7 Carson City NV	14:07	01:24	14:21	01:04	14:40	00:47	14:55	00:38	15:10	00:37	15:17	00:43
W7 Cheyenne WY	13:09	00:22	13:25	00:02	13:45	23:42	14:01	23:33	14:16	23:31	14:24	23:37
W7 Olympia WA	14:27	01:28	14:47	01:04	15:12	00:39	15:32	00:27	15:49	00:22	15:57	00:29
W7 Phoenix AZ	13:31	00:57	13:43	00:41	13:58	00:27	14:11	00:21	14:24	00:21	14:31	00:28
W7 Salem OR	14:25	01:31	14:44	01:08	15:07	00:46	15:25	00:35	15:42	00:31	15:49	00:37
W7 Salt Lake City UT	13:37	00:51	13:52	00:30	14:12	00:11	14:28	00:02	14:43	00:00	14:51	00:07
W8 Charleston WV	11:33	22:52	11:47	22:33	12:06	22:16	12:20	22:08	12:35	22:06	12:42	22:13
W8 Columbus OH	11:40	22:56	11:56	22:36	12:15	22:18	12:31	22:09	12:45	22:07	12:53	22:14
W8 Lansing MI	11:49	23:00	12:06	22:38	12:28	22:18	12:45	22:07	13:00	22:05	13:08	22:11
W9 Indianapolis IN	11:53	23:09	12:08	22:49	12:27	22:31	12:42	22:22	12:57	22:21	13:05	22:27
W9 Madison WI	12:09	23:19	12:26	22:57	12:48	22:36	13:05	22:26	13:21	22:23	13:28	22:29
W9 Springfield IL	12:07	23:23	12:22	23:03	12:41	22:45	12:56	22:36	13:11	22:34	13:19	22:41
XE1 MEXICO CITY MEXICO	12:30	00:15	12:35	00:06	12:44	23:58	12:52	23:57	13:02	00:00	13:09	00:07
XW VIENTIANE LOAS	23:03	10:50	23:07	10:41	23:15	10:34	23:22	10:32	23:32	10:36	23:39	10:42
YA KABUL AFGANISTAN	01:27	12:52	01:39	12:35	01:55	12:20	02:08	12:13	02:21	12:13	02:29	12:19
YB JAKARTA INDONESIA	22:33	10:46	22:28	10:46	22:26	10:49	22:27	10:54	22:33	11:02	22:39	11:09
ZD8 ASCENSION IS.	06:30	18:46	06:25	18:47	06:22	18:50	06:23	18:56	06:29	19:04	06:35	19:11
ZD9 TRISTAN DA CUNHA	06:09	19:02	05:51	19:15	05:35	19:31	05:28	19:45	05:27	19:59	05:34	20:06
ZF CAYMAN IS.	11:16	23:01	11:21	22:51	11:29	22:44	11:37	22:42	11:47	22:45	11:54	22:52
ZL1 AUCKLAND NZ	17:40	06:35	17:22	06:48	17:05	07:06	16:57	07:20	16:55	07:35	17:01	07:42
ZL4 DUNEDIN NZ	17:48	07:01	17:23	07:20	16:59	07:45	16:46	08:04	16:42	08:22	16:47	08:30
ZP ASUNCION PARAGUAY	09:17	21:55	09:05	22:03	08:55	22:14	08:51	22:24	08:53	22:35	09:00	22:43
ZS1 CAPETOWN S.AFRICA	04:07	16:58	03:51	17:09	03:36	17:25	03:29	17:38	03:29	17:51	03:35	17:59
ZS8 Pr.EDWARD & MARION	02:36	15:53	02:11	16:13	01:47	16:38	01:34	16:58	01:29	17:16	01:35	17:24
1S SPRATLY IS.	22:20	10:18	22:21	10:12	22:24	10:09	22:30	10:10	22:38	10:15	22:45	10:22
3V TUNIS TUNISIA	05:25	16:47	05:38	16:29	05:56	16:12	06:10	16:04	06:24	16:04	06:32	16:10
3Y BOUVET IS.	04:43	18:23	04:10	18:51	03:36	19:26	03:16	19:52	03:06	20:15	03:11	20:24
3Y PETER 1ST IS.	10:20	01:20	09:18	02:17	07:56	03:40	NO:SR	NO:SS	NO:SR	NO:SS	NO:SR	NO:SS
4J BAKU AZERBAIJAN	02:49	14:05	03:05	13:45	03:25	13:26	03:40	13:17	03:55	13:15	04:03	13:21
4U UNITED NATIONS HQ	11:05	22:19	11:20	21:59	11:40	21:40	11:56	21:31	12:11	21:29	12:19	21:35
4X ISRAEL	03:46	15:14	03:57	14:59	04:12	14:45	04:24	14:39	04:37	14:39	04:45	14:45
5R TANANARIVE MALAGASY	02:21	14:50	02:12	14:55	02:04	15:03	02:03	15:11	02:06	15:21	02:13	15:28
5W WESTERN SAMOA	17:02	05:26	16:55	05:29	16:49	05:35	16:49	05:43	16:53	05:52	17:00	05:59
5Z NAIROBI KENYA	03:15	15:23	03:12	15:21	03:12	15:22	03:15	15:26	03:21	15:33	03:28	15:40
6W DAKAR SENEGAL	07:01	18:51	07:04	18:44	07:10	18:38	07:17	18:38	07:26	18:42	07:33	18:49
7O ADEN YEMAN P.DEM.	02:50	14:43	02:52	14:36	02:57	14:31	03:04	14:31	03:13	14:36	03:20	14:42
7P LESOTHO	03:32	16:16	03:18	16:25	03:05	16:39	03:00	16:50	03:01	17:03	03:07	17:10
8Q MALDIVE IS	00:59	13:01	00:59	12:57	01:01	12:56	01:05	12:58	01:12	13:04	01:19	13:11
9G ACCRA GHANA	05:47	17:48	05:46	17:43	05:49	17:42	05:53	17:44	06:01	17:49	06:08	17:56
9J LUSAKA ZAMBIA	03:41	16:05	03:33	16:09	03:27	16:15	03:26	16:23	03:30	16:32	03:37	16:39
9K KUWAIT	02:45	14:18	02:54	14:04	03:07	13:52	03:18	13:47	03:30	13:49	03:37	13:55
9N KATMANDU NEPAL	00:17	11:52	00:26	11:39	00:38	11:28	00:49	11:23	01:00	11:25	01:08	11:31
9Q KINSHASA ZAIRE	04:39	16:51	04:35	16:50	04:34	16:52	04:36	16:57	04:42	17:04	04:49	17:11
9V SINGAPORE	22:49	10:54	22:47	10:51	22:47	10:51	22:51	10:54	22:58	11:00	23:05	11:07
9X RWANDA	03:43	15:52	03:40	15:51	03:39	15:52	03:42	15:56	03:48	16:03	03:55	16:10
9Y TRINIDAD IS.	09:52	21:48	09:54	21:42	09:58	21:39	10:04	21:40	10:12	21:45	10:19	21:51

NCDXF/IARU INTERNATIONAL BEACON PROJECT. . . HOW IT ALL BEGAN AND WHERE IT IS TODAY"

Shortly after the formation of the Northern California DX Foundation (NCDXF) in October 1972, the newly formed Board of Directors decided that the organization should expand its horizons to include something of a scientific nature in which all amateurs—DXers and non-DXers alike—could participate. NCDXF support for DXpeditions and overseas operations would be a priority because DXers of the world would be the major contributors to the Foundation. But we hoped to do more than just send radios and DXpeditions and print QSL cards.

We consulted our Scientific Advisor, Dr. O.G. "Mike" Villard, Jr., W6QYT, Professor of Electrical Engineering at Stanford University and Senior Research Scientist at Stanford Research Institute (now SRI International). Mike had an idea. He was concerned about the disappearance of fishing and other boats in Alaskan waters every year. He believed that if the circulation of the Arctic currents were better understood, searching rescue ships would have a clearer idea where to look, thus increasing the chance for finding small lost boats. To help solve this problem, Mike suggested the possibility that a floating beacon be dropped into these Alaskan currents. This could be tracked by amateurs around the world to monitor the drifting course of the beacons.

Mike had a friend with just such a drifting beacon which would transmit on 20 meters with 1 watt or 25 watts, and it was being tested in Washington, D.C. Mike arranged for his friend to turn on the beacon one Saturday morning, and we organized a listening group on 20 meters via the Northern California DX Club 2-meter network. The beacon was easily readable with 25 watts, and pretty good with just one watt. So even the 1-watt QRP transmission could be monitored and useful to amateurs some distance away.

There was One Big Problem with the drifting beacons: their cost of $25,000 each was about 25 times greater than NCDXF had in the treasury. In addition, the beacons were non-recoverable. The idea of a beacon remained appealing, however, and with further thought, the development of a series of relatively low-cost stationary beacons world-wide appeared possible. It would be much cheaper and would still work!

We set up a series of brown bag lunch meetings at SRI to explore possibilities. To do the heavy thinking, we recruited some of the fellows who had worked on Oscar I–IV: Chuck Towns, K6LFH, who was president of Project Oscar at the time; Lance Ginner; and Board Member Jim Maxwell, W6CF, who was always full of imagination and creative thoughts.

After a month or so of meetings, we agreed it would be possible to develop a world-wide beacon network. It would feature all beacons transmitting the same message, each about one minute in length, all on the same frequency, one after the other, and going around the world. At Mike's suggestion, we also planned to step down the power output of the beacons in 10 dB steps, beginning at 100 watts. The beacon would come on the air with 100 watts, sign the beacon callsign, then step down to each of four power levels, 100, 10, 1, 0.1 watt, and finally back to 100 watts for the sign-off call. Each power level would last about 10 seconds before automatically switching to the next level.

All this planning and daydreaming was just fine, but who was going to design and build this thing? We turned to Chuck Towns and he looked deep into the engineering talent of Project Oscar and came up with an enthusiastic,

knowledgeable designer and builder, Jim Ouimet, K6OPO.

One seemingly small matter had to be addressed now, but it was potentially the biggest hurdle. We would need an FCC license! So we wrote a letter to Mr. A. Prose Walker, W4BW, then Chief of the Amateur Branch of the FCC. We received a prompt answer from him saying, essentially, he thought this was a good idea. The plan showed the creative ingenuity that amateurs had used in creating Oscar-1. It was a program that would be for the benefit and interest of all amateurs world-wide, and thus of interest to the WARC-79 planners (he expressed this sentiment personally later). He invited us to join as a member of the WARC-79 group which was beginning to develop the amateur agenda for that important international conference. The NCDXF/IARU World-Wide Beacon Network owes its existence to the early encouragement of Mr. A. Prose Walker.

Attending WARC-79 meetings in Washington offered the all-important opportunity to discuss with the FCC Amateur Branch engineers what requirements we would have to meet before submitting the proper application for an unmanned, automatic beacon on 14 MHz. One requirement was that we submit a contour map showing the beacon location, as well as the location of all amateur station operators in the San Francisco Bay Area who would be monitoring the beacon 24 hours a day. This was a precaution in case the beacon drifted off frequency, the keying mechanism failed, or anything else went wrong (the map requirement was the same as for early 2-meter repeaters). A lot of Northern California DX Club members did not realize they were now expected to have a receiver on 14.1 MHz day and night, and listen to it...continuously!

Meanwhile, back in Palo Alto, CA, Jim Ouimet was busy designing and building the beacon. He became so busy at work that he had to turn beacon construction over to a colleague. But the work proceeded and, on bench tests, did exactly what it was supposed to do.

The license arrived and we were assigned the call WB6ZNL, not exactly a nice, crisp, short beacon-type call, but it was a license and we were elated and grateful!

In 1979, the beacon was put in operation from a trailer on a low hill overlooking the Stanford University campus. It worked remarkably well, transmitting a one-minute message every 10 minutes for about two years. We received reports from all over the world telling of its reception. So the beacon was doing what it was supposed to do.

Now all we needed was to build eight or nine beacons and distribute them around the world. But there was another problem. Our beacon transmitter was very complicated to build and, we had to admit it, a real boat anchor. Also, Jim Ouimet was being sent world-wide by his company for extended periods. We definitely had a problem in manufacturing those other beacons.

About this time, Dave Lesson, W6QHS, came on the Board of Directors of NCDXF. We described the problem to Dave and he went to work in his lab. He came up with a solution: use a Kenwood TS-120 as the beacon transmitter and build a black box to control the entire system. Dave built the control unit and hooked it up to the TS-120 and, Voila!, we had a beacon transmitter that an amateur with an average build and strength could lift.

We now needed eight more beacons. Who was going to build them? Fortunately, the late Cam Pierce, K6RU, another NCDXF Board Member, took on the project with great enthusiasm. He had the control circuit boards designed and built, cabinets designed and made, cables fabricated, and the units tested. He turned on a real engineering production line.

The new beacons worked beautifully. We put up two

quad loops at right angles, complete with a phasing box, designed by Mike Stahl, K6MYC, then at KLM Electronics. At about the same time, we received the call W6WX/B for the beacon. NCDXF had acquired the call after the untimely death of a well-known local DXer, Dave Baker, W6WX. That beacon was on the air almost continuously until 1990—when it was stolen from the trailer!

As Cam Pierce was building beacons, we began to contact potential beacon station operators spaced around the world. At the United Nations, we talked to Dr. Max de Hensler, HB9RS, "Mr. U.N. Amateur Radio." Max immediately said yes, he would like to operate a beacon there. Martti Laine, OH2BH, arranged for a beacon at the University of Helsinki and also in Madeira. Local DX Club friend Bruno Bienenfeld, AA6AD, introduced us to an astronomy professor, Dr. Ahron Slonim, 4X4FQ, at his alma mater, Tel Aviv University. Kan Mizoguchi, JA1BK, introduced the beacon idea to the JARL. We also contacted old DX friend ZS6DN for a good location in the Southern Hemisphere. And at Honolulu City College, we spoke with Professor Bob Jones, KH6O. Later we received approval from Radio Club Argentina to put a beacon in Buenos Aires. Here were eight groups ready to operate a beacon and join W6WX/B at Stanford to complete the first World-Wide Beacon Network.

These new beacons were unique. Each would transmit the same one-minute message in sequence one after the other on 14.1 MHz. The message was the same as before: callsign at 100 watts, then four 9-second dashes at power levels descending from 100 watts to 10 watts, to 1 watt, to 0.1 watt—then back to 100 watts to sign-off. This same message has been transmitted on 14.1 MHz by beacons for almost 14 years.

Beacons were distributed to the operators as they were completed and tested. They have all been in almost continuous operation since being put on the air. We have had two thefts, one at W6WX/B, the other at JA2IGY. Lightning struck a tree which crashed into the antenna at ZS6DN/B. A hurricane flattened the vertical at KH6O/B. Once in a great while something did go wrong with the beacon or power supply, but was repaired locally. But the TS-120s, in general, have been remarkably free of problems. This is a very good record, considering that the beacons have been on the air for between 10 to 12 years each.

The International Amateur Radio Union (IARU) had been interested in beacons on a world-wide basis for many years. In 1984, at an IARU Advisory Council meeting, Alberto Shaio, HK3DEU, then Secretary of Region 2 of IARU, had an idea. He suggested that a frequency and time-sharing network, as used in the NCDXF system, would be the best way to present beacons on a world-wide basis. We talked it over and have been working together ever since.

About four years ago, it was decided that the network should be expanded and up-graded to a multiband network. Also, we had ideas about expanding the number of beacons and making the entire system multiband. But back to the old problem—who would do the work?

Quite fortuitously, at a meeting of the Northern California Contest Club, this writer met Bob Fabry, N6EK, retired Professor of Computer Science at the University of California, Berkeley. Somehow we drifted into a conversation about beacons and Bob said he would be interested in building rather than programming for a while. I suggested he talk to Dave Lesson, W6QHS, who designed the present generation of beacons, and who just happened to be sitting at the next table. They immediately started drawing pictures on the table and closely collabo-

rated over many months until Bob had a prototype. Thus was born the new generation of beacons.

We wanted to shorten the time of each beacon's transmission so we could increase the number of beacons without stretching listening time beyond listener attention span. So, Bob recorded beacon messages at various speeds, from about 10 to 20 seconds; that is, beacon call, then four short, power-stepping dashes only. He played the tape for the Directors of NCDXF, and we voted unanimously that 12 seconds would be about right. The same tape was played for a meeting of the Executive Committee of IARU Region 2. They agreed that 12-15 seconds was good. However, as a practical matter, Bob used a 10 second transmission for each beacon. This allows six beacons per minute, or 18 per three minutes, which is the number of beacons we wished to use.

Originally it was planned to use the Kenwood TS-140 transceiver for the beacon transmitter. However, various technical factors pointed to the Kenwood TS-50 as the better transmitter to use. It should be stated that Kenwood Corporation donated 16 of these transceivers to the International Beacon Project, for which we are sincerely grateful. Kenwood requested that a plaque be affixed to each TS-50 that states that the unit is dedicated to the memory of Jim Rafferty, N6RJ.

Bob constructed a control unit to control the functions of bandswitching and power stepping the beacon. He also used the Trimble Global Positioning System (GPS) receiver as a time control unit to assure accurate functions. Everything's state-of-the-art in this new system.

As this new beacon is being crafted, we went to work with the IARU, with their worldwide associations, to secure additional beacon locations, principally in the Southern Hemisphere. To date, Radio Club of Kenya, Radio Club Peruano, Radio ClubVenezuelano, Radio Amateurs of Canada, New Zealand Amateur Radio Transmitters, Wireless Institute of Australia, Chinese Radio Sports Association, and Radio Club of Sri Lanka have accepted our invitation to join the network. One location is being held for Central Russia. These additions will bring the total number of beacons in the network to 18.

Frequencies were chosen for the five new bands after a survey of several months by Bob Knowles, ZL1BAD, and his worldwide crew at the IARU Monitoring Service. Frequencies chosen were 14.100, 18.110, 21.150, 24930, and 28.200 MHz. We are quite aware that 14.100 MHz is in the middle of packet station QRM. W6WX is limited to that frequency because of our FCC license. Actually 14.100 has been designated on the IARU band plan as a "guarded" frequency for the beacon network for many years. We are hopeful that packeteers will give a little up and down to keep 14.100 clear for the network.

Distribution of the new five-band "Phase 3" beacons begins in the fall of 1995 and at least 16 of the beacons should be in place before summer of 1996. Then the listener can listen to 16 (later 18) beacons transmitting around the world every three minutes. OR he/she can monitor a single beacon, switching through the bands in 50 seconds to check the MUF.

We are grateful to the Universities, National Societies, and individuals who have volunteered to operate the beacons in this expanded network. It will be interesting to monitor the beacons during the increase in HF band activity as the sunspot cycle passes through its minimum and begins its climb back to DX glory!

(Reference: "The NCDXF/IARU International Beacon Project", by John Troster, W6ISQ, and Robert Fabry, N6EK, QST, October and November 1994)

Northern California DX Foundation/International Amateur Radio Union
14.100 MHz World-Wide Beacon Network

Nr.	Time (hh:mm:ss)	Callsign	Country	Operator	Bands	Status*
1	00:00:00	4U1UN/B	United Nations, NY	UN Radio Club	14	14.1
2	00:10	VE8?	Canada	Radio Amateurs of Canada		new
3	00:20	W6WX/B	USA	NCDXF	14/21/28	14.1
4	00:30	KH6?	Hawaii	U of Hawaii/Honolulu RC	14	14.1
5	00:40	ZL?	New Zealand	NZART		new
6	00:50	VK6?	Australia	Wireless Inst of Australia		new
7	01:00	JA2IGY	Japan	JARL	14	14.1
8	01:10	BY?	China	Chinese Radio Sports Assn.		new
9	01:20	UA	Russia	?		new
10	01:30	4S7?	Sri Lanka	Radio Society of Sri Lanka		new
11	01:40	ZS6DN/B	South Africa	ZS6DN	14	14.1
12	01:50	5Z4?	Kenya	Radio Society of Kenya		new
13	02:00	4X6TU	Israel	Univ. of Tel Aviv	14	14.1
14	02:10	OH2B	Finland	Univ. of Helsinki	14	14.1
15	02:20	CT3B	Madeira Isl.	ARRM	14	14.1
16	02:30	LU4AA/B	Argentina	Radio Club of Argentina	14	14.1
17	02:40	OA4?	Peru	Radio Club Peruano		new
18	02:50	YV5?	Venezuela	Radio Club Venezuelano		new

* "14.1" indicates beacon is part of original network, operating now.
  "new" indicates responsibility accepted for operating a beacon in the Phase 3 Network

The 10-Second "Phase 3" Message Format

"W6WX dah-dah-dah-dah"

-each "dah" lasts a little more than a second
-W6WX is transmitted at 100 watts, then each "dah" is attenuated in order, beginning at 100 watts, then 10 watts, then 1 watt, and finally 0.1 watt

10 METER BEACONS
September 15, 1995

FREQ.	CALL	OPERATION	LOCATION	NOTES
28.175	VE3TEN	C	OTTAWA,ONT.	10W,GP
28.1825	SV3AQR/B	C	AMALIAS, GREECE	4W,GP
28.191	VE6YF		EDMONTON,ALTA.	10W
28.195	IY4M	ROBOT	BOLOGNA,ITALY	20W,5/8 GP
28.2	GB3SX	C	CROWBOROUGH,ENGLAND	8W,DIPOLE
28.201	LU8ED		ARGENTINA	5W
28.202	KE5GY	I	ARLINGTON,TX	5W,VERTICAL
28.2025	ZS5VHF		NATAL,RSA	5W,GP
28.204	DL1GI	C	GERMANY	100W,VERT DIPOLE
28.205	KA3OEM		MEADVILLE,PA	27W,YAGI/WEST
28.206	KJ4X		PICKENS,SC	2W,VERTICAL
28.2075	N8ZKP	C	BROOKPARK OH	10W,VERT
28.208	WA1IOB	C	MARLBORO,MA	75W,VERT
28.209	NX2O	C	STATEN ISLAND,NY	10W,GP
28.21	3B8MS	C	MAURITIUS	GP
28.21	K4KMZ	I	ELIZABETHTOWN,KY	20W,VERT
28.21	KC4DPC	C	WILMINGTON,NC	4W,DIPOLE
28.212	EA6RCM		PALMA DE MALLORCA	4W,5 EL YAGI NNE
28.2125	ZD9GI	C	GOUGH ISLAND	GP
28.213	PT7BCN	C	CEARA, BRAZIL	
28.216	GB3RAL	C	SLOUGH,ENGLAND	20W,GP
28.2175	WB9VMY	C	CALUMET,OK	2W,DIPOLE
28.218	W8UR	C	MACKINAC ISLAND,MI	0.5W,VERT
28.2185	PT8AA		RIO BRANCO,BRAZIL	5W,GP
28.2195	LU4XS		CAPE HORN,ARGENTINA	
28.22	5B4CY	C	CYPRUS	26W,GP
28.221	PY2GOB		SAN PAULO,BRAZIL	15W,VERT
28.2215	K5PF	I	APEX NC	16W,VERT
28.222	W9UXO	C	NR CHICAGO,IL	10W,GP
28.2225	HG2BHA	C	TAPOLCA,HUNGARY	10W,GP
28.225	N6TWX	I	GRASSVALLEY,CA	30W,3 EL YAGI

28.225	KW7Y		EVERETT,WA	4W,OMNI
28.225	PY2AMI	C	SAO PAULO,BRAZIL	5W,DIOPOLE
28.2275	EA6AU	C	MALLORCA,BALEARIC IS	10W,5/8 GP
28.23	ZL2MHF	C	MT.CLMIE,NEW ZEALAND	50W,VERT DIPOLE
28.232	W7JPI	C	SONOITA,AZ	5W,3 EL YAGI NE
28.233	KD4EC	C	JUPITER,FL	7W,GP
28.235	VP9BA	C	HAMILTON,BERMUDA	10W,GP
28.237	NV6A	C	SAN DIEGO, CA	0.5W, VERT
28.2375	LA5TEN	C	OSLO,NORWAY	10W,5/8 GP
28.2405	5Z4ERR	C	KIAMBU,KENYA	
28.244	WA6APQ	C	LONG BEACH, CA	30W,Vert
28.245	A92C		BAHRAIN	DIPOLE NW/SE
28.2455	ZS1CTB	C	CAPETOWN,RSA	20W,1/4 VERT
28.246	N8KHE	C	MACKINAW,MI	0.050W,VERT
28.2475	EA2HB	I	SPAIN	6W,GP
28.248	K1BZ	C	BELFAST,ME	5W,VERT DIPOLE
28.2495	EA3JA		BARCELCONA,SPAIN	
28.25	W3SV	C	ELVERSON,PA	10W,VERT
28.25	KØHTF	C	DES MOINES,IA	2W,GP
28.25	Z21ANB	C	BULAWAYO,ZIMBABWE	8W,GP
28.250	WA4SLT	I	HASTINGS, FL	20W, Vert
28.2505	4N3ZHK	C	MT.KUM,YUGOSLAVIA	1W,VERT
28.252	WJ7X	C	SEATTLE,WA	5W,RINGO
28.2525	OH2TEN		FINLAND	
28.255	LU1UG		G RAL PICO,ARGENTINA	5W,GP
28.2575	DKØTEN	C	ARBEITSGEN,GERMANY	40W,GP
28.258	WB4JHS	I	KISSIMMEE, FL	5W,VERT
28.259	WB9FVR	C	PEMBROKE PINES,FL	1W,DIPOLE
28.26	VK5WI	C	ADELAIDE,AUSTRALIA	10W,GP
28.262	VK2RSY	C	SYDNEY,NSW,AUSTRALIA	25W,GP
28.263	N6PEQ	C ?	TUSTIN, CA	2W, HORZ DIPOLE
28.264	VK6RWA	C	PERTH,WA,AUSTRALIA	
28.265	VK4RIK		CAIRNS,AUSTRALIA	
28.266	VK6RTW	C	ALBANY,WA,AUSTRALIA	
28.266	KB4UPI	C	BIRMINGTON,AL	20W,1/4 VERT
28.268	VK8VF		DARWIN,AUSTRALIA	
28.2685	W9KFO	I	EATON,IL	0.750W,VERT
28.27	ZS6PW	C	PRETORIA,RSA	10W,3 EL YAGI
28.27	VK4RTL	C	TOWNSVILLE,QLD,AUSTRALIA	
28.2705	KF4MS	C	ST PETERSBURG,FL	5W
28.2725	9L1FTN	I	FREETOWN,SIERRA LEONE	10W,VERT DIPOLE
28.2745	ZS1LA		STILLBAY,RSA	20W,3 EL YAGI NW
28.275	AL7GQ	C	DENVER,CO	1W,LOOP
28.2755	N6RDX	I	STOCKTON,CA	20W,3 EL YAGI
28.277	DFØAAB	C	KIEL,GERMANY	10W,GP
28.28	LU8EB		ARGENTINA	5W
28.282	VE1MUF	C	FREDRICKTON,NB,CANADA	0.5W,DIPOLE
28.282	VE2HOT	C	BEACONSFIELD,QUE,CANADA	5W,VERT DIPOLE
28.2825	OKØEG	C	HRADEC KRALOVE,CZECH	10W,DIPOLE
28.284	VP8ADE	C	ADELAIDE IS,NR ANTARCTICA	8W,V BEAM TO ENGLAND
28.285	N2JNT	C	TROY, NY	1W, GP
28.286	KE2DI		NR ROCHESTER,NY	2W,VERT DIPOLE
28.286	KK4M	C	LAS VEGAS,NV	5W,VERT
28.2865	N5AQM	C	ARIZONA	2W, VERTICAL
28.287	W8OMV		NR ASHVILLE,NC	5W,GP
28.287	H44SI	C	SOLOMON IS	15W
28.288	W2NZH	I	MOORESTOWN,NJ	3W GP
28.29	SK5TEN		SWEDEN	5W
28.29	VS6TEN	C	HONG KONG	10W,VERT
28.292	ZD8HF		ASCENSION ISLAND	
28.2925	LU2FFV		SAN JORGE,ARGENTINA	5W,GP
28.284	KEØUL		GREELY, CO	5W Omni vert
28.295	WC8E	I	CINCINNATI,OHIO	10W,RINGO
28.295	W3VD	C	LAUREL,MD	1,5W,VERT DIPOLE
28.297	WA4DJS	I	FT LAUDERDALE,FL	30W,GP
28.3025	PT7AAC		FORTALEZA,BRAZIL	5W,GP
28.993	DFØANN	C	NUREMBERG, GERMANY	0.30W, 1-EL DELTA LOOP

C=Continuous; I=Intermittent
Copy supplied compliments of The DX Bulletin, Box 50 Fulton CA 95439 USA. Phone (707) 523-1001.

VHF/UHF BEACON STATIONS
Sources: DUBUS/K2LME

Freq	Callsign	Grid	Power	Antenna
50.000	GB3BUX	IO93	15	TURNSTILE
50.003	BV2FG	PL05	3	5/8 VERTICAL
50.003	7Q7SIX	KH74	5	
50.004	PJ2SIX	FK52	22	OMNI
50.004	VE8KM	DP79	50	5LBEAM
50.005	VK9RNI	RG30	25	
50.005	ZS2SIX	KF25	25	DIPOLE
50.007	SR5SIX	KO02	10	
50.008	DX1HB	PK04	20	J.POLE
50.008	KØGUV	EN26	8	HALO
50.008	VE8SIX	CP38?	80	COLINEAR
50.008	XE2HWB	DL44	01	6LBEAM
50.010	VE7SIX	DN09	130	QUAD
50.010	SV9SIX	KM25	30	VERTICAL DIPOLE
50.010	JA2IGY	PM84	10	5/8 G.P.
50.011	VP2EA	FK88	50	
50.013	CU3URA	HM68	05	5/8 VERTICAL
50.013	JD1ADP	QL17	01	DIPOLE
50.013	S55ZRS	JN76	8	G.P.
50.014	9M6SMC	OJ85	05	G.P.
50.015	PJ4B	FK52	15	5/8 G.P.
50.0155	LU9EHF	FF95	15	INV. VEE
50.017	JA6YBR	PM51	50	TURNSTILE
50.018	V51VHF	JG87	50	G.P.
50.019	P29BPL	QI30	30	G.P.
50.019	CX1CCC	GF15	05	G.P.
50.020	GB3SIX	IO73	25	5L BEAM@58m
50.021	OZ7IGY	JO55	20	TURNSTILE
50.0215	FR5SIX	LG78	02	HALO
50.023	4NØSIX	KN04	01	
50.023	LXØSIX	JN39	05	DIPOLE
50.023	SR5SIX	KO02	05	
50.0245	ZP5AA	GG14	05	G.P.
50.025	OH1SIX	KP11	40	OMNI
50.025	YV4AB	FK50	15	RINGO
50.025	9H1SIX	JM75	07	5/8 G.P.
50.027	ZS6PW	KG44	30	
50.027	JA7ZMA	QM07	50	2-TURNSTILE
50.028	SR6SIX	JO81	10	G.P.
50.028	XE2UZL	DM10	25	LOOPS
50.030	CTØWW	IN61	40	DIPOLE 700M
50.031	VE6XIS	DO21	25	4L BEAM
50.032	VE6MTR	DO33		
50.032	JRØYEE	PM97	02	LOOP
50.0325	ZD8VHF	II22	50	5/8
50.0335	LU8YYO	FF50	1.5	1/2 VERTICAL
50.035	V31SMC	EK57	10	VERTICAL
50.035	ZB2VHF	IM76	30	5L BEAM
50.037	ESØSIX	KO18	15	X/DIPOLES
50.037	JR6YAG	PL36	08	5/8 G.P.
50.038	FP5EK			
50.039	FY7THF	GJ35	100	G.P.
50.040	SV1SIX	KM17	25	VERTICAL
50.040	VO1ZA	GN37	10	G.P.
50.042	GB3MCB	IO70	40	DIPOLE
50.043	ZL3MHF	RE66	20	VERTICAL
50.045	OX3VHF	GP60	15	G.P.
50.045	YV5ZZ	FK60	10	VERTICAL
50.046	VK8RAS	PG66	15	X/DIPOLE
50.047	JW7SIX	JQ88	10	4L BEAM
50.047	CX8BE	GF15	08	BEAM
50.047	4N1SIX	KN04	10	VEE
50.050	FO5DR	BH52	50	

50.050	ZS6DN	KG44	100	5L BEAM
50.050	GB3NHQ	IO91	15	TURNSTILE
50.051	LA7SIX	JP99	30	4L BEAM 190 DEG.
50.052	Z21SIX	KH52	08	G.P.
50.053	VE1PZ		15	OMNI
50.054	VK3SIX	QF02	15	2 X 9L BEAM
50.054	OZ6VHF	JO57	50	TURNSTILE
50.055	VE2TWO	FN08	18	DIPOLE
50.056	V44K	FK87	03	5/8 VERTICAL
50.0567	VK7RNW	QE38	25	X/DIPOLES
50.057	TF3SIX	HP94	08	G.P.
50.057	VK8VF	PH57	20	VERTICAL
50.058	VK4RGG	QG62	06	
50.059	VE3UBL	FN03	10	TURNSTILE
50.059	PY2AA	GG66	5	G.P.
50.060	K4TQR	EM63	4	DIPOLE
50.060	WA8ONQ	EM79	2	X/DIPOLE
50.060	W5VAS	EM40	25	QUAD
50.060	GB3RMK	IO77	40	DIPOLE
50.061	KH6HME	BK29	20	DIPOLE
50.061	WBØRMO	EN10	25	SQUALO
50.062	GB3NGI	IO65	20	DIPOLE
50.062	WA8R	EM79	1	LOOP
50.062	WA8HTL	EN82	2	OMNI
50.063	W3VD	FM19	7	SQUALO
50.063	VE5US	DO61		
50.063	KB6BCN	CM88	3	3L BEAM
50.064	WD7Z	DM75	75	SQUALO
50.064	KH6HI	BL01	60	TURNSTILE
50.064	GB3LER	IP90	30	DIPOLE
50.065	WØIJR	DM79	20	HALO
50.065	WØMTK	DM59	2	4 Vee Dipoles
50.0655	GB3IOJ	IN89	10	VERTICAL
50.066	WA1OJB	FN54	10	J.POLE
50.067	N7DB	CN85	5	HALO
50.067	KD4LP	EM86	75	G.P.
50.067	W4RFR	EM66	2	HALO
50.067	OH9SIX	KP36	35	OMNI
50.068	W7US	DM42	2	G.P.
50.069	K6FV	CM87	100	
50.070	N4LTA	EM94	10	HALO
50.070	W2CAP/1	FN41	15	VERTICAL DIPOLE
50.070	KØHTF	EN31	2	Inverted Vee
50.070	EA3VHF	JN01	0.25	VERTICAL
50.070	SK3SIX	JP71	10	X/DIPOLES
50.070	ZS1SES			
50.071	WB9STR	EN61	5	
50.071	WØVD	EM27	10	G.P.
50.071	WB5LUA	EM12	1.5	HALO
50.072	KS2T	FM29	10	G.P.
50.072	KW2T	FN13	.5	SQUALO
50.072	WA4NTF	EN81		
50.073	KH6HI	BL01		OMNI
50.073	WB4WTC	FM06	10	LOOPS
50.073	ES6SIX	KO37	10	G.P.
50.073	VE9MS	FN65	2	DIPOLE
50.075	JY6ZZ	KM71	8	5/8 VERTICAL
50.075	NL7XM/2	FN20	1	SQUALO
50.075	EA8SIX	IL29	10	
50.075	K7IHZ	DM43	20	SQUALO
50.0755	PY2AMI	GG67	10	G.P.
50.076	KL7GLK/3	FM18	4	TURNSTILE
50.076	W6SKC/7	DM41	20	HALO
50.077	NØLL	EM09	20	HALO
50.077	W8UR	EN75	2	
50.077	WB2CUS	EL98	1	LOOP

50.0775	VK4BRG	QG48	3	TURNSTILE
50.078	PT7BCN	HI06	5	G.P.
50.078	OD5SIX	KM74	8	VERTICAL
50.079	TI2NA	EJ79	20	DIPOLE
50.080	WB4OOJ	EL87	10	VERTICAL
50.080	SK6SIX	JO57	10	G.P.
50.080	ZS1SIX	JF96JG	10	HALO
50.082	HC8SIX	EI59	4	VERTICAL
50.0825	LU8DCH	GF05		
50.084	3D2FJ	????	20	2L BEAM 20 DEG
50.086	VP2MO	FK86	10	6L BEAM
50.0865	LU1MA	FF57	8	G.P.
50.087	PBØALN	JO22		
50.087	VK4RTL	QH30	10	
50.0873	YU1SIX	KN03	15	DIPOLE
50.089	VE2TWO	FO13	18	DIPOLE
50.095	PT5XX	GG54	50	DIPOLE
50.098	LU2MFO	FF97	4	BEAM
50.200	VKØIX	OC53	50	3L BEAM
50.315	FX4SIX	JN06	25	X/DIPOLES
50.480	JH8ZND	QN02	10	DISCONE
50.490	JG1ZGW	PM95	1	DIPOLE
50.499	5B4CY	KM64	15	G.P.
50.521	SZ2DF	KM25	1000	4 X 16L
51.022	ZL1UHF	RF73	25	VERTICAL
51.029	ZL2MHB	RF80	1/10	VERTICAL
52.320	VK6RTT	OG89	10	
52.326	VK2RHV	QF57	10	VERTICAL DIPOLE
52.330	VK3RGL	QF22	20	X/DIPOLES
52.347	VK4ABP	QG26	10	VERTICAL
52.350	VK6RTU	PF09		
52.370	VK7RST	QE37		
52.410	VK1RCC	QF44		
52.418	VKØMA	QD95		
52.420	VK2RSY	QF56	25	TURNSTILE
52.425	VK2RGB	QF59	5	OMNI
52.435	VK3RMV	QF12		
52.445	VK4RIK	QH23	15	DIPOLE
52.445	VK4RBM	QG48		
52.450	VK5VF	PF95	10	TURNSTILE
52.460	VK6RPH	OF78		
52.465	VK6RTW	OF85		
52.470	VK7RNT	QE38		
52.510	ZL2MHF	RE78	4	DIPOLE
70.000	GB3BUX	IO93		
70.010	GB3REB	JO01		2L BEAM N.E.
70.020	GB3ANG	IO86		3L BEAM S.S.E.
70.025	GB3MCB	IO70		2L BEAM N.E.
70.120	ZB2VHF	IM76		4L QUAD N.
70.130	EI4RF	IO63		5L BEAM N.E./BEAM S.E.
144.275	KS2D	FN20		
144.275	W4RFR	EM66		
144.275	WB4OOJ	EL87		
144.276	W2RTB	FN12		
144.276	VE1SMU	GN03		
144.277	VE2FUT	FN25		
144.277	VE2TWO	FN08		
144.280	N4MW	FM17		
144.280	VE3TBX	EN58		
144.281	KW2T	FN13		
144.282	WD4LWG	EL87	75W	OMNI
144.287	KB8JI	EN64		
144.287	WØVD	EM27		
144.287	WØVB	EN34		
144.289	KL7GLK/3	FM18		
144.290	VE1SMU	FN84		10L BEAM W.N.W.

144.290	WA2UMX	FN23		
144.292	W9IP/2	FN24		
144.295	WA8R	EM79		
144.296	WD4KPD	FM15	20W	OMNI
144.296	W3VD	FM19		
144.297	VE9MS	FN65		10L BEAM N.W./S.W.
144.297	WB2CUS	EL98	.5W	OMNI
144.299	WZ8D	EM79		
144.867	EA1VHF	IN53		5L BEAM N.E.
144.905	FX3THF	IN88		9L BEAM E.
144.915	GB3MCB	IO70		3L BEAM N.E.
144.920	EI2WRB	IO62		5L BEAM E./S.
144.925	GB3VHF	JO01		6L BEAM S.W.
144.940	TF8THF	HP84		VERT
144.950	CTØSAT	IM59		
144.955	FX4VHF	JN05		OMNI
144.965	GB3LER	IO90		3L BEAM N.N.E.
144.975	GB3ABG	IO86		4L BEAM S.S.E.
222.051	WB2IEY	FN31		
222.055	N4MW	EM55		
222.055	VE1SMU	GN03		5L BEAM W.
222.060	VE1SMU	FN84		OMNI
432.070	N2CYM	FM29	10W	
432.074	WA4PGI	FM18		
432.289	W3VD	FM19		
432.300	VE1SMU	GN03		
432.303	KW2T	FN13		
432.308	WA4ZTK	EM75		
432.315	W1UHE	FN41		
432.340	KB3NQ	FN20		
432.365	K8AXU	EM99		
432.395	W4HHK	EM55		
903.073	KB3NQ	FN20		
903.075	K4EJQ	EM86		
903.080	N3CX	FN20		
1296.001	K4EJQ	EM86		
1296.046	W3VD	FM19		
1296.050	K3IVO	FM18		
1296.080	N3CX	FN20		
1296.180	WD5AGO	EM26		
1296.200	WA4PGI	FM07		
1296.260	KD5RO	FN12		
1296.350	VE1SMU	GN03		
2304.018	N3CX	FN20		
2304.050	K4EJQ	EM86		
2304.070	N4MW	EM55		
3456.040	N4MW	EM55		
3456.080	K4EJQ	EM86		
3456.140	N3CX	FN20		
10368.275	KS2D	FN20		

EUROPEAN 48/49MHz TV TRANSMITTERS

Six-meter band openings can often be predicted in the short-term by monitoring European television transmitter frequencies. This list of TV Transmitters in Europe may prove useful.

Freq (MHz)	Prefix	Loc.	ERP kW
48.239.6	DL	JN39	100
48.239.6	SM	JO79	60
48.241.4	EA	????	??
48.242.2	CT	IN51	40
48.246.1	LA	JP20	30
48.247.4	DL	JO40	100
48.249.7	LA	KP59	30
48.250.0	HB9	JN36	50
48.251.4	EA	IN80	250
48.250.0	EA	IN52	40
48.250.0	OE	JN78	60
48.250.0	SM	JP93	60

48.252.6	LA	JO38	60
48.256.1	LA	JP53	100
48.260.4	DL	JN57	100
48.260.4	LA	JP77	60
49.224.0	F	JN25	100
49.739.6	HG	JN97	150
49.739.6	UR	KN29	150
49.739.6	OK	JO70	150
49.739.6	UR	KN74	50
49.739.6	UA3	KO89	35
49.739.6	UR	KO62	35
49.740.9	UA1	KP78	10
49.744.8	HG	JN86	50
49.747.6	UA3	KO85	240
49.750.0	UA6	KN95	50
49.750.0	UR	KN67	35
49.750.0	UA1	KP75	10
49.750.0	UA3	KO59	240
49.750.0	EU	KO33	150
49.750.0	UA6	KN93	35
49.751.3	SP	JO93	100
49.757.8	UA3	KO56	90
49.757.8	UA1	KP63	10
49.760.4	YL	KO06	50
49.760.4	EU	KO41	50
49.760.4	UR	KN79	35
49.760.4	UA3	KO91	35
53.739.6	I5	JN53	34
53.760.4	IT9	JM67	35
53.760.4	I7	JN81	34
53.760.4	I3	JN65	1600
53.757.8	EI	IO52	100

(Provided by GJ4ICD, G3USF & UKSMG. Last update 5/7/95)

OTHER HF BEACONS

The International Telecommunication Union (ITU) has activated a new HF beacon transmitter in Sveio, Norway, as part of a worldwide HF field strength measurement campaign. The new beacon was provided by the Norwegian Telecommunications Regulatory Authority and Norwegian Telecom, and joins a similar beacon station on the air from Australia since 1990.

LN2A operates 24 hours a day, 1 kW on CW to a 5-band trap vertical antenna. VK4IPS in Brisbane, Australia also operates 24 hours a day, running 1 kW to an omni-directional spiral antenna.

The schedule of frequencies (in kHz) for both is as follows:

Time (min past hr)	VK4IPS	LN2A
00, 20, 40	5470	14405
04, 24, 44	7870	20945
08, 28, 48	10407	5470
12, 32, 52	14405	7870
16, 36, 56	20945	10407

A new beacon began operation May 13, 1993 from Cape Prince of Wales, Alaska (67N, 168W). Transmissions are narrowband CW and FSK. Its call sign is NAF and it runs 100 watts to a 3-band fan dipole on the following schedule (all frequencies in kHz):

Time (min past hr)	Frequency
00, 01	5604
20, 21	11004
40, 41	16804

The beacon will be monitored by government facili-ties in Fairbanks, Alaska; Seattle, Washington; State College, Pennsylvania; and San Diego, California. The project's sponsor is the Naval Security Group Command in Washington, D.C.

A government spokesman said the project is "purely scientific." The purpose of the project is described as to provide a "rigorous verification" of field strength models done by HF propagation prediction programs like ION-CAP which are used, for example, to generate the predictions published each month in *QST*.

Reception reports are encouraged and will be acknowledged with a colorful QSL card. Send them to:

Bob Rose, K6GKU, Code 54, NRaD Division, NCCOSC, 271 Catalina Blvd., San Diego CA 92152; or...

Dr. Gus Lott, KR4K, Code GX, COMNAVSEC-GRU, 3801 Nebraska Ave NW, Washington DC 20393

TIME BROADCAST STATIONS
AROUND THE WORLD

Station	Location	Transmit Frequencies (MHz.)
CHU	Ottawa, Canada	3.330, 7.335, 14.670
FFH	Chevannes, France	2.5
JJY	Tokyo, Japan	2.5, 5.0, 10.0, 15.0
LOL	Buenos Aires, Argentina	5.0, 10.0, 15.0
OMA	Liblice, Czech Republic	2.5
RCH	Tashkent, Uzbek	2.5, 5.0, 10.0 (Call sign should change to UK series.)
RID	Irkutsk, Russia	5.004, 10.004, 15.004
RTA	Novosibirsk, Russia	10.0, 15.0
RWM	Moscow, Russia	4.996, 9.996, 14.996
VNG	Llandilo, NSW, Australia	5.0, 8.638, 12.984, 16.0
ZUO	Johannesburg, South Africa	2.5, 5.0

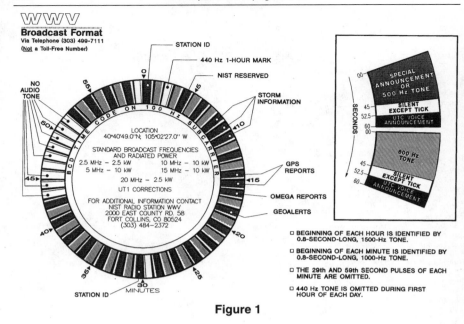

Figure 1

Figure 2

Summary of Radio Broadcast Services

Characteristics & Services:	**WWV**		**WWVH**		**WWVB**
Date Service Began	March 1923		November 1948		July 1956
Geographical Coordinates	40° 40' 49.0" N		21° 59' 26.0" N		40° 40' 28.3" N
	105° 02' 27.0" W		159° 46' 00.0" W		105° 02' 39.5" W
Standard Carrier Frequencies	2.5 & 20 MHz	5, 10, & 15 MHz	2.5 MHz	5, 10, & 15 MHz	60 kHz
Power	2500 W	10,000 W	5000 W	10,000 W	13,000 W
Standard Audio Frequencies	440 (A above middle C), 500, & 600 Hz				——
Time Intervals	1 pulse/s; minute mark; hour mark				s; min.
Time Signals: Voice	Once per minute				——
Time Signals: Code	BCD code on 100-Hz subcarrier, 1 puls/s				BCD code
UT1 Corrections	UT1 corrections are broadcast with an accuracy of ±0.1 s				
Special Announcements	Omega Reports, Geoalerts, Marine Storm Warnings, Global Positioning System Status Reports				

WWV/WWVH/WWVB INFORMATION

The following is excerpted from NIST Time and Frequency Services [Special Publication 432 (Revised 1990)]. A complete copy is available from NIST.

Abstract

NIST Time and Frequency Services [Special Publication 432 (Revised 1990)] is a revision of SP 432, last published in 1979. It describes services available, as of December 1990, from NIST radio stations WWV, WWVH, and WWVB; from GOES satellites; from Loran-C; by telephone (voice and modem); and from the NIST Frequency Measurement Service.

Key words: broadcast of standard frequencies; computer time setting; frequency calibrations; GOES satellite; high frequency; low frequency; satellite time code; shortwave; standard frequencies; time calibrations; time signals.

Introduction

Precise time and frequency information is needed by electric power companies, radio and television stations, telephone companies, air traffic control systems, participants in space exploration, computer networks, scientists monitoring data of all kinds, and navigators of ships and planes. These users need to compare their own timing equipment to a reliable, internationally recognized standard. The National Institute of Standards and Technology (NIST), formerly the National Bureau of Standards, provides this standard for most users in the United States.

NIST began broadcasting time and frequency information from radio station WWV in 1923. Since then, NIST has expanded its time and frequency services to meet the needs of a growing number of users. NIST time and frequency services are convenient, accurate, and easy to use. They contribute greatly to the nation's space and defense programs, to manufacturers, and to transportation and communications. In addition, NIST services are widely used by the general public.

Broadcast services include radio signals from NIST radio stations WWV, WWVH, and WWBV; the GOES satellites, and Loran-C. Services are also available using telephone voice and data lines. This is a guide to these services.

Shortwave Services—WWV and WWVH

NIST operates two high-frequency (shortwave) radio stations, WWV and WWVH. WWV is in Ft. Collins, CO, and WWVH is in Kauai, HI. Both stations broadcast continuous time and frequency signals on 2.5, 5, 10, and 15 MHz. WWV also broadcasts on 20 MHz. All frequencies provide the same information. Although radio reception conditions in the high-frequency band vary greatly with factors such as location, time of year, time of day, the particular frequency being used, atmospheric and ionospheric propagation conditions, and the type of receiving equipment used, at least one frequency should be usable at all times. As a general rule, frequencies above 10 MHz work best in the daytime, and the lower frequencies work best at night.

Services provided by WWV and WWVH include:

Time announcements
Standard time intervals
Standard frequencies
UT1 time corrections
BCD time code
Geophysical alerts
Marine storm warnings
OMEGA Navigation System status reports
Global Positioning System (GPS) status reports

Figures 1 and 2 show the hourly broadcast schedules of these services along with station location, radiated power, and details of the modulation.

Accuracy and Stability

WWV and WWVH are referred to the primary NIST Frequency Standard and related NIST atomic time scales in Boulder, CO. The frequencies as transmitted are accurate to about 1 part in 100 billion (1×10^{-11}) for frequency and about 0.01 ms for timing. The day-to-day deviations are normally less than 1 part in 1,000 billion (1×10^{-12}). However, the received accuracy is far less due to various propagation effects. The usable received accuracy is about 1 part in 10 million for frequency (1×10^{-7}) and about 1 ms for timing.

Radiated Power, Antennas, and Modulation

WWV and WWVH radiate 10, 000 W on 5, 10, and 15 MHz. The radiated power is lower on the other frequencies: WWV radiates 2500 W on 2.5 and 20 MHz while WWVH radiates 5000 W on 2.5 MHz and does not broadcast on 20 MHz.

The WWV antennas are half-wave dipoles that radiate omnidirectional patterns. The 2.5-MHz antenna at WWVH is also of this type. The other antennas at WWVH are phased vertical half-wave dipole arrays. They radiate a cardioid pattern with the maximum gain pointed toward the west.

Both stations use double sideband amplitude modula-

tion. The modulation level is 50 percent for the steady tones, 25 percent for the BCD time code, 100 percent for the seconds pulses and the minute and hour markers, and 75 percent for the voice announcements.

Time Announcements

Voice announcements are made from WWV and WWVH once every minute. Since both stations can be heard in some locations, a man's voice is used on WWV, and a woman's voice is used on WWVH to reduce confusion. The WWVH announcement occurs first, at about 15 s before the minute. The WWV announcement follows at about 7.5 s before the minute. Though the announcements occur at different times, the tone mark-

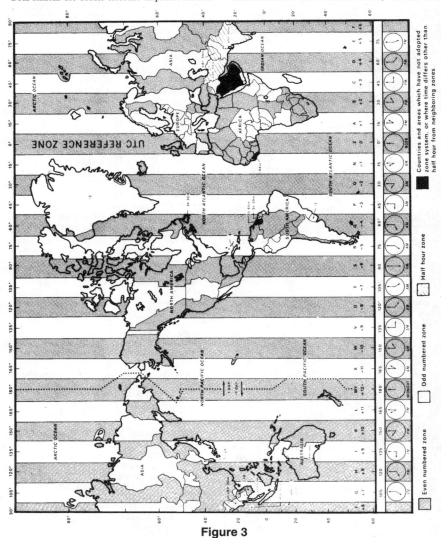

Figure 3

ers are transmitted at the exact same time from both sta-tions. However, they may not be received at exactly the same instant due to differences in the propagation delays from the two station sites.

The announced time is "Coordinated Universal Time" (UTC). UTC was established by international agreement in 1972, and is governed by the International Bureau of Weights and Measures (BIPM) in Paris, France. Coordination with the international UTC time scale keeps NIST time signals in close agreement with signals from other time and frequency stations throughout the world.

UTC differs from your local time by a specific num-ber of hours. The number of hours depends on the num-ber of time zones between your location and the location of the zero meridian (which passes through Greenwich, England). When local time changes from Daylight Saving to Standard Time, or vice versa, UTC does not change. However, the difference between UTC and local time does change—by 1 hour. Use the chart of world time zones (figure 3) to find out how many hours to add to or subtract from UTC to obtain your local standard time. If DST is in effect at your location, subtract 1 hour less in the U.S. than shown on the chart. Thus, Eastern Daylight Time (EDT) is only 4 hours behind UTC, not 5 as shown on the chart for EST.

UTC is a 24-hour clock system. The hours are num-bered beginning with 00 hours at midnight through 12 hours at noon to 23 hours and 59 minutes just before the next midnight.

The international agreement that established UTC in 1972 also specified that occasional adjustments of exact-ly 1 s will be made to UTC so that UTC should never dif-fer from a particular astronomical time scale, UT1, by more than 0.9 s. This was done as a convenience for some time-broadcast users, such as boaters using celes-tial navigation, who need to know time that is based on the rotation of the Earth. These occasional 1-s adjust-ments are known as "leap seconds." When deemed nec-essary by the International Earth Rotation Service in Paris, France, the leap seconds are inserted into UTC, usually at the end of June or at the end of December, making that month 1 s longer than usual. Typically, a leap second has been inserted at intervals of 1 to 2 years.

Standard Time Intervals

The most frequent sounds heard on WWV and WWVH are the seconds pulses. These pulses are heard every second except on the 29th and 59th seconds of each minute. The first pulse of each hour is an 800-ms pulse of 1500 Hz. The first pulse of each minute is an 800-ms pulse of 1000 Hz at WWV and 1200 Hz at WWVH. The remaining seconds pulses are short audio bursts (5-ms pulses of 1000 Hz at WWV and 1200 Hz at WWVH) that sound like the ticking of a clock.

Each seconds pulse is preceded by 10 ms of silence and followed by 25 ms of silence. The silence makes it easier to pick out the pulse. The total 40-ms protected zone around each seconds pulse is shown in figure 4.

Standard Audio Frequencies

In alternate minutes during most of each hour, 500-Hz or 600-Hz audio tones are broadcast. A 440-Hz tone (the musical note A above middle C) is broadcast once each hour. In addition to being a musical standard, the 440-Hz tone provides an hourly marker for chart

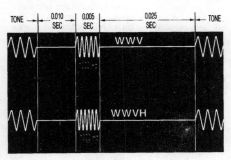

Figure 4

recorders and other automated devices. The 440-Hz tone is omitted, however, during the first hour of each UTC day. See figures 1 and 2 for further details.

Silent Periods

The silent periods are without tone modulation. However, the carrier frequency, seconds pulses, time announcements, and the 100-Hz BCD time code contin-ue during the silent periods. In general, one station will not broadcast an audio tone while the other station is broadcasting a voice message.

On WWV, the silent period extends from 43 to 46 and from 47 to 52 minutes after the hour. WWVH has two silent periods; from 8 to 11 minutes after the hour, and from 14 to 20 minutes after the hour. Minutes 29 and 59 on WWV and minutes 00 and 30 on WWVH are also silent.

BCD Time Code

WWV and WWVH continuously broadcast a binary coded decimal (BCD) time code on a 100-Hz subcarrier. The time code presents UTC information in serial fashion at a rate of 1 pulse per second. The information carried by the time code includes the current minute, hour, and day of the year. The time code also contains the 100-Hz frequency from the subcarrier. The 100-Hz frequency may be used as a standard with the same accuracy as the audio frequencies.

At the time of publication of this revision of Special Publication 432 (late 1990), further changes to the content of the WWV and WWVH time codes are in the planning stage. The proposed changes will not affect currently encoded information, but will add information in the form of the last two digits of the current year, improved indica-tors for when Daylight Saving Time is in effect, and a warning for the insertion of a leap second at the end of the current month.

UT1 Time Corrections

The UTC time scale broadcast by WWV and WWVH meets the needs of most users. UTC runs at an almost per-fectly constant rate, since its rate is based on cesium atomic frequency standards. Somewhat surprisingly, some users need time less stable than UTC but related to the rotation of the Earth. Applications such as celestial navigation, satellite observations of the Earth, and some types of surveying require time referenced to the rotation-al position of the Earth. These users rely on the UT1 time scale. UT1 is derived by astronomers who monitor the speed of the Earth's rotation.

You can obtain UT1 time by applying a correction to the UTC time signals broadcast from WWV and WWVH. UT1 time corrections are included in the WWV and WWVH broadcasts at two levels of accuracy. First, for those users only needing UT1 to within 1 s, occasional corrections of exactly 1 s are inserted into the UTC time scale. These corrections, called leap seconds, keep UTC within ±0.9 s of UT1. Leap seconds are coordinated under international agreement by the International Earth Rotation Service in Paris, France. Leap seconds can be either positive or negative, but so far, only positive leap seconds have been needed. A positive leap second is normally added every 1 or 2 years, usually on June 30 or December 31.

The second level of correction is for the small number of users needing UT1 accurate to within 0.1 s. These corrections are encoded on the broadcasts by using doubled ticks during the first 16 s of each minute. The amount of correction (in tenths of 1 s) is determined by counting the number of successive doubled ticks heard each minute. The sign of the correction depends on whether the doubled ticks are in the first 8 s of the minute or in the second 8 s. If the doubled ticks are in the first 8 s (1–8) the sign is positive, and if they are in the second 8 s (9–16) the sign is negative. For example, if ticks 1, 2, and 3 are doubled, the correction is "plus" 0.3 s. This means that UT1 equals UTC + 0.3 s. If UTC is 8:45:17, then UT1 is 8:45:17.3. If ticks 9, 10, 11, and 12 are doubled, the correction is "minus" 0.4 s. If UTC is 8:45:17, then UT1 is 8:45:16.6. An absence of doubled ticks indicates that the current correction is 0.

Official Announcements

Announcement segments 45 s long are available by subscription to other Federal agencies (see figures 1 & 2). These segments are used for public service messages. The accuracy and content of these messages is the responsibility of the originating agency.

For information about the availability of these segments, contact the NIST Time and Frequency Division.

The segments currently in use (late 1990) are described below. Since these are subject to change from time to time, contact NIST for more current status information.

OMEGA Navigation System Status Reports—The OMEGA Navigation System status reports are voice announcements broadcast on WWV at 16 minutes after the hour, and on WWVH at 47 minutes after the hour. The OMEGA Navigation System consists of eight radio stations transmitting in the 10- to 14-kHz frequency band. These stations serve as international aids to navigation. The status reports are updated as necessary by the U.S. Coast Guard.

For more information about the OMEGA Navigation System or these announcements, contact: Commanding Officer, U.S. Coast Guard OMEGA Navigation System Center, 7323 Telegraph Road, Alexandria, VA 22310-3998; telephone (703) 866-3800.

Geophysical Alerts—Current geophysical alerts (Geo-alerts) are broadcast in voice from WWV at 18 minutes after the hour and from WWVH at 45 minutes after the hour. The messages are less than 45 s in length and are updated every 3 hours (typically at 0000, 0300, 0600, 0900, 1200, 1500, 1800, and 2100 UTC). Hourly updates are made when necessary.

Part A of the message gives the solar-terrestrial indices for the day: specifically the 1700 UTC solar flux from Ottowa, Canada, at 2800 MHz, the estimated A-index for Boulder, CO, and the current Boulder K-index.

Part B gives the solar-terrestrial conditions for the previous 24 hours.

Part C gives optional information on current conditions that may exist (that is, major flares, proton or polar cap absorption [PCA] events, or stratwarm conditions).

Part D gives the expected conditions for the next 24 hours. For example:

A) "Solar-terrestrial indices for 26 October follow: Solar flux 173 and estimated Boulder A-index 20; repeat: Solar flux one-seven-three and estimated Boulder A-index two-zero.The Boulder K-index at 1800 UTC on 26 October was four; Repeat: four."

B) "Solar-terrestrial conditions for the last 24 hours follow:
Solar activity was high.
Geomagnetic field was unsettled to active."

C) "A major flare occurred at 1648 UTC on 26 October. A satellite proton event and PCA are in progress."

D) "The forecast for the next 24 hours follows:
Solar activity will be moderate to high. The geomagnetic field will be active."

Definitions

1. Solar Activity is defined as transient perturbations of the solar atmosphere as measured by enhanced x-ray emission, typically associated with flares. Five standard terms are used to describe solar activity:

Very Low:	X-ray events less than C-class.
Low:	C-class x-ray events.
Moderate:	isolated (1 to 4) M-class x-ray events.
High	several (5 or more) M-class x-ray events, or isolated (1 to 4) M5 or greater x-ray events.
Very High	several M5 or greater x-ray events.

2. The geomagnetic field experiences natural variations classified quantitatively into six standard categories depending upon the amplitude of the disturbance. The Boulder K- and estimated A-indices determine the category according to the following table:

Condition	Range of A-index	Typical K-indices
Quiet	$0 \leq A < 8$	usually no K-indices > 2
Unsettled	$8 \leq A < 16$	usually no K-indices > 3
Active	$16 \leq A < 30$	a few K-indices of 4
Minor storm	$30 \leq A \leq 50$	K-indices mostly 4 and 5
Major storm	$50 \leq A < 100$	some K-indices 6 or greater
Severe storm	$100 \leq A$	some K-indices 7 or greater.

3. Solar Flares are classified by their x-ray emission as:

	Peak Flux Range (1–8 Angstroms)	
Class	SI system (W m^{-2})	cgs system (erg cm^{-2} s^{-1})
A	$\emptyset < 10^{-7}$	$\emptyset < 10^{-4}$
B	$10^{-7} \leq \emptyset < 10^{-6}$	$10^{-4} \leq \emptyset < 10^{-3}$
C	$10^{-6} \leq \emptyset < 10^{-5}$	$10^{-3} \leq \emptyset < 10^{-1}$
M	$10^{-5} \leq \emptyset < 10^{-4}$	$10^{-2} \leq \emptyset < 10^{-1}$
X	$10^{-4} \leq \emptyset$	$10^{-1} \leq \emptyset$

The letter designates the order of magnitude of the peak value. Following the letter the measured peak value is given. For descriptive purposes, a number from 1.0 to 9.9 is appended to the letter designation. The number acts as a multiplier. For example, a C3.2 event indicates an x-ray burst with peak flux of 3.2 x 10^{-6} Wm^{-2}.

Forecasts are usually issued only in terms of the broad C, M, and X categories. Since x-ray bursts are observed as a full-Sun value, bursts below the x-ray background level are not discernible. The background drops to class A level during solar minimum; only bursts that exceed B1.0 are classified as x-ray events. During solar maximum the background is often at the class M level, and therefore class A, B, and C x-ray bursts cannot be seen. Data are from the NOAA GOES satellites, monitored in real time at the NOAA Space Environment Services Center. Bursts greater than 1.2 x 10^{-3} Wm^{-2} may saturate the GOES detectors. If saturation occurs estimated peak flux values are reported.

4. The remainder of the report is as follows:

Major Solar Flare = flare which produces some geophysical effect, usually flares that have x-rays ≥ M5 class.
Proton Flare = protons by satellite detectors (or polar cap observed in time association with H-alpha flare.
Satellite Level Proton Event = proton enhancement detected by Earth-orbiting satellites with measured particle flux of at least 10 protons cm^{-2}s^{-1}sr^{-1} at ≥ 10 MeV.
Polar Cap Absorption = proton-induced absorption ≥ 2 dB daytime, 0.5 dB night, as measured by a 20-MHz riometer located within the polar cap.
Stratwarm = reports of stratospheric warmings in the high latitude regions of the winter hemisphere of the Earth associated with gross distortions of the normal circulation associated with the winter season.

To hear these Geophysical Alert messages by telephone (at any minute of the hour, but without time information), dial (303) 497-3235.

Inquiries regarding these messages should be addressed to:
Space Environment Services Center, NOAA R/E/SE2, 325 Broadway, Boulder, CO 80303-3328. Or call (303) 497-5127.

Marine Storm Warnings—Marine storm warnings are broadcast for the marine areas that the United States has warning responsibility for under international agreement. The storm warning information is provided by the National Weather Service. Storm warnings for the Atlantic and eastern North Pacific are broadcast by voice on WWV at 8, 9, and 10 minutes after the hour. Storm warnings for the western, eastern, southern, and north Pacific are broadcast by WWVH at 48, 49, 50, and 51 minutes after the hour. An additional segment (at 11 minutes after the hour on WWV and at 52 minutes after the hour on WWVH) is used occasionally if there are unusually widespread storm conditions. The brief voice messages warn mariners of storm threats present in their areas.

The storm warnings are based on the most recent forecasts. Updated forecasts are issued by the National Weather Service at 0500, 1100, 1700, and 2300 UTC for WWV; and at 0000, 0600, 1200, and 1800 UTC for WWVH.

A typical storm warning announcement text is as follows:
> North Atlantic weather West of 35 West at 1700 UTC; Hurricane Donna, intensifying, 24 North, 60 West, moving northwest, 20 knots, winds 75 knots; storm, 65 North, 35 West, moving east, 10 knots; winds 50 knots, seas 15 feet.

For more information about marine storm warnings, write to: The Director, National Weather Service, Silver Spring, MD 20910.

Global Positioning System (GPS) Status Announcements—Since March 1990 the U.S. Coast guard has sponsored two voice announcements per hour on WWV and WWVH, giving current status information about the GPS satellites and related operations. The 45-s announcements begin at 14 and 15 minutes after each hour on WWV and at 43 and 44 minutes after each hour on WWVH. For further information, contact the Commanding Officer, U.S. Coast Guard Center, 7323 Telegraph Road, Alexandria, VA 22310-3998.

WWV and WWVH Audio Signals by Telephone
The audio portions of the WWV and WWVH broadcasts can be heard by telephone. The accuracy of the telephone time signals is normally 20 ms or better in the continental United States. In rare instances when the telephone connection is made by satellite, there is an additional delay of 0.25 to 0.5 s.

To hear these broadcasts, dial (303) 499-7111 for WWV, and (808) 335-4363 for WWVH. Callers are disconnected after three minutes. These are not toll-free numbers; callers outside the local calling area are charged for the call at regular long-distance rates.

Low-Frequency Service—WWVB
Radio station WWVB is located on the WWV site near Ft. Collins, CO. WWVB continuously broadcasts time and frequency signals at 60 kHz, primarily for the continental United States. WWVB does not broadcast voice announcements, but provides standard time information, including the year; time intervals; Daylight Saving Time, leap second, and leap-year indicators; and UT1 corrections by means of a BCD time code. In addition, the 60-kHz carrier frequency provides an accurate frequency standard which is referenced to the NIST Frequency Standard.

Accuracy and Stability
The transmitted accuracy of WWVB is normally better than 1 part in 100 billion (1 x 10^{-11}). Day-to-day deviations are less than 5 parts in 1000 billion (5 x 10^{-12}). The BCD time code can be received and used with an accuracy of approximately 0.1 ms. Propagation effects are minor compared to those of WWV and WWVH. When proper receiving and averaging techniques are used, the received accuracy of WWVB should be nearly as good as the transmitted accuracy.

Station Identification
WWVB identifies itself by advancing its carrier phase 45° at 10 minutes after the hour and returning to normal phase at 15 minutes after the hour. WWVB is also identified by its unique time code.

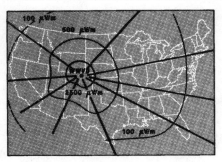

Figure 5

Radiated Power, Antenna, and Coverage

The effective radiated power from WWVB is 13,000 watts. The antenna is a 122-m, top-loaded vertical, installed over a radial ground screen. Some measured field intensity contours are shown in the coverage map in figure 5.

WWVB Time Code

The WWVB time code is synchronized with the 60-kHz carrier and is broadcast continuously at a rate of 1 pulse per second using pulse-width modulation. Each pulse is generated by reducing the carrier power 10 dB at the start of the second, so that the leading edge of every negative-going pulse is on time. Full power is restored either 0.2, 0.5, or 0.8 s later to convey either a binary "0," "1," or a position marker, respectively. Details of the time code are in appendix C.

The WWVB code contains information on the current year (since early 1990), day of year, hour, minute, second, status of Daylight Saving Time, leap year, and a leap-second warning (planned for mid-1991 implementation). Since the WWVB code is undergoing some revision, users are encouraged to contact the NIST Time and Frequency Division for the most current information.

How NIST Controls the Transmitted Frequencies

Figure 6 shows the relationship between the NIST broadcasts, the primary NIST Frequency Standard, and the NIST atomic time scale. The NIST Frequency Standard and atomic time scale systems are located in Boulder, CO. They include:

• The primary NIST Frequency Standard. This standard is a laboratory cesium beam device, built and maintained by NIST. It provides a frequency and time interval reference, based on the international definition of the second.

• A group of commercial cesium standards, hydrogen

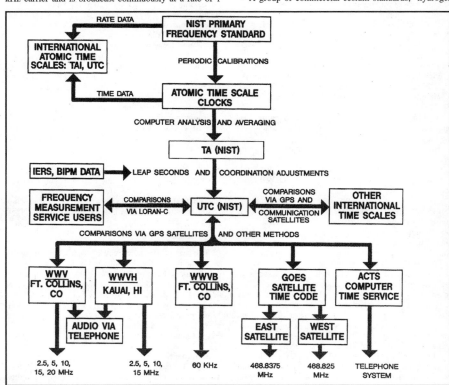

Figure 6

maser frequency standards, and possibly other devices. These are kept in controlled environments, and serve as the continuously operating "working" standards.

- Sophisticated time comparison equipment, computer systems, and computer software. This equipment generates a composite time scale that is better than any of the individual standards. The composite time scale, TA(NIST), is based on (approximately) annual calibrations of the working standards using the primary NIST Frequency Standard.

- Provisions for inserting small adjustments and leap seconds into the time scale. These adjustments let NIST generate UTC(NIST), an internationally coordinated UTC time scale. UTC(NIST) is distributed to users of the services described in this booklet, including WWH, WWVH, WWVB, the GOES satellites, the Frequency Measurement Service, and the telephone services.

The NIST radio station sites (Fort Collins, CO, and Kauai, HI) each have three commercial atomic standards that emulate UTC(NIST). Each standard can provide the frequency input to the station's time-code generators and transmitters.

The local standards at each station are compared to the UTC(NIST) time scale in Boulder, CO. These comparisons are made using GPS (Global Positioning System) satellites. GPS consists of orbiting navigation satellites that broadcast precise timing signals. These signals are received simultaneously at Boulder and at both radio station sites. The results obtained at the radio stations are then compared to the results obtained in Boulder. The timing differences (accurate to less than 100 ns), are then used to calibrate the local standards.

Other time and frequency resources are also used to check the local standards. These include portable clocks (physically carried to the radio station sites), Loran-C, and GOES satellite broadcasts.

GOES Satellite Time Services

In 1974, NIST began broadcasting a time code from the GOES (Geostationary Operational Environmental Satellite) satellites of the National Oceanic and Atmospheric Administration (NOAA). This cooperative arrangement between NIST and NOAA was formalized by a renewable agreement, the latest version of which extends until 1997. The primary purpose of the GOES satellites is to collect environmental data from thousands of sensing platforms located throughout the Western Hemisphere and to relay this information to a central processing facility from where it is made available to interested organizations, such as the World Meteorological Organization, radio and TV stations, and various government agencies. The time code is referenced to UTC(NIST) and is broadcast continuously to the entire Western Hemisphere from two satellites (GOES/East and GOES/West).

GOES time-code receivers are commercially available. Some of them provide timing signals accurate to within 100 μs over periods of hours, months, or years. Other versions are accurate to within 1–2 ms over the same time periods.

Inquiries about NIST Time and Frequency Services

If you have specific questions about the operations of NIST radio stations, contact:

Engineer-in-Charge, NIST Radio Stations WWV and WWVB, 2000 East County Road 58, Fort Collins, CO 80524; (303) 484-2372. Or...

Engineer-in-Charge, NIST Radio Station WWVH, P.O. Box 417, Kekaha, Kauai, HI 96752; (808) 335-4361.

If you have specific questions about the other time and frequency services, contact:

NIST Time and Frequency Services, 847.40, National Institute of Standards and Technology, 325 Broadway, Boulder, CO 80303-3328; (303) 497-3294.

Tours of NIST Facilities

Public guided tours of the NIST Laboratories in Boulder are held twice a week from Memorial Day to Labor Day, and once a week the rest of the year. They offer a chance to see the NIST Atomic Clock that provides the basis for the time and frequency services, as well as visiting other laboratories of NIST and the National Oceanic and Atmospheric Administration.

Contact the Tour Program Office, Division 360.06, NIST, 325 Broadway, Boulder, CO 80303-3328; telephone (303) 497-5507; for information about when tours are scheduled or to arrange special tours for groups of 15 to 30 people.

There are no public tours available to the radio stations.

A Brief History of NIST Time and Frequency Services

March	1923	First scheduled broadcasts of WWV, Washington, D.C.
April	1933	WWV gets first 20-kW transmitter, Beltsville, MD.
January	1943	WWV relocated to Greenbelt, MD.
November	1948	WWVH commenced broadcasts, Maui, HI.
January	1950	WWV added voice announcements.
July	1956	WWVB began 60-kHz broadcasts (as KK2XEI), Sunset, CO.
April	1960	WWVL began 20-kHz experimental broadcasts, Sunset, CO.
July	1963	WWVB began high power broadcasts, Ft. Collins, CO.
August	1963	WWVL began high power broadcasts, Ft. Collins, CO.
July	1964	WWVH added voice announcements.
December	1966	WWV relocated to Ft. Collins, CO.
July	1971	WWVH relocated to Kauai, HI.
June	1972	First leap second in history was added to UTC time scale.
July	1972	WWVL went off the air.
January	1974	Voice announcements changed from GMT to UTC (WWV/WWVH).
July	1974	GOES satellite time code initiated.
March	1975	Frequency calibration using network color TV became a nationwide service.
August	1975	Line-10 time comparisons using TV synchronization pulses became a nationwide service.
February	1977	20- and 25-MHz broadcasts from WWV and 20-MHz broadcasts from WWVH were discontinued.
December	1978	20-MHz broadcasts from WWV were reinstated.
February	1984	Frequency Measurement Service began.
March	1988	Automated Computer Time Service (ACTS) began on experimental basis.

THE SUN, THE IONOSPHERE, AND THE RADIO AMATEUR

Most amateurs know that the sun plays an important role in radio propagation. Numerous books and papers have been written on the subject. The most popular is probably *The New Shortwave Propagation Handbook,* by George Jacobs, W3ASK, Robert B. Rose, K6GKU and Theodore Cohen, N4XX. It provides detailed explanations of the science of propagation forecasting.

Sunspot Numbers

In 1748, the Swiss astronomer Johann Rudolph Wolf introduced a daily measurement of sunspot number. His method, which is still used today, counts the total number of spots visible on the face of the sun and the number of groups into which they cluster, because neither quantity alone satisfactorily measures sunspot activity.

An observer computes a daily sunspot number by multiplying the number of groups he sees by 10 and then adding this product to the total count of individual spots. Results, however, vary greatly since the measurement strongly depends on observer interpretation and experience and on the stability of the Earth's atmosphere above the observation site. Moreover, the use of Earth as a platform from which to record these numbers contributes to their variability, too, because the sun rotates and the evolving spot groups are distributed unevenly across solar longitudes. To compensate for these limitations, each daily international number is computed as a weighted average of measurements made from a network of cooperating observatories.

Two sets of monthly sunspot numbers are recorded: a "mean" number and a "smoothed" number (both are included here). The smoothed count best approximates the sunspot cycle, but due to the data required for the smoothing, a given month's smoothed number is not available until six months later. The monthly mean number is available immediately.

How do sunspot numbers in these tables compare with the largest values ever recorded? The highest daily count on record occurred December 24 and 25, 1957. On both of those days the sunspot number totaled 355. In contrast, during years near the minimum of the spot cycle, the count can fall to zero. Today, much more sophisticated measurements of solar activity are made routinely, but none has the link with the past that sunspot numbers have.

Monthly Mean Sunspot Numbers

Year	Jan	Feb	Mar	Apr	May	Jun	Jul	Aug	Sep	Oct	Nov	Dec
1749	58.0	62.6	70.0	55.7	85.0	83.5	94.8	66.3	75.9	75.5	158.6	85.2
1750	73.3	75.9	89.2	88.3	90.0	100.0	85.4	103.0	91.2	65.7	63.3	75.4
1751	70.0	43.5	45.3	56.4	60.7	50.7	66.3	59.8	23.5	23.2	28.5	44.0
1752	35.0	50.0	71.0	59.3	59.7	39.6	78.4	29.3	27.1	46.6	37.6	40.0
1753	44.0	32.0	45.7	38.0	36.0	31.7	22.0	39.0	28.0	25.0	20.0	6.7
1754	0.0	3.0	1.7	13.7	20.7	26.7	18.8	12.3	8.2	24.1	13.2	4.2
1755	10.2	11.2	6.8	6.5	0.0	0.0	8.6	3.2	17.8	23.7	6.8	20.0
1756	12.5	7.1	5.4	9.4	12.5	12.9	3.6	6.4	11.8	14.3	17.0	9.4
1757	14.1	21.2	26.2	30.0	38.1	12.8	25.0	51.3	39.7	32.5	64.7	33.5
1758	37.6	52.0	49.0	72.3	46.4	45.0	44.0	38.7	62.5	37.7	43.0	43.0
1759	48.3	44.0	46.8	47.0	49.0	50.0	51.0	71.3	77.2	59.7	46.3	57.0
1760	67.3	59.5	74.7	58.3	72.0	48.3	66.0	75.6	61.3	50.6	59.7	61.0
1761	70.0	91.0	80.7	71.7	107.2	99.3	94.1	91.1	100.7	88.7	89.7	46.0
1762	43.8	72.8	45.7	60.2	39.9	77.1	33.8	67.7	68.5	69.3	77.8	77.2
1763	56.5	31.9	34.2	32.9	32.7	35.8	54.2	26.5	68.1	46.3	60.9	61.4
1764	59.7	59.7	40.2	34.4	44.3	30.0	30.0	30.0	28.2	28.0	26.0	25.7
1765	24.0	26.0	25.0	22.0	20.2	20.0	27.0	29.7	16.0	14.0	14.0	13.0
1766	12.0	11.0	36.6	6.0	26.8	3.0	3.3	4.0	4.3	5.0	5.7	19.2
1767	27.4	30.0	43.0	32.9	29.8	33.3	21.9	40.8	42.7	44.1	54.7	53.3
1768	53.5	66.1	46.3	42.7	77.7	77.4	52.6	66.8	74.8	77.8	90.6	111.8
1769	73.9	64.2	64.3	96.7	73.6	94.4	118.6	120.3	148.8	158.2	148.1	112.0
1770	104.0	142.5	80.1	51.0	70.1	83.3	109.8	126.3	104.4	103.6	132.2	102.3
1771	36.0	46.2	46.7	64.9	152.7	119.5	67.7	58.5	101.4	90.0	99.7	95.7
1772	100.9	90.8	31.1	92.2	38.0	57.0	77.3	56.2	50.5	78.6	61.3	64.0
1773	54.6	29.0	51.2	32.9	41.1	28.4	27.7	12.7	29.3	26.3	40.9	43.2
1774	46.8	65.4	55.7	43.8	51.3	28.5	17.5	6.6	7.9	14.0	17.7	12.2
1775	4.4	0.0	11.6	11.2	3.9	12.3	1.0	7.9	3.2	5.6	15.1	7.9
1776	21.7	11.6	6.3	21.8	11.2	19.0	1.0	24.2	16.0	30.0	35.0	40.0
1777	45.0	36.5	39.0	95.5	80.3	80.7	95.0	112.0	116.2	106.5	146.0	157.3
1778	177.3	109.3	134.0	145.0	238.9	171.6	153.0	140.0	171.7	156.3	150.3	105.0
1779	114.7	165.7	118.0	145.0	140.0	113.7	143.0	112.0	111.0	124.0	114.0	110.0
1780	70.0	98.0	98.0	95.0	107.2	88.0	86.0	86.0	93.7	77.0	60.0	58.7

Year	Jan	Feb	Mar	Apr	May	Jun	Jul	Aug	Sep	Oct	Nov	Dec
1781	98.7	74.7	53.0	68.3	104.7	97.7	73.5	66.0	51.0	27.3	67.0	35.2
1782	54.0	37.5	37.0	41.0	54.3	38.0	37.0	44.0	34.0	23.2	31.5	30.0
1783	28.0	38.7	26.7	28.3	23.0	25.2	32.2	20.0	18.0	8.0	15.0	10.5
1784	13.0	8.0	11.0	10.0	6.0	9.0	6.0	10.0	10.0	8.0	17.0	14.0
1785	6.5	8.0	9.0	15.7	20.7	26.3	36.3	20.0	32.0	47.2	40.2	27.3
1786	37.2	47.6	47.7	85.4	92.3	59.0	83.0	89.7	111.5	112.3	116.0	112.7
1787	134.7	106.0	87.4	127.2	134.8	99.2	128.0	137.2	157.3	157.0	141.5	174.0
1788	138.0	129.2	143.3	108.5	113.0	154.2	141.5	136.0	141.0	142.0	94.7	129.5
1789	114.0	125.3	120.0	123.3	123.5	120.0	117.0	103.0	112.0	89.7	134.0	135.5
1790	103.0	127.5	96.3	94.0	93.0	91.0	69.3	87.0	77.3	84.3	82.0	74.0
1791	72.7	62.0	74.0	77.2	73.7	64.2	71.0	43.0	66.5	61.7	67.0	66.0
1792	58.0	64.0	63.0	75.7	62.0	61.0	45.8	60.0	59.0	59.0	57.0	56.0
1793	56.0	55.0	55.5	53.0	52.3	51.0	50.0	29.3	24.0	47.0	44.0	45.7
1794	45.0	44.0	38.0	28.4	55.7	41.5	41.0	40.0	11.1	28.5	67.4	51.4
1795	21.4	39.9	12.6	18.6	31.0	17.1	12.9	25.7	13.5	19.5	25.0	18.0
1796	22.0	23.8	15.7	31.7	21.0	6.7	26.9	1.5	18.4	11.0	8.4	5.1
1797	14.4	4.2	4.0	4.0	7.3	11.1	4.3	6.0	5.7	6.9	5.8	3.0
1798	2.0	4.0	12.4	1.1	0.0	0.0	0.0	3.0	2.4	1.5	12.5	9.9
1799	1.6	12.6	21.7	8.4	8.2	10.6	2.1	0.0	0.0	4.6	2.7	8.6
1800	6.9	9.3	13.9	0.0	5.0	23.7	21.0	19.5	11.5	12.3	10.5	40.1
1801	27.0	29.0	30.0	31.0	32.0	31.2	35.0	38.7	33.5	32.6	39.8	48.2
1802	47.8	47.0	40.8	42.0	44.0	46.0	48.0	50.0	51.8	38.5	34.5	50.0
1803	50.0	50.8	29.5	25.0	44.3	36.0	48.3	34.1	45.3	54.3	51.0	48.0
1804	45.3	48.3	48.0	50.6	33.4	34.8	29.8	43.1	53.0	62.3	61.0	60.0
1805	61.0	44.1	51.4	37.5	39.0	40.5	37.6	42.7	44.4	29.4	41.0	38.3
1806	39.0	29.6	32.7	27.7	26.4	25.6	30.0	26.3	24.0	27.0	25.0	24.0
1807	12.0	12.2	9.6	23.8	10.0	12.0	12.7	12.0	5.7	8.0	2.6	0.0
1808	0.0	4.5	0.0	12.3	13.5	13.5	6.7	8.0	11.7	4.7	10.5	12.3
1809	7.2	9.2	0.9	2.5	2.0	7.7	0.3	0.2	0.4	0.0	0.0	0.0
1810	0.0	0.0	0.0	0.0	0.0	0.0	0.0	0.0	0.0	0.0	0.0	0.0
1811	0.0	0.0	0.0	0.0	0.0	0.0	6.6	0.0	2.4	6.1	0.8	1.1
1812	11.3	1.9	0.7	0.0	1.0	1.3	0.5	15.6	5.2	3.9	7.9	10.1
1813	0.0	10.3	1.9	16.6	5.5	11.2	18.3	8.4	15.3	27.8	16.7	14.3
1814	22.2	12.0	5.7	23.8	5.8	14.9	18.5	2.3	8.1	19.3	14.5	20.1
1815	19.2	32.2	26.2	31.6	9.8	55.9	35.5	47.2	31.5	33.5	37.2	65.0
1816	26.3	68.8	73.7	58.8	44.3	43.6	38.8	23.2	47.8	56.4	38.1	29.9
1817	36.4	57.9	96.2	26.4	21.2	40.0	50.0	45.0	36.7	25.6	28.9	28.4
1818	34.9	22.4	25.4	34.5	53.1	36.4	28.0	31.5	26.1	31.6	10.9	25.8
1819	32.8	20.7	3.7	20.2	19.6	35.0	31.4	26.1	14.9	27.5	25.1	30.6
1820	19.2	26.6	4.5	19.4	29.3	10.8	20.6	25.9	5.2	8.9	7.9	9.1
1821	21.5	4.2	5.7	9.2	1.7	1.8	2.5	4.8	4.4	18.8	4.4	0.2
1822	0.0	0.9	16.1	13.5	1.5	5.6	7.9	2.1	0.0	0.4	0.0	0.0
1823	0.0	0.0	0.6	0.0	0.0	0.0	0.5	0.0	0.0	0.0	0.0	20.4
1824	21.7	10.8	0.0	19.4	2.8	0.0	0.0	1.4	20.5	25.2	0.0	0.8
1825	5.0	15.5	22.4	3.8	15.5	15.4	30.9	25.7	15.7	15.6	11.7	22.0
1826	17.7	18.2	36.7	24.0	32.4	37.1	52.5	39.6	18.9	50.6	39.5	68.1
1827	34.6	47.4	57.8	46.0	56.3	56.7	42.3	53.7	49.6	56.1	48.2	46.1
1828	52.8	64.4	65.0	61.1	89.1	98.0	54.2	76.4	50.4	54.7	57.0	46.9
1829	43.0	49.4	72.3	95.0	67.4	73.9	90.8	77.6	52.8	57.2	67.6	56.5
1830	52.2	72.1	84.6	106.3	66.3	65.1	43.9	50.7	62.1	84.4	81.2	82.1
1831	47.5	50.1	93.4	54.5	38.1	33.4	45.2	55.0	37.9	46.3	43.5	28.9
1832	30.9	55.6	55.1	26.9	41.3	26.7	14.0	8.9	8.2	21.1	14.3	27.5
1833	11.3	14.9	11.8	2.8	12.9	1.0	7.0	5.7	11.6	7.5	5.9	9.9
1834	4.9	18.1	3.9	1.4	8.8	7.8	8.7	4.0	11.5	24.8	30.5	34.5
1835	7.5	24.5	19.7	61.5	43.6	33.2	59.8	59.0	100.8	95.2	100.0	77.5

Year	Jan	Feb	Mar	Apr	May	Jun	Jul	Aug	Sep	Oct	Nov	Dec
1836	88.6	107.6	98.2	142.9	111.4	124.7	116.7	107.8	95.1	137.4	120.9	206.2
1837	188.0	175.6	134.6	138.2	111.7	158.0	162.8	134.0	96.3	123.7	107.0	129.8
1838	144.9	84.8	140.8	126.6	137.6	94.5	108.2	78.8	73.6	90.8	77.4	79.8
1839	105.6	102.5	77.7	61.8	53.8	54.6	84.8	131.2	132.7	90.9	68.8	63.7
1840	81.2	87.7	67.8	65.9	69.2	48.5	60.7	57.8	74.0	55.0	54.3	53.7
1841	24.1	29.9	29.7	40.2	67.5	55.7	30.8	39.3	36.5	28.5	19.8	38.8
1842	20.4	22.1	21.7	26.9	24.9	20.5	12.6	26.6	18.4	38.1	40.5	17.6
1843	13.3	3.5	8.3	9.5	21.1	10.5	9.5	11.8	4.2	5.3	19.1	12.7
1844	9.4	14.7	13.6	20.8	11.6	3.7	21.2	23.9	7.0	21.5	10.7	21.6
1845	25.7	43.6	43.3	57.0	47.8	31.1	30.6	32.3	29.6	40.7	39.4	59.7
1846	38.7	51.0	63.9	69.3	59.9	65.1	46.5	54.8	107.1	55.9	60.4	65.5
1847	62.6	44.9	85.7	44.7	75.4	85.3	52.2	140.6	160.9	180.4	138.9	109.6
1848	159.1	111.8	108.6	107.1	102.2	129.0	139.2	132.6	100.3	132.4	114.6	159.5
1849	157.0	131.7	96.2	102.5	80.6	81.1	78.0	67.7	93.7	71.5	99.0	97.0
1850	78.0	89.4	82.6	44.1	61.6	70.0	39.1	61.6	86.2	71.0	54.8	61.0
1851	75.5	105.4	64.6	56.5	62.6	63.2	36.1	57.4	67.9	62.5	51.0	71.4
1852	68.4	66.4	61.2	65.4	54.9	46.9	42.1	39.7	37.5	67.3	54.3	45.4
1853	41.1	42.9	37.7	47.6	34.7	40.0	45.9	50.4	33.5	42.3	28.8	23.4
1854	15.4	20.0	20.7	26.5	24.0	21.1	18.7	15.8	22.4	12.6	28.2	21.6
1855	12.3	11.4	17.4	4.4	9.1	5.3	0.4	3.1	0.0	9.6	4.2	3.1
1856	0.5	4.9	0.4	6.5	0.0	5.2	4.6	5.9	4.4	4.5	7.7	7.2
1857	13.7	7.4	5.2	11.1	28.6	16.0	22.2	16.9	42.4	40.6	31.4	37.2
1858	39.0	34.9	57.5	38.3	41.4	44.5	56.7	55.3	80.1	91.2	51.9	66.9
1859	83.7	87.6	90.3	85.7	91.0	87.1	95.2	106.8	105.8	114.6	97.2	81.0
1860	82.4	88.3	98.9	71.4	107.1	108.6	116.7	100.3	92.2	90.1	97.9	95.6
1861	62.3	77.7	101.0	98.5	56.8	88.1	78.0	82.5	79.9	67.2	53.7	80.5
1862	63.1	64.5	43.6	53.7	64.4	84.0	73.4	62.5	66.6	41.9	50.6	40.9
1863	48.3	56.7	66.4	40.6	53.8	40.8	32.7	48.1	22.0	39.9	37.7	41.2
1864	57.7	47.1	66.3	35.8	40.6	57.8	54.7	54.8	28.5	33.9	57.6	28.6
1865	48.7	39.3	39.5	29.4	34.5	33.6	26.8	37.8	21.6	17.1	24.6	12.8
1866	31.6	38.4	24.6	17.6	12.9	16.5	9.3	12.7	7.3	14.1	9.0	1.5
1867	0.0	0.7	9.2	5.1	2.9	1.5	5.0	4.8	9.8	13.5	9.6	25.2
1868	15.6	15.7	26.5	36.6	26.7	31.1	29.0	34.4	47.2	61.6	59.1	67.6
1869	60.9	59.9	52.7	41.0	103.9	108.4	59.2	79.6	80.6	59.3	78.1	104.3
1870	77.3	114.9	157.6	160.0	176.0	135.6	132.4	153.8	136.0	146.4	147.5	130.0
1871	88.3	125.3	143.2	162.4	145.5	91.7	103.0	110.1	80.3	89.0	105.4	90.4
1872	79.5	120.1	88.4	102.1	107.6	109.9	105.5	92.9	114.6	102.6	112.0	83.9
1873	86.7	107.0	98.3	76.2	47.9	44.8	66.9	68.2	47.1	47.1	55.4	49.2
1874	60.8	64.2	46.4	32.0	44.6	38.2	67.8	61.3	28.0	34.3	28.9	29.3
1875	14.6	21.5	33.8	29.1	11.5	23.9	12.5	14.6	2.4	12.7	17.7	9.9
1876	14.3	15.0	30.6	2.3	5.1	1.6	15.2	8.8	9.9	14.3	9.9	8.2
1877	24.4	8.7	11.9	15.8	21.6	14.2	6.0	6.3	16.9	6.7	14.2	2.2
1878	3.3	6.6	7.8	0.1	5.9	6.4	0.1	0.0	5.3	1.1	4.1	0.5
1879	1.0	0.6	0.0	6.2	2.4	4.8	7.5	10.7	6.1	12.3	13.1	7.3
1880	24.0	27.2	19.3	19.5	23.5	34.1	21.9	48.1	66.0	43.0	30.7	29.6
1881	36.4	53.2	51.5	51.6	43.5	60.5	76.9	58.4	53.2	64.4	54.8	47.3
1882	45.0	69.5	66.8	95.8	64.1	45.2	45.4	40.4	57.7	59.2	84.4	41.8
1883	60.6	46.9	42.8	82.1	31.5	76.3	80.6	46.0	52.6	83.8	84.5	75.9
1884	91.5	86.9	87.5	76.1	66.5	51.2	53.1	55.8	61.9	47.8	36.6	47.2
1885	42.8	71.8	49.8	55.0	73.0	83.7	66.5	50.0	39.6	38.7	30.9	21.7
1886	29.9	25.9	57.3	43.7	30.7	27.1	30.3	16.9	21.4	8.6	0.3	13.0
1887	10.3	13.2	4.2	6.9	20.0	15.7	23.3	21.4	7.4	6.6	6.9	20.7
1888	12.7	7.1	7.8	5.1	7.0	7.1	3.1	2.8	8.8	2.1	10.7	6.7
1889	0.8	8.5	6.7	4.3	2.4	6.4	9.4	20.6	6.5	2.1	0.2	6.7
1890	5.3	0.6	5.1	1.6	4.8	1.3	11.6	8.5	17.2	11.2	9.6	7.8

Year	Jan	Feb	Mar	Apr	May	Jun	Jul	Aug	Sep	Oct	Nov	Dec
1891	13.5	22.2	10.4	20.5	41.1	48.3	58.8	33.0	53.8	51.5	41.9	32.5
1892	69.1	75.6	49.9	69.6	79.6	76.3	76.5	101.4	62.8	70.5	65.4	78.6
1893	75.0	73.0	65.7	88.1	84.7	89.9	88.6	129.2	77.9	80.0	75.1	93.8
1894	83.2	84.6	52.3	81.6	101.2	98.9	106.0	70.3	65.9	75.5	56.6	60.0
1895	63.3	67.2	61.0	76.9	67.5	71.5	47.8	68.9	57.7	67.9	47.2	70.7
1896	29.0	57.4	52.0	43.8	27.7	49.0	45.0	27.2	61.3	28.7	38.0	42.6
1897	40.6	29.4	29.1	31.0	20.0	11.3	27.6	21.8	48.1	14.3	8.4	33.3
1898	30.2	36.4	38.3	14.5	25.8	22.3	9.0	31.4	34.8	34.4	30.9	12.6
1899	19.5	9.2	18.1	14.2	7.7	20.5	13.5	2.9	8.4	13.0	7.8	10.5
1900	9.4	13.6	8.6	16.0	15.2	12.1	8.3	4.3	8.3	12.9	4.5	0.3
1901	0.2	2.4	4.5	0.0	10.2	5.8	0.7	1.0	0.6	3.7	3.8	0.0
1902	5.5	0.0	12.4	0.0	2.8	1.4	0.9	2.3	7.6	16.3	10.3	1.1
1903	8.3	17.0	13.5	26.1	14.6	16.3	27.9	28.8	11.1	38.9	44.5	45.6
1904	31.6	24.5	37.2	43.0	39.5	41.9	50.6	58.2	30.1	54.2	38.0	54.6
1905	54.8	85.8	56.5	39.3	48.0	49.0	73.0	58.8	55.0	78.7	107.2	55.5
1906	45.5	31.3	64.5	55.3	57.7	63.2	103.6	47.7	56.1	17.8	38.9	64.7
1907	76.4	108.2	60.7	52.6	42.9	40.4	49.7	54.3	85.0	65.4	61.5	47.3
1908	39.2	33.9	28.7	57.6	40.8	48.1	39.5	90.5	86.9	32.3	45.5	39.5
1909	56.7	46.6	66.3	32.3	36.0	22.6	35.8	23.1	38.8	58.4	55.8	54.2
1910	26.4	31.5	21.4	8.4	22.2	12.3	14.1	11.5	26.2	38.3	4.9	5.8
1911	3.4	9.0	7.8	16.5	9.0	2.2	3.5	4.0	4.0	2.6	4.2	2.2
1912	0.3	0.0	4.9	4.5	4.4	4.1	3.0	0.3	9.5	4.6	1.1	6.4
1913	2.3	2.9	0.5	0.9	0.0	0.0	1.7	0.2	1.2	3.1	0.7	3.8
1914	2.8	2.6	3.1	17.3	5.2	11.4	5.4	7.7	12.7	8.2	16.4	22.3
1915	23.0	42.3	38.8	41.3	33.0	68.8	71.6	69.6	49.5	53.5	42.5	34.5
1916	45.3	55.4	67.0	71.8	74.5	67.7	53.5	35.2	45.1	50.7	65.6	53.0
1917	74.7	71.9	94.8	74.7	114.1	114.9	119.8	154.5	129.4	72.2	96.4	129.3
1918	96.0	65.3	72.2	80.5	76.7	59.4	107.6	101.7	79.9	85.0	83.4	59.2
1919	48.1	79.5	66.5	51.8	88.1	111.2	64.7	69.0	54.7	52.8	42.0	34.9
1920	51.1	53.9	70.2	14.8	33.3	38.7	27.5	19.2	36.3	49.6	27.2	29.9
1921	31.5	28.3	26.7	32.4	22.2	33.7	41.9	22.8	17.8	18.2	17.8	20.3
1922	11.8	26.4	54.7	11.0	8.0	5.8	10.9	6.5	4.7	6.2	7.4	17.5
1923	4.5	1.5	3.3	6.1	3.2	9.1	3.5	0.5	13.2	11.6	10.0	2.8
1924	0.5	5.1	1.8	11.3	20.8	24.0	28.1	19.3	25.1	25.6	22.5	16.5
1925	5.5	23.2	18.0	31.7	42.8	47.5	38.5	37.9	60.2	69.2	58.6	98.6
1926	71.8	69.9	62.5	38.5	64.3	73.5	52.3	61.6	60.8	71.5	60.5	79.4
1927	81.6	93.0	69.6	93.5	79.1	59.1	54.9	53.8	68.4	63.1	67.2	45.2
1928	83.5	73.5	85.4	80.6	77.0	91.4	98.0	83.8	89.7	61.4	50.3	54.9
1929	68.9	62.8	50.2	52.8	58.2	71.9	70.2	65.8	34.4	54.0	81.1	108.0
1930	65.3	49.9	35.0	38.2	36.8	28.8	21.9	24.9	32.1	34.4	35.6	25.8
1931	14.6	43.1	30.0	31.2	24.6	15.3	17.4	13.0	19.0	10.0	18.7	17.8
1932	12.1	10.6	11.2	11.2	17.9	22.2	9.6	6.8	4.0	8.9	8.2	11.0
1933	12.3	22.2	10.1	2.9	3.2	5.2	2.8	0.2	5.1	3.0	0.6	0.3
1934	3.4	7.8	4.3	11.3	19.7	6.7	9.3	8.3	4.0	5.7	8.7	15.4
1935	18.6	20.5	23.1	12.2	27.3	45.7	33.9	30.1	42.1	53.2	64.2	61.5
1936	62.8	74.3	77.1	74.9	54.6	70.0	52.3	87.0	76.0	89.0	115.4	123.4
1937	132.5	128.5	83.9	109.3	116.7	130.3	145.1	137.7	100.7	124.9	74.4	88.8
1938	98.4	119.2	86.5	101.0	127.4	97.5	165.3	115.7	89.6	99.1	122.2	92.7
1939	80.3	77.4	64.6	109.1	118.3	101.0	97.6	105.8	112.6	88.1	68.1	42.1
1940	50.5	59.4	83.3	60.7	54.4	83.9	67.5	105.5	66.5	55.0	58.4	68.3
1941	45.6	44.5	46.4	32.8	29.5	59.8	66.9	60.0	65.9	46.3	38.4	33.7
1942	35.6	52.8	54.2	60.7	25.0	11.4	17.7	20.2	17.2	19.2	30.7	22.5
1943	12.4	28.9	27.4	26.1	14.1	7.6	13.2	19.4	10.0	7.8	10.2	18.8
1944	3.7	0.5	11.0	0.3	2.5	5.0	5.0	16.7	14.3	16.9	10.8	28.4
1945	18.5	12.7	21.5	32.0	30.6	36.2	42.6	25.9	34.9	68.8	46.0	27.4

Year	Jan	Feb	Mar	Apr	May	Jun	Jul	Aug	Sep	Oct	Nov	Dec
1946	47.6	86.2	76.6	75.7	84.9	73.5	116.2	107.2	94.4	102.3	123.8	121.7
1947	115.7	133.4	129.8	149.8	201.3	163.9	157.9	188.8	169.4	163.6	128.0	116.5
1948	108.5	86.1	94.8	189.7	174.0	167.8	142.2	157.9	143.3	136.3	95.8	138.0
1949	119.1	182.3	157.5	147.0	106.2	121.7	125.8	123.8	145.3	131.6	143.5	117.6
1950	101.6	94.8	109.7	113.4	106.2	83.6	91.0	85.2	51.3	61.4	54.8	54.1
1951	59.9	59.9	55.9	92.9	108.5	100.6	61.5	61.0	83.1	51.6	52.4	45.8
1952	40.7	22.7	22.0	29.1	23.4	36.4	39.3	54.9	28.2	23.8	22.1	34.3
1953	26.5	3.9	10.0	27.8	12.5	21.8	8.6	23.5	19.3	8.2	1.6	2.5
1954	0.2	0.5	10.9	1.8	0.8	0.2	4.8	8.4	1.5	7.0	9.2	7.6
1955	23.1	20.8	4.9	11.3	28.9	31.7	26.7	40.7	42.7	58.5	89.2	76.9
1956	73.6	124.0	118.4	110.7	136.6	116.6	129.1	169.6	173.2	155.3	201.3	192.1
1957	165.0	130.2	157.4	175.2	164.6	200.7	187.2	158.0	235.8	253.8	210.9	239.4
1958	202.5	164.9	190.7	196.0	175.3	171.5	191.4	200.2	201.2	181.5	152.3	187.6
1959	217.4	143.1	185.7	163.3	172.0	168.7	149.6	199.6	145.2	111.4	124.0	125.0
1960	146.3	106.0	102.2	122.0	119.6	110.2	121.7	134.1	127.2	82.8	89.6	85.6
1961	57.9	46.1	53.0	61.4	51.0	77.4	70.2	55.8	63.6	37.7	32.6	39.9
1962	38.7	50.3	45.6	46.4	43.7	42.0	21.8	21.8	51.3	39.5	26.9	23.2
1963	19.8	24.4	17.1	29.3	43.0	35.9	19.6	33.2	38.8	35.3	23.4	14.9
1964	15.3	17.7	16.5	8.6	9.5	9.1	3.1	9.3	4.7	6.1	7.4	15.1
1965	17.5	14.2	11.7	6.8	24.1	15.9	11.9	8.9	16.8	20.1	15.8	17.0
1966	28.2	24.4	25.3	48.7	45.3	47.7	56.7	51.2	50.2	57.2	57.2	70.4
1967	110.9	93.6	111.8	69.5	86.5	67.3	91.5	107.2	76.8	88.2	94.3	126.4
1968	121.8	111.9	92.2	81.2	127.2	110.3	96.1	109.3	117.2	107.7	86.0	109.8
1969	104.4	120.5	135.8	106.8	120.0	106.0	96.8	98.0	91.3	95.7	93.5	97.9
1970	111.5	127.8	102.9	109.5	127.5	106.8	112.5	93.0	99.5	86.6	95.2	83.5
1971	91.3	79.0	60.7	71.8	57.5	49.8	81.0	61.4	50.2	51.7	63.2	82.2
1972	61.5	88.4	80.1	63.2	80.5	88.0	76.5	76.8	64.0	61.3	41.6	45.3
1973	43.4	42.9	46.0	57.7	42.4	39.5	23.1	25.6	59.3	30.7	23.9	23.3
1974	27.6	26.0	21.3	40.3	39.5	36.0	55.8	33.6	40.2	47.1	25.0	20.5
1975	18.9	11.5	11.5	5.1	9.0	11.4	28.2	39.7	13.9	9.1	19.4	7.8
1976	8.1	4.3	21.9	18.8	12.4	12.2	1.9	16.4	13.5	20.6	5.2	15.3
1977	16.4	23.1	8.7	12.9	18.6	38.5	21.4	30.1	44.0	43.8	29.1	43.2
1978	51.9	93.6	76.5	99.7	82.7	95.1	70.4	58.1	138.2	125.1	97.9	122.7
1979	166.6	137.5	138.0	101.5	134.4	149.5	159.4	142.2	188.4	186.2	183.3	176.3
1980	159.6	155.0	126.2	164.1	179.9	157.3	136.3	135.4	155.0	164.7	147.9	174.4
1981	114.0	141.3	135.5	156.4	127.5	90.9	143.8	158.7	167.3	162.4	137.5	150.1
1982	111.2	163.6	153.8	122.0	82.2	110.4	106.1	107.6	118.8	94.7	98.1	127.0
1983	84.3	51.0	66.5	80.7	99.2	91.1	82.2	71.8	50.3	55.8	33.3	33.4
1984	57.0	85.4	83.5	69.7	76.4	46.1	37.4	25.5	15.7	12.0	22.8	18.7
1985	16.5	15.9	17.2	16.2	27.5	24.2	30.7	11.1	3.9	18.6	16.2	17.3
1986	2.5	23.2	15.1	18.5	13.7	1.1	18.1	7.4	3.8	35.4	15.2	6.8
1987	10.4	2.4	14.7	39.6	33.0	17.4	33.0	38.7	33.9	60.6	39.9	27.1
1988	59.0	40.0	76.2	88.0	60.1	101.8	113.8	111.6	120.1	125.1	125.1	179.2
1989	161.3	165.1	131.4	130.6	138.5	196.2	126.9	168.9	176.7	159.4	173.0	165.5
1990	177.3	130.5	140.3	140.3	132.2	105.4	149.4	200.3	125.2	145.5	131.4	129.7
1991	136.9	167.5	141.9	140.0	121.3	169.7	173.7	176.3	125.3	144.1	108.2	144.4
1992	150.0	161.1	106.7	99.8	73.8	65.2	85.7	64.5	63.9	88.7	91.8	82.6
1993	59.3	91.0	69.8	62.2	61.3	49.8	57.9	42.2	22.4	56.4	35.6	48.9
1994	57.8	35.5	31.7	16.1	17.8	28.0	35.1	22.5	25.7	44.0	18.0	26.2
1995	24.2	29.9	31.1	14.6	14.7	15.8						

Note: Data are preliminary after March 95.

No observations were available during February 1824. The value shown was interpolated from the January and March monthly means of that year.

Smoothed Monthly Mean Sunspot Numbers

Year	Jan	Feb	Mar	Apr	May	Jun	Jul	Aug	Sep	Oct	Nov	Dec
1749							81.6	82.8	84.1	86.3	87.8	88.7
1750	89.0	90.2	92.3	92.6	88.2	83.8	83.3	81.8	78.6	75.4	72.9	69.6
1751	66.8	64.2	59.5	54.9	51.7	49.0	46.2	45.0	46.4	47.5	47.6	47.1
1752	47.2	46.4	45.3	46.4	47.8	48.0	48.2	47.8	46.0	44.1	42.2	40.9
1753	38.2	36.2	36.7	35.8	34.2	32.1	28.8	25.8	22.8	19.9	18.3	17.4
1754	17.1	15.8	13.9	13.0	12.7	12.3	12.6	13.4	14.0	13.9	12.7	10.7
1755	9.2	8.4	8.4	8.8	8.5	8.9	9.7	9.6	9.4	9.4	10.1	11.1
1756	11.5	11.4	11.3	10.6	10.7	10.6	10.3	10.9	12.4	14.1	16.0	17.1
1757	18.0	20.7	23.8	25.7	28.4	31.4	33.4	35.7	37.9	40.6	42.7	44.4
1758	46.5	46.8	47.2	48.4	47.7	47.2	48.0	48.2	47.7	46.6	45.6	46.0
1759	46.5	48.1	50.1	51.5	52.7	53.4	54.8	56.2	58.0	59.6	61.1	62.0
1760	62.5	63.3	62.8	61.8	62.0	62.7	63.0	64.4	66.0	66.8	68.8	72.4
1761	75.7	77.5	79.8	83.0	85.9	86.5	84.8	82.9	80.7	78.8	75.5	71.7
1762	68.3	64.8	62.5	60.4	59.0	59.9	61.7	60.5	58.3	56.7	55.3	53.2
1763	52.4	51.5	49.8	48.8	47.1	45.8	45.3	46.5	48.0	48.3	48.8	49.1
1764	47.8	46.9	45.4	43.0	40.8	37.8	34.9	32.0	29.9	28.8	27.3	25.8
1765	25.3	25.2	24.6	23.6	22.5	21.4	20.4	19.3	19.1	19.0	18.6	18.1
1766	16.4	14.4	12.8	12.0	11.2	11.2	12.1	13.5	14.5	15.9	17.2	18.6
1767	20.6	22.9	26.0	29.3	32.9	36.4	38.9	41.5	43.1	43.7	46.1	49.9
1768	53.0	55.4	57.8	60.6	63.5	67.4	70.7	71.5	72.1	75.1	77.2	77.8
1769	81.2	86.2	91.5	97.9	103.7	106.1	107.3	111.9	115.8	114.5	112.5	111.9
1770	111.1	110.9	109.3	105.2	102.3	101.2	98.0	91.1	85.7	84.9	88.9	93.9
1771	93.6	89.1	86.1	85.4	83.5	81.9	84.3	88.9	90.1	90.5	86.9	79.5
1772	77.3	77.6	75.4	72.8	70.7	67.8	64.6	60.1	58.3	56.7	54.3	53.3
1773	50.0	46.1	43.5	40.4	37.4	35.6	34.5	35.6	37.3	38.0	38.9	39.3
1774	38.9	38.2	37.1	35.6	34.2	31.9	28.9	24.4	19.8	16.6	13.3	10.6
1775	9.3	8.6	8.5	7.9	7.5	7.2	7.7	8.9	9.2	9.4	10.2	10.7
1776	11.0	11.7	12.9	14.5	16.3	18.5	20.8	22.8	25.2	29.6	35.6	41.0
1777	47.5	55.1	62.9	70.3	78.1	87.6	98.0	106.6	113.6	119.6	128.2	138.6
1778	144.8	148.4	151.9	156.3	158.5	156.5	151.8	151.5	153.2	152.5	148.4	141.9
1779	139.0	137.5	133.8	129.9	127.0	125.7	124.1	119.4	115.7	112.8	109.4	106.9
1780	103.5	100.0	98.2	95.5	91.3	86.9	86.0	86.2	83.4	80.4	79.2	79.5
1781	79.4	78.0	75.4	71.5	69.8	69.1	66.2	62.8	60.6	58.8	55.6	51.0
1782	47.0	44.5	42.9	42.0	40.4	38.7	37.4	36.3	36.0	35.0	33.2	31.3
1783	30.6	29.4	27.7	26.4	25.1	23.6	22.2	20.3	18.3	16.9	15.5	14.1
1784	12.3	10.8	10.0	9.7	9.8	10.0	9.9	9.6	9.5	9.7	10.5	11.9
1785	13.9	15.5	16.9	19.4	22.0	23.5	25.4	28.3	31.6	36.1	42.0	46.3
1786	49.6	54.5	60.7	66.7	72.6	79.3	86.9	93.4	97.5	100.9	104	107.9
1787	111.4	115.3	119.2	123.0	125.9	129.5	132.2	133.3	136.6	138.0	136.4	137.8
1788	140.7	141.2	140.4	139.1	136.6	132.8	129.9	128.7	127.6	127.3	128.3	127.3
1789	124.9	122.5	119.9	116.5	116.0	117.9	117.7	117.3	116.4	114.2	111.7	109.2
1790	106.0	103.4	101.2	99.6	97.2	92.5	88.6	84.6	81.0	79.4	77.8	75.9
1791	74.9	73.1	70.8	69.4	67.9	66.9	66.0	65.4	65.1	64.5	64.0	63.4
1792	62.2	61.9	62.2	61.8	61.3	60.5	60.0	59.5	58.8	57.6	56.2	55.4
1793	55.1	54.0	51.3	49.3	48.3	47.3	46.4	45.5	44.3	42.6	41.7	41.4
1794	40.7	40.7	40.7	39.3	39.6	40.8	40.0	38.9	37.6	36.2	34.7	32.7
1795	30.5	28.7	28.2	28.0	25.8	22.7	21.3	20.6	20.1	20.8	20.9	20.1
1796	20.2	19.8	19.0	18.8	17.8	16.6	15.7	14.6	13.3	11.6	9.9	9.5
1797	8.8	8.0	7.7	7.0	6.7	6.5	5.9	5.4	5.7	5.9	5.5	4.7
1798	4.1	3.8	3.5	3.2	3.2	3.8	4.1	4.4	5.1	5.8	6.5	7.3
1799	7.8	7.8	7.5	7.6	7.3	6.8	7.0	7.1	6.6	5.9	5.4	5.9
1800	7.2	8.8	10.1	10.9	11.5	13.2	15.3	17.0	18.5	20.4	22.8	24.3

Year	Jan	Feb	Mar	Apr	May	Jun	Jul	Aug	Sep	Oct	Nov	Dec
1801	25.2	26.6	28.3	30.0	32.1	33.7	34.9	36.5	37.7	38.6	39.6	40.7
1802	41.8	42.8	44.1	45.1	45.1	45.0	45.1	45.4	45.1	43.9	43.2	42.8
1803	42.4	41.7	40.8	41.2	42.5	43.1	42.9	42.6	43.2	45.1	45.7	45.2
1804	44.3	44.0	44.6	45.3	46.1	47.0	48.1	48.6	48.6	48.2	47.9	48.3
1805	48.9	49.2	48.8	47.1	44.9	43.1	41.3	39.8	38.4	37.2	36.3	35.2
1806	34.2	33.2	31.7	30.7	30.0	28.7	27.0	25.1	23.0	22.3	21.5	20.2
1807	18.9	17.6	16.3	14.7	13.0	11.1	9.6	8.7	8.0	7.1	6.8	7.0
1808	6.8	6.4	6.5	6.6	6.8	7.6	8.4	8.9	9.2	8.8	7.9	7.2
1809	6.7	6.1	5.3	4.6	4.0	3.0	2.2	1.6	1.1	1.0	0.8	0.4
1810	0.1	0.0	0.0	0.0	0.0	0.0	0.0	0.0	0.0	0.0	0.0	0.0
1811	0.3	0.6	0.7	1.0	1.3	1.4	1.9	2.4	2.5	2.6	2.6	2.7
1812	2.5	2.9	3.7	3.7	3.9	4.6	4.5	4.4	4.8	5.5	6.4	7.0
1813	8.1	8.6	8.7	10.1	11.5	12.0	13.1	14.1	14.3	14.8	15.1	15.3
1814	15.4	15.2	14.6	14.0	13.5	13.7	13.8	14.5	16.2	17.4	17.9	19.8
1815	22.2	24.8	27.6	29.2	30.7	33.5	35.7	37.5	41.0	44.1	46.7	47.6
1816	47.3	43.4	46.1	47.7	48.7	47.3	46.2	46.2	46.7	46.3	44.0	42.8
1817	43.2	44.5	45.0	43.2	41.6	41.1	41.0	39.5	35.0	32.4	34.1	35.2
1818	34.2	32.7	31.7	31.5	31.0	30.2	30.0	29.8	28.8	27.3	25.3	23.9
1819	24.0	23.9	23.2	22.5	23.0	23.7	23.4	23.1	23.4	23.4	23.7	23.1
1820	21.7	21.2	20.8	19.6	18.1	16.5	15.8	14.9	14.1	13.7	12.1	10.6
1821	9.5	7.8	6.9	7.3	7.5	7.0	5.7	4.7	5.0	5.6	5.7	5.9
1822	6.3	6.4	6.1	5.1	4.2	4.0	4.0	4.0	3.3	2.1	1.4	1.2
1823	0.6	0.2	0.1	0.1	0.1	0.9	2.7	4.0	4.5	5.3	6.2	6.3
1824	6.3	6.3	7.2	9.1	10.2	9.4	7.9	7.4	8.5	8.8	8.6	9.8
1825	11.7	14.0	14.8	14.2	14.3	15.7	17.1	17.7	18.4	19.9	21.4	23.0
1826	24.9	26.3	27.1	28.7	31.3	34.4	37.0	38.9	41.0	42.8	44.7	46.5
1827	46.9	47.1	49.0	50.5	51.2	50.6	50.5	51.9	52.9	53.9	55.9	59.0
1828	61.2	62.6	63.6	63.5	63.8	64.2	63.8	62.8	62.4	64.1	64.7	62.8
1829	63.3	64.9	65.1	65.3	65.8	66.7	67.4	68.7	70.2	71.2	71.7	71.3
1830	68.9	65.8	65.1	66.6	68.3	69.9	70.8	69.7	69.1	67.3	63.9	61.4
1831	60.2	60.4	59.6	57.0	53.8	50.0	47.1	46.7	45.3	42.5	41.5	41.4
1832	39.8	36.6	33.4	31.1	28.9	27.6	26.7	24.2	20.7	17.9	15.7	13.5
1833	12.1	11.7	11.7	11.3	10.3	9.3	8.3	8.1	7.9	7.5	7.3	7.4
1834	7.8	7.8	7.7	8.4	10.2	12.2	13.4	13.7	14.7	17.8	21.8	24.3
1835	27.5	31.9	37.9	44.6	50.4	55.1	60.2	67.1	73.8	80.5	86.7	93.3
1836	99.5	103.9	105.7	107.2	109.9	116.1	125.6	132.6	136.9	138.2	138.0	139.4
1837	142.7	145.8	146.9	146.4	145.2	141.5	136.5	130.9	127.4	127.2	127.8	126.2
1838	121.3	116.7	113.5	111.2	108.6	105.2	101.6	100.8	98.9	93.6	87.4	82.2
1839	79.6	80.8	85.4	87.9	87.5	86.5	84.7	83.0	81.5	80.7	81.5	81.9
1840	80.7	76.6	71.1	66.9	64.6	63.6	60.8	56.0	52.5	50.5	49.4	49.7
1841	48.7	46.7	44.3	41.8	39.5	37.4	36.7	36.2	35.5	34.5	32.1	28.9
1842	26.6	25.4	24.1	23.8	25.1	25.1	23.9	22.8	21.5	20.2	19.3	18.7
1843	18.1	17.4	16.2	14.2	12.0	10.9	10.5	10.8	11.5	12.2	12.3	11.7
1844	11.9	12.9	13.5	14.3	14.6	14.6	15.7	17.6	20.0	22.7	25.7	28.4
1845	29.9	30.7	31.9	33.7	35.7	38.5	40.6	41.5	42.6	44.0	45.0	46.9
1846	49.0	50.6	54.8	58.6	60.1	61.3	62.5	63.2	63.9	63.8	63.4	64.9
1847	66.0	69.8	75.6	83.1	91.5	96.6	102.5	109.3	113.0	116.6	120.3	123.0
1848	128.3	131.6	128.7	124.2	121.1	122.2	124.2	124.9	125.3	124.6	123.5	120.8
1849	116.5	110.9	107.7	104.9	101.7	98.5	92.6	87.5	85.2	82.2	79.0	77.7
1850	75.6	74.0	73.7	73.4	71.5	68.1	66.4	67.0	66.9	66.7	67.2	67.0
1851	66.6	66.3	65.4	64.2	63.7	64.0	64.2	62.3	60.6	60.8	60.9	59.9
1852	59.5	59.0	57.0	55.9	56.2	55.3	53.1	50.9	48.9	47.2	45.6	44.5
1853	44.3	45.0	45.2	44.0	41.9	39.9	38.0	35.9	34.3	32.7	31.3	30.1
1854	28.2	25.6	23.7	22.0	20.8	20.7	20.4	20.0	19.5	18.4	16.9	15.6
1855	14.2	12.9	11.4	10.4	9.2	7.5	6.2	5.4	4.5	3.8	3.6	3.2

Year	Jan	Feb	Mar	Apr	May	Jun	Jul	Aug	Sep	Oct	Nov	Dec
1856	3.3	3.6	3.9	3.9	3.8	4.1	4.9	5.5	5.8	6.2	7.6	9.3
1857	10.5	11.7	13.7	16.8	19.3	21.5	23.8	26.0	29.4	32.7	34.3	36.0
1858	38.6	41.7	44.8	48.5	51.5	53.6	56.7	60.7	64.3	67.6	71.7	75.5
1859	78.9	82.6	85.9	87.9	90.8	93.2	93.7	93.7	94.0	93.8	93.9	95.4
1860	97.2	97.9	97.0	95.4	94.4	95.1	94.9	93.7	93.3	94.5	93.6	90.6
1861	88.1	85.8	84.5	83.1	80.3	77.8	77.2	76.7	73.7	69.5	67.9	68.1
1862	67.7	66.7	65.3	63.7	62.5	60.8	58.5	57.6	58.2	58.6	57.6	55.4
1863	51.9	49.6	47.1	45.2	44.5	44.0	44.4	44.4	44.0	43.8	43.0	43.2
1864	44.8	46.0	46.8	46.6	47.2	47.5	46.6	45.9	44.4	43.1	42.5	41.3
1865	39.1	37.2	36.2	35.2	33.2	31.1	29.8	29.0	28.4	27.2	25.9	24.2
1866	22.8	21.0	19.4	18.7	17.9	16.8	15.0	12.1	9.9	8.7	7.8	6.7
1867	5.9	5.4	5.2	5.3	5.3	6.3	7.9	9.2	10.5	12.6	14.9	17.1
1868	19.3	21.5	24.2	27.6	31.7	35.5	39.2	42.9	45.8	47.1	50.5	56.9
1869	61.4	64.6	68.0	69.4	70.1	72.4	74.6	77.6	84.3	93.8	101.7	105.8
1870	110.0	116.2	121.6	127.5	134.0	138.0	139.6	140.5	140.2	139.6	138.5	135.4
1871	132.3	129.3	125.1	120.4	116.3	112.9	110.8	110.3	107.8	103.0	98.9	98.0
1872	98.9	98.3	99.0	101.0	101.9	101.9	102.0	101.7	101.6	100.9	97.4	92.2
1873	87.8	85.2	81.4	76.2	71.5	67.7	65.2	62.4	58.4	54.4	52.4	52.0
1874	51.8	51.5	50.4	49.1	47.4	45.5	42.7	39.1	36.8	36.1	34.6	32.7
1875	29.8	25.5	22.5	20.5	19.2	17.9	17.1	16.8	16.3	15.1	13.7	12.5
1876	11.7	11.6	11.7	12.0	11.8	11.4	11.7	11.9	10.8	10.6	11.8	13.0
1877	13.1	12.6	12.7	12.7	12.6	12.5	11.4	10.4	10.1	9.3	8.0	7.1
1878	6.6	6.0	5.3	4.6	4.0	3.5	3.3	3.9	2.4	2.3	2.4	2.2
1879	2.5	3.2	3.7	4.2	5.0	5.7	6.9	9.0	10.9	12.3	13.7	15.8
1880	17.7	19.8	23.9	27.6	29.7	31.3	32.8	34.4	36.8	39.5	41.6	43.6
1881	47.0	49.7	49.6	49.9	51.8	53.5	54.6	55.6	57.0	59.5	62.2	62.4
1882	60.4	58.4	57.9	57.8	58.9	59.9	60.3	60.0	58.1	56.5	54.6	54.5
1883	57.3	59.0	59.0	59.8	60.9	62.3	65.0	67.9	71.4	73.0	74.2	74.6
1884	72.4	71.7	72.4	71.3	67.8	64.6	61.4	58.8	56.6	54.2	53.6	55.2
1885	57.1	57.4	56.2	54.9	54.4	53.2	51.6	49.2	47.6	47.4	45.2	41.1
1886	37.2	34.3	32.2	30.2	27.5	25.8	24.6	23.2	20.5	16.7	14.7	13.8
1887	13.1	13.0	12.6	11.9	12.1	12.7	13.2	13.0	12.9	13.0	12.4	11.5
1888	10.3	8.6	7.9	7.8	7.8	7.3	6.3	5.8	5.8	5.8	5.6	5.3
1889	5.6	6.6	7.2	7.1	6.7	6.3	6.5	6.3	5.9	5.7	5.7	5.6
1890	5.5	5.0	5.0	5.8	6.6	7.0	7.4	8.6	9.8	10.8	13.1	16.5
1891	20.5	23.5	26.0	29.2	32.2	34.6	37.9	42.5	46.3	50.0	53.7	56.5
1892	58.4	62.0	65.2	66.4	68.1	71.0	73.2	73.4	73.9	75.3	76.3	77.0
1893	78.0	79.7	81.5	82.5	83.3	84.3	85.3	86.1	86.0	85.2	85.6	86.7
1894	87.9	86.2	83.2	82.5	81.6	79.4	77.2	75.6	75.3	75.4	73.8	71.3
1895	67.7	65.2	64.8	64.2	63.5	63.5	62.5	60.7	59.9	58.2	55.1	52.5
1896	51.5	49.6	48.0	46.5	44.5	43.0	42.3	41.6	39.5	38.0	37.1	35.2
1897	32.9	32.0	31.2	30.1	28.3	26.6	25.8	25.7	26.3	26.0	25.6	26.3
1898	26.0	25.6	25.4	25.7	27.5	27.6	26.3	24.7	22.7	21.9	21.1	20.3
1899	20.4	19.4	17.1	15.1	13.2	12.2	11.7	11.5	11.2	10.9	11.3	11.3
1900	10.7	10.5	10.6	10.6	10.4	9.9	9.1	8.2	7.6	6.8	5.9	5.4
1901	4.8	4.4	3.9	3.2	2.8	2.8	3.0	3.1	3.3	3.6	3.3	2.8
1902	2.6	2.7	3.1	3.9	4.7	5.0	5.2	6.0	6.8	7.9	9.5	10.6
1903	12.3	14.6	15.8	16.9	19.3	22.5	25.4	26.6	27.9	29.6	31.4	33.5
1904	35.5	37.7	39.7	41.1	41.5	41.6	42.9	46.4	49.8	50.5	50.7	51.3
1905	52.5	53.5	54.6	56.6	60.5	63.4	63.1	60.4	58.5	59.5	60.6	61.6
1906	63.4	64.2	63.8	61.3	55.9	53.5	55.1	59.6	62.7	62.4	61.7	60.1
1907	56.9	55.0	56.4	59.6	62.6	62.8	60.5	55.9	51.4	50.3	50.4	50.6
1908	50.5	51.6	53.2	51.9	49.9	48.9	49.3	50.5	52.6	53.1	51.9	50.6
1909	49.4	46.4	41.6	40.7	42.2	43.3	42.6	40.7	38.2	35.4	33.8	32.8
1910	31.5	30.1	29.1	27.7	24.7	20.6	17.6	15.7	14.2	14.0	13.8	12.8

Year	Jan	Feb	Mar	Apr	May	Jun	Jul	Aug	Sep	Oct	Nov	Dec
1911	12.0	11.2	10.0	7.6	6.0	5.9	5.6	5.1	4.6	4.0	3.3	3.2
1912	3.2	3.0	3.1	3.4	3.4	3.4	3.7	3.9	3.8	3.5	3.2	2.8
1913	2.6	2.5	2.2	1.8	1.7	1.6	1.5	1.5	1.6	2.4	3.3	4.0
1914	4.6	5.1	5.8	6.5	7.4	8.8	10.4	12.9	16.1	18.6	20.7	24.3
1915	29.4	34.8	38.9	42.3	45.3	46.9	48.3	49.8	51.5	53.9	56.9	58.6
1916	57.8	55.6	54.0	53.7	54.6	56.3	58.3	60.2	62.1	63.3	65.1	68.7
1917	73.4	81.2	89.7	94.1	96.3	100.7	104.8	105.4	104.2	103.5	102.2	98.3
1918	95.5	92.8	88.5	87.0	87.0	83.5	78.6	77.2	77.5	76.1	75.4	78.0
1919	78.4	75.2	72.8	70.4	67.4	64.6	63.7	62.8	61.9	60.5	56.7	51.4
1920	46.8	43.2	40.3	39.4	38.7	37.9	36.8	34.9	32.1	31.0	31.3	30.6
1921	31.0	31.7	31.1	29.0	27.3	26.5	25.3	24.4	25.5	25.8	24.3	22.5
1922	20.1	18.1	16.9	15.8	14.9	14.4	13.9	12.6	9.4	7.1	6.7	6.6
1923	6.4	5.9	6.0	6.6	6.9	6.4	5.6	5.6	5.7	5.8	6.8	8.1
1924	9.8	11.6	12.9	14.0	15.1	16.1	16.9	17.9	19.3	20.9	22.6	24.5
1925	25.9	27.1	29.3	32.6	35.9	40.9	47.2	51.8	55.6	57.7	58.9	60.9
1926	62.6	64.1	65.1	65.2	65.4	64.7	64.3	65.7	66.9	69.5	72.4	72.4
1927	72.0	71.8	71.7	71.7	71.6	70.5	69.1	68.4	68.3	68.4	67.7	69.0
1928	72.1	75.1	77.3	78.1	77.3	77.2	77.1	76.1	74.2	71.6	69.2	67.7
1929	66.2	64.3	61.3	58.6	59.6	63.0	64.8	64.0	62.8	61.1	60.6	57.5
1930	53.6	49.8	48.0	47.1	44.2	39.0	33.5	31.2	30.7	30.2	29.4	28.3
1931	27.6	26.9	25.9	24.2	22.6	21.6	21.1	19.7	17.8	16.3	14.8	14.8
1932	14.8	14.2	13.3	12.6	12.2	11.4	11.2	11.7	12.0	11.7	10.7	9.4
1933	8.4	7.9	7.7	7.5	6.9	6.2	5.4	4.3	3.4	3.6	4.6	5.4
1934	5.7	6.3	6.6	6.7	7.2	8.1	9.4	10.6	11.9	12.7	13.0	15.0
1935	17.6	19.6	22.0	25.6	29.9	34.2	37.9	42.0	46.5	51.3	55.0	57.2
1936	59.0	62.2	65.9	68.8	72.5	77.2	82.6	87.8	90.3	92.1	96.1	101.2
1937	107.6	113.5	116.7	119.2	119.0	115.8	113.0	111.2	110.9	110.6	110.8	109.8
1938	109.3	109.2	107.9	106.3	107.1	109.4	108.8	106.3	103.6	103.0	103.0	102.8
1939	101.1	96.9	97.4	97.9	95.2	90.9	87.6	85.5	85.5	84.3	79.6	76.3
1940	74.2	73.0	71.1	67.8	66.0	66.7	67.6	66.8	64.6	61.9	59.7	57.6
1941	56.6	54.7	52.8	52.4	51.2	49.0	47.0	47.0	47.6	49.1	50.2	47.8
1942	43.7	41.1	36.5	33.3	31.8	31.0	29.6	27.7	25.6	23.0	21.1	20.5
1943	20.1	19.9	19.6	18.8	17.5	16.5	16.0	14.4	12.6	10.8	9.2	8.6
1944	8.2	7.7	7.8	8.4	8.8	9.2	10.2	11.3	12.3	14.0	16.5	19.0
1945	21.9	23.8	25.1	28.1	31.7	33.1	34.3	38.6	43.9	48.1	52.1	56.0
1946	60.6	67.0	72.9	76.8	81.4	88.6	95.3	100.2	104.3	109.6	117.6	126.2
1947	131.7	136.8	143.4	149.0	151.8	151.7	151.2	148.9	145.5	145.7	146.2	145.3
1948	144.8	142.8	140.5	138.2	135.8	135.3	136.6	141.1	147.7	148.5	143.9	139.2
1949	136.6	134.5	133.2	133.0	134.8	136.0	134.4	130.0	124.4	121.0	119.6	118.0
1950	115.0	111.9	106.4	99.5	92.9	86.6	82.2	79.0	75.3	72.2	71.4	72.2
1951	71.7	69.5	69.8	70.7	70.2	69.8	68.6	66.3	63.3	59.2	53.0	46.8
1952	43.2	42.0	39.5	36.0	33.6	31.9	30.8	29.4	28.2	27.6	27.1	26.0
1953	24.1	21.6	19.9	18.9	17.4	15.2	12.8	11.5	11.4	10.4	8.8	7.4
1954	6.4	5.6	4.2	3.4	3.7	4.2	5.4	7.2	7.8	7.9	9.5	12.0
1955	14.2	16.4	19.5	23.4	28.8	35.1	40.1	46.5	55.5	64.4	73.0	81.0
1956	88.8	98.5	109.3	118.7	127.4	136.9	145.5	149.6	151.5	155.8	159.6	164.3
1957	170.2	172.2	174.3	181.0	185.5	187.9	191.4	194.4	197.3	199.5	200.8	200.1
1958	199.0	200.9	201.3	196.8	191.4	186.8	185.2	184.9	183.8	182.2	180.7	180.5
1959	178.6	176.9	174.5	169.2	165.1	161.4	155.8	151.3	146.3	141.1	137.2	132.5
1960	128.9	125.0	121.6	119.6	117.0	113.9	108.4	101.9	97.2	92.6	87.2	82.9
1961	80.2	74.8	68.9	64.3	60.1	55.8	53.1	52.5	52.3	51.4	50.5	48.7
1962	45.2	41.8	39.8	39.4	39.2	38.3	36.8	34.9	32.7	30.8	30.0	29.8
1963	29.4	29.8	29.7	29.0	28.7	28.2	27.7	27.2	26.9	26.0	23.8	21.3
1964	19.5	17.8	15.4	12.7	10.9	10.2	10.3	10.2	9.9	9.6	10.2	11.0
1965	11.7	12.0	12.5	13.6	14.6	15.0	15.5	16.4	17.4	19.7	22.3	24.5

Year	Jan	Feb	Mar	Apr	May	Jun	Jul	Aug	Sep	Oct	Nov	Dec
1966	27.7	31.3	34.5	37.4	40.7	44.6	50.3	56.6	63.1	67.6	70.2	72.7
1967	75.0	78.8	82.2	84.6	87.4	91.3	94.1	95.3	95.3	95.0	97.1	100.6
1968	102.6	102.9	104.7	107.2	107.6	106.6	105.2	104.8	107.0	109.9	110.6	110.1
1969	110.0	109.6	108.0	106.4	106.2	106.1	105.8	106.4	105.4	104.1	104.6	104.9
1970	105.6	106.0	106.2	106.1	105.8	105.3	103.8	101.0	97.2	93.9	89.4	84.1
1971	80.4	77.8	74.4	70.9	68.1	66.7	65.4	64.6	65.8	66.2	66.8	69.4
1972	70.8	71.2	72.4	73.4	72.9	70.5	68.2	65.5	62.2	60.6	58.7	55.1
1973	50.9	46.5	44.2	42.7	40.7	39.1	37.5	36.1	34.4	32.6	31.8	31.5
1974	32.7	34.4	34.0	33.9	34.6	34.5	34.0	33.1	32.1	30.3	27.6	25.2
1975	23.0	22.1	21.3	18.6	16.8	16.0	15.0	14.2	14.4	15.4	16.1	16.2
1976	15.2	13.2	12.2	12.6	12.5	12.2	12.9	14.0	14.3	13.4	13.5	14.8
1977	16.7	18.1	20.0	22.2	24.2	26.3	29.0	33.4	39.1	45.6	51.9	56.9
1978	61.3	64.5	69.6	76.9	83.2	89.3	97.4	104.0	108.4	111.1	113.3	117.7
1979	123.7	130.9	136.5	141.1	147.2	153.0	155.0	155.4	155.7	157.8	162.3	164.5
1980	163.9	162.6	160.9	158.7	156.3	154.7	152.8	150.3	150.1	150.2	147.7	142.7
1981	140.3	141.5	143.0	143.4	142.9	141.5	140.3	141.1	142.8	142.2	138.9	137.8
1982	137.0	133.3	129.2	124.3	119.9	117.3	115.2	109.4	101.0	95.7	94.7	94.6
1983	92.8	90.3	85.9	81.5	77.1	70.5	65.5	65.8	67.9	68.2	66.8	64.0
1984	60.2	56.4	53.0	49.8	47.5	46.5	44.2	39.6	33.9	28.9	24.7	21.7
1985	20.5	19.6	18.6	18.3	18.3	18.0	17.4	17.1	17.3	17.3	16.8	15.3
1986	13.8	13.1	13.0	13.7	14.3	13.8	13.7	13.2	12.3	13.2	14.9	16.3
1987	17.6	19.6	22.1	24.4	26.5	28.4	31.2	34.8	39.0	43.6	46.7	51.3
1988	58.2	64.6	71.3	77.5	83.8	93.7	104.3	113.7	121.2	125.3	130.4	137.6
1989	142.0	145.0	149.7	153.5	156.9	158.4	158.5	157.7	156.6	157.4	157.5	153.5
1990	150.6	152.9	152.0	149.3	147.0	143.8	140.6	140.5	142.1	142.1	141.7	143.9
1991	147.6	147.6	146.6	146.5	145.5	145.2	146.3	146.6	144.9	141.7	138.1	131.7
1992	123.7	115.4	108.2	103.3	100.3	97.1	90.7	84.0	79.5	76.4	74.4	73.2
1993	71.4	69.3	66.6	63.6	59.9	56.1	54.7	52.3	48.4	44.9	41.2	38.4
1994	36.6	34.8	34.1	33.7	32.5	30.8	28.5	26.8	26.6	26.5	26.3	25.6

NOTE: Values after Sep 94 are preliminary.

Each smoothed number represents the average of two adjacent 12-month running means of monthly means. Consider, for example, the June 1980 value. We built this number by first computing two 12-month running means: one that averaged monthly means from December 1979 through November 1980 and a second that averaged means from January 1980 through December 1980. The June 1980 smoothed number equals the arithmetic average of these two running means.

(Source: Solar-Terrestrial Physics Division, National Geophysical Data Center, 325 Broadway, Mail Code E/GC2, Boulder, CO 80303-3328 USA)

SMOOTHED SUNSPOT NUMBER RECENT HISTORY AND PREDICTIONS

Year	Jan	Feb	Mar	Apr	May	Jun	Jul	Aug	Sep	Oct	Nov	Dec
1992	123.7	115.4	108.2	103.3	100.3	97.1	90.7	84.0	79.5	76.4	74.4	73.2
1993	71.4	69.3	66.6	63.6	59.9	56.1	54.6	52.3	48.5	45.0	41.3	38.6
1994	36.6(0)	34.8(0)	34.1(0)	33.8(0)	32.6(0)	30.9(0)	28.5(0)	26.8(0)	26.6(0)	26.5(0)	26.3(0)	25.6(0)
1995	24.7(4)	23.8(5)	22.6(7)	21.5(8)	20.4(8)	19.5(9)	18.9(10)	18.4(11)	17.9(11)	17.3(11)	16.6(11)	15.7(10)
1996	14.6(10)	13.4(11)	12.7(11)	12.3(12)	12.0(12)	11.3(12)	10.7(12)	10.1(12)	9.6(12)	9.0(12)	8.6(12)	8.4(11)
1997	8.5(11)	8.8(11)	9.1(10)	9.3(10)	9.8(11)	10.5(12)	11.3(13)	11.9(15)	12.6(16)	13.3(17)	14.1(19)	15.2(22)

Note: Numbers in parentheses indicate 90% confidence interval.
(Source: National Geophysical Data Center)

WORLDWIDE INDICES

Daily Solar and Geomagnetic Terrestrial Data, July 1994 through June 1995

The following 13 observatories, which lie between 46 and 63 degrees north and south geomagnetic latitude, now contribute to the planetary indices: Lerwick (UK), Eskdalemuir (UK), Hartland (UK), Ottawa (Canada), Fredericksburg (USA), Meannook (Canada), Sitka (USA), Eyrewell (New Zealand), Canberra (Australia), Lovo (Sweden), Rude Skov (Denmark), Wingst (Germany), and Witteveen (The Netherlands).

Three-Hour-Range K-Index

K indices isolate solar particle effects on the earth's magnetic field; over a three-hour period, they classify into disturbance levels the range of variation of the more unsettled horizontal field component. Each activity level relates almost logarithmically to its corresponding disturbance amplitude. Three-hour indices discriminate conservatively between true magnetic field perturbations and the quiet-day variations produced by ionospheric currents.

K indices range in 28 steps from 0 (quiet) to 9 (greatly disturbed) with fractional parts expressed in thirds of a unit. The indices reported in the table below have been rounded to resemble the reports announced on WWV at 18 minutes after each hour.

Equivalent Amplitude, or A-Index

The A-index ranges from 0 to 400 and represents a K-value converted to a linear scale in gammas (nanoTeslas). A scale that measures equivalent disturbance amplitude of a station at which K=9 has a lower limit of 400 gammas.

Daily Sunspot Number, "SN"

"SN" in the table is the daily sunspot number for that date, unsmoothed, from the Royal Observatory in Brussels.

SFI—Solar Flux Index

The SFI column represents the Ottawa 10.7-cm (2800 MHz) Solar Radio Flux ADJUSTED TO 1 AU. Solar flux is measured at 1700 UT daily and expressed in units of 10 to the -22 Watts/meter^2/hertz. The numbers here are the "observed" and "adjusted" values. Since the Solar Flux Index is closely followed by radio amateurs, it is instructive to know how this number is derived.

The sun emits radio energy with a slowly varying intensity. This radio flux, which originates from atmospheric layers high in the sun's chromosphere and low in its corona, changes gradually from day-to-day, in response to the number of spot groups on the disk. Radio intensity levels consist of emission from three sources: from the undisturbed solar surface, from developing active regions, and from short-lived enhancements above the daily level. Solar flux density at 2800 MHz has been recorded routinely by radio telescope near Ottawa since February 14, 1947. Each day, levels are determined at local noon (1700 GMT) and then corrected to within a few percent for factors such as antenna gain, atmospheric absorption, bursts in progress, and background sky temperature.

Three sets of flux values are recorded: the observed, the adjusted, and the absolute. Of the three, the observed numbers are the least refined, since they contain fluctuations as large as 7% that arise from the changing sun-earth distance, but they are perhaps more relevant in determining the sun's effect on radio wave propagation. In contrast, adjusted fluxes have this variation removed; these numbers reflect the energy flux received by a detector located at the mean distance (1AU) between sun and earth. Finally, the absolute levels carry the error reduction one step further; here each adjusted value is multiplied by 0.90 to compensate for uncertainties in antenna gain and in waves reflected from the ground.

Month	day	|K-Index| 03Z	06Z	09Z	12Z	15Z	18Z	21Z	24Z	|A-index| 03Z	06Z	09Z	12Z	15Z	18Z	21Z	24Z	SN	adj SFI
Jul-94	1	4	4	4	4	3	3	4	5	32	22	27	22	15	15	22	39	32	89.7
	2	3	5	4	4	3	4	3	4	18	48	32	27	18	27	15	27	43	85.6
	3	3	2	3	3	2	2	2	3	12	9	15	15	9	6	7	15	39	89.3
	4	2	2	2	1	3	3	3	3	6	9	7	5	15	12	15	12	41	87.4
	5	2	1	1	1	1	1	1	2	6	4	5	3	3	4	3	9	27	86.2
	6	2	2	2	2	2	2	2	4	7	6	9	6	7	6	9	32	27	87.1
	7	3	4	3	2	3	3	4	3	15	27	15	9	18	15	22	12	47	91.1
	8	1	1	1	1	2	1	1	1	5	5	4	4	6	4	3	3	60	88.7
	9	1	1	1	0	2	3	1	1	5	3	3	2	7	15	5	3	49	88.8
	10	0	0	1	1	1	1	1	1	2	2	5	3	5	5	5	4	60	88.9
	11	1	1	0	0	2	2	1	1	4	4	0	2	6	7	4	5	72	88.5
	12	2	1	0	0	0	1	0	0	6	4	0	0	2	4	2	0	68	85.7
	13	0	1	2	0	1	1	1	1	2	4	6	2	3	4	5	5	59	84.1
	14	1	2	3	4	5	6	4	5	5	6	6	27	56	67	27	48	45	84.6
	15	5	3	3	2	3	3	3	5	48	12	12	7	12	12	12	56	43	85.7
	16	6	4	3	5	4	4	4	4	80	27	15	39	22	22	22	32	54	85.1
	17	4	3	2	3	3	2	2	3	22	12	9	18	15	9	9	12	53	83
	18	3	3	2	2	3	2	2	1	18	15	9	7	15	6	7	5	48	82.7
	19	3	3	3	2	2	1	2	3	18	12	12	7	9	4	7	18	27	80.1
	20	2	1	1	1	2	1	1	1	6	3	5	5	6	4	4	4	29	79.4
	21	2	3	2	2	3	2	2	2	7	12	9	9	15	6	6	7	27	79.8

Month	Day																			
	22	2	2	1	1	1	1	1	0	9	7	5	5	5	3	4	2	19	80.1	
	23	1	2	1	1	2	2	2	2	4	7	4	5	6	9	6	7	16	78.2	
	24	2	1	1	2	1	1	2	3	6	5	5	7	5	5	6	15	14	77.7	
	25	3	2	2	2	3	2	2	2	15	9	7	7	15	9	6	9	12	77.8	
	26	1	1	1	1	1	1	1	1	4	5	3	4	4	3	5	5	17	76.3	
	27	2	5	2	1	1	1	2	3	9	48	7	4	4	5	9	18	14	76.6	
	28	3	4	3	3	3	2	4	3	18	22	15	18	15	7	22	18	16	77.1	
	29	3	3	3	2	2	3	2	3	15	12	12	7	7	12	9	15	7	78	
	30	4	3	2	1	2	2	2	3	22	12	6	5	7	6	7	15	12	77.6	
	31	2	2	1	2	2	1	1	4	9	7	5	9	7	5	5	27	10	76.8	
Aug-94	1	3	2	2	1	1	1	1	2	12	9	6	4	4	4	5	6	11	76.2	
	2	2	1	0	1	0	1	1	1	6	3	2	4	2	3	3	5	10	77.2	
	3	1	1	1	1	1	1	1	1	4	4	4	5	3	4	5	5	13	78.7	
	4	1	0	0	0	1	1	0	1	4	2	2	2	4	3	2	3	14	77.4	
	5	1	2	1	1	1	1	1	1	5	7	5	5	3	4	5	3	18	77.8	
	6	1	1	2	1	1	1	1	1	3	5	6	4	4	3	3	5	17	77.5	
	7	0	0	0	1	2	1	1	2	2	2	2	3	6	5	3	6	13	77.9	
	8	1	0	0	0	0	1	0	1	4	2	0	0	2	3	2	3	12	75.9	
	9	1	1	1	0	1	1	2	2	3	4	4	2	3	4	9	6	12	77.4	
	10	2	1	3	2	4	3	3	2	7	5	12	7	22	12	15	9	15	79.7	
	11	3	3	2	3	3	3	3	3	18	18	9	12	12	18	12	15	20	78.7	
	12	4	4	3	3	3	3	4	3	22	22	15	12	15	12	32	15	36	82.9	
	13	3	5	4	3	5	4	4	3	18	39	32	12	39	27	32	18	39	86.7	
	14	4	4	5	4	3	3	4	3	27	27	48	32	12	18	22	18	44	91.2	
	15	4	2	2	2	3	2	3	3	27	7	7	7	15	9	15	18	42	83.5	
	16	2	2	1	2	2	1	3	3	9	6	5	9	6	5	15	15	43	78.8	
	17	2	2	3	2	2	2	1	1	6	7	12	6	6	7	5	4	41	79.5	
	18	1	1	2	2	2	2	1	2	3	5	6	9	9	6	4	6	39	79.3	
	19	2	1	2	1	1	1	2	1	6	5	6	3	4	5	6	3	43	77.1	
	20	1	2	2	2	1	2	2	2	4	9	9	6	5	6	7	7	30	74	
	21	2	2	0	1	3	3	1	1	9	6	2	3	15	15	4	3	19	72.7	
	22	2	2	1	1	2	2	2	4	7	6	3	5	9	9	9	32	10	72.2	
	23	3	2	1	1	1	1	1	1	12	9	4	4	4	4	5	3	25	73.5	
	24	1	2	1	1	1	2	2	2	3	6	4	3	4	9	9	9	23	73.5	
	25	2	3	2	3	3	2	2	1	7	15	9	15	15	9	9	3	12	72.5	
	26	0	2	2	2	1	1	2	3	2	6	7	6	3	5	6	12	12	73.7	
	27	3	2	2	2	1	2	3	2	18	9	9	6	5	7	12	7	11	72.5	
	28	2	3	2	2	1	1	1	1	9	18	7	6	3	3	5	3	13	72.5	
	29	1	1	2	1	1	1	0	2	4	3	7	5	5	5	2	6	16	79.1	
	30	1	1	1	1	1	1	1	1	3	3	4	4	4	5	4	5	22	84.7	
	31	1	1	0	1	2	2	1	3	4	4	2	3	9	9	4	12	35	83.8	
Sep-94	1	4	4	1	2	1	1	3	3	27	22	5	6	4	4	15	12	38	87.3	
	2	2	3	1	1	1	1	0	1	9	15	3	3	4	3	2	3	49	94.9	
	3	2	2	2	1	2	1	1	2	6	7	7	5	6	5	3	6	59	98.2	
	4	2	0	1	1	1	1	1	1	7	2	4	4	4	5	3	4	53	95.8	
	5	2	2	2	3	3	2	2	2	6	7	9	15	12	6	6	7	65	95.1	
	6	1	3	2	5	3	3	2	2	5	18	9	39	12	12	9	9	57	95	
	7	2	4	4	5	5	4	6	5	6	22	32	56	48	27	80	48	57	92.1	
	8	3	5	5	5	3	4	4	3	18	39	39	48	12	22	27	18	47	89.6	
	9	4	4	5	4	5	5	5	4	22	32	39	22	56	39	39	32	41	87.3	
	10	4	4	4	3	3	3	2	2	32	22	22	18	18	12	7	7	31	82.8	
	11	3	4	4	3	3	2	3	2	18	27	22	12	15	7	15	7	19	82	
	12	2	1	2	4	2	3	3	2	7	4	7	22	9	18	18	7	10	78.2	
	13	1	4	3	1	2	2	2	4	4	22	15	5	6	7	7	27	9	76.5	
	14	3	3	1	1	3	1	1	1	12	12	5	4	18	5	3	3	9	74.8	
	15	2	1	1	2	2	2	2	3	7	5	5	6	6	6	6	12	8	72.3	
	16	3	3	3	2	2	3	2	2	12	15	12	7	7	12	7	7	8	71.6	
	17	2	3	3	3	1	2	2	2	7	12	12	12	5	9	6	9	12	71	
	18	2	2	1	1	2	1	1	2	9	6	5	4	7	5	3	7	22	72.5	
	19	3	1	2	0	1	2	2	1	18	4	6	0	4	6	6	5	14	70.8	
	20	2	2	1	1	1	1	2	2	6	7	3	4	4	5	6	7	0	70.3	
	21	2	1	1	1	2	2	2	3	7	5	4	4	7	6	6	12	0	70.4	
	22	3	2	1	1	1	1	1	1	15	6	4	3	3	3	5	5	11	71.2	
	23	1	2	0	1	0	1	1	2	5	7	2	3	2	4	4	6	12	71.8	
	24	3	3	1	1	0	2	1	1	12	15	5	4	2	6	3	5	14	73.3	
	25	2	1	1	0	1	2	5	5	9	4	5	2	5	9	56	56	16	76.6	
	26	4	5	4	4	3	3	2	4	22	39	27	22	15	18	7	6	26	77.4	
	27	4	4	3	3	2	3	2	4	27	22	15	18	7	18	9	27	28	75	
	28	4	4	3	4	1	2	1	2	32	22	15	22	5	7	5	6	21	74.2	
	29	3	2	2	1	1	1	1	2	12	6	9	5	5	4	4	7	17	74.7	
	30	1	1	1	1	1	1	1	2	3	4	5	4	4	5	3	6	17	74.7	
Oct-94	1	2	1	1	1	1	0	1	1	6	5	5	4	3	2	3	3	16	74.7	
	2	0	1	1	1	2	4	4	5	2	4	4	5	6	27	22	56	18	74.6	
	3	6	7	7	6	7	6	6	4	94	132	132	67	111	80	67	22	16	74.3	

Date																		
4	5	4	5	4	4	5	4	4	39	32	39	22	32	39	27	22	27	74.7
5	5	4	4	5	5	5	5	4	39	32	27	56	56	39	39	32	48	79.4
6	4	4	5	5	5	4	4	5	27	32	39	39	39	27	22	48	53	84.2
7	5	5	5	3	5	5	5	5	39	39	39	18	39	48	56	48	46	83.7
8	4	4	4	4	2	3	4	2	22	27	22	27	6	15	22	9	44	86
9	3	3	3	1	2	2	4	2	15	18	15	5	6	9	22	7	50	86.9
10	4	3	3	3	2	3	4	4	22	18	12	12	7	15	27	32	49	86.6
11	3	4	3	4	3	3	4	3	15	27	18	22	18	15	32	12	48	87.2
12	3	4	4	2	1	2	4	4	12	27	22	9	5	6	27	22	45	87.7
13	3	3	3	3	2	3	3	3	12	15	12	18	7	12	12	15	48	92.7
14	3	2	1	2	2	3	2	3	15	9	3	7	6	18	7	15	60	92.1
15	4	4	2	3	2	1	2	2	22	27	9	12	9	5	7	7	57	92.2
16	2	1	2	2	1	1	1	1	7	5	6	7	5	4	4	4	51	90.9
17	1	1	1	1	1	0	1	2	5	4	5	5	4	2	5	7	39	91
18	2	0	0	0	0	1	2	3	7	2	2	2	2	5	7	12	55	89.8
19	1	1	1	0	2	3	3	3	5	3	3	0	7	12	15	18	56	89.9
20	3	1	1	1	2	3	3	2	18	4	4	4	6	12	12	9	49	89.3
21	2	1	1	1	0	0	1	1	6	3	3	4	2	2	3	3	41	87
22	1	1	2	4	4	4	4	6	3	3	7	22	27	22	22	94	29	84.7
23	6	5	6	5	5	4	5	4	80	56	94	48	56	27	39	27	28	83.3
24	4	5	5	4	4	5	4	4	27	48	39	22	22	56	27	22	28	81.3
25	3	5	4	2	1	1	2	1	12	39	32	6	4	3	6	3	27	88.5
26	3	2	2	2	2	1	2	2	18	9	6	7	6	4	9	6	56	91.5
27	1	2	1	2	2	1	1	1	3	6	5	6	6	4	3	3	57	91.9
28	1	1	2	1	1	0	1	2	5	3	6	4	5	2	4	7	57	96.1
29	4	3	4	6	7	7	3	3	27	15	32	80	132	111	18	12	55	96.9
30	3	5	7	6	5	4	6	4	18	56	111	67	48	27	94	32	59	96.4
31	5	4	4	6	5	4	5	4	48	22	32	67	39	27	39	22	51	95.2
Nov-94 1	4	4	3	3	3	4	2	3	27	32	15	12	15	22	9	12	45	90.9
2	4	4	3	3	4	3	2	3	22	22	15	18	27	18	9	18	48	90
3	2	3	3	2	4	3	3	2	7	12	12	9	22	15	18	9	34	85.9
4	3	5	3	3	2	3	4	4	18	56	18	18	9	12	27	22	28	82.6
5	4	3	2	3	3	3	4	5	27	15	6	12	12	12	27	39	31	82
6	5	6	6	5	4	5	4	2	39	67	67	56	27	48	32	6	24	79.7
7	2	3	3	2	1	2	3	1	9	15	12	9	4	6	12	4	17	80.4
8	1	1	1	1	2	1	1	3	5	3	3	5	6	4	3	12	16	78.6
9	2	2	2	2	4	4	4	4	9	7	6	7	22	27	27	32	23	77.4
10	3	5	4	3	2	2	3	2	18	39	27	15	7	9	15	9	24	78.5
11	2	3	2	3	2	1	1	2	7	12	9	12	6	4	5	7	16	77.5
12	2	1	1	1	1	1	2	1	7	5	5	4	3	5	6	4	17	78.5
13	1	2	1	2	2	2	2	1	4	7	5	6	6	9	6	5	8	79.5
14	2	2	2	2	1	2	3	3	9	9	7	6	5	6	15	18	9	77
15	3	2	2	2	2	2	3	3	12	6	7	9	7	7	15	18	8	77.2
16	3	2	1	1	2	1	2	0	12	6	5	3	4	7	4	2	8	77.2
17	1	3	2	1	1	1	2	2	4	15	6	5	3	3	9	6	10	77.3
18	1	1	1	2	2	2	2	2	3	4	3	9	7	9	6	6	19	78.1
19	3	5	4	3	3	4	3	3	18	39	27	15	15	27	15	18	15	76.6
20	3	4	5	4	5	3	2	2	15	27	48	32	39	15	7	9	9	76.9
21	3	3	2	2	1	1	1	1	12	12	7	6	5	4	4	5	7	76
22	1	2	1	2	1	2	2	2	4	9	4	6	4	6	9	6	8	74.5
23	1	1	1	1	1	1	0	2	4	5	3	3	4	4	2	6	8	75.6
24	1	2	2	1	1	2	1	1	3	6	7	3	3	6	4	4	8	76.6
25	1	1	0	0	1	0	1	1	3	5	2	2	3	2	4	4	18	79.4
26	1	3	5	6	6	4	3	2	4	12	39	94	80	32	18	9	17	81
27	3	4	4	5	5	4	4	3	18	27	32	39	39	32	22	12	20	78.3
28	2	2	4	4	3	3	3	1	9	7	22	22	15	15	12	5	23	77.6
29	1	2	2	2	2	3	3	3	4	9	7	7	7	12	12	12	11	77.4
30	3	4	4	5	3	3	3	3	18	22	27	39	18	12	12	18	10	76.1
Dec-94 1	3	4	3	3	4	3	2	4	12	22	18	18	22	18	6	27	11	76.9
2	3	4	4	3	5	4	4	4	18	22	22	15	39	22	22	32	9	76.8
3	5	3	3	2	3	1	2	2	39	12	18	9	18	5	9	9	0	79.5
4	3	2	1	2	1	1	3	1	12	6	5	6	5	3	12	3	10	79.5
5	1	1	2	2	2	2	2	3	4	5	7	7	9	9	6	12	15	78.4
6	3	3	5	3	4	4	4	4	15	15	56	18	22	32	27	27	12	76.2
7	4	3	3	2	2	4	3	4	27	18	18	9	9	22	12	22	20	77.7
8	3	3	3	3	3	3	3	3	18	12	15	15	12	12	15	15	25	78.8
9	3	2	2	2	2	2	2	3	18	6	9	9	9	7	9	15	42	84
10	3	2	2	2	1	3	3	3	18	9	6	6	4	12	12	12	44	83.8
11	2	2	1	1	2	3	4	4	7	6	5	5	6	15	27	22	57	92.3
12	4	3	3	2	2	3	4	4	22	15	12	9	9	12	22	22	51	96.4
13	4	4	4	3	3	3	2	2	22	27	27	12	15	18	6	6	40	93.8
14	2	2	0	2	2	1	3	3	9	6	2	6	6	5	12	15	35	91.7
15	2	3	3	3	5	4	4	4	9	12	18	12	39	27	22	32	32	90.3
16	3	4	4	3	4	3	2	2	18	22	22	18	27	15	6	6	26	90.1

	Day																		
	17	2	2	3	2	1	2	2	2	6	7	12	7	5	7	9	6	35	88.9
	18	2	2	2	1	2	2	2	2	6	6	6	3	6	9	7	7	34	84.6
	19	1	3	2	2	2	1	0	1	5	18	6	7	7	3	2	4	33	82.3
	20	2	2	2	2	2	3	3	3	9	7	9	9	9	12	12	15	21	79.8
	21	2	2	2	1	2	1	1	1	9	6	9	4	6	5	4	5	18	79.6
	22	2	3	0	0	1	0	0	0	7	12	2	2	3	2	2	2	30	79.5
	23	3	2	2	3	1	2	1	3	15	9	9	12	3	9	5	12	21	77.2
	24	4	4	5	4	4	5	5	5	27	27	39	32	32	39	39	39	30	77
	25	4	4	3	3	3	3	2	3	32	22	15	12	18	18	9	18	36	78
	26	4	3	3	3	3	4	3	4	27	15	12	12	15	22	15	22	43	77.2
	27	2	2	2	2	2	2	5	4	9	9	6	6	9	6	39	27	29	73.4
	28	3	1	1	2	3	3	1	3	18	5	3	6	12	15	5	12	17	76.3
	29	4	3	4	3	3	3	1	1	22	15	27	18	12	12	5	4	17	77
	30	2	3	2	2	1	1	1	1	9	18	7	6	5	5	5	3	10	74
	31	1	1	2	1	1	1	1	2	3	5	6	4	4	4	5	7	9	74.8
Jan-95	1	1	0	1	1	0	1	3	1	4	0	3	5	2	4	12	3	10	72.9
	2	1	2	3	3	2	4	5	4	5	9	12	18	9	22	39	27	8	74.8
	3	5	4	4	4	3	5	4	4	56	27	32	22	18	39	22	32	11	74.3
	4	4	4	3	3	4	3	3	3	27	22	18	18	22	12	15	12	14	74.4
	5	3	3	3	4	4	4	4	4	18	15	18	27	27	27	27	22	11	73.8
	6	3	3	3	3	3	3	3	4	12	12	12	18	18	12	15	32	15	72
	7	3	2	3	3	3	2	2	3	12	9	12	15	15	9	9	12	7	72.2
	8	3	2	1	1	2	2	2	1	18	9	5	4	7	7	9	4	9	71.3
	9	2	2	3	1	1	1	1	2	7	7	18	4	3	4	5	7	9	71.4
	10	1	1	1	2	2	2	3	3	3	3	5	6	9	6	12	18	8	70.9
	11	3	3	3	3	4	3	2	2	18	18	12	18	32	15	9	9	9	72.8
	12	3	1	1	1	1	1	2	2	12	3	3	3	3	4	6	9	12	73.7
	13	1	3	3	2	2	1	1	1	4	15	18	9	7	4	5	5	8	72.5
	14	2	1	1	1	0	1	1	2	6	4	5	3	2	3	5	7	11	74.7
	15	2	1	1	1	1	1	1	1	6	3	5	3	3	3	3	3	12	78
	16	2	2	2	4	4	3	4	3	9	7	6	32	27	12	27	18	17	80.1
	17	3	4	4	5	3	4	5	6	15	22	27	39	15	27	56	94	23	80.8
	18	6	6	5	4	3	2	2	3	80	94	56	22	15	7	7	12	35	83.9
	19	2	1	0	2	2	1	1	1	7	4	2	6	7	4	4	3	42	87.3
	20	1	1	0	1	1	3	3	3	3	3	2	3	5	5	18	12	51	89.6
	21	2	2	2	1	1	2	3	3	9	7	6	4	3	7	15	12	55	93.4
	22	2	1	1	1	3	2	2	2	9	5	4	5	15	9	6	9	53	92.5
	23	3	3	1	1	1	0	1	1	12	18	5	4	4	2	5	5	50	93
	24	2	1	1	1	1	0	1	0	6	5	4	5	4	2	3	2	53	94.1
	25	1	2	2	2	1	1	1	1	5	9	6	6	5	5	4	3	49	86.8
	26	1	0	0	1	1	1	0	1	5	2	2	4	4	4	2	4	30	83.1
	27	1	0	1	1	2	1	1	0	3	2	3	5	6	5	3	2	21	85.8
	28	0	1	1	1	1	0	1	1	2	4	4	4	3	2	4	4	25	80.2
	29	2	5	3	4	4	4	4	6	7	39	12	22	27	32	32	94	16	82.1
	30	5	5	4	4	4	5	5	5	56	48	27	27	22	39	48	48	31	83.9
	31	5	4	4	2	4	4	5	3	56	32	27	7	22	22	48	18	32	84.2
Feb-95	1	3	3	3	3	3	3	3	2	15	18	15	15	15	15	18	7	35	84.6
	2	3	4	4	4	4	5	4	4	18	27	27	22	27	39	22	27	41	83.5
	3	4	4	3	3	4	4	4	4	22	27	18	18	27	22	32	27	38	80.4
	4	5	4	3	3	4	4	3	3	56	22	18	12	22	27	15	12	35	83.6
	5	2	1	2	1	1	2	1	1	7	4	6	3	4	7	5	3	35	78.8
	6	3	3	1	2	3	3	1	3	15	12	5	7	12	15	5	12	34	82
	7	4	3	3	1	1	1	1	5	22	15	12	5	4	5	5	39	16	81.5
	8	4	5	5	4	4	3	1	2	27	56	48	32	32	12	5	7	12	83.2
	9	1	2	2	0	0	0	0	0	4	9	6	2	0	2	2	2	17	81.5
	10	0	0	0	0	0	1	1	2	0	0	0	2	2	5	3	6	23	79.3
	11	3	3	5	3	4	4	4	5	12	12	56	18	27	27	22	48	13	79
	12	4	4	3	4	3	5	5	4	32	27	15	22	15	39	56	32	14	79.1
	13	5	4	5	5	6	5	3	4	56	32	39	39	67	39	18	32	14	83.5
	14	5	4	3	3	5	4	4	3	48	27	12	18	48	32	27	12	17	80.4
	15	3	3	3	3	2	4	3	4	18	18	12	15	7	22	18	22	25	82.9
	16	2	2	2	2	1	3	1	2	6	7	7	6	5	18	5	7	26	83.5
	17	1	1	2	2	2	2	1	1	4	4	9	9	7	7	5	5	33	86.9
	18	3	3	2	2	2	2	3	1	15	15	7	9	7	9	15	5	31	86.7
	19	2	3	2	2	1	1	2	2	6	12	7	6	4	5	6	9	46	92.4
	20	2	2	1	1	1	1	1	1	6	9	5	5	4	5	5	3	46	88.8
	21	2	1	1	2	1	0	1	2	6	4	4	7	4	0	4	7	45	87.1
	22	0	1	1	1	0	0	0	0	2	3	4	4	2	0	2	2	47	83.4
	23	1	1	2	2	2	1	1	1	4	4	7	6	6	4	3	4	42	82.9
	24	1	0	1	1	0	0	1	1	4	2	3	3	0	2	5	4	30	81.7
	25	0	0	0	0	1	0	0	0	0	0	0	2	3	2	2	2	26	81.6
	26	2	3	4	2	3	3	3	3	6	15	27	9	12	15	18	15	26	84.5
	27	2	4	3	4	4	4	5	4	9	27	15	27	32	22	48	22	32	85.9
	28	4	4	4	5	3	3	6	5	32	27	32	48	12	15	67	39	38	89.1

Date	Day																			
Mar-95	1	4	5	4	6	3	4	5	4	27	39	32	67	12	32	39	32	51	88.3
	2	4	5	5	4	4	5	3	3	27	39	39	22	22	39	12	12	51	88.7
	3	3	1	1	1	1	1	1	2	18	5	5	4	4	3	5	6	55	89.4
	4	2	1	2	3	3	4	5	6	9	5	7	12	18	22	39	94	65	87.6
	5	5	2	2	3	3	3	4	3	39	9	9	18	15	15	22	15	61	82.9
	6	1	3	1	0	1	1	0	0	5	15	4	2	3	4	0	0	38	82.2
	7	0	1	1	1	1	1	1	1	0	4	3	3	3	5	4	3	36	82.7
	8	1	1	1	0	0	0	1	1	5	5	5	2	2	0	5	3	20	79.6
	9	1	3	3	2	2	2	5	4	3	12	18	6	7	9	48	32	10	75.8
	10	5	4	2	2	3	3	3	3	56	22	9	7	12	12	15	18	0	78.1
	11	5	3	4	1	3	4	5	6	39	18	22	4	15	32	39	67	8	75.1
	12	6	6	4	3	3	4	5	5	80	80	27	18	15	27	39	56	12	75.3
	13	4	4	4	5	5	6	5	4	32	22	27	39	56	80	39	22	10	76.6
	14	4	4	3	3	4	4	4	3	27	27	18	18	22	22	22	15	10	78.1
	15	2	3	3	3	2	1	3	3	9	15	18	15	9	4	15	18	17	80.2
	16	3	3	3	2	3	3	3	2	18	18	15	9	15	12	12	9	13	83.3
	17	2	3	2	2	2	3	2	2	6	12	7	6	7	12	9	7	16	82.5
	18	2	1	2	1	1	0	1	1	7	5	6	4	4	2	5	3	35	90.9
	19	2	1	1	0	1	1	1	1	6	4	4	2	3	4	5	5	38	83.4
	20	1	0	1	1	2	2	1	0	4	0	3	5	7	7	5	2	41	88.3
	21	0	1	1	0	0	0	0	0	0	3	4	2	2	2	2	2	42	89.3
	22	0	1	1	1	1	1	0	1	2	4	4	3	3	3	2	3	45	92.9
	23	1	0	0	3	2	2	2	1	3	0	0	12	9	9	9	5	47	93.6
	24	2	1	2	1	1	2	1	1	9	4	7	3	5	6	4	4	44	94.4
	25	2	2	1	2	1	1	1	2	6	9	4	6	4	3	5	6	42	91.8
	26	2	1	3	4	6	5	4	4	7	5	18	32	67	56	32.	27	41	89.9
	27	3	2	4	3	4	3	4	3	18	7	27	18	27	18	22	18	28	88.6
	28	4	3	3	2	4	4	3	2	22	18	18	7	22	32	18	6	19	83.4
	29	3	5	4	3	3	3	2	3	15	39	27	18	18	12	9	12	23	80.7
	30	3	3	1	2	3	2	1	1	15	12	5	6	15	6	4	3	23	79.5
	31	1	1	2	3	2	2	2	2	5	5	7	12	6	7	7	7	22	76.7
Apr-95	1	1	0	0	0	1	4	5	4	3	2	2	2	5	22	39	27	16	75.3
	2	4	6	2	4	4	2	1	1	32	67	9	27	27	6	5	4	17	76.2
	3	0	0	0	0	0	0	0	0	2	2	2	2	2	0	2	2	19	75.2
	4	1	1	1	0	0	1	0	1	5	3	4	2	2	4	2	5	10	73.1
	5	3	2	1	3	2	2	3	2	12	7	5	18	7	6	15	9	0	72.2
	6	3	1	0	0	1	0	2	3	12	5	2	2	3	2	7	12	0	71.9
	7	4	6	6	5	7	8	7	6	27	67	80	56	132	207	154	80	0	70.7
	8	4	5	5	5	4	5	3	3	27	39	39	48	27	56	18	18	0	72
	9	5	4	4	4	4	4	3	3	39	32	32	27	22	27	12	12	8	72.9
	10	3	5	5	4	5	4	3	3	15	39	27	39	27	27	15	15	7	74.8
	11	2	4	4	3	3	3	5	4	9	32	22	15	12	15	48	22	10	78
	12	4	4	3	2	2	3	3	2	32	27	12	9	9	15	15	7	18	82.3
	13	3	3	2	1	1	2	1	2	12	18	7	4	3	6	5	9	31	83.3
	14	3	1	1	1	1	1	1	2	12	5	3	3	4	4	5	6	38	88.9
	15	1	1	0	0	1	1	1	0	4	4	2	2	5	5	4	2	42	92
	16	1	0	1	1	0	0	2	2	3	2	5	4	2	2	6	6	46	89.7
	17	3	2	1	1	0	1	1	1	15	6	4	4	2	4	3	4	48	89.6
	18	2	1	1	2	2	2	2	1	9	5	5	9	6	7	7	5	37	90.7
	19	2	2	2	1	1	1	2	2	7	9	9	4	3	4	6	7	34	92.6
	20	2	3	2	2	2	2	2	1	6	15	9	6	7	9	6	4	25	87.2
	21	0	0	0	0	0	1	1	1	2	2	2	0	2	3	3	3	13	85.5
	22	1	0	0	2	4	4	4	3	4	2	2	9	22	27	32	15	11	84.8
	23	4	3	3	4	3	4	3	5	22	15	18	22	18	27	15	39	0	77.6
	24	4	5	4	2	3	4	4	3	27	48	32	9	12	27	32	18	0	73.4
	25	5	5	3	2	3	3	3	2	39	39	12	7	12	12	12	6	0	71.3
	26	3	3	3	3	3	4	3	4	15	18	12	12	18	22	15	32	0	70.2
	27	4	3	6	4	4	3	3	1	27	18	94	32	32	15	15	5	8	69.1
	28	2	3	3	4	3	2	2	1	7	12	15	22	18	9	6	5	0	68.6
	29	2	2	2	1	3	2	2	2	9	7	6	5	12	6	7	7	0	69
	30	1	1	1	1	1	0	1	2	5	3	3	3	3	2	5	7	0	68.2
May-95	1	0	1	0	0	1	.1	1	2	0	3	2	2	3	3	3	6	9	69.7
	2	3	5	4	3	5	6	5	6	12	48	27	15	48	94	56	94	0	69.9
	3	6	5	5	6	5	4	4	4	80	56	48	67	39	32	32	32	8	70.2
	4	5	6	5	3	4	3	4	4	48	67	48	15	27	18	22	32	8	71.8
	5	4	5	5	5	6	5	4	4	32	56	39	67	48	39	32	32	10	74.2
	6	5	3	5	5	3	2	3	3	56	18	48	39	15	9	15	12	9	77.8
	7	2	3	4	4	4	3	4	5	9	18	27	27	27	15	27	39	8	79
	8	4	4	4	3	4	3	3	3	32	32	9	12	27	12	15	18	9	79.4
	9	2	2	2	2	2	1	3	3	9	9	9	9	6	5	18	12	10	79.2
	10	2	2	1	1	2	1	2	1	7	7	3	5	6	5	6	5	11	79.3
	11	2	2	2	2	1	2	1	1	9	6	6	5	6	5	6	5	12	78.4
	12	2	2	2	1	2	2	2	1	9	6	6	5	9	6	7	4	26	82.4
	13	1	2	2	2	3	3	1	2	5	9	7	9	18	12	5	7	26	82.2

14	2	2	2	1	1	1	1	1	6	9	9	5	4	5	5	3	27	81.6
15	0	1	1	1	2	1	1	2	2	3	3	4	7	4	5	9	36	87.8
16	3	4	4	5	6	5	5	5	12	27	32	39	80	56	56	56	40	95.9
17	5	6	5	4	3	3	4	3	39	67	39	27	12	18	22	15	43	97.3
18	2	1	2	2	3	2	3	4	9	5	6	7	15	6	12	22	40	93.4
19	2	3	4	2	2	2	3	2	9	12	22	9	7	7	15	9	35	88.3
20	3	3	3	3	2	2	4	3	15	12	12	15	7	9	22	15	25	82.8
21	3	2	1	1	1	1	1	1	12	6	3	3	4	3	4	3	13	77
22	2	2	1	1	2	1	1	2	6	9	5	5	6	4	3	9	10	72.6
23	2	1	1	1	2	3	5	6	7	4	4	5	9	18	48	67	0	70.4
24	3	5	5	4	3	4	3	3	18	56	39	32	15	22	12	12	0	68.8
25	3	4	2	1	3	2	3	2	15	22	7	5	12	9	15	6	0	68.9
26	2	2	2	3	4	3	4	3	7	6	6	12	22	12	22	15	0	68.1
27	1	2	1	2	1	2	2	0	4	6	5	6	5	7	9	2	0	67.5
28	2	2	2	1	2	2	1	2	6	7	6	5	9	6	5	7	9	68.4
29	1	2	1	1	3	2	2	2	4	6	4	5	12	7	9	6	12	68.7
30	2	4	5	6	6	4	5	4	9	32	39	67	67	27	48	22	10	69.4
31	4	3	4	4	4	5	5	3	32	18	22	27	32	39	48	18	9	70.9

Jun-95

1	4	4	4	3	4	5	4	3	32	27	22	18	27	39	22	15	8	72.5
2	3	4	3	3	3	4	3	3	18	27	15	15	18	22	12	15	9	75.4
3	4	5	4	4	3	3	4	3	22	39	32	22	15	15	22	15	12	76.8
4	2	1	2	2	2	2	2	2	7	4	6	6	6	9	7	6	14	81.4
5	2	2	1	1	1	2	2	3	7	6	5	3	4	7	6	12	30	84.2
6	3	1	1	2	2	1	2	2	12	5	5	6	9	5	7	6	24	81.8
7	3	1	1	1	1	1	2	2	12	3	5	5	4	5	6	9	28	85
8	1	2	1	1	0	1	1	0	5	6	3	3	2	3	3	2	28	86.8
9	1	1	1	1	1	2	2	2	3	3	4	5	5	6	6	6	24	89
10	3	3	2	2	1	2	1	1	12	18	9	9	5	6	3	4	30	87
11	2	2	2	1	1	1	1	1	9	9	6	3	3	3	3	3	24	85.1
12	1	1	1	1	1	1	1	0	3	3	4	4	3	4	3	2	20	83.3
13	0	0	1	1	0	0	1	1	2	2	3	3	2	2	3	3	17	79.2
14	0	0	1	1	2	2	3	2	2	0	3	5	6	6	12	9	12	78.1
15	1	2	2	1	2	1	2	1	5	6	6	5	6	5	6	5	0	75.6
16	4	3	1	2	3	2	4	2	22	12	4	9	15	7	27	7	0	73.1
17	2	1	1	2	1	1	1	1	7	5	5	6	4	4	4	5	7	72.1
18	1	3	5	3	3	2	1	3	4	12	39	18	18	6	4	12	9	71.8
19	3	4	4	4	6	3	4	3	15	27	27	32	67	18	32	15	18	73.7
20	2	5	4	3	3	4	3	2	9	48	32	12	15	27	12	9	13	74.2
21	3	3	2	2	2	2	2	2	12	18	9	9	9	9	7	7	15	74.7
22	2	2	1	2	3	2	1	2	9	6	5	6	12	7	5	6	17	75.5
23	2	2	3	2	2	2	1	1	6	7	12	9	6	9	4	4	16	74.3
24	1	1	2	1	1	1	1	1	4	5	6	4	4	4	5	4	13	73.4
25	0	1	1	2	4	4	5	3	0	4	5	6	27	32	39	18	13	73.2
26	4	4	3	3	2	2	3	3	32	27	18	12	7	9	15	15	12	73.2
27	2	2	2	1	1	1	1	2	7	6	9	5	3	4	5	7	10	74.7
28	2	1	2	2	3	3	3	4	6	5	6	6	12	12	12	22	9	77
29	1	1	1	1	2	3	2	1	5	4	5	4	7	15	6	5	14	80.1
30	3	2	2	4	4	4	5	5	15	9	7	22	27	22	48	56	27	80.8

This listing gives valuable information on manufactures and/or importers of amateur radio transmitting equipment and supplies.

Beyond being a handy source of phone numbers and addresses, it also gives you insight to a company's business longevity and size, the latter by number of employees. Moreover, you'll learn which companies sell their own products directly to end users, usually through mail order, rather than only through dealers for resale.

The listing that follows presents retail operations that sell transmitting products made or imported by companies listed here.

While every effort has been made to create accurate and useful listings, we recognize errors and omissions may occur. Should any be observed, please inform us so they may be corrected in the next edition of the Almanac.

A

Advanced Electronic Applications, Inc. (AEA)
2006 196th St. SW
PO Box C-2160
Lynnwood, WA 98036
Phone: 206 774-5554
FAX: 206 775-2340
Tech. Support: 206 775-7373
Literature: 800 432-8873
Established: 1977
Sells direct and through dealers.
Major Lines: Packet controllers, multimode data controllers, terminal software, HF/VHF/UHF antennas, UHF/ VHF handheld antennas, high speed radio modems, remote radio controllers, handheld antenna analysts, weather FAX demodulator, CW keyers and trainers.

Advanced Specialties Inc.
P.O. Box 1099R
Lodi, NJ 07644
Phone: 201 VHF-2067
Sells direct and through dealers.
Major Lines: U.S. Importer and distributor of RMS Communications accessories, SWR/power meters, VHF amplifiers, switches, etc.

Alinco Electronics, Inc.
438 Amapola Ave., Suite 130
Torrance, CA 90501
Phone: 310 618-8616
FAX: 310 618-8758
Established: 1977; Employs 203
Sells through dealers.
Major Lines: HF, VHF & UHF mobile and HT transceivers.

Ameritron
116 Willow Road
Starkville, MS 39759
Phone: 800 647-1800, 601 323-8211
FAX: 601 323-6551
Sells through dealers.
Major Lines: HF amplifiers, antenna tuners, remote antenna switches, dummy loads, in-rush current protectors, QSK switch.

Atlas Radio Company
722-G Genevieve Street
Solana Beach,CA 92075
Phone: 619 259-7321
FAX: 619 259-7392
Established: 1992
Sells factory direct.
Major Lines: Transceivers, Desk Top Power Consoles, Mobile Mounts.

Azden Corporation
147 New Hyde Park Road
Franklin Square, NY 11010
Phone: 516 328-7501
FAX: 516 328-7506
Sells direct and through dealers.
Major Lines: VHF/UHF mobile & handheld radios, communications headphones with/without boom mic.

C

Communications Concepts
508 Millstone Drive
Beavercreek, OH 45434
Phone: 513 426-8600
FAX: 513 429-3811
Established: 1978
Sells direct and through dealers.
Major Lines: HF and 2-meter amplifiers, ATV Downconverters & hard to find components, metal clad, chip and arco caps, low-pass filters, transistors(ferrite).

D

Down East Microwave
954 Rt. 519
Frenchtown, NJ 08825
Phone: 908 996-3584
FAX: 908 996-3702
Major Lines: VHF/UHF/Microwave transverters, Downconverters, LNA's, Power Amps and RF components. Distributor of Directive Systems VHF/UHF/Microwave antennas and accessories.

E

Ehrhorn Technological Operations, Inc. (ETO)
4975 N. 30th St.
Colorado Springs, CO 80919-4101
Phone: 719 260-1191
FAX: 719 260-0395
Established: 1970
Sells direct in U.S.
Major Line: RF power amplifiers.

Electro Automatic Corporation
599 Canal St.
Lawrence, MA 01840
Phone: 508 687-6411
FAX: 508 687-6493
Established: 1994; Employs 8
Sells direct and through dealers.
Major Lines: 2 Meter and 70cm Handhelds, Power Supplies, Inverters, Tower Installation, Accessories for HF, VHF, UHF Transceivers.

F

Fair Radio Sales Co., Inc.
PO Box 1105
Lima, OH 45802
Phone: 419 227-6573
FAX: 419 227-1313
Established: 1947; Employs 12
Sells direct.
Major Lines: Government surplus radio electronics.

G

G.L.B. Electronics, Inc.
151 North America Drive
Buffalo, NY 14224
Phone: 716 675-6740
Established: 1972; Employs 10
Sells direct.
Major Lines: NETLINK 220 high speed data transceiver, preselector preamps, automatic CW identifiers, transmitter & receiver strips.

GLS Co.
2506 Caddy Lane
Joliet, IL 60435
Phone: 800 240-3307
Major Lines: VHF/UHF Base/Mobile Transceivers, Astron power supplies/ all models.

H

Hamtronics, Inc.
65-C Moul Road
Hilton, NY 14468-9535
Phone: 716 392-9430
FAX: 716 392-9420
Established: 1961
Sells direct.
Major Lines: VHF/UHF FM exciters, receivers, weather fax receivers, digital voice recorders, repeaters, PA's, preamps, RCVG & XMTG converters, autopatches & other repeater accessories, computer data links.

J

Japan Radio Company, Ltd. (JRC)
430 Park Avenue, 2nd Floor
New York, NY 10022
Phone: 212 355-1180
FAX: 212 319-5227
Established: 1915
Sells through dealers.
Major Lines: HF transceivers, general-coverage receivers, linear amplifiers.

K

Kenwood Communications Corporation
2201 E. Dominguez Street
PO Box 22745
Long Beach, CA 90801-5745
Phone: 310 639-4200
Tech: 310 639-5300
BBS: 310 761-8284 (8Nl up to 14.4K baud)
Established: 1975
Sells through dealers.
Major Lines: HF/VHF/UHF base, mobile, portable transceivers and receivers, power supplies, automatic antenna tuners, external speakers for base and mobile use, SWR and RF power meters, HF mobile antenna, dual band (2m/70cm) mobile antenna, headphones, microphones, and accessories.

M

MFJ Enterprises, Inc.
PO Box 494
Mississippi State, MS 39762
Phone: 800 647-1800; 601 323-5869
Tech: 800 647-8324; FAX: 601 323-6551
Established: 1972; Employs 160
Sells direct and through dealers.
Major Lines: Antenna tuners, keyers, wattmeters, packet controllers, dummy loads, antenna bridge, noise bridge, antenna current probe, clocks, coaxial switches, filters, speaker mics, mobile speaker, telescoping antennas, VHF and UHF antennas, interfaces, code oscillators, books, licensing, code and theory programs.

Maggiore Electronic Lab
600 Westtown Road
West Chester, PA 19382
Phone: 610 436-6051
FAX: 610 436-6268
Established: 1968; Employs 8
Sells direct and through dealers.
Major Lines: VHF & UHF Repeaters, Transmitters, Receivers, Power Amplifiers, C.O.R.'s. Microprocessor Repeater Controllers and Auto Patches. Miniature, Marine Packages Including Duplexers, Antennas and High Power Amplifiers.

Micro Control Specialties
23 Elm Park
Groveland, MA 01834
Phone: 508 372-3442
Established: 1978; Employs 5
Sells direct.
Major Lines: Receivers, transmitters, controllers, complete repeaters, voter systems.

Midland Consumer Radio Co., Inc.
1670 N. Topping
Kansas City, MO 64120-1224
P.O. Box 33865
Kansas City, MO 64120-3865
Phone: 800 669-4567 x1165
Tech: 800-669-4567 x 57
Established: 1995; Employs 100
Sells through dealers.
Major Lines: VHF, UHF transceivers,and antennas.

Mirage Communications Equipment
PO Box 494
Mississipi State, MS 39762
Phone: 800 647-1800
Tech: 408-779-7363
FAX: 601 323-6551
Sells through dealers.
Major Lines: VHF/UHF Solid State Amplifiers, commercial power amplifiers, wattmeters, and SWR meters.

P

PATCOMM CORP.
Mills Pond Rd., Bldg #7
St. James, NY 11780
Phone: 516 862-6511
FAX: 516 862-6529
Established: 1992; Employs 10
Sells direct.
Major Line: HF transceivers.

P.C. Electronics
2522 Paxson Lane
Arcadia, CA 91007
Phone: 818 447-4565
FAX: 818 447-0489
Established: 1965; Employs 10
Sells direct and through dealers.
Major Lines: Fast scan amateur television equipment, mini video cameras, transmitters, transceivers, down converters, antennas, linear amps,accessories.

Pauldon
210 Utica Street
Tonawanda, NY 14150
Phone/FAX: 716 692-5451
Major Line: VHF/UHF amplifiers.

Pro•Am (Div. of Valor Enterprises Inc.)
1711 N. Commerce Drive
Piqua, OH 45356
Phone: 513 778-0074
FAX: 513 778-0259
Established: 1974; Employs 180
Sells through dealers.
Major Lines: Two meter base/mobile, multi-band base/mobile, 450 portables, HF yagis, scanner mobile/base, two-meter portable, HF mobile antenna, cellular, commercial, marine antennas and accessories, RV antennas, UHF/VHF amplifiers.
Branch Office:
185 W. Hamilton Street
West Milton, OH 45383
Phone: 513 698-4194

R

RF Concepts, Div. of Kantronics
1202 E. 23rd Street
Lawrence, KS 66046
Phone: 913 842-7745
FAX: 913 842-2031
Service: 913 842-4476
Established: 1987
Sells through dealers.
Major Lines: VHF all mode RF amplifiers, UHF all mode RF amplifiers, dual-band amplifiers.

RF Limited/Clear Channel Corporation
PO Box 1124
Issaquah, WA 98027
Phone: 206 222-4295
FAX: 206 222-4294
Established: 1977
Sells through dealers.
Major Lines: Amplifiers, microphones, antennas, handheld transceiver accessories.

RF Technologies, Inc.
6055 Fairmount Ave.
San Diego,CA 92120
Phone: 619 282-4947
FAX: 619 283-3402
Established: 1990; Employs 15
Sells direct and through dealers.
Major Line: GaAsFET Pre-Amp, VHF amplifiers and repeater amps.

Radio Adventures Corp.
Main Street
Seneca, PA 16346
Phone: 814 677-7221
FAX: 814 437-5432
Established: 1993; Employs 10
Sells direct.
Major Lines: HF Receiver Kits, HF Transmitter Kits, HF Transceiver Kits, Accessories, Test Equipment Kits, Classic Radio Accessories.

Radio Shack
1500 One Tandy Center
Ft. Worth, TX 76102
Phone: 1 800 843-7422
Established: 1921; Employs 26,000
More than 6,600 stores in the U.S.
Sells direct and through dealers.
Major Lines: Computers, scanners, antennas, transceivers, coax, plugs, jacks, parts and supplies.

S

SGC, Inc.
The SGC Building
13737 S.E. 26th St.
Bellevue, WA 98005
PO Box 3526
Bellevue, WA 98009
Phone: 206 746-6310
FAX: 206 746-6384
Established: 1972; Employs 100
Sells direct and through dealers.
Major Lines: Manufactures HF SSB radios, antenna couplers, power supplies, and antennas.

S&S Engineering
14102 Brown Road
Smithsburg, MD 21783
Phone: 301 416-0661
FAX: 301 416-0963
Established: 1989
Sells direct.
Major Lines: HF Transceiver Kits, QRP, CW; Accessories (Speed Keyers, Frequency Counters).

Spectrum Communications Corp.
1055 W. Germantown Pike
Norristown, PA 19403
Phone: 610 631-1710
FAX: 610 631-5017
Established: 1974
Sells direct.
Major Lines: VHF/UHF repeaters and amplifiers, repeater control boards, RF link and repeater receivers/transmitters and accessories. Receiver preselectors. 900 MHz TXs.

Standard Amateur Radio Products, Inc.
PO Box 48480
Niles, IL 60714
Phone: 312 763-0081
Established: 1990
Sells through dealers.
Major Lines: VHF/UHF base/mobile & handheld transceivers; transceiver accessories.

T

TE Systems
PO Box 25845
Los Angeles, CA 90025
Phone: 310 478-0591
FAX: 310 473-4038
Established: 1982; Employs 28
Sells direct and through dealers.
Major Lines: RF power amplifiers (VHF, UHF, Microwave) low noise preamplifiers.

Ten-Tec, Inc.
1185 Dolly Parton Parkway
Sevierville, TN 37862
Phone: 423 453-7172
Tech: 423 428-0364
FAX: 423 428-4483
Established: 1968; Employs 100
Sells direct.
Major Lines: HF transceivers, mobile and base. HF amplifiers to 1500 watts. Manual and automatic antenna tuners. Line of kits including VHF transceiver. Metal and plastic equipment enclosures. Custom tools and dies.

V

Vectronics Corporation Inc.
Division of Valor Enterprises, Inc.
1711 Commerce Drive
Piqua, OH 45356
Phone: 513 778-0074
FAX: 513 778-0259
Established: 1981; Employs 10
Sells through dealers.
Major Lines: HF amplifiers, HF antenna tuners, power supplies, HP/LP filters, dummy loads amplifier components.

Y

Yaesu
17210 Edwards Road
Cerritos, CA 90703
Phone: 310 404-2700
FAX: 310 404-1210
Established: 1956
Sells through dealers.
Major Lines: HF, VHF, UHF all-mode transceiver, VHF and UHF FM mobiles and handhelds, HF amplifiers, antenna tuners, portable HF transceiver, HF receivers, headphones, microphones, power/SWR meters, antenna rotors, power supplies, speakers, and other related accessories.

The following listing consists of retailers who sell other makers' brands of radio products either through a retail store(s) or by mail order, or both. (Manufacturers or importers who sell only their own brands to end users are listed in the previous section).

Top lines they carry are noted, as well as information concerning equipment trade-ins and on-site repairs. Additionally, the year the company was established, the number of employees, and branch store information, if any, are also indicated. Furthermore, key employee names and call signs are shown, enabling you to personalize your buying contacts.

Note that toll-free "800" telephone numbers should only be used to place a purchase order, not to gather general information, for repairs, etc. This reduces the time that other callers have to wait (it might be you!) for someone to take their order.

8Radio Products
6198 Marlo Drive
Concord, OH 44077
Phone: 216 946-6889

A

AA6EE—Callbook Distributor
16832 Whirlwind #B
Ramona, CA 92065
Phone: 619 789-3674
Established: 1982
Sells via mail order.
Key Employee: Duane Heise, AA6EE
Top Line: Callbooks, ARRL publications, HamCalls on CDROM, CQ books.

A-B-C Communications
17550 15th Avenue NE
Seattle, WA 98155
Phone: 206 364-8300
Sells via showroom and mail order.
Top Lines: ICOM, books, and all major antenna lines.

A.M.C. Sales, Inc.
193 Vaquero Drive
Boulder, CO 80303
Phone: 303 499-5405; 800 926-2488
Established: 1970; Employs 3
Sells via mail order.
Key Employee: Jim Walshe, Pres.
Top Lines: Long play recorders, VOX switches, telephone recording adapters, 3-hour micro recorders, telephone voice changers, microphones, RF Bug Detectors.

A.S.A./Antenna Sales & Accessories
PO Box 3461
4551 Highway 17 Bypass South
Myrtle Beach, SC 29578
Phone: 803 293-7888; 800-722-2681
FAX: 803 293-7888
Established: 1991; Employs 3
Sells via dealer, mail order, hamfests.
Key Employees: Jim Wood
Top Lines: A.S.A., Pro-Am, Midland, Maldol, Anli, Fiberwhips, Workman, Maxell Alkaline batteries.

AXM Enterprises
11791 Loara St., Suite B
Garden Grove, CA 92640-2321
Phone toll-free: 1 800 755-7169
Local phone: 714 638-8807
FAX: 714 638-9556
Established: 1976; Employs 3
Sells via mail order and direct.
Key Employees: Susan, N6ORA; Gar, W6AXM.
Top Lines: Patriot, Ranger and Uniden type-accepted wide-band mobile and handheld transceivers; Centurion batteries and antennas, Larsen and Maxrad VHF/UHF antennas; Mobile-Mark single and multiband HF and VHF mobile antennas; CES phone patches.

Ack Radio Supply Company
3101 4th Avenue
South Birmingham, AL 35233
Phone: 205 322-0588; 800-338-4218
FAX: 205-322-0580
Established: 1947; Employs 5
Sells via showroom and mail order.
Key Employees: Larry N4HYX; Mike, KC4OIT; Bill; Kirk, N4WYC.
Top Lines: ICOM, MFJ, ARRL, Vibroplex, Bencher, Hustler, DAIWA.
Accepts Trade-Ins.
Branch: 554 Deering Road
Atlanta, GA 30367
Phone: 404 351 -6340
FAX: 404 351-1879
Key Employees: Tommy W4RRW; Jim WA4APG.

Advanced Specialties
114 Essex Street
Lodi, NJ 07644
Phone: 201 VHF-2067
Sells via showroom and mail order.
Top Lines: Alinco, A.O.R., ANLI, Larsen, A.D.I., Daiwa, Valor Pro-am, Nevada, TS Antennas, CTE, K-COM, Ranger, Para-Dynamics, Midland, Anttron, RMS, Ramsey Kits, J-com, Ameco, CQ, NARA, Uniden-Bearcat, Vectronics, Van Gorden—Amateur radio equipment, power supplies, antennas, GMRS, Scanners, CB & Marine radio equipment, books, cable, microphones, filters, meters.

Affordable Electronic Repair
7110 E. Thomas Road
Scottsdale, AZ 85251
Phone: 602 945-3908
Top Lines: Service and repair of HF Transceivers, VHF/UHF Multimode & Base/Mobile Transceivers, Handhelds.

Alabama Radio
8907 Alabama Hwy #9
Anniston, AL 36207
Phone: 205 235-1917
Key Employees: Charlie, N6IFL; Carol, N4YGI

Alfa Electronics, Inc.
741 Alexander Road
Princeton, NJ 08540
Phone: 609 520-2002; 800 526-2532
Established: 1988; Employs 7
Sells via mail order.
Key Employee: Jeff Kao
Top Lines: Digital meters, oscilloscope, signal generator, frequencycounter, power supply, dip meter, SWR meter.
Provides repairs.

All Electronics
PO Box 567
Van Nuys, CA 91408
Phone: 800 826-5432
Tech: 818 904-0524
FAX: 818 781-2653
Established: 1967; Employs 35
Sells via showroom and mail order.
Top Lines: Full line of parts—rechargeable batteries, capacitors, semi-conductors, fuses, switches, fasteners, lamps, optoelectronics, etc.
Free Catalog sent 6 times per year.
Branches: 905 S. Vermont Avenue
Los Angeles, CA 90006
Phone: 213 380-8000
14928 Oxnard Street
Van Nuys, CA 91411
Phone: 818 997-1806

AlphaLab Inc.
1272 Alameda Avenue
Salt Lake City, UT 84102
Phone: 801 359-0204; 801 487-3866
Established: 1989; Employs 11
Sells via mail order.
Key Employee: William, John, Veronica
Top Line: Magnetic/Electric/Radio field meters.
Provides repairs.

Amateur & Advanced Communications
3208 Concord Pike, Rt 202
Wilmington, DE 19803
Phone: 302 478-2757
Established: 1977; Employs 2
Sells via showroom and mail order.
Key Employees: Gisele, K3WAJ
Top Lines: ARRL, Alinco, Ameritron, Cushcraft, MFJ, Ten-Tec, Alpha Delta, Hustler, Butternut, Larsen, Rohn, ProAm, Bencher, Henry, Kantronics, Astron, Shortwave Equipment, Ham Classes.
Provides repairs.

Amateur Electronic Supply, Inc.
5710 W. Good Hope Road
Milwaukee, WI 53223
Phone: 800 558-0411, 414 358-0333 BBS: 414 358-3472 8-n-1, up to 14.4 kb
Established: 1957; Employs 39
Sells via showroom and mail order.
Key Employee: Dale, N9BRX.
Top Lines: Azden, Kenwood, Yaesu, ICOM, Cushcraft, Hy-Gain, AEA, Kantronics, Nye-Viking, MFJ, Ten-Tec, Hustler, Mirage/KLM, Standard, Alinco. Large used gear inventory.
Accepts trade-ins, provides repairs and warranty service.

Amateur Electronic Supply, Inc.
28940 Euclid Avenue
Wickliffe, OH 44092
Phone: 800 321-3594; 216 585-7388
Employs 8
Sells via showroom and mail order.
Key Employees: Dave, WB8BAG.
Top Lines: Kenwood, Yaesu, Azden, ICOM, Cushcraft, Hy-Gain, AEA, Kantronics, Nye-Viking, MFJ, Ten-Tec, Hustler. Large used gear inventory.
Accepts trade-ins, provides repair and warranty service.

Amateur Electronic Supply, Inc.
621 Commonwealth Avenue
Orlando, FL 32803
Phone: 800 327-1917; 407 894-3238
Employs 11
Sells via showroom and mail order.
Key employees: Grant, K41QW.
Top Lines: Kenwood, Yaesu, Azden, ICOM, Cushcraft, Hy-Gain, AEA, Kantronics, Nye-Viking, MFJ, Ten-Tec, Hustler. Large used gear inventory.
Accepts trade-ins, provides repairs and warranty service.

Amateur Electronic Supply, Inc.
1898 Drew Street
Clearwater, FL 34625
Phone: 813 461-4267
FAX: 813 443-7893
Employs 4
Sells via showroom and mail order.
Key Employee: Len, K4BDP.
Top Lines: Azden, Kenwood, Yaesu, ICOM, Cushcraft, Hy-Gain, AEA, Kantronics, Nye-Viking, MFJ, Mirage/KLM. Large used gear inventory.
Accepts trade-ins and provides repairs.

Amateur Electronic Supply, Inc.
1072 N. Rancho Drive
Las Vegas, NV 89106
Phone: 800 634-6227; 702 647-311 4
FAX: 702 647-3412
Employs 8
Sells via showroom and mail order.
Key Employees: Squeak, AD7K.
Top Lines: Azden, Kenwood, Yaesu, ICOM, Cushcraft, Hy-Gain, AEA, Kantronics, Nye-Viking, Mirage/KLM, Standard, MFJ, Ten-Tec. Large used gear inventory.
Accepts trade-ins, provides repairs and warranty service.

Amateur Radio Supply, Co.
5963 Corson Avenue S. Suite #140
Seattle, WA 98108-2707
Phone: 206 767-3222
FAX: 206 763-8176
Established: 1956; Employs 6
Sells via showroom and mail order.
Key Employees: Marlo, N7TQU; Floyd, KB7PLW; Eric, N7DLV; Casey, W6PKW, George, K7HZ.
Top Lines: Kenwood, ICOM, Rohn, AEA, Cushcraft, Bird Wattmeters, Kantronics, Yaesu, New-Tronics, NYE-Viking, Alinco, Comet, Valor, Larsen, Maldol, Standard, Timewave Technology, Diamond.

American Electronics
164 Southpark Blvd.
PO Box 301
Greenwood, IN 46142
Phone: 317 888-7265; 800 872-1373
FAX: 317 888-7368
Established: 1965; Employs 8
Sells via showroom and phone/mail order to whole sale dealers.
Key Employees: Win, K9EMV; Mike, N9LPJ; Jon, N9KZB; John, N9KZG.
Top Lines: Kenwood, Dick Smith, Cushcraft, DAIWA, Hustler, Telex/Hy-Gain, ARRL, Uniden, AEI, Nye-Viking, Antenna Specialists, Valor, Maxon, Radio Amateur Callbook, Sangean,B & W, Trippe, W5YI, Heights, Wintenna, Tekk, Relm.
Provides repairs and custom installations.

Antennas Etc.
PO Box 4215
Andover, MA 01810-4215
Phone: 508 475-7831
FAX: 508 474-8949
Established: 1986; Employs 6
Sells direct to Int'l & OEM dist.
Top Lines: Unadilla antennas, James Millen components, RF groundingaccessories.

Antique Electronic Supply
6221 South Maple
Tempe, AZ 85283
Phone: 602 820-5411
FAX: 602 820-4643
Employs: 26
Sells via mail order.
Key Employee: Noreen Cravener.
Top Lines: Vacuum tubes and parts for tube equipment.

Antique Radio Classified
P.O. Box 2-C15A
Carlisle, MA 01741
Phone: 508 371-0512
FAX: 508 371-7129

Arnold Company
PO Box 512
Commerce, TX 75429
Phone: 903 395-2922
FAX: 903 395-2340.
Established: 1981; Employs 5.
Sells via showroom and mail order.
Key Employees: Roger; Brenda.
Top Lines: Commander, Citizen, Amphenol, J&I, Pyramid, pre-owned computers.
Accepts trade-ins and provides repairs.

Associated Radio Comm.
PO Box 4327 8012 Conser
Overland Park, KS 66204
Phone: 913 381-5900, 800 497-1457
FAX: 913 648-3020.
Established: 1945.
Sells via showroom and mail order.
Top Lines: Kenwood, ICOM, Yaesu, Comet, Alinco, Kantronics, R.F. Concepts, Cushcraft, Maxrad, Hustler, B&W, NYE-Viking, Panasonic, Sangean, Uniden-Bearcat, Drake, Collins, AEA, Daiwa, Vectronics, MFJ, W5YI, Valor, Midland, Butternut, Astron, MFJ.
Accepts trade-ins and provides repairs.

Austin Amateur Radio Supply
5325 North IH-35
Austin, TX 78723
Phone: 512 454-2994; 800 423-2604. FAX: 512 454-3069
Established: 1986.
Sells via showroom and mail order.
Top Lines: Kenwood, ICOM, Yaesu, Alinco, AEA, MFJ, Astron, Cushcraft, Hustler, Butternut, Larsen, Telex-Hy-Gain, Periphex, ARRL, Kantronics, Comet, Diamond, Alpha Delta, Anli, Bencher, Heil, Daiwa, Outbacker, RF Concepts, Van Gorden, W5YI, Time-wave.

B

B.C. Communications, Inc.
The 211 Bldg.-Depot Road
Huntington Station, NY 11746
Phone: 516 549-8833
Tri-State Area: 800 924-9884
FAX: 516 549-8820
Established: 1972; *Employs* 12
Sells via showroom and mail order.
Key Employees: William, W2WBY.
Top Lines: Yaesu, MFJ, Larsen, William Nye-Viking, DAIWA, Kantronics, Astatic, Hustler, Newtronics, Valor, Mobile Mark, Bearcat, Uniden, Maxon, Mirage, B&W, Callbook, ARRL, KLM, Ameco, Bencher, etc.
Accepts trade-ins and provides repairs.
Two-way radio systems and repeater service.

B & H Sales
707 North Baltimore
Derby, KS 67037
Phone: 316 788-4225
Top Lines: AEA, ARRL, Astron, Kantronics, MFJ, Radio Shack, Valor, Larsen, W5YI.
Accepts trade-ins.

B & S Sales
51756 Van Dyke, Suite 330
Shelby Township, MI 48316
Phone: 810 566-7248

Bamcom Communications
PO Box 557
Arabi, LA 70023
Phone: 504 277-6815; 800 283-8696
Established: 1987; Employs 2
Sells via showroom and mail order.
Key Employees: Bert, WD5HFC, Mary Ellen, KB5EUM.
Top Lines: Yaesu, ICOM, Alinco, AEA, MFJ, Daiwa, ANLI, ARRL, Mirage, TE Systems, Ameritron, Valor.

Barry Electronics, Corp.
540 Broadway
New York, NY 10012
Phone: 212 925-7000
FAX: 212 925-7001
Established: 1950; Employs 11
Sells via showroom and mail order "world wide".
Key Employees: Kitty, WA2BAP; Arnie, WB2YXB; Deanna, KB2JYL, Toni, Anil.
Top Lines: Motorola, ICOM, Kenwood, Yaesu, Alinco, Standard, Hy-Gain, AEA, MFJ, Astron, B&W, Sony, Grundig, Sangean, Larsen, Comet, Diamond, Maxon, Relm, Mirage, Cushcraft, Hustler, Bencher, Astatic, ARRL Books, Tapes, ANLI Antennas, Belden coaxial cable, connectors, rotor cable. Books on Ham radio, modifications, theory, antennas, etc. Ranger, Ameco, RF Concepts, Kantronics, Scanners, Ace, ICOM, Uniden & scanner books.
Provides full-service repairs, technical assistance.

Meet Jim Vogler, WA7CJO. Here he sits at his station holding a 1,000-watt, 10-GHz tube.

Impressive antennas adorn the landscape at the Mt. Greylock, MA, VHF/UHF site of the W2SZ/1 crew.

Jerry Seefeld, WA9O, of Milwaukee, WI, brings mobiling to a new level with this set-up.

The sunset only adds to this beautiful antenna system owned by Larry Smith, WA4QQV, in Alvaton, KY.

Here's the crack operating team from South Georgia Island, VP8BSGP. Left to right: Vince, K5VT, Al, WA3YVN, Bob, VP8BFH, and Jon, WA4VQD.

The Hashemite Kingdom of Jordan

Confirming with Pleasure The Contact with

John Dow – K1AR

Warm Regards and Best 73's

Hussein 1.

A royal QSL confirmation by one of the world's most famous hams, King Hussein, JY1, of Jordan.

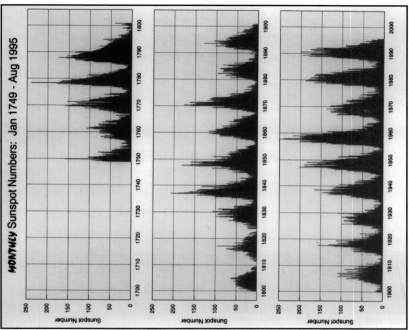

Monthly mean sunspot numbers: January 1749–August 1995.

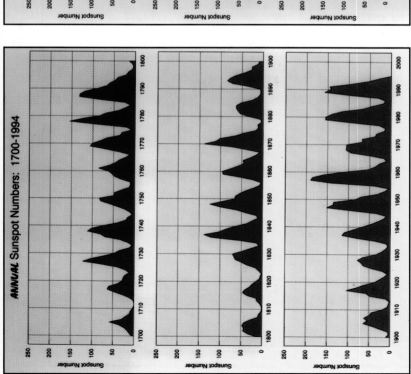

Annual mean sunspot numbers: 1700–1994.

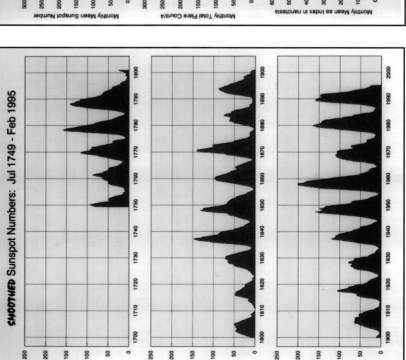

Sunspots, solar flares, and aa Indices.

Smoothed sunspot numbers: July 1749–February 1995.

Base Station, Inc.
1839 East Street
Concord, CA 94520
Phone: 510 685-7388
Established: 1976; Employs 5
Sells via showroom and mail order.
Key Employees: Art, AB6HB; Charley, N5TBV.
Top Lines: Yaesu, Cushcraft, Larsen, MFJ, AEA, Nye-Viking, B&
W, Bearcat, Uniden, Astron, Sony, Hustler, Bencher,
Astatic, Ameco, Tab Books, Gordon West Books & Tapes, W5YI
Books and Software, ANLI Antennas, Diamond Antennas, Call-
books, ARRL publications, QRZ CD- ROM's.
Accepts trade-ins and provides repairs.

Brownville Sales Co., Inc.
Rt. 2, Box 104
Stanley, WI 54768-9418
Phone: 715 644-2112
Key Employee: Richard, N9MEH

Burghardt Amateur Center, Inc.
PO Box 73,182 N. Maple St. Watertown, SD 57201-0073
Phone: 605 886-7314; 800 927-4261
FAX: 605 886-3444
Established: 1937; Employs 12
Sells via showroom and mail order.
Key Employees: Darrell, WDØGDF; Tim, WDØFKC; Stan,
WØJT; Jim, WBØMJY, David, KAØJDN, Jason, KBØIUS, Marty,
KBØION.
Top Lines: Kenwood, ICOM, Yaesu, AEA, Kantronics, Astron,
Ten-Tec,B&W, Cushcraft, Telex, MFJ.
Accepts trade-ins and provides repairs.

Burk Electronics
35 N. Kensington
LaGrange, IL 60525
Phone: 708 482-9310
Established: 1983; Employs 7
Sells via mail order, Hamfests.
Key Employees: Wayne, NA9B; Ann, AA9FN; Jeff, N9MLU;
Maria, N9QWQ; James, N9MSX, Laura, N9SXD, Steve N9TVQ.
Top Lines: AEA, Alinco, Antenex, Ameritron, Ameco, Antenna
Specialist, Antler, ARRL Publications, Bencher, Butternut, Comet,
Diamond, Hustler, ICOM, Larsen, Lunar, Maxrad, Mobile Mark,
MFJ, Maldol, NARA, Radio Amateur Callbook, Radio
Publications, Pyramid, Sams, Sangean, Smiley Antenna, Standard
TAB Books, Valor, Wintenna, W5YI, Yaesu,.
Accepts trade-ins, provides repairs and technical assistance.

C

CABLE X-PERTS
113 Mc Henry Road, Suite 240
Buffalo Grove, IL 60089-1797
Questions? Please Call or FAX
Phone/Tech: 708 506-1886
Orders Only: 800 828-3340
FAX: 708 506-1970
Established: 1989; Employs 7
Sells via phone, mail order, hamfests & dealers.
Key Employees: Marc, KC9VW and Chuck, KB9BQN.
Top Lines: We specialize in coax cable, rotor cable, antenna wire,
baluns, connectors, and related products. We do custom connector
work too. Mail a #10 SASE for complete literature.

C.A.T.S.
7368 S. R. 105
Pemberville, OH 43450
Phone: 419352-4465
Established: 1980; Employs 2
Sells via mail order.
Key Employee: Craig, N8DJB.
Major Lines: Parts/service on rotators, Hy-Gain/CDE, Alliance,
Channel Master, cable wire, rotator accessories. Hy-Gain, Create
models for sale.
Accepts trade-ins.

CBC International, Inc.
PO Box 31500
Phoenix, AZ 85046
FAX: 602 996-8700
Established: 1976; Employs 3
Sells via mail order.
Key Employees: Lou, K6NH.
Top Lines: CB-to-ham radio modification, plans and hardware;
FM conversion kits, books, plans, accessories.

C & S Sales
150 West Carpenter Avenue
Wheeling, I L 60090
Phone: 800 292-7711; 708 541-0710
FAX: 708 520-0085
Established: 1985; Employs 5
Sells via mail order.
Key Employees: Jim.
Top Lines: John Fluke, B&K Precision, Elenco, Hitachi test equip-
ment and standard amateur radio products, and robotic kits.

Communication Headquarters, Inc.
3830 Oleander Drive
Wilmington, NC 28403
Orders Only: 800 688-0073
Phone: 910 791-8885
Established: 1988
Sells via showroom, Hamfests, and mail order.
Key Employees: Dwight, K4YNY; Dennis, KC4YOU, Mike,
N3DJT, Linwood, WA4LEQ; Stanley, KC4DPC; Cindy,
KC4DUK; Brandie, KE4LJY.
Top Lines: ICOM, Kenwood, Yaesu, Diamond, Cushcraft, AEA,
MFJ, Larsen, Hustler, Kantronics, RF Concepts, B & W, ARRL,
Bencher, Astron, and Cellular.
Accepts trade-ins and provides repairs.
Full-time service department.

Communication Source, The
2713 Galleria Drive
Arlington, TX 76011
Phone: 817 640-1663
FAX: 817 649-1255
Key Employee: Lee Wilmeth, N5RAT

Communications City
175 SE 3rd Avenue
Miami, FL 33131
Phone: 305 579-9709
FAX: 305 579-9805
Top Lines: ICOM, Yaesu, Kenwood. Special export prices.

Communications Data Corporation
1051 Main Street
St. Joseph, Ml 49085
Phone: 616 982-0404; 800 382-2562 FAX: 616 982-0433
Established: 1981; Employs 6
Sells via showroom and mail order.
Key Employees: Duane, KX8D; Barb, N8JML; Jim, N8LXQ.
Top Lines: Yaesu, ICOM, Alinco, Standard, AEA, Kantronics,
Vibroplex, Bencher, RF Concepts, Comet, Hustler, Ameritron,
ARRL, Astron, MFJ, Various computer products.

Communications Electronics Inc.
Emergency Operations Center
PO Box 1045 -CQBG
Ann Arbor, Ml 48106
Phone: 313 996-8888
Established: 1969; Employs 38
Sells via showroom and mail order.
Top Lines: ICOM, Uniden, Relm, Mo-torola, Super Disk, Cobra,
Sangean, Grundig, Ranger, Davis weather.

Comm-Pute, Inc.
7946 S. State Street
Midvale, UT 84047
Phone: 801 567-9944; 800 942-8873
Established: 1988; Employs 5
Sells via showroom and mail-order.
Key Employees: Preben, K7KMZ.
Top Lines: Kenwood, Yaesu, ICOM, Cushcraft, Larsen, B&W,
AEA, Kantronics, Hustler, Bencher, Diamond.

Connectors Unlimited
PO Box 5973
Manchester, NH 03108-5973
Phone: 603 668-5926; 800 549-5955
FAX: 603 641-1179
Established: 1984; Employs 3
Sells via direct and mail order.
Top Lines: Amphenol, Kings, Gilbert, Flexwell, Andrew,
Cablewave.

Consolidated Electronics
705 Watervliet Avenue
Dayton, OH 45420
Phone: 800 543-3568; 513 252-5662
FAX: 513 252-4066
Key Employee: Steve KB8UHY.
Top Lines: B&K, Wavetek, Goldstar, Weller, Ungar, Amphenol,
Chemtronics, Rawn, Japanese Semiconductors.

Contact East, Inc.
335 Willow St. So.
North Andover, MA 01845
Phone: 508 682-2000, 800 225-5370 FAX: 508 688-7829
Established: 1964; Employs 75
Sells via phone and mail order.
Top Lines: Tektronix, Leader, Fluke, and Kenwood precision hand
tools and tool kits.

Copper Electronics, Inc.
3315 Gilmore Industrial Blvd.
Louisville, KY 40213
Phone: 502 968-8500; 800 626-6343
FAX: 502 968-0449
Established: 1974; Employs 36
Sells via showroom , phone and mail order.
Top Lines: Uniden, Ranger, Cobra, Astatic, Dosy, Firestick,
Antenna Specialist, Valor, Wilson, Telex, CEI Computers and spe-
cial orders on tubes.

D

De La Hunt Electronics
Highway 34E
Park Rapids, MN 56470
Phone: 218 732-3306
Top Line: ICOM.
Accepts trade-ins.

Dentronics
6102 Deland Road
Flushing, MI 48433
Phone: 810 659-1776; 800 722-5488 FAX: 810 659-1280
Established: 1978; Employs 2
Sells via showroom and mail order.
Key Employee: Dennis, WB8QWL.
Top Lines: Kantronics, AEA, MFJ, R.F. Concepts, Pac Comm,
Diamond, Sierra Computers.
Accepts trade-ins.
Packet Radio Specialists.

Doc's Communications
702 Chickamauga Avenue
Rossville, GA 30741
Phone: 706 866-2302; 706 861-5610 FAX: 706 866-6113
Established: 1975; Employs 6
Sells via showroom and mail order.
Key Employees: Maxine, N4ECA; Doc, KC4EV; John,
WD4AQO; Pat; Jim, Darrel.
Top Lines: ICOM, Kenwood, Yaesu, Alinco, MFJ, Larsen, Pro-
Am, Astron.
Accepts trade-ins.

E

El Original Electronics
1257 East Levee
Brownsville, TX 78520
Phone: 512 546-9846; 512 542-8507 Established: 1979;
Employs 4
Sells via showroom.
Key Employee: Emilio, XE2XES.
Top Lines: Alinco, Azden, Kenwood, ICOM, Cushcraft, Yaesu,
Larsen, Comet, Maxrad.

El Paso Communications System
1630 Paisano
El Paso, TX 79901
Phone: 915 533-5119

Electronic Equipment Bank (EEB)
323 Mill Street NW
Vienna,VA 22180
Phone: 800 368-3270, 703 938-3350 FAX: 703 938-6911
Established: 1971; Employs 37
Sells via showroom and mail order.
Key Employees: Bob, K7RDH, Mgr.; Alan KD4CGR, Gary
N3EAH, Darrin, N4WTN, Ken, KC4YMD.
Top Lines: ICOM, Kenwood, Yaesu, Alinco, AOR, Sony,
DAIWA, Astron, AEA, Bird, Emoto, Create, Datong, Novex,
Antennas, Cable, Accessories, Shortwave, Scanners and more.
Accepts trade-ins and provides repairs.

Eli's Amateur Radio, Inc.
2513 SW Ninth Avenue
Ft. Lauderdale, FL 33315
Phone: 305 525-0103, 305 944-3383; 800 780-0103; FAX: 305
944-3383
Established: 1980; Employs 5
Sells via showroom and mail order.
Key Employees: Eli, AA4BK; Al, N4AXQ; Jose, WT4G.
Top Lines: ICOM, Yaesu, Kenwood, Cushcraft, Hustler, AEA,
Kantronics, Bird, DAIWA, MFJ, Antenna Specialists, Larsen,
Diamond, SGC.
Accepts trade-ins and provides repairs.

Erickson Communications
5456 N. Milwaukee Avenue
Chicago, I L 60630
Phone: 800 621-5802; 312 631-5181 Established: 1969;
Employs 7
Sells via showroom and mail order.
Key Employees: Vince, KA9ZDM; Verne, K9TMR.
Top Lines: Kenwood, Yaesu, ICOM, Cushcraft, Hy-Gain, AEA,
Kantronics, Nye-Viking, MFJ, Ten-Tec, Hustler.
Accepts trade-ins.

F

Fair Radio Sales
1016 E. Eureka Street
Lima, OH 45802
Phone: 419 227-6573, 419 223-2196 FAX: 419 227-1313
Established: 1947; Employs 12
Sells via showroom and mail order.
Top Lines: Military surplus, receivers, test equipment, vacuum tubes, electronic parts.

For Hams Only Inc.
4309 Northern Pike Blvd.
Monroeville, PA 15146
Phone: 412 374-9744
Orders Only: 800 854-0815
Key Employee: Robb Weiss

G

G and G Electronics of Maryland
8524 Dakota Drive
Gaithersburg, MD 20877
Phone: 301 258-7373
Established: 1975; Employs 1
Sells via showroom and mail order.
Key Employee: Jeff, K3DUA
Accepts trade-ins and provides repairs.

G&W
524 Independence Blvd.
Virginia Beach, VA 23462
Phone: 804 499-9217
Key Employee: Bob

Galaxy Electronics
PO Box 1202, 67 Eber Avenue
Akron, OH 44309
Phone: 216 376-2402
Established: 1982; Employs 6
Sells via mail order.
Key Employees: Robert, KA8VWR; Al, WD8KTR; Mary, Rose, Steve.
Top Lines: Uniden, AOR, Regency, Cobra, ICOM, Kenwood, Yaesu, Sony, Grundig, Sangean.
Accepts trade-ins.

Gateway Electronics Corp.
8123 Page Blvd.
St. Louis, MO 63130
Phone: 314 427-6116; 800 669-5810
FAX: 314 427-3147
Established: 1960
Sells via showroom and mail order.
Key Employee: Stu, WØIGU; Lou, WØIBX; Lisa, KBØBBB.
Top Lines: ICOM, Alinco, MFJ, Larsen, Hustler, Maldol, Protek, Sangean, Heil, ARRL, Radio Amateur Callbook, Van Gorden, W5YIU.

Gilfer Shortwave
52 Park Avenue
Park Ridge, NJ 07656
Phone: 201 391-7887
FAX: 201 391-7433
E-Mail: GilferSW@aol.com; gilfer@haven.ios.com
Established: 1953; Employs 5
Sells via showroom and mail order.
Key Employees: Paul, N2HIE; Tom, N2ONK.
Top Lines: Kenwood, Yaesu, ICOM, Japan Radio, Drake, Sony, Grundig, Lowe, Datong, Dressler, Optoelectronics, JPS, AOR.

H

H.R. Electronics
722-24 Evanston Avenue
Muskegon, MI 49442
Phone: 616 722-2246
FAX: 616 722-4993
Established: 1978; Employs 3
Sells via showroom and mail order.
Key Employees: Sue, N8MMH; Dan, N8PPQ, Chuck, KG8CF.
Top Lines: ICOM, Alinco, Cushcraft, DAIWA, Hustler, Mirage/KLM, Van Gorden, Butternut, Rohn, B&W, Astron, Larsen, Spi-ro, MFJ, A/S, Grundig, ARRL, Callbooks, Bencher, Vibroplex, AEA, KPC, Ameco, Yaesu, Kenwood, Diamond, Comet, Pouch, and M2.
Accepts trade-ins and provides repairs.

H.S.C. Electronic Supply
6819 Redwood Blvd.
Cotati, CA 94931
Phone: 707 792-2277; 707 792-2357 FAX: 707 792-0146
BBS: 707 792-1042 8N1 (300-14,400) Established: 1963; Employs 7
Sells via showroom and mail order.
Key Employee: George, KS6W.
Top Lines: MFJ, Larsen, Antenna Specialists, ARRL, Ameco, Call Books, software.

Hal-Tronix, Inc.
12671 Dix Toledo Hwy.
Southgate, MI 48195
Phone: 313 281-7773
Established: 1968; Employs 4
Sells via showroom and mail order.
Key Employees: Hal, W8ZXH; Greg; Dick; Betty.
Top Lines: Ten-Tec, MFJ, IBM Clone Computers, Laser Computers.
Accepts trade-ins and provides repairs.

"Ham" Buerger, Inc.
417 Davisville Road
Willow Grove, PA 19090
Phone: 215 659-5900
Established: 1956; Employs 4
Sells via showroom and mail order.
Key Employees: Bob, WA3ZID; Jeff, WB3HOF; Dave.
Top Lines: ICOM, Azden, MFJ, Astatic, Hustler, Antenna Specialist, Uniden Bearcat, B&W, W2AW Baluns,Bencher, Larsen, Astron, Sangean, Code Alarm, ARRL, AOR, Daiwa, Nutone, Ademco, Cushcraft, ECG, Genesys.
Accepts trade-ins and provides repairs.

Ham Contact, The
PO Box 3624, Dept. CQ
Long Beach, CA 90803
Phone: 310 433-5860
Order Line: 800 933-4264
Established: 1987; Employs: 4
Key Employees: Joe, N6YYO; Joan N3SAA, David, KD6JBN; Don, KD6QHZ; Mike, KC6PZB; Lee and Mark.
Top Lines: Importers of The Power Station, Major dealer for SGC Smartuner, QM5II Quick Mount System, and SG2000 Commercial Quality HF radio.
Accepts trade-ins and provides repairs.

Ham Radio Outlet
1509 N. Dupont Hwy., #7
New Castle, DE 19720
Phone: 302 322-7092; 800 644-4476
FAX: 302 322-8808
Established: 1971; Employs: 5
Key Employee: John, N1IFL
Top Lines: Kenwood, ICOM, Yaesu, Alinco, US Tower, RFC, Rohn, AEA, Cushcraft, Hy-Gain, Butternut, Bird, Kantronics, Larsen, Diamond, NCG, MFJ, Ameritron, Standard, Bencher, Mirage/KLM, Sony, Astron, Hustler, M2, Outbacker.
Accepts trade-ins and provides repairs.

Ham Radio Outlet
933 N. Euclid Street
Anaheim, CA 92801
Phone: 800 854-6046
Local: 714 533-7373
FAX: 714 533-9485
Established: 1971; Employs 10
Sells via showroom and mail order.
Key Employee: Janet, WA7WMB/A35YL.
Top Lines: Kenwood, ICOM, Yaesu, Alinco, US Tower, RFC,
Rohn, AEA, Cushcraft, Hy-Gain, Butternut, Bird, Kantronics,
Larsen, Diamond, NCG, MFJ, Ameritron, Standard, Bencher,
Mirage/KLM, Sony, Astron, Hustler, M2, Outbacker.
Accepts trade-ins.

Ham Radio Outlet
6071 Buford Highway
Atlanta, GA 30340
Phone: 404 263-0700; 800 444-7927
FAX: 404 263-9548
Established: 1971; Employs 8
Sells via showroom and mail order.
Key Employee: John, KB4NUC.
Top Lines: Kenwood, ICOM, Yaesu, Alinco, US Tower, RFC,
Rohn, AEA, Cushcraft, Hy-Gain, Butternut, Bird, Kantronics,
Larsen, Diamond, NCG, MFJ, Ameritron, Standard, Bencher,
Mirage/KLM, Sony, Astron, Hustler, M2, Outbacker.
Accepts trade-ins.

Ham Radio Outlet
8400 E. Iliff Avenue, #9
Denver, CO 80231
Phone: 303 745-7373; 800 444-9476
FAX: 303 745-7394
Employs: 5
Sells via showroom and mail order. Exports.
Key Employee: Joe, KDØGA.
Top Lines: Kenwood, ICOM, Yaesu, Alinco, US Tower, RFC,
Rohn, AEA, Cushcraft, Hy-Gain, Butternut, Bird, Kantronics,
Larsen, Diamond, NCG, MFJ, Ameritron, Standard, Bencher,
Mirage/KLM, Sony, Astron, Hustler, M2, Outbacker.
Accepts trade-ins.

Ham Radio Outlet
2210 Livingston Street
Oakland, CA 94606
Phone: 415 534-5757, 800 854-6046 FAX: 415 534-0729
Established: 1971; Employs 4
Sells via showroom and mail order.
Key Employee: Mach, K6KAP.
Top Lines: Kenwood, ICOM, Yaesu, Alinco, US Tower, RFC,
Rohn, AEA, Cushcraft, Hy-Gain, Butternut, Bird, Kantronics,
Larsen, Diamond, NCG, MFJ, Ameritron, Standard, Bencher,
Mirage/KLM, Sony, Astron, Hustler, M2, Outbacker.
Accepts trade-ins.

Ham Radio Outlet
11705 SW Pacific Highway
Suite Z
Portland, OR 97223
Phone: 503 598-0555; 800 854-6046
FAX: 503 684-0469
Established: 1991; Employs 4
Sells via showroom and mail order.
Key Employee: Earl, KE7OA.
Top Lines: Kenwood, ICOM, Yaesu, Alinco, US Tower, RFC,
AEA, Cushcraft, Hy-Gain, Butternut, Bird, Kantronics, Larsen,
Diamond, NCG, MFJ, Ameritron, Standard, Bencher, Mirage/
KLM, Sony, Astron, Hustler, M2, Outbacker.
Accepts trade-ins.

Ham Radio Outlet
1702 W. Camelback Road, Suite 4
Phoenix, AZ 85015
Phone: 602 242-3515, 800 854-6046 FAX: 602 242-3481
Established: 1971; Employs 5
Sells via showroom and mail order.
Key Employee: Gary, WB7SLY.
Top Lines: Kenwood, ICOM, Yaesu, Alinco, US Tower, RFC,
Rohn, AEA, Cushcraft, Hy-Gain, Butternut, Bird, Kantronics,
Larsen, Diamond, NCG, MFJ, Ameritron, Standard, Bencher,
Mirage/KLM, Sony, Astron, Hustler, M2, Outbacker.
Accepts trade-ins.

Ham Radio Outlet
224 N. Broadway D-12
Salem, NH 03079
Phone: 603 898-3750, 800 444-0047 FAX: 603 898-1041
Established: 1986; Employs 10
Sells via showroom and mail order.
Key Employee: Chuck, KM4NZ.
Top Lines: Kenwood, ICOM, Yaesu, Alinco, US Tower, RFC,
Rohn, AEA, Cushcraft, Hy-Gain, Butternut, Bird, Kantronics,
Larsen, Diamond, NCG, MFJ, Ameritron, Standard, Bencher,
Mirage/KLM, Sony, Astron, Hustler, M2, Outbacker.
Accepts trade-ins.

Ham Radio Outlet
5375 Kearny Villa Road
San Diego, CA 92123
Phone: 619 560-4928, 800 854-6046 FAX: 619 560-1705
Established: 1971; Employs 5
Sells via showroom and mail order.
Key Employee: Tom, KM6K.
Top Lines: Kenwood, ICOM, Yaesu, Alinco, US Tower, RFC,
Rohn, AEA, Cushcraft, Hy-Gain, Butternut, Bird, Kantronics,
Larsen, Diamond, NCG, MFJ, Ameritron, Standard, Bencher,
Mirage/KLM, Sony, Astron, Hustler, M2, Outbacker.
Accepts trade-ins.

Ham Radio Outlet
510 Lawrence Expwy #102
Sunnyvale, CA 94086
Phone: 408 736-9496, 800 854-6046
FAX: 408 736-9499
Established: 1971; Employs 8
Sells via showroom and mail order. Exports.
Key Employee: Bill, K6YDQ.
Top Lines: Kenwood, ICOM, Yaesu, Alinco, US Tower, RFC,
Rohn, AEA, Cushcraft, Hy-Gain, Butternut, Bird, Kantronics,
Larsen, Diamond, NCG, MFJ, Ameritron, Standard, Bencher,
Mirage/KLM, Sony, Astron, Hustler, M2, Outbacker.
Accepts trade-ins.

Ham Radio Outlet
2492 W. Victory Blvd.
Burbank, CA 91506
Phone: 818 842-1786
FAX: 818 842-5283
Established: 1971; Employs 5
Sells via showroom and mail order.
Key Employee: Jon, KB6ZBI .
Top Lines: Kenwood, ICOM, Yaesu, Alinco, US Tower, RFC,
Rohn, AEA, Cushcraft, Hy-Gain, Butternut, Bird, Kantronics,
Larsen, Diamond, NCG, MFJ, Ameritron, Standard, Bencher,
Mirage/KLM, Sony, Astron, Hustler. M2, Outbacker.
Accepts trade-ins.

Ham Radio Outlet
14803 Build America Drive, Bldg B Woodbridge, VA 22191
Phone: 800 444-4799, 703 643-1063 FAX: 703 494-3679
Established: 1977; Employs 9
Sells via showroom and mail order.
Key Employee: Curtis, WB4KZL.
Top Lines: Kenwood, ICOM, Yaesu, Alinco, US Tower, RFC,
AEA, Cushcraft, Hy-Gain, Butternut, Bird, Kantronics, Larsen,
Diamond, NCG, MFJ, Ameritron, Standard, Bencher, Mirage/
KLM, Sony, Astron, Hustler, M2, Outbacker.
Accepts trade-ins andprovides repairs.

Ham Radio Toy Store, Inc.
117 West Wesley Street
Wheaton, IL 60187
Phone: 708 668-9577
FAX: 703 690-9335
Established: 1989; Employs 5
Sells via showroom and mail order.
Key Employee: Roberto, WA9E; Wayne, NA9B.
Top Lines: ICOM, Yaesu, Ameritron, AEA, Ameco, Antenna Specialist, Antenex, Antler, ARRL Publications, Belden, Bencher, Butternut, Comet, Diamond, Gordon West, Hustler, Larsen, Lunar, Maxrad, MFJ, Mobile Mark, Radio Amateur Callbook, Radio Publications, Pyramid, Sams, Smiley Antenna, Valor, Wintenna, Alinco, W5YI, Alpha Delta, Radio School, Standard, Cushcraft, NCG, Sangean, Kantronics, RFConcepts, Mirage, Pro Am.
Accepts trade-ins and provides repairs.

Ham Station, Inc.
220 N. Fulton Avenue
Evansville, IN 47710
Phone: 800 729-4373, 812 422-0231
Service Line: 812 422-0252
FAX: 812 422-4253
BBS: 812 424-3614
Sales and Service
Sells via showroom, mail order/phone.
Key Employees: Dan, N9APA owner; Rick, WB9SAN sales; Jeff, KA9YKA sales and shipping; Russ, KB9GJ used equipment sales; Steve, KA9OKH service dept.; Ruth, N9ZCA Office and sales.
Hours: Mon-Fri 8AM-5PM CST. Sat. 9AM-3PM CST.
Top Lines: Kenwood, Yaesu, ICOM, Standard, MFJ, AEA, RFC, Comet, Diamond, Cushcraft, Kantronics, ARRL, many others.
Accepts trade-ins andprovides repairs.

Hamtronics/Trevose
4033 Brownsville Road
Trevose, PA 19053
Phone: 215 357-1400; 800 426-2820 FAX: 215 355-8958
BBS: 215 355-8830
Established: 1954; Employs 15
Sells via showroom and mail order.
Key Employees: Dave, N3TS; Mike, KA3WVZ; Jim, KA3JSO, Dave Jr., KA3BKG; Gene, WA3STW, Eva, WA3USJ; Marrion, WA3VEP; Sam, N3DFV Jim, N3119; Jim N3SIR.
Top Lines: Kenwood, Yaesu, ICOM, Alinco, AEA, MFJ, Cushcraft, Hustler, JRC, Ameritron, Larsen, Kantronics, Rohn, Bencher, Heil, B&W, RF Concepts, Daiwa, Diamond, Comet, Optoelectronics, Genesys, Astron, Outbacker.
Accepts trade-ins and provides repairs.

Hardin Electronics
5635 E. Rosedale Street
Ft. Worth, TX 76112
Phone: 800 433-3203; 817 429-9761 FAX: 817 457-2429
Established: 1965; Employs 20
Sells via showroom and mail order.
Key Employees: Lee, N5RAT; Richard, K5ZIM; Rick, N5QKB.
Top Lines: Kenwood, Standard, ICOM, Yaesu, Alinco, Larsen, Comet, Diamond, Daiwa, ARRL.
Accepts trade-ins and provides repairs.

Hatry Electronics
500 Ledyard Street
Hartford, CT 06114
Phone: 203 296-1881
FAX: 203 296-7110
Established: 1928; Employs 20
Sells via showroom and mail order.
Key Employees: Lenny, WA1 VPT; Spiro, KJ1R.
Top Lines: ICOM, MFJ, B&W, Alinco, Ameco, Sangean, Van Gorden, Cushcraft, Larsen, Hustler.
Accepts trade-ins.

Heaster, Inc., Harold
84 North Tymber Creek Road
Ormond Beach, FL 32174
Phone: 904 673-4066
Established: 1984; Employs 5
Sells via showroom and mail order.
Key Employees: Harold, KE8MR; Vern, K3FOA; Barb, KA8RRD; Lewis, KC4DSQ; Jeff, KA8RRE.
Top Lines: Kenwood, ICOM, Larsen.
Accepts trade-ins.

Henry Radio Inc.
2050 South Bundy Drive
Los Angeles, CA 90025
Phone: 800 877-7979, 310 820-1234
FAX: 310 826-7790
Established: 1941; Employs 35
Sells via showroom.
Key Employees: Ray, WW6X; Leo, KJ6HI; Nate, KI6IK; Ted, W6YEY; Paul, N6VLV; Joe, WA6FST.
Top Lines: Kenwood, ICOM, Yaesu, Henry Amps, Tempo, Astron, Bird, Hy-Gain, Cushcraft, MFJ, Alinco, Tohtsu.
Accepts trade-in and provides repairs.

Hialeah Communications
801 Hialeah Drive
Hialeah, FL 33010
Phone: 305 885-9929
FAX: 305 888-8768
Established: 1978; Employs 20
Sells via showroom and mail order.
Key Employees: Sara, Ruben, Jose, Miguel, Juvenal.
Top Lines: Motorola, ICOM, Kenwood, Yaesu, Bird, CES, Larsen, Cushcraft, Antenna Specialists, Hustler.
Provides repairs.

HighText Publications, Inc.
PO Box 1489
Solana Beach, CA 92075
Phone: 619 793-4142
Orders Only: 1 800 247-6553
Established: 1990; Employs 4
Sells via mail order.
Key Employees: Harry, AA6FW and Carol.
Top Lines: Publisher of All About Ham Radio.

Hirsch Sales Corporation
219 California Drive
Williamsville, NY 14221
Phone: 716 632-1189
FAX: 716 632-6304
Established: 1961; Employs 13
Sells via showroom and mail order.
Key Employees: Jerry, WA2ZFA; Joe, N2FEO; Rick; Mike, Ken.
Top Lines: All popular brands stocked.
Accepts trade-ins and provides repairs.

Hooper Electronics
1702 Pass Road
Biloxi, MS 39531
Phone: 601 432-1100; 601 432-0584 FAX: 601 432-7651
Established: 1960; Employs 10
Sells via showroom and mail order.
Key Employees: Dave, WB5KDV; Bill, W5WWJ; Joyce, WB5LKC.
Top Lines: ICOM, Kenwood, MFJ, Cushcraft, Hustler, Astron, Bird, Larsen, Antenna Specialist.
Accepts trade-ins.
Branch: 1700 Terry Road
Jackson, MS 38204
Phone: 601 353-0922, 601 354-4531 FAX: 601 948-3807
Key Employee: Wayne, KB5JCI.

Hosfelt Electronics, Inc.
2700 Sunset Blvd.
Steubenville, OH 43952-1158
Phone: 800 524-6464
FAX: 800 524-5414
Established: 1969; Employs 28
Sells via mail order.
Top Lines: Bourns, Chemtronics, Kester, Lyton, Diodes Inc., LMB Heeger, Fans, Qualtek, NTE Replacement Line, Xicon, Mueller, Speco.
Call or FAX toll free for FREE 146-page Catalog.

Houston Amateur Radio Supply
181 Cypresswood Drive
Spring, TX 77388
Phone: 713 355-7373; 800 471-7373
FAX: 713 355-8007
Key Employee: George, KDØRW.
Top Lines: Yaesu, ICOM, Alinco, Cushcraft, Butternut, Maldol, MFJ, Valor, Timewave, Kantronics, AEA, RF Concepts, Bencher, Daiwa.
Authorized repair for Yaesu and ICOM.
Accepts trade-ins.

I

IRI Amateur Electronics
918 Plantation Farms Road
Greensboro, NC 27409-9257
Phone: 910 299-1725
Key Employee: James, KD4IRI.
Top Lines: Anttron Antenna Products, Genesys Products Group, Maldol USA, Periphex Batteries, Oak Bay Technologies, TVI Filters & Etc., G5RV's (Omega), No-Tenna (Grove).

International Radio & Computer, Inc.
3804 South US #1
Fort Pierce, FL 34982-6620
Phone: 407 489-0956
FAX: 407 464-6386
Established: 1979; Employs 6
Sells via showroom and mail order.
Key Employees: Rob, N8RT; Peg, KB4LRD; Robert, KB4JMQ, Mark, N4FMC; Michelle.
Top Lines: High-performance 8 pole crystal filters for popular transceivers, newsletters, enhancement kits for popular transceivers. Kenwood Service Center. Authorized ICOM dealer. Catalog SASE.

International Radio Systems
5001 NW 72nd Avenue
Miami, FL 33166
Phone: 305 594-4313
Key Employee: John

J

JWO Services
12 Hickory Place
Camp Hill, PA 17011
Phone: 717 731-4747
FAX: 717 730-9373
Key Employee: John, W3IS
Top Lines: World wide mail order specialists of Amateur radio publications, software, audio and video tapes. Free Discount Catalog. Special services to clubs, classes and instructors. AMECO/ARRL/CQ publications, Gordon West, Callbook, QRZ Ham Radio CD ROM, RSGB.

Jones & Associates, Marlin P.
PO Box 12685
Lake Park, FL 33403-0685
Phone: 407 848-8236
FAX: 407 844-8764
Established: 1976
Sells via mail order catalogs.
Top Lines: Connectors, fans, motors, power supplies, meters, switches, knobs, LED's, semiconductors, tools, relays, lens, lasers, valves, nicads.

Juneau Electronics
8111 Glacier Hwy
Juneau, AK 99801-8035
Phone: 907 586-2260

Jun's Electronics
5563 Sepulveda Blvd.
Culver City, CA 90230
Phone: 310 390-8003, 800 882-1343 FAX: 310 390-4393
Established: 1975; Employs 8
Sells via showroom and mail order.
Key Employees: PJ, WA6IBY; Raul, KB6GMR; Rudy, KC6LYY; Jack, WB6BBZ; Steve, KE6HYL; James, N6KBV, and Kyung.
Top Lines: ICOM, Kenwood, Yaesu, Alinco, Grundig, Mirage/KLM, RF Concepts, Astron, Ameritron, AEA.
Provides repairs.

K

KComm, Inc.
5730 Mobud
San Antonio, TX 78238
Phone: 210 680-6110; 800 344-3144
Established: 1988; Employs 10
Sells via showroom and mail order.
Key Employees: Craig, KB5BI; Oscar, AA5DB; Ed, KS5V; Bill, KB5COU, Edwin, WC6X.
Top Lines: Kenwood, Yaesu, ICOM, Alinco, Genesys, MFJ, AEA, Kantronics, Cushcraft, KLM, Ramsey.
Accepts trade-ins and provides repairs.
Si Habla Espanol.

KJI Electronics
66 Skytop Road
Cedar Grove, NJ 07009
Phone: 201 239-4389
Established: 1975; Employs 4
Sells via showroom and mail order.
Key Employees: Gene, K2KJI; Maryann, K2RVH, Chris.
Top Lines: ICOM, Kenwood, Yaesu, Azden, Alinco, KLM/Mirage, RF Concepts, Bencher, AEA, Kantronics, Larsen, Astron, Diamond, MFJ.
Accepts trade-ins and provides repairs.
Polish spoken/Polska Mowa

L

LaCue Communications, Inc.
132 Village Street
Johnstown, PA 15902
Phone: 814 536-5500

Ladd Electronics Co.
111 N. 41st St.
Omaha, NE 68131
Phone: 402 556-3023

LaRue Electronics
1112 Grandview Street
Scranton, PA 18509
Phone: 717 343-2124
Established: 1976; Employs 3
Sells via showroom and mail order.
Key Employees: Gene, K3HAM; Les, W3LPZ.
Top Lines: ICOM, AEA, Amphenol, Astron, Beckman, Belden, Bencher, Bird, Cushcraft, Larsen, Vibroplex, Weller, Xcelite, 3M. Provides repairs.

Lentini Communications
21 Garfield Street
Newington, CT 06111
Phone: 203 666-6227, 800 666-0908
FAX: 203 667-3561
Established: l954; Employs 7
Sells via showroom and mail order.
Key Employees: Alex, N1EBU; Larry, N1RGI; Martin, N1FOC; Bill, N1JBS; Joe, AA1GW; Mike, W1VLA; Kris.
Top Lines: Kenwood, Yaesu, ICOM, Alinco, Standard, Drake, Sony, Comet, Bencher, DAIWA, Hustler, Larsen, Uniden, Astatic, RF Concepts, NCG, MFJ, Austin Custom Antennas, Kantronics, Diamond, AEA, Cushcraft, Standard.
Accepts trade-ins and provides repairs.

M

M.D. Electronics
875 South 72nd Street
Omaha,NE 68114
Phone: 402 392-2284; 402 392-0810
Established: 1984; Employs 50
Sells via mail order.
Key Employees: Joe; Curt.
Top Lines: Video cable TV hardware: Everquest, Jerrold, Zenith, Panasonic, Hamlin, Oak brands.

MECI
340 E. First Street
Dayton, OH 45402-1257
Phone: 800 344-4465
Top Lines: Mail-order components including fans, capacitors, connectors, relays, switches relays, motors, hard-to-find electronic items.
Free Catalog.

MacFarlane Electronic, Ltd., H.C.
(See Canada Listings)

Maryland Radio Center
8576 Laureldale Drive
Laurel, MD 20707
Phone: 301 725-1212,800 447-7489; FAX: 301 725-1198
Established: 1986; Employs 9
Sells via showroom.
Key Employees: Jerry, WA3WZF; Steve, KD3EH; Mike, WA8MCQ; Ike, WB3LRM; John, N3FNG.
Top Lines: Kenwood, ICOM.
Accepts trade-ins and provides repairs.

McCarthy, N6CI0, Loraine
2775 Mesa Verde Dr. E., Ste E101
Costa Mesa, CA 92626
Phone: 714 979-CODE
Established: 1979; Employs 3
Sells via showroom and mail order.
Key Employee: Loraine, N6CIO.
Top Lines: Gordon West Radio School training materials, ARRL training materials, AMECO material.

Medicine Man CB
PO Box 37
Clarksville, AR 72830
Phone: 501 754-2076
Established: 1980
Sells via mail order.
Key Employee: J.L. Richardson
Major Line: Books

Memphis Amateur Electronics, Inc.
1465 Wells Station Road
Memphis, TN 38108
Phone: 800 238-6168; 901 683-9125
Established: 1966; Employs 8
Sells via showroom and mail order.
Key Employees: Bill, W4TNP; Marshall, KU40; Stan, W4RMW; Chic, WB4KXN.
Top Lines: Kenwood, ICOM, MFJ, Cushcraft, Hustler, Larsen, Van Gorden, Astron, Hy-Gain, Alinco, Daiwa, Valor, Yaesu.
Accepts trade-ins and provides repairs.

Miami Radio Center
5590 W. Flagler Street
Miami, FL 33134
Phone: 305 264-8406
Established: 1980; Employs 6
Sells via showroom.
Top Lines: ICOM, Kenwood, Yaesu, Motorola, Cobra, Astron, Hustler, Mirage.

Michigan Radio
23040 Shoenherr
Warren, Ml 48089
Phone: 313 771-4711; 800 878-4266
Service: 313 771-4712
FAX: 313 771-6546
Established: 1980; Employs 8
Sells via showroom and mail order.
Key Employees: Jerry, W8MR; Vicky, XYL W8MR; Thomas, KA8LSU; Dave, AA8DH; Chris, N8FZT; Rose; John.
Top Lines: Kenwood, ICOM, Yaesu, Standard, Alinco, Drake, Sangean, Rohn, Cushcraft, KLM, Hustler, Butternut, Diamond, Comet, Larsen, AEA, Alpha Delta, Astron, Bencher, Daiwa, Diamond, Heil, MFJ, RF Concepts, Kantronics, VanGorden, ARRL, W5YI, Callbook, and more.
Accepts trade-ins and provides repairs.

Mike's Electronics
1001 NW 52nd Street
Ft. Lauderdale, FL 33309
Phone: 305 491-7110, 800 427-3066
Established 1980; Employs 4
Sells via showroom and mail order.
Key Employees: Walt, KN4SL; Mike, WB4RFC.
Top Lines: ICOM, Yaesu, Spirit, JPS, ANLI, Ham Stick, Pouch, ARRL, Radio Amateur Callbooks, Valor, Alpha Delta, VGE, NCG, Maxcom, Bearcat, SONY, Rohn, Butternut, Astron, Daiwa, AOR, Nye Viking, Cushcraft, Bencher, Hustler, AEA, Kantronics, RF Products, MFJ, Ritron, Jabro, Uniden, Comet, RF Concepts, Belden, Periphex, QRZ, Mirage.
Accepts trade-ins.

Morgan & Associates, C.T.
2300 E. 34thSt.
Chattanooga, TN 37407
Phone: 615 629-7911
Key Employee: Tom

Mouser Electronics
2401 Hwy. 287 North
Mansfield, TX 76063
Phone: 800 346-6873; 817 483-4422
FAX: 817 483-0931
Established: 1964; Employs 350
Sells via showroom and mail order.
Top Lines: SGS-Thomson, Amp, NTE, Carol Cable International Power, 3M, Keystone, Beckman Industrial, Cornell Dubilier, Continental Industries, Thompson Passive Components.

Musero, K3UKW, Anthony
1609 S. Iseminger Street
Philadelphia, PA 19148-1010
Phone: 215 271-8898
Established: 1972; Employs 1
Sells via mail order.
Top Lines: All used ham & SWL gear.
Buy/Sell/Trade/Consign

N

N & G Electronics
1950 NW 94th Avenue
Miami, FL 33172
Phone: 305 592-9685
Top Lines: ICOM, Kenwood, Yaesu, Interconnect Spec., Astron, Hy-Gain, Cushcraft, AEA, C.E.S., Hustler, Alinco, Motorola, and many more.
Provides repairs.

National Tower Company
PO Box 15417
Shawnee Mission, KS 66215
Phone: 913 888-8864; 800 762-5049
FAX: 913 894-2136
Key Employees: Lee, KBØDE; Mark, NØRWJ; Gary, NØRWG.
Top Lines: Rohn Towers, Cushcraft, Hustler, Midland, Astron, Diamond, Larsen, Telex.

North Olmsted Amateur Radio Depot
29462 Lorain Road
N. Olmsted, OH 44070
Phone: 216 777-9460
Established: 1988; Employs 2
Sells via showroom and mail order.
Key Employees: Rick, K8SCI; Pauline, KA8FOE.
Top Lines: Ameritron, ARRL, DAIWA, Larsen, MFJ, Heath, Rohn Towers.
Accepts trade-ins; provides repairs.

O

Ocean State Electronics
PO Box 1458
Westerly, RI 02891
Phone: 401 596-3080
Sells via mail order.
Top Lines: Electronic components, kits, books, test equipment and amateur radio accessories.
120-page Catalog $2.00.

Oklahoma Comm Center
13424 Railway Drive
Oklahoma City, OK 73114
Phone: 800 765-4267; 405 748-3066
FAX: 405 748-3077
Sells via showroom and mail order.
Key Employees: Craig, AH9B; Tony, KG5SA; Mike, V73OK; Clay, WD4O, Troy, K5CBL; Lee, N5KXI; Marcia.
Top Lines: Yaesu, ICOM, Alinco, Cushcraft, Comet, Diamond, AEA, Kantronics, Kenwood.
Accepts trade-ins.

Omar Electronics
2130 GA Hwy. 81 SW
Loganville, GA 30249
Phone: 404 466-3241
FAX: 404 466-9013
Established: 1974; Employs 4
Sells via showroom and mail order.
Key Employees: Omar, WA8FON; Mark, N4YHM, Alex; Christopher.
Top Lines: MFJ, Van Gorden, Daiwa, J&I, Bencher, Uniden, ARRL AEA, Drake, Kantronics, Azden, Diamond, ANLI, Pyramid, Ameritron, Genesys, Pro-Am, Smiley, ADI.
Accepts trade-ins.

OMEGA Electronics
101-D Railroad Street
PO Box 579
Knightdale, NC 27545
Phone: 919 231-7373
Orders Only: 800 900-7388
FAX: 919 250-0073
Established: 1988; Employs 2
Sells via showroom and mail order.
Key Employees: Bill, K4BWC; WG, KB4ZMB.
Top Lines: Omega "G5RV" Antenna, JPS "DSP" Audio Filters, Maxrad Antennas, Astron P/S, Pride Tubes, Peter Dahl, Consignment Equipment Sales.
Specialize in "HF" Amplifier Repair.

Omni Electronics
1007 San Dario
Laredo, TX 78040
Phone: 210 725-OMNI
Established: 1984; Employs 8
Sells via showroom and mail order.
Key Employee: Eduardo, XE2HHC.
Top Lines: ICOM, Kenwood, Yaesu, Antenna Specialists, Telemobile, Tempo, CES, CSI, Anteco, Astron, Arco-Solar, Azden, Centurion, Hustler, Neutec, RF, Motorola, Standard.
Provides repairs.

P

P.A.C.E.
1870 W. Prince, Ste. 2
Tucson, AZ 85705
Phone: 602 888-3333

PAS Ham Enterprises
2215 C Ingalls Ave.
Pascagoula, MS 39567
Phone: 601 762-8070
Key Employee: Winston

Paramount Communications Elec.
PO Box 506
Dalton, OH 44618
Phone: 800 431-7777; 216 828-2071
FAX: 216 828-8308

Phillips- Tech Electronics
5420 E. Sahuro Drive
Scottsdale, AZ 85252
Phone: 602-947-7700
Top Lines: Wireless cable, amateur television, and microwave television antennas.

Portland Radio Supply
234 SE Grand Avenue
Portland, OR 97214-1115
Phone: 503 233-4904
Established: 1932; Employs 8
Sells via showroom and mail order.
Key Employees: Ken, WR7D; Stan, WB7OKN; Stephanie, Office Manager.
Top Lines: NTE, Larsen, CB Distributing, Jim-Pak, ICOM, Yaesu, ARRL, Mouser, MFJ, Caltronics.
Large selection of used gear.
Accepts trade-ins.

Q

QRV Electronics
PO Box 330
Crawford, GA 30630
Phone: 706 743-3344
Top Lines: Azden, Alinco, Valor, Vectronics, Pyramid, Kantronics, RF Concepts, Wintenna.
Provides repairs.
ADI Factory Service Center.

R

R.C. Distributing Company
PO Box 552
South Bend, IN 46624
Phone: 219-236-5776
Established: 1992; Employs 5
Sells via mail order.
Key Employees: Ron; Patti; Robert; Max; Lucy.
Top Lines: Video Sync Generator, Specialty microwave equipment; high performance antennas (parabolic & yagi), portable satellite antennas.

R&D Electronics
10511 Phelps Streets
New Orleans, LA 70123
Phone: 206 364-8300
Sells via mail order.
Top Lines: Elsie Faser L/C Phase, Unit Broad-band Dipoles.

R.F. Connection, The
213 N. Frederick Ave. Suite 11
Gaithersburg, MD 20877
Phone: 301 840-5477; 800 783-2666
FAX: 301 869-3680
Established: 1979; Employs 2
Sells via showroom and mail order.
Key Employee: Joel, KA3QPG; Don.
Top Lines: Amphenol, Kings, Delta, J&I, Andrew, International Wire & Cable, Certified Quality, Times, Intercomp, Custom Connectors and wire assemblies.

R & L Electronics
1315 Maple Avenue
Hamilton, OH 45011
Phone: 800 221-7735, 513 868-6399
FAX: 513 868-6574
Established: 1980; Employs 13
Sells via showroom and mail order.
Key Employees: Larry, N8CHL; Rita, WD8POC; Troy, N8ASZ; Roger, N8EKG.
Top Lines: Yaesu, Kenwood, ICOM, MFJ, Ten-Tec, Rohn, AEA, DAIWA, Cushcraft, Hustler.
Accepts trade-ins.
Branch: 8524 E. Washington St. Indianapolis, IN 46219
Phone: 800 524-4889; 317 897-7362 FAX: 317 898-3027
Established : 1992; Employs 5
Key Employee: Roger

R & S Electronics Ltd.
(See Canada Listings)

RT Systems Amateur Radio Supply
8207 Stephanie Drive
Huntsville, AL 35802
Phone: 205 882-9292; 1 800 723-6922
FAX: 205 880-0332
Established: 1993; Employs 3
Sells via showroom, mail order, and hamfests.
Top Lines: Yaesu, Alinco, Kantronics, AEA, MFJ, Pride Tubes, Comet, Daiwa.

RAD-COMM RADIO
3300 82nd Street, Suite E
Lubbock, TX 79423
Phone: 806 792-3669
Orders Only: 1 800 588-2426
Top Lines: Alinco, MFJ, Mirage, Kantronics, Comet, Maldol, Larsen, Genesys, ARRL Publications, CQ Publications, Artsci Publication, W5YI Publica- tions, Startek Freq Counters, ProAm Valor, Ameco.

Radio Bookstore
(Formerly Ham Radio Bookstore)
PO Box 209
Rindge, NH 03461-0209
Phone: 800 457-7373; 603 899-6957 FAX: 603 899-6826
Established: 1991
Sells via mail/telephone order.
Key Employee: J. Craig Clark, Jr.
Top Lines: Amateur Radio, SWL, CB & Scanner related books and software.

Radio City Inc.
2663 County Road I
Mounds View, MN 55112
Phone: 612 786-4475; 800 426-2891
Established: 1982; Employs 10
Sells via showroom and mail order.
Key Employees: Dan, KBØXC; Phil, AAØTR, John, NØISL, Maline and Denise.
Top Lines: Yaesu, ICOM, Kenwood, Alinco, Cushcraft, Comet, Telex, ARRL, Kantronics, Timewave, Midland, Fritzel, Bencher, MFJ, Mirage.
Accepts trade-ins; provides repairs.

Radio Club of J.H.S. 22, NYC, Inc.
PO Box 1052
New York, NY 10002
Phone: 516 674-4072
FAX: 516 674-9600
Established: 1980; Employs 5
Key Employees: Joe, WB2JKJ.
Non-Profit Educational Organization seeking used equipment.

Radio Communications
Box 212
Royal Oak, MI 48068-0212
Phone: 800 551-1955; 810 546-2010
Established: 1971; Employs 4
Sells via showroom and mail order.
Key Employees: Gil, K8CQM; Wally, WD8PSN; Gloria; TC.
Top Lines: Larsen, Hustler, Valor, Transel Bearcat, Alinco, Comet, Multiplier, Motorola, ARRL Book.
Accepts trade-ins and provides repairs.

Radio Comm. of Charleston, Inc.
102 Farm Road
Goose Creek, SC 29445
Phone: 803 553-4101
FAX: 803 553-3564

Radio Depot
16 W. 36th St.
New York, NY 10018
Phone: 212 714-9194
FAX: 212 714-1543
Sells via showroom and mail order.
Top Lines: Alinco, Motorola, Kenwood, Yaesu, IFR, Revex, Maxon, Lancer, Superfone, ICOM, International World Exporters

Radio Depot
2135 Sheridan Road, Suite A
Bremerton, WA 98310
Phone: 360 377-9067
Established: 1992; Employs 2
Sells via showroom and mail order.
Key Employee: Dave, N7KZN.
Top Lines: AEA, Alinco, Astron, Bencher, Cushcraft, Heil, Larsen, Maldol, MFJ, Periphex, Standard.

Radio Place, The
5675A Power Inn Road
Sacramento, CA 95824
Phone: 916 387-0730
FAX: 916 387-0744
Established: 1976; Employs 6
Sells via showroom and mail order.
Key Employees: Glenn, WR60; Gary, KI6T; Rudy, KC6MUR;
Roger, WD6CLZ, Justin, KE6TWS.
Top Lines: Yaesu, ICOM, Cushcraft, Larsen, Hustler, ARRL,
MFJ, Kantronics, Astron, Diamond, DAIWA, AEA and Comet.
Accepts trade-ins and provides repairs.

Radio Products
6198 Marlo Drive
Concord, OH 44077
Phone: 216 946-6889

The Radio Room
898 N. Broadway
N. Massapequa, NY 11758
Phone: 516 795-9371
Established: 1992; Employs 1
Sells direct and through dealers.
Major Lines: G5RV antennas and single band dipoles; Pro-Am,
Sirio, Taiwan Serene, Daiwa, Maldol Antennas and accessories.

Radio Works
PO Box 6159
Portsmouth, VA 23703
Phone: 804 484-0140
Established: 1984; Employs 4
Sells via mail order.
Key Employees: Jim, W4THU; Judy.
Top Lines: MFJ, B&W, Heil, Metz, ARRL and Radio Publications
books, Spi-Ro, Ameco, Alpha-Delta, ProAm, Van Gorden, Smiley.

Radio World
1656 Nevada Hwy.
Boulder City, NV 89005
Phone: 702 294-2666
FAX: 702 294-2668
Established: 1988
Key Employee: Dave, W9MPD; Karl, WB7DRJ; Dave, N7RNV.
Top Lines: Alinco, MFJ, Sangean, AOR, Maxon, Relm—Amateur,
Commercial, Marine, Shortwave Scanners.
New/Used—Sales/Service—Trades

RadioKit
Box 973
Pelham, NH 03076
Phone: 603 635-2235
FAX: 603 635-2943
Established: 1975; Employs 4
Sells via mail order.
Key Employee: Carl, KM1H.
Top Lines: Svetlana Tubes, Pride Tubes, LMB, Hammond, Ten-
Tec Enclosures, Jackson, JW Miller, Cardwell, Multronics,
RadioSwitch, Electroswitch/Centralab, Van Gorden, James Millen.
RADIOKIT plate and filament transformers, variable capacitors/
amplifier parts. QRP Kits & supplies.
International Sales: ICOM, Kenwood, Yaesu, Telex/HyGain,
Cushcraft.
Provides HF-VHF-UHF amplifier repairs, modifications.

Radioware Corp.
PO Box 1478
Westford, MA 01886
Phone: 800 950-9273; 508 452-5555
FAX: 508 251-0515
Established: 1992
Sells via mail/phone orders.
Key Employees: John, WA1KYH.
Top Lines: Austin Antenna, ICE, Davis RF-Flex-Weave, Unadilla,
M2 Enterprises, Antennaco, Pasokon SSTV, Lakeview, Palomar,
Radioware, Comet, Timewave, JPS, Digital Vision, Daiwa, Sigma,
Create, Solder-It.

Ray's Amateur Electronics
1701-G N. Main St.
High Point, NC 27262
Phone: 910 883-6038; 800 854-5338
FAX: 910 883-1464
Key Employee: Ray, N4LXW
Sells via showroom, mail order and hamfest.
Top Lines: Alinco, Standard, Azden, Comet, Genesys, Anttron,
MFJ, Daiwa, Uniden, Cobra, Hy-Gain, Mirage, Ameritron, used
ham radios & amplifiers.
Accepts trade-ins.

Rio Radio Supply, Inc.
515 S. 12th St., Box 1808
McAllen, TX 78501
Phone: 512 682-5224

Rivendell Electronics
8 Londonderry Road
Derry, NH 03038
Phone: 603 434-5371
Established 1982; Employs 5
Sells via showroom and mail order.
Key Employees: Joe, KC1D; Peter, KI1M; Nancy, N1CXC; Herb,
AK1V; Mario, WA1EZE.
Top Lines: ICOM, AEA, B&W, Cushcraft, Hy-Gain, Alinco,
Larsen, MFJ, Rohn, Bencher, Kantronics, ARRL.
Accepts trade-ins and provides repairs.

Rogus Electronics
250 Meriden-Waterbury Turnpike Southington, CT 06489
Phone: 203 621-2252
Established: 1979
Key Employees: John, WA1JKR; Frank, W1FD; Joe, N1ECB; Jan,
KA1NXX.
Sells via showroom and mail order.
Top Lines: Rohn Towers, Alinco, Kantronics, MFJ, Cushcraft,
Butternut, B&W, Diamond, RF Concepts, Astron, Bencher,
Ameritron, VanGorden, ProAm, Comtelco.

Rosen's Electronics, Inc.
104 E 2nd Avenue
Williamson, WV 25661
Phone: 304 235-3677
FAX: 304 235-8038
Established: 1980; Employs 2
Sells via showroom and mail order.
Key Employees: Larry, WR8M; and Liz, KB8GDG.
Top Lines: MFJ, Tandy Computers, JDR Microdevices, Relm,
Shinwa, TE Systems.
Accepts trade-ins and provides repairs.

Rosewood Company
PO Box 229, Hwy 78 East
Elko, SC 29826
Phone: 803 266-7900; 800 875-7762
Established: 1991; Employs 2
Sells via showroom, mail order, hamfests.
Key Employees: Ray, WA4OMM; Kathleen, KB4HWC.
Top Lines: Amateur Radio, SWL, CB & Scanner related books,
videos and software. Also carry used radios, antennas, and acces-
sories.

Ross Distributing, Co.
PO Box 234
78 S. State Street
Preston, ID 83263
Phone: 208 852-0830
FAX: 208 852-0833
Established: 1953; Employs 6
Sells via showroom and mail order.
Key Employees: Ross, WB7BYZ; Karen, KA7BLB; Kathy, Gae,
Shannon, David.
Top Lines: Kenwood, ICOM, Yaesu, AEA, Larsen, Kantronics,
Cushcraft, Telex, MFJ, Astron, Butternut, Diamond.
Accepts trade-ins and used gear.

S

S & S Amateur Radio Supply
16 Kelly
Cabot, AR 72023
Phone: 501 843-8262
Key Employees: John, Brenda

Satman, Inc.
6310 N University No. 3798
Peoria, IL 61612
Phone: 1 800 472-8626
Sells via mail order.
Top Line: Satellite Systems.

Scanner World USA
10 New Scotland Avenue
Albany, NY 12208
Phone: 518 436-9606

Slep Electronics
PO Box 100, Hwy.441
Otto, NC 28763-0100
Phone: 704 524-7519
Established: 1955; Employs 6
Sells via mail order.
Key Employee: Bill, W4FHY.
Top Lines: Military surplus radios and test equipment, technical manuals, receiver/transmitter tubes, early American amateur, military & antique radios.
Accepts trade-ins and provides repairs.

Soundnorth Electronics
1802 Highway 53
International Falls, MN 56679
Phone: 218 283-9290, 800 932-3337
Established: 1976; Employs 7
Sells via showroom and mail order.
Key Employees: Terry, WV0G: Gale, WV0O; Dan, WZ0A; Jami, WY0D; Mike, KB0DXW.
Top Lines: Alinco, DAIWA, NCG, Valor, ANTECO, Maxrad, JRC, T.E. Systems Amps, MFJ and all used gear.
Accepts trade-ins and provides repairs.

Spectronics Inc.
1009 Garfield Street
Oak Park, l L 60304
Phone: 708 848-6777
FAX: 708 848-3398
Established: 1967; Employs 10
Sells via showroom and mail order.
Top Lines: Surplus two-way equipment and SW receivers, accessories for both.

Spokane Radio
S. 25 Girad
Spokane, WA 99212
Phone: 509 928-3073

Sports Communication
PO Box 36
Scotts Mills, OR 97375
Phone: 503 873-2256
FAX: 503 873-2051
Established: 1993; Employs 3
Sells via mail order.
Key Employee: John, KB7ZUI
Top Lines: ICOM, Vertex (Yaesu), Uniden, Maxon (GMRS Radios).

Star Fire Tec
1679 Hollow Creek Court
San Jose, CA 95121
Phone: 408 274-7396
Key Employee: Doug, KC6FRY, Susan.
Established: 1990
Sells via mail order, conventions, swaps.
Top Lines: Anli, Daiwa, Genesys, Lakeview, Larsen, Maldol, The Pouch, ProAm, Tune Belt, Smiley, Transel, Vibroplex. Specialize in antennas and accessories.
Call for Free Mail Order Catalog.

Surplus Sales of Nebraska
1502 Jones Street
Omaha, NE 68102
Phone: 402 346-4750
FAX: 402 346-2939
Established: 1978 Employs 9
Sells via showroom and mail order.
Key Employee: Bob, WDØFDE.
Top Lines: B&W, SGC, RF transmitting components, connectors, Collins parts, Jackson Brothers, MicroMetals, Vectronics, Diamond, Vacuum caps/relays.

T

TNR Technical, Inc.
279 Douglas Avenue #1112
Altamonte Springs, FL 32714
Phone: 800 346-0601
FAX: 407 682-4469
Top Line:Replacement batteries and inserts for popular ham handhelds, gel cells, custom batteries.

Texas Radio Products
5 East Upshaw
Temple, TX 76501
Phone: 817 771-1188

Texas Towers
1108 Summit Avenue Suite 4
Plano, TX 75074
Phone: 800 272-3467, 214 422-7306
FAX: 214 881-0776
Established: 1977; Employs 9
Sells via showroom and mail order.
Key Employees: Gerald, K5GW; Matt, N5VQT; Neil, AA6TR; Cheryl, N5JSS; Steve, KC5BTX; Mark, N5XDI; Charlie, KB5GMG.
Top Lines: AEA, Alinco, Alpha Delta, Aluminum Tubing, Ameritron, Amphenol, Andrew, Anli, Artsci, ARRL, Astron, B&W, Bencher, Butternut, Callbook, Certified Quality, Comet, Create, Cushcraft, Daiwa, Diamond, Glenn Martin, Heil, Hustler, Hygain, Icom, Kantronics, Kenwood, Klein, KLM, Larsen, M2, Maldol, MFJ, Mirage, NyeViking, Outbacker, Palomar, Phillystran, Radio Publications, RF Concepts, Rohn, SAM Database, Standard, TE Systems, Timewave, Uni- versal Tower, US Tower, Van Gordon, Vibroplex, W5YI, Yaesu.
Send for Free Mail Order Catalog.

Tiare Publications
P.O. Box 493
Lake Geneva, WI 53147
Phone: 414 248-4845
Orders Only: 1 800 420-0579
Established: 1986; Employs 2
Sells via mail order.
Key Employees: Gerry L. Dexter, Pres.
Top Lines: Books for ham/radio monitor includes: Luchi's Amateur Radio License Guides, LowPower Communications (3 vol), The Code Book (CW), Ham Radio Contesting, HamSat Handbook, World Ham Net Directory, Pse QSL!, Hidden Ham Antennas, Weather Radio.
Full Catalog $1.

Transel Technologies/LJ Electronic Inc.
123 E. South Street
Harveysburg, OH 45032
Phone: 800 829-8321; 513 897-3442 FAX: 513 897-0738
Established: 1981; Employs 5
Sells via showroom and mail order.
Key Employee: Darrell, N801B
Top Lines: Transel Tech., Transel Corp., Centurion, Pack-lt,
Uniden, Maxon.
Accepts trade-ins and provides repairs.

Traxit , Inc.
2002 S. Main Street
McAllen, TX 78503-5416
Phone: 210 682-6559
FAX: 210 682-1658
Established: 1988; Employs 3
Sells via showroom and mail order.
Key Employees: Felipe, Alfredo,Lupita, Ratael.
Top Lines: Yaesu, Motorola, Hustler, Larsen, Cushcraft, Alinco,
ASP,Astron.
Branch Office: Traxit Comunicaciones
6144 Ave DelBosque
Pueba, Mexico
Phone: (22) 445193
Key Employees: Mario, Miguel, Ricardo.

Tucker Electronics
1717 Reserve Street
Garland, TX 75042
Phone: 800 527-4642; 214 340-0631
Established: 1967; Employs 116
Sells via showroom and mail order.
Key Employees: Alan, WQ5W; Hazen, KC5DX; Dick, K5WOR;
Wayne, N5LDD; Richard, KB5RFO; Mark, KC5EWT; Shane,
KC5EWU.
Top Lines: ICOM, Alinco, MFJ, AEA, Kantronics, Bearcat, Sony,
Daiwa, Hy-Gain, Hustler, Grundig, Edco, Create, Anli, Datong,
AOR, JPS, DRSI, Emoto, Jim, Sigma, Novex, Astron, Sangean,
Pro•am, Heil Sound, Tucker Line of Antennas, Antenna Tuners,
and dummy loads.
Accepts trade-ins and provides repairs.

Tune Belt, Inc.
2601 Arbor Place
Cincinnati, OH 45209
Phone: 513 531-1712
Established: 1985; Employs 8
Sells via mail order.
Key Employees: Richard, Doug.
Top Lines: Handheld radio belt.

U

US Radio
377 Plaza
Granbury, TX 76048
Phone: 817 573-0220
Orders Only: 800 433-SAVE
Established: 1981; Employs 4
Sells via showroom and mail order.
Key Employees: Stan and Steve.
Top Lines: Radio Shack/Realistic: Scanner, S.W. Radios, Ham
Radios.

US Scanner Publications
234 SE Grand Avenue
Portland, OR 97214
Phone: 503 233-4904

Universal Radio Inc.
6830 Americana Parkway
Reynoldsburg, OH 43068
Phone: 614 866-4267, 800 431-3939
FAX: 614 866-2339
Established: 1942; Employs 18
Sells via showroom and mail order.
Key Employees: Steve, NI8F; Fred, N8EKU; Dave, N8FVL; Scott,
KB2ARL; Jim, KC8XZ, Jerry, N8XMV.
Top Lines: Kenwood, ICOM, Yaesu, Japan Radio, AEA, Info-
Tech, Kantronics, Sony, Telex/Hy-Gain, Alpha-Delta, Sangean,
Alinco, Standard, MFJ, Ameritron, Cushcraft, Hustler, Lowe,
Grundig, ARRL, Diamond, Astron, Bearcat, AOR, Heil, Daiwa,
Maldol, Nye-Viking.
Accepts trade-ins and provides repairs

V

Valley Radio Center
1522 N. 77 Sunshine Strip
Harlingen, TX 78550
Phone: 210 423-6407; 800 869-6439
FAX: 210 423-1705
Established: 1959; Employs 8
Sells via showroom and mail order.
Key Employees: Bob, WD5KBZ; Rick, WD5ADC.
Top Lines: Kenwood, ICOM, Yaesu, Alinco, Larsen, Antenna
Specialists, Hustler, B&W, Mirage/KLM, RF Concepts, AEA,
Kantronics, Astron, CES, Connect System, Comm Systems.
Accepts trade-ins and provides repairs.

Van Valzah Company, H.C.
38W 111 Horseshoe Drive
Batavia, IL 60510
Phone: 708 406-9210
Established: 1981; Employs 1
Sells via mail order.
Key Employees: Howard, WB9IPG.
Top Line: Antenna parts.

W

W5YI Group, Inc.
2000 E. Randol Mill Road, Suite 608-A Arlington, TX 76011
Phone: 817 461-6443, 817 274-0400, 800 669-9594; FAX: 817
548-9594
Established: 1979; Employs 7
Sells via mail order. Accepts checks/ VISA/MasterCard.
Key Employees: Fred, W5YI; Larry, N5XBM; Arlene, N5YLT,
Steve, NS51, Sherry, KB5YUP.
Top Lines: Amateur Radio License Preparation materials, code
tapes, license study manuals, videos, educational software, part 97
rulebooks, logbooks, Commercial operator license preparation
material etc.

Williams Radio Sales
600 Lakedale Road
Colfax, NC 27235
Phone: 910 993-5881
Established: 1974; Employs 3
Sells via showroom, hamfests and mail order.
Key Employees: Wayne, K4MOB; Gerry, KB4SEL; Chris,
KE4FUP.
Top Lines: Alinco, Hustler, Larsen, ARRL books, Paccomm,
Pyramid, Startec, Factory Direct Batteries.
Accepts trade-ins.

The Wireman, Inc.
261 Pittman Road
Landrum, SC 29356
Phone: 800 727-WIRE, 803 895-4195
FAX: 803 895-5811
Established: 1975; Employs 8
Sells via mail order and dealers.
Key Employee: Press, N8UG.
Top Lines: Wire & cable, wire antennas and accessories, baluns, kits, noise bridges.
Provides technical assistance.

Woody's Antenna & Tower Service
PO Box 222
Levittown, NY 11756
Phone: 516 462-2524
Established: 1979; Employs 3
Sells via mail order.
Key Employees: Woody, K2UU; Joe, KA2UYV.
Top Lines: Tri-Ex Tower, US Tower.
Accepts trade-ins.

Canada

Atlantic Ham Radio, Ltd.
368 Wilson Avenue
Downsview, ONT M3H 1S9
Phone: 416 636-3636
FAX: 416 631 -0747
Established: 1979; Employs 12
Sells via showroom and mail order.
Key Employees: Lutz; Mike, VA3MW; Nick, VE3SEC; Mario, VE2MBZ; Howard, VE2HNL; Steve, VE3OOS; Alan, VE3XAG.
Top Lines: Kenwood, ICOM, Yaesu, Sony, JRC, AEA, Kantronics, Hy-Gain, Cushcraft, MFJ, Astron, Nye-Viking, Uniden, Butternut, DAIWA, Hustler, Larsen, Mirage/KLM, Diamond, Yupiteru. We have full 800 MHz scanners. (Yupiteru, AOR, ICOM, etc.)
Accepts trade-ins and provides repairs.

Burnaby Radio Communications
4257 E. Hastings St.
Burnaby, BC V5C 2J5
Phone: 604 298-5444
FAX: 604 298-5455
Established: 1992; Employs 4
Sells via showroom and mail order.
Key Employees: Bill, VE7CIM; John, VE7AYP; Ron, VE7HRE; Don, VE7CBT.
Top Lines: Kenwood, Yaesu, Alinco, ICOM, Comet, Astron, PacComm, Larsen, Bencher, Valor.
Accepts trade-ins and provides repairs.

Caltronics Communications
#9 7139-40th St. S.E.
Calgary, Alberta T2C 2H7
Phone: 403 279-2242
Key Employee: Claude

Com-West Radio Systems, Ltd.
48 East 69th Avenue
Vancouver, BC V5X 4K6
Phone: 604 321-3200
FAX: 604 321-6560
Established: 1970; Employs 6
Sells via showroom and mail order.
Key Employees: Ron, VE7XR; Stan, VE7STN; Al, VE7DL; Fred, VE7EE.
Service: Doug, VE7HDL.
Top Lines: ICOM, Kenwood, Yaesu, Cushcraft, Mirage/KLM, Larsen, AEA, Butternut, Hustler, Unadilla, Kantronics, ARRL, Astron, B&W, MFJ, Telex/ HyGain, Alinco, Maldol, Paccomm, Bencher, Daiwa, Falcon, Optoelectronics, Outbacker, Vectronics.
Provides repairs.

MacFarlane Electronics, Ltd., H.C.
R.R. #2
Battersea, ONT KOH 1H0
Phone: 613 353-2800
FAX: 613 353-1294
Established: 1958; Employs 4
Sells via showroom and mail order.
Key Employee: Harold, VE3BPM, Tom, VE3UXP .
Top Lines: Kenwood, Yaesu, ICOM, Alinco, Telex/HyGain, Cushcraft, Larsen, Hustler, Diamond, ARRL Publications, Vibroplex, Uniden, MFJ, Ameritron, Nye-Viking, Bencher, Alpha-Delta, Uniden, Delhi and Trylon Towers. Antenna systems designed and installed within a radius of 300 KM.
Accepts trade-ins and provides repairs.

Norham Radio, Inc.
4373 Steeles Ave. W
North York, ON M3N 1V7
Phone: 416 667-1000
Established: 1990; Employs 8
Sells direct
Top Lines: Kenwood, ICOM, Yaesu, AEA, Kantronics, Cushcraft, Telex-HyGain, Astron, Timewave, JPS, MFJ, JRC, Comet, Diamond, Heil, Larson, Outbacker. Also AOR & Yupiteru (800 MHz scanners).

R & S Electronics Ltd.
306 Prince Albert Road
Dartmouth, Nova Scotia B2Y 1N2
Phone: 902 464-0464
FAX: 902 464-0090
E-Mail: dgrantha@fox.nstn.ca
Established: 1976; Employs 4
Sells via showroom and mail order.
Key Employee: Dick, VE1AI; Mike, VE1MTG, Jim, VE1JMM, Sandra.
Top Lines: Kenwood, ICOM, Ten-Tec, Hy-Gain, Yaesu, Cushcraft, KLM, Alinco, MFJ, Uniden, AEA, M2, Comet;Heil, GAP, X-10, Timewave, cellular scanners, etc.
Accepts trade-ins and provides repairs.

Radio Progressive Inc.
8104-A Trans Canada Highway
Ville St. Laurent, Quebec H4S 1M5
Phone: 514 336-2423
FAX: 514 336-5929
Key Employees: Jean-Claude, VE2DRL; Joe, VE2ALE; Bruno, VE2JFX; Julio, VE2NTO; Patrick, SW2.
Top Lines: ICOM, Kenwood, Yaesu, Kantronics, AEA, Cushcraft, Telex/Hy-Gain, Daiwa, Diamond, Anli, Valor, MFJ, Timewave, Yupiteru, Ace, & more.
Free Extended Warranty.
Accepts trade-ins and factory authorized repair centre.

Skyware Communication Technology
64590-1942 Como Lake Ave.
Coquitlam, B.C. V3J 7V7
Phone: 604 936-1790
FAX: 604 936-1740
Established: 1991; Employs 5
Sells via mail order.
Top Lines: Antenna & communication accessory specialist. Full 800 MHz scanners available.
Free Catalog.

Wholesale Connexion
c/o Target Importers
620 Supertest Road
Downview, ON M3J 2C6
Phone: 416 665-6239
FAX: 416 739-5423
Key Employee: Peter Klein
Top Lines: Cobra, Maco, Paradynamics, Uniden.

LIST OF PRODUCT REVIEWS
(From CQ and QST)

A & A Engineering, 20 Meter QRP CW Transceiver Kit, CQ May-94, Carr, Paul, N4PC

A & A Engineering, Deluxe Memory Keyer Kit, QST Nov-88, Hale, Bruce KB1MW

A & A Engineering, Spectrum Analyzer, CQ Feb-91, Bertini, Peter, K1ZJH

A & A Engineering, View Port VGA Slow-Scan TV System, QST Feb-93, Taggart, Ralph WB8DQT

A P Products, Powerface Solderless Breadboard, CQ Feb-81, Weinstein, Martin Bradley, WB8LBV

A.R.E., Silencer External Speaker and DTMF Decoder, CQ Jul-92, Ingram, Dave, K4TWJ

Accu-Circuits, Accu-Memory II Keyer, QST Jul-79, Westbrook, Jim K1FD

Advanced Computer Controls, RC-850 Repeater Controller, QST Feb-84, O'Dell, Pete KB1N

Advanced Computer Controls, Shackmaster 100, CQ Nov-86, McCoy, Lew, W1ICP

Advanced Radio Devices, 230A HF/MF Linear Amplifier, QST May-89, Wilson, Mark AA2Z

Advanced Receiver Research, MM144VDG Mast-Mounted Preamp, QST Feb-87, Wilson, Mark AA2Z

Advanced Receiver Research, TRS04VD TR Sequencer, QST Feb-87, Wilson, Mark AA2Z

AEA, 430-16 Wideband 70-CM Yagi Antenna, QST May-92, Taggart, Ralph WB8DQT

AEA, AD-1 Auto Dialer, QST May-79, Barker, George WB8PBC

AEA, AEA-FAX HF-Facsimile Receiving System, QST Apr-92, Kleinschmidt, Kirk NTØZ

AEA, AEASOFT Morse University, CQ Jan-87, McCoy, Lew, W1ICP

AEA, AMT-1 AMTOR Terminal Unit, QST Nov-83, Bender, Chuck W1WPR

AEA, AT-300 Antenna Tuner, CQ Apr-90, McCoy, Lew, W1ICP

AEA, AT-300 Antenna Tuner, QST Aug-90, DeMaw, Doug W1FB

AEA, ATU-1000 Advanced Terminal Unit, CQ Nov-86, Sternberg, Norm, W2JUP

AEA, BT-1P Code Trainer, QST Feb-83, Anderson, Marian WB1FSB

AEA, CK-1 Memory Keyer, QST Aug-81, Aurick, Lee W1SE

AEA, CP-1 and AEAsoft RTTY System, CQ Nov-83, Ingram, Lew, K4TWJ

AEA, CP-1 Computer Patch Interface, QST Apr-84, Pagel, Paul N1FB

AEA, Doctor DX C64 Morse Trainer, QST Dec-84, Ward, Jeff K8KA

AEA, Doctor DX Morse Code DX and Contest Simulator, CQ Oct-84, Lochner, Bob, W9KNI

AEA, DSP-2232 Multimode Communications Processor, QST Apr-93, Ford, Steve WB8IMY

AEA, Fast-Scan Television System, QST May-92, Taggart, Ralph WB8DQT

AEA, Fax WEFAX (Weather Facsimile), CQ Nov-92, Rogers, Buck, K4ABT

AEA, Hamlink Radio Controller, CQ Aug-93, Lynch, Joe, N6CL

AEA, Hot Rod Antenna for 2-M, QST Nov-83, O'Dell, Peter KB1N

AEA, HR1 Half Wave 2 Meter HT Antenna, CQ Jul-83, Ingram, Dave, K4TWJ

AEA, Isoloop 14- to 30-MHz Antenna, QST Apr-91, DeMaw, Doug W1FB

AEA, Isoloop Antenna, CQ Apr-93, McCoy, Lew, W1ICP

AEA, Isoloop HF Antenna, CQ Jul-90, Rogers, Buck, K4ABT

AEA, Isopole 2-M Antenna, QST Apr-80, Aurick, Lee W1SE

AEA, Isopole 220-MHz Vertical Gain Antennas, QST Jun-82, O'Dell, Pete KB1N

AEA, KT-2 Keyer/Trainer, QST Dec-83, Place, Steve WB2EYI

AEA, KT-Keyer, QST Jan-81, O'Dell, Pete KB1N

AEA, Map 64/2 Micropatch and TI-1 Tuning Indicator, CQ May-85, Gode, Bill, KB9IY

AEA, MBA RC Code Reader, QST Aug-83, DeMaw, Doug W1FB

AEA, MBA RO Code Reader, QST Aug-83, Pagel, Paul N1FB

AEA, MBA-RC Model-BAudot-ASCII Reader/Code Converter, CQ Jul-83, Schultz, John J., W4FA

AEA, MK-1 Memory Keyer, QST Oct-80, Collins, George KC1V

AEA, MM-3 Morse Machine, QST Jul-90, Kilgore, Jeff KC1MK and Wolfgang, Larry WR1B

AEA, Morsematic MM1, QST Oct-80, Collins, George KC1V

AEA, Moscow Muffler-A Russian Woodpecker Blanker, CQ Feb-83, McCoy, Lew, W1ICP

AEA, MPS-100 Power Supply, QST May-92, Taggart, Ralph WB8DQT

AEA, Pakratt Model PK-64, QST Jun-86, Palm, Rick K1CE

AEA, PK-232 Multi Mode Data Controller, CQ Nov-87, Mayo, Jonathan L., KR3T

AEA, PK-232 Multi-Mode Digital Communications Terminal, QST Jan-88, Hale, Bruce KB1MW

AEA, PK-64 Pakratt and HFM-64 Modem, CQ May-86, McCoy, Lew, W1ICP

AEA, PK-88 Packet Controller, QST Dec-93, Ford, Steve WB8IMY

AEA, PK-900 Multimode Communications Processor, QST Oct-93, Ford, Steve WB8IMY

AEA, PK-96 1200/9600 Packet TNC, QST Sep-94, Ford, Steve WB8IMY

AEA, PKT-1 Terminal Node Controller, QST Nov-85, Ward, Jeff K8KA and Wilson, Mark AA2Z

AEA, RLA-70 Power Amplifier/Preamplifier, QST May-92, Taggart, Ralph, WB8DQT

AEA, SWR 121 Graphical HF Antenna Analyst, CQ Jun-94, Carr, Paul, N4PC

AEA, VSB-70 Video/Audio Transceiver, QST May-92, Taggart, Ralph WB8DQT

Aftronics, Superscaf Audio Filter, CQ Nov-87, McCoy, Lew, W1ICP

Alda, 103 Transceiver, QST Dec-78, Gerli, Sandy AC1Y

Alinco, ALM-203T 2-M FM HH Transceiver, June-86 , Williams, Bruce WA6IVC

Alinco, DJ-100T HH 2-M FM Transceiver, QST Mar-89, Norton, Glen KC1MM

Alinco, DJ-180 2 Meter FM Handheld, CQ May-93, Thompson, David L., K4JRB

Alinco, DJ-560T Dual-Band HH FM Transceiver, June-91, Healy, Rus NJ2L

Alinco, DJ-580T Dual Band FM Handheld, CQ Dec-92, Ingram, Dave, K4TWJ

Alinco, DJ-580T Dual-Band HH FM Transceiver, QST Mar-94, Ford, Steve WB8IMY

Alinco, DJ-582T Dual-Band Hand-Held Transceiver, QST Jul-95, Ford, Steve WB8IMY

Alinco, DJ-G1 2 Meter Handheld, CQ Nov-94, Lynch, Joe N6CL

Alinco, DR-110T 2-M FM Transceiver, QST Jul-89, Kleinschmidt, Kirk NTØZ

Alinco, DR-110T Transceiver, CQ Dec-90, Steer, Chuck, WA3IAC

Alinco, DR-112T Mobile 2-M FM Transceiver, QST Dec-91, Healy, Rus NJ2L

Alinco, DR-130 2-M FM Transceiver, QST Jan-95, Ford, Steve WB8IMY

Alinco, DR-600T Dual-Band Mobile FM Transceiver, QST Jun-93, Healy, Rus NJ2L

Alinco, ELH230D 2 Meter RF Amplifier, CQ Dec-85, Ingram, Dave, K4TWJ

Alinco, ETS-210 Roof Tower, CQ Aug-87, Katz, Steve, WB2WIK

Alliance, HD-73 Heavy-Duty Rotator, QST Dec-80, Gerli, Sandy AC1Y

Allied, A-2516 Receiver, QST Jan-70, Lange, Walter W1YDS

Allied, A-2517 Transceiver, QST Nov-70, DeMaw, Doug W1CER

Allied, A-2587 146- to 175-MHz FM Receiver, QST Mar-70, Lange, Walter W1YDS

Allied Radio Shack, Series 190 Receivers, QST May-72, Scherer, Wilfred M., W2AEF

Alpha-Delta, Delta 4 Coax Switch, CQ Sep-87, McCoy, Lew, W1ICP

Alpha-Delta, DX-A "Twin Sloper" Antenna, CQ Aug-85, McCoy, Lew, W1ICP

AlphaLab, TriField Meter, QST May-93, Overbeck, Wayne N6NB

AMCOMM, S225 2-M Transceiver, QST Jan-78, Kearman, Jim W1XZ

Ameco Equipment Co., Ameco PT-3 1.8-54 MHz Preamplifier, QST Apr-88, Wilson, Mark AA2Z

American Circuits and Systems, MK1 Function Generator, QST Dec-74, DeMaw, Doug W1CER

Ameritron, AL-1200 HF Linear Amplifier, QST Dec-85, Wilson, Mark AA2Z

Ameritron, AL-1200 HF Linear Amplifier, CQ Oct-90, Schultz, John J., W4FA

Ameritron, AL-80A Linear Amplifier, CQ Apr-90, O'Dell, Peter, WB2D

Ameritron, AL-80B HF Amplifier, CQ Aug-93, McCoy, Lew, W1ICP

Ameritron, AL-811 HF Linear Amplifier, CQ Jul-92, Schultz, John J., W4FA

Ameritron, AL-811 MF/HF Linear Amplifier, QST Feb-92, Jahnke, Bart KB9NM

Ameritron, AL-82 Amplifier, July-92, Wilson, Mark AA2Z and Healy, Rus NJ2L

Amperex, 110 2-M Amplifier, QST Mar-79, Bender, Chuck W1WPR

Angle Linear, VHF/UHF Receiving Preamplifiers, QST Aug-78, Kearman, Jim W1XZ

Antec, Universal Transmatch Model UT-1, QST Feb-72, Nelson, John WØDRE/1

Antenna, Inc., Model 10043 Power/SWR Meter, QST Nov-76, Hall, Jerry K1TD

Antenna Mart, AMQ-2-5 Two-Element Five Band Quad Antenna, CQ Apr-95 McCoy, Lew W1ICP

Antenna Specialists, HM-224 220-MHz Mobile Antenna, QST Apr-77, Unknown

Antennas West, Product Line, CQ Apr-89, Ingram, Dave, K4TWJ

AOR, AR2500 Scanning Receiver, QST Sep-91, Kleinschmidt, Kirk NTØZ

Apollo(Village Ring), 2000X-2 Antenna Tuner, QST May-80, Pagel, Paul N1FB

Archer, 12-V. Air Compressor, CQ Aug-79, Dorhoffer, Alan M., K2EEK

Archer, Micronta 12V./8 Amp Power Supply, CQ Aug-79, Dorhoffer, Alan M., K2EEK

ARCOS, 432-MHz Trans. Conv./Amps., QST Aug-76, McMullen, Tom W1SL

Astrolite, 436B Headset, QST Jan-79, Aurick, Lee W1SE

Astron, RS-50M and RM-50M Power Supplies, CQ Nov-90, Bertini, Peter J., K1ZJH

Atlas, 110 Series QRP Transceiver, CQ Dec-79, Weiss, Adrian, K8EEG/WØRSP

Atlas, 210 SSB Transceiver, CQ May-75, Schultz, John, W2EEY

Atlas, 215 SSB Transceiver, CQ May-75, Schultz, John, W2EEY

Austin, Dual Band (144 and 440 MHz) Antennas, CQ Feb-87, McCoy, Lew, W1ICP

Austin, Metropolitan Triband VHF/UHF Antenna, QST Jan-89, Wolfgang, Larry WA3VIL

Austin, OMNI 2-M Antenna, QST Dec-83, Tilton, Ed W1HDQ

Austin, Superlite 2 Meter Beam, CQ Jan-84, McCoy, Lew, W1ICP

Autek Research, MK-1 Keyer, June-77, DeMaw, Doug W1FB

Autek Research, QF-1 RC Active Filter, QST Mar-77 , DeMaw, Doug W1FB

Autek Research, QF-1A Audio Filter, QST Jul-80, DeMaw, Doug W1FB

Autek Research, RF-1 R.F. Analyst, QST May-95, Kennamer, Bill K5FUV

Autek Research, WM1 SWR/Wattmeter, QST Nov-89, Kleinschmidt, Kirk NTØZ

Avanti, AH 151.3G Window-Mount Antenna, QST Dec-79, Lindholm, John W1XX

Avatar Magnetics, Transformers, QST Nov-82, Woodward, George W1RN

Azden, AZ-21A 2 Meter Handheld, CQ Feb-95, McCoy, Lew W1ICP

Azden, AZ-61 50 MHz FM Handheld Transceiver, CQ Jan-94, Lynch, Joe, N6CL

Azden, AZ-61 6-Meter FM Hand-Held Transceiver, QST Mar-95, Ford, Steve WB8IMY

Azden, PCS 2000 2-M FM Transceiver, QST Aug-80, Kleinman, Joel N1BKE

Azden, PCS-300 2-M Hand Held, QST Sep-82, Colvin, Carol AJ2I

Azden, PCS-7000H 2-M FM Transceiver, QST Jan-95, Ford, Steve WB8IMY

Azden, PCS-7000H Mobile 2-M FM Transceiver, QST Dec-91, Healy, Rus NJ2L

Azden, PCS-9600D 440 Mhz Voice/Data Tranceiver, QST May-95, Ford, Steve WB8IMY

B & K , Model 1820 Frequency Counter, QST Nov-78, Kearman, Jim W1XZ

Barker and Williamson, AS-40/AS-20 Trap Dipoles, CQ Mar-85, Schultz, John J., W4FA

Barker and Williamson, Model 593 Coaxial Switch, CQ Mar-85, Schultz, John J., W4FA

Barker and Williamson, RF Clipper, QST Oct-79, Halprin, Bob K1XA

Barlow-Wadley, XCR-30 Receiver, QST Jan-77, McCoy, Lew W1ICP

Bearcat, Model 100 Scanner, QST Mar-83, Steinman, Hal K1ET

Bearcat, Model 210 Scanner, QST Jul-78, Karpiej, Dave K1THP

Bencher, Paddle, QST May-78, Cain, Jim K1TN

Bencher, Paddle (Improved model), QST Dec-80, Pelham, John W1JA

Bencher, Paddle and MFJ Keyers, CQ Dec-81, Schultz, John J., W4FA

Bencher, ZA-1 and ZA-2 Baluns, QST Oct-80, DeMaw, Doug W1FB

Berk-Tek, RG-8X Coaxial Cable, QST Dec-79, Woodward, George W1RN

Bird, 4360 and 4362 Ham-Mate Wattmeters, QST Aug-79, Nelson, John W1GNC

Bird, 4381 Power Analyst, QST Jul-80, Collins, George KC1V

Bird, 4381 RF Power Analyst, CQ Jul-83, Schultz, John J., W4FA

Bird, 43P Peak-reading Directional Wattmeter, QST Dec-89, Healy, Rus NJ2L

Bird, 4410 Thruline Wattmeter, QST Oct-83, Wilson, Mark AA2Z

Bird, Bird Ham-Mate Directional Wattmeter, QST May-72, Wooten, Walter W1NTH

Bird, Ham-Mate Directional RF Wattmeters, CQ Jul-72, Scherer, Wilfred M., W2AEF

Bird, Series 4410 RF Directional Wattmeter, CQ Feb-91, McCoy, Lew, W1ICP

Bitcil Systems, Inc., Magnum Six RF Processor, QST Nov-72, Myers, Robert W1FBY

Blacksburg Group, Fist Fighter Keyer, QST Jan-83, Pagel, Paul N1FB

Bowmar, MX-100 Electronic Calculator, QST Aug-74, DeMaw, Doug W1CER

Braun, TTV 1270 Transverter, QST Jul-71, Blakeslee, Douglas W1KLK

Brown & Simpson Eng., MK-75 Electronic Keyer, QST Aug-76, Carrol, Charles W1GQO

Buckmaster, CD ROM Amateur Listing, CQ Mar-92, McCoy, Lew W1ICP

Buckmaster, Hamcall BBS Service, CQ Apr-93, Lynch, Joe, N6CL

Bunker Ramo, Solderless Coaxial-Cable Connectors, QST Mar-77, Watts, Chuck WA6GVC

Butternut, 2MCV 2 Meter Collinear Antenna, CQ Apr-83, Schultz, John J., W4FA

Butternut, HF2V Vertical Antenna, CQ Jul-85, Hagen, Jerry, N6AV

Butternut, HF4B Butterfly Beam, CQ Aug-86, McCoy, Lew, W1ICP

Butternut, HF5V-II Multiband Vertical, QST May-79, Clark, Craig Jr., N1ACH

Butternut, HF6V Multiband Vertical Antenna, CQ Mar-85, McCoy, Lew, W1ICP

Butternut Electronics, HF5V-III Multiband Verrtical Antenna, CQ Feb-82, Schultz, John J., W4FA

CEPCO, T10-C Touch-Call Decoder, CQ Jul-74, Sternberg, Norm, W2JUP

CES, 510SA Smart Patch, QST Apr-84, Hall, Jerry K1TD

Clear Channel Comm., AR-3000 Ranger 10-M All-Mode Trans., QST Jun-87, Bartoloth, Leslie KA1MJP

Clegg, AB-144 All-Bander Receiving Converter, QST Oct-80, Glassmeyer, Bernie W9KDR

Clegg, FM-27B FM Transceiver, QST May-73, McCoy, Lewis W1ICP

Clegg, FM-28 2-M FM Transceiver, QST Jun-78, Leipper, Bryan K1CD

Clegg, FM-DX 2 Meter FM Transceiver, CQ Jul-76, Paul, Hugh R., W6POK

Clegg, FM-DX 2-M FM Transceiver, QST Sep-76, McCoy, Lew W1ICP

Coax-Seal, Antenna Protection, CQ Mar-83, McCoy, Lew, W1ICP

Coaxial Dynamics, 81000-A Directional Wattmeter, CQ Dec-87, O'Dell, Peter R., KB1N

Coaxial Dynamics, 83000-A MF/HF Wattmeter, QST Feb-91, Healy, Rus NJ2L

Collins, 30L-1 Linear Amplifier, CQ Oct-70, Scherer, WilfredM., W2AEF

Collins, KWM-380 HF Transceiver, QST Oct-82, Collins, George KC1V

Collins, Low-Cost Mechanical Filter, QST Jun-78, DeMaw, Doug W1FB

Comcraft, CTR-144 2-Meter Transceiver, QST May-72, Tilton, Edward W1HDQ

Comet, CD-160H MF/HF Wattmeter, QST Feb-91, Healy, Rus NJ2L

Comet, Dual Band Base/Repeater Vertical Antenna, CQ Aug-92, McCoy, Lew W1ICP

Comm Center, Bantam Dipole, CQ May-79, Scwartz, Irwin, K2VG

Command Technologies, Commander HF-1250 Amp, July-92, Wilson, Mark AA2Z and Healy, Rus NJ2L

Command Technologies, Commander HF-2500 HF Amplifier, CQ May-90, McCoy, Lew, W1ICP

Command Technologies, Commander HF-2500 Linear Amplifier, QST May-91, Wilson, Mark AA2Z

Communications Associates, Inc., CF-8 FSK Converter/Keyer, QST May-70, Hall, Gerald K1PLP

Communications Elec. Spec., CES 200 & CES 210 Tone Pads, QST Feb-76, Watts, Chuck WA6GVC

Communications Elec. Spec., Model 100 Digital Display, QST Apr-76, Hall, Jerry K1TD

Communications Power, WM-7000 Wattmeter, QST Sep-78, Gerli, Sandy AC1Y

Communications Specialists, SS-32 M CTCSS Encoder, QST Mar-83, O'Dell, Pete KB1N

Communications Specialists, TE-64 Tone Encoder, QST Sep-80, O'Dell, Pete KB1N

Computer Warehouse Store, Video Monitor, QST Apr-77, McCoy, Lew W1ICP

Comtronix, FM-80 10-M FM Transceiver, QST Dec-80, O'Dell, Pete KB1N

Continental Specialties, Mini-Max 50 MHz Frequency Counter, CQ Jan-79, Schwartz, Irwin, K2VG

Control Signal Company, Automatic Keyer Module AKM-25, QST Nov-74, McMullen, Thomas W1SL

CPP1, Microcomputer Code Practice Chip, CQ Jul-83, McCoy, Lew, W1ICP

Create, 730V-1 A Multiband V Dipole, CQ Aug-88, Schultz, John J., W4FA/SVØDX

Create, CLP 5130-1 Log Periodic Antenna, CQ Jun-94, Lynch, Joe, N6CL

Create, CV48 80/40 Meter Vertical Antenna, CQ Jul-88, O'Dell, Peter R., WB2D

Create, RC Rotators, CQ Jan-93, McCoy, Lew, W1ICP

Create, RC5A Antenna Rotator, CQ Aug-87, McCoy, Lew, W1ICP

Create, VHF/UHF Log-Periodic Antenna (CLP5130-1, 2, &3), QST Aug-88, Jahnke, Bart KB9NM

Crosby, Jerry, LOGBOOK Software (TRS-80), QST Dec-83, Horzepa, Stan WA1LOU

Crown Microproducts, ROM-116, CQ Nov-83, McCoy, Lew, W1ICP

Cubic, Astro 102 BXA HF Transceiver, QST Dec-81, Pelham, John W1JA

Curtis, EK-39M Mnemonic Keyer, QST Mar-71, Hall, Gerald K1PLP and Myers, Robert W1FBY

Curtis, EK-402 Electronic Keyer, QST Mar-72, Niswander, Rick WA1PID and Unknown, WA8VRB

Curtis, EK-420 Programmable Electronic Keyer, QST Oct-73, Myers, Robert W1FBY

Curtis, EK-430 Keyer & 8044-2 Kit, QST Feb-76, DeMaw, Doug W1FB

Curtis, EK-480M CMOS Deluxe Keyer, QST Jun-80, Pagel, Paul N1FB

Curtis, Electro Devices EK-404 Delux Keyer, CQ Mar-73, Ross, Richard A., K2MGA

Curtis, Electro Devices Electronic Fists, CQ Jan-71, Scherer, WilfredM., W2AEF

Curtis, IK-440 Instructokeyer, QST Mar-76, DeMaw, Doug W1FB

Curtis, KB-4200 Morse Keyboard, QST Oct-74, Hall, Gerald K1PLP

Curtis, KM-420 Programmable Electronic Keyer, QST Oct-73, Myers, Robert W1FBY

Curtis Devices, EK-402 Programmable Electronic Keyer, CQ Jan-72, Scherer, Wilfred M., W2AEF

Cushcraft, 13B2 2 Meter Boomer Antenna, CQ Aug-93, Rogers, Buck, K4ABT

Cushcraft, 13B2 2-Meter "Boomer" Yagi Antenna, QST May-92, Lau, Zack KH6CP

Cushcraft, 17B2 2 Meter (Long Boom) Yagi, CQ Sep-92, Lynch, Joe, N6CL

Cushcraft, 2-M Collinear Array, QST Feb-78, DeMaw, Doug W1FB

Cushcraft, 20-3CD 20 Meter Beam, CQ Jan-86, Schultz, John J., W4FA

Cushcraft, 220B 220-MHz Boomer, QST Aug-83, Wilson, Mark AA2Z

Cushcraft, 225WB 220-MHz Yagi, QST Dec-89, Wilson, Mark AA2Z

Cushcraft, 26B2 2 Meter Boomer Antenna, CQ Aug-93, Rogers, Buck, K4ABT

Cushcraft, 32-19 Boomer & 324K Stacking Kit, QST Nov-80, Sumner, Dave K1ZZ

Cushcraft, 40-2CD 40 Meter Yagi Antenna, CQ May-87, Dorr, John, K1AR

Cushcraft, 40-2CD 40-M Skywalker Yagi, QST Jul-83, Wilson, Mark AA2Z

Cushcraft, 617-6B Boomer 6-M Yagi, QST Sep-82, Hull, Gerry AK4L

Cushcraft, A3WS 12 and 17 Meter Beam, CQ Apr-91, McCoy, Lew, W1ICP

Cushcraft, A4 Triband Beam, CQ Aug-84, Schultz, John J., W4FA

Cushcraft, A4 Yagi Antenna, QST Jan-83, DeMaw, Doug W1FB

Cushcraft, A50-6S 6-M Beam, QST Feb-92, Wilson, Mark AA2Z

Cushcraft, AP8, 8 Band Vertical Antenna, CQ Apr-88, Schultz, John J. W4FA

Cushcraft, AR-270 144 MHz/440 MHz Dual Band Vertical, CQ Jan-91, Schultz, John J., W4FA

Cushcraft, ARX-220B Ringo Ranger Antenna, QST Nov-87, Hale, Bruce KB1MW

Cushcraft, ASL 2010 Log-Periodic Antenna, CQ Aug-95, Carr, Paul N4PC

Cushcraft, ATB-34 Tribander, QST Jun-79, McCoy, Lew W1ICP

Cushcraft, D3W World Ranger 12, 17 and 30-Meter Rotatable Dipole, QST Oct-90, Newkirk, Dave AK7M

Cushcraft, R3 3-Band Vertical, QST Mar-83, Wolfgang, Larry WA3VIL

Cushcraft, R3 Vertical Multiband Antenna, CQ Apr-83, Schultz, John J., W4FA

Cushcraft, R4 Four Band Vertical Antenna, CQ Apr-89, McCoy, Lew, W1ICP

Cushcraft, R45K A 17 Meter Conversion Kit for the R4, CQ Nov-89, McCoy, Lew, W1ICP

Cushcraft, R5 Multiband Vertical Antenna, QST Oct-90, Sumner, Dave K1ZZ

Cushcraft, R7 HF Vertcal Antenna, CQ Apr-92, McCoy, Lew W1ICP

Daiwa, AF-606K Audio Filter, QST Jan-83, DeMaw, Doug W1FB

Daiwa, Automatic Antenna Tuners, CQ Apr-82, Schultz, John J., W4FA

Daiwa, CN-720 SWR and Power Meter, QST Jan-79, Aurick, Lee W1SE

Daiwa, Cross Needle Power/SWR Meters, CQ Apr-82, Schultz, John J., W4FA

Daiwa, CS-201 and CS-401 Coaxial Switches, QST May-79, Bartels, Garry WB1CPM

Daiwa, PS-304 Power Supply, CQ Nov-92, McCoy, Lew, W1ICP

Daiwa, RF 440 RF Speech Processor, QST Apr-79, Bartlett, Jim K1TX

Data Precision, 938 Digital Capacitance Meter, QST Nov-79, DeMaw, Doug W1FB

Datong Electronics, Ltd., PC1 General-Coverage RCVR Adptr., QST Apr-83, Glassmeyer, Bernie W9KDR

Datong Electronics, Ltd., Datest 1 IC/Transistor Tester, QST Apr-77, Unknown

Datong Electronics, Ltd., FL1 Audio Filter, QST Aug-79, Bartels, Garry WB1CPM

Davis, CTR-2-500 Frequency Counter, QST Apr-78, Leland, Stu W1JEC

Daytronics, Mimic Programmable-Memory Keyer, QST Dec-78, Bartlett, Jim K1TX

Debco, Rapid Mobile Charger, CQ Nov-80, Dorhoffer, Alan M., K2EEK

Decibel Products, DB-4048 Duplexer, QST Jun-75, McCoy, Lewis W1ICP and Unknown, WR1ABH

Decibel Products, DB-702/DB-702T Antennas, QST Oct-76, Watts, Chuck WA6GVC

Decibel Products, Vapor-Bloc Coaxial Cable, QST Nov-78, DeMaw, Doug W1FB

Delta Loop Antennas, Three Band Big Horn Beams, CQ Jun-88, McCoy, Lew, W1ICP

Dentron, 160-V Skyclaw Antenna, QST Nov-76, Price, Larry W4RA

Dentron, MLA 2500 Amplifier, QST Mar-78, Rusgrove, Jay W1VD

Dentron, MLA-2500 Linear Amplifier, CQ Sep-77, Paul, Hugh R., W6POK

Design Electronics, QSK 1500 High Power RF Switch, QST Sep-85, Towle, Jon WB1DNL

Design Electronics, QSK-1500, CQ Apr-85, McCoy, Lew, W1ICP

DGM Electronics, Inc., SRT 3000 Terminal, QST Dec-83, Kluger, Leo WB2TRN

Diamond, SX-600 Wattmeter/SWR Bridge, CQ Apr-89, Ingram, Dave, K4TWJ

Diamond Antenna, Antenna Line, CQ Apr-91, McCoy, Lew, W1ICP

Diamond Antenna, SX-100 MF/HF Wattmeter, QST Feb-91, Healy, Rus NJ2L

Diawa, NS-660PA MF/HF Wattmeter, QST Feb-91, Healy, Rus NJ2L

Dielectric Communications, Model 1000A RF Wattmeter, QST Dec-79, Glassmeyer, Bernie W9KDR

Digi-Field, Field Strength Meter, CQ Jul-94, DeMaw, Doug, W1FB

Digipet, Digipet-60 Frequency Counter, QST Apr-73, Dorbuck, Tony W1YNC

Digipet, Digitpet-160 Converter, QST Apr-73, Dorbuck, Tony W1YNC

Digital Radio System, PCPacket Adapter, CQ Jan-89, McCoy, Lew, W1ICP

Digital Radio Systems, PC Packet Adapter, QST Feb-89, Bloom, Jon KE3Z

Digitrex, PFC-4500 Frequency Counter, CQ Jan-90, Ingram, Dave, K4TWJ

Direct Conversion Technique, DC-10A Receiver Module/ VV-10 VFO, QST Jan-79, Bartlett, Jim K1TX

Douglas Randall, Scrubber, QST Nov-71, Myers, Robert W1FBY

Down East Microwave, 23-cm & 13-cm Loop Yagis, QST Jul-90, Wilson, Mark AA2Z

Down East Microwave, 3333LY 33-CM Loop Yagi, QST Apr-87, Wilson, Mark AA2Z

Down East Microwave, 432PA 432-MHz Amplifier Kit, QST Mar-93, Healy, Rus NJ2L

Down East Microwave, DEM432 No-Tune 432-MHz Transverter, QST Mar-93, Healy, Rus NJ2L

Down East Microwave, SHF-2400 2.4 Ghz Satellite Downconverter, QST Feb-94, Ford, Steve WB8IMY

Down East Microwave, Transverter IF Switch, June-92, Healy, Rus NJ2L

Drake, Marker Luxury 2M FM Transceiver, CQ May-71, Zook, GlenE., K9STH/5

Drake, ML-2 Marker Luxury FM Transceiver, QST Sep-71, Campbell, Laird W1CUT

Drake, R-4C Communications Receiver, QST Jan-74, Myers, Robert W1FBY

Drake, R7 Receiver, QST Jan-80, DeMaw, Doug W1FB

Drake, R8 Communications Receiver, CQ Feb-92, Schultz, John J., W4FA

Drake, R8 Shortwave Receiver, QST Mar-92, Kearman, Jim KR1S

Drake, RCS-4 Remote Coax Switch, QST Dec-76, Cain, Jim K1TN

Drake, SPR-4 Communications Receiver, CQ Nov-70, Scherer, WilfredM., W2AEF

Drake, SPR-4 Receiver, QST Dec-70, Blakeslee, Douglas W1KLK

Drake, SW8 General-Coverage Receiver, QST Oct-94, Kearman, Jim KR1S

Drake, T-4XC Transmitter, QST Feb-74, Sumner, Dave K1ZND

Drake, TC-6 Six Meter Transmitting Converter, Sept-70, Scherer, WilfredM., W2AEF

Drake, TR-22C 2-Meter Transceiver, CQ Apr-76, Paul, Hugh R., W6POK

Drake, TR-6 50-MHz Transceiver, QST Jul-70, Tilton, Edward W1HDQ

Drake, TR-6 Six-Meter Transceiver, CQ Feb-70, Scherer, Wilfred M., W2AEF

Drake, TR-7 HF Transceiver, QST May-79, Sumner, Dave K1ZZ

Drake, UV-3 VHF FM System, QST Aug-78, Kearman, Jim W1XZ

DRSI, DPK-2 Packet Controller, QST Dec-93, Ford, Steve WB8IMY

DSI Instruments, 3600A Frequency Counter, QST Feb-79, Kearman, Jim W1XZ

Dunestar, Model 600 Multiband Bandpass Filter, QST Mar-95, Straw, R.Dean N6BV

Dycom, PSU-13 VHF Scaler, QST Jan-72, Blakeslee, Douglas W1KLK

Dycomm, PSU-13 VHF Scaler, CQ Nov-71, Scherer, Wilfred M., W2AEF

E-Tek, FR4TR Frequency Counter, QST Oct-80, Halprin, Robert K1XA

E. F. Johnson, Fleetcom 550 UHF FM Transceiver, QST Sep-73, McMullen, Thomas W1SL

Egbert, RTTY Program, QST Jun-82, Kaczynski, Mike W1OD

Eldorado Electrodata, 225 Frequency Meter, QST Feb-71, Hall, Gerald K1PLP

Electronic Research, SL-55 Actiive Audio Filter, CQ Oct-78, Schwartz, Irwin, K2VG

Electronic Research, SL-65 VSWR/Net Power Indicator, CQ Jul-79, Schwartz, Irwin, K2VG

Electronic Signal Products, VHF 144-5A 2-M FM Receiver Kit, QST Nov-79, Bartlett, Jim K1TX

Electrospace Systems, HP-2 160-M Matching Network, QST Mar-79, Bartlett, Jim K1TX

Electrospace Systems, HV-580 10-M Dual-Mode Antenna, QST Mar-79, Bartlett, Jim K1TX

ELNEC, The Smart Mininec-Based Antenna Analysis Program, CQ Aug-91, Carr, Paul, N4PC

Emergency Beacon Corporation, EBC-144 Jr 2M Transceiver, CQ Jun-75, Sternberg, Norman, W2JUP

ENCOMM, Tokyo Hy-Power HL-1K HF Linear Amplifier, CQ Oct-87, Schultz, John J., W4FA/SVØDX

ENCON, Solar Array - A Study in Photovoltaic Power, CQ Apr-83, McCoy, Lew, W1ICP

ETO, Alpha 374 Bandpass Linear Amplifier, QST Apr-75, Unknown, WA1JZC

ETO, Alpha 76 Linear Amplifier, QST Jan-78, Cain, Jim K1TN

ETO, Alpha 77 Linear Power Amplifier, QST Mar-73, Myers, Robert W1FBY

ETO, Alpha 86 Linear Amplifier, QST Apr-89, Sumner, David K1ZZ

ETO, Alpha 87A HF Amplifier, CQ Aug-92, McCoy, Lew W1ICP

ETO, Alpha 87A MF/HF Linear Amplifier, QST Jun-92, Wilson, Mark AA2Z

ETO, Alpha 89 Linear Power Amplifier, QST Jul-94, Sumner, David K1ZZ

ETO, Alpha/Vomax SBP-3 Split Band Speech Processor, CQ Oct-77, Paul, Hugh R., W6POK

ETO, VOMAX, Speech Processor, QST Aug-77, LeLand, Stu W1JEC

ETO/Alpha, 78 High Frequency Linear Power Amplifier, CQ Dec-82, McCoy, Lew, W1ICP

Farr Technologies, 450 MHZ Corner Reflector, CQ July-95, McCoy, Lew W1ICP

Flesher, TR-128 Baud-Rate Converter, QST Apr-80 , Horzepa, Stan WA1LOU

Flesher, TU-170 RTTY Terminal, QST Mar-79, Horzepa, Stan WA1LOU

Flesher, TU-170A RTTY Terminal Unit, CQ Aug-83, McCoy, Lew, W1ICP

Flesher, TU-300 & TU-400 RTTY Modem, QST Jun-83, Pagel, Paul N1FB

Flesher Corp., TU-1170 RTTY Terminal Unit, CQ Oct-78, Schwartz, Irwin, K2VG

Fluke, 8000A 3 1/2 Digital Multimeter, QST Jun-73, McMullen, Thomas W1SL

Fluke, 8020A Multimeter, QST Sep-78, Greene, Clarke K1JX

Forbes Group, Ventenna, CQ May-92, Ingram, Dave, K4TWJ

Fox-Tango, 2.1-kHz TS-830 Filter, QST Sep-83, Kaczynski, Michael W1OD

Fox-Tango, YF-90H1.8 Crystal Filter, QST Apr-82, Woodward, George W1RN

GAP, Challenger DX-VI Vertical Antenna, CQ Mar-90, McCoy, Lew W1ICP

GAP, Challenger DX-VIII Vertical Antenna, QST Jan-95, Kennamer, Bill K5FUV

Garant Enterprises, GD-8 "Windom" Antenna, QST Sep-90, Hall, Jerry K1TD

GEM, Quad Antenna, QST Jan-78, Aurick, Lee W1SE

Genave, GTX-1 2-M Handheld Transceiver, QST Dec-76, Tilton, Ed W1HDQ

Genave, GTX-100 220-MHz Transceiver, QST May-76, Carroll, Charles W1GQO

Genave, GTX-2 FM Transceiver, QST Mar-74, Williams, Perry W1UED

Genave, GTX-200 FM Transceiver, QST Mar-74, Williams, Perry W1UED

Genave, GTX-600 6-Meter FM Transceiver, QST May-75, Hall, Gerald K1PLP

Genave, GTX-800 2-M FM Transceiver, QST Nov-78, LaPorta, Jim N1CC

Genwest Engineering, Voicebox, CQ Nov-90, McCoy, Lew, W1ICP

GLA Systems, Texas Bug Catcher, CQ Jul-90, McCoy, Lew, W1ICP

Gladding, 25 FM Transceiver, QST Dec-71, DeMaw, Doug W1CER

GLB Electronics, 400B Channelizer, QST Oct-73, McMullen, Thomas W1SL

Glen Martin Engineering, Hazer Elevator System, CQ Apr-89, Wilson, Steve, KØJW

GRF Computer Services, N6RJ 2ND OP Computerized Operating Aid, CQ Feb-92, McCoy, Lew, W1ICP

Grove Enterprises, Scanner Beam, CQ Jun-81, Romney, David

HAL, 2550/ID Keyer, QST Sep-76, Cain, Jim K1TN

HAL, CWR 6850 Telereader RTTY/CW Terminal, QST May-83, Hull, Gerry VE1CER

HAL, DKB-2010 Dual-Mode Keyboard, QST Jan-75, Hart, George W1NJM

HAL, DS-3000 Video Display Terminal, QST May-77, Bloom, Alan WA3JSU

HAL, DS-3100 ASR Video Display Terminal, QST Apr-80, Pagel, Paul N1FB

HAL, DS3200 ASR Video Terminal For CW and RTTY, CQ Nov-81, Robinson, H.B. "Robby", W2SR

HAL, FYO Key, QST Dec-76, Cain, Jim K1TN

HAL, HAL Devices 1550 Keyer w/Station Identifier, QST Dec-72, Dorbuck, Tony W1YNC

HAL, HAL System II RTTY/Morse Communications Terminal, CQ Nov-82, Schultz, John J., W4FA

HAL, ID-1A Repeater Identifier, QST Jun-73, Myers, Robert W1FBY

HAL, MCEM-8080 Microcomputer, QST Dec-76, Watts, Chuck WA6GVC

HAL, MKB-1 Morse Keyboard, QST Nov-73, Hall, Gerald K1PLP

HAL, P38 Communications Modem, QSY Aug-95, Ford, Steve WB8IMY

HAL, PCI-3000 Multimode HF Data Modem, QST Dec-92, Kleinschmidt, Kirk NTØZ

HAL, PCI-3000 System, CQ Apr-91, McCoy, Lew, W1ICP

HAL, PCI-4000 CLOVER-III Data Controller, QST May-93, Ford, Steve WB8IMY

HAL, RKB-1 TTY Keyboard, QST Apr-73, Hall, Gerald K1PLP

HAL, RTTY Keyboard, Video Display and Demodulator, CQ Jan-75, Robinson, Harry B., W2SR

HAL, RVD-1002 RTTY Video Display Unit , QST Apr-73, Hall, Gerald K1PLP

HAL, RVM-100 Radio Voice Mail Controller, CQ Jun-94, Rogers, Buck, K4ABT

HAL, SPT-1 Tuning Indicator for RTTY and SSTV, CQ Jul-85, Ingram, Dave, K4TWJ

HAL, ST-6 RTTY Demodulator, QST Apr-73, Hall, Gerald K1PLP

HAL, ST-6000 Demodulator, QST May-77, Bloom, Alan WA3JSU

HAL, ST-7000 Modem, CQ Nov-89, Rogers, Buck, K4ABT

HAL, ST-8000 HF Modem (modulator Demodulator), CQ Nov-86, Griffith, Jim S., WA5RAX

HAL Communictations, CRI-200 Computer RTTY Interface, CQ Nov-84, Ingram, Dave, K4TWJ

Hal-Tronix, 5312 12/24 Hour 6-Digit Clock Kit, CQ Sep-82, Swearengin, Bob, W5HJV

Hallicrafters, FPM-300 "SAFARI" SSB Transceiver, CQ May-73, Ross, Richard A., K2MGA

Hallicrafters, FPM-300 SSB Transceiver, QST Aug-73, Dorbuck, Tony W1YNC

Hallicrafters, HC-100 2-M FM Transceiver, QST Nov-71, McCoy, Lewis W1ICP

Hallicrafters, SX-122A Receiver, Jul/Aug-70, Scherer, WilfredM., W2AEF

Hallicrafters, SX-122A Receiver, QST Aug-70, Blakeslee, Douglas W1KLK

Ham Contact, Power Station, CQ Nov-94, Dorhoffer, Alan K2EEK

Ham Radio Center, Inc., Adjustable key, QST Jul-76, Rusgrove, Jay W1VD

Ham-Pro, 15 Meter Beam, CQ Jan-92, McCoy, Lew, W1ICP

Ham-Pro, H144-15 2 Meter Beam, CQ Dec-92, Lynch, Joe, N6CL

Hamco, Scotia Paddle, QST Dec-78, Bartlett, Jim K1TX

Hameg, HM-203 Dual Trace Oscilloscope, CQ Jun-83, Prentiss, Stan

Hameg, HM-204 Dual Trace Oscilloscope, CQ Jun-83, Prentiss, Stan

Hamlog/Applecoder, Apple II Software, QST Jan-83, Steinman, Hal K1ET

Hamtronics, 432-435 MHz Converter Kits, QST Jul-78, Harris, Chod WB2CHO

Hamtronics, P8 VHF Preamplifier, QST May-77, Kearman, Jim W1XZ

Hamtronics, XV-4 Transmitting Converter, QST Jan-82, Place, Steve WB1EYI

Harp, H. Alan, CW Sendin' Machine, QST Jul-76, Kearman, Jim W1XZ

Heathkit, Auto-Tune Antenna Tuner SA-2500, CQ Jun-85, McCoy, Lew, W1ICP

Heathkit, Color TV Set, CQ Mar-72, Scherer, Wilfred M., W2AEF

Heathkit, Computer Based License Preparation Courses, CQ Mar-89, O'Dell, Peter R., WB2D

Heathkit, EE-3404 6809 uP Course, QST Jul-84, O'Dell, Pete KB1N

Heathkit, ETS-3401 Microcomputer Training System, QST Sep-82, O'Dell, Pete KB1N

Heathkit, GC-1000 "Most Accurate Clock", QST Jan-86, Kaczynski, Mike W1OD

Heathkit, GC-1005 Electronic Clock, QST Dec-73, Hall, Gerald K1PLP

Heathkit, General Class License Course, CQ Jul-82, Schultz, John J., W4FA, Schultz, E., N2BVN

Heathkit, GH-17A Soldering Iron Kit, CQ Sep-77, Dorhoffer, Alan M., K2EEK

Heathkit, GR-110 Scanning Monitor, QST Aug-73, McCoy, Lewis W1ICP

Heathkit, GR-740 Scanner, QST Jan-85, Pagel, Paul N1FB

Heathkit, GR-78 Receiver, QST Oct-70, Huntoon, John W1LVQ

Heathkit, GU-1820 AC Power System, QST Dec-82, DeMaw, Doug W1FB

Heathkit, HA-202 2-Meter FM Amplifier, QST Aug-73, Sumner, Dave K1ZND

Heathkit, HD-1250 Dip Meter, QST Jan-76, DeMaw, Doug W1FB

Heathkit, HD-1250 Dip Meter, CQ Jul-81, Schultz, John J., W4FA

Heathkit, HD-1250 Solid State Dip Meter, CQ Dec-75, Math, Irwin, WA2NDM

Heathkit, HD-1410 Electronic Keyer, QST Mar-78, LeLand, Stu W1JEC

Heathkit, HD-1416 Code Oscillator, CQ Sep-77, Smith, Kim

Heathkit, HD-1418 Active Audio Filter, QST Mar-84, Wolfgang, Larry WA3VIL

Heathkit, HD-1420 VLF Converter, QST Nov-86, Williams, Bruce WA6IVC

Heathkit, HD-1422 Antenna Noise Bridge, QST Nov-86, Williams, Bruce WA6IVC

Heathkit, HD-1982 Micoder Microphone, QST Nov-76, Margolin, Bob K1BM

Heathkit, HD-3030 Computer Interface, QST Feb-85, Schetgen, Bob KU7G

Heathkit, HD-3030 RTTY Terminal Interface, CQ Nov-84, Rash, Wayne Jr., N4HCR

Heathkit, HD-4040 Terminal Node Controller, QST Nov-85, Ward, Jeff K8KA and Wilson, Mark AA2Z

Heathkit, HD-8999 Ultra Pro CW Keyboard, QST Apr-84, Kaczynski, Mike W1OD

Heathkit, HFT-9 Antenna Tuner, QST Jul-84, Wilson, Mark AA2Z

Heathkit, HG-10B V.F.O., CQ Mar-70, Scherer, Wilfred M., W2AEF

Heathkit, HK-232 Packkit Multi-Mode Digital, QST Jan-88, Hutchinson, Chuck K8CH

Heathkit, HL-2200 Amplifier, QST Nov-83, Pagel, Paul N1FB

Heathkit, HM-102 RF Power Meter, CQ Nov-71, Scherer, Wilfred M., W2AEF

Heathkit, HM-102 RF Power Meter, QST Dec-71, Blakeslee, Douglas W1KLK

Heathkit, HM-2103 RF-Load Wattmeter, QST Sep-73, Dorbuck, Tony W1YNC

Heathkit, HM-2140 Dual HF Wattmeter, QST Feb-80, Leland, Stu W1JEC

Heathkit, HM-2140-A MF/HF Wattmeter, QST Feb-91, Healy, Rus NJ2L

Heathkit, HM-2141 VHF Wattmeter, QST Sep-80, Pelham, John W1JA

Heathkit, HN-31A Cantenna, QST May-84, Hutchinson, Chuck K8CH

Heathkit, HO-5404 Station Monitor, QST Jan-87, Pagel, Paul N1FB

Heathkit, HR-1680 Receiver, CQ Oct-76, Paul, Hugh R., W6POK

Heathkit, HR-1680 Receiver, QST Jan-77, Kearman, Jim W1XZ

Heathkit, HV-2000 Voice Synthesizer, QST Dec-87, Hale, Bruce KB1MW

Heathkit, HW-101 SSB Transceiver, QST Jan-72, Unknown, WN1LZQ

Heathkit, HW-104 HF Transceiver, QST Dec-76, McMullen, Tom W1SL

Heathkit, HW-202 2 Meter FM Transceiver, CQ Jun-74, Math, Irwin, WA2NDM

Heathkit, HW-202 2-Meter FM Transceiver, QST Jul-74, Pride, Mark WA1ABV

Heathkit, HW-2021 2-M FM Hand-Held Transceiver, QST Jan-77, McCoy, Lew W1ICP

Heathkit, HW-2021 Hand-Held 2-Meter Transceiver, CQ Jun-76, Paul, Hugh R., W6POK

Heathkit, HW-2036 Synthesized 2-Meter Transceiver, CQ Mar-78, Paul, Hugh R., W6POK

Heathkit, HW-5400 HF Transceiver, QST Oct-84, Wolfgang, Larry WA3VIL

Heathkit, HW-5400 Transceiver, CQ Dec-83, McCoy, Lew, W1ICP

Heathkit, HW-6502 2 Meter HT and HWA 6502-2 Mobile Console, CQ Mar-87, McCoy, Lew, W1ICP

Heathkit, HW-7 CW QRP Transceiver, QST Jan-73, DeMaw, Doug W1CER

Heathkit, HW-7 Low-Power CW Transceiver, CQ Apr-73, Scherer, Wilfred M., W2AEF

Heathkit, HW-8 QRP Transceiver, QST Apr-76, DeMaw, Doug W1FB

Heathkit, HW-8 QRPP Transceiver, CQ May-77, Weiss, Adrian, K8EEG/Ø

Heathkit, HW-9 Deluxe QRP CW Transceiver, QST Jul-85, Hutchinson, Chuck K8CH

Heathkit, HW-99 Novice CW Transceiver, QST Mar-86 , Holsopple, Curt K9CH

Heathkit, HX-1681 CW Transmitter, QST Mar-81, Hull, Gerry AK4L

Heathkit, IB-101 Frequency Counter, QST May-71, DeMaw, Doug W1CER

Heathkit, IB-101 Frequency Counter, CQ Jul-71, Scherer, Wilfred M., W2AEF

Heathkit, IB-102 Frequency Scaler, QST Feb-72, Wooten, Walter W1NTH

Heathkit, IB-1100 Frequency Counter, CQ Jan-74, Scherer, Wilfred M., W2AEF

Heathkit, IB-1102 Frequency Counter, CQ Jan-74, Scherer, Wilfred M., W2AEF

Heathkit, IB-5281 RLC Bridge, QST Dec-80, Pagel, Paul N1FB

Heathkit, IC-2100 Elec. Slide Rule, QST Aug-76, McCoy, Lew W1ICP

Heathkit, ID-4101 Electronic Switch, CQ Sep-82, Schultz, John J., W4FA

Heathkit, ID-4801 EPROM Programmer, QST Aug-86, Bloom, Jonathan KE3Z

Heathkit, IG-4505 Oscilloscope Calibrator, CQ Sep-82, Schultz, John J., W4FA

Heathkit, IM-102 Digital Multimeter, CQ Jun-72, Scherer, Wilfred M., W2AEF

Heathkit, IM-102 Digital Multimeter, QST Oct-72, Blakeslee, Douglas W1KLK

Heathkit, IM-103 Line Voltage Monitor, QST Sep-72, Hall, Gerald K1PLP

Heathkit, IM-1202 Digital Multimeter, QST May-74, McMullen, Thomas W1SL

Heathkit, IM-2215 Digital Multimeter, QST Jun-80, Pagel, Paul N1FB

Heathkit, IM-2320 Digital Multimeter, QST May-87, Williams, Bruce WA6IVC

Heathkit, IMM-105 VOM, CQ Dec-71, Scherer, Wilfred M., W2AEF

Heathkit, IO-102 Oscilloscope, QST Jun-72, DeMaw, Doug W1CER

Heathkit, IO-4235 Dual Trace 35 MHz Oscsilloscope, CQ May-83, McCoy, Lew, W1ICP

Heathkit, IP-18 Regulated Power Supply, QST Dec-71, Hall, Gerald K1PLP

Heathkit, IP-2715 Battery Eliminator, QST Jun-77, Watts, Chuck WA6GVC

Heathkit, IP-2718 Tri-Power Supply, QST May-77, Hall, Jerry K1TD

Heathkit, IP-28 Regulated DC Supply, QST May-70, Blakeslee, Douglas W1KLK

Heathkit, IT-1121 Semiconductor Curve Tracer, QST Nov-74, Dorbuck, Tony W1YNC

Heathkit, M-4190 Bi-directional RF Wattmeter Kit, CQ Nov-78, Schwartz, Irwin, K2VG

Heathkit, Packkit HK-232 Multi-Mode Packet Controller, CQ Nov-88, O'Dell, Peter, WB2D

Heathkit, SA-1480 Remote Antenna Switch, QST Jul-80, Frenaye, Tom K1KI

Heathkit, SA-2040 Antenna Tuner, QST Nov-80, Pagel, Paul N1FB

Heathkit, SA-2040 Antenna Tuner Kit, CQ Jul-80, Dorhoffer, Alan M., K2EEK

Heathkit, SA-2060 Transmatch, QST Jul-82, Collins, George KC1V

Heathkit, SA-2500 Antenna Tuner, QST Mar-85, Pagel, Paul N1FB

Heathkit, SA-2550 Remote Antenna Matcher, QST Aug-88, DeMaw, Doug W1FB

Heathkit, SA-7010 HF Tribander, QST Aug-80, DeMaw, Doug W1FB

Heathkit, SB-1000 HF Linear Amplifier, QST Feb-88, Pagel, Paul N1FB

Heathkit, SB-102 Transceiver, QST Feb-71, Blakeslee, Douglas W1KLK

Heathkit, SB-104 SSB Transceiver, QST Oct-75, Dorbuck, Tony W1YNC

Heathkit, SB-104 SSB/CW Transceiver Kit, CQ Aug-75, Ross, Richard A., K2GMA

Heathkit, SB-104A 80-10 Meter SSB Transceiver, CQ Mar-80, Paul, Hugh R., W6POK

Heathkit, SB-104A HF Transceiver, QST Oct-75, Unknown

Heathkit, SB-1400 MF/HF Transceiver, QST Oct-89, Kleinschmidt, Kirk NTØZ

Heathkit, SB-201 Linear Amplifier, CQ Feb-83, Schultz, John J., W4FA

Heathkit, SB-220 Linear Amplifier, CQ Jun-70, Scherer, WilfredM., W2AEF

Heathkit, SB-220 Linear Amplifier, QST Aug-70, Blakeslee, Doug N1RM

Heathkit, SB-221 Linear Amplifier, QST Mar-80, DeMaw, Doug W1FB

Heathkit, SB-221 Linear Amplifier, CQ Feb-83, Schultz, John J., W4FA

Heathkit, SB-230 1KW Conduction-Cooled Linear, CQ Apr-77, Paul, Hugh R., W6POK

Heathkit, SB-230 HF Amplifier, QST Feb-76, Myers, Bob W1XT

Heathkit, SB-303 Receiver, QST Jul-71, Hall, Gerald K1PLP

Heathkit, SB-303 Solid-State Receiver, CQ Apr-71, Scherer, WilfredM., W2AEF

Heathkit, SB-500 2-M Transverter, QST Sep-70, Campbell, Laird W1CUT

Heathkit, SB-610 Monitorscope, QST Jul-72, Myers, Robert W1FBY

Heathkit, SB-614 Monitorscope, QST Jun-76, Rusgrove, Jay W1VD

Heathkit, SB-614 Station Monitor, CQ Aug-76, Paul, Hugh R., W6POK

Heathkit, SB-650 Digital Frequency Display, CQ Oct-72, Scherer, Wilfred M., W2AEF

Heathkit, SB-650 Frequency Display, QST Aug-72, Baldwin, Dick W1RU

Heathkit, SB.COM: Computer-Control Software for the SB-1400, QST Oct-89, Kleinschmidt, Kirk NTØZ

Heathkit, Solid State Triggered Sweep Oscilloscopes, CQ Aug-72, Scherer, Wilfred M., W2AEF

Heathkit, SS-9000 HF Synthesized Transceiver, CQ Feb-84, Schultz, John J., W4FA

Heathkit, SS-9000 HF Transceiver, QST Feb-84 , Raso, Ed WA2FTC

Heathkit, SW-7800 Gen. Cov. Receiver, QST Apr-85, Schetgen, Bob KU7G

Heathkit, uMatic Memory Keyer SA-5010, QST May-82, O'Dell, Pete KB1N

Heathkit, VF-2031 2-M FM Hand-Held Transceiver, QST Oct-79, Bartlett, Jim K1TX

Heathkit, VF-7401 2-M Transceiver, QST Nov-81, O'Dell, Sally KB1O

Heathkit, VL-1180 2-M Mobile Amplifier, QST May-82, Pagel, Paul N1FB

Heathkit, VL-2280 2-M Base Station Amplifier, QST Jun-82, Hull, Gerry AK4L

Heil, BM-10 Contester Headset, QST Feb-92, Wilson, Mark AA2Z

Heil, Concept 2000 Audio Products, CQ Dec-89, Schultz, John J., W4FA

Heil, EQ-200 Microphone Equalizer, CQ Jul-82, Schultz, John J., W4FA

Heil, HC-3 Microphone Cartridge, CQ May-83, Schultz, John J., W4FA

Heil, HM-5 Microphone, QST Apr-84, DeMaw, Doug W1FB

Heil, Pro-Set Headset, Wilson, Mark AA2Z

Henry Radio, 2K Ultra Amplifier, QST Jun-72, Nelson, John WØDRE/1

Henry Radio, 2K-4 Amplifier, QST May-74, Myers, Robert W1FBY

Henry Radio, Kenwood Pair, QST Oct-74, Myers, Robert W1FBY

Henry Radio, Kenwood TS-511S Transceiver, QST May-73, Myers, Robert W1FBY

Henry Radio, Kenwood TS-520 Transceiver, QST Sep-74, Pride, Mark WA1ABV

Henry Radio, Kenwood TS-900 Transceiver, QST Jul-73, Myers, Robert W1FBY

Henry Radio, R-599 and T-599 (Pair), QST Jun-71, Myers, Robert W1FBY

Henry Radio, Tempo 3002A 2-Meter Linear, QST Nov-89, Wilson, Mark AA2Z

Henry Radio, Tempo CL-146 FM Transceiver, QST May-73, Unknown, WA1FCM

Henry Radio, Tempo CL-220 FM Transceiver, QST May-73, Unknown, WA1FCM

Hewlett-Packard, 3476A Digital Multimeter, QST Apr-77, Hall, Jerry K1TD

Hewlett-Packard, HP-25 Programmable Calculator, QST Oct-76, Hall, Jerry K1TD

High Sierra, HS-100 Mobile Antenna, CQ Apr-94, McCoy, Lew, W1ICP

Horizon, 10-FM 2-M Vertical Antenna, QST Jan-76, Hall, Jerry K1TD

Horizon, Model 15-147 Vertical Antenna, QST Jan-79, Hall, Jerry K1TD

HUA Electronics, 1BC-1A Frequency Counter, QST Apr-72, Hall, Gerald K1PLP

Hustler, 2-Meter Mag-Mount Antenna, QST Feb-89, Kleinschmidt, Kirk NTØZ

Hustler, 6-BTV Vertical Antenna, QST Jan-84, Schetgen, Bob KU7G

Hy-Gain, 155CA 15-M Yagi Antenna, QST Apr-92, Sumner, Dave K1ZZ

Hy-Gain, 214 2-M Yagi, QST Oct-78, Sumner, Dave K1ZZ

Hy-Gain, 400 Antenna Rotator, CQ Aug-71, Scherer, WilfredM., W2AEF

Hy-Gain, 400 Rotator, QST Feb-71, McCoy, Lewis W1ICP

Hy-Gain, OMNI DX-88 Eight-Band Vertical Antenna, CQ Sep-90, Schultz, John J., W4FA

Hy-Gain, TH7DX Antenna, QST Feb-83, O'Dell, Pete KB1N

Hy-Gain, V-2 2-M Antenna, QST May-82, Lusis, Dennis W1LG

ICOM, AH 2 HF Mobile Antenna System, CQ Apr-87, McCoy, Lew, W1ICP

ICOM, AT-160 Automatic Antenna Tuner , QST Feb-93, Wolfgang, Larry WR1B Healy, Rus NJ2L

ICOM, IC-02A/T 2-Meter Handheld, CQ Jun-84, McCoy, Lew, W1ICP

ICOM, IC-175 HF All Band and General Coverage Transceiver, CQ Sep-84, Schultz, John J., W4FA

ICOM, IC-21 2-Meter FM Transceiver, QST Feb-73, Baldwin, Dick W1RU

ICOM, IC-211 Multimode 2-M Transceiver, QST Dec-78, Sumner, David K1ZZ

ICOM, IC-228H 2-Meter FM Transceiver, QST Jan-89, Palm, Rick K1CE

ICOM, IC-229A/H 2 Meter FM Transceiver, CQ May-91, Ingram, Dave, K4TWJ

ICOM, IC-229H Mobile 2-M FM Transceiver, QST Dec-91, Healy, Rus NJ2L

ICOM, IC-22S 2-M FM Transceiver, QST Dec-77, O'Dell, Pete KB1N

ICOM, IC-230, QST Jul-74, Unknown, W1GRE

ICOM, IC-2410H Dual-Band Mobile FM Transceiver, QST Jun-93, Healy, Rus NJ2L

ICOM, IC-245 2-M Transceiver, QST Sep-77, DeMaw, Doug W1FB

ICOM, IC-25A 2 Meter Transceiver, CQ Sep-82, Gordon, Al, WD6HAK

ICOM, IC-25A 2-M Transceiver, QST Jul-82, Accardi, Phil AJ1N

ICOM, IC-271 2-M Transceiver, QST May-85, Wilson, Mark AA2Z

ICOM, IC-271A 2-Meter Transceiver, CQ Jul-84, Ingram, Dave, K4TWJ

ICOM, IC-275 All-Mode 2 Meter Transceiver, CQ Jul-87, Ingram, Dave, K4TWJ

ICOM, IC-275A 2-Meter Multimode Transceiver, QST Oct-87, Wilson, Mark AA2Z

ICOM, IC-27H, CQ Feb-85, McCoy, Lew, W1ICP

ICOM, IC-281H 2-M FM Transceiver, QST Jan-95, Ford, Steve WB8IMY

ICOM, IC-290H 2-M All-Mode Transceiver, QST May-83, Kleinman, Joel N1BKE

ICOM, IC-2A 2-M FM Hand-Held Transceiver, QST Jan-81, Clift, Dale WA3NLO

ICOM, IC-2GAT Hand-Held 2-Meter FM Transceiver, QST Jun-90, Battles, Brian WA1YUA

ICOM, IC-3230H Dual-Band Mobile FM Transceiver, QST Jun-93, Healy, Rus NJ2L

ICOM, IC-32AT Dual-Band Hand-Held FM Transceivers, QST Jun-91, Healy, Rus NJ2L

ICOM, IC-32AT Two Band VHF/UHF HT, CQ Jul-89, McCoy, Lew, W1ICP

ICOM, IC-375A 220-MHz Multimode Transceiver, QST Mar-88, Wilson, Mark AA2Z

ICOM, IC-3AT 220-MHz FM Transceiver, QST Feb-83, O'Dell, Pete KB1N

ICOM, IC-45A 450-MHz FM Transceiver, QST Nov-83, Palm, Rick K1CE

ICOM, IC-471 70-cm Transceiver, QST Aug-85, Lindholm, John W1XX

ICOM, IC-47A, CQ Feb-85, McCoy, Lew, W1ICP

ICOM, IC-551 6-M Multimode Transceiver, QST Jun-81, Hull, Gerry AK4L

ICOM, IC-575A 50/28-MHz Transceiver, QST Nov-88, Gamble, M. N1FOZ and Kleinschmidt, K. NTØZ

ICOM, IC-701 HF Transceiver, QST Apr-79, Rusgrove, Jay W1VD

ICOM, IC-707 MF/HF Trnasceiver, QST Apr-94, Ford, Steve WB8IMY

ICOM, IC-720A All Band/General Coverage HF Transceiver, CQ Nov-81, Schultz, John J., W4FA

ICOM, IC-720A HF Transceiver, QST Aug-82, Wilson, Mark AA2Z

ICOM, IC-725 MF/HF Transceiver, QST Mar-90, Kleinschmidt, Kirk NTØZ

ICOM, IC-728 Transceiver, QST Feb-93, Wolfgang, Larry WR1B and Healy, Rus NJ2L

ICOM, IC-729 Transceiver, QST Feb-93, Wolfgang, Larry WR1B and Healy, Rus NJ2L

ICOM, IC-730 HF Transceiver, QST Dec-82, Wolfgang, Larry WA3VIL

ICOM, IC-730 HF Transceiver, CQ Jan-83, Schultz, John J., W4FA

ICOM, IC-735 HF Transceiver, CQ Jan-86, Ingram, Dave, K4TWJ

ICOM, IC-735 HF Transceiver, QST Jan-86, Schetgen, Bob KU7G

ICOM, IC-737 MF/HF Transceiver, QST Aug-93, Healy, Rus NJ2L

ICOM, IC-740 HF Transceiver, QST Sep-83, Hull, Gerry AK4L

ICOM, IC-745 HF Transceiver, QST Sep-85, O'Dell, Pete KB1N

ICOM, IC-745 Transceiver, CQ Sep-85, McCoy, Lew, W1ICP

ICOM, IC-751 HF Transceiver, QST Jan-85, Wilson, Mark AA2Z

ICOM, IC-761 160- to 10-Meter Transceiver, QST Sep-88, Miller, Tom NK1P

ICOM, IC-765 160- to 10-Meter Transceiver, QST Dec-90, Wilson, Mark AA2Z

ICOM, IC-781 160- to 10-Meter Transceiver, QST Jan-90, Wilson, Mark AA2Z

ICOM, IC-781 HF Super Transceiver Part I, CQ Nov-88, Ingram, Dave, K4TWJ

ICOM, IC-781 HF Super Transceiver Part II, CQ Dec-88, Ingram, Dave, K4TWJ

ICOM, IC-820H VHF/UHF Multi-mode Transceiver, QST Mar-95, Ford, Steve WB8IMY

ICOM, IC-900 A/E Super Multi-Bander System, CQ May-88, McCoy, Lew, W1ICP

ICOM, IC-900 Multiband VHF/UHF FM Mobile Transceiver, QST Dec-88, Wolfgang, Larry WA3VIL

ICOM, IC-R7100 VHF/VHF/UHF Comm. Receiver, QST Apr-93, Newkirk, Dave WJ1Z, Healy, Rus NJ2L

ICOM, IC-u2AT 2 Meter Handheld Pocket Transceiver, CQ Jun-87, Ingram, Dave, K4TWJ

ICOM, IC-u2AT 2-Meter FM Hand-Held Transceiver, QST May-87, Hale, Bruce KB1MW

ICOM, IC-W21A Dual-Band HH FM Transceiver, QST Mar-94, Ford, Steve WB8IMY

ICOM, IC-W2A Dual Band 144/450 MHz Handheld, CQ Jan-92, McCoy, Lew, W1ICP

ICOM, IC-Z1A Dual-Band Hand-Held Transceiver, QST Jul-95, Ford, Steve WB8IMY

ICOM, IC-Z1A Dual-Band HT, CQ Sept-95, McCoy, Lew W1ICP

ICOM, R-70 General Coverage Receiver, QST Jun-83, DeMaw, Doug W1FB

ICOM, R-9000 All-Band All-Mode Communications Receiver, CQ Feb-90, Ingram, Dave, K4TWJ

ICOM, R71A General Coverage Receiver, CQ Aug-84, McCoy, Lew, W1ICP

ICOM, SM-10 Graphic Equalized Compressor Desk Mike, CQ Nov-86, Ingram, Dave, K4TWJ

Info-Tech, M-150 and M-75 RTTY Units, QST Apr-78, Bartlett, Jim K1TX

Info-Tech, M-44 AMTOR Converter, QST Aug-85, Newland, Paul AD7I

Info-Tech, Model 300 Keyboard, QST Apr-79, Gibilisco, Stan W1GV

Info-Tech, Model 30C CW-to-Video Converter, QST Nov-78, Bartlett, Jim K1TX

Inline Instruments, Coaxial Relays and Couplers, QST Apr-76, McMullen, Tom W1SL

Innova, Cordless Battery Charger and Portable DC Supply, CQ May-93, Ingram, Dave, K4TWJ

Inque, IC-21, CQ May-72, Zook, Glen E., K9STH/5

Instant Software, Electronic Breadboard, QST Dec-82, Hutchinson, Chuck K8CH

International Instrumentation, Digital Capacitance Converter, QST Mar-79, Gibilisco, Stan W1GV

IPS Radio and Space Services, ASAPS HF Prop. Prediction Software, QST Dec-94, Straw, R. Dean N6BV

IRL, FSK-1000 RTTY Demodulator, QST Jun-80, Schenck, Chris W1EH

ITT , Mackay Marine 3020A Receiver, QST Jan-74, Dorbuck, Tony W1YNC

j-Com, Magic Notch Automatic Audio Notch Filter, QST Oct-91, Kearman, Jim KR1S

Jade Products, 160 Meter Twin-Lead Marconi Antenna, QST Aug-94, Ford, Steve WB8IMY

Janal Laboratories, 432CA Converter, QST Dec-75, McMullen, Thomas W1SL

Japan Radio Company, JST-135HP MF/HF Transceiver, QST Mar-92, Newkirk, Dave WJ1Z

Japan Radio Company, JST-245 MF/HF/6-Meter Transceiver, QST Sep-95, Swanson, Glenn KB1GW

Japan Radio Company, NRD 515 All-Wave Receiver, QST Nov-81, Hull, Gerry AK4L

Japan Radio Company, NRD-525 General-Coverage Receiver, QST Jul-88, Newkirk, Dave AK7M

Japan Radio Company, NSD 515 HF Transmitter, QST Nov-82, Hull, Gerry AK4L

Japan Radio Corp., JST-135HP HF Transceiver, CQ Jul-91, McCoy, Lew, W1ICP

Jini Micro Systems Inc., Mini Jini Record Keeper, CQ Jan-84, Thurber, Karl T. Jr., W8FX

Johnson, 504 2-Meter Transceiver, QST Jun-71, Blakeslee, Douglas W1KLK

JPS Communications, NF-60 Automatic DSP Notch Filter, QST Feb-94, Healy, Rus NJ2L

JPS Communications, NF-60 Notch Filter, CQ Mar-93, McCoy, Lew, W1ICP

JPS Communications, NIR-10 Noise/Interference-Reduction Filter, QST Oct-91, Kearman, Jim KR1S

JPS Communications, NRF-7 and NIR-10 Audio Processing Units, CQ Jul-94, Carr, Paul, N4PC

JPS Communications, NRF-7 Multimode DSP Audio Filter, QST Feb-94, Healy, Rus NJ2L

JRC, JRL-2000F HF Linear Amplifier, CQ May-93, Carr, Paul, N4PC

K1EA Software, Digital Voice Processor, QST Jan-93, Healy, Rus NJ2L

Kampp Electronics, Autobrak Kits, QST Aug-78, Aurick, Lee W1SE

Kangaroo Tabor Software, CAPMAN HF Prop. Prediction Software, QST Dec-94, Straw, R. Dean N6BV

Kansas City Keyer, KC-1, CQ Oct-85, Locher, Bob, W9KNI

Kantronics, 8040-B Receiver, QST Jun-78, Bartlett, Jim K1TX

Kantronics, All Mode Communicator , QST Jun-89, Wolfgang, Larry WA3VIL

Kantronics, All-Mode Software Ver.4.0 Upgrade, QST Jan-92, Wolfgang, Larry WR1B

Kantronics, CW Training System, QST Mar-83, Kaczynski, Mike W1OD

Kantronics, Host Master II Terminal Software, QST Jan-92, Wolfgang, Larry WR1B

Kantronics, Interface I & Software, QST Feb-84, Clift, Dale WA3NLO

Kantronics, Interface II & Software, QST Sep-84, Pagel, Paul N1FB

Kantronics, Interface II and Hamsoft, CQ Nov-84, McCoy, Lew, W1ICP

Kantronics, Kam Plus, CQ Jan-94, Rogers, Buck, K4ABT

Kantronics, KAM Plus Multimode TNC, QST Jun-94, Ford, Steve WB8IMY

Kantronics, KPC-2400 Packet Communicator, QST Nov-87, Hall, Jerry K1TD

Kantronics, KPC-3 Mini Packet TNC, CQ Nov-92, Rogers, Buck K4ABT

Kantronics, KPC-3 Packet Communicator 3 TNC, QST Dec-93, Ford, Steve WB8IMY

Kantronics, KPC-4 Multi-Function Packet Controller, CQ Nov-88, Rogers, Buck, K4ABT

Kantronics, KPC-9612 1200/9600 Packet TNC, QST Sep-94, Ford, Steve WB8IMY

Kantronics, KT-130 Single Band 30 Meter Transceiver, CQ May-89, Ingram, Dave, K4TWJ

Kantronics, Packet Communicator, CQ Nov-85, McCoy, Lew, W1ICP

Kantronics, Version 3.0 Firmware, CQ Dec-90, Rogers, Buck, K4ABT

Kantroniics, Feild Day 2 Morse/RTTY Reader, CQ Mar-81, Wertz, Arthur H., N5AEN

KDK, FM-2051 2-M Transceiver, QST Oct-78, O'Dell, Pete KB1N

Kenwood, A Trio of Accessories for the TS-850S HF Transceiver, CQ Nov-91, Schultz, John J., W4FA

Kenwood, AT-250 Automatic Antenna Tuner, QST Oct-88, Jahnke, Bart KB9NM

Kenwood, DM-81 Dip Meter, QST Dec-80, DeMaw, Doug W1FB

Kenwood, MC-55 Mobile Microphone, CQ Jan-87, Schultz, John J., W4FA/SVØDX

Kenwood, MC-85 Communication Microphone and Control Unit, CQ Mar-86, Schultz, John J., W4FA

Kenwood, R-1000 General-Coverage Receiver, QST Dec-80, Rusgrove, Jay W1VD

Kenwood, R-5000 General-Coverage Receiver, QST Feb-88, Newkirk, Dave AK7M

Kenwood, R-820 Receiver, QST Jul-79, Rusgrove, Jay W1VD

Kenwood, R599D Receiver and the T599D Transmitter, CQ Mar-78, Paul, Hugh R., W6POK

Kenwood, RC-10 Remote Control Headset, QST Feb-89, Healy, Rus NJ2L

Kenwood, SM-230 Station Monitor, CQ Jul-90, Schultz, John J., W4FA

Kenwood, TH 25AT/45AT and Accessories, CQ Jul-88, O'Dell, Peter R., WB2D

Kenwood, TH-21AT 2 Meter HT, CQ Feb-85, Ingram, Dave, K4TWJ

Kenwood, TH-21AT 2-M Transceiver, QST Apr-85, Wilson, Mark AA2Z

Kenwood, TH-22AT 2 Meter Handheld Transceiver, CQ May-94, McCoy, Lew W1ICP

Kenwood, TH-25AT 2 Meter HT, CQ Mar-88, Clarke, Bill, WA4BLC

Kenwood, TH-27A and TH-47A UHF Handheld Transceivers, CQ Mar-91, Ingram, Dave, K4TWJ

Kenwood, TH-31BT Hand-Held Transceiver, QST Nov-87, Hale, Bruce KB1MW

Kenwood, TH-77A Dual-Band Hand-Held FM Transceiver, QST Jun-91, Healy, Rus NJ2L

Kenwood, TH-78A Dual-Band HH FM Transceiver, QST Mar-94, Ford, Steve WB8IMY

Kenwood, TH-79A Dual-Band Hand-Held FM Transceiver, QST Jul-95, Ford, Steve WB8IMY

Kenwood, TL-922A Linear Amplifier, QST Sep-80, Pelham, John W1JA

Kenwood, TL-922A Linear Amplifier, July-92, Wilson, Mark AA2Z and Healy, Rus NJ2L

Kenwood, TM-221A/321A/421A VHF/UHF FM Transceivers, QST Jul-88, Hale, Bruce KB1MW

Kenwood, TM-241A, QST Dec-91, Healy, Rus NJ2L

Kenwood, TM-241A 2-M FM Transceiver, QST Jan-95, Ford, Steve WB8IMY

Kenwood, TM-251A 2-M FM Transceiver, QST Jan-95, Ford, Steve WB8IMY

Kenwood, TM-2570A 2-M Transceiver, QST Oct-86, Schetgen, Bob KU7G

Kenwood, TM-353A 220 MHz FM Transceiver, CQ Feb-88, O'Dell, Peter R., WB2D

Kenwood, TM-701 Dual-Band VHF/UHF FM Transceiver, QST Feb-89, Healy, Rus NJ2L

Kenwood, TM-732A Dual-Band Mobile FM Transceiver, QST Jun-93, Healy, Rus NJ2L

Kenwood, TM-941A 144/440/1200 MHz FM Triband Transceiver, CQ Feb-91, Schultz, John J., W4FA

Kenwood, TR-2200A 2 Meter Portable Transceiver, CQ Jan-78, Paul, Hugh R., W6POK

Kenwood, TR-2200A 2-M FM Transceiver, QST Nov-76, McMullen, Tom W1SL

Kenwood, TR-2400 2-M FM Transceiver, QST Apr-81, Place, Steve WB1EYI

Kenwood, TR-2400 Two Meter HT, CQ Jan-81, Schultz, John J., W4FA

Kenwood, TR-2600 Two Meter HT, CQ May-85, Ingram, Dave, K4TWJ

Kenwood, TR-7400A 2 Meter Transceiver, CQ Jun-77, Paul, Hugh R., W6POK

Kenwood, TR-7400A 2-M Transceiver, QST Sep-77, Arnold, Max W4WHN

Kenwood, TR-751A 144-MHz All-Mode Transceiver, QST Mar-87, Healy, Rus NJ2L

Kenwood, TR-7730 2-M FM Transceiver, QST May-82, DeMaw, Doug W1FB

Kenwood, TR-7800 2-M FM Transceiver, QST Sep-81, Palm, Rick K1CE

Kenwood, TR-9000 2-M Multimode Transceiver, QST Dec-81, Sumner, David K1ZZ

Kenwood, TS-120S HF Transceiver, QST Feb-80, Pagel, Paul N1FB

Kenwood, TS-130S HF Transceiver, QST Jul-81, O'Dell, Pete KB1N

Kenwood, TS-140S 160- to 10-M Transceiver, QST Jun-88, Wolfgang, Larry WA3VIL

Kenwood, TS-140S Compact HF Transceiver, CQ Sep-88, Ingram, Dave, K4TWJ

Kenwood, TS-180S HF Transceiver, QST May-80, DeMaw, Doug W1FB

Kenwood, TS-430S HF Transceiver, QST Mar-84, Kluger, Leo WB2TRN

Kenwood, TS-440S HF Transceiver, QST Dec-86, Miller, Thomas KA1JQW

Kenwood, TS-440S HF Transceiver Part I, CQ Dec-87, Schultz, John J., W4FA

Kenwood, TS-450S HF Transceiver, CQ May-92, Schultz, John J., W4FA

Kenwood, TS-450S Transceiver, QST Apr-92, Healy, Rus NJ2L

Kenwood, TS-50S MF/HF Transceiver, QST Sep-93, Healy, Rus NJ2L

Kenwood, TS-50S Transceiver, CQ Jun-93, McCoy, Lew, W1ICP

Kenwood, TS-520S HF Transceiver, QST May-78, DeMaw, Doug W1FB

Kenwood, TS-530S HF Transceiver, QST Mar-82, Hull, Gerry AK4L

Kenwood, TS-60S 6-M All-Mode Transceiver, QST Sep-94, Wilson, Mark AA2Z

Kenwood, TS-680S 160- to 6-M Transceiver, QST Oct-88, Jahnke, Bart KB9NM

Kenwood, TS-690S Transceiver, QST Apr-92, Healy, Rus NJ2L

Kenwood, TS-700A 2-M Transceiver, QST Mar-76, McMullen, Tom W1SL

Kenwood, TS-700A 2-Meter All-Mode Transceiver, CQ Mar-76, Paul, Hugh R., W6POK

Kenwood, TS-700S 2-M Transceiver, QST Feb-78, DeMaw, Doug W1FB

Kenwood, TS-790A, All Mode VHF/UHF/OSCAR Transceiver, CQ Oct-89, Ingram, Dave K4TWJ

Kenwood, TS-790A VHF/UHF Transceiver, QST Apr-91, Healy, Rus NJ2L

Kenwood, TS-820 HF Transceiver, QST Sep-76, DeMaw, Doug W1FB

Kenwood, TS-820 Transceiver, CQ Feb-77, Paul, Hugh R., W6POK

Kenwood, TS-830S HF Transceiver, QST May-81, Pagel, Paul N1FB

Kenwood, TS-8400 UHF FM Transceiver, QST Jan-82, O'Dell, Pete KB1N and Hall, Jerry AK4L

Kenwood, TS-850S 160-10 M Transceiver, QST Jul-91, Healy, Rus NJ2L

Kenwood, TS-850S HF Transceiver, CQ Jun-91, Schultz, John J., W4FA

Kenwood, TS-900 SSB Transceiver, CQ Feb-75, Ross, Richard A., K2GMA

Kenwood, TS-930S HF Transceiver, QST Jan-84, Wilson, Mark AA2Z

Kenwood, TS-940S HF Transceiver, QST Feb-86, Hutchinson, Chuck K8CH

Kenwood, TS-940S HF Transceiver, CQ May-86, Schultz, John J., W4FA

Kenwood, TS-950S/TS-950S-Digital HF Transceiver, CQ Jul-90, Schultz, John J., W4FA

Kenwood, TS-950SD MF/HF Transceiver, QST Jan-91, Healy, Rus NJ2L

Kenwood, TS-950SDX MF/HF Transceiver, QST Dec-92, Healy, Rus NJ2L

Kenwood, TV-502 2 Meter Transducer, CQ Sep-76, Paul, Hugh R., W6POK

Kenwood, TV-502 2-M Transceiver, QST Aug-77, Rusgrove, Jay W1VD

Kenwood, TW-4000A 2-M/70-CM FM Dual Bander, QST Aug-84, Yoshida, Wayne T. KH6WZ

Klitzing, 70CM10W60 UHF Amplifier, QST Jun-79, Bartlett, Jim K1TX

Klitzing, SSB-1 Squelch Board, QST Jun-79, Bartlett, Jim K1TX

KLM, 144-148-13LBA 2-M Yagi, QST Feb-85, Hutchinson, Chuck K8CH

KLM, 16-Element 2-M Yagi, QST Aug-79, Kearman, Jim W1XZ

KLM, 2-M-16LBX 2-M Yagi, QST Mar-85, Wilson, Mark AA2Z

KLM, 21.0-21.5-6A Big Sticker Antenna, QST Dec-83, Wilson, Mark AA2Z

KLM, 220-22LBX 220-MHz Yagi, QST Sep-86, Wilson, Mark AA2Z

KLM, 2M-22C and 435-49CX Yagis, QST Oct-85, Jansson, Dick WD4FAB

KLM, 40-M Yagi, QST Nov-77, Cain, Jim K1TN

KLM, 7.2-2 40-M Yagi, QST Jul-84, Kaczynski, Michael W1OD

KLM, AP-144DIII Base Stn. Antenna, QST Sep-83, Gerli, Sandy AC1Y

KLM, JV-2 "J" 2 Meter Vertical Antenna, CQ Apr-83, Ingram, Dave, K4TWJ

KLM, KT34A Beam, CQ Sep-84, McCoy, Lew, W1ICP

KLM, KT34XA Beam, CQ Sep-84, McCoy, Lew, W1ICP

KLM, Log-Periodic Antenna, QST Jan-74, Myers, Robert W1FBY

KLM, Multi-2000 2-Meter FM/SSB/CW Transceiver, CQ Jan-76, Paul, Hugh R., W6POK

KLM, PA 15-80BL 2-M Linear Amplifier, QST Sep-79, Glassmeyer, Bernie W9KDR

Knight-Kit, R-195 Receiver, QST Oct-70, Unknown, WN1LZQ

Kronotek, RT-1 RF-Actuated Timer, QST Jan-77, Watts, Chuck WA6GVC

Kurt Fritzel, FB-DX 506 Poly Beam System, CQ May-95, Carr, Paul N4PC

KW Electronics, KW-103 SWR/Power Meter, QST Sep-72, Godwin, Morgan W4WFL/1

KW Electronics, KW107 Supermatch, QST Jul-72, Godwin, Morgan W4WFL/1

L-Tronics, Little L-Per VHF Direction Finder, QST Apr-78, Kearman, Jim W1XZ

Lab Science, Econotrace Curve Tracer, QST Oct-76, Watts, Chuck WA6GVC

Lafayette, HA-146 2-M FM Transceiver, QST Jul-76, Williams, Perry W1UED

Lambda, Coaxial Portal Unit, QST Nov-82, Collins, George KC1V

Lance Johnson Engineering, D-Lay-5 Rotator Control, QST Apr-83, Pagel, Paul N1FB

Lance Johnson Engineering, GP-1 Vertical Antenna Mount, CQ Aug-84, McCoy, Lew, W1ICP

Larsen, JM Mobile Antenna Mount, QST Apr-76, Watts, Chuck WA6GVC

Larson Antennas, Dual Band Coupler for 2 Meter and 450 MHz, CQ Feb-88, McCoy, Lew W1ICP

Layfayette, BCR-101 Communications Receiver, QST Jan-79, Gibilisco, Stan W1GV

Layfayette, HA-750 6-Meter Transceiver, QST Apr-71, Tilton, Edward W1HDQ

Layfeyette, 99-35313L 146- to 175- MHz FM Receiver, QST Mar-70, Lange, Walter W1YDS

Layfeyette, HA-800 Receiver, QST Feb-70, Lange, Walter W1YDS

Leader, LAC-895 Antenna Coupler, QST Jul-76, DeMaw, Doug W1FB

Leader, LBO-310 Oscilloscope, QST Aug-74, McMullen, Thomas W1SL

Leader, LDM-810 Grid-dip Meter, CQ Feb-73, Scherer, Wilfred M., W2AEF

LEL, Dynamic Serviset E-C, QST Mar-71, Scherer, WilfredM., W2AEF

Lighting Bolt, Multiband Quad, CQ Aug-90, McCoy, Lew, W1ICP

Lightning Bolt, Five-Band Quad, CQ Apr-93, McCoy, Lew, W1ICP

LMW Electronics, 2304TRV2 2304-MHz Transverter, QST Dec-87, Wilson, Mark AA2Z

LMW Electronics, LMW 1296TRV1K 23-cm Transverter Kit, QST Dec-87, Lau, Zack KH6CP

Logikey, K-1 Electronic Keyer, CQ Feb-93, Dorr, John, K1AR

Lowe, HF-150 LF/MF/HF Communications Receiver, QST Aug-93, Newkirk, David WJ1Z

Luke Company, S40 A 40 Amp 13.8 Volt Power Supply, CQ Jan-94, DeMaw, Doug, W1FB

M & M Electronics, Model MSB-1 Audio Filter, QST Jun-82 , Kaczynski, Mike W1OD

M2 Enterprises, EB-144 Eggbeater Antenna, QST Sep-93, Ford, Steve WB8IMY

Macrotronics, Code Class, QST Mar-83, Horzepa, Stan WA1LOU

Macrotronics, M800 RTTY System, QST Nov-79, Horzepa, Stan WA1LOU

Macrotronics, Ritty Riter, QST Dec-80, Horzepa, Stan WA1LOU

Macrotronics, RM1000 Modem, QST Jun-84, Pagel, Paul N1FB

Macrotronics, Terminal RTTY Modem, QST Jun-82, Horzepa, Stan WA1LOU

Maggiore Elec Lab, Hi Pro MK I 2-M Repeater, QST Aug-84, Kaczynski, Mike W1OD

Maggiore Elec Lab, Hi Pro MK1 220-MHz Repeater, QST Feb-82, Kaczynski, Mike W1OD

Maldol, 28HS2HB 2-Element 10-Meter Beam, QST Dec-92, Jahnke, Bart KB9NM

MAXCOM, Antenna Matcher & Dipole, QST Nov-84, Hall, Jerry K1TD

McKay Dymek, DR33C All-Wave Receiver, QST Sep-79, Rusgrove, Jay W1VD

MET, NBS 144/7T Two Meter Yagi, CQ Aug-86, McCoy, Lew, W1ICP

MFJ, 259 HF/VHF SWR Analyzer, CQ June-95, Carr, Paul N4PC

MFJ, 784 Super DSP Filter, CQ Mar-95, DeMaw, Doug W1FB/8

MFJ, A Trio of Station Clocks, CQ Dec-83, Schultz, John J., W4FA

MFJ, CMOS-440RS Electronic Keyer, QST May-76, Godwin, Morgan W4WFL

MFJ, Complete Line of Accessory Audio Filters, CQ Mar-88, Schultz, John J., W4FA

MFJ, Gray Line DX Advantage, CQ Jul-88, O'Dell, Peter R., WB2D

MFJ, MFJ-1224 RTTY Interface and Mating Software Packages, CQ Nov-84, Ingram, Dave, K4TWJ

MFJ, MFJ-1270 Terminal Node Controller, QST Sep-86, Hale, Bruce KB1MW

MFJ, MFJ-1270B TNC2 Packet Controller, QST Dec-93, Ford, Steve WB8IMY

MFJ, MFJ-1274/T 2400 BPS Packet Radio Controller, CQ Nov-90, Rogers, Buck, K4ABT

MFJ, MFJ-1278 Multi-Mode Data Controller, CQ May-89, McCoy, Lew, W1ICP

MFJ, MFJ-1278 Multi-Mode Data Controller, QST Jul-89, Kilgore, Jeff KC1MK

MFJ, MFJ-1278 Multi-Mode Data Controller-Revisited, QST Sep-89, Kilgore, Jeff KC1MK

MFJ, MFJ-1289 Multicom.EXE Terminal Software Package, CQ Aug-90, Rogers, Buck, K4ABT

MFJ, MFJ-16010 ST Super Antenna Tuner, CQ Nov-77, Paul, Hugh R., W6POK

MFJ, MFJ-1786 High-Q Loop Antenna for 10-30 Mhz., QST Aug-94, Kleinschmidt, Kirk NTØZ

MFJ, MFJ-202B RF Noise Bridge, CQ Aug-84, Schultz, John J., W4FA

MFJ, MFJ-207 HF SWR Analyzer, CQ Jun-91, Schultz, John J., W4FA

MFJ, MFJ-207 SWR Analyzer, QST Nov-93, Gruber, Mike WA1SVF

MFJ, MFJ-208 VHF SWR Analyzers, CQ Jun-91, Schultz, John J, W4FA

MFJ, MFJ-249 HF/VHF SWR Analyzer, CQ Mar-94, McCoy, Lew W1ICP

MFJ, MFJ-249 SWR Analyzer, QST Nov-93, Gruber, Mike WA1SVF

MFJ, MFJ-264 HF/UHF Dry Dummy Load, CQ Sep-89, Schultz, John J., W4FA/SVØDX

MFJ, MFJ-308 World Explorer II Shortwave Converter, CQ Apr-82, Dorhoffer, Alan M., K2EEK

MFJ, MFJ-312 Dual Band VHF Converter, CQ Jul-82, Ingram, Dave, K4TWJ

MFJ, MFJ-422 Keyer/Paddle, CQ Jan-90, Schultz, John J., W4FA

MFJ, MFJ-432 Voice Memory Keyer, QST May-95, Swanson, Glenn KB1GW

MFJ, MFJ-484 Grandmaster Keyer, QST Aug-80, DeMaw, Doug W1FB

MFJ, MFJ-484 Grandmaster Memory Keyer, CQ Feb-79, Scwartz, Irwin, K2VG

MFJ, MFJ-486 Super Grand Master Memory Keyer, CQ Jan-90, Schultz, John J., W4FA

MFJ, MFJ-496 Keyboard, QST Jul-82, Hall, Jerry AK4L

MFJ, MFJ-8043 Electronic Keyer, CQ Nov-77, Paul, Hugh R., W6POK

MFJ, MFJ-8100 Shortwave Regenerative Receiver Kit, Newkirk, David WJ1Z

MFJ, MFJ-8100 World Band Receiver Kit, CQ Jun-94, Carr, Paul, N4PC

MFJ, MFJ-815B MF/HF Wattmeter, QST Feb-91, Healy, Rus NJ2L

MFJ, MFJ-9017 18-Mhz QRP CW Transceiver, QST Jul-93, Hale, Bruce KB1MW

MFJ, MFJ-931 Artificial Ground, CQ Feb-88, Thurber, Karl T. Jr., W8FX

MFJ, MFJ-931 Artificial RF Ground, QST Apr-88, DeMaw, Doug W1FB

MFJ, MFJ-945-C and MFJ-989B HF Antenna Tuners, CQ Aug-88, Schultz, John J., SVØDX

MFJ, MFJ-948 "Deluxe Versa Tuner II" Antenna Tuner, QST Aug-91, Wilson, Mark AA2Z

MFJ, MFJ-986 Differential-T Tuner, CQ Sep-89, Schultz, John J., W4FA/SVØDX

MFJ, MFJ-989 Versa Tuner V Anatenna Tuner, CQ Mar-83, Schultz, John J., W4FA

MFJ, Super Keyboards, MFJ-494 and MFJ-496, for CW and RTTY, CQ Nov-81, Schultz, John J., W4FA

MFJ, Versa-Tuners, CQ Feb-81, Schultz, John J., W4FA

MFJ Enterprises, MFJ-949C Versa II, CQ Jul-87, McCoy, Lew, W1ICP

Micro Pro Systems, MPS CW Machine II, QST Jul-84, Towle, Jonathan WB1DNL

Micro-80, Inc., Morse Code Trainer II, QST Oct-83, Place, Steve WB1EYI

Microcomm, UHF Modules, QST Aug-76, McMullen, Tom W1SL

Microcraft, Code*Star Reader Kit, QST Jul-83 , O'Dell, Pete KB1N

Microlog, "SWL Cartridge Unit for RTTY and CW Reception, CQ Nov-85, Ingram, Dave, K4TWJ

Microlog, Air 1 RTTY/Morse System, CQ Nov-84, McCoy, Lew, W1ICP

Microlog, Morse Coach, Aug 86, Ingram, Dave, K4TWJ

Microtronics, M-80 Ham Interface, QST May-79, Horzepa, Stan WA1LOU

Microwave Associates, 89127 10-GHz Transceiver, QST Sep-77, Kearman, Jim W1XZ

Microwave Modules, MMC 1296 Receiving Converter, QST Dec-77, Kearman, Jim W1XZ

Microwave Modules, MMD050 Counter, QST Nov-76, McMullen, Tom W1SL

Microwave Modules, MMD500P Prescaler, QST Nov-76, McMullen, Tom W1SL

Microwave Modules, MMS1 & MMS2 Code Trainers, QST Jun-84, Place, Steve WB1EYI

Microwave Modules, MMT432 Transverters, QST Sep-77, Kearman, Jim W1XZ

Microwave Modules, MMV 1296 Varactor Tripler, QST Dec-77, Kearman, Jim W1XZ

Microwave Modules LTD., MML432/100 Power Amplifier, CQ Jun-86, Katz, Steve, WB2WIK

Mida, 6354 Mini-Multi-Meter, CQ Sep-73, Scherer, Wilfred M., W2AEF

Mida, Digipet 60 Digital Frequency Counter, CQ Nov-72, Scherer, Wilfred M., W2AEF

Midland, 13-509 220-MHz Transceiver, QST Oct-77, Gerli, Sandy AC1Y

Polaroid, Polapulse Battery Designer's Kit, CQ Jun-81, Weinstein, Martin Bradley, WB8LBV

Processor Technology, 8KRA Static Memory Module, QST May-77, Watts, Chuck WA6GVC

Processor Technology, SOL-20 uComputer, QST Jul-77, Watts, Chuck WA6GVC

Processor Technology, VDM-1 Video Display Module, QST Mar-77, Watts, Chuck WA6GVC

Propagation Products, Insulators and Quad Kit, QST Mar-78, McCoy, Lew W1ICP

QM70 Products, FMT-440 Transverter, QST Nov-77, Kearman, Jim W1XZ

Quantics, W9GR DSP-3 Audio Filter, QST Aug-95 Swanson, Glenn KB1GW

Quantum, Ham Battery, CQ May-92, Ingram, Dave, K4TWJ

Radio Amateur Callbook, A Tradition for 63 Years, CQ Dec-83, Dorhoffer, Alan M., K2EEK

Radio Shack, 12-159 Timekube, CQ Sep-77, Dorhoffer, Alan M., K2EEK

Radio Shack, DSP Communication Noise Reduction System, QST Jul-94, Kearman, Jim KR1S

Radio Shack, DX-302 Communications Receiver, QST Aug-81, Kampe, Bruce WA1POI

Radio Shack, HTX-100 10-M Mobile Transceiver, QST Feb-92, Hale, Bruce KB1MW and Healy, Rus NJ2L

Radio Shack, HTX-404 440 MHz FM Handheld Transceiver, CQ Feb-94, Lynch, Joe, N6CL

Radio Shack, LCD DMM, CQ Dec-81, Grove, Robert B., WA4PYQ

Radio Shack, Micro-80 Morse Trainer II, CQ Sep-81, Gorsky, Buzz, K8BG

Radio Shack, RG-8M Coaxial Cable, QST Dec-80, Hall, Jerry K1TD

Radio Shack, SCT-11 Stereo Cassette Tape Deck, CQ Apr-77, DeWitt, Bill, W2DD

Radio Shack, TRS-80 Micrcomputer, CQ Jun-78, Stites, Robert L.

Radio Shack, TRS-80 Microcomputer, QST Jun-78, O'Dell, Peter KB1N

Radio Ware, Ladder-Loc Center Insulator, QST Aug-94, Ford, Steve WB8IMY

Radio Works, Line Isolators, CQ July-95, DeMaw, Doug W1FB/8

Radiokit, UHF Converter Kit, CQ Oct-84, McCoy, Lew, W1ICP

Radioware, SSTV Explorer, QST Apr-94, Pagel, Paul N1FB

Ramsey, FX-146 2-Meter Transceiver Kit, QST Oct-93, Bloom, Jon KE3Z

Ranger Comm., RCI-2950 10-M Transceiver, QST Feb-92, Hale, Bruce KB1MW and Healy, Rus NJ2L

Raytrack, 6-Meter and 10-80-Meter 2Kw Linear Amplifiers, CQ Jan-70, Scherer, Wilfred M., W2AEF

RCA, COSMAC VIP Microcomputer, QST Feb-79, Place, Steve WB1EYI

Realistic, DX-150A Receiver, QST Sep-70, Myers, Robert W1FBY

Realistic, HTX-100 10 Meter SSB/CW Transceiver, CQ May-89, McCoy, Lew, W1ICP

Regency, AR-2 Two Meter Amplifier, CQ Mar-73, Zook, Glen E., K9STH/5

Regency, HR-2 FM Transceiver, CQ Feb-71, Zook, GlenE., K9STH/5

Regency Electronics, AR-2 FM Power Amplifier, QST Aug-75, Tilton, Edward W1HDQ

Regency Electronics, HR-2 FM Transceiver, QST Aug-71, DeMaw, Doug W1CER

Regency Electronics, HR-220 Transceiver, QST Mar-75, Pride, Mark WA1ABV

Regency Electronics, HR-6 FM Transceiver, QST Jan-75, McMullen, Thomas W1SL

Regency Electronics, HR2MS FM Transceiver, QST Jun-73, Williams, Perry W1UED

Regency Electronics, HR2S FM Transceiver, QST Jun-73, Williams, Perry W1UED

Regency Electronics, HRT-2 FM Transceiver, QST Oct-74, McMullen, Thomas W1SL

RF Applications, P-3000 HF Digital Power and SWR Bridge, CQ Oct-94, McCoy, Lew, W1ICP

RF Concepts, MM-1 Mobile Mount, CQ May-88, Dorhoffer, Alan M., K2EEK

RF Concepts, RFC 2-23 Solid-State 144-Mhz Amplifier, QST Mar-88, Hale, Bruce KB1MW

RF Concepts, RFC 2-317 2-M Amplifier , QST Oct-87, Wilson, Mark AA2Z

RF Concepts, RFC 3-22 Solid-State 220-MHz Amplifier, QST Mar-88, Hale, Bruce KB1MW

RF Concepts, RFC 3-312 220-MHz Amplifier, QST Apr-88, Hale, Bruce KB1MW and Wilson, Mark AA2Z

RF Concepts, RFC 8-RC Repeater Controller, QST Apr-89, Francis, Tom NM1Q

RF Concepts, VHF1-60 and RFC-2/70H RF Amplifiers, CQ Mar-94, Lynch, Joe, N6CL

RF Engineering Corp., RFE-100 Digital Frequency Display, CQ Jan-78, Paul, Hugh R., W6POK

RF Products, 5/8-Wave 220-MHz & 450-MHz Antennas, QST Aug-83, O'Dell, Pete KB1N

RIW, 432-19 19-Element 432-MHz Yagi, QST Dec-78, Bartlett, Jim K1TX

Robot Research, 70 SSTV Monitor, QST Nov-71 , Hall, Gerald K1PLP

Robot Research, 80 Camera, QST Nov-71, Hall, Gerald K1PLP

Robot Research, 800 Terminal, QST Apr-82, Pagel, Paul N1FB

Robot Research, Slow-Scan TV Equipment, CQ Aug-74, MacDonald, Copthorne, W0ORX

Robot Research Inc., 1200C Colors Slow Scan Converter, CQ Jun-88, DeWitt, William H., W2DD

Robyn, Digital 500 Transceiver, QST Apr-72, Scherer, Wilfred M., W2AEF

Rockwell-Collins, KWM-380 Transceiver, CQ Nov-82, Schultz, John J., W4FA

Rohn, Rotating Tower Systems Rohn 25 Rotation System, QST Sep-91, Healy, Rus NJ2L

Ross & White, RW-BND 2 Meter Transceiver, CQ Jun-72, Zook, Glen E., K9STH/5

RP Electronics, MFA-22 Frequency Synthesizer, QST Feb-75, Dorbuck, Tony W1YNC

RUPP Enterprises, TR-100 Power Bridge, CQ Jun-92, McCoy, Lew, W1ICP

Rutland Arrays, FO-22 432-Mhz Yagi Antenna, QST Apr-91, Healy, Rus NJ2L and Wilson, Mark AA2Z

Rutland Arrays, FO-25 432-MHz Yagi Antenna, QST Apr-91, Healy, Rus NJ2L and Wilson, Mark AA2Z

S & S Engineering, ARK 40 CW QRP Transceiver Kit, QST May-94, Gold, Jeff AC4HF

S&S Engineering, ARK 40 - QRP 40M CW Transceiver Kit, CQ Jan-94, Carr, Paul, N4PC

S&S Engineering, FC-X Frequency Counter Kit, CQ Aug-94, Carr, Paul, N4PC

S&S Engineering, PC1 Programmable Counter, CQ Aug-95, Carr, Paul N4PC

Sabtronics, 2000 Digital Multimeter, QST Feb-79, Bartlett, Jim K1TX

Sabtronics, 2000DMM Digital Multimeter, CQ May-79, Weiss, Adrian, K8EEG/WØRSP

Santec, LS-202A 2-M SSB/FM Transceiver, QST Dec-85, Wolfgang, Larry WA3VIL

Santec, ST-142 Two Meter HT, CQ May-84, McCoy, Lew, W1ICP

Santec, ST20T Two Meter Handheld Transceiver, CQ May-87, Ingram, Dave, K4TWJ

Santec, W720 Wattmeter, CQ Mar-89, McCoy, Lew, W1ICP

SAY, SPS-20M Power Supply, QST Oct-78, Harris, Chod WB2CHO

SBE Linear Systems, SB-450 UHF-FM Transceiver, QST Nov-73, Unknown, W1GRE

Sem Con, HA-2 2-M Mobile Antenna, QST May-79, Gerli, Sandy AC1Y

SGC Inc., Model SG-230 "Smartuner" Automatic Antenna Tuner, QST Nov-93, Bauer, Jeff WA1MBK

Sherwood, Crystal Filter, QST Feb-77, Myers, Bob W1XT

Sherwood, Se-1 Microphone Equalizer, QST Jan-83, O'Dell, Pete KB1N

SHF Systems, SHF 1240K 1296-MHz Transverter, QST Feb-90, Lau, Zack KH6CP

Sierra Technologies, 80386SX Computer System, CQ Nov-90, Rogers, Buck, K4ABT

Simpson, A2 Meter FM Transceiver, CQ Jul-71, Zook, Glen E., K9STH/5

Simpson, Model A FM Transceiver, QST Oct-71, Williams, Perry W1UED

Sinclabs, SBP 144-4 Cavity Filter, CQ Nov-93, Juge, Ed, W5TOO

Sinclair, D0236H 2-M Antenna, QST Apr-77, McCoy, Lew W1ICP

Sinclair, IC-10 Audio Amplifier, QST Feb-71, Blakeslee, Douglas W1KLK

Sky Lane Products, Cubical Quad Antenna, QST Nov-75, Unknown, WA1JZC

Smith Electronics, Dick, K-6345 Radio Direction Finder, QST Aug-86, Williams, Bruce WA6IVC

Snyder Antenna Corporation, 80 Meter Brooad-Band Dipole, CQ Jun-82, McCoy, Lew, W1ICP

Software Systems, Consulting PC Slow Scan TV System, CQ Jan-93, Rogers, Buck, K4ABT

Solar Power Corp., Series E Solar Electric Generator, QST Aug-77, DeMaw, Doug W1FB

Solder-It, Slodering Kit, CQ Jan-93, McCoy, Lew, W1ICP

Solder-It, Soldering Kit, QST Apr-94, Gruber, Mike WA1SVF

Solid State Sales, CCD Camera Kit, QST Feb-77, Watts, Chuck WA6GVC

Sommer, Trapless Multiband Beam, CQ Apr-86, McCoy, Lew, W1ICP

Sonar, 2307 Transistorized Portable Radio Telephone, QST Mar-72, McMullen, Thomas W1SL

Sonar, FM-3601 2 Meter FM Transceiver, CQ Jan-72, Zook, Glen E., K9STH/5

Sonar, FM-3601 Amateur FM Transmitter-Receiver, QST Aug-72, Myers, Robert W1FBY

Sonar, Sonar Sentry FM/A-M Monitor Receiver Model FR-1035A, QST Oct-72, DeMaw, Doug W1CER

SONY, ICF-2001 Receiver and the MFJ 1020 Active Antenna, CQ Mar-82, Schultz, John J., W4FA

Soundpower, SP100 Audio Processor, QST Jan-80, Frenaye, Tom K1KI

Southwest Tech. Prod. Corp., 6800 Computer System, QST Apr-77, McCoy, Lew W1ICP

Southwest Tech. Prod. Corp., AC-30 & MF-68 Interface & Disc System, QST Aug-78, McCoy, Lew W1ICP

Southwest Tech. Prod. Corp., CT-1024 Terminal Systems Kit, QST Mar-77, McCoy, Lew W1ICP

Spectronics, DD-1 Digital Display, QST Jul-74, DeMaw, Doug W1CER

Spectronics, DD-1K Digital Frequency Display, CQ Mar-77, Paul, Hugh R., W6POK

Spectrum Communications, SCR 1000 2-M FM Repeater, QST Jul-83, O'Dell, Pete KB1N

Spectrum International, 1296-MHz Loop Yagi, QST Jun-78, Kearman, Jim W1XZ

Spectrum International, 70/MBM28 Jaybeam Antenna, CQ Apr-89, McCoy, Lew, W1ICP

Spectrum International, JMF 432 & JMF 1296 UHF Filters, QST Apr-76, McMullen, Tom W1SL

Spider, HF Mobile Antenna, QST Jul-86, Wolfgang, Larry WA3VIL

SSB Electronics, LT 33S 902-MHz Transverter, QST Apr-87, Wilson, Mark AA2Z

SSB Electronics, SP-70 Mast-Mount Preamplifier, QST Mar-93, Jansson, Dick WD FAB

SSB Electronics, UEK-2000S 2.4 Ghz Satellite Downconverter, QST Feb-94, Ford, Steve WB8IMY

SST, T-1 Random Wire Antenna Tuner, QST Oct-77, Paul, Hugh R., W6POK

Standard, C1208DA 2-M FM Transceiver, QST Jan-95, Ford, Steve WB8IMY

Standard, C158A 2 Meter HandHeld, CQ Nov-94, Schroeder, Joe W9JUV

Standard, C228A Dual Band HT, CQ Jan-93, McCoy, Lew, W1ICP

Standard, C228A Dual-Band Hand-Held FM Transceiver, QST Jun-91, Healy, Rus NJ2L

Standard, C558A Dual-Band HH FM Transceiver, QST Mar-94, Ford, Steve WB8IMY

Standard, C5608DA Dual-Band Mobile FM Transceiver, QST Jun-93, Healy, Rus NJ2L

Standard, C568A Dual-Band Hand-Held Transceiver, QST Jul-95, Ford, Steve WB8IMY

Standard, C5718DA Twin Band Transceiver, CQ July-95, Schroeder, Joseph W9JUV

Standard, Horizon 2 2-Meter FM Transceiver, CQ May-76, Paul, Hugh R., W6POK

Standard, RPT-1 Repeater, QST Jun-74, McCoy, Lewis W1ICP and Unknown, WR1ABH

Standard, SR-C146 FM Transceiver, QST Mar-73, DeMaw, Doug W1CER

Standard, SR-C806MA FM Transceiver, CQ Feb-71, Zook, GlenE., K9STH/5

Standard, SR-C826M 2M Transceiver, CQ Nov-71, Zook, Glen E., K9STH/5

Stewart, Quad "Battery-Beater" HT DC Power Source, CQ Sep-81, Stern, Mark, N1BLH

Superflex, Cordless Headphones, CQ Feb-72, Scherer, Wilfred M., W2AEF

Superspanner, A 75-10 Meter Mobile Antenna, CQ Dec-93, McCoy, Lew W1ICP

Swan, 100 MX HF Transceiver & Acc., QST Jun-79, Woodward, George W1RN

Swan, 600-R, QST Jan-73, Nelson, John W1GNC and Niswander, Rick WA1PID

Swan, 600-T, QST Jan-73, Nelson, John W1GNC and Niswander, Rick WA1PID

Swan, Astro 102 BXA, QST Dec-81, DeMaw, Doug W1FB

Swan, Astro 150 HF Transceiver, QST Jul-80, Pelham, John W1JA

Swan, FM 1210-A 2 Meter FM Transceiver, CQ Dec-72, Zook, Glen E., K9STH/5

Swan, FM-2X Meter FM Transceiver, CQ Oct-71, Zook, Glen E., K9STH/5

Swan, Twins 600-R Receiver, and 600-T Transmitter, CQ Oct-71, Scherer, Wilfred M., W2AEF

Swan, WM-1500 RF Wattmeter, QST Feb-73, Myers, Robert W1FBY

Symtek, SABA-5 Signal Intensifier, QST May-74, Pride, Mark WA1ABV

TAPR, Remote Control and Telemetry System, QST Jan-93, new-land, Paul AD7I

TAPR, TAPR-1 Terminal Node Controller, QST Nov-85, Ward, Jeff K8KA & Wilson, Mark AA2Z

Technico, TEC-9900-SS Computer Kit, QST Jul-78, Schueckler, James WB2YZL

TEDCO, Model 1 QRP Transceiver, QST Nov-80, Pagel, Paul N1FB

TEDCO, Model-1 80 Meter QRPP Transceiver, CQ Aug-80, Weiss, Adrian, K8EEG/WØRSP

Tejas, Backpacker I-A QRP Transceiver, CQ Feb-93, Ingram, Dave, K4TWJ

Tejas RF Technologies, TRFT-550 Backpacker II, QST Nov-93, Bauer, John, MA1MBK

Tektronix, T922 Dual-Trace Scope, QST Nov-76, McGivern, James III WA1QZH

TELCO, 125 2-M Class-C Amplifier, QST Mar-78, O'Dell, Pete KB1N

Teletron, Slinky Dipole, QST Feb-74, Godwin, Morgan W4WFL

Telex/Hy-Gain, Contester Headset, QST Feb-92, Wilson, Mark AA2Z

Telex/Hy-Gain, T2X "Tail Twister" Rotator, CQ Apr-87, Schultz, John J., W4FA

Telex/Hy-Gain, TH11DX 5-Band Super Thunderbird Antenna, CQ Apr-94, McCoy, Lew, W1ICP

Telex/Hy-Gain, TH2MK3S 2 Element 10/15/20M Trap Beam, CQ Aug-88, Schultz, John J., SVØDX

Telex/Hy-Gain, TH7DX Super Thunderbird Triband Beam, CQ Apr-86, Schultz, John J., W4FA

Telex/HyGain, Explorer 14 Triband Beam, CQ Apr-85, Schultz, John J., W4FA/SV

Telex/HyGain, Ham-IV Rotor System, CQ Dec-85, Schultz, John J., W4FA

Tempo, 1002-3 2M Amplifier, CQ Jul-72, Zook, Glen E., K9STH/5

Tempo, K6FZ 20-M Loop Antenna, QST Sep-79, Horzepa, Stan WA1LOU

Tempo, MR-2 VHF Monitor Receiver, CQ Sep-76, Paul, Hugh R., W6POK

Tempo, S1 2-M FM Transceiver, QST Jun-79, DeMaw, Doug W1FB

Tempo, Tempo-ONE Transceiver, CQ May-71, Scherer, WilfredM., W2AEF

Tempo, VHF/ONE 2 Meter FM/SSB Transceiver, CQ Feb-76, Paul, Hugh R., W6POK

Ten Tec, 2591 2 Meter Talkie, CQ Feb-84, Ingram, Dave, K4TWJ

Ten-Tec, 229 2 KW Antenna Tuner, CQ May-83, Schultz, John J., W4FA

Ten-Tec, 247 & 277 Antenna Tuners, QST Apr-80, Leland, Stu W1JEC

Ten-Tec, 2510 Mode-B Satellite Station, QST Oct-85, Hutchinson, Chuck K8CH

Ten-Tec, 2510 Oscar Satellite Station/Converter, CQ Feb-85, Ingram, Dave, K4TWJ

Ten-Tec, 253 Automatic Antenna Tuner, CQ Aug-91, McCoy, Lew, W1ICP

Ten-Tec, 315 Receiver, CQ Feb-74, Scherer, Wilfred M., W2AEF

Ten-Tec, 425 Titan HF Linear Amplifier, QST Apr-86, Wilson, Mark AA2Z

Ten-Tec, 544 HF Transceiver, QST Jul-79, DeMaw, Doug W1FB

Ten-Tec, 561 Corsair II HF Transceiver, QST Aug-87, Hare, Ed KA1CV

Ten-Tec, 562 OMNI V HF Transceiver, CQ Jan-90, Schultz, John J., W4FA

Ten-Tec, 585 Paragon 160-10 M Transceiver, QST May-88, Wilson, Mark AA2Z

Ten-Tec, Argonaut 505 , QST Nov-72, Tilton, Edward W1HDQ

Ten-Tec, Argonaut 509 QRPp SSB/CW Transceiver, CQ Jul-78, Weiss, Adrian, K8EEG/Ø

Ten-Tec, Argonaut 515 Transceiver, CQ Mar-82, Weiss, Adrian, WØRSP

Ten-Tec, Argonaut II & Delta II MF/HF Transceivers, QST Jan-92, Newkirk, Dave WJ1Z

Ten-Tec, Argonaut II QRP Transceiver, CQ Nov-92, Carr, Paul, N4PC

Ten-Tec, Argosy HF Transceiver, QST Oct-82, Hutchinson, Chuck K8CH

Ten-Tec, Argosy Transceiver, CQ Dec-82, Schultz, John J., W4FA

Ten-Tec, Centurion MF/HF Linear Amplifier, July-92, Wilson, Mark AA2Z and Healy, Rus NJ2L

Ten-Tec, Century 21 HF Transceiver, QST Dec-77, Bartlett, Jim K1TX

Ten-Tec, Century 22 HF CW Transceiver, QST May-85, DeMaw, Doug W1FB

Ten-Tec, Century 22 Transceiver, CQ May-85, Ingram, Dave, K4TWJ

Ten-Tec, Corsair HF Transceiver, CQ Jan-84, Schultz, John J., W4FA

Ten-Tec, Corsair II Transceiver Part I, CQ Sep-86, Schultz, John J., W4FA

Ten-Tec, Delta II HF Transceiver, CQ Jul-92, McCoy, Lew, W1ICP

Ten-Tec, Equipment Enclosures, CQ Oct-88, Schultz, John J., W4FA/SVØDX

Ten-Tec, Hercules II HF Linear Amplifier and 9420 Power Supply, CQ Aug-90, Schultz, John J., W4FA

Ten-Tec, KR-40 Electronic Keyer, QST Dec-73, Godwin, Morgan W4WFL

Ten-Tec, KR-5 Electronic Keyer, QST Dec-73, Godwin, Morgan W4WFL

Ten-Tec, OMNI D HF Transceiver, QST Jan-80, Pagel, Paul N1FB

Ten-Tec, OMNI V Model 562 160-10 M Transceiver, QST Nov-90, Kleinschmidt, Kirk NTØZ

Ten-Tec, OMNI VI HF Transceiver, CQ Jul-93, McCoy, Lew, W1ICP

Ten-Tec, OMNI VI MF/HF Transceiver, QST Jan-93, Healy, Rus NJ2L

Ten-Tec, Paragon HF Transceiver, Part II, CQ May-88, Schultz John J., W4FA/SVØDX

Ten-Tec, Paragon HF Transceiver Part III, CQ Jun-88, Schultz, John J., W4FA/SVØDX

Ten-Tec, Paragon HF Trnsceiver, Part I, CQ Apr-88, Schultz, John J., W4FA

Ten-Tec, PM-2 Transmitter, QST Jun-70, DeMaw, Doug W1CER

Ten-Tec, Power-Mite Solid-State C.W. Transceivers, CQ Apr-70, Scherer, Wilfred M., W2AEF

Ten-Tec, RX10 Communications Receiver, QST Aug-71, Godwin, Morgan W4WFL/1

Ten-Tec, Scout 555 HF Transceiver, CQ Nov-93, McCoy, Lew, W1ICP

Ten-Tec, Scout Model 555 MF/HF Transceiver, QST Dec-93, Newkirk, David WJ1Z

Ten-Tec, Series of "Ultimate" HF Mobile Antennas, CQ Aug-87, Schultz, John J., W4FA

Ten-Tec, Titan Legal Limit Power Amplifier, CQ Feb-86, Schultz, John J., W4FA

Terlin, Outbacker JR8 HF Mobile Antenna, QST Apr-93, Bauer, Jeff WA1MBK

TET, 3F35DX Triband Antenna, QST Apr-80, Aurick, Lee W1SE

TET, HB-35 Triband Antenna, QST Dec-82, Aurick, Lee W1SE

Texas Instruments, SR-10 Electronic Slide Rule Calculator, QST Oct-73, Dorbuck, Tony W1YNC

Texas Instruments, SR-50 Electronic Slide Rule Calculator, QST Jun-75, Steinman, Hal K1FHN

TimeKit, Blinky SSTV/RTTY Tuner, CQ Nov-83, Ingram, Dave, K4TWJ

Timewave, DSP-59+ Digital Signal Processor, QST Oct-94, Healy, Rus NJ2L

Timewave, DSP-9 Audio Noise Reduction Filter, CQ May-94, Cohen, Ted, N4XX

Timewave, DSP-9+ Digital Signal Processor, QST Oct-94, Healy, Rus NJ2L

Tokyo Hy-Power Labs, HC-200 Transmatch, QST May-83, Hutchinson, Chuck K8CH

Tonna, F9FT 144/16 2-M Yagi, QST Jul-79, Sumner, Dave K1ZZ

Tonna, F9FT Tonna 50/5, 5 Element Six Meter Beam, CQ Dec-87, Katz, Steve, WB2WIK

Tono, EXL-5000E RTTY/CW/AMTOR Term, QST Jul-85, Bloom, Jon KE3Z

Tono, Theta-777 Communications Terminal, QST Apr-87, Hale, Bruce KB1MW

Trimble, Scout HH GPS Receiver, QST Mar-94, Wilson, Mark AA2Z

Triplett, 7000 Universal Counter, CQ Aug-81, Weiss, Adrian, K8EEG/WØRSP

TUCH-COM, 1215 Tone Encoder Microphone, QST Feb-79, Morris, Jim K1UJ

Twin Oaks, Morse Code Training Program, QST Aug-83, Colvin, Carol AJ2I

Uni-Hat, CTSVR Antenna, CQ Dec-94, McCoy, Lew W1ICP

Uniden, HR-2510 10 Meter SSB/CW/FM Transceiver, CQ Feb-89, Ingram, Dave, K4TWJ

Uniden Corp of America, President HR2510 10-M Transceiver, QST May-89, Kleinschmidt, Kirk NTØZ

Unique Products Company, Identiminder, QST Nov-70, Hall, Gerald K1PLP

Universal Software, Inc., Super-RATT RTTY/CW Software, QST Nov-83, Pagel, Paul N1FB

UPI Communications Systems, Climbing-Safety Belt, QST Dec-89, Healy, Rus NJ2L

Varitronics, InqueIC-2F 2 Meter FM Transceiver, CQ Mar-71, Zook, GlenE., K9STH/5

Varitronics-Inoue, AS2G Antenna, QST Feb-71, Blakeslee, Douglas W1KLK

Varitronics-Inoue, IC-2F FM Transceiver, QST Jan-71, DeMaw, Doug W1CER

Vectronics, VC-300 and HFT-1500 Antenna Tuners, CQ Apr-92, Schultz, John J., W4FA

Yaesu, FT-757GX HF Transceiver, QST Dec-84, Kluger, Leo WB2TRN

Yaesu, FT-767GX All Mode HF Transceiver, QST Sep-87, McGrath, Don KZ1A

Yaesu, FT-77 Compact HF Transceiver, CQ Dec-84, Schultz, John J., W4FA

Yaesu, FT-77 HF Transceiver, QST Nov-83, Schetgen, Bob KU7G

Yaesu, FT-7B HF Transceiver, QST Mar-80, Woodward, George W1RN

Yaesu, FT-840 MF/HF Transceiver, QST May-94, Ford, Steve WB8IMY

Yaesu, FT-890 HF Transceiver, CQ Dec-92, Schultz, John J., W4FA

Yaesu, FT-900A MF/HF Transceiver, QST Feb-95, Swanson, Glenn KB1GW

Yaesu, FT-901DM HF Transceiver, QST Nov-78, DeMaw, Doug W1FB

Yaesu, FT-980 Cat, CQ Jan-85, Schultz, John J., W4FA

Yaesu, FT-980 HF Transceiver, QST Nov-84, Ward, Jeff K8KA

Yaesu, FT-990 160-10 M Transceiver, QST Nov-91, Healy, Rus NJ2L

Yaesu, FT-990 HF Transceiver, CQ Jun-92, Schultz, John J., W4FA

Yaesu, FT-One All-Mode General Coverage Transceiver Part I, CQ Sep-83, Schultz, John J., W4FA

Yaesu, FT-One All-Mode General Coverage Transciver Part II, CQ Oct-83, Schultz, John J., W4FA

Yaesu, FT-ONE HF Transceiver, QST Aug-83, Hutchinson, Chuck K8CH

Yaesu, FTDX 570 SSB/CW Transceiver, CQ Sep-72, Scherer, Wilfred M., W2AEF

Yaesu, FTdx560 Transceiver, CQ May-70, Scherer, WilfredM., W2AEF

Yaesu, FTV-901R Transverter, QST Feb-82, Sumner, Dave K1ZZ

Yaesu, YC-355D Frequency Counter, CQ Jan-77, Paul, Hugh R., W6POK

Yaesu, YC-7 Frequency Display, QST Mar-80, Woodward, George W1RN

Yaesu, YS-60 MF/HF Wattmeter, QST Feb-91, Healy, Rus NJ2L

Z.R.C., Cold Galvanizing Compound, QST Oct-80, DeMaw, Doug W1FB

COMPARISON CHART OF HF TRANSCEIVERS

Model	Frequency Range	Power Required	Current Drain TX	Power Output	Receiver Circuitry	Sensitivity 1.8-30 MHz
Alinco						
DX-70T	15-30,50-54MHz	13.8V DC		100w		
Atlas Radio						
Model 400-X	1.8-30 MHz	13.8V DC		150w	Single	
Azden						
DX-70T	.5MHz-54MHz	13.8V DC		100w/10w		
Icom						
IC-706	.03kHz-200 MHz	13.8V DC		100/10w		
IC-707	500kHz-30 MHz	13.8V DC	20A	100w	Double	<.16uv @ 10 db
IC-725	500kHz-30 MHz	13.8V DC	20A	100w	Double	<.16uv @ 10 db
IC-728	500kHz-30 MHz	13.8V DC	20A	100w	Triple	<.16uv @ 10 db
IC-735	100kHz-30 MHz	13.8V DC	20A	100w	Triple	<.15uv @ 10db
IC-736	.1-30,50-54 MHz	120V AC	500VA	100w	Triple	<.15uv @ 10db
IC-737A	100kHz-30 MHz	13.8V DC	20A	100w	Triple	<.15uv @ 10db
IC-738	100kHz-30 MHz	13.8V DC		100w	Triple	<.15uv @ 10db
IC-765	100kHz-30 MHz	120V AC	650VA	100w	Quad	<.15uv @ 10db
IC-775	100kHz-30 MHz	120V AC		200w		
IC-781	100kHz-30 MHz	120V AC	760VA	150w	Quad	<.16uv @ 10 db
JRC						
JST-245	.1-30,48-54MHz	115V AC		150w	Quad	
Kenwood						
TS-950S	100kHz-30 MHz	120V AC	700VA	150w	Quad	<.2 uv
TS-850S	100kHz-30 MHz	13.8V DC	205A	100w	Quad	<.2 uv
TS-450S	500kHz-30 MHz	13.8V DC	205A	100w	Triple	<.25uv
TS-140S	500kHz-30 MHz	13.8V DC	20A	100w	Double	<.25uv @ 10db
TS-690S	.5-30,50-54 MHz	13.8V DC	205A	100w	Triple	<.25uv @ 10db
TS-50S	100kHz-30 MHz	13.8V DC	205A	100w	Double	<.25uv
Ten-Tec						
OMNI-VI	.1-30 MHz	13.8V DC	20A	100w	Triple	
Paragon II	10-160m	13.8V DC	20A	100w	Triple	
Scout 555	10-160m	13.8V DC	10A	50w		
Argonaut II	10-160m	13.8V DC	2A	5w		
Yaesu						
FT-1000MP	100kHz-30 MHz	AC or DC		100w		
FT-1000D	100kHz-30 MHz	117V AC	1050VA	200w	Quad	<.25
FT-990	100kHz-30 MHz	117V AC	470VA	100w	Triple	<.25
FT-890	100kHz-30 MHz	13.8V DC	20A	100w	Double	<.25
FT-840	100kHz-30 MHz	13.8V DC	20A	100w	Double	<.25
FT-650	24.5-56 MHz	117V AC	500VA	100w	Triple	<.125uv

COMPARISON CHART OF HANDHELD RADIOS

Model	Type	Frequency Range	Power Required	Current Drain TX	Current Drain RX	Power Out	Memories	Sensitivity	Height Inches	Width Inches	Depth Inches	Weight Ounces
ADI												
AT-200	5W H.T.	2M+	7.2V DC			3w/5w	20					
AT-400	5W H.T.	440 MHz+	7.2V DC	1.6 A	120 ma	3w/5w	20	<.15uv @ 12db	4.5	1.9	1.4	0.79lbs.
Alinco												
DJ-G5T	5W H.T.	2M/ 440 MHz+	13.8V DC			5W	80	<.15uv @ 12db	5.3	2.14	1.4	0.79lbs.
DJ-G1T	5W H.T.	2M/ 440 MHz+				4.5W	100/100	<.15uv @ 12db				0.77lbs.
DJ-16ZT	5W H.T.	2M+	13.8V DC			5W	20	100/Opt100 <.15uv @ 12db	5.2	2.3	1.3	0.77lbs.
DJ-180T	5W H.T.	2M+	13.8V DC	1.05A	130 ma	5W		<.15uv @ 12db			1.5	0.78lbs.
DJ-180T-HP	5W H.T.	2M+	13.8V DC		130 ma	2W		<.15uv @ 12db	4.3	2.1		0.78lbs.
DJ-F1T-TH	5W H.T.	2M+	13.8V DC	1.4A	130 ma	5W	40	<.15uv @ 12db			1.23	0.97lbs.
DJ-F1T-HP	5W H.T.	2M+	13.8V DC		130 ma	2W		<.15uv @ 12db		2.24		0.97lbs.
DJ-580T	5W H.T.	2M/ 440 MHz+	13.8V DC			2W/Opt5W	40/40	<.15uv @ 12db	6.65	2.24	1.23	
DJ-580T	5W H.T.	2 M+/ 440 MHz+	13.8V DC	1.2 A	120 ma	5W /5W	40	<.15uv @ 12db	6.65	2.24	1.23	
Azden												
AZ-11	5W H.T.	10M	12V DC	1A	150ma	0.5w/5w	40	<.16uv @ 12db	6.85	2.6	1.3	1.00lbs.
AZ-61	5W H.T.	6M	12V DC	1A	150ma	0.5w/5w	40	<.16uv @ 12db	6.85	2.6	1.3	1.00lbs.
AZ-21A	5W H.T.	2M+	12V DC	1A	150ma	0.5w/5w	40	<.16uv @ 12db	6.85	2.6	1.3	1.00lbs.
Icom												
IC-Z1A	5W H.T.	2M+/440	4.5V DC+	1.5 A max	170 ma	5w/5w	104	<.16uv @ 12db	4.9	2.6	1.4	13.4lbs.
IC-2AT	5W H.T.	144-147.995 MHz	8.4V DC	550 ma	200 ma	2w/.5w		5uv @ 20db	6.6	2.6	1.4	19lbs.
IC-2iA	5W H.T.	144-148 MHz	13.8V DC	1.2 A	200 ma	2.5w/.02w	10	<.16uv @ 12db	3.6	2.3	1.2	9.1lbs.
IC-D1A	5W H.T.	2m+440MHz/1.24G	13.8V DC	1.2 A	210 ma	2.5w/.02w	26/26/26	<.18uv @ 12db	3.6	2.3	1.2	6.0lbs.
IC-4A	5W H.T.	440-450 MHz	13.8V DC	1.8 A	250 ma	2w/2w/1w	10	<.18uv @ 12db	4.9	2.3	1.4	9.0lbs.
IC-4SAT	5W H.T.	144-148 MHz RX	13.8V DC	1.8 A	250 ma	3w/1w	41	<.18uv @ 12db	4.9	2.3	1.4	13lbs.
IC-T21A	6W H.T.	2M / with440 RX	4.5-16 V	1.8 A	160 ma	6w/.15w	100	<.16uv @ 12db	4.4	2.1	1.4	11.1lbs.
IC-T22A	5W H.T.	2M/ 220	13.8V DC			5w/	100					11.1lbs.
IC-T41A	Handheld	440 / with 2M RX	4.5-16 V			5w/	40					
IC-T42A	5W H.T.	2m / 220	7.2V DC				40					
IC-V21	5W H.T.	2m +/ 440	13.8V DC	1.5 A	200 ma	1.5w	70	<.16uv @ 12db	4.9	2.2	1.4	14lbs.
IC-W2A	5W H.T.	2m +/ 440	13.8V DC	1.3 A	200 ma	2w/.5w	70	<.25uv @ 12db	6.7	2.1	1.4	19lbs.
IC-W21AT	5W H.T.	Dual Band	4.5V DC+	1.4 A	150 ma	2w/.5w	70	<.16uv @ 12db	6.8	2.3	1.4	13lbs.
IC-W31A	5W H.T.	2M	13.8V DC+			1.5w/.5w	104					13lbs.
IC-2SRA	5W H.T.	440 MHz	13.8V DC	1.3 A	250 ma	1.5w/.5w	96	<.16uv @ 12db	6.7	2.1	1.4	13lbs.
IC-4SRA	5W H.T.	440 / 1.2Ghz	13.8V DC	1.6 A	250 ma	5w/1w	96	<.18uv @ 12db	6.7	2.1	1.4	13lbs.
IC-X21AT	5W H.T.	1.2 Ghz	13.8V DC	1.6 A/.9A	180 ma	1w	70	<.22uv @ 12db		2.6	1.4	18lbs.
IC-21GAT	Handheld	222 MHz	13.8V DC	990 ma	250 ma	2.5w	20	<.16uv @ 12db	5.1			
IC-03AT	5W H.T.	222 MHz				2.5w						
IC-3SAT	5W H.T.	222 MHz	13.8V DC	1.6 A	250 ma	1.5w/.5w	100	<.16uv @ 12db	4.1	2.1	1.5	10lbs.
IC-P3AT	5W H.T.	222 MHz	13.8V DC			1.5w/.5w						
Kenwood												
TH-22AT	5W H.T.	144-148 MHz	6V DC	1.6 A	45 ma	3w	40	<.18uv	4.6	2.2	0.96	10.2lbs.
TH-28A	5W H.T.	438-450 MHz	7.2V DC	1.6 A	45 ma	2.5w	40	<.18uv	4.6	2.2	0.96	9lbs.
TH-48A	5W H.T.	144-148 MHz	7.2V DC	1.5 A	60 ma	1.5w	41/Opt240	<.18uv	4.6	1.9	1.5	12lbs.
TH-78A	5W H.T.	438-450 MHz	7.2V DC	1.4 A	60 ma	2.5w	57/Opt252	<.18uv	5.3	1.9	1.5	11lbs.
TH-79A	5W H.T.	2M+/440 MHz+	7.2V DC	1.3A	60 ma	2.5w/2w	41/41	<.16uv	5.1	2.2	1.6	11lbs.
Radio Shack												
HTX-202	5W H.T.	2M	7.2V DC	0.8 A	35 ma	2.5w	12	<2 uv	4.6	2.7	1.9	19lbs.
HTX-404	5W H.T.	440 MHz	13.8V DC	1.8A		5w	16		4.6		1.9	
Standard												
C108A	Handheld	144-148 MHz+	3V DC	0.8 A	220 ma	200 mw	20	<.112uv @ 12db	3.15	2.28	0.98	4.9lbs.
C156A	5W H.T.	144-148 MHz+	7.2V DC	700 ma	40 ma	2w	20	<.158uv @ 12db	5.38	2.8	1.2	12.5lbs.
C168A	5W H.T.	144-148 MHz+	7.2V DC	700 ma	38 ma	2.8w	40/Opt200	<.158uv @ 12db	4.9	1.8	1.2	10lbs.
C188A	5W H.T.	2m+/440 MHz+	7.2V DC	700 ma	38 ma	2w/50mw	40/Opt200	<.158uv @ 12db	4.9	2.3	1.2	10lbs.
C288A	5W H.T.	220-224 MHz	7.2V DC	1.1A	38 ma	5w	40/Opt200	<.158uv @ 12db	4.9	2.3	1.2	10lbs.
C468A	5W H.T.	438-450 MHz	7.2V DC	700 ma	25/70 ma	2.8w		<.158uv @ 12db	4.9	1.8	1.8	10lbs.
C528A	5W H.T.	2m+440 MHz+	7.2V DC	700 ma	25/70 ma	2.8w/2.5w	40/Opt200	<.158uv @ 12db	5.8	1.8	1.2	17lbs.
C628A	5W H.T.	440MHz+/1.2Ghz+	7.2V DC	700 ma	25/70 ma	2.8w/1w	20/20	<.158uv @ 12db	6.2	2.2	1.2	17lbs.
C558A	5W H.T.	2m+/440 MHz+	7.2V DC	700 ma	25/70 ma	2.8w/2.5w	40/Opt200	<.158uv @ 12db	6.2	2.2	1.2	17lbs.
C228A	5W H.T.	2m+/220 MHz	7.2V DC	700 ma	25/70 ma	2.8w/2.5w	20/20	<.158uv @ 12db	6.2	2.2	1.2	17lbs.
Yaesu												
FT-11	5W H.T.	144-148 MHz	4.8V DC	900 ma	19 ma	1.5w	150	<.25uv @ 12db	3.9	2.3	1	15lbs.
FT-23R	5W H.T.	144-225 MHz			19 ma	2w	10	<.25uv 12db	4.8	2.3	1.3	15lbs.
FT-33R	5W H.T.	220-225 MHz	4.8V DC	1.3 A		5w	10		3.9	2.3	1.3	
FT-41R	5W H.T.	440-450 MHz				1.5w	150					
FT-411E	5W H.T.	144-148 MHz	7.2V DC	900 ma	150 ma	2.5w	120	<.158uv @ 12db	4.8	2.3	1.3	13lbs.
FT-416	5W H.T.	144-148 MHz	7.2V DC	900 ma	150ma	2.5w	49	<.158uv @ 12db	5.7	2.3	1.3	15lbs.
FT-811	5W H.T.	430-450 MHz	7.2V DC	900 ma	150ma	2.5w	49	<.158uv @ 12db				
FT-816	5W H.T.	430-450 MHz	7.2V DC	900 ma	150 ma	2.5w	41	<.158uv @ 12db	5.7	2.3	1.3	30lbs.
FT-911	Handheld	1.24-1.3 Ghz	7.2V DC	900 ma	150 ma	1w	49	<2uv @ 12db	4.8	2.3	1.3	
FT-530	5W H.T.	2m440 MHz	7.2V DC	900 ma	190 ma	2w	82	<.18uv @ 12db	5.3	2.3	1.3	18lbs.

COMPARISON CHART OF VHF/UHF BASE MOBILE TRANSCEIVERS

Model	Frequency Range	Power Req'd	Current Drain TX	Current Drain RX	Power Output	Memories	Sensitivity
ADI							
AR-146	2m+	13.8V DC			50w	40	
AR-446	440 MHz+	13.8V DC			35w	40	
Alinco							
DR-130T	2m	13.8V DC	11A	500ma	50w	20/Opt100	<.16uv @ 12db
DR-430T	440 MHz	13.8V DC	11A	500ma	35w	20/Opt100	<.16uv @ 12db
DR-119T	2m	13.8V DC	11A	500ma	50w	14	<.16uv @ 12db
DR-1200T	2m @ 1.2-2.4Kb	13.8V DC	5.5A	500ma	25w	14	<.16uv @ 12db
DR-1200TH2	2m @ 9.6Kb	13.8V DC	5.5A	500ma	25w	14	<.16uv @ 12db
DR-570T	2m/440 MHz	13.8V DC	11A	500ma	45/35w	28	<.16uv @ 12db
DR-592T	2m/440 MHz	13.8V DC	11A	500ma	45/35w	28	<.16uv @ 12db
DR-600T	2m+/440 MHz	13.8V DC	10A	500ma	45/35w	28	<.16uv @ 12db
DR-M06T	6m	13.8V DC		500ma	1/10w	100	
DR-150T	2M+/440 RX	13.8V DC		500ma	50w	100	
DR-610T	2M+/440 MHz	13.8V DC		500ma	50/35w	120	
Azden							
PCS-7000H	2m	13.8V DC	9.0A	600ma	50w	20	<.19uv @ 12db
PCS-7200	220 MHz	13.8V DC	9.0A	600ma	25w	20	<.19uv @ 12db
PCS-7300H	440 MHz	13.8V DC	9.0A	600ma	35w	20	<.19uv @ 12db
PCS-7500H	6m	13.8V DC	9.0A	600ma	50w/5w	20	<.19uv @ 12db
PCS-7800H	10m	13.8V DC	9.0A	600ma	50w	20	<.19uv @ 12db
PCS-9600D	440MHz	13.8V DC	9.0A	600ma	35w	20	<.19uv @ 12db
Icom							
IC-28H	2m	13.8V DC	105A	800ma	45w	21	<.18uv @ 12db
IC-275H *	2m	13.8V DC	20A	1A	100w	99	<.1 uv @ 12db
IC-281H	2m/440 MHz RX	13.8V DC	105A	800ma	50w	50	<.16uv @ 12db
IC-449A	440 MHz	13.8V DC	9.5A	800ma	35w	20	<.16uv @ 12db
IC-475H *	440 MHz	13.8V DC	20A	1A	80w	99	<.18uv @ 12db
IC-575H *	6m/10m	13.8V DC	20A	1A	100w	99	<.25uv @ 12db
IC-820H *	2m/440 MHz	13.8V DC	16A	2A	45/35w	50	<.11uv @ 12db
IC-970A *	2m/440/1.2GHz	110V AC	400VA	60VA	25w	396	<.18uv @ 12db
IC-970H *	2m/440/1.2GHz	13.8V DC	16A	2.5A	45/35w	396	<.18uv @ 12db
IC-2000H	2m+	13.8V DC			50w	50	
IC-2330	2m/220 MHz	13.8V DC	10.5A	1.8A	25w	36	<.16uv @ 12db
IC-2340H	2m/440 MHz	13.8V DC	10.5A	1.8A	45/35w	110	<.18uv @ 12db
IC-2700H	2m/440 MHz	13.8V DC	12A	1.8A	50/35w	100	<.16uv @ 12db
IC-901	10m-1.2GHz	13.8V DC	105A	600ma	10/45w		<.25uv @ 12db
IC-D100	2m/440/1.2GHz	13.8V DC	12A	1.8A	50/10w	642	<.16uv @ 12db
Kenwood							
TM-251A	2m+/440MHzRX	13.8V DC	11A		50w	41	<.16uv @ 12db
TM-241A	2m	13.8V DC	11A		50w	20	<.16uv @ 12db
TM-331A	220 MHz	13.8V DC			25w	20	<.16uv @ 12db
TM-441A	440 MHz	13.8V DC	11A			20	<.16uv @ 12db
TM-451A	440 MHz/2m RX	13.8V DC	11A		35w	41	<.16uv @ 12db
TM-541A	1.2 GHz	13.8V DC	11A			20	<.16uv @ 12db
TM-551A	1.2 GHz/440 RX	13.8V DC			10w	41	<.16uv @ 12db
TM-641A	2m/220 MHz	13.8V DC	115A	1.2A	50/25w	100	<.16uv @ 12db
TM-642A	2m/220 MHz	13.8V DC	115A		50/25w	101	<.16uv @ 12db
TM-732A	2m/440 MHz	13.8V DC	115A	0.8A	50/25w	100	<.16uv @ 12db
TM-733A	2m/440 MHz	13.8V DC	115A		50/25w	72	<.16uv @ 12db
TM-742A	2m/440 MHz	13.8V DC	115A	1.2A	50/10w	101	<.16uv @ 12db
TM-255A *	2m	13.8V DC	13A		40w	41	<.16uv
TM-455A *	440 MHz	13.8V DC	15A		35w	41	<.16uv
TS-790A *	2m/440/1.2GHz	13.8V DC	15A	2.5A	45/10w	59	<.22uv
TM-942A	2m/440/1.2 GHz	13.8V DC	11SA	1.2A	50/10w	101	<.16uv
TS-60S *	6m All-Mode	13.8V DC			90w	100	
RadioShack							
HTX-212	2m	13.8V DC			45w	30	
Standard							
C1208DA	144-148 MHz+	13.8V DC	105A		50w	100	<.158uv@12db
C5718A	2m+/440 MHz+	13.8V DC	105A		50/40w	40	<.158uv@12db
Yaesu							
FT-736 *	2m/440AllMode	110V AC			25/10w	2	<.15uv @ 12db
FT-712RH	440 MHz	13.8V DC			35w	21	
FT-912RH	1.2 GHz	13.8V DC			10w	21	
FT-290R *	2m All-Mode	13.8V DC	8A		25/2.5w	10	<.20uv
FT-690R *	6m All-Mode	13.8V DC	8A		10w	10	<.20uv
FT-790R *	440 All-Mode	13.8V DC	8A		25/2.5w	10	<.20uv
FT-2200	2m+	13.8V DC	11A		50w	50	<.158uv@12db
FT-2400H	2m	13.8V DC			50w		
FT-2500M	2m+	13.8V DC	12A		50w	31	<.19uv @ 12db
FT-5100	2m+/440 MHz	13.8V DC	9A		50/35w	94	<.158uv@12db
FT-5200	2m/440 MHz	13.8V DC	115A		50/35w	32	<.158uv@12db
FT-6200	440 MHz/1.2GHz	13.8V DC	9A		35/10w	32	<.158uv@12db
FT-7200	440 MHz	13.8V DC	11A		50w	50	<.158uv@12db
FT-7400H	440 MHz	13.8V DC	11A		35w	31	<.158uv@12db
FT-8500	2M / 440 MHz	13.8V DC	11A		50/35w		

* = All-Mode
\+ = Extended RX range

COMPARISON CHART OF HF AMPLIFIERS

Model	Frequency Range	A.C. Power Required	Current Drain TX	Drive Power	Power Output	Weight (lbs.)	Final Tubes/ Transistors
Ameritron							
AL-1500	160-10m	220V AC	15A	65w	1500w+	85	1 x 8877
AL-1200	160-10m	220V AC	15A	100w	1500w	85	1 x 3CX1200A7
ALS-500M	160-10m	13.8V DC	80A	100w	600w	7	2SC2879
ALS-600	160-10m No Tune	110/220V AC	12A/6A	100w	700w	48	MRF150
AL-80B	160-10m No Tune	110/220V AC	15A/10A	100w	850w	54	1 x 3-500Z
AL-811	160-10m	110/220V AC	8A/4A	75w	600w	40	3 x 811A
AL-811H	160-10m	110/220V AC	8A/4A	75w	800w	41	4 x 811A
AL-82	160-10m	220V AC	15A	100w	1300w	87	2 x 3-500Z
B & W							
PT-2500A	160-10m			100w	1300w		2 x 3-500Z
Commander							
HF-2500	160-10m	220V AC	20A	60w	1500w+	75	2 x 3CX800A7
HF-1250	160-10m	110/220V AC	20A/10A	60w	1250w	75	1 x 3CX800A7
E.T.O.							
Alpha 91b	160-10m			50w	1500w		2 x 4CX800A7
Alpha 89	160-10m			80-100w	1500w+		2 x 3CX800A7
Alpha 87	160-10m Auto-Tune	220V AC		80-100w	1500w+		2 x 3CX800A7
HENRY							
2KD Standard	80-10m			85w	700w		1 x 3-500Z
2KD Classic	80-10m			100w	1300w		2 x 3-500Z
2K Classic	80-10m			100w	1300w		2 x 3-500Z
3KD Classic	80-10m			75w	1500w		1 x 3CX1200
3KD Premier	160-10m			75w	1500w		1 x 3CX1200
3K Classic MKII	80-10m			75w	1500w		1 x 3CX1200
3K Premier	160-10m			75w	1500w		1 x 3CX1200
Icom							
IC-2KL	160-10m	110/220V AC	20A/10A	50-80w	500w	45	4 x 2SC2652
IC-4KL	160-10m	110/220V AC	35A/18A	100w	1000w	80	8 x 2SC2652
JRC							
JRL-2000F	160-10m	110/220V AC	2,500VA	100w	1000w	62	48 x 2SK408/9
Kenwood							
TL-922A	160-10m	120/220V AC	20A/10A	80w	1300w	83	2 x 3-500Z
QRO Technologies							
HF-1000	160-10m	110/220V AC	20A/10A	80w	800w	75	1 x 3-500Z
HF-2000	160-10m	110/220V AC	20A/10A	110w	1200w	76	2 x 3-500Z
Ten-Tec							
Titan	160-10m	220V AC		80-100w	1500w+	41	2 x 3CX800A7
Centurion	160-10m	110/220V AC	30A/15A	100w	1300w	47	2 x 3-500Z
Hercules II	160-10m NoTune	13.8V DC	80A	65w	500w+		8 x MRF-458
Yaesu							
FL-7700	160-10m	110/220V AC	20A/10A	100w	600w	60	Transistors

This directory gives valuable information on manufacturers and/or importers of amateur radio transmitting equipment.

Beyond being a handy source of phone numbers and addresses, it also gives you insight to a company's business longevity and size, the latter by number of employees. Moreover, you'll learn which companies sell their own products directly to end users, usually through mail order, rather than only to dealers for resale.

The directory that follows, "Who's Who in Dealers," presents retail operations that sell radio products made or imported by companies listed here.

A

AAE/Bandmaster
3164 Cahaba Heights Road
Birmingham, AL 35243
Phone: 205 967-6122
FAX: 205 970-0622
E-mail: wa4fat@scott.net
Worrld Wide Web: http://wweb.net/comart/marcom/bmstr.html
Established: 1988
Sells direct and through dealers
Major Lines: Bandmaster quad antenna systems, Codeck—The original Morse code training flash cards.

ANLI Corp.
20277 Valley Blvd #J
Walnut, CA 91789
Phone: 800 666-2654
Tech: 909 869-5711
FAX: 909 869-5710
Established: 1963
Sells through dealers.
Major Lines: Antennas and radio accessories.

Advanced Electronic Applications, Inc. (AEA)
2006 196th St. SW
PO Box C-2160
Lynnwood, WA 98036
Phone: 206 774-5554
FAX: 206 775-2340
Tech. Support: 206 775-7373
Literature: 800 432-8873
Established: 1977
Sells direct and through dealers.
Major Lines: Packet controllers, multimode data controllers, terminal software, HF/VHF/UHF antennas, UHF/ VHF handheld antennas, high speed radio modems, remote radio controllers, handheld antenna analysts,weather FAX demodulator, CW keyers and trainers.

Alinco Electronics, Inc.
438 Amapola Ave., Suite 130
Torrance, CA 90501
Phone: 310 618-8616
FAX: 310 618-8758
Established: 1977; *Employs* 203
Sells through dealers.
Major Lines: HF, VHF & UHF mobile and HT transceivers.

Alpha Delta Communications, Inc.
PO Box 620
Manchester, KY 40962
Phone: 606 598-2029
FAX: 606 598-4413
Established: 1981
Sells direct and through dealers.
Major Lines: Lightning/EMP ceramic gas tube Transi-Trap surge protectors, lightning/emp-protected DELTA-4 precision coaxial switches, HF (160–10 meter) NO-TRAP spacelimited sloper, dipole DX SERIES wire antennas, and DELTA-C antenna hardware kit.

American Antenna Corp.
1500 Executive Drive
Elgin, IL 60123
Phone: 800 323-6768; 708 888-7200
FAX: 708 888-7094
Established: 1977; *Employs* 125
Sells direct and through dealers.
Major Lines: Manufacturers of HF & VHF mobile antennas and accessories.

Antenna Mart
PO Box 699
Loganville, GA 30249
Phone: 770 466-4353
FAX: 770 466-3095
Established: 1984; *Employs* 3
Major Lines: Quad antennas and parts. Heavy duty gin poles. HD mast pipes RF switches—remote type, dist, fiberglass tubes and solid fiberglass, Rotator capacitor kits & more.

Antenna Sales & Accessories (ASA)
4551 Highway 17 Bypass South
Myrtle Beach, SC 29577
Phone: 800 722-2681
Tech: 803 293-7888
Established: 1991; *Employs* 3
Sells direct and through dealers.
Major Lines: Specializes in antennas: HF mobile, UHF & VHF mobiles, Rubberducks, UHF-VHF Base Antennas, Cellular, Scanner, C.B., Business & Marine, adaptors, connectors, accessories.

Antenna Specialists Division
Allen Telecom Group Inc.
30500 Bruce Industrial Pkway.
Cleveland, OH 44139
Phone: 216 349-8400
FAX: 216 349-8407
Established: 1953
Sells through dealers.
Major Lines: Mobile antennas for all wireless applications.

Antenna Supermarket
PO Box 563
Palatine, IL 60078
Phone: 708 359-7092
FAX: 708 359-8161
Established: 1974
Sells: direct and through dealers.
Major Lines: SWL antennas, multi-band SWL antennas, lightning surge arrestors.

Antennaco, Inc.
PO Box 218
Milford, NH 03055-0218
Phone: 603 673-3153
FAX: 603 673-4347
Established: 1992
Sells through dealers.
Major Lines: Commercial and amateur antennas (Yagis, omnis and dipoles) and kits. Custom hardware including power divides, stacking frames, mounting clamps and cable kits.

Antennas West
PO Box 50062
1500 North 150 West
Provo, UT 84605
Phone: 801 373-8425
FAX: 801 373-8426
Order Line: 800 926-7373
Established: 1987; *Employs* 10
Catalog: Radio Adventure $1
Sells direct and through dealers.
Major Lines: HF antennas, VHF/UHF Antennas, Masts & Towers, SolarPower, Publications.

Austin Antenna Ltd.
10 Main Street
Gonic, NH 03839
Phone: 603 335-6339
FAX: 603 335-1756
Established: 1973; *Employs* 19
Sells direct and through dealers.
Major Lines: VHF/UHF/GHz fixed and mobile, single and multiband antennas, including the five band Pentenna version.

B

Barker and Williamson
10 Canal Street
Bristol, PA 19007
Phone: 215 788-5581
FAX: 215 788-9577
Established: 1932; *Employs* 25
Sells direct and through dealers.
Major Lines: HF antennas, air-wound inductors, baluns, coax switches, filters, coils, insulators, center connectors.

Bilal Company
137 Manchester Drive
Florissant, CO 80816
Phone: 719 687-0650
Established: 1980
Sells direct and through dealers.
Major Line: Compact HF Antennas.

Broadcast Technical Services
11 Walnut Street
Marshfield, MA 02050
Phone: 617 837-2880
Sells direct.
Major Line: Sidekick VHF/UHF antennas.

Butternut Manufacturing
831 N.Central Avenue
Wood Dale, IL 60191
Phone: 708 238-1854
Fax: 708-238-1186
Established: 1995; *Employs* 6
Sells through dealers.
Major Line: HF Antennas.

C

Cellular Security Group
4 Gerring Road
Gloucester, MA 01930
Phone: 508 281-8892
Established: 1990; *Employs* 3
Sells direct.

Centurion International, Inc.
PO Box 82846
Lincoln, NE 68501-2846
Phone: 402 467-4491
FAX: 800 848-3825
Major Lines: Replacement antennas and batteries for two-way portable radios, cordless telephones and specialized OEM applications.

Com-Rad Industries
PO Box 88
4230 East Lake Road
Wilson, NY 14172
Phone: 716 751-9945
FAX: 716 751-9879
Established: 1979; *Employs* 2
Sells direct and through dealers.
Major Lines: Un-tenna line of low profile VHF and UHF antennas for amateur and commercial use.

ComTek Systems
PO Box 470565
Charlotte, NC 28247
Phone: 704 542-4808
FAX: 704 542-9652
Established: 1988
Sells direct.
Major Lines: Original dual hybrid mfg. for phased arrays.

Create Design Co., LTD
Phone: 703 938-8105
FAX: 703 938-4525
Major Lines: Amateur and Commercial Antennas, HF, VHF, UHF, Rotators, Highpower Baluns, Towers and RF Generator for Plasma Processing, etc.
Amateur Radio Sales:
Electronic Distributors Corp.
325 Mill Street
Vienna, VA 22180
Phone: 703 938-8105
FAX: 703 938-4525
Henry Radio, Inc. 2050 S. Bundy Drive
Los Angeles, CA 90025
Phone: Toll Free: 800 877-7979
Local: 213 820-1234
FAX: 310 826-7790

Cubex Co.
2761 Saturn St. Unit E
Brea, CA 92621
Phone: 714 577-9009
FAX: 714 577-9124
Established: 1956; *Employs* 3
Sells direct.
Major Lines: Exclusive manufacturer of cubical quad beam antennas and accessories.

Cushcraft, Inc.
PO Box 4680
48 Perimeter Road
Manchester, NH 03108
Phone: 603 627-7877
FAX: 603 627-1764
E-mail: sales@cushcraft.com
Established: 1950
Sells through dealers.
Major Lines: HF/VHF/UHF antennas and accessories.

D

DC Sales
1602 Chestnut Ridge
Kingwood, TX 77339
Phone: 713 358-0051
Established: 1985; *Employs* 3
Sells direct and through dealers.
Major Lines: Antenna brackets (mobile), antennas (mobile).

DX Engineering
618 Spaulding Avenue
Brownsville, OR 97327
Phone: 503 466-3138
FAX: 503 466-5453
Established: 1989 *Employs* 3
Sells direct.
Major Lines: High-powered remote coax switches, choice of ETO for their amps. HF vertical phased array boxes, stacking boxes for mono or multi band Yagis. High performance 40 through 6 meter monoband Yagis, the high performance 10-20M log periodic antenna.

Davis RF Co.
PO Box 230-G
Carlisle, MA 01741
Phone: 508 371-1356, 800 328-4773
Tech: 508 369-1738
FAX : 508 369-3484
Established: 1988
Sells direct.
Major Lines: Dacron rope, DSP audio noise reduction filters (JPS Co.), B&W, Vibroplex, Polyphaser, commercial wire/cable, wire antenna parts, commercial custom wire/cable design.

Diamond Antennas
435 So. Pacific
San Marcos, CA 92069
Phone: 619 744-0900
Sells through authorized dealers.
Major Line: Complete line of VHF/UHF base and mobile antennas for amateur radio.

Down East Microwave
954 Rt. 519
Frenchtown, NJ 08825
Phone: 908 996-3584
FAX: 908 996-3702
Major Lines: VHF/UHF/Microwave transverters, Down-converters, LNA's, Power Amps and RF components. Distributor of Directive Systems VHF/UHF/Microwave antennas and accessories.

E

EUR-AM Electronics
PO Box 990
Meredith, NH 03253-0990
Phone: 603 279-1393
FAX: 603 279-1394
Established: 1992; *Employs* 4
Sells direct, catalog ordering.
Major Lines: European Made Communications Antennas. Amateur, CB, Commercial, Base, Mobile, Scanner & Cellular. Mounting accessories, tools, connectors & Dup-Triplexers, Microphone with DTMF pad for HT's, Communications programs (AMIGA, PC & ATARI), Electronic Kits.

Electron Processing, Inc.
PO Box 68
Cedar, MI 49621
Phone: 616 228-7020
Established: 1984; *Employs* 10
Sells direct.
Major Lines: Receiver preamplifiers, antennas, assorted accessories for Hams and SWL's.

Electronic Distributors Corp. (EDCO)
325 Mill St. NE
Vienna, VA 22180
Phone: 703 938-8105
FAX: 703 938-4525
Established: 1989; *Employs* 8
Sells through dealers.
Major Lines: Create antennas and roof top towers and rotors; DAIWA coax switches, SWR/power meters, power supplies, amplifiers; Datong audio filters; Emoto rotators; Novex SWR/power meters, speaker mics, Nevada discone and scanner antennas. J.I.M. receiver preamps, handheld scanner stands and power supplies. AOR receivers and wide range scanners and communications receivers, loop antenna systems, DSP audio filters and compact SSTV converters, Sirio commercial and marine, mono and dual band amateur antennas, Lowe communications receivers and SASI satellite tracking systems, computer control software for ICOM & AOR.

Electronic Switch Co., Inc.
4343 Shallowford Road, Suite E-6
Marietta, GA 30062
Phone: 404 518-4634
FAX: 404 642-9035
Established: 1990
Sells direct and through dealers.
Major Lines: North American Distributors for Fritzel Antennas, Versatower Antenna Tower Equipment, Schurr Morse Keys & Paddles, Hofi Switches and Lightning Protection Equipment.

F

Forbes Group, The
PO Box 445
Rocklin, CA 95677
Phone: 916 624-7069, 800 551-5156
FAX: 916 624-5186
Sells direct and through dealers.
Major Lines: Ventenna™ concealed antennas.

Force 12
3015-B Copper Road
Santa Clara,CA 95051-0701
Phone: 408 720-9073; 800 248-1985
FAX: 408 720-9055
Major Lines: Large variety of HF antennas.

G

GAP Antenna Products Inc.
6010 Bldg. B, N. Old Dixie Hwy.
Vero Beach, FL 32967
Phone: 407 778-3728
Established: 1989; *Employs* 6
Sells direct.
Major Line: HF vertical and beam antennas utilizing GAP technology.

GRE America, Inc.
425 Harbor Blvd.
Belmont, CA 94002
Phone: 415 591-1400; 800 233-5973
FAX: 415 591-2001
Established: 1979; *Employs* 25
Sells direct and through dealers.
Major Lines: Scanner accessories, pre-amplifiers, frequency converters, all band antennas, 900 MHz spread spectrum products and other wireless telecommunications products.

Genesys Products Group
5730 Mobud
San Antonio, TX 78238
Phone: 210 521-6868
FAX: 210 647-8007
Established: 1992
Sells through dealers.
Major Lines: Dual band antennas, mount, accessories, wattmeters, spkr/ mics, eartalk microphone, throat microphone.

H

High Sierra Antennas
Box 2389
Nevada City, CA 95959
Phone: 916 273-3415
Major Line: HF antennas.

Hustler Antenna
One Newtronics Place
Mineral Wells, TX 76067
Phone: 817 325-1386; 800 949-9490
FAX: 817 328-1409
Established: 1968; *Employs* 100
Sells through dealers.
Major Line: Amateur, land-mobile, CB antennas and scanner antennas.

I

IVIE Technologies
1366 W. Center Street
Orem, UT 84057
Phone: 801 224-1800
FAX: 801 224-7526
Major Line: Ground cooperative antennas that can be laid on the ground.

J

J•Com
793 Canning Pkwy.
Victor, NY 14560
Orders: 1 800 446-2295
FAX: 716 924-4555
Established: 1990
Sells direct and through dealers.
Major Lines: MagiNotch automatic notch audio filter; Ventriloquist audio memory; Ham-Base software and databases; Stealth Antennas; SofTNC and packet ;modems. Other hardware andsoftware accessories for the radio amateur.

Jade Products, Inc.
PO Box 368
East Hampstead, NH 03826-0368
Phone: 603 329-6995
Orders Only: 800 JADEPRO
FAX: 603 329-4499
Sells direct.
Major Lines: Specializing in twin-lead ladder-line antennas: multiband dipole, marconi, windham "J" pole. Battery controller kits (including solar), power supply kits, prototyping kits, QRP transceivers, Curtis Keyer Kit, accessories.

Jo Gunn Enterprises
Route 1, Box 32C
Ethelsville AL 35461
Phone: 205 658-2229
FAX: 205 658-2259
Established: 1976; *Employs* 5
Sells direct through distributors/dealers.
Major Lines: Mobile and base CB antennas, 10M amateur antennas, coax and accessories.

K

KLM Antennas,Inc.
14792 172nd Dr. SE#1
PO Box 694
Monroe, WA 98272
Phone: 360 794-2923
FAX: 360 794-0294
Established: 1974
Sells through dealers.
Major Lines: UHF/VHF/HF Yagis, monoband and triband; UHF/VHF/HF verticals; satellite circular yagis, baluns, antenna couplers and power dividers, fiberglass masts.

Kenwood Communications Corporation
2201 E. Dominguez Street
PO Box 22745
Long Beach, CA 90801-5745
Phone: 310 639-4200
Tech: 310 639-5300
BBS: 310 761-8284 (8N1 up to 14.4K baud)
Established: 1975
Sells through dealers.
Major Lines: HF/VHF/UHF base, mobile, portable transceivers and receivers, power supplies, automatic antenna tuners, external speakers for base and mobile use, SWR and RF power meters, HF mobile antenna, dual band (2m/70cm) mobile antenna, headphones, microphones, and accessories.

Kilo-Tec
PO Box 10
Oak View, CA 93022
Phone: 805 646-9645
FAX: 805 646-9645
Established: 1982
Sells direct and through dealers.
Major Lines: Antennas HF/VHF, ATU components & tuners, coax and antenna accessories, HV variable capacitors, roller inductors, Hi-Q mobile loading coil and turns counter.

L

Lakeview Co. Inc.
3620-9A Whitehall Road
Anderson, SC 29624
Phone: 803 226-6990
FAX: 803 225-4565
Established: 1979; *Employs* 7
Sells direct and through dealers.
Major Lines: WD4BUM mobile antennas and mounts for HF, VHF and UHF. Fixed station ground plane antennas for 10M, UHF and VHF.

Larsen Electronics
3611 NE 112th Avenue
PO Box 1799
Vancouver, WA 98668-1799
Phone: 360 944-7551; 800 426-1656
Established: 1965; *Employs* 100 +
Sells direct and through dealers.
Major Lines: Mobile, portable and base antennas from 27 MHz to 1.2+ GHz and accessories.
Branch: Canadian Larsen Electronics Ltd.
5049 Still Creek Avenue
Burnaby, BC V5C 5V1
Phone: 604 299-8517; 800 663-6734

Lightning Bolt Antennas
RD #2, Rt. 19
Volant, PA 16156
Phone: 412 530-7396
FAX: 412 530-6796
Established: 1984; *Employs* 2
Sells direct.
Major Lines: Fiberglass quad antennas and parts, custom quads.

M

M2 Enterprises
7560 N. Del Mar Avenue
Fresno, CA 93711
Phone: 209 432-8873
FAX: 209 432-3059
Established: 1985; *Employs* 16
Sells direct and through dealers.
Major Lines: HF, VHF, UHF and microwave antennas, antenna accessories, rotators and elevation rotators.

MAX System Antennas
4 Gerring Road
Gloucester, MA 01930
Orders: 1 800 487-7539
Phone: 508 281-8892
FAX: 1 800 487-7539 (24 hr.)
Established: 1990; *Employs* 5
Sells through dealers only.
Major Line: UHF & VHF ground plane antennas.

MFJ Enterprises, Inc.
PO Box 494
Mississippi State, MS 39762
Phone: 800 647-1800; 601 323-5869
Tech: 800 647-8324; *FAX:* 601 323-6551
Established: 1972; *Employs* 160
Sells direct and through dealers.
Major Lines: Antenna tuners, keyers, wattmeters, packet controllers, dummy loads, antenna bridge, noise bridge, antenna current probe, clocks, coaxial switches, filters, speaker mics, mobile speaker, telescoping antennas, VHF and UHF antennas, interfaces, code oscillators, books, licensing, code and theory programs.

Maggiore Electronic Lab
600 Westtown Road
West Chester, PA 19382
Phone: 610 436-6051
FAX: 610 436-6268
Established: 1968; *Employs* 8
Sells direct and through dealers.
Major Lines: VHF & UHF Repeaters, Transmitters, Receivers, Power Amplifiers, C.O.R.'s. Microprocessor Repeater Controllers and Auto Patches. Miniature, Marine Packages Including Duplexers, Antennas and High Power Amplifiers.

Maldol Antennas
4711 NE 50 Street
Seattle, WA 98105
Answer FAX: 206 525-1896
FAX: 206 524-7826
Established: 1976; *Employs* 6
Sells through dealers
Major Lines: Full line Mobile VHF, UHF & dual band antennas; monoband and dual band HT antennas; VHF/UHF, Dual Band Yagi portable antennas; 30 different mobile mounts/power mount; HF/VHF limited space yagis; towers and accessories.

Metal & Cable Corp., Inc.
9241 Ravenna Rd., Unit C-10
PO Box 117
Twinsburg, OH 44087
Phone: 216 425-8455
Established: 1985; *Employs* 3
Sells direct and through dealers.
Major Lines: W3BMW Impossible Dream Whip HF Mag Mount; Copper Ground System Strap; Tubing and Rod Aluminum 6061-T6 for masts & antennas.

Midland Consumer Radio Co., Inc.
1670 N. Topping
Kansas City, MO 64120-1224
P.O. Box 33865
Kansas City, MO 64120-3865
Phone: 800 669-4567 x1165
Tech: 800-669-4567 x 57
Established: 1995; *Employs* 100
Sells through dealers.
Major Lines: VHF, UHF transceivers,and antennas.

Mosley Electronics
10812 Ambassador Blvd.
St. Louis, MO 63132
Orders: 800 966-7539; 800 325-4016 *Phone:* 314 994-7872
FAX: 314 994-7873
Established: 1940
Sells direct.
Major Lines: Beam antennas, verticals, dipoles, and mobiles for HF, VHF, and UHF use.

Myers, Communications, R.
PO Box 17108
Fountain Hills, AZ 85269-7108
Phone: 602 837-6492
FAX: 602 837-6872
Established: 1979
Sells direct.
Major Lines: Complete line of satellite products for beginner/advanced at discount prices, newsletters and magazines. MODE S ANTENNA, dealer for AEA, M^2 and YAESU Products. Cata-log and samples $3.00.

Myers Engineering International, Inc.
PO Box 15908
Ft. Lauderdale, FL 33318-5908
Phone: 305 345-5000
FAX: 305 345-5005
Major Lines: Antennas, custom and specialty antennas (commercial, amateur, military, broadcast). Standard "off-the-shelf" antennas available in 1996.

N

N.C.G. Companies/Comet Antenna Distr.
1275 N. Grove Street
Anaheim, CA 92806-2114
Phone: 714 630-4541; 800 962-2611
FAX: 714 630-7024
Established: 1968; *Employs* 7
Sells through dealers.
Major Lines: US and Canadian distributor of the COMET Multi-Band Antenna Line. Exclusive distributor for Comet Co. Ltd. antennas.

Nil-Jon Antennas
1898 JoAnn Dr.
Parma, OH 44134
Phone: 1 216 777-9460
Orders Only: 1 800 277-9460
Dealer Inquiries: 1 216 327-5300
Established: 1993; *Employs* 5
Sells through dealers.
Major Lines: VHF/UHF Yagis (single & stacked arrays); Television (single channel & all channel models); FM-Broadcast yagi; Commercial (Land-Mobile); VHF/UHF yagis.

O

OFS WeatherFAX
6404 Lakerest Court
Raleigh, NC 27612
Phone/FAX: 919 847-4545
E-mail: jdahl@cybernetics.net
Established: 1990
Sells direct and through dealers.
Major Line: Satellite products, PCMCIA cards, ISA adapters, interactive software, turnstyle VHF antennas, Marine quadri-filier helix antennas.

OMEGA Electronics
101-D Railrod Street
PO Box 579
Knightdale, NC 27545
Phone: 800 900-7388; 919 266-7373
FAX: 919 250-0073
Established: 1988
Sells direct/showroom and dealers.
Major Line: Manufactures the OMEGA "G5RV" "Signal Enhancer" Antenna. Dealer Inquiries Invited!

Outbacker Antenna Sales
410 Cyndica Drive
Chattanooga, TN 37421
Phone: 423 899-3390
FAX: 423 899-6536
CompuServe: 74127,2725
Sells direct.
Major Lines: Importers of mobile HF, multiband vertical antennas manufactured by Terlin Aerials of Australia.

P

P.C. Electronics
2522 Paxson Lane
Arcadia, CA 91007
Phone: 818 447-4565
FAX: 818 447-0489
Established: 1965; *Employs* 10
Sells direct and through dealers.
Major Lines: Fast scan amateur television equipment, mini video cameras, transmitters, transceivers, down converters, antennas, linear amps,accessories.

Palomar Engineers
PO Box 462222
Escondido, CA 92046
Phone: 619 747-3343
FAX: 619 747-3346
Established: 1965; *Employs* 7
Sells direct and through dealers.
Major Lines: Noise bridge, tuner-tuner SWR/power meters, HF preamplifiers, loop antennas, antenna baluns, VLF converters, RFI kit, iron powder, ferrite toroids and beads, audio filter, keys and keyers, digital frequency display.

Productivity Resources
PO Box 813
Bellville, TX 77418
Phone: 409 865-2727
FAX: 409 865-9800
E-mail: tomk5rc@aol.com
Established: 1990; *Employs* 2
Sells direct.
Major Lines: Antenna Masts, Rotators, Towers, Antennas.

Pro•Am (Div. of Valor Enterprises Inc.)
1711 N. Commerce Drive
Piqua, OH 45356
Phone: 513 778-0074
FAX: 513 778-0259
Established: 1974; *Employs* 180
Sells through dealers.
Major Lines: Two meter base/mobile, multi-band base/mobile, 450 portables, HF yagis, scanner mobile/base, two-meter portable, HF mobile antenna, cellular, commercial, marine antennas and accessories, RV antennas, UHF/VHF amplifiers.
Branch Office:
185 W. Hamilton Street
West Milton, OH 45383
Phone: 513 698-4194

Q

Quorum Communications, Inc.
8304 Esters Blvd., Suite 850
Irving, TX 75063
Phone: 214 915-0256
FAX: 214 915-0270
BBS: 214 915-0346
Established : 1988; *Employs* 5
Sells direct and through dealers.
Major Lines: Weather facsimile products—scan converters, receivers, down converters, preamps, antennas, feed, satellite tracking systems.

R

RAI Enterprises
5638 West Alice Avenue
Glendale, AZ 85302
Phone: 602 435-9523
Established: 1989
Sells direct.
Major Lines: Raider™ Antennas, the Quickyagi, and EZ Log Plus (full function logging program).

R.C. Distributing Company
PO Box 552
South Bend, IN 46615
Phone: 219 236-5776
Established: 1992; *Employs* 5
Sells direct and through dealers.
Major Lines: Video sync generator, specialty microwave equipment, High performance parabolic and yagi antennas.

RF Limited/Clear Channel Corporation
PO Box 1124
Issaquah, WA 98027
Phone: 206 222-4295
FAX: 206 222-4294
Established: 1977
Sells through dealers.
Major Lines: Amplifiers, microphones, antennas, handheld transceiver accessories.

Radio Engineers/Technitron
7969 Engineer Road, Suite 102
San Diego, CA 92111
Phone: 619 565-1319
FAX: 619 571-5909
Established: 1979; *Employs* 4
Sells direct and through dealers.
Major Lines; Communication equipment, antennas, direction finding systems, Weather Satellite Antennas and Receivers, voice amplifiers, RF Engineering, State registered professional engineers.

Radio Shack
1500 One Tandy Center
Ft. Worth, TX 76102
Phone: 1 800 843-7422
Established: 1921; *Employs* 26,000
More than 6,600 stores in the U.S.
Sells direct and through dealers.
Major Lines: Computers, scanners, antennas, transceivers, coax, plugs, jacks, parts and supplies.

Radio Works, Inc.
PO Box 6159
Portsmouth, VA 23703
Phone: 804 484-0140
Established: 1984; *Employs* 4
Sells direct.
Major Lines: HF antennas, baluns, VHF base antennas, portable antennas.

Radioware Corporation
PO Box 1478
Westford, MA 01886
Phone: 800 950-9273; 508 452-5555
FAX: 508 251-0515
Established: 1992
Sells via mail/phone orders.
Major Lines: VHF/UHF Antennas, Scanner Antennas, SSTV equipment, Flex-Weave antenna wire, Bury and Flex RF cable, grounding items.

S

SGC, Inc.
The SGC Building
13737 S.E. 26th St.
Bellevue, WA 98005
PO Box 3526
Bellevue, WA 98009
Phone: 206 746-6310
FAX: 206 746-6384
Established: 1972; *Employs* 100
Sells direct and through dealers.
Major Lines: Manufactures HF SSB radios, antenna couplers, power supplies, and antennas.

Sentech, Inc.
P.O. Box 2136
Riverview, FL 33569
Phone: 813 677-4410
Established: 1989; *Employs* 8
Sells direct and through dealers.
Major Lines: VHF Quagi antennas, 160M active receiving loop, aluminum crank-up towers, HF Quads, VHF transverters.

Sinclabs Amateur Radio Products
675 Ensminger Road
Tonawanda, NY 14150
Phone: 716 874-3682
FAX: 716 874-4007
Major Lines: Base and repeater antennas, repeater duplexers, mobile duplexer and band pass filters for intermod.

Smiley Antenna Co.
408 La Cresta Heights Road
El Cajon, CA 92021
Phone: 619 579-8916
FAX: 619 579-8598
Established: 1985; *Employs* 4
Sells through dealers.
Major Line: Antennas for handheld radios for amateur commercial and business band use.

Solorcon Corporation
7134 Railroad St., Box 176
Holland, OH 43611
Phone: 800 445-3991
FAX: 419 865-9449
Established: 1979; *Employs* 30
Sells through dealers.
Major Lines: Mobile and baseantennas.

Sommer Antenna Systems
395 Osceola Road
PO Box 710
Geneva, FL 32732
Phone: 407 349-9114
FAX: 407 349-2485
Sells direct.
Major Line: HF antennas.

Spectrum International Inc.
PO Box 1084
Concord, MA 01742
Phone: 508 263-2145
FAX: 508 263-7008
Established: 1970
Sells direct and through dealers.
Major Lines: Antennas, filters, crystal filters, receivers, preamps and weather satellite systems.

Spider Antenna
7131 Owensmouth Avenue, Suite 663
Canoga Park, CA 91303
Phone: 818 341-5460
Established: 1979
Sells direct.
Major Lines: Spider 4-band HF mobile antenna, maritimer antenna, 4-band dipole 10, 15, 20 & 40 meters, ideal for apartments, condominiums and CC&Rs.

Spi-Ro Manufacturing, Inc.
PO Box 2800
Hendersonville, NC 28793-2800
Phone: 704 693-1001
FAX: 704 693-3002
Established: 1970; *Employs* 6
Sells direct and through dealers.
Major Lines: Antennas, antenna traps, antenna connectors, insulators, transmission lines, antenna wire, antenna shorteners, limited space antennas, lightning-surge protectors, baluns, custom inter-connect coax cables, multi-band & mono-band antennas.

Stewart Designs, H.
PO Box 643
Oregon City, OR 97045
Phone: 503 654-3350
Established: 1985
Sells direct.
Major Lines: HF/VHF directionals and omnidirectionals; small-space and hidden antennas; VHF/UHF vertical dipoles, mobile, belt-pack and base;the BIG DK-DX, remote tuning 3.5-30 MHz mobile antenna designed by W6AAQ; and various mobile antenna mounts.

T

Telex Communications Inc.
Hygain Division
8601 E. Cornhusker Hwy
Lincoln, NE 68505
Phone: 402 467-5321 Ext. 287
Established: 1936
Sells through dealers.
Major Lines: Multiband antennas, HF monobanders, vertical antennas, VHF/ UHF beam antennas,VHF/UHF omnidirectional antennas, rotators, crank-up towers.

Telrex
PO Box 879
Asbury Park, NJ 07712
Phone: 908 775-7252
Established: 1921; *Employs* 15
Sells direct.
Major Lines: HF,VHF & UHF antennas.

Tennadyne Corporation
PO Box 1894
Rockport, TX 78381
Phone/FAX: 512 790-7745
Established: 1989
Sells direct.
Major Lines: Log-Periodic dipole ar-ray antennas, 1.8-1300 MHz (Beamantennas).

Texas Radio Products
5 East Upshaw
Temple, TX 76501
Phone: 817 771-1188
Major Lines: Bugcatcher mobile antennas, mobile antenna accessories, VHF/ UHF antennas.

U

Unadilla Antenna Manufacturing Co.
PO Box 4215
Andover, MA 01810-4215
Phone: 508 475-7831
FAX: 508 474-8949
Established: 1986
Sells through dealers.
Major Lines: Baluns, antenna traps for 10-160 meters, remote coaxial switches, antenna kits for amateur, CAP and MARS antenna accessories, wire, end insulators and center insulators, mobile slot antennas, grounding systems.

UniHat Corp. Engineering & Marketing Group
3816 Royal Lane, Suite 100
Dallas, TX 75229
Phone: 214 352-4623
Established: 1993; *Employs* 5
Sells direct.
Major Lines: Short, highly-efficient vertical antennas (CTSVR).

V

Van Gorden Engineering
PO Box 21305
S. Euclid, OH 44121
Phone: 216 481-6590
FAX: 216 481-8329
Established: 1964; *Employs* 7
Sells direct and through dealers.
Major Lines: Dipole antennas, shortened antennas, baluns, insulators, coils, traps, wire, feedline, and cable for amateur, SWL and commercial use.

Van Valzah Company, H.C.
38 W. 111 Horseshoe Drive
Batavia, IL 60510-9730
Phone: 708 406-9210
Established: 1981; *Employs* 1
Major Line: Antenna parts.

W

W9INN Antennas
PO Box 393, 811 Cathy Lane
Mt. Prospect, IL 60056
Phone: 708 394-3414
Established: 1982; *Employs* 2
Sells direct.
Major Lines: Space-restricted antennas, HF multi-band dipoles, $1/2$ and full slopers, MARS & marine wire antennas, multi-antenna systems, remotely-tuned antenna systems, and accessories.

Wallen USA, Les
19 Aero Drive
Amherst, NY 14225
Phone: 716 634-0634
FAX: 716 632-6304
Established: 1961; *Employs* 14
Sells direct and through dealers.
Major Lines: Antennas, voltage converters, zap traps.

Wilson Antenna, Inc.
1181 Grier Dr., Suite A
Las Vegas, NV 89119
Phone: 800 541-6116, 702 896-0399
FAX: 702 896-0409
Established: 1985
Sells through dealers.
Major Lines: Amateur and CB mobile antennas.

Wintenna,Inc.
911 Amity Road
Anderson, SC 29621-7097
Phone: 803 261-3965, 800 845-9724
FAX: 803 224-7920
Major Lines: Manufactures wide line of base and mobile antennas.

Woodhouse Communication
PO Box 73
Plainwell, MI 49080-0073
Phone: 616 226-8873
Sells direct.
Major Line: VHF antennas - amateur 50/144/220/440 MHz; VHF antennas - Apt satellite; Custom VHF antennas.

CANADA

Antennas Connectors Coax & Surplus (A.C.C.S.)
2115 Ridgeway St.
Thunder Bay, Ont., Canada P7E 5J8
Phone: 807 622-2700
Established: 1992
Sells direct and through dealers.
Major Line: VHF/HF/UHF and Handheld Antennas, Batteries, Connectors, Wire & Coax Antenna supplies, Rotor Cable, Heliax Connectors, Duplexers, Extension Speakers, Mag Mounts, Antenna Mounting Kits. Manufactures and Distributor of Connectors and MoeTech Antennas.

Databank Technical
55 Queen Street East, Suite 505
Toronto, Ontario, Canada M5C 1R6
Phone: 416 376-9045
FAX: 416 759-4988
Internet: databank.technical@canrem.com
Established: 1977; *Employs:* 27
Sells direct and through dealers.
Major Line: HF/VHF/UHF Antennas for Base, Mobile, Portable Transceivers and Communications Receivers. Antenna Manufacturers & Distributors.
Branch Offices:
1170 Bay Street, Suite 102
Toronto, Ontario, Canada M5S 2B4
Phone: 416 361-3306
181 Cocksfield Avenue, Unit 1
Downsview, Ont., Canada M3H 3T4
Phone: 416 606-6359

Gem Quad
PO Box 291
Boissevain, Manitoba R0K 0E0, Canada
Phone: 204 534-6184
Established: 1968; *Employs* 4
Sells direct.
Major Line: Cubical quad antennas.

Gunfleet Produc ts Inc.
Box 102
Coombs, B.C., V0R 1M0, Canada
Established: 1991; *Employs* 4
Sells to dealers.
Major Line: Manufactures antennas and associated components.

Sinclabs Amateur Radio Products
Div. of Sinclair Technologies, Inc.
85 Mary Street
Aurora, ON L4G 6X5, Canada
Phone: 905 727-0163
FAX: 905 841-6255
Sells direct and through dealers.
Major Lines: VHF and UHF repeater duplexers and antennas, VHF and UHF mobile antennas, mobile diplexers, 12v DC power supplies, and intermod filters.

Skyware Communication Technology
64590 1942 Como Lake Ave.
Coquitlam, B.C. V3J 7V7
Phone: 604 936-1790
FAX: 604 936-1740
Established: 1991; *Employs* 5
Sells via mail order.
Major Lines: HF/VHF/UHF antennas & communication accessories. Specializes in antenna consultancy. Full 800 MHz scanners available.

HF ANTENNAS

Manufacturer	Model	Band(s)	Gain	F/B	# Elements	Boom Length (ft.)	Price
AAE/ Bandmaster	Q-28-2	28	8.0	24.0	2	4.00	$170
Antenna Mart	AMQ-2-5-8**	10/12/ 15/17/20	7.0	19	2*	8.00	599
	AMQ-5-5-24**	10/12/ 15/17/20	9.3/10.2	26	5 -10, 12 m* 4 -15, 17, 20 m*	24.00	980
	AMQ-5-5-31**	10/12/ 15/17/20	10.3/11.2	27	5 (10, 12 m)* 4 -15, 17, 20 m*	31.00	1295
	AMQ-4-30-40-48	30/40	10.2/10.7	25	4*	48.00	2995
	AMQ-6-6-12	6	13.8	30	6*	12.00	799
	AMQ-8-10-31	10	14.7	30	8*	31.00	1295
	AMQ-8-15-41	15	14.5	30	8*	41.00	1495
	AMQ-6-20-48	20	13.8	30	5*	48.00	1550
	AMQ-4-40-48	40	10.2	25	4*	48.00	2995
Butternut	HF5B	10/12/15/17/20	5	20	2	6.00	376
Cushcraft	TEN-3	10	7.8	25	3	8.00	150
	10-3CD	10	8.0	30	3	12.00	232
	10-4CD	10	10.0	30	4	17.00	299
	15-3CD	15	8.0	30	3	14.00	305
	15-4CD	15	10.0	30	4	20.00	354
	20-3CD	20	8.0	30	3	18.00	438
	20-4CD	20	10.0	30	4	32.00	626
	D40	40	—	—	1	—	296
	40-2CD	40	5.5	20	2	23.00	626
	D3W	12/17/30	—	—	1	—	249
	D3	10/15/20	—	—	1	—	244
	D4	10/15/20/40	—	—	1	—	331
	A3S	10/15/20	8.0	25	3	14.00	497
	A4S	10/15/20	8.9	25	4	18.00	597
	A3WS	12/17	8.0	25	3	14.00	390
	ASL2010	10/12/15/17/20	6.4	15-20	8	18.00	828
Force 12	C-3	10/15/20(+12/17)	4.2-4.5	14-18	7	18.00	489
	C-3S	10/15/20(+12/17)	4.1-4.4	14-18	6	12.00	399
	C-4	10/15/20/40(+12/17)	0,4.2-4.5	0,14-18	8	18.00	649
	C-4S	10/15/20/40(+12/17)	0,4.1-4.4	0,14-18	7	12.00	559
	C-4XL	10/15/20/40(+12/17)	4.1-4.5	12-18	9	30.00	849
	4BA	10/12/15/17	5.4-5.8	14-20	12	24.00	799
	5BA	10/12/15/17/20	5.4-5.9	14-23	15	33.00	1049
	DXer	15/17/20	5.2-5.8	14-23	9	24.00	779
	DXer/S	15/17/20	4.3-4.6	14	6	12.00	449
	EF-606	6	7.9	24	6	12.00	209
	EF-240	40	4.2	12	2	18.00	519
	EF-420	20	6.4	22	4	30.00	459
	EF-420/240	20-40	6.4,4.3	22,13	6	30.00	849
	MAGNUM 620	20	7.8	23	6	44.00	1050
	MAGNUM 620/340	20-40	7.8,5.2	23,16	9	44.00	1895
	MAGNUM 280B	80/75	4.2	12	2	36.00	1640
	MAGNUM 2/2	80/75-40	4.2,4.2	14,12	4	35.00	2495
	MAGNUM 3/4	80/75-40	4.8,5.2	16,16	7	62.00	3950
GAP Antenna	DXII	10/11	7.0	25	3	10.00	130
Gem Quad	2EL	10/15/20	7.0	25	2*	9.5(Ant.)	220

HF ANTENNAS (cont.)

Manufacturer	Model	Band(s)	Gain	F/B	# Elements	Boom Length (ft.)	Price
Gem Quad	3EL	10/15/20	8.9	30	3*	16.0(Ant.)	340
(Cont.)	4EL	10/15/20	10.0	30	4*	22.5(Ant.)	450
Hy-Gain	LP-1009	10/12/15/17/20	5.1	22	12	27.00	1232
	LP-1010	10/12/15/17/20/30	4.8	22	14	36.00	1779
	TH11DX	10/12/15/17/20	7.1	27	11	24.00	1027
	TH7DX	10/15/20	7.4	27	7	24.00	756
	TH5Mk2	10/15/20	6.0	27	5	19.00	649
	TH3JR	10/15/20	5.9	25	3	12.00	317
	TH3Mk4	10/15/20	5.9	25	3	14.00	381
	TH2Mk3	10/15/20	3.4	20	2	6.00	321
	EXP 14	10/15/20	6.7	27	4	14.10	499
	DISC7-1	30 or 40	—	35†	1	2.70	251
	DISC7-2	40	4.4	15	2††	22.60	561
	105CA	10	8.6	36	5	24.00	234
	155CA	15	7.5	40	5	26.00	355
	205CA	20	7.3	23	5	34.00	629
	204CA	20	6.0	28	4	26.00	458
	103BA	10	5.7	25	3	8.50	123
	153BA	15	5.7	25	3	12.00	170
	203BA	20	5.0	25	3	16.00	275
KLM	10M-4	10	7.7	25	4	10.00	257
Antennas	10M-6	10	11.0	30	6	27.00	406
	15M-4	15	7.7	25	4	14.00	282
	15M-6LD	15	10.5	30	6	36.00	498
	15M-6	15	11.0	30	6	36.00	663
	20M-4	20	7.7	25	4	21.00	530
	20M-5	20	9.7	30	5	42.00	821
	20M-6	20	11.0	30	6	57.00	1180
	40M-1	40	—	—	1	46.50	255
	40M-2A	40	4.9	12	2	16.00	626
	40M-3A	40	6.5	20	3	32.00	856
	40M-4	40	7.2	20	4	42.00	1187
	80M-1	80	—	—	1	45.00	967
	80M-2	80	4.0	12	2	36.00	3185
	80M-3	80	7.0	18	3	60.00	4792
	KT-31	10/15/20	—	—	1	24.40	310
	KT34A	10/15/20	8.2	20	4	16.00	616
	KT34XA	10/15/20	10.3	20	6	32.00	884
	12-17-30D	17	—	—	1	39.7	310
	12-17-30V	17	—	—	1	21.25	163
	17M-3	17	6.5	20	3	17.00	472
	10-30-7LPA	10-20MHz	7.0	15	7	30.00	1124
	20-30-6LPA	20-30MHz	7.0	20	6	24.00	383
	6-12-8LPA	6-12MHz	6.0	15	8	46.00	2414
	6-30—15LPA	6-30MHz	6.0	15	15	46.00	3992
	7.2/10-30LPA	7-30MHz	3/7	15	8	42.00	1325
Lightning	10MCQ	10	7.0	25	2*	6.00	90
Bolt	10MCQ-4	10	10.0	25	4*	16.00	170
	22MCQ	10/12/15	7.0	25	2*	8.00	99
	17MCQ	10/12/15/17	7.0	25	2*	9.00	225
	32MCQ	10/15/20	7.0	25	2*	10.00	265
	34MCQ	10/15/20	10.0	30	4*	24.00	675
	32MCQ/WB	10/12/15/ 17/20	7.0	26	2*	10.00	289
	34MCQ/WB	10/12/15/ 17/20	10.0	25	4*	24.00	725
	42MCQ	17/20 40	7.0	25	2	16.00	995
M²	10M4	10	8.7	25	4	23.00	309
	10M7	10	10.3	25	7	45.00	579
	15M4	15	8.5	25	4	34.00	429
	15M6	15	9.4	25	6	45.00	579
	17M3	18.0-18.0	6.3	25	3	18.00	319

HF ANTENNAS (cont.)

Manufacturer	Model	Band(s)	Gain	F/B	# Elements	Boom Length (ft.)	Price
M² (cont.)	17M5	18.05-18.0	8.6	24	5	36.00	549
	20M4	20	8.7	25	4	45.00	749
	20M6	20	9.0	25	6	60.00	1149
	40M1L	40/30	na	na	1	46.00	339
	40M2L	40	4.2	12	2	20.00	569
	40M3L	40	5.6	20	3	30.00	799
	40M4L	40	6.2	22	4	42.00	1149
	80M1	80	na	na	1	85.00	999
	80M2	80	4.2	15-20	2	30.00	2600
	80M3	80	6.3	20	3	58.00	3400
	10-30LP8	10–30 MHz	6	15	8	32.00	1159
	17-30LP7	17–30 MHz	6.5	20	7	24.00	799
	7-1030LP8	7, 10–30 MHz	1/7,6/10–30	3	8	30.00	1289
	6-10LP5	6–10 MHz	5	15	5	30.00	1289
	6M5	6	9.4	12	5	16.00	199
	6M7	6	10.5	23	7	27.00	289
	6M2WLC	6	11.9	25	6	39.50	439
	6M2.5WLC	6	12.6	23	11	50.00	549
Mosley	TA-31	10/15/20	—	—	1	—	152
	TA-32	10/15/20	5.0	20	2	7.00	280
	TA-33	10/15/20	8.0	20	3	14.00	359
	TA-53-M	10/12/15/17/20	varies	varies	3	14.00	572
	CL-33	10/15/20	8.4	23	3	18.00	463
	TA-34-XL	10/15/20	9.1	21	4	21.00	633
	CL-36	10/15/20	9.1	24	6	24.00	655
	PRO-57-B	10/12/15/ 17/20	na	20-25	7	24.00	718
	PRO-67-B	10/12/15/ 17/20/40	n/a	10–25	7	24.00	888
	PRO-67-C	10/12/15/ 17/20/40	n/a	n/a	7	24.00	1120
Sommer	XP403	10/15/20	6.0	20-25	4	8	450
	XP504	10/12/15/20	7.0	25	7	15	665
	XP507	10/12/15/17/20/ 30/40	7.0 / 0-2	25 / 0–3	8	15	950
	XP704	10/12/15/20	9.0	30	9	20	820
	XP707	10/12/15/17/20/ 30/40	9.0 / 0-3	15–30 / 0–10	10	20	1060
	XP804	10/12/15/20	10–11	15–30	11	26	1140
	XP807	10/12/15/20 30/40	10–11 / 0-4	15–30 / 0	12	26	1380
Tennadyne	T5	13–30**	4.5	14–24	5	12.00	335
	T6	13–30**	5.0	14–24	6	12.00	380
	T7	13–30**	5.6	14–24	7	18.00	465
	T8	13–30**	5.8	15–24	8	18.00	500
	T10	13–30**	6.1	15–25	10	24.00	605
	T12	13–30**	6.3	15–24	12	30.00	750
	T18	6–30	5.8	17–24	18	48.80	8500
	T18WL (non-rotatable)	4–22	5.8	17–24	18	100.00	1800
	T21	3–30	5.7	14–24	21	58.80	16000
	T31	50–1300**	6.5	15–24	31	12.00	225

* elements per band
** UPS shippable models
† denotes front to side ratio
†† add on kit available

VHF/UHF ANTENNAS

Manufacturer	Model	Freq.	Gain	F/B	# Elements	Boom Length (ft.)	Price
AAE/	Q-144-2	144	8.0	24.0	2	1.20	$35
Bandmaster	Q-144-3	144	9.0	24.0	3	2.50	50
	Q-144-4	144	10.0	24.0	4	3.30	60
	Q-144-6	144	12.0	24.0	6	5.00	80
	Q-220-6	220	12.0	24.0	6	3.30	70
	Q-440-6	440	12.0	24.0	6	3.00	60
	Q-50-2	50	8.0	24.0	2	3.00	75
	Q-50-4	50	10.0	24.0	4	8.5	170
Create	CL6DX	50	13.0	22.0	6	19.00	161
Design	X209	144	11.5	16.0	9	12.08	155
	2X209	144	17.0	18.0	18(9X2)	12.17	285
	4X209	144	19.5	20.0	36(9X4)	12.17	769
	X713	430	16.0	18.0	13	7.58	113
	2X713	430	18.5	18.0	26(13X2)	7.58	277
	CLP130-1	50-1300	11.0	15.0	21	6.8	450
	CLP5130-2	105-1300	12.0	15.0	20	4.7	280
Comet	CY-1205	1280	11.0	na	5	na	41
	CYA-1216E	1260-1300	16.6	na	16	4.50	140
	PYA-913	900-930	15.8	na	13	na	145
Cushcraft	738XB	435	15.5	25.0	38	14.33	268
	22XB	146	14.0	25.0	22	19.33	290
	A50-3S	50	8.0	20.0	3	6.00	110
	A50-5S	50	10.5	24.0	5	12.00	182
	A50-6S	50	11.6	26.0	6	20.00	296
	617-6B	50	14.0	30.0	6	34.00	423
	A148-3S	144	9.0	18.0	4	3.67	46
	124WB	144	10.2	24.0	4	4.00	73
	A148-10S	144	13.5	20.0	11	12.00	85
	13B2	144	15.8	26.0	13	15.00	142
	A148-20S	144	16.2	24.0	20(2X10)	12.00	233
	17B2	144	18.0	26.0	17	31.00	246
	26B2	144	18.8	26.0	26	15.00	406
	224WB	220	10.2	24.0	4	3.00	67
	225WB	220	15.5	24.0	15	10.00	149
	220B	220	17.2	30.0	17	19.00	192
	A449-6S	450	10.0	18.0	6	2.90	59
	A449-11S	450	13.2	20.0	11	5.00	83
	A430-11S	430	13.2	20.0	11	4.70	87
	424B	430	18.2	30.0	24	17.00	161
	A270-10S	146/440	10/10	20/18	5/5	6.17	99
	A270-6S	146/440	7.8/7.9	20/18	3/3	2.8	73
	A148-20T	146	11.1	20.0	20.0	10.80	126
	719B	430/450	15.5	25.0	19	13.5	142
	729B	430/440	17.8	25.0	29	22.2	220
Hy-Gain	64DX	50	8.2	25.0	4	12.00	115
	66DX	50	10.3	25.0	6	24.50	216
	23FM	144	6.1	20.0	3	3.60	35
	25FM	144	9.1	20.0	5	6.25	45
	28FM	144	11.8	20.0	8	12.30	59
	214FM	144	13.0	20.0	14	15.50	69
	215DX	144	14.2	30.0	15	28.00	196
	216SAT	145	11.5	25.0	16	14.00	193
	215SAT	440	14.0	25.0	30	11.20	186
	218SAT	145/435	215SAT	AND	216SAT	Relays	432
	70-31DX	440	17.6	28.0	31	24.00	190
KLM	6M-5	50	9.7	30.0	5	11.75	255
Antennas	6M-7LD	50	10.5	30.0	7	20.00	294
	6M-7LB	50	11.5	30.0	7	25.75	372
	6M-10	50	11.7	25.0	10	34.20	494
	6M-14	50	14.0	26.0	14	61.00	685
	2M-4X	144	8.5	20.0	4	4.20	79

VHF/UHF ANTENNAS (cont.)

Manufacturer	Model	Freq.	Gain	F/B	# Elements	Boom Length (ft.)	Price
KLM	2M-8	144	10.3	30.0	8	7.25	124
Antennas.	2M-14C	144	11.0	20.0	14	12.75	207
(cont.)	2M-11X	144	12.5	20.0	11	15.30	107
	2M-22C	144	13.0	20.0	22	19.10	256
	2M-13LBA	144	13.3	20.0	13	21.50	121
	2M-16LBXM	144	14.5	20.0	16	28.00	235
	2M-20LBX	144	15.5	20.0	20	36.50	286
	220-7	220	8.8	20.0	7	4.75	96
	220-14X	220	13.5	20.0	14	14.67	139
	220-22LBX	220	15.6	20.0	22	29.75	175
	432-20LBX	430	15.3	20.0	20	12.30	185
	432-30LBX	430	17.3	20.0	30	21.90	217
	435-18C	435	12.0	20.0	18	7.30	257
	435-40CX	435	15.2	20.0	40	14.63	278
	440-10X	440	10.0	20.0	10	4.80	76
	440-16X	440	14.0	20.0	16	12.00	139
	440-6X	440	8.0	20.0	6	7.30	63
	1.2-15LBX	1260	13.6	39.0	15	3.50	116
	1.2-24LBX	1260	16.2	39.0	24	6.30	140
	1.2-44LBX	1260	18.2	39.0	44	12.30	195
Lightning	6MCQ	50	7.0	25.0	2	4.00	80
Bolt	6MCQ-4	50	10.0	26.0	4	12.00	135
	2MCQ-2	144	7.0	25.0	2	1.00	30
	2MCQ-4	144	10.0	25.0	4	4.00	40
	2MCQ-6	144	12.0	25.0	6	6.00	60
	2MCQ-10	144	14.0	25.0	10	12.00	100
	70MCQ-4	440	9.0	25.0	4	1.00	30
	70MCQ-8	440	12.5	25.0	8	4.00	50
	270MCQ-4	144/440	7.5/9.0	25.0	2/4	1.00	50
	270MCQ-8	144/440	9.0/12.5	25.0	4/8	4.00	80
	OPP-1	145/435	2MCQ-4	AND	70MCQ-7	CrossBoom	115
	OPP-2	145/435	2MCQ-6	AND	70MCQ-7.5	CrossBoom	155
M2	6M5	50-50.2	9.4	12.0	5	16.00	199
	6M7	49.5-50.5	10.5	23.0	7	27.00	289
	6M2WLC	49.7-50.5	11.9	25.0	6	39.50	439
	6M2.5WLC	49.5-50.350	12.6	23.0	11	50.00	549
	2M4	144-148	7.5	20.00	4	4.00	89
	2M7	144-148	10.3	20.0	7	9.00	109
	2M9FM	145-146	12	24.0	9	14.50	119
	2M9SSB	144-146	12	24.0	9	14.50	119
	2M12	144-148	12.8	25.0	12	16.50	149
	2M5WL	144-148	14.8	22.0	16	33.00	189
	2M18XXX	144-146	15.3	25.0	18	36.00	229
	2M8WL	144-146	16.7	23.0	25	53.00	339
	2MCP14	143-148	10.3*	20.0	7/7	10.50	169
	2MCP22	144-148	12.5*	25.0	11/11	18.50	229
	2MXP28	144-146**	15.1	24.0	14/14	34.50	205
	2M5-440XP	144/440	9.0/12.0	12/25	5/10	6.00	159
	222-7EZ	220-226	9.8	22.0	7	6.00	89
	222-10EZ	222-226	12.0	23.0	10	10.00	99
	222-7WL	222-226	16.4	25.0	23	32.50	189
	420-50-5HD	420–450	7.8	20.0	5	2.00	99
	420-50-11	420–450	11.3	20.0	11.0	5.00	79
	440-18	420–453	14.5	23.0	18.0	11.50	109
	43630CP	430–440	14.5*	22.0	15/15	10.00	229
	432-9WL	420–440	17.3	24.0	28	21.00	159
	432-13WLA	430–434	18.6	22.0	38	31.00	229
	902-10EZ	900–928	12.0	23.0	10	3.50	89
	902-16EZ	900–928	14.7	25.0	16	5.50	129
	902-14WL	900–910	19.0	30.0	41	15.50	199
	23CM22EZ	1250–1300	16.0	26.0	22	5.50	89
	22CM35	1250–1300	18.4	28.0	35	10.00	129

VHF/UHF ANTENNAS (cont.)

Manufacturer	Model	Freq.	Gain	F/B	# Elements	Boom Length (ft.)	Price
M²	S22EZ	2300–2500	15.5	26.0	22	3.00	call
(Cont.)	S40EZ	2375–2640	18.4	26.0	40	5.50	call
MFJ	MFJ-1763	144	6.9	17.0	3	2.75	40
Mosley	MY144-5	144	10.0	20.0	5	4.50	50
	AM-14-2	144	13.0	20.0	14	12.00	170
	MY144-9	144	14.0	20.0	9	9.00	70
	MY220-9	220	14.0	20.0	9	8.17	70
	AM-2N6	50/144	11/9	11.0	5/4	14.00	450
Radioware	R24	144	7.0	16.0	4	3.70	74
	R28	144	11.0	20.0	8	12.00	139
	R211	144	12.5	25.0	11	17.80	199
	R2234	220	6.75	16.0	4	2.60	68
	R2236	220	8.75	20.0	6	5.00	79
	R22311	220	12.5	25.0	11	12.00	149
	R4406	440	8.5	17.0	6	2.75	83
	R43010	430	11.0	20.0	10	5.50	93
	R44010	440	11.0	20.0	10	5.50	149
	R43218	432	15.0	25.0	18	12.0	149
	R91510	915	12.0	20.0	10	3.50	94
Spectrum	CY137-2	137	3.0	OMNI	2+2	3.75	99
International	CY137-4	137	7.8	GAIN	5+5	5.50	150
	1268-LY	1268	20.0	20.0	29 (loop)	8.00	99
	1296LY	1296	20.0	20.0	29 (loop)	8.00	99
	1691LY (N)	1691	20.0	20.0	29 (loop)	6.00	99

* circular
** dual polarity

ANTENNA BASICS

An antenna is the part of a radio station responsible for the conversion of the transmitter power into electromagnetic waves, and for the collecting of power from electromagnetic waves for the receiver to process. The efficiency of the antenna in performing these two jobs is one of the biggest factors in how well a station hears and how well it is heard by others.

Since the first days of radio, transmission antennas have been evolving to fill various needs and restrictions. Each type of antenna was created to fit someone's particular need at the time. Many of these have been modified, optimized, bent, folded, until someone gives them a new name and a new antenna is born. Some antennas have been created to fit in a limited space, some to provide particular radiation patterns, and others just because they were there.

Hams love to build and experiment with antennas. It's one of the few areas in amateur radio where you can build something yourself that works as well or better than one you can buy off the shelf. First, get a good book and do a bit of reading. There are several books that give excellent background on different types of amateur antennas. They also show construction techniques and installation methods. Next, find a source of materials. This will depend on what type of antenna you decide to build. Wire and insulators can often be found in hardware stores, while aluminum tubing can be bought from various suppliers as well as from advertisers in amateur radio magazines. You may also find flea markets a good source for wire and parts.

Antenna performance is often misunderstood. Many beginners believe that a good SWR means an antenna is good. Actually, the SWR reading shows only how much power is going in each direction on the feedline. A lossy feedline will make the reading lower even though less power is actually getting to the antenna. Likewise a high SWR can be caused by bad connectors or cable as well as a damaged or misadjusted antenna. But in any case it is never a measure of how well the antenna is actually radiating.

Measuring or predicting gain on antennas is subject to many variables that make comparisons between claims by manufacturers almost impossible. Not only is there a difference of about 2 dB depending on the reference chosen (dBi referenced gains are about 2 dB higher than dBd referenced gains), but the height of the antenna and the vertical angle measured can result in many dB difference in reported gain.

A poorly designed Yagi, or one at the wrong height may not perform as well as a smaller one, or even a dipole.

The best height for an antenna depends on lots of factors. Generally a low antenna is better for local coverage, and a higher one better for DX. But there are times when lower antennas outperform high ones for DX on the HF bands. And on VHF and UHF sometimes a high antenna is needed to clear local obstructions. There are also some applications where height doesn't matter much at all, like moonbounce, aurora, meteor scatter, and satellites.

Choosing which type of antenna to use depends on what you want to do. Do you want to compete in 160 meter contests, or just listen on the local repeater? Do you live in an apartment, or do you have a thousand acre back yard? Can you put up a tower to support an antenna, or is hanging a wire out a window as much as you can do? The list below describes some of the more popular antenna types and what they are normally used for. This list is not all inclusive by any means, but should give enough options to get you started on picking a type of antenna.

Half-Wave Dipole
Description: Most basic antenna, just two pieces of wire or tubing about 1/4 wavelength long, fed in the middle.
Bands: All
Common Uses: HF, VHF
Radiation Pattern: If wire is horizontal, the pattern is bidirectional with nulls off the ends of the wires. If the center is higher than the ends, the configuration is called an "Inverted V, " and the nulls off the ends are less pronounced. If the wire is vertical, the pattern is circular with a null straight up.
Advantages: Easy to build and put up, materials are cheap and easy to find.
Disadvantages: They are quite large for the lower frequency HF bands.

Multiband/Trap Dipole
Description: A dipole or inverted V with traps and/or loading coils that help it operate on several bands.
Bands: HF
Common Uses: HF, where space doesn't permit multiple full size dipoles.
Radiation Pattern: Same as 1/2 wave dipole.
Advantages: Easy to put up, saves space
Disadvantages: Traps or coils may limit maximum transmitter power, harder to design and tune than basic dipole.

Long Wire, Random Wire
Description: Usually a piece of wire as long as practical, fed at one end.
Bands: HF
Common Uses: HF, portable or base when quick and easy antenna needed.
Radiation Pattern: Hard to predict, varies with length and frequency
Advantages: Easy to put up, cheap, can be put just about anywhere.
Disadvantages: Normally needs tuner to keep transmitter happy, needs good ground to keep RF out of the shack. Performance hard to predict.

Windom, Zepp, Double Zepp, Extended Double Zepp, G5RV
Description: These are variations on dipoles and longwires that have evolved to fit certain situations. The G5RV is the most popular version as it is designed to cover several HF bands.
Bands: HF, some work on certain combinations of HF bands
Common Uses: HF
Radiation Pattern: Hard to predict; varies by band
Advantages: Cover multiple bands
Disadvantages: Some rather large, some need certain combination of feedlines, tuners, or baluns to work as intended.

Quarter-Wave Vertical
Description: This is essentially a 1/2 wave dipole where the bottom half has been replaced by either a conductive plate (i.e., a car roof) or a "ground-plane" consisting of several wires or pieces of tubing. A common variation is the "Inverted L" where the top is bent over to reduce the height.
Bands: HF, VHF, UHF

Common Uses: HF/VHF/UHF mobile or base
Radiation Pattern: Omnidirectional
Advantages: Omnidirectional pattern, easy to build for most bands, easy to tune. With proper feed system, several verticals can be combined into arrays for higher gain and directing beams.
Disadvantages: Rather large for lower HF bands.

Multiband/Trap Vertical
Description: These are just 1/4 or 1/2 wave verticals with traps and/or loading coils to make them work on multiple bands.
Bands: HF, VHF, UHF
Common Uses: HF, VHF/UHF Mobile
Radiation Pattern: Omnidirectional
Advantages: Omnidirectional pattern, covers multiple bands with single vertical radiator, relatively small compared to full size vertical, some don't need radials or ground plane.
Disadvantages: More expensive than a single band vertical, not as efficient due to losses in traps and coils. Harder to design and build.

Full-Wave Loop
Description: This is a full wave length of wire or tubing bent into a loop and fed where the ends come together.
Bands: HF
Common Uses: HF
Radiation Pattern: Bidirectional to omnidirectional depending on orientation and feed point location.
Advantages: Gain is slightly better than a dipole. Material usually easy to find and cheap.
Disadvantages: Much bigger than a dipole, harder to put up, harder to match to 50-ohm feed line.

Parasitic Arrays (Yagi, Quad, Quagi)
Description: These are arrays of either approximately 1/2 wavelength wires or tubes (Yagis), or near full wavelength loops (Quads), or combinations of loops and straight elements (Quagis). Normally only one of the elements is driven from the feedline, the others do their job by picking up the waves from the driven element and re-radiating them.
Bands: HF, VHF, UHF
Common Uses: Any time a directional pattern is desired
Radiation Pattern: A single directed beam
Advantages: Can be designed to give quite a bit of gain over most other antennas for HF and VHF. Are small enough to be rotated to concentrate power where you want it. Can reduce interference by receiving only from one direction. Can be built from wire for low cost to beam a specific direction.
Disadvantages: Harder to build, especially for lower frequency HF bands. Larger ones get expensive. Materials harder to find to build your own.

Multiband/Trap Yagi (e.g., Tribander)
Description: Similar to the Yagi above, but with traps and/or loading coils in some or all of the elements to let them cover multiple bands.
Bands: HF(40 through 10 meters mostly)
Common Uses: These are usually used on HF where you don't have room or money for single band Yagis for all the bands you want to use.
Radiation Pattern: Single main lobe.
Advantages: Cover multiple HF bands from one rotateable antenna. Much smaller and simpler than multiple single band antennas.

Disadvantages: Not as efficient as single band antennas. Much harder to design and build yourself.

Log Periodic Dipole Array

Description: This is an array of approximately 1/2 wave dipoles that get progressively shorter along the length of the antenna. All the dipoles are fed power from the feedline, with an alternating phasing arrangement. The idea is to have two or more elements working together on any particular frequency to create a beam in the direction the antenna is pointing.

Bands: HF, VHF, UHF

Common Uses: To cover multiple bands where you don't have room for multiple single band antennas.

Radiation Pattern: Single main lobe

Advantages: Continuous frequency coverage over the design range. More efficient than trapped Yagi antennas.

Disadvantages: Very hard to design and build yourself. Usually bigger than trapped yagi antenna for same amateur band coverage.

Beverage

Description: Named for Dr. Harold Beverage who first designed it, this antenna has become very popular as a low noise receiving antenna for the lower HF bands. It is a low long wire, usually a wavelength or more long and under 3-4 meters off the ground. Many variations of this exist including running insulated wire down the streets in cities.

Bands: HF (40-160 meters)

Radiation Pattern: Dependent on length and termination methods, but normally a series of lobes along the length of wire with much smaller lobes off the sides.

Common Uses: Receive only on lower HF bands (160 through 40 meters)

Advantages: Very directional, very low noise pickup, easy to put up, lots of variations possible. Materials are cheap and easy to find.

Disadvantages: Long, 200 meters and up fairly common. Relatively inefficient, some users add a preamplifier to boost signal levels.

Feedhorns, Dishes, Slotted Waveguides, etc.

Description: These are all specialty UHF and Microwave antennas. Bands: UHF and up

Radiation Pattern: From single narrow beam to multiple directed beams.

Common Uses: Any UHF application.

Advantages: Very high gain possible. Some are very small and light weight.

Disadvantages: Can be hard to build since accuracy is extremely important as the wavelengths get smaller. Larger dishes take special care for mounting and aiming.

Contributed by David Robbins, KY1H

Antenna Books

General

The ARRL Antenna Book, ARRL
The ARRL Antenna Compendium Series, ARRL
The Handbook of Antenna Design, IEEE Press
HF Antennas for All Locations, Moxon
Lew McCoy on Antennas, CQ Communications
RSGB HF Antenna Collection

Specialty Antenna Books

Yagi Antenna Design, Lawson
Physical Design of Yagi Antennas, Leeson
Low Band DXing, Devoldere
The Quad Antenna, Haviland
The Vertical Antenna Handbook, Lee

DIPOLE LENGTHS

Table of Half-Wave Dipole Lengths for Wire Antennas
(assuming 5% correction for wire diameter & end effect)

Band	Freq. kHz	Length Feet	Inches	Meters	Band	Freq. kHz	Length Feet	Inches	Meters
160 Meters	1750	267	1.0	81.43	30 Meters	10050	46	6.1	14.18
	1775	263	3.9	80.28		10075	46	4.7	14.14
	1800	259	8.0	79.17		10100	46	3.3	14.11
	1825	256	1.3	78.08		10125	46	2.0	14.07
	1850	252	7.8	77.03		10150	46	0.6	14.04
	1875	249	3.4	76.00		10175	45	11.2	14.00
	1900	246	0.0	75.00		10200	45	9.9	13.97
	1925	242	9.7	74.03					
	1950	239	8.3	73.08	20 Meters	13950	33	6.1	10.22
	1975	236	7.9	72.15		14000	33	4.6	10.18
	2000	233	8.4	71.25		14050	33	3.2	10.14
	2025	230	9.8	70.37		14100	33	1.8	10.11
	2050	228	0.0	69.51		14150	33	0.4	10.07
						14200	32	11.0	10.04
80 Meters	3450	135	5.7	41.30		14250	32	9.6	10.00
	3475	134	6.0	41.01		14300	32	8.2	9.97
	3500	133	6.5	40.71		14350	32	6.9	9.93
	3525	132	7.1	40.43		14400	32	5.5	9.90
	3550	131	7.9	40.14					
	3575	130	8.9	39.86	17 Meters	18018	25	11.3	7.91
	3600	129	10.0	39.58		18068	25	10.4	7.89
	3625	128	11.3	39.31		18118	25	9.6	7.87
	3650	128	0.7	39.04		18168	25	8.7	7.84
	3675	127	2.2	38.78		18218	25	7.9	7.82
	3700	126	3.9	38.51					
	3725	125	5.7	38.26	15 Meters	20950	22	3.7	6.80
	3750	124	7.7	38.00		21000	22	3.1	6.79
	3775	123	9.8	37.75		21050	22	2.5	6.77
	3800	123	0.0	37.50		21100	22	1.8	6.75
	3825	122	2.4	37.25		21150	22	1.2	6.74
	3850	121	4.8	37.01		21200	22	0.6	6.72
	3875	120	7.4	36.77		21250	21	11.9	6.71
	3900	119	10.2	36.54		21300	21	11.3	6.69
	3925	119	1.0	36.31		21350	21	10.7	6.67
	3950	118	3.9	36.08		21400	21	10.1	6.66
	3975	117	7.0	35.85		21450	21	9.5	6.64
	4000	116	10.2	35.63		21500	21	8.9	6.63
	4025	116	1.5	35.40					
	4050	115	4.9	35.19	12 Meters	24840	18	9.8	5.74
						24890	18	9.3	5.73
40 Meters	6950	67	3.0	20.50		24940	18	8.9	5.71
	6975	67	0.1	20.43		24990	18	8.4	5.70
	7000	66	9.3	20.36		25040	18	8.0	5.69
	7025	66	6.4	20.28					
	7050	66	3.6	20.21	10 Meters	27800	16	9.8	5.13
	7075	66	0.8	20.14		28000	16	8.3	5.09
	7100	65	10.0	20.07		28200	16	6.9	5.05
	7125	65	7.2	20.00		28400	16	5.5	5.02
	7150	65	4.4	19.93		28600	16	4.1	4.98
	7175	65	1.7	19.86		28800	16	2.8	4.95
	7200	64	11.0	19.79		29000	16	1.4	4.91
	7225	64	8.3	19.72		29200	16	0.1	4.88
	7250	64	5.6	19.66		29400	15	10.8	4.85
	7275	64	3.0	19.59		29600	15	9.5	4.81
	7300	64	0.3	19.52		29800	15	8.2	4.78
	7325	63	9.7	19.45		30000	15	7.0	4.75
	7350	63	7.1	19.39					

ANTENNA LENGTHS FOR AMATEUR BANDS

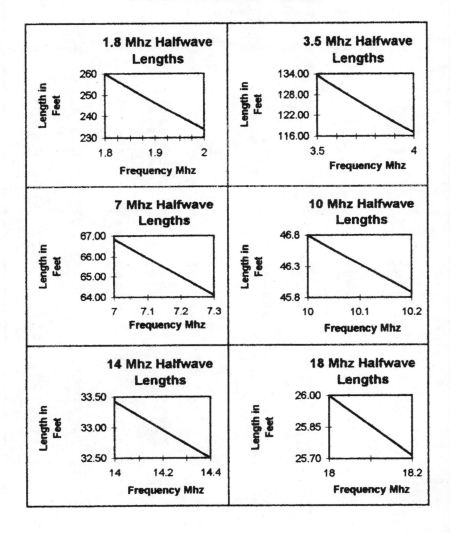

ANTENNA LENGTHS FOR AMATEUR BANDS

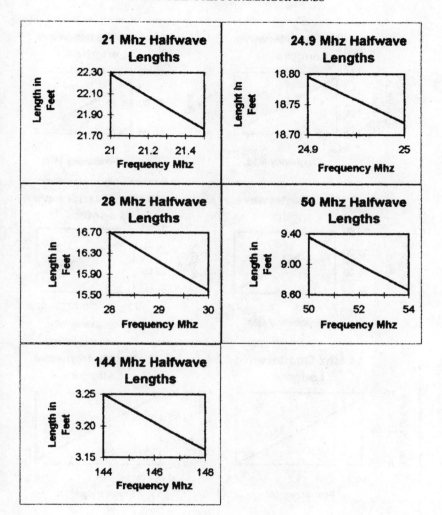

ANTENNA LENGTHS FOR AMATEUR BANDS

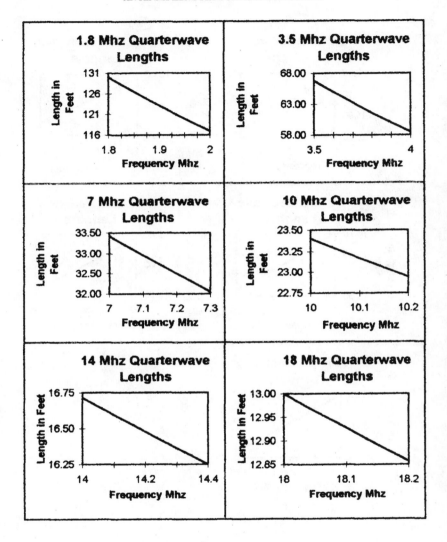

ANTENNA LENGTHS FOR AMATEUR BANDS

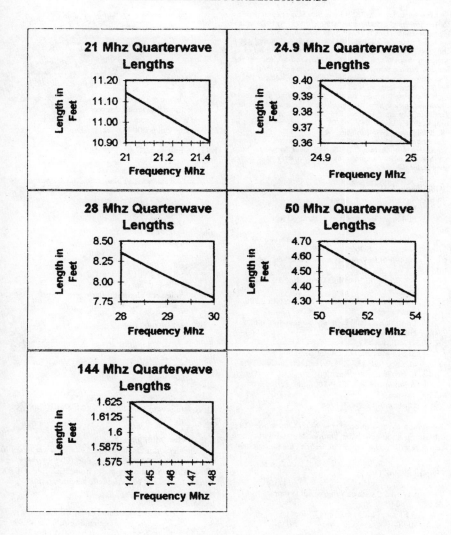

THE BIGGEST AMATEUR ANTENNAS IN THE WORLD

Most hams are satisfied with a triband Yagi and dipoles for the low bands. Some amateurs are driven to build much bigger antennas. This list of exceptionally large antennas was collected through informal surveys conducted during 1994 and 1995. Photographs of some of these arrays appear elsewhere in the Almanac. Only antennas used exclusively for amateur radio are included. The 1000-foot radio astronomy dish at Arecibo, PR is occasionally used for amateur EME, but is not considered an amateur antenna.

If you have (or know of) any antennas larger than those listed here, please forward a detailed description and a photograph to Almanac Editor, CQ Communications, 76 N. Broadway, Hicksville, NY 11801. They will be included in this listing in the 1996 Almanac.

Since many of these stations are using arrays of stacked Yagis, the nomenclature "N/N/N/N" will be used where appropriate and is interpreted as a stacked array of 4 N-element Yagis.

NO8D, Stow, OH:
160M
Full-size ground-plane, feedpoint at 40 feet
Full-size Delta Loop, bottom at 30 feet
80M
2 full-size ground-planes, feedpoints at 35 feet, phased end-fire or broadside
1000-foot Beverage receiving antenna
40M
3/3/3 at 66/132/198 feet on rotating 200 foot tower
2/2/2 at 66/132/198 on a separate rotating 200 foot tower
(antennas selectable in any combination on either tower)
20/17/15/12/10M
Six 13-element DX Engineering Log Periodics at 33/66/99/132/165/198 feet, selectable in sets of two in any combination
Additonal Hy-Gain LP-1007 Log Periodic on 135 foot tower
Cushcraft R3 at 140 Feet

This station exists at present, but NO8D has moved, and is building another similar station. He claims to have ideas that will "make it much better than the present one"

N7AVK, Salem, OR
160M
Inverted L with elevated radial system
80M
3 Element full-size Yagi (70-foot boom) on 30-to-90-foot crankup tower
40M
3 Element full-size Yagi (42-foot boom) on 30-to-75 foot crankup tower
20M
5/5 (44-foot booms) at 110/55 feet on rotating Rohn 55 tower
15M
6/6 (36-foot booms) at 100/60 feet on rotating Rohn 55 tower
10M
6/6 (30-foot booms) at 90/40 feet on rotating Rohn 55 tower

6M
10/10 (48-foot booms) at 55/25 feet
2M
23/23/23/23 at 60 feet az-el mount
11 elements at 65 feet
222 MHz
10 elements at 55 feet
432 MHz
21/21/21/21 at 60 feet
1296 MHz
55 elements at 58 feet
Also - 13 element Log-Periodic Yagi (36-foot boom) on 30-90 foot crankup tower
Station is situated on a hill with rapid dropoff to Willamette Valley 600-800 feet below.

K4JPD, Fairburn, GA
160M
4 Half-wave slopers from 150 feet
Elevated ground plane with 10 elevated radials
4 1200-foot Beverages
80M
2-element Yagi at 160 feet
Dipole at 100 feet
40M
3 element Yagi at 140 feet
2 element Yagi at 70 feet
20M
5/5/5 at 150/100/50 feet
6 element Yagi at 125 feet
5 element Yagi at 80 feet
3 element Yagi at 60 feet
15M
4/4 at 90/60 feet
6 element Yagi at 85 feet
4 element Yagi at 60 feet
10M
4/4 at 60/30 feet
6 element Yagi at 85 feet
5 element Yagi at 65 feet
8 element Yagi at 100 feet
All on 8 towers of 160, 140, 125, 105, 100, 95, 85, 65 feet.

K3LR, West Middlesex, PA
The antenna system includes one 190 foot tower, one 170 foot tower, two 120 foot towers, one 100 foot tower, one four-square vertical array made from four Rohn HDBX towers, dipoles and beverage wire antennas.

160M
120 foot base insulated tower with four inverted "L" parasitic wires. Ground system has 45,000 feet of #16 wire. Antenna has 30 dB of front/back and 4 or 5 dB of gain that can be beamed in 4 directions. Single wire Beverage, 580 foot long NE for receiving.
80M
2 element W2PV quad with the top at 185 feet. switchable NE/SW. W1FC 4-square made from 24 foot Rohn HDBX self supporting towers and aluminum masts making each vertical 75 feet tall. 30,000 feet of #16 for the ground system. Single wire Beverage, 290 feet long NE for receiving. Dipole at 75 feet

40M

4/4, 50 ft booms at 190/100 ft
2 element Yagi, 24 ft boom at 140 feet
2 element wire Yagi aimed straight up. Driven ele at 50 feet, reflector at 22 feet.
Dipole at 50 feet

20M

5/5/5, 48 ft booms at 170/110/50 feet
5 element Yagi, 40' boom at 120' on another tower
3/3, 22' booms at 80/40 feet fixed south

15M

6/6/6, 48 ft booms at 120/80/40 feet
5 element Yagi, 32' boom at 55' on another tower
4 element Yagi, 22' boom at 80' fixed NE

10M

7/7/7, 48 ft booms at 100/66/33 feet
6 element Yagi, 36 foot boom at 200 feet
5 element Yagi, 26 foot boom at 60' on another tower

W3LPL Glenwood, MD

The antenna system includes four 195 foot towers, three 100 foot towers, two vertical arrays and numerous phased Beverage-pair receiving arrays. An unusual capability allows reception of weak DX signals within a few kHz of the transmitting frequency on each band. This capability derives from phased Beverage-pair arrays specifically designed for very high front-to-side ratio, located at least 500 ft southeast of the towers and verticals.

160M

130 ft vertical ground plane with 4 elevated radials 10 ft high. Phased Beverage-pair arrays, 580 ft long, spaced 270 ft: NE & West

80M

2 element horizontally polarized Quads, 50 ft booms, top at 170 ft: NE, SE, SW, W & NW. 4-square vertical array. Phased Beverage-pair arrays, 290 ft long, spaced 135 ft: NE, South & West

40M

3/3, 48 ft booms at 100/195 ft. 4-square vertical array. Phased Beverage-pair arrays, 145 ft long, spaced 70 ft: NE, South & West

20M

5/5, 48 ft booms at 50/100 ft fixed on Europe
5/5, 48 ft booms at 50/100 ft, independently 360 deg rotatable. 5 element Yagi, 48 ft boom at 195 ft. Phased Beverage-pair array, 110 ft long, spaced 35 ft, NE

15M

6/6, 48 ft booms at 45/90 ft, fixed on Europe.
6/6, 48 ft booms at 50/100 ft, independently 360 deg rotatable
6 element Yagi, 48 ft boom, at 195 ft. Phased Beverage-pair array, 70 ft long, spaced 23 ft, NE

10M

7 element Yagi, 48 ft boom, at 195 ft 5 element Yagi, 24 ft boom, at 60 ft fixed south (more 10M antennas to come if Sunspots ever return!)

W5WMU, Lafayette, LA

160M

Slopers NE and SE at 190 feet

80M

4-element wire array at 150 feet
3-element Vertical array at NE with a 300 radial ground system
Dipole at 190 feet

15M

6-element Yagi at 190 feet
4-element Yagi at 110 feet
4-element Yagi at 70 feet
4-element Yagi at 50 feet
2/2 at 65 feet/35 feet (these are 2-element quads)

40M

4-element Yagi at 160 feet
3-element Yagi at 130 feet
2-element Yagi at 90 feet

20M

3-element Yagi at 200 feet
6-element Yagi at 160 feet
4-element Yagi at 120 feet
4-element Yagi at 90 feet
4-element Yagi at 70 feet
4-element Yagi at 40 feet

15M

6-element Yagi at 190 feet
4-element Yagi at 110 feet
4-element Yagi at 70 feet
4-element Yagi at 50 feet
2/2 at 65 feet/35 feet (these are 2-element quads)

10M

6-element Yagi at 170 feet
6-element Yagi at 110 feet
4/4 at 60 feet/45 feet
4/4 at 45 feet/30 feet
4-element Yagi at 40 feet

W6KPC, McFarland, CA

20M

36-element array of six 6-element/60-foot boom KLM Yagis at 200 feet
Array is arranged as 12/12/12, with each "12" consisting of unit 6-element Yagis spaced 20 meters apart.
Vertical spacing is 21 meters. Array is electronically-steered, with vertical angle variable from 4.8 to 21 degrees. Azimuth can be varied +/- 30 degrees electronically without rotating the array.
Gain is approximately 18 dB referred to a dipole.
Details appeared in June 1980 QST.

15M

A similar array, but using 8-element Yagis on 60-foot booms is used on 15M at W6KPC

1996 - 1997 CONVENTIONS and HAMFESTS CALENDAR

1996

January 20	+	MO Valley, Green Hills, & Ray-Clay ARCs, St. Joseph, MO John Winkler, WBØVRA, Rt. 1, Box 53A, Gower, MO 64454, (816)-424-6484
January 21	x	Metro 70cm Network, Yonkers, NY Otto Supliski, WB2SLQ, 53 Hayward St., Yonkers, NY 10704, (914)-969-1053
January 28	+	Maryland Mobileers ARC, Odenton, MD Bob Andersen, W3HEI, 601 Elizabeth Rd., Glen Burnie, MD 21061, (410)-766-3481
January 28	x	WCRA Hamfest, Villa Park, IL WCRA, P.O. Box QSL, Wheaton, IL 60189, (708)-545-9950
February 10	+	Pensacola Area Hamfest Association, Pensacola, FL Bill Behrends, WA4YRN, 1050 West Carlton Rd., Pensacola,FL 32534-1130, (904)-476-8537
February 10	+	Charleston ARS, North Charleston, SC Jenny Myers, WA4NGV, 2630 Dellwood Ave, North Charleston, SC 29405-6814 , (803)-747-2324
February 16-18	**	Florida State Convention, Orlando, FL John Lenkerd, W4DNU, 1046 Turner Rd., Winter Park, FL 32789, (407)-645-2026
February 24-25	**	Great Lakes Division Convention, Cincinnati, OH Stanley Cohen, WD8QDQ, 2301 Royal Oak Ct., Cincinnati, OH 45237, (513)-531-1011
March 8-10	**	Nebraska State Convention, Norfolk, NE Patrick Adams, NØAZC, 2002 Sunset Ave., Norfolk, NE 68701, (402)-371-7295
March 14-15	*	Arkansas State Convention, Little Rock, AR Jim Blackmon, KB5IFV, 1008 Pine Street, Arkadelphia, AR 71923-4919, (501)-246-7833
March 16-17	+	Midland ARC, Midland, TX Larry Nix, N5TQU, 3900 Douglas Ave., Midland, TX 79703, (915)-699-5441
March 30-31	**	Maryland State Convention/Greater Baltimore Hamboree & Computerfest, Timonium, MD William Dobson, WA3ZER, 482 Shirley Manor Rd., Reisterstown, MD 21136, (410)-426-3378
April 6	+	Wichita ARS, Wichita Falls, TX Steven Guerra, WB5LCN, 5100 Edgecliff Dr., Wichita Falls, TX 76302-4814, (817)-723-6500
April 12-13	**	Oklahoma State Convention, Lawton, OK Robert Morford, KA5YED, 1415 NW 33rd, Lawton, OK 73505, (405)-353-8074
April 13	+	Kentucky Colonels ARC, Bowling Green, KY Leon Garrett, K4CIT, 2901 Smallhouse Rd., Bowling Green, KY 42104, (502)-842-5307
April 13	+	Inland NW Hamfest Association, Spokane, WA Al Lafky, K7YY, PO Box 14643, Spokane, WA 99214, (509)-924-3475
April 14	+	Madison Area Repeater Association, Madison, WI Jim Waldorf, KB9AQQ, PO Box 8890, Madison, WI 53708-8890, (608)-249-7579

April 21	x	MIT RS & Harvard Wireless Club, Cambridge, MA Steve Finberg, W1GSL, PO Box 397082, MIT Branch, Cambridge, MA 02139, Nick Alternburnd, KA1MQX, (617)-253-3776
April 27	+	Owensboro ARC, Owensboro, KY Suzzie Young, N9LJF, 2427 Latrobe Ave., Owensboro, KY 42301, (502)-684-5157
April 27	+	Umpqua Valley ARC, Roseburg, OR Ed Pahl, W5PII, 1440 Wild Iris Ln., Roseburg, OR 97470-9469, (503)-673-1310
May 5	x	Metro 70cm Network, Yonkers, NY Otto Supliski, WB2SLQ, 53 Hayward St., Yonkers, NY 10704, (914)-969-1053
May 5	+	Warminster ARC, Wrightstown, PA Woody Woodside, N6XES, 665 St. Davids Ave., Werminster, PA 18974, (215)-672-8482
May 17-19	x	Dayton Hamvention, Dayton, OH Dayton ARA, P.O. Box 964, Dayton, OH 45401-0964, (513)-276-6930
May 19	x	MIT RS & Harvard Wireless Club, Cambridge, MA Steve Finberg, W1GSL, PO Box 397082, MIT Branch, Cambridge, MA 02139, Nick Alternburnd, KA1MQX, (617)-253-3776
May 31-June 2	*	Atlantic Division/New York State Convention, Rochester, NY Harold Smith, K2HC, 300 White Spruce Blvd., Rochester, NY 14623, (716)-424-7184
June 14-15	x	Albany GA ARC, Albany, GA John Crosby, K4XA, 308 Residence Ave, Albany, GA 31701, (912)-883-7373
June 16	x	MIT RS & Harvard Wireless Club, Cambridge, MA Steve Finberg, W1GSL, PO Box 397082, MIT Branch, Cambridge, MA 02139, Nick Alternburnd, KA1MQX, (617)-253-3776
June 30	+	Kankakee Area Radio Society, Peotone, IL Willis Bowser, K9IFO, 1210 North Riverside Dr., Momence, IL 60954-3452, (815)-472-2079
July 7	+	Murgas ARC, Wilkes-Barre, PA James Post, KA3A, 15 Monarch Rd., Wilkes-Barre, PA 18702, (717)-825-3940
July 21	x	MIT RS & Harvard Wireless Club, Cambridge, MA Steve Finberg, W1GSL, PO Box 397082, MIT Branch, Cambridge, MA 02139, Nick Alternburnd, KA1MQX, (617)-253-3776
August 18	x	MIT RS & Harvard Wireless Club, Cambridge, MA Steve Finberg, W1GSL, PO Box 397082, MIT Branch, Cambridge, MA 02139, Nick Alternburnd, KA1MQX, (617)-253-3776
September 7-8	**	Kentucky State Convention, Louisville, KY Herbert Rowe, W4WQD, 5612 Highway 160, Charlestown, IN 47111, (812)-294-4905
September 8	+	Putnam Emergency & Repeater League, Brewster, NY Shirley Dahlgren, N2SKP, Box 677, 152 Broadway, Verplanck, NY 10596, (914)-736-0717
September 8	+	Butler County ARA, Butler, PA Gerald H. Wetzel, W3DMB, 784 Mercer Rd., Butler, PA 16001, (412)-282-6777
September 13-15	*	ARRL National Convention, Peoria, IL Ron Morgan, KB9NW, PO Box 3508, Peoria, IL 61612-3508, (309)-694-2649
September 15	x	MIT RS & Harvard Wireless Club, Cambridge, MA Steve Finberg, W1GSL, PO Box 397082, MIT Branch, Cambridge, MA 02139, Nick Alternburnd, KA1MQX, (617)-253-3776

September 29	x	Metro 70cm Network, Yonkers, NY Otto Supliski, WB2SLQ, 53 Hayward St., Yonkers, NY 10704, (914)-969-1053
October 5-6	**	New England Division Convention, Boxboro, MA Anthony Penta, WA1MWN, 66 Pleasant Ave., Lynnfield, MA 01940, (617)-334-3945
October 11-13	*	Southwestern Division Convention, Mesa, AZ Robert Myers, W1XT, PO Box 17108, Fountain Hills, AZ 85269, (602)-837-6492
October 20	x	MIT RS & Harvard Wireless Club, Cambridge, MA Steve Finberg, W1GSL, PO Box 397082, MIT Branch, Cambridge, MA 02139, Nick Alternburnd, KA1MQX, (617)-253-3776
November 9	+	Twin City Ham Club, West Monroe, LA Lynn Tiller, 2907 Fort Miro, Monroe, LA 71201, (318)-322-3129
November 16-17	*	Indiana State Convention, Fort Wayne, IN Don Gagnon, WB8HQS, 2805 Nordholme Ave., Fort Wayne, IN 46805, (219)-484-3317
November 23-24	*	Southeastern Division Convention, Tampa, FL Charlotte Frazier, WB4PEL, 617 Highland Ave., Dunedin, FL 34698, (813)-733-6937

1997

March 14-15	*	Arkansas State Convention, Little Rock, AR Jim Blackmon, KB5IFV, 1008 Pine Street, Arkadelphia, AR 71923-4919, (501)-246-7833
April 5-6	**	Maryland State Convention, Timonium, MD William Dobson, WA3ZER, 482 Shirley Manor Rd., Reisterstown, MD 21136, (410)-HAM-FEST
September 7	+	Butler County ARA, Butler, PA Gerald Wetzel, W3DMB, 784 Mercer Rd., Butler, PA 16001, (412)-282-6777
November 15-16	*	Indiana State Convention, Fort Wayne, IN Don Gagnon, WB8HQS, 2805 Nordholme Ave., Fort Wayne, IN 46805, (219)-484-3317

*= Conventions, + = ARRL Hamfest, x = Non-ARRL Hamfest, ** = Convention Pending ARRL Executive Committee Action*

INTRODUCTION TO AWARDS HUNTING

One of the best features of Amateur Radio is the number of ways there are to enjoy this hobby. Ragchewing, experimentation, VHF/UHF, Packet, Traffic Handling, DX, Contests and Awards Hunting. Old timers will remember Clif Evans, K6BX, and his Certificate Hunters Club which was formed in the early 1960s and continued until about 1980. Clif published an Awards Directory listing the requirements for fascinating and sometimes obscure awards.

Most countries are divided into smaller political areas, similar to our states and counties here in the U.S. Best of all, there are many awards for working them. These were to be even a tougher challenge than WAS. Many DX cards have this location information printed on them. Oblasts, Laens and DOKs slowly become familiar old friends to Award Hunters. K6BX died a few years ago leaving a void which was only sporadically filled by a few others who published directories for a short time and then disappeared.

In 1981, K1BV started to collect information about some of the lesser known awards. This became a minor passion. Award information was gathered from DX newsletters, DX columns in several magazines and even info from the backs of QSL cards where the rules were either printed or were attached to a gossamer piece of paper glued to the back. In late 1986, the scope of the data collection was broadened to start gathering data from the national organizations and known major sponsors all around the world.

The Scope of Ham DX Awards

The first tier of awards includes the major international awards like DXCC (DX Century Club), WAZ (Worked All Zones). In addition, the national organizations of most larger countries seem to have an awards series emphasizing the geographical or political organization of their country. These awards form the core of the well known ones: WAS and WACA from the U.S., DUF from France, CCC from England, R-150 from Russia. If you have earned DXCC, you have a fine start in meeting the requirements for more than several dozen colorful DX awards.

Next down the chain come local and regional awards. They may require the contacting of specific numbers of stations in stated areas of their country. Examples include WAGM for Scotland, WAPY for Brazil, etc. The DXer will need a bigger card collection for most of these. It's very likely that you have earned many of these right now!

Club and individual awards form the third category. They almost always require the hunter to contact certain numbers of members or even one station a number of times. Some clubs have tens of thousands of easy-to-find members, like the 10-X International Club. Others, like the Freebooters Club (of Hassleholm, Sweden), have somewhat fewer members (but the Freebooter Statuette is one of the nicer Awards in Amateur Radio). In most cases, though, the farther away you are from them, the fewer contacts are needed. Even so, your DX card collection will have to be fairly extensive, or you will have to do some hunting. The odds are that you've got some of these earned right now and don't know it. In this category, sponsors will often make lists of members available.

List of Countries Offering Amateur Radio Awards

Alaska	Guam	Norfolk Island
Andorra	Guantanamo Bay	Northern Ireland
Anguilla	Guernsey	Norway
Argentina	Honduras	Oman
Aruba	Hong Kong	Panama
Ascension Isl.	Hungary	Papua New
Australia	Iceland	Guinea
Austria	India	Paraguay
Balearic Isl.	Indonesia	Peru
Belarus	Ireland	Philippines
Belgium	Israel	Poland
Bermuda	Italy	Portugal
Bolivia	Japan	Puerto Rico
Brazil	Jersey	Rhodes
Brunei	Jordan	Romania
Bulgaria	Kaliningrad	Russia
Canada	Kazakhstan	Sardinia
Canary Isl.	Kenya	Saudi Arabia
Chile	Kirghizia	Scotland
Colombia	Korea, South	Slovakia
Corsica	Kuwait	Slovenia
Costa Rica	Kwajalein	Solomon Isl.
Croatia	Kyrghystan	South Africa
Cuba	Latvia	Spain
Cyprus	Lebanon	Sri Lanka
Czech Republic	Liechtenstein	St. Pierre & M.
Denmark	Lithuania	Sweden
Djibouti	Lord Howe Isl.	Switzerland
Dominican Rep.	Luxembourg	Tahiti
Ecuador	Macedonia	Taiwan
England	Malaysia	Thailand
Estonia	Malta	Turkey
Faroe Isl.	Man, Isle of	Ukraine
Finland	Martinique	Uruguay
France	Mexico	USA
Gabon	Monaco	Vanuatu
Germany	Mongolia	Venezuela
Gibraltar	Morocco	Vietnam
Greece	Netherlands	Wales
Greenland	New Zealand	Yugoslavia
Guadeloupe	Nigeria	

Some Awards Tips From The K1BV DX Awards Directory

The newcomer to awards hunting should read this section to keep his or her expenses down and success ratio high. Your submissions should follow these general guidelines.

Applications:

Most major awards require that their specific application forms be used. The second and third-tier awards usually accept home-made applications, as long as they contain all required information. A sample form that K1BV has used to apply with great success follows this section. It includes all the data generally required by Award Sponsors. Rubber stamp or print your name and address in the top section and add more pages as required. You should write in advance for less known awards and ask the sponsor if it is still offered. Enclose a SASE/IRC as necessary. There is a fairly high mortality rate for awards, and if you don't get a reply in 3-4 months, it's probable that the award is defunct.

Don't forget to get the necessary signatures if the General Certification Rule (see below) is acceptable by the sponsor (99% allow GCR).

Include the fee as indicated. If money has to be sent, a piece of carbon paper folded around currency may prevent tampering. In some countries, mail which may contain money is always a target for pilfering. If you live in a large city, check the International Department of a bank for availability of foriegn currency. It may be a whole lot less expensive than IRCs.

IRCs are an almost universal medium of currency among awards hunters. Many awards specify 10-15 IRCs as the fee. The U.S. Post Office charges $.95 each. You can get them for about half that price from DX QSL managers. Keep an eye out in the various DX newsletters for notices. IRCs change "editions" every few years, so don't hoard large quantities. Always send valid ones of the current edition which are postmarked on the LEFT side.

GCR (General Certification Rule)

Most award sponsors allow GCR in lieu of actually wanting to see your cards. You need to have the cards! GCR usually means getting the signatures of two witnesses who certify that you possess the cards and that the information you state on the application is correct. If the award rules specify club officials, you should make sure their title follows their signature; include the name of the club just to make sure.

Levels

If the award is issued in several different levels, always specify the one you are seeking. If you have the basic level, indicate the serial number, if any, of the basic award and the correct designation of the one now being applied for. If all your contacts for a particular award are SSB and you don't say you want a special mode endorsement, you probably won't get it.

Number of Amateur Radio Awards Listed in K1BV Directory of DX Awards

Year	Number
1987	680
1988	830
1989	1055
1990	1380
1991	1729
1992	1980
1993	2240
1994	2215
1995	2315
1996	2450

(Not including awards offered for a single year)

The K1BV Directory of DX Awards is issued annually as a service to amateur radio awards hunters throughout the world. All correspondence and inquiries should be directed to: Ted Melinosky K1BV HCR 10 © Box 837A Spofford, New Hampshire 03462 U.S.A. (SASE Appreciated).

Contributed by Ted Melinosky, K1BV.

Ted Melinosky, K1BV, at his station.

April, 1995 Edition
(with additions to Section 2, Point 2, as approved by the ARRL Awards Committee in April 1995 indicated in *italics*)
Current Countries Total: 327
(Includes P5, North Korea, effective May 14, 1995)

INTRODUCTION

"...the number of countries worked is increasingly becoming the criterion of excellence among outstanding DX stations."—

Clinton B. DeSoto, W1CBD, October 1935 *QST*

From its simple beginnings, culminating in the announcement of the new DX award, The DX Century Club, in September 1937 *QST* (which was itself based on the "ARRL List of Countries" published in January 1937 *QST*), membership in the ARRL DX Century Club (DXCC) has been the mark of distinction among radio amateurs the world over. That it is regarded with such prestige by DXers is a testament to its integrity and level of achievement. The high standards of DXCC are intensely defended and supported by its membership. The rules established by the founders of DXCC were consistent with the art of Amateur Radio as it existed at the time. As technology improved the ability to communicate, the rules were progressively changed to maintain a competitive environment and complement the gaining popularity of DXCC.

Because of the vast changes in the international scene brought about by World War II, it logically followed that DXCC needed to be recast, as indicated in December 1945 *QST*. Ultimately, after a great deal of study, the first postwar DXCC Countries List emerged as published in February 1947 *QST*. The new DXCC Rules appeared in March 1947 *QST*. Contacts were valid from November 15, 1945, the date US amateurs were authorized by the FCC to return to the air.

The DXCC rules today represent the aggregate of experience gained from administering postwar DXCC. Some countries on the DXCC Countries List do not, of course, meet the present criteria. This includes countries "grandfathered" from the WWII era or those that met the criteria as it existed at the time and are not subject to deletion (see Section III for the appropriate grounds for deletion). Changes are announced under DXCC Notes in *QST*.

SECTION I. BASIC RULES

1) The DX Century Club Award, with certificate and lapel pin (there is a nominal fee of $3 for the DXCC lapel pin) is available to Amateur Radio operators throughout the world (see #15 below for the DXCC Award Fee Schedule). ARRL membership is required for DXCC applicants in the US and possessions, and Puerto Rico. ARRL membership is not required for foreign applicants.

All DXCCs are endorsable (see Rule 5). There are 12 separate DXCC awards available, plus the DXCC Honor Roll:

(a) Mixed (general type): Contacts may be made using any mode since November 15, 1945.

(b) Phone: Contacts must be made using radiotelephone since November 15, 1945. Confirmations for cross-mode contacts for this award must be dated September 30, 1981, or earlier.

(c) CW: Contacts must be made using CW since January 1, 1975. Confirmations for cross-mode contacts for this award must be dated September 30, 1981, or earlier.

(d) RTTY: Contacts must be made using radioteletype since November 15, 1945. (Baudot, ASCII, AMTOR and packet count as RTTY.) Confirmations for cross-mode contacts for this award must be dated September 30, 1981, or earlier.

(e) 160 Meter: Contacts must be made on 160 meters since November 15, 1945.

(f) 80 Meter: Contacts must be made on 80 meters since November 15, 1945.

(g) 40 Meter: Contacts must be made on 40 meters since November 15, 1945.

(h) 10 Meter: Contacts must be made on 10 meters since November 15, 1945.

(i) 6 Meter: Contacts must be made on 6 meters since November 15, 1945

(j) 2 Meter: Contacts must be made on 2 meters since November 15, 1945.

(k) Satellite: Contacts must be made using satellites since March 1, 1965. Confirmations must indicate satellite QSO.

(l) Five-Band DXCC (5BDXCC): The 5BDXCC certificate is available for working and confirming 100 current DXCC countries (deleted countries don't count for this award) on each of the following five bands: 80, 40, 20, 15, and 10 Meters. Contacts are valid from November 15, 1945.

The 5BDXCC is endorsable for these additional bands: 160, 17, 12, 6, and 2 meters. 5BDXCC qualifiers are eligible for an individually engraved plaque (at a charge of $25.00 US). (m) Honor Roll: Attaining the DXCC Honor Roll represents the pinnacle of DX achievement:

*Mixed-To qualify, you must have a total confirmed country count that places you among the numerical top ten DXCC countries total on the current DXCC Countries List (example: if there are 327 current DXCC countries, you must have at least 318 countries confirmed).

*Phone-same as Mixed.

*CW-same as Mixed.

To establish the number of DXCC country credits needed to qualify for the Honor Roll, the maximum possible number of current countries available for credit is published monthly in *QST*. First-time Honor Roll members are recognized monthly in *QST*. Complete Honor Roll standings are published annually in *QST*, usually in the July issue. See DXCC notes in *QST* for specific information on qualifying for this Honor Roll standings list. Once recognized on this list or in a subsequent monthly update of new members, you retain your Honor Roll standing until the next standings list is published. In addition, Honor Roll members are recognized in the DXCC Annual List for those who have been listed in the previous Honor Roll listings or have gained Honor Roll status in a subsequent monthly listing. Honor Roll qualifiers receive an Honor Roll endorsement sticker for their DXCC certificate and are eligible for an Honor Roll lapel pin ($2) and an Honor Roll plaque ($25 plus shipping). Write the DXCC Desk for details.

#1 Honor Roll: To qualify for a Mixed, Phone or CW Number One plaque, you must have worked every country on the current DXCC Countries List. Write the DXCC Desk for details.

2) Written proof (confirmations, ie, QSL cards) of having made two-way communication must be submitted directly to ARRL Headquarters for all DXCC countries claimed. Photocopies and electronically transmitted confirmations (including, but not limited to, FAX, telex and telegram) are not acceptable for DXCC purposes. Applicants for their first DXCC award may have the cards checked by ARRL DXCC Field Representatives-see Section V for details. The use of the official DXCC application forms or an approved facsimile (eg, produced by a computer program) is required. Complete application materials are available from ARRL Headquarters. Confirmations for a total of 100 or more countries must be included with your first application. By ARRL Board of Directors action, 10-MHz confirmations are creditable to the Mixed, CW and RTTY awards only.

3) The ARRL DXCC Countries List criteria will be used in determining what constitutes a DXCC country.

4) Confirmation data for two-way communications (ie, contacts) must include the call signs of both stations, the country, mode, and date, time and frequency band.

5) Endorsement stickers for affixing to certificates or pins will be awarded as additional DXCC credits are granted. For the Mixed, Phone, CW, RTTY and 10-Meter DXCC, these stickers are in exact multiples of 25, ie 125, 150, etc, between 100 and 250 DXCC countries; in multiples of 10 between 250 and 300, and in multiples of 5 above 300 DXCC countries. For 160-Meter, 80-Meter, 40-Meter, 6-Meter, 2-Meter and Satellite DXCC, the stickers are in exact multiples of 10 starting at 100 and multiples of 5 above 200. Confirmations for DXCC countries may be submitted for credits in any increment. (See #15 for applicable fees, if any.)

6) All contacts must be made with amateur stations working in the authorized amateur bands or with other stations licensed or authorized to work amateurs. Contacts made through "repeater" devices or any other power relay method (aside from Satellite DXCC) are invalid for DXCC credit.

7) Any Amateur Radio operation should take place only with the complete approval and understanding of appropriate administration officials. In countries where amateurs are licensed in the normal manner, credit may be claimed only for stations using regular government-assigned call signs or portable call signs where reciprocal agreements exist or the host government has so authorized portable operation. No credit may be claimed for contacts with stations in any country that has temporarily or permanently closed down Amateur Radio operations by special government edict where amateur licenses were formerly issued in the normal manner. Some countries, in spite of such prohibitions, issue authorizations which are acceptable.

8) All stations contacted must be "land stations." Contacts with ships and boats, anchored or under way, and airborne aircraft, cannot be counted.

9) All stations must be contacted from the same DXCC country.

10) All contacts must be made by the same station licensee. However, contacts may have been made under different call signs in the same country if the licensee for all was the same. That is, you may simultaneously feed one DXCC from several call signs held, as long as the provisions of Rule 9 are met.

11) Any altered, forged, or otherwise invalid confirmations submitted by an applicant for DXCC credit may result in disqualification of the applicant. Any holder of a

DXCC award submitting altered, forged or otherwise invalid confirmations may forfeit the right to continued DXCC membership. The ARRL Awards Committee shall rule in these matters and may also determine the eligibility of any DXCC applicant who was ever barred from DXCC to reapply and the conditions of such application.

12) Operations Ethics:

(a) Fair play and good sportsmanship in operating are required of all DXCC members. In the event of specific objections relative to continued poor operating ethics, an individual may be disqualified from DXCC by action on the ARRL Awards Committee.

(b) Credit for contacts with individuals who have displayed continued poor operating ethics may be disallowed by action of the ARRL Awards Committee.

(c) For (a) and (b) above, "operating" includes confirmation procedures and/or documentation submitted for DXCC accreditation.

13) Each DXCC applicant must stipulate that he/she has observed all DXCC rules as well as all pertinent governmental regulations established for Amateur Radio in the country or countries concerned, and agrees to be bound by the decisions of the ARRL Awards Committee. Decisions of the ARRL Awards Committee regarding interpretations of the rules here printed or later amended shall be final.

14) All DXCC applications (both new and endorsements) must include sufficient funds to cover the cost of returning all confirmations (QSL cards) via the method chosen. Funds must be in US dollars, utilizing US currency, check or money order made payable to the ARRL, or International Reply Coupons (IRCs). Address all correspondence and inquiries relating to the various DXCC awards and all applications to: ARRL Headquarters, DXCC Desk, 225 Main St, Newington, CT 06111, USA.

(15) Effective January 1, 1994, all amateurs applying for their very first DXCC Award will be charged a one-time Registration Fee of $10.00. This same fee applies to both ARRL members and foreign non-members, and both will receive one DXCC certificate and a DXCC pin. Applicants must provide funds for Postage charges for QSL return.

(b) A $5.00 Shipping and handling fee will be charged for each additional DXCC certificate issued, whether new or replacement. A DXCC pin will be included with each certificate.

(c) Endorsements and new applications may be presented at ARRL HQ, and at certain ARRL conventions. When presented in this manner, such applications shall be limited to 110 cards maximum, and a $2.00 handling charge will apply.

(d) Each ARRL member will be allowed one submission in each calendar year at no cost (except as in (c) above, or return postage). This annual submission may include any number of QSL cards for any number of DXCC Awards, and may be a combination of new and endorsement applications. Fees as in (b) above will apply for additional new DXCC Awards.

(e) Foreign non-ARRL members will be allowed the same annual submission as ARRL members, however, they will be charged a $10.00 DXCC Award fee, in addition to return postage charges. Fees in (b) and (c) may also apply.

(f) DXCC participants who wish to submit more than once per year will be charged a DXCC fee for each additional submission made during the remainder of the calendar year. These fees are dependent upon membership status: ARRL Members: $10.00 Foreign non-members: $20.00. Additionally, return postage must be provided by applicant, and charges from (b) and (c) above may be applied.

16) The ARRL DX Advisory Committee (DXAC) requests your comments and suggestions for improving DXCC. Address correspondence, including petitions for new country consideration, to ARRL Headquarters, DXAC, 225 Main St, Newington, CT 06111, USA.

SECTION II. COUNTRIES LIST CRITERIA

The ARRL DXCC Countries List is the result of progressive changes in DXing since 1945. The full list will not necessarily conform completely with current criteria since some of the listings were recognized from pre-WWII or were accredited from earlier versions of the criteria. While the general policy has remained the same, specific mileages in Point 2(a) and Point 3, mentioned in the following criteria, have been used in considerations made April 1960 and after. The specific mileage in Point 2(b) has been used in considerations made April 1963 and after.

When an area in question meets at least one of the following three points, it is eligible as a separate country listing for the DXCC Countries List. These criteria address considerations by virtue of Government [Point 1] or geographical separation [Points 2 and 3], while Point 4 addresses ineligible areas. All distances are given in statute miles.

Point 1, GOVERNMENT

An independent country or nation-state having sovereignty (that is, a body politic or society united together, occupying a definite territory and having a definite population, politically organized and controlled under one exclusive regime, and engaging in foreign relations-including the capacity to carry out obligations of international law and applicable international agreements) constitutes a separate DXCC country by reason of Government. This may be indicated by membership in the United Nations (UN). However, some nations that possess the attributes of sovereignty are not members of the UN, although these nations may have been recognized by a number of UN-member nations. Recognition is the formal act of one nation committing itself to treat an entity as a sovereign state. There are some entities that have been admitted to the UN that lack the requisite attributes of sovereignty and, as a result, are not recognized by a number of UN-member nations.

Other entities which are not totally independent may also be considered for separate DXCC country status by reason of Government. Included are Territories, Protectorates, Dependencies, Associated States, and so on. Such an entity may delegate to another country or international organization a measure of its authority (such as the conduct of its foreign relations in whole or in part, or other functions such as customs, communications or diplomatic protection) without surrendering its sovereign status. DXCC country status for such an entity is individually considered, based on all the available facts in the particular case. In making a reasonable determination as to whether a sufficient degree of sovereignty exists for DXCC purposes, the following characteristics (list not necessarily all-inclusive) are taken into consideration:

(a) Membership in specialized agencies of the UN, such as the International Telecommunications Union (ITU).

(b) Authorized use of ITU-assigned call sign prefixes.

(c) Diplomatic relations (entering into international agreements and/or supporting embassies and consulates), and maintaining a standing army.

(d) Regulation of foreign trade and commerce, customs, immigration and licensing (including landing and operating permits), and the issuance of currency and stamps.

An entity that qualifies under Point 1, but consists of two or more separate land areas, will be considered a single DXCC country (since none of these areas alone retains an independent capacity to carry out the obligations of sovereignty), unless the areas can qualify under Points 2 or 3.

Point 2, SEPARATION BY WATER

An island or a group of islands which is part of a DXCC country established by reason of Government, Point 1, is considered as a separate DXCC country under the following conditions:

(a) The island or islands are situated off shore, geographically separated by a minimum of 225 miles of open water from a continent, another island or group of islands that make up any part of the "parent" DXCC country.

For any additional island or islands to qualify as an additional separate DXCC country or countries, such must qualify under Point 2(b).

(b) This point applies to the "second" island or island grouping geographically separated from the "first" DXCC country created under Point 2(a). For the second island or island grouping to qualify, at least a 500-mile separation of open water from the first is required, as well as meeting the 225-mile requirement of (a) from the "parent". For any subsequent island(s) to qualify, the 500-mile separation would again have to be met. This precludes, for example, using the 225-mile measurement for each of several islands from the parent country to make several DXCC countries.

(c) An island is defined as a naturally formed area of land surrounded by water, the surface of which is above water at high tide. Rocks which cannot sustain human habitation shall not be considered for DXCC country status.

(d) An island must meet or exceed size standards. To be eligible for consideration, the island must be visible, and named, on a chart with a scale of not less than 1:1,000,000. Charts used must be from recognized national mapping agencies. The island must consist of a single unbroken piece of land not less than 10,000 square feet in area, which is above water at high tide. The area requirements shall be demonstrated by the chart.

Point 3, SEPARATION BY ANOTHER DXCC COUNTRY

(a) Where a Point 1 DXCC country, composed of one or more continental land areas or of continental land areas and islands, is totally separated by an intervening DXCC country into two land areas which are at least 75 miles apart, two DXCC countries result. This distance is measured along the great circle between the two closest points of the two areas divided. The measured distance

may include inland lakes and seas which are part of the intervening DXCC country. The test for total separation into two areas requires that a great circle cannot be drawn from any point on the continental land and/or islands of one area to any point on the continental land and/or islands of the other area without intersecting any land of the intervening DXCC country.

(b) Where a Point 1 DXCC country, composed entirely of islands, is totally separated by an intervening DXCC country into two areas, then two DXCC countries result. No minimal distance is required for the separation. The test for total separation into two areas requires that a great circle cannot be drawn from any point on any island of one area to any point on any island of the other area without intersecting any land of the intervening DXCC country.

Point 4, INELIGIBLE AREAS

(a) Any area which is unclaimed or unowned by any recognized government does not count as a separate DXCC country.

(b) Any area which is classified as a Demilitarized Zone, Neutral Zone or Buffer Zone does not count as a separate DXCC country.

(c) The following do not count as a separate DXCC country from the host country: Embassies, consulates and extra-territorial legal entities of any nature, including, but not limited to, monuments, offices of the United Nations agencies or related organizations, other inter-governmental organizations or diplomatic missions.

SECTION III. DELETION CRITERIA

A DXCC country is subject to deletion from the ARRL DXCC Countries List if political change causes it to cease to meet Point 1 of the Countries List Criteria (a derivative of such change may cause it to cease to meet Points 2 or 3) or if it falls into Point 4 of the criteria. Additions to and deletions from the DXCC Countries List come about as a result of a myriad of such political changes. Reviewing the nature of the changes which have occurred since 1945 as they affect DXCC, these changes can be grouped into categories as follows:

(a) Annexation. When an area that has been recognized as a separate country under Point 1 is annexed or absorbed by an adjacent Point 1 country, the annexed area becomes a deleted country. Examples: India annexed Sikkim (AC3); China annexed Tibet (AC4); Indonesia annexed Portuguese Timor (CR8).

(b) Unification. When two or more entities that have been separate DXCC countries under Point 1 unite or combine into a single entity under a common administration, one new DXCC country is created and two or more DXCC countries have been deleted. Example: Italian Somaliland (I5) plus British Somaliland (VQ6) became Somalia (60/T5).

(c) Partition. When one country is divided or partitioned into two or more countries, one DXCC country is deleted and two or more DXCC countries are created. Example: French Equatorial Africa (FQ) was deleted and replaced by Central Africa (TL), Congo (TN), Gabon (TR) and Chad (TT). The partition category is not employed when the original political entity continues in some form. That is, if part of country A splits off to form country B, the original DXCC country (A) is retained and one new DXCC country (B) is added. Examples: the

British Sovereign Bases on Cyprus (ZC4); Aruba (P4).

(d) Independence. Mere independence does not result in a Countries List deletion. Examples: the Tonga Islands, then a British protectorate (VR5), is the same country as the present listing of the Kingdom of Tonga (A3). Further, an entity already recognized as a separate DXCC country is not deleted because of a change in its independent status. Bangladesh (S2) is the same listing as East Pakistan (AP), which was already separate from West Pakistan by virtue of Point 3. Also, a country that merely changes its name (such as when Upper Volta became Burkina Faso) does not change its basic status as a DXCC country on the DXCC Countries List.

SECTION IV. ACCREDITATION CRITERIA

1) The many vagaries of how each nation manages its telecommunications matters does not lend itself to a hard set of rules that can be applied across the board in accrediting all Amateur Radio DX operations. However, during the course of more than 40 years of DXCC administration, basic standards have evolved in determining whether a DX operation meets the test of legitimate operation. The intent is to assure that DXCC credit is given only for contacts with operations that are conducted appropriately in two respects: (1) proper licensing; and (2) physical presence in the country to be credited.

2) The following points should be of particular interest to those seeking accreditation for a DX operation:

(a) The vast majority of operations are accredited routinely without any requirement for submitting authenticating documentation.

(b) In countries where Amateur Radio operation has not been permitted or has been suspended or where some reluctance to license amateur stations has been evidenced, authenticating documents may be required prior to accrediting an operation.

(c) Some DXCC countries, even though part of a country with no Amateur Radio restrictions, nevertheless require the permission of a governmental agency or private party prior to conducting Amateur Radio operations on territory within their jurisdiction. Examples: Desecheo I. (KP5); Palmyra I. (KH5); Kingman Reef (KH5K).

3) In those cases where supporting documentation is required, the following should be used as a guide as to what information may be necessary to make a reasonable determination of the validity of the operation:

(a) Photocopy of license or operating authorization.

(b) Photocopy of passport entry and exit stamps.

(c) For islands, a landing permit and/or signed statement of the transporting ship's, boat's, or aircraft's captain, showing all pertinent data, such as date, place of landing, etc.

(d) For some locations where special permission is known to be required to gain access, evidence of this permission having been given is required.

4) These accreditation requirements are intended to preserve the DXCC program's integrity and to ensure that the program does not encourage amateurs to "bend the rules" in their enthusiasm, thus jeopardizing the future development of Amateur Radio. Every effort will be made to apply these criteria in a uniform manner in conformity with these objectives.

SECTION V. FIELD CHECKING OF QSL CARDS

QSL cards for new DXCC awards may be checked by two DXCC Field Representatives. This program applies to any DXCC award for an individual or station. Specifically excluded from this program are additional new DXCC awards and endorsements of existing awards. Also excluded are 5BDXCC, 6 meter, 2 meter and Satellite DXCC and any 160-meter QSL.

1) Countries Eligible for Field Checking:

(a) Eligible countries will be indicated in the ARRL DXCC Countries List, and are subject to change. Only cards from these eligible countries may be checked by DXCC Field Representatives. QSLs for other DXCC countries must be submitted directly to ARRL Headquarters.

(b) The ARRL Awards Committee determines which countries are eligible for Field Checking.

2) DXCC Field Representatives:

(a) DXCC Field Representatives must be ARRL members who have a DXCC award endorsed for at least 300 countries.

(b) To become a DXCC Field Representative, a person must be nominated by a DX club. (A DX club is an ARRL affiliated club with at least 25 members who are DXCC members and which has, as its primary interest, DX. If there are any questions regarding the validity of a DX club, the issue shall be determined by the Division Director where the DX club is located.) A person does not have to be a member to be nominated by a DX club.

(c) DXCC Field Representatives are approved by the Director of the ARRL Division in which they reside and appointed by the President of the ARRL.

(d) DXCC Field Representative appointments must be renewed annually.

3) Card Checking Process:

(a) Only cards from the list of eligible countries can be checked by DXCC Field Representatives. An application shall contain a minimum of 100 QSL confirmations from the list and shall not contain any QSLs from countries that are not on the list of eligible countries. Additional cards should not be sent with field checked applications. The application may contain the maximum number of countries that appear on the list of eligible countries. That is, if there are 245 countries on the list, the initial application for a field-checked DXCC award could contain 245 countries.

(b) It is the applicant's responsibility to get cards to and from the DXCC Field Representatives.

(c) Field Representatives may, at their own discretion, handle members' cards by mail.

(d) The ARRL is not responsible for cards handled by DXCC Field Representatives and will not honor any claims.

(e) The QSL cards must be checked by two DXCC Field Representatives.

(f) The applicant and both DXCC Field Representatives must sign the application form. (See SECTION I no. 11 regarding altered, forged or otherwise invalid confirmations.)

(g) The applicant shall provide a stamped no. 10 envelope (business size) addressed to ARRL HQ to the DXCC Field Representatives. The applicant shall also provide the application fee (check or money order payable to ARRL—no cash) for the initial DXCC award.

(h) The DXCC Field Representatives will forward

completed applications and appropriate fee(s) to ARRL HQ.

4) ARRL HQ involvement in the card checking process: a) ARRL HQ staff will receive field-checked applications, enter application data into DXCC records and issue DXCC credits and awards as appropriate.

(b) ARRL HQ staff will perform random audits of applications. Applicants or members may be requested to forward cards to HQ for checking before or after credit is issued.

c) The applicant and both DXCC Field Representatives will be advised of any errors or discrepancies encountered by ARRL staff.

(d) ARRL HQ staff provides instructions and guidelines to DXCC Field Representatives.

5) Applicants and DXCC members may send cards to ARRL Headquarters at any time for review or recheck if the individual feels that an incorrect determination has been made.

ARRL DXCC FIELD REPRESENTATIVES
(as of July 1, 1995)

Atlantic Division

EPA
M. Stokley Benson, AJ3H, 1192 Divot Drive, Wescosville, PA 18106
Charles E.Moraller, K2CM, RD 1, Box 99-C, New Ringgold, PA 17960
Norman Zoltack, K3NZ, 4333 Locust Drive, Schnecksville, PA 18078

MDC
Murray Green, K3BEQ, 5730 Lockwood Road, Cheverly, MD 20785
Michael C. Cizek, KO7V, Box 227, Oxon Hill, MD 20750
Everett C. Bollin, WA3DVO, 8000 Ray Leonard Court, Palmer Park, MD 20785

SNJ
John M. Fisher, K2JF, 538 Wesley Avenue, Pitman, NJ 08071-1924
James M. Mollica, Sr., K2OWE, 22 Wright Loop, Williamstown, NJ 08094
Steve Branca, K2SB, 202 Minnetonka Road, Somerdale, NJ 08083
John D. Imhof, N2VW, 3 Seeley Drive, Mt. Holly, NJ 08060

WNY
John C. Yodis, K2VV, P. O. Box 460, Hagaman, NY 07642
Eugene W. Nadolny, W2FXA, 21 Hidden Valley Drive, Elma, NY 14059
Bob Rossi, W2HG, 75 Olde Erie Trail, Rochester, NY 14626
Elmer Wagner, WB2BNJ, 23 Red Barn Circle, Pittsford, NY 14534
Robert Dow, WB2CJL, 115 Puritan Road, Tonawanda, NY 14150
James R. Ciurczak, WB2IVO, 10404 Cayuga Drive, Niagara Falls, NY 14304
Robert E. Nadolny, WB2YQH, 135 Wetherstone Drive, West Seneca, NY 14224

Central Division

IL
Tim Wright, KØBFR, 1131 Warren, Alton, IL 62002
Jerry Brunning, K9BG, 15307 Shamrock Lane, Woodstock, IL 60098
Andrew White, K9CW, RR 1 Box 115, Thomasboro, IL 61878-9642
Merv Schweigert, K9FD, Rt 2, Box 138A, Red Bud, IL 62278
R. Eugene Duncan, K9IKP, 6032 N. Knoll Road, Monroe Center, IL 61052
Robert E. Nielsen, K9RN, 1372 Skyridge Drive, Crystal Lake, IL 60014

James J. Coleman, KA6A, 250 County Road 700N, Ivesdale, IL 61851
Greg Wilson, N4CC, 3865 Staunton Road, Edwardsville, IL 62025
Ed Doubek, N9RF, 25W063 Wood Court, Naperville, IL 60563
Jim Dawson, WW9Q, 257 S. 9th Street, East Alton, IL 62024
Gary H. Hilker, K9LJN, 804 Otto Road, Machesney Park, IL 61115
Joseph L. Pontek, Sr., K8JP, P.O. Box 80262, Indianapolis, IN 46280-0262
David Bunte, K9FN, 129 Ivy Hill Drive, West Lafayette, IN 47906
John C. Goller, K9UWA, 4836 Ranch Road, Leo, IN 46765
James R. Weigand , N9BW, 1422 Buckskin Drive, Fort Wayne, IN 46804
Victor Keller, N9GK, 4011 Daner Drive, Fort Wayne, IN 46815
Mark Reese, NB9F, 3511 Windlass Court, Fort Wayne, IN 46815
Vernon Seitz, W9HLY, 45 Homestead, Decatur, IN 46733
Bill Gibbons, W9KBV, 535 Grapevine Lane, Fort Wayne, IN 46825
Robert C. Webb, W9OCL, 8208 Valley Estates Drive, Indianapolis, IN 46227
William D. DeGeer, W9TY, 3601 Tyler Street, Gary, IN 46408

WI
Peter Byfield, K9VAL, 4534 S. Hill Ct., Deforest, WI 53532
Ron Gorski, N9AU, 3241 S. Clement Avenue, Milwaukee, WI 53207
Gerald Scherkenbach, N9AW, 5952 So. Elaine Avenue, Cudahy, WI 53110
Herb Jordan, W9LA, 6318 Putnam Road, Madison, WI 53711
Richard F. Roll, W9TA, 14880 W. Maple Ridge Court, New Berlin, WI 53151
Frank Holliday, WB9NOV, 23 Walworth Ct., Madison, WI 53705
Ed Toal, WI9L, 5141 Sunrise Ridge Trail, Middleton, WI 53562

Dakota Division

MN
Curt Swenson, KØCVD, 4821 Westminster Road, Minnetonka, MN 55345
Dave Wester, KØIEA, 10205 217th Street North, Forest Lake, MN 55205
Gary Reichow, KNØV, 15615 Harmony Way, Apple Valley, MN 55124
Keith H. Gilbertson , KSØZ, HC10, Box 44 , Detroit Lakes, MN 56501
John Hill, NJØM, 3353-34th Avenue South, Minneapolis, MN 55406
Jack Falker, W8KR, 5716 View Lane, Edina, MN 55436

ND
Ron L. Roche, KØALL, 1437 No. University Drive, Fargo, ND 58105

Delta Division

AR
Paul E. Wynne, AF5M, 5306 Marion Street, N. Little Rock, AR 72118
Stanley W. Krueger, K5AS, 3169 Skillern Road, Fayetteville, AR 72703
J. Sherwood Charlton, K5GOE, 306 N. Willow Avenue, Fayetteville, AR 72701
Leonard Mendel, K5OVC, 309 Rolling Acres Drive, Pearcy, AR 71964
Oliver A. Gade, K9GPN, 156 Wildcat Estates Lane, Hot Springs, AR 71913
Earl F. Smith, KD5ZM, 818 Green Oak Lane, White Hall, AR 71602
Stewart W. Long, KE5PO, 28 Nob View Circle, Little Rock, AR 72205
F. Martin Hankins, W5HTY, 406 East Cedar Street, Warren, AR 71671
Andrew Toth , W5LQN, 117 Camellia Drive, Hot Springs, AR 71913
Ben Lockerd, W5NF, 3231 S. Cliff Drive, Fort Smith, AR 72903

LA

Howard N. Schmidt, K5BLV, PO Box 675-Elm Street, Clinton, LA 70722

John Wondergem, K5KR, 600 Smith Drive, Metairie, LA 70005

Jim O'Brien, K5NV, P.O. Box 403, Harvey, LA 70059

Silvano Amenta, KB5GL, 5028 Hearst Street, Metairie, LA 70001

Paul Azar, N5AN, 412 Live Oak, Lafayette, LA 70503

Shirley Wondergem, N5GGO, 600 Smith Drive, Metairie, LA 70005

Wes Attaway, N5WA, 2048 Pepper Ridge, Shreveport, LA 71115

Floyd Teetson, W5MUG, Route 1, Box 261, Heflin, LA 71039

Willis J. Robert, W5RRK, Rt 2 Box 538, Erath, LA 70533

Michael Mayer III, W5ZPA, 5836 Marcia Avenue, New Orleans, LA 70124

Roy Bonvillian, WD5DBV, P.O. Box 350, Cade, LA 70519

Cornell C. Bodensteiner, WD5GJB, 129 Patricia Ann Place, Lafayette, LA 70508

Randy Hollier, WX5L, 1421 North Atlanta, Metairie, LA 70003

MS

Steven Alexander, KX5V, 22 Yorkshire Pkwy, Gulfport, MS 39501

Floyd Gerald, N5FG, 4706 Washington Avenue, Gulfport, MS 39507-4043

Vic West, WA5SUE, 113 Ben Drive, Gulfport, MS 39503

Michael J. Parisey, WDØGML, 251 Eisenhower Drive Apt 337, Biloxi, MS 39531

George G. Bethea, WN5IJZ, 609 Woodward Avenue, Gulfport, MS 39501

TN

Gary S. Pitts, AA4DO, 2023 Kimberly Drive, Mt. Juliet, TN 37122

John L. Anderson, AG4M, 1041 West Outer Drive, Oak Ridge, TN 37830

Don Payne, K4ID, 117 Sam Davis Drive, Springfield, TN 37172

Robert M. May II, K4SE, P.O. Box 453, Jonesboro, TN 37659

Don Moore, KB4HU, 309 Brookhaven Trail, Smyrna, TN 37167

Vernon T. Underwood, WA4NIB, P.O. Box 3771, Knoxville, TN 37927

Lawrence D. Strader, WA4ZWH, 11915 Turpin Lane, Knoxville, TN 37932

William B. Jones, WB4PUD, 5081-HY 49 West, Springfield, TN 37172

Great Lakes Division

KY

James C. Vaughan, K4TXJ, 5504 Datura Lane, Louisville, KY 40258

Robert K. Ledford, KC4MK, 4533 Highway 155, Fisherville, KY 40023

Timothy B. Totten, KJ4VH, 10400 Broad Run Road, Louisville, KY 40299

Roy M. Dobbs, W4KHL, 516 Dorsey Way, Louisville, KY 40223

Gary Hext, WB4FLB, 4953 Westgate Drive, Bowling Green, KY 42101

MI

Wayne W. Wiltse, K8BTH, 14468 Bassett Avenue, Livonia, MI 48154

Merlin D. Anderson, K8EFS, 4300 South Cochran, Charlotte, MI 48813-9109

Donald L. Defeyter, KC8CY, 5061 East Clark Road, Bath, MI 48808

Edmund Gurney, Jr., WD8Y, 18712 Westbrook Drive, Livonia, MI 41852

OH

Pete Michaelis, N8ATR, 12224 East River Road, Columbia Station, OH 44028

John M. Hugentober, Sr., N8FU, 4441 Andreas Avenue, Cincinnati, OH 45211

Ron Moorefield, W8ILC, 6531 Le Mans Lane, Huber Heights, OH 45424

Harry T. Flasher, W8KKF, 7425 Barr Circle, Dayton, OH 45459

Alfred V. Altomari, W8QWI, 2009 W. 36 Street, Lorain, OH 44053

Robert A. Frey, WA6EZV, 11938 Belgreen Lane, Cincinnati, OH 45240

Jerome Kurucz, WB8LFO, 5338 Edgewood Drive, Lorain, OH 44053

John Udvari, WB8VPA, 2896 Pine Lake Road, Uniontown, OH 44685

Reno M. Tonsi, WT8C, 9248 Woodvale Court, Mentor, OH 44060

Hudson Division

ENY

Jay Musikar, AF2C, 6 Cherry Lane, Putnam Valley, NJ 10579

Saul M. Abrams, K2XA, RD #1, Maple Road, Slingerlands, NY 12159

G. William Hellman, NA2M, 3713 Valleyview Street, Mohegan Lake, NY 10547

Steven L. Weinstein, NI2C, 45 Esterwood Avenue, Dobbs Ferry, NY 10522

NLI

Richard J. Tygar, AC2P, 5 Clelmsford Drive, Wheatley Heights, NY 11798-1504

Arthur M. Albert, K2ENT, 2476 Fortesque Avenue, Oceanside, NY 11572

Larry Strasser, K2LS, 1404 Amend Drive, North Merrick, NY 11566-1301

Frank Kiefer, K2PWG, 1-Sherrill Lane, Port Jefferson Station, NY 11776

Leonard Zuckerman, KB2HK, 2444 Seebode Court, Bellmore, NY 11710-4522

Edward J. Manheimer, KD2OD 14 Seneca Trail, Ridge, NY 11961

Allen Singer, N2KW, Box 382, Village Station, New York, NY 10014-0382

Martin P. Miller, NN2C, 24 Earl Road , Melville, NY 11747

L. E. Dietrich, NN2G, 32 Beach Road, Massapequa, NY 11758

Frank E. Gut, NY2A, 129 North Side Road, Wading River, NY 11792

Jim Stiles, W2NJN, 21 Branch Drive, Smithtown, NY 11787

Jean Chittenden, WA2BGE, RR2, Box 2105, Syosset, NY 11791

NNJ

Robert B. Larson, K2TK, P.O. Box 53, Stockholm, NJ 07460

Warren Hager, K2UFM, 31 Forest Drive, Hillsdale, NJ 07642

Ronald Loneker, KA2BZS, 2092 Nicholl Avenue, Scotch Plains, NJ 07076

Derry Galbreath, KB2HZ, 207 Gordon Place, Neshanic Station, NJ 08853

Guy Glaser, KE2CG, 240 Grant Avenue, Piscataway, NJ 08854

Harry Johnson, KS2O, 424 Jefferson Avenue, Hasbrouck Heights, NJ 07604

Eugene Ingraham, N2BIM, 79 Stillwater Road, Newton, NJ 07860

Michael Pagan, N2GBH, 1 Woodview Avenue, Long Beach, NJ 07740

Harry Westervelt, NA2K, 72 Kuhlthau Avenue, Milltown, NJ 08850

John Sawina, NA2R, 61 Gallmeier Road, Frenchtown, NJ 08825

Peter Pellack, NO2R, 77 Money Street, Lodi, NJ 07644

Ted Marks, W2FG, 81 Oakey Drive, Kendall Park, NJ 08824

Orion Arnold, W2HN, 2 Sleepy Hollow Drive, Ho-Ho-Kus, NJ 07423

T. Edward Berzin, W2MIG, 47 Palisade Road, Elizabeth, NJ 07208

Stan Dicks, W8YA, 9 Settlers Court, Neshanic Station, NJ 08855

William W. Hudzik, WA2UDT, 111 Preston Drive, Gillette, NJ 07933

Angel M. Garcia, WA2VUY, 7 Markham Drive, Long Valley, NJ 07853

Mario Karcich, WB2CZB, 311 Liberty Avenue, Hillsdale, NJ 07642

Andrew Birmingham, WB2RQX, 235 Van Emburgh Avenue, Ridgewood, NJ 07450

Midwest Division

IA
Thomas White, KØVZR, 2027 Carter Avenue, Jessup, IA 50648

Thomas Noel Vinson, NYØV, 10211 Hall Road , Cedar Rapids, IA 52400

Robert W. Walstrom, WØEJ, 7431 Macon Drive NE, Cedar Rapids, IA 52402

Dale E. Repp, WØIZ, 1618 Texas Avenue, NE, Cedar Rapids, IA 52402

James L. Spencer, WØSR, 3712 Tanager Drive, NE, Cedar Rapids, IA 52402

Thomas G. Vavra, WB8ZRL, 682 Palisades Access Road, Ely, IA 52227

KS
Rick Barnett, KBØU, P.O. Box 4798, 8909 W. 81, Overland Park, KS 66204

Richard G. Tucker, WØRT, P. O. Box 875 , Parsons, KS 67357

Dean Lewis, WAØTKJ, 609 Otto Avenue, Salina, KS 67401-7261

John Shoultys, WDØBNC, 2157 Edward, Salina, KS 67401

MO
Joseph F. Nemecek, KØJN, 1208 N.E. 77th Street, Kansas City, MO 64118

Thomas L. Bishop, KØTLM, 4936 N. Kansas Avenue, Kansas City, MO 64119

Ken Kreski, KWØA, 6645 St. John's Lane, House Springs, MO 63051

Jim Higgins, NDØF, 219 Birchwood Drive, St. Peters, MO 63376

Udo Heinze, NIØG, #2 Wildwood Circle, Crestwood, MO 63126

Jim Glasscock, WØFF, 3416 Manhattan Avenue, St. Louis, MO 63143

Bill Wiese, WØHBH, 9520 Old Bonhomme, Olivette, MO 63132

John L. Chass, WØJLC, 12167 N.W. Hwy. 45, Parkville, MO 64152

Don Gaikins, WDØCHW, 344 Hillside, Webster Groves, MO 63119

New England Division

CT
Joel Wilks, AK1N, 27 Champion Hill Road, East Hampton, CT 06424

William C. Stapleford, K1DII, Bird Road, P. O. Box 1093, Bristol, CT 06010

Ralph M. Hirsch, K1RH, 172 Newton Road, Woodbridge, CT 06525

Peter Budnik, KB1HY, 40 Bartlett Street, Plainville, CT 06062

Raoul Elton, NI2B, 60 Padanaram Road #18, Danbury, CT 06811

John P. Larson, NQ1K, 34 Donna Drive, Burlington, CT 06013

EMA
William G. Bithell, Jr., K1JKS, 96 Greenwood Avenue, Swampscott, MA 01907

Gary Young, K2AJY, 1 Sutton Place, Swampscott, MA 01907

Clifford O. Thomson, N1EOA, 5 Lakeshore Drive, Beverly, MA 01915

Ray Sylvester, Jr., NR1R, 20 Gardner Road, Reading, MA 01867

Russell E. Small, W1JJ, 224 Pleasant Street, Marblehead, MA 01945

Melrose R. Cole, WZ1Q, P. O. Box 8, Prides Crossing, MA 01965

NH
W. David Gerns, KA1CB, 3 Rolling Hill Avenue, Plaistow, NH 03865

Carl Huether, KM1H, 169 Jeremy Hill Road, Pelham NH 03076

Paul I. Cleveland, W1OHA, PO Box 89, Craighill, North Sutton, NH 03260-0089

William A. Dodge, WA1PEL, 78 Littleworth Road, Dover, NH 03820

Ann M. Santos, WA1S, 245 Colburn Road, Milford, NH 03055

VT
David A. Wilson, N4DW, Rt 1, Box 11A, East Burke, VT 05832

Gareth B. Gaudette, W1GG, 21 Westview Road, Lanesboro, MA 01237

WMA
Edward Landry, WA1ZAM, 140 Cliff Street, North Adams, MA 01247

Northwestern Division

EWA
William Martinek, K7EFB, N 522 Moore Road, Veradale, WA 99037

Michael J. Klem, K7ZBV, 804 Cedar Avenue, Richland, WA 99352

Lester S. Morgan, NR7B, N 6811 Monroe Street, Spokane, WA 99208-4129

Donald W. E. Calbick, W7GB, 447 Knolls Vista Drive, Moses Lake, WA 98837

Warren L. Triebwasser, W7YEM, 4624 NW Blvd, Spokane, WA 99205

Jay W. Townsend, WS7I, P.O. Box 644, Spokane, WA 99210

MT
George E. Martin, K7ABV, 3608 5th Avenue S., Great Falls, MT 59405

Robert E. Leo, W7LR, 6790 S. 3rd Road, Bozeman, MT 59715

Doug Laubach, WQ7B, Box 19, Carter, MT 59420

OR
Gerald D. Branson, AA6BB, 93787 Dorsey Lane, Junction City, OR 97448

Robert S. Wruble, AI7B, SW Pleasant Valley Road, Beaverton, OR 97007

Larry R. Johnson, K7LJ, 10036 SW 50th Avenue, Portland, OR 97219

Vincent J. Varnas, K8REG, 3229 SW Luradel Street, Portland, OR 97219

Joan E. Branson, KA6V, 93787 Dorsey Lane, Junction City, OR 97448

Hugh C. Dean, KF7SH, 420 W. Rose, Lebanon, OR 97355

Benjamin F. Shearer, Jr., N7GMN, 15364 S. Clackamas River Drive, Oregon City, OR 97045

Sanford T. Weinstein, NK7Y, 2945 SW 4th Avenue, Portlnd, OR 97201

Robert A. Moore, W7JNC, P.O. Box 33197, Portland, OR 97233

Michael J. Huey, W7ZI, 12385 SW Stillwell Lane, Beaverton, OR 97005

Richard A. Zalewski, W7ZR, 10735 SW 175th Avenue, Beaverton, OR 97007

WWA
David C. Norton, AB9O, 2612 NW 18th Avenue, Camas, WA 98607

Herbert Anderson, K7GEX, 20148 6th NE, Seattle, WA 98155

Jim Fenstermaker, K9JF, 1032 NE 161st Avenue, Vancouver, WA 98682

Thomas H. Renfro, N7OT, 8703 NE 94th Avenue, Vancouver, WA 98662

Dick Moen, N7RO, 2935 Plymouth Drive, Bellingham, WA 98225

John Gohndrone, N7TT, PO Box 863, Bothell, WA 98041

Brenda G. Murphy, NS7F, 2612 NW 18th Avenue, Camas, WA 98607

Kermit W. Raaen, W7BG, 15509 E. Mill Plain Blvd 42, Vancouver, WA 98664

Al Johnson, W7EKM, 8186 Stein Road, Custer, WA 98240

Randy Stegemeyer, W7HR, P.O. Box 1590, Port Orchard, WA 98366

Leonard A. Westbo, Jr., W7MCU, 10528 S.E. 323nd Street, Auburn, WA 98002

Alan N. Rovner, WA2TMP, 18809 N.E. 21st Streetk, Vancouver, WA 98684

Pacific Division

EB
Charles E. McHenry, W6BSY, 1612 Via Escondido, San Lorenzo, CA 94580

Rubin Hughes, WA6AHF, 17494 Via Alamitos, San Lorenzo, CA 94580

Randy Wright, WB6CUA, 18432 Milmar Road, Castro Valley, CA 94546

NV

Frank Dziurda, K7SFN, 225 W. Coyote Drive, Carson City, NV 89704

JB Coats, KB7YX, 1839 Deep Creek Drive,Sparks, NV 89434

James R. Frye, NW7O, 4120 Oakhill Avenue, Las Vegas, NV 89121

William L. Dawson, 7TVF, HCR 65 Box 71623, Pahrump, NV 89041-9678

PAC

Richard I. Senones, KH6JEB, 95-161 Kauopae Place, Mililani Town, HI 96789

Charles S. Y. Yee, KH6WU, 27 Moe Moe Place, Wahiawa, HI 96786

SCV

Larry Souza, KG6GF, 11000 Cienega Road, Hollister, CA 95023

Hillar Raamat, N6HR, P. O. Box 60611, Sunnyvale, CA 94088

Stan Goldstein, N6ULU, 791 Calabasas, Watsonville, CA 95076

J. Edward Muns, WØYK PO Box 1877, Los Gatos, CA 95031-1877

James A. Maxwell, W6CF, P. O. Box 473, Redwood Estates, CA9 5044

John G. Troster, W6ISQ 82 Belbrook Way, Atherton, CA 94025

William J. Stevens, W6ZM, 2074 Foxworthy Avenue, San Jose, CA 95124

Gerald D. Griffin, W8MEP, 123 Forest Avenue, Pacific Grove, CA 93950

Richard G. Whisler, WA6SLO, 716 Hill Avenue, South San Francisco, CA 94080

Richard M. Letrich, WB6WKM, 3686 Kirk Road, San Jose, CA 95124

SF

Chuck Ternes, N6OJ, 535 Cherry Street, Petaluma, CA 94952

Dave Stocham, W6NPY, 106 Locust Avenue, Mill Valley, CA 94941

Al Lotze, W6RQ, 46 Cragmont Avenue San Francisco, CA 94116-1308

Jerry Foster, WA6BXV, 16 San Ramon, Novato, CA 94947

SJV

Perry Foster, K6XJ, 10575 E. Bullard, Clovis, CA 93612

Charles McConnell, W6DPD, 1658 W. Mesa, Fresno, CA 93711-1944

Robert Craft, W6FAH, 8136 Grenoble Way, Stockton, CA 95210

Robert W. Smith, W6GR, 13447 Road 23, Madera, CA 93637

Charles Allessi, W6IEG, PO Box 1244, Oakhurst, CA 93644

Stanley R. Ostrom, W6XP, 4921 E. Townsend , Fresno, CA 93727

Jules Wenglare, W6YO, 1416 Seventh Avenue, Delano,CA 93215

Carl Boone, WB6VIN, 1642 Seventh Place, Delano, CA 93215

Dennis J. DuGal, WG6P, 2008 Sharilyn Drive, Modesto, CA 95355

SV

Larry Murdoch, K6AAW, 14370 Brian Road, Red Bluff, CA 96080

Ken Anderson, K6PU, P. O. Box 853, Pine Grove, CA 95665

John D. Brand , K6WC, 9655 Tanglewood Circle, Orangevale, CA 95662

Phil Sanders, KF6A, 8580 Krogh Ct., Orangevale, CA 95662

Ted Davis, W6BJH, 11595 Ridgewood Road , Redding, CA 96003

Jerry Fuller, W6JRY, PO Box 363, Forest Ranch, CA 95942

Jettie B. Hill, W6RFF, 306 St. Charles Ct., Roseville, CA 95661

Roanoke Division

NC

Bill Parris, AA4R, 16741 100 Norman Place, Huntersville, NC 28078

Ron Oates, AA4VK, 9908 Waterview Road, Raleigh, NC 27615

Bruce M. Gragg, AG4L, Route 2, Box 329, Conover, NC 28613

Frank Dowd, Jr., K4BVQ, PO Box 35430, Charlotte, NC 28235

Larry Sossoman, K4CEB, 4838 Ponderosa Lane, Concord, NC 28025

Bill McDowell, K4CIA, 2317 Norwood Road, Raleigh, NC 27614

Les Murphy, K4DY, PO Box 626, Hickory, NC 28603-0626

Norman Pendleton, Jr., K4NYV, 3617 Country Cove Lane, Raleigh, NC 27606

Mike Jackson, N4AYO, 2568 Devon Drive, Dallas, NC 28034

Joe Young, N4DAZ, 1912 Miles Chapel Road, Mebane, NC 27302

Roger Burt, N4ZC, RFD 1 Box 246, Mt. Holly, NC 28120

Ewell M. Brown, W4CZU, PO Box 447, Taylorsville, NC 28681

Robert H. McNeill, II, W4MBD, PO Box 843, Morehead City, NC 28557

Mark McIntyre , WA4FFW, 2903 Maple Avenue, Burlington, NC 27215

Robert C. Cranford, WA4SSI, 772 Fernwood Road, Lincolnton, NC 28092

Joe Simplins, WD4R, 2400 Flintwood Lane, Charlotte, NC 28226

Gene Turner, WJ4T, 159 Windsor Drive, Graham, NC 27253

SC

Rick Porter, AA4SC, Box 2731, Rock Hill, SC 29731

Gary Dixon, K4MQG, 1606 Crescent Ridge, Ford Mill, SC 29715

Jack Jackson, N4JJ, PO Box 2612, Florence, SC 29504

Charles Johnson, N4TJ, 109 Kimberly Lane, Florence, SC 29505

Bill Jennings, W4UNP, 630 Whitepine Drive, Catawba, SC 29704

Murphy Ratterree, W4WMQ, 264 Wayland Drive , Rock Hill, SC 29732

C. W. Eldridge, WA4PLR, 220 Pinewood Lane, Rock Hill, SC 29730

Ted F. Goldthorpe, WA4VCC, 209 Swamp Fox Drive, Fort Mill, SC 29715

VA

M. David Gaskins, AA4AV, 228 Unser Drive, Chesapeake, VA 23320

Ralph King , K1KOB, 3409 Morninton Drive, Chesapeake, VA 23321

John N. Kirkham, KC4B, 10920 Byrd Drive, Fairfax, VA 22030

Gordon Garrett, KC4DY, 4528 North Fork Road, Elliston, VA 24087

Emmett Stine, KD4OS, 1269 Parkside Place, Virginia Beach, VA 23454

Fran Sledge, N4CRU, 3004 Oakley Hall Road, Portsmouth, VA 23702

Clay Partin, N4VG, 6642 Shingle Ridge Road, SW Roanoke, VA 24018

Karl H. Oyster, Jr., NQ1W, 524 Vicksdell Cres., Chesapeake, VA 23320-3551

Nap Perry, W4DHZ, 9317 Marlow Avenue, Norfolk, VA 23503

Art Westmont, W4EEU, 4717 Little John Road, Virginia Beach, VA 23455

Grover Brewer, W4FPW, 1359 Eagle Avenue , Norfolk, VA 23518

Howell G. Gwaltney, W4IF, 209 Pennington, Portsworkt, VA 23701

Charlie Sledge, W4JVU, 3004 Oakley Hall Road, Portsmouth, VA 23703

Pradyumna Rana, WB4NFO, 29 East Chapman Street, Alexandria, VA 22301

James A. Sladek, WB4UBD, 215 Delaney Drive, Suffolk, VA 23434

Richard T. Williams, WE6H, 7902 Viola Street, Springfield, VA 22152

Rocky Mountain Division

CO

Ed Wood , KØAB,PO Box 863, Elizabeth, CO 80107-0863

Steve P. Gecewicz, KØCS, P. O. Box 1048, Elizabeth, CO 80107

John D. Peters, K1ER, 2439 Sherman Street, Longmont, CO 80501

Karen Schultz, KAØCDN, 15643 East 35th Place, Aurora, CO 80011

Glenn Schultz, WØIJR, 15643 East 35th Place, Aurora, CO 80011

Jim Spaulding, WØUO, 6277 S. Niagara Court, Englewood, CO 80111

NM

Robert L. Norton, N5EPA, 116 Pinon Heights Road, Sandia Park, NM 87047

Arne Gjerning, N7KA, PO Box 1485, Corrales, NM 87048

Fred F. Seifert, W5FS, 3106 Florida Street NE Albuquerque, NM 87110-2617

Eugene F. Carter, W5OLN, 320 Camino Tres, SW Albuquerque, NM 87105

Paul Rubinfield, WF5T, Rt 19 Box 91-JR, Sante Fe, NM 87505

UT

Steven C. Salmon, K7OXB, 2184 East 6070 South Salt Lake City, UT 84121

Southeastern Division

AL

Bill Christian , K4IKR, 2800 Cave Avenue, Huntsville, AL 35810

Tim Pearson, KU4J, 6214 Rime Village Drive #205, Huntsville, AL 35806

GA

Neil R. Foster, KC4MJ, 3185 Friar Tuck Way, Atlanta, GA 30340

Robert Varone, WA4ETN, 5177 Holly Springs Drive, Douglasville, GA 30135

NFL

William R. Hicks, K4UTE, 7002 Deauville Road, Jacksonville, FL 33205

Jim Buth, K9CJK, Rt 4 Box 7216-1, Crawfordville, FL 32327

Richard A. Knox, WR4K, 5172 Pine Avenue, Orange Park, FL 32072

PR

Telesforo Figueroa, KP4P, PO Box 1651, Yabucoa, PR 00767-1651

Eduardo Negron, KP4EQF, Urb. Valle Real AG-78, Ponce, PR 00731

SFL

Harry D. Belock, 8108 N. W. 102 Terr, Tamarac, FL 33321

Michael M. Raskin, K4KUZ, 561 West Tropical Way, Plantation, FL 33317

Gary Fowks, K4MF, 635 W.64 Drive, Hialeah, FL 33012

Peter Rimmel, K8UNP, 4209 Madison Street, Hollywood, FL 33021

Marv Westerdahl, KC2KU, 1934 Hawaii Avenue N.E., St. Petersburg, FL 33703

Left Boggess, KE4VU, 11636 Grove Street North, Seminole, FL 34642

Ray Riker, NY2E, 360 Ponte Vedra Road, Palm Springs, FL 33461

Donald B. Search, W3AZD, 10550 State Road 84 #147 Park City,Fort Lauderdale, FL 33324

William E. Fells, W5OG, 6857 Broadmoor, No. Lauderdale, FL 33068

Mark Horowitz, WA2YMX, 6831 SW 16th Street, Plantation, FL 33317

Joseph L. Picior, WB4OSN, 1485 NW66 Avenue, Margate, FL 33063

Harry T. Henley, WK9Z, 2013 Massachusetts Avenue NE, St.Petersburg, FL 33703

Edward Grogan, WW1N, 902 Bay Point Drive, Madeira Beach, FL 33708

Southwestern Division

AZ

Ned Stearns, AA7A, 7038 E. Aster Drive , Scottsdale, AZ 85254

Mike Fulcher, KC7V, 6545 E. Montgomery Road, Cave Creek, AZ 85331

Lee Finkel, KY7M, 6928 E. Ludlow, Scottsdale, AZ 85254

Hardy Landskov, N7RT, 15814 N 44th Street, Phoenix, AZ 85032

Jim McDonald, N7US, 9235 N 32nd Place, Phoenix, AZ 85028

Steve Towne, NN7X, 12002 N. Oakhurst Way, Scottsdale, AZ 85254

Frank Schottke, W2UE, 5049 E. Laurel Lane, Scottsdale, AZ 85254

Stuart Greene, WA2MOE, 7537 N. 28th Avenue, Phoenix, AZ 85051

LAX

John Alexander, K6SVL, 28403 Covecrest Drive, Palos Verdes, CA 90274

Leonard Svidor, W6AUG, 17760 Alonzo Place, Encino, CA 91316

Sheldon C. Shallon, W6EL, 11058 Queensland Street , Los Angeles, CA 90034

ORG

Rick Samoian, WB6OKK, 5302 Cedarlawn Drive, Placentia, CA 92670

Milt Bramer, N6MB, 4161 Shadyglade Drive , Santa Maria, CA 93455

SB

John O. Norback, W6KFV, 133 Pino Solo Court, Nipomo, CA 93444

Neil B. Taylor, W6MUS, PO Box 392, Atascadero, CA 93423

Jim Robb, W6OUL, 501 N. Poppy, Lompoc, CA 93436

Steve Nolan, WA6ARG, 904 N. Western Avenue, Santa Maria, CA 93454

SDG

Bruce Clark, K6JYO, 508 Washingtonia Drive, San Marcos, CA 92069

Cliff Smith, K6TWU, 10565 Rancho Road, La Mesa, CA 91941

Art Charette, K6XT, 16604 Adrienne Way, Ramona, CA 92065

Rick Craig, N6ND, PO Box 741, Ramona, CA 92065

Ed Andress, W6KUT, 12821 Corte Dorotea, Poway, CA 92064

Jim McCook, W6YA, 1029 Passiflora Place, Leucadia, CA 92024

George Pugsley, W6ZZ, 1362 Via Rancho Prky, Escondido, CA 92029

Pat Bunsold, WA6MHZ, 14291 Rios Canyon Road #33, El Cajon, CA 92021

Charles L. Roy, WS6F, 6231 Lake Shore Drive, San Diego, CA 92110

West Gulf Division

NTX

Phil Clements, K5PC, 1313 Applegate Lane, Lewisville, TX 75069

Mike Krzystyniak, K9MK, 249 Bayne Road, Haslet, TX 76052

Joel Rubenstein, KA5W, 3601 Larkin Lane, Rowlett, TX 75088

Vernon Brunson, KD5HO, 2717 Whispering Trail, Arlington, TX 76013

Mickey Heimlich, N5AJW, 9207 Mill Hollow, Dallas, TX 75243

Orin Helvey, N5ORT, 969 Tupelo, Coppell, TX 75019

John Clifford, NK5K, 125 Alta Mesa Drive, Fort Worth, TX 76108

David Jaksa, W0VX, 626 Torrey Pines Lane, Garland, TX 75044

Thomas Little, W5GVP, 2109 Miriam Lane, Arlington, TX 76010

Ken Knudson, W5PLN, 5725 Wales Avenue, Fort Worth, TX 76133

Richard Pruitt, WB5TED, 2232 Jensen, Fort Worth, TX 76112

STX

Robert C. Walworth, AK5B, 3210 Chaparral Way, Spring, TX 77380

Don Hall, K5AQ, 4870 Enchanted Oaks Drive, College Station, TX 77845

William E. Barnes, K5GE, 434 Windcrest Drive, San Antonio, TX 78239

Charles R. Crutchfield, KC5SC, 3031 Rose Lane, La Marque, TX 77568

Linda G. Walworth, KE5TF, 3210 Chaparral Way, Spring, TX 77380

Evonne M. Lane, KF5MY, 1602 Chestnut Ridge, Kingwood, TX 77339

G.B. "Jim" Lane II, N5DC, 1602 Chestnut Ridge, Kingwood, TX 77339

Joseph A. Castorina, N5FJY, 4437 F. M. 646 N. Sante Fe, NM 77510

Harley L. Dillon, N5HB, 4902 Lambeth, San Antonio, TX 78228

Robert A. Wood, WB5CRG, 1013 Lewis Drive, Kemah, TX 77565

Wayne Wyatt, WB5QBV, 7114 Quail Landing, San Antonio, TX 78250

Randy Light, WC5Q, 2100 Bent Oak Street, College Station, TX 77840

WTX

David Zulawski, KA5TQF, 2808 Catnip Street, El Paso, TX 79925

(Courtesy ARRL)

RECENT CHANGES TO THE DXCC LIST

1993 Changes

Added Croatia 9A as of June 26, 1991
Added Slovenia S5 as of June 26, 1991
Added Bosnia-Hercegovina T9 as of October 15, 1991
Added Republic of Macedonia Z3 as of September 8, 1991
Note: Yugoslavia continues to hold the YT-YU,4N-4O ITU allocations.

OK Czechoslovakia Deleted December 31,1992
Added OK,OL Czech Republic as of January 1,1993
Added OM Slovakia as of January 1,1993
A15 Abu Ail Deleted as of March 31, 1991

January 1994

E3 Eritrea moved from deleted to active (5/24/91)

April 1994

Deleted ZS1 Penguin Islands effective March 1, 1994
Deleted ZS9 Walvis Bay effective March 1, 1994

July 1995

Added P5, North Korea (Democratic People's Republic of Korea), effective May 14, 1995.

(Note: At press time, no decision on the pending application for country status for Scarborough Shoal has been made)

ARRL DXCC COUNTRIES LIST
October 1995 Edition

The ARRL DXCC Countries List is also available (in full-page format) from the ARRL Publication Sales Department for $2. In addition to the Countries List, that version includes the DXCC rules, DXCC Award Application, ARRL DXCC Field Representatives, Incoming and Outgoing QSL Bureau information and more.

NOTES:

@ Indicates eligible for field checking.

* Indicates current list of countries for which QSLs may be forwarded by the ARRL membership outgoing QSL service.

#Indicates countries with which U.S. amateurs may legally handle third-party message traffic.

& Indicates third-party traffic permitted with special-events stations in the United Kingdom having the prefix GB only, with the exception that GB3 stations are not included in this agreement.

Prefix	Country	Cont.	Zone ITU	CQ
1AØ(1)	Sov. Mil. Order of Malta	EU	28	15
1S(1)	Spratly Is.	AS	50	26
@3A*	Monaco	EU	27	14
3B6,7*	Agalega & St. Brandon	AF	53	39
@3B8*	Mauritius	AF	53	39
@3B9*	Rodriguez I.	AF	53	39
@3C	Equatorial Guinea	AF	47	36
3CØ	Pagalu I.	AF	52	36
@3D2*	Fiji	OC	56	32
@3D2*	Conway Reef	OC	56	32
@3D2*	Rotuma I.	OC	56	32
@3DA#*	Swaziland	AF	57	38
3V	Tunisia	AF	37	33
3W,XV	Vietnam	AS	49	26
3X	Guinea	AF	46	35
3Y*	Bouvet	AF	67	38
3Y*	Peter I I.	AN	72	12
4J, 4K*	Azerbaijan	AS	29	21
4J1, R1MV	Malyj Vysotskij I.	EU	29	16
@4K2,UA1, R1FJ*	Franz Josef Land	EU	75	40
4L YU5(2)*	Georgia	AS	29	16
@4P-4S*	Sri Lanka	AS	41	22
@4U#*	ITUHQ	EU	28	14
@4U*	United Nations HQ	NA	08	05
@4X,4Z#*	Israel	AS	39	20
5A	Libya	AF	38	34
@5B*	Cyprus	AS	39	20
@5H-5I	Tanzania	AF	53	37
@5N-5O*	Nigeria	AF	46	35
5R-5S	Madagascar	AF	53	39
@5T(2)	Mauritania	AF	46	35
5U(3)	Niger	AF	46	35
@5V*	Togo	AF	46	35
@5W*	Western Samoa	OC	62	32
5X	Uganda	AF	48	37
@5Y-5Z*	Kenya	AF	48	37
@6V-6W(4)*	Senegal	AF	46	35
@6Y#*	Jamaica	NA	11	08
7O(5)	Yemen	AS	39	21
@7P*	Lesotho	AF	57	38
@7Q	Malawi	AF	53	37
@7T-7Y*	Algeria	AF	37	33
@8P*	Barbados	NA	11	08
@8Q	Maldives	AS/AF	41	22
@8R#*	Guyana	SA	12	09
9A,YU2(6)*	Croatia	EU	28	15
9G(7)#	Ghana	AF	46	35
@9H*	Malta	EU	28	15
@9I-9J*	Zambia	AF	53	36
@9K*	Kuwait	AS	39	21
@9L#*	Sierra Leone	AF	46	35
@9M2,4(8)*	West Malaysia	AS	54	28
@9M6,8(8)*	East Malaysia	OC	54	28
@9N	Nepal	AS	42	22
@9Q-9T	Zaire	AF	52	36
9U(9)	Burundi	AF	52	36
@9V(10)	Singapore	AS	54	28
@9X(11)	Rwanda	AF	52	36
@9Y-9Z#*	Trinidad & Tobago	SA	11	09
@A2*	Botswana	AF	57	38
@A3*	Tonga	OC	62	32
@A4*	Oman	AS	39	21
A5	Bhutan	AS	41	22
@A6	United Arab Emirates	AS	39	21
@A7*	Qatar	AS	39	21
@A9*	Bahrain	AS	39	21
@AP-AS*	Pakistan	AS	41	21
@BV*	Taiwan	AS	44	24
@BY,BT*	China	AS	(A)	23,24
@C2*	Nauru	OC	65	31

Prefix	Entity	Cont.		
@C3*	Andorra	EU	27	14
@C5#*	The Gambia	AF	46	35
@C6*	Bahamas	NA	11	08
C8-9	Mozambique	AF	53	37
@CA-CE#*	Chile	SA	14,16	12
@CEØ#*	Easter I.	SA	63	12
@CEØ#*	Juan Fernandez Is.	SA	14	12
CEØ#*	San Felix & San Ambrosio	SA	14	12
@CE9/KC4^*	Antarctica	AN	(B)	(C)
@CM,CO#*	Cuba	NA	11	08
@CN*	Morocco	AF	37	33
@CP#*	Bolivia	SA	12,14	10
@CT*	Portugal	EU	37	14
@CT3*	Madeira Is.	AF	36	33
@CU*	Azores	EU	36	14
@CV-CX#*	Uruguay	SA	14	13
@CYØ*	Sable I.	NA	09	05
@CY9*	St. Paul I.	NA	09	05
D2-3	Angola	AF	52	36
@D4*	Cape Verde	AF	46	35
@D6#(11)*	Comoros	AF	53	39
@DA-DL, Y2-Y9(12)*	Fed. Rep. of Germany	EU	28	14
@DU-DZ#*	Philippines	OC	50	27
E3(13)	Eritrea	AF	48	37
@EA-EH*	Spain	EU	37	14
@EA6-EH6*	Balearic Is.	EU	37	14
@EA8-EH8*	Canary Is.	AF	36	33
@EA9-EH9*	Ceuta & Melilla	AF	37	33
@EI-EJ*	Ireland	EU	27	14
EK	Aremenia	AS	29	21
@EL#*	Liberia	AF	46	35
EP-EQ	Iran	AS	40	21
ER*	Moldavia	EU	29	16
@ES*	Estonia	EU	29	15
ET	Ethiopia	AF	48	37
EU,EV,EW*	Belarus	EU	29	16
EX*	Kyrgystan	AS	30,31	17
EY*	Tajikistan	AS	30	17
EZ	Turkmenistan	AS	30	17
@F*	France	EU	27	14
@FG*	Guadeloupe	NA	11	08
@FJ,FS(1)*	Saint Martin	NA	11	08
@FH(11)*	Mayotte	AF	53	39
@FK*	New Caledonia	OC	56	32
@FM*	Martinique	NA	11	08
FO*	Clipperton I.	NA	10	07
@FO*	French Polynesia	OC	63	32
@FP*	St. Pierre & Miquelon	NA	09	05
FR/G(14)*	Glorioso Is.	AF	53	39
FR/J,E(14)*	Juande Nova, Europa	AF	53	39
@FR*	Reunion	AF	53	39
FR/T*	Tromelin I.	AF	53	39
@FT8W*	Crozet I.	AF	68	39
@FT8X*	Kerguelen Is.	AF	68	39
@FT8Z*	Amsterdam & St. Paul Is.	AF	68	39
@FW*	Wallis & Futuna Is.	OC	62	32
@FY*	French Guiana	SA	12	09
@G,GX*	England	EU	27	14
@GD,GT*	Isle of Man	EU	27	14
@GI,GN*	Northern Ireland	EU	27	14
@GJ,GH*	Jersey	EU	27	14
@GM,GS*	Scotland	EU	27	14
@GU,GP*	Guernsey	EU	27	14
@GW,GC*	Wales	EU	27	14
H4*	Solomon Is.	OC	51	28
@HA,HG*	Hungary	EU	28	15
@HB*	Switzerland	EU	28	14
@HB*	Liechtenstein	EU	28	14
@HC-HD#*	Ecuador	SA	12	10
@HC8-HD8#*	Galapagos Is.	SA	12	10
@HH#*	Haiti	NA	11	08
@HI#*	Dominican Republic	NA	11	08
@HJ-HK#*	Colombia	SA	12	09
@HKØ#*	Malpelo I.	SA	12	09
@HKØ#*	San Andres & Providencia	NA	11	07
@HL*	South Korea	AS	44	25
@HO-HP#*	Panama	NA	11	07
@HQ-HR#*	Honduras	NA	11	07
@HS*	Thailand	AS	49	26
@HV*	Vatican	EU	28	15
@HZ*	Saudi Arabia	AS	39	21
@I*	Italy	EU	28,37	15,33
@IS,IM*	Sardinia	EU	28	15
@J2*	Djibouti	AF	48	37
@J3#*	Grenada	NA	11	08
@J5	Guinea-Bissau	AF	46	35
@J6#*	St. Lucia	NA	11	08
@J7#*	Dominica	NA	11	08
@J8#*	St. Vincent	NA	11	08
@JA-JS*	Japan	AS	45	25
@JD1(15)*	Minami Torishima	OC	90	27
@JD1(16)*	Ogasawara	AS	45	27
@JT-JV*	Mongolia	AS	32,33	23
@JW*	Svalbard	EU	18	40
@JX*	Jan Mayen	EU	18	40
@JY#*	Jordan	AS	39	20
@K,W,N, AA-AK#	United States of America	NA	6,7,83,4,5	
@KC6(17)	Belau (W. Caroline Is.)	OC	64	27
@KG4#*	Guantanamo Bay	NA	11	08
@KHØ#	Mariana Is.	OC	64	27
KH1#	Baker & Howland Is.	OC	61	31
@KH2#*	Guam	OC	64	27
@KH3#*	Johnston I.	OC	61	31
@KH4#*	Midway I.	OC	61	31
KH5#	Palmyra & Jarvis Is.	OC	61,62	31
KH5K#	Kingman Reef	OC	61	31
@KH6#*	Hawaii	OC	61	31
@KH7#	Kure I.	OC	61	31
@KH8#	American Samoa	OC	62	32
@KH9#	Wake I.	OC	65	31
@KL7#*	Alaska	NA	1,2	1
KP1#	Navassa I.	NA	11	08
@KP2#*	Virgin Is.	NA	11	08
@KP4#*	Puerto Rico	NA	11	08
KP5(18)#	Desecheo I.	NA	11	08
@LA-LN*	Norway	EU	18	14
@LO-LW#*	Argentina	SA	14,16	13
@LX*	Luxembourg	EU	27	14
@LY,UP*	Lithuania	EU	29	15
@LZ*	Bulgaria	EU	28	20
@OA-OC#*	Peru	SA	12	10
@OD	Lebanon	AS	39	20
@OE*	Austria	EU	28	15
@OF-OI*	Finland	EU	18	15
@OHØ*	Aland Is.	EU	18	15
@OJØ*	Market Reef	EU	18	15
OK-OL(19)*	Czech Republic	EU	28	15
OM(19)*	Slovak Republic	EU	28	15
@ON-OT*	Belgium	EU	27	14
@OX*	Greenland	NA	5,7,5	40
@OY*	Faroe Is.	EU	18	14
@OZ*	Denmark	EU	18	14
@P2(20)*	Papua New Guinea	OC	51	28
@P4(21)*	Aruba	SA	11	09
P5(38)	North Korea	AS	44	25
@PA-PI*	Netherlands	EU	27	14
@PJ2,4,9*	Bonaire, Curacao (Neth. Antilles)	SA	11	09
@PJ5-8*	St. Maarten, Saba, St. Eustatius	NA	11	08
@PP-PY#*	Brazil	SA	(D)	11
@PPØ-PYØ#	Fernando de Noronha	SA	13	11
PPØ-PYØ#*	St. Peter & St. Paul Rocks	SA	13	11
@PPØ-PYØ#*	Trindade & Martim Vaz Is.	SA	15	11
@PZ*	Suriname	SA	12	09
SØ(1),(22)*	Western Sahara	AF	46	33
S2	Bangladesh	AS	41	22
S5,YU3(7)*	Slovenia	EU	28	15
@S7	Seychelles	AF	53	39

Prefix	Entity	Cont.		
@S9*	Sao Tome & Principe	AF	47	36
@SA-SM*	Sweden	EU	18	14
@SN-SR*	Poland	EU	28	15
@ST*	Sudan	AF	48	34
@STØ*	Southern Sudan	AF	47,48	34
@SU*	Egypt	AF	38	34
@SV-SZ*	Greece	EU	28	20
SV/A*	Mount Athos	EU	28	20
@SV5*	Dodecanese	EU	28	20
@SV9*	Crete	EU	28	20
@T2(23)	Tuvalu	OC	65	31
@T3Ø	W. Kiribati (Gilbert Is.)	OC	65	31
@T31	C. Kiribati (Brit. Phoenix Is.)	OC	62	31
@T32	E. Kiribati (Line Is.)	OC	61,63	31
@T33	Banaba I.(Ocean I.)	OC	65	31
T5	Somalia	AF	48	37
@T7*	San Marino	EU	28	15
T9,4N4 4O4,YU4(24)*	Bosnia-Herzegovina	EU	28	15
@TA-TC*	Turkey	EU/AS	39	20
@TF*	Iceland	EU	17	40
@TG,TD#*	Guatemala	NA	11	07
@TI,TE#*	Costa Rica	NA	11	07
@TI9#*	Cocos I.	NA	11	07
@TJ	Cameroon	AF	47	36
@TK*	Corsica	EU	28	15
@TL(25)	Central Africa	AF	47	36
TN(26)	Congo	AF	52	36
@TR(27)*	Gabon	AF	52	36
TT(28)	Chad	AF	47	36
@TU(29)*	Cote d'Ivoire	AF	46	35
@TY(30)	Benin	AF	46	35
@TZ(31)	Mali	AF	46	35
UA1,3,4,6*	European Russia	EU	(E)	16
UA2*	Kaliningrad	EU	29	15
UA8,9,Ø*	Asiatic Russia	AS	(F)	(G)
UB,UT,UY*	Ukraine	EU	29	16
UC*	Belarus	EU	29	16
UD*	Azerbaijan	AS	29	21
UF*	Georgia	AS	29	21
UG*	Armenia	AS	29	21
UH*	Turkmenistan	AS	30	17
UI*	Uzbekistan	AS	30	17
UJ*	Tajikistan	AS	30	17
UL*	Kazakhstan	AS	29-31	17
UM*	Kyrgyzstan	AS	30,31	17
UO*	Moldova	EU	29	16
@V2#*	Antigua & Barbuda	NA	11	08
@V3#*	Belize	NA	11	07
@V4(32)#	St. Kitts & Nevis	NA	11	08
@V5*	Namibia	AF	57	38
@V6(33)	Micronesia (E. Caroline Is.)	OC	65	27
@V7#*	Marshall Is.	OC	65	31
@V8*	Brunei	OC	54	28
@VE,VO,VY#*	Canada	NA	(H)	1-5
@VK#*	Australia	OC	(I)	29,30
VKØ#*	Heard I.	AF	68	39
@VKØ#*	Macquarie I.	OC	60	30
@VK9C#*	Cocos-Keeling Is.	OC	54	29
@VK9L#*	Lord Howe I.	OC	60	30
@VK9M#*	Mellish Reef	OC	56	30
@VK9N#*	Norfolk I.	OC	60	32
@VK9W#*	Willis I.	OC	55	30
@VK9X#*	Christmas I.	OC	54	29
@VP2E(32)	Anguilla	NA	11	08
@VP2M(32)	Montserrat	NA	11	08
@VP2V(32)*	British Virgin Is.	NA	11	08
@VP5*	Turks & Caicos Is.	NA	11	08
@VP8*	Falkland Is.	SA	16	13
@VP8,LU*	South Georgia I.	SA	73	13
@VP8,LU*	South Orkney Is.	SA	73	13
@VP8,LU*	South Sandwich Is.	SA	73	13
@VP8,LU, CE9,HFØ,4K1*	South Shetland Is.	SA	73	13
@VP9*	Bermuda	NA	11	05
@VQ9*	Chagos Is.	AF	41	39
@VR6#	Pitcairn I.	OC	63	32
@VS6,VR2*	Hong Kong	AS	44	24
@VU*	India	AS	41	22
@VU*	Andaman & Nicobar Is.	AS	49	26
@VU*	Laccadive Is.	AS	41	22
@XA-XI#*	Mexico	NA	10	06
XA4-XI4#*	Revilla Gigedo	NA	10	06
@XT(34)	Burkina Faso	AF	46	35
XU	Cambodia	AS	49	26
XW	Laos	AS	49	26
@XX9	Macao	AS	44	24
XY-XZ	Myanmar	AS	49	26
YA	Afghanistan	AS	40	21
@YB-YH(33)*	Indonesia	OC	51,54	28
YI	Iraq	AS	39	21
@YJ*	Vanuatu	OC	56	32
YK*	Syria	AS	39	20
@YL,UQ*	Latvia	EU	29	15
@YN#*	Nicaragua	NA	11	07
@YO-YR*	Romania	EU	28	20
@YS#*	El Salvador	NA	11	07
@YT-YU,YZ*	Yugoslavia	EU	28	15
@YV-YY#*	Venezuela	SA	12	09
YVØ#*	Aves I.	NA	11	08
@Z2*	Zimbabwe	AF	53	38
Z3,4N5, YU5(36)*	Macedonia	EU	28	15
ZA*	Albania	EU	28	15
@ZB2*	Gibraltar	EU	37	14
@ZC4(37)*	UK Sov. Base Areas on Cyprus	AS	39	20
@ZD7	St. Helena	AF	66	36
@ZD8*	Ascension I.	AF	66	36
@ZD9	Tristan da Cunha & Gough I.	AF	66	38
@ZF*	Cayman Is.	NA	11	08
ZK1*	N. Cook Is.	OC	62	32
@ZK1*	S. Cook Is.	OC	62	32
@ZK2*	Niue	OC	62	32
ZK3	Tokelau Is.	OC	62	31
@ZL-ZM*	New Zealand	OC	60	32
@ZL7*	Chatham Is.	OC	60	32
ZL8*	Kermadec Is.	OC	60	32
ZL9*	Auckland & Campbell Is.	OC	60	32
@ZP#*	Paraguay	SA	14	11
@ZR-ZU*	South Africa	AF	57	38
@ZS8*	Prince Edward & Marion Is.	AF	57	38

NOTES:
1. Unofficial prefix.
2. (5T) Only contacts made June 20, 1960, and after, count for this country.
3. (5U) Only contacts made August 3, 1960, and after, count for this country.
4. (6W) Only contact made June 20, 1960, and after, count for this country.
5. (70) Only contacts made May 22, 1990, and after, count for this country.
6. (9A, YU2, S5, YU3) Only contacts made June 26, 1991, and after, count for this country.
7. (9G) Only contacts made March 5, 1957, and after, count for this country.
8. (9M2, 4, 6, 8) Only contacts made September 16, 1963, and after count for this country.
9. (9U, 9X) Only contacts made July 1, 1962, and after, count for this country.
10. (9V) Contacts made from September 16, 1963 to August 8, 1965, count for West Malaysia.
11. (D6, FH8) Only contacts made July 15, 1975, and after, count for this country.
12. (DA-DL, Y2-Y9) Only contacts made with DA-DL stations September 17, 1973, and after, and contacts made with Y2-Y9 stations October 3, 1990 and after, count for this country.
13. (E3) Only contacts made November 14, 1962, and before, or May 24, 1991, and after, count for this country.

14. (FR) Only contacts made June 25, 1960, and after, count for this country.
15. (JD) Formerly Marcus Island.
16. (KC6) Formerly Bonin and Volcano Islands
17. (KC6) Includes Yap Islands December 31, 1980, and before.
18. (KP5, KP4) Only contacts made March 1, 1979, and after, count for this country.
19. (OK-OL, OM) Only contacts made January 1, 1993, and after, count for this country.
20. (P2) Only contacts made September 16, 1975, and after, count for this country.
21. (P4) Only contacts made January 1, 1986, and after, count for this country.
22. (SØ) Contacts with Rio de Oro (Spanish Sahara), EA9, also count for this country.
23. (T2) Only contacts made January 1, 1976, and after, count for this country.
24. (T9, 4N4, 4O4, YU4) Only contacts made October 15, 1991, and after, count for this country.
25. (TL) Only contacts made August 13, 1960, and after, count for this country.
26. (TN) Only contacts made August 15, 1960, and after, count for this country.
27. (TR) Only contacts made August 17, 1960, and after, count for this country.
28. (TT) Only contacts made August 11, 1960, and after, count for this country.
29. (TU) Only contacts made August 7, 1960, and after, count for this country.
30. (TY) Only contacts made August 1, 1960, and after, count for this country.
31. (TZ) Only contacts made June 20, 1960, and after, count for this country.
32. (V4, VP2) For DXCC credit for contacts made May 31, 1958, and before, see page 97, June 1958 *QST*.
33. (V6, KC6) Includes Yap Islands January 1, 1981, and after.
34. (XT) Only contacts made August 16, 1960, and after, count for this country.
35. (YB) Only contacts made May 1, 1963, and after, count for this country.
36. (Z3,4N5,YU5) Only contacts made September 8, 1991, and after, count for this country.
37. (ZC4) Only contacts made August 16, 1960, and after, count for this country.
38. (P5) Only contacts made May 14, 1995, and after, count for this country.
^ Also 3Y, 8J1, ATØ, DPØ, FT8Y, LU, OR4, VKØ, R1AW, VP8, ZL5, ZS1, ZXØ, etc. QSL via country under whose auspices the particular station is operating. The availability of a third- party traffic agreement and a QSL Bureau applies to the country under whose auspices the particular station is operating.
Zone notes can be found with Prefix Cross References.

Deleted Countries (Total: 58)

Credit for any of these countries can be given if the date of contact in question agrees with the date(s) shown in the corresponding footnote.

Prefix	Country	Cont.	Zone	
			ITU	CQ
(2)	Blenheim Reef	AF	41	39
(3)	Geysey Reef	AF	53	39
(4)	Abu Ail Is.	AS	39	21
1M(1), (5)	Minerva Reef	OC	62	32
4W(6)	Yemen Arab Rep.	AS	39	21
7J1(7)	Okino Tori-shima	AS	45	27
8Z4(8)	Saudi Arabia/ Iraq Neut. Zone	AS	39	21
8Z5,9K3(9)	Kuwait/Saudi Arabia Neut. Zone	AS	39	21
9S4(10)	Saar	EU	28	14
9U5(11)	Ruanda-Urundi	AF	52	36
AC3(1),(12)	Sikkim	AS	41	22
AC4(1),(13)	Tibet	AS	41	23
C9(14)	Manchuria	AS	33	24
CN2(15)	Tangier	AF	37	33

CR8(16)	Damao, Diu	AS	41	22
CR8(16)	Goa	AS	41	22
CR8, CRØ(17)	Portuguese Timor	OC	54	28
DA-DM(18)	Germany	EU	28	14
DM,Y2-9(19)	German Dem. Rep.	EU	28	14
EA9(20)	Ifni	AF	37	33
FF(21)	French West Africa	AF	46	35
FH,FB8(22)	Comoros	AF	53	39
FI8(23)	French Indo-China	AS	49	26
FN8(24)	French India	AS	41	22
FQ8(25)	Fr. Equatorial Africa	AF	47,52	36
HKØ(26)	Bajo Nuevo	NA	11	08
HKØ,KP3, KS4(26)	Serrana Bank & Roncador Cay	NA	11	07
I1(27)	Trieste	EU	28	15
I5(28)	Italian Somaliland	AF	48	37
JZ0(29)	Netherlands N. Guinea	OC	51	28
KR6,8,JR6, KA6(30)	Okinawa (Ryukyu Is.)	AS	45	25
KS4(31)	Swan Is.	NA	11	07
KZ5(32)	Canal Zone	NA	11	07
OK-OM(33)	Checkoslovakia	EU	28	15
P2,VK9(34)	Papua Territory	OC	51	28
P2,VK9(34)	Terr. New Guinea	OC	51	28
PK1-3(35)	Java	OC	54	28
PK4(35)	Sumatra	OC	54	28
PK5(35)	Netherlands Borneo	OC	54	28
PK6(35)	Celebe & Molucca Is.	OC	54	28
UN1(36)	Karelo-Finnish Rep.	EU	19	16
VO(37)	Newfoundland, Labrador	NA	09	02,05
VQ1,5H1(38)	Zanzibar	AF	53	37
VQ6(39)	British Somaliland	AF	48	37
VQ9(40)	Aldabra	AF	53	39
VQ9(40)	Desroches	AF	53	39
VQ9(40)	Farquhar	AF	53	39
VS2,9M2(41)	Malaya	AS	54	28
VS4(41)	Sarawak	OC	54	28
VS9A,P,S(42)	People's Dem. Rep of Yemen	AS	39	21
VS9H(43)	Kuria Muria I.	AS	39	21
VS9K(44)	Kamaran Is.	AS	39	21
ZC5(42)	British North Borneo	OC	54	28
ZC6,4X1(45)	Palestine	AS	39	20
ZD4(46)	Gold Coast, Togoland	AF	46	35
ZSØ,1*	Penguin Is.	AF	57	38
ZS9(37)*	Walvis Bay	AF	57	38

NOTES:

1. Unofficial prefix.
2 . (Blenheim Reef) Only contacts made from May 4, 1967, to June 30, 1975, count for this country. Contacts made July 1, 1975, and after, count as Chagos (VQ9).
3. (Geysey Reef) Only contacts made from May 4, 1967, to February 28, 1978, count for this country.
4. (Abu Ail Is.) Only contacts made March 30, 1991, and before, count for this country.
5. (1M) Only contacts made from July 15, 1972, and before, count for this country. Contacts made July 16, 1972, and after, count as Tonga (A3).
6. (4W) Only contacts made before May 21, 1990, and before, count for this country.
7. (7J1) Only contacts made May 30, 1976, to November 30, 1980, count for this country. Contacts made December 1, 1980, and after country as Ogasawara (JD1).
8. (8Z4) Only contacts made December 25, 1981, and before, count for this country.
9. (8Z5,9K3) Only contacts made December 14, 1969, and before, count for this country.
10. (9S4) Only contacts made March 31, 1957, and before, count for this country.
11. (9U5) Only contacts made from July 1, 1960, to June 30, 1962, count for this country. Contact made July 1, 1962, and after, count as Burundi (9U) or Rwanda (9X).

12. (AC3) Only contacts made April 30, 1975, and before, count for this country. Contacts made May 1, 1975, and after, count as India (VU).

13. (AC4) Only contacts made May 30, 1974, and before, count for this country. Contact made May 31, 1974, and after, count as China (BY).

14. (C9) Only contacts made September 15, 1963, and before, count for this country. Contact made September 16, 1963, and after, count as China (BY).

15. (CN2) Only contacts made June 30, 1960, and before, count for this country. Contact made July 1, 1960, and after, count as Morocco (CN).

16. (CR8)Only contacts made December 31, 1961, and before, count for this country.

17. (CR8, CRØ) Only contacts made September 14, 1976, and before, count for this country.

18. (DA-DM) Only contacts made September 16, 1973, and before, count for this country. Contacts made September 17, 1973, and after, count as either FRG (DA-DL) or GDR (Y2-Y9).

19. (DM, Y2-Y9) Only contacts made from September 17, 1973, to October 2, 1990 count for this country. On October 3, 1990 the GDR became part of the FRG.

20. (EA9) Only contacts made May 13, 1969, and before, count for this country.

21. (FF) Only contacts made August 6, 1960, and before, count for this country.

22. (FH, FB8) Only contacts made July 5, 1975, and before, count for this country. Contacts made July 6, 1975, and after, count as Comoros (D6) or Mayotte (FH).

23. (FI8) Only contacts made December 20, 1950, and before, count for this country.

24. (FN8) Only contacts made October 31, 1954, and before, count for this country.

25. (FQ8) Only contacts made August 16, 1960, and before count for this country.

26. (HK, KP3, KS4) Only contacts made September 16, 1981, and before, count for this country. Contacts made September 17, 1981, and after, count as San Andres (HK).

27. (I1) Only contacts made March 31, 1957, and before, count for this country.

28. (I5) Only contacts made June 30, 1960, and before, count for this country.

29. (JZ) Only contacts made April 30, 1963, and before, count for this country.

30. (KR6, 8, JR6, KA6) Only contacts made May 14, 1972, and before, count for this country. Contacts made May 15, 1972, and after, count as Japan (JA).

31. (KS4) Only contacts made August 31, 1972, and before, count for this country. Contacts made September 1, 1972, and after, count as Honduras (HR).

32. (KZ5) Only contacts made September 30, 1979, and before, count for this country.

33. (OK-OM) Only contacts made December 31, 1992, and before, count for this country.

34. (P2, VK9) Only contacts made September 15, 1975, and before, count for this country. Contacts made September 16, 1975, and after, count as Papua new Guinea (P2).

35. (PK1-6) Only contacts made April 30, 1963, and before, count for this country. Contact made May 1, 1963, and after, count as Indonesia.

36. (UN1) Only contacts made June 30, 1960, and before, count for this country. Contacts made July 1, 1960, and after count as European RSFSR (UA).

37. (VO) Only contacts made March 31, 1949, and before, count for this country. Contacts made April 1, 1949, and after, count as Canada (VE).

38. (VQ1, 5H1) Only contacts made May 31, 1974, and before, count for this country. Contacts made June 1, 1974, and after, count as Tanzania (5H).

39. (VQ6) Only contacts made June 30, 1960, and before, count for this country.

40. (VQ9) Only contacts made June 28, 1976, and before, count for this country. Contacts made June 29, 1976, and after, count as Seychelles (S7).

41. (VS2, VS4, ZC5, 9M2) Only contacts made September 15, 1963, and before, count for this country. Contacts made September 16, 1963, and after, count as West malaysia (9M2) or East Malaysia (9M6, 8).

42. (VS9A, P, S) Only contacts made before May 22, 1990, and before, count for this country.

43. (VS9H) Only contacts made November 29, 1967, and before, count for this country.

44. (VS9K) Only contacts made on March 10, 1982, and before, count for this country.

45. (ZC6, 4X1) Only contacts made June 30, 1968, and before, count for this country. Contacts made July 1, 1968, and after, count as Israel (4X).

46. (ZD4) Only contacts made March 5, 1957, and before, count for this country.

47. (ZSØ,1) Only contacts made February 29, 1994, and before, count for this country.

48. (ZS9) Only contacts made from September 1, 1977 to February 28, 1994, count for this country.

Courtesy of the ARRL

THE 1994 *DX MAGAZINE* 100 MOST-WANTED COUNTRIES SURVEY

Each year since the late 1960s, the *DX Magazine* had surveyed its subscribers to determine which DXCC countries are most needed. This information is used by DXpedition organizers to plan major operations. The 1994 survey results are shown below:

95 Rank	Country		% Need	94 Rank	Change	80 Rank	Change
1	Bhutan	A5	65.5	1		55	54
2	Andaman	VU4	63.4	2		10	8
3	Heard Island	VKØ	60.8	4	1	62	59
4	Yemen	7O	55.5	5	1		4
5	Libya	5A	55.0	3	2	22	17
6	Macquarie Island	VKØ	48.7	10	4	52	46
7	Tromelin	FR/T	47.6	7		38	31
8	Laccadive Island	VU7	47.3	9	1	55	47
9	Mount Athos	SV/A	46.7	12	3	45	36
10	Burma	XZ	45.8	8	2	3	7
11	Bouvet	3Y	42.4	14	3	12	1
12	Eritrea	E3	40.3	15	3		12
13	Glorioso Island	FR/G	40.3	13		15	2
14	Kermadec I.	ZL8	40.0	11	3	44	30
15	Tunisia	3V	39.1	6	9	71	56
16	AucklandCampbell Islands	ZL9	39.1	16		58	42
17	Amsterdam	FT/Z	38.6	17			17
18	Agalega & St. Brandon Is	3B6	37.0	21	3		18
19	Laos	XW	34.9	18	1	26	7
20	Juan de Nova	FR/J	34.5	20		14	6
21	Iran	EP	34.3	23	2		21
22	Marion Island	ZS8	34.1	22		74	52
23	Crozet Island	FT/W	33.7	24	1	9	14

24	Chad	TT	32.8	33	9	33	9
25	Annobon Island	3CØ	32.4	26	1	39	14
26	Malpelo Island	HKØ	31.2	39	13	18	8
27	Kerguelen	FT/X	31.1	31	4	80	53
28	Spratly Island	1S	30.0	30	2	53	25
29	CocosKeeling Island	VK9C	29.8	28	1	52	23
30	Maldives	8Q	28.2	42	12		30
31	Afghanistan	YA	28.2	40	9	19	12
32	South Sandwich Islands	VP8S	28.1	34	2	48	16
33	Christmas Island	VK9X	28.0	37	4		33
34	SMOM	1AØ	27.5	17	17		34
35	Vietnam	3W	26.9	46	11	17	18
36	Willis Island	VK9W	26.6	52	16	37	1
37	Cambodia	XU	26.6	40	3	81	44
38	South Sudan	STØ	26.5	44	6	57	19
39	Nepal	9N	26.4	50	11		39
40	United Arab Emirates	A6	26.1	54	14	30	10
41	St. Peter & Paul Rocks	PYØS	25.7	39	2	36	5
42	Syria	YK	25.7	42			42
43	Iraq	YI	25.6	34	9	41	2
44	Burundi	9U	25.2	21	23	16	28
45	Bangladesh	S2	24.5	51	6	49	4
46	Rodriguez I.	3B9	24.3	58	12	82	36
47	Kingman Reef	KH5K	22.8	47			47
48	Congo	TN	22.8	22	26	28	20
49	Monaco	3A	22.3	58	9	79	30
50	Tristan da CunhaGough Is	ZD9	21.7	33	17	35	15
51	UK Sovereign Bases	ZC4	21.6	64	13		51
52	Benin	TY	21.5	55	3	34	18
53	Madagascar	5R	21.1	48	5	31	22
54	Equatorial Guinea	3C	21.0	62	8		54
55	Sudan	ST	20.7	52	3	85	30
56	Mellish Reef	VK9M	20.5	61	5	50	6
57	Ethiopia	ET	20.2	49	8		57
58	Comoros Is.	D6	20.1	57	1	43	15
59	Western Sahara	SØ	19.9	51	8		59
60	Minami Torishima	JD1	19.3	63	3	93	33
61	Trinidade	PYØT	19.1	60	1	62	1
62	South Georgia Island	VP8G	19.1	24	38	53	9
63	Somalia	T5	19.0	72	9	21	42
64	Midway I.	KH4	18.6	80	16		64
65	San Felix	CEØX	18.5		65		65
66	Bahrain	A9	18.5	59	7	65	1
67	Jan Mayen	JX	18.5	66	1	77	10
68	Palmyra Island	KH5	18.4	71	3		68
69	Conway Reef	3D2	18.1	25	44		69
70	Malyj Vystoskij Island	4J	17.6	83	13		70
71	Banaba (Ocean Island)	T33	17.1	69	2		71
72	South Orkney Islands	VP8O	17.0	73	1		72
73	Belau	KC6	16.3	74	1	83	10
74	Sable I.	CY	16.1	84	10		74
75	Central Kiribati	T31	16.1	67	8		75
76	Baker, Howland Islands	KH1	15.6	87	11		76
77	Macao	XX	15.4	76	1		77
78	Kure Island	KH7K	15.2		78		78
79	Market Reef	OJØ	15.1	88	9		79
80	Qatar	A7	15.0	75	5		80
81	Juan Fernandez	CEØZ	15.0	81		40	41
82	Burkino Faso	XT	15.0	65	17	88	6
83	Aves Island	YVØ	15.0	90	7	70	13
84	Cocos I.	TI9	14.4	79	5		84
85	Uganda	5X	14.0	89	4	23	62
86	Togo	5V	13.6	70	16	45	41
87	Clipperton Island	FO/X	13.3	93	6	64	23
88	Wake I.	KH9	13.2	86	2		88
89	Cameroon	TJ	13.2	68	21	68	21
90	Peter 1 Island	3Y1	13.1	94	4		90
91	Egypt	SU	12.7	82	9		91
92	Nauru	C2	12.2	85	7		92
93	Guinea	3X	11.9	97	4	13	80
94	Tokelau Island	ZK3	11.9	77	17	29	65
95	Djibouti	J2	11.8	91	4		95
96	GuineaBissau	J5	11.7	96		61	35
97	Rwanda	9X	11.5	78	19		97
98	Mauritania	5T	10.9	92	6		98
99	St. Paul I.	CY9	8.6	98	1		99
100	North Cook Is.	ZK1	7.5	99	1		100

THE 1994 OHIO/PENN DX CLUSTER/NORTHERN OHIO DX ASSOCIATION DX SURVEY

The following is the percentage of votes from different areas received (percentages are rounded off):

U.S. Call Areas   **Foreign**
W1 - 4%   Canada - 3%
W2 - 3%   (E/W)Europe - 29%
W3 - 5%   Russia/Asia - 2%
W4 - 7%   VK/Pacific - 2%
W5 - 8%
W6 - 6%
W7 - 3%
W8 - 6%
W9 - 2%
WØ - 11%

1994 DXer of the Year - Rudi, DK7PE

Rudi, DK7PE, not only received the most amount of votes for DXer of the Year, but there were many who expressed their comments about how Rudi's skills on CW had supplied them with a new country on that mode. Rudi was mostly noted for his outstanding operation from TNØCW in 1994. There were 33 DXers nominated in all. 29% voted for DK7PE. Out of the 33 DXers that

were mentioned, here are some that should receive "Honorable Mention" for their efforts and hard work:

—Rodger/G3SXW, Jim/VK9NS and each member of the 3YØPI team.

1994 DXpedition of the Year (by a group) - 3YØPI

There were 6 nominated. The 3YØPI Peter I Island DXpedition was voted, by far, the best operation of 1994 with 89% of the vote. This DXpedition netted around 70,000 QSOs and will go down in DX history as one of the best DXpeditions. Here are a few DXpeditions that should receive "Honorable Mention" for their efforts and hard work: —1AØKM and YKØA.

1994 DXpedition of the Year (by an individual) - Rudi, DK7PE/TNØCW

Rudi's DXpedition to the Congo has netted him this second honor. There were 28 individuals nominated in all with Rudi receiving 38% of the vote. Here are some of the individuals that should receive "Honorable Mention" for their efforts and hard work:

—Roger-G3SXW/XX9/others and Jacques-FR5ZU/G.

THE OPDX/NODXA 1994 MOST-WANTED COUNTRIES (200 COUNTRIES WERE SUBMITTED)

Last Year's Rating (xx)	Country	Needed		
		Any mode	CW	SSB
1(4)	5A Libya	96%	40%	43%
2(5)	VU4 Andaman	91%	39%	39%
3(2)	VKØ/H Heard Is.	90%	36%	43%
4(3)	A5 Bhutan	84%	44%	36%
5(10)	ZL8 Kermadec Is.	64%	46%	38%
6(7)	7O Yemen	62%	51%	32%
7(6)	VKØ/M Macquarie Is.	61%	51%	37%
8(8)	3V Tunisia	52%	44%	41%
9(18)	FT8Z Amsterdam/StPaul	49%	43%	37%
10(11)	SV/A Mount Athos	48%	48%	40%
11(13)	3Y/B Bouvet	47%	41%	38%
12(9)	FR/T Tromelin	40%	53%	31%
13(**)	ZL9 Auckland/Campbell	39%	41%	43%
14(**)	3D2/CR Conway Reef	38%	39%	47%
15(12)	XZ Burma (Myanmar)	31%	33%	44%
16(17)	FR/G Glorioso	30%	46%	27%
17(**)	HKØ/M Malpelo Is.	29%	42%	42%
18(14)	VU7 Laccadive Is.	26%	41%	28%
19(**)	FT8/X Kerguelen Is.	25%	55%	23%
20(**)	FT8/W Crozet	24%	48%	31%

(**) Means new to this year's list.

Editor's Analyzation: To start, the survey was taken between December of 1994 and January of 1995. As you can see the list has changed a bit since last year. Five of the countries that were on last year's list are no longer in the "Top 20". These are 1AØ, 3YØ/P, FR/J, TN and XW.

All these countries were fairly active in 1994, except XW. There may be a possible reason why 5A jumped to the top of the list. The reasoning may be that individuals who voted last year might have felt Romeo's 5AØRR DXpedition would count for DXCC credit and chose not to mark it needed. This year I was pleased to see that more DXers are beginning to realize that P5 (North Korea) should be counted in the list.

Even though P5 will not be added to the DXCC Countries List until after the first accredited operation, DXers should show the need for it to become active. P5 ranked 22nd with 22% needing it (compared to 15% in 1993). The only predicted changes I see (so far) for 1995's list may be Conway Reef (3D2/CR) dropping from the "Top 20". Also, there is a possibility Tunisia (3V) may move down the need list. This is due to the promising activity by individuals and groups in trying to get this country active again on the air.

In closing, I want to thank everyone who took the time to participate in the survey. The InterNet balloting had a great response. I also want to thank those who listed their needs by prefixes and in order. You made my job alot easier. 73 de Tedd KB8NW

(Contributed by Tedd Mirgliotta, KB8NW, editor, OPDX Bulletin. To contribute DX info to the OPDX Bulletin, call BARF-80 BBS online at 216-237-8208 14400/9600/2400/1200/300 and leave a message with the Sysop or send InterNet Mail to: aq474@cleveland.freenet.edu)

HOLDERS OF THE TOP TEN DXCC COUNTRY TOTALS
(based on total countries, including deleted countries)

MIXED	K6OJ	W8AH	W4LYV	W6PT	**374**	WØBW	**339**
383	WØBW	W8LKH	W5KGX	W7CG	W2OKM	**371**	DJ2BW
W1GKK	**378**	W9FKC	WØCM	W7GN	W5IO	ZP5CF	**338**
381	N4SU	W9JUV	4X4JU	W8KPL	W9WHM	W3CWG	DL6EN
W2AGW	K5UC	**375**	**373**	WØPGI	WØCM	W4SKO	JA3FYC
W2BXA	WØMLY	W1FZ	DL1BO		**373**	**370**	OZ1LO
K6ZO	**377**	W5IO	ON4DM	**PHONE**	GW3AHN	DL6EN	SMØAJU
W7MB	W2AG	K6CH	W1JR	**379**	ON4DM	W4EEE	K4PI
LU6DJX	W3EVW	W7IR	K2FL	W2BXA	SM3BIZ	W5KGX	K4XO
380	W3VT	W8JBI	W2SAW	VK6RU	K5UC	W9YSX	K6GA
VK6RU	**376**	**374**	W2SSC	**378**	W8AH		W6PT
W6HX	G3AAE	DJ2BW	W3CWG	W6GVM	**372**	**CW**	W8AH
W7KH	GW3AHN	SM3BIZ	W4BQY	**377**	DJ2YI	**341**	**337**
W9ZM	KH6CD	K4EZ	W4DR	VK5MS	PAØHBO	W9KNI	MANY
379	W2LV	K4KQ	W6BS	**375**	TI2HP	**340**	
W1AX	W2OKM	K4PDV	W6BSY	IØAMU	W2HTI	N4WW	
W1BIH	W6KUT	W4AIT	W6FSJ				

(Courtesy ARRL)

FIRST 10 WINNERS OF THE 5BDXCC AWARD

1. W4QCW (now W4DR)	3. W1EVT (now NY1N)	5. W8BT (now W8AH)	7. W1AX	9. K2BZT
2. DL7AA	4. W8GZ	6. W4IC	8. W4BRB	10. LA7Y

THE CQ DX AWARDS PROGRAM

1. The CQ DX Award is issued in three categories. The CQ DX CW Award is issued to an amateur radio station submitting proof of contact with 100 or more countries. The CQ DX SSB Award is issued to an amateur radio station submitting proof of contact with 100 or more countries. The CQ DX RTTY Award is issued to an amateur radio station submitting proof of contact with 100 countries using two-way RTTY. Applications should be submitted on the official CQ DX Award application (form 1067B). Reasonable facsimiles or computer printouts are also acceptable.

2. All contacts must be two-way in the mode for which the application is made. Cross-mode or one-way contacts are not valid. QSLs must be listed in alphanumeric order (A to Z and 1 to Ø) by prefix. All contacts must have been made after November 15, 1945. Deleted countries do not count. Only currently active countries are acceptable.

3. QSL cards must be verified by one of the authorized check points for the CQ DX Awards, or must be included with the application. Return postage must be included.

4. Country endorsement stickers are issued for 150, 200, 250, 275, 300, 310, and 320 active countries. A fee of $1.00 per sticker (where stickers are issued) is charged.

5. Special endorsements, as follows, are available for a fee of $1.00 each:

(a) 28 MHz endorsement—for 100 or more countries confirmed in the 10 meter band.

(b) 3.5/7 MHz endorsement—for 100 or more countries confirmed using any combination of the 40 and 80 meter bands.

(c) 1.8 MHz endorsement—for 50 or more countries confirmed using the 160 meter band.

(d) QRPp endorsement—for 50 or more countries confirmed using 5 watts output or less.

(e) Mobile endorsement—for 50 or more countries confirmed with the applicant operating mobile.

(f) Slow Scan TV endorsement—for 50 or more countries confirmed using two-way SSTV.

(g) OSCAR endorsement—for 50 or more countries confirmed via amateur satellite.

6. Any altered or forged confirmations will result in permanent disqualification of the applicant.

7. Fair play and good sportsmanship in operating are required for all amateurs working toward CQ DX Awards. Continued use

HISTORY OF THE CQ WAZ AWARD

The Worked All Zones (WAZ) Award program actually began long before CQ magazine was founded. The original sponsor was R/9 magazine, published by K.V.R. Lansingh, W6QX. In the November 1934 issue of R/9 magazine, the WAZ award was announced for the first time. The award was called "R/9 DX Zones of the World."

The reasons for the WAZ award were given in the November 1934 issue of R/9. The same reasons were later repeated by W6QD in an article in Radio magazine. The WAZ was conceived in response to the increasing ease of working the WAC Award. DX enthusiasts needed something more difficult to strive to achieve. Several attempts had been made to define a comprehensive "countries" list, but there was no DXCC nor CQ DX award available in 1934. In the hope of providing a more challenging goal, the WAZ was born.

The original WAZ map indicated at the bottom that it was drafted by Mark M. Bowelle, W4CXY. The 1934 "R/9 DX Zones of the World" map was nearly identical to the modern CQ WAZ map, except for the boundaries of zones 17, 18, 19 and 25. The exact criteria for defining the zone boundaries (and who made the decisions) is not known with certainty. An educated guess would be that it was W6QX, a man of many ideas, who thought it up and assigned the project to others. The map was then drawn by Mr. Bowelle. The boundaries were supposedly defined with consideration given to topographical maps, "calls heard" lists, and similar factors. In drawing the boundaries, zone lines were "made to coincide with political or call area boundaries, even where slight departures from natural geographic boundaries were necessitated." It was noted in the R/9 article that "portions of the boundaries of zones 17, 18, 19, and 22 are incorrect on the map," and that a corrected, more detailed, and larger version would "soon be available at a nominal cost."

In January 1936 R/9 magazine bought out Radio magazine and published the new Radio magazine from Southern California. The name Radio was retained because it had the earlier origin of the two. In its February 1936 issue, Radio magazine announced the WAZ award. The WAZ map was the same one that was first published in the November 1934 R/9. It had Mr. Bowelle's name on it and contained the original errors. It was now called "Radio Magazine's DX Zones of the World."

The WAZ map appeared again in the January 1939 issue of Radio. It was still incorrect because it seems no one checked the original for errors. The first station to accomplish the WAZ Award under the Radio magazine program was ON4AU, in March 1937. He was followed by G2ZQ, W3EVT, and J5CC. W3EVT's award appears to have been revoked, and his total reduced to 39 zones, probably due to uncertainties about some stations' exact location. Since amateur radio contact between U.S. amateurs and most foreign countries was prohibited for the duration of World War II, the WAZ award was suspended in 1941.

The new CQ magazine revived the WAZ program in 1947. The first to achieve post-war WAZ was Ben Stevenson, W2BXA, in September 1947, followed by W6VFR that same month. The first 10 qualifiers were:

1. W2BXA
2. W6VFR
3. W6PFD
4. W6SAI
5. W6ITA (now W6RR)
6. W6MJB
7. W6SA
8. ZS2X
9. W6LEE
10. W8HGW

The only really significant changes to the WAZ map were made in the 1970s when UAØ stations on the South Sakhalin Islands and the Kuril Islands were changed from Zone 25 to zone 19. In the early 1980s, a major effort to accurately define the boundaries of Zones 23 and 24 was undertaken, since stations in China had become active once again.

THE CQ WAZ AWARDS PROGRAM

The CQ Worked All Zones (WAZ) Award and its variations are issued to any licensed amateur station presenting proof of contact with the appropriate number of CQ zones of the world. This proof consists of proper QSL cards which in many cases may be checked by any of the authorized check points or sent directly to the WAZ Award Manager, Jim Dionne, K1MEM, 31 DeMarco Road, Sudbury, MA 01776. Many of the major DX clubs in the USA and Canada and most national amateur radio societies are authorized CQ check points. If in doubt, consult the WAZ Award Manager. Any legal type of emission may be used, providing communication was established after November 14, 1945.

1. The Official CQ WAZ Zone Map and the printed zone list which follows these rules are used to determine the zone in which a station is located.

2. Confirmations must be accompanied by a list of claimed zones, using CQ form 1479 or a facsimile, showing the callsign of the station contacted within each zone, with the date, time, band, and mode of each contact. The list must also clearly show the applicant's name, callsign, and complete mailing address. The applicant must show the type of award for which he or she is applying, such as Mixed Mode, All SSB, or All CW (see #7 below). A hand-written list may be submitted and will be accepted for processing, provided that the above information is shown.

3. All contacts must be made with licensed, land-based, amateur service stations operating in authorized (on the date of contact) amateur bands. Contacts with non-land-based stations, such as ships at sea, aeronautical mobile, or polar ice stations are not acceptable. Awards obtained by any mobile station will be endorsed as such, as will awards obtained by QRP stations, providing that the QSL cards submitted specify these modifiers.

4. All contacts submitted by the applicant must be made from within the same DXCC country. It is recommended that each submitted QSL clearly show the contacted station's zone number, if possible. When the applicant submits cards made out to different callsigns, evidence should be provided to show that he or she also holds or has held those callsigns at the time of the contact.

5. Any altered or forged confirmations submitted by an applicant for WAZ credit may result in permanent disqualification. The WAZ Manager may request the resubmission of certain confirmations as required. While a QSL card is normally accepted as proof of a contact, the final proof is an entry in the DX station's logbook

for the listed QSO. Failure to resubmit QSLs in a timely manner when request-ed by the WAZ Manager may result in the recall of the award in question. Submission of an application for any WAZ award acknowledges consent to abide by the decisions of the CQ WAZ Manager and the CQ Awards Committee.

6. An application must be accompanied by the processing fee (subscribers must include a recent CQ mailing label or copy with the application fee of $4.00; non-subscribers $10.00) and a self-addressed envelope with sufficient postage or International Reply Coupons (IRCs) to return the QSL cards by the class of mail desired and indicated. International Reply Coupons equal in redemption value to the processing fee are acceptable. The 1993 redemption value of IRCs is $.50 each. Checks should be made out to the name of the WAZ Award Manager.

7. In addition to the basic certificate for which any and all HF bands may be used, specially endorsed and numbered certificates are available for Phone (including AM), Single Sideband, and CW operation. The Phone certificate requires that all contacts be Two-Way Phone, the SSB certificate requires that all contact be Two-Way SSB, and the CW certificate requires that all contacts be Two-Way CW.[1]

8. If at the time of the original application a note is made stating the possibility of a subsequent application for an endorsement or special certificate, only the missing confirmations required for that endorsement need be submitted with the later application, providing a copy of the original authorization signed by the WAZ Manager is enclosed.

9. Decisions of the CQ DX Advisory Committee on any matter pertaining to the administration of this award are final.

10. All applications should be sent to the WAZ Award Manager after the QSL cards have been checked by an authorized CQ check point. If the application is for 160 meters or the 5 Band WAZ, all QSL cards must be checked by the WAZ Award Manager. Photocopies are not acceptable.

11. Zone maps, printed rules, and application forms are available from the WAZ Award Manager. Send a large self-addressed envelope with 2 units of postage, or $1.00 and an address label. For DX stations, send an address label and 2 IRCs.

The following list of zones is presented as a guide. Any questions will be decided by the CQ Zone Map. For rulings on borderline areas, consult the WAZ Award Manager.

Zone 1. Northwestern Zone of North America: KL7, VY1/VE8 Yukon, the Northwest Territories west of 102 degrees. (Includes the islands of Victoria, Banks, Melville, and Prince Patrick.)

Zone 2. Northeastern Zone of North America: VO2 Labrador, the portion of VE2 Quebec north of the 50th parallel, the VE8 Northwest Territories east of 102 degrees. (Includes the islands of King Christian, King William, Prince of Wales, Somerset, Bathurst, Devon, Ellesmere, Baffin, and the Melville and Boothia Peninsulas, excluding Akimiski Island.)

Zone 3. Western Zone of North America: VE7, W6, and the W7 states of Arizona, Idaho, Nevada, Oregon, Utah, and Washington.

Zone 4. Central Zone of North America: VE3, VE4, VE5, VE6, VE8 Akimiski Island, and W7 states of Montana and Wyoming. WØ, W9, W8 (except West Virginia), W5, and the W4 states of Alabama, Tennessee, and Kentucky.

Zone 5. Eastern Zone of North America: 4U1UN, CY9, CYØ, FP, VE1/VY2, VO1, the portion of VE2 Quebec south of the 50th parallel, VP9, W1, W2, W3, and the W4 states of Florida, Georgia, South Carolina, North Carolina, Virginia, and the W8 state of West Virginia.

Zone 6. Southern Zone of North America: XE/XF, XF4 (Revilla Gigedo).

Zone 7. Central American Zone: FO (Clipperton), HKØ (San Andres), HP, HR, TG, TI, TI9, V3, YN, and YS.

Zone 8. West Indies Zone: C6, CO, FG, FJ, FM, FS, HH, HI, J3, J6, J7, J8, KG4 (Guantanamo), KP1, KP2, KP4, KP5, PJ (Saba, St. Maarten, St. Eustatius), V2, V4, VP2, VP5, YVØ (Aves Is.), ZF, 6Y, and 8P.

Zone 9. Northern Zone of South America: FY, HK, HKØ (Malpelo), P4, PJ (Bonaire, Curacao), PZ, YV, 8R, and 9Y.

Zone 10. Western Zone of South America: CP, HC, HC8, and OA.

Zone 11. Central Zone of South America: PY, PYØ, and ZP.

Zone 12. Southwest Zone of South America: 3Y (Peter I), CE, CEØ (Easter Is., Juan Fernandez Is., San Felix Is.), and some Antarctic stations.[2]

Zone 13. Southeast Zone of South America: CX, LU, VP8 Islands, and some Antarctic stations.[2]

Zone 14. Western Zone of Europe: C3, CT, CU, DL, EA, EA6, EI, F, G, GD, GI, GJ, GM, GU, GW, HB, HBØ, LA, LX, ON, OY, OZ, PA, SM, ZB, 3A, and 4U1ITU.

Zone 15. Central European Zone: ES (UR), HA, HV, I, ISØ, LY (UP), OE, OH, OHØ, OJØ, OK, OM, S5, SP, T7, T9, TK, UA2, YL (UQ), YU, ZA, 1AØ, Z3, 9A, 9H.

Zone 16. Eastern Zone of Europe: UA1, UA3, UA4, UA6, UA9 (S,W), UR-UZ (UB) EU-EW (UC), ER (UO), and R1M (M.V. Island).

Zone 17. Western Zone of Siberia: UA9 (A, C, F, G, J, K, L, M, Q, X) and EZ (UH), UK (UI), EY (UJ), UN-UQ (UL), EX (UM).

Zone 18. Central Siberian Zone: UA8 (T,V), UA9 (H, O, U, V, Y, Z), and UAØ (A, B, H, S, U, W).

Zone 19. Eastern Siberian Zone: UAØ (C, D, F, I, J, K, L, Q, X, Z).

Zone 20. Balkan Zone: JY, LZ, OD, SV, TA, YK, YO, ZC4, 4X, and 5B.

Zone 21. Southwestern Zone of Asia: A4, A6, A7, A9, AP, EP, HZ, 4J (UD), 4L (UF), EK (UG), YA, YI, 4W, 7O, and 9K.

Zone 22. Southern Zone of Asia: A5, S2, VU, VU (Laccadive Is.), 4S, 8Q, and 9N.

Zone 23. Central Zone of Asia: JT, UAØY, BY3G-L, BY9A-L, BY9T-Z, and BYØ.

Zone 24. Eastern Zone of Asia: BV, BY1, BY2, BY3A-F, BY3M-S, BY3T-Z, BY4, BY5, BY6, BY7, BY8, BY9M-S, VS6, and XX.

Zone 25. Japanese Zone: HL and JA.

Zone 26. Southeastern Zone of Asia: HS, VU (Andaman and Nicobar Islands), XV, XU, XW, XZ, and 1S (Spratly Islands).

Zone 27. Philippine Zone: DU (Philippines), JD1 (Minami Torishima), JD1 (Ogasawara), KC6 (Republic of Belau), KH2 (Guam), KHØ (Marianas Is.), V6 (Fed. States of Micronesia).

Zone 28. Indonesian Zone: H4, P2, V8, YB, 9M, and 9V.

Zone 29. Western Zone of Australia: VK6, VK8, VK9X (Christmas Is.), VK9Y (Cocos-Keeling Is.), and some Antarctic stations.[2]

Zone 30. Eastern Zone of Australia: VK1, VK5, VK7, VK9L (Lord Howe Is.), VK9 (Willis Is.), VK9 (Mellish Reef), VKØ (Macquarie Is.), and some Antarctic stations.[2]

Zone 31. Central Pacific Zone: C2, FO (Marquises), KH1, KH3, KH4, KH5, KH6, KH7, KH9, T2, T3, V7, and ZK3.

Zone 32. New Zealand Zone: A3, FK, FO (except Marquises and Clipperton), FW, KH8, VK9 (Norfolk Is.), VR6, YJ, ZK1, ZK2, ZL, 3D2, 5W, and some Antarctic stations.[2]

Zone 33. Northwestern Zone of Africa: CN, CT3, EA8, EA9, IG9, IH9 (Pantellaria Is.), SØ, 3V, and 7X.

Zone 34. Northeastern Zone of Africa: ST, STØ, SU, and 5A.

Zone 35. Central Zone of Africa: C5, D4, EL, J5, TU, TY, TZ, XT, 3X, 5N, 5T, 5U, 5V, 6W, 9G, and 9L.

Zone 36. Equatorial Zone of Africa: D2, TJ, TL, TN, S9, TR, TT, ZD7, ZD8, 3C, 9J, 9Q, 9U, and 9X.

Zone 37. Eastern Zone of Africa: C9, ET, E3, J2, T5, 5H, 5X, 5Z, and 7Q.

Zone 38. South African Zone: A2, ZD9, Z2, ZS1-9, 3DAØ, 3Y (Bouvet Is.), 7P, and some Antarctic stations.[2]

Zone 39. Madagascar Zone: D6, FT-W, FT-X, FT-Z, FH, FR, S7, VKØ (Heard Is.), VQ9, 3B6/7, 3B8, 3B9, 5R8, and some Antarctic stations.[2]

Zone 40. North Atlantic Zone: JW, JX, OX, TF, and R1F (Franz Josef Land).

[1]Previously all Mixed WAZ Awards were numbered consecutively, with ALL CW noted. This will continue for mixed mode applications. As of January 1, 1991, the ALL CW WAZ Award is issued. To qualify, the QSL card from each zone must be dated after January 1, 1991.

[2]Antarctic notes. The boundries of CQ zones 12, 13, 29, 30, 32, 38, and 39 coverage at the South Pole. Stations KC4AAA and KC4USN are at the South Pole, ansd will count for any one of the listed zones. Most Antarctic stations indicate their zone on the QSL card. A few stations and their zones are: 4K1A 39, 4K1B 29, 4K1C 29, 4K1D 38, 4K1E 29, 4K1F 13, 4K1G 30, 4K1H 32, 4K1J 13, 8J1RL 39, CE9 13, DPØ 38, FT-Y 30, HFØPOL 13, HL5BDS 13, KC4AAC 13, KC4AAD 13, KC4AAE 29, KC4USB 32, KC4USV 30, LU-Z 13, VKØGM 29, VP8ME 38, Y38ANT 38, AND ZL5AA 30. This list changes frequently. Questions regarding the zone location of a particular Antarctic station should be directed to the WAZ Award Manager.

Single Band WAZ

Awards will be issued to amateur radio stations presenting proof of contact with the 40 zones of the world on one of the bands: 80, 40, 20, 15, 10. Contacts for a single band WAZ must have been made after January 1, 1973. Single band certificates will be awarded for ALL SSB or for ALL CW only.

5 Band WAZ

Applicants who succeed in presenting proof of contact with the 40 zones of the world on the 80, 40, 20, 15, and 10 meter bands (for a total of 200) receive a special certificate in recognition of this achievement. A prerequisite for 5 Band WAZ is that the applicant must already be a holder of any 40-zone WAZ.

The 5 Band WAZ Award is offered for any combination of CW, SSB, RTTY, or other mode contact, MIXED MODE only. Separate awards will not be offered for the different modes. Contacts must have been made after 0000Z January 1, 1979. Proof of contact shall consist only of QSL cards checked by the WAZ Manager, and all provisions of Rule 5 of the WAZ rules are strictly enforced. The first plateau is a total of 150 zones on a combination of 5 bands. Applicants should use a separate sheet for each band, using CQ form 1479 or a facsimile. See rule 6 for fee and postage information.

After the 150 zone point, each 10 zones requires the submission of the QSL cards and a $1.00 fee. At the 200 zone point the applicant will be awarded a gold sticker for the certificate and the opportunity to purchase a hand-engraved plaque to commemorate the achievement.

All applications should be sent to the WAZ Award Manager. The 5 Band Award is governed by the same rules as the regular WAZ Award and uses the same boundaries.

WARC Bands WAZ

Effective January 1, 1991 single band WAZ Awards will be issued to amateur radio stations presenting proof of contact with the 40 zones of the world on any one of the WARC bands: 30, 17, or 12 meters. (Each band constitutes a separate award, and each may be applied for separately.) This award is available for MIXED MODE, ALL SSB, ALL CW, or ALL RTTY. Contacts for each one of these WARC WAZ Awards must have been made after each station involved in the contact had permission from its licensing authority to operate on the band and mode.

RTTY WAZ

Special WAZ Awards are issued to amateur radio stations presenting proof of contact with the 40 zones of the world using RTTY. For the mixed band award QSL cards must show a date of November 15, 1945 or later. The RTTY WAZ is also available with a single band endorsement. For a single band endorsement for 80, 40, 20, 15, or 10 meters the QSL cards must show a date of January 1, 1973 or later.

WNZ

WNZ stands for "Worked Novice Zones" and is available only to holders of a U.S. Novice or Technician class license. Proof of contact with at least 25 of the 40 CQ zones as defined by the WAZ rules is required. All contacts must be made using the 80, 40, 15, and 10 meter Novice bands. All contacts must be made while holding a Novice or Technician class license, although the application may be submitted at a later date. Contacts must be made prior to receiving authorization to operate with higher class privileges. The WNZ is available as a MIXED MODE, CW ONLY, or SSB

ONLY award. It may also be endorsed for a single band. The WNZ Award may be used to fulfill part of the application requirement for the WAZ Award when all 40 zones are confirmed.

1. The basic award may be secured by submitting QSL cards for 25 zones. The processing fee is $5.00 for all applicants.

2. All QSL cards must show a date of January 1, 1952 or later.

3. Use CQ Form 1479 or a facsimile to make application for this award.

160 Meter WAZ

The WAZ Award for 160 Meters requires that the applicant submit directly to the WAZ Manager QSL cards from at least 30 zones.

All QSL cards must be dated January 1, 1975 or later, and a $5.00 application fee must accompany all applications. The 160 WAZ is a mixed mode award only. The basic 160 WAZ Award may be secured by submitting QSL cards from 30 zones. Stickers for 35, 36, 37, 38, 39, and 40 zones may be secured from the WAZ Manager upon the submission of the QSL cards and $2.00 for each sticker.

Satellite WAZ

The Satellite WAZ Award is issued to amateur radio stations submitting proof of contact with all 40 CQ zones through any amateur radio satellite. The award is available for MIXED MODE only. All QSL cards must show a date of January 1, 1989 or later.

HOLDERS OF THE CQ 5-BAND WORKED-ALL-ZONES AWARD
(All 200 Zones Worked and Confirmed)
The First Ten

1. ON4UN	3. SM4CAN	5. W8AH	7. EA8AK	9. EA3SF
2. K4MQG	4. AA6AA	6. W6KUT	8. LA7JO	10. OH1XX

Alphabetical List of All Holders

4N7ZZ	DL8AN	HA8XX	JA0DWY	K0CS	LA9GV	OH8KN
4X6DK	DL8MAG	HA9RE	JA1BWA	K0ZZ	LU8DPM	OH8SR
4Z4DX	EA1OD	HB9AHL	JA1ELY	K1MEM	LU9FFA	OK1ADM
5B4TI	EA2IA	HB9AMO	JA1FNA	K1MM	LY2BR	OK1AWZ
A71AD	EA2KL	HB9ATA	JA1GTF	K1VKO	LZ1GC	OK1DDS
A92BE	EA3AD	HB9CIP	JA1GV	K2ENT	LZ1HA	OK1MG
AA4LU	EA3SF	HB9RG	JA1IFP	K2TQC	LZ1NG	OK1MP
AA4V	EA4BVE	HC5EA	JA1KQX	K3TW	LZ2CC	OK1RD
AA6AA	EA4DO	HG19HB	JA1SDV	K4CEB	LZ2DF	OK2DB
CT1BH	EA5SP	HK3DDD	JA1SVP	K4MQG	LZ2JF	OK3CGP
CT1FL	EA6ET	HL1IUA	JA1UQP	K5CTG	LZ2KK	OK3CSC
CT1TM	EA6NB	I0RIZ	JA2AAQ	K5OVC	NX7K	OK3DG
CT2AK	EA8AK	I1APQ	JA2BL	K5TSQ	N0XA	OK3TCA
CT3BM	EA8OZ	I1BSN	JA2DSY	K5UC	N2MF	ON4ACG
CT4BD	EA8QL	I1HAG	JA2ODS	K5UR	N4CC	ON4ADZ
CT4NH	EA8XS	I1JQJ	JA3CSZ	K6EID	N4JJ	ON4QX
DF3GY	EA9IE	I1POR	JA3DY	K6SIK	N4KE	ON4UN
DF4DI	F5OHS	I1ZEU	JA3EMU	K6SSS	N4KG	ON5NT
DF6CY	F5VU	I2EOW	JA3FYC	K6YRA	N4RR	ON5WQ
DF7NM	F6BEE	I2JQ	JA3MNP	K7EJ	N4VZ	ON6HE
DF9ZP	F6BKI	I2TZK	JA4IKD	K8EJ	N4WJ	ON6OS
DJ2YA	F6BLP	I2UIY	JA4MRL	K8NA	N5CB	ON7EM
DJ4PI	F6DZU	I2ZGC	JA4VUQ	K9AJ	N5FG	ON7PQ
DJ5JH	F6EXV	I3MAU	JA5DQH	K9EL	N6AR	OZ3PZ
DJ6RX	FM5WD	I4EAT	JA6JPS	K9FD	N6DX	OZ7OP
DJ7RD	G3GIQ	I4EWH	JA6LCJ	K9JF	N7MC	OZ7YY
DJ8NK	G3MCS	I4IKW	JA6NQT	K9RHY	N7RT	PA0CLN
DJ9RQ	G3MXJ	I4RYC	JA6VU	K9UWA	NR1R	PA0LVB
DJ9ZB	G3LNS	I4USC	JA7FS	KB0G	NS7Z	PA0XPQ
DK2OY	G3NLY	I4YNO	JA7FWR	KB0U	NW5K	PA0ZH
DK5AD	G3TJW	I6FLD	JA7GLB	KB8DB	OA4OS	PA3BUD
DK6WL	G3UML	I8IGS	JA7HMZ	KC7EM	OE1ZJ	PY1APS
DK7PE	G3XTT	I8SAT	JA7IL	KC8PG	OE2VEL	PY1OL
DL1SDN	G4BUE	IK0AGU	JA7JWF	KD7P/KH2	OE3WWB	PY7Z
DL1YD	G4BWP	IK0IOL	JA8EAT	KG4W	OE6IMD	RB5GW
DL3RK	G4GED	IK4ALM	JH1XYR	KM1H	OH1XX	RB7GG
DL4YAH	G4GIR	IK4GME	JH7FMJ	KS1L	OH2EE	RT4UA
DL6EN	G4IUF	IK6BOB	JH7QXL	LA1VFA	OH3JF	RT5UN
DL6WD	G4LJF	IT9TQH	JR3HZW	LA2GV	OH3RF	RT5UO
DL7AA	GW4OFQ	IT9ZGY	JR1FYS	LA4HW	OH3TQ	RT5UY
DL7AFV	HA0DU	IV3PRK	JR3IIR	LA5YJ	OH3TY	S59VM
DL7HZ	HA0MM	IV3YYK	JR6PGB	LA6OT	OH3YI	SM0AJU
DL7PR	HA3NU	JA0CWZ	JT1BG	LA7JO	OH6JW	SM0BZH
DL7XS	HA5WA	JA0DAI	JW5NM	LA7ZO	OH7KI	SM0DJZ

SM3BIZ	SM7FIG	UA6JWW	VE7IG	W4YV	XE1OX	YU7DX
SM3EVR	SP3BQD	UA9CBO	VK3QI	W5EU	XE1VIC	YZ1MB
SM3GSK	SP3IBS	UA9FAR	VK6HD	W6GO	Y22JD	ZL1BIL
SM4CAN	SP3IOE	UK2RDX	VK9NS	W6KUT	YBØRX	ZL1BOQ
SM4CTT	SP5AA	UP1BZZ	WØJLC	W6TC	YBØWR	ZL1BQD
SM5AKT	SP6CDK	UQ1GXX	WØMLY	W7OM	YO3AC	ZL3GQ
SM5AQD	SP7KTE	UQ2MU	WØSD	W8AH	YO3CD	ZL4BO
SM5CLE	SP8EMO	UR2QD	WØZV	W8SEY	YT3AA	ZP5JCY
SM6BGG	SV1ADG	UWØMF	W1NG	W8UVZ	YT7DX	ZS4TX
SM6CST	SV1JA	UZ2FWA	W1NW	W9ZR	YU1EXY	ZS5BK
SM6CTQ	SV1JG	UZ9CWA	W2UE/7	WA1AER	YU2AA	ZS5LB
SM6CVX	TG9NX	VE1AST	W3AP	WA2TMP/7	YU2CBM	ZS5MY
SM6DYK	TG9VT	VE1NG	W3GG	WB2P	YU2HDE	ZS6BCR
SM7BYP	UA3TT	VE3EJ	W3UM	WB9Z	YU2TW	
SM7CRW	UA4RZ	VE3ICR	W4DR	WS9V	YU2WV	
SM7DZZ	UA6JD	VE7DX	W4VQ	XE1J	YU3AN	

How Hard is the 5BWAZ Award from Different parts of the U.S.?

As of May, 1994, 74 U.S. amateurs had earned the 5BWAZ Award, or about 20% of the worldwide total. The breakdown by call area is as follows:

W1 9
W2 3
W3 4
W4 14
W5 6
W6 8
W7 6
W8 6
W9 9
WØ 9

Dividing the above numbers by the number of General through Advanced class licensees in each call area yields the following data:

W1	9/18305	4.9%	W6	8/38810	2.1%
W2	3/23922	1.25%	W7	6/28024	2.1%
W3	4/18257	2.2%	W8	6/24450	2.5%
W4	14/55671	2.5%	W9	9/22259	4.0%
W5	6/32368	1.9%	WØ	9/26442	3.4%

(Courtesy K1MEM)

WAZ AWARDS
Worked-All-Zones Awards from 1 Aug 94 to 30 July 95

CW/Phone

7487	DL6YH	7518	OH7MKR	7549	K8BN	7580 CW	K6HMS
7488 CW	DK8NM	7519	KF2C	7550	OD5PL	7581 CW	HA5AF
7489 CW	OK1JTR	7520	SM6MCX	7551	KQ9O	7582	SM3KOR
7490 CW	IK1OWC	7521	KZØX	7552	SP5GRM	7583	KD9WK
7491 CW	DL1SDN	7522 CW	K1MBO	7553 CW	DL4FAD	7584	KM6HB
7492	N3CAM	7523	K1ACL	7554	DL9GCF	7585 CW	IK1GPK
7493 CW	Z31RQ	7524	JI1CZK	7555	N2PLE	7586	UT3UY
7494 CW	OE7JLI	7525	JA1IFB	7556 CW	OE5HIL	7587	AC4JO
7495	JG1UZD	7526 CW	IK4DLW	7557	DL6LBD	7588 CW	N4MHQ
7496	JJ1CZR	7527 CW	JO1MQV	7558	DL6LA	7589	IK8UDP
7497	OH3MKH	7528	K3KO	7559 CW	KC2AN	7590	4X6UU
7498	JH1LPZ	7529	KØDEW	7560 CW	EU7SA	7591	IK2GWY
7499 CW	KB2FD	7530	HA7UW	7561 CW	K4IKM	7592	I2JHF
7500 CW	JA7ECT	7531	KN4UI	7562 CW	DL9VDQ	7593 CW	EA6AA
7501	KF2KT	7532	OE3EPW	7563 CW	EA2CLL	7594	W3GRW
7502	KJ6ZH	7533	AD1E	7564	JFØSGW	7595	DK7LA
7503	K8JJC	7534	DL1HBB	7565 CW	DL2NY	7596	HA8IB
7504	PT2AZ	7535 CW	K8JJC	7566 CW	IK1GNC	7597	OH6NVC
7505	EA3DRE	7536 CW	YO3YC	7567	GM4AGL	7598	W7SE
7506	KØJUH	7537	SMØBNK	7568 CW	SM7MPM	7599	AA8CH
7507	I6BDS	7538 CW	SMØBNK	7569	AD4KW	7600	DL7UGU
7508 CW	JA4MIF	7539	EA3FBO	7570	F6IGF	7601 CW	KA5GMN
7509 CW	YU1ZD	7540	I4JGE	7571	K8MBH	7602 CW	W1JA
7510	HA6NF	7541	DL1VDL	7572	YU1KJ	7603 CW	JI1DHY
7511 CW	AA5AU	7542 CW	9A2AJ	7573 CW	IK1PQT	7604	WA4QDM
7512 CW	WK7Z	7543	WK1F	7574 CW	NQ6N	7605 CW	JH1BAM
7513	5B4ES	7544	KD7E	7575	N4CBR	7606	JH6JMN
7514	W1HIC	7545	KN6DV	7576	JM2DRM	7607	N2BYM
7515	IT9JPS	7546 CW	WM9X	7577	WAØGOZ	7608	HA1DAE
7516	SV8ZC	7547 CW	WG9L	7578 CW	JG1HQI		
7517 CW	DL8TQ	7548	G3LNS	7579	JR2KJJ		

SSB

4208	JH1FVE	4229	KB7M	4250	AB4IQ	4271	VK6APW
4209	DF3IN	4230	OZ3FS	4251	N5WNG	4272	W6UPI
4210	DL1SDN	4231	W1HIC	4252	WU1F	4273	KD9WK
4211	IK6PBX	4232	KR4DA	4253	KN6DV	4274	WO2N
4212	N5QDE	4233	IK3LGC	4254	TU2QW	4275	W5EU
4213	GM3EDZ	4234	KG6KW	4255	N9JVC	4276	WZ3E
4214	WDØGLF	4235	W1WTG	4256	CT1AUO	4277	AC4JO
4215	DF2IS	4236	K9MIE	4257	N5BLK	4278	DL3EAY
4216	JJ1CZR	4237	K2PBP	4258	K4IKM	4279	HB9DMQ
4217	K6ICS	4238	N1BUA	4259	HL9FC	4280	YV3BAP
4218	JR1KAG	4239	W5OXA	4260	EA3GJW	4281	LX1BG
4219	JG1EGG	4240	YO3YO	4261	EA3BT	4282	KM6HG
4220	W2OW	4241	OE3EPW	4262	F5PAC	4283	VK4SJ
4221	IK2JYK	4242	DF1IC	4263	LU8MCO	4284	IK2GWY
4222	K2UVG	4243	KB8ZM	4264	DL6DBF	4285	JR6SVM
4223	AH9B/W5	4244	KK6EX	4265	F/HH2HM	4286	LU3HBO
4224	KQ4YI	4245	HB9DDW	4266	JL1UFW	4287	KI6BN
4225	EA1FEK	4246	EA2CLK	4267	CT1KT	4288	WO1P
4226	WA5TUD	4247	EA5DL	4268	KM6CB	4289	LU5EWD
4227	N8KOL	4248	EA7CRL	4269	IK7PUD	4290	EA3NB
4228	N8KUS	4249	EA5GMB	4270	GM4VWV		

CW

60	SM5BUH	65	NØABA	70	I7ALE	75	AD1E
61	JA6NQT	66	WB4DBB	71	K2AZ	76	KB5OHT
62	NV1C	67	VE1UK	72	WA2UUK	77	N5FG
63	K2TK	68	DJ9SO	73	JM1VKW	78	JL3SBE
64	OE3EPW	69	KJ6HO	74	I3JTE		

Single-Band/Mode WAZ

10 METERS

SSB				CW			
480	IK2DUW	482	DK4TA	146	NZØR	148	IK3QAR
481	F5NBX	483	WP4U	147	N6DX	149	UB3EA

SSB		MIXED		CW	
8	I2JQ	17	I1JQJ	NONE	
9	OH3TY				

15 METERS

SSB						CW	
472	AA2GQ	476	WBØTVY	481	JF1UVJ	258	JAØHC
473	IK2DUW	477	F5OHS	482	JI6URU	259	LU2BRG
474	JJ2QXI	478	EA5BY	483	JRØAMD	261	JA1AS
475	YZ7AA	479	KC7V	484	JA8XDA	260	N6DX
		480	JA7MGP	485	JF5APX		

17 METERS

SSB				MIXED		CW	
9	DJ9ZB	29	I1JQJ	27	I2JQ	NONE	
28	W3KHQ	30	WA2UUK				

20 METERS

SSB							
955	AA2GQ	963	W3KHZ	972	EA1AYN	455	VE3IAY
956	IK8MVH	964	KD1HN	973	EA3CB	456	HB9BGV
957	KF8UN	965	EA3BT	CW		457	N6DX
958	JA5OP	966	YZ7AA	449	UA3ECJ	458	JA2IU
959	N5QDE	967	WBØTVY	450	LU2BRG	459	VE6JO
960	VK6LC	968	PT7CB	451	KØJUH	460	JA1AS
961	AA2SZ	969	KD9WK	452	ON4ACG	461	WO2N
962	KN6DV	970	AC4JO	453	K2TK	462	W5USM
		971	KEØRR	454	9A3SM		

30 METERS

MIXED						CW	
2	G4BWP	13	4Z4DX	15	N4MM	11	I1JQJ
12	G4BWP	14	DL1PM	16	N6DX		

40 METERS

SSB						CW	
81	XE1VIC	177	K2TK	180	OE5JDL	175	N9AU
82	EA4KD	178	DK4TA	181	DK6WL	176	LU2BRG
		179	HB9BGV	182	SM5JE		

80 METERS

SSB				CW			
63	4X6DK	66	VE7DX	42	DK5AD	45	JE7MQB
64	XE1VIC	67	SM5GZ	43	KC8PG	46	JA1PMN
65	I2EOW	68	K6YRA	44	N6DX		

160 METERS

76	I4EAT	30	ZONES
77	W1JZ	34	"
78	OK1DQT	31	"
79	DJ7RD	31	"

RTTY

MIXED   **20 METERS**

93	IØKKY	42	JA1BWA
94	K6EID		

WNZ

10 SSB		SATELLITE	
60	KD4HXT	14	OE3JIS

THE CQ WPX AWARDS PROGRAM

The CQ WPX Award recognizes the accomplishments of confirmed QSOs with the many prefixes used by amateurs throughout the world. Separate, distinctively marked certificates are available for 2× SSB, CW, and MIXED (CW and SSB/Phone) modes, as well as the VPX Award for shortwave listeners and the WPNX Award for Novices.

1. Applications

A. All applications for WPX certificates (and endorsements) must be submitted on the official application form (CQ 1051A). This form can be obtained by sending a self-addressed, stamped, business-size (4 × 9 inch) envelope to the WPX Award Manager, Norm Koch, K6ZDL, P.O. Box 593, Clovis, NM 88101.

B. All QSOs must be made from the same country.

C. All call letters must be in strict alphabetical order and the entire call letters must be shown.

D. All entries must be clearly legible.

E. Certificates are issued for HF (160–10) for the following modes and number of prefixes (VHF contacts do not count for the WPX Award program). Cross-mode QSOs are not valid for the CW or 2× SSB certificates. Mixed (CW and SSB/Phone only): 400 prefixes confirmed. CW: 300 prefixes confirmed. 2× SSB: 300 prefixes confirmed. Separate applications are required for each mode.

F. Cards need not be sent, but they must be in the possession of the applicant. Any and all cards may be requested by the WPX Award Manager or by the CQ DX Committee.

G. The application fee for each certificate is $4.00 for *CQ* subscribers (subscribers must include a recent *CQ* mailing label, or a photocopy of it) and $10.00 for non-subscribers, or the equivalent in IRCs.

H. All applications and endorsement requests should be sent to the WPX Award Manager.

2. Endorsements

A. Prefix endorsements are issued for each 50 additional prefixes submitted. Minimum submission at any one time is 50 prefixes.

B. Band endorsements are available for working the following numbers of prefixes on the various bands: 1.8 MHz—50, 3.5 MHz—175, 7 MHz—250, 14 MHz—300, 21 MHz—300, 28 MHz—300.

C. Continental endorsements are given for working the following numbers of prefixes in the respective continents: North America 160, South America 95, Europe 160, Africa 90, Asia 75, and Oceania 60.

D. Endorsement applications must be submitted on CQ form 1051A. Use a separate application for each mode of your endorsement application.

E. For prefix endorsements list only additional call letters confirmed since the last endorsement application.

F. A self-addressed, stamped envelope or proper IRCs for surface or airmail return is required, and $1.00 or 5 IRCs for *each* endorsement sticker.

3. Prefixes

A. The letter/numeral combinations which form the first part of the amateur call will be considered the prefix. Examples: K6, N6, Y22, Y23, WD4, HG1, HG19, WB2, WB2ØØ, KC2, KC2ØØ, OE2, OE25, U3, GB75, ZS66, NG84, etc. Any difference in the numbering, lettering, or order of same shall constitute a separate prefix.

B. Any prefix will be considered legitimate if its use was licensed or permitted by the governing authority in the country of operation after November 15, 1945.

C. In cases of portable operation in another country or call area, the portable designator would then become the prefix. Example: K6ZDL/7 would count as K7, J6/K6ZDL would count as J6, KH6/K6ZDL would count as KH6, etc. Portable designators without numbers will be assigned a zero (Ø) at the end of the designator to form the prefix. Example: LX/K6ZDL would count as LXØ. When claiming a prefix which has been sent as K6ZDL/XV5, for example, and you are claiming the XV5 for credit, it is requested that the *claimed prefix* be listed in the proper alphabetical position, such as XV5/K6ZDL, if for XV5; or K6ZDL/XV5 if for K6. The portable prefix must be an authorized prefix in the country/area of operation. Maritime mobile, mobile, /A, /E, /J, /P, or interim license class identifiers do not count as prefixes.

D. All calls without numbers will be assigned a zero (Ø) plus the first two letters to form a prefix. Examples: XEFTJW would count as XEØ, RAEM would count as RAØ, AIR as AIØ, etc.

VPX

The VPX (Verified Prefixes) Award may be earned by shortwave listeners (SWLs) who possess QSL cards confirming reception of at least 300 different amateur prefixes. No mode endorsements are available. Applications must be submitted to the WPX Award Manager in accordance with the WPX rules.

WPNX

The WPNX Award can be earned by USA Novices who work 100 different prefixes prior to receiving a higher class license. The application may be submitted after receiving the higher license, providing the actual contacts were made as a Novice. Prefixes worked for the WPNX Award may be used later for credit toward the WPX Award.

The rules for the WPNX Award are the same as for the WPX Award except that only 100 prefixes must be confirmed. Applications are to be sent to the WPX Award Manager.

WPX Honor Roll

The WPX Honor Roll recognizes the operators and stations that maintain a high standing in confirmed, current prefixes. The rules, therefore, reflect the belief that Honor Roll membership should be accessible to all active radio amateurs and not be unduly advantageous to the "old timers." With the exceptions listed below, all general rules for WPX apply toward Honor Roll credit.

A minimum of 600 prefixes is required to be eligible for the WPX Honor Roll. No certificates are issued, but a listing of members appears in *CQ* every other month.

A. Only current prefixes will be counted toward the WPX Honor Roll standings. A list of prefixes to be deleted from the Honor Roll is published annually in *CQ* and is available from the WPX Award Manager. Prefixes will be deleted from the Honor Roll listing two years after they are no longer authorized for use by the governing authority or by the ITU.

B. Special-issue prefixes (i.e., OF, OS, 4A, etc.) will be considered current for as long as they are assigned to a particular country and deducted as credit for Honor Roll standings after cessation of their use or assignment.

C. Honor Roll applicants must submit their list of current prefixes (entire call required) separate from their regular WPX applications. Use regular form 1051 and indicate "Honor Roll" at the top of the form. Forms may be obtained by sending a business-size, self-addressed, stamped envelope or 1 IRC (foreign stations send extra postage or IRC if airmail is desired) to the WPX Award Manager. A separate application must be made for each mode. Lifetime Honor Roll fee *for each mode* is $4.00. A computer printout of your individual Honor Roll file may be obtained from the WPX Award Manager for $5.00.

D. Endorsements for the Honor Roll may be made for 10 or more prefixes. An SASE or IRC should be included. For prefixes by countries see the *Callbook* listings.

WPX Award of Excellence

This is the ultimate award for the prefix DXer. The requirements are 1000 prefixes Mixed mode, 600 prefixes SSB, 600 prefixes CW, all 6 continental endorsements, and the 5 band endorsements 80–10 meters. A special 160 meter endorsement bar is also available.

The WPX Award of Excellence plaque fee is $60.00. The 160 meter bar is $5.25.

Only plaque holders will be listed in the WPX program in *CQ* magazine.

WORKED ALL CONTINENTS (WAC) AWARD RULES

In recognition of international two-way amateur radio communication, the International Amateur Radio Union (IARU) issues Worked-All-Continents certificates to amateur radio stations of the world.

Qualification for the WAC award is based on an examination by the International Secretariat, or a member-society, of the IARU of QSL cards that the applicant has received from other amateur stations in each of the six continental areas of the world. All contacts must be made from the same country or separate territory within the same continental area of the world. All QSL cards (no photocopies) must show the mode and/or band for any endorsement applied for.

Contacts made on 10/18/24 MHz or via satellites are void for the 5-band certificate and 6-band sticker. All contacts for the QRP sticker must be made on or after January 1, 1985 while running a maximum power of 5 watts output or 10 watts input.

The following information should be helpful in determining the continental area of a station located adjacent to a continental boundary. North America includes Greenland (OX) and Panama (HP). South America includes Trinidad & Tobago (9Y), Aruba (P4), Curacao & Bonaire (PJ2-4) and Easter Island (CEØ). Oceania includes Minami Tori-shima (JD1), Philippines (DU), East Malaysia (9M6-8) and Indonesia (YB). Asia includes Ogasawara Islands (JD1), Maldives (8Q), Socotra Island (7O), Abu Ail Island (J2/A), Cyprus (5B, ZC4), Eastern Turkey (TA2-8) and Georgia (UF). Europe includes the fourth and sixth call areas of RSFSR (UA4-6), Istanbul (TA1), all Italian islands(I) and Azores (CU). Africa includes Ceuta & Melilla (EA9), Madeira (CT3), Gan Island (VS9M), French Austral Territory (FT) and Heard Island (VKØ).

For amateurs in the United States or countries without IARU representation, applications should be directed to the below address. After verification, the cards will be returned and the award sent soon afterward. There is no application fee. However, sufficient return postage for the cards, in the form of a self-addressed stamped enveloped, or IRCs, is required. US amateurs must have current ARRL membership.

(For other amateurs)—Applicants must be members of their national amateur radio societies affiliated with IARU, apply through the society.

International Amateur Radio Union, P.O. Box AAA, Newington, CT 06111, U.S.A.

THE FIRST WINNERS OF THE WORKED ALL CONTINENTS AWARD

The Worked All Continents (WAC) award began as the "WAC Club," and was announced in April 1926 *QST*. The first certificate was issued on April 13, 1926, to Brandon Wentworth, u6OI. Call signs of the era need some explanation to today's hams. The official call signs issued in 1926 were generally in the format of one numeral followed by two or three letters, with no official indicator of country. The lower-case first letter or letters were called "intermediates," and indicated the station's country. The intermediate "u" for example, desgnated the United States. The other initial qualifiers were u6HM, u1AAO, c4GT (Canada), pr4SA (Puerto Rico), u9ZT-9XAX, b4YZ (Belgium), and gi5NJ (Ireland). The WAC Club membership certificates were signed by the "Grand Wacker" of the IARU, perhaps a distant relative of the "Old Sock" who signs the Rag Chewer's Club certificates.

CQ USA-CA AWARD PROGRAM

The United States of America Counties Award sponsored by *CQ* is issued for confirmed two-way radio contacts with specified numbers of U.S. counties under rules and conditions hereafter stated.

A. Award Classes

The USA-CA Award is issued in seven different classes, each a separate achievement as endorsed on the basic certificate by use of special seals for higher class. Also, special endorsements are made for all one band or mode questions subject to the rules.

Class	Counties Required	States Required
USA-500	500	any
USA-1000	1000	25
USA-1500	1500	45
USA-2000	2000	50
USA-2500	2500	50
USA-3000	3000	50

USA 3076-CA for ALL counties and Special Honors Plaque $50.00

B. Conditions

1. USA-CA is available to all licensed amateurs everywhere in the world and is issued to them as individuals for all county contacts made, regardless of calls held, operating QTHs, or dates.

Special USA-CA Awards are also available to SWLs on a heard basis.

2. All contacts must be confirmed by QSL, and such QSLs must be in the applicant's possession for identification by certification officials.

3. Any QSL card found to be altered in any way disqualifies the applicant.

4. QSOs via repeaters, satellites, moon bounce, and phone patches are **not** valid for USA-CA.

5. So-called "team" contacts, wherein one person acknowledges a signal report and another returns a signal report, while both amateur callsigns are logged, are **not** valid for USA-CA. Acceptable contact may be made with only one station at a time.

C. County Identity

1. Unless otherwise indicated on QSL cards, the QTH printed on cards will determine county identity.

2. The National Zip Code & Directory of Post Offices will be helpful in some cases in determining the identity of counties of contact as ascertained by name or nearest municipality. Publication No. 65 is available from the Superintendent of Documents, U.S. Government Printing Office, Washington, D.C. 20402, Stock No. 039-000-00264-7, shipped one, to U.S.A. or Canada.

3. For mobile and portable operations the postmark shall identify the county unless information stated on QSL cards makes other positive identity.

4. In the case of cities, parks, or reservations not within counties proper, applicants may claim any one of the adjoining counties for credit (once).

5. QSOs via repeaters, satellites, moon bounce, and phone patches are *not* valid for USA-CA.

D. Administration of USA-CA Program

1. The USA-CA program is administered by a *CQ* staff member acting as USA-CA Custodian, and all applications and related correspondence should be sent directly to this person's QTH.

2. Decisions of the Custodian in administering these rules and their interpretation, including future amendments, are final.

E. Record Book and Bookkeeping

1. The scope of the USA-CA Award makes it mandatory that special Record Books be used for application. For this purpose, *CQ* has provided a 64-page 4 1/4" x 11" Record Book which contains application and certification forms and provides record-log space meeting the conditions of any class award and/or endorsement requested.

2. A completed USA-CA Record Book constitutes the medium of basic application and becomes the property of *CQ* for record purposes. On subsequent applications for either higher classes or for special endorsements, the applicant may use additional Record Books to list required data, or he or she may make up alphabetical lists conforming to requirements.

3. Record Books are to be obtained directly from CQ, 76 North Broadway, Hicksville, NY 11801, for $2.00 each. We recommend that two be obtained—one for application use and one for a personal file copy.

F. Application

1. Make Record Book entries necessary for county identity and enter other log data necessary to satisfy any special endorsements (band/mode) requested.

2. Have the certification form provided signed by two licensed amateurs (General Class or higher) or an official of a national-level radio organization or affiliated club verifying that QSL cards for all contacts as listed have been seen. The USA-CA Custodian reserves the right to request any specific cards to satisfy any doubt whatsoever. In such cases the applicant should send sufficient postage for the return of cards by registered mail.

3. Send the *original* completed Record Book (*not* a copy) and certification forms and handling fee. Fee for *CQ* subscribers is $4.00 or 12 IRCs (subscribers must include a recent *CQ* mailing label), and for non-subscribers it is $10.00 or 40 IRCs. Send to Norm Van Raay, WA3RTY, Box 76, Pleasant Mount, PA 18453-0076. For later applications for higher class seals, send Record Book or self-prepared list per rules and $1.25 or 6 IRCs handling charge. For application for later special endorsements (band/mode) where certificates must be returned for endorsement, send certificate and $1.50 or 8 IRCs for handling charges. *Note:* At the time any USA-CA Award certificate is being processed, there are no charges other than the basic fee, regardless of the number of endorsements or seals; likewise, one may skip lower classes of USA-CA and get higher classes without losing any lower awards credits or paying any fee for them. *Also note:* IRCs are *not* accepted from U.S. stations.

UNITED STATES COUNTIES

ALASKA
First
Second
Third
Fourth

ALABAMA
Autauga
Baldwin
Barbour
Bibb
Blount
Bullock
Butler
Calhoun
Chambers
Cherokee
Chilton
Choctaw
Clarke
Clay
Cleburne
Coffee
Colbert
Conecuh
Coosa
Covington
Crenshaw
Cullman
Dale
Dallas
DeKalb
Elmore
Escambia
Etowah
Fayette
Franklin
Geneva
Greene
Hale
Henry
Houston
Jackson
Jefferson
Lamar
Lauderdale
Lawrence
Lee
Limestone
Lowndes
Macon
Madison
Marengo
Marion
Marshall
Mobile
Monroe
Montgomery
Morgan
Perry
Pickens
Pike
Randolph
Russell
SaintClair
Shelby
Sumter
Talladega
Tallapoosa
Tuscaloosa
Walker
Washington
Wilcox
Winston

ARKANSAS

Arkansas
Ashley
Baxter
Benton
Boone
Bradley
Calhoun
Carroll
Chicot
Clark
Clay
Cleburne
Cleveland
Columbia
Conway
Craighead
Crawford
Crittenden
Cross
Dallas
Desha
Drew
Faulkner
Franklin
Fulton
Garland
Grant
Greene
Hempstead
HotSpring
Howard
Independence
Izard
Jackson
Jefferson
Johnson
Lafayette
Lawrence
Lee
Lincoln
LittleRiver
Logan
Lonoke
Madison
Marion
Miller
Mississippi
Monroe
Montgomery
Nevada
Newton
Ouachita
Perry
Phillips
Pike
Poinsett
Polk
Pope
Prairie
Pulaski
Randolph
Saline
Scott
Searcy
Sebastian
Sevier
Sharp
St.Francis
Stone
Union
VanBuren
Washington
White
Woodruff
Yell

ARIZONA
Apache
Cochise
Coconino
Gila
Graham
Greenlee
LaPaz
Maricopa
Mohave
Navajo
Pima
Pinal
SantaCruz
Yavapai
Yuma

CALIFORNIA
Alameda
Alpine
Amador
Butte
Calaveras
Colusa
ContraCosta
DelNorte
ElDorado
Fresno
Glenn
Humboldt
Imperial
Inyo
Kern
Kings
Lake
Lassen
LosAngeles
Madera
Marin
Mariposa
Mendocino
Merced
Modoc
Mono
Monterey
Napa
Nevada
Orange
Placer
Plumas
Riverside
Sacramento
SanBenito
SanBernadino
SanDiego
SanFrancisco
SanJoaquin
SanLuisObispo
SanMateo
SantaBarbara
SantaClara
SantaCruz
Shasta
Sierra
Siskiyou
Solano
Sonoma
Stanislaus
Sutter
Tehama
Trinity
Tulare
Tuolumne
Ventura
Yolo
Yuba

COLORADO
Adams
Alamosa
Arapahoe
Archuleta
Baca
Bent
Boulder
Chaffee
Cheyenne
ClearCreek
Conejos
Costilla   ·
Crowley
Custer
Delta
Denver
Dolores
Douglas
Eagle
ElPaso
Elbert
Fremont
Garfield
Gilpin
Grand
Gunnison
Hinsdale
Huerfano
Jackson
Jefferson
Kiowa
KitCarson
LaPlata
Lake
Larimer
LasAnimas
Lincoln
Logan
Mesa
Mineral
Moffat
Montezuma
Montrose
Morgan
Otero
Ouray
Park
Phillips
Pitkin
Prowers
Pueblo
RioBlanco
RioGrande
Routt
Saguache
SanJuan
SanMiguel
Sedgwick
Summit
Teller
Washington
Weld
Yuma

CONNECTICUT
Fairfield
Hartford
Litchfield
Middlesex
NewHaven
NewLondon
Tolland
Windham

DELAWARE
Kent
NewCastle
Sussex

FLORIDA
Alachua
Baker
Bay
Bradford
Brevard
Broward
Calhoun
Charlotte
Citrus
Clay
Collier
Columbia
Dade
DeSoto
Dixie
Duval
Escambia
Flagler
Franklin
Gadsden
Gilchrist
Glades
Gulf
Hamilton
Hardee
Hendry
Hernando
Highlands
Hillsborough
Holmes
IndianRiver
Jackson
Jefferson
Lafayette
Lake
Lee
Leon
Levy
Liberty
Madison
Manatee
Marion
Martin
Monroe
Nassau
Okaloosa
Okeechobee
Orange
Osceola
PalmBeach
Pasco
Pinellas
Polk
Putnam
SantaRosa
Sarasota
Seminole
St.Johns
St.Lucie
Sumter
Suwannee
Taylor
Union
Volusia
Wakulla
Walton
Washington

GEORGIA
Appling
Atkinson
Bacon
Baker
Baldwin
Banks
Barrow
Bartow
BenHill
Berrien
Bibb
Bleckley
Brantley
Brooks
Bryan
Bulloch
Burke
Butts
Calhoun
Camden
Candler
Carroll
Catoosa
Charlton
Chatham
Chattahoochee
Chattooga
Cherokee
Clarke
Clay
Clayton
Clinch
Cobb
Coffee
Colquitt
Columbia
Cook
Coweta
Crawford
Crisp
Dade
Dawson
Decatur
Dodge
Dooly
Dougherty
Douglas
Early
Echols
Effingham
Elbert
Emanuel
Evans
Fannin
Fayette
Floyd
Forsyth
Franklin
Fulton
Gilmer
Glascock
Glynn
Gordon
Grady
Greene
Gwinnett
Habersham
Hall
Hancock
Haralson
Harris
Hart
Heard

Henry
Houston
Irwin
Jackson
Jasper
JeffDavis
Jefferson
Jenkins
Johnson
Jones
Lamar
Lanier
Laurens
Lee
Liberty
Lincoln
Long
Lowndes
Lumpkin
Macon
Madison
Marion
McDuffie
McIntosh
Meriwether
Miller
Mitchell
Monroe
Montgomery
Morgan
Murray
Muscogee
Newton
Oconee
Oglethorpe
Paulding
Peach
Pickens
Pierce
Pike
Polk
Pulaski
Putnam
Quitman
Rabun
Randolph
Richmond
Rockdale
Schley
Screven
Seminole
Spalding
Stephens
Stewart
Sumter
Talbot
Taliaferro
Tattnall
Taylor
Telfair
Terrell
Thomas
Tift
Toombs
Towns
Treutlen
Troup
Turner
Twiggs
Union
Upson
Walker
Walton
Ware
Warren
Washington
Wayne

Webster
Wheeler
White
Whitfield
Wilcox
Wilkes
Wilkinson
Worth

HAWAII

Hawaii
Honolulu
Kalawao
Kauai
Maui

IOWA

Adair
Adams
Allamakee
Appanoose
Audubon
Benton
BlackHawk
Boone
Bremer
Buchanan
BuenaVista
Butler
Calhoun
Carroll
Cass
Cedar
CerroGordo
Cherokee
Chickasaw
Clarke
Clay
Clayton
Clinton
Crawford
Dallas
Davis
Decatur
Delaware
DesMoines
Dickinson
Dubuque
Emmet
Fayette
Floyd
Franklin
Fremont
Greene
Grundy
Guthrie
Hamilton
Hancock
Hardin
Harrison
Henry
Howard
Humboldt
Ida
Iowa
Jackson
Jasper
Jefferson
Johnson
Jones
Keokuk
Kossuth
Lee
Linn
Louisa
Lucas
Lyon

Madison
Mahaska
Marion
Marshall
Mills
Mitchell
Monona
Monroe
Montgomery
Muscatine
O'Brien
Osceola
Page
PaloAlto
Plymouth
Pocahontas
Polk
Pottawattamie
Poweshiek
Ringgold
Sac
Scott
Shelby
Sioux
Story
Tama
Taylor
Union
VanBuren
Wapello
Warren
Washington
Wayne
Webster
Winnebago
Winneshiek
Woodbury
Worth
Wright

IDAHO

Ada
Adams
Bannock
BearLake
Benewah
Bingham
Blaine
Boise
Bonner
Bonneville
Boundary
Butte
Camas
Canyon
Caribou
Cassia
Clark
Clearwater
Custer
Elmore
Franklin
Fremont
Gem
Gooding
Idaho
Jefferson
Jerome
Kootenai
Latah
Lemhi
Lewis
Lincoln
Madison
Minidoka
NezPerce
Oneida

Owyhee
Payette
Power
Shoshone
Teton
TwinFalls
Valley
Washington

ILLINOIS

Adams
Alexander
Bond
Boone
Brown
Bureau
Calhoun
Carroll
Cass
Champaign
Christian
Clark
Clay
Clinton
Coles
Cook
Crawford
Cumberland
DeKalb
DeWitt
Douglas
DuPage
Edgar
Edwards
Effingham
Fayette
Ford
Franklin
Fulton
Gallatin
Greene
Grundy
Hamilton
Hancock
Hardin
Henderson
Henry
Iroquois
Jackson
Jasper
Jefferson
Jersey
JoDaviess
Johnson
Kane
Kankakee
Kendall
Knox
LaSalle
Lake
Lawrence
Lee
Livingston
Logan
Macon
Macoupin
Madison
Marion
Marshall
Mason
Massac
McDonough
McHenry
McLean
Menard
Mercer
Monroe

Montgomery
Morgan
Moultrie
Ogle
Peoria
Perry
Piatt
Pike
Pope
Pulaski
Putnam
Randolph
Richland
RockIsland
Saline
Sangamon
Schuyler
Scott
Shelby
St.Clair
Stark
Stephenson
Tazewell
Union
Vermilion
Wabash
Warren
Washington
Wayne
White
Whiteside
Will
Williamson
Winnebago
Woodford

INDIANA

Adams
Allen
Bartholomew
Benton
Blackford
Boone
Brown
Carroll
Cass
Clark
Clay
Clinton
Crawford
Daviess
DeKalb
Dearborn
Decatur
Delaware
Dubois
Elkhart
Fayette
Floyd
Fountain
Franklin
Fulton
Gibson
Grant
Greene
Hamilton
Hancock
Harrison
Hendricks
Henry
Howard
Huntington
Jackson
Jasper
Jay
Jefferson
Jennings

Johnson
Knox
Kosciusko
LaPorte
Lagrange
Lake
Lawrence
Madison
Marion
Marshall
Martin
Miami
Monroe
Montgomery
Morgan
Newton
Noble
Ohio
Orange
Owen
Parke
Perry
Pike
Porter
Posey
Pulaski
Putnam
Randolph
Ripley
Rush
Scott
Shelby
Spencer
St.Joseph
Starke
Steuben
Sullivan
Switzerland
Tippecanoe
Tipton
Union
Vanderburgh
Vermillion
Vigo
Wabash
Warren
Warrick
Washington
Wayne
Wells
White
Whitley

KANSAS

Allen
Anderson
Atchison
Barber
Barton
Bourbon
Brown
Butler
Chase
Chautauqua
Cherokee
Cheyenne
Clark
Clay
Cloud
Coffey
Comanche
Cowley
Crawford
Decatur
Dickinson
Doniphan
Douglas

Edwards
Elk
Ellis
Ellsworth
Finney
Ford
Franklin
Geary
Gove
Graham
Grant
Gray
Greeley
Greenwood
Hamilton
Harper
Harvey
Haskell
Hodgeman
Jackson
Jefferson
Jewell
Johnson
Kearny
Kingman
Kiowa
Labette
Lane
Leavenworth
Lincoln
Linn
Logan
Lyon
Marion
Marshall
McPherson
Meade
Miami
Mitchell
Montgomery
Morris
Morton
Nemaha
Neosho
Ness
Norton
Osage
Osborne
Ottawa
Pawnee
Phillips
Pottawatomie
Pratt
Rawlins
Reno
Republic
Rice
Riley
Rooks
Rush
Russell
Saline
Scott
Sedgwick
Seward
Shawnee
Sheridan
Sherman
Smith
Stafford
Stanton
Stevens
Sumner
Thomas
Trego
Wabaunsee
Wallace

Washington
Wichita
Wilson
Woodson
Wyandotte

KENTUCKY
Adair
Allen
Anderson
Ballard
Barren
Bath
Bell
Boone
Bourbon
Boyd
Boyle
Bracken
Breathitt
Breckenridge
Bullitt
Butler
Caldwell
Calloway
Campbell
Carlisle
Carroll
Carter
Casey
Christian
Clark
Clay
Clinton
Crittenden
Cumberland
Daviess
Edmonson
Elliott
Estill
Fayette
Fleming
Floyd
Franklin
Fulton
Gallatin
Garrard
Grant
Graves
Grayson
Green
Greenup
Hancock
Hardin
Harlan
Harrison
Hart
Henderson
Henry
Hickman
Hopkins
Jackson
Jefferson
Jessamine
Johnson
Kenton
Knott
Knox
Larue
Laurel
Lawrence
Lee
Leslie
Letcher
Lewis
Lincoln
Livingston

Logan
Lyon
Madison
Magoffin
Marion
Marshall
Martin
Mason
McCracken
McCreary
McLean
Meade
Menifee
Mercer
Metcalfe
Monroe
Montgomery
Morgan
Muhlenberg
Nelson
Nicholas
Ohio
Oldham
Owen
Owsley
Pendleton
Perry
Pike
Powell
Pulaski
Robertson
Rockcastle
Rowan
Russell
Scott
Shelby
Simpson
Spencer
Taylor
Todd
Trigg
Trimble
Union
Warren
Washington
Wayne
Webster
Whitley
Wolfe
Woodford

LOUISIANA
Acadia
Allen
Ascension
Assumption
Avoyelles
Beauregard
Bienville
Bossier
Caddo
Calcasieu
Caldwell
Cameron
Catahoula
Claiborne
Concordia
DeSoto
E.BatonRouge
EastCarroll
EastFeliciana
Evangeline
Franklin
Grant
Iberia
Iberville
Jackson

Jefferson
JeffersonDavis
LaSalle
Lafayette
Lafourche
Lincoln
Livingston
Madison
Morehouse
Natchitoches
Orleans
Ouachita
Plaquemines
PointeCoupee
Rapides
RedRiver
Richland
Sabine
St.Bernard
St.Charles
St.Helena
St.James
St.JohntheBts
St.Landry
St.Martin
St.Mary
St.Tammany
Tangipahoa
Tensas
Terrebonne
Union
Vermilion
Vernon
W.BatonRouge
Washington
Webster
WestCarroll
WestFeliciana
Winn

MASSACHUSETTS
Barnstable
Berkshire
Bristol
Dukes
Essex
Franklin
Hampden
Hampshire
Middlesex
Nantucket
Norfolk
Plymouth
Suffolk
Worcester

MARYLAND
Allegany
AnneArundel
Baltimore
BaltimoreCity
Calvert
Caroline
Carroll
Cecil
Charles
Dorchester
Frederick
Garrett
Harford
Howard
Kent
Montgomery
PrinceGeorges
QueenAnnes
Somerset
St.Marys

Talbot
Washington
Wicomico
Worcester

MAINE
Androscoggin
Aroostook
Cumberland
Franklin
Hancock
Kennebec
Knox
Lincoln
Oxford
Penobscot
Piscataquis
Sagadahoc
Somerset
Waldo
Washington
York

MICHIGAN
Alcona
Alger
Allegan
Alpena
Antrim
Arenac
Baraga
Barry
Bay
Benzie
Berrien
Branch
Calhoun
Cass
Charlevoix
Cheboygan
Chippewa
Clare
Clinton
Crawford
Delta
Dickinson
Eaton
Emmet
Genesee
Gladwin
Gogebic
GrandTraverse
Gratiot
Hillsdale
Houghton
Huron
Ingham
Ionia
Iosco
Iron
Isabella
Jackson
Kalamazoo
Kalkaska
Kent
Keweenaw
Lake
Lapeer
Leelanau
Lenawee
Livingston
Luce
Mackinac
Macomb
Manistee
Marquette
Mason

Mecosta
Menominee
Midland
Missaukee
Monroe
Montcalm
Montmorency
Muskegon
Newaygo
Oakland
Oceana
Ogemaw
Ontonagon
Osceola
Oscoda
Otsego
Ottawa
PresqueIsle
Roscommon
Saginaw
Sanilac
Schoolcraft
Shiawassee
St.Clair
St.Joseph
Tuscola
VanBuren
Washtenaw
Wayne
Wexford

MINNESOTA
Aitkin
Anoka
Becker
Beltrami
Benton
BigStone
BlueEarth
Brown
Carlton
Carver
Cass
Chippewa
Chisago
Clay
Clearwater
Cook
Cottonwood
CrowWing
Dakota
Dodge
Douglas
Faribault
Fillmore
Freeborn
Goodhue
Grant
Hennepin
Houston
Hubbard
Isanti
Itasca
Jackson
Kanabec
Kandiyohi
Kittson
Koochiching
L.oftheWoods
LacQuiParle
Lake
LeSueur
Lincoln
Lyon
Mahnomen
Marshall
Martin

McLeod
Meeker
MilleLacs
Morrison
Mower
Murray
Nicollet
Nobles
Norman
Olmsted
OtterTail
Pennington
Pine
Pipestone
Polk
Pope
Ramsey
RedLake
Redwood
Renville
Rice
Rock
Roseau
Scott
Sherburne
Sibley
St.Louis
Stearns
Steele
Stevens
Swift
Todd
Traverse
Wabasha
Wadena
Waseca
Washington
Watonwan
Wilkin
Winona
Wright
YellowMedicine

MISSOURI
Adair
Andrew
Atchison
Audrain
Barry
Barton
Bates
Benton
Bollinger
Boone
Buchanan
Butler
Caldwell
Callaway
Camden
CapeGirardeau
Carroll
Carter
Cass
Cedar
Chariton
Christian
Clark
Clay
Clinton
Cole
Cooper
Crawford
Dade
Dallas
Daviess
DeKalb
Dent

Douglas
Dunklin
Franklin
Gasconade
Gentry
Greene
Grundy
Harrison
Henry
Hickory
Holt
Howard
Howell
Iron
Jackson
Jasper
Jefferson
Johnson
Knox
Laclede
Lafayette
Lawrence
Lewis
Lincoln
Linn
Livingston
Macon
Madison
Maries
Marion
McDonald
Mercer
Miller
Mississippi
Moniteau
Monroe
Montgomery
Morgan
NewMadrid
Newton
Nodaway
Oregon
Osage
Ozark
Pemiscot
Perry
Pettis
Phelps
Pike
Platte
Polk
Pulaski
Putnam
Ralls
Randolph
Ray
Reynolds
Ripley
Saline
Schuyler
Scotland
Scott
Shannon
Shelby
St.Charles
St.Clair
St.Francois
St.Louis
St.LouisCity
Ste.Genevieve
Stoddard
Stone
Sullivan
Taney
Texas
Vernon
Warren

Washington
Wayne
Webster
Worth
Wright

MISSISSIPPI
Adams
Alcorn
Amite
Attala
Benton
Bolivar
Calhoun
Carroll
Chickasaw
Choctaw
Claiborne
Clarke
Clay
Coahoma
Copiah
Covington
DeSoto
Forrest
Franklin
George
Greene
Grenada
Hancock
Harrison
Hinds
Holmes
Humphreys
Issaquena
Itawamba
Jackson
Jasper
Jefferson
JeffersonDavis
Jones
Kemper
Lafayette
Lamar
Lauderdale
Lawrence
Leake
Lee
Leflore
Lincoln
Lowndes
Madison
Marion
Marshall
Monroe
Montgomery
Neshoba
Newton
Noxubee
Oktibbeha
Panola
PearlRiver
Perry
Pike
Pontotoc
Prentiss
Quitman
Rankin
Scott
Sharkey
Simpson
Smith
Stone
Sunflower
Tallahatchie
Tate

Tippah
Tishomingo
Tunica
Union
Walthall
Warren
Washington
Wayne
Webster
Wilkinson
Winston
Yalobusha
Yazoo

MONTANA
Beaverhead
BigHorn
Blaine
Broadwater
Carbon
Carter
Cascade
Chouteau
Custer
Daniels
Dawson
DeerLodge
Fallon
Fergus
Flathead
Gallatin
Garfield
Glacier
GoldenValley
Granite
Hill
Jefferson
JudithBasin
Lake
LewisandClark
Liberty
Lincoln
Madison
McCone
Meagher
Mineral
Missoula
Musselshell
Park
Petroleum
Phillips
Pondera
PowderRiver
Powell
Prairie
Ravalli
Richland
Roosevelt
Rosebud
Sanders
Sheridan
SilverBow
Stillwater
SweetGrass
Teton
Toole
Treasure
Valley
Wheatland
Wibaux
Yellowstone

NORTH CAROLINA
Alamance
Alexander
Alleghany

Anson
Ashe
Avery
Beaufort
Bertie
Bladen
Brunswick
Buncombe
Burke
Cabarrus
Caldwell
Camden
Carteret
Caswell
Catawba
Chatham
Cherokee
Chowan
Clay
Cleveland
Columbus
Craven
Cumberland
Currituck
Dare
Davidson
Davie
Duplin
Durham
Edgecombe
Forsyth
Franklin
Gaston
Gates
Graham
Granville
Greene
Guilford
Halifax
Harnett
Haywood
Henderson
Hertford
Hoke
Hyde
Iredell
Jackson
Johnston
Jones
Lee
Lenoir
Lincoln
Macon
Madison
Martin
McDowell
Mecklenburg
Mitchell
Montgomery
Moore
Nash
NewHanover
Northampton
Onslow
Orange
Pamlico
Pasquotank
Pender
Perquimans
Person
Pitt
Polk
Randolph
Richmond
Robeson
Rockingham
Rowan

Rutherford
Sampson
Scotland
Stanly
Stokes
Surry
Swain
Transylvania
Tyrrell
Union
Vance
Wake
Warren
Washington
Watauga
Wayne
Wilkes
Wilson
Yadkin
Yancy

NORTH DAKOTA
Adams
Barnes
Benson
Billings
Bottineau
Bowman
Burke
Burleigh
Cass
Cavalier
Dickey
Divide
Dunn
Eddy
Emmons
Foster
GoldenValley
GrandForks
Grant
Griggs
Hettinger
Kidder
LaMoure
Logan
McHenry
McIntosh
McKenzie
McLean
Mercer
Morton
Mountrail
Nelson
Oliver
Pembina
Pierce
Ramsey
Ransom
Renville
Richland
Rolette
Sargent
Sheridan
Sioux
Slope
Stark
Steele
Stutsman
Towner
Traill
Walsh
Ward
Wells
Williams

NEBRASKA
Adams
Antelope
Arthur
Banner
Blaine
Boone
BoxButte
Boyd
Brown
Buffalo
Burt
Butler
Cass
Cedar
Chase
Cherry
Cheyenne
Clay
Colfax
Cuming
Custer
Dakota
Dawes
Dawson
Deuel
Dixon
Dodge
Douglas
Dundy
Fillmore
Franklin
Frontier
Furnas
Gage
Garden
Garfield
Gosper
Grant
Greeley
Hall
Hamilton
Harlan
Hayes
Hitchcock
Holt
Hooker
Howard
Jefferson
Johnson
Kearney
Keith
KeyaPaha
Kimball
Knox
Lancaster
Lincoln
Logan
Loup
Madison
McPherson
Merrick
Morrill
Nance
Nemaha
Nuckolls
Otoe
Pawnee
Perkins
Phelps
Pierce
Platte
Polk
RedWillow
Richardson
Rock
Saline

Sarpy
Saunders
ScottsBluff
Seward
Sheridan
Sherman
Sioux
Stanton
Thayer
Thomas
Thurston
Valley
Washington
Wayne
Webster
Wheeler
York

NEW HAMPSHIRE
Belknap
Carroll
Cheshire
Coos
Grafton
Hillsboro
Merrimack
Rockingham
Strafford
Sullivan

NEW JERSEY
Atlantic
Bergen
Burlington
Camden
CapeMay
Cumberland
Essex
Gloucester
Hudson
Hunterdon
Mercer
Middlesex
Monmouth
Morris
Ocean
Passaic
Salem
Somerset
Sussex
Union
Warren

NEW MEXICO
Bernalillo
Catron
Chaves
Cibola
Colfax
Curry
DeBaca
DonaAna
Eddy
Grant
Guadalupe
Harding
Hidalgo
Lea
Lincoln
LosAlamos
Luna
McKinley
Mora
Otero
Quay
RioArriba

Roosevelt
SanJuan
SanMiguel
Sandoval
SantaFe
Sierra
Socorro
Taos
Torrance
Union
Valencia

NEVADA
Churchill
Clark
Douglas
Elko
Esmeralda
Eureka
Humboldt
Lander
Lincoln
Lyon
Mineral
Nye
Pershing
Storey
Washoe
WhitePine

NEW YORK
Albany
Allegany
Bronx
Broome
Cattaraugus
Cayuga
Chautauqua
Chemung
Chenango
Clinton
Columbia
Cortland
Delaware
Dutchess
Erie
Essex
Franklin
Fulton
Genesee
Greene
Hamilton
Herkimer
Jefferson
Kings
Lewis
Livingston
Madison
Monroe
Montgomery
Nassau
NewYork
Niagara
Oneida
Onondaga
Ontario
Orange
Orleans
Oswego
Otsego
Putnam
Queens
Rensselaer
Rockland
Saratoga
Schenectady
Schoharie

Schuyler
Seneca
St.Lawrence
StatenIsland
Steuben
Suffolk
Sullivan
Tioga
Tompkins
Ulster
Warren
Washington
Wayne
Westchester
Wyoming
Yates

OHIO
Adams
Allen
Ashland
Ashtabula
Athens
Auglaize
Belmont
Brown
Butler
Carroll
Champaign
Clark
Clermont
Clinton
Columbiana
Coshocton
Crawford
Cuyahoga
Darke
Defiance
Delaware
Erie
Fairfield
Fayette
Franklin
Fulton
Gallia
Geauga
Greene
Guernsey
Hamilton
Hancock
Hardin
Harrison
Henry
Highland
Hocking
Holmes
Huron
Jackson
Jefferson
Knox
Lake
Lawrence
Licking
Logan
Lorain
Lucas
Madison
Mahoning
Marion
Medina
Meigs
Mercer
Miami
Monroe
Montgomery
Morgan
Morrow

Muskingum
Noble
Ottawa
Paulding
Perry
Pickaway
Pike
Portage
Preble
Putnam
Richland
Ross
Sandusky
Scioto
Seneca
Shelby
Stark
Summit
Trumbull
Tuscarawas
Union
VanWert
Vinton
Warren
Washington
Wayne
Williams
Wood
Wyandot

OKLAHOMA
Adair
Alfalfa
Atoka
Beaver
Beckham
Blaine
Bryan
Caddo
Canadian
Carter
Cherokee
Choctaw
Cimarron
Cleveland
Coal
Comanche
Cotton
Craig
Creek
Custer
Delaware
Dewey
Ellis
Garfield
Garvin
Grady
Grant
Greer
Harmon
Harper
Haskell
Hughes
Jackson
Jefferson
Johnston
Kay
Kingfisher
Kiowa
Latimer
LeFlore
Lincoln
Logan
Love
Major
Marshall
Mayes

McClain
McCurtain
McIntosh
Murray
Muskogee
Noble
Nowata
Okfuskee
Oklahoma
Okmulgee
Osage
Ottawa
Pawnee
Payne
Pittsburg
Pontotoc
Pottawatomie
Pushmataha
RogerMills
Rogers
Seminole
Sequoyah
Stephens
Texas
Tillman
Tulsa
Wagoner
Washington
Washita
Woods
Woodward

OREGON
Baker
Benton
Clackamas
Clatsop
Columbia
Coos
Crook
Curry
Deschutes
Douglas
Gilliam
Grant
Harney
HoodRiver
Jackson
Jefferson
Josephine
Klamath
Lake
Lane
Lincoln
Linn
Malheur
Marion
Morrow
Multnomah
Polk
Sherman
Tillamook
Umatilla
Union
Wallowa
Wasco
Washington
Wheeler
Yamhill

PENNSYLVANIA
Adams
Allegheny
Armstrong
Beaver
Bedford
Berks

Blair	Barnwell	Hyde	Humphreys	Briscoe	Hardin
Bradford	Beaufort	Jackson	Jackson	Brooks	Harris
Bucks	Berkeley	Jerauld	Jefferson	Brown	Harrison
Butler	Calhoun	Jones	Johnson	Burleson	Hartley
Cambria	Charleston	Kingsbury	Knox	Burnet	Haskell
Cameron	Cherokee	Lake	Lake	Caldwell	Hays
Carbon	Chester	Lawrence	Lauderdale	Calhoun	Hemphill
Centre	Chesterfield	Lincoln	Lawrence	Callahan	Henderson
Chester	Clarendon	Lyman	Lewis	Cameron	Hidalgo
Clarion	Colleton	Marshall	Lincoln	Camp	Hill
Clearfield	Darlington	McCook	Loudon	Carson	Hockley
Clinton	Dillon	McPherson	Macon	Cass	Hood
Columbia	Dorchester	Meade	Madison	Castro	Hopkins
Crawford	Edgefield	Mellette	Marion	Chambers	Houston
Cumberland	Fairfield	Miner	Marshall	Cherokee	Howard
Dauphin	Florence	Minnehaha	Maury	Childress	Hudspeth
Delaware	Georgetown	Moody	McMinn	Clay	Hunt
Elk	Greenville	Pennington	McNairy	Cochran	Hutchinson
Erie	Greenwood	Perkins	Meigs	Coke	Irion
Fayette	Hampton	Potter	Monroe	Coleman	Jack
Forest	Horry	Roberts	Montgomery	Collin	Jackson
Franklin	Jasper	Sanborn	Moore	Collingsworth	Jasper
Fulton	Kershaw	Shannon	Morgan	Colorado	JeffDavis
Greene	Lancaster	Spink	Obion	Comal	Jefferson
Huntingdon	Laurens	Stanley	Overton	Comanche	JimHogg
Indiana	Lee	Sully	Perry	Concho	JimWells
Jefferson	Lexington	Todd	Pickett	Cooke	Johnson
Juniata	Marion	Tripp	Polk	Coryell	Jones
Lackawanna	Marlboro	Turner	Putnam	Cottle	Karnes
Lancaster	McCormick	Union	Rhea	Crane	Kaufman
Lawrence	Newberry	Walworth	Roane	Crockett	Kendall
Lebanon	Oconee	Yankton	Robertson	Crosby	Kenedy
Lehigh	Orangeburg	Ziebach	Rutherford	Culberson	Kent
Luzerne	Pickens		Scott	Dallam	Kerr
Lycoming	Richland	**TENNESSEE**	Sequatchie	Dallas	Kimble
McKean	Saluda	Anderson	Sevier	Dawson	King
Mercer	Spartanburg	Bedford	Shelby	DeWitt	Kinney
Mifflin	Sumter	Benton	Smith	DeafSmith	Kleberg
Monroe	Union	Bledsoe	Stewart	Delta	Knox
Montgomery	Williamsburg	Blount	Sullivan	Denton	LaSalle
Montour	York	Bradley	Sumner	Dickens	Lamar
Northampton		Campbell	Tipton	Dimmit	Lamb
Northumberland	**SOUTH**	Cannon	Trousdale	Donley	Lampasas
Perry	**DAKOTA**	Carroll	Unicoi	Duval	Lavaca
Philadelphia	Aurora	Carter	Union	Eastland	Lee
Pike	Beadle	Cheatham	VanBuren	Ector	Leon
Potter	Bennett	Chester	Warren	Edwards	Liberty
Schuylkill	BonHomme	Claiborne	Washington	ElPaso	Limestone
Snyder	Brookings	Clay	Wayne	Ellis	Lipscomb
Somerset	Brown	Cocke	Weakley	Erath	LiveOak
Sullivan	Brule	Coffee	White	Falls	Llano
Susquehanna	Buffalo	Crockett	Williamson	Fannin	Loving
Tioga	Butte	Cumberland	Wilson	Fayette	Lubbock
Union	Campbell	Davidson		Fisher	Lynn
Venango	CharlesMix	DeKalb	**TEXAS**	Floyd	Madison
Warren	Clark	Decatur	Anderson	Foard	Marion
Washington	Clay	Dickson	Andrews	FortBend	Martin
Wayne	Codington	Dyer	Angelina	Franklin	Mason
Westmoreland	Corson	Fayette	Aransas	Freestone	Matagorda
Wyoming	Custer	Fentress	Archer	Frio	Maverick
York	Davison	Franklin	Armstrong	Gaines	McCulloch
	Day	Gibson	Atascosa	Galveston	McLennan
RHODE	Deuel	Giles	Austin	Garza	McMullen
ISLAND	Dewey	Grainger	Bailey	Gillespie	Medina
Bristol	Douglas	Greene	Bandera	Glasscock	Menard
Kent	Edmunds	Grundy	Bastrop	Goliad	Midland
Newport	FallRiver	Hamblen	Baylor	Gonzales	Milam
Providence	Faulk	Hamilton	Bee	Gray	Mills
Washington	Grant	Hancock	Bell	Grayson	Mitchell
	Gregory	Hardeman	Bexar	Gregg	Montague
SOUTH	Haakon	Hardin	Blanco	Grimes	Montgomery
CAROLINA	Hamlin	Hawkins	Borden	Guadalupe	Moore
Abbeville	Hand	Haywood	Bosque	Hale	Morris
Aiken	Hanson	Henderson	Bowie	Hall	Motley
Allendale	Harding	Henry	Brazoria	Hamilton	Nacogdoches
Anderson	Hughes	Hickman	Brazos	Hansford	Navarro
Bamberg	Hutchinson	Houston	Brewster	Hardeman	Newton

Nolan	Wilson	Fauquier	Caledonia	Douglas	Gilmer
Nueces	Winkler	Floyd	Chittendon	Dunn	Grant
Ochiltree	Wise	Fluvanna	Essex	EauClaire	Greenbrier
Oldham	Wood	Franklin	Franklin	Florence	Hampshire
Orange	Yoakum	Frederick	GrandIsle	FondduLac	Hancock
PaloPinto	Young	Giles	Lamoille	Forest	Hardy
Panola	Zapata	Gloucester	Orange	Grant	Harrison
Parker	Zavala	Goochland	Orleans	Green	Jackson
Parmer		Grayson	Rutland	GreenLake	Jefferson
Pecos	**UTAH**	Greene	Washington	Iowa	Kanawha
Polk	Beaver	Greensville	Windham	Iron	Lewis
Potter	BoxElder	Halifax	Windsor	Jackson	Lincoln
Presidio	Cache	Hanover		Jefferson	Logan
Rains	Carbon	Henrico	**WASHINGTON**	Juneau	Marion
Randall	Daggett	Henry	Adams	Kenosha	Marshall
Reagan	Davis	Highland	Asotin	Kewaunee	Mason
Real	Duchesne	IsleofWight	Benton	LaCrosse	McDowell
RedRiver	Emery	JamesCity	Chelan	Lafayette	Mercer
Reeves	Garfield	KingandQueen	Clallam	Langlade	Mineral
Refugio	Grand	KingGeorge	Clark	Lincoln	Mingo
Roberts	Iron	KingWilliam	Columbia	Manitowoc	Monongalia
Robertson	Juab	Lancaster	Cowlitz	Marathon	Monroe
Rockwall	Kane	Lee	Douglas	Marinette	Morgan
Runnels	Millard	Loudoun	Ferry	Marquette	Nicholas
Rusk	Morgan	Louisa	Franklin	Menominee	Ohio
Sabine	Piute	Lunenburg	Garfield	Milwaukee	Pendleton
SanAugustine	Rich	Madison	Grant	Monroe	Pleasants
SanJacinto	SaltLake	Mathews	GraysHarbor	Oconto	Pocahontas
SanPatricio	SanJuan	Mecklenburg	Island	Oneida	Preston
SanSaba	Sanpete	Middlesex	Jefferson	Outagamie	Putnam
Schleicher	Sevier	Montgomery	King	Ozaukee	Raleigh
Scurry	Summit	Nelson	Kitsap	Pepin	Randolph
Shackelford	Tooele	NewKent	Kittitas	Pierce	Ritchie
Shelby	Uintah	Northampton	Klickitat	Polk	Roane
Sherman	Utah	Northumberland	Lewis	Portage	Summers
Smith	Wasatch	Nottoway	Lincoln	Price	Taylor
Somervell	Washington	Orange	Mason	Racine	Tucker
Starr	Wayne	Page	Okanogan	Richland	Tyler
Stephens	Weber	Patrick	Pacific	Rock	Upshur
Sterling		Pittsylvania	PendOreille	Rusk	Wayne
Stonewall	**VIRGINIA**	Powhatan	Pierce	Sauk	Webster
Sutton	Accomack	PrinceEdward	SanJuan	Sawyer	Wetzel
Swisher	Albemarle	PrinceGeorge	Skagit	Shawano	Wirt
Tarrant	Alleghany	PrinceWilliam	Skamania	Sheboygan	Wood
Taylor	Amelia	Pulaski	Skamania	St.Croix	Wyoming
Terrell	Amherst	Rappahannock	Snohomish	Taylor	
Terry	Appomattox	Richmond	Spokane	Trempealeau	**WYOMING**
Throckmorton	Arlington	Roanoke	Stevens	Vernon	Albany
Titus	Augusta	Rockbridge	Thurston	Vilas	BigHorn
TomGreen	Bath	Rockingham	Wahkiakum	Walworth	Campbell
Travis	Bedford	Russell	WallaWalla	Washburn	Carbon
Trinity	Bland	Scott	Whatcom	Washington	Converse
Tyler	Botetourt	Shenandoah	Whitman	Waukesha	Crook
Upshur	Brunswick	Smyth	Yakima	Waupaca	Fremont
Upton	Buchanan	Southampton		Waushara	Goshen
Uvalde	Buckingham	Spotsylvania	**WISCONSiN**	Winnebago	HotSprings
ValVerde	Campbell	Stafford	Adams	Wood	Johnson
VanZandt	Caroline	Surry	Ashland		Laramie
Victoria	Carroll	Sussex	Barron	**WEST**	Lincoln
Walker	CharlesCity	Tazewell	Bayfield	**VIRGINIA**	Natrona
Waller	Charlotte	Warren	Brown	Barbour	Niobrara
Ward	Chesterfield	Washington	Buffalo	Berkeley	Park
Washington	Clarke	Westmoreland	Burnett	Boone	Platte
Webb	Craig	Wise	Calumet	Braxton	Sheridan
Wharton	Culpeper	Wythe	Chippewa	Brooke	Sublette
Wheeler	Cumberland	York	Clark	Cabell	Sweetwater
Wichita	Dickenson		Columbia	Calhoun	Teton
Wilbarger	Dinwiddie	**VERMONT**	Crawford	Clay	Uinta
Willacy	Essex	Addison	Dane	Doddridge	Washakie
Williamson	Fairfax	Bennington	Dodge	Fayette	Weston
			Door		

List Courtesy of N6OQZ

HOLDERS OF THE CQ USA-CA WORKED ALL U.S.A. COUNTIES AWARD
(All 3076 Counties)

Number	Callsign	Number	Callsign	Number	Callsign	Number	Callsign	Number	Callsign
1	K9EAB	70	N2CW	141	W6CCM	211	W4OWY	281	W4JUJ
1	KFØLZ	71	WA4LMR	142	K8EUX	212	KØMT	282	K4ZT
2	WØBK	72	N4YY	143	VE4QZ	213	WB4UPW	283	W6NAT
3	K8CIR	73	N5DGQ	144	WA7GMX	214	K8WXJ	284	WB2NFB
4	W2QHH	74	WB6CPE	145	WAØUPL	215	W5FS	285	K2UVG
5	WØBL	75	W4HA	146	W7PXA	216	AD8W	286	N9ER
6	WØGYM	76	W6TCD	147	W8ZCV	217	WB6VRR	287	KC7EI
7	K1QZV	77	WAØLRQ	148	W9ABM	218	K9DZG	288	K6RLR
8	K8IWI	78	K3LXN	149	WAØKQQ	219	WO4L	289	WØDSY
9	K8KOM	79	WAØGZA	150	WØGQR	220	WB9SPD	290	W5VNW
10	NS9Y	80	W2KXL	151	K5JBC	221	W7CB	291	WB4ZXP
11	WAØEVO	81	K1OAZ	152	W6QPF	222	WØUM	292	WA3ZTY
12	W2JWK	82	W7OK	153	K1IK	223	WA9GOH	293	KBØXB
13	K4LSP	83	VE3CBY	154	W2CUC	224	VE4XN	294	WAØRJJ
14	W8UMR	84	W4IZR	155	KB4IF	225	WØHNV	295	W7ULC
15	N4PN	85	K5BTM	156	WA3TUC	226	KC9A	296	W5UMD
16	W7KOI	86	K9DCJ	157	VE4EL	227	K1KPS	297	W4EHN
17	K1CXP	87	WB4FBS	158	W8CXS	228	K3LK	298	WB7AYN
18	K5DRF	88	WDX6ETT	159	W4JVN	229	W8NJC	299	KE6RN
19	WA4BMC	89	K7JWZ	160	W8RSW	230	WB9OOE	300	KL7MF
20	K4AUL	90	WA4MGC	161	WA9CNV	231	K5GC	301	WA3ZMY
21	WA3HGV	91	WB6EXT	162	K8NQP	232	WA9OBR	302	K9DAF
22	W6HVU	92	W8WT	163	W1LQQ	233	WØIU	303	N2RT
23	W1EQ	93	K8ODY	164	KBØKS	234	KB5KM	304	WDØEMS
24	K8BHG	94	W3FNT	164	WAØYJL	235	WD4HRN	305	W2MEI
25	W8DCD	95	WB5DVT	165	AB5C	236	WØDG	306	N7CLA
26	W8DCH	96	WPE9ETT	166	W6HDV	237	WAØUHC	307	VE4ZX
27	WØKZZ	97	WA9OFF	167	WØFBB	238	KK6DU	308	WA2JFL
28	W7LUQ	98	K8IQB	168	KW3F	238	WA6JJC	309	WD9BCG
29	WA7IRD	99	W6UNP	169	W5AWT	239	N6QS	310	VE3BFJ
30	W4NXD	100	K8DCR	170	WA6ZUD	240	K7WUR	311	K4ZA
31	K7WQJ	101	W4LK	171	W2QKJ	241	KC4AE	312	WA6GQY
32	WØSJE	102	WA4LSU	172	W4YWV	242	W9LMT	313	WD4HVZ
33	NJØC	103	WB6RMZ	173	K1UNM	243	WA6LBO	314	AH6IP
34	WB2SJQ	104	K9HRC	174	W9CNG	244	W4ISF	315	WB5YDH
35	W6DIX	105	W8UOQ	175	AK4N	245	NFØX	316	W8UPH
36	K7ZJP	106	KØAYO	176	WB9DCZ	246	WB8SNO	317	W5QEM
37	N4XE	107	K7SQD	177	AE5B	247	WA2GLU	318	KCØUJ
38	W2OST	108	WA3GLJ	178	WB6EGQ	248	KD9Q	319	WA3UQR
39	W5TQE	109	W5HDK	179	WA1UVX	249	WØRP	320	WB9RCY
40	N4GLV	110	W1AQE	180	W2MCY	250	K9GTQ	321	WA5ZDZ
41	WA4YQC	111	K4ELK	181	WA2WCW	251	W7NXZ	322	WBØAXN
42	WAØSHE	112	K5VYT	182	WBØCQO	252	VE3IR	323	W4LQF
43	W4KA	113	W3LDD	183	NI7B	253	WA6AQR	324	K6XZ
44	W7KBC	114	K9UTI	184	WB5C	254	WB4RVW	325	W4XT
45	K5KDG	115	W8WUT	185	W7GHT	255	WA3QVJ	326	W7LQT
46	K4ISE	116	WA6OTV	186	KØITP	256	WB9TKR	327	WB7OHB
47	WA8NDL	117	K5HKG	187	K2PBU	257	W1UYL	328	WB7WBZ
48	W4GGU	118	G4JZ	188	W9JR	258	KD6PP	329	K9CSL
49	W5VD	119	KØARS	189	KØPFV	259	K2QK	330	WA5DTK
50	K1WQU	120	N6UW	190	VE1RQ	260	W7DXN	331	6Y5RS
51	W6JHV	121	AC2O	191	N7TT	261	KØDJC	332	WA4AUL
52	ZL1KG	122	W5VPV	192	W2CUE	262	K7TM	333	NZØA
53	NFØN	123	WAØCEL	193	W4SWW	263	KD6HZ	334	WA2TJL
54	K7NN	124	WA4WQG	194	WB2NHP	264	K7GNC	335	W9FD
55	WAØSBR	125	KØEQY	195	WA9ZRP	265	VE3RN	336	W9VPE
56	K2OO	126	W9ZD	196	WB4TNY	266	W1FAB	337	AC8F
57	W4LXI	127	W1DIT	197	K7CLO	267	K3GOO	338	K1NWE
58	K5RPC	128	WA6MAR	198	WAØBPE	268	AC2J	339	WD8AYN
59	W9UZC	129	WA5YSC	199	WG9A	269	KØXT	340	WB9ELH
60	W8MJG	130	K9QGR	200	WØKMH	270	N4ANV	341	WB6SRK
61	WAØJRZ	131	W5UUM	201	K4IUO	271	WØLRH	342	K1YPR
62	N2BL	132	K1QFD	202	VE1DI	272	W1ORV	343	WB9YZE
63	WA9FZR	133	VE7ATI	203	W5LXG	273	K7JJ	344	WB5KEA
64	W3JZY	134	N7SU	204	K2LFG	274	K7SE	345	WW1N
65	WØAYL	135	WB5AKI	205	W2UP	275	K5IW	346	K9EHP
66	WA1RAN	136	W9CRN	206	K7LQI	276	WA4UNS	347	NGØT
67	WAØWOB	137	WA2GPT	207	W4SSU	277	W5VDW	348	WB5UJO
68	WA1CXE	138	GW3NWV	208	WN5MBS	278	SM4EAC	349	WD7N
69	W9DRL	139	W7WVD	209	WB4AIL	279	KE9O	349	WBØGRN
		140	WØGV	210	W7CUJ	280	WA9WGJ	350	KA5A

No.	Call	No.	Call	No.	Call	No.	Call	No.	Call
351	WBØMNE	425	WAØMQM	497	WA9QNI	571	N8EMV	645	W5RJH
352	N9ATA	426	K7XN	498	N9DEH	572	K6SLP	646	K1QPV
353	N8BNI	427	W5QLD	499	WA1KPJ	573	WB6TJW	647	KD9ZF
354	W5RBO	428	K2HVN	500	WA6VJP	574	K1CGI	648	KF4BU
355	WN4M	429	WA6UFY	501	W4RKV	575	K3GWA	649	WA6MUK
356	W1VJ	430	WØRSR	502	WB3IQJ	576	W6TKV	650	KD8HA
357	N5QQ	431	WDØCFZ	503	N9CLZ	577	W6PXE	651	WA4NBC
358	VE3GCO	432	KS7T	504	WAØLMK	578	W2FXA	652	WB9STT
358	VE3XN	432	WV5S	505	VE7OR	579	W7GQK	653	WAØZBK
359	KB7W	433	KKØL	506	K5HT	580	KY9Y	654	WA8RSQ
360	W1III	434	N1GMU	507	W3XE	581	NB8R	655	WA4KER
361	KC9DD	434	N1GMU	508	KA1CKX	582	NF8G	656	WD7X
362	NØBHO	435	K5CKQ	509	K5OUK	583	K7EQ	657	KA9ZRW
363	WAØYFQ	436	KX2W	510	W2DWO	584	WBØDPD	658	WB4UHN
364	KDØU	437	N5BDY	511	N9BDM	585	W6YMV	659	I2PHN
365	WA2AKJ	438	AB4HR	512	KB5FU	586	NK8P	660	WB7QID
366	WAØSGJ	439	KCØJG	513	KX1A	587	N9CVP	661	K5IID
367	AG9S	440	KCØVB	514	KC4OV	587	N8CVP	662	KA4IFF
368	KI4W	441	WBØTVL	515	W2CUK	588	W2EZ	663	WB9NUL
369	K1VSJ	442	NØCOL	516	AI9Y	589	W6NNV	664	K9ETB
370	W4ARH	443	KC2RS	517	KA4SAX	590	KAØHJR	665	KA9JOL
371	N7AKG	444	W3ARK	518	WA1JYO	591	WA2VQW	666	CT1TZ
372	N9CHU	445	N5DUQ	519	W3NB	592	K9KKX	667	W9MYY
373	K7GTK	446	W6KAW	520	WA4CHI	593	K5UTH	668	KØGEN
374	WØGOR	447	KC5UO	521	NG9L	594	K9MGF	669	N4UMR
375	N7BKW	448	W1CRL	522	G2AFQ	595	WØMHK	670	NT9V
376	K9BX	449	K8MW	523	KF6CN	596	DJ3OE	671	W2RPZ
377	K4QFK	450	WØEWH	524	AK8A	597	K2CTJ	672	K7OQZ
378	N4CCJ	451	K2YIY	525	W4IGW	598	WA6OCI	673	KØOJG
379	W1WHQ	452	WA2JTY	526	N9AUV	599	KJ4EJ	674	N6QA
380	W1SXX	453	W2IN	527	N4EED	600	KC3YT	675	K2POF
381	KS5A	454	VE7AIO	528	NØDPX	601	KA5RNH	676	N2CWG
382	K5MOF	455	KB3SN	529	KD4ON	602	WØIZV	677	KA1LSD
383	WB6ALC	456	KA2K	530	WA4WIN	603	W8DZL	678	W2EMW
384	KB7QO	457	AI1Q	531	W7IEU	604	WA6CQW	679	WU4S
385	WB6GMM	458	KB5DM	532	N4IWY	605	KF5HY	680	KC5P
386	KKØV	459	WØBXM	533	W4WXJ	606	KA6BTU	681	K9ZWH
387	AI5P	460	W7EFO	534	YV5AGD	607	W9ET	682	N8HAM
388	WA3HMJ	461	WD9ITF	535	WD4RAF	608	W7BKM	683	KAØNVT
389	K5WQM	462	W1EKZ	536	G4KHG	609	KF9FU	684	KX8Z
390	WBØLOU	463	KT4U	537	KA1NX	610	N8ESR	685	K5XY
391	KK5P	464	AA4FF	538	KC3X	611	N4OA	686	N5AWE
392	KM4W	465	K2POA	539	K8CW	612	NT7R	687	KD2NN
393	W5PWG	466	KU7F	540	K2UPD	613	WD9GSU	688	W9DC
394	WAØLKL	467	WA9EZT	541	WB7VIZ	614	W7HRD	689	AB4OI
395	WB5CWI	468	KB4XK	542	WØGOQ	615	ZL2ACP	690	K8MDU
396	WA9IWM	469	WB3DWH	543	W5ILR	616	NF9A	691	W7GVF
397	KD4PY	470	WA4PGM	544	KC5CV	617	WA1YZV	692	W3SQA
398	W8ILC	471	K6HZI	545	WBØVNN	618	W7HZL	693	K7IOO
399	N8BLO	472	N7CYQ	546	W2PDM	619	K8DTO	694	KG8I
400	VE3BHZ	473	WBØJYB	547	WA9ROU	620	W7KEU	695	AA4LY
401	N8BGF	474	N3ANX	548	N9DR	621	KD4ZJ	696	K8BXT
402	KU9G	475	WA7NNH	549	K1ZIT	622	WB2ABD	697	NU9M
403	W1JR	476	WBØODS	550	K8OHC	623	KF5AT	698	WA6KHK
404	W4KFA	477	N8GEQ	551	KA2CNG	624	N3DRO	699	WDØEAM
405	KCØMB	478	KC4IF	552	WØFF	625	WØOWY	700	K8GPC
406	W1APU	479	KCØVA	553	W5UJO	626	WT4S	701	KW3H
407	KN5I	480	K8OOK	554	W1TEE	627	AG2K	702	KCØZU
408	KD6PY	481	W3HQU	555	N4FSZ	628	K8IXU	703	OE2EGL
409	VE4SK	482	KZ2P	556	KØGSV	629	ZL2BCX	704	NV6L
410	N7AKT	483	N8CIJ	557	W4LHP	630	WB2HXZ	705	NA5F
411	KF5F	484	WA7YID	558	4X4JU	631	N5KGY	706	KF4FP
412	WA3QNT	485	WB2ZSO	559	KD8GL	631	K5KGY	707	VE2YM
413	WD9GBH	486	KG5J	560	WØNNH	632	N4KE	708	WI9C
414	WB2WZE	487	K4CCW	561	KC7JC	633	KJ5W	709	K8KIR
415	W8GZF	488	K4BZV	562	WB6FJU	634	HB9AFI	710	KB3WN
416	N9TN	489	NØCKC	563	KYØE	635	WPE6YL	711	W3RWJ
417	ON4UN	490	W7ULA	564	N6EBU	636	KB9ER	712	W5MW
418	NØCKN	491	W2CC	565	WA5INV	637	NØCYB	713	KA3DRO
419	W5EHY	492	NØAKC	566	N3AHA	638	N5JRH	714	NZ8Q
420	VE7ATH	493	KØIFL	567	W4UYC	639	NV4Z	715	K1BM
421	W2XQ	494	WDX9DCJ	568	W1WLW	640	VE2MS	716	KD9OT
422	N8AIL	495	WB9YCO	569	WB2RCJ	641	KB1GN	717	WA4IMC
423	KD4DJ	496	WDX4KEF	570	KC3AD	642	KD9KK	718	KA5VWD
						643	AK2H	719	W5UGD
						644	VE3OEE		

720	KJ4LG	751	HR1KAS	782	N4SMH	813	WB3HTK	846	KJ5PQ
721	N6ERM	752	K5AAY	783	WD4NEG	814	ND3T	847	KA3MMM
722	K4JFI	753	NWØF	784	W3DYA	815	N7OTR	848	KEØAY
723	ND1H	754	WD9HEB	785	KD2Q	816	PT2TF	849	N4UGH
724	KQ1Z	755	AC4MP	786	KM8U	817	W5VRA	850	KE2EA
725	WØAWP	756	K2NJ	787	N8FEB	818	AJ3X	851	K1DFO
726	WB9QNX	757	KN4Y	788	AA9CW	819	WD6CKT	852	N1FJR
727	W4HSA	758	G5PQ	789	WA2CNJ	820	K3IMC	853	W4XQ
728	VE1GU	759	KB3GN	790	N7LWX	821	KFØYF	854	KD4NFE
729	W6TMD	760	WD9HAW	791	HB9RG	822	WA2MUA	855	KBØFQC
730	N6PLQ	761	WA5OPO	792	KS3I	823	W2BUO	856	WB4HIN
731	VK5AQZ	762	N8ELQ	793	NV6I	824	WB3IET	857	NX4Z
732	N9HRX	763	N4IXV	794	WA4HXG	825	KA7JAS	858	WAØRKQ
733	W5RIT	764	WØWYX	795	NØLDT	826	KK7X	859	W6ISQ
734	WØWYJ	765	N6OKX	796	N2JNE	827	KG5UZ	860	N5BLK
735	K5TVC	766	I6FLD	797	AA4HD	828	AB4OQ	861	WS3F
736	VE3FNM	767	WB4CCT	798	WA8KIW	829	KA6SWI	862	KG5E
737	KC4DUP	768	WM9F	799	N6HJY	830	SM5BHW	863	KA4BHL
738	KA5ZXF	769	N7POK	800	AB4QD	831	N5UR	864	KS4Q
739	N8AJC	770	AA2AV	801	KO4QZ	834	NC6M	865	NØDRX
740	KC4SF	771	KD8HB	802	KI6YX	835	N6PYN	866	KM4ES
741	VE1AIT	772	W8PN	803	KA5PVB	836	W5FHL	867	KK6QW
742	K4BBF	773	WV2B	804	NO2W	837	WØDFK	868	GM3BCL
743	KJ4JC	774	KK6BB	805	N2ARE	838	WA2UJH	869	KE2FZ
744	WD8AGC	775	KN4JR	806	N1API	839	PS8YL	870	WU3H
745	KA1JPR	776	K6PQA	807	AA7CP	840	WØLLU	871	K1ER
746	KF7RU	777	SM4BNZ	808	KV1M	841	W9GPC	872	KE2C
747	WA1FNS	778	AA4LB	809	W3FG	842	AA6PI	873	AB5SL
748	W9OP	779	NW1O	810	KA9PZS	843	N5QOS		
749	K9AGB	780	N9AC	811	WU8Q	844	N4MYZ		
750	KE9CA	781	NC2O	812	WD9EJK	845	KM6QF		*(Courtesy N6OQZ)*

ARRL VHF/UHF CENTURY CLUB AWARD RULES

1. (a) The VHF/UHF Century Club Award (VUCC) is awarded for contact with a minimum number of "Maidenhead" 2 degree X 1 degree grid square locators per band as indicated below. Grid squares are designated by a combination of two letters and two numbers. More information on grid squares can be found in January 1983 QST, pp. 49-51 (reprint available on request). The ARRL World Grid Locator Atlas and the ARRL Grid Locator for North America (see latest QST for prices and ordering information). The VUCC certificate and endorsements are available to ARRL members in Canada, the U.S. and possessions, and Puerto Rico, and other amateurs worldwide.

(b) The minimum number of squares needed to initially qualify for each individual band award is as follows: 50 MHz—100; 144 MHz—100; Satellite—100; 220 MHz—50; 432 MHz—50; 902 MHz—25; 1296 MHz—25; 2.3 GHz—10; 3.4 GHz—5; 5.7 GHz—5; 10 GHz—5; and Laser—5. Certificates for 220 and 432 MHz are designed as Half Century, 902 and 1296 MHz as Quarter Century, and those above SHF Awards.

2. Only those contacts dated January 1, 1983, and later are creditable for VUCC purposes.

3. Individual band awards are endorsable in the following increments: (a) 50 and 144 MHz and Satellite—25 credits; (b) 220 and 432 MHz—10 credits; (c) 902 MHz and above—5 credits.

4. (a) Separate bands are considered as separate awards. (b) No crossband contacts permitted except for Satellite award. (c) No contacts through active "repeater" or satellite devices, or any other power relay methods are permitted for nonsatellite awards. (d) Contacts with aeronautical mobiles (in the air) do NOT count. (e) Stations who claim to operate from more than one grid square simultaneously (such as from the intersection of 4 grid squares) must be physically present in more than one square to give multiple square credit with a single contact. This requires the operator to know precisely where the intersection lines are located and placing the station exactly on the boundary to meet this test. To achieve this precision work requires either current markers permanently in place, or the precision work of a professional surveyor. Operators of such stations should be prepared to provide some evidence of meeting this test if called upon to do so. Multiple QSL cards are not required.

5. There are no specialty endorsements such as "CW only," etc.

6. For VUCC awards on 50 through 1296 MHz and Satellite, all contacts must be made from a location or locations within the same grid square or locations in different grid squares no more than 50 miles apart. For SHF awards, contacts must be made from a single location, defined as within a 300-meter diameter circle.

7. APPLICATION PROCEDURE (please follow carefully):

(a) Confirmations (QSLs) and application format (MSD-259 and MSD-260) must be submitted to an approved VHF Awards Manager for certification. ARRL Special Service Clubs appoint VHF Managers whose names are on file at Hq. If you do not know of an Awards Manager in your area, Hq. will give you the name of the closest manager. DO NOT SEND cards to Hq.

(b) For the convenience of the Awards Manager in checking cards, applicants may indicate in pencil (pencil only) the grid square locator on the address side of the cards that DO NOT clearly indicate the grid locator. The applicant affirms that he/she has accurately determined the proper locator from the address information given on the card by signing the affirmation statement on the application.

(c) Cards must be sorted by: (1) Alphabetical by field; (2) Numerical from 00 to 99 within that field.

(d) Where it is necessary to mail cards for certification, postage equal to the amount needed to send them will be included for return of cards along with a separate self-addressed mailing label. In addition, U.S. and Canadian applicants will enclose $1.00 U.S. for any initial application to cover postage and packaging of the certificate. For endorsements, enclose an S.A.S.E. with appropriate postage for the weight of your field sheets. When mailing cards, REGISTERED or CERTIFIED mail is recommended.

(e) Enclosed with the initial VUCC certificate from Hq. will be a photocopy of the original list of grid squares for which the applicant has received credit (MSD-259). When applying for endorsement, the applicant will indicate in RED on that photocopy those new grid squares for which credit is sought, and submit cards for certification to an Awards Manager. A new updated photocopy listing will be returned with the endorsement sticker. Thus, a current list of grid squares worked is always in the hands of the VUCC award holder, available to the VHF manager during certification, and a permanent historical record always maintained at Hq. Reminder: for initial application, enclose $1.00, and for endorsements, enclose an S.A.S.E with appropriate postage for the weight of your field sheets. For endorsement applications, it is only necessary to submit those MSD-259s that indicate new grid squares worked since the previous submission (indicated in RED).

8. DISQUALIFICATION. Altered/forged confirmations or fraudulent applications submitted may result in disqualification of the applicant from VUCC participation by action of the ARRL Awards Committee. The applicant affirms he/she has abided by all the rules of membership in the VUCC and agrees to be bound by the decisions of the ARRL Awards Committee.

9. Decisions of the ARRL Awards Committee regarding interpretation of the rules here printed or later amended shall be final.

10. Operating Ethics: Fair play and good sportsmanship in operating are required of all VUCC members.

(Courtesy of ARRL)

WORKED ALL STATES (WAS) AWARD RULES

1. The WAS (Worked All States) award is available to all amateurs worldwide who submit proof with written confirmation of having contacted each of the 50 states of the United States of America. The WAS awards program includes 10 different and separately numbered awards as listed below. In addition, ENDORSEMENT stickers are available as listed below.

2. Two-way communications must be established on amateur bands with each state. Specialty awards and endorsements must be two-way (2X) on that band and/or mode. There is no minimum signal report required. Any or all bands may be used for general WAS. The District of Columbia may be counted for Maryland.

3. Contacts must all be made from the same location, or from locations no two of which are more than 50 miles apart which is affirmed by signature of the applicant on the application. Club station applicants, please include clearly the club name and callsign of the club station (or trustee).

4. Contacts may be made over any period or years. Contacts must be confirmed in writing, preferably in the form of QSL cards. Written confirmations must be submitted (no photocopies). Confirmations must show your call and indicate that two-way communications was established. Applications for SPECIALTY awards or ENDORSEMENTS must submit confirmations that clearly confirm two-way contact on the specialty mode/band. Contacts made with Alaska must be dated January 3, 1959 or later and with Hawaii date August 21, 1959, or after.

5. Specialty awards (numbered separately) are available for OSCAR satellite, SSTV, RTTY, 432 MHz., 220 MHz., 144 MHz., 50 MHz., and 160- meters. In addition, the "75-Meter, 2-letter Extra Class" award is available to any Extra Class amateur who has worked all states consisting of 1X2, 2X1, or 2X2 callsigns on the Extra Class portion of 75-meters. Acceptable callsigns for the "75-Meter 2-letter Extra Class" specialty award include one of the following:

  a. All 1X2s
  b. All 2X1s
  c. 2X2s with AA through AL prefixes only in all districts (including AH6 and AL7 in Alaska and Hawaii)
  d. 2X2s with KH6 and KL7 prefixes only

ENDORSEMENT stickers for the basic mixed mode/band award and any of the SPECIALTY awards are available for SSB, CW, Novices, QRP, Packet, EME, and any Single Band. The Novice endorsement is available for the applicant who has worked all states as a Novice licensee. QRP is defined as 10 watts input (or 5 watts output) of the applicant only and is affirmed by signature of the applicant on the application.

6. Contacts made through "repeater" devices or any other power relay method cannot be used for WAS confirmation. A separate WAS is available for OSCAR contacts. All stations contacted must be "land stations." Contact with ships, anchored or otherwise, and aircraft, cannot be counted.

7. A W/VE applicant must be an ARRL member to participate in the WAS program. DX stations are exempt from this requirement.

8. Hq. reserves the right to spot call for inspection of cards (at

ARRL expense) of applications verified by an HF Awards Manager. The purpose of this is not to call into question the integrity of any individual, but rather to ensure the overall integrity of the program. More difficult-to-attain specialty awards (such as 220 MHz WAS for example) are more likely to be so called. Failure of the applicant to respond to such spot check will result in non-issuance of the WAS certificate.

9. DISQUALIFICATION. False statements on this application or submission of forged or altered cards may result in disqualification. ARRL does not submit any marked over cards. The decision of the ARRL Awards Committee in such cases is final.

10. APPLICATION PROCEDURE (Please follow carefully):

(a) Confirmations (QSLs) and applications for (MCS-217) may be submitted to an approved ARRL Special Service Club HF Awards Manager. ARRL Special Service Clubs appoint HF Awards Managers whose names are on file at Hq. If you do not know of HF Awards Managers in your local area, call a club officer to see if one has bee appointed. If you can have your application so verified locally, you need not submit your cards to HQ., as indicated on the application form.

(b) Be sure that when cards are presented for verification (either locally or to Hq.) sort them alphabetically by state, as listed on the back of the application form, MCS-217.

(c) All QSL cards sent to Hq. must be accompanied by sufficient postage for their safe return.

Courtesy of the ARRL

FIRST 10 WINNERS OF THE 5BWAS AWARD

1. W1AX
2. W4IC
3. K9LBQ/7
4. W6ISQ
5. W8YEK
6. KØGJD
7. KH6SP
8. K4GHR (now AA4R)
9. W4YWX (now N4PN)
10. XE1WS

W1JR AND HIS AWARDS

W1JR, Joseph H. Reisert Jr., Amherst, NH. Born 1936 - Extra Class

His first WAS award was issued on July 15, 1975 as #25083. Next he qualified for the Bicentennial W.A.S. in 1976.

5 Band Worked-All-States was # 505 issued in February 1979.
6 Meter WAS, #374 in February 1980.
2 Meter WAS, #26 in August 1980.
160 Meter WAS, #139 in March 1981.
432 MHZ WAS, #6 in April 1982.
Oscar satellite WAS, #82 in December 1982.
12 Meter Endorsement of his 5 Band WAS, in November 1987.
E.M.E. (Moonbounce) Endorsement in October 1988.
17 Meter Endorsement of his 5 Band WAS in March 1989.

Joe has completed his 30 Meter WAS but, no awards are issued for that band. Joe also has 47 states confirmed on 220 MHz!

Presently, Joe has been Awarded WAS on 11 Bands plus OSCAR and E.M.E !

Bill Kennamer, K5FUV, has confirmed that Joe has completed the DXCC requirements on 160M, 80M, 40M, 20M, 17M, 15M, 12M, 10M, 6 M. That's 9 Band DXCC. Additionally, Joe has completed 30M, for which there is no award offered.

Joe has completed the Worked-All-Continents requirements on 160M, 80M, 40M, 20M, 17M, 15M, 12M, 10M, 6M, 2M, and 432 MHz. Additionally, Joe has completed the requirements for 30M WAC but this award is not offered.

Joe's callsign is listed with the other operators who are waiting to work their last ones for 5 Band WAZ. He has 199 Band Zones submitted out of the 200 and has yet to work Zone #23 on 80M.

Joe is also a member of the Yankee Clipper Contest Club.

ISLANDS ON THE AIR (IOTA)

The IOTA program was created by Geoff Watts, a leading British short wave listener, in the mid-1960s. When it was taken over by the RSGB in 1985, it had already become for some a favorite award. Its popularity grows each year, not only among ever-increasing numbers of island chasers, but also among a growing band of amateurs attracted by the possibilities for operating portable from islands. Under an agreement signed in September 1994, Yaesu has become principal sponsor of the program.

The program consists of 18 separate awards which may be claimed by any licensed radio amateur (or SWL on a heard basis) who has had contacts with the required number of islands/groups listed in the IOTA Directory. Many of the islands are DXCC countries in their own right. Others are not, but, by meeting particular eligibility criteria, also count for credit. Part of the fun of IOTA is that it is an evolving program with new islands/groups on the Directory list being activated for the first time. Currently, over 880 of the 1,175 listed have seen activity, and, therefore, have reference numbers, for example. EU-005.

The basic award is for working stations located on 100 islands/groups. There are higher achievement awards for working 200, 300, 400, 500, 600 and 700 islands/groups. In addition, there are seven continental awards (including Antarctica) and three regional awards—Arctic Islands, British Islands and West Indies—for contacting a specified number of islands/groups listed in each area. The IOTA World Diploma is for working a set number in all seven continents. A Plaque of Excellence is available for confirmed contacts with at least 750 islands/groups, with shields for every 25 thereafter.

All contacts must be made by the applicant from the same DXCC country using a call sign or call signs issued personally to him/her by the licensing authority. This means that a mainland Stateside station may claim credit for contacts made from any mainland State but has to start again if he/she moves to Hawaii or Alaska. Club stations may not apply for certificates.

All contacts must have been made after 15 November 1945, on any amateur band between 1.8 and 30 MHz, including all three WARC bands, and on a single mode or mixed mode. Credit will not be given for crossmode, crossband or satellite-aided contacts. Certificate endorsements for single mode and/or single band transmission may be made upon submission of cards with the mode and/or frequency clearly confirmed. The request must be made at the time of the first submission. Only one record is maintained per applicant, so, if he/she chooses a particular mode or band, updates will only be accepted on the same basis.

The IOTA Directory details the rules—they are comprehensive and need careful attention—as well as the island list. It is a key requirement that each applicant should have his own numbered copy of the Directory. This ensures that he/she has the application details including forms, procedures and charges. It is available in the United States from Dewitt L. Jones, W4BAA, P.O. Box 379, Glen Arbor, MI 49636, price $10; and in disk form, personal use only, from Datamatrix, 5560 Jackson Loop NE., Rio Rancho, NM 87124, price $8.50. Both prices include domestic postage.

The rules require that, in order for credit to be given, QSL cards be submitted to nominated IOTA checkpoints for checking. Three checkpoints have been appointed in the U.S. to handle North American applications. Details are given in the Directory.

A feature of the IOTA program is the Honour Roll and Annual Listing which encourages continual updating of scores. This is published in the RSGB's monthly journal, RadCom (May or June issue), and also in the Society's DX News Sheet. The Honour Roll lists the call signs of stations who have a checked score which equals or exceeds 50 percent of the activated islands/groups at the time of the preparation of the Honour Roll. The Annual Listing lists those who have a checked score of 100 or more islands but less than the qualifying threshold for entry into the Honour Roll.

Official IOTA information, for example, new reference numbers, etc, is published regularly in the DX News Sheet (contact RSGB headquarters for details of DXNS subscriptions) and, in the U.S. in QRZ-DX. News of current and forthcoming activity is also published in the DX Bulletin, with reports on recent island DXpeditions featured in the DX Magazine. For up-to-the-minute information, monitor the IOTA frequencies: 14.260 and 21.260 MHz.

Intending island activators should be aware that, as not all islands qualify for IOTA, they should check carefully the Directory rules before incurring expenditure on a visit. The key requirement in the case of a coastal island is that it is separated from the mainland by 200 meters of water at low tide. Some of the Intracoastal Waterway islands cause problems on this count.

The IOTA program is managed by the RSGB's IOTA Committee. The North American representative on that committee, Dewitt L. Jones, W4BAA, is happy to deal with questions about the program from the U.S. and Canada. The enquirer can also write to the RSGB IOTA Director, Roger Balister, G3KMA, La Quinta, Mimbridge, Chobham, Woking, Surrey GU24 8AR, ENGLAND. In both cases, enclose a SASE or an SAE with 2 IRCs for a reply.

IOTA Honor Roll 1995

1	F9RM	834
2	I1ZL	822
3	I1HYW	814
4	I1SNW	807
5	I1JQJ	806
5	VE3XN	806
5	9A2AA	806
8	W9DC	803
8	EA4MY	803
10	ON5KL	798
11	W9DWQ	796
12	GM3ITN	795
12	I8XTX	795
12	W4BAA	795
15	VE7IG	794
16	IK1AIG	792
17	G3AAE	790
18	ON5NT	787
18	DL8NU	787
20	OH2QQ	786
21	IK1GPG	785
22	ON6HE	784
22	G4WFZ	784
24	I8KNT	780
25	G3GIQ	778
26	IØOLK	777
27	K9PPY	775
28	IK1JJB	773
29	I2MWZ	771
30	IT9GAI	768

(Courtesy of G3KMA)

W.A.E. COUNTRIES LIST

Prefix	Country
CT1,4-8,Ø	Portugal
CU	Azores
C3	Andorra
DA-DL	Germany
EA1-5,7,Ø	Spain
EA6	Balearic Island
EI	Ireland
ES	Estonia
F,TP	France
G	England
GD	Isle of Man
GI	N. Ireland
GJ	Jersey
GM	Scotland
GM	Shetland Islands
GU	Guernsey (along with Aldernesy, Brechou, Great Sark, Herm, Jethou, Lihou, and Little Sark.)
GW	Wales
HA,HG	Hungary
HB4,9	Switzerland
HBØ	Liechtenstein

HV	Vatican City
I	Italy
IS,IM,IWØ	Sardinia
IT,IW9	Sicily
JW	Spitzbergen
JW/B	Bear Island
JX	Jan Mayen
LA	Norway
LX	Luxembourg
LY	Lithuania
LZ	Bulgaria
OE	Austria
OH1-9	Finland
OHØ	Aland Island
OJØ	Market Reef
OK,OL	Czech Republic
OM	Slovakia
ON	Belgium
OY	Faroes Islands
OZ	Denmark
PA	Netherlands
SM	Sweden
SP	Poland
SV1-4,6-8	Greece
SV5	Dodecanese
SV9	Crete
SY	Mount Athos
S5	Slovenia
TA1	Turkey (European)
TF	Iceland
TK	Corsica
T7	San Marino
T9	Bosnia and Herzegovina
R,UA-UI1,3,4,6	Russian Federation
R,UA-UI-UA2F	Kaliningrad
UR-UZ,EM-EO	Ukraine
EU-EW	Belarus
ER	Moldavia
YL	Latvia
YO	Romania
YT,YU,YZ,4N,4O	Yugoslavia
ZA	Albania
ZB	Gibraltar
Z3	Macedonia
1A	Sovereign Military Order of Malta
3A	Monaco
4J1	Malyj Vysotskij Island
4K2	Franz Josef-land
4U1ITU	International Telecommunications Union HQ Station
4U1VIC	Vienna International Center
9A	Croatia
9H	Malta

72 Countries

Provided by DL2DN and D.A.R.C.

CQ CONTEST HALL OF FAME

Inductee	Date Inducted
1. Hazzard "Buzz" Reeves, K2GL	Sept 1986
2. Katashi Nose, KH6IJ	Apr 1987
3. Al J. Slater, G3FXB	Apr 1988
4. Martti Laine, OH2BH	Apr 1989
5. Bernie W. Welch, W8IMZ	Jun 1989
6. Lenord Chertok, W3GRF	Dec 1991
7. W. Gerry Mathis, W3GM	Dec 1991
8. Frank Anzalone, W1WY	Apr 1993
9. Jim Lawson, W2PV	Apr 1993
10. Ed Bissell, W3AU	Apr 1993
11. Fred Laun, K3ZO	Apr 1993
12. Vic Clark, W4KFC	Apr 1993

13. Rush Drake, W7RM	Apr 1993
14. John Thompson, W1BIH/PJ9JT	Apr 1994
15. Atilano de Oms, PY5EG	Apr 1994
16. Herb Becker, W6QD	Apr 1994
17. Ken Wolff, K1EA	Apr. 1995
18. Dick Norton, N6AA	Apr. 1995
19. Jim Neiger, N6TJ	Apr. 1995
20 Tine Brajnik, S52AA	Apr. 1995

CQ DX HALL OF FAME

Inductee	Date Inducted
1. Gus M. Browing, W4BPD	Nov 1967
2. John M. Cummings, W2CTN	Mar 1968
3. Stewart S. Perry, W1BB	Aug 1968
4. Richard C. Spenceley, KV4AA	Mar 1969
5. Danny Weil, VP2VB	Sep 1969
6. H. Dale Strieter, W4DQS	May 1970
7. Stuart Meyer, W2GHK	Oct 1970
8. Martti Laine, OH2BH	Jan 1972
9. Ted Thorpe, ZL2AWJ	Aug 1972
10. Chuck Swain, K7LMU	Aug 1972
11. C. J. (Joe) Hiller, W4OPM	Mar 1973
12. Ernst Krenkel, RAEM	Apr 1974
13. Frank Anzalone, W1WY	Jun 1976
14. Lloyd and Iris Colvin, W6KG & W6QL	Nov 1976
15. Geoff Watts, Editor and Publisher	Jun 1977
16. Don C. Wallace, W6AM	1978
17. Joe Arcure, W3HNK	Dec 1979
18. Hugh Cassidy, WA6AUD	Apr 1980
19. Eric Sjolund, SMØAGD	Apr 1981
20. Franz Langner, DJ9ZB	1982
21. Dr. Sanford Hutson, K5YY	Jan 1983
22. Rodney Newkirk, W9BRD	Feb 1984
23. Ronald Wright, ZL1AMO	Apr 1985
24. Herb Becker, W6QD	Apr 1985
25. Jim Smith, P29JS/VK9NS	Apr 1986
26. Kan Mizoguchi, JA1BK	Apr 1987
27. John Troster, W6ISQ	Apr 1988
28. Charlie Mellen, W1FH	Apr 1994
29. Carl Henson, WB4ZNH	Apr. 1995

RECIPIENTS OF SCHOLARSHIP AWARDS OF THE ARRL FOUNDATION SCHOLARSHIP PROGRAM

1980–81

James M. Zambik, WB2QYG—Long Island Scholarship—$250
Gregory D. Jay, WA2EDY—Long Island Scholarship—$250
Larry E. Smith, Jr., WB9UKE—YL ISSB Memorial Scholarship—$300

1982–83

Paul M. Silverman, KA2DSP—Long Island Scholarship—$250
Larry E. Smith, WB9UKE—YL ISSB Memorial Scholarship—$709

1983–84

Pamela S. Hayward, WBØMUS—Paul & Helen L. Grauer Scholarship—$500

1985

John Allcorn, KAØEWS—Paul and Helen L. Grauer Scholarship—$500
David J. Schmocker, KJ9I—Perry F. Hadlock Memorial Scholarship—$500

1986

William J. Hulka, KA9AKI—The ARRL Scholarship to Honor Barry Goldwater—$5000
John Allcorn, KAØEWS—Paul and Helen L. Grauer Scholarship—$500
Michael R. Dargel, N1AMR—Perry F. Hadlock Memorial Scholarship—$500

1987
William H. Sands, IV, KA3FXX—The ARRL Scholarship to Honor Barry Goldwater—$5000
Raymond J. Gomez, Jr., NØGNA—Paul and Helen L. Grauer Scholarship—$500
Peter S. Jaworski, KC2KK—Perry F. Hadlock Memorial Scholarship—$500
Stephanie A. Dougherty, N8FIT—You've Got A Friend In Pennsylvania Scholarship—$500
Robert S. Hulka, Jr., KA9AKJ—Edmond A. Metzger Scholarship—$500

1988
Shawn A. Wakefield, WK5P—The ARRL Scholarship to Honor Barry Goldwater—$5000
Raymond J. Gomez, Jr., NØGNA—Paul and Helen L. Grauer Scholarship—$500
Robert N. Keenan, WU6L—Perry F. Hadlock Memorial Scholarship—$500
Douglas N. Benish, N3CXB—You've Got A Friend In Pennsylvania Scholarship—$500
Raymond P. Klump, KA9WFR—Edmond A. Metzger Scholarship—$500
Paul W. Hoffman, NK3M—L. Phil and Alice J. Wicker Scholarship—$500
David A. Clemons, K1VUT—Dr. James A. Lawson Memorial Scholarship—$500
Stephanie A. Dougherty, N8FIT—Edward D. Jaikins Memorial Scholarship—$500

1989
David A. Clemons, K1VUT—The ARRL Scholarship to Honor Barry Goldwater—$5000
Cary E. Watkins, KA5YRK—Paul and Helen L. Grauer Scholarship—$500 (1)
Robert N. Keenan, WU6L—Paul and Helen L. Grauer Scholarship—$500 (2)
William A. Kjontvedt, NØHZB—Perry F. Hadlock Memorial Scholarship—$1000
Jonathan J. Vidal, N3GUI—You've Got A Friend In Pennsylvania—$1000
David A. Hulka, KD9UA—Edmond A. Metzger Scholarship—$500
Andrew J. Jackson, N4FUF—L. Phil and Alice J. Wicker Scholarship—$1000
David A. Stein, KD2ZE—Dr. James A. Lawson Memorial Scholarship—$500
Troy A. Campbell, N7KWY—Edward A. Jaikins Memorial Scholarship—$500

1990
Kurt D. Schwehr, N6XWB—ARRL Scholarship to Honor Barry Goldwater—$5000
Michelle M. Lehman, KBØFNE—Paul and Helen L. Grauer Scholarship—$500 (1)
Christopher N. Haddan, NØGXB—Paul and Helen L. Grauer Scholarship—$500(2)
Daniel C. Lawry, KA1PNC—Perry F. Hadlock Memorial Scholarship—$1000
Douglas M. Benish, N3CXB—You've Got A Friend In Pennsylvania—$1000
Scott L. Young, N9FZS—Edmond A. Metzger Scholarship—$500
Christopher C. Peters, KB4MRH—L. Phil and Alice J. Wicker Scholarship—$1000
Gregory H. Laufman, N6GPA—Dr. James L. Lawson Memorial Scholarship—$500
Dennis P. Ward, KI8X—Edward D. Jaikins Memorial Scholarship—$500
Robert J. Inderbitzen, NQ1R—New England FEMARA Scholarship—$600

1991
James R. Harper, KB5CTQ—ARRL Scholarship to Honor Barry Goldwater—$5000

Jason N. Efken, NØKOQ—Paul and Helen L. Grauer Scholarship—$1000
Daniel C. Lawry, KA1PNE—Perry F. Hadlock Memorial Scholarships—$1000
George P. Gumbrell, KA3RLZ—You've Got A Friend In Pennsylvania—$1000
Michael W. Barton, KB9AQR—Edmond A. Metzger Scholarship—$500
John T. Crago, KB8DAN—L. Phil and Alice J. Wicker Scholarship—$1000
Michael D. Ambrose, KC1UK—Dr. James L. Lawson Memorial Scholarship—$500
Dennis P. Ward, KT8X—Edward D. Jaikins Memorial Scholarship—$500
Jeffrey A. Brower, N6XZX—Charles N. Fisher Memorial Scholarship—$500
William G. Rubin, N1HWC—Irving W. Cook, WAØCGS Scholarship—$500
Christos N. Conner, N5MZD—Mississippi Scholarship—$500
Michael T. Decerbo, N1FYO—New England FEMARA Scholarship—$600
Jason G. Lovett, N1EJD—New England FEMARA Scholarship—$600
Joel F. Kluender, NF9G—New England FEMARA Scholarship—$600
Frederick J. Darling, Jr.—New England FEMARA Scholarship—$600
John R. Labenski, KA1KRL—New England FEMARA Scholarship—$600

1992
Harry Litaker, Jr., AC4BO—ARRL Scholarship to Honor Barry Goldwater—$5000
Kristin Poduska, WOØH—Paul and Helen L. Grauer Scholarship—$1000
Daniel Lawry, KA1PNE—Perry F. Hadlock Memorial Scholarship—$1000
John Smertneck, KA3SFX—You've Got A Friend In Pennsylvania—$1000
James P. Harper, KB5CTQ—General Fund Scholarship—$1000 (1)
Dennis Washington, N4YFL—General Fund Scholarship—$1000 (2)
Andrew Ross, KC6OHS—Charles N. Fisher Memorial Scholarship—$1000
Brian J. Kuebert, N4UEZ—L. Phil and Alice J. Wicker Scholarship—$1000
Stephen Byrd, N5WGE—Mississippi Scholarship—$500
Rachel Fremmer, KB2GOT—Dr. James L. Lawson Memorial Scholarship—$500
Robert Goemans, N9HAD—Edmond A. Metzger Scholarship—$500
Chad Seuser, KBØCTL—Irving W. Cook, WAØCGS Scholarship—$500
Richard E. McLaughlin, Jr.—Edward D. Jaikins Memorial Scholarship—$500
Mark Gilbert, N5MG/1—New England FEMARA Scholarship—$600
Laurie Mann, KA1KXP—New England FEMARA Scholarship—$600
Deborah Stein, KB2BM—New England FEMARA Scholarship—$600
David Perrin, Jr., KC1TS—New England FEMARA Scholarship—$600
Allan Hallberg, N1KBE—New England FEMARA Scholarship—$600

1993
Phillip Rowe, KA3TOK—ARRL Scholarship to Honor Barry Goldwater—$5000
Grant Kesselring, NØICI—Paul and Helen L. Grauer Scholarship—$1000
Ravi Hariprasad, AA2CR—Perry F. Hadlock Memorial Scholarship—$1000
Melissa L. Benish, N3FAC—You've Got A Friend In Pennsylvania—$1000
Brian J. Kuebert, N4UEZ—L. Phil and Alice J. Wicker Scholarship—$1000

Peter C. Laing, N7TXI—Charles N. Fisher Memorial
    Scholarship—$1000
Craig E. Flavin, AAØFF—K2TEO Martin J. Green, Sr. Memorial
    Scholarship—$1000
Vern J. Wirka, WBØGQM—General Fund Scholarship—$1000
Steven D. Kraft, KE9RW—General Fund Scholarship—$1000
John C. Evanson, WJ1U—New England FEMARA Scholarship—
    $600
Scott H. Ledder, KA1RLT—New England FEMARA
    Scholarship—$600
David B. Perrin, Jr., KC1TS—New England FEMARA
    Scholarship—$600
Michael T. Decerbo, N1FYO—New England FEMARA
    Scholarship—$600
Cristofor M. Cataudella, N1FWC—New England FEMARA
    Scholarship—$600
Elizabeth A. Skolaut, KAØYSP—PHD Scholarship—$500
Michael B. Madden, N9OHW—Edmond A. Metzger
    Scholarship—$500
Matthew J. Minney, N8PGI—Edward D. Jaikins Memorial
    Scholarship—$500
Michael D. Ambrose, KC1UK—Dr. James L. Lawson Memorial
    Scholarship—$500
Martin L. Kollman, Jr., NØRQP—Irving W. Cook, WAØCGS
    Scholarship—$500
Mansel P. Bell, KB5HVV—Mississippi Scholarship—$500

1994

The ARRL Foundation Scholarship Committee has announced
recipients of its 1994 academic scholarship awards:

Stephen K. Gee, AA2GE, Great Neck, NY, The ARRL
    Scholarship to Honor Barry Goldwater of $5000.
Brian T. Kuehn, KBØETT, Emporia, KS, The Paul and Helen L.
    Grauer Scholarship of $1000.
Michael G. Taylor, KA3SVX, Drexel Hill, PA, The You've Got A
    Friend In Pennsylvania Scholarship of $1000.
Elliot J. Bernstein, AA2KR, New Hyde Park, NY, The Perry F.
    Hadlock Memorial Scholarship of $1000.
Jeffrey H. Johnson, N4YRC, Cary, NC, The L. Phil and Alice J.
    Wicker Scholarship of $1000.
James H. Muiter, N6TDC, San Mateo, CA, The Charles N. Fisher
    Memorial Scholarship of $1000.
Amy L. Tlachac, N9TLN, Wisconsin Rapids, WI, The K2TEO
    Martin J. Green Sr. Memorial Scholarship of $1000.
Michael R. Hawk, NØOSY, Omaha, NE, The PHD Scholarship of
    $500.
Eric N. Johnson, N9TAR, Danville, IL, The Edmond A. Metzger
    Scholarship of $500.
John T. Crago, KB8DAN, Liberty, WV, The Edward D. Jaikins
    Memorial Scholarship of $500.
Owen Debowy, N2WPO, Stony Brook, NY, The Dr. James L.
    Lawson Memorial Scholarship of $500.
Leanna R. Gordon, KFØZL, Augusta, KS, The Irving W. Cook,
    WAØ$CGS Scholarship of $500.

The General Fund Scholarships of $1000 each to:
Todd A. Kramer, N4WOR, Longwood, FL
Kenneth R. Leitch, KB5OKI, El Paso, TX
James R. Phillips, KA3WSZ, Newark, DE
James W. Gregory, AD4GN, Paducah, KY
Tamara D. Britain, KB5RYE, Grand Prairie, TX

The New England FEMARA Scholarships of $600 each to:
Justin K. Munger, AA1AS, Pittsfield, MA
Andrew K. Freeston, KA1VYX, Windham, NH
Jason G. Lovett, N1EJD, Kennebunk, ME
Brian Montmarquet, N1NKH, Pembroke, NH
William G. Nelson, KB1AWA, Madison, NH
Michael D. Ambrose, KC1UK, Monroe, CT
Kevin T. Jensen, N1KCG, Cheshire, CT
Michael T. Decerbo, N1FYO, Trumbull, CT

1995

Benjie Chen, KE6BCU, Sunnyvale, CA, The ARRL
Scholarship to Honor Barry Goldwater—$5000
    Steven J. Sandlin, KBØLJV, Lamoni, IA, The Paul and Helen
L. Grauer Scholarship—$1000
    Suzanne L. Chimel, KD3GG, Clarks Summit, PA, The You've
Got A Friend In Pennsylvania Scholarship—$1000
    Brian D. Kuebert, N4UEZ, Warrenton, VA, The L. Phil and
Alice J. Wicker Scholarship—$1000
    Jonathan M. Becker, KC6QOQ, Lancaster, CA, The Charles N.
Fisher Memorial Scholarship—$1000
    Mary F. Alestra, KB2IGG, Staten Island, NY, The K2TEO
Martin J. Green Sr. Memorial Scholarship—$1000

The General Fund Scholarships—$1000 each to:
Thomas P. Schwabel, N2WLG, Clarence, NY
Leann P. Foss, KB9HHB, Washington, IN
David A. Case, KA1NCN, Hampton, CT
Richard E. Kutter, KB8LOE, New Madison, OH
Mike Orlando, KC6QYA, Costa Mesa, CA

Sarah R. Hughes, NØQGE, Shellsburg, IA 52332—The PHD
Scholarship—$1000
    Adam T. Tate, AB5PO, Baton Rouge, LA— The F. Charles
Ruling, N6FR Memorial Scholarship—$1000

The New England FEMARA Scholarships—$ 600 each to:
Alison R. Burns, N1QPK, Carmel, ME
Hendrik M. Gruteke, KA1LHC, Derby Line, VT
Nathan J. Goyette, N1OHE, Bennington, VT
Adam J. Stachelek, N1IJJ, Dudley, MA
Elizabeth M. Pelczar, KA1SLD, Rocky Hill, CT
Scott H. Ledder, KA1RLT, Medway, MA
Melina A. Roy, KA1UDA, Amesbury, MA

Riva C. Robeson, N9ESZ, Chicago, IL—The Edmond A.
Metzger Scholarship—$500
    Stephen A. Hopper, AD4CT, Pickerington, OH—The NEMAL
Electronics Scholarship—$500
    Keith J. Leitch, KB5JVM, El Paso, TX—The Tom and Judith
Comstock Scholarship—$500
    Eric W. Belloma, KB8PNU, North Ridgeville, OH—The
Edward D. Jaikins Memorial Scholarship—$500
    Michael J. Delman, AK2N, Kew Gardens, NY—The Dr. James
L. Lawson Memorial Scholarship—$500
    Steven D. Hicks, AB5ZW, Hernando, MS—The Mississippi
Scholarship—$500
    Steven J. Izell, KB5JWN, Choctaw, OK—The Fred R.
McDaniel Memorial Scholarship—$500
    Timothy S. Mosher, WX9I, Woodridge, IL—The Six Meter
Club of Chicago Scholarship—$500

To apply for 1996 scholarships, write to: The ARRL
Foundation, Inc., 225 Main Street, Newington, CT 06111.
Deadline for applications and transcripts is February 1, 1996.

HISTORY OF THE ARRL TECHNICAL EXCELLENCE AWARD

The ARRL Technical Excellence Award was established in
1975 to honor the author(s) whose article (or series of articles) published in QST for that year is judged to have the highest degree of
technical merit (Minute #65, May 1975 Board Meeting). In 1988,
the scope of the eligible candidates was widened to include all peri-
odicals published by the ARRL. The award is presented annually,
and has been earned by the following authors (list is by year,
author(s), article):
    1975, Wes Hayward, W7ZOI, "Defining and Measuring
        Receiver Dynamic Range," QST, July 1975
    1976, P. D. Rhodes, K4EWG, J. R. Painter, W4BBP, "The
        Log-Yag Array," QST, December 1976

1977, Wayne Overbeck, N6NB/K6YNB, "The VHF Quagi," *QST*, April 1977

1978, Richard W. Harris, J. F. Cleveland, WB6CZX, "A Baseband Communications System," *QST*, November and December 1978

1979, Ed Oxner, KB6QJ, "Build a Broadband Ultralinear VMOS Amplifier," *QST*, May 1979

1980, David T. Geiser, WA2ANU, "The Impedance-Match Indicator," *QST*, July 1980

1981, Ray Cracknell, ZE2JV, Fred Anderson, ZS6PW, Costas Fimerellis, SV1DH, "The Euro-Asia to Africa VHF Transequatorial Circuit During Solar Cycle 21," *QST*, November and December 1981

1982, Wes Hayward, W7ZOI, "A Unified Approach to the Design of Crystal Ladder Filters," *QST*, May 1982

1983, William Sabin, WØIYH, "Spread-Spectrum Applications in Amateur Radio," *QST*, July 1983

1984, James C. Rautio, AJ3K, "The Effect of Real Ground on Antennas," *QST*, February, April, June, August and November 1984

1985, E. R. "Chip" Angle, N6CA, "A Quarter-Kilowatt 23-cm Amplifier," *QST*, March and April 1985

1986, Rich Arndt, WB4TLM, Joe Fikes, KB4KVE, "SuperSCAF and Son—A Pair of Switched-Capacitor Audio Filters," *QST*, April 1986

1987, Charles J. Michaels, W7XC, "Some Reflections on Vertical Antennas," *QST*, July 1987

1988, John Grebenkemper, KI6WX, "Phase Noise and its Effects on Amateur Communications," *QST*, March and April 1988

1989, Andre Kesteloot, N4ICK, "A Practical Direct-Sequence Spread-Spectrum UHF Link," *QST*, May 1989.

1990, Roy Lewallen, W7EL, "A Simple and Accurate QRP Directional Wattmeter," *QST*, February 1990.

1991, Rick Campbell, KK7B, "A Single-Board, No-Tune 902-MHz Transverter," *QST*, July 1991.

1991, Bruce S. Hale, KB1MW, "An Introduction to Digital Signal Processing," *QST*, July 1991.

1991, Roy Lewallen, W7EL, "MININEC: The Other Edge of the Sword," *QST*, February 1991.

1992, Ken Macleish, W7TX, "Why an Antenna Radiates," *QST*, November 1992.

1993, Rick Campbell, KK7B, "High-Performance, Single-Signal, Direct-Conversion Receivers," *QST*, January 1993.

1994 Dr. Ulrich L. Rohde, KA2WEU, "Key Components of Modern Receiver Design" *QST*, May, June, July, and December.

In the case of joint authors of an award-winning article, each author is presented with an engraved cup. All winners (except those for 1981) lived in the U.S., and each of those awards was presented through Division Directors. The 1981 award presentations were made through sister societies of the ARRL.

In 1992, the Volunteer Resources Committee recommended that the Board of Directors present three (3) Technical Excellence Awards for 1991—a first. We ask Technical Advisors to vote for one (1) candidate as we never have given second- or third-place awards.

After the close of a calendar year, the process for determining the winner for that year begins. As established by the Board, the judging is done by a panel of Technical Advisor appointees, with liaison through the Technical Department at Hq. The decision of the judging panel is ratified at the second meeting of the Board in the year following the date of the award, after which the award is engraved and presented. The award itself is a 12-inch pewter loving cup and base (the 1983 award was a bowl rather than a cup). The engraving identifies the award and includes the ARRL diamond logo, the year, and the name and call of the author(s).

(Courtesy of ARRL)

ARRL TEACHER-OF-THE-YEAR AWARDS

The ARRL annually recognizes teachers for their efforts. One award (the Herb S. Brier, W9EGQ, award) is for volunteer teachers. The other award is for professional educators using ham radio in the classroom.

Herb S. Brier, W9EGQ, Volunteer Instructor Award

1991	Tom Hammond, NØSS
1992	James Stafford, W4QD
1993	James Dalley, WØNAP
1994	Chris Townsend, NU7V

Professional Teacher of the Year Award

1991	Patricia Hensley, N4ROS
1992	Bob Maurais, KC1IV
1993	Sheila Perry, NØUOP
1994	Philip A. Downes, N1IFP—Professional Educator (paid teacher)
1994	Charles S. Ward, KJ4RV—Professional Instructor (non-paid teacher)

DAYTON HAMVENTION "AMATEUR OF THE YEAR" AWARD

Each year, the Dayton Hamvention Committee recognizes "the special person who has made a long-term committment to the advancement of amateur radio...a well-rounded individual who has contributed to our hobby in some outstanding way" with the "Amateur of the Year" Award. This award is presented at the Dayton Hamvention in April.

The winner is selected from all those nominated by the Hamvention Awards Committee. Information on the nomination process is available from: Hamvention Awards Chairman, Box 964, Dayton, OH 45401-0964

1995 Dayton HamVention Award Winners

Technical Excellence Award
Philip Ferrell, K7PF, developer of the FingerPrinting method of transmitter identification using the turn-on transient of a transmitter as it comes on a frequency.

DARA Special Achievement Award
Ed Briner, WA3TVG, for work with the "Flying Dentists," and other Medical Amateur Radio Council (MARCO) activities.

Dayton HamVention Ham of the Year.
Rosalie White, WA1STO, head of ARRL's Educational Activities Department and member of SAREX Working Group.

Dayton Hams of the Year

1955	William C. Jenney, W8FYW; Westlake, Ohio Benjamin S. Zieg, W9EHU, K4OQK; Atlanta, Georgia
1956	Edmund C. Ryan, W8LRR; Mansfield, Ohio
1957	Ralph Crammer, W8VHO; Columbus, Ohio
1958	Rev. C. Lynn White, K4CC; Tavarres, Florida
1959	Harlow Lucas, W8QQ; Columbus, Ohio

1960	Paul Wolfe, W8IVE; Cincinnati, Ohio	1978	Frank Schwab, W8OK; Dayton, Ohio
1961	Ed Bonnet, W8OVG; Dayton, Ohio	1979	George B. Batterson, W2GB; Rochester, New York
1962	Dana Cartwright, W8UPB; Cincinnati, Ohio	1980	Wayne Overbeck, N6NB; Woodland Hills,
1963	Chester Funk, K8EUF; Phoenix, Arizona		California
1964	H. Ruble, W8PTF; Dayton, Ohio	1981	Eric C. Shalkhauser, W9CI; Washington, Illinois
1965	Carl B. Snyder, W8ARW; Greenville, Ohio	1982	Robert G. Heil, Jr., K9EID; Marissa, Illinois
1966	Robert K. Caskey, W9DNQ; Indianapolis, Indiana	1983	Katashi Nose, KH6IJ; Honolulu, Hawaii
1967	Jack Gray, W8JDV; Mason, Ohio	1984	Dave L. Bell, W6AQ; Los Angeles, California
1968	Elmer Schubert, W8ALW; Cincinnati, Ohio	1985	John J. Willig, W8ACE; Sarasota, Florida
1969	Wayne Walters, W9DOG; Plainfield, Indiana	1986	Roy Neal, K6DUE; Woodlawn Hill, California
1970	Kay Anderson, W8DUV; Huntington, West	1987	Carole Perry, WB2MGP; Staten Island, New York
	Virginia	1988	Bill Bennett, W7PHO; Seattle, Washington
1971	Al Michel, W8WC; Cincinnati, Ohio	1989	Bill Pasternak, WA6ITF; Saugus, California
1972	Don C. Miller, W9NTP; Waldron, Indiana	1990	Stephen Mendelsohn; WA2DHF; Dumont,
1973	Ray E. Myers, W6MLZ; San Gabriel, California		New Jersey
1974	Barry Goldwater, K7UGA; Scottsdale, Arizona	1991	John B. Johnston, W3BE; Derwood, Maryland
1975	Richard A. Daniels, WA4DGU; Arlington, Virginia	1992	Richard Baldwin, W1RU; Waldoboro, Maine
1976	Joseph M. Hertzberg, N3EA; Bryn Mawr, Pennsylvania	1993	Harry Dannals, W2HD, Charlottesville, Virginia
1977	Rafael M. Estevez, WA4ZZG; Hialeah, Florida	1994	Perry Williams, W1UED, Unionville, Connecticut

QSL CARDS

"The final courtesy of a QSO is a QSL"—one of the oldest quotes in amateur radio. A QSL card is a reminder, souvenir , or proof of a contact. Many of the awards pursued by amateurs require submission of QSL cards as proof of contact.

Hams take pride in their own QSLs and their collections. QSL card contests are held at many hamfests, with prizes for the most original design. Many clubs offer their members the opportunity to purchase club QSL cards, and awards are available for obtaining a number of QSLs from club members.

Exchanging QSL cards can often be a challenge, particularly when dealing with stations in remote locations where postal service is unreliable. It is also a great expense for stations who make large numbers of contacts to verify every contact directly. To simplify things, QSL bureaus and QSL managers are often used by the more active stations.

QSL Bureaus

QSL Bureaus serve as "clearinghouses" for QSL cards for hams in a particular geographic area. In the United States, each call area has a QSL bureau for incoming cards. Cards are sorted and distributed by volunteers to individual hams, usually by means of envelopes provided to the bureau. These bureaus are listed below.

The ARRL Incoming QSL Bureau System

Purpose

Within the U.S. and Canada, the ARRL DX QSL Bureau System is made up of numerous call area bureaus that act as central clearing houses for QSLs arriving from foreign countries. These "incoming" bureaus are staffed by volunteers. The service is free and ARRL membership is not required.

How it Works

Most countries have "outgoing" QSL bureaus that operate in much the same manner as the ARRL Outgoing QSL Service. The member sends his cards to his outgoing bureau where they are packaged and shipped to the appropriate countries.

A majority of the DX QSLs are shipped directly to the individual incoming bureaus where volunteers sort the incoming QSLs by the first letter of the call sign suffix. One individual may be assigned the responsibility of handling from one or more letters of the alphabet. Operating costs are funded from ARRL membership dues.

Claiming your QSLs

Send a 5 x 7-1/2 or 6 x 9 inch self-addressed, stamped envelope (SASE) to the bureau serving your callsign district. Neatly print your call-sign in the upper left corner of the envelope. Place your mailing address on the front of the envelope. A suggested way to send envelopes is to affix a first class stamp and clip extra postage to the envelope. Then, if you receive more than 1 oz. of cards, they can be sent in the single package.

Some incoming bureaus sell envelopes or postage credits in addition to the normal SASE handling. They provide the proper envelope and postage upon the prepayment of a certain fee. The exact arrangements can be obtained by sending your inquiry with a SASE to your area bureau. A list of bureaus appears below.

Helpful Hints

Good cooperation between the DXer and the bureau is important to ensure a smooth flow of cards. Remember that the people who work in the area bureaus are volunteers. They are providing you with a valuable service. With that thought in mind, please pay close attention to the following DOs and DON'Ts.

DOs

• DO keep self-addressed 5 x 7-1/2 or 6 x 9 inch envelopes on file at your bureau, with your call in the upper left corner, and affix at least one unit of first-class postage.

• DO send the bureau enough postage to cover SASEs on file and enough to take care of possible postage rate increases.

• DO respond quickly to any bureau request for SASEs, stamps or money. Unclaimed card backlogs are the bureau's biggest problem.

• DO notify the bureau of your new call as you upgrade. Please send SASEs with new call, in addition to SASEs with old call.

• DO include a SASE with any information request to the bureau.

• DO notify the bureau in writing if you don't want your cards.

DON'Ts

• DON'T send domestic US to US cards to your call-area bureau.

• DON'T expect DX cards to arrive for several months after the QSO. Overseas delivery is very slow. Many cards coming from overseas bureaus are over a year old.

• DON'T send your outgoing DX cards to your call-area bureau.

• DON'T send SASEs to your "portable" bureau. For example, AA2Z/1 sends SASEs to the W2 bureau, not the W1 bureau.

• DON'T send SASEs to the ARRL Outgoing QSL Service.

• Don't send SASEs larger than 6 x 9 inches. SASEs larger than 6 x 9 inches require additional postage surcharges.

ARRL INCOMING DX QSL BUREAU ADDRESSES

First Call Area: All calls*, W1 QSL Bureau, Y.C.C.C., P.O. Box 80216, Springfield, MA 01138-0216

Second Call Area: All calls*, ARRL 2nd District QSL Bureau, N.J.D.X.A., P.O. Box 599, Morris Plains, NJ 07950.

Third Call Area: All calls, C-CARS, P.O. Box 448, New Kingstown, PA 17072 - 0448

Fourth Call Area: All single-letter prefixes (K4, N4, W4), Mecklenburg Amateur Radio Club, P.O. Box DX, Charlotte, NC 28220

Fourth Call Area: All two-letter prefixes (AA4, KB4, NC4,

WD4, etc.), Sterling Park Amateur Radio Club, Call Box 599, Sterling, VA 20167

Fifth Call Area - All calls*, ARRL W5 Incoming QSL Bureau, P.O. Box 50625, Midland, TX 79710
Sixth Call Area: All calls* 1, ARRL Sixth (6th) District DX QSL Bureau, P.O. Box 1460, Sun Valley, CA 91352
Seventh Call Area: All calls*, Willamette Valley DX Club, Inc., P.O. Box 555, Portland, OR 97207
Eighth Call Area: All calls, 8th Area QSL Bureau, P.O. Box 182165, Columbus, OH 43218-2165
Ninth Call Area: All calls*, Northern Illinois DX Assn., Box 519, Elmhurst, IL 60126
Zero Call Area: All calls*, WØ QSL Bureau, P.O. Box 4798
Overland Park, KS 66204
Puerto Rico: All calls*, KP4 QSL Bureau, P.O. Box 1061, San Juan, PR 00902
U.S. Virgin Islands: All calls, Virgin Islands ARC, GPO Box 11360, Charlotte, Amalie, Virgin Islands 00801
Hawaiian Islands: All calls*, Wayne Jones, NH6GJ, P.O. Box 788, Wahiawa, HI 96786
Alaska: All calls*, Alaska QSL Bureau, 4304 Garfield St., Anchorage, AK 99503
Guam: MARC, Box 445, Agana, Guam 96910
SWL: Mike Witkowski, WDX9JFT, 4206 Nebel St., Stevens Point, WI 54481

QSL Cards for Canada may be sent to:
RAC National Incoming QSL Bureau, Loyalist City Amateur Radio Club, P. O. Box 51, Saint John NB E2L 3X1, Canada

QSL cards for Canada may also be sent to the individual bureaus:
VE1, VE9, VEØ, VY2*
VE2—J. Dube, VE2QK, Brit Fader Memorial Bureau, 875 St. Severe St., P. O. Box 8895, Trois-Rivieres, PQ G9A **4G4**, Halifax, NS B3K 5M5
VE3—The Ontario Trilliums
VE4—Adam Romanchuck, VE4SN, PO Box 157, 26 Morrison St., Downsview ON M3M 3A3, Winnipeg, MB R2V 3B4
VE5*- Bj. Madsen, VE5FX
VE6*—Larry Langston, VE6LLL, 739 Washington Dr., PO Box 3364, Weyburn, SK S4H 2S4, Saskatchewan AB T8L 2T3
VE7*—Dennis Livesey, VE7DK
VE8*—Rolf Ziemann, VE8RZ, 8309 112th St. 2 Taylor Road, Delta, BC, V4C 4W7, Yellowknife, NWT X1A 2K9
VY1—W.L. Champagne, VY1AU.
VO1, VO2—Roland Peddle, VO1BD, P.O. Box 4597, P.O. Box 6, Whitehorse, YU Y1A 2RB, St. John's, NF A1C 5H5

* These bureaus sell envelopes or postage credits. Send an SASE to the bureau for further information.

1 These bureaus can only accept specific sized envelopes. Send an SASE to the bureau for further information.

THE ARRL OUTGOING QSL SERVICE

Note: The ARRL QSL Service should not be used to exchange QSL cards within the 48 contiguous states.

One of the greatest bargains of League membership is being able to use the ARRL Outgoing QSL Service to conveniently send your DX QSL cards overseas to foreign QSL Bureaus. Your ticket for using this service is your QST address label and just $3.00 per pound. For those not quite so DX active (sending 10 cards or fewer), enclose $1.00. You can't even get a deal like that at your local warehouse supermarket! And the potential savings over the substantial cost of individual QSLing is equal to many times the price of your annual dues. Your cards are sorted promptly by the Outgoing Service staff, and cards are on their way overseas usually within a week of arrival at ARRL HQ. Approximately two million cards are handled by the Service each year!

QSL cards are shipped to QSL Bureaus throughout the world, which are typically maintained by the national Amateur Radio Society of each country. While no cards are sent to individuals or individual QSL managers, keep in mind that what you might lose in speed is more than made up in the convenience and savings of not having to address and mail QSL cards separately. (In the case of DXpeditions and/or active DX stations that use US QSL managers, a better approach is to QSL directly to the QSL manager. The various DX newsletters, the W6GO QSL manager directory, and other publications, are good sources of up-to-date QSL manager information.)

As postage costs become increasingly prohibitive, don't go broke before you're even halfway towards making DXCC. There's a better and cheaper way—"QSL VIA BURO" through the ARRL Outgoing QSL Service!

How To Use The ARRL Outgoing QSL Service

1) Presort your DX QSLs alphabetically by parent call-sign prefix (AP, C6, CE, DL, ES, F, G, JA, LY, PY, YL, 5N, 9Y and so on). NOTE: Some countries have a parent prefix and use additional prefixes, i.e., CE (parent prefix) = XQ, 3G, When sorting countries that have multiple prefixes, keep that country's prefixes grouped together in your alphabetical stack. Addresses are not required. DO NOT separate the country prefix by use of paper clips, rubber bands, slips of paper or envelopes.

2) Enclose the address label from your current copy of QST. The label shows that you are a current ARRL member.

3) Members (including honorary and QSL managers) should enclose payment of $3.00 per each pound of cards or portion thereof—approximately 150 cards weigh one pound. A package of ten (10) cards or fewer costs only $1.00. Please pay by check (or money order) and write your callsign on the check. Send "green stamps" (cash) at your own risk. DO NOT send postage stamps or IRCs. (DXCC credit CANNOT be used towards the QSL Service fee.)

4) Include only the cards, address label and check in the package. Wrap the package securely and address it to the ARRL Outgoing QSL Service, 225 Main Street, Newington CT 06111.

5) Family members may also use the service by enclosing their QSLs with those of the primary member. Include the appropriate fee with each individual's cards and indicate "family membership" on the primary member's QST address label.

6) Blind members who do not receive QST need only include the appropriate fee along with a note indicating the cards are from a blind member.

7) ARRL affiliated-club stations may use the service when submitting club QSLs by indicating the club name.

Club secretaries should check affiliation papers to ensure that affiliation is current. In addition to sending club station QSLs through this service, affiliated clubs may also "pool" their members' individual QSL cards to effect an even greater savings. Each club member using this service must also be a League member. Cards should be sorted "en masse" by prefix, and a *QST* label enclosed for each ARRL member.

Recommended QSL-Card Dimensions

The efficient operation of the worldwide system of QSL Bureau requires that cards be easy to handle and sort. Cards of unusual dimensions, either much larger or much smaller than normal, slow the work of the Bureaus, most of which is done by unpaid volunteers. A review of the cards received by the ARRL Outgoing QSL Service indicates that most fall in the following range: Height = 2-3/4 to 4-1/4 in. (70 to 110 mm), Width = 4-3/4 to 6-1/4 in. (120 to 160 mm). Cards in this range can be easily sorted, stacked and packaged. Cards outside this range create problems; in particular, the larger cards often cannot be handled without folding or otherwise damaging them. In the interest of efficient operation of the worldwide QSL Bureau system, it is recommended that cards entering the system be limited to the range of dimensions given. [Note: IARU Region 2 has suggested the following dimensions as optimum: Height 3 1/2 in. (90 mm), Width 5 1/2 in. (140 mm).]

Countries Not Served By The Outgoing QSL Service

Approximately 260 DXCC countries are served by the ARRL Outgoing QSL Service, as detailed in the ARRL DXCC Countries List. This includes nearly every active country. As noted previously, cards are forwarded from the ARRL Outgoing Service to a counterpart Bureau in each of these countries. In some cases, there is no Incoming Bureau in a particular country and cards therefore cannot be forwarded. However, QSL cards can be forwarded to a QSL manager, i.e.; 3C1MB via (EA7KF). The ARRL Outgoing Service cannot forward cards to the following countries:

A5	Bhutan
A6	United Arab Emirates
D2	Angola
EP	Iran
J5	Guinea-Bissau
KC6	Belau
KHØ	Mariana Is.
KH1	Baker and Howland Is.
KH4	Midway I.
KH5	Palmyra and Jarvis Is.
KH7	Kure I.
KH8	Am. Samoa
KH9	Wake I.
KP1	Navassa I.
KP5	Desecheo I.
P5	North Korea
S2	Bangladesh
S7	Seychelles
T2	Tuvalu
T3	Kiribati
T5	Somalia
TJ	Cameroon
TL	Central African Republic
TN	Congo
TT	Chad
TY	Benin
TZ	Mali
V6 (KC6)	Micronesia
VP2M	Montserrat
XU	Kampuchea
XW	Laos
XX9	Macao
XZ (1Z)	Myanmar (Burma)
YA	Afghanistan
ZD9	Tristan da Cunha
3CØ	Pagalu I.
3C	Equatorial Guinea
3V	Tunisia
3W, XV	Vietnam
3X	Guinea
5A	Libya
5H	Tanzania
5R	Madagascar
5T	Mauritania
5U	Niger
5X	Uganda
70,4W	Yemen
7Q	Malawi
8Q	Maldives
9G	Ghana
9N	Nepal
9Q	Zaire
9U	Burundi
9X	Rwanda

Additional information:

—SWL cards can be forwarded through the QSL Service.

—We no longer hold cards for countries with no Incoming Bureau. Only cards indicating a QSL manager for a station in these particular countries will be forwarded.

—The Outgoing QSL Service CANNOT forward stamps, IRCs or "green stamps" (cash) to the foreign QSL bureaus.

(Courtesy ARRL)

IARU LIST OF WORLDWIDE BUREAUS

3A MONACO: Association des Radio-Amateurs de Monaco, Box 2, MC-98001 Monaco Cedex

3B MAURITIUS: Mauritius Amateur Radio Society, Box 467, Port Louis

3D2 FIJI: Fiji Association of Radio Amateurs, Box 184, Suva

3DA SWAZILAND: Radio Society of Swaziland, Box 3744, Manzini

4P-4S SRI LANKA: Radio Society of Sri Lanka, Box 907, Colombo

4X,4Z ISRAEL: IARC QSL Bureau, Box 17600, Tel Aviv 61176

5B CYPRUS: Cyprus Amateur Radio Society, Box 1267, Limassol

5N-5O NIGERIA: Nigerian Amateur Radio Society, Box 2873, GPO, Marina, Lagos

5W WESTERN SAMOA: WSARC QSL Bureau, Box 1069, Apia

5Y-5Z KENYA: Radio Society of Kenya, Box 45681, Nairobi

6V-6W SENEGAL: Association des Radio-Amateurs du Senegal, Box 971, Dakar

6Y JAMAICA: Jamaica Amateur Radio Association, 75 Arnold Rd., Kingston 5

7P LESOTHO: Lesotho Amateur Radio Society, Box 949, Maseru 100

7T-7Y ALGERIA: Amateurs Radio Algeriens, Box 2, Alger Gare

8P BARBADOS: Amateur Radio Society of Barbados, Box 814E, Bridgetown

8R: GUYANA: c/o S. D'Ornellas, 8R1R, 110 Barrack St., Kingston, Georgetown

9A CROATIA: Hrvatski Radio-Amaterski Savez, Box 564, HR-41000 Zagreb

9G GHANA: Ghana Amateur Radio Society

9H MALTA: Malta Amateur Radio League, Box 575, Valletta

9I-9J ZAMBIA: Radio Society of Zambia, Box 20332, Kitwe

9K KUWAIT: Kuwait Amateur Radio Society, Box 5240, Safat 13053

9L SIERRA LEONE: Sierra Leone Amateur Radio Society, Box 10, Freetown

9M MALAYSIA: Malaysian Amateur Radio Transmitters' Society, Box 10777, 50724 Kuala Lumpur

9O-9T ZAIRE: Union Zairoise des Radio-Amateurs (closed)

9V SINGAPORE: Singapore Amateur Radio Transmitting Society, Box 2728, GPO, Singapore 9047

9Y-9Z TRINIDAD AND TOBAGO: Trinidad and Tobago Amateur Radio Society, Box 1167, Port of Spain

A2,8O BOTSWANA: Botswana Amateur Radio Society, Box 1873, Gaborone

A3 TONGA: Amateur Radio Club of Tonga, c/o M. Schuster, Box 1078, Nuku'alofa

A4 OMAN: Royal Omani Amateur Radio Society, Box 981, Muscat 113

A7 Qatar: Qatar Amateur Radio Society, Box 22122, Doha

A9 BAHRAIN: ARAB QSL Bureau, Box 22,381, Muharraq

AP-AS PAKISTAN: Pakistan Amateur Radio Society, Box 1450, Islamabad 44000

BA-BZ CHINA: Chinese Radio Sports Association, Box 6106, Beijing 1000061

BV TAIWAN: Chinese Taipei Amateur Radio League, Box 73, Taipei 100

C3 ANDORRA: Unio de Radioaficionats Andorrans, Box 150, La Vella

C5 GAMBIA: Radio Society of The Gambia, c/o J.-M. Voinot, PMB 120, Banjul

C6 BAHAMAS: Bahamas Amateur Radio Society, Box SS-6004, Nassau NP

C8-C9 MOZAMBIQUE: Liga dos Radio Emissores de Moambique, Box 25, Maputo

CA-CE, XQ-XR CHILE: Radio Club de Chile, Box 13630, Santiago 21

CM,CO,T4 CUBA: Federacion de Radioaficionados de Cuba, Box 1, Havana 10100

CN MOROCCO: Association Royale des Radio-Amateurs du Maroc, Box 299, Rabat

CP BOLIVIA: Radio Club Boliviano, Box 2800, Cochabamba

CQ-CU PORTUGAL: Rede dos Emissores Portugueses, Rua D. Pedro V 7-4, P-1200 Lisboa

CV-CX URUGUAY: Radio Club Uruguayo, Box 37, Montevideo

D2-D3 ANGOLA: Liga dos Amadores de Radio de Angola

DA-DR GERMANY: Deutscher Amateur-Radio-Club, Box 1155, 34216 Baunatal

DU-DZ, 4D-4I PHILIPPINES: Philippine Amateur Radio Association, Box 4083, Manila

EA-EH, AM-AO SPAIN: Union de Radioaficionados Espanoles, Box 220, E-28080 Madrid

EI-EJ IRELAND: Irish Radio Tramsmitters Society, Box 462, Dublin 9

EL,5L-5M LIBERIA: Liberia Radio Amateur Association, Box 10-1477, 1000 Monrovia 10

ES ESTONIA: Eesti Raadioamatooride Uhing, Box 125, EE-0090 Tallinn

F,HW-HY,TK TM, TO-TQ FRANCE: Reseau des Emetteurs Francais, Box 2129, F-37021 Tours Cedex

FO FRENCH POLYNESIA: CORA QSL Bureau, Box 5006, Pirae, Tahiti

G,2A-2Z EN6 UNITED KINGDOM: RSGB QSL Bureau, Box 1773, Potters Bar 3EP, England

H4 SOLOMON ISLANDS: Solomon Islands Radio Society, Box 418, Honiara

HA,HG HUNGARY: MRASZ QSL Bureau, Box 214, H-1368 Budapest 5

HB,HE SWITZERLAND: USKA QSL Service, Box 15, CH-4705 Wangen an der Aare

HBØ 501, LIECHTENSTEIN: AFVL QSL Bureau, Box 9494 Schaan

HC-HD ECUADOR: Guayaquil Radio Club, Box 5757, Guayaquil

HH,4V HAITI: Radio Club d'Haiti, Box 1484, Port-au-Prince

HI DOMINICANA: Radio Club Dominicana, Box 1157, Santo Domingo

HJ-HK, 5J-5K COLOMBIA: Liga Colombiana de Radioaficionados, Box 584, Bogota

HL KOREA (Republic of): Korean Amateur Radio League, Box 162, CPO, Seoul 100-601

HO-HP,H3 PANAMA: Liga Paname a de Radioaficionados, Box 175, Panama 9A

HQ-HR HONDURAS: Radio Club de Honduras, Box 273, San Pedro Sula

HS,E2 THAILAND: RAST QSL Bureau, Box 2008, GPO, Bangkok 10501

I ITALY: Associazione Radioamatori Italiani, Via Scarlatti 31, 20124 Milano

J2 DJIBOUTI: Association des Radioamateurs de Djibouti, Box 1076, Djibouti

J3 GRENADA: Grenada Amateur Radio Club, Box 737, St. George's

J7 DOMINICA: Dominica Amateur Radio Club, Box 389, Roseau

JA-JS, 7J-7N,8J-8N JAPAN: Japan Amateur Radio League, 1-14-2 Sugamo, Toshima, Tokyo 170

JT-JV MONGOLIA: Mongolian Radio Sports Federation, Box 639, Ulaan Baatar 13

JY	JORDAN: Royal Jordanian Radio Amateur Society, Box 2353, Amman
LA-LN,	NORWAY: Norsk Radio Relae
JW-JX,3Y	Liga, Box 21, Refstad, N-0513 Oslo 5
LO-LW,AY-	ARGENTINA: Radio Club
AZ,L2-L9	Argentino, Box 97, 1000 Buenos Aires
LX	LUXEMBOURG: c/o A. Rischette, LX1AR, 25 Rue A. Munchen, L-2172 Luxembourg
LY	LITHUANIA: Lietuvos Radijo Megeju Draugija, Box 1000, 2001 Vilnius
LZ	BULGARIA: Bulgarian Federation of Radio Amateurs, Box 830, 1000 Sofia
OA-OC,4T	PERU: Radio Club Peruano, Box 538, Lima 100
OD	LEBANON: Association des Radio-Amateurs Libanais, Box 118888, Beirut
OE	AUSTRIA: Oesterreichischer Versuchssenderverband, Theresiengasse 11, A-1180 Vienna
OF-OJ	FINLAND: SRAL QSL Bureau, Box 30, SF-00381 Helsinki
OK-OL	CZECH REPUBLIC: Cesky Radioklub, Box 69, 11327 Praha 1
OM	SLOVAKIA: Slovak Amateur Radio Association, Box 1, 85299 Bratislava 5
ON-OT	BELGIUM: UBA QSL Bureau, Box 400, B-8400 Ostend WV
OU-OZ	DENMARK: c/o B.W. Nielsen, OZ7BW, Solbjerghedevej 76, DK-8355 Ny Solbjerg
OY 1358,	FAROE ISLANDS: FRA QSL Bureau, Box FR-110 Torshavn
P2-P3	PAPUA NEW GUINEA: PNGARS QSL Bureau, Box 141, Port Moresby
P4	ARUBA: Aruba Amateur Radio Club, Box 2273, San Nicolas
PA-PI	NETHERLANDS: Dutch QSL Bureau, Box 330, NL-6800 AH Arnhem
PJ	NETHERLANDS ANTILLES: Vereniging voor Experimenteel Radio, Onderzoek in Nederlandse Antillen, Box 3383, Curacao
PP-PY,	BRAZIL: Liga de Amadores Brasileiros de
ZV-ZZ	Radio Emissao, Box 07-0004, 70.359 Brasilia DF
PZ	SURINAME: VRAS QSL Bureau, Box 566, Paramaribo
S2-S3	BANGLADESH: Bangladesh Amateur Radio League, Box 3512, GPO, Dhaka
S5	SLOVENIA: Zveza Radioamaterjev Slovenije, Box 180, 61001 Ljubljana
SA-SM,7S,	SWEDEN: Foreningen Sveriges
8S	Sandareamatorer, Ostmarksgatan 43, S-12342 Farsta
SN-SR,HF,	POLAND: PZK QSL Bureau, Box 320, PL-
3Z	00950 Warszawa 1
SU	EGYPT: EAWC QSL Bureau, c/o Wireless Officers Club, Ramsis Bldg. Floor 13 Flat 10, No. 6 Ramsis Sq., Cairo 11111
SV-SZ,J4	GREECE: Radio Amateur Association of Greece, Box 3564, GR-10210 Athens
T7	SAN MARINO: Associazione Radioamatori della Repubblica di San Marino, Box 77, RSM-47031 San Marino
T9	BOSNIA AND HERZEGOVINA: Savez Radio-amatera Bosne i Hercegovine (mail service suspended)
TA-TC,YM	TURKEY: Telsiz Radyo Amatorleri Cemiyeti, Box 109, Istanbul
TF	ICELAND: Islenzkir Radioamatorar, Box 1058, IS-121 Reykjavik
TG,TD	GUATEMALA: Club de Radioaficionados de Guatemala, Box 115, Guatemala City
TI,TE	COSTA RICA: Radio Club de Costa Rica, Box 2412, San Jose 1000
TR	GABON: Association Gabonaise des Radio-Amateurs, Box 1826, Libreville

TU	IVORY COAST: Association des Radio-Amateurs Ivoiriens, Box 2946, Abidjan 01
UR-UZ,	
EM-EO	UKRAINE: Ukrainian Amateur Radio League, Box 56, Kiev 252001
V2	ANTIGUA AND BARBUDA: Antigua and Barbuda Amateur Radio Society, Box 1111, St. John's
V3	BELIZE: c/o B. Leonard, V31HK, Box 168, Belmopan
V5	NAMIBIA Namibian Amateur Radio League, Box 1100, Windhoek 9000
V8	BRUNEI: Negara Brunei Darussalam Amateur Radio Association, Box 73, Gadong, Bandar Seri Begawan 3100
VA-VG,VO,	CANADA: Kennebcasis Valley A.R.C., Box 51,
VX-VY,CF-	St. John, NB E2L 3X1
CK,CY-CZ,	
XJ-XO	
VE,VO,VY	
VEØ-1,VY2	c/o J. Wade, VE1DH, Box 141, Petitcodiac, NB E0A 2H0
VE2	c/o A.G. Daemen, VE2IJ, 2960 Douglas Ave., Montreal, PQ H3R 2E3
VE3	c/o G. Hammond, VE3XN, 5 McLaren Ave., Listowel, ON N4W 3K1
VE4	c/o A. Romanchuk, VE4SN, 26 Morrison St., Winnipeg, MB R2V 3B4
VE5	c/o Bj. Madsen, VE5FX, 739 Washington Dr., Weyburn, SK S4H 3C7
VE6	c/o N.F. Waltho, VE6VW, Box 1890, Morinvile, AB T0G 1P0
VE7	c/o D. Livesay, VE7DK, 8309 112th St., Delta, BC V4C 4W7
VE8	c/o R. Ziemann, VE8RZ, 2 Taylor Rd., Yellowknife, NT X1A 2K9
VO	c/o R. Peddle, VO1BD, Box 6, St. John's, NF A1C 5H5
VY1	c/o W. Champagne, VY1AU, Box 4597, Whitehorse, YT Y1A 2R8
VH-VN,AX	AUSTRALIA:
VK1	Box E46, Queen Victoria Terrace, ACT 2600
VK2	Box 73, Teralba, NSW 2284
VK3	Box 757G, GPO, Melbourne, Victoria 3001
VK4	Box 638, GPO, Brisbane, QLD 4001
VK5	Box 10092, Gouger Street, Adelaide, SA 5000
VK6	c/o J. Rumble, VK6RU, Box F319, GPO, Perth, WA 6001
VK7	Box 371D, GPO, Hobart, TAS 7001
VK8	c/o H.G. Andersson, VK8HA, Box 619, Humpty Doo, NT 0836
VK9-Ø	c/o N. Penfold, VK6NE, 2 Moss Court, Kingsley, WA 6026
VP2E	ANGUILLA: Anguilla Amateur Radio Society, Box DX, The Valley
VP2M	MONTSERRAT: Montserrat Amateur Radio Society, Box 448, Plymouth
VP2V	BRITISH VIRGIN ISLANDS: British Virgin Islands QSL Bureau, c/o Dirk J. de Jong, PO Box 137, Road Town, Tortola, British Virgin Islands. Only cards for calls starting with VP2V are handled by the service. Cards for calls starting with VP2E, VP2M, or /VP2V reciprocal licensed stations will be discarded.
VP5	TURKS AND CAICOS ISLANDS: Turks and Caicos Amateur Radio Society, c/o J. Millspaugh, VP5JM, Box 218, Providenciales
VP9	BERMUDA: Radio Society of Bermuda, Box HM 275, Hamilton HM AX
VS6	HONG KONG: Hong Kong Amateur Radio Transmitting Society, Box 541, Hong Kong
VT-VW	INDIA: Amateur Radio Society of India, Box 6538, Bombay 400026 ARSI Madras Branch, Box 6143, Madras 600017
W,AA-AL,	U.S.A.:
K,N	

W1	Box 80216, Springfield, MA 01138-0216
W2	North Jersey DX Association, Box 599, Morris Plains, NJ 07950
W3	Cumberland County Amateur Radio Service, Box 448, New Kingstown, PA 17072-0448
K4,N4,W4	Mecklenburg Amateur Radio Society, Box DX, Charlotte, NC 28220
AA-AK4,KA-KZ4,NA-NZ4, WA-WZ4	Sterling Park Amateur Radio Club, Call Box 599, Sterling, VA 20167
W5	Box 50625, Midland, TX 79710
W6	Box 1460, Sun Valley, CA 91352
W7	Willamette Valley DX Club, Box 555, Portland, OR 97207
W8	Box 182165, Columbus, OH 43218-2165
W9	Northern Illinois DX Association, Box 519, Elmhurst, IL 60126
WØ	Box 4798, Overland Park, KS 66204
KA2-5**,	Far East Amateur Radio League:
KA7-9**	c/o D. Arthur, KA8DA, or J. Arthur, KA8JA, 6920 ESG, PSC, Box 2253, APO, AP 93319-2253
KA6**	Radio Club of Okinawa, Box 217, Torii Station, APO AP 96331
KG4**	Guantanamo Amateur Radio Club, Box 73, FPO, AE 09593
KH2	Marianas Amateur Radio Club, Box 445, Agana, GU 96910
KH3	Box 73, APO, AP 96558
KH4	U.S. Naval Air Facility, FPO, AP 96614
KH6	Box 788, Wahiawa, HI 96786
KL7	4304 Garfield St., Anchorage, AK 99503
KP2	Virgin Islands Amateur Radio Club, Box 11360, GPO, Charlotte Amalie, VI 00801
KP4	Radio Club de Puerto Rico, Box 1061, San Juan, PR 00902
SWL	c/o M. Witkowski, 4206 Nebel St., Stevens Point, WI 54481
XA-XI,4A-4C,6D-6J	MEXICO: Federacion Mexicana de Radio Experimentadores, Box 907, 06000 Mexico D.F.
XY-XZ	MYANMAR: Burma Amateur Radio Transmitting Society (closed)
YB-YE,8A-8I	INDONESIA: ORARI QSL Bureau, Box 96, Jakarta 10002
YJ	VANUATU: Vanuatu Amateur Radio Society, Box 665, Port Villa
YK,6C	SYRIA: Technical Institute of Radio, Box 245, Damascus
YN,HT	NICARAGUA: Club de Radioexperimentadores de Nicaragua, Box 925, Managua
YO-YR	ROMANIA: Federatia Romana de Radioamatorism, Box 22-50, R-71100 Bucuresti
YS	El SALVADOR: Club de Radio Aficionados de El Salvador, Box 517, San Salvador
YT-YU,YZ, 4N-4O	YUGOSLAVIA: Savez Radio-Amatera Jugoslavije, Box 48, YU-11001 Beograd
YV-YY,4M	VENEZUELA: Radio Club Venezolano, Box 2285, Caracas 1010A
Z2	ZIMBABWE: Zimbabwe Amateur Radio Society, Box 2377, Harare
Z3	FORMER YUGOSLAV REPUBLIC OF MACEDONIA: Association of Amateur Radio Operators of Macedonia, Box 14, 91000 Skopje
ZA	ALBANIA: Albanian Amateur Radio Association, Box 66, Tirana
ZB2	GILBRALTAR: Gibraltar Amateur Radio Society, Box 292, Gibraltar
ZF	CAYMAN ISLANDS: Cayman Amateur Radio Society, Box 1029, Grand Cayman
ZL-ZM 857	NEW ZEALAND: NZART QSL Bureau, Box Wanganui 5000
ZP	PARAGUAY: Radio Club Paraguayo, Box 512, Asuncion
ZR-ZU	SOUTH AFRICA: South African Radio League, Box 807, Houghton 2041

In addition to those listed above, the following QSL bureaus are in operation in countries or territories where there is no IARU member-society. This listing, however, neither confirms nor denies the possibility of their being affiliated with the IARU in the future.

4J-4K	AZERBAIJAN: Box 165 ROSTK DVPSTO, 4K7DWA, Baku 370000 Azerbaidjan
4L	GEORGIA: Box 1, Tbilisi 380002 Georgia
4U1ITU	SWITZERLAND: International Amateur Radio Club: Box 6, CH-1211 Geneva 20
EK	ARMENIA: Box 22, Yerevan 375000 Armenia
ER	MOLDOVA: Box 6637, Kishinev-50, 277050 Moldova
ET	ETHIOPIA: Ethiopian Amateur Radio Society, Box 7447, Addis Ababa
EU-EW	BELARUS: Box 469, 220050 Minsk
EX	KYRGYZSTAN: Union of Radioamateurs of Kirghizstan, Box 1100, 720020 Bishkek 20
EY	TAJIKISTAN: Tajik Amateur Radio League, Box 303, Glavpochtamt Dushanbe 734025
EZ	TURKMENIA: Box 555 (T.R.A.L) Ashgabat 744020, Turkmenia
FK	NEW CALEDONIA: Association des Radio-Amateurs de Nouvelle-Caledonie, Box 3956, Noumea
HL9	US PERSONNEL IN THE REPUBLIC OF KOREA: American Amateur Radio Club of Korea, Dependent Mail Section, Box 153, APO, AP 96206
UK	UZBEKISTAN: Box 0, Tashkent, 700000, Uzbekistan
UN-UQ	KAZAKHSTAN: Kazakhstan Amateur Radio Union, Box 112, 470055 Karaganda
V7	MARSHALL ISLANDS: Kwajalein Amateur Radio Club, Box 444, APO, AP 96555, USA
VP8	FALKLAND ISLANDS: Falkland Islands Radio Club, Box 260, Mount Pleasant Airport
VQ9 AP	BRITISH INDIAN OCEAN TERRITORY: c/o R. Shaw, VQ9RS/KAØMXI, NSF, Box 16, FPO, 96464, USA
YI	IRAQ: Iraqi Radio Amateur Club, Box 55072, Baghdad 12001
ZC4 Board	BRITISH FORCES CYPRUS: Joint Signal Hq., BFC, BFPO 53, London GPO, UNITED KINGDOM
ZD8	ASCENSION ISLAND: Ascension Amateur Radio Relay League, Box 4127, Patrick AFB, FL 32925-0127, USA

(Sources: IARU, RA6YR)

QSL MANAGERS

A QSL manager is an individual who volunteers to handle the QSL chores for a DX station, freeing the rare DX to operate more and spend less time filling out cards. The DX station sends a copy of his log periodically to his manager, who takes care of answering cards and mailing replies. Usually, DX operators will announce their manager during a contact, but sometimes a pileup's volume may make it difficult for the DX station to announce his manager to every station he works.

Several QSL Manager Lists have been compiled over the years to assist DXers in obtaining the cards they need for various awards. Managers are also routinely listed in the DX columns of the major amateur magazines, on-line databases, and PacketCluster (tm) systems.

The best-known QSL manager list is the "GO LIST," now published by AE4AP/KE4RGW, and available in both paper and electronic formats. An interesting database has been compiled over the last few years on the Copenhagen DX Cluster (OZ2DXC). It contains more than 80,000 QSL references. The input sources come from all known DX magazines, packet worldwide DX files and from DX Spots on the DX clusters in Europe. The incoming rate is about 500–700 new entries per month.

Some QSL managers handle QSLs for several stations. A select few managers provide QSL services for so many stations that their calls are almost synonymous with QSL Manager. The following are the Top Ten QSL managers, in terms of number of calls for which they manage cards.

1.	W3HNK	305
2.	F6FNU	220
3.	WA3HUP	106
4.	DJ9ZB	89
5.	YASME	82
6.	F6AJA	76
7.	LU8DPM	73
8.	W4FRU	54
9.	AA6BB	46
10.	K8LJG	43

(Compiled by AE4AP/KE4RGW)

QSL PRINTERS AND PRINTING SERVICES

There are many sources of QSL cards. Most ham magazines contain advertisements every month. Many will provide a few samples of their work and a price list for a self-addressed stamped envelope (SASE). Some of the QSL card printers who advertise regularly are listed below:

WX9X, 161 W. Lincolnway, Valparaiso, IN 46383, Phone: (219) 465-7128, Fax: (219) 464-7333
QSLs by W4MPY, Wayne Carroll, W4MPY, 682 Mt. Pleasant Road, Monetts, SC 29105, Fax: (803) 685-7117
Shell Printing, KD9KW, Box 50, Rockton, IL 61072, Phone: (815) 629-2193
Colorful QSLs, WA7LNW, P.O. Box 5358, Glendale, AZ 85312-5358
Marcum's QSLs, KA6GND, 4645 Pine Street, Riverside, CA 92501
KD6EUT Perryprints, 12812 Shadowline Drive, Poway, CA 92064, Phone/FAX: (619) 748-8315
Mac's Shop, P.O. Box 43175, Seven Points, TX 75143
Harry A. Hamlen, K2QFL, P.O. Box 1, Stewartsville, NJ 08886, Fax:1-(800)-AIR-FAXX
Olde Press, WB9MPP, Box 1252, Kankakee, IL 60901
Mahre & Sons Print Shop, 2095 Prosperity Avenue, Maplewood, MN 55109-3621
WorldWise Services, 107 Giles Court, Newark, DE 19702
Little Print Shop, Box 1160, Pflugerville, TX 78660, Phone: (512) 990-1192

Jerry Fitz-Randolph, K5KRN, P.O. Box 3473 , Jackson, TN 38303
Brownies QSL Cards, 3035 Lehigh Street, Allentown, PA 18103
Charlie Hansen , NØTT, RR 1, Box 108-B, Napolean, MO 64074
Samcards, 48 Monte Carlo Drive, Pittsburgh, PA 15239
Visual Conception, Vi-Con International, P.O. Box 10013, Kansas City, MO 64111, Phone:1 (800) 869-7527, Phone: (816) 531-3939
K-K-Labels, Box 412 , Troy, NY 12181-0412
K2MK QSLs, 551 Norwood Road, Mt. Laurel, NJ 08054
Bud Smith, Box 1948 Blain, WA 98231
Spangler X-Pressions, P.O. Box 6262, Kansas City, MO 66106, Phone:1 (800) 466-1616
W5YI Group (QSLs), Box 565101, Dallas, TX 75356, Phone: (817) 461-6443
Chester QSLs, 310 Commercial, Emphoria, KS 66801, Phone: (316) 342-8792, Fax: (316) 342-4705
Raum's, 8617 Orchard Road, Coopersburg, PA 18036, Phone/Fax: 1 (215) 679-7238
Ebbert Graphics D-3, Box 70, Westerville, OH 43081
New Dimension QSLs, 6600 Lucia Lane, Minneapolis, MN 55432, Phone: (612) 571-5881
Network QSLs, P.O. Box 13200, Alexandria, VA 71315 Phone:1 (800) 354-0830, Phone: (318) 443-7261, Fax: (318) 445-9940
Bert P. van der Berg, N6ID, BVE Professional Printing, 2023 Chicago Ave #b13, Riverside, CA 92507, Phone: (714) 781-0283
Pro-Print, WA5LKS, 5301 Junction Road, Norman, OK 73701, Phone: 1 (405) 364-6676
Rusprint, 26037 220th Terr., Spring Hill, KS 66083
Gazebo Press, 4148 Mimosa Lane , La Plata, MD 20646
WD5ADH, 4209 McConnell, El Paso, TX 79904-6224
K8NUJ, Louis Craycraft, 1502 53rd Ave. W., Bradenton, FL 34207, Phone/Fax: (813) 758-9758
Creative Imprinting Co.,WB4FIH, 6522 Chesterfield Ave., McLean, VA 22101
AACO, 1639 Fordham Way, Mountain View, CA 94040
W0LQV, Box 4133, Overland Park, KS 66204
Wilkins, Box 787, Atascadero, CA 93423
K7HLR Ray, 25 South Terrace Drive, Clearfield, UT 84015

CANADA
M. Smith, VE7FI, 18610 62nd Avenue, Surrey, B.C. CANADA , V3S 7P1

THE WORLD'S BIGGEST QSL CARD COLLECTION

The biggest collection of QSL cards in the world is believed to be that of the YASME Foundation. "Yasme" is a Japanese word meaning "good luck," and was the name of a boat sailed around the world by Danny Weil, VP2VB, who activated many rare island groups in the 1960s. The YASME Foundation was formed in 1961 by KV4AA and VP2VB. Lloyd and Iris Colvin, W6KG and W6QL, joined in 1965.

Lloyd and Iris, married over 50 years, may be the most-traveled DXers in the world. They have visited over 200 DXCC countries and operated from over half of them, including 14 of the 15 Republics of the former U.S.S.R. The QSL cards from the many hams who have sent their "QSL via YASME" total close to a million, all neatly stored in organized file drawers.

The YASME address has become one of the best-known addresses in the world among DXers: YASME Foundation, Box 2025, Castro Valley, CA 94546.

POSTAL RATES

Hams in search of QSL cards from overseas stations need to know the latest postal rates. The following information is from *Publication 51* - International Postal Rates and Fees, July 1995 Edition, published by the United States Postal Service. This publication is a condensed version of general information included in the *International Mail Manual*. The international Mail Manual is the definitive reference book on international mailing. A copy of the IMM may be ordered from:

Superintendent of Documents, P.O. BOX 371954, Pittsburgh, PA 15250-7954

The postal rates listed here were in effect at press time and are subject to change. If rates have changed, details should be available from your local post office

SOURCES OF FOREIGN POSTAGE

DX QSL Associates, 434 Blair Road NW, Vienna, VA 22180
Mackey DX Postage, 187 Ridgewood Road / Box 270569, West Hartford, CT 06107
Plum DX Supplies, 12 Glenn Road, Flemington, NJ 08822

Post Cards/Postal Cards	
Canada	$0.40
Mexico	0.35
All other countries	0.50

Letters and Letter Packages—Airmail rates
Countries other than Canada & Mexico

Weight not over	Rate	Weight not over	Rate	Weight not over	Rate
0.5 ozs.	$0.60	16.5 ozs.	$13.40	33 ozs.	$26.20
1.0	1.00	17.0	13.80	34	26.60
1.5	1.40	17.5	14.20	35	27.00
2.0	1.80	18.0	14.60	36	27.40
2.5	2.20	18.5	15.00	37	27.80
3.0	2.60	19.0	15.40	38	28.20
3.5	3.00	19.5	15.80	39	28.60
4.0	3.40	20.0	16.20	40	29.00
4.5	3.80	20.5	16.60	41	29.40
5.0	4.20	21.0	17.00	42	29.80
5.5	4.60	21.5	17.40	43	30.20
6.0	5.00	22.0	17.80	44	30.60
6.5	5.40	22.5	18.20	45	31.00
7.0	5.80	23.0	18.60	46	31.40
7.5	6.20	23.5	19.00	47	31.80
8.0	6.60	24.0	19.40	48	32.20
8.5	7.00	24.5	19.80	49	32.60
9.0	7.40	25.0	20.20	50	33.00
9.5	7.80	25.5	20.60	51	33.40
10.0	8.20	26.0	21.00	52	33.80
10.5	8.60	26.5	21.40	53	34.20
11.0	9.00	27.0	21.80	54	34.60
11.5	9.40	27.5	22.20	55	35.00
12.0	9.80	28.0	22.60	56	35.40
12.5	10.20	28.5	23.00	57	35.80
13.0	10.60	29.0	23.40	58	36.20
13.5	11.00	29.5	23.80	59	36.60
14.0	11.40	30.0	24.20	60	37.00
14.5	11.80	30.5	24.60	61	37.40
15.0	12.20	31.0	25.00	62	37.80
15.5	12.60	31.5	25.40	63	38.20
16.0	13.00	32.0	25.80	64	38.60

Weight limit: 64 ozs. (4lbs.)

Letters and Letter Packages (Canada & Mexico)

Weight not over		Canada	Mexico	Weight not over		Canada	Mexico
0 lbs.	0.5 ozs.	$0.46	$0.40	0	10	2.28	4.06
0	1.0	.52	.46	0	11	2.47	4.46
0	1.5	.64	.66	0	12	2.66	4.86
0	2	.72	.86	1	0	3.42	6.46
0	3	.95	1.26	1	8	4.30	9.66
0	4	1.14	1.66	2	0	5.18	12.86
0	5	1.33	2.06	2	8	6.06	16.06
0	6	1.52	2.46	3	0	6.94	19.26
0	7	1.71	2.86	3	8	7.82	22.46
0	8	1.90	3.26	4	0	8.70	25.66
0	9	2.09	3.66				

Maximum weight: 4 pounds, except that registered items to Canada may weigh up to 66 pounds. For registered items weighing more than 4 pounds, the rate is $1.76 for each additional pound up to the 66-pound limit.

International Reply Coupons— The REAL Story

International Reply Coupons (IRCs) provide a mechanism for sending return postage to a person in another country. They are often sent by hams with their QSL cards as a courtesy when requesting a QSL card from a rare station. They are exchangeable in any country of the Universal Postal Union (which includes most countries) for the minimum postage for a foreign letter. The minimum postage varies from country to country. In the U.S., for example, the "first unit" is a half-ounce letter. In other countries, the minimum varies from a quarter-ounce to an ounce.

IRCs are issued by the UPU, and sold on request to the various postal administrations in the member countries, who then resell them to their customers. When a customer presents an IRC to the local post office for exchange, that customer is entitled to stamps equal to the cost of the first unit of airmail postage. The local post office forwards the exchanged IRC to the national postal administration, and at the end of the year, all exchanged coupons are sent to the UPU office in Bern, Switzerland for compensation.

Historically, IRCs have been exchangeable for the first unit of surface postage. However, in 1989, the UPU member countries met in Washington, DC to discuss and update numerous international mail regulations, including IRCs.

It was decided that since surface mail is rarely used for letters, that IRCs should be upgraded to be exchangeable for airmail postage, effective February 1991. All post-1975 IRCs are now exchangeable for airmail postage, regardless of what is printed on them. IRCs printed after 1991 indicate that they are valid for airmail postage in the last few word of the French text on the front of the coupon ("...par voie aerienne"). IRCs printed between 1975 and 1991 refer to surface-mail validity at the end of the text ("...par voie de surface").

Many amateurs purchase IRCs from other amateurs (especially active DXers who receive them frequently). Many hams are confused about the difference between the airmail and surface IRCs —the fact is that there is no difference! Both are equally valid for airmail postage. If you bring a valid IRC originally purchased in another country to a U.S. Post office, you are entitled to exchange it for 60 cents worth of stamps (postal rates in effect as of September 1995). If you buy an IRC in the U.S. and decide to cash it in at a later date, you are entitled to $1.04 worth of stamps (for an IRC purchased for $1.05).

Amateurs who have tried to exchange IRCs at their local post office have often been frustrated by postal employees who have never seen IRCs before. Most smaller post offices don't stock them, and don't know how to exchange them for postage. The revised rules make it even more difficult to convince a postal employee that an IRC marked "surface" is actually valid for airmail postage.

If this happens to you, refer the postal employee to the International Mail Manual, Section 392, reproduced here in its entirety:

International Reply Coupons

392.1 Description

a. The sender of a letter may prepay a reply by purchasing reply coupons which are sold and exchangeable for postage stamps at post offices in member countries of the Universal Postal Union. The period of exchange of international reply coupons issued by the Universal Postal Union on or after January 1, 1975, is unlimited.

b. International reply coupons (in French, Coupons-Reponse Internationaux) are printed in blue ink on paper which has the letters UPU in large characters in the watermark. The front of each coupon is printed in French. The reverse side of the coupon shows the text relating to its use in German, English, Arabic, Chinese, Spanish, and Russian.

c. Coupons sold in the United States have the selling price printed on them, while coupons in other countries may not.

392.2 Availability

Reply coupons may be requisitioned by post offices in the same manner as postage stamps. The coupons should be stocked at post offices which have a demand for them.

393.3 Selling Price and Rate of Exchange

a. The selling price of a reply coupon in the United States is $1.05. One coupon is exchangeable in any other member country for a stamp or stamps representing the minimum postage on an unregistered air letter. Unused U.S. coupons (that is, those with the U.S. selling price stamped on them) may be exchanged only for United States postage stamps by the original purchaser at a discount of 1 cent below the purchase price.

b. International reply coupons purchased in foreign countries are exchangeable in U.S. post offices toward the purchase of postage stamps, postage meter stamps, and embossed stamped envelopes (including aerogrammes) at the rate of 60 cents per coupon, irrespective of the country where it was purchased.

392.4 Processing Requirements

a. U.S. post offices must postmark coupons in the lower left circle at the time of sale. Coupons issued by foreign countries will contain a control stamp of the country of origin.

b. A post office redeeming an unused U.S. coupon must postmark it in the right circle. Post offices must not accept foreign coupons which already bear a USPS postmark.

c. Reply coupons issued by foreign countries prior to January 1, 1975, are no longer redeemable at U.S. post offices. These old-style coupons are distinguishable from the newer coupons printed by the International Bureau of the Universal Postal Union, because the name of the country of origin is always present on the old-style coupons. Customers processing pre-1975 coupons should be advised to return them to their correspondents in the country of origin for replacement or redemption through the selling post office.

d. Reply coupons formerly issued by the Postal Union of the Americas and Spain are no longer valid. These coupons are printed in green ink and bear the caption Cupons Respuesta America-Espanol. Customers possessing any of these coupons should return them to their correspondents in the country of issue for redemption through the selling post office.

e. Postmasters must process exchanged foreign and redeemed U.S. coupons as prescribed in 458 International Reply Coupons (IRCs) Handbook F-1, Post Office Accounting Procedures.

AMATEUR RADIO IN STAMPS

Argentina
  Date of Issue: Nov 1,1980, Title: The Radio Club of Argentina, Value: 7 pesos.
  Date of Issue: Dec 28,1991, Title: LUSAT 1 Amateur Radio satellite

Ascension Island
  Date of Issue: Feb 22,1982, Title: Boy Scout Jamboree, Values: 10 pence, 15 pence, 25 pence, 40 pence

Australia
  Date of Issue: May 22,1985, Title: 75th Anniversary of the Wireless Institute of Australia, Value: 33 cents

Bolivia
  Date of Issue: Mar 26,1979, Title: 38th Anniversary of the Radio Club of Bolivia, Value: 3 pesos
  Date of Issue: Mar 1,1991, Title: 50th Anniversary of the Radio Club of Bolivia, Value: 2.40 Bs

Brazil
  Date of Issue: Nov 5,1977, Title: Brazilian Amateur Radio Day, Value: 1.30 cruzei

Bulgaria
  Date of Issue: Dec 10,1986, Title: 60th Anniversary of the Bulgarian Amateur Radio Club, Value: 13 stontinki

Chile
  Date of Issue: Dec 29,1982, Title: 60th Anniversary of the Radio Club of Chile, Value: 7 peseos

Colombia
  Date of Issue: Jun 25,1959, Title: Miss Universe, Luz Marina Zuruaga HK6LT, Value: 10 centavos
  Date of Issue: May 10,1973, Title: 40th Anniversary of the Radio Amateurs League, Value: 60 centavos
  Date of Issue: Jun 11,1983, Title: 50th Anniversary of the Colombia Radio Amateurs League, Value: 12 pesos

Costa Rica
  Date of Issue: Apr 16,1975, Title: 16th Convention of Federation Radio Amateur Clubs, Values: 1.00 colones, 1.10 colones, 2.00 colones

Czechoslovakia
  Date of Issue: Mar 28,1959, Title: 10th Anniversary of Radiosport, Value: 60 heleru

Djibouti
  Date of Issue: Jun 25,1981, Title: Djibouti Radio Club, Value: 250 francs

Dominican Republic
  Date of Issue: Oct 8,1976, Title: 50th Anniversary of the Dominican Republic Radio Club, Values: 6 centavos, 10 centavos
  Date of Issue: Jan 25,1979, Title: Beate Island DXpedition, Value: 10 centavos
  Date of Issue: Oct 3,1980, Title: Catilina Island DXpedition, Value: 7 centavos

East Germany
  Date of Issue: Aug 8,1972, Title: Society for Sports and Technology, Value: 25 pfennigs

West Germany
  Date of Issue: Jul 12, 1979, Title: W.A.R.C. (shows a Collins KWM-2 tuned to 21.275 kHz., Value: 60 pfennigs
  Date of Issue: May 25,1973, Title: O. Maksymillian Kolbe, SP3RN, Value: 40 pfennigs

Indonesia
  Date of Issue: Oct 1991, Title: 8th I.A.R.U. Region III Conference, Value: 30 Rp

Israel
  Date of Issue: Jun 14,1987, Title: 40th Anniversary of the Israel Radio Amateurs; club station 4X4Z, Value: 2.50 nis

Japan
  Date of Issue: Sept 24,1977, Title: 50th anniversary of the Japanese Amateur Radio League, Value: 50 yen

Jordan
  Date of Issue: Aug 11,1983, Title: Royal Jordanian Amateur Radio Society and King Hussein, JY1, Value: 10 fils, 25 fils, 40 fils, 50 fils, 100 fils

Liberia
  Date of Issue: Nov 23,1987, Title: 25th anniversary of the Liberian Radio Amateur Society , Value: 10 c (Jubilee Emblem), 10 c (Village), 35 c (Jubilee Award), 35 c (Flag and Award)

Luxembourg
> *Date of Issue*: Mar 9,1987, *Title*: 50th anniversary of Amateur Radio in Luxembourg, *Value*: 12 francs

Morocco
> *Date of Issue*: Jul 9,1957, *Title*: King Hassan, CN8AA, *Value*: 15 francs

New Caledonia
> *Date of Issue*: Jan 7,1987, *Title*: 25th anniversary of the New Caledonia Radio Amateur Association, *Value*: 64 francs

Nicaragua
> *Date of Issue*: Oct 7,1983, *Title*: Radio Amateur Federations of Central American and Panama, *Value*: 1 cordoba, 4 cordoba

Norfolk Island
> *Date of Issue*: Apr 9,1991, *Title*: Amateur Radio; call signs of five operators, *Value*: 43 cents, 1 dollar, 1.20 dollars

Oman
> *Date of Issue*: 1985, *Title*: Sultan Qaboos Bin Said, A4XAA, *Value*: 2.50 baisa
> *Date of Issue*: Dec 23,1987, *Title*: 15th anniversary of the Royal Omani Amateur Radio Society, *Value*: 1.30 baisa

Peru
> *Title*: 55th anniversary of the Amateur Radio Service, *Value*: 1300 sol

Poland
> *Date of Issue*: Jun 26,1961, *Title*: Conference of Communications Ministries in Communist, *Value*: 40 gr, 60 gr, 2.30 zlotys
> *Date of Issue*: Apr 15,1975, *Title*: IARU Region I Conference, *Value*: 1.50 zlotys
> *Date of Issue*: 1980, *Title*: Postal card; 50th anniversary of Polish amateur radio society, *Value*: 2.00 zlotys
> *Date of Issue*: Oct 10,1982, *Title*: O. Maksymillian Kolbe, *Value*: 27 zlotys

Russia
> *Date of Issue*: May 20,1973, *Title*: Ernst Krenkel, RAEM, *Value*: 4 kopecks
> *Date of Issue*: Feb 23,1979, *Title*: RS-1 and RS-2 Amateur Radio satellites, *Value*: 4 kopecks

> *Date of Issue*: Dec 25,1979, *Title*: Ernst Krenkel, RAEM, *Value*: 4 kopecks
> *Date of Issue*: Mar 12,1981, *Title*: 30th All Unions Amateur Radio Exhibition, *Value*: 4 kopecks
> *Date of Issue*: Sep 1,1983, *Title*: Radiotelegraphy Championship, *Value*: 6 kopecks

San Marino
> *Date of Issue*: Apr 28,1983, *Title*: World Communications Year honors Amateur Radio Society of San Marino, *Value*: 400 lira

Solomon Island
> *Date of Issue*: Dec 19,1983, *Title*: Radio Society, H44SI, *Value*: 18 cents

Spain
> *Title*: King Juan Carlos, EAØJC, *Value*: 2 peseta, 3 peseta, 12 peseta

Sri Lanka
> *Date of Issue*: Jan 17,1983, *Title*: 55th anniversary of Amateur Radio, *Value*: 2.50 rupee

Switzerland
> *Date of Issue*: Sep 6,1979, *Title*: 50th anniversary of the Swiss Amateur Radio Union, *Value*: 70 centimes

Thailand
> *Date of Issue*: Dec 5,1980, *Title*: King Bhumibol Adulyadeji, HS1A, *Value*: 25 satangs

United States
> *Date of Issue*: Dec 15,1964, *Title*: Amateur Radio; issued on the 50th anniversary of ARRL, *Value*: 5 cents

Uruguay
> *Date of Issue*: Apr 16,1984, *Title*: 50th anniversary of the Radio Club of Uruguay, *Value*: 7 pesos

Venezuela
> *Date of Issue*: Nov 18,1983, *Title*: 50th anniversary of the Radio Club of Venezuela, *Value*: 2.70 bolivars

Yugoslavia
> *Date of Issue*: May 23,1966, *Title*: 20th anniversary of Amateur Radio in Yugoslavia, *Value*: 85 paras

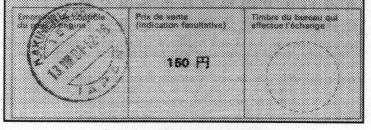

Comparison of Surface and Airmail IRCs

(Both were purchased for 150 yen in Japan in 1991, and both are currently valid for airmail postage!)

U.S. AMATEUR RADIO CLUBS

ALABAMA

AUBURN UNIVERSITY RADIO CLUB, RYAN WELTY KD4VZX TREAS, FOY UNION BLDG, BOX AUBURN UNIVERSITY, AUBURN, AL 36849

BIRMINGHAM AMATEUR RADIO CLUB, BOB RUSSELL WD4DZZ PRES, P O BOX 603, BIRMINGHAM, AL 35201

CALHOUN COUNTY ARA, STEWART GARRETT KA4PSE PRES, PO BOX 1624, ANNISTON, AL 36202

CLARKE COUNTY ARS, JOHN K DAVIS KJ4RU PRES PO BOX 36, WHATLEY, AL 36482

CULLMAN ARC, HERMAN GARDNER N4TUN 6824 COUNTY RD 310, CRANE HILL, AL 35053

DECATUR ARC, R W MCKENZIE K4MLR SECY PO BOX 9, DECATUR, AL 35602

EAST ALABAMA ARC, CHERYL WHITLOCK N4JFV SCTR 1101 STALEY AVE, OPELIKA, AL 36801

ENTERPRISE ARS, MIKE ALLEN N4WIG PRES PO BOX 34, ENTERPRISE, AL 36331

HUNTSVILLE ARC, JARED CASSIDY KQ4VT SCTR PO BOX 423,HUNTSVILLE, AL 35804

HUNTSVILLE AREA YOUNG LADIES, RORY STRATTON KD4FJJ PRES, 2614 BONITA CIR SW, HUNTSVILLE, AL 35801

LAKE MARTIN ARC, WAYNE BANKS KD4DLH PRES PO BOX 33, JACKSONS GAP, AL 36861

MONROE ARC, HOWARD C MCKINLEY JR KD4EJA 752 N MOUNT PLEASANT AVE, MONROEVILLE, AL 36460

MONTGOMERY ARC, STEVE SCOGGIN N4WVW SECY PO BOX 3141, MONTGOMERY, AL 36109

MORGAN COUNTY ARC, ROGER D COOK KO4O SECY 512 MEADOWVIEW DR NW, DECATUR, AL 35601

MUSCLE SHOALS ARC, KENNETH K MOOR WA4ZDW TREAS, 1634 CULLMAN ST, FLORENCE, AL 35630

NORTH ALABAMA DX CLUB, WILLIAM G HULL W4GBF SCTR, 405 GWENDOLYN AVE NW, HUNTSVILLE, AL 35811

RAINBOW CITY ARC, EUGENE MCGLAUGHN KC4TFF SECY, 315 ALLEN ST, GADSDEN, AL 35903

SOC. FOR THE PROMOTION OF ARC, RAYMOND C DODGE III N4KFM, PO BOX 2423, OPELIKA, AL 36801

SOUTH BALDWIN ARC, CHARLES VASUT AF4I SECY, PO BOX 6, ROBERTSDALE, AL 36567

SUMTER COUNTY ARC, JEFF MANUEL N4WLZ SCTR PO BOX 361, LIVINGSTON, AL 35470

TALLADEGA RADIO AMATEUR CLUB, ANNITA MARTIN KD4PJC SCTR, PO BOX 636, TALLADEGA, AL 35160

UNIVERSITY OF ALABAMA ARC, BRIAN BENDER KD4FGL VPRS, 11307 MAPLECREST DR SE, HUNTSVILLE, AL 35803

WEST ALABAMA ARS, TOM HENDERSON K4CIH SCTR PO BOX 1741, TUSCALOOSA, AL 35403

WIREGRASS ARC, WALTER C HAYMON WA6MWS TREAS PO BOX 958, DOTHAN, AL 36302

ALASKA

ANCHORAGE ARC INC, ROBERT WUKSON KL7AA PRES P O BOX 101987, ANCHORAGE, AK 99510

COAST GUARD ARC OF KODIAK, KEITH FONCREE AL7MP PRES, PO BOX 190421, KODIAK, AK 99619

MATANUSKA AMATEUR RADIO ASSN, ERIC SANFORD NL7OL SCTR, PO BOX 873131, WASILLA, AK 99687

PETERSBURG ARC, MILDRED FUGLVOG WL7ALG SCTR PO BOX 781, PETERSBURG, AK 99833

SITKA ARC, MARGARET L DANGEL KL7BYA SCTR 1324 SAWMILL CREEK RD, SITKA, AK 99835

SOUTH CENTRAL RADIO CLUB, JIM WILEY KL7CC EDITOR 8023 E 11TH CT, ANCHORAGE, AK 99504

ARIZONA

AMATEUR RADIO COUNCIL OF AZ, CLIFFORD HAUSER KD6XH PRES, PO BOX 32756, PHOENIX, AZ 85064

ARIZONA ARC, TOM IVAN KF7GC PRES 3323 N 79TH AVE, PHOENIX, AZ 85033

ARIZONA NETWORK INTERTIE GROUP, DANIEL J MEREDITH N7MRP PRES, PO BOX 44563, PHOENIX, AZ 85064

ARIZONA REPEATER ASSN, ROY HEJHALL K7QWR PO BOX 35758, PHOENIX, AZ 85069

CARL HAYDEN HIGH SCHOOL ARC, ALLAN CAMERON N7UJJ, 3333 W ROOSEVELT AVE, PHOENIX, AZ 85009

CATALINA RADIO CLUB, CHARLES MICHELS KB7RFI PRES, HUGHES MISSILE SYSTEMS CO, PO BOX 11337, TUCSON, AZ 85734

COCHISE AMATEUR RADIO ASSN, DALE GREEN K7RDG PRES, PO BOX 1855, SIERRA VISTA, AZ 85636

COCONINO COUNTY ARC, KENNETH D GARDNER N7LQS PRES, PO BOX 1695, FLAGSTAFF, AZ 86002

GREEN VALLEY ARC, LLOYD MILLER K8AVH PRES 601 N LA CANADA DR, GREEN VALLEY, AZ 85614

HUALAPAI ARC, PEGGY AKE KB7YGZ SECY, PO BOX 4364, KINGMAN, AZ 86402

KACHINA RADIO CLUB, PAUL REED KE7KO PRES PO BOX 2996, SHOW LOW, AZ 85901

LONDON BRIDGE RADIO ASSN, GERALD R FRANCE KD0KZ PRES, PO BOX 984, LAKE HAVASU CITY, AZ 86405

MINGUS MOUNTAIN RPTR GROUP, ERNIE HARWAN WA7VWG PRES, 4907 W VERDE LN, PHOENIX, AZ 85031

MOTOROLA ARC OF ARIZONA, PAUL BOYCE N7ZQA SECY, 2100 E ELLIOT RD # EL316, TEMPE, AZ 85284

NORTHERN ARIZONA DX ASSN., ARTHUR M PHILLIPS III NN7A, PO BOX 201, FLAGSTAFF, AZ 86002

OLD PUEBLO RADIO CLUB, SCOTT KETCHER KF7XD PRES, 9005 E 8TH ST, TUCSON, AZ 85710

PAGE HIGH SCHOOL ARC, LEHAMAN BURROW KA7ZLE, PO BOX 1927, PAGE, AZ 86040

SCOTTSDALE ARC, RON AVERY WB6PEB PRES PO BOX 10878, SCOTTSDALE, AZ 85271

SUPERSTITION ARC, RICHARD CHECKETTS KA0KZB PRES, PO BOX 1551, APACHE JUNCTION, AZ 85217

TRICITY ARC, JOHN ROBINETT N7RAT PRES, PO BOX 8662, MESA, AZ 85214

TUCSON REPEATER ASSN, TED WILLIS AA7HX 5649 E 32ND ST, TUCSON, AZ 85711

UNIVERSITY OF ARIZONA ARC, CHARLES E ABERNETHY WB5VHZ, UNIVERSITY OF ARIZONA, NEE/SIE BLDG 20, TUCSON, AZ 85721

WEST VALLEY ARC, RAYMOND W SUMNER NW7R EDIT 11037 W FARGO DR, SUN CITY, AZ 85351

YAVAPAI ARC, JERRY SAGER KG7ZF PRES, PO BOX 11572, PRESCOTT, AZ 86304

ARKANSAS

ARC OF THE UNIVERSITY OF ARK, JEFF MCFARLAND KB5RUI PRES, DEPT OF ELECTRIAL ENGINEERING, BELL 3217 UNIV OF ARK,FAYETTEVILLE, AR 72701

ARKANSAS DX ASSOCIATION, DENNIS SCHAEFER W5VOX SCTR, 181 SCHAEFER DR, DOVER, AR 72837

ARKANSAS RIVER VALLEY AR FOUN, TOM NEUMEIER WB5VOX PRES, PO BOX 582, RUSSELLVILLE, AR 72811

BATESVILLE AREA ARC, DAVID A NORRIS AA5GY PRES PO BOX 2846, BATESVILLE, AR 72503

BENTON ARS, LARRY PORTERFIELD WD5HJC SECY 4426 CONGO RD, BENTON, AR 72015

CENTRAL ARKANSAS RADIO EMERG, BOB HANCOCK KB5IDB, 6116 NICOLE DR, NORTH LITTLE ROCK, AR 72118

CLINTON ARC, JOE EVANS KD5GC PRES, PO BOX 143, CLINTON, AR 72031

DRIVEN ELEMENTS A R GROUP, CARL RICHARDSON KB5FJX VPRS, RR 1 BOX 55, MARION, AR 72364

FAULKNER COUNTY ARC, INC, BILL THOMAS AA5YZ PRES, PO BOX 324, CONWAY, AR 72033

FORT SMITH AREA ARC, MARGARET BURKS KI5OC SECY PO BOX 32, FORT SMITH, AR 72902

GREERS FERRY ARC, WILLIAM A HARPER KB4JKQ PRES
   650 CLIFFVIEW DR, QUITMAN, AR 72131
GREERS FERRY LAKE AREA ARS, THOMAS F KORN,
   W9ZJZ EDIT, 150 RIDGECREST RD, HEBER SPRINGS, AR
   72543
HOT SPRING CTY AR EMERG NET, JACK EDWARDS
   KA5OVQ PRES, RR 1 BOX 908, MALVERN, AR 72104
HOT SPRINGS ARC, JIMMY BALLEW N5ZI SECY, 612
   MCCLENDON, HOT SPRINGS, AR 71901
J D LEFTWICH HIGH SCHOOL ARC, JOHN C BELL SR
   KB5ZGV, RR 1 BOX 112, MAGAZINE, AR 72943
JONESBORO ARC, EVELYN CASTLEBERRY N5DSY SECY
   3362 CR 333, BONO, AR 72416
MALVERN ARC, MARGARETT FINLEY KC5FNN SECY
   RT 1 BOX 115, MALVERN, AR 72104
MARION COUNTY ARC, ROBERT F RICHTER N5UYS PRES
   PO BOX 35, PEEL, AR 72668
METROPOLITAN ARC, DORA ANNA GRAZIANI NI5D
   5712 ALTA VISTA DR, NORTH LITTLE ROCK, AR 72118
NORTH CENTRAL ARK AR SERVICE, ROGER GRAY
   KB5REE PRES, PO BOX 911, JUDSONIA, AR 72081
NORTHWEST ARKANSAS ARC, CHARLES TILLOTSON
   N5NXH PRES, 4117 W WEIR RD, FAYETTEVILLE, AR
   72704
OUACHITA ARA, LANDON J BREWER JR. KG5QO
   268 POLK ROAD 36, HATFIELD, AR 71945
OZARK ARC, GEORGE LEHMKUHL K0SXQ PRES.
   956 TANGLEWOOD DR, MOUNTAIN HOME, AR 72653
PINE BLUFF ARC, JACK JEHLEN N5STJ PRES
   604 TRACY RD, PINE BLUFF, AR 71602
RESPOND OF ARKANSAS, DAVID MOORE N5MOT PRES
   PO BOX 1672, STATE UNIVERSITY, AR 72467
RIDGECREST HIGH SCHOOL ARC, DR JAY SPRINGMAN
   KB5KAA, 83 GREENE ROAD 743, PARAGOULD, AR
   72450
SALINE COUNTY ARC, LESIA A GRIFFIN N5VPU SECY
   PO BOX 573, BENTON, AR 72018
SHARP COUNTY ARC, BILL GRIFFIN KE5HD SECY
   8 TESSUNTEE DR, CHEROKEE VILLAGE, AR 72529
SMALL TOWN AMATEUR RADIO SERV, V ELDON
   BRYANT K7ZQR PRES, RR 2 BOX 123, WARD, AR 72176
SOUTH ARKANSAS ARC, JEFFERSON D BURCHFIELD
   N5AXE, 4243 CALION HWY, EL DORADO, AR 71730
SOUTHWEST ARKANSAS RADIO CLUB, HOWARD
   HOLMES KF5TK PRES, RR 3 BOX 458, DE QUEEN, AR
   71832
SPA ARA, DON NOBLES KC5QY PRES, 523 ROCKDALE RD,
   HOT SPRINGS NATIONAL PARK, AR 71901
TWIN LAKES ARC INC, PAM J LAFFERTY KB5SIV SECY
   206 W 8TH ST, MOUNTAIN HOME, AR 72653
VILLAGE ARC, HAROLD SWITZER W9YCE PRES
   110 ARIAS WAY, HOT SPRINGS VILLAGE, AR 71909

CALIFORNIA

220 CLUB OF SAN DIEGO, STEVE ADAMS K6PD PRES
   43699 MORAGA AVE, SAN DIEGO, CA 92117
AEROJET RADIO AMATEURS CLUB, WILL MAXTON
   AB6RT PRES, 15750 REED DR, FONTANA, CA 92336
ALAMEDA COUNTY RADIO CLUB, RICH ANDERSON
   N6LIM, 5082 ABBOTFORD CT, NEWARK, CA 94560
ALLAN HANCOCK COLLEGE BULLDOGS, ROBERT ALL
   DREDGE KE6BXR ADVIS, 525 E ORCHARD ST, SANTA
   MARIA, CA 93454
AMADOR COUNTY ARC, BARBARA BROWN WZ6Y
   TREAS, PO BOX 1094, PINE GROVE, CA 95665
AMATEUR COMMUNICATIONS SOC INC, IRVIN R WOLD
   W6JEU SECY, 16 ANGELA AVE, SAN ANSELMO, CA
   94960
AMATEUR TELEVISION NETWORK, MICHAEL V COLLIS
   WA6SVT VPRS, PO BOX 1594, CRESTLINE, CA 92325
AMBASSADOR ARA, CLARK MILLER W6T}PQ PRES
   590 N OAKLAND AVE, PASADENA, CA 91101
ANAHEIM AMATEUR RADIO ASSN, BARRY D IVES K6QJ
   801 E DONNY BROOK AVE, LA HABRA, CA 90631

ANTELOPE VALLEY ARC, IVAN E HINKLE JR. N6PQB
   PRES, 2121 E AVENUE I SPC 70, LANCASTER, CA 93535
ANZA VALLEY RC, PAUL R LARSON KN6JL SCTR
   38395 MCDONALD LN, ANZA, CA 92539
ARA OF LONG BEACH, ROBERT A BUHBE AA6HV PRES
   PO BOX 7493, LONG BEACH, CA 90807
ARC AT UNIV OF CA SANTA BARBA, STEPHEN I LONG
   AC6T ADVISOR, 895 N PATTERSON AVE, SANTA BAR
   BARA, CA 93111
ARC OF EL CAJON, AL GENTZ KA6RLX PRES, 9627 RAMS
   GATE WAY, SANTEE, CA 92071
AUTONETICS RADIO CLUB, DAN R VIOLETTE KI6X PRES
   M/C HD01, 3370 MIRALOMA, ANAHEIM, CA 92803
BALDWIN HILLS ARC, EDWARD L WALKER WA6MDJ
   PO BOX 43639, LOS ANGELES, CA 90043
BARSTOW ARC, JOHN SEGESMAN N6TAP SCTR, PO BOX
   451, BARSTOW, CA 92312
BEACH CITIES WIRELESS SOCIETY, DAVE TRUPKIN
   KD6KLZ PRES, PO BOX 4016, SAN CLEMENTE, CA 92674
BENICIA ARC, JACK B COOPER JR. KG6LV, PO BOX 1881,
   BENICIA, CA 94510
BERRYESSA AMATEUR RADIO KLUB, GARY FORD N6GF
   SCTR, 226 DIABLO AVE, DAVIS, CA 95616
BIG BEAR ARC, JAY DOWN KC6JTJ PRES, PO BOX 790,
   BIG BEAR LAKE, CA 92315
BISHOP ARC, JON PATZER NW6C EDIT, 730 KEOUGH ST,
   BISHOP, CA 93514
BLACKBIRD AIRPARK RADIO CLUB, WILLIAM WELSH
   W6DDB PRES, 45527 3RD ST E, LANCASTER, CA 93535
BUENA PARK ARC, COLLIER MCDERMON KC6WJM PRES
   4792 SUNNYBROOK AVE, BUENA PARK, CA 90621
CA CENTRAL COAST DX CLUB, SHERRY ROBB N7LTO
   SECY, 501 N POPPY ST, LOMPOC, CA 93436
CA STATE UNIV-SACRAMENTO STUD, GARY WEBBEN
   HURST KC6URB ADVIS, 8430 LA RIVIERA DR, SACRA
   MENTO, CA 95826
CALAVERAS ARS, JAMES M CLARK KC6ZWQ SCTR
   PO BOX 4453, CAMP CONNELL, CA 95223
CALIF POLYTECHNIC STATE UNIV, DAN MALONE
   KC6WOT ADVISOR, CAL POLY STATE UNIVERSITY, UU
   BOX 53, SAN LUIS OBISPO, CA 93407
CALTECH ARC, DAVID B RITCHIE N6DLU SCTR
   3521 YORKSHIRE RD, PASADENA, CA 91107
CASTLE ARC, JANET SIEGEL KB6SBH SECY
   1145 JULIE DR, MERCED, CA 95348
CENTRAL CA AMATEUR COMMUNICAT, WALTER
   WILLMS WB6FWO, 2605 E SUSSEX WAY, FRESNO, CA
   93726
CENTRAL CALIFORNIA DX CLUB, GILBERT DE LA LAING
   W6BJI SECY, 1260 W SAN RAMON AVE, FRESNO, CA
   93711
CENTRAL VALLEY INTER SCH RADIO, WILLIAM P ADDI
   SON N6GLL ADVIS, 4411 MOUNTAIN LAKES BLVD,
   REDDING, CA 96003
CHALLENGER JR HIGH SCHOOL ARC, FRANK FOR
   RESTER KI6YG PRES, CHALLENGER JHS, 10810 PARK
   DALE AVE, SAN DIEGO, CA 92126
CITRUS BELT ARC, ANDREW T HALE N6IQK PRES
   PO BOX 3788, SAN BERNARDINO, CA 92413
CLAIREMONT REPEATER ASSN, JACKIE CADOTTE
   KD6FHY SECY, PO BOX 7675, HUNTINGTON BEACH, CA
   92615
CLOVIS AMATEUR RADIO PIONEERS, RAY HARKINS
   KB6LQV SECY, 6114 E SHIELDS, FRESNO, CA 93727
COASTAL AR EMERG. SERVICES, MARCIA R BRUNO
   N6ISW SECY, 15849 LOS REYES ST, FOUNTAIN VAL
   LEY, CA 92708
COASTSIDE ARC, ROGER G SPINDLER WA6AFT TREA
   977 PARK PACIFICA AVE, PACIFICA, CA 94044
CONEJO VALLEY ARC INC, ED PIERCE WB6DFW SECY,
   PO BOX 2093, THOUSAND OAKS, CA 91358
CORONA NORCO ARC, MIKEL FRASIER KN6KH PRES
   PO BOX 253, CORONA, CA 91718
CRESCENTA VALLEY RADIO CLUB, DONALD HANSON
   KD6DIJ PRES, 600 W BROADWAY STE 235, GLENDALE,
   CA 91204

DEL NORTE ARC, GERRY FREDRICKSON KA6WYX SCTR
388 7TH ST, CRESCENT CITY, CA 95531

DOWNEY ARC, DOUGLAS E LYON N6WZI PRES
PO BOX 207, DOWNEY, CA 90241

EAST BAY ARC INC, JOHN S PERCIVAL WI6O SECY
PO BOX 1393, EL CERRITO, CA 94530

EDISON AMATEUR RADIO NET, CHARLES F SPETNAGEL
N7QQ PRES, 5327 CAROL AVE, ALTA LOMA, CA 91701

EDISON HIGH SCHOOL ARC, JOHN MORRICE WB6ITM
4768 N HARRISON AVE, FRESNO, CA 93704

EL DORADO COUNTY ARC, MICHAEL PICCO KD6BMV
PRES, PO BOX 451, PLACERVILLE, CA 95667

ELECTRONICS MUSEUM ARC, SHELDON EDELMAN
KM6GV TREAS, 3708 STARR KING CIR, PALO ALTO, CA
94306

ENGINEERING RADIO GROUP, JIM PRATT N6IG
3603 RIDGEVIEW DR, EL DORADO HILLS, CA 95762

ESCONDIDO ARS, KONRAD F WALLENDA N6UKO PRES
3040 QUAIL RD, ESCONDIDO, CA 92026

ESTERO RADIO CLUB, NORM WAGNOR KD6SAKZ SCTR
PO BOX 1813, MORRO BAY, CA 93443

FIFTY CLUB OF CALIF INC, HUGH S ALLEN JR W6MFC
SECY, 3601 ROYAL WOODS DR, SHERMAN OAKS, CA
91403

FRESNO ARC, FRED ATTOIAN JR KN6TA SECY
PO BOX 783, FRESNO, CA 93712

FULLERTON RADIO CLUB INC, CHUCK HUNGATE K6TAT
SECY, PO BOX 545, FULLERTON, CA 92632

G S LADD PIONEER RADIO CLUB, GENE HILDEMAN
W6KQZ SECY, 346 PERKINS DR, HAYWARD, CA 94541

GABILAN ARC, DON TRIGUEIRO KA6AUR PRES
1729 FALLBROOK AVE, GILROY, CA 95021

GENENTECH ARC, RON SIDELL KM6BH
GENENTECH INC, 460 POINT SAN BRUNO BLVD # 31,
SOUTH SAN FRANCISCO, CA 94080

GOLDEN EMPIRE ARS, INC., STUART KING KN6ME PRES
136 W FRANCIS WILLARD AVE, CHICO, CA 95926

GOLDEN STATE ARC, ANANIAS G GASACAO KM6EF PRES
PO BOX 1324, UNION CITY, CA 94587

GOLDEN TRIANGLE ARC, CLIFF SHINN AB6PB PRES
PO BOX 1335, WILDOMAR, CA 92595

HENRY RADIO OF ORANGE CTY RADI, STEVEN L WALLIS
WA6PYE PRES, 25108-B MARGUERITE PKWY #76, MIS
SION VIEJO, CA 92692

HILLTOP AMATEUR MASTERTIE SYST, FREDERICK E
DEEG N6FD PRES, 13234A FIJI WAY, MARINA DEL REY,
CA 90292

HUGHES-FULLERTON EMPLOYEES, STEVE DAVIES
N6MNG SECY, BLDG 604 MS B260, P O BOX 3310,
FULLERTON, CA 92634

HUMBOLDT ARC, SCOTT BINDER AB6TR SECY
PO BOX 5251, EUREKA, CA 95502

INLAND EMPIRE A R C, KEN D WALSTON SR WA6ZEF
1248 N CYPRESS AVE, ONTARIO, CA 91762

JAPANESE AMERICAN ARS, CHARLOTTE ISEDA KB6FXS
SECY, 6781 VERDE RIDGE RD, RANCHO PALOS
VERDES, CA 90275

KELLER PEAK REPEATER ASSOC, GEORGE W POULSEN
KB6AWJ SECY, PO BOX 8051, RIVERSIDE, CA 92515

KENNEDY HIGH SCHOOL ARC, ROBERT HAZARD N6NBF
ADVISOR, KENNEDY HIGH SCHOOL, 11254 GOTHIC
AVE, GRANADA HILLS, CA 91344

KERN COUNTY RC INC, DR ROBERT GARDINER KE6FFN
4219 GLENCANNON ST, BAKERSFIELD, CA 93308

KERN RIVER VALLEY ARC, TOM H CORSO KN6TS PRES
PO BOX 3645, LAKE ISABELLA, CA 93240

KINGS ARC, INC., CARLETON REED AA6GZ EDITOR
PO BOX 548, ARMONA, CA 93202

LA AREA COUNCIL OF ARC, DICK RITTERBAND AA6BC
PRES, 3823 CODY RD, SHERMAN OAKS, CA 91403

LADIES AID OF ORANGE COUNTY, JERI HAINES KB6USX
PRES, 16139 OCASO AVE, LA MIRADA, CA 90638

LAKE COUNTY ARS, PHIL MCCLEEARY W6OZT PRES
PO BOX 814, KELSEYVILLE, CA 95451

LAMBDA ARC GOLDEN GATE CHAPTER, KITCHELL
BROWN WB6QVU PRES, PO BOX 14073, SAN FRANCIS
CO, CA 94114

LAS CUMBRES ARC INC, DAVE FERGUSON WA6BXP
PRES, PO BOX 160577, CUPERTINO, CA 95016

LEE DEFOREST RC OF HEMET, LANDON BROWN KK6BC
PRES, 25108 ROSEBRUGH LN, HEMET, CA 92544

LIVERMORE AMATEUR RADIO KLUB, LEE ZALAZNIK
KI6OY PRES, PO BOX 3190, LIVERMORE, CA 94551

LOCKHEED ARC (LERA), JAMES N WOODS KC7FG SECY
1042 ROBIN WAY, SUNNYVALE, CA 94087

LODI ARC, DALE KRETZER K6PJV SCTR, PO BOX 2088,
LODI, CA 95241

LOMPOC ARC, DALE J LEISTRO WA4ANW SCTR
313 SOMERSET PL, LOMPOC, CA 93436

LOYOLA MARYMOUNT UNIV ARC, CLIFF D'AUTREMONT
KI6TM, 7101 W 80TH ST, BOX 555, LOS ANGELES, CA
90045

MADERA COUNTY ARC, DAN STANFORD KNGMV PRES
PO BOX 251, MADERA, CA 93639

MARIN ARC, GRANT PRITCHARD KK6JJ PRES, PO BOX
151231, SAN RAFAEL, CA 94915

MENDOCINO COUNTY ARS, JOE PARK WB6AGR PRES
PO BOX 1396, UKIAH, CA 95482

MILE HIGH MOUNTAIN RADIO CLUB, ROBERT G HAY
TON KC6GQC PRES, PO BOX 1776, IDYLLWILD, CA
92549

MONTEREY PARK ARC, RON GROSS N6KNF, PO BOX 403,
MONTEREY PARK, CA 91754

MORENO VALLEY ARA, LOIS GOODINE WB6PLR PRES
PO BOX 7642, MORENO VALLEY, CA 92552

MORONGO BASIN ARC, GLENN MILLER N6GIW PRES
P O BOX 1995, YUCCA VALLEY, CA 92286

MOUNT DIABLO ARC, RICHARD SCHULZE AA6DL PRES
1599 LA CASA CT, WALNUT CREEK, CA 94598

MOUNT SHASTA ARC, CHRISTINE CROCKER KD6LNH
SECY, PO BOX 73, MOUNT SHASTA, CA 96067

MOUNTAIN ARC, JOE BENTSON K6STR, 48947 QUAIL DR,
OAKHURST, CA 93644

MOUNTAIN REPEATER ASSN, ELLIOTT BLOCH K6ELX
SCTR, 3049 DONA NENITA PL, STUDIO CITY, CA 91604

NEVADA COUNTY ARC, A FRANK DOTING W6NKU PRES
PO BOX 2923, GRASS VALLEY, CA 95945

NICHOLS ELEM SCHOOL ARC, THOMAS M ZANE K6URI
ADVISOR, 2310 W TOKAY ST, LODI, CA 95242

NORTH BAY AMATEUR RADIO ASSN, DOUG DYKES
KC6UET PRES, PO BOX 1468, VALLEJO, CA 94590

NORTH HILLS RADIO CLUB, KEN SNYDER AB6BD PRES
PO BOX 41635, SACRAMENTO, CA 95841

NORTH SHORES ARC, MARK CARROLL KB6OSP PRES
P O BOX 639005-127, SAN DIEGO, CA 92163

NORTHERN AMATEUR RELAY COUNCIL, JOHN
KERNKAMP WB4YJT PRES, PO BOX 60531, SUNNY
VALE, CA 94088

NORTHERN CALIF CONTEST CLUB, BRUCE B SAWYER
AA6KX SCTR, 15430 BOHLMAN RD, SARATOGA, CA
95070

NORTHERN CALIFORNIA DX CLUB, GEORGE ALLAN
WA6O PRES, PO BOX 608, MENLO PARK, CA 94026

NORTHROP RADIO CLUB, GREGORY T LANE K7SDW
PRES, 3970 CORONADO CIR, NEWBURY PARK, CA
91320

NUCLEAR ARC, SHAWN DIENHART WB6JWB PRES
317 PRIMROSE LN, PASO ROBLES, CA 93446

ORANGE COUNTY ARC, KEN KONECHI W6HHC SECY
340 S CRAIG DR, ORANGE, CA 92669

ORANGE COUNTY COUNCIL OF ARO, LORRAINE HAST
INGS KK6CG, 854 E BERNARD DR, FULLERTON, CA
92635

PAL-HAM KIDS, PAUL BOWMAN N6WZR ADVISOR
LONG BEACH POLICE DEPT, 400 W BROADWAY JUVE
NILE DIV, LONG BEACH, CA 90802

PALO ALTO ARA, LILY ANNE HILLIS N6PGM PRES
PO BOX 911, MENLO PARK, CA 94026

PALOMAR ARC INC, STAN ROHRER W9FQN EDITOR
  30311 CIRCLE R LN, VALLEY CENTER, CA 92082
PALOS VERDES ARC, HERBERT E CLARKSON KM6DD
  POB 2316, PALOS VERDES PENINSULA, CA 90274
PASADENA RADIO CLUB, MERRIE SUYDAM AB6LR PRES
  321 CHERRY HILLS LN, AZUSA, CA 91702
PASO ROBLES ARC, L S MOORE KI6LA, 1628 POPPY LN,
  PASO ROBLES, CA 93446
PETALUMA DX & EXPERIMENTERS, FRITZ O MAASS
  WB6EGE PRES, 1291 ELYSIAN AVE, PENNGROVE, CA
  94951
PIEDMONT ARC, DERALD SUE KD6HJX PRES, 8 GLEN
  ALPINE RD, PIEDMONT, CA 94611
PLACENTIA RADIO WATCH, CHARLES TIELROOY
  KD6DON, PLACENTIA POLICE DEPT, 401 E CHAPMAN
  AVE, PLACENTIA, CA 92670
POINSETTIA ARC, RUSSELL REID KD6QOH PRES
  PO BOX 268, VENTURA, CA 93002
PORTERVILLE AMATEUR RPTR ASSN, EDWARD LAMB
  K6LSB TREAS, 23433 AVENUE 184, PORTERVILLE, CA
  93257
PREPARED RADIO IN MIGHTY ORAN, KATHERINE
  SCHAFFSTEIN WABFAH, 731 W HOUSTON AVE,
  FULLERTON, CA 92632
QSO CLUB OF PASADENA AREA C.C., HARVEY D HET
  LAND N6MM ADVISOR, PO BOX 73, ALTADENA, CA
  91003
RADIO ACTIVE TEENAGERS, GAYLE OLSON KC6YDQ
  21400 MAGNOLIA ST, HUNTINGTON BEACH, CA 92646
REDWOOD EMPIRE DX ASSN, STEVE LUND WA8LLY
  SCTR, PO BOX 476, GRATON, CA 95444
RELAY REPEATER CLUB, DONALD WERNER WA6KKR
  PRES, PO BOX 660081, ARCADIA, CA 91066
RHOADES SCHOOL HAM RADIO CLUB, RON MIELY
  KC6YLA PRES, 3030 GARBOSO ST, CARLSBAD, CA
  92009
RIVER CITY AR COMMUNICATIONS, LYLE B AUFRANC
  AA6DJ LIAS, PO BOX 215073, SACRAMENTO, CA 95821
RIVER CITY CONTESTERS, TIM DOLAN KM6AS SCTR
  PO BOX 51, FOLSOM, CA 95763
RIVERSIDE COUNTY ARA, JOANIE VERDUFT KC6NXK
  PRES, 5505 GOLDEN AVE, RIVERSIDE, CA 92505
SABIN PIONEER RADIO CLUB, LES COBB W6TEE SECY
  4124 PASADENA AVE, SACRAMENTO, CA 95821
SACRAMENTO ARC, INC, GARY BRYANT KB6KZZ PRES
  PO BOX 161903, SACRAMENTO, CA 95816
SAM'S RADIO HAMS, BOBBY MOONEYHAM WB6BRU
  PRES, 9481 N WILLOW AVE, CLOVIS, CA 93611
SAN BRUNO CIVIL DEFENSE RC, GEORGE P DELICH
  WA6IZB PRES, 321 FERNWOOD DR, SAN BRUNO, CA
  94066
SAN DIEGO COUNTY RADIO COUNCIL, PATRICK C BUN
  SOLD WA6MHZ, 14291 RIOS CANYON RD SPC 33, EL
  CAJON, CA 92021
SAN DIEGO DX CLUB, SAMUEL E JOHNSON JR W6BS
  1647 CORTE LADERA, ESCONDIDO, CA 92025
SAN DIEGO METROPOLITAN ARC, WILLIAM CHAPMAN
  AA6JZ PRES, 3951 THE HILL RD, BONITA, CA 91902
SAN FERNANDO VALLEY ARC, IRVING SLITZKY N6PTI
  PRES, PO BOX 280517, NORTHRIDGE, CA 91328
SAN FRANCISCO ARC, ART SAMUELSON W6VV PRES
  PO BOX 420741, SAN FRANCISCO, CA 94142
SAN GABRIEL VALLEY RC, TONY CHILLEMI KD6IFC
  5023 N WILLOW AVE, COVINA, CA 91724
SAN LEANDRO RADIO CLUB, BOB RETT W6MEI PRES
  301 DOWLING BLVD, SAN LEANDRO, CA 94577
SAN MATEO RADIO CLUB INC, JIM MUITER W6KXG PRES
  PO BOX 751, SAN MATEO, CA 94401
SANDRA INC, BOB HELT NY6J PRES, P O BOX 81103, SAN
  DIEGO, CA 92138
SANTA BARBARA ARC, RONALD D FUGATE N6ANF PRES
  PO BOX 3232, SANTA BARBARA, CA 93130
SANTA CLARA COUNTY ARA, DOUG EATON WN6U PRES
  PO BOX 6, SAN JOSE, CA 95103

SANTA CLARA VALLEY RPTR SOC, LARRY FLETCHER
  KC6VRK PRES, 17162 TASSAJARA CIR, MORGAN HILL,
  CA 95037
SANTA CRUZ COUNTY ARC INC, ALLAN HANDFORTH
  KC6VJL PRES, PO BOX 238, SANTA CRUZ, CA 95061
SANTA MARIA AR TELE & TELE ARC, CHARLES H
  DANIEL KD6AUY PRES, 3958 LOCH LOMOND DR,
  SANTA MARIA, CA 93455
SATELLITE ARC, SHARI RICH KD6KZY SECY
  PO BOX 5117, LOMPOC, CA 93437
SCIENCE & TECH CTR AR DIV, KEVIN WILLIAMS KN6UO
  PRES, P O BOX 895, LUCERNE VALLEY, CA 92356
SHAFTER HIGH RADIO CLUB, RALPH IRONS AA6UL
  ADVISOR, 526 MANNEL AVE, SHAFTER, CA 93263
SHASTA CASCADE ARS, DON CHURCHILL K6EIX PRES
  21815 SHADY OAK LN, PALO CEDRO, CA 96073
SHASTA DX & CONTEST CLUB, JON CASAMAJOR KN6EL
  PRES, 24 GRACELAND CT, CHICO, CA 95926
SIERRA ARC, BOB LENTZ K6PYH PRES, PO BOX 236,
  CHESTER, CA 96020
SIERRA ARC OF THE HIGH MOJAVE, LLOYD BRUBAKER
  WA6KZV SECY, 235 E KENDALL AVE, RIDGECREST, CA
  93555
SIERRA FOOTHILLS ARC, ROGER HEDGPETH N7EUQ
  PRES, 460 FLOWER DR, FOLSOM, CA 95630
SILICON GULCH ARC, DOYLE E SOUDERS KG6MY PRES
  1342 FIELDFAIR CT, SUNNYVALE, CA 94087
SIMI SETTLERS ARC, RON MCCLURE KD6VLM PRES
  PO BOX 3035, SIMI VALLEY, CA 93093
SISKIYOU REPEATER ASSN, AL KIEP WA6IHK TREAS
  PO BOX 41, ETNA, CA 96027
SOLANO COUNTY ARS, DARWIN THOMPSON K6USW
  PRES, 2050 DORLAND DR., FAIRFIELD, CA 94533
SONOMA COUNTY RADIO AMATEURS, ALAN BLOOM
  N1AL SECY, PO BOX 116, SANTA ROSA, CA 95402
SONORA PASS AR KLUB, GERTRUDE ANDERSON
  KA6VFO TREA, 17300 MONTE GRANDE DR, SOULS
  BYVILLE, CA 95372
SONS IN RETIREMENT ARC, CHARLES UNFRIED K6IGJ
  1642 BALBOA CT, PLEASANT HILL, CA 94523
SOUTH BAY ARA, PHIL KELLER N6MWC TREAS
  48775 BIG HORN CT, FREMONT, CA 94539
SOUTH BAY ARC, JOE LANPHEN WB6MYD SCTR
  PO BOX 536, TORRANCE, CA 90508
SOUTH BAY ARS, JEROME ROEBUCK N6ZQC PRES
  1003 MELROSE AVE, CHULA VISTA, CA 91911
SOUTH CTY AR EMERG SERVICE INC, HARRY R COLLINS
  K6ANN PRES, 1921 BIRCH AVE, SAN CARLOS, CA 94070
SOUTH ORANGE COUNTY ARA, GREGORY A PASQUA
  KB6YOM, 12 WEST AVENIDA JUNIPERO, SAN
  CLEMENTE, CA 92672
SOUTHERN CA SIX METER CLUB, BOB HASTINGS K6PHE
  PRES, 854 E BERNARD DR., FULLERTON, CA 92635
SOUTHERN CALIF CONTEST CLUB, MARK BECKWITH
  WA6OTU EDITOR, PO BOX 4, SIERRA MADRE, CA 91025
SOUTHERN CALIF DX CLUB, STEVE LOCKS W6FRZ SECY
  2324 N HEATHER AVE., LONG BEACH, CA 90815
SOUTHERN CALIFORNIA ATS, PAT MCNULTY N6GXZ
  SECY, PO BOX 1770, COVINA, CA 91722
SOUTHERN HUMBOLDT ARC, JACK FOSTER KM6TE PRES
  PO BOX 193, GARBERVILLE, CA 95542
SOUTHERN SIERRA ARS, CAROLINE PARSONS
  20277 BACKES LN., TEHACHAPI, CA 93561
SOUTHWESTERN REACT ARC, ERIC CROSSER N6SUB
  PRES, PO BOX 4554, SAN DIEGO, CA 92164
SS LANE VICTORY ARC, JENNY LUKENBILL K6JCL SCTR
  C/O SS LANE VICTORY, PO BOX 629, SAN PEDRO, CA
  90733
ST JUDE HOSP & REHAB CTR ARA, APRIL MOELL
  WA6OPS CHRM, PO BOX 2508, FULLERTON, CA 92633
STANISLAUS ARA, SANDRA INGRAM KC6TBK PRES
  PO BOX 4601, MODESTO, CA 95352
STOCKTON-DELTA ARC, JOHN KESTER KD6FVA PRES
  PO BOX 690271, STOCKTON, CA 95269

SULPHUR MOUNTAIN REPEATER ASSN., DAVID R
WILKINS KM6UU PRES, PO BOX 5131, VENTURA, CA
93005
SUN CITY ARC, HERB LIPSON W8FBH PRES
26960 PINEHURST RD., SUN CITY, CA 92586
TAHOE AMATEUR RADIO ASSN, PAUL GULBRO WA6EWV
PRES, PO BOX 624772, SOUTH LAKE TAHOE, CA 96154
TANDEM RADIO AMATEURS CLUB, PAUL WESLING
KM6LH SECY, TANDEM COMPUTERS INC., 10400
RIDGEVIEW CT M/S 208-57, CUPERTINO, CA 95014
TECH RAC & EXPERIMENTAL SYSTEM, JAN F STICHA
WA6HWT TRUSTEE, 846 PLATT CT., MILPITAS, CA
95035
TEEN ARA, ANDREAS CSEPELY KD6HNE SECY
1246 SANDIA AVE., SUNNYVALE, CA 94089
TRI-COUNTY ARA, JOHN T PHILLIPP N6ZAE SECY
PO BOX 142, POMONA, CA 91769
TULARE COUNTY ARC, DAVE MILLHOUSE N6YMM SCTR
PO BOX 723, VISALIA, CA 93279
TURLOCK ARC, GEORGE STEVANS K6SNA VPRS
1804 E RUMBLE RD, MODESTO, CA 95355
U.H.F. ASSOCIATES, WILLIAM R SEYMOUR WA6MOD
SCTR, PO BOX 609, GLENDORA, CA 91740
UNITED RADIO AMATEUR CLUB, BEVERLY PITMAN
WA6TIU SECY, 2902 ONRADO ST., TORRANCE, CA
90503
UNIVERSITY OF CALIFORNIA ARC, JAMES CHESKO
AB6YH PRES, UNIVERSITY OF CALIFORNIA, 609 E.
SHLEMAN HALL, BERKELEY, CA 94720
VACA VALLEY RADIO CLUB, GLENN D BISSELL N6WVF
PRES, 419 MASON ST SUITE 126, PO BOX 521, VACAV
ILLE, CA 95696
VALLEY OF THE MOON ARC, STEVE TODD K26KVY SECY
175 FIRST ST W, SONOMA, CA 95476
VENTURA COUNTY ARC, DICK GEORGE WA6JOX PRES
603 EVERGREEN SQ, PORT HUENEME, CA 93041
VICTOR VALLEY ARC, JAMES A CHANDLER N5COT PRES
PO BOX 869, VICTORVILLE, CA 92393
VILLAGES ARC, EMIL MARTIN K6MJ PRES, 7652 FALKIRK
DR., SAN JOSE, CA 95135
WEST COAST ARC, CAM HARRIOT KI6WF PRES
PO BOX 2617, COSTA MESA, CA 92628
WEST VALLEY ARA, TRAVIS WISE KB8FOU PRES
P O BOX 6544, SAN JOSE, CA 95150
WESTERN AMATEUR RADIO ASSN, CHRIS PREWITT
WA6OQC, PO BOX 5403, HUNTINGTON BEACH, CA
92615
WESTSIDE ARC, RAY W SPECKMAN KB6PBM SCTR
1046 ROSE AVE., VENICE, CA 90291
WILDCAT ARS OF WAWONA MIDDLE S., CHARLES P MC
CONNELL W6DPD ADV, 1658 W MESA AVE., FRESNO,
CA 93711
WILLITS ARS, BILL PERAC W6JCG SECY, PO BOX 73,
WILLITS, CA 95490
WORLDRADIO STAFF ARC, NORMAN BROOKS K6FO
PRES, 5901 ADANA CIR., CARMICHAEL, CA 95608
YOLO ARS, KYLE NODERER KB6OLL PRES
PO BOX 659, DAVIS, CA 95617
YOUNG LADIES RADIO LEAGUE, MARGARET PITT
KC6PRB TREAS, 2379 LEEDWARD CIRCLE, THOUSAND
OAKS, CA 91361
YOUNG LADIES RC OF LOS ANGELES, IRMA W WEBER
K6KCI PRES, 165 N MARCELLO AVE., THOUSAND
OAKS, CA 91320
YUBA-SUTTER ARC, CLARA M ANSLEY KC6JPP TREAS
PO BOX 1169, YUBA CITY, CA 95992

COLORADO
ARKANSAS VALLEY ARC, HOWARD LUEHRING W0KEV
123 SEELEY AVE, LA JUNTA, CO 81050
ASSN FOR RESPONSIBLE COMMUNIC., PENDELL
PITTMAN N0DZA, 14700 E PENWOOD PL, AURORA, CO
80015
AURORA REPEATER ASSOCIATION, JANICE CHRISTO
PHERSON KA7TYU, PO BOX 39666, DENVER, CO 80239

BEGINNERS AR GROUP OF STEAMBOA, GEORGE R
WEBER KC0ZQ ADVISOR, STRAWBERRY PARK, P O
BOX 774368, STEAMBOAT SPRINGS, CO 80477
BOULDER ARC, KEITH BOBO N0VNX PRES, PO BOX 2033,
BOULDER, CO 80306
COLORADO REACT AMATEUR EMERG, WALTER GREEN
III N0PSB PRES, 3624 CITADEL DR N STE 309, COL
ORADO SPRINGS, CO 80909
DENVER RADIO CLUB INC., JOE DELWICHE KF0OD PRES
PO BOX 44173, DENVER, CO 80201
DENVER RADIO LEAGUE, GEORGE STOLL WA0KBT PRES
2055 S MADISON ST, DENVER, CO 80210
DURANGO ARC, WENDELL D MCCOY NX0S
29308 HIGHWAY 160, DURANGO, CO 81301
EMPIRE RADIO CLUB, MERILYN MORRIS KB0CMN SCTR
PO BOX 32, COMMERCE CITY, CO 80037
EXPLORER POST 59 RADIO CLUB, KATHY HAYS N0ZIP
PRES, 704 RIVERBEND DR, FORT COLLINS, CO 80524
LONGMONT ARC, GARY BAILEY KA0ABK SCTR
2156 DALEY DR, LONGMONT, CO 80501
LOVELAND REPEATER ASSN, BRIAN WOOD W0DZ PRES
PO BOX 1733, LOVELAND, CO 80539
MILE HIGH DX ASSN, STEVE BOONE WA8SWM PRES
110 PINEWOOD LOOP, MONUMENT, CO 80132
MONTROSE ARC, AUDREY G LOWTHER N0JAB PRES
15866 6200 RD, MONTROSE, CO 81401
MOUNTAIN ARC, DON CHAMBERLAIN AA0NW PRES
P O BOX 1012, WOODLAND PARK, CO 80866
PIKES PEAK RADIO AMATEUR ASSN, RON & LIZ
DEUTSCH NK0P EDITORS, PO BOX 16521, COLORADO
SPRINGS, CO 80935
PUEBLO HAM CLUB, PAULINE BUFFINGTON KA0YXB
SECY, PO BOX 92, PUEBLO, CO 81002
REMINGTON ELEM SCHOOL ARC, GLENN DURANT
KB0BHN ADVISOR, 4735 PECOS ST, DENVER, CO 80211
ROYAL GORGE HAM CLUB, MARILYN REDMAN WD5AAR
PRES, PO BOX 2044, CANON CITY, CO 81215
WATERTON ARS, PETER HILLS W0HXB
2048 HUDSON ST, DENVER, CO 80207
WESTERN COLORADO ARC, LARRY BALL W0IOL PRES
PO BOX 3422, GRAND JUNCTION, CO 81502

CONNECTICUT
BETHEL EDUCATIONAL ARS, PETER W KEMP KZ1Z ADVI
SOR, EDUCATIONAL PARK, BETHEL MIDDLE SCHOOL,
BETHEL, CT 06801
BETTER EMERG AR SERVICE, WILLIAM DOWNARD
N1MQR PRES, 39 GREENWOOD DR, MANCHESTER, CT
06040
BLOOMFIELD ARC, DICK BAGNALL N1BMC SECY
616 BLOOMFIELD AVE., BLOOMFIELD, CT 06002
CANDLEWOOD ARA, JOHN M AHLE N2DVX PRES
POB 3441, DANBURY, CT 06813
CHIPPENS REPEATER ASSN, NORMAND RIQUIER W1GNS
TREAS, 78 NORWOOD RD., BRISTOL, CT 06010
COASTLINE ARA, ROBERT W FUCHS N1GIF TREAS
35 PASTORS WALK, MONROE, CT 06468
CQ RADIO CLUB, TIMOTHY FERRAROTTI,N1BPD
86 WILDWOOD RD, TORRINGTON, CT 06790
CRICKET WIRELESS ASSN, ROBERT A PULITO, WB1DWO
SECY, 32 MAPLE ST, GLASTONBURY, CT 06033
EASTERN CONNECTICUT ARA, MICHAEL THERRIEN
N1OPZ PRES, 476 RIVER RD, PUTNAM, CT 06260
GREATER BRIDGEPORT ARC, DOUGLAS P WATERHOUSE
KB1SS, 187 SUNFLOWER AVE, STRATFORD, CT 06497
GREATER FAIRFIELD ARA, JOHN BISACK KD1SX PRES
340 HALF MILE RD, SOUTHPORT, CT 06490
GREATER NORWALK ARC, ED ASHWAY W9KTH PRES
88 HILLBROOK RD, WILTON, CT 06897
HARTFORD COUNTY ARA, PETER KOHANASKI, WB1DWR
SECY, 16 BERKELEY CIR., NEWINGTON, CT 06111
HEN HOUSE GANG, ROBERT J O'NEIL, W1FHP PRES
283 HARD HILL RD N., BETHLEHEM, CT 06751
INSURANCE CITY REPEATER CLUB, CHARLES I MOTES
JR K1DFS TREAS, 22 WOODSIDE LN, PLAINVILLE, CT
06062

MERIDEN ARC, BILL WAWRZENIAK W1KKF PRES
PO BOX 583, MERIDEN, CT 06450
MIDDLESEX ARS, BRIAN BATTLES WS1O PRES, PRODUC
TION DEPT, 225 MAIN ST., NEWINGTON, CT 06111
MILFORD AMATEUR REPEATER ASSN, EARL W DUGAN
KA1DCL PRES, 147 HARRISON AVE, MILFORD, CT 06460
MURPHY'S MARAUDERS, ALEXANDER J MELEG, JR N1JW
SCT 224 OAK ST., EAST HARTFORD, CT 06118
NAKED CHICKEN CONTEST CLUB, AL BROGDON K3KMO
VPRS, 114 LYONS ST, NEW BRITAIN, CT 06052
NEWINGTON AMATEUR RADIO LEAGUE, WILLIAM G
BESENYEI WB1ESR, PO BOX 310282, NEWINGTON, CT
06131
NORTH EAST WEAK SIGNAL GROUP, RON KLIMAS WZ1V
SECY, 458 ALLENTOWN RD, BRISTOL, CT 06010
PIONEER VALLEY RADIO ASSN, MARK PEARL N1NYN
PRES, 138 REGINA DR, EAST WINDSOR, CT 06088
RADIO AMATEUR SOC OF NORWICH, DAVID W SMITH
KA1LK PRES, PO BOX 329, NORWICH, CT 06360
SHORELINE ARC INC, DON KRAHL N1KSY PRES
348 CLARK AVE, BRANFORD, CT 06405
SOUTH CENTRAL CONNECTICUT ARA, ROBERT HAUSER
JR KA1PCG PRES, 40 YOUNGS APPLE ORCHARD RD,
NORTHFORD, CT 06472
SOUTHEASTERN CT RA MOBILE SYS., DARRYL
DELGROSSO WA1WY, PO BOX 219, GROTON, CT 06340
SOUTHERN BERKSHIRES ARC, EDWARD H WILBUR
WB1CEI, RT 41 BOX 547, SHARON, CT 06069
SOUTHINGTON ARA, STEVEN MOSS N1GCV PRES
PO BOX 873, SOUTHINGTON, CT 06489
STAMFORD ARA, WILLIAM LUSH KC2CF PRES
PO BOX 4225, STAMFORD, CT 06907
STRATFORD ARC, MAY BLAKLEY WA1EHK
17 CORAM RD APT. 4F, SHELTON, CT 06484
SUBMARINE BASE ARC, GERALD J SCARANO W1ZM
NAVAL SUBMARINE BASE, BOX 200, GROTON, CT
06349
TRI-CITY ARC - CT, MICHAEL SULLIVAN N1OKK PRES
471 BEACH POND RD, VOLUNTOWN, CT 06384
VALLEY AMATEUR RADIO ASSN., JON KRIJGSMAN
N1BDF PRES, PO BOX 184, DERBY, CT 06418
WATERBURY ARC, DEREK HOOK WM1U PRES
24 CARRIAGE LANE RD, ROXBURY, CT 06783
WATERTOWN HIGH SCHOOL ARC, TOM FIX N1GPQ
ADVISOR, 324 FRENCH ST, WATERTOWN, CT 06795
WIRELESS OPERATORS OF WINSTED, PERRY T GREEN
WY1O SECY, 234 E WAKEFIELD BLVD, WINSTED, CT
06098
YALE UNIVERSITY ARC, ANTHONY J NIESZ KA1HBP
ADVISOR, YALE LANGUAGE LAB, 111 GROVE ST.,
NEW HAVEN, CT 06511
ZYGO ARC, GARY FOSKETT W1ECH SECY
ZYGO CORP LAUREL BROOK RD, P O BOX 448, MID
DLEFIELD, CT 06455

DELAWARE

ASSOCIATION OF WILMINGTON AMAT, WILLIAM J MAR
TIN KD3GB PRES, 719 BURNLEY RD, WILMINGTON, DE
19803
DELAWARE ARC, JOSEPH M GRIB JR. KI3B SCTR
42 ANDERSON CT, BEAR, DE 19701
DELAWARE REPEATER ASSN., INC., BOB MADER N3LYM
SECY, PO BOX 304, ROCKLAND, DE 19732
DUPONT EXPERIMENTAL STATION, HERBERT L GROH
KA3CDB PRES, 323/164, DUPONT EXPERIMENTAL STA
TION, WILMINGTON, DE 19880
FIRST STATE ARC, ROBERT A RODGERS WA3HDS PRES
29 DEMPSEY DR., NEWARK, DE 19713
KENT COUNTY ARC, EDWARD D BITER JR., NS3E
PO BOX 1000, DOVER, DE 19903
NANTICOKE ARC, WILLIAM L ZOULEK KA3ZOO SCTR
PO BOX 584, SEAFORD, DE 19973
NEWARK EXTRAS FROM RURAL DE, BOB PENNEYS
WN3K PRES, 12 E MILL STATION DR., NEWARK, DE
19711

PENN-DEL ARC, DEBRA A FRANTZ KA3TZG SCTR
950 RIDGE RD STE C27, CLAYMONT, DE 19703
SUSSEX ARC, LARS SPENCER KA3CDF TREAS
RR 1 BOX 437T, ELLENDALE, DE 19941
UNIVERSITY OF DELAWARE ARA, DOUGLAS RAMBO
KA3KHZ PRES, UNIVERSITY OF DELAWARE, 135
DUPONT HALL, NEWARK, DE 19716

DISTRICT OF COLUMBIA

CAPITAL CITY ARS, CHESTER R MARTIN W3EIL
1719 FRANKLIN ST NE, WASHINGTON, DC 20018
NAVAL RESEARCH LAB ARC, CODE 5541 NRL
RUTH E PHILLIPS K3AGR LIAS, WASHINGTON, DC
20375
PENTAGON ARC, DENNIS B DOLLE NX5W SECY
PO BOX 47063, WASHINGTON, DC 20050

FLORIDA

GEORGE THURSTON W4MLE
PO BOX 37127, TALLAHASSEE, FL 32315
AMERICA RADIO CLUB, INC., IVAN PRADA KP4EDL LIAS
7931 NW 190TH TER, HIALEAH, FL 33015
ARA OF SOUTHWEST FLORIDA, JORDAN E MASH
WB2QLP PRES, 2033 42ND ST SW., NAPLES, FL 33999
BEACHES AMATEUR RADIO SOCIETY, MICHAEL D COR
WIN AD4CV SECY, 2906 CANYON FALLS DR., JACK
SONVILLE, FL 32224
BELL TOWER PIONEER RADIO CLUB, VERNON H FERRIS
KB4VPU, PO BOX 390, JACKSONVILLE, FL 32201
BRANDON ARS, ROY G BENTLEY WD8RBB SECY
2823 TIMBERWAY PL., BRANDON, FL 33511
BROWARD ARC, JAN LORAH WB4ROC SECY
2407 FLAMINGO LN, FORT LAUDERDALE, FL 33312
CALOOSA MIDDLE SCHOOL ARC, PHYLLISAN WEST
KA4FZI, 1410 SHELBY PKY, CAPE CORAL, FL 33904
CENTRAL FLORIDA DX ASSN, LESLIE E SMALLWOOD
KW4V SCTR, 24 APPLE HILL HOLW, CASSELBERRY, FL
32707
CHARLOTTE ARC, RONALD D OBLINGER KD4AMS PRES
PO BOX 415, PUNTA GORDA, FL 33951
CLEARWATER ARS, ELI NANNIS K4JMH SECY.
14996 IMPERIAL POINT DR. N., LARGO, FL 34644
CLOVER LEAF ARC, INC., RICHARD D.TOWNSEND AC4HR
TRUS, 900 N BROAD ST.,PO BOX 144, BROOKSVILLE, FL
34605
DADE RADIO CLUB OF MIAMI INC., EVELYN D GAUZENS
W4WYR, 2780 NW 3RD ST., MIAMI, FL 33125
DAYTONA BEACH ARA, STEPHEN W SZABO WB4OMM
PRES, 3736 HUGH ST., PORT ORANGE, FL 32119
EAST PASCO ARS, ROBERT W OWENS AA4RU SECY
PO BOX 942, DADE CITY, FL 33526
EGLIN ARS, FRANK M BUTLER JR. W4RH SECY
PO BOX 1773, EGLIN AFB, FL 32542
ENGLEWOOD ARS, BRUCE ROBIDEAU K2OY PRES
PO BOX 572, ENGLEWOOD, FL 34295
EVERGLADES ARC, ARRELL DAY KB4ONU PRES
P O BOX 113, HOMESTEAD, FL 33090
FLORIDA ARS, INC., TIM ROBINETTE WD8MVU
PO BOX 145, NEW PORT RICHEY, FL 34656
FLORIDA CW CONTEST GROUP, JIM WHITE K1ZX PRES
5605 E 127TH AVE., TAMPA, FL 33617
FLORIDA GULF COAST AR COUNCIL, PATRICIA M BARBI
ERE WB1GZW, PO BOX 2423, CLEARWATER, FL 34617
FLORIDA KEYS ARC, LURENE LEWIS WN1L TREAS
P O BOX 545, BIG PINE KEY, FL 33043
FLORIDA WESTCOAST DX-RING, RENNY BARRETO
WA1DHM SECY, 11160 130TH AVE., LARGO, FL 34648
FORT MYERS ARC, JERRY DEUTSCHER KQ4UW PRES
802 ELINOR WAY, SANIBEL, FL 33957
FORT PIERCE RADIO CLUB, PETER AMAR KD4SPW SECY
1046 TRINIDAD AVE., FORT PIERCE, FL 34982
GAINESVILLE ARS, PAUL R BENNETT N4EGO SECY
4000 SW 47TH ST. LOT K13, GAINESVILLE, FL 32608

GATOR ARC, DR LEON COUCH K4GWQ TRUSTEE
 UNIVERSITY OF FLORIDA, 277 LARSEN HALL,
 GAINESVILLE, FL 32611
GULF AMATEUR RADIO SOCIETY, BERNARD PRIDGEON
 KD4PDU PRES, PO BOX 104, PORT SAINT JOE, FL 32456
GULF COAST ARC, ELLIS L GRAVES WA4PHL PRES
 214 WESTWINDS DR., PALM HARBOR, FL 34683
GULFCOAST AMATEUR SOCIETY, VITO TRUPIANO IV
 KD4FRA, 104 FLAME VINE DR., NAPLES, FL 33942
HAMM/RAMM ARC, C E SHOWALTER W4UJL PRES
 1810 LORENA LN, ORLANDO, FL 32806
HARRIS ARC, RAY LILES WA4VME PRES
 957 PELLAM AVE NE, PALM BAY, FL 32907
HERNANDO COUNTY ARA, GLENN BROWN AC4QH PRES
 31188 PARK RIDGE DR., BROOKSVILLE, FL 34602
HIGHLANDS COUNTY ARC, GERALD EICHHORN
 WB4WDK LIAS, PO BOX 993, LAKE PLACID, FL 33862
HOLLYWOOD ARC, ROBERT YOUNG KC4KME PRES
 P O BOX 6306, HOLLYWOOD, FL 33021
HONEYWELL EMERG AR TEAM, JIM HANSEN WD0DIA
 PRES, 10433 SHADY OAK LANE, LARGO, FL 34647
IBM ARC OF BOCA RATON, NEAL OSBORN N4PYB PRES
 IBM, INTERNAL ZIP 1302, BOCA RATON, FL 33432
INDIAN RIVER ARC, RICHARD B MCKLVEEN W4YWA
 597 CAPRI RD, COCOA BEACH, FL 32931
JACKSONVILLE RANGE ASSN, DON HARGIS KI4ZL
 PO BOX 10623, JACKSONVILLE, FL 32247
LAKE MONROE ARS INC., ALFRED LA PETER WB4DRF
 PRES, 7564 GLENMOOR LN, WINTER PARK, FL 32792
LAKELAND ARC INC., G PAUL LETTS KD4EFL SECY
 PO BOX 792, EATON PARK, FL 33840
MANATEE ARC, FRANK MORTON AC4MK PRES
 PO BOX 10038, BRADENTON, FL 34282
MARTIN COUNTY ARA, ROBERT J HESS KA3EDLPRES
 PO BOX 1901, STUART, FL 34995
METROPOLITAN REPEATER ASSN, LANCE LESTER
 N4VMZ TRUSTEE, PO BOX 2735, PINELLAS PARK, FL
 34664
MILTON ARC, WILLIAM S COUCH WA4MYK PRES
 PO BOX 7112, MILTON, FL 32570
MOTOROLA ARC, BILL PENCE KI4US PRES
 8000 W SUNRISE BLVD., FORT LAUDERDALE, FL 33322
NORTH DADE REPEATER ASSN, NORM BORENSTEIN
 WA4ILQ PRES, 5406 HAYES ST., HOLLYWOOD, FL 33021
NORTH FLORIDA ARS, BILLY F WILLIAMS N4UF
 TRUSTEE, PO BOX 9673, JACKSONVILLE, FL 32208
NORTH FLORIDA DX ASSN, BILL MARTIN WB4KSP SECY
 2544 BOTTOMRIDGE DR., ORANGE PARK, FL 32065
ORANGE PARK ARC, JAMES E GRAVES III WD4NXY TRE
 PO BOX 27033, JACKSONVILLE, FL 32205
PALATKA ARC, BILL RAYMOND KC4FMYZ PRES
 RR 2 BOX 2920, PALATKA, FL 32177
PALM BEACH ARS, INC., DIANE BAKER KC4TIA
 PO BOX 5454, LAKE WORTH, FL 33466
PALMETTO ARC, BERNARD STERNBERG AA4EE
 9790 NW 24TH ST., SUNRISE, FL 33322
PASCO HILLS DX ASSN., RON MCLEAN KK4CR SECY
 32036 TALLY HO LN, ZEPHYRHILLS, FL 33543
PENSACOLA AREA HAMS ASSN. ,BILL BREHRENDS
 WA4YRN PRES, 1050 CARLTON RD, PENSACOLA, FL
 32534
PLATINUM COAST ARC, GARRY WENTZ KC4EHT EDITOR
 672 INDIAN RIVER DR., MELBOURNE, FL 32935
PLAYGROUND ARC, FRANK BUTLER W4RH
 PO BOX 873, FORT WALTON BEACH, FL 32549
POLK HAM CLUB, CHET CARRUTH AA4XK
 P O BOX 66, EAGLE LAKE, FL 33839
PORT ST LUCIE ARA, BRUCE BUCKEY KD4HQJ SECY
 PO BOX 7461, PORT SAINT LUCIE, FL 34985
RACAL-MILGO ARC, JOE LOEWY KB4FO
 MS E2133, 1601 NW 136TH AVE, SUNRISE, FL 33323
ROYAL PALM ELEM SCHOOL ARC, ALAN E WOLFE
 WB4DYU SPONSOR
 4200 SW 112TH CT, MIAMI, FL 33165

SANTA FE COMMUNITY COLLEGE ARS, BOB LIGHTNER
 WA4PWF P-132
 3000 NW 83RD ST, GAINESVILLE, FL 32606
SARASOTA ARA, EDDIE MARTIN KI4ZJ PRES
 PO BOX 3182, SARASOTA, FL 34230
SARASOTA EMERGENCY RADIO CLUB, A J WARNOCK
 KC4LNU PRES, PO BOX 1175, SARASOTA, FL 34230
SERIOUS HAMS ARC, JIM PINES AD4BU SECY
 10697 BRIDGE CREEK DR., PENSACOLA, FL 32506
SILVER SPRINGS RC INC., WILLIAM C BRITT JR. KA9CQU
 PO BOX 787, SILVER SPRINGS, FL 34489
SKY HIGH ARC, RICK RINGEL N7KAA PRES
 PO BOX 572, LECANTO, FL 34460
SOUTH BREVARD ARC, LAWRENCE H LUNDY KD4JRJ
 1785 VIA ROMA, MERRITT ISLAND, FL 32953
SOUTH FLORIDA FM ASSN, SAM HARTE KQ4MR
 PO BOX 430025, MIAMI, FL 33243
SOUTH HILLSBOROUGH ARC, CHARLES R MILLER N4GOI
 SECY, 1522 HARTWICK DR., SUN CITY CENTER, FL
 33573
SPRING HILL ARC, ROBERT G BOCK KD4QLQ PRES
 PO BOX 5773, SPRING HILL, FL 34606
ST AUGUSTINE ARS, MARY ALBRECHT WD4MJS LIAS
 1 PARK TERRACE DR., SAINT AUGUSTINE, FL 32084
ST LUCIE REPEATER ASSN INC., JAMES LASETER N4ZYX
 SECY, 2716 ROBIN ST., FORT PIERCE, FL 34982
ST PETERSBURG ARC, ROBERT H RUSSELL N4ZMQ PRES
 2014 7TH ST N., SAINT PETERSBURG, FL 33704
SUN CITY CENTER ARC, JOHN D BOWKER WA2WEN
 SECY, 1811 FORT DUQUESNA DR., SUN CITY CENTER,
 FL 33573
SUNCOAST ARC, RON WRIGHT N9EE SECY
 PO BOX 1992, NEW PORT RICHEY, FL 34656
TAMARAC ARA, BERNIE MANN N4JBH EDITOR
 4917 NW 58TH ST., TAMARAC, FL 33319
TAMIAMI RADIO CLUB, LARRY YACOBELLI KD4THJ
 TREAS, PO BOX 6034, VENICE, FL 34292
TAMPA ARC, ARTHUR GIBSON KA4WNZ PRES
 PO BOX 10453, TAMPA, FL 33679
THUNDER BAY ARA, WILLIAM H HOLCOMB KC4YTP
 SUITE 204-168, PO BOX 103, LARGO, FL 34649
TITUSVILLE ARC, HORACE T HUGHES K4CWG PRES
 PO BOX 73, TITUSVILLE, FL 32781
TRAVELER'S REST ARC, EDNA ZIEGLER KB0KAM SCTR
 29129 JOHNSTON RD LOT 16-23, DADE CITY, FL 33525
TREASURE COASTERS RPTR ASSN, JACK NOVOTNY
 K3CEW SCTR, 2580 84TH TER, VERO BEACH, FL 32966
TRI-COUNTY ARS, ROGER KELLEY ND4E PRES
 9075 NW 24TH CT, CORAL SPRINGS, FL 33065
UNIVERSITY OF CENTRAL FL ARC, MITCH WINKLE
 AC4IY PRES, P O BOX 168052, ORLANDO, FL 32816
VERO BEACH ARC INC., RICHARD JACKSON AB4AZ SECY
 PO BOX 2082, VERO BEACH, FL 32961
WALTON COUNTY ARC, DARRYL I DAVIS K4DVL PRES
 RT 4 BOX 195-F, DEFUNIAK SPRINGS, FL 32433
WASHINGTON/HOLMES ARC, JONATHAN MCKEOWN
 KQ4EQ PRES, 704 S 2ND ST, CHIPLEY, FL 32428
WEST PALM BEACH ARC INC., BURCK GROSSE KC4UEV
 PRES, 11 HUNTLY CIR, PALM BEACH, FL 33418
WR4AYC REPEATER GROUP, BRUCE STANLEY KD4GHV
 SECY, 3505 W ATLANTIC BLVD APT 810, POMPANO
 BEACH, FL 33069
ZEPHYRHILLS AREA AMATEUR RADIO, ERNA
 VANSELOW KD4VRV PRES, PO BOX 1534,
 ZEPHYRHILLS, FL 33539

GEORGIA
ACADEMY ELEMENTARY RADIO CLUB, BEN D ROY
 KB4IXL PRES, 1274 REEVES STATION RD SW., CAL
 HOUN, GA 30701
ALBANY AMATEUR RADIO CLUB, F LEON PERRETT JR
 K4GCR, PO BOX 70601, ALBANY, GA 31708
ALFORD MEMORIAL RADIO CLUB INC., AL HUGHES
 N4SIR PRES, PO BOX 1282, STONE MOUNTAIN, GA
 30086

AMATEUR RADIO CLUB OF AUGUSTA, DONALD ANDERSON K4PSW PRES, PO BOX 3072, AUGUSTA, GA 30914

ARC OF SAVANNAH INC., ANDREW J BLACKBURN III WD4AFY, PO BOX 13342, SAVANNAH, GA 31416

ATHENS RADIO CLUB, SCOTT H SIKES KD4MSR SECY 251 WAKEFIELD TRCE, ATHENS, GA 30605

ATLANTA CHAPT 49 QCWA, JUD WHATLEY W4NZJ PRES 2156 WINDSOR DR., SNELLVILLE, GA 30278

ATLANTA RADIO CLUB, VERM FOWLER W8BLA PRES PO BOX 77171, ATLANTA, GA 30357

BILL GREMILLION MEMORIAL RC, TED HATFIELD KQ4IC SCTR, PO BOX 2327, NEWNAN, GA 30264

CEDAR VALLEY ARC, JAMES T. SCHLIESTETT W4IMQ SCT, PO BOX 93, CEDARTOWN, GA 30125

CENTRAL GEORGIA ARC, GENE MCKINLEY KD4NGB PRES, PO BOX 2585, WARNER ROBINS, GA 31099

CHARLES E NEWTON RADIO CLUB, GENE RAY WD4GUA SCTR, 2388 HIGHWAY 36 E., MILNER, GA 30257

COASTAL ARS, PEG DAVIS KC4BAZ SECY, 619 COLUMBUS DR., PO BOX 23972, SAVANNAH, GA 31403

COASTAL PLAINS ARC, INC., WAYNE HARRELL WD4LYV PRES, 2716 DENHAM RD, SYCAMORE, GA 31790

COLQUITT CTY HAM RADIO SOCIETY, JOEL D GOINGS AA4P SCTR, PO BOX 813, MOULTRIE, GA 31776

COLUMBUS ARC, JOE M OWEN KO4RR PRES PO BOX 6336, COLUMBUS, GA 31907

CONFEDERATE SIGNAL CORPS INC., F L WOMACK W4BJT SECY, 2545 PLANTATION DR., EAST POINT, GA 30344

CONYERS AMATEUR RADIO GROUP, KEVIN WOOD KQ4LE PRES, PO BOX 80721, CONYERS, GA 30208

DALTON ARC, INC., HAROLD W JONES N4OTC PRES PO BOX 143, DALTON, GA 30722

DOUGLAS AREA ARC, ROBERT K MERRITT KE4BMT SECY, 516 FOREST CIR., DOUGLAS, GA 31533

DUBLIN ARC, RAY DUCK KC4FKD PRES 307 RIDGECREST RD, DUBLIN, GA 31021

GEORGIA R A PACKET ENTHUSIASTS, WM TURRENTINE JR WA4GAI TREAS, 4602 WESTHAMPTON CIR, TUCKER, GA 30084

GEORGIA SINGLE SIDE BAND ASSN, ELLIE WATERS WB4CJB PRES, PO BOX 231, PEMBROKE, GA 31321

GEORGIA TECH RADIO CLUB, DR PAUL STEFFES W8ZI GA TECH STATION, BOX 32705, ATLANTA, GA 30332

GLYNN ARA INC., T K TOUW N4AKD SECY 121 MILITARY RD, SAINT SIMONS ISLAND, GA 31522

GWINNETT ARS, WILLIAM CHAPMAN KE4GYM SECY PO BOX 88, LILBURN, GA 30226

IBM RADIO CLUB-ATLANTA, LORI VANDEGRIFT KD4SXI 4111 NORTHSIDE PKY NW, ATLANTA, GA 30327

KENNEHOOCHEE ARC INC., JANE D BRUNE KB4QKX SECY, PO BOX 1245, MARIETTA, GA 30061

KIBBEE RA PRESERVATION SOCIETY, ROBERT B WHEELER K4HVK PRES, PO BOX 47939, ATLANTA, GA 30362

LIBERTY CTY EMERG COMM. ARC, DOUGLAS HARRIS KM4NE PRES., 814 MANDARIN DR., HINESVILLE, GA 31313

MACON ARC, ROBERT B WRIGHT W4OZF PRES PO BOX 4862, MACON, GA 31208

METRO ATLANTA TELEPHONE PIONEE, JIM LACEY N4ZXR PRES., 1711 PARLIAMENT DR., DUNWOODY, GA 30338

MIDDLE GEORGIA RADIO ASSN., DAVID L SHIPLETT WL7ACY PRES, 107 MOSSY LAKE RD, PERRY, GA 31069

NORTH FULTON A R LEAGUE, JEFFREY J WEINBERG AA4RC PRES, 2714 FIELDSTONE PATH, MARIETTA, GA 30062

NORTHEAST GEORGIA ARC., DR JOSEPH A TILLER KD4VHX, 8806 JEFFERSON RD, COMMERCE, GA 30529

NORTHWEST GEORGIA ARC, SHEILA BARRETT KA4WCW SCTR, 11 DOGWOOD ST., ROME, GA 30161

PICKENS ARC, HOWELL J PARRY W4OOM SECY 175 SWEETGUM CIR, JASPER, GA 30143

PINE LOG MOUNTAIN RPTR GROUP, WILLIAM N LITTLE KE4PN PRES, PO BOX 237, WALESKA, GA 30183

SKINT CHESTNUT ARS, RALPH UBRAS N4ZCV PRES 4090 OAK STONE DR., DOUGLASVILLE, GA 30135

SOUTH GEORGIA ARC, WILLIAM O LEY WA4NKL SCTR RR 2 BOX 3950, QUITMAN, GA 31643

SOUTHEASTERN DX CLUB, JOHN TRAMONTANIS N4TOL TREA, PO BOX 19871, ATLANTA, GA 30325

STATESBORO ARS, STANLEY G YARBER W2HXW PRES 208 S EDGEWOOD DR., STATESBORO, GA 30458

SUMTER-COUNTY ARC, CHARLES T ROYAL WD4EIK SECY, PO BOX 195, AMERICUS, GA 31709

THOMASVILLE ARC, MICHAEL GREENE KE4BZG SCTR PO BOX 251, THOMASVILLE, GA 31799

VALDOSTA ARC, WESLEY HESTERS KD4QYI PRES PO BOX 3846, VALDOSTA, GA 31604

VIDALIA EMERG COORDINATING RPT, DAVE THEURER W8ARB SCTR, PO BOX 750, VIDALIA, GA 30474

HAWAII

BIG ISLAND ARC, MYRTLE BARTRON NH6WX SECY PO BOX 1938, HILO, HI 96721

EMERGENCY ARC, ROBERT JONES NH6GJ PRES PO BOX 30315, HONOLULU, HI 96820

HONOLULU DX CLUB, PATRICK CORRIGAN KH6DD PO BOX 67, HONOLULU, HI 96810

KALAWAO COUNTY ARC, SHERMAN NAPOLEON WH6IT PRES, PO BOX 95, KUALAPUU, HI 96757

KAUAI ARC, KEVIN DEVITT WH6GK PRES PO BOX 458, KALAHEO, HI 96741

MAUI ARC, WILLIAM HEYDE KH6UU SECY PO BOX 1791, KAHULUI, HI 96732

PACIFIC RA TRANSMITTING SOC., L ROGER WICAL KH6BZF PRES, KOOLAUPOKO, 45-601 LULUKU RD CRT #44-25, KANE'OHE, HI 96744

IDAHO

CLEARWATER VALLEY ARC, DENNIS BURGESS KA7FAH LIAS, PO BOX 1263, OROFINO, ID 83544

KOOTENAI ARS, HARVEY ZION KI7EG PRES PO BOX 5222, COEUR D ALENE, ID 83814

MAGIC VALLEY CHAPTER ISRA, DEXTER ROGERS N7GKJ SCTR, RR 5 BOX 5144, BUHL, ID 83316

NORTH IDAHO AMATEUR RPTR ASSN, ED GROGER KA7DIS PRES, PO BOX 967, COEUR D ALENE, ID 83816

POCATELLO ARC, JOHN WILSON WA0DYU PRES BOX 2722, POCATELLO, ID 83201

SNAKE RIVER CHAPTER ISRA, MIKE JOHNSON WA7NRP SECY, 850 ALMO, PO BOX 833, BURLEY, ID 83318

ILLINOIS

AMATEUR CROSS LINK REPEATER, BILL DAVIDSON KA9SWW PRES, PO BOX 34446, CHICAGO, IL 60634

AMATEUR RADIO CLUB-POLONIA, JOHN PARTYKA KB9IDQ, 7597 CHURCHILL DR., HANOVER PARK, IL 60103

ARC of MT VERNON, CHERYL HERTENSTEIN AA9EM SECY, 12 EVERGREEN, PO BOX 1342, MOUNT VERNON, IL 62864

BIG THUNDER ARC INC., JAMES G GRIMSBY W9HRF PRES 210 OAKLAWN LN, POPLAR GROVE, IL 61065

BLACKHAWK DX & CONTEST CLUB, CLAY DEWITT N9HUB SECY, PO BOX 7343, ROCKFORD, IL 61126

BOLINGBROOK ARS, MARK J O'DONNELL N9OCL PRES 121 W BRIARCLIFF RD, BOLINGBROOK, IL 60440

CENOIS ARC, JEFFREY W WELLS KB9HVA EDITOR PO BOX 4595, DECATUR, IL 62525

CENTRAL IL/ST LOUIS AREA TELE, SCOTT MILLICK K9SM PRES, 907 BIG 4 AVE., HILLSBORO, IL 62049

CENTRALIA WIRELESS ASSN, ALVA (BUD) KING JR WA9U, 776 BETHEL RD, SANDOVAL, IL 62882

CHICAGO FM CLUB, RICHARD HERSH K9FFY SECY P O BOX 1532, EVANSTON, IL 60204

CHICAGO RADIO TRAFFIC ASSN, WILLIAM E RUSS W9REC SCTR, 1501 CUYLER AVE., BERWYN, IL 60402

CHRISTIAN COUNTY ARC, PATRICK J GRAFTON AA9II
PO BOX 93, TAYLORVILLE, IL 62568
COMMONWEALTH EDISON EMPLOYEES, GERRY SWAN
SON N9MEP PRES, 124 TAMARACK AVE., NAPERVILLE,
IL 60540
DUPAGE ARC, JACK J CARR NV9S PRES
PO BOX 71, CLARENDON HILLS, IL 60514
ECHO REPEATER ASSN, KEITH HOOVER WA9VGX SECY
5020 THORNBARK DR., BARRINGTON, IL 60010
EGYPTIAN RADIO CLUB INC., DENNIS MCCANN W9UH
SECY, 2432 HEMLOCK AVE., GRANITE CITY, IL 62040
ELGIN ARS, WAYNE STACH KA9HJC PRES
1216 HIAWATHA DR., ELGIN, IL 60120
ELK GROVE ARC, LARRY HOFFMAN KB9L PRES
115 HASTINGS AVE., ELK GROVE VILLAGE, IL 60007
FEDERATION OF AR OPERATORS, KEVIN KAUFHOLD
WB9GKA PRES, 910 N CHURCH ST., BELLEVILLE, IL
62220
FOX RIVER RADIO LEAGUE, WARREN GEARY KI9H PRES
PO BOX 673, BATAVIA, IL 60510
HAMFESTERS RADIO CLUB, WALTER R GESELL
WD9DYR, 9142 PEMBROKE LN, BRIDGEVIEW, IL 60455
HIGHLAND PK HS ARC, ROBERT H WEGNER W9MON
SPONSO, HIGHLAND PARK H S., 433 VINE AVE., HIGH
LAND PARK, IL 60035
ILLINOIS VALLEY ARC INC., JAMES DUNHAM KA9UFX
BOX 232, BROWNING, IL 62624
ILLINOIS VALLEY RADIO ASSN., DAN DE MATTIA N9OBB
SCTR, P O BOX 322, HENNEPIN, IL 61327
IROQUOIS COUNTY ARC, SAM RIPPLE WB9QKF SCTR
420 E LOCUST ST., WATSEKA, IL 60970
IROQUOIS-FORD ARS, ALLEN J GHARST K9UXC
P O BOX 151, 205 S ELM ST., LODA, IL 60948
JBARS GROUP, JIM BRADY W9JB PRES
PO BOX 79, DOWNERS GROVE, IL 60515
KISHWAUKEE RADIO CLUB, HOWARD R NEWQUIST
WA9TXW TREAS, 128 MANOR DR., DE KALB, IL 60115
KNOX COUNTY ARC, BRENT ZHORNE WB9FHI PRES
PO BOX 305, GALESBURG, IL 61402
LAMOINE EMERGENCY ARC, BRUCE BOSTON KD9UL
EDITOR, 815 E 3RD ST., BEARDSTOWN, IL 62618
LEWIS & CLARK RADIO CLUB, LARRY H ROBERTS
W9MXC, 5319 DOVER DR., GODFREY, IL 62035
LIBERTYVILLE AND MUNDELEIN, FLOYD VLASAK NE9L
SCTR, PO BOX 751, LIBERTYVILLE, IL 60048
METRO DX CLUB, JOHN W NIENHAUS WA9NJB SCTR
PO BOX 239, OAK FOREST, IL 60452
MOULTRIE ARK, RALPH ZANCHA WC9V SCTR
PO BOX 91, LOVINGTON, IL 61937
NATIONAL TRAIL ARC, DAVE JEFFRIES N9KDJ PRES
111 MARTIN, EFFINGHAM, IL 62401
NODOT DX'ERS, BRUCE FORD KS9U PRES
23042 N APPLE HILL LN, PRAIRIE VIEW, IL 60069
NORTH SHORE RADIO CLUB, ALAN MARCUS PRES
PO BOX 1066, HIGHLAND PARK, IL 60035
NORTHERN ILLINOIS DX ASSN, RAY HIBNICK WA9YYY
SECY, 920 THOMPSON BLVD, BUFFALO GROVE, IL
60089
NORTHERN LAKE COUNTY ARC, ANTHONY J VIVERITO
WR9D SECY, 465 POPLAR AVE., ANTIOCH, IL 60002
NORTHWEST ARC, PAUL BOVLACONTI WD9DJD
PO BOX 121, ARLINGTON HEIGHTS, IL 60006
NORTHWESTERN UNIV ARS, PETER ANVIN N9ITP PRES
2145 SHERIDAN RD, EEGS DEPT, EVANSTON, IL 60208
P.R.I.M.E. ARA, JOHN J KOZOWSKI KN9G PRES
164 TWIN LAKE DR., BELLEVILLE, IL 62221
PEKIN ARC, ARTHUR L OATES K9GBN PRES
122 ARROW ST., PEKIN, IL 61554
PEORIA AREA AMATEUR RADIO CLUB, JIM CUSEY NT9C
PRES, PO BOX 3508, PEORIA, IL 61612
RADIO AMATEUR MEGACYCLE SOC, ELMER P FRO
HARDT JR W9DY TREA, 3620 N OLEANDER AVE.,
CHICAGO, IL 60634
ROCK RIVER ARC, BRION C GILBERT N0AE PRES
PO BOX 283, DIXON, IL 61021

ROCKFORD ARA, GENE YOUNG N9MXQ PRES
P O BOX 8465, ROCKFORD, IL 61126
S ILLINOIS UNIV ARC, GARY R SMITH AA9JS PRES
SIU CARBONDALE, ENG & TECHNOLOGY BLDG, CAR
BONDALE, IL 62901
SALT CREEK RADIO CLUB, WILFRID F BERG KR9A PRES
17W050 HAWTHORNE AVE., BENSENVILLE, IL 60106
SCHAUMBURG ARC, HAROLD ROBIN KB9DJD PRES
PO BOX 68251, SCHAUMBURG, IL 60168
SHAWNEE AMATEUR RADIO ASSN, BRUCE TALLEY
WA9APQ PRES, RR 1 BOX 143F, CARTERVILLE, IL 62918
SIX METER CLUB OF CHICAGO INC., JOSEPH GUTWEIN
WA9RIJ SECY, 7109 BLACKBURN AVE., DOWNERS
GROVE, IL 60516
SOCIETY OF MIDWEST CONTESTERS, KEITH MORE
HOUSE WB9TIY, 14N082 FRENCH RD, HAMPSHIRE, IL
60140
SOUTHERN ILLINOIS ARS, LES BILDERBACK KF9TZ PRES
PO BOX 2323, CARBONDALE, IL 62902
ST CLAIR ARC, TOD A WEST SR KB9AIL PRES
709 GAWAIN DR., TROY, IL 62294
STERLING ROCKFALLS ARS, JOE BALDWIN N9ORQ PRES
PO BOX 521, STERLING, IL 61081
STREATOR ARC, EDWIN HAUSAMAN WB9CKS PRES
1820 CARR ST, STREATOR, IL 61364
SUBURBAN AMATEUR RPTR ASSN., ROBERT SCHIFF
KA9HHH PRES, 6628 N RICHMOND ST, CHICAGO, IL
60645
TRI-TOWN RADIO AMATEUR CLUB, BRIAN BEDOE
WD9HSY PRES, 11248 MARILYN WAY, MOKENA, IL
60448
VERMILION COUNTY ARA, GARY S DENISON KA9SKS
PRES, RR 2 BOX 89, DANVILLE, IL 61832
WESTERN ILLINOIS ARC, DARELL TAYLOR KA9RTV
PRES, 1815 N HIGHLAND DR., QUINCY, IL 62301
WHEATON COMM RADIO AMATEURS, PAT BYRNE
K9JAU SECY, P O BOX QSL, WHEATON, IL 60189
YORK RADIO CLUB, JOHN KAMM WA9OHU
PO BOX 1201, ELMHURST, IL 60126

INDIANA
JOHN MORRIS KA9LPN SPONSOR
8112 N 200 W., FORTVILLE, IN 46040
21 REPEATER GROUP, BENNIE L GRIMM KA9KOG SCTR
10294 E-100 S., AVILLA, IN 46710
807 CLUB OF KOKOMO, DONALD J GROSS W9KQD LIAS
9068 E 400 N., GREENTOWN, IN 46936
ADAMS COUNTY ARC, FRED EYANSON KA9NJT PRES
1538 CHERRY LN, DECATUR, IN 46733
AMERICAN RED CROSS ARC, WESLEY SCHAEFER N9IXI
5440 W HIGH ST., SILVER LAKE, IN 46982
ANDERSON REPEATER CLUB, STEVE RILEY WA9GWE
SECY, RR 2 BOX 225A, ALEXANDRIA, IN 46001
BLOOMINGTON ARC INC., MILLARD H QUALLS K9DIY
PRES, 2129 S ROGERS ST, BLOOMINGTON, IN 47403
BOONE COUNTY AR, CAROL MARTIN JR N9JSS PRES
603 E WALNUT ST, LEBANON, IN 46052
BOZO AND THE LIDS, DANIEL SHUDICK JR WB9YRT
SCTR, 1908 MARTHA ST, MUNSTER, IN 46321
CALUMET AMATEUR RADIO ENTHUSIA, MICHAEL A
KASRICH AJ9C SECY, 3035 GRAND BLVD, LAKE STA
TION, IN 46405
CASS COUNTY ARC, ED NORRIS K9FSR
PO BOX 1092, LOGANSPORT, IN 46947
CLARK COUNTY ARC INC., HERBERT E ROWE W4WQD
SECY, 5612 HIGHWAY 160, CHARLESTOWN, IN 47111
CLIFTY ARS, GARY MAAS WN9X PRES
517 CLIFTY DR., MADISON, IN 47250
CLINTON COUNTY ARC, JUDY E SMITH KA9SVG SCTR
267 S COUNTY ROAD 380 E., FRANKFORT, IN 46041
COLUMBUS ARC, DENNIS ARNHOLT WD9DWE SECY
10655 N CEDARS RD, SEYMOUR, IN 47274
DELAWARE ARA, GILBERT T RAGER W9BZI TRUSTEE
1407 S MAY AVE., MUNCIE, IN 47302

ELEC APPLICATIONS OF RADIO SOC, MARTIN L HENS
LEY KA9PCT PRES, 1506 S PARKER DR., EVANSVILLE, I
N 47714

ELKHART COUNTY RADIO ASSN INC., JOHN H ALLYN
KA9SYE SECY, 1117 S 8TH ST, GOSHEN, IN 46526

ELKHART RED CROSS ARC, ANDY KEIL KB9FMZ SECY
10935W W 805 N., SHIPSHEWANA, IN 46565

EVANS MIDDLE SCHOOL RC, PAUL A RICE WA9BYZ
ADVISOR, 17 N ALVORD BLVD, EVANSVILLE, IN 47711

FORT WAYNE RC INC., CAROLE BURKE WB9RUS SECY
PO BOX 15127, FORT WAYNE, IN 46885

FULTON COUNTY ARC, JAMES G BROWN WA9PKL SCTR
3331 S 1300 E., AKRON, IN 46910

GOSHEN ARC, JOHN H ALLYN KA9SYE SCTR
1117 S 8TH ST, GOSHEN, IN 46526

GOSHEN COLLEGE ARC, GENE CRUSIE KA9QVM
1700 S MAIN ST, GOSHEN, IN 46526

GRANT COUNTY ARC, P EDWIN BROWN DDS N9GTL
1511 IRONWOOD DR., MARION, IN 46952

HANCOCK ARC, TOM DONALDSON N9LFU PRES
PO BOX 7033, GREENFIELD, IN 46140

HENDRICKS COUNTY ARS, MIKE WALKER N9LGH SECY
9036 E COUNTY ROAD 100 N., INDIANAPOLIS, IN 46234

HOOSIER CONTESTERS, MICHAEL R O'CULL WO9Z SCTR
3650 W 141ST ST, WESTFIELD, IN 46074

HUNTINGTON COUNTY ARS, MICHAEL BROOKER
WD9JFC, 15035 FEICHNER RD, ROANOKE, IN 46783

INDIANAPOLIS HAMFEST ASSN, RICK L OGAN N9LRR
PRES, PO BOX 11776, INDIANAPOLIS, IN 46201

INDIANAPOLIS RADIO CLUB, RICKIE L SEXTON N9NCH
VPRS, PO BOX 5201, INDIANAPOLIS, IN 46251

INDIANAPOLIS RED CROSS ARC, JOHN DICKERSON
KA9YPM VPRS, 7248 TARRAGON LN, INDIANAPOLIS,
IN 46237

IVY TECH ARC, MIKE HALL N9TW}OT 316, ONE WEST
26TH ST, INDIANAPOLIS, IN 46206

KEY AND MIKE CLUB, LOUIS KRUIZINGA N7BBW LIAS
3807 CARVER ST, NEW ALBANY, IN 47150

KOKOMO ARC, CHARLES J SPONAUGLE N9LYY
1631 OSAGE DR S., KOKOMO, IN 46902

LAKE COUNTY ARC, LUCILLE M SCHENDERA N9DTG
LIAS, 812 E 40TH PL., GRIFFITH, IN 46319

LAUGHERY VALLEY ARC, DAVID M MAYER WD8NMZ
TREAS, 3689 N DEARBORN RD, WEST HARRISON, IN
47060

MARSHALL COUNTY ARC, RICHARD BASHAM K9ILU
PRES, PO BOX 151, PLYMOUTH, IN 46563

MIAMI COUNTY ARC, ANGIE FLORY KF9QW SCTR, PO
BOX 1011, PERU, IN 46970

MICHIANA ARC, PATRICIA WOLF N8XOA SECY
3220 E JEFFERSON BLVD, SOUTH BEND, IN 46615

MICHIGAN CITY ARC, ROY G JACKSON NY9B TRUSTEE
PO BOX 2013, MICHIGAN CITY, IN 46361

MID-STATE ARC, ROY BARNES N9PFZ
PO BOX 836, FRANKLIN, IN 46131

MILLTOWN ARS, KEN CUNDIFF WB9ZHL PRES, RR 1 BOX
328A, MILLTOWN, IN 47145

MIZPAH TEMPLE RADIO CLUB, RONALD L STABLER
N9PTQ SECY, 5016 FIRWOOD DR., FORT WAYNE, IN
46835

MUNCIE AREA ARC, SONNY MCCOY WA9DOL EDITOR
2004 S BATAVIA AVE., MUNCIE, IN 47302

NORTHEASTERN INDIANA ARC, VIRGIL ED DUNKIN
WA6OIZ PRES, PO BOX 745, AUBURN, IN 46706

OLD POST ARS, BRIAN FILLINGIM WD9ELZ EDITOR
PO BOX 834, VINCENNES, IN 47591

OWEN COUNTY ARA, KATHRYN SMITH KB9INU SECY
RR 1 BOX 368D, POLAND, IN 47868

PIKE COUNTY ARC, MAXINE HAGEMEYER SECY
PO BOX 12, STENDAL, IN 47585

RANDOLPH ARA, LLOYD M SPENCER KA9MNR SCTR
5 WOODCREST AVE., WINCHESTER, IN 47394

RICHMOND ARA INC., CHARLES M PHILHOWER WA9KZC
SCT, 1437 HUNT ST, RICHMOND, IN 47374

RIPLEY COUNTY REPEATER ASSN, STEVE KRISTOFF
N9PSG LIAS, PO BOX 44, OLDENBURG, IN 47036

ROSE TECH RADIO CLUB, DAVID PETTIT WO9M PRES
5500 WABASH AVE., PO BOX 1584, TERRE HAUTE, IN
47808

SEYMOUR ARC, GREGORY TATLOCK N9PUG TREAS
PO BOX 1274, SEYMOUR, IN 47274

STEUBEN COUNTY RADIO AMATEUR C, BECKY HILL
N9FAK SCTR, PO BOX 352, ANGOLA, IN 46703

THOMSON ARC, JIM RINEHART WB9CEP
TECHNICAL CENTER INH600, 101 WEST 103RD ST,
INDIANAPOLIS, IN 46290

TRI-STATE ARS, CHARLES APFELSTADT N9GWS PRES
PO BOX 4521, EVANSVILLE, IN 47724

U S I AMATEUR RADIO CLUB, D NEIL RAPP WB9VP PRES
8600 UNIVERSITY BLVD, EVANSVILLE, IN 47712

WABASH COUNTY ARC, LARRY D MANNING N9AFI PRES
5199 E STATE ROAD 218, LA FONTAINE, IN 46940

WABASH VALLEY ARA INC., GARY WHEELER WB9SWG
SECY, PO BOX 81, TERRE HAUTE, IN 47808

WHITEWATER VALLEY ARC, KEVIN M FESSLER N9TIF
PRES, 507 W SCHOOL ST, CENTERVILLE, IN 47330

WHITLEY COUNTY ARC, HENRY R MACKEY KB9CQO
SECY, 333 N WALNUT ST, COLUMBIA CITY, IN 46725

IOWA

3900 CLUB, RICHARD W PITNER W0FZO SCTR
2931 PIERCE ST., SIOUX CITY, IA 51104

BELMOND ARC, HAROLD BARNES N0YWG SCTR
403 5TH AVE NE, BELMOND, IA 50421

BOONE AMATEUR RADIO KLUB, CALVIN ROSSMAN
K0VRA SCTR, 1717 BOONE ST., BOONE, IA 50036

CEDAR VALLEY ARC INC., RON BREITWISCH KC0OX
SECY, PO BOX 994, CEDAR RAPIDS, IA 52406

CENTRAL IOWA RADIO AMATEUR SOC., BRIAN KRUMM
N0MXK TREAS, 911 SOUTH EIGHTH AVE., RT #6, MAR
SHALLTOWN, IA 50158

CENTRAL IOWA TECHNICAL SOCIETY, RALPH GANDY
N0NOU SCTR, 1876 E PARK AVE., DES MOINES, IA
50320

COON VALLEY ARC, JIM DOWD N0GVJ TREAS
1822 EVELYN ST., PERRY, IA 50220

CYCLONE ARC-IA STATE UNIV., TODD WHEELER N0NUM
PRES, FRILEY 5567, AMES, IA 50012

DES MOINES RADIO AMATEUR ASSN, DAN BUREMAN
WB0QAM PRES, PO BOX 88, DES MOINES, IA 50301

EASTERN IOWA DX ASSOCIATION, GARY TOOMSEN
K0GT SCTR, 2730 TOWER DR., CEDAR RAPIDS, IA 52411

FORT DODGE ARC, JOHN R CLARK NQ7G SECY
PO BOX 111, FORT DODGE, IA 50501

FORT MADISON ARC, W T STANSBERY N0MDX PRES
3310 AVENUE L, FORT MADISON, IA 52627

GREAT RIVER ARC, JERRY EHLERS N0NLU PRES
3115 BRUNSWICK ST., DUBUQUE, IA 52001

HEARTLAND HAMS ARC, DALE SARGENT N0WKF
RR 1 BOX 290, PACIFIC JUNCTION, IA 51561

HUMBOLDT AMATEUR RADIO KLUB, PHILL MCLARNAN
N0RXX, 206 1ST AVE S., HUMBOLDT, IA 50548

I M A RADIO CLUB, BOB JACKSON WA0AUF SECY
RR 1, BLOOMFIELD, IA 52537

IOWA CITY ARC, MARK ATHERTON N0RXD PRES
POB 4, IOWA CITY, IA 52240

IOWA GREAT LAKES ARC, RON HARRIS K0GVC SECY
R B 9269, SPIRIT LAKE, IA 51360

IOWA-ILLINOIS ARC, CHUCK GYSI N2DUP VPRS
PO BOX 911, BURLINGTON, IA 52601

JONES COUNTY ARC, JAMES R MCCLINTOCK N0CWP
BOX 462, 301 VINE ST., MORLEY, IA 52312

JUNIOR CRESCO ARC, SCOTT VOYNA KB0DCU PRES
417 5TH AVE W., CRESCO, IA 52136

KOSSOUTH A R OPERATORS, CHRIS VAN OSBREE N0WHI
P O BOX 388, EMMETSBURG, IA 50536

MEGAHERTZ MANOR MANIACS, DONALD L SCHMIDT
W0ANZ SECY, 2161 NW 80TH PL., DES MOINES, IA 50325

NEWTON ARA, JOHN C BRITSON K0ZAL SECY
   203 W 11TH ST S., NEWTON, IA 50208
NORTH IOWA ARC, CHRIS RYE N0OKH PRES
   1009 FAIR MEADOW DR., MASON CITY, IA 50401
O'BRIEN COUNTY ARA, GARY L STEINBECK N0UYJ SECY
   3893 TANAGER AVE., PRIMGHAR, IA 51245
QUAD CITY AMATEUR TELEVISION, DON SCHNEIDER
   WD0AMA TREAS, 518 W LOCUST ST., DAVENPORT, IA
   52803
QUAD CITY ARC, JOHN T HOENSHELL N0BFJ SECY
   2331 N LINWOOD AVE., DAVENPORT, IA 52804
SOOLAND ARA, HARVARD A BAMGAARS KA0KUA
   1328 26TH ST., SIOUX CITY, IA 51104
TURKEY ISLAND DX CLUB, BILL BISHOP NQ0W PRES
   1106 W FINLEY AVE., OTTUMWA, IA 52501
VALLEY EMERG COMMUNICATIONS, CHUCK GYSI,
   N2DUP, P O BOX 911, BURLINGTON, IA 52601

KANSAS
AMATEUR RADIO ADVANCED TECH, HAROLD E CHIL
   DRESS WB0LFH, 2925 EVANS, WICHITA, KS 67216
APOSTOLIC ACADEMY RC, STEVE CARRIER N0JVS
   SPONSOR, 1525 MCFARLAND RD, JUNCTION CITY, KS
   66441
BOEING EMPLOYEES ARS, MICHAEL THORNTON WC0L
   PRES, 1106 PAIGE ST., WICHITA, KS 67207
CENTRAL KANSAS ARC, ROB KELLY N0SMR SECY
   PO BOX 2493, SALINA, KS 67402
CHANUTE AREA ARC, PAUL D SMITH N0NBD
   RR 1 BOX 208, HUMBOLDT, KS 66748
COWLEY COUNTY ARC, BEVIN RUSSELL KB0GBB SCTR
   PO BOX 227, WINFIELD, KS 67156
DOUGLAS COUNTY ARC, TRUMAN C WAUGH N0APJ
   SECY, 1916 MELHOLLAND RD, LAWRENCE, KS 66044
EMPORIA ARS, RAY POLLEY K0JDB SCTR
   276 COUNTRY RD 180, EMPORIA, KS 66801
FLINT HILLS ARC, JEFFREY DREW N0TIN SCTR
   PO BOX 173, AUGUSTA, KS 67010
GOLDEN BELT ARC, JAY KNUDSON AA0DP PRES
   PO BOX 287, GREAT BEND, KS 67530
HAYS ARC, DOUG YOUNKER N0LKK PRES
   PO BOX 622, HAYS, KS 67601
HEARTLAND HF ASSOCIATION, TOM LAPPIN W0UY
   PRES, 44 TOMAHAWK RD, HUTCHINSON, KS 67502
HIAWATHA ARC, DEAN BAILEY KF0UB SCTR
   106 SIOUX AVE., HIAWATHA, KS 66434
JACKSON ARC, ROSS S MCCLAIN AA0MM SCTR
   620 WISCONSIN AVE., HOLTON, KS 66436
JAYHAWK ARS, BERT VOTH WA0PWE PRES
   PO BOX 4282, KANSAS CITY, KS 66104
JOHNSON COUNTY RADIO AMATEURS, CALVIN GOOD
   MAN KA0LFT PRES, 8201 W 87TH ST, OVERLAND
   PARK, KS 66212
KANSAS CITY DX CLUB, RICK BARNETT KB0U SCTR
   PO BOX 4798, OVERLAND PARK, KS 66204
KANSAS STATE UNIV ARC, NORM DILLMAN N0JCC
   ADVISOR, EE DURLAND HALL, KANSAS STATE UNI
   VERSITY, MANHATTAN, KS 66506
KANSAS-NEBRASKA RC, WENDELL D WILSON W0TQ
   PRES, 717 2ND AVE., CONCORDIA, KS 66901
KAW VALLEY ARC, STEVE HAMILTON KB0JYL PRES
   1221 SW 17TH ST., TOPEKA, KS 66604
MANHATTAN AREA ARS, PAUL SCHLIFFKE N0UZN PRES
   P O BOX 613, MANHATTAN, KS 66502
NEOSHO VALLEY ARC, MIKE CLEMENS N0RLR SECY
   521 NIAGARA ST., BURLINGTON, KS 66839
NEWTON ARC, MICHAEL BRUNGARDT WA0SXR PRES
   P O BOX 224, NORTH NEWTON, KS 67117
NORTH EAST KANSAS ARC, ROBERT NALL WV0S PRES
   5707 SW 28TH TER, TOPEKA, KS 66614
OSAGE COUNTY ARC, JAMES R WORSLEY N0HXU PRES
   RR 2 BOX 383A, OVERBROOK, KS 66524
PILOT KNOB ARC, DAVID BRICE N0NOM PRES
   416 OAKBROOK, LANSING, KS 66043

PITTSBURG RPTR ORGANIZATION, CHARLES CHANCEY
   KA0EGE TREAS, PO BOX 1303, PITTSBURG, KS 66762
SANDHILLS AMATEUR RADIO CLUB, RODNEY HOGG
   K0EQH TRUSTEE, 1104 MAIN ST, SCOTT CITY, KS 67871
SHAWNEE RADIO AMATEUR COMMUNIC, WILBUR E
   GOLL W0DEL PRES, PO BOX 3754, SHAWNEE, KS 66203
SOUTH WEST KANSAS ARC, BOB JONES N6USP PRES
   9105 N JENNIE BARKER RD, GARDEN CITY, KS 67846
TROJANS ARC, BRUCE FRAHM K0BJ ADVISOR
   P O BOX DX, COLBY, KS 67701
U S CENTER ARC, LARRY LAMBERT N0LL
   405 SHELTON DR., SMITH CENTER, KS 66967
UNIVERSITY OF KANSAS ARC, MIKE MARMOR N0MGE
   PRES, KANSAS UNIVERSITY, 1013 LEARNED HALL,
   LAWRENCE, KS 66045
WHEAT STATE WIRELESS ASSOCIATION, STEPHEN D
   HILL AA0GC SECY, 13 MORNINGSIDE DR., PAOLA, KS
   66071
WICHITA ARC, INC., LEO SPENCER KG0BZ SECY, 853 N
   WESTLINK AVE., WICHITA, KS 67212

KENTUCKY
AMATEUR RADIO TRANSMITTING SOC, VERNON C
   NUNN N4UL PRES, PO BOX 22917, LOUISVILLE, KY
   40252
BLUEGRASS ARS, GREGORY A CROSS WA8FJK PRES
   PO BOX 4411, LEXINGTON, KY 40544
BULLITT ARS, GEORGE A LEAVER JR WA4AGH
   1020 LARKSPUR AVE., LOUISVILLE, KY 40213
CAPITAL ARS, MARIE RASSENFOSS KA4TFC SCTR
   PO BOX 3066, FRANKFORT, KY 40603
DERBY CITY DX ASSOCIATION, JAMES C VAUGHAN
   K4TXJ PRES, 5504 DATURA LANE, LOUISVILLE, KY
   40258
EASTERN ARS, TERRY L HOLMAN AC4PY SECY
   156 NORTON DR., RICHMOND, KY 40475
GREATER MASON COUNTY ARA, ROBERT A CLARKE
   KO4GM PRES, PO BOX 73, MAYSVILLE, KY 41056
HENDERSON ARC, RICK POWELL KC4MRP PRES
   PO BOX 2058, HENDERSON, KY 42420
KENTUCKY COLONEL ARC, WARREN T HARRIS W4PKX
   362 CEDAR RIDGE RD, BOWLING GREEN, KY 42101
KENTUCKY CONTEST GROUP, TIM TOTTEN KJ4VH PRES
   10400 BROAD RUN RD, LOUISVILLE, KY 40299
KENTUCKY MOUNTAINS ARC, JOHN W FARLER K4AVX
   SCTR, 109 HALL ST, HAZARD, KY 41701
KOSAIR ARC, JAMES C VAUGHAN K4TXJ LIAS
   5504 DATURA LN, LOUISVILLE, KY 40258
LAUREL ARS-KY, STANLEY BAKER KD4ERF SECY
   112 PIPER WAY, CORBIN, KY 40701
LINCOLN TRAIL ARC INC., WALT BOWMAN WD4RAK
   LIAS, 336 SYCAMORE DR., RADCLIFF, KY 40160
MAMMOTH CAVE AMATEUR RADIO CL, JOE TAYLOR
   N4NAS SCTY, PO BOX 1852, GLASGOW, KY 42142
MARSHALL COUNTY ARA, JOYJCE SHEMWELL N4KKW
   PRES, 1486 PHELPS RD, BENTON, KY 42025
MOUNTAIN ARC, ANDREW PITT WB8WEZ SCTR
   PO BOX 2164, MIDDLESBORO, KY 40965
MURRAY STATE UNIV ARC, WILLIAM L CALL KJ4W
   ADVIS, P O. BOX 2580, UNIVERSITY STATION, MUR
   RAY, KY 42071
NORTHERN KENTUCKY ARC, BOB ILLIG WA4YVW PRES
   2312 SUMMERSET CIR, FLORENCE, KY 41042
OWENSBORO AMATEUR RADIO CLUB, WIN SCAR
   BROUGH WA4MXD EDITOR, 3417 S GRIFFITH AVE.,
   OWENSBORO, KY 42301
PADUCAH ARC, CRAIG MARTINDALE WA4WBU PRES
   2509 TRIMBLE ST, PADUCAH, KY 42001
RIVER CITIES ARA INC., PAUL ADAMS KC4ZSV PRES
   PO BOX 612, ASHLAND, KY 41105
WOODFORD ARS, JEROME E MUELLER KC4WZO SECY
   PO BOX 734, VERSAILLES, KY 40383

LOUISIANA

ACADIANA ARA INC., JAMES DUGAL N5KNX VPRS
PO BOX 51174, LAFAYETTE, LA 70505
ACADIANA DX ASSOCIATION, ROY KELLER NG5X SCTR
412 CHEYENNE CIR, SCOTT, LA 70583
ARK-LA-TEX RADIO ASSOCIATION, WINFRED TEETSON
WN5YTR PRES, 1914 FRANKLIN RD, HEFLIN, LA 71039
ARKLA ARA, LOUISE EDWARDS N5NSX SCTR
606 5TH ST NE, SPRINGHILL, LA 71075
ASCENSION ARC INC., WAYNE RUSSELL SR N5VWM
SECY, 40390 SYCAMORE AVE., GONZALES, LA 70737
BATON ROUGE ARC, NORMA F RAMEY WD5GFD SECY
PO BOX 68, GREENWELL SPRINGS, LA 70739
CATHOLIC HIGH ARC, JAY HARMON KB5YZH
855 HEARTHSTONE DR., BATON ROUGE, LA 70806
CENTRAL LOUISIANA ARC, JON KISSICK KB5NOB TREAS
9223 HWY SOUTH, ALEXANDRIA, LA 71302
DELTA DX ASSOCIATION, FRANK SICURO, W5UP PRES
5136 RANDOLPH ST, MARRERO, LA 70072
FIST & MOUTH CONTEST COMPANY, WILLIAM B BLAN
TON WC5N PRES, 202 TWIN OAKS DR., WEST MONROE,
LA 71291
IBERIA ARC, CHARLES V LANZA K5PR PRES
804 VICTORY DR., NEW IBERIA, LA 70560
JEFFERSON ARC INC., MARK UNLAND WB9VTN PRES
PO BOX 73665, METAIRIE, LA 70033
JONESBORO AREA AR OPERATORS, QUINCY HEMPHILL
N5ZWO PRES, RR 1 BOX 152, DODSON, LA 71422
LOUISIANA COUNCIL OF ARCS INC., PO BOX 182,
ROSEDALE, LA 70772
MINDEN ARC, JAMES D WHITE KB5SUE PRES
RR 1 BOX 590, DUBBERLY, LA 71024
MOREHOUSE ARC, RAYMOND HALEY N5YEE PRES
9544 CUTOFF RD, BASTROP, LA 71220
OPELOUSAS AREA ARC, CLAUDIA BERNARD N5ZBT
SCTR, 304 TORONTO DR., LAFAYETTE, LA 70507
OZONE ARC, G VERNON HENNLEIN N5PEV TREAS
203 SCOTT DR., SLIDELL, LA 70458
RADIO AMATEUR SERVICE CLUB, SHELTON MC ANELLY
KD5SL PRES, PO BOX 4393, BATON ROUGE, LA 70821
SHREVEPORT ARA, BRIAN LEWIS N5OCD TREAS
PO BOX 37632, SHREVEPORT, LA 71133
SOUTHEAST LOUISIANA ARC, J ERNEST BUSH N5NIB
PRES, PO BOX 1324, HAMMOND, LA 70404
SOUTHEASTERN LA UNIV ARC, RALPH W SHAW JR.
K5CAV TRUS, BOX 402, SLU, HAMMOND, LA 70402
SPRINGHILL ARC, JAMES CHEATHAM N5NLX SECY
RR 1 BOX 485, SAREPTA, LA 71071
UNITED RADIO AMATEUR CLUB, LLOYD D JACKSON
KA5DDA TRUSTEE, P O BOX 190, 508 REA ST, MANS
FIELD, LA 71052
WINN ARC, QUINCY HEMPHILL N5ZWO PRES
PO BOX 239, JOYCE, LA 71440

MAINE

AUGUSTA ARA, ARNOLD SMITH, KA1LPW, PRES
RR 1 BOX 475, AUGUSTA, ME 04330
BAGLEY ARC, JOSEPHINE LOUPIN KA1ZAV SECY
14 WASHINGTON ST, LINCOLN, ME 04457
BEAN SCHOOL ARC, PHILIP DOWNES N1IFP
BEAN SCHOOL, RFD 3, AUGUSTA, ME 04330
ELLSWORTH AMATEUR WIRELESS ASS, WILLIAM C
TOWNSEND SECY, 12 SPRING ST., BAR HARBOR, ME
04609
MAINE MARITIME ACADEMY ARC, JOHN F HACKNEY
WA4SWM PRES, MAINE MARITIME ACADEMY, BOX E-
15, CASTINE, ME 04420
MERRYMEETING ARA, ROBERT ROWLAND KA1IQA SECY
PO BOX 115, BOWDOINHAM, ME 04008
MID-COAST A R REPEATER CLUB, DAVID A HAWKE
KQ1L TRUSTEE, 198 CONY STREET EXT, AUGUSTA, ME
04330
PEN BAY ARC, INC., EDWARD C ROTCH KC1CG
PO BOX 255, WASHINGTON, ME 04574

PINE STATE ARC, ROGER DOLE KA1TKS PRES
PO BOX 6304, BANGOR, ME 04402
PORTLAND AMATEUR WIRELESS ASSN, RONALD
LEVERE KA1FI SECY, PO BOX 1605, PORTLAND, ME
04096
SANDY RIVER ARC, JOHN A MESSEDER JR N1HKZ PRES
PO BOX 114, NEW VINEYARD, ME 04956
WALDO COUNTY ARA, LIONEL L MERRILL KB1OF SECY
BOX 298, TURNPIKE RD, SEARSPORT, ME 04974
YANKEE RADIO CLUB INC., BERNARD R LANGLEY
W1EZR SECY, RR 1 BOX 575-17, POLAND, ME 04273
YARMOUTH JR.SR. HIGH ARC, BOB MAURAIS KC1IV
ADVISOR, 46 WILLIAM KNIGHT RD, WINDHAM, ME
04062
YARMOUTH RADIO CLUB INC., JEFF WEINSTEIN K1JW
PRES, PO BOX 373, YARMOUTH, ME 04096

MARYLAND

ANNE ARUNDEL RC, ROBERT ISRAEL WC3I PRES
PO BOX 308, DAVIDSONVILLE, MD 21035
ANTIETAM RADIO ASSOCIATION, ELLEN HAMMONS
N3MZA SECY, 1745 EDGEWOOD HILL DR APT 101,
HAGERSTOWN, MD 21740
APPALACHIAN RADIO AMATEUR ASSN, GREG LATTA
AA8V PRES, PO BOX 635, FROSTBURG, MD 21532
APPLIED PHYSICS LAB ARC, JOHN GRIMES PRES,
APPLIED PHYSICS LAB, JOHNS HOPKINS ROAD, LAUREL,
MD 21073
ARINC ARC, WARREN R OSTERLOH W0FRS SECY
MAIL STOP 4-232, 2551 RIVA RD, ANNAPOLIS, MD
21401
BALTIMORE AMATEUR RADIO CLUB, WILLIAM A DOB
SON III WA3ZER, 482 SHIRLEY MANOR RD, REISTER
STOWN, MD 21136
BALTIMORE RADIO AMATEUR TV SO., MAYER D ZIM
MERMAN W3GXK SECY, PO BOX 5915, BALTIMORE,
MD 21208
BAY AREA ARS, ALAN RAICHEL N3IKI SECY
263 GINA CT., PASADENA, MD 21122
BOWIE HIGH ARC, THOMAS L DOVE K3ORC ADVISOR
BOWIE HIGH SCHOOL, 15200 ANNAPOLIS RD, BOWIE,
MD 20715
BUREAU RADIO AMATEUR SIGNAL SO., JAMES C OWEN
III K4CGY PRES, NIST BLDG 225/B360, GAITHERS
BURG, MD 20899
CAPITOL HILL ARS, GEORGE R STEPHENS WB3DAC
STAR ROUTE BOX 30, AVENUE, MD 20609
CARROLL COUNTY ARC, DAVID V FURMAN KG6TU SECY
PO BOX 2099, WESTMINSTER, MD 21158
CHESAPEAKE BAY RADIO ASSN, LOUIS J BUSBY KK3D
PRES, PO BOX 222, ABERDEEN, MD 21001
COLUMBIA ARA, KEN WILLIAMS N3CCI PRES
3321 ROSCOMMON DR., GLENELG, MD 21737
EASTON ARS, O BROOK TODD K3PEN SECY
PO BOX 311, EASTON, MD 21601
FREDERICK ARC, NORVAL G KENNEDY N3JXC PRES
PO BOX 1260, FREDERICK, MD 21702
FREE STATE ARC, JOHN S V WEISS KD3YU SCTR
PO BOX 19, ANNAPOLIS JUNCTION, MD 20701
GODDARD ARC, JAMES H BLACKWELL JR N3KWU
8507 GRUBB RD., SILVER SPRING, MD 20910
GORETTI HIGH SCHOOL ARC, CHRIS MOLESKIE KA3WVC
PRES, 20102 CHERRY HILL DR., HAGERSTOWN, MD
21742
GREEN MOUNTAIN REPEATER ASSN., JOHN NUNEMAK
ER KD3VR PRES, PO BOX 429, RIVERDALE, MD 20738
KENT ARS, CINDY S HAMM KA3TLF SECY
5829 MAIN ST, ROCK HALL, MD 21661
LAUREL ARC-MD, JIM CROSS WI3N PRES
PO BOX 3039, LAUREL, MD 20709
LEISURE WORLD ARC, ROBERT BECK KA3IHQ PRES
3400 GLENEAGLES DR., SILVER SPRING, MD 20906
MARYLAND MOBILEERS ARC INC., ELISA COLBURN
N3PCR SECY, PO BOX 935, SEVERN, MD 21144

MD APPLE DUMPLING RADIO AMATEUR, LARRY ZIECHECK N3GKZ PRES, PO BOX 2468, WHEATON, MD 20915

METROVISION INC., RUTH E PHILLIPS K3AGR LIAS 2901 ACCOKEEK RD W., ACCOKEEK, MD 20607

MONTGOMERY ARC, ALBERT RABASSA NW2M PRES 17100 AMITY DR., ROCKVILLE, MD 20855

MONTGOMERY AUX COMM SERVICE, THOMAS E MINDTE WB3DIO VPRS, PO BOX 4243, ROCKVILLE, MD 20849

MOOSE ARC, TOM MOLYNEAUX KE3GK PRES 266 HAMMERLEE RD, GLEN BURNIE, MD 21060

MOUNTAIN ARC, SHANNON PETERSON KB8JTQ PRES PO BOX 234, CUMBERLAND, MD 21501

POTOMAC AREA VHF SOCIETY, PAUL H ROSE WA3NZL SCTR, 25116 OAK DR., DAMASCUS, MD 20872

POTOMAC VALLEY RADIO CLUB, FRED LAUN K3ZO EDI TOR, 5801 HUNTLAND RD, TEMPLE HILLS, MD 20748

PRINCE GEORGES WIRELESS, E FRANKLIN HOLTON WA3ZMW
4109 KENNEDY ST., HYATTSVILLE, MD 20781

ROCK CREEK ARA, JOHN C HALWEG N3ETD PRES 2611 NEWTON ST., SILVER SPRING, MD 20902

SOMERSET COUNTY ARC, BETTY BALL AK8O SECY. PO BOX 417, GRANTSVILLE, MD 21536

SOUTHERN MARYLAND ARC, ALLEN STEVENSON KA3ZPA PRES
5289 W BONIWOOD TURN, CLINTON, MD 20735

SOUTHERN PATUXENT ARC, DWAYNE KINCAID WD8OYG PRES, 1445 PARRAN RD, SAINT LEONARD, MD 20685

UNIVERSITY OF MARYLAND ARA, BENJAMIN SCHULTZ N3RUR PRES, UNIVERSITY OF MARYLAND, P O BOX 88, COLLEGE PARK, MD 20742

US NAVAL ACADEMY, ANDREW A PARKER WV1B C/O SATELLITE EARTH STA FACILITY, 590 HOL LOWAY RD, ANNAPOLIS, MD 21402

WORCESTER CAREER & TECH CTR, WILLIAM HAMMOND JR. N3IOD, 6268 WORCESTER HWY, NEWARK, MD 21841

MASSACHUSETTS

ACTON-BOXBORO ARC, BILL CHURCHILL AA1O SECY PO BOX 20, CARLISLE, MA 01741

ALGONQUIN ARC, WALTER ZILONIS KC1XE PRES PO BOX 258, MARLBOROUGH, MA 01752

BARNSTABLE RADIO CLUB, CHARLES E KITSON W1DPN TREAS, 22 POPLAR DR., OSTERVILLE, MA 02655

BILLERICA ARS, MARY BETH CORKUM N1FER PRES 7 ELM ST #4, ACTON, MA 01720

BOSTON ARC, G JOHN GARRETT WN9T PRES BOX 15585, BOSTON, MA 02215

CAPEWAY RADIO CLUB OF MA, RAYMOND WITT WA1OWQ C/SEC, 62 CALDWELL ST, NORTH WEY MOUTH, MA 02191

CENTRAL MASSACHUSETTS ARA INC., HOWARD MOUL TON KB1ARQ PRES, 10 VIEW ST, WORCESTER, MA 01610

COLONIAL WIRELESS, WILLIAM R WEISS W1QGL PRES 16 ESTABROOK RD, LEXINGTON, MA 02173

COUNCIL OF EASTERN MA ARC'S, ELAINE CHASE N1GTB PO BOX 3773, NATICK, MA 01760

EAST COAST AMATEUR TELEVISON, DAVID CRAIG WB1CEA TREAS, PO BOX 1219, RANDOLPH, MA 02368

FALMOUTH ARA INC., JOANNE REID N1LNE SECY 29 HIAWATHA ST, TEATICKET, MA 02536

FEDERATION OF EASTERN MA ARA, EUGENE H HAST INGS W1VRK SCTR, 18 CHURCHILL RD, MARBLEHEAD, MA 01945

FRAMINGHAM AMATEUR RADIO ASSN, LEW NYMAN K1AZE, SAXONVILLE STATION, P O BOX 3005, FRAM INGHAM, MA 01701

FRANKLIN COUNTY REPEATER CLUB, BERT PHILLIPS N1IUM PRES, PO BOX 3172, GREENFIELD, MA 01302

GENESIS ARS INC., RICHARD EATON KA1VVS VPRS 549 ELLISVILLE DR., PLYMOUTH, MA 02360

GREATER LAWRENCE AR FELLOWSHIP, ROBERT H SMITH KA1VF PRES, 128 WINONA AVE., HAVERHILL, MA 01830

GREEN MEADOW SCHOOL ARC, JUDITH JOHNSON KA1WZM PRES, GREEN MEADOW SCHOOL, GREAT RD, MAYNARD, MA 01754

HAMPDEN CO RADIO ASSN., INC., JAMES SEBOLT N1DUY SECY, PO BOX 482, WEST SPRINGFIELD, MA 01090

HARVARD WIRELESS, JOE SHELTON N5NKY PRES 6 LINDEN ST., CAMBRIDGE, MA 02138

HEWLETT-PACKARD ANDOVER RC, DOUG YATES KB1APT SECY, HEWLETT PACKARD CO., 3000 MINUTE MAN RD, ANDOVER, MA 01810

MARCONI RADIO CLUB, ROBERT J DOHERTY K1VV PRES 153 COUNTY ST, LAKEVILLE, MA 02347

MASHPEE ARA, PHILIP H CHOATE N1GLC TREAS 46 PRINCE HENRY DR., EAST FALMOUTH, MA 02536

MASSASOIT ARA INC., CARL AVENI N1FYZ PRES PO BOX 428, BRIDGEWATER, MA 02324

MAYFLOWER AMATEUR RADIO CLUB, ROBERT ALLAN WF1M, 17 PLEASANT ST., PLYMOUTH, MA 02360

MIDDLESEX ARC, LAWRENCE OBER KC1VS PRES 51 SCHOOL ST., ACTON, MA 01720

MINUTEMAN REPEATER ASSN, FRANK P MORRISON KB1FZ SECY, PO BOX 2282, LEXINGTON, MA 02173

MOHAWK ARC, WILLIAM CURTIS WJ1Y PRES PO BOX 532, ATHOL, MA 01331

MONTACHUSETT ARA, RON COOK KA1VPK SECY PO BOX 95, LEOMINSTER, MA 01453

MOUNT TOM AMATEUR REPEATER ASN, CHRISTINE BERGERON N1JSK SECY, PO BOX 3494, SPRINGFIELD, MA 01101

MYSTIC VALLEY AMATEUR GROUP, NICHOLAS MAGLIANO KC1MA PRES, 12 KILSYTH RD, MEDFORD, MA 02155

NASHOBA VALLEY ARC, STANLEY POZERSKI KD1LE PRES, PO BOX 900, PEPPERELL, MA 01463

NATICK HIGH SCHOOL RADIO CLUB, GERALD E ASH N1DGC ADVIS, 15 WEST ST., NATICK HIGH SCHOOL, NATICK, MA 01760

NE PROFESSIONAL BROADCASTERS, PETER W KODIS N1EXA PRES, PO BOX 79187, NORTH DARTMOUTH, MA 02747

NORFOLK COUNTY RADIO ASSN, RICHARD C SIMONSON KA1INO, 15 HIGHVIEW AVE., W ROXBURY, MA 02132

NORTH SHORE REPEATER ASSN, KENNETH SMITH AA1DR PRES, PO BOX 3724, PEABODY, MA 01961

NORTHEASTERN UNIV RC, CHRISTOPHER E PERKINS N1LSX, 255 ELL CENTER, 360 HUNTINGTON AVE., BOSTON, MA 02115

NORTHERN BERKSHIRE ARC, CHUCK LOWERY NZ1Z PRES, P O BOX 2097, PITTSFIELD, MA 01202

NORWOOD ARC, EDWARD V LAJOIE K1CB 20 HEMLOCK ST., NORWOOD, MA 02062

PATRIOT DX ASSOCIATION, TONY PENTA WA1MWN PRES, 66 PLEASANT AVE., LYNNFIELD, MA 01940

PILGRIM ARC, EARLE H CHADDOCK WA1KZT SCTR PO BOX 947, PROVINCETOWN, MA 02657

POLICE AR TEAM OF WESTFORD MA, TERRY M STADER KA8SCP PRES, 8 CHRISTOPHER RD, WESTFORD, MA 01886

QUABOAG VALLEY ARC, LARRY SMITH N1BDH PRES 130 SCANTIC RD, HAMPDEN, MA 01036

QUANNAPOWITT RADIO ASSN., E W BILL ARNOLD JR KA1CNU, 6 VILLAGE LN, HAVERHILL, MA 01832

RADIO OP. ASSN OF NEW BEDFORD, HENRY OLDEN NW1W PRES, 20 PHOENIX ST., FAIRHAVEN, MA 02719

SOUTH SHORE ARC, DONALD B ANDERSON W1KPX SECY 6 MAY AVE., BRAINTREE, MA 02184

SOUTHEASTERN MASS ARA, PETER J CARREIRO KA1WOJ PRES, PO BOX 80007, SOUTH DARTMOUTH, MA 02748

STURDY MEMORIAL HOSPITAL ARC, RAY CORD K2TGX SECY, 316 S WORCESTER ST., NORTON, MA 02766

THREE AMIGOS RADIO ASSN, KEVIN LYNCH KQ1N PRES
GENERAL DELIVERY, WHEELWRIGHT, MA 01094

TRI-COUNTY VOCATIONAL H.S. ARC, DONALD B COE
N1LRL PRES, 147 POND ST., FRANKLIN, MA 02038

WALTHAM ARA, ANDREW DONOVAN WA1GEP TREAS
PO BOX 411, WALTHAM, MA 02254

WAREHAM ARC, BARRY S KENNEDY N1EZH SECY
24 BUNGALOW LN, RR #3, BUZZARDS BAY, MA 02532

WELLESLEY ARS, ARTHUR W ANDERSON NT1M PRES
200 H LINDEN ST, WELLESLEY, MA 02181

WHITMAN ARC, WILLIAM F HAYDEN N1FRE SECY
118 WASHINGTON ST., WHITMAN, MA 02382

YANKEE CLIPPER CONTEST CLUB, CHARLOTTE L
RICHARDSON KQ1F SC, 11 MICHIGAN DR., HUDSON,
MA 01749

MICHIGAN

ADRIAN ARC, TOM PARSONS N8QEW PRES
4436 EVERGREEN DR., ADRIAN, MI 49221

ALLEGAN COUNTY ARC, DON MINER N8IGO PRES
PO BOX 21, ALLEGAN, MI 49010

AMATEUR RADIO & YOUTH, DEBRA ALLEN N8TXA SECY
PO BOX 7136, FLINT, MI 48507

AMERICAN RED CROSS AR SERVICE, DARRYL SMITH
KA8UKO PRES, 1818 GRISWOLD ST., PORT HURON, MI
48060

ARA OF HANSON HILLS, MELVIN MOORE NM8L SECY
RR 5 BOX 5084, GRAYLING, MI 49738

ARROW COMMUNICATIONS ASSOCIATION, MARK
O'BRIEN N8PQJ PRES, PO BOX 1572, ANN ARBOR, MI
48106

AUSABLE VALLEY ARC, GERRY VOLZ WT8G PRES
510 KNEPP RD, FAIRVIEW, MI 48621

BAY AREA ARC, BOB CUTHBERT N8BBR EDIT
1122 16TH ST, BAY CITY, MI 48708

BIG RAPIDS AREA ARC, WALTER R NEVILL N8HDD SECY
PO BOX 41, RODNEY, MI 49342

BLOSSOMLAND ARA, DUANE L DURFLINGER KX8D
1051 MAIN ST, SAINT JOSEPH, MI 49085

BRANCH COUNTY ARC, LARRY CAMP WB8R PRES
774 CENTRAL RD, QUINCY, MI 49082

CASCADES ARS, TERRY OSBORN KD8B PRES
508 DALTON RD, JACKSON, MI 49203

CATALPA ARS, PAUL HATFIELD KE8UD PRES
PO BOX 721422, BERKLEY, MI 48072

CENTRAL MICHIGAN ARC, SCOTT ROWE KB8HTM SECY
2613 MONTEGO DR., LANSING, MI 48912

CHAIN O LAKES ARC, MARTI ROETH N8MKY SECY
504 W UPRIGHT ST, CHARLEVOIX, MI 49720

CHELSEA COMMUNICATIONS CLUB, MARJORIE BIES
N8UNK SECY, 104 E MIDDLE ST, CHELSEA, MI 48118

CHERRYLAND ARC, WARD KUHN N8UVD TREAS
PO BOX 987, TRAVERSE CITY, MI 49685

CHIPPEWA HILLS HS ARC, FRANK E BENN WD8LUV SCTR
6445 W AIRLINE RD, WEIDMAN, MI 48893

COPPER COUNTRY RADIO AMATEUR A, RICHARD E
BROOKS KB8FCX, PO BOX 217, DOLLAR BAY, MI 49922

CRAWFORD-ROSCOMMON ARC, EDWARD H BASSETT
N8FVM PRES, 11734 MARBER DR., ROSCOMMON, MI
48653

DELTA COUNTY ARS, RANDY PARR N8CKT PRES
4491 26TH RD, GLADSTONE, MI 49837

DETROIT MIKE & KEY CLUB, DAVID W WHEELER
KB8OLF PRES, 11401 GRANDVILLE AVE., DETROIT, MI
48228

EASTERN MICHIGAN ARC, BRUCE CAMPBELL WA8OJR
SECY, 3126 GRATIOT AVE., PORT HURON, MI 48060

EATON COUNTY ARC, ROBERT C BACHMAN KG8HQ
SECY, 5775 N CLINTON TRL, CHARLOTTE, MI 48813

EDISON RADIO AMATEUR'S ASSN, MICHAEL T
KORALEWSKI K8VA, 4726 JACKSON ST., TRENTON, MI
48183

FARMINGTON ARC, RAY KASUBOSKY K8RAY PRES
23404 LARKSHIRE ST, FARMINGTON HILLS, MI 48336

FENTON AREA ARA, TOM PARSHALL KB8PBJ PRES
502 SHERMAN, HOLLY, MI 48442

FORD AR LEAGUE TIN LIZZY CLUB, CHARLES THOMAS
K3SGS PRES, 16745 SUNDERLAND RD, DETROIT, MI
48219

FOUR FLAGS ARC, JOHN M SCHLAMERSDORF WD8IZG
66788 CONRAD RD, EDWARDSBURG, MI 49112

GARDEN CITY ARC, TIM GOVIE KE8XY EDITOR
7295 ROSEMONT AVE., DETROIT, MI 48228

GENERAL MOTORS ARC, VAL BREAULT N8OEF
8101 WARREN BLVD, CENTER LINE, MI 48015

GENESEE COUNTY RC, BILL COALE KB8MBJ PRES
PO BOX 485, FLINT, MI 48501

GLADWIN AREA ARC, TOM SANDBORN KB8OTH
2500 E ROCK RD, CLARE, MI 48617

GRAND RAPIDS ARA, JEFF BELKNAP N8RWE VPRS
PO BOX 1248, GRAND RAPIDS, MI 49501

GRATIOT COUNTY ARA, NORMAN B KEON WA8AEG
10732 RIVERSIDE DR., SAINT LOUIS, MI 48880

GREATER LANSING DX GROUP, KENNETH KRUGER
WB8AAX SCTR, 1539 OHIO AVE., LANSING, MI 48906

HAZEL PARK ARC, DON STOCKTON AA8EG PRES
PO BOX 368, HAZEL PARK, MI 48030

HENRY FORD COMMUNITY COLLEGE, MICHAEL RUDZKI
KF8BE PRES, PATTERSON TECH BLDG, 5101 EVER
GREEN RD, DEARBORN, MI 48128

HIAWATHA ARC, RONALD S SIHTALA KO8U TREAS
PO BOX 1183, MARQUETTE, MI 49855

HUSKY ARC OF MICHIGAN TECH, DAVID GILLAHAN
KB8POR PRES, 1802C WOODMAR, HOUGHTON, MI
49931

KALAMAZOO ARC, JAMES H GORKA N8MCF PRES
7914 FOXWOOD, RICHLAND, MI 49083

KEY-CLICKERS ARC, DEBORAH MACAULEY N8DPH PRES
826 S PUTNAM ST, WILLIAMSTON, MI 48895

L'ANSE CREUSE ARC, ALLAN C KOCH KA8JJN LIAS
23682 KIM DR., CLINTON TOWNSHIP, MI 48035

LAPEER COUNTY ARA, WARD MCGINNIS KB8QDI SECY
112 N COURT ST, LAPEER, MI 48446

LIVINGSTON AMATEUR RADIO KLUB, PETER ROBAK
KF8YC SECY, 3328 CEMETARY RD, FOWLERVILLE, MI
48836

LIVINGSTON COUNTY ARES, JOSEPH E GALIPEAU JR.
WA1LRL, 11200 YOUNG DR., BRIGHTON, MI 48116

LIVONIA ARC, JAMES HARVEY WB8NBS SECY
P O BOX 8404, LIVONIA, MI 48151

LOWELL AMATEUR RADIO YOUTH, ALLAN L ECKMAN
KD8FU, PO BOX 5, LOWELL, MI 49331

LOWELL ARC, JACK AMELAR NY8D PRES
1554 LINCOLN LAKE AVE N., LOWELL, MI 49331

MAD RIVER RC, DAVE PRUETT K8CC PRES
2727 N HARRIS RD, YPSILANTI, MI 48198

MASON COUNTY RADIO CLUB, JOSEPH A JURKOWSKI
W2VGW SCTR, PO BOX 426, LUDINGTON, MI 49431

MICHIGAN A R ALLIANCE, RICHARDM RANTA WB8HJX
PRES, 812 GRACELAND NE, GRAND RAPIDS, MI 49505

MICHIGAN QRP CLUB, LOWELL D CORBIN KD8FR PRES
3315 SHEFFER AVE., LANSING, MI 48906

MICHIGAN STATE UNIV. ARC, VANCE J STRINGHAM
N8UGV, 1617 F SPARTAN VILLAGE, E. LANSING, MI
48823

MIDLAND ARC, RALEIGH L WERT W8QOI
309 E GORDONVILLE RD, MIDLAND, MI 48640

MONROE COUNTY RADIO COMM ASSN, JOHN FUNK
N8GVG PRES, PO BOX 237, MONROE, MI 48161

MONTCALM AREA ARC, LARRY ALMAN WA8QCW
PO BOX 312, GREENVILLE, MI 48838

MOTOR CITY RC INC., ANNE TRAVIS KB8HGM SECY
PO BOX 337, WYANDOTTE, MI 48192

MUSKEGON AREA AMATEUR RADIO CO., BRADLEY R
KING N8PFC PRES, PO BOX 691, MUSKEGON, MI 49443

NORTH OTTAWA ARC, BOB VAN RHEE N8LAS PRES
PO BOX 44, FERRYSBURG, MI 49409

NOVI AMATEUR RADIO CLUB, MIKE HEWITT N8TDD
PRES, 47266 CIDER MILL DR., NOVI, MI 48374

OAK PARK AMATEUR RADIO CLUB, DIANE PENN N8KBA
  SECY, 28771 FAIRFAX ST, SOUTHFIELD, MI 48076
OAKLAND COUNTY ARS, HARRY BEDARD W8RRE PRES
  P O BOX 431-244, PONTIAC, MI 48343
RADIO ACTIVE COMM CL OF SE MI, DAVE LASKOWSKI
  N8TMN, 1767 HEMPSTEAD DR., TROY, MI 48083
SHIAWASSEE ARA, NORMAN HIGGINS KA8YVT SECY.
  5550 CHURCH RD, PERRY, MI 48872
SOUTH MICHIGAN DX ASSN, ROBERT MACAULEY
  WB8G PRES, 826 S PUTNAM, WILLIAMSTON, MI 48895
SOUTHERN MICHIGAN ARS, JAMES R HOLLOWAY
  KB8JGD SECY, 30 E LANGELY RD, BATTLE CREEK, MI
  49015
STRAITS AREA RC, HARRY E LEIBER JR. N8OIV PRES
  5300 PICKEREL LAKE RD, PETOSKEY, MI 49770
STU ROCKAFELLOW ARS, DICK MARKS W8UTB PRES
  45075 N TERRITORIAL RD, PLYMOUTH, MI 48170
THUMB ARC, ELWYN B RAYSIN W8IDT SCTR
  105 E SPEAKER ST, SANDUSKY, MI 48471
TOP OF MICHIGAN ARC, RENEE MENCH N8VKS SECY
  4823 MURNER ROAD, PO BOX 930, GAYLORD, MI 49735
TUSCOLA COUNTY ARC, DONALD STUART N8RZQ SECY
  159 MILLWOOD ST, CARO, MI 48723
UNIVERSITY OF MICHIGAN ARC, ADAM THODEY
  KB6NMI PRES, 4021 MICHIGAN UNION, ANN ARBOR,
  MI 48109
UTICA SHELBY EMERG COMM ASSN, DAVID C MARTIN
  KF5CT, 37342 MARION DR., STERLING HEIGHTS, MI
  48312
WEXAUKEE AMATEUR RADIO ASSN, DAN SCHMIDT
  KE8KU SECY, PO BOX 163, CADILLAC, MI 49601

MINNESOTA

ALBERT LEA ARC, DEXTER HENSCHEL W0DH TRUSTEE
  1806 GRAND, ALBERT LEA, MN 56007
AMATEUR RADIO ASSN OF BLOOMINGTON, BILLY C
  STONE N0HEV PRES, PO BOX 20174, BLOOMINGTON,
  MN 55420
ARROWHEAD RADIO AMATEURS CLUB, DOUGLAS NEL
  SON AA0AW PRES, PO BOX 7164, DULUTH, MN 55807
BOUNDARY WATERS ARC, OSCAR E SANDEN K4EH
  PRES, HCR 1 BOX 485, GRAND MARAIS, MN 55604
BRAINERD AREA ARC, CRAIG BUCHHOLZ N0SRU TREAS
  732 PINE BEACH RD W., BRAINERD, MN 56401
CARLETON COLLEGE ARC, MARK K DIETERICH N2PGD
  PRES, CARLETON COLLEGE, NORTHFIELD, MN 55057
COLLEGE CITY RADIO CLUB, TIM ISOM WD9HDQ PRES
  1021 MAPLE ST., NORTHFIELD, MN 55057
COURAGE CENTER, ATTN: PAT TICE WA0TDA
  3915 GOLDEN VALLEY RD, GOLDEN VALLEY, MN
  55422, LAKE REGION ARC, DR BILL MORGAN AA0AX
  SECY, RT 1 BOX 46, BATTLE LAKE, MN 56515
MANKATO AREA RC INC., KEN E BERG KB0LPC PRES
  PO BOX 1961, MANKATO, MN 56002
MARSHALL ARC, SHERLEE A GREGG WD0BZU PRES
  1007 PINE ST., MARSHALL, MN 56258
MESABI WIRELESS ASSOCIATION, GORDON DAHL
  N0RPX SECY, 604 9TH ST S., VIRGINIA, MN 55792
MINNEAPOLIS RADIO CLUB, STEVEN FILEK N0OWL
  PRES, PO BOX 583281, MINNEAPOLIS, MN 55458
MINNETONKA MINNESOTA ARC, JOHN A ROBERTSEN
  KA0OSC LIAS, 17273 HAMPTON CT, MINNETONKA, MN
  55345
NEW ULM ARC, DAVE MATTKE N0XPM SECY
  P O BOX 851, NEW ULM, MN 56073
NORTHERN LAKES ARC (NLARC), ROBERT ROSS
  KA0FME SCTR, PO BOX 525, GRAND RAPIDS, MN 55744
NORTHERN LIGHTS RADIO SOCIETY, RICH WESTERBERG
  N0HJZ PRES, 17500 CHERRY DR., EDEN PRAIRIE, MN
  55346
PAYNESVILLE ARA, ALLEN A HERTZBERG WD0DEH
  PRES, 18114 263RD AVE., PAYNESVILLE, MN 56362
ROBBINSDALE ARC, DENNIS ANDERSON K0YVZ PRES
  PO BOX 22613, ROBBINSDALE, MN 55422

ROCHESTER ARC, MARK BAILEY N0QZL PRES
  PO BOX 1, ROCHESTER, MN 55903
RUNESTONE RADIO CLUB, NORMAN D YEUTTER W0KCJ
  2301 S LE HOMME DIEU DR NE, ALEXANDRIA, MN
  56308
SOUTH EAST METRO ARC, DAVID HARRELL N0IPN PRES
  8772 GREYSTONE AVE S. COTTAGE GROVE, MN 55016
SOUTHWEST METRO AR TRANSMITTIN, AUDREY ZELL
  MAN N0OKX SECY, PO BOX 144, CHASKA, MN 55318
ST CLOUD RADIO CLUB, CHUCK SMITH KB0IQE PRES
  21914 OAK HEIGHTS CIR, COLD SPRING, MN 56320
ST CLOUD STATE UNIV ARC, MATT HOLDEN N0QFM
  PRES, ST CLOUD UNIVERSITY, 720 SOUTH 4TH AVE AC
  117, ST CLOUD, MN 56301
ST PAUL RADIO CLUB, CHARLES ESCH W0KZM SECY
  PO BOX 9375, NORTH SAINT PAUL, MN 55109
THE 33'S, ANN FOSTER N0LLC PRES
  6654 E RIVER RD, FRIDLEY, MN 55432
TRI-STATE ARC, GARY STOAKES N0IHX SCTR
  PO BOX 308, LUVERNE, MN 56156
TWIN CITIES REPEATER CLUB, STEVEN J FILEK N0OWL
  PRES, PO BOX 20274, BLOOMINGTON, MN 55420
TWIN CITY DX ASSOCIATION, HENRY KNOLL WA0GOZ
  PRES, 10081 103RD ST N., STILLWATER, MN 55082
TWIN CITY FM CLUB, JOHN K BISPALA N0EGG TREAS
  PO BOX 580555, MINNEAPOLIS, MN 55458
VIKING ARS,. WILLIAM J KNISH WB0KEK PRES
  PO BOX 3, WASECA, MN 56093
WILLMAR AREA E A R
  PO BOX 882, WILLMAR, MN 56201
WORTHINGTON ARC, ARNOLD SEXE WB0OPZ SECY
  422 GALENA ST, WORTHINGTON, MN 56187

MISSISSIPPI

BOONEVILLE ARC, KENNETH CHRISTIAN KB5PBG
  312 WASHINGTON ST., BOONEVILLE, MS 38829
HATTIESBURG ARC INC., ARTHUR C ALLEN KG5XA
  PRES, PO BOX 15025, HATTIESBURG, MS 39404
JACKSON ARC, STEVE MILLER KB5RZP EDITOR
  424 HOLLY HEDGE DR., MADISON, MS 39110
JACKSON COUNTY ARC, MARILYN HARVEY N5RPV
  SECY, PO BOX 845, GAUTIER, MS 39553
KEESLER ARC, BOB MATHIS N5MRN PRES
  81 MWRS, KESSLER AFB, MS 39534
LAUREL ARC-MS, BRUCE BARNES KB5RQL PRES
  PO BOX 6252. LAUREL, MS 39441
LOWNDES COUNTY ARC, STAN KOZLOWITZ AA5XO
  PRES, PO BOX 9291, COLUMBUS, MS 39705
MAGNOLIA DX ASSOCIATION, FLOYD GERALD N5FG
  4706 WASHINGTON AVE., GULFPORT, MS 39507
MISSISSIPPI COAST ARA, KENNETH BURTON KI5LV PRES
  PO BOX 1785, GULFPORT, MS 39502
MISSISSIPPI STATE UNIVERSITY A, CRAIG LINDSEY
  KC5AUG PRES, POB 591, MISSISSIPPI STATE, MS 39762
OLD SOUTH ARC, MELVIN RATZLAFF JR KA5BEE PRES
  PO BOX 773, GLOSTER, MS 39638
PEARL RIVER COUNTY ARC INC., HUGH RUSSELL K5MFN
  VPRS, 700 N CURRAN AVE., PICAYUNE, MS 39466
PETAL AMATEUR WIRELESS SOCIETY, BILLY RICHARD
  SON N5VEI PRES, USM BOX 6273, HATTIESBURG, MS
  39406
ST ANDREWS SCHOOL RADIO SOC., JOHN C UNDER
  WOOD JR AG5Y, 17 HIGHLAND MEADOWS DR., JACK
  SON, MS 39211
ST STANISL AUS HS ARC, EDUARDO BALDIOCEDA WJ5C,
  P O BOX 8001, 304 S BEACH BLVD, BAY SAINT LOUIS,
  MS 39520
SUMRALL ARC, MAX D BALL K5WUX PRES
  15-2ND AVE., PO BOX C, SUMRALL, MS 39482
TUPELO ARC, KEN TAYLOR KC5IX
  PO BOX 3401, TUPELO, MS 38803
VICKSBURG ARC, MELINDA LAMB N5EZX SCTR
  PO BOX 821612, VICKSBURG, MS 39182
WEST JACKSON COUNTY ARC, ERNEST E ORMAN JR
  W5OXA PRES, P O BOX 1822, OCEAN SPRINGS, MS
  39564

MISSOURI

ARC OF THE OZARKS, RICHARD B HOWE K7QCS PRES
PO BOX 369, ISABELLA, MO 65676

BLUE SPRINGS ARC, PAUL THOMSON KC0VG TRUSTEE
1501 W JEFFERSON ST., BLUE SPRINGS, MO 64015

BLUFF ARC, JERRY MOORE W0HMA SECY
2438 KATY LANE, POPLAR BLUFF, MO 63901

BOOTHEEL ARC, DAVID WILKINS WC0U SECY
RR 1 BOX 177, HORNERSVILLE, MO 63855

CALLAWAY AMATEUR RADIO LEAGUE, RICHARD C
WHITE KS0M SCTR, 1405 KENWOOD DR., FULTON, MO
65251

CENTRAL MISSOURI RADIO ASSN, WM J STROUD N0NMB
PRES, RR 6 BOX 231A, COLUMBIA, MO 65202

FLORISSANT VALLEY COMM COLLEGE, THOMAS J BING
HAM JR K9ZYW, 3400 PERSHALL RD, FERGUSON, MO
63135

FRANCIS HOWELL NORTH H.S. ARC, RON OCHU JR KO0Z
ADVISOR, 2549 HACKMANN RD, SAINT CHARLES, MO
63303

HANNIBAL ARC INC., CLIFFORD H AHRENS KI0W SECY
PO BOX 1522, HANNIBAL, MO 63401

JEFFERSON BARRACKS ARC, HARRISON B SCOTT
KA0FJA, 4121 FABIAN DR., SAINT LOUIS, MO 63125

JEFFERSON COUNTY ARC INC., JAMES J BERGER
WA0FQK TRUSTEE, 4895 BROOKS DR., HOUSE
SPRINGS, MO 63051

KANSAS CITY ARC INC., J K ENENBACH KC0WX SECY
PO BOX 30352, KANSAS CITY, MO 64112

KIMBERLING ARC, JIM DAVIS NQ0G TREAS
PO BOX 1171, KIMBERLING CITY, MO 65686

LEBANON ARC, GREG TYRE WX0E EDITOR
RT 17 BOX 81, LEBANON, MO 65536

MACON COUNTY ARC, DALE C BAGLEY NZ0S PRES
PO BOX 13, MACON, MO 63552

MACON HIGH SCHOOL ARC, DALE BAGLEY NZ0S SPON
SOR, HWY 63 NORTH, MACON, MO 63552

MARLBOROUGH ELEM SCHOOL ARC, CHARLES BRYAN
KB0CUS ADV, 1300 E 75TH ST, KANSAS CITY, MO 64131

MCDONNELL DOUGLAS ARC, J FRED PETERS WQ0Z
SECY, 3567 BOSWELL AVE., SAINT LOUIS, MO 63114

MID MO ARC, RON BRIZENDINE KA0ZIY SCTR
7209 RED BIRD LN, JEFFERSON CITY, MO 65101

MISSISSIPPI VALLEY DX CONTEST, UDO A HEINZE NI0G
VPRS, 2 WILLOUGH CIR, SAINT LOUIS, MO 63126

MISSOURI VALLEY ARC INC., WAYNE HOLMES AA0ML
PRES, PO BOX 1533, SAINT JOSEPH, MO 64502

MONSANTO ARA, ROGER H VOLK K0GOB
4773 OAKBRIER DR., SAINT LOUIS, MO 63128

MOOLAH ARC, MICHAEL ROLLA N0NEC PRES
3107 EMERALD DR., HIGH RIDGE, MO 63049

NORTHWEST AR ELECTRONICS ASSN, GORDON G
WILLARD N0OG TRUSTEE, 11846 ADMIRALTON DR.,
BRIDGETON, MO 63044

NORTHWEST ST LOUIS ARC, RICHARD L ZYSK K0GSV
SCTR, 3457 A HUMPHREY ST, ST LOUIS, MO 63118

OBP #1 ARC, JAMES R BURKHART W0BA
408 GEYER FOREST DR., KIRKWOOD, MO 63122

PHD AMATEUR RADIO ASSN INC., LYNDELL C MILLER
WA0KUH TREAS, PO BOX 11, LIBERTY, MO 64068

RAY-CLAY RADIO CLUB, ELIZABETH L MILLER N0HKH
15816 OAKMONT PL., KEARNEY, MO 64060

SACRED HEART SCHOOL RADIO CLUB, REV DAVID
NOVAK N0DN TRUSTEE, 10 ANN AVE., VALLEY PARK,
MO 63088

SEDALIA/PETTIS AR KLUB, MARK S JACKSON N0OWZ
PRES, 1121 E 15TH ST, SEDALIA, MO 65301

SOUTHSIDE ARC INC., JERRY GORRELL W0CLR PRES
PO BOX 1142, GRANDVIEW, MO 64030

SOUTHWEST MISSOURI ARC, JIM E JEFFRIES K0ICB PRES
PO BOX 11363, SPRINGFIELD, MO 65808

ST CHARLES ARC, ERIC N KOCH NF0Q LIAS
2805 WESTMINISTER DR., SAINT CHARLES, MO 63301

ST LOUIS ARC INC., DAVID P SERRA NY0S SECY
415 SPRING AVE., WEBSTER GROVES, MO 63119

ST LOUIS REPEATER INC., JAMES E WELBY WB0ZJW
PRES, PO BOX 50202, SAINT LOUIS, MO 63105

ST PETERS ARC, ALLEN J UNDERDOWN N0GOM PRES
4136 TOWERS RD, SAINT CHARLES, MO 63304

SUBURBAN RADIO CLUB, ROBERT BENNETT N0RQI PRES
12271 CREVE COEUR RIDGE, ST LOUIS, MO 63043

TELEPHONE EMPLOYEES ARC, JAMES B HARPER
WD0EWR SCTR, 515 WINDING TRAIL LN, DES PERES,
MO 63131

TRANSWORLD AIRLINES ARC, PAUL A SHAW WA0SLR
SCTR, 5624 ASH AVE., RAYTOWN, MO 64133

TRI-COUNTY ARC, MONTIE G BARCUS N0AUY PRES
PO BOX 341, MOBERLY, MO 65270

TRI-LAKES ARC, ROBERT G SIMONS N0IDF PRES
PO BOX 1234, FORSYTH, MO 65653

TRICO RADIO CLUB, JOSEPH A MUNGER JR WB0RLI SEC
RR 2 BOX 2090, SIKESTON, MO 63801

WARRENSBURG AREA ARC INC., KEITH HAYE WE0G
PRES, PO BOX 1364, WARRENSBURG, MO 64093

WASHINGTON UNIVERSITY ARC, TEFFORD REED N0WYE
PRES, 6515 WYDOWN, PO BOX 5471, SAINT LOUIS, MO
63147

MONTANA

ANACONDA ARC, KENNETH G KOPP K0PP T/SECY
PO BOX 848, ANACONDA, MT 59711

BUTTE ARC, BOB EVANS AA7LU SECY
ROUTE 1, 260 MEADOW VIEW DR., BUTTE, MT 59701

CAPITAL CITY RC, ALLEN KNUTH WA9BFL PRES
PO BOX 1112, HELENA, MT 59624

FLATHEAD VALLEY ARC, ED MAHLUM AA7TN EDITOR
300 LEISURE DR., KALISPELL, MT 59901

GALLATIN HAM RADIO CLUB, DON REGLI KI7OJ PRES
PO BOX 4381, BOZEMAN, MT 59772

GREAT FALLS AREA ARC, DARRELL THOMAS N7KOR
PRES, P O BOX 1763, GREAT FALLS, MT 59403

HELLGATE ARC, ERIC SEDGWICK KI7BR PRES
1825 ARLINGTON DR., MISSOULA, MT 59801

HI-PLAINS ARC, DENNIS PAULBECK KB7SDF
COURTHOUSE, 100 W LAUREL AVE., PLENTYWOOD,
MT 59254

HILINE RADIO CLUB INC., LLOYD STALLKAMP KA0ADZ
PRES, 817 17 ST, HAVRE, MT 59501

LOWER YELLOWSTONE AR SYSTEM, LEIF ANDERSON
N7KCH PRES, 1153 S CENTRAL AVE., SIDNEY, MT 59270

NORTHERN LIGHTS AR GROUP, RANDY J HANRAHAN
K7PGL SCTR, BOX 73, WHITETAIL, MT 59276

SOUTHEASTERN MONTANA ARC, DR JAMES ELLIOT
WB7WBA SECY, PO BOX 1030, MILES CITY, MT 59301

VALLEY ARC, GEORGE A ASLESON KA7BFU PRES
BOX 286, OPHEIM, MT 59250

YELLOWSTONE RADIO CLUB, MIKE HARDTKE N7LEQ
SECY, 4427 KING AVE EAST, BILLINGS, MT 59101

NEBRASKA

AK-SAR-BEN RC INC., TODD LEMENSE KG0EJ
PO BOX 291, OMAHA, NE 68101

BELLEVUE ARC, THOMAS C HUBER WD0BFO PRES
7518 CHANDLER HILLS DR., BELLEVUE, NE 68147

BLUE VALLEY ARC, BEVERLY NABER KA0VSO SECY
BOX 116, 609 CHAPIN ST, WACO, NE 68460

CRETE ARC, B J FICTUM N0YNC SECY
1201 LINDEN AVE., CRETE, NE 68333

ELKHORN ARC, RICK KROPF KG0IX PRES
P O BOX 1033, 3RD & WILLIS ST, NORFOLK, NE 68702

GRAND ISLAND ARS, DENNIS WING KC0GF TRUSTEE
3012 W 16TH ST, GRAND ISLAND, NE 68803

HASTINGS ARC, THOMAS C VAUGHAN WA0UJZ
926 N BELLEVUE AVE., HASTINGS, NE 68901

JEFFERSON COUNTY ARS, LOUIE HANSON JR. WB0RMO
PRES, 1321 ELM ST, FAIRBURY, NE 68352

LINCOLN ARC INC., ROY BURGESS WB0WWA PRES
PO BOX 5006, LINCOLN, NE 68505

MIDWAY ARC, JERRY RAMSEY W0PXD PRES
PO BOX 1231, KEARNEY, NE 68848

NORTH EAST NEBRASKA HF CLUB, WALTER HILKE
MANN JR. N0MWU, RT 1 BOX 5, 1405 FIR, STANTON,
NE 68779
OREGON TRAIL ARC, FRANK JORDAN KB0HFJ PRES
807 10TH ST, FAIRBURY, NE 68352
PINE RIDGE ARC, JIM MCCAFFERTY K0YIY SCTR
PO BOX 68, WHITNEY, NE 69367
PIONEER ARC, DAVID C BAUER N0LGU
326 S BIRCHWOOD DR APT A, FREMONT, NE 68025
TRI CITY ARC, CHARLES HAGGARD N0EUF SECY
PO BOX 925, SCOTTSBLUFF, NE 69363
WEST NEBRASKA ARC,JAMES R PAXTON WA0NKC VPRS
1320 COUNTRY CLUB DR., SIDNEY, NE 69162

NEVADA
EASTERN NEVADA ARS, JAMES FAIRCHILD N7VLD PRES
PO BOX 12, ELY, NV 89301
ELKO ARC, GARY HASKETT KD0TA PRES
PO BOX 5607, ELKO, NV 89802
FRONTIER ARS, LEONA WALLACE WA6OHB PRES
5805 W HARMON AVE SPACE 159, LAS VEGAS, NV
89103
LAS VEGAS RADIO AMATEUR CLUB, LEE SCHRAM
N7YBE PRES, PO BOX 27342, LAS VEGAS, NV 89126
LAS VEGAS REPEATER ASSN, JOE LAMBERT W8IXD PRES
PO BOX 61201, BOULDER CITY, NV 89006
NEVADA ARA, GORDON HARRIS W7UIZ SECY
PO BOX 73, RENO, NV 89504
NORTHERN NEVADA DX & CONTEST, RICHARD HALL
MAN KI3V PRES, 11870 HEARTPINE ST, RENO, NV
89506
SIERRA INTERMOUNTAIN EMERG RAD, GEORGE UEBELE
WW7E TREAS, 875 BOLLEN CIR, GARDNERVILLE, NV
89410
SIERRA NEVADA ARSINC, JOHN A PETERSON N7SEC
SECY, PO BOX 7727, RENO, NV 89510
WIDE AREA DATA GROUP, RICHARD H WARREN JR.
N7RH, 680 TASKER WAY, SPARKS, NV 89431

NEW HAMPSHIRE
AMOSKEAG RADIO CLUB INC., REESE FOWLER N1KIM
PRES, PO BOX 996, MANCHESTER, NH 03105
CENTRAL EMERG MGT A ASSOC GROUP, HERBERT
CALVITTO WA1WOK, PO BOX 551, CONCORD, NH 03302
CENTRAL NEW HAMPSHIRE ARC, LORRAINE DESCH
ENES WA4LWB, PO BOX 1112, LACONIA, NH 03247
CONTOOCOOK VALLEY RADIO CLUB, PAUL LACLAIR
SECY, 82 HIGH ST, PENACOOK, NH 03303
CROTCHED MOUNTAIN SCHOOL, CHRIS EDSCORN
N0CUH ADVISOR, 1 VERNEY DR., GREENFIELD, NH
03047
GOSHEN-LEMPSTER COOP SCH ARC, CONRAD EKSTROM
WB1GXM ADVISOR, PO BOX 1076, CLAREMONT, NH
03743
GRANITE STATE ARA, GARDNER PAGE W1UGV PRES
PO BOX 5439, MANCHESTER, NH 03108
GREAT BAY RADIO ASSN, WILLIAM A DODGE WA1PEL
LIAS, 78 LITTLEWORTH RD, DOVER, NH 03820
INTERNATIONAL POLICE ASSN RC, WILLIAM A DENNIS
W1WA SCTR. PO BOX 463, SANBORNVILLE, NH 03872
INTERSTATE REPEATER SOC., INC., JOHN BRUNELLE
KA1FYB PRES, 1 PAGET DR., HUDSON, NH 03051
MANCHESTER BOYS & GIRLS AIRWA, RICHARD ZAMOI
DA N1QZA, 555 UNION ST, MANCHESTER, NH 03104
MASCOMA VALLEY REG HS, DAN HUNTLEY K1QLK
ADVISOR, RR 1 BOX 168A, CANAAN, NH 03741
NASHUA AREA RADIO CLUB, DONALD K DILLABY
KA1GOZ SECY, PO BOX 248, NASHUA, NH 03061
NB CONTEST CLUB, DONALD KIELBASA WW1G PRES
PO BOX 1111, DERRY, NH 03038
NEW HAMPSHIRE ARA COUNCIL OF C, DON KIELBASA
WW1G PRES, P O BOX 599, DERRY, NH 03038
NORTH COUNTRY ARC, RICHARD C FORCE WB1ASL
PRES, 12 COTTAGE ST, LANCASTER, NH 03584

NORTH EAST CONVENTION ASSN., AL SHUMAN N1FIK
PRES, PO BOX 475, GOFFSTOWN, NH 03045
PORT CITY ARC, BARRY SHORE NR1P PRES
PO BOX 1587, PORTSMOUTH, NH 03802
SOUHEGAN VALLEY ARC, SEAN P MCINERNEY N1JDM
SECY, 83 MASON RD, NEW IPSWICH, NH 03071
TRITON REGIONAL SCHOOL ARC, JOHN LOVERING
KC1XG ADVISOR, 70 PARK AVE., HAMPTON, NH 03842
TWIN STATE RADIO CLUB, INC., MIKE SCHMITT N1JYT
PRES, PO BOX 5078, HANOVER, NH 03755
WOMEN RADIO OPERATORS OF NE, DAWN CUMMINGS
K1TQY, 23 MARGUERITE ST, KEENE, NH 03431

NEW JERSEY
10-70 REPEATER ASSN INC., ANDREW G BIRMINGHAM
WB2RQX, 235 VAN EMBURGH AVE., RIDGEWOOD, NJ
07450
AMATEUR RADIO EXCPERIMENTERS, ROBERT REED
WB2DIN TRUSTEE, 596 BREWERS BRIDGE RD, JACK
SON, NJ 08527
APPLE ARC, CLAIRE ROSENBAUM W2KQL SECY
170 OAK AVE., RIVER EDGE, NJ 07661
BERGEN ARA, BRUCE LEMKEN WG2Y PRES
47 FURMAN DR., EMERSON, NJ 07630
BURLINGTON COUNTY RADIO CLUB, THOMAS B DEMEIS
K2TD SECY, 121 KATHLEEN AVE., DELRAN, NJ 08075
CAMDEN COUNTY AUTOPATCH RA, JOE FISHER KC2TN
PRES, PO BOX 24, VOORHEES, NJ 08043
CAPE MAY COUNTY ARC, JAMES H CHADWICK KB2JMJ
PRES, PO BOX 352, RIO GRANDE, NJ 08242
CENTRAL JERSEY VHF SOCIETY, HENRY G EVELAND
KC2TA PRES, 24 ADELPHI DR., JACKSON, NJ 08527
CHAVERIM OF NEW JERSEY, ROBERT M FELLER N2JWK
PRES, 340 N 4TH AVE., HIGHLAND PARK, NJ 08904
CHERRYVILLE REPEATER ASSOCIATION, DUNCAN
MACRAE KE2HG PRES, O BOX 308, QUAKERTOWN, NJ
08868
CHESTNUT RIDGE RADIO CLUB, RICHARD COLTEN
K1JMI VPRS, 36 HAMPSHIRE HILL RD, UPPER SADDLE
RIVER, NJ 07458
CLIFTON ARS, ED MARCHESE N2TE PRES
PO BOX 4842, CLIFTON, NJ 07015
CLIFTON HIGH SCHOOL ARC, JAMES J LYON N2FDM
PRES, 333 COLFAX AVE., CLIFTON, NJ 07013
COUNTY LINE ARA OF NW NJ, CHRISTOPHER D LINNE
N2OPO PRES, PO BOX 291, HACKETTSTOWN, NJ 07840
ELECTRONIC TECHNOLOGY OF NJ, AL BLASUCCI
W2KOG TREAS, 2087 WEST FIELD AVE., SCOTCH
PLAINS, NJ 07076
ENGLEWOOD ARA INC., DAVID B POPKIN W2CC PRES
PO BOX 528, ENGLEWOOD, NJ 07631
FAIRLAWN ARC, CHRIS PAPPAS N2MFH
69 CEDARHURST AVE., WEST PATERSON, NJ 07424
FORT MONMOUTH ARC, GERALD SILVERMAN WB2GYS
15 PARTRIDGE LN, EATONTOWN, NJ 07724
GARDEN STATE ARA INC., BRUCE A MCLEOD K2QXW
SECY, PO BOX 34, FAIR HAVEN, NJ 07704
GLOUCESTER COUNTY ARC, MARLA BOZARTH N2DWR
PRES, P O BOX 370, PITMAN, NJ 08071
HOLIDAY CITY ARC, MURRAY GOLDBERG KD2IN SCTR
2 TROPICANA CT, TOMS RIVER, NJ 08757
HUDSON AMATEUR RADIO COUNCIL, DAVID B POPKIN
W2CC TREAS, PO BOX 528, ENGLEWOOD, NJ 07631
INTERVALE SCHOOL ARC, RICH HIBBARD AA2FY ADVI
SOR, 60 PITT RD, P O BOX 52, PARSIPPANY, NJ 07054
IRVINGTON-ROSELAND ARC, JIM HOWE N2TDI
#5 IROQUOIS AVE., LAKE HIAWATHA, NJ 07034
JERSEY SHORE ARS, BARRY V KEAVENY N2NVP PRES
38 DOUBLE TROUBLE RD, TOMS RIVER, NJ 08757
JERSEY SHORE CHAVERIM, GERALD G SILVERMAN
WB2GYS, 15 PARTRIDGE LN, EATONTOWN, NJ 07724
KNIGHT RAIDERS VHF CLUB INC., JACK D WILK N2DXP
PRES, 120 BROOK AVE., PASSAIC, NJ 07055

LAKEWOOD HIGH SCHOOL, NORMAN PALMER WE2J
ADVISOR, 15 STUYUESANT PLACE, ELBERON, NJ 07740
LAND ROVERS ARC, ROBERT E KAKASCIK WB2IHI PRES
175 PASSAIC ST, GARFIELD, NJ 07026
MAJOR ARMSTRONG MEMORIAL ARC I, SUSAN VESEY
KB2KGG SECY, 78 PALISADE AVE., CLIFFSIDE PARK,
NJ 07010
MAPLE SHADE ARC. FRANK COCHRANE KA2DGF PRES
11 WAGON BRIDGE RUN, MOORESTOWN, NJ 08057
MONMOUTH COUNTY REPEATER ASSN, KENNETH R
HAMPTON JR KY2S PRES, 1119 GRASSMERE AVE.,
ASBURY PARK, NJ 07712
MORRIS RADIO CLUB, BENJAMIN J FRIEDLAND K2PBP
9 KNOLLWOOD TRL W. MENDHAM, NJ 07945
NEPTUNE ARC, GARY D GOODMAN N2XNN PRES
PO BOX 2181, NEPTUNE, NJ 07754
NEW LISBON DEVELOPMENTAL CTR, TRENT A DAVIS SR
KB2KGW PRES, 209 ATLANTIS AVE., MANA
HAWKIN, NJ 08050
NEW PROVIDENCE ARC INC., JOHN I SHEETZ K2AGI
TREAS, PO BOX 813, NEW PROVIDENCE, NJ 07974
NJ INSTITUTE OF TECH ARC, FREDDY A BALADY N2MTA
PRES, NJIT SENATE, 323 KING BLVD, NEWARK, NJ
07102
NORTH JERSEY DX ASSOCIATION, WILLIAM HUDZIK
WA2UDT SECY, PO BOX 599, MORRIS PLAINS, NJ 07950
NUTLEY ARS, JOHN M WALCH KA2QLR
166 HIGHFIELD LN, NUTLEY, NJ 07110
OCEAN-MONMOUTH ARC, IRA GOLDSTEIN AA2RZ PRES
157 FLINTLOCK DR., LAKEWOOD, NJ 08701
OLD BARNEY ARC, JAMES HEPBURN W2IIC SECY
1711 MILL CREEK RD, MANAHAWKIN, NJ 08050
OLD BRIDGE RADIO ASSOCIATION, MARVIN SHULDMAN
WB2SZI SECY, 74 DOE CT, MONMOUTH JUNCTION, NJ
08852
PANASONIC ARC, TIM ORTON N2UAQ SECY
PANA ZIO 4D-8, ONE PANASONIC WAY, SECAUCUS,
NJ 07094
PENN-JERSEY ARC, ROBERT ESCOTT N2VWM SECY
P IO BOX 7617, BLOOMSBURG, NJ 08804
PINE BARRENS RADIO CLUB, CHARLES PORCH NG2S
SECY, 14 BURTONS DR., SOUTHAMPTON, NJ 08088
PISCATAWAY ARC, BILL KOETH KF2DD PRES
PO BOX 1233, PISCATAWAY, NJ 08855
RAMAPO MOUNTAIN ARC, WARREN WALSH N2BCC
PRES, 163 SKYLINE LAKE DR., RINGWOOD, NJ 07456
RARITAN BAY RADIO AMATEURS, LARRY MAKOSKI
N2ELW, PO BOX 173, SAYREVILLE, NJ 08871
RARITAN VALLEY RADIO CLUB, ROGER SCHROEDER
N2LAQ PRES, 506 COUNTRY CLUB RD, BRIDGEWATER,
NJ 08807
RUTGERS ARC, JOHN BUNGS N2YLR PRES, 613 GEORGE
ST, SAC BOX 114, NEW BRUNSWICK, NJ 08903
SHORE POINTS ARC, JOHN G BARBIERI KB2HZU SECY 1
1002 BARTLETT AVE., PO BOX 142, LINWOOD, NJ 08221
SINGER EMPLOYEES ARC, JOHN SANTILLO N2HMM PRES
RR 2 BOX 310, HIGHLAND LAKES, NJ 07422
SOMERSET COUNTY ARS, WILLIAM DETTELBACK
N2ASK SECY, PO BOX 742, MANVILLE, NJ 08835
SOMERSET HILLS ARC, DENIS J DOOLEY II WB2HDB
PRES, 17 OLCOTT AVE., BERNARDSVILLE, NJ 07924
SOURLAND MOUNTAIN ARC, RON HAUSER KE2JR SECY
59 DE HART DR., BELLE MEAD, NJ 08502
SOUTH CENTRAL JERSEY REPEATER, JOSEPH NUZZO
N2BEI PRES, 14 SURREY DR., OLD BRIDGE, NJ 08857
SOUTH JERSEY DX ASSOCIATION, BOB SCHENCK N2OO
PRES, PO BOX 345, TUCKERTON, NJ 08087
SOUTH JERSEY RADIO ASSN, EDWARD RAMMING AB2Y
PRES, 4500 WESTFIELD AVE., PENNSAUKEN, NJ 08110
SOUTHERN COUNTIES ARA, JACOB G KAFERLE W2IFI
PRES, PO BOX 121, LINWOOD, NJ 08221
SPLITROCK ARA INC., JEFFREY M FRIEDMAN KB2OBU
PO BOX 610, ROCKAWAY, NJ 07866
ST BARNABAS ARC, HOWARD M EISENSTODT N2VDH
43 EDGEWOOD RD, PO BOX 786, SUMMIT, NJ 07902

STATELINE RADIO CLUB, CHRIS FAGAS WB2VVV PRES
68 KINDERKAMACK RD, PARK RIDGE, NJ 07656
SUSSEX COUNTY ARC, BILL SLACK NX2P SECY
321 E SHORE TRL, SPARTA, NJ 07871
TRI-COUNTY ARA, DANIEL FERGUSON AA2PT SECY
P O BOX 412, SCOTCH PLAINS, NJ 07076
UNDERGROUND DISCHARGE ARC, DAVID R KANITRA
WB2AZE PRES, 74 PORT READING AVE., WOOD
BRIDGE, NJ 07095
VICTOR ARA, PETER GREENE N2LVI PRES
MARTIN MARIETTA CORP, 1 FEDERAL ST A&E -2W,
CAMDEN, NJ 07112
WEST JERSEY DX GROUP, JOHN HULTS KF2BH PRES
714 MOUNTAIN RD, ASBURY, NJ 08802

NEW MEXICO

ALAMOGORDO ARC INC., OLE JORGENSEN WA5IPS PRES
PO BOX 1191, ALAMOGORDO, NM 88311
ALBUQUERQUE ARC, ART PRIEBE N5OQJ SECY
10225 KAREN AVE NE, ALBUQUERQUE, NM 87111
ALBUQUERQUE DX ASSOC., EUGENE F CARTER W5OLN
SCTR, 320 CAMINO TRES SW, ALBUQUERQUE, NM
87105
AMATEUR RADIO CARAVAN CLUB, DAN KIRBY KC5DIV
SECY, P O BOX 36365, ALBUQUERQUE, NM 87110
DEMING AMATEUR RADIO CLUB, MARGARET GRIFFY
KB5OYU SECY, 1401 S PLATINUM, DEMING, NM 88030
EASTERN NEW MEXICO ARC, HAROLD LANDSPERG
KA5BAT PRES, 1001 W PLAZA DR., CLOVIS, NM 88101
GILA ARS, ROLAND SHOOK KB5SYZ SCTR
PO BOX 1874, SILVER CITY, NM 88062
MESILLA VALLEY RADIO CLUB, JACKIE LEMONS KJ5FW
PRES, PO BOX 1443, LAS CRUCES, NM 88004
NEW MEXICO BIG RIVER CONTESTER, BRUCE L DRAPER
AA5B PRES, 7827 PIONEER TRL NE, ALBUQUERQUE,
NM 87109
PECOS VALLEY ARC, TOMMY DOW W5BLO PRES
PO BOX 162, ROSWELL, NM 88202
TOTAH ARC, DOUGLAS K ARNOLD KC5CHL SCTR
PO BOX 1991, FARMINGTON, NM 87499
UPPER RIO FM SOC INC., LESLIE M RIVLIN KA5BEM
TREAS, 3004 MATADOR DR NE, ALBUQUERQUE, NM
87111

NEW YORK

ALBANY ARA, BOB RAFFAELE W2XM PRES
5 GADSEN CT, ALBANY, NY 12205
ALLEGANY HIGHLANDS ARC, RAY STEVENS W2BYO
PRES, 65 N MAIN ST, WELLSVILLE, NY 14895
AMERICAN RED CROSS EMERGENCY R, GEORGE SAU
WB2ZTH TRUS, 25 RODSFIELD CT, HUNTINGTON, NY
11743
ARA OF THE SOUTHERN TIER INC., DONALD CAMPANEL
LI WB2ABK LIAS, PO BOX 388, ELMIRA, NY 14902
ARA OF THE TONAWANDAS INC., RICHARD A STEIN
K2ZR PRES, PO BOX 430. NORTH TONAWANDA, NY
14120
BINGHAMTON AMATEUR RADIO ASSOC, RON CLOTHIER
AA2EQ PRES, PO BOX 853, BINGHAMTON, NY 13902
BLACK RIVER VALLEY ARC, LA VELL MILLER N2PFN
PRES, 5324 SUNSET DR., LOWVILLE, NY 13367
BLUE MOUNTAIN MIDDLE SCH ARC, BILL MACHONIS
AA2NU PRES, PO BOX 14, MONTROSE, NY 10548
BOONVILLE ARC, HANDLEY E JACKSON JR KA2SJG S
RR 1 BOX 79, BOONVILLE, NY 13309
BROADCAST EMPLOYEES ARS INC., HOWARD PRICE
KA2QPJ PRES, ANSONIA STATION, P O BOX 995, NEW
YORK, NY 10023
BROCKPORT AMATEUR RADIO KLUB, PAUL MACKANOS
K2DB PRES, 86 CLOSE HOLLOW DR., HAMLIN, NY
14464
BROOKHAVEN ARC, CHRIS NEUBERGER KA2GAV SECY
PO BOX 5000-0201, UPTON, NY 11973

BROOKLYN TECH HS ARC, LEWIS A MALCHICK N2RQ ADVS, 29 FORT GREENE PL., BROOKLYN, NY 11217

BROOKLYN UHF AMATEUR SERVICE, ARNOLD SCHIFF MAN WB2YXB PRES, 8122 250TH ST, BELLEROSE, NY 11426

CAPITAL DISTRICT AR COUNCIL, JACK DONNELLY WA2YBM, 21 RED FOX DR., ALBANY, NY 12205

CHAMPLAIN VALLEY ARC, ALVAH B HAGGETT KB2LML PRES, PO BOX 313, MORRISONVILLE, NY 12962

CHAUTAUQUA COUNTY AMATEUR FM A, JAMES H BEL LUZ NY2Z PRES, 529 DODGE RD, FREWSBURG, NY 14738

CHENANGO VALLEY ARA, KIM EDWARD MILLER N2PAO SECY, RR 1 BOX 584, NORWICH, NY 13815

CLARKSON UNIV ARC, DR DAVID BRAY K2LMG ADVIS CLARKSON UNIVERSITY, BOX 8550, POTSDAM, NY 13699

COLUMBIA UNIVERSITY ARC, HARRY Y XU AA2NO PRES, COLUMBIA UNIVERSITY, 530 MUDD COLUMBIA UNIV., NEW YORK, NY 10027

COMMUNICATIONS CLUB OF NEW ROC, RICHARD A SANDELL WK6R, 2250 BOSTON POST RD, LARCH MONT, NY 10538

CORNELL UNIVERSITY ARC, SANJAY HIRANANDANI N2MRZ, CORNELL UNIVERSITY, 401 BARTON HALL, ITHACA, NY 14853

COUNCIL FOR ADVANCEMENT OF AR, CAROLE PERRY WB2MGP LIAS, 33 FERNDALE AVE., STATEN ISLAND, NY 10314

CROSS ISLAND RADIO RAGCHEW NET, TONY DIBIASE WA2RHY PRES, PO BOX 16, RIDGE, NY 11961

CRYSTAL RADIO CLUB, ABE GLITTLER K2YHM PRES 126 W MAPLE AVE., MONSEY, NY 10952

DRUMLIN ARC, STEVE ELLIOTT N2COJ TRUSTEE 1609 MARYLAND STREET RD, PHELPS, NY 14532

EASTERN AMATEUR RADIO SERVICE, CHARLES GAL LANTI SR AA2DB, 289 FERRY RD, SAG HARBOR, NY 11963

EBONAIRE ARS, GWEN (VIKKI) NEAL KA2VKW STAPLETON STATION, P O BOX 040113, STATEN ISLAND, NY 10304

ELEC & RADIO COMM CTR OF S.M.L., SIDNEY WOLIN K2LJH PRES, 3072 CLUBHOUSE RD, MERRICK, NY 11566

ELECTCHESTER VHF CLUB INC., ROGER JACOBS KB2HB PRES, 22 WOOD AVE., ALBERTSON, NY 11507

ENCON ARS, CHARLES T COGSWELL N2RKC PRES 50 WOLF RD RM 617, ALBANY, NY 12233

ERIC TROJAHN MEMORIAL ARC, HENRY BOOKOUT N2FCZ ADVISOR, 167 SOUND AVE., RIVERHEAD, NY 11901

FIVE COUNTY ARC, ROCCO J CONTE WU2M PRES PO BOX 547, FONDA, NY 12068

FULTON ARC, BRUCE ZELLAR KA2GJV EDITOR 366 S 5TH ST, FULTON, NY 13069

GENESEE RADIO AMATEURS, CHERYL YAXLEY SECY PO BOX 228, BYRON, NY 14422

GENESEE REPEATER ASSN, OTTO BLUNTZER WB2RJB VPRS, 95 MARIPOSA DR., ROCHESTER, NY 14624

GREAT NECK SOUTH H.S. ARC, DANIEL REISMAN N2UAC PRES, 341 LAKEVILLE RD, GREAT NECK, NY 11020

GREAT SOUTH BAY ARC, JAY EICHNER N2PIK PO BOX 491, BABYLON, NY 11702

GRUMMAN ARC, PATRICK MASTERSON KE2LJ 6 WEATHERVANE WAY, DIX HILLS, NY 11746

HAITIAN AMATEUR RADIO YOUTH CL, PAUL CAMEAU KE2LM, 35 NASSAU PL., HEMPSTEAD, NY 11550

HALL OF SCIENCE ARC, CHARLES BECKER WA2JUJ PRES 25 FARRAGUT RD, OLD BETHPAGE, NY 11804

HARBORFIELDS HIGH SCHOOL ARC, SACHA J BERN STEIN N2NSZ PRES, PO BOX 15, GREENLAWN, NY 11740

HELLENIC ARA, HIPPOCRATES KOUTSOUPAKIS PO BOX 6307, LONG ISLAND CITY, NY 11106

HUDSON VALLEY CONTESTERS & DX, ROBERT SCHWENK W2XL PRES, 133 CLIFTON AVE., KINGSTON, NY 12401

HUDSON WATERSHED VHF SOCIETY, TOM RICHMOND WB2IEY, PO BOX 7, MARYKNOLL, NY 10545

HUNTINGTON VHF FM ASSOC INC., CHUCK LAUFMAN K2JLD PRES, PO BOX 234, EAST NORTHPORT, NY 11731

IBM EXPLORER POST 204 BSA, BILL COLEMAN WF2A PRES, P O BOX 312, ENDICOTT, NY 13760

INTERMEDIATE SCHOOL 72, CAROLE J PERRY WB2MGP PRES, 33 FERNDALE AVE., STATEN ISLAND, NY 10314

INTERNATIONAL MISSION RADIO A, REV MICHAEL MULLEN WB2GQW, ST JOHN'S UNIVERSITY, JAMAICA, NY 11439

KEUKA LAKE ARA, JAMES C WHITE KV2W 8653 CORYELL RD, HAMMONDSPORT, NY 14840

KINGS COUNTY RADIO CLUB, GEORGE M DONAHUE KD2AU PRES, 616 E 24TH ST., BROOKLYN, NY 11210

KINGS CTY RPTR ASSN., HARVEY BIRD KB2EA PRES 34 ROBINSON AVE., STATEN ISLAND, NY 10312

KINGS PARK SCHOOLS ARC, MARSHALL L CARLOZZI KB2LYS, WILLIAM T ROGERS MIDDLE SCHOOL, OLD DOCK RD, KINGS PARK, NY 11754

KNICKERBOCKER ARC, RICHARD FARBER W2KXB PRES 147 ROBBY LN, NEW HYDE PARK, NY 11040

LANCASTER ARC, LUKE CALIANNO N2GDU PRES 1105 RANSOM RD, LANCASTER, NY 14086

LARKFIELD ARC, WILLIAM MORRISSEY N2QZF SECY PO BOX 1450, HUNTINGTON, NY 11743

LIVERPOOL AMATEUR REPEATER CL, CHARLES SILVIA KB2DIO PRES, P O BOX 103, NORTH SYRACUSE, NY 13212

LOCKPORT ARA, EDMUND P BUSCH N2MYM PRES 5316 ERNEST RD, LOCKPORT, NY 14094

LONG ISLAND DX ASSN, EDWARD A WHITMAN K2MFY SECY, 2 NUTLEY CT, PLAINVIEW, NY 11803

LONG ISLAND MOBILE ARC, KEN GUNTHER WB2KWC SECY, 219 NORMANDY RD, MASSAPEQUA, NY 11758

MADISON-ONEIDA ARC, CHRIS HANSEN N3DXJ PRES PO BOX 241, VERONA, NY 13478

MANHATTAN-AVE OF THE AMERICAS, HENRY SCHICK LER W2ICW, 1618 163RD ST, WHITESTONE, NY 11357

MAPLE HILL HIGH SCHOOL ARC, JOHN F KIENZLE WA2UON TRUSTEE, 1216 MAPLE HILL RD, CASTLE TON ON HUDSON, NY 12033

MIDLAKES H S R A TRANSMITTING, JEFF JENSEN N2MKT PRES, 3190 BIRD RD, CLIFTON SPRINGS, NY 14432

NASSAU ARC, HOWARD S LIEBMAN W2QUV 1750 BELMONT AVE., NEW HYDE PARK, NY 11040

NASSAU COUNTY POLICE ARC, J HACKETT KF2MQ VPRS PO BOX 8001, HICKSVILLE, NY 11802

NASSAU COUNTY WIRELESS ASSN, MARK NADEL NK2T PRES, 22 SPRINGTIME LN, LEVITTOWN, NY 11756

NEW YORK CITY REPEATER ASSOCIA, STEVE ZUVICICH KA2HXU TREAS, PO BOX 140819, STATEN ISLAND, NY 10314

NIAGARA RADIO CLUB INC., HAROLD O KLINGELE K2JDF, PO BOX 104 LASALLE, NIAGARA FALLS, NY 14304

NORTH FRANKLIN ARC, RICHARD C SHERMAN WZ2T PRES, HC 1 BOX 85A, MALONE, NY 12953

NORTHERN CATSKILLS VHF ASSN, THOMAS G VALOSIN WB2KLD PRES, 30 WARRIOR WAY, MIDDLEBURGH, NY 12122

NORTHERN CHAUTAUQUA ARC, FRANK M ROSSI N2RGS PRES, 23 BIRD AND TREE RD, MAYVILLE, NY 14757

OGDENSBURG ARC, LOIS G IERLAN WA2RXO SCTR 725 PROCTOR AVE., OGDENSBURG, NY 13669

ONEONTA ARC, JEROME J BATIK K2TNN PRES RR 1 BOX 213, STAMFORD, NY 12167

ORANGE COUNTY ARC, INGE BECKENBACH WA1WTG SECY, BOX 753, GREENWOOD LAKE, NY 10925

ORDER OF BOILED OWLS, JACK L SCHULTZ W2GGE SECY, 2 HUXLEY DR., HUNTINGTON, NY 11743

ORLEANS COUNTY ARC, ROBERT L HAZEL WA2QDV
    564 EAST AVE., MEDINA, NY 14103
OTSEGO COUNTY ARA, FRED ST JOHN KA2EOP SCTR
    13 EAGLE ST, COOPERSTOWN, NY 13326
OVERLOOK MOUNTAIN ARC, BRIAN MARTIN N2NMF
    SECY, 115 PETTICOAT LANE, HURLEY, NY 12443
PECONIC ARC, WILLIAM NORRIS N2YKH SECY
    PO BOX 113, PECONIC, NY 11958
PIONEER RADIO OPERATORS SOC, MR GAIL LEWIS
    W2CRY PRES, 9765 S PROTECTION RD, HOLLAND, NY
    14080
PLAINEDGE PUBLIC SCHOOLS ARC, WALTER M WENZEL
    KA2RGI PRES, C/O AV DEPT, 241 WYNGATE DR., MASS
    APEQUA, NY 11758
POLYTECHNIC RADIO CLUB OR, PETER G UNFRIED
    KA2NYR PRES, 8009 78TH AVE., GLENDALE, NY 11385
POUGHKEEPSIE ARC, ANDREW D SCHMIDT N2FTR
    19 GERRY RD, POUGHKEEPSIE, NY 12603
PUTNAM EMERGENCY & ARL, JAMES F WARD N2EGS
    PRES, 22 TROUT PLACE RD 4, MAHOPAC, NY 10541
RADIO AMATEURS OF GREATER SYRA, VIVIAN DOU
    GLAS WA2AE, PO BOX 88, LIVERPOOL, NY 13088
RADIO ASSN OF WESTERN NY, BURCHARD ROYCE
    N2QWX SECY, 194 SMALLWOOD DR., NY 14226
RADIO CENTRAL ARC, NEIL HEFT KC2KY PRES
    PO BOX 680, MILLER PLACE, NY 11764
RADIO CLUB OF J.H.S. 22 NYC, JOSEPH J FAIRCLOUGH
    WB2JKJ, PO BOX 1052, NEW YORK, NY 10002
RIP VAN WINKLE ARS, SHAILER EVANS AA2Y PRES
    PO BOX 365, CLAVERACK, NY 12513
ROBERT MOSES MIDDLE SCHOOL ARC, WILLIAM L
    JANSEN K2HVN TRUSTEE, MILLER AVE., N BABYLON,
    NY 11703
ROCHESTER AMATEUR RADIO ASSN, LLOYD CAVES
    WB2EFU PRES
    PO BOX 93333, ROCHESTER, NY 14692
ROCHESTER DX ASSN, RUTH HOFFMAN AA2IO SCTR
    PO BOX 274, WEST HENRIETTA, NY 14586
ROCHESTER RADIO REPEATER ASSN, RAYMOND R PICK
    ENS WA2MYG, PO BOX 92031, ROCHESTER, NY 14692
ROCHESTER VHF GROUP, JOHN M GILLY WD4RDZ PRES
    PO BOX 92122, ROCHESTER, NY 14692
ROCKAWAY ARC, BRO ROBERT LINDEMANN KD2PB
    144 BEACH 111TH ST, ROCKAWAY PARK, NY 11694
ROCKAWAY BEACH JHS ARC, GERALD SKLOOT KE2N
    ADVISOR, 2923 MANDALAY BEACH RD, WANTAGH,
    NY 11793
ROCKLAND WIRELESS RADIO CLUB, THOMAS BITTNER
    N2ONK PRES, 442 SOMERSET DR., PEARL RIVER, NY
    10965
ROME RADIO CLUB, JAMES BUTLER KA2JXA SECY
    1021 WOOD ST, ROME, NY 13440
RUSTY POLECATS VHF SOCIETY, JOHN C FANDL
    WA2FUZ PRES, HC 1 BOX 9F, SURPRISE, NY 12176
SACHEM HIGH SCHOOL ARC, ROGER QUINN N2JGK
    PO BOX 26, FARMINGVILLE, NY 11738
SALT CITY DX ASSOCIATION, WILSON B PARKER KB2G
    SECY, 210 PLEASANT DR., EAST SYRACUSE, NY 13057
SARATOGA COUNTY RACES ASSN, MILDRED FEENY
    KV2A SECY, 5 COESA DR.,SARATOGA SPRINGS, NY
    12866
SCHENECTADY MUSEUM ARA, JOEL JOHNSON N2QII
    SECY, 3 WHITE ST, SCHENECTADY, NY 12308
SECOND AREA YOUNG LADIES ARC, MIRIAM D LAMB
    AA2DX SCTR, 7 DRURY LANE, HIGHLAND FALLS, NY
    10928
SKYLINE ARC, REGINA CANFIELD KB2AED SECY
    PO BOX 5241, CORTLAND, NY 13045
SNOWBELT AMATEUR GROUP, WILLIAM MCCARNS
    N2PBX SECY, 1243 BRAY RD, ARCADE, NY 14009
SOUTH SHORE ARC OF SUFFOLK, SY SIEGEL WB2PBG
    VPRS, 40 W 9TH ST., DEER PARK, NY 11729
SOUTH TOWNS ARS, JOHN H LEITTEN KA2RFT VPRS
    6120 MCKINLEY PKY, HAMBURG, NY 14075

SOUTHERN CATSKILL ARC, CHARLIE HAAS WB2JDT
    PRES, 3083 RT 42, MONTICELLO, NY 12701
SPRING VALLEY REPEATER SERVICE, STANLEY J TEICH
    K2IRS TRUS, 34 BRIDLE RD, SPRING VALLEY, NY 10977
SQUAW ISLAND ARC, ERNIE BROWN K2BWK TREAS
    1 COVILLE ST., VICTOR, NY 14564
STARC, MICHAEL P GRUSZKA N2NW PRES
    PO BOX 7082 , ENDICOTT, NY 13761
STATE UNIV OF NY AT ALBANY ARC, BOB RAFFAELE
    W2XM, 5 GADSEN CT, ALBANY, NY 12205
STATEN ISLAND ARA, RICH DYRACK K2LUQ
    27 MANOR CT, STATEN ISLAND, NY 10306
SUFFOLK COUNTY RADIO CLUB, BARNEY LINDEN
    N2DQR PRES, 59 COMERFORD ST, PORT JEFFERSON
    STATION, NY 11776
SUFFOLK COUNTY VHF/UHF ASSN, LEONARD A BUON
    AIUTO SR KE2LE, PO BOX 212, ISLIP TERRACE, NY
    11752
SUFFOLK POLICE ARC, JEFFREY SAVASTA KB4JKL
    9 PEPPERMINT RD, COMMACK, NY 11725
TIOGA AMATEUR REPEATER INC., SUE PORTER N2DBK
    SCTR, PO BOX 113, OWEGO, NY 13827
TOMPKINS COUNTY ARC, KEITH MORGAN-DAVIE
    N2MUU SECY, 848 JEWETT HILL RD, BERKSHIRE, NY
    13736
TROY ARA, WILLIAM J EDDY NY2U PRES
    2404 22ND ST, TROY, NY 12180
TRYON ARC, GARY D ASHE K2RKW PRES
    9 S CLARK AVE., GLOVERSVILLE, NY 12078
TU-BORO RADIO CLUB INC., NELSON SIDMAN N2CKK
    VPRS, 14205 ROOSEVELT AVE APT 330, FLUSHING, NY
    11354
UNATEGO JR SR HIGH SCHOOL ARC, PAUL AGOGLIA
    WN2K ADVISOR, UNATEGO JR SR HIGH SCHOOL, RD
    #1, OTEGO, NY 13825
UTICA ARC, HOWARD COHEN WA2TVE SECY
    PO BOX 71, UTICA, NY 13503
WALTON RADIO ASSOCIATION, ALLAN J MONGARDI
    K2EZK SCTR, 24 CONCORD ST, SIDNEY, NY 13838
WANTAGH ARC, HERMAN MILATZ W2TLC PRES
    1836 GARDENIA AVE., MERRICK, NY 11566
WARD MELVILLE HS ARC, JOHN F POMFRET W2AAF
    TRUS, WARD MELVILLE HIGH SCHOOL, OLD TOWN
    RD, SETAUKET, NY 11733
WARREN COUNTY RACES ASSN., THOMAS J HARIG
    KT2M PRES, 1 SWEETBRIAR LN, GLENS FALLS, NY
    12804
WESTCHESTER ARA, LOUIS GARY LEONARD WA2UIJ
    SECY, 13 MONTCLAIR RD, YONKERS, NY 10710
WESTCHESTER EMERG COMM ASSN, PAUL S VYDARENY
    WB2VUK, 259 N WASHINGTON ST, NORTH
    TARRYTOWN, NY 10591
WESTCHESTER FM REPEATER ASSN, GERSON A LEVY
    W2LAP, PO BOX 27, BRONX, NY 10457
WESTERN NEW YORK DX ASSN., BOB NADOLNY,
    WB2YQH SCTR, PO BOX 73, SPRING BROOK, NY 14140
XEROX ARC, DAVESWEET N2BHL VPRS
    2136 PENFIELD RD, PENFIELD, NY 14526
YATES ARC, RICHARD A AYERS KB2DMK PRES
    2590 AYERS RD, PENN YAN, NY 14527
YONKERS ARC, JOHN A COSTA WB2AUL
    195 WOODLANDS AVE., YONKERS, NY 10703

NORTH CAROLINA

ALAMANCE ARC, PAUL KEARNS N4ZDZ PRES
    PO BOX 3064, BURLINGTON, NC 27215
ASHE COUNTY ARC, LORY WHITEHEAD KE4CYO
    PO BOX 545, JEFFERSON, NC 28640
BLUE RIDGE ARC, STEPHEN M GROSE KE4CLE PRES
    POB 183, FLAT ROCK, NC 28731
BRIGHTLEAF ARC, BERNARD NOBLES WA4MOK PRES
    PO BOX 8387, GREENVILLE, NC 27835
CABARRUS AMATEUR RADIO SOCIETY, GEORGE PAT
    TERSON KD4YSJ SECY, PO BOX 1290, CONCORD, NC
    28026

CAPE FEAR ARS, ROBERT DUDLEY KC4RSZ PRES
113 COCHRAN AVE., FAYETTEVILLE, NC 28301
CARTERET COUNTY ARS, ANDREW GRIFFITH W4ULD
SECY, PO BOX 1302, NEWPORT, NC 28570
CARY ARC INC., HERBERT L LACEY JR N4UE SECY
1022 MEDLIN DR., CARY, NC 27511
CENTRALINA ARC, JOHN SETZLER KD4YJO
P O BOX 1513, HICKORY, NC 28603
CHICORA AR GROUP, TOM WOOD N4CID TRUSTEE
PO BOX 116, DUNN, NC 28335
DURHAM FM ASSN, JIM LANGLEY WB4YYY PRES
11800 NORWOOD RD, RALEIGH, NC 27613
EAST WAKE SCHOOLS ARC, MARY ALICE RAMSEY
KD4OKQ, 2700 OLD MILBURNIE RD, RALEIGH, NC
27604
FORSYTH ARC, MELISSA HALL KD4VED
PO BOX 11361, WINSTON SALEM, NC 27116
FRANKLIN ARC, KAY CORIELL KB4USN PRES
50 SANDERS RD, FRANKLIN, NC 28734
GREENSBORO ARA, CHRIS SPIVEY KB4UYP PRES
PO BOX 7054, GREENSBORO, NC 27417
GREENSBORO DAY SCHOOL ARC, CARL FENSKE
KC4WGA ADVISOR, 5401 LAWNDALE DR., PO BOX
26805, GREENSBORO, NC 27429,
HEALING SPRINGS MTN VHF SOC IN, RAEFORD EVER
HART K4SWN LIAS, PO BOX 41, LEXINGTON, NC 27293
HIGH POINT ARC, BARRY GARNER N4UOH PRES
PO BOX 4941, HIGH POINT, NC 27263
JOHNSTON AMATEUR RADIO SOCIETY, JIM CRESWELL
KJ4SW TREAS, 4567 BENSON HARDEE RD, BENSON, NC
27504
KINSTON ARS, STEVE PITTMAN KC4EJU VPRS
1206 LOCKWOOD RD, KINSTON, NC 28501
MECKLENBURG ARS INC., MARY S HUNT KA4EXP SECY
3213 BRIDGEMERE TER, MATTHEWS, NC 28105
NEW BERN ARC, ARTHUR M SLAUGHTER KE4BRV
P O BOX 2483, NEW BERN, NC 28560
NORTH CAROLINA STATE UNIV ARC, ADAM HARRIMAN
N4ZSM PRES, 131 TUCKER, P O BOX 4642, RALEIGH, NC
27607
ORANGE COUNTY RADIO AMATEURS, JOHN W HUGHES
N0RXK PRES, 1914 NEW HOPE CHURCH RD, CHAPEL
HILL, NC 27514
OUTER BANKS REPEATER ASSN., BOB MAY N4LBR PRES
PO BOX 1329, KILL DEVIL HILLS, NC 27948
PAMLICO ARC, ROBERT E RICHARDSON KR4TW
RT 5 BOX 12, WASHINGTON, NC 27889
PIEDMONT AMATEUR RADIO KLUB, GARY HENDRICKS
KD4NW PRES, RR 2 BOX 144, HURDLE MILLS, NC 27541
POLK COUNTY H.S. ARS, THOMAS D TAYLOR KC4QPR
ADVIS, P O BOX 5003, MORGANTON, NC 28655
RADIO AMATEUR GROUP INC., HENRY HORNE W4MZP
PRES, PO BOX 688, RED SPRINGS, NC 28377
RALEIGH ARS INC., CHARLES R LITTLEWOOD K4HF LIAS
PO BOX 17124, RALEIGH, NC 27619
ROBESON COUNTY ARS, INC., SUZANNE DEDRICK
W4PET, PO BOX 692, LUMBERTON, NC 28359
ROWAN ARS, RALPH K BROWN WB4AQK SCTR
1621 EMERALD AVE., SALISBURY, NC 28144
SAMPSON CTY AMATEUR RADIO SERV, JOHN NAYLOR
WB4OMN SCTR, PO BOX 64, CLINTON, NC 28328
SHELBY RADIO CLUB, JUNE MELVIN WA4JNJ SECY
902 HENRY ST, KINGS MOUNTAIN, NC 28086
TAR RIVER ARC, ROD BRIDGERS
6363 NC 42 HWY EAST, ELM CITY, NC 27822
TRIANGLE EAST ARA, HOWARD THORNTON AC4UD
PRES, P O BOX 480, ZEBULON, NC 27597
TRIODE ARC, DAVID LEWIS KB4YSX PRES
PO BOX 1721, ANDREWS, NC 28901
WAKE TECHNICAL COLLEGE ARC, MARTIN CLARK
W2IRQ ADVISOR, 9101 FAYETTEVILLE RD, RALEIGH,
NC 27603
WATAUGA ARC, CHRIS E PENICK N4YGY VPRS
RR 2 BOX 785, BOONE, NC 28607

WAYNE COUNTY ARC, LLOYD E DAVIS K4EHZ PRES
3431 N 117 HWY, GOLDSBORO, NC 27530
WESTERN CAROLINA ARS, DAISY J GREEN KD4NZN
SECY, PO BOX 1488, ASHEVILLE, NC 28802
WILSON ARC, BOWIE GRAY WD4PZY EDITOR
PO BOX 3033, WILSON, NC 27895
YADKIN VALLEY ARC, LEE GROCE N4AAD VPRS
PO BOX 213, YADKINVILLE, NC 27055

NORTH DAKOTA

CENTRAL DAKOTA ARC, DEE TRACY KB0CGK TREAS
PO BOX 7162, BISMARCK, ND 58507
FORX ARC, PAUL SLUSSAR KA0CAF TRUSTEE
1816 1ST AVE N., GRAND FORKS, ND 58203
JAMESTOWN ARC, ARTHUR F JENSEN W0EOZ PRES
411-3RD ST SE, JAMESTOWN, ND 58401
MINOT ARA, LYNN NELSON WA0WBU
2700 23RD ST SW, MINOT, ND 58701
RED RIVER RADIO AMATEURS, MARK B KERKVLIET
KG0FR PRES, 507 WALLY ST., HARWOOD, ND 58042
THEODORE ROOSEVELT ARC, JOSEPH CALLAHAN
N0QAU VPRS, 1051 3RD AVE W., DICKINSON, ND 58601

OHIO

ALLIANCE ARC, PAMELA MYERS N8IAK PRES
510 W HARRISON ST, ALLIANCE, OH 44601
AMATEUR RADIO FELLOWSHIP, EDWARD T CLARK
K8ZM SECY, 1105 JESSIE AVE., KENT, OH 44240
ARC OF WALLY BYAM CARVAN CLUB, MARION C SNY
DER WB8MSE LIAS, PO BOX 30078, EAST CANTON, OH
44730
ASHLAND AREA ARC, ERIC WEBNER KA8FAN PRES
435 SNADER AVE., ASHLAND, OH 44805
ASHTABULA COUNTY RADIO CLUB, KENNETH L STEN
BACK AI8S PRES, 722 LYNDON AVE., ASHTABULA, OH
44004
ATHENS COUNTY ARA, CARL J DENBOW KA8JXG SECY
63 MORRIS AVE., ATHENS, OH 45701
BETHEL ARA, LYNN M ELBE KY8R PRES
3377 MUSGROVE RD, WILLIAMSBURG, OH 45176
BUTLER COUNTY VHF ASSN., THOMAS M BUCHANAN
AA8GC, 2587 UNMSTON AVE., HAMILTON, OH 45011
CAMBRIDGE ARC, FRED BARTON KB8MBI PRES
PO BOX 1545, CAMBRIDGE, OH 43725
CANTON ARC, SAUNDRA DAY N8TZB SECY
1607 19TH ST NE, CANTON, OH 44714
CASE ARC, ROBERT LESKEVEC K8DTS
118 ROCKEFELLER BLDG., 10900 EUCLID AVE.,
CLEVELAND, OH 44106
CENTRAL OHIO ARES, JOHN CHAPMAN WB8INY PRES
743 FLEETRUN AVE., GAHANNA, OH 43230
CHAMPAIGN LOGAN ARC INC., JERRY L TEMPLE N8MTZ
PRES, PO BOX 343, BELLEFONTAINE, OH 43311
CHIPPEWA ARC, DINO POCARO WD8DBQ PRES
25150 COLUMBUS RD, BEDFORD HEIGHTS, OH 44146
CLEVELAND STATE UNIV. ARC, GREGORY ROMANIAK
N8XOS PRES, 2121 EUCLID AVE., UNIVERSITY CTR
BOX 75, CLEVELAND, OH 44115
CLEVELAND WIRELESS ASSN INC., GLENN L WILLIAMS
AF8C SCTR, 513 KENILWORTH RD, BAY VILLAGE, OH 44140
CLINTON COUNTY ARA, HARRY PAULEY KB8UJN TREAS
641 LYTLE PLACE, WILMINGTON, OH 45177
CLYDE ARS, ERVIN REMALEY KA8CAS SCTR
333 BELLE AVE., BELLEVUE, OH 44811
COLUMBUS ARA, ERIC BABER KB8KOQ PRES
66 W COMO AVE., COLUMBUS, OH 43202
CONNEAUT AMATEUR RADIO CLUB, ANDREW SMITH
N8QBP EDITOR, 5 BISCOFF AVE., CONNEAUT, OH 44030
COSHOCTON COUNTY RADIO AMATEUR, THOMAS
CORDES KB8HEA SECY, PO BOX 501, COSHOCTON,
OH 43812
CUYAHOGA FALLS A.R.C., TED WANDS WD8CVH MEMB.
CHMN., 1930 NORTON RD, STOW, OH 44224
DAYTON AMATEUR RADIO ASSN, JIM SCHOETTINGER
WA8SVV SECY, PO BOX 44, DAYTON, OH 45401

DEFIANCE COUNTY ARC INC., BEN POLASEK KB8ISC
PRES, 8463 INDEPENDENCE RD, DEFIANCE, OH 43512
DEFOREST ARC, RICHARD NEWBAUER KA8EKC PRES
2832 UNITY RD, WEST UNION, OH 45693
DELAWARE ARA, PAUL L DAMON N8TPY SCTR
5683 US 42 N., DELAWARE, OH 43015
DIAL RADIO CLUB, GERALD B LOONEY N8WUY SECY
4718 SEBALD DR.,FRANKLIN, OH 45005
DRAKE ARC, ERNEST M HELTON W8MVN VPRS
36 WALNUT ST., FRANKLIN, OH 45005
EVENDALE ARS, JOHN P HAUNGS WA8STX SCTR
10615 THORNVIEW DR., CINCINNATI, OH 45241
FAROUT ARC, CHARLES D BERRY WD8NVY
4150 RONDEAU RIOGE, DAYTON, OH 45429
FAYETTE ARA, MAX ALEXANDER KA8ITE SCTR
6800 WHITE OAK RD NE, BLOOMINGBURG, OH 43106
FINDLAY AMATEUR RADIO CLUB, RON GRIFFIN N8AEH
EDITOR, PO BOX 587, FINDLAY, OH 45839
FIRELANDS AMATEUR RPTR ASSN., TOM O'CONNER
NI8G PRES, PO BOX 442, HURON, OH 44839
FULTON COUNTY ARC, ADAM T CATELY KB8MDF SCTR
319 W LEGGETT ST., WAUSEON, OH 43567
GEAUGA ARA, BETTY SCHOLZ KC8FF SCTR
14653 RUSSELL LN, NOVELTY, OH 44072
GRANT ARC, GARRY HOWARD N8SVW PRES
3363 CARPENTER RD, MT ORAB, OH 45154
GREATER CINCINNATI ARA, J EUGENE REILLY NU8U
PRES, 5856 WESTON CT, CINCINNATI, OH 45248
GREATER TOLEDO ARA INC., THOMAS P SPETZ WA8WZX
TRUSTEE, 2303 MURRAY DR., TOLEDO, OH 43613
HIGHLAND ARA, JEFF MARTIN NS8G SCTR
PO BOX 203, HILLSBORO, OH 45133
HUBER HEIGHTS ARC, RON MOOREFIELD W8ILC PRES.
6531 LEMANS LN, HUBER HEIGHTS, OH 45424
INTERCITY AMATEUR RADIO CLUB, JODI HARRINGTON
SECY, 1166 KELLER DR., MANSFIELD, OH 44905
JACKSON COUNTY ARC, BARRY L GREENE KA8PYH
TREAS, 1242 STANDPIPE RD, JACKSON, OH 45640
LAKE COUNTY ARA INC., ROBERT G BROADY KE8PW
LIAS, 5777 FENWOOD CT, MENTOR, OH 44060
LANCASTER & FAIRFIELD CTY ARC, EDWARD L CAMP
BELL WD8PGO, 1243 QUARRY RD SE, LANCASTER, OH
43130
LORAIN COUNTY ARA, WILBUR L WILSON WA8HED
PRES, 15494 WHEELER RD, LAGRANGE, OH 44050
MAHONING VALLEY ARA, JAMES T VIELE N8IRL PRES
161 FOX ST, HUBBARD, OH 44425
MARION ARC, KAREN K ECKARD N8JDH SCTR
6583 SOUTH STREET MEEKER, MARION, OH 43302
MASSILLON ARC, JERRY LAROCCA KF8EB SCTR
PO BOX 73, MASSILLON, OH 44648
MERCURY MIDWEST CHAPTER, PAUL FORGRAVE K8ES
PRES, 1850 CARRIAGE RD, POWELL, OH 43065
MIAMI COUNTY ARC, ED LATTA KA8CBE SECY
BOX 214, TROY, OH 45373
MID OHIO VALLEY ARC, MICHAEL MCKEAN N8PRX
PRES, 50 CHILLICOTHE RD, GALLIPOLIS, OH 45631
MILFORD ARC, PETE ENGEL N8OGW PRES
PO BOX 100, MILFORD, OH 45150
MOUNT VERNON ARC, BOB BRUFF N8PCE PRES
PO BOX 372, MOUNT VERNON, OH 43050
NORTH COAST ARC, H HAUSMANN WB8RNI PRES
PO BOX 30529, CLEVELAND, OH 44130
NORTH COAST CONTEST CLUB, BOB MATE AA8BV PRES
24132 ELM RD, NORTH OLMSTED, OH 44070
NORTHERN OHIO ARS, THOMAS W PORTER W8KYZ PRES
777 WASHINGTON AVE., ELYRIA, OH 44035
NORTHERN OHIO DX ASSN, TEDD MIRGLIOTTA KB8NW
PRES, 16806 W 130TH ST, STRONGSVILLE, OH 44136
OH-KY-IN ARS, R DANA LAURIE WA8M PRES
1031 OVERLOOK AVE., CINCINNATI, OH 45238
OH-KY-IN DXERS ARS, DANA LAURIE WA8M PRES
1031 OVERLOOK AVE., CINCINNATI, OH 45238
OHIO VALLEY ARA, JACK YERIGAN K8SW SECY
4693 MCCORMICK LN, FAIRFIELD, OH 45014

OHIO VALLEY REPEATER CLUB INC., RALPH L KEGLEY
WD8ICU PRES, 2554 DOGWOOD RIDGE RD, WHEELERS
BURG, OH 45694
OMAR H MUNG ARS, SANDY SHRIGLEY KB8FZX SECY
1725 39TH ST NW, CANTON, OH 44709
OTTAWA AREA RC, RICHARD MAAS WB8YHO SCTR
10267 CO RD M, OTTAWA, OH 45875
OTTAWA COUNTY ARC, THOMAS SCHEMMER WJ8V
PRES, AMERICAN RED CROSS, 109 MADISON ST, PORT
CLINTON, OH 43452
PARMA RADIO CLUB INC., THEODORE F HINZ NQ8I
7610 MARLBOROUGH AVE., PARMA, OH 44129
PIONEER A R FELLOWSHIP, DAVE RAINES KE8HA SECY
5081 ELNO ST., AKRON, OH 44319
PORTAGE ARC INC., JOANNE SOLAK KJ3O SECY
9971 DIAGONAL RD, MANTUA, OH 44255
QUEEN CITY EMERGENCY NET, DEAN WINKELMAN
KB8GFN PRES, 720 SYCAMORE DT, CINCINNATI, OH
45202
RESERVOIR ARA, WILLIAM L WHITE KB8LYI SECY
215 JEFFERSON ST., MENDON, OH 45862
SALEM AREA ARA, KEVIN BRANDT N8UEX PRES
2267 STATE ROUTE 45 S., SALEM, OH 44460
SANDUSKY RADIO EXPERIMENTAL LE, JEFF LEE
KB8DFV PRES, 2909 W PERKINS AVE., SANDUSKY, OH
44870
SANDUSKY VALLEY ARC, NINA BRAZZELL WY8B SECY
3316 SR 20, LINDSEY, OH 43442
SCIOTO VALLEY ARC, WAYNE E COOK KA8PQC SCTR
112 ALPINE DR., WAVERLY, OH 45690
SENECA RC, MATT PLOTTS N8WOC PRES
74 N TECUMSEH TRAILI, TIFFIN, OH 44883
SENECA REPEATER ASSN, GARY G FLECHTNER WJ8C
SCTR, 188 N STATE ROUTE 101, TIFFIN, OH 44883
SIMON PERKINS MIDDLE SCH ARC, CHARLES P FITZSIM
MONS N8LGE, SIMON PERKINS MIDDLE SCHOOL, 630
MULL AVE., AKRON, OH 44313
SOUTHWEST OHIO DX ASSN., BOB ESLAIRE W9UI PRES
8334 PLEASANT PLAIN RD, BROOKVILLE, OH 45309
STARK DX ASSN, JOHN H SCHAFFNER JR NM8K
6079 MEESE RD, LOUISVILLE, OH 44641
STEUBENVILLE-WEIRTON ARC, JOSEPH M PLESICH
W8DYF SECY, 173 BROCKTON RD, STEUBENVILLE, OH
43952
SUBURBAN ARA, JOHN T COLBURN SR. KF8LB PRES
1100 MAPLEWOOD ST LOT 24, DELTA, OH 43515
SUNDAY CREEK AR FEDERATION, RUSSELL ELLIS
N8MWK PRES, 8051 OREGON RIDGE, GLOUSTER, OH
45732
SYRIAN TEMPLE ARC, RALPH T WEHKING W8YDC PRES
1145 WITTSHIRE LN, CINCINNATI, OH 45255
TEAYS ARC, DON TINCH N8SIY SECY
PO BOX 13137, CIRCLEVILLE, OH 43113
TOLEDO MOBILE RADIO ASSN., BRENDA KRUKOWSKI
KB8IUP PRES, 9408 SALISBURY RD, MONCLOVA, OH
43542
TRI-STATE AMATEUR RPTR ASSN., ROBERT E HEATON
N8MCT, RR 1 BOX 545, TORONTO, OH 43964
TRIPLE STATES RADIO AMATEUR CL., RALPH A MC
DONOUGH K8AN PRES, BOX 240 RD 1, ADENA, OH
43901
TUSCO ARC, D ROBERT JACOBS N8RNL PRES
PO BOX 725, NEW PHILADELPHIA, OH 44663
TWENTY OVER NINE ARC INC., SHARON SPENCER
N8VPQ SECY, 424 PEFFER, NILES, OH 44446
UNION COUNTY ARC, R ALAN TOOPS N8PTD PRES
PO BOX 72, MILFORD CENTER, OH 43045
UNIV OF CINCINNATI ARC, RICK RIESS N8NVF PRES
340 TANGEMAN CENTER, CINCINNATI, OH 45221
UNIVERSITY SCHOOL ARC, ROBERT MORGAN K8RBV
20701 BRANTLEY RD, SHAKER HEIGHTS, OH 44122
VAN WERT ARC, LOUIE THOMAS WD8LLO PRES
PO BOX 602, VAN WERT, OH 45891
VOICE OF ALADDIN ARC, WILLIAM STEBLETON N8PPJ
SCTR, 7580 S OAKBROOK DR., REYNOLDSBURG, OH
43068

WARREN AMATEUR RADIO ASSN., DON GMUCS WD8OZN
PRES, PO BOX 809, WARREN, OH 44482
WAYNE AMATEUR RADIO TECH SOC, STAN KINNEY
N8RNK SCTR, 1039 E WAYNE AVE., WOOSTER, OH
44691
WEST PARK RADIOPS ARC, GLENN WILLIAMS AF8C
SECY, 513 KENILWORTH RD, BAY VILLAGE, OH 44140
WESTERN RESERVE RADIO ASSN., BERT W BUGANSKI
WA8TTTZ PRES, 7501 HUDSON PARK DR., HUDSON, OH
44236
WILLIAMS COUNTY ARC, LAMARR G RUPP K8MZY SECY
358 E HIGH ST, BRYAN, OH 43506
WOODCHUCK ARC, ROGER T WENNER N8WEH SECY
8547 SAYBROOK DR., BROOKLYN, OH 44144

OKLAHOMA

ADA ARC, THOMAS GILLIAM KB5STS
PO BOX 1333, ADA, OK 74821
ALTUS AREA ARA, ROBERT HERON KE4BN SCTR
PO BOX 73, ALTUS, OK 73522
AMERICAN AIRLINES ARC, RICHARD EICHELBERGER
KJ5NA, 8920 S BRADEN AVE., TULSA, OK 74137
BARTLESVILLE ARC, INC., EARL E VAUGHN AA5ZU SECY
1405 CEDAR ST, BARTLESVILLE, OK 74006
BROKEN ARROW ARC, BERNIE COOPER N5PCZ PRES
PO BOX 552, BROKEN ARROW, OK 74013
CANADIAN VALLEY ARC, CLYDE L WARD WB5CPA
SCTR, PO BOX 207, WASHINGTON, OK 73093
CHISHOLM TRAIL ARC INC., BETTY CANFIELD KC5CYZ
PRES, RR 5 BOX 510B, DUNCAN, OK 73533
ENID ARC, TOM WORTH N5LWT SECY
P O BOX 261, ENID, OK 73702
KAY COUNTY ARC, MIKE MORRIS N5JJR PRES
PO BOX 2750, PONCA CITY, OK 74602
LAWTON FT SILL ARC, PAUL GOULET KC5CYY SCTR
PO BOX 892, LAWTON, OK 73502
MUSKOGEE ARC, DAVID L HAMMOND KA5IIS SCTR
1101 BARRON RD, FORT GIBSON, OK 74434
OKLAHOMA DX ASSOCIATION, COY C DAY N5OK PRES
PO BOX 73, OWASSO, OK 74055
OKLAHOMA INDEPENDENT ARC, MICKEY PHILLIPS
N5VEA PRES, PO BOX 1051, BLACKWELL, OK 74631
SOUTHWEST OK REPEATER ASSN, LOREN C SIMMS
WA5CBF SECY, 209 MOCKINGBIRD DR N., ALTUS, OK
73521
TULSA ARC, INC., RICHARD MORGAN WB5IQS PRES
PO BOX 4283, TULSA, OK 74159
WHEATSTRAW ARC, JOSEPH GARLAND WA5FLT SCTR
P O BOX 204, 241 N FREEHOME AVE., CALUMET, OK
73014

OREGON

AMATEUR RADIO RELAY GROUP, WARREN WINNER
W7JDT TREAS, PO BOX 10031, PORTLAND, OR 97210
ARC SILVERTON, RICHARD G AHLEFELD KD7X PRES
405 WHITTIER ST, SILVERTON, OR 97381
CENTRAL OREGON DX CLUB, WILLIAM J SAWDERS
K7ZM PRES, 19821 PONDEROSA ST, BEND, OR 97702
CENTRAL OREGON RADIO AMATEURS, NANCY A SMITH
N7RPE SECY, PO BOX 723, BEND, OR 97709
CLACKAMAS ARC, HENRY STEWART N7WMW SECY
13610 SE BREEN ST, MILWAUKIE, OR 97222
COOS COUNTY RADIO CLUB, ZANE C ALBERTSON
WA7OXM, RR 1 BOX 199-D6, COQUILLE, OR 97423
CRESCENT VALLEY HS, ARTHUR E POWELL N7IAR ADVI
SOR, 4444 NW HIGHLAND DR., CORVALLIS, OR 97330
EMERALD ARS, SIDNEY A STRONG KA7C PRES
1560 SCANDIA ST, EUGENE, OR 97402
GRANDE RONDE RADIO AMATEURS, LOIS E ROGERS
KA7BBE SCTR, PO BOX 171, LA GRANDE, OR 97850
HERMISTON ARC, GARY COOPER N7ZHG PRES
PO BOX 962, HERMISTON, OR 97838
HOODVIEW ARC, PATSY BARMORE KA7MZZ LIAS
PO BOX 459, SANDY, OR 97055

KENO ARC, THOMAS A HAMILTON WD6EAW
PO BOX 678, KENO, OR 97627
KLAMATH BASIN ARA, FRED BECHDOLDT ND7V SECY
JEFFERSON MALL, P O BOX 8106, KALMATH FALLS,
OR 97601
LINCOLN COUNTY ARC, WILLIAM WELLMAN ND7I PRES
11525 SE ASH ST, SOUTH BEACH, OR 97366
MCMINNVILLE ARC, TOM WHITELAW N7OHH TREAS
PO BOX 891, MCMINNVILLE, OR 97128
MID-WILLAMETTE ARC, FRED DICKSON W7LBH SECY
38566 PARKSIDE RD NE, ALBANY, OR 97321
NORTHWEST AR COUNCIL, PO BOX 5097, ALOHA, OR
97006
OREGON TUALATIN VALLEY ARC, BILL MERWIN N7VZF
PRES, 10740 SW SUNNYHILL LN, BEAVERTON, OR
97005
PENDLETON ARC, DENTON SPRAGUE WB7TDG SCTR
709 SW 13TH ST, PENDLETON, OR 97801
PORTLAND ARC, RON MAYER K7BT TRUSTEE
6115 SE 13TH AVE., PORTLAND, OR 97202
ROGUE VALLEY ARC, JOSH MOULIN N7VTM PRES
PO BOX 1021, JACKSONVILLE, OR 97530
SALEM ARC INC., WILLIAM BERRY KG7OU
PO BOX 61, SALEM, OR 97308
SANTIAM RADIO CLUB, PAUL M ZIMICK WB9HZT PRES
2395 CASCADE DR., LEBANON, OR 97355
SOUTHERN OREGON RADIO CLUB, PAUL C JOHNSTON
WM7K PRES, PO BOX 1164, GRANTS PASS, OR 97526
STEWART LAKE ARC, DAVID HACKLEMAN WA7IQH
PRES, C/O HEWLETT-PARKARD, 1000 NE CIRCLE BLVD,
CORVALLIS, OR 97330
SUNSET EMPIRE RC, E M MARRIOTT N7LWQ PRES
PO BOX 264, ASTORIA, OR 97103
TILLAMOOK EMERG. AR SERVICE, F A HALVERSON
N7ILB, 401 MCCORMICK LOOP RD, TILLAMOOK, OR
97141
UMPQUA VALLEY ARC, ED PAHL W5PII TREAS
1440 WILD IRIS LN, ROSEBURG, OR 97470
VALLEY RADIO CLUB OF EUGENE, JAMES WALSH
W7LVN, 159 E 16TH AVE., EUGENE, OR 97401
WESTERN OREGON RADIO CLUB INC., FRANK L HOFF
MAN WB7DZG PRES, P O BOX 2259, BEAVERTON, OR
97075
WILLAMETTE VALLEY DX INC. CLUB, BOB NORIN
W7YAQ SECY, PO BOX 555, PORTLAND, OR 97207

PENNSYLVANIA

96 OVER THE HILL GANG/METRO, P O BOX 1041, LIN
WOOD, PA 19061
ADAMS COUNTY ARS, ROBERT B LAUDER JR SCTR
40 GRASSGIOOER LANE, FAIRFIELD, PA 17320
ALLEGHENY COLLEGE ARA, KAREN COX KB8PRP
BOX 1154, ALLEGHENY COLLEGE, MEADVILLE, PA
16335
ALLENTOWN WORKS ARC, STEVEN E STRAUSS NY3BY
PRES, AT&T BELL LAB RM 23R-135GA, 555 UNION
BLVD, ALLENTOWN, PA 18103
ANTHRACITE REPEATER ASSN INC., JOSEPH R KOVAL
W3IVG SECY, 705 W 12TH ST, HAZLETON, PA 18201
BEACON RADIO AMATEURS, NATHAN SHUMAN W3ATR
6628 HOLLIS ST, PHILADELPHIA, PA 19138
BEAVER VALLEY ARA, WILLIAM R LAUDERBAUGH
KE3BZ, RR 2 BOX 2392, WAMPUM, PA 16157
BUTLER COUNTY ARA, GWEN BEATTY N3EMT SECY
316 HOON RD, BUTLER, PA 16001
CAMERON COUNTY ARC, FRANK METZLER KD3LM SECY
1562 BUCKTAIL RD, SAINT MARYS, PA 15857
CARBON ARC, LARRY LILLY KA3AFY PRES
PO BOX 4162, JIM THORPE, PA 18229
CARNEGIE TECH RADIO CLUB, THOMAS M STRICKER
N3JNK SCTR, STUDENT ACTIVITY INFO DESK,
CARNIEGIE MELLON UNIVERSITY, PITTSBURGH, PA
15213
CENTRAL PA REPEATER ASSN., KEN SUTTON WA3HDP
PRES, P O BOX 60101, HARRISBURG, PA 17112

COLUMBIA AREA ARC, JAMES E BEAR   WB3FQY PRES
    PO BOX 126, LANCASTER, PA 17608
COLUMBIA-MONTOUR ARC, DAVE SCHACK WC3A
    TREAS, 6020 FORT JENKINS LN, BLOOMSBURG, PA
    17815
CONEMAUGH VALLEY ARC, JOHN H LENZ WA3BIX VPRS
    117 ROCKWELL AVE., JOHNSTOWN, PA 15905
CRAWFORD ARS, JAMES J CIHON K3TLP SECY
    743 BEVERLY DRIVE RD #7, RR 7, MEADVILLE, PA
    16335
CUMBERLAND ARC, GARY L FASICK   K3EYK SCTR
    107 HILLTOP RD, BOILING SPRINGS, PA 17007
CUMBERLAND COUNTY ARS, BRIAN R GREENWAY
    K3WKK, 28 PINE ST., CARLISLE, PA 17013
CUMBERLAND VALLEY ARC, PAUL BEVARD KA3RVM
    PRES, PO BOX 172, CHAMBERSBURG, PA 17201
DAUBERVILLE DX ASSN, DAN DEVINE WB3FYL PRES
    PO BOX 73, DAUBERVILLE, PA 19517
DELAWARE VALLEY OMIK ELEC COMM, VINCENT C
    BRAXTON K3SNZ PRES, 6047 N 21ST STREET,
    PHILADELPHIA, PA 19138
DELAWARE VALLEY VHF SOC, DAVID MICHAEL N3FUJ
    VPRS, 689 IVERS LN, WARMINSTER, PA 18974
DELAWARE-LEHIGH ARC INC., ROBERT A GREEN KE3AW
    PRES, GREYSTONE BUILDING, RD 4, NAZARETH, PA
    18064
DELMONT RADIO CLUB, ROBERT R SMITH KK3KP PRES
    1215 CHURCH RD, ORELAND, PA 19075
DIGITAL COMMUNICATIONS OF BEAV, DAVID HEIM
    KA3SMF SECY, PO BOX 9, SOUTH HEIGHTS, PA 15081
DREXEL UNIV ARC, WILLIAM F SORETHN3QVA PRES
    3915 LAWNDALE ST, PHILADELPHIA, PA 19124
DUQUESNE UNIV ARC, J CLIFTON HILL W3GLA ADVS
    PHYSICS DEPT, DUQUESNE UNIV, PITTSBURGH, PA
    15282
EASTERN PA VHF SOC INC., GARY H WEISS N3ECW SCTR
    1098 OAKHURST DR., SLATINGTON, PA 18080
ELK COUNTY ARA, ROSIE SCHNEIDER KE3JE SECY
    PO BOX 175, SAINT MARYS, PA 15857
ENDLESS MOUNTAINS ARC, DAN WILLIAMS   KA3TOV
    PRES, BOX 261   RD #2, DALTON, PA 18414
EPHRATA AREA REPEATER SOC.,INC., MARY HAMILL
    N3EPT, 435 NETZLEY DR.,DENVER, PA 17517
FOOTHILLS ARC INC., JIM YEX WB3CQA PRES
    PO BOX 236, GREENSBURG, PA 15601
FORT ARMSTRONG WIRELESS ASSN, ROBERT TRAUTER
    MAN W3YEY PRES, 664 ROSS AVE., FORD CITY, PA
    16226
FORT VENANGO MIKE & KEY CLUB, DEBRA CRISE   NV3B
    PRES, RR 1 BOX 591, CRANBERRY, PA 16319
FRANKFORD RC, JACK IMHOF N2VW PRES
    PO BOX 431, ALBURTIS, PA 18011
GANNON UNIV WIRELESS SOC, MATTHEW STEGER
    N3NTJ PRES, GANNON UNIVERSITY, BOX 1031, ERIE,
    PA 16541
GREATER PITTSBURGH VHF SOC, JODY NELIS K3JZD
    TRES, 132 AUTUMN DR., TRAFFORD, PA 15085
GROVE CITY COLLEGE ARC, AL CHRISTMAN KB8I ADVI
    SOR, 1101 W MAIN ST., GROVE CITY, PA 16127
HANOVER AREA HAMMING ASSN, KEN PARVIS KA3RAI
    SECY, 1054 BALTIMORE ST, HANOVER, PA 17331
HARFORD AR EMERGENCY MGT TEAM, DAVE HAHN
    N3CNJ VPRS, RR 2 BOX 459J, DELTA, PA 17314
HARRISBURG RADIO AMATEUR CLUB, THOMAS B HALE
    WU3X PRES, PO BOX 418, HALIFAX, PA 17032
HAVERFORD TOWNSHIP EMERGENCY R, HAROLD P
    GRACE W3HFY SECY, 108 N CONCORD AVE., HAVER
    TOWN, PA 19083
HEADWATERS ARC, ALONZO W BUNCH N3BUM PRES
    PO BOX 131, COUDERSPORT, PA 16915
HILLTOP REPEATER ASSN, DONALD L WALKER WA3IMX
    4049 ELTON RD, JOHNSTOWN, PA 15904
HOLMESBURG ARC, JAY KUPERMAN WA3IFY TREAS
    1934 DEVEREAUX AVE., PHILADELPHIA, PA 19149

HORSESHOE ARC, TOM COONEY JR KD3SA SECY
    4006 CORTLAND AVE., ALTOONA, PA 16601
HUNTINGDON COUNTY ARC, BUTCH HINTON KD3YT
    SECY, P O BOX 462, HUNTINGDON, PA 16652
INDIANA COUNTY ARC, DICK CHRISTENSEN N3EHO
    SCTR, 2660 MELLONEY LN, INDIANA, PA 15701
IRWIN AREA ARA, JODY NELIS K3JZD TREAS
    132 AUTUMN DR., TRAFFORD, PA 15085
KENSINGTON HS SOC PROP WAVES, ALBERT MEINSTER
    K3EAX, 8940 KREWSTOWN RD APT 301, PHILADEL
    PHIA, PA 19115
KEYSTONE ARC, RICHARD A MOLL W3RM SECY
    POB 88, ABINGTON, PA 19001
KEYSTONE VHF CLUB INC., GEORGE W KUTCHER JR
    N3GKP PRES, 1333 W MARKET ST, YORK, PA 17404
LAWRENCE COUNTY ARA, JOSEPHINE CUNNINGHAM
    WV3Q PRE, RR 4 BOX 242, NEW CASTLE, PA 16101
LEBANON VALLEY SOC OF RADIO AM, RANDY MILLER
    N3FOG PRES, 303 W CUMBERLAND ST, LEBANON, PA
    17042
LEHIGH VALLEY ARC INC., PAUL RYAN N0KIA
    6359 OPPOSSUM LN, SLATINGTON, PA 18080
LOG COLLEGE MIDDLE SCHOOL ARC, THOMAS E
    MICHAUD WA3TQJ ADVIS, 720 NORRISTOWN RD,
    WARMINSTER, PA 18974
MCKEAN COUNTY ARC, MERLE JOHNSON   KF2JB SCTR
    PO BOX 107, BRADFORD, PA 16701
MEGA CONNECTION, MELVIN C MARTZ N3IPW   PRES
    408 GLENHAYES DR., ALIQUIPPA, PA 15001
MERCER COUNTY ARC, TIM DUFFY K3LR PRES
    44 ELLIOT RD, WEST MIDDLESEX, PA 16159
MID-ATLANTIC ARC, TOM PORETT N3JMA PRES
    673 AUBREY AVE., ARDMORE, PA 19003
MOBILE SIXERS RADIO CLUB INC., JOHN C DYCKMAN
    WA3KFT, 117 RICHARD RD, ASTON, PA 19014
MT AIRY VHF RADIO CLUB INC., GEOFFREY H KRAUSS
    WA2GFP, 1927 AUDUBON DR., DRESHER, PA 19025
MURGAS AMATEUR RADIO CLUB, CAROL NYGREN
    KA3EEO PRES, PO BOX 1094, WILKES BARRE, PA 18703
NITTANY ARC, WILBER D FILES W3SAY TREAS
    PO BOX 614, STATE COLLEGE, PA 16804
NORTH CLARION SCHOOL ARC, GLORIA BARLETT N3IOP
    ADVISOR, 242 NORTH MAIN ST., PO BOX 12, KNOX, PA
    16232
NORTH COAST CONTESTERS, TIM DUFFY K3LR PRES
    PO BOX 59, NEW BEDFORD, PA 16140
NORTH HILLS ARC, C BUD FAULHABER N3DOS   TREAS
    1059 BALMORAL DR., PITTSBURGH, PA 15237
OLYMPIA RADIO AMATEUR CLUB, WILLIAM E RHOADS
    SR ND3Q, 4726 SHELDON ST, PHILADELPHIA, PA 19127
PA EMERG COMMUNICATIONS COUNCIL, STEVE GOBAT
    KA3PDQ PRES. PO BOX 8797, ALLENTOWN, PA 18105
PENN STATE ARC, DAN RANERI N2OQN PRES
    HUB INFORMATION DESK, UNIVERSITY PARK, PA
    16802
PENN WIRELESS ASSN INC., CHARLES H SCHULTZ
    KE3CH, PO BOX 734, LANGHORNE, PA 19047
PENN-MAR RADIO CLUB, GARY VIANDS KE3FN PRES
    PO BOX 763, HANOVER, PA 17331
PHIL-MONT MOBILE RADIO CLUB, RICHARD A MOLL
    W3RM, POB 88, ABINGTON, PA 19001
PHILADELPHIA AREA REPEATER ASSOC., RICHARD E
    STEWART K3ITH PRES, PO BOX 954, VALLEY FORGE,
    PA 19482
POCONO AMATEUR RADIO KLUB, DEBBIE PORTER
    KE3FY PRES, PO BOX 985, BRODHEADSVILLE, PA 18322
PUNXSUTAWNEY ARC, ROGER HACKENBERRY KA3TWB
    SECY, RR 1 BOX 325, SUMMERVILLE, PA 15864
PYMATUNING ARC, BRIAN E DICK NI3B SCTR
    38 LOPER RD, HADLEY, PA 16130
QUAD COUNTY ARC, JAMES F BYRNE KA3WSX PRES
    RT 410 G D, TROUTVILLE, PA 15866
R F HILL ARC INC., JOHN LAWSON   W3ZC   TREAS
    PO BOX 29, COLMAR, PA 18915

RADIO AMATEURS OF CORRY, NORMA R VANDERHOFF
W3CG SCTR, 713 W CHURCH ST, CORRY, PA 16407
RADIO ASSOCIATION OF ERIE, DAN MILLER K3UFG PRES
130 W 21ST ST, ERIE, PA 16502
READING RADIO CLUB INC., HARRY HOFFMAN W3VBY
PRES, PO BOX 13777, READING, PA 19612
RED ROSE REPEATER ASSN, CARLOS M RIVERA WV3Y
PRES, PO BOX 8316, LANCASTER, PA 17604
SKYVIEW RADIO SOCIETY, JOHN E THOMPSON WB3FYP
1014 CABLE AVE., PITTSBURGH, PA 15238
SOUTH HILLS ARC, JIM MOUNTS KA3EBX PRES
211 GERRIE DR., PITTSBURGH, PA 15241
SOUTH MOUNTAIN REPEATER ASSN, JOHN R YUPATOFF
KD3QZ SCTR, 9 GRANT CV, EAST BERLIN, PA 17316
STEEL CITY ARC, INC., MARK STRAYER KE3JA
5976 MCPHERSON AVE., BETHEL PARK, PA 15102
SULLIVAN COUNTY ARA, ROBERT D MONTGOMERY
N3ERE, 316 PENNSYLVANIA AVE., WATSONTOWN, PA
17777
SUSQUEHANNA COUNTY ARC, ELLEN D THOMPSON
N3EMI LIAS, RR 1 BOX 1M, GREAT BEND, PA 18821
SUSQUEHANNA VALLEY ARC, DAVID WELKER AA3BO
SECY, PO BOX 73, HUMMELS WHARF, PA 17831
SWARTHMORE COLLEGE ARC, SAM WEILER N3NSR
SWARTHMORE COLLEGE, SWARTHMORE, PA 19081
TAMAQUA AREA SIDE BAND ARA, ALLEN R BREINER
W3TI SEC, 212 RACE ST, TAMAQUA, PA 18252
TEMPLE UNIVERSITY ARC, DENNIS SILAGE WB2LGJ
PRES, DEPT OF EE TEMPLE UNIVERSITY, 12TH & NOR
RIS ST, PHILADELPHIA, PA 19122
TIOGA COUNTY ARC, DEBORAH CRAWFORD N3FVQ
PRES, RR 5 BOX 112, WELLSBORO, PA 16901
TRI STATE ARC, RUDY SWEISFURTH N3EBG
1579 KELLY ANN DR., WEST CHESTER, PA 19380
TRI-COUNTY AMATEUR RADIO GROUP, DENNIS SHEGDA
N3KRE PRES, 1565 ARLINE AVE., ABINGTON, PA 19001
TRI-COUNTY TRI-BANDERS, T VAN HAMEL N2HTF SECY
243 LEEDOM WAY, NEWTOWN, PA 18940
TRI-STATE ARC, RON SCHUCK WA2SNL PRES
PO BOX 292, MATAMORAS, PA 18336
TRIAC, BOB REYMOS KA3SMV PRES
526 DOGWOOD LN, COATESVILLE, PA 19320
TRIPLE A ARA INC., DAVID J LEISER K3NPX PRES
9 LEXINGTON DR., FREEDOM, PA 15042
TWO RIVERS AMATEUR RADIO, LOU ZIMMWERMAN,
WV3Z PRES, P O BOX 225, GREENOCK, PA 15047
UNIONTOWN ARC, D, HERB SHAFFER KA3WSO PRES
RR 2 BOX 50, PERRYOPOLIS, PA 15473
UNISYS ARC-BLUE BELL, JOHN H TAYLOR W3YSB SECY
324 MEADOWBROOK RD, NORTH WALES, PA 19454
UNIV OF PITTSBURGH PANTHER ARC, APRIL RODGERS
N3PPU, UNIVERSITY OF PITTSBURGH, 501 WILLIAM
PITT UNION, PITTSBURGH, PA 15260
UNIVERSITY OF PA RADIO CLUB, ERIC REITER WI2N
PRES, 101 S 39TH ST APT C202, 200 S 33RD ST,
PHILADELPHIA, PA 19104
WARMINSTER AMATEUR RADIO CLUB, AL FOLSOM
KY3T SECY, P O BOX 113, WARMINSTER, PA 18974
WASHINGTON AMATEUR COMM INC., STEVEN G
ELLIOTT KA3UDR, RR 2 BOX 402, EIGHTY FOUR, PA
15330
WEST BRANCH ARA, KIM DOCKEY KA3TTH PRES
PO BOX 3002, WILLIAMSPORT, PA 17701
WESTERN PA DX ASSOCIATION, GEORGE GROSS KA3JWJ
SECY, 108 EASTERN DR., LOWER BURRELL, PA 15068
WESTERN PA HILLTOPPERS, DONALD FLETCHER
KA3WZO SCTR, 62 LINCOLN AVE., PITTSBURGH, PA
15205
WYNDMOOR REPEATER CLUB, ROBERT W AGANS SR
WA3EPA SECY, 99 S SHADY RETREAT RD,
DOYLESTOWN, PA 18901

PUERTO RICO

PUERTO RICO AR LEAGUE, VICTOR M MADERA KP4PQ
VPRS, PO BOX 191917, SAN JUAN, PR 00919

PUERTO RICO DX CLUB, INC., SERGIO RUBIO KP4L PRES
PO BOX 191917, SAN JUAN, PR 00919

RHODE ISLAND

ASSOCIATED RA OF SOUTHERN NE, MICHAEL ARMINO
NO1U PRES, 54 KELLY AVE., E PROVIDENCE, RI 02916
BLACKSTONE VALLEY ARC, RAYMOND ST ONGE W1HW
TRUSTEE, 18 BLANCHE AVE., CUMBERLAND, RI 02864
FIDELITY ARC, BOB RITOLI NE1E PRES
31 MARCY ST, CRANSTON, RI 02905
JAMESTOWN SCHOOL ELEC GROUP, JAMES I SAMMONS
KA1ZOU ADVISOR, 55 LAWN AVE, PO BOX 318,
JAMESTOWN, RI 02835
N RHODE ISLAND RC, JOHN VOTA WB1FDY TRUSTEE
41 BROOKSIDE AVE, NORTH PROVIDENCE, RI 02911
NEWPORT COUNTY RADIO CLUB, ROBERT DAY ND1CL
PRES, SEAMAN'S CHURCH INSTITUTE, 1 MARKET SQ.,
NEWPORT, RI 02840
OCEAN STATE AR GROUP INC., CHARLES HATHAWAY
WJ1K PRES, PO BOX 8238, CRANSTON, RI 02920
PROVIDENCE RADIO ASSN INC., DR RICHARD ROSEN
K1DS SECY, 1 LUDLOW ST, JOHNSTON, RI 02919
ROGER WILLIAMS VHF SOCIETY, MICHAEL G CAR
DARELLI JR N1FKI, PO BOX 40001, PROVIDENCE, RI
02940
SOUTH COAST WIRELESS SOCIETY, MATTHEW T
ATWOOD N1RJF SECY, PO BOX 1516, WESTERLY, RI
02891
UNIVERSITY OF RHODE ISLAND ARC, WILLIAM J HAM
N1SMY PRES, MEMORIAL UNION BLDG RM 138, URI
STUDENT CENTER, KINGSTON, RI 02881
VIKING AMATEUR RADIO SERVICE, EDNA D BARTRAM
WA1WKK PRES, 94 KANE AVE., MIDDLETOWN, RI
02842
WASHINGTON COUNTY ARA, MIKE BILOW N1BEE PRES
40 PLANTATION DR., CRANSTON, RI 02920
WOONSOCKET HIGH SCHOOL ARC, JOSEPH DESROSIERS
N1JFY PRES, 777 CASS AVE., WOONSOCKET, RI 02895

SOUTH CAROLINA

ANDERSON RC, JAMES KAUFMAN N4NWB SECY
PO BOX 1525, ANDERSON, SC 29622
BLUE RIDGE ARS INC., SUE CHISM N4ENX SECY
PO BOX 6751, GREENVILLE, SC 29606
CAROLINA DX ASSOCIATION, WILLIAM L JENNINGS
W4UNP SCTR, 630 WHITE PINE RD, CATAWBA, SC
29704
CHARLESTON ARS, SHEILA FRANK KC4UDD SECY
614 LONG STREET CIR, SUMMERVILLE, SC 29483
CHARLESTON SOUTHERN UNIV -CSA, MICHAEL B CONE
KB4REI PRES, 9200 UNIVERSITY BLVD, CSU #298 P O
BOX 118087, CHARLESTON, SC 29423
COLUMBIA ARC, ERIC ZIMMERMAN WB9OLE
PO BOX 595, COLUMBIA, SC 29202
CRESCENT HILL AMATEUR OP SOC., GENE STANFIELD
KC4ZZI PRES, PO BOX 2343, ORANGEBURG, SC 29116
FLORENCE ARC, DAVID W TILLEY WD4KAT SCTR
PO BOX 5062, FLORENCE, SC 29502
GRAND STRAND ARC, BRAD FRANCE-KELLY WA4YBD
SCTR, 640 TIGER PAW RD, ORIS, SC 29569
ISLANDER ARA, JAMES GASEN KR4DF PRES
31 DEER RUN LN, HILTON HEAD ISLAND, SC 29928
LAURENS ARS, CHARLES H GUERRY WD4CWY PRES
PO BOX 1394, LAURENS, SC 29360
MCDUFFIE HIGH SCHOOL ARC, MICHAEL A EPSTEIN
KD1DS, 108 KANARD RD, STARR, SC 29684
NORTH AUGUSTA-BELVEDERE RADIO, MICHAEL D FUL
FORD KC4YLD PRES, 304 LAURENS ST, AIKEN, SC
29801
NORTH MYRTLE BEACH VHF CLUB IN, JOSEPH JONES
W3TMO SCTR, PO BOX 532, NORTH MYRTLE BEACH,
SC 29597
PALMETTO ARC, JOHN MOOD KD4HTX SECY
242 PAR CT, GASTON, SC 29053

RIDGE ARC, W A BILL ROBISON W4GIV PRES
P O BOX 1046, LEESVILLE, SC 29070
SALKEHATCHIE ARS, EDGAR A HARTZOG KE4FIC SCTR
P O BOX 454, BLACKVILLE, SC 29817
SPARTANBURG ARC, JOSEPH G SAYRE W4OJZ SCTR
104 GARNER RD, SPARTANBURG, SC 29303
TRIDENT ARC, SCOTT CHIPPENDALE WB3EFS PRES
PO BOX 73, SUMMERVILLE, SC 29484

SOUTH DAKOTA

BLACK HILLS ARC, GARY PETERSON K0CX SECY
PO BOX 294, RAPID CITY, SD 57709
HOT SPRINGS ARC, SUSAN KNAPP KA0IGV SECY
PO BOX 173, HOT SPRINGS, SD 57747
HUB CITY ARC, BOB KING KA0QVK PRES
821 12TH AVE. SE, ABERDEEN, SD 57401
HURON ARC INC., LLOYD TIMPERLEY WB0ULX SCTR
PO BOX 205, HURON, SD 57350
LAKE AREA RADIO KLUB, DENNY WARRICK WB0MWJ
SCTR, PO BOX 642, WATERTOWN, SD 57201
MOBRIDGE AREA ARC, DAVID DEKKER K0ERM SECY
617 2ND AVE W., MOBRIDGE, SD 57601
PRAIRIE DOG ARC, ROY JORGENSEN W0MMQ TRUSTEE
803 E LEWIS ST., VERMILLION, SD 57069
SD SCHOOL OF MINES & TECH ARC, NEAL H HODGES II
KA0SEZ TRUS, MIS DEPT/SD SCH OF MINES & TE, 501 E
SAINT JOSEPH ST., RAPID CITY, SD 57701
SIOUX EMPIRE ARC, ROBERT MCCAFFREY KB0GNG
PRES, PO BOX 91, SIOUX FALLS, SD 57101

TENNESSEE

ANDREW JOHNSON ARC, FORREST L PRIDE WB4KEG
SCTR, 408 SUNSET BLVD, GREENVILLE, TN 37743
ARS OF TENNESSEE TECH, BRYAN D YOUNG KD4IIC
PRES, TTU BOX 5262, COOKEVILLE, TN 38505
BIG SOUTH FORK ARC, RANDALL GILREATH KC4ZSQ
SCTR, P O BOX 175, WINFIELD, TN 37892
BRISTOL AMATEUR RADIO CLUB, JAMES T OXENDINE
KB4EX, 276 ALHAMBRA DR., BRISTOL, TN 37620
CAMPBELL COUNTY ARC, JIMMY ELKINS KQ4AB TREAS
RR 1 BOX 1199, JACKSBORO, TN 37757
CAMPBELL COUNTY H.S. ARC, JERRY L STOUT KB4VRE
SPONSOR, 125 MAPLE DR., LA FOLLETTE, TN 37766
CHATTANOOGA ARC, WILLIAM F WIGGINS N4BMR
SECY, PO BOX 23121, CHATTANOOGA, TN 37422
CLARKSVILLE AMATEUR TRANSMITTING, MICHAEL R
WARNER NX7T SCTR, 447 WINDING WAY RD,
CLARKSVILLE, TN 37043
DELTA ARC, STEVE FELTMAN KC4ZOV PRES
6034 PEBBLE BEACH AVE., MEMPHIS, TN 38115
EAST TENNESSEE DX ASSN., JOHN ANDERSON AG4M
1041 W OUTER DR., OAK RIDGE, TN 37830
HAWKINS COUNTY ARC, MAURICE DE WITTE N4GQU
PRES, TIMBERLAKE SUBDIVISION, ROGERSVILLE, TN
37857
HUMBOLDT ARC, WILLIAM E HOLMES W4IGW SCTR
501 N 18TH AVE., HUMBOLDT, TN 38343
JACKSON RADIO CLUB, INC., KENNY JOHNS AB4EG PRES
PO BOX 3382, JACKSON, TN 38303
JOHNSON CITY RADIO ASSN INC., ED INGRAHAM WX4S
SECY, 377 A A DEAKINS RD, JONESBOROUGH, TN
37659
KERBELA AMATEUR RADIO SERVICE, CARL MICHAELS
WA4TKN SCTR, 231 TIPTON STATION RD, KNOXVILLE,
TN 37920
LAFOLLETTE MIDDLE SCHOOL ARC, VIC KING N4RFV
SPONSOR, 217 COUNTRY CLUB RD, LA FOLLETTE, TN
37766
MAURY ARC, GEORGE T RUSSELL WB43JCR SECY
PO BOX 832, COLUMBIA, TN 38402
MID SOUTH ARA, STAN MOORE AC4CQ TREAS
PO BOX 751841, MEMPHIS, TN 38175
MID SOUTH VHF ASSOCIATION, J DALTON MCCRARY
N4OYS SECY, 2111 BROOKWOOD LN, MURFREESBORO,
TN 37129

MID-SOUTH DX ASSOCIATION, JIM H MCLEAN K4XO
SECY, 1527 W CHURCHILL DOWNS, GERMANTOWN,
TN 38138
MIDDLE TENNESSEE ARS, TERRY BARTHOLOMEW NQ4Y
PRES, PO BOX 932, TULLAHOMA, TN 37388
MIDDLE TENNESSEE DX ARC, BERT NOLL K4UVH
TRUSTEE, 6935 BETHEL RD, GREENBRIER, TN 37073
MORGAN COUNTY ARC INC., PHILIP MEHLHORN
WD4ORB SCTR, RR 1 BOX 129A, OLIVER SPRINGS, TN
37840
MUSIC CITY ARA, ARIEL M ELAM K4AAL SCTR
1065 BARNES RD, ANTIOCH, TN 37013
NASHVILLE ARC, COL MURRAY G JONES K4ANH
1044 FOREST HARBOR DR., HENDERSONVILLE, TN
37075
OAK RIDGE ARC INC., JAMES E WHITTLESEY KC4RHW
PO BOX 4291, OAK RIDGE, TN 37831
RAC OF KNOXVILLE, SHIRLEY RUSSELL KC4FGE PRES
PO BOX 124, KNOXVILLE, TN 37901
RADIO AMATEUR TRANSMITTING SOC, MILTON H FAN
NING WA4GZZ, 4936 DANBY DR., NASHVILLE, TN 37211
REELFOOT ARC, GLENN SNOW N4MJ PRES
8959 BRUNDIGE RD, SOUTH FULTON, TN 38257
ROANE COUNTY ARC INC., JOHNNY W COLE KA4QYI
TREAS, 309 WASHINGTON AVE., ROCKWOOD, TN 37854
SHORT MTN. REPEATER CLUB, INC., MILTON H FANNING
WA4GZZ TREA, 4936 DANBY DR., NASHVILLE, TN
37211
SMOKY MOUNTAIN ARC, CARROLL W PEABODY W4PCA
SCTR, 2054 INDEPENDENCE DR., MARYVILLE, TN 37803
STONES RIVER ARC, JIM BURNETTE KD4BAM SECY
106 EAST MCNIGHT DR., MURFREESBORO, TN 37130
SUMNER COUNTY ARA INC., STEVE JOHNSON KR4MC
SCTR, 1016 CHEYENNE BLVD, MADISON, TN 37115
TENNESSEE VALLEY AR NETWORK, WM R FERRELL
N4SSB PRES, 1120 DOUGLAS BEND RD, GALLATIN, TN
37066
TRI-COUNTY ARC, LUCILE P SHARROCK KB4CJS PRES
205 SUMMER ST, MARTIN, TN 38237
TRI-COUNTY ARC-SPARTA TN, BILL L ENGLAND W4PCT
RR 7 BOX 24, SPARTA, TN 38583
UNIVERSITY ARC, ROBERT M MAY II K4SE ADVISOR
E TENN STATE UNIVERSITY, BOX 70552, JOHNSON
CITY, TN 37614
UNIVERSITY OF TENNESSEE ARC OR, BEN GAMBLE
KD4UMT PRES, 422 OAK VALLEY DR., KNOXVILLE, TN
37918
WILSON ARC, NICK RAMOS N4VSM PRES
PO BOX 88, GLADEVILLE, TN 37071

TEXAS

ALAMO AREA RADIO ORGANIZATION, RAY CALLAHAN
KB4YGO PRES, 2730 VILLAGE PKY, SAN ANTONIO, TX
78251
ALAMO DX AMIGOS, HARRY DILLON N5HB SECY
4902 LAMBETH DR., SAN ANTONIO, TX 78228
ARC OF PARKER COUNTY, TX, DARLENE AHLEFELD
KC5DCW SCTR, PO BOX 1795, WEATHERFORD, TX
76086
ARLINGTON RADIO CLUB INC., PAUL BAUMGARDNER III
KB5FJ PRE, 4600 WINDSTONE DR APT 1511,
ARLINGTON, TX 76018
ARMADILLO GANG, RICHARD MARTIN N5IED PRES
PO BOX 72094, CORPUS CHRISTI, TX 78472
ASHCRAFT REPEATER CLUB INC., GARY W MOSS AA5IK
PRES, 5620 FONDREN DR., DALLAS, TX 75206
ATHENS ARC INC., RONNIE E DAY N5CKJ SECY
PO BOX 1641, ATHENS, TX 75751
AUSTIN ARC INC., JIM NEELY WA5LHS PRES
PO BOX 13413, AUSTIN, TX 78711
BAY AREA ARC, J B OVERMEYER N5SCY SECY
718 HEATHGATE DR., HOUSTON, TX 77062
BEAUMONT ARC, JOHN D ALLEN JR. KB5RNL PRES
PO BOX 7073, BEAUMONT, TX 77726
BELLAIRE HIGH SCHOOL ARC, B F SECOR JR. KB5VSY

SPONSOR, BELLAIRE HIGH SCHOOL, 5100 MAPLE ST.,
BELLAIRE, TX 77401
BIG BEND ARC, BOB C WARD WA5ROE LIAS
1402 N 5TH ST, ALPINE, TX 79830
BOLES JUNIOR HIGH SCHOOL ARC, MORRIE PICKLER
KB5UNX PRES, 3900 SW GREEN OAKS BLVD, ARLING
TON, TX 76017
BRAZOS VALLEY ARC INC., DOUGLAS W HOLLEY KE5SR
PRES, PO BOX 1630, MISSOURI CITY, TX 77459
BRYAN ARC, LEO OWENS KB5VAY SECY
3311 BIG HORN DR., BRYAN, TX 77803
BURLESON ARS, STEVE SOWDER N5SDZ SECY
PO BOX 561, KEENE, TX 76059
CEDAR CREEK ARC, HERB BRYDON KC5AMP PRES
PO BOX 388, MABANK, TX 75147
CENTRAL TX DX & CONTEST CLUB, EDWARD L LINDE
N5DDT PRES, 3900 SORREL COVE, AUSTIN, TX 78730
CLEAR LAKE ARC, MATT BORDELON KC5BTL SECY
PO BOX 57714, WEBSTER, TX 77598
COMM HOBBY AR RELAY ORGANIZATION, ARTHUR R
ROSS W5KR SCTR, 132 SALLY LN, BROWNSVILLE, TX
78521
COMPAQ RADIO CLUB, WALTER HOLMES WD5GAZ
20555 SH 249, HOUSTON, TX 77070
COOKE COUNTY ARC, INC., JAMES T FLOYD N5FBG PRES
PO BOX 100, GAINESVILLE, TX 76241
COPPELL AMATEUR CLUB, DON THOMAS KA1CWM PRES
PO BOX 112, COPPELL, TX 75019
COVE REPEATER ASSN, JACK H SIMS KG5PM PRES
3310 LAKE INKS, BRADY BLVD, KILLEEN, TX 76543
DALLAS ARC, JAY MILLER KB5YXG PRES
PO BOX 744266, DALLAS, TX 75374
DALLAS COUNTY REACT ARC, CHARLES A THOMPSON
N5IAG, 2909 ROSEDALE AVE., DALLAS, TX 75205
DALLAS/FORT WORTH ARC, DONALD THOMAS
KA1CWM PRES, PO BOX 112, COPPELL, TX 75019
DENTON COUNTY ARA, JERRY E SMITH AB5GF PRES
PO BOX 50433, DENTON, TX 76206
DISASTER & COMM ACT TEAM OF TX, ROGER RUNNELS
WA5WCY SECY, 11706 WOLF RUN LN, HOUSTON, TX
77065
DUMAS ARC, ROBERT MUSICK WB5QMZ PRES
PO BOX 955, DUMAS, TX 79029
EAST TEXAS VHF-FM SOCIETY, INC., AL SPRINGER
WM5V PRES, 411 HIGHLAND DR., KILGORE, TX 75662
EL PASO AMATEUR RADIO, MILLY WISE W5OVH TREAS
2100 SAN DIEGO AVE., EL PASO, TX 79930
ELLIS COUNTY ARC, DANNY WOODRUFF KA5RDB
TREAS, PO BOX 373, WAXAHACHIE, TX 75165
FOUR STATES ARC INC., LAVERNE H CLARK JR KF5YO
PRES, 18 HIGHLAND HILLS, TEXARKANA, TX 75502
GARLAND AMATEUR CLUB, JOHN R GALVIN N5TIM
PRES, 1027B AUSTIN ST, GARLAND, TX 75040
GOLDEN CRESCENT ARC, RONALD E NICHOLAS KB5DWF
SCTR, 211 FAHRENTHOLD ST, EL CAMPO, TX 77437
GRAPEVINE ARC, DAVID PEEL K4HIX LIAS
4133 HARVESTWOOD DR., GRAPEVINE, TX 76051
GULF AREA YOUNG LADIES ARK, MARGARET H PEARRE
K5MXO, 7011 SHARPVIEW DR., HOUSTON, TX 77074
HAM ASSOCIATION OF MESQUITE, EDDIE WAGONER
KA5RWT TREAS, PO BOX 851835, MESQUITE, TX 75185
HEART O'TEXAS ARC INC., ROGER MILES WB5MBO PRES
800 N OLD DALLAS RD, ELM MOTT, TX 76640
HILL COUNTRY ARC, HARVEY WEST AK5M PRES
115 LOMA VISTA DR., KERRVILLE, TX 78028
HOCKADAY SCHOOL AR KLUB, DR DAVID KOCH W8LNJ
ADVISOR, THE HOCKADAY SCHOOL, 11600 WELCH
RD, DALLAS, TX 75229
HOCKLEY COUNTY ARC, CALLEN GILBERT AB5JY
PO BOX 338, LEVELLAND, TX 79336
HOOD COUNTY ARC, DELTON WOOLSEY KB5LMG PRES
192 GLADYS DR., GRANBURY, TX 76049
HOU-TEXINS ARC, BILL NOVAK KA9IKK SECY
11735 S GLEN DR APT 603, HOUSTON, TX 77099

HOUSTON AMATEUR MOBILE SOC, TOM ONCKEN
KB5CLK SECY, PO BOX 34822, HOUSTON, TX 77234
HOUSTON AMATEUR RADIO HELPLINE, JAMES D HEIL
KB5AWM PRES, 16410 HAVENHURST DR., HOUSTON,
TX 77059
HOUSTON ECHO SOCIETY INC., WILLIAM K BRUNE
N5XWU PRES, POB 270540, HOUSTON, TX 77277
HUNTSVILLE ARS, DAVID RATHKE KI5NG PRES
103 NESTOR RD, HUNTSVILLE, TX 77340
HURST ARC, FRANK L SCHNELL W5EFZ PRES
1811 GREENBRIAR DR., EULESS, TX 76040
INTERMEDICS INC ARC, PATRICK PAUL N4SSL PRES
PO BOX 4000, ANGLETON, TX 77516
INTERTIE, INC., LAWRENCE S HIGGINS W5QMU PRES
1222 N MAIN AVE STE 1016, SAN ANTONIO, TX 78212
IRVING ARC, C J BEAUPRE WA5MJB PRES
P O BOX 153333, IRVING, TX 75015
JAMISON SCHOOL ARC, JERRY VENABLE KI5MB PRES
RR 1 BOX 1086, PEARLAND, TX 77584
JEFFERSON COUNTY ARC, STEVE GOMEZ KE5O SCTR
3130 BERRY AVE., GROVES, TX 77619
JOHNSON SPACE CENTER, LARRY DIETRICH WD8KUJ
PRES, 17511 HERITAGE COVE DR., WEBSTER, TX 77598
KATY ARS, RANDY ROBBINS N5SXF SECY
PO BOX 200, KATY, TX 77492
KENDALL ARS, RAWLINS M MORRIS KB5TX
RT 2 BOX 2324, BOERNE, TX 78006
KEY CITY ARC, PEGGY A RICHARD KA4UPA TREAS
1442 LAKESIDE DR., ABILENE, TX 79602
KILOCYCLE CLUB OF FT WORTH, POLLIE FOBBS
KB5UQM, PO BOX 6910, FORT WORTH, TX 76115
LAKE HOUSTON REPEATER ASSN., CHARLES WHEELER
WA5PJD PRES, 13210 FERRY HILL LN, HOUSTON, TX
77015
LAKES AREA ARC, ROBERT C MC WHORTER K5PFE SCTR
PO BOX 461, JASPER, TX 75951
LAREDO ARC, JOSE CAMPOS KB5TXC PRES
1010 EDEN LN, LAREDO, TX 78041
LONE STAR DX ASSOCIATION
RAY MOYER WD8JKV, 303 WESTOVER DR., EULESS,
TX 76039
LUBBOCK ARC, JERRY GARRY KB5ARQ SECY
RR 1 BOX 168, WOLFFORTH, TX 79382
MCKINNEY ARC, JOHN A FRANCIS WB5EXM PRES
P O BOX 759, 217 N 4TH ST, PRINCETON, TX 75407
MENASCO ARC, JAMES M KEGLEY KB5QFU PRES
4000 S HIGHWAY 157, EULESS, TX 76040
MIDLAND ARC, TOM SEAWRIGHT WA5WQC PRES
4405 WILSHIRE DR., MIDLAND, TX 79703
MOBILE A R AWARDS CLUB, DON MAGERS KE5WL SECY
406 CHERRY PARK DR., SHERMAN, TX 75090
NACOGDOCHES MEDICAL CENTER ARC, ALBERT L FISH
ER AC5Z SECY, 1637 SHELTON DR., NACOGDOCHES,
TX 75961
NORTH TEXAS CONTEST CLUB, JOHN HAWKINS K5NW
SECY, 1723 SHUFORDS CT, LEWISVILLE, TX 75067
NORTH TEXAS MICROWAVE SOC, WES ATCHISON
WA5TKU SCTR, RR 4 BOX 565, SANGER, TX 76266
NORTH TX SYNCHRONIZATION CLUB, ANDREW D
CARSTARPHEN WY5V PRES, 1409 WESLEY DR.,
MESQUITE, TX 75149
NORTHWEST ARS, RUDOLF J NOVOTNY KB5ZXO PRES
12914 WINCREST CRT, CYPRESS, TX 77429
ORANGE ARC INC., IRENE THOMAS N5GNF SCTR
PO BOX 232, ORANGE, TX 77631
PALESTINE/ANDERSON COUNTY ARC, LARRY DAVIS
KB5JHW PRES, 206 MOUND PRAIRIE DR., PALESTINE,
TX 75801
PAMPA ARC, HERMAN WHATLEY W5IJQ TRUSTEE
521 N WEST ST., PAMPA, TX 79065
PANHANDLE ARC, GLENN D DALLKE KB5VVM PRES
4525 GOODNIGHT TRL, AMARILLO, TX 79109
PEARLAND ARC, MARTY HALEY AB5GU PRES
PO BOX 2654, PEARLAND, TX 77588

PLAINVIEW ARC, STANLEY J FOSTER N5ZXC TREAS
PO BOX 313, PLAINVIEW, TX 79073
PLANO AMATEUR RADIO KLUB INC., WES MARSHALL
N5ODC SECY, PO BOX 860435, PLANO, TX 75086
PRAIRIE DOG ARC, JIMMIE TRAMMEL K5OVO PRES
PO BOX 485, CHILDRESS, TX 79201
QUARTER CENTURY W ASSN CHAP 27, WILLIAM H
PEARRE K5MMP SCTR, 7011 SHARPVIEW DR., HOUS
TON, TX 77074
RADIO EAST TEXAS STATE UNIV., DAVID FOX KB5ULK
PRES, C/O DEPT OF PHYSICS, EAST TEXAS STATE UNI
VERSITY, COMMERCE, TX 75429
RED RIVER VALLEY ARC, HARRY H BRIGHT KB5CER
SCTR, PO BOX 772, PARIS, TX 75461
RICHARDSON WIRELESS KLUB, BILL LOESSBERG JR.
N5RWQ, 739 BRENTWOOD LN, RICHARDSON, TX 75080
ROCKWELL-COLLINS ARC, DAVID M JAKSA W0VX
626 TORREY PINES LN, GARLAND, TX 75044
SAN ANGELO ARC, GLENN MILLER AA5PK PRES
PO BOX 4002, SAN ANGELO, TX 76902
SAN ANTONIO RADIO CLUB, STEVE CERWIN WA5FRF
PRES, 10227 MOUNT CROSBY, SAN ANTONIO, TX 78251
SAN BENITO ARC, FRED WASIELEWSKI WA2VJL
RR 8 BOX 20, SAN BENITO, TX 78586
SOUTH TX AMATEUR REPEATER CLUB, ELLA PERRY
N5XCD TREAS, PO BOX 2182, CORPUS
CHRISTI, TX 78403
SOUTHERN METHODIST UNIVERSITY, W MILTON GOS
NEY JR KG5RO, 315 CARUTH HALL, SOUTHERN
METHODIST UNIV., DALLAS, TX 75275
SOUTHWEST DALLAS COUNTY ARC, RON ERHART
KB5RON PRES, PO BOX 381023, DUNCANVILLE, TX
75138
STEPHEN F AUSTIN RADIO CLUB, TOM TAORMINA K5RC
PRES, PO BOX 813, BELLVILLE, TX 77418
TEMPLE ARC INC., FAITH PETERSON N5XIY SECY
P O BOX 616, TEMPLE, TX 76503
TEXAS DX SOCIETY, BOB PERRING N5RP PRES
12715 WESTMERE DR., HOUSTON, TX 77077
TEXAS INSTRUMENTS ARC, W F MORRISON KB5WQE
PRES, 6104 JENNINGS DR., THE COLONY, TX 75056
TEXAS SOUTHMOST ARC INC., CHARLES BRABHAM JR
WA5ZRP SEC, RR 2 BOX 46, LYFORD, TX 78569
TEXAS TECH ARS, BRIAN GALLIMORE KB5WON EDITOR
8014 RICHMOND AVE., LUBBOCK, TX 79424
TEXAS VHF-FM SOCIETY, ROBERT MCWHORTER K5PFE
SCTR, PO BOX 461, JASPER, TX 75951
TEXOMA ARC, P O BOX 295, ,SHERMAN, TX 75091
TIDELANDS ARS, JAMES R REX K5YYC SECY
2114 EVERGREEN, LA MARQUE, TX 77568
TOP OF THE PANHANDLE ARC, JEROME DOERRIE K5IS
TRUSTEE, RR 2 BOX 72, BOOKER, TX 79005
TRINITY VALLEY ARC, GEORGE PURNELL K5SXE PRES
RR 2 BOX 608, TERRELL, TX 75160
TYLER ARC, INC., EDWARD WILKINSON N5YIH PRES
P O BOX 6393, TYLER, TX 75711
UNIV OF TX BROWNSVILLE ARC, RICHARD SAMMONS
N5DNV PRES, PO BOX 1872, PORT ISABEL, TX 78578
UNIVERSITY OF TEXAS ARC, MIKE BECKER KW5F
ADVISOR, UNIVERSITY OF TEXAS, BOX 170 TEXAS
UNION, AUSTIN, TX 78713
UPPER LAKE LIVINGSTON WIRELESS, WILLIAM VANCE
WA5DQF, RR 1 BOX 113K, CROCKETT, TX 75835
VICTORIA ARC, ROBERT B WHITAKER KI5PG SCTR
121 S MAIN ST STE 205, VICTORIA, TX 77901
WICHITA AMATEUR RADIO SOCIETY, DAVID H GAINES
KC4WVK EDITOR, PO BOX 4363, WICHITA FALLS, TX
76308
YL ROSES OF TEXAS ARC, JUDI JAKSA N0IDR SECY
626 TORREY PINES LN, GARLAND, TX 75044

UTAH

DAVIS COUNTY ARC, BILL SPATZ KB7OKD PRES
1622 S 1000 W., SYRACUSE, UT 84075

OGDEN ARC INC., RICHARD N COWLEY KB7KYC PRES
PO BOX 3353, OGDEN, UT 84409
PAYSON HIGH SCHOOL ARC, ROBERT E STRANGE
K7VVU ADVISOR, 1050 S MAIN ST., PAYSON, UT 84651
UTAH ARC, GERALD G BENNION WR7N PRES
3666 S STATE ST., SALT LAKE CITY, UT 84115

VERMONT

AMATEUR RADIO ASSOCIATES, ALLAN RAY MACHELL
KC1BT, RR 3 BOX 7762, BARRE, VT 05641
BORDER ARC, CENA GALBRAITH N1HJR SECY
13 ARLINGTON ST., ESSEX JUNCTION, VT 05452
BRATTLEBORO ARC, JOSEPH B ARMSTRONG KA1YLN
80 CLARK ST APT 1, BRATTLEBORO, VT 05301
BURLINGTON AMATEUR RADIO CLUB, RALPH STETSON
KD1R PRES, OSGOOD HILL RD, RR1 BOX 185, WEST
FORD, VT 05494
CENTRAL VERMONT ARC, TOM GIRARDI WA1YNU PRES
PO BOX 261, WATERBURY, VT 05676
CONNECTICUT VALLEY FM ASSN, CARL SNYDER N1JRA
PRES, RFD 1 BOX 57, CAVENDISH, VT 05142
RADIO AMATEURS OF NORTHERN VT, RON STERN
KA1NRR PRES, PO BOX 9392, SOUTH BURLINGTON, VT
05407
RANDOLPH HIGH SCHOOL ARC, ROGER INNIS KA1TNC
FOREST ST, RANDOLPH, VT 05060
SILICON JUNCTION RADIO CLUB, FRANK ALWINE
N1GPY PRES, 422/965-2 IBM BURLINGTON, ESSEX JCT,
VT 05452
SOUTHERN VERMONT ARC, RANDALL H GATES N1GWL
PRES, P O BOX 309, RTE 7A, ARLINGTON, VT 05250
VERMONT TECH COLLEGE ARC, DONALD E NEVIN KK1U
ADVISOR, VT TECH COLLEGE, RANDOLPH CENTER,
VT 05061

VIRGIN ISLANDS

ST CROIX ARC, MATTHEW E RODINA NP2FK
PO BOX 25543, CHRISTIANSTED, VI 00824
ST JOHN ARC, MALCOLM M PRESTON NP2L PRES
PO BOX 1318, CRUZ BAY, VI 00831
VIRGIN ISLANDS ARC, DEBORAH M THOMAS NP2DJ
SCTR, PO BOX 9280, ST THOMAS, VI 00801

VIRGINIA

ALBEMARLE ARC INC., PAT WILSON N0RDQ LIAS
4260 PINE GROVE, EARLYSVILLE, VA 22936
AMATEUR RADIO RESEARCH DEVEL A, DAVID ROGERS
N4JGQ SECY, PO BOX 6148, MC LEAN, VA 22106
ARLINGTON ARC, DENNIS BODSON W4PWF VPRS
233 N COLUMBUS ST., ARLINGTON, VA 22203
BEDFORD ARC, AUGUST MEIDLING JR N4PVU
RR 1 BOX 110, HUDDLESTON, VA 24104
CENTRAL VIRGINIA CONTEST CLUB, THOMAS P
OGBURN WB4BVY SECY
3211 WHITEHORSE RD, RICHMOND, VA 23235
CHESAPEAKE AR SERVICE, LYMAN BYRD WA4YSE PRES
748 SADDLEHORN DR., CHESAPEAKE, VA 23322
CULPEPER ARA, CYNTHIA P. GRAY N4GJW SECY
9442 BARNES RD, MINE RUN, VA 22568
DANVILLE ARS, HAROLD MANASCO KM4AU PRES
RR 1 BOX 1542, RINGGOLD, VA 24586
EASTERN MENNONITE COLLEGE ARC, PHIL HARDER
WC0Q ADVISOR EASTERN MENNONITE COLLEGE, 1200
PARK RD, HARRISONBURG, VA 22801
EASTERN SHORE ARC, DIXIE J GRINNALDS WA4ACL S/T
PO BOX 305, MELFA, VA 23410
FOUNDATION FOR AMATEUR RADIO INC., ETHEL SMITH
K4LMB, 2012 ROCKINGHAM ST, MC LEAN, VA 22101
GILES ARA, INC., DONALD L WILLIAMS JR WA4K
412 RIDGEWAY DR., BLUEFIELD, VA 24605
GREAT DISMAL SWAMP DX ASSN., EMMETT STINE
KD4OS, 1269 PARKSIDE PL., VIRGINIA BEACH, VA
23454
HALIFAX CTY SR H.S. ARC, CHARLES LOWERY N4QGW
ADVISOR, PO BOX 310, SOUTH BOSTON, VA 24592

LYNCHBURG ARC, G P HOWELL JR. WA4RTS TREAS
   PO BOX 4242, LYNCHBURG, VA 24502
MASSANUTTEN ARA, MARSHALL COOPER N4ZKH PRES
   PO BOX 219, BASYE, VA 22810
MIDDLE PENINSULA ARC, WARREN W BOWERS
   WB4MRH SCTR, PO BOX 1121, GLOUCESTER POINT, VA
   23062
MOUNTAIN EMPIRE ARS, TED DINGLER N4KSO PRES
   PO BOX 33, ABINGDON, VA 24212
MT VERNON ARC, WILLIAM A SCHMITT KD4CNS PRES
   4434 FLINTSTONE RD, ALEXANDRIA, VA 22306
NEW RIVER VALLEY ARC, JERRY TAYLOR KC4ENM
   PRES, 127 15TH ST NW, PULASKI, VA 24301
NORTH POINT FM ASSOCIATION, ALAN RONALD MOECK
   WA2RPX, 113 WARBLER DR., STEPHENS CITY, VA
   22655
NORTH SHENANDOAH DX ASSN, JOHN C KANODE N4MM
   PRES, RR 1 BOX 73A, BOYCE, VA 22620
NORTHERN VA FM ASSN INC., DAN GROPPER KC4OCG
   SECY, 9908 DALE RIDGE CT., VIENNA, VA 22181
OLE VIRGINIA HAMS ARC INC., BUTCH TONOLETE
   N6NMSM PRES, PO BOX 1255, MANASSAS, VA 22110
PENINSULA ARC, D L STEBERL KR4DS PRES
   116 POSEIDON DR., NEWPORT NEWS, VA 23602
PORTSMOUTH ARC, WILLIAM G ANDERSON KE4IPS
   3803 TOWNE POINT RD, PORTSMOUTH, VA 23703
RAPPAHANNOCK ARA, RAY MASSIE K3RZR
   RT 1 BOX 697, TAPPAHANNOCK, VA 22560
RAPPAHANNOCK VALLEY RADIO CLUB, JOHN STONE
   KE4DCQ SECY, 1445 CLOVER DR., FREDERICKSBURG,
   VA 22407
RICHMOND AMATEUR TELEC SOC, WARREN WINNER
   N4NCL TREAS, PO BOX 14828, RICHMOND, VA 23221
RICHMOND ARC, HELEN RATCLIFF KC4KHH SECY
   PO BOX 73, RICHMOND, VA 23201
ROANOKE VALLEY ARC, DANNY W PENDLETON N4NPD
   PRES, 4763 JACKLIN DR NE, ROANOKE, VA 24019
SHENANDOAH VALLEY ARC, IRV BARB KD4BHY PRES
   RR 3 BOX 5385, BERRYVILLE, VA 22611
SOUTH WESTERN VA WIRELESS ASSN, WES MYERS JR.,
   KB4PW SCTR, 2921 OAK CREST AVE SW, ROANOKE,
   VA 24015
SOUTHERN PENINSULA ARK, NANCY C PIERPONT
   KB4MIF TREAS, 204 CEDAR POINT CRES., YORKTOWN,
   VA 23692
SOUTHSIDE ARA, GENE LYLES KB4TLZ PRES
   RR 2 BOX 135, KEYSVILLE, VA 23947
SPRINGFIELD ESTATES ELEM SCH, WILLIAM W SCHMITT
   KD4CNX, 4434 FLINTSTONE RD, ALEXANDRIA, VA
   22306
STERLING PARK ARC, EVAN MANN KM4SK EDITOR
   107 ENVIRONS RD, STERLING, VA 20165
VIENNA WIRELESS SOCIETY, JORGE A THEVENET
   KD4DGQ PRES, PO BOX 418, VIENNA, VA 22183
VIRGINIA BEACH ARC, CLIFF IRELAND KN4DV PRES
   PO BOX 62003, VIRGINIA BEACH, VA 23466
VIRGINIA DX CENTURY CLUB, O B CORNING N4AIG
   SCTR, 528 WATER OAK RD, VIRGINIA BEACH, VA 23452
VIRGINIA TECH ARA, JIM DIXON NL7HI PRES
   E EGGLESTON HALL VPI, 347 SQUIRES STUDENT CEN
   TER, BLACKSBURG, VA 24003
WILLIAMSBURG AREA ARC, CLAUDE FEIGLEY W3ATQ
   TREAS, 135 THE MAINE, WILLIAMSBURG, VA 23185
WOODBRIDGE WIRELESS INC., EDWARD QUINTO
   KD4KBT, PO BOX 112, WOODBRIDGE, VA 22194

WASHINGTON

ACADEMY ARC, RICHARD HEADRICK WA7QCC SECY
   3513 ORCHARD PL SE, AUBURN, WA 98092
APPLE CITY RADIO CLUB, GREGG JOHNSON WA7TSP
   PRES, 806 11TH ST NE, EAST WENATCHEE, WA 98802
ARA OF BREMERTON, WEIFORD WELLS KA7LBG SECY
   3710 HARBEL DR., BREMERTON, WA 98310
BOEING EMPLOYEES AR OPERATORS, HOWARD SELMER
   N7RTT PRES, 19716 FILBERT RD, BOTHELL, WA 98012

BOEING EMPLOYEES ARS, DENE LEACH N7RZC PRES
   MAIL STOP 8L-35, PO BOX 3707, SEATTLE, WA 98124
BURLEY ARC, FRANK MOORE N7LPS, PO BOX 262, BUR
   LEY, WA 98322
CASCADE RADIO CLUB, AUDREY SPARLIN WB7ECH
   SECY, 5125 SEAHURST AVE., EVERETT, WA 98203
CHEHALIS VALLEY ARS, RON EDWARDS N7REM PRES
   PO BOX 775, CHEHALIS, WA 98532
CLALLAM COUNTY ARC INC., LOUIS STEVENSON AA7FZ
   SECY, PO BOX 661, PORT ANGELES, WA 98362
CLARK COUNTY ARC-WA, DAN BAULIG N7PKB PRES
   12507 NE 71ST ST., VANCOUVER, WA 98682
EASTERN WASHINGTON ARG, JAY W TOWNSEND WS7I
   PO BOX 644, SPOKANE, WA 99210
EASTERN WASHINGTON DX CLUB, PAM FOLLANSBEE
   WM7R SCTR, 3641 FRONTIER RD, PASCO, WA 99301
GONZAGA PREP HAM RADIO CLUB, KURT L
   KROMHOLTZ KC7FJ ADVS, GONZAGA PREP SCHOOL,
   1224 E EUCLID AVE., SPOKANE, WA 99207
GRAYS HARBOR ARC, TOM COOK KA7EXP
   PO BOX 2250, ABERDEEN, WA 98520
INLAND EMPIRE VHF RADIO AMATEUR, GARY CASEY
   N7BFJ SECY, 4704 N VISTA RD, SPOKANE, WA 99212
INLAND NW HAMFEST ASSOCIATION, WARREN KELSEY
   N7KYH PRES, 1405 S CRESTLINE ST, SPOKANE, WA
   99203
ISLAND COUNTY ARC, TERRY PERMENMTER KF7OI PRES
   PO BOX 352, COUPEVILLE, WA 98239
ISSAQUAH ARC, ROD JOHNSON KA7OU PRES
   15819 266TH AVE. SE, ISSAQUAH, WA 98027
JEFFERSON COUNTY ARC, JACK WEST W7LD PRES
   PO BOX 88, CHIMACUM, WA 98325
LOWER YAKIMA VALLEY ARC, DENNIS SIMPSON N7JNV
   SCTR, 921 UPLAND DR., SUNNYSIDE, WA 98944
MAPLE VALLEY WIRELESS SOCIETY, DAVID BARBER
   K7LHB PRES, PO BOX 586, MAPLE VALLEY, WA 98038
MASON COUNTY ARC, FRANCIS SCHROEDER KN7D SECY
   SE 401 ARCADIA SHORES, SHELTON, WA 98584
MID-COLUMBIA ARC, JANE ALLAN KB7IGL SECY
   PO BOX 25, GOLDENDALE, WA 98620
MIKE & KEY AMATEUR RADIO CLUB, NILS HALLSTROM
   WB7TJK PRES, PO BOX 2121, KIRKLAND, WA 98083
MT BAKER ARC, GARY PROWSE KB7IGR PRES
   7646 TERRACE ST., FERNDALE, WA 98248
NORTH KITSAP ARC, TOM SANDERS W6QJI PRES
   PO BOX 450, PORT ORCHARD, WA 98366
NORTH MASON AR EMERGENCY SERV, ROBERT H
   DEWEY N7SSP PRES, 3610 S MISSION RD, BREMERTON,
   WA 98312
OLYMPIA ARS, LEE CHAMBERS WB7UEU PRES
   4015 11TH AVE. NW, OLYMPIA, WA 98502
PANASONIC ARC MASCA CHAPTER, LESLIE SMITH
   KC7BYN, M/S 2C36, 1111 39TH AVE SE, PUYALLUP, WA
   98374
PORT LUDLOW ARC, ORVILLE BENNETT N7MOZ SECY
   50 SEAFARER LANE, PT LUDLOW, WA 98365
PUGET ARS, STUART WHITING KE7UUX PRES
   16307 TIGER MOUNTAIN RD SE, ISSAQUAH, WA 98027
RADIO CLUB OF TACOMA, JERRY SELIGMAN W7BUN
   PO BOX 11188, TACOMA, WA 98411
RHO-EPSILON RADIO FRAT ALPHA C, TODD THOMPSEN
   KF7LX PRES, RHO EPSILON RM 429B, COMPTON
   UNION BLDG, PULLMAN, WA 99164
SEATAC REPEATER ASSN, MARY E LEWIS W7QGP V-
   PRES, 10352 SAND POINT WAY NE, SEATTLE, WA 98125
SKAGIT ARC INC., JACK L CARLSON W7GHO SCTR
   121 15TH AVE., KIRKLAND, WA 98033
SNO-ISLE SKILLS CTR TEKTRONS, LARRY R LUCHI
   W7KZE ADVISOR, 9001 AIRPORT RD, EVERETT, WA
   98204
SNOHOMISH COUNTY HAMS CLUB, CARL COWL
   WA6DPL SECY, PO BOX 911, MARYSVILLE, WA 98270
SOPER HILL ARC, TOM BRUHNS K7ITM PRES
   8600 SOPER HILL RD STE 330, EVERETT, WA 98205

SPOKANE RADIO AMATEURS INC., ART COWAN KA7FQB
PRES, S 2420 MYRTLE, SPOKANE, WA 99223
STANWOOD-CAMANO ARC, VIC HENRY N7KRE PRES
PO BOX 941, STANWOOD, WA 98292
TRI-CITY ARC - WA, PAM FOLLANSBEE WM7R TREAS
3641 FRONTIER RD, PASCO, WA 99301
VALLEY ARC INC., RICHARD R ANDERSEN W7TYI SCTR
PO BOX 12, PUYALLUP, WA 98371
WALLA WALLA VALLEY ARC INC., CRISTY STIMMEL
KB7PTM SECY, PO BOX 321, WALLA WALLA, WA 99362
WEST SEATTLE ARC, BOB LUBIN N7UOR VPRS
5042 PUGET BLVD SW, SEATTLE, WA 98106
YAKIMA ARC, RICHARD UMBERGER N7HHU EDITOR
PO BOX 9211, YAKIMA, WA 98909

WEST VIRGINIA

CHARLESTON AREA HAMFEST & COMP, WILLIAM H
KIBLER JR. K8WMX, PO BOX 916, SAINT ALBANS, WV
25177
EAST RIVER RADIO CLUB, ROBERT HOGE N8RIR PRES
PO BOX 1362, BLUEFIELD, WV 24701
JACKSON COUNTY ARC INC., TED JACOBSON W8KVK
SECY, PO BOX 598, RIPLEY, WV 25271
KANAWHA ARC, W DOUGLAS SWEENEY N8AJC PRES
PO BOX 1694, CHARLESTON, WV 25326
KANAWHA ELEMENTARY ARC, DAVID R STONE KD8YY
ADVISOR, RR 1 BOX 38A, DAVISVILLE, WV 26142
MONONGALIA WIRELESS ASSN., NORTON P SMITH
WD8AFJ SCTR, PO BOX 4263, MORGANTOWN, WV
26504
MOUNTAIN ARA INC., RICHARD D GILLESPIE N8HON
PO BOX 471, FRANKLIN, WV 26807
MOUNTAINEER ARA, CHARLES T MCCLAIN K8UQY PRES
PO BOX 571, FAIRMONT, WV 26555
OPEQUON RADIO SOCIETY, FREDERICK S SCHLICHTING
K8SDG, 513 S RALEIGH ST, MARTINSBURG, WV 25401
PARKERSBURG AMATEUR RADIO KLUB, ROY S MAULL
N8YYS PRES, P O BOX 2112, PARKERSBURG, WV 26101
PLATEAU AMATEUR RADIO ASSN., MICHAEL W VARGO
WB8WKO, 211 KELLY AVE., MOUNT HOPE, WV 25880
PLATEAU ARA, MICHAEL W VARGO WB8WKO SECY
328 MAIN ST, OAK HILL, WV 25901
SOUTHERN APPALACHIAN WIRELESS, BRUCE E ROSEN
KB4RXO SECY, 208 LOGAN ST., WILLIAMSON, WV
25661
ST ALBANS EMERG SERV AR COMMUN., WILLIAM H
KIBLER JR K8WMX, 182 MONTEREY DR., SAINT
ALBANS, WV 25177
STONEWALL JACKSON ARA, STEVE MARTIN N8NPP
PRES, PO BOX 752, CLARKSBURG, WV 26302
TRI-STATE AMATEUR RADIO ASSN., FRED M COX N8TVP
PRES, PO BOX 4120, HUNTINGTON, WV 25727
WEST VIRGINIA STATE ARC, L ANN RINEHART KA8ZGY
SECY, 1256 RIDGE DR., SOUTH CHARLESTON, WV
25309
WEST VIRGINIA UNIV. ARC, DR ROY NUTTER N8BHI
ADVISOR, WEST VA UNIV ECE DEPT, P O BOX 6101,
MORGANTOWN, WV 26505
WV WESLEYAN COLLEGE ARC, RICHARD C CLEMENS
KB8AOB, WV WESLEYAN COLLEGE, BOX 18, BUCK
HANNON, WV 26201

WISCONSIN

BADGER CONTESTERS, BRUCE RICHARDSON KE9QT
PRES, PO BOX 233, HILLSBORO, WI 54634
BARRON COUNTY ARES, MICHAEL A PERSSON W9MP
SECY, 694 WISCONSIN AVE., CHETEK, WI 54728
BELDENVILLE AMATEUR TELEVISION, SCOT THOMPSON
WB0WOT PRES, W7137 770TH AVE., BELDENVILLE, WI
54003
BELOIT ARC INC., GARY A COOK W9JSN SECY
737 MILWAUKEE RD, BELOIT, WI 53511
BSA PACK 139 ARC, FREDERICK GRAY JR N9MTT
4335 N 49TH ST, MILWAUKEE, WI 53216

CALUMET AR EMERG SERVICE, RICHARD SCHMIDT
KA9OJV SCTR, W2049 CTY H, NEW HOLSTEIN, WI 53061
CENTRAL WI AMATEURS LTD, ART WYSOCKI N9BCA
PRES, 3356 APRIL LN, STEVENS POINT, WI 54481
CHETEK HS ARC, MICHAEL A PERSSON W9MP ADVS
1001 KNAPP ST, CHETEK, WI 54728
DISASTER EMERG RESPONSE ASSN, JAY WILSON N3DAK
PRES, PO BOX 37324, MILWAUKEE, WI 53237
EAU CLAIRE ARC, J BRENNAN FITZL N9LIJ SECY
1418 ALTOONA AVE., EAU CLAIRE, WI 54701
FALLS ARC INC., PHILLIP REBENSBURG KC9CI
W 159 N 9737 BUTTERNUT RD, GERMANTOWN, WI
53022
FOND DU LAC ARC, EDWARD J BELTZ N9PJQ PRES
W4820 MAIN ST., TAYCHEEDAH, WI 54935
FOUR LAKES ARC, MARYLYNN FRANZEN N9PQN TREA
1710 SACHTJEN ST, MADISON, WI 53704
FOX CITIES ARC, MYRON JACKSON WB9BBI PRES
2020 S CARPENTER ST, APPLETON, WI 54915
GOODRICH SR HS ARC, JEFFREY D HOEFT K9NC ADVS
481 WILLOW DR., FOND DU LAC, WI 54935
GREATER MILWAUKEE DX ASSN, CHARLIE CHIARIELLO
NK9I, 3806 30TH ST., KENOSHA, WI 53144
GREEN BAY MIKE & KEY CLUB INC., JACOB DERENNE
N9DKH TREA, E1190 CTY RD X, LUXEMBURG, WI 54217
GREEN COUNTY ARA, SCOTT W FELDT KC9YI TREAS
N5299 FELDT RD, MONTICELLO, WI 53570
GREEN FOX ARC, JACK BREMER KB9WC SECY
RT 1 BOX 142, MARKESAN, WI 53946
LAKES AREA ARC, NORMAN GORE N9SKL SECY
225 VALENCIA DR., DELAVAN, WI 53115
LOCAL EMERG FIELD RADIO OPERAT, WILLIAM HOWE
KA9WRL PRES, 1781 1ST AVE., GRAFTON, WI 53024
MADISON AREA RPTR ASSN, INC., JIM WALDORF
WB9AER PRES, P O BOX 8890, MADISON, WI 53708
MADISON DX CLUB, PETER BYFIELD K9VAL SECY
4534 S HILL CT, DE FOREST, WI 53532
MANCORAD ARC, ERIC RUGOWSKI WD9IAB SECY
POB 204, MANITOWOC, WI 54221
MARINETTE/MENOMNEE RADIO CLUB, LYNNE F RYNISH
N8OSK PRES, PO BOX 1082, MARINETTE, WI 54143
MARSHFIELD AREA ARS, KEITH BORNBACH N9KPC
VP/SECY, 818 S WEBER AVE., STRATFORD, WI 54484
MILWAUKEE RADIO AMATEURS CLUB, FRED LINN
W9NZF SECY, PO BOX 25707, MILWAUKEE, WI 53225
NAMEKAGON VALLEY WIRELESS ASSN., RAY KET
TLEWELL KA9SSQ SCTR, PO BOX 333, HAYWARD, WI
54843
NORTHEAST WI RADIO LEAGUE, MARK OLSON KE9PQ
PRES, P O BOX 22130, PO BOX 22130, GREEN BAY, WI
54305
NORTHEAST WISCONSIN DX ASSN., GEORGE CROY
W9MDP PRES, 2113 W TWIN WILLOWS DR., APPLETON,
WI 54914
NORTHERN WISCONSIN RADIO CLUB, STEVE HART
KA9OMC SCTR, RR 1 BOX 54, KNAPP, WI 54749
NORTHLAND ARC, R CHARLES PERRY N9CZM PRES
STAR ROUTE BOX 13-A, CORNUCOPIA, WI 54827
NORTHWOODS ARC, MARY L ANDERSON N9PFN PRES
35 LAKE CREEK RD, RHINELANDER, WI 54501
OSHKOSH ARC, JIM ECKSTROM WB9VKD SECY
127 SUNSET CT, OMRO, WI 54963
OZAUKEE RADIO CLUB, GABE CHIDO N9QQA PRES
W 58 N 985 ESSEX DR., CEDERBURG, WI 53010
RACINE MEGACYCLE CLUB, ARNE MOXNESS KA9ZPY
SECY, 3345 NEWMAN RD, RACINE, WI 53406
RED CEDAR REPEATER ASSN., INC., STEVE HART
KA9OMC SCTR, RR 1 BOX 54, KNAPP, WI 54749
RIVERLAND ARC, ROBERT K WILSON N9LZK PRES
PO BOX 621, ONALASKA, WI 54650
ROCK RIVER RADIO CLUB, JOHN P KLEIN W9YOA SECY
615 ONEIDA ST, BEAVER DAM, WI 53916
SHEBOYGAN COUNTY ARC INC., STEVE MEIFERT KF9KZ
PRES, 1525 NEW JERSEY, SHEBOYAN, WI 53081

SOUTH MILWAUKEE ARC, ROBERT J KASTELIC WB9TIK
SECY, 7410 S CLEMENT AVE., OAK CREEK, WI 53154
ST CROIX VALLEY ARC, BOB EMBERGER KB9BTB PRES
126 W VINE ST, RIVER FALLS, WI 54022
W/K ARC OF GREATER MILWAUKEE, TERRY HUBBARD
KF9HI SECY, 5278 S ELAINE AVE., CUDAHY, WI 53110
WASHINGTON COUNTY ARC, DAVID PARKINSON
KA9RNU SCTR, 708 ROBIN ST, WEST BEND, WI 53095
WATERTOWN ARC, WILLIAM KIMBLE WA9OAY SECY
502 CARL SCHURZ DR., WATERTOWN, WI 53098
WEST ALLIS RADIO A C INC., PHILIP NOFTZ N9FEW PRES
2155 S 83RD ST, MILWAUKEE, WI 53219
WI AMATEUR CONTEST ORGANIZATION, GARY HOEHNE
KB9AIT PRES
W8244 COUNTY ROAD BB, NEENAH, WI 54956
WINNEBAGO ARC, TERRY LYNN WITT AA9BA SECY
9145 IVY LN, BERLIN, WI 54923
YELLOW THUNDER ARC INC., RAYMOND J MATLOSZ
SECY, 450 W 2ND ST, REEDSBURG, WI 53959

WYOMING

CAMPBELL COUNTY ARC, LEE TELKAMP N7XUB PRES
1109 ALMON CIR, GILLETTE, WY 82718
CASPER ARC INC., STACY BOYER N7VLM PRES
P O BOX 2802, CASPER, WY 82602
CEDAR MOUNTAIN ARC, IVAN Z CHRISTOPHERS WA7NZI
P O BOX 2732, 61 ROAD 2CD, CODY, WY 82414
CONVERSE COUNTY ARC, KEITH C FRANCE W7OGT
SECY, 719 S 10TH ST., DOUGLAS, WY 82633
FREMONT COUNTY ARS, LARRY HUDSON KD7BN
HC 33 BOX 25013, RIVERTON, WY 82501
HIGH PLAINS ARC, LEE MILNER KT7VC SECY
111 CAMINO DEL REY, TORRINGTON, WY 82240
SHY-WY RADIO CLUB, GREG RIX KB7MUP SECY
PO BOX 6262, CHEYENNE, WY 82003
UNIVERSITY ARC, J D BURKE KB7SGS VPRS
1118 RENSHAW ST., LARAMIE, WY 82070
WYOMING DX CONTEST CLUB, DALE PUTNAM WC7S
5503 N COLLEGE DR., CHEYENNE, WY 82009

SPECIAL-INTEREST RADIO CLUBS AND ORGANIZATIONS WITH LARGE MEMBERSHIPS

The Quarter Century Wireless Association (QCWA)

The Quarter Century Wireless Association is one of the largest independent amateur radio organizations in the world. Founded in the Northeastern U.S. in November 1947, members must have been first licensed at least 25 years ago to be eligible to join. With membership numbers issued in the range of 27,000 there are currently about 10,000 active members in 185 Chapters in the U.S., Canada, Austria and Germany. QCWA has established a Memorial Scholarship program to honor Silent Key members. It is administered by the Foundation for Amateur Radio in Washington, DC. As of September 1995, the officers are:

Lew McCoy, W1ICP President
Jack Kelleher, W4ZC Vice President
John Swafford, W4HU   Secretary
Wes Randles, W4COW   ·   Treasurer
The address for QCWA headquarters is: QCWA, Inc., 159 East 16th Avenue, Eugene, OR 97401-4017.

The 10-X International Net

10-X is an organization to sponsor the use of the 10 meter band. It was first conceived when propagation was poor on 10 meters, as a method to get more hams to use 10 meters. The past few years have been poor for 10 meters worldwide, and the band has been limited primarily to local communications, except for sporadic band openings, until the next sunspot cycle. At the peak of the sunspot cycle, 10 meter propagation supports worldwide DX QSOs.

To join 10-X, work ten 10-X members and LOG each 10-X number, callsign, operator's name and location. Send the list to your numeric call sign or DX area manager (as shown below), with $5.00 U.S. new membership registration fee ($6.00 for foreign addresses). Members receive a 10-X membership number valid for life. Numerous awards are available for contacting net members.

10-X International Area Managers:
USA 1 - Al Kaiser N1API, 194 Glen Hills Rd, Meriden, CT 06450
USA 2 - Larry Berger WA2SUH, 9 Nancy Blvd., Merrick, NY 11566
USA 3 - Chester Gardner N3GZE, 9028 Overhill Dr, Ellicott Cty, MD 21042
USA 4 - KY, TN, FL, VA, NC, SC only Rick Roberts N4KCC, 7106 Ridgestone Dr,Ooltewah, TN 37363
USA 4 - GA, AL, Puerto Rico only, Jim Beswick W4YHF, 112 Owl Town Farm, Ellijay, GA 30540
USA 5 - Grace Dunlap K5MRU,* Box 445, LaFeria, TX 78559, *summer addr Jun-Oct Box 13, Rand, CO 80473
USA 6 - Dick Rauschler W6ANK, 4371 Cambria St, Fremont, CA 94538
USA 7 - Willie Madison WB7VZI, 10512 W Butler Dr, Peoria, AZ 85345
USA 8 - John Hugentober N8FU, 4441 Andreas Ave, Cincinatti, OH 45211
USA 9 - Jim Williams N9HHU, 240 Park Rd, Creve Coeur, IL 61611
USA Ø - Debbie Peterson KFØNV, RR 1 Box 35, Duncombe, IA 50532
All DX- Carol Hugentober K8DHK,4441 Andreas Ave, Cincinatti, OH 45211

The Military Affiliate Radio System (MARS)

To join MARS, you have to be 14 years or older (parental consent required under age 17), be a US citizen or resident alien, possess a valid Amateur Radio license, possess a station capable of operating on MARS HF frequencies, and be able to operate the minimum amount of time for each quarter (12 hours for Army and Air Force; 18 hours for Navy-Marines. Novices must upgrade to Technician within 6 months, else be dropped from MARS. No-Code Techs can apply, provided they have transmit and receive HF capability for MARS frequencies (they don't need transmit capability for Amateur HF frequencies). For application forms contact:

Chief, Air Force MARS, HQ AFC4A/SYXR, 203 W. Losey St. Room 1020, Scott AFB, IL 62225-5219; (618)256-5552, Fax: (618) 256-5126
Chief, US Army MARS, HQ USA Information Systems Command, ATTN: ASOP-HF, Ft. Huachuca, AZ 85613-5000; 800-633-1128
Chief, Navy-Marine Corps MARS, Naval Communication Unit, Washington, DC 20397-5161

Radio Amateur Civil Emergency Service (RACES)

The Radio Amateur Civil Emergency Service is a part of a municipal, county, or state government. This does not mean, however, that every such government has a RACES program. Contact your nearest Civil Defense or Emergency Management Agency. If your government does not have a RACES, ask them to refer you to the nearest jurisdiction that does have a RACES program.

International Amateur Radio Union
List of Member Societies

ALBANIA:
  Albanian Amateur Radio Association [AARA]
  Address: P.O. Box 66, Tirana
  Telephone: +355 (42) 33767 <ZA1Z>, (42) 34738 <ZA1B>
  Chairman: Dajlan Omeri, ZA1Z
  Secretary: Jovan Bojdani, ZA1H
  IARU liaison: Marenglen Mema, ZA1B

ALGERIA:
  Amateurs Radio Algeriens [ARA]
  Address: P.O. Box 2, Alger Gare
  Location: 7 Square Port Said, Alger
  Telephone: +213 (2) 644432 <HQ>, (2) 561657 <7X2SX>
  +213 (6) 213849 <7X4MD>
  President: Mohamed Yacoubi, 7X2SX
  Secretary: Malek Messadi
  IARU liaison: Driss Bendani, 7X4MD

ANDORRA:
  Unio de Radioaficionats Andorrans [URA]
  Address: P.O. Box 150, La Vella
  Telephone: +33 (628) 25380 <C31US>
  Telefax: +33 (628) 25380 <C31US>
  President: Juan M. Sauri, C31US
  Secretary: Victor Nunez, C31NH
  IARU liaison: President

ANGOLA:
  Liga dos Amadores de Radio de Angola [LARA]

ANGUILLA:
  Anguilla Amateur Radio Society [AARS]
  Address: P.O. Box DX, The Valley
  Telephone: +1 (809) 497-2219/3659 <VP2EM>
  +1 (809) 497-2150 <VP2EHF>
  President: Langford Morton, VP2EM
  Secretary: Dorothea Mann, VP2EE
  IARU liaison: Dave Mann, VP2EHF

ANTIGUA & BARBUDA:
  Antigua and Barbuda Amateur Radio Society [ABARS]
  Address: P.O. Box 1111, St. John's
  Telephone: +1 (809) 461-1693 <V21AO>
  Telex: 2158 KNIT [Ramez Hadeed, V2AU]
  President: Randolph Prescod, V21AO
  Secretary: Carol Thomas
  IARU liaison: Norris Mendes, V21CH

ARGENTINA:
  Radio Club Argentino [RCA]
  Address: P.O. Box 97, 1000 Buenos Aires
  Location: Carlos Calvo 1420/24, 1102 Buenos Aires
  Telephone: +54 (1) 304-0555 & 26-0505 <HQ>
  Telefax: +54 (1) 304-0555 <HQ>, (1) 613-3004 <LU2AH>
  Telex: 18506 MILIA
  President: Alberto Sixto Grandoli, LU1AG
  Secretary: Antonio Dasso, LU4AU
  IARU liaison: President

ARUBA:
  Aruba Amateur Radio Club [AARC]
  Address: P.O. Box 2273, San Nicolas
  Telephone: +297 (8) 25504 <P43IDP>
  Telefax: +297 (8) 24647 <P43IDP>
  President:
  Secretary: Irwin D. Provence, P43IDP
  IARU liaison: Secretary

AUSTRALIA:
  Wirelesss Institute of Australia [WIA]
  Address: P.O. Box 2175, Caulfield Junction, Victoria 3161
  Location: 3/105 Hawthorn Rd., North Caulfield, Vic. 3161
  Telephone: +61 (3) 528-5962 <HQ>, (6) 254-7129 <VK1OK>
  Telefax: +61 (3) 523-8191 <HQ>
  President: Neil Penfold, VK6NE
  Secretary: Bruce Thorne
  IARU liaison: Kevin Olds, VK1OK

AUSTRIA:
  Oesterreichischer Versuchssenderverband [OEVSV]
  Address: Theresiengasse 11, A-1180 Vienna
  Telephone: +43 (1) 4085535 <HQ, Wed. 4-8 p.m.>
  +43 (2622) 71853 <OE3REB>
  Telefax: +43 (1) 4031830 <HQ>
  President: Dr. Ronald Eisenwagner, OE3REB
  Secretary: Mrs. Beatrix Eisenwagner
  IARU liaison: President

BAHAMAS:
  Bahamas Amateur Radio Society [BARS]
  Address: P.O. Box SS-6004, Nassau, N.P.
  President: Rhinehart Pearson, C6ANO
  Secretary: Philip Dawkins, C6ACN
  IARU liaison: Secretary

BAHRAIN:
  Amateur Radio Association Bahrain [ARAB]
  Address: P.O. Box 22381, Muharraq
  Telephone: +973 740426 <A92BE>, 756034 <A92BW>
  Telex: 8214 BAPCO <Ian Cable>
  President: H.E. Tariq Al-Moayyed
  Chairman: Sheridon Street, A92BE
  Secretary: Ian Cable, A92BW

BANGLADESH:
  Bangladesh Amateur Radio League [BARL]
  Address: G.P.O. Box 3512, Dhaka 1000
  Telephone: +880 (2) 811097 <S21A>
  Telefax: +880 (2) 832915 <S21A>, (2) 891177 <S21B>
  Telex: 65848 BXIM <S. Shahid>
  President: Saif D. Shahid, S21A
  Secretary: Nizam A. Chowdhury, S21B
  IARU liaison: President

BARBADOS:
    Amateur Radio Society of Barbados [ARSB]
    Address: P.O. Box 814E, Bridgetown
    Telephone: +1 (809) 426-2502
    President: Decarlo Howell, 8P6RY
    Secretary: Arthur Farmer, 8P6AA
    IARIU liaison: Secretary
BELGIUM:
    Union Belge des Amateurs-Emetteurs
    Unie van de Belgische Amateur-Zenders [UBA]
    Union der Belgischen Amateurfunker
    Address: c/o Etienne David, ON5IA
    A. Vermeylenstraat 49, B-8400 Oostende
    Telephone: +32 (2) 246-8585 <ON5IA before 4:30 p.m.>
    +32 (59) 803816 <ON5IA after 7:30 p.m.>
    Telefax: +32 (59) 514047 <ON5IA>
    Honorary President: Rene A. Vanmuysen, ON4VY
    President: Gaston Bertels, ON4WF
    Secretaries: August Vollemaere, ON5VA (Flemish), Pierre
    Cornelis, ON7PC (French)
    IARU liaison: Etienne David, ON5IA
BELIZE:
    Belize Amateur Radio Club [BARC]
BERMUDA:
    Radio Society of Bermuda [RSB]
    Address: P.O. Box HM 275, Hamilton HM AX
    Telephone: +1 (809) 236-6907 <VP9HK>
    +1 (809) 292-0754 & 295-5881 <VP9IM>
    Telefax: +1 (809) 295-0871 <VP9HK>, (809) 292-6600
    <VP9IM>
    President: Walter Carlington, VP9KD
    Secretary: Dianna Doe, VP9LZ
    IARU liaison: Tony Siese, VP9HK, Box HM 1060, Hamilton
    HM JX
BOLIVIA:
    Radio Club Boliviano [RCB]
    Address: P.O. Box 2800, Cochabamba
    Telephone: +591 (42) 31469 <HQ>, (42) 32076
    Telefax: +591 (42) 29905
    Telex: 4442 VIMAC
    President: Guillermo Sanabria, CP5NO
    Secretary: Osvaldo Coca M., CP5BP
    IARU liaison: President
BOSNIA AND HERZEGOVINA:
    Savez Radioamatera Bosne i Hercegovine [SRABiH]
    Address: P.O. Box 61, 71001 Sarajevo, (via 58000 Split,
    Croatia)
    Location: Daniela Ozme 7, 71000 Sarajevo
    Telephone: +49 (89) 6805983 <DJØJV>
    Telefax: +49 (89) 6805983 <DJ$$JV>
    President: Sead Celjo, T94AC
    Secretary: Salih Hukelic, T94A
    IARU liaison: Nusret Abadzic, DJØØJV/T93N, Erminoldstr.
    189, W-8000 Muenchen 83, Germany
BOTSWANA:
    Botswana Amateur Radio Society [BARS]
    Address: P.O. Box 1873, Gaborone
    Telephone: +267 314765 <A22MN>, 353982 <David Heil>
    Telefax: +267 347956 <A22MN>, 356947 <David Heil>
    President: David E. Heil, A22MN
    Secretary: Udo Stuesser, A22ST
    IARU liaison: President
BRAZIL:
    Liga Brasileira de Radioamadores [LABRE]
    Address: P.O. Box 07-0004, 70359 Brasilia D.F.
    Location: Setor de Clubes Esportivos Sul, Trecho 4, Lote 1/A,
        70359-970 Brasilia, D.F.
    Telephone: +55 (61) 223-1157 <HQ>, (84) 228-2151
    <PY2BJO>
    Telefax: +55 (61) 223-1157 <HQ>
    President: Mauricio Carrilho Barreto, PS7RK
    Secretary: Americo Barbosa Fortes, PT2ABF
    IARU liaison: Junior Torres de Castro, PY2BJO

BRITISH VIRGIN ISLANDS:
    British Virgin Islands Radio League [BVIRL]
    Address: P.O. Box 4, West End, Tortola
    President: Robert W. Denniston, VP2VI
    Secretary: Edward W. White, VP2VBK
    IARU liaison: Secretary
BRUNEI:
    Brunei Darussalam Amateur Radio Association [BDARA]
    Address: P.O. Box 73, Gadong, Bandar Seri Begawan 3100
    Telephone: +673 (2) 42903 X438 <V85HG>
    Telex: 2346 & 2575
    President: Hj. Hassan Bin Hj. Ghani, V85HG
    Secretary: Hasnan Bin Hj. Sha'ari, V85MH
    IARU liaison: President
BULGARIA:
    ulgarian Federation of Radio Amateurs [BFRA]
    Address: P.O. Box 830, 1000 Sofia
    Location: Arso Pandurski St. No. 1, 1100 Sofia
    Telephone: +359 (2) 623022 X216 <HQ>
    +359 (2) 467211 <LZ1CZ>, (2) 238222 <LZ1ZQ>
    Telefax: +359 (2) 467553 <HQ>
    Telex: 23741 COMPTR
    President: Dimitar Zvezdev, LZ1CZ
    Secretary: Miss Zdravka Buchkova, LZ1ZQ
    IARIU liaison: Secretary
CANADA:
    Radio Amateurs of Canada [RAC]
    Address: P.O. Box 356, Kingston, Ontario K7L 4W2
    Location: 614 Norris Court - Unit 6, Kingston, Ontario K7P
    2R9
    Telephone: +1 (613) 634-4184 <HQ>, (604) 985-1267
    <VE7RD>
    +1 (905) 562-4891 <VE3AGS>, (416) 763-1761
    <VE3DSS>
    Telefax: +1 (613) 634-7118 <HQ>, (604) 985-1267 <VE7RD>
    President: J. Farrell Hopwood, VE7RD
    Secretary: James Kenneth Pulfer, VE3PU
    General Manager: Deborah Norman
    IARU liaison: President
CAYMAN ISLANDS:
    Cayman Amateur Radio Society [CARS]
    Address: P.O. Box 1029, Grand Cayman
    Telephone: +1 (809) 949-9406/0111 <ZF1RC>
    Telefax: +1 (809) 949-8163 <ZF1RC>
    President: Joseph Jackman, ZF1MA
    Secretary: Richard Harris, ZF2JN
    IARU liaison: Jeff Parker, ZF1JF
CHILE:
    Radio Club de Chile [RCCH]
    Address: P.O. Box 13630, Santiago 21
    Location: Nataniel Cox 1054, Santiago
    Telephone: +56 (2) 696-4707 <HQ>
    Telefax: +56 (2) 672-2623 <HQ>
    President: Eduardo Ibaceta Veas, CE3BOC
    Secretary: Felipe Bray Navarrete, CE3HYW
    IARU liaison: President
CHINA:
    Chinese Radio Sports Association [CRSA]
    Address: P.O. Box 6106, Beijing 100061
    Location: 9 Tiyuguan Road, Beijing
    Telephone: +86 (1) 701-1177 <HQ>, (1) 751313 <BZ1HAM>
    Telefax: +86 (1) 701-5858 <State Sports Commission>
    Telex: 22323 CHOC <State Sports Commission>
    President: Qin Duxun
    Secretary: Wang Xun, BZ1WX
    IARU liaison: Chen Ping, BZ1HAM
COLOMBIA:
    Liga Colombiana de Radioaficionados [LCRA]
    Address: P.O. Box 584, Bogota
    Location: Carrera 32 No. 94-67, Bogota
    Telephone: +57 (1) 610-8499 <HQ>
    Telefax: +57 (1) 610-9877 <HQ>
    President: Alvaro Martinez Salcedo, HK3AVA
    Secretary: Gloria Velez Henao, HK3IOU
    IARU liaison: President

COSTA RICA:
Radio Club de Costa Rica [RCCR]
Address: P.O. Box 2412, San Jose 1000
Telephone: +506 216903 <HQ>, 233701 <TI2MCL>
Telefax: +506 535044 <TI2YO>, 336896 <TI2MCL>
President: Minor Barrantes, TI2YO
Secretary: Luis Barrantes, TI2BLL
IARU liaison: Mario Cordero, TI2MCL

CROATIA:
Hrvatski Radio-Amaterski Savez [HRS]
Address: P.O. Box 564, HR-41000 Zagreb
Location: Dalmatinska 12, HR-41000 Zagreb
Telephone: +385 (1) 433025 <HQ>
Telefax: +385 (1) 274391 <HQ>
President: Zvonimir Jakobovic, 9A2RQ
Secretary: Tomislav Cosic, YT2MQ
IARU liaison: Tonci Majica

CUBA:
Federacion de Radioaficionados de Cuba [FRC]
Address: P.O. Box 1, Habana 10100
Location: Paseo #611 e/ 25 y 27, Vidado, Habana 10400
Telephone: +53 (7) 34811 <HQ>, (7) 302223
President: Pedro Rodriguez Perez, CO2RP
Secretary: Angel Labori Urgelles, CM2LU
IARU liaison: President

CYPRUS:
Cyprus Amateur Radio Society [CARS]
Address: P.O. Box 1267, Limassol
Telephone: +357 (5) 372997/369529 <5B4AP>, (5) 362792
<5B4JE>
Telefax: +357 (5) 367033 <Totos Theodossiou>
Telex: 4154 TOTO THEO <Totos Theodossiou>
President: Totos Theodossiou, 5B4AP
Secretary: Aris Kaponides, 5B4JE
IARU liaison: Secretary

CZECH REPUBLIC::
Cesky Radioklub [CRK], Czech Radio Club [CRC]
Address: P.O. Box 69, 11327 Praha 1
Location: U. Pergamenky 3, 17000 Praha 7
Telephone: +42 (2) 876989 <HQ>, (2) 704620 <OK1MP>
+42 (2) 7992205 <M. Prosteky>
Telefax: +42 (2) 809587 <HQ>, (2) 7992318 <M. Prostecky>
President: Milos Prostecky, OK1MP
Secretary: Jiri Blaha, OK1VIT
IARU liaison: President

DENMARK:
Eksperimenterende Danske Radioamatorer [EDR]
Address: P.O. Box 172, DK-5100 Odense C
Telephone: +45 (66) 137700 <HQ>, (75) 941066 <OZ1DHQ>
+45 (86) 579242 <OZ5KM>, (54) 858844
<OZ5DX>
Telefax: +45 (54) 860979 <H.O.Pyndt>
President: Per Wellin, OZ1DHQ
Secretary: Kjeld Majland, OZ5KM
IARU liaison: Hans Otto Pyndt, OZ5DX, Kirstinebergparken
25, DK-4800 Nykobing F

DJIBOUTI:
Association des Radioamateurs de Djibouti [ARAD]
Address: P.O. Box 1076, Djibouti
Telephone: 350027/352849 <J28DQ>
Telex: 5808/5868 <Sylvain Affinito>
President: Mohamed Omar Moussa, J28AP
Secretary: Michel Caillaux, J28DN
IARU liaison: Sylvain Affinito, J28DQ

DOMINICA:
Dominica Amateur Radio Club [DARC]
Address: P.O. Box 389, Roseau
Location: 69 Cork Street, Roseau
Telephone: +1 (809) 448-8533 <HQ>
Telefax: +1 (809) 448-7708 <HQ>
President: Clement James, J73CI
Secretary: Olwyn Norris, J73NO
IARU liaison: Secretary

DOMINICAN REPUBLIC:
Radio Club Dominicano [RCD]
Address: P.O. Box 1157, Santo Domingo
Telephone: +1 (809) 533-2211
President: Manuel E. Gomez L., HI8MEL
Secretary: Tomas Lambertus F., HI8TLF
IARU liaison: William Read, HI8WA

ECUADOR:
Guayaquil Radio Club [GRC]
Address: P.O. Box 5757, Guayaquil
Telephone: +593 (4) 392671 <HQ>, (4) 294671/392860
<HC2EE>
Telefax: +593 (4) 307252 <HC2EE>, (4) 200650 <HC2ENM>
President: Cesar Rodriguez, HC2RB
Secretary: Eduardo Estrada, HC2EE
IARU liaison: Secretary

EGYPT:
Egypt Amateurs Wireless Club [EAWC]
Address: c/o E.S. Ramadan, Box 78, Heliopolis, Cairo 11341
Telephone: +20 (2) 257-4270 <SU1ER>
Telefax: +20 (2) 260-9792 <Ezzat S. Ramadan>
President: Ezzat Sayed Ramadan, SU1ER
Secretary: Said Kamel, SU1SK
IARU liaison: President

EL SALVADOR:
Club de Radio Aficionados de El Salvador [CRAS]
Address: P.O. Box 517, San Salvaodr
Location: Centro Comercial El Rosal, Local 12 Pte., Calle
Progreso 2823, San Salvador
Telephone: +503 246714/245975 <HQ>, 267424/299740
<YS1FAF>
Telefax: +503 236234 <YS1FAF>
Telex: 20534 QUINTEGRA <F. Fischnaler>
President: Francisco A. Fischnaler, YS1FAF
Secretary: Jose Enrique Celis, YS1CZ
IARU liaison: President

ESTONIA:
Eesti Raadioamatooride Uhing [ERAU]
Address: P.O. Box 125, EE-0090 Tallinn
Telephone: +372 (2) 449312 <HQ>, (2) 599990/429699
<ES1AR>
Telefax: +372 (2) 4238082 <ES1AR>
President: Enn Lohk, ES1AR
Secretary: Laine Kallaste, ES1YL
IARU liaison: President

FAROE ISLANDS:
Foroyskir Radioamatorar [FRA]
Address: P.O. Box 343, FR-110 Torshavn
Telephone: 009-298-10644 <HQ>, 009-298-11740 or 1169
<OY1A>
President: Annfinn Hansen, OY4AH
Secretary: Jan Egholm, OY3JE
IARU liaison: Arne Juul Arnskov, OY1A

FIJI:
Fiji Association of Radio Amateurs [FARA]
Address: P.O. Box 184, Suva
Telephone: +679 321605 <3D2CM>
President: Richard L. Northcott, 3D2CM
Secretary: Raj Singh, 3D2ER
IARU liaison: President

FINLAND:
Suomen Radioamatooriliitto [SRAL]
Address: P.O. Box 44, SF-00441 Helsinki
Location: Kaupinmaenpolku 9, SF-00440 Helsinki
Telephone: +358 (0) 562-5973 <HQ>
+358 (0) 298-5060 <OH2BU>, (0) 418-462
<OH2BQZ>
Telefax: +358 (0) 562-3987 <HQ>
President: Jari Jussila, OH2BU
Secretary: Hannu Lattu, OH2BKQ
IARU liaison: Vice President Markku Toijala, OH2BQZ

FORMER YUGOSLAV REPUBLIC OF MACEDONIA:
Radioamaterski Sojuz na Makedonija [RSM]
Radioamateur Society of Macedonia

Address: P.O. Box 14, 91000 Skopje
Location: Gradski zid blok 5, 91000 Skopje
Telephone: +389 (1) 237371/118339 <HQ>
Telefax: +389 (1) 238257 <HQ>
President: Venco Stojcev, Z32JA
Secretary: Mitko Kikerekov, Z34XMA
IARU liaison: President

FRANCE:
Reseau des Emetteurs Francais [REF]
Address: P.O. Box 2129, F-37021 Tours Cedex
Location: 32 Rue de Suede, Tours
Telephone: +33 (47) 418873 <HQ>, (32) 533695 <F3YP>
Telefax: +33 (47) 418888 <HQ>, (32) 523871 <F3YP>
President: Jean-Marie Gaucheron, F3YP
Secretary: Serge Phalippou, F5HX
IARU liaison: Vincent Magrou, F5JFT

FRENCH POLYNESIA:
Club Oceanien de Radio et d'Astronomie [CORA]
Address: P.O. Box 2139, Papeete, Tahiti
Telephone: +689 533121/438500 <FO5IW>, 436258/412525
<FO5EC>
+689 412923/425025 <FO4NR>
Telefax: +689 435230
President: Stan Wisniewski, FO5IW
Secretary: Alain Portal, FO5EC
IARU liaison: Richard Slavov, FO4NR

GABON:
Association Gabonaise des Radio-Amateurs [AGRA]
Address: P.O. Box 1826, Libreville
Telephone: +241 762602 <TR8CC>
President: Jean-Claude Villard, TR8JCV
Secretary:
IARU liaison: President

GAMBIA:
Radio Society of The Gambia [RSTG]
Address: c/o Jean-Michel Voinot, C53GB, P.M.B. 120, Banjul

GERMANY:
Deutscher Amateur-Radio-Club [DARC]
Address: P.O. Box 1155, D-34216 Baunatal
Location: Lindenallee 6, D-34216 Baunatal
Telephone: +49 (561) 949880 <HQ>, (451) 7070891 <DJ6TJ>
Telefax: +49 (561) 9498850 <HQ>, (451) 73630 <DJ6TJ>
President: Dr. Horst Ellgering, DL9MH
Secretary: Mr. Bernd W. Haefner, DB4DL
IARU liaison: Mr. Hans Berg, DJ6TJ

GHANA:
Ghana Amateur Radio Society [GARS]
Address: P.O. Box 3936, Accra
President: Kofi A. Jackson, 9G1AJ
Secretary: Samir Nassar, 9G1NS

GIBRALTAR:
Gibraltar Amateur Radio Society [GARS]
Address: P.O. Box 292, Gibraltar
Location: 3 Hargreaves Court, Gibraltar
Telephone: +350 73285 <ZBØØD>
Telefax: +350 73385 <ZBØØD>
President: Jimmy Bruzon, ZB2BL
Secretary: Wilfred Guerrero, ZB2IB
IARU liaison: Secretary

GREECE:
Radio Amateur Association of Greece [RAAG]
Address: P.O. Box 3564, GR-10210 Athens
Location: 2 Lenorman Street, Athens
Telephone: +30 (1) 522-6516 <HQ>
Telefax: +30 (1) 522-6505 <HQ>
President: Kostas Thanopoulos, SV1DC
Secretary: Dimitri Tzelatidis SV1RL
IARU liaison: Secretary

GRENADA:
Grenada Amateur Radio Club [GARC]
Address: P.O. Box 737, St. George's
Telephone: +1 (809) 443-2662 <J39DF>
President: John Phillip, J39CR
Secretary: Jerry Aberdeen, J39DF
IARU liaison: Secretary

GUATEMALA:
Club de Radioaficionados de Guatemala [CRAG]
Address: P.O. Box 115, Guatemala City 01901
Location: 5o. Nivel, Local 517, 12 Calle 2-04, Zona 9,
Guatemala City
Telephone: +502 (2) 314683 <HQ>
President: Eduardo Estrada, TG9CL
Secretary: Raul Guerra Rivera, TG4ZR
IARU liaison: Secretary

GUYANA:
Guyana Amateur Radio Association [GARA]
Address: c/o 221 Aubrey Barker Rd., So. Ruimveldt Gardens,
Georgetown
President: Pastor Cleophas Quashie, 8R1CJ
Secretary: Rev. George Richmond, 8R1AR

HAITI:
Radio Club d'Haiti [RCH]
Address: P.O. Box 1484, Port-au-Prince
President: Victor Lemoine, HH2V
Secretary: Joseph Baguidy Jr., HH2JO
IARU liaison: President

HONDURAS:
Radio Club de Honduras [RCH]
Address: P.O. Box 273, San Pedro Sula
Location: Colonia El Altiplano, San Pedro Sula
Telephone: +504 532035 <HQ>, 520877 <HR2HM>
President: Eduardo MacCullagh, HR2EMC
Secretary: Hugo Hernandez
IARU liaison: President

HONG KONG:
Hong Kong Amateur Radio Transmitting Society [HARTS]
Address: G.P.O. Box 541, Hong Kong
Telephone: +852 356-5606 & 875-2144 <VS6EY>
Telefax: +852 328-2672 <VS6BG>, 551-4364 <VS6EY>
President: Brett Graham, VS6BG
Secretary: Steven Beesley, VS6XMQ
IARU liaison: Robert Palitz, VS6EY

HUNGARY:
Magyar Radioamator Szovetseg [MRASZ]
Hungarian Radioamateur Society
Address: P.O. Box 11, H-1400 Budapest
Location: Zoltan u. 6 III/315, H-1054 Budapest
Telephone: +36 (1) 112-1616
Telefax: +36 (1) 111-6204
President: Bela Berzsenyi, HA5EB
Secretary: Imre Gajarszki, HA4YD
IARU liaison: Laszlo Berzsenyi, HA5EA

ICELAND:
Islenzkir Radioamatorar [IRA]
Address: P.O. Box 1058, IS-121 Reykjavik
Chairman: Saemundur Thorsteinsson, TF3UA
Secretary: Gisli Ofeigsson, TF3US
IARU liaison: Secretary

INDIA:
Amateur Radio Society of India [ARSI]
Address: P.O. Box 6538, Bombay 400026
Telephone: +91 (11) 668929 <VU2HV>, (11) 727857
<VU2SDN>
+91 (22) 361682 <VU2IN>, (22) 640136
<VU2ST>
President: Maj. Haveli Ram, VU2HV
Secretary: N.D. Mulla, VU2IN
IARU liaison: Saad Ali, VU2ST

INDONESIA:
Organisasi Amatir Radio Indonesia [ORARI]
Address: P.O. Box 6797 JKSRB, Jakarta 12067
Location: Jl. Danau Tondano T 10, Pejompongan Jakarta
10210
Telephone: +62 (21) 582226 <HQ>
Telefax: +62 (21) 5705034 <YB$$BNB>
President: Sugito, YF$$AL
Secretary: Sriwijaya Mertonegoro, YB$$BNB
IARU liaison: Erlangga Suryadarma, YB$$BZZ

IRAQ:
  Iraqi Radio Amateur Club [IRAC]
  Address: P.O. Box 55072, Baghdad 12001
  Location: Planetorium Building, Al-Zawra Park, Damascus
Square, Baghdad
  Telephone: +964 (1) 8843521
  President: Majid Abdul Hameid Rasheid, YI1MH
  Vice President: Saad Abdul Karim Al-Tai, YI1AB
  IARU liaison: Vice President
IRELAND:
  Irish Radio Transmitters Society [IRTS]
  Address: P.O. Box 462, Dublin 9
  Telephone: +353 (1) 896134 <EI7CX>
  President: Tom Rea, EI2GP
  Secretary: Martin F. O'Dea, EI3FI
  IARU liaison: Ian McStay, EI7CX, 37 Clonkeen Dr., Foxrock,
      Dublin 18
ISRAEL:
  Israel Amateur Radio Club [IARC]
  Address: P.O. Box 17600, Tel-Aviv 61176
  Telephone: +972 (3) 5658203 <HQ>
  +972 (3) 6349049 <4X6KJ>, (3) 5043253 <4X1AT>
  Telefax: +972 (3) 6349049 <4X6KJ>, (3) 6828335
  <4X1AT>
  Chairman: Joseph Obstfeld, 4X6KJ
  Secretary: Ami Rozenberg, 4Z9GCB
  IARU liaison: Chairman
ITALY:
  Associazione Radioamatori Italiani [ARI]
  Address: Via Scarlatti 31, I-20124 Milano
  Telephone: +39 (2) 6692192 <HQ>
  +39 (10) 564975 <I1BYH>, (534) 43104 <I4SN>
  Telefax: +39 (2) 66714809 <HQ>
  President: Alessio Ortona, I1BYH
  Secretary: Mario Ambrosi, I2MQP
  IARU liaison: arino Miceli, I4SN, Via Santo 192/1, I-40030
      Badi
IVORY COAST:
  Association des Radio-Amateurs Ivoiriens [ARAI]
  Address: P.O. Box 2946, Abidjan 01
  President: Jean-Jacques Niava, TU2OP
  Secretary: Bakayoko Soumaila, TU2XA
  IARU liaison: President
JAMAICA:
  Jamaica Amateur Radio Association [JARA]
  Address: 76 Arnold Road, Kingston 5
  Telephone: +1 (809) 926-7861 <HQ>, (809) 927-6977
  <6Y5MM>
  +1 (809) 928-1248/5139/5753 <6Y5LA>
  Telefax: +1 (809) 928-5632 <Lloyd Alberga>
  Telex: 2188 CITRAD <Lloyd Alberga>
  President: Michael Matalon, 6Y5MM
  Secretary: Thelma Findlay, 6Y5TG
  IARU liaison: Lloyd Alberga, 6Y5LA, c/o Palace Amusement
      Co., Box 8009, CSO Kingston
JAPAN:
  Japan Amateur Radio League [JARL]
  Address: C.P.O. Box 377, Tokyo 100-91
  Location: 1-14-2 Sugamo, Toshima, Tokyo 170
  Telephone: +81 (3) 5395-3106 <HQ>, (3) 5395-3100
  <JA1DM>
  Telefax: +81 (3) 3943-8282 <HQ>
  Telex: 23868 JAPRETAR
  President: Shozo Hara, JA1AN
  Secretary: Masayoshi Ebisawa, JA1DM
  IARU liaison: Yoshiji Sekido, JJ1OEY
JORDAN:
  Royal Jordanian Radio Amateur Society [RJRAS]
  Address: P.O. Box 2353, Amman
  Telephone: +962 (6) 666235 <HQ>
  President: H.R.H. Prince Raad Bin Zeid, JY2RZ
  Secretary: Mohammad Balbisi, JY4MB
  IARU liaison: Secretary

KENYA:
  Radio Society of Kenya [RSK]
  Address: P.O. Box 45681, Nairobi
  Telephone: +254 (2) 749667/729669 <5Z4FN>
  Telefax: +254 (2) 337978 <5Z4FN>
  Telex: 22297 NAIROBI <5Z4FN>
  Chairman: Leonard H. Raburn Jr., 5Z4DU
  Secretary: Paul Wyse, 5Z4FO
  IARU liaison: Max Raicha, 5Z4MR
REPUBLIC OF KOREA:
  Korean Amateur Radio League [KARL]
  Address: C.P.O. Box 162, Seoul 100-601
  Telephone: +82 (2) 817-7493 <HQ>
  Telefax: +82 (2) 817-7494 <HQ>
  Telex: 22065
  President: Joong-Geun Rhee, HL1AQQ
  Secretary: Ill-Boo Kim, HL5BBG
  IARU liaison: J.S. Kim, HL1KBW
KUWAIT:
  Kuwait Amateur Radio Society [KARS]
  Address: P.O. Box 5240, Safat 13053
  Location: House No. 2, Subst. No. 12, St. No. 1, Block No. 2,
  Al Surra
  Telephone: +965 533-3762 <HQ>
  Telefax: +965 531-1188 <HQ>
  Telex: 23357 KALIFCO
  Chairman: Dr. Abdel Rahman Al-Awadi, 9K2FF
  Secretary: Mr. M.J. Al-Amiri, 9K2MJ
  IARU liaison: Mr. Ahmed K. Al-Jassim, 9K2DQ
LEBANON:
  Association des Radio-Amateurs Libanais [RAL]
  Address: P.O. Box 8888, Beirut
  Telephone: +961 (9) 949346 <OD5CN>, (1) 581912
  <OD5KU>
  Telefax: +961 (1) 602817 <Elie Kadi>
  Telex: 42579 CEBEX <Elie Kadi>
  President: Aref Mansour, OD5CN
  Secretary: Elie Kadi, OD5KU
  IARU liaison: Secretary
LESOTHO:
  Lesotho Amateur Radio Society [LARS]
  Address: P.O. Box 949, Maseru 100
  Telephone: +266 312585 <7P8CI>, 340603 <7P8EB>
  Telefax: +266 310081 <Gunter Barak>
  President: Gunter Barak, 7P8CI
  Secretary: Rick Atherton, 7P8EB
  IARU liaison: President
LIBERIA:
  Liberia Radio Amateur Association [LRAA]
  Address: P.O. Box 10-1477, 1000 Monrovia 10
  Telephone: +231 225855 <HQ>
  +1 (718) 727-7080 <W2/EL2BA>
  Telefax: +231 224153
  President: Kamal T. Hamzi, EL2AY
  Secretary: Fr. Joe Brown, EL2FM
  IARU liaison: H. Walcott Benjamin, W2/EL2BA, 55 Austin Pl.
      #6T, Staten Island, NY 10304, U.S.A.
LIECHTENSTEIN:
  Amateurfunk Verein Liechtenstein [AFVL]
  Address: P.O. Box 629, FL-9495 Triesen
  Telephone: +41 (75) 392-1665 <HQ>
  Telefax: +41 (75) 373-5960 <HBØHTA>
  President: Alois Buechel, HBØMUO
  Secretary: Guenter Marogg, HB$$UTA
  IARU liaison: Leo Marxer, HBØØDMI
LITHUANIA:
  Lietuvos Radijo Megeju Draugija [LRMD]
  Address: P.O. Box 1000, Vilnius 2001
  Location: 8 Pilies, Vilnius
  Telephone: +370 (2) 221-836 <Tue. evening and Sun. morn
      ing>
  Telefax: +370 (2) 662-470 <HQ>
  President: Vytautas Kudelis, LY3BG
  Secretary: Antanas Zdramys, LY1DL

LUXEMBOURG:
  Reseau Luxembourgeois des Amateurs d'Ondes Courtes [RL]
  Address: c/o J. Kirsch, LX1DK, 23 Route de Noertzange, L-
      3530 Dudelange
  Telephone: +352 511133 <HQ>, 92434 <LX1JX>
  President: Josy Kirsch, LX1DK
  Secretary: Jacquot Junck, LX1JX
  IARU liaison: President
MALAYSIA:
  Malaysian Amateur Radio Transmitters' Society [MARTS]
  Address: P.O. Box 10777, 50724 Kuala Lumpur
  Telephone: +60 (3) 636-5299 <9M2CJ>, (3) 408-0260
      <9M2RS>
  Telefax: +60 (3) 262-1479 <9M2CJ>, (3) 775-2098 <9M2RS>
  President: Mohamad Anuar Bin Nordin, 9M2MO
  Secretary: Thiam Chee Ming, 9M2CJ
  IARU liaison: Abdul Rashid Sultan, 9M2RS
MALTA:
  Malta Amateur Radio League [MARL]
  Address: P.O. Box 575, Valletta
  Telephone: +356 492258 <9H1AQ>
  President: Anthony Vella, 9H1FG
  Secretary: Tony Bugeja, 9H1FM
  IARU liaison: Carmel A. Fenech, 9H1AQ, 35 Main St., Attard
MAURITIUS:
  Mauritius Amateur Radio Society [MARS]
  Address: P.O. Box 467, Port Louis
  Telephone: +230 4245866 <3B8CF>, 6761598 <3B8FP>
  Telefax: +230 2122102 <R. Karroo>
  President: S. Mandary, 3B8CF
  Secretary: Rashid Karroo, 3B8FP
  IARU liaison: President
MEXICO:
  Federacion Mexicana de Radio Experimentadores [FMRE]
  Address: P.O. Box 907, 06000 Mexico, D.F.
  Location: Av. Molinos 51, Desp. 307/308, Mexico 19, D.F.
  Telephone: +52 (5) 563-1405/2264 <5-8 p.m.>
  Telefax: +52 (5) 563-2264
  President: Oscar Oropeza, XE1ZOG
  Secretary: Hector Rodriguez, XE1RHZ
  IARU liaison: Carlos F. Narvaez, XE1FOX
MONACO:
  Association des Radio-Amateurs de Monaco [ARM]
  Address: P.O. Box 2, MC-98001 Monaco Cedex
  Telephone: +33 (93) 303498 <3A2LF>
  Telefax: +33 (93) 506034 <Claude Passet>
  President: Dr. Robert Scarlot, 3A2CR
  Secretary: Mr. Serge Salganik, 3A2HH
  IARU liaison: Mr. Henri Van Klaveren, 3A2AH
MONGOLIA:
  Mongolian Radio Sports Federation [MRSF]
  Address: P.O. Box 639, Ulaanbaatar 13
  Telephone: +976 (1) 320058 <HQ>
  Chairman: D. Garam-Ochir
  Secretary: G. Ulziysaikhan, JT1CG
  IARU liaison: Secretary
MONTSERRAT:
  Montserrat Amateur Radio Society [MARS]
  Address: P.O. Box 448, Plymouth
  Telephone: +1 (809) 492-5451 <VP2MQ>, (809) 491-3073
      <VP2MAG>
  President: Victor James, VP2MQ
  Secretary: Joseph Galloway, VP2MAG
  IARU liaison: Secretary
MOROCCO:
  Association Royale des Radio-Amateurs du Maroc [ARRAM]
  Address: P.O. Box 299, Rabat
  Location: 12 Rue Ahmed Arabi, Agdal, Rabat
  Telephone: +212 (7) 673703 <HQ>
  +212 (7) 55152/54060 <CN8BC>
  Telefax: +212 (7) 674757
  President: Col. Maj. Housni Benslimane, CN8BE
  Secretary: Said Boulhimez, CN8BL
  IARU liaison: Brahim Sidate, CN8BC

MOZAMBIQUE:
  Liga dos Radio Emissores de Mocambique [LREM]
  Address: P.O. Box 25, Maputo
MYANMAR:
  Burma Amateur Radio Transmitting Society [BARTS]
NAMIBIA:
  Namibian Amateur Radio League [NARL]
  Address: P.O. Box 1100, Windhoek 9000
  Telephone: +264 (61) 44114
  Telefax: +264 (61) 2012519
  President: Derek Moore, V51DM
  Secretary: Mike Alberts, V51MA
  IARU liaison: Secretary
NETHERLANDS:
  Vereniging voor Experimenteel Radio Onderzoek, in
      Nederland [VERON]
  Address: P.O. Box 1166, NL-6801 BD Arnhem
  Telephone: +31 (85) 426760 <HQ>
  +31 (4990) 72191 <PA3AVV>, (71) 761871
      <PAØTO>
  Telefax: +31 (4990) 72191 <PA3AVV>
  President: Thomas I. Sprenger, PA3AVV
  Secretary: J. Hoek, PAØJNH
  IARU liaison: A.J. Dijkshoorn, PAØTO, Jan van Gelderdreef
      11, NL-2253 VH Voorschoten
NETHERLANDS ANTILLES:
  Vereniging voor Experimenteel Radio Onderzoek, in de
      Nederlandse Antillen [VERONA]
  Address: P.O. Box 3383, Curacao
  Telephone: +599 (9) 375491 <PJ2WG>, (9) 656217/614300
      <PJ2HB>
  Telefax: +599 (9) 614819
  President: Mr. W.J. Gravenhorst, PJ2WG
  Secretary: Dr. J.H.R. Beaujon, PJ2HB
  IARU liaison: Dr. J.H.R. Beaujon, PJ2HB, P.O. Box 3052,
      Curacao
NEW ZEALAND:
  New Zealand Association of Radio Transmitters [NZART]
  Address: P.O. Box 40-525, Upper Hutt 6415
  Location: Astral Tower Bldg. 5F, 88-90 Main St., Upper Hutt
      6007
  Telephone: +64 (4) 528-2170 <HQ>, (4) 527-8694 <ZL2BHF>
  +64 (3) 351-5630 <ZL3QL>, (4) 528-8313 <ZL2AMJ>
  Telefax: +64 (4) 528-2170 <HQ>, (3) 351-5630
      <ZL3QL>
  President: Jim Meachen, ZL2BHF
  Secretary: Neville Copeland, ZL2AKV
  IARU liaison: Terry Carrell, ZL3QL
NICARAGUA:
  Club de Radioexperimentadores de Nicaragua [CREN]
  Address: P.O. Box 925, Managua
  Telephone: +505 (2) 71274/74937 <HQ>, (2) 283103
      <YN1KW>
  Telefax: +505 (2) 665879 <HQ>
  President: Reinerio Montiel B., YN1RM
  Secretary: William Karam Z., YN1KW
  IARU liaison: Secretary
NIGERIA:
  Nigeria Amateur Radio Society [NARS]
  Address: P.O. Box 2873, G.P.O. Marina, Lagos
  Location: 13 Orimolade Rd. (off Aina St.), Egbe (Alimosho
      Local Government), Lagos State
  Telephone: +234 (1) 884145 <HQ>, (1) 860974/964912
      <5NØOBA>
  +234 (60) 232330/232019 <5N9MBT>
  Telefax: +234 (1) 863237 <HQ>
  President: Muhammed Bello Tunau, 5N9MBT
  Secretary: Oyekunle B. Ajayi, 5N$$OBA
  IARU liaison: Secretary
NORWAY:
  Norsk Radio Relae Liga [NRRL]
  Address: P.O. Box 21, Refstad, N-0513 Oslo 5
  Location: Nedre Rommen 5 E, N-0988 Oslo 9
  Telephone: +47 (2) 2213790 <HQ>, (2) 2146860 <LA1ZH>

+47 (5) 1655721 <LA5QK>, (4) 951826/638800
<LA2RR>
Telefax: +47 (2) 2213791 <HQ>
President: Victor Hvistendahl, LA1ZH
Secretary: Alf Almedal, LA5QK
IARU liaison: Ole Garpestad, LA2RR, Brages vei 14, N-1540
Vestby

OMAN:
Royal Omani Amateur Radio Society [ROARS]
Address: P.O. Box 981, Muscat 113
Telephone: +968 600407 <HQ>, 537777/950999 <A41JT>
Telefax: +968 698558 <HQ>
Telex: 5556 POSTAL <HQ>
Chairman: H.E. Ahmed Suwaidan Al-Balushi, A41FK
Secretary: Abdul Razak Al-Shahwarzi, A41JT
IARU liaison: Secretary

PAKISTAN:
Pakistan Amateur Radio Society [PARS]
Address: P.O. Box 1450, Islamabad 44000
Telephone: +92 (51) 810755/213755 <Nasir Khan>
+92 (51) 253782 <AP2MYC>, (42) 301630 <AP2DM>
Telefax: +92 (51) 250912 <AP2MYC>
President: Nasir H. Khan, AP2NK
Secretary: M. Yunus Chaudry, AP2MYC
IARU liaison: Idress Mohsin, AP2DM

PANAMA:
Liga Panamena de Radioaficionados [LPRA]
Address: P.O. Box 9A-175, Panama
Telephone: +507 26-3160 <HQ>
President: Jose Moreira, HP1BUM
Secretary: Bolivar Vega, HP1CEV
IARU liaison: Dario Jurado, HP1DJ

PAPUA NEW GUINEA:
Papua New Guinea Amateur Radio Society [PNGARS]
Address: P.O. Box 204, Port Moresby, N.C.D.
Telephone: +675 274701 <P29ZGD>, 281297 <P29DX>
Telefax: +675 212199 <P29ZGD>
Telex: 22220 <mark attention Gordon Darling>
President: Gordon Darling P29ZGD
Secretary: Norman Beasley, P29NB
IARU liaison: President

PARAGUAY:
Radio Club Paraguayo [RCP]
Address: P.O. Box 512, Asuncion
Location: Humaita 1057, Asuncion
Telephone: +595 (21) 446124
+595 (21) 441975 <ZP5PX>, (21) 661484
<ZP5HSB>
Telefax: +595 (21) 440762
President: Francisco Schubeius, ZP5FGS
Secretary: Hernando Bertoni, ZP5HSB
IARU liaison: Secretary

PERU:
Radio Club Peruano [RCP]
Address: P.O. Box 538, Lima 100
Telephone: +51 (14) 414837 <HQ>, (14) 458134 <OA4PQ>
+51 (14) 489566/333592 <OA4ACK>
+51 (14) 354573 <OA4AVM>, (14) 618238
<OA4GL>
Telefax: +51 (14) 338401 <OA4ACK>
+51 (14) 462408 <Alfonso Alvarez-Calderon,
OA4PQ>
President: Eduardo Hoyle de Rivero, OA4ACK
Secretary: Luis Aguirre Tejada, OA4BQO
IARU liaison: Mario Caballero Olivera, OA4AVM

PHILIPPINES:
Philippine Amateur Radio Association [PARA]
Address: P.O. Box 4083, Manila
Location: Room 207, Remedios Bldg., #55 A. Rices Ave.,
Quezon City
Telephone: +63 (2) 964069 <HQ>
Telefax: +63 (2) 984705 <HQ>
President: Rene O. So, DU2CG
Secretary: Reynaldo G. Ang, DU2EF
IARU liaison: Tomas Hashim, DU1TH

POLAND:
Polski Zwiazek Krotkofalowcow [PZK]
Address: P.O. Box 61, PL-64100 Leszno 1
Telephone: +48 (65) 209529 <HQ>
Telefax: +48 (65) 209529 <HQ>
President: Ryszard Grabowski, SP3CUG
Secretary: Pawel Szmyd, SP7RJK
IARU liaison: President

PORTUGAL:
Rede dos Emissores Portugueses [REP]
Address: Rua D. Pedro V 7-4, P-1200 Lisboa
Telephone: +351 (1) 361186
President: Manuel Augusto Esteves, CT1LC
Secretary: Antonio Mendes, CT1DTC
IARU liaison: President

QATAR:
Qatar Amateur Radio Society [QARS]
Address: P.O. Box 22122, Doha
Location: 82 Suhaim Bin Hamad Rd., Doha
Telephone: +974 439191 <HQ>
Telefax: +974 439595 <HQ>
President: H.E. Abdulla Bin Hamad Al-Attyah
Secretary: Eng. Hashim Mustafavi Al-Hashimi
IARU liaison: Samir M. El Battah

ROMANIA:
Federatia Romana de Radioamatorism [FRR]
Address: P.O. Box 22-50, R-71100 Bucuresti
Location: Str. W. Maracineanu 1, Bucuresti
Telephone: +40 (01) 211-9787
Telex: 11180 SPORTROM
President: Vasile Oceanu, YO3NL
Secretary: Vasile Ciobanita, YO3APG
IARU liaison: Secretary

RUSSIA:
Soyuz Radiolyubitelej Rossii [SRR]
Union of Radio Amateurs of Russia
Address: P.O. Box 59, Moscow 105122
Telephone: +7 (095) 1664487 <HQ>, (86534) 53040
<UA6HZ>
Telefax: +7 (095) 1665856 <HQ>, (86534) 53040 <UA6HZ>
President: Valery Agabekov, UA6HZ
Vice President: Boris Stepanov, RU3AX
IARU liaison: Vice President

SAN MARINO:
Associazione Radioamatori della Repubblica di San Marino
[ARRSM]
Address: P.O. Box 77, RSM-47031 San Marino
Telephone: +378 906790 <HQ>, 997851 <T77C>, 944200
<T77J>
Telefax: +378 906790 <HQ>
President: Tony Ceccoli, T77C
Secretary: Rick Guirgetti, T72EB
IARU liaison: Julian Giacomoni, T77J

SENEGAL:
Association des Radio-Amateurs du Senegal [ARAS]
Address: P.O. Box 971, Dakar
Location: Immeuble des Colis Postaux, Avenue El-Hadj
Malick Sy, Dakar
Telephone: +221 223653/221643 <6W1KI>
Telefax: +221 229726 <Mustafa Diop>
Telex: 21430 <mark attention Mustafa Diop>
President: Mustafa Diop, 6W1KI
Secretary: Daniel Borowiec, 6W7OG
IARU liaison: President

SIERRA LEONE:
Sierra Leone Amateur Radio Society [SLARS]
Address: P.O. Box 10, Freetown
Telephone: +232 223335
President: Mrs. Cassandra Davies, 9L1YL
Secretary: William Sawyer, 9L1WS
IARU liaison: Alfred Koroma, 9L1AK

SINGAPORE:
Singapore Amateur Radio Transmitting Society [SARTS]
Address: G.P.O. Box 2728, Singapore 9047

Telephone: +65 281-9227 <9V1XC>
Telefax: +65 281-5898 <9V1XC>
President: K.C. Selvadurai, 9V1UV
Secretary: Chee Phuay Kit, 9V1SX
IARU liaison: President

SLOVAKIA:
Slovensky Zvaz Radioamaterov [SZR]
Slovak Amateur Radio Association [SARA]
Address: P.O. Box 1, 85299 Bratislava 5
Location: Wolkrova 4, 85101 Bratislava
Telephone: +42 (7) 847501 <HQ>, (7) 033093 <OM3LU>
Telefax: +42 (7) 845138 <HQ>
President: Anton Mraz, OM3LU
Secretary: Stefan Horecky, OM3JW
IARU liaison: President

SLOVENIA:
Zveza Radioamaterjev Slovenije [ZRS]
Address: Lepi Pot 6, 61000 Ljubljana
Telephone: +386 (1) 222459 <HQ>
Chairman: Anton Stipanic, S53BH
Secretary: Drago Grabensek, S59AR
IARU Liaison: Joze Vehovc, S51EJ

SOLOMON ISLANDS:
Solomon Islands Radio Society [SIRS]
Address: P.O. Box 418, Honiara
Telephone: +677 30417 <H44GR>
Telefax: +677 30051 <G. Richardson>
President: Greg Pearson, H44GP
Secretary: Graham Richardson, H44GR
IARU liaison: Secretary

SOUTH AFRICA:
South African Radio League [SARL]
Address: P.O. Box 807, Houghton 2041
Location: Cor. Duff Rd. & Louis Botha Ave., Houghton 2198
Telephone: +27 (11) 484-2830 <HQ>, (31) 765-6334 <ZS6AKV>
Telefax: +27 (11) 484-2831 <HQ>, (31) 765-6456 <ZS6AKV>
President: Hans v.d. Groenendaal, ZS6AKV
Secretary: Gerald Klatzko, ZS6BTD
IARU liaison: President

SPAIN:
Union de Radioaficionados Espanoles [URE]
Address: P.O. Box 220, E-28080 Madrid
Location: Monte Igueldo 102, E-28018 Madrid
Telephone: +34 (1) 477-1413 <HQ>, (41) 24-4045 <EA1QF>
Telefax: +34 (1) 477-2071 <HQ>, (41) 24-4045 <EA1QF>
President: Gonzalo Belay Pumares, EA1RF
Secretary: Pablo Barahona Aires, EA2NO
IARU liaison: Angel Padin de Pazos, EA1QF

SRI LANKA:
Radio Society of Sri Lanka [RSSL]
Address: P.O. Box 907, Colombo
Telephone: +94 (1) 505420
Telefax: +94 (1) 698315/695602 <Ranjit Gunawardena, 4S7RR>
+94 (1) 587891 <Anthony Jayaranjan, 4S7AJ>
President: Ernest Amarasinghe, 4S7EA
Secretary: K.K.G. Kulasekara, 4S7KG
IARU liaison: Vice President Paddy Gunasekera, 4S7PB

SURINAME:
Vereniging van Radio Amateurs in Suriname [VRAS]
Address: P.O. Box 1153, Paramaribo
Telephone: +597 472883/475300 <PZ1AC>
President: H. Bechan, PZ1EE
Secretary: A.D. Van Wijk, PZ5OC
IARU liaison: President

SWAZILAND:
Radio Society of Swaziland [RSS]
Address: P.O. Box 3744, Manzini
Telephone: +268 36511 <3DAØAJ>
Telefax: +268 36330 <Robin Seal>
Chairman: Willie Long, 3DAØBD
Secretary: Robin Seal, 3DAØAJ
IARU liaison: Secretary

SWEDEN:
Foreningen Sveriges Sandareamatorer [SSA]
Address: Ostmarksgatan 43, S-12342 Farsta
Telephone: +46 (8) 604-4006 <HQ>, (8) 552-482-70 <SMØCOP>
+46 (8) 581-737-66 <SMØSMK>
Telefax: +46 (8) 604-4007 <HQ>
President: Rune Wande, SMØCOP
Secretary: Gunnar Ahl, SM5CWV
IARU liaison: Gunnar Kvarnefalk, SMØSMK

SWITZERLAND:
Union Schweizerischer Kurzwellen-Amateure [USKA]
Address: P.O. Box 9, CH-4539 Rumisberg
Telephone: +41 (65) 763676 <HB9BTT>, (61) 7115391 <HB9DX>
President: Armin Wyss, HB9BOX
Secretary: Silvia Klaus, HB9BTT
IARU liaison: Etienne Heritier, HB9DX
P.O. Box 906, CH-4153 Reinach BL1

SYRIA:
Technical Institute of Radio [TIR]
Address: P.O. Box 245, Damascus
Telephone: +963 (11) 714540 <YK1AO>, (11) 717570 <YK1AM>
+963 (11) 215829 <YK1AN>
Telex: 412923 MAHASN <Dr. Omar Shabsigh>
President: Dr. Omar Shabsigh, YK1AO
Secretary: Mr. Hikmat Zuhdi, YK1AM
IARU liaison: President

TAIWAN:
Chinese Taipei Amateur Radio League [CTARL]
Address: P.O. Box 39, Changhua 500
Location: 17 Changnan Rd., Sec. 2, Changhua 500
Telephone: +886 (4) 738-8746 <HQ>, (2) 282-6957 <BV2FB>
+886 (2) 282-6957 <BV2FB>
Telefax: +886 (4) 738-5441 <HQ>
President: Bolon Lin, BV5AF
Secretary: Anthony Li, BV4OB
IARU liaison: Ralph Yang, BV2FB

THAILAND:
Radio Amateur Society of Thailand [RAST]
Address: G.P.O. Box 2008, Bangkok 10501
Telephone: +66 (2) 277-9453 <HQ>
Telefax: +66 (2) 275-7288 <HQ>
President: Mrs. Mayuree Chotikul, HS1YL
Secretary: Sombat Tharincharoen, HS1BV
IARU liaison: Tony Waltham, HSØ/G4UAV

TONGA:
Amateur Radio Club of Tonga [ARCOT]
Address: c/o Manfred Schuster, P.O. Box 1078, Nuku'alofa
President: Hama Na'ati, A35HN
Secretary: Manfred Schuster, A35MS
IARU liaison: Secretary

TRINIDAD AND TOBAGO:
Trinidad and Tobago Amateur Radio Society [TTARS]
Address: P.O. Box 1167, Port of Spain
Telephone: +1 (809) 637-4773 <9Y4NED>, 645-6933 <9Y4SS>
Telefax: +1 (809) 645-3352 <9Y4NED>
President: Noel Donawa, 9Y4NED
Secretary: Stephen Sheppard, 9Y4SS
IARU liaison: President

TURKEY:
Telsiz Radyo Amatorleri Cemiyeti [TRAC]
Address: P.O. Box 109, Istanbul
Telephone: +90 (1) 571-1651 <HQ>
Telefax: +90 (1) 266-4531 <TA1E>
President: Aziz Sasa, TA1E
Secretary: Kamil Aksoylu, TA2FL
IARU liaison: Bahri Kacan, TA2BK, Camlica Cad. 26, Uskudar, Istanbul

TURKS AND CAICOS ISLANDS:
Turks and Caicos Amateur Radio Society [TACARS]
Address: P.O. Box 218, Providenciales

Telephone: +1 (809) 946-4436 <VP5JM>
Telefax: +1 (809) 946-4998 <Jody Millspaugh>
President: Frederick Braithwaite, VP5FEB
Secretary: Jody Millspaugh, VP5JM
IARU liaison: Jody Millspaugh, P.O. Box 350567, Ft.
    Lauderdale, FL 33335, U.S.A.
UKRAINE:
    Ukrainian Amateur Radio League [UARL]
    Address: P.O. Box 57, Kiev 252001
    Location: Yangelia Str. 1/39, Kiev 252056
    Telephone: +7 (044) 446-2239 <HQ>
    Telefax: +7 (044) 488-3968 <N. Gostry>
    President: Nickolai V. Gostry, UT5UT
    Secretary: George A. Chlijanc, UY5XE
    IARU liaison: President
UNITED KINGDOM:
    Radio Society of Great Britain [RSGB]
    Address: Lambda House, Cranborne Rd., Potters Bar, Herts.
    EN6 3JE
    Telephone: +44 (707) 659015 <HQ>, (732) 353360 <G3GVV>
    Telefax: +44 (707) 645105 <HQ>
    Telex: 9312-130923 RSGB <HQ>
    President: I.D. Suart, GM4AUP
    Executive Vice President: C. Trotman, GW4YKL
    Secretary: J.C. Hall, G3KVA
    General Manager: Peter Kirby, GØØTWW
    IARU liaison: Tim Hughes, G3GVV, 10 Farm Lane,
    Tonbridge, Kent TN10 3DG
U.S.A.:
    American Radio Relay League [ARRL]
    Address: 225 Main Street, Newington, CT 06111
    Telephone: +1 (203) 594-0200 <HQ>
    Telefax: +1 (203) 594-0259 <HQ>
    Telex: 650215-5052 MCI <HQ>
    President: Rodney Stafford, KB6ZV
    Secretary: David Sumner, K1ZZ
    IARU liaison: Vice President Larry E. Price, W4RA
URUGUAY:
    Radio Club Uruguayo [RCU]
    Address: P.O. Box 37, Montevideo
    Location: Simon Blivar 1195, Montevideo
    Telephone: +598 (2) 787523/787879 <HQ>, (2) 787887
        <CX4AA>
    President: Prof. Yamandu Amen P., CX4AA
    Secretary: Mr. Jorge De Castro, CX8BE
    IARU liaison: President
VANUATU:
    Vanuatu Amateur Radio Society [VARS]
    Address: P.O. Box 665, Port Vila
    Telephone: Port Vila 3092
    President: Rodney S. Newel, YJ8RN
    Secretary: Colin W. Joly, YJ8CW
    IARU liaison: Secretary
VENEZUELA:
    Radio Club Venezolano [RCV]
    Address: P.O. Box 2285, Caracas 1010-A
    Location: Av. Lima, con Av. La Salle, Los Caboros, Caracas
    Telephone: +58 (2) 793-5404 <HQ>
    +58 (2) 261-3321 & (2) 227798 <YV5LTR>
    +58 (14) 217145 <YV5NLQ>, (2) 781-4878
    <YV5KAJ>
    Telefax: +58 (2) 793-6883 <HQ>
    President: Daniel J. Mancin R., YV5LTR
    Secretary: Juan M. Tesoro P., YV5NLQ
    IARU liaison: Pasquale Casale V., YV5KAJ
WESTERN SAMOA:
    Western Samoa Amateur Radio Club [WSARC]
    Address: P.O. Box 2015, Apia
    Telephone: +685 23980 <5W1AU>, 23055/21297 <5W1AT>
    Telefax: +685 23173 <Marty Maessen>
    Telex: 213 HIDETOUR <Phil Williams>
    President: Phil Williams, 5W1AU
    Secretary: Marty Maessen, 5W1AT
    IARU liaison: Secretary

YUGOSLAVIA:
    Savez Radio-Amatera Jugoslavije [SRJ]
    Address: P.O. Box 48, YU-11001 Beograd
    Telephone: +381 (1) 332216 <HQ>
    +381 (3) 46444 & (1) 344414 <YT7MM>
    Telex: 12405 UPRAD <M.S. Mandrino, Int'l Affairs Sec.>
    President: Ismet Karsniqi, YU8IK
    Secretary: Ivan Filipovic, YU1FF
    IARU liaison: Mirko Mandrino, YT7MM, Box 14, 26001
    Pancevo
ZAIRE:
    Union Zairoise des Radio-Amateurs [UZRA]
ZAMBIA:
    Radio Society of Zambia [RSZ]
    Address: P.O. Box 20332, Kitwe
    Telephone: 260 (2) 227627 <9J2FB>
    Telefax: +260 (2) 226219 <mark attention F. Bunce>
    Telex: 51230 <mark attention F. Bunce>
    Chairman: Chris Cotton, 9J2CP
    Secretary: Fred Buncem 9J2FB
    IARU liaison: Secretary
ZIMBABWE:
    Zimbabwe Amateur Radio Society [ZARS]
    Address: P.O. Box 2377, Harare
    Telephone: +263 (4) 735187 <Z21EK>, (4) 46341 <Z21JE>
    President: Howard Kramer, Z21EK
    Secretary: Molly E. Henderson, Z21JE
    IARU liaison: ?

ADMINISTRATIVE COUNCIL
    Richard L. Baldwin, W1RU - President, IARU
    Michael J. Owen, VK3KI - Vice President, IARU
    Larry E. Price, W4RA - Secretary IARU
    L.v.d. Nadort, PAØLOU - Chairman, IARU Region 1
    John Allaway, G3FKM - Secretary, IARU Region 1
    Alberto Shaio, HK3DEU - President, IARU Region 2
    Pedro Seidemann, YV5BPG - Secretary, IARU Region 2
    David Rankin, 9V1RH - Chairman, IARU Region 3
    Fred Johnson, ZL2AMJ - Director, IARU Region 3

INTERNATIONAL SECRETARIAT
    Address: P.O. Box 310905, Newington, CT 06131-0905,
    U.S.A.
    Location: 225 Main Street, Newington, CT 06111-1494,
    U.S.A.
    Telephone: +1 (203) 594-0200
    Telefax: +1 (203) 594-0259
    Telex: 650215-5052 MCI

OFFICE OF THE PRESIDENT:
    H.C. 60, Box 60, Waldoboro, ME 04572, U.S.A.
    Office of the Vice President:
    3 Gordon Rd., Mount Waverley, Vic. 3149, AUSTRALIA
    Office of the Secretary:
    P.O. Box 2067, Statesboro, GA 30459-2067, U.S.A.

PRESIDENT EMERITUS
    Noel B. Eaton, VE3CJ, P.O. Box 660, Waterdown, Ontario
    L0R 2H0, Canada

REGIONAL ORGANIZATIONS
IARU Region 1
    Address: 10 Knightlow Road, Birmingham B17 8QB, United
    Kingdom
    Telephone: +44 (21) 429-3200
    Telefax: +44 (21) 429-4800
    Chairman: L.v.d. Nadort, PAØLOU, Laarpark 34, NL-4881
    Ed Zundert, The Netherlands
    Vice Chairman: W. Nietyksza, SP5FM, Mazurska 11, PL-
    05806 Komorow, Poland
    Secretary: John Allaway, G3FKM, 10 Knightlow Road,
    Birmingham B17 8QB, United Kingdom
    Treasurer: Rossella Strom, I1RYS, P.O. Box 597, NL-2130
    AN Hoofddorp, The Netherlands

Executive Committee Members: A. Razak Al-Sharwarzi, A41JT, P.O. Box 933, Seeb 121, Oman; Vincent Magron, F5JFT, 36 Ave. G. Clemenceau, F-69230 Saint Genis Laval, France; Jari Jussila, OH2BU, Pilvijarvi, SF-02400 Kirkkonummi, Finland; Hans van de Groenendaal, ZS6AKV, P.O. Box 1842, Hillcrest 3650, South Africa; Mustafa Diop, 6W1KI, P.O. Box 2276, Dakar, Senegal

IARU Region 2

Address: P.O. Box 2253, Caracas 1010A, Venezuela

Telephone: +58 (2) 986-8759

Telefax: +58 (2) 986-8759

President: Alberto Shaio, HK3DEU, 9 Sidney Lanier Lane, Greenwich, CT 06831-3735, U.S.A.

Vice President:

Thomas B.J. Atkins, VE3CDM, 55 Havenbrook Blvd., Willowdale, Ont. M2J 1A7, Canada

Secretary: Pedro Seidemann, YV5BPG, P.O. Box 2253, Caracas 1010A, Venezuela

Treasurer: Steve Dunkerley, VP9IM, P.O. Box HM 2215, Hamilton HM JX, Bermuda

Executive Committee Members: Frank M. Butler, Jr., W4RH, 323 Elliott Rd. SE, Fort Walton Beach, FL 32548, U.S.A.; Alfonso Alvarez Calderon, OA4PQ, Av. Santa Cruz 937, Miraflores, Lima 18, Peru; Guillermo Nunez, XE1NJ, P.O. Box 21-386, 04000 Coyoacan, D.F., Mexico; Reinaldo J. Szama, LU2AH, Gorostiaga 2320 P. 15 A, 1426 Buenos Aires, C.F.,Argentina (Telefax: +54 1-613-3004); Fabian Zarrabe, YS1FI, Box 294, San Salvador, El Salvador

IARU Region 3

Address: P.O. Box 73, Toshima, Tokyo 170-91, Japan

Location: Matsuoka Bldg. No. 2, 1-14-5 Sugamo, Toshima, Tokyo 170, JPN

Telephone: +81 (3) 3944-3322

Telefax: +81 (3) 3943-8282

Telex: 23868 JAPRETAR

Chairman: David Rankin, 9V1RH/VK3QV, P.O. Box 14, Pasir Panjang, Singapore 9111, Singapore

Secretary: Masayoshi Fujioka, JM1UXU, P.O. Box 73, Toshima, Tokyo 170-91, Japan

Directors: Fred Johnson, ZL2AMJ, 15 Field Street, Upper Hutt, New Zealand; Keigo Komuro, JA1KAB, 1-6-13 Kokubunjidai, Ebina, Kanagawa 243, Japan; Sangat Singh, 9M2SS, 111 Jalan Terasik Lapan, Bangsar Baru, 59100 Kuala Lumpur, Malaysia (Telefax: +603 256-1571); David A. Wardlaw, VK3ADW, 21 Tormey Street, North Balwyn, Victoria 3104, Australia

Coordinators

Beacon Project: John G. Troster, W6ISQ, 82 Belbrook Way, Atherton, CA 94025, U.S.A.

CISPR: Tom Sprenger, PA3AVV, Dolomietenlaan 3, NL-5691 JP Son, The Netherlands

Monitoring System: International - R.E. Knowles, ZL1BAD, Onewhero, R.D. 2, Tuakau, New Zealand

Region 1 - Ronald Roden, G4GKO, 27 Wilmington Close, Hassocks, West Sussex BN6 8QB, United Kingdom

Region 2 - Malcolm Hamon, VE3KXH, 5 Eastbank Road, Newcastle, Ontario L0A 1H0, Canada

Region 3 - R.M. Wahrlich, ZL1CVK, Whangamarino Road, R.D. 2, Te Kauwhata, New Zealand

ARRL OFFICERS AND DIRECTORS
(As of September 1995)

Stafford, Rodney KB6ZV— President
Holladay, Jay A. W6EJJ—First Vice President
Frenaye, Tom K1KI—Vice President
Price, Larry W4RA—Vice President, International Affairs

Atlantic Division:
Hugh Turnbull, W3ABC—Director
Kay C. Craigie, WT3P—Vice Director

Central Division:
Edmond A. Metzger, W9PRN—Director
Howard Huntington, K9KM—Vice Director

Dakota Division:
Tod Olson, KØTO—Director
Hans Brakob, KØHB—Vice Director

Delta Division:
Joel Harrison, WB5IGF—Director
Rick Roderick, K5UR—Vice Director

George E. Race, WB8BGY—Director
John Thernes, WM4T—Vice Director

Hudson Division:
Stephen Mendelsohn, WA2DHF—Director
Paul Vydareny, WB2VUK—Vice Director

Midwest Division:
Lew Gordon, K4VX—Director
Bruce Frahm, KØBJ—Vice Director

New England Division:
Bill Burden, WB1BRE—Director
Warren Rothberg, WB1HBB —Vice Director

Northwestern Division:
Mary Lou Brown, NM7N—Director
Greg Milnes, W7AGQ—Vice Director

Pacific Division:
Brad Wyatt, K6WR—Director
Jim Maxwell, W6CF—Vice Director

Roanoke Division:
John Kanode, N4MM—Director
Dennis Bodson, W4PWF—Vice Director

Rocky Mountain Division:
Marshall Quiat, AGØX—Director
Walt Stinson, WØCP—Vice Director

Southeastern Division:
Frank Butler Jr., W4RH —Director
Evelyn Gauzens, W4WYR—Vice Director

Southwestern Division:
Fried Heyn, WA6WZO—Director
Art Goddard, W6XD—Vice Director

West Gulf Division:
Thomas Comstock, N5TC—Director
Jim Haynie, WB5JBP—Vice Director

ARRL Section Managers

Atlantic Division:
Delaware—Randall K. Carlson, WBØJJX
Eastern Pennsylvania—Bob Stanhope, KB3YS
Maryland-D.C.—William Howard, WB3V
Northern New York- *(new section)*
Southern New Jersey—Bruce Eichmann, KE2OP
Western New York—William Thompson, W2MTA
Western Pennsylvania—Bernie Fuller, N3EFN

Central Division:
Illinois—Bruce Boston, KD9UL
Indiana—Peggy Coulter, W9JUJ
Wisconsin—Richard R. Regent, K9GDF

Dakota Division:
Minnesota—Randy "Max" Wendel, NØFKU
North Dakota—Roger "Bill" Kurtti, WCØM
South Dakota—Roland Cory, WØYMB

Delta Division:
Arkansas—George Mitchell, KI5BV
Louisiana—Lionel A. "Al" Oubre, K5DPG
Mississippi—Ernest Orman Jr., W5OXA
Tennessee—O.D. Keaton, WA4GLS

Great Lakes Division:
Kentucky—Steve Morgan, WB4NHO
Michigan—Dale Williams, WA8EFK
Ohio—David Kersten, N8AUH

Hudson Division:
Eastern New York—Paul S. Vydarney, WB2VUK
New York City-Long Island—Richard Ramhap, N2GQR
Northern New Jersey—Richard S. Moseson, NW2L

Midwest Division:
Iowa— Jim Lasley, NØJL
Kansas—Robert M. Summers, KØBXF
Missouri—Roger Volk, KØGOB
Nebraska—Bill McCollum, KEØXQ

New England:
Connecticut—Betsey Doane, K1EIC
Eastern Massachusetts—Phil Temples, K9HI
Maine—Michelle Mann, WM1C
New Hampshire—Alan Shuman, N1FIK
Rhode Island—Rick Fairweather, K1KYI
Vermont—Justin Barton, WA1ITZ
Western Massachusetts—Daniel Senie, N1JEB

Northwestern Division:
Alaska—Larry Flanagan, NL7XG
Eastern Washington—Kyle Pugh, KA7CSP
Idaho—Don Clower, KA7T
Montana—Darrell Thomas, N7KOR
Oregon—Randy Stimson, KZ7T
Western Washington—Harry Lewis, W7JWJ

Pacific Division:
East Bay—Bob Vallio, W6RGG
Nevada—Bill Smith Jr., W4HMV
Pacific—Robert Schneider, AH6J
Sacramento Valley—Jettie Hill, W6RFF
San Francisco—Arthur Samuelson, W6VV
San Joaquin Valley—Mike Siegel, KI6PR
Santa Clara Valley—Kit Blanke, WA6PWW

Roanoke Division:
North Carlonia—W. Reed Whitten, AB4W
South Carolina—Michael Epstein, KD1DS
Virginia—Edward Dingler, N4KSO
West Virginia—O. N. (Olie) Rinehart, WD8V

Rocky Mountain Division:
Colorado—Tim Armagost, WBØTUB
New Mexico—Joe Knight, W5PDY
Utah—Jim Rudnicki, NZ7T
Wyoming—Warren "Rev" Morton, WS7W

Southeastern Division:
Alabama—Tom Moore Jr., KL7Q
Georgia—Jim Altman, N4UCK
Nothern Florida—Rudy Hubbard, WA4PUP
Southern Florida—Robert (Rip) Van Winkle, AA4HT
Puerto Rico—Guillermo Schwarz, KP4DDB
Virgin Islands—Ronald Hall, Sr., KP2N

Southwestern Division:
Arizona—Clifford Hauser, KD6XH
Los Angeles—Phineas J. Icenbice Jr.,W6BF
Orange—Joe H. Brown, W6UBQ
San Diego—Patrick Bunsold, WA6MHZ
Santa Barbara—Jennifer Roe, AA6MX

West Gulf Division:
North Texas—Robert Adler, NZ2T
Oklahoma—Joseph Lynch, N6CL
South Texas—Alan Cross, WA5UZB
West Texas—Amelia "Milly" Wise, W5OVH

ARRL ADVISORY COMMITTEES

Contest Advisory Committee
(As of July 1995)

Atlantic: Tim Duffy, K3LR
44 Elliott Road, West Middlesex, PA 16159

Central: Joe Sego, KJ9D
5336 Elmwood Ave., Indianapolis, IN 46203

Dakota: John Baungarten, KØIJL
2107 Vermilion Road, Duluth, MN 55803

Delta: John Day, W4XJ
289 Taraview, Collierville, TN 38107

Great Lakes: Gary Hext, WB4FLB
4953 Westgate Drive, Bowling Green, KY 42101

Hudson: J.P. Kleinhaus, AA2DU-Chair
29 Dirubbo Drive, Cortland, NY 10566

Midwest: Larry Lambert, NØLL
405 Shelton Drive, Smith Center, KS 66967

New England: Kurt Pauer, W1PH
PO Box 754, Amherst, NH 03031

Northwestern: Larry Tyree, N6TR
15125 SE Bartel Road, Boring, OR 97009

Pacific: Bob Wilson, N6TV
51 Cheltenham Way, San Jose, CA 95139

Roanoke: Don Daso, WZ3Q
1260 College Avenue, Wilkesboro, NC 28697

Rocky Mountain: Doug Allen, W2CRS
PO Box 5646, Woodland Park, KS 80866

Southeastern: Steve McElroy, K4JPD
5587 Five Notch Trail, Douglasville, GA 30135

Southwestern: Ned Stearns, AA7A
7038 E. Aster Dr., Scottsdale, AZ 85254

West Gulf: James Eppright, K5RX
3123 Tower Trail, Dallas, TX 75229

RAC: Tim Ellam, VE6SH
307 Christie Knoll Point SW Calgary, AB T3H 1R5

Board Liaison: Rick Roderick, K5UR
PO Box 1463, Little Rock, AR 72203

Administrative Liaison: Lisa DeLude
225 Main Street, Newington, CT 06111

DX Advisory Committee
(As of July 11, 1995)

Atlantic: Tony Gargano, N2SS
26 Winchester Drive, Sewell, NJ 08080

Central: Jim O'Connell, W9WU
512 West Elm Avenue, La Grange, IL 60525

Dakota: Thomas Lutz, WØKZV
3677 Steele Street, Minnetonka, MN 55345

Delta: Rick Roderick, K5UR
PO Box 1463, Little Rock, AR 72203

Great Lakes: Theodore Mirgliotta, KB8NW
16806 W 130th Street, Strongville, OH 44136

Hudson: **Not Yet Elected**

Midwest: James Spencer, WØSR
3712 Tanager Drive NE, Cedar Rapids, IA 52402

New England: James Dionne, K1MEM
31 De Marco Road, Sudbury, MA 01776

Northwestern: Dick Moen, N7RO
2935 Plymouth Drive, Bellingham, WA 98225

Pacific: John Troster, W6ISQ
82 Belbrook Way, Atherton, CA 94025

Roanoke: Gary Dixon, K4MQG
Crescent Ridge, Fort Mill, SC 29715

Rocky Mountain: Charlie Summers, KYØA
6392 S Yellowstone Way, Aurora, CO 80016

Southeastern: Robert R. Beatty III, W4VQ
11 Heritage Cove Court, Casselberry, FL 32707

Southwestern: John Alexander, K6SVL
28403 Covecrest Dr, Rancho Palos Verdes, CA 90274

West Gulf: Jim Lane, N5DC
1602 Chestnut Ridge Road, Kingwood, TX 77339

RAC: Garth Hamilton, VE3HO
PO Box 2641, Niagara Falls, NY 14302

Board Liaison: Bruce Frahm, KØBJ
PO Box DX, Colby, KS 67701

Staff Liaison: Chuck Hutchinson, K8CH
225 Main Street, Newington, CT 06111

Administrative Liaison: Lisa DeLude
225 Main Street, Newington, CT 06111

*New Member

Public Service Advisory Committee

Pacific: Steve Wilson, KA6S, CHAIR
813 Berryessa Street
Milpitas, CA 95035

Board Liaison: Howard Huntington, K9KM
25350 N Marilyn Ln.
Hawthorn Woods, IL 60047

Staff Liaison: Jay Mabey, NUØX
225 Main St.
Newington, CT 06111

Great Lakes: David L. Kersten, N8AUH
2197 McKinley Av.
Lakewood, OH 44107-5432

Midwest: Les Myers, KØSCM
5001 Normal Blvd.
Lincoln, NE 68506

Atlantic: Bob Josuweit, WA3PZO
3341 Sheffield Av.
Philadelphia, PA 19136

Central: Richard R. Regent, K9GDF
5003 S 26th St.
Milwaukee, WI 53221

Dakota: R. H. Munger, KAØARP
8386 Pequaywan Lake Rd.
Duluth, MN 55803

Delta: Jim Leist, KB5W
2632 Valley Wood Dr.
Gautier, MS 39553

Hudson: Joe Schimmel, W2HPM
41 Midvale Av.
Farmingville, NY 11738

New England: Bob Salow, WA1IDA
Box 3773
Natick, MA 01760-0030

Northwestern: John White, K7RUN
PO Box 13274
Portland, OR 972113

West Gulf: Sam Sitton, KV5X
417 Ridge Road
Edmond, OK 73034

Roanoke: Carl E. Smith, N4AA
PO Box 6843
Asheville, NC 28816

Rocky Mtn: Joe T. Knight, W5PDY
10408 Snow Hgts NE
Albuquerque, NM 87112-3057

Southeastern: Joel I. Kandel, KI4T
9370 SW 87 Ave S-18
Miami, FL 33176

Southwestern: Richard Rudman, W6TIA
KFWB-6230 Yucca Street
Hollywood, CA 90028

Canada: William Parkes, VE7PAR
1211 Greenbriar Way
N Vancouver, BC V7R 1L8 Canada

SPECTRUM COMMITTEE
(As of July 10, 1995)

Atlantic Devision: *Robert Bennett, W3WCQ (CHAIR)
1006 Green Acre Road, Towson, MD 21204

Willem Van Aller, K3CZ
7623 Old Washington Road, Woodbine, MD 21797

John Hansen, WAØPTV
49 Maple Avenue, Fredonia, NY 14063

Central Division: Joe Schroeder, W9JUV
2120 Fir Street, Glenview, IL 60025

Dakota Division: *Paul Emeott, KØLAV
3960 Schuneman Road, St. Paul, MN 55110

Stan Kittelson, WDØDAJ
261 10th Street E, Dickinson, ND 58601

Paul Ramey, WGØG
16266 Finland Avenue, Rosemount, MN 55068

Delta Division: Dalton McCrary, N4OYS
2111 Brookwood Lane, Murfreesboro, TN 37129

Bob Taylor, WB5LBT
10715 Waverland, Baton Rouge, LA 70815

Dr. Jim Akers, W5VZF
21 Whispering Pines, Starkville, MS 39759

Great Lakes Division: *David Prestel, W8AJR
525 Coy Lane, Chagrin Falls, OH 44022

David Smith, W8YZ
530 Hollywood Drive, Monroe, MI 48161

John Thernes, WM4T
60 Locust Avenue, Covington, KY 41017

Hudson Division: *Chris Peckman, WG2W
31 Della Avenue, Pompton Plains, NJ 07444

Gordon Beattie, N2DSY
206 N Vivyen Street, Bergenfield, NJ 07621

Midwest: *Jim McKim, WØCY
1721 Glen Avenue, Salina, KS 67401

Donald Hoehne, KFØRE
1900 Elm Street, Box 632, Stanton, NE 68779
Larry Reeves,WAØGKZ
12936 Virginia, Kansas City, MO 64146

New England Division: Mitch Stern, WB2JSJ
14 Kimberly Drive, Essex Junction, VT 05452

David Upton, WB1CMG
25 Harwood Road, Mt. Vernon, NH 03057

Northwestern Division: Clay Freinwald, K7CR
32521 107th Ave SE, Auburn, WA 98002

Merle Cox, W7YOZ
12411 86 Place NE, Kirkland, WA 98034

Pacific Division: *Jim Von Striver, W6ASL
3712 Merridan Drive, Concord, CA 94518

Carl Guastaferro, WH6OP
11790 Kathryn Lane, Nevada City, CA 95959

Eric Williams, WD6CMU
5860 Clinton Avenue, Richmond, CA 94805

Roanoke Division: Michael A. Baker, W8CM
306 Woodbury Lane, Lynchburg, VA 24502

Russell Platt, Jr., WJ9F
4226 Peppertree Avenue, Concord, NC 28027

Rocky Mountain Division: Michael Barnell, WDØFVV
PO Box 621, Parker, CO 80134

Lauren Libby, KXØO
6166 Del Paz Drive, Colorado Springs, CO 80918

John Lloyd, K7JL
11560 S Sandy Creek Drive, Sandy, UT 84094

Southeastern Division: Joel Kandel, KI4T
9370 SW 87th Avenue, Apt. S18, Miami, FL 33176

Robert Schafer, KA4PKB
PO Box 57, Loachapoka, AL 36865

Walter Maxwell, W2DU
243 N Cranor Avenue, Deland, FL 32720

Southwestern Division: *Karl Pagel, N6BVU
PO Box 6080, Anaheim, CA 92816

David Gutierrez, WA6PMX
5221 Del Norte Circle, La Palma, CA 90623

James Fortney, K6IYK
11865 E Pradera Road, Rancho Santa Rosa, CA 93012

West Gulf Division: Kent Britain, WA5VJB
1626 Vineyard, Grand Prairie, TX 75052

Joe Jarrett, K5FOG
13411 Overland Pass, Austin, TX 78736

Robert Diersing, N5AHD
4129 Montego, Corpus Christi, TX 78411

Radio Amateurs of Canada: Dana Shtun, VE3DSS
500 Willard Avenue, Toronto, ON M6S 3R6 Canada

Board Liaison: George Race, WB8BGY
Director, Great Lakes, 3856 Gibbs Road, Albion, MI 49224
Staff Liaison: Jay Mabey, NUØX
225 Main Street, Newington, CT 06111
Administrative Liaison: Lisa DeLude
225 Main Street, Newington, CT 06111

Board Liasion: Fried Heyn
* Spokesman for Division

DIGITAL COMMITTEE LIST
(As of July 24, 1995)

Dave Speltz, KB1PJ
9 Heather Lane, Amherst, NH 03031

Craig McCartney, WA8DRZ
160 Montalvo Road, Redwood City, CA 94062

Paul Newland, AD7I
PO Box 205, Holmdel, NJ 07733-0205

Perry (Bo) McLean, WØXK
9955 NW Windover Drive, Kansas City, MO 64153

Vic Poor, W5SMM
401 Riverside Drive, Melbourne Beach, FL 32951

Dale Sinner, W6IWO
1904 Carolton Lane, Fallbrook, CA 92028-4614

Kay Craigie, WT3P
5 Faggs Manor Lane, Paoli, PA 19301-1905

AMERICAN RADIO RELAY LEAGUE FIELD ORGANIZATION

The United States is divided into 15 ARRL Divisions. Every two years, the ARRL full members in each of these divisions elect a director and a vice director to represent them on the League's Board of Directors. The Board determines the policies of the League, which are carried out by the Headquarters staff. A director's function is principally policymaking at the highest level, but the Board of Directors is all-powerful in the conduct of League affairs.

The 15 divisions are further broken down into 70 sections and the ARRL full members in each section elect a Section Manager (SM). The SM is the senior elected ARRL official in the section, and in cooperation with the director, fosters and encourages all ARRL activities within the section. A breakdown of sections within each division (and counties within each split-state section) follows:

Atlantic Division:
Delaware;
Eastern Pennsylvania;
    (Adams, Berks, Bucks, Carbon, Chester, Columbia, Cumberland, Dauphin, Delaware, Juniata, Lackawanna, Lancaster, Lebanon, Lehigh, Luzerne, Lycoming, Monroe, Montgomery, Montour, Northampton, Northumberland, Perry, Philadelphia, Pike, Schuylkill, Snyder, Sullivan, Susquehanna, Tioga, Union, Wayne, Wyoming, York.)
Maryland-D.C.;
Northern New York (effective, Jan. 96);
    (Clinton, Essex, Franklin, Fulton, Hamilton, Herkimer, Jefferson, Lewis, Montgomery, Otsego, St. Lawrence, Schoharie.)
Southern New Jersey;
    (Atlantic, Burlington, Camden, Cape May, Cumberland, Gloucester, Mercer, Ocean, Salem.)
Western New York;
    (Allegany, Broome, Cattaraugus, Cayuga, Chautauqua, Chemung, Chenango, Cortland, Delaware, Erie, Genesee, Livingston, Madison, Monroe, Niagara, Oneida, Onondaga, Ontario, Orleans, Oswego, Schuyler, Seneca, Steuben, Tioga, Tompkins, Wayne, Wyoming, Yates.)
Western Pennsylvania;
    (Those counties not listed under Eastern Pennsylvania.)

Central Division:
Illinois;
Indiana;
Wisconsin.

Dakota Division:
Minnesota;
North Dakota;
South Dakota.

Delta Division:
Arkansas;
Louisiana;
Mississippi;
Tennessee.

Great Lakes Division:
Kentucky;
Michigan;
Ohio.

Hudson Division:
Eastern New York;
   (Albany, Columbia, Dutchess, Greene, Orange, Putnam,
   Rensselaer, Rockland, Saratoga, Schenectady, Sullivan, Ulster,
   Warren, Washington, Westchester.)
New York City, Long Island;
   (Bronx, Kings, Nassau, New York, Queens, Staten Island,
   Suffolk.)
Northern New Jersey;
   (Bergen, Essex, Hudson, Hunterdon, Middlesex, Monmouth,
   Morris, Passaic, Somerset, Sussex, Union, Warren.)

Midwest Division:
Iowa;
Kansas;
Missouri;
Nebraska.

New England Division:
Connecticut;
Maine;
Eastern Massachusetts;
   (Barnstable, Bristol, Essex, Middlesex, Nantucket, Norfolk,
   Plymouth, Suffolk.)
New Hampshire;
Rhode Island;
Vermont;
Western Massachusetts;
   (Those counties not listed under Eastern Massachusetts.)

Northwestern Division:
Alaska;
Idaho;
Montana;
Oregon;
Eastern Washington;
   (Okanogan, Chelan, Douglas, Kittitas, Yakima, Klickitat,
   Benton, Franklin, Walla Walla, Adams, Grant, Lincoln, Ferry,
   Stevens, Pend Oreille, Spokane, Whitman, Columbia, Garfield,
   Asotin.)
Western Washington;
   (Those counties not listed under Eastern Washington.)

Pacific Division:
East Bay;
   (Alameda, Contra Costa, Napa, Solano.)
Nevada;
Pacific;
   (Hawaii and U.S. possessions in the Pacific.)
Sacramento Valley;
   (Alpine, Amador, Butte, Colusa, El Dorado, Glenn, Lassen,
   Modoc, Nevada, Placer, Plumas, Sacramento, Shasta, Sierra,
   Siskiyou, Sutter, Tehama, Trinity, Yolo, Yuba.)
San Francisco;
   (Del Norte, Humboldt, Lake, Marin, Mendocino, San
   Francisco, Sonoma.)
San Joaquin Valley;
   (Calaveras, Fresno, Kern, Kings, Madera, Mariposa, Merced,
   Mono, San Joaquin, Stanislaus, Tulare, Tuolumne.)
Santa Clara Valley;
   (Monterey, San Benito, San Mateo, Santa Clara, Santa Cruz.)

Roanoke Division:
North Carolina;
South Carolina;
Virginia;
West Virginia.

Rocky Mountain Division:
Colorado;
Utah;
New Mexico;
Wyoming.

Southeastern Division:
Alabama;
Georgia;
Northern Florida;
   (Alachua, Baker, Bay, Bradford, Calhoun, Citrus, Clay,
   Columbia, Dixie, Duval, Escambia, Flagler, Franklin, Gadsden,
   Gilchrist, Gulf, Hamilton, Hernando, Holmes, Jackson,
   Jefferson, Lafayette, Lake, Leon, Levy, Liberty, Madison,
   Marion, Nassau, Okaloosa, Orange, Pasco, Putnam, Santa
   Rosa, Seminole, St. Johns, Sumter, Suwanee, Taylor, Union,
   Volusia, Wakulla, Walton, Washington.)
South Florida;
   (Those counties not listed under Northern Florida.)
Puerto Rico;
U.S. Virgin Islands;
   (including Guantanamo Bay.)

Western Gulf Division:
Northern Texas;
   (Anderson, Archer, Baylor, Bell, Bosque, Bowie, Brown, Camp;
   Cass, Cherokee, Clay, Collin, Comanche, Cooke, Coryell,
   Dallas, Delta, Denton, Eastland, Ellis, Erath, Falls, Fannin,
   Franklin, Freestone, Grayson, Gregg, Hamilton, Harrison,
   Henderson, Hill, Hood, Hopkins, Hunt, Jack, Johnson,
   Kaufman, Lamar, Lampasas, Limestone, Mc Lennan, Marion,
   Mills, Maontague, Morris, Nacogdoches, Navarro, Palo Pinto,
   Panola, Parker, Rains, Red River, Rockwall, Rusk, Shelby,
   Smith, Somervell, Stephens, Tarrant, Throckmorton, Titus,
   Upshur, Van Zandt, Wichita, Wilbarger, Wise, Wood, Young.)
Oklahoma;
Southern Texas;
   (Angelina, Aransas, Atacosa, Austin, Bandera, Bastrop, Bee,
   Bexar, Blanco, Brazoria, Brazos, Brooks, Burleson, Burnet,
   Caldwell, Calhoun, Cameron, Chambers, Colorado, Comal,
   Concho, DeWitt, Dimmitt, Duval, Edwards, Fayette, Fort
   Bend, Frio, Galveston, Gillespie, Goliad, Gonzales, Grimes,
   Guadalupe, Hardin, Harris, Hays, Hidalgo, Houston, Jackson,
   Jasper, Jefferson, Jim Hogg, Jim Wells, Karnes, Kendall,
   Kenedy, Kerr, Kimble, Kinney, Kleberg, LaSalle, Lavaca, Lee,
   Leon, Liberty, Live Oak, Llano, Madison, Mason, Matagorda,
   Maverick, McCulloch, McMullen, Medina, Menard, Milam,
   Montgomery, Newton, Nueces, Orange, Polk, Real, Refugio,
   Robertson, Sabine, San Augustine, San Jacinto, San Patricio,
   Victoria, Walker, Waller, Washington, Webb, Wharton,
   Willacy, Williamson, Wilson, Zapata, Zavala.)
Western Texas;
   (Those Counties not already listed under Northern and
   Southern Texas.)

The following entities were formerly part of the Canadian
Division of the ARRL, and are often used along with the sections as
multipliers in contests such as the ARRL November Sweepstakes:

Alberta;
British Columbia;
Manitoba;
Maritime;
   (Nova Scotia, New Brunswick, Prince Edward Island,
   Labrador, Newfoundland.)
Northwest Territories;
(and Yukon.)
Ontario;
Quebec;
Saskatchewan.

(Courtesy of ARRL)

HAM ASTRONAUTS IN SPACE

Name	Callsign	Name	Callsign
Viktor Afanasiev	U9MIR	Ernst Messerschmid	DG2KM/DPØSL
Jay Apt	N5QWL	Mamoru Mohri	7L2NJY
Anatoli P. Artsebaski	U7MIR	Steven "Steve" R. Nagel	N5RAW
Sergei Avdeyev	RV3DW	Ellen Ochoa	KB5TZZ
Ellen S. Baker	KB5SIX	Wubbo Ockels	PE1LFO/DPØSL
Alexander Balandin	U7MIR	Steve Oswald	KB5YSR
John David F. Bartoe	W4NYZ	Ronald "Ron" A. Parise	WA4SIR
Charles Bolden	KE4IQB	Valery Polyakov	U3MIR
Kenneth "Ken" D. Cameron	KB5AWP	Charles Precourt	KB5YSQ
Robert "Bob" D. Cabana	KC5HBV	Richard "Dick" N. Richards	KB5SIW
Kenneth "Ken" D. Cockrell	KB5UAH	Jerry L. Ross	N5SCW
Sherlock Curie	KC5OZX	Hans-Wilhelm Schlegel	DG1KIH
Brian Duffy	N5WQW	Rick Searfoss	KC5CKM
Sam Durrance	N3TQA	Ronald Sega	KC5ETH
Anthony "Tony" W. England	WØORE	Helen Sharman	GB1MIR
Martin Fettman	KC5AXA	Anatoly Solovov	U6MIR
Klau-Dietrich Flade	DPØMIR	Kathryn "Kathy" D. Sullivan	N5YYV
Colin "Mike" M. Foale	KB5UAC	Donald Thomas	KC5FVF
Dirk D. Frimount	ON1AFD	Vladimir Titov	U1MIR
Reinhard Furrer	DD6CF/DPØSL	Michael Tognini	F5MIR
Owen K. Garriott	W5LFL	Yuri Usachev	R3MIR
Linda M. Godwin	N5RAX	Alexander Viktorenko	U9MIR
L. Blaine Hammond, Jr.	KC5HBS	Alexander Volkov	U4MIR
Alexander Kaleri	U8MIR	Janice Voss	KC5BTK
Sergei Krikalev	U5MIR	Ulrich Walter	DG1KIM
Wendy Lawrence	KC5KII		
Dave C. Leetsma	N5WQC	*** It has long been rumored that the first man in space,	
Jerry M. Linenger	KC5HBR	Yuri Gagarin, was at some point in time licensed as a	
Bill McArthur	KC5ACR	Radio Amateur. This has been determined to be false.	
Musa Manarov	RV3AM/U2MIR	****	

Amateur Radio Satellites

Satellite	Launch	Status
OSCAR-1	DEC12-61	FUNCTIONED FOR 21 DAYS-BATTERY EXHAUSTED.
OSCAR-2	JUN02-62	FUNCTIONED FOR 19 DAYS-BATTERY EXHAUSTED.
OSCAR-3	MAR09-65	FUNCTIONED FOR 18 DAYS-BATTERY EXHAUSTED/FIRST WITH SOLAR CELLS.
OSCAR-4	DEC21-65	FUNCTIONED FOR 85 DAYS-FULLY SOLAR POWERED-BAD ORBIT.
Australis/OSCAR-5	JAN23-70	FUNCTIONED FOR 52 DAYS-CONTAINED BEACON ONLY-BATTERY EXHAUSTED.
PHASE-2a/OSCAR-6	OCT15-72	FUNCTIONED FOR 4 1/2 YRS/FIRST TRANSPONDER/SOLARCELLS+BATTERIES.
PHASE-2b/OSCAR-7	NOV15-74	FUNCTIONED FOR 6 1/2 YRS/TWO TRANSPONDERS.
PHASE-2c/OSCAR-8	MAR05-78	FUNCTIONED FOR 5 1/2 YRS/TWO TRANSPONDERS.
RS-1	OCT26-78	FUNCTIONED FOR SEVERAL MONTHS/ONE TRANSPONDER.
RS-2	OCT26-78	FUNCTIONED FOR 3 YRS/ONE TRANSPONDER.
Phase-3a/OSCAR-10	MAY23-80	LAUNCH FAILURE.
ISKRA-1	JUL10-81	FUNCTIONED 89 DAYS.
UO-9	OCT06-81	FUNCTIONED 8 YRS/FIRST WITH CAMERA.
RS-3	DEC17-81	FUNCTIONED FOR SEVERAL MONTHS/BATTERIES EXHAUSTED.
RS-4	DEC17-81	FUNCTIONED FOR SEVERAL MONTHS/BATTERIES EXHAUSTED.
RS-5	DEC17-81	FUNCTIONED 7 YRS/BATTERIES EXHAUSTED.
RS-6	DEC17-81	FUNCTIONED FOR SEVERAL YRS/BATTERIES EXHAUSTED.
RS-7	DEC17-81	FUNCTIONED FOR 8YRS/BATTERIES EXHAUSTED.
RS-8	DEC17-81	FUNCTIONED FOR SEVERAL YRS/BATTERIES EXHAUSTED.
ISKRA-2	MAY17-82	FUNCTIONED FOR 53 DAYS/LAUNCHED BY HAND FRO SALYUT-7.
ISKRA-3	NOV18-82	FUNCTIONED FOR 28 DAYS/LAUNCHED BY HAND FROM SALYUT-7.
PHASE-3B/AO-10	JUN16-83	PRESENTLY FUNCTIONING/BAD ORBIT - Only operates while in Sunlight.
UoSAT-B/UO-11	MAR02-84	FAILED SEPT09-93
NUSAT-1	APR29-85	FUNCTIONED 2 YRS/SHUTTLE LAUNCHED/COMMERCIAL-AMATEUR BUILT.
JAS-1a/FO-12	AUG12-86	FUNCTIONED FOR 3 YRS/ELECTRICAL FAILURE.
RS-10/11	JUN23-87	PRESENTLY OPERATING.
PHASE-3C/AO-13	JUN15-88	PRESENTLY OPERATING. Orbit Decay Dec-1996
UoSAT-D/UO-14	JAN22-90	PRESENTLY OPERATING.
UoSAT-E/UO-15	JAN22-90	FAILED IMMEDIATELY AFTER LAUNCH.
PACSAT/AO-16	JAN22-90	PRESENTLY FUNCTIONING.
DOVE/DO-17	JAN22-90	PRESENTLY FUNCTIONING.

NUSAT-2/WEBER SAT/WO-18	JAN22-90	PRESENTLY FUNCTIONING-CAMERA.
LUSAT/LO-19	JAN22-90	PRESENTLY FUNCTIONING.
8J1JBS/JAS-1b/FO-20	FEB07-90	PRESENTLY FUNCTIONING.
BADR-A	JUL16-90	FUNCTIONED FOR 145 DAYS.
AO-21/RS-14	JAN29-91	TURNED OFF OCT10-94.
RS-12/13	FEB05-91	PRESENTLY FUNCTIONING.
MICROSAT-1>7	JUL17-91	FUNCTIONED FOR 6 MONTHS/BAD ORBIT/NON-AMATEUR.
UoSAT5/UO-22	JUL17-91	PRESENTLY FUNCTIONING.
KITSAT-A/HL01/KO-23	AUG10-92	PRESENTLY FUNCTIONING.
ARSENE/AO-24	MAY13-93	NO TELEMETRY SINCE SEP09-93
KITSAT-B/HL02/KO-25	SEP26-93	PRESENTLY FUNCTIONING.
ITMSAT/IO-26	SEP26-93	PRESENTLY FUNCTIONING.
EYESAT-1/AO-27	SEP26-93	PRESENTLY FUNCTIONING.
RS-15	DEC26-94	PRESENTLY FUNCTIONING.
UNAMSAT	MAR28-95	LAUNCH FAILURE.
TECHSAT	MAR28-95 .	LAUNCH FAILURE

O.S.C.A.R.: Orbital Satellite Carrying Amateur Radio.
R.S.: Radiosputnik
AO: AMSAT OSCAR
DO: DOVE OSCAR
FO: Fuji OSCAR
IO: ITAMSAT
AM: AMRAD
KO: KITSAT OSCAR
PO: Portuguese OSCAR
Proposed satellites or waiting for launch:
FASat-Alfa August 1995
SEDSAT-1
SUNSAT January 1996
NANOSAT

AMATEUR RADIO SATELLITE FREQUENCIES

Designation	Frequencies	Transponder/ Beacon	Mode
AO-10			
Downlinks	145.810	B	B
	145.825-.975	T	B
	145.987	B	B (Usually off)
Uplinks	435.027-.179	T	B
RS-10			
Downlinks	29.357	B	A
	29.360-.400	T	A
	29.403	B (Robot)	A
	145.857	B	T/KT
	145.903	B (Robot)	T/KT
Uplinks	145.860-.900	T	T/KT
RS-12			
Downlinks	29.408	B	K
	29.410-.450	T	K
	29.454	B (Robot)	K
	145.913	B	T/KT
	145.959	B (Robot)	T/KT
Uplinks	21.210-.250	T	K
AO-13			
Downlinks	145.812	B	B
	145.825-.975	T	B
	145.985	B	B (Usually off)
	435.651	B	L/JL
	435.677		
RUDAK	435.715-6.005	T	L/JL
	2400.664	B	S
	2400.711-.749	T	S
Uplinks	435.423-.573	T	B/S
	435.601-.637	T	B/S
AO-16 (PACSAT)			
Downlinks	437.02625	T/B	J Dig. (1200b SSB) (secondary)
	437.05130	T/B	J Dig. (1200b Rai. Cos SSB) (pri)
	2401.14280	B	1200 bps SSB (Usually off)

Uplinks	145.900	T	1200 bps AFSK FM Digital
	145.920	T	1200 bps AFSK FM Digital
	145.940	T	1200 bps AFSK FM Digital
	145.960	T	1200 bps AFSK FM Digital
DO-17			
Downlinks	145.82438	B	1200 bps AFSK FM or Dig Voice
	145.82516	B	1200 bps AFSK FM or Dig Voice
	2401.22050	B	1200 bps BPSK (SSB) (usually off)
Uplinks	None		
WO-18			
Downlink	437.10200	B	1200 bps BPSK, J Dig (Telem, Image)
Uplink	None		
LO-19 (LUSAT)			
Downlinks	437.125	T/B	J Digital (secondary)
	437.127	B	CW
	437.154	T/B	J Digital (primary)
Uplinks	145.840	T	1200 bps AFSK FM Digital
	145.860	T	1200 bps AFSK·FM Digital
	145.880	T	1200 bps AFSK FM Digital
	145.900	T	1200 bps AFSK FM Digital
FO-20 (8J1JBS)			
Downlinks	435.795	B	J Analog
	435.800-.900	T	J Analog (See below)
	435.910	T/B	1200 bps BPSK (SSB), J Digital
Uplinks	145.850	T	1200 bps AFSK FM Digital
	145.870	T	1200 bps AFSK FM Digital
	145.890	T	1200 bps AFSK FM Digital
	145.910	T	1200 bps AFSK FM Digital
OR	145.900-6.00	T	CW/SSB (Alternates with above every other week. Changes on Wednesdays)
AO-21			
Downlinks	145.852-.932	T	CW/SSB
	145.866-.946	T	CW/SSB
	145.985	Repeater	FM (Alternates with voice bulletins and telemetry)
Uplinks	435.022-.102	T	CW/SSB
	435.601-.637	T	CW/SSB
	435.015	Repeater	FM (See above)
UO-22 (UOSAT5)			
Downlink	435.120	T	9600 bps FM Digital
Uplinks	145.900	T	9600 bps FM Digital
	145.975	T	9600 bps FM Digital
KO-23 (KITSAT, HLØ1)			
Downlink	435.175	T	9600 bps FM Digital
Uplinks	145.850	T	9600 bps FM Digital
	145.900	T	9600 bps FM Digital
KO-25 (KITSAT-B, HLØ2)			
Downlink	435.175/436.500 MHz		9600 bps FSK FM Digital
Uplink	145.870/145.980 MHz		9600 bps FSK FM Digital
IO-26 (ITMSAT)			
Downlink	435.867 MHz		1200 bps PSK Digital
Uplinks	145.875 MHz		1200 bps FM Digital
	145.900 MHz		1200 bps FM Digital
	145.925 MHz		1200 bps FM Digital
	145.950 MHz		1200 bps FM Digital
PO-28 (POSAT1)			
Uplink	145.975 MHz		JD 9600 bps FSK (Primary)
	145.925 MHZ		JD 9600 bps FSK (Secondary)
Downlink	435.075 MHz		JD 9600 bps FSK (Primary)
	435.050 MHz		JD 9600 bps FSK (Secondary)

AMATEUR RADIO-RELATED TELEPHONE BULLETIN-BOARD SYSTEMS
(N2UTO Ham-Radio Land-Line BBS List - 08/01/95 revision)

BOARDNAME & ABBREVIATIONS	PHONE NUMBER/ST	CALLSIGN	ADDRESS & MODEMS
The Roy Hobbs BBS hobbs.com	201-641-7307 NJ	N2UTO	Cono 1:2604/122 HST8 V.34
PCB /FI /INET /Email	201-641-3126 NJ		Node 2 1:2604/121 HST8 V.34
Franklin's Tower BBS	201-794-8437 NJ		Fido 1:2604/402 VFC
CyberNet BBS	201-939-6986 NJ	KE2SL	David
BFWK BBS	201-941-3302 NJ	KB2JXK	Fido 1:2604/143
KA2HHB Ham BBS	201-481-4108 NJ	KA2HHB	
Trumbull-Mini	203-261-6434 CT	WA1QKS	Fido 1:141/370
Info-Link	203-295-8384 CT		Fido 1:320/73
Windsor Amateur Radio Union	203-298-9989 CT	WB1CBY	Fido 1:142/100
The Planet Earth BBS	203-335-7742 CT	WA1UCQ	Fido 1:141/455
The Dark Side BBS	203-438-4721 CT		Fido 1:141/745
Gemstone BBS	203-449-8747 CT	N1GZA	Fido 1:320/157
Rod's Maximus	203-621-9936 CT	N1FNE	Fido 1:141/1840
ARRL BBS	203-666-0578 CT	W1AW	
Teletalk	203-799-7454 CT		Fido 1:141/328
Bill VE4UB	204-785-8518 Manitoba		
Electro BBS	205-491-8402 AL		Fido 1:3602/8402
The Data Connection	205-601-0917 AL	N4SZO	Fido 1:3625/465
The Bulletin Board BBS	205-758-5017 AL	W4WYP/WD4DAT	
Bears BBS	206-237-3472 WA	WM7O	
The AA6ED Packet Gateway	206-271-4657WA	AA6ED	
Boone's Farm	206-282-2851 WA		
Bill & Moe's BBS	206-328-4522 WA	KD0JU	Fido 1:343/159
The Precedent	206-355-1295 WA	N7NIP	Fido 1:343/9
KA1SVC BBS	207-247-6225 ME	KA1VSC	
Cobra BBS	207-725-0899 ME	N1MQY	
The Rebel BBS	208-997-3937 ID		
The Ninety 6 Quarts of Gin	209-625-0618 CA	N6QOG	
WCPB	209-661-5355 CA	KM6HK	Fido 1:10/45
Communication Specialties	212-645-8673 NY	KF2JO	Fido 1:278/712
Signal Intelligence	212-864-0112 NY	David Torres SWL BBS	
Kenwood Factory BBS	213-761-8284 CA		
COM Port One	214-226-1181 TX	WA5EHA	Fido 1:124/7009
The One Stop BBS	214-240-2069 TX	N5NMA	
Datalink BBS AMSAT	214-394-7438 TX	N5ITU	
Smoked Armadillos	214-669-9645 TX	AA5SA	Fido 1:124/5118
Better Amateur Radio	216-237-8208 OH	KB8NW	
Bill	216-526-9482 OH		
The Radio Room BBS	216-686-8800 OH	KD8GC	Fido 1:157/607.0
The Cleveland Hamnet BBS	216-942-6382 OH	WB8APD	14.4 v32 v42b
Radio Daze	219-256-2255		
PC-HAM	301-593 9067 MD	G3ZCZ	
Around and About	301-621-9669 MD	WA3TKW	
Diamond Jims	301-645-7964 MD		
The Dungeon BBS	301-737-0543 MD	N3RDP	Fido 1:2612/666
WJ3P Ham Exchange BBS	301-831-5954 MD	WJ3P	
First Due BBS	301-949-1927 MD	WA3ZKE	
WB3IKP BBS	302-798-8186 DE	WB3IKP	Fido 1:150/220.0
Mile High	303-431-1404 CO	N0RSE	Fido 1:104/223
The S.T.A.R BBS	305-238-8851 FL		Fido 1:135/17
The Right Connections BBS	305-382-6687 FL	N4LDG	Fido 1:135/63
Brass Ponder	305-472-7715 FL	N4BP	Fido 1:369/120
Telcom Central	305-828-7909 FL	KD4BBM	Fido 1:135/23
FTW	307-328-1923 WY	N7PQZ	(C-64)
TP PACKET BBS(Twisted pair)	309-822-8781 IL	WD9HYY	
PC Haven	310-374-7929 CA	AB6GA	Fido 1:102/137
The LINK BBS	310-459-1264 CA		
GFRN BBS	310-541-2503 CA	WB6YMH	
On The Air	313-522-5349 MI	N8FKV	
Air Studio	313-546-7045 MI	KA8NCR	Fido 1:120/216
The M.A.R.C. BBS	313-669-4279 MI	N8NSX	
Arthur BBS	313-879-2318 MI	KB8IQW	Fido 1:120/120
The Black Hole BBS	313-879-7387 MI	N8MAX/N8NMX	Fido 1:120/36
Fun City USA	314-893-9166 MO	N0IWK	Fido 1:289/18
Dots-N-Dashes BBS	315-642-1220 NY		Fido 1:2609/811
Ham-Net BBS	315-682-1824 IN	N2PMM	Fido 1:260/375
N2JEU BBS	315-697-5123 NY	N2JEU	Fido 1:260/304
Southside BBS	317-535-9097 IN	KB9BVN	Fido 1:231/30
American Silver Dollar BBS	318-443-0271 LA	WB5ASD	
Tri-State Data Exchange	319-556-4536 IA		Fido 1:283/610

BBS Name	Phone	Callsign	Network
The Roadrunner BBS	401-821-1457 RI		
Aksarben ARC BBS	402-289-4658 NE	WBØQPP	
KipNet	402-597-1740 NE	KBØKRV	
The Cameo Gateway BBS	403-234-9837 AB	VE6TAK	Fido 1:134/63
The Foothills Ham BBS	403-283-1107 AB		
Alta Packet Info Net	403-464-5069 AB		Fido 1:342/7
Terraplex	403-743-4696 AB	VE6CYR	Fido 1:3402/4
Thunder Bay Trading Post	404-516-1282 GA	WA3ZLB	
WA4BRO'S Ham Board BBS	404-552-0868 GA	WA4BRO	
WB4MZO AT&T Radio Club BBS	404-573-6048 GA	WB4MZO	
Top of the Rock BBS	404-921-8687 GA	KB4ZTN	
Retriever's Retreat	405-359-1540 OK	KB5WIF	Fido 1:147/2001
AVALON	405-636-4022 OK	KB5CNG	
N7GXP Hamshack	406-458-9379 MT		
HAM BYTES	407-575-9680 FL	KD4VBS	
Pclogic	407-879-4823 FL	W4NVC	
Atlanta Amateur Radio Club	404-393-3083 GA		
Lockheed Arc BBS	404-949-0687 GA		
NOHO BBS	408-259-3864 CA	KD6SBW	
Mt. Retreat	408-335-4595 CA		Fido 1:216/506.0
The House of Ill Compute	408-338-6860 CA		Fido 1:216/21
Saratoga Clone	408-395-1402 CA	WA6LYZ/WD5ICZ	
WBBS	409-447-4267 TX	WB5ITT	Fido 1:106/4267
BorderTech BBS	410-239-4247 MD	WA3UTC	Fido 1:261/1355
The Amateur Radio BBS	410-661-2475 MD	WB3FFV	
NB3P BBS	410-750-6403 MD	NB3P	
Allegheny-Kiski	412-226-7357 PA	KA3NVP	
Beaver Valley Ara BBS	412-775-7536 PA	KA3CYW	
Joe's Garage	414-453-5145 WI	N9JR	Fido 1:154/222
Milwaukee Heath Users Grp	414-548-9866 WI	KA9TGN	
No-Name RBBS	415-481-0252 CA	N6MON	RBBS 8:914/101
BBS-JC	415-961-7250 CA	K6LLK	
Nortown ARC BBS	416-223-2186 Ontario		
The Black Hole BBS	419-228-7236 OH	KB8BMQ	Fido 1:234/16
The Paragould Ham Net	501-239 2397 AR	NØCHP	
The Crossing	501-239-2969 AR	KB5VTL	
The Ether Net	501-455-0628 AR	WDØGRC	Fido 1:3821/7
Freedom One HST	501-932-7932 AR	KB5SEJ	Fido 1:389/6
Deckman's Exchange	502-267-7422 KY	N4VEH	
The Rose	503-286-3855 OR		Fido 1:105/7
MacHam	503-656-0510 OR	KI7FG	
Com-Dat BBS	503-681-0543 OR	KJ4TX	Fido 1:105/314
The Wireless BBS	503-692-7097 OR	WB7VHB	
Helpnet of Baton Rouge	504-273-3116 LA	W5KGG	Fido 1:396/101
The Digital Cottage	504-897-6614 LA	K8VDU	Fido 1:396/65
Salt Air	508-385-3427 MA	KQ1K	
The Cul-de-Sac BBS	508-429-8385 MA	WA1YDL	Fido 1:322:360
Waystar BBS	508-481-7147 MA		
Ham Shack BBS	508-949-3590 MA		
Secret BBS	508-998-6434 MA		Fido 1:101/873
Think Tank II	509-244-6446 WA		Fido 1:346/8
Code Three Outfitters BBS	510-799-2921 CA	WA6QQF	kleong@ctobbs.com
Time & Eternity	510-886-5467 CA	KB2NSX	Fido 1:215/17
The Antenna Farm BBS	512-444-1052 TX		
Hams on Health-Link	512-444-9908 TX	KB5IN	Fido 1:382/5
The Thirst for Knowledge	512-454-8065 TX	K5ARS	Fido 1:382/74
The 128 P.C.	512-827-1025 TX		Fido 1:387/1101
The Armadillo BBS	512-837-2003 TX	N5OWD	Fido 1:382/32
Computer Data Services	512-887-0787 TX		
The Listening Post	513-474-3719 OH		
Light in the Darke BBS	513-547-3313 OH	WA8RUO	Fido 1:108/210
Bad Element BBS	513-625-4303 OH	N8MST	USR 28.8
KIC-BBS	513-762-1115 OH	N8KTW	Fido 1:108/89
VE2MMM Amateur Radio BBS	514-624-5651 CANADA		Fido 1:167/230
TTGCITN SuperBBS	515-265-0164 IA	NØPBS	Fido 1:290/10
The Ham Tech BBS	515-981-9116 IA	NØEPU	
The Long Island AR BBS	516-368-4979 NY	N2MCS	
Amateur Radio Selling Post	516-581-1896 NY		
No Frills Plus Ham BBS	516-661-3643 NY		
The Silicon Garden	516-736-7010 NY	N2HAA	Fido 1:2619/211
Datalink BBS	516-862-8764 NY	KB2KUR	Fido 1:107/241
Ham It Up	516-878-4906 NY	KE2NK	
AA8ES Amateur Radio BBS	517-939-0169 MI	AA8ES	Fido 1:239/1020
RADIO FREQS	518-782-0507 NY		Fido 1:267/103

Name	Phone	Callsign	Fido
Kitchener-Waterloo AR BBS	519-578-9314 ON	VE3MTS	Fido 1:221/177
The Voice of Windsor	519-969-0747 ON	VE3TIZ	Fido 1:246/36
Neighborhood Net	602-495-1797 AZ	KB7DJE	Fido 1:114/24
The Pembroke Connection	603-485-8549 NH	N1MEN	Fido 1:132/169
The Legal Beagle	603-883-4466 NH	K1TCD	Fido 1:132/115
Fantasy Lights Software	604-583-3919 BC	VE7HTP	Fido 1:153/919
The Grapevine BBS	604-764-4672 BC		Fido 1:353/220
Pics Online(sm)	609-654-0999 NJ	N2LQH	Fido 1:266/21
The Radio Wave BBS	609-764-0812 NJ		
Pinelands RBBS	609-859-1910 NJ	W2XQ	Fido 1:266/32
Jersey Devil Citadel	609-893-2152 NJ	K2NE	Fido 1:266/33
MadMax BBS	610-792-9667 PA	N3MAX	Fido 1:2614/736
The 5th Dimension BBS	610-827-7689 PA	N3QLZ/K3DSM	Fido 1:273/304
Ham>Link<	612-HAM-0000 MN	KØTG	Fido 1:282/100
NFB	612-729-0538 MN		
The Lion's Den BBS	613-392-8294 Canada	VE3JFD	Fido 1:249/303
Drl BBS	613-548-3691 ON		Fido 1:249/124
CHANNEL-23 BBS	613-830-5391 OTTAWA CANADA	VE3CUZ	FIDO 1:243/23
Cross-Fire BBS	614-294-5336 OH		Fido 1:226/170
South Parking Lot	614-351-2274 OH		Fido 1:226/330
Ham BBS	614-895-2553 OH	N8EMR	
The Sounding Board	615-281-2104 TN	WB4HQN	
Lebanon Link	615-443-2237 TN	N4SCT	Fido 1:116/26
The Paragon Data Serv. BBS	615-892-1031 TN	WA8GKH	Fido 1:362/575
Sunrise BBS	616-795-7886 MI	N/H	
De Shop	617-233-0297 MA		
Tom's BBS	617-356-3538 MA	KA1TOX	Fido 1:101/470
The Garden Spot	617-545-6239 MA	NS1N	
DX ONLINE	617-592-8404 MA	NV1L	
Baystate BBS	617-598-6646 MA	WB1ERG	Fido 1:101/370
Also Cheers BBS	617-770-9451 MA		Fido 1:101/2306
JAFO! BBS	618-797-1675 IL	N9LBE	Fido 1:2250/26
Radio Sport BBS	619-279-3921 CA	WB6BDY	
Lakeside Wildcat! BBS	619-390-7328 CA	N6CQW	Fido 1:135/63
Ham/Shortwave BBS	702-368-0846 NV		Fido 1:209/730
Frequency Forum	703-207-9622 VA	N4ULS	Fido 1:109/239
Jack's Emporium	703-373-8215 VA	W4VQX	Fido 1:274/30
The Technicians Exchange BBS	703-387-1780 VA	KA4YUY	
Max's Doghouse	703-548-7849 VA		FIDO 1:109/136
KC3OL BBS	703-689-7156 VA	KC3OL	
AMRAD BBS	703-734-1387 VA		
Das Spitzen Sparken Board	703-791-6198 VA	WD4AZG	
WB4YZA BBS	704-284-4851 NC	WB4YZA	
N7IJI	704-372-8925 NC	N7IJI	
NC EMS BBS	704-637-6906 NC	W4HG/KD4DBH	FIDO 1:379/308
Radio Hobby Online	708-238-1901 IL		Fido 1:115/747
WB4QOJ Network 57 BBS	708-362-8673 IL	WB4QOJ	Fido 1:2602/530
Micro Overflow	708-355-6942 IL	AF9M	
Samson BBS	708-394-0071 IL	KB9DIP	Fido 1:115/108
Elk Grove Repeater	708-529-1586 IL		
The Precision Board	708-980-9544 IL	N9MHT	1:115/410
The Far Point Relay	713-463-8324 TX		Fido 1:106/8324
The Politically Incorrect	713-550-1070 TX	KC5CJA	Fido 1:106/1010
USS Pegasus - Houston node	713-777-0821 TX	KB5NFN	Fido 1:106/9637.0
Cell Bio BBS	713-798-4955 TX	K2TNO/KB5NFN	Fido 1:106/9636.1
Acom II	713-879-1448 TX		
The Comport	713-947-9866 TX	KE5WJ	Fido 1:106/18
InterFlex Systems Design	714-497-5860 CA	WB6UUT	
T.E.L. Net Systems #2	714-597-7858 CA	KC6ZOL	Fido 1:207/107
F.O.G.	714-638-2298 CA	N6GIS	MAX 1:103/100
The Byte Stops Here!	715-341-9723 WI	Mike Witkowski N/H	
The Shack Too	716-288-5848 NY	WD4SGU	Fido 1:260/232
Vector Board	716-544-1863 NY		
NF2G Online	716-663-8478 NY	NF2G	Fido 1:260/218
Highland BBS	716-761-6460 NY	N2JYG	
HamCenter BBS	716-885-6604 NY	KB2KRB	
Vaccumn Valley BBS	717-323-1645 PA	N3DQC	
Tec-Board	717-561-8145 PA	KA3ADU	
Ham Shack BBS	717-652-6014 PA	NZ3U	Fido 1:270/117
KB7UV Roserver/PRMBS	718-956-7133 NY	KB7UV	
The Mav Plus BBS	801-634-3655 UT	KB4YHB	Fido 1:15/10
We Serve Your Drives	802-453-6074 VT	N1PEC	Fido 1:325/101
N1HUM's HAM BBS	802-479-2159 VT	N1HUM	Fido 1:32504
The HAM Connection	802-868-5274 VT	KA2UZC	

Name	Number/Location	Call	Address
WV4B	803-223-5185 NC		
KC4YJC	803-269-7899 NC		
The Byte Bucket BBS	803-871-3076 SC	N4PGN	
Ben & Cheryl BBS	804-261-1819 VA		Fido 1:264/166.6
The Computer Forum BBS	804-471-3360 VA	KF4GL	
HamdiNet BBS	804-496-3320 VA		Fido 1:275/429
Relatively Speaking	804-566-0219 VA	KE4GWV	Fido 1:271/300
HamBones Fish House	804-665-7906 VA	KD4NFI	
The Listening Post	804-850-8586 VA	KD4IEN	Fido 1:271/263
Believers Bounty BBS	804-855-8421 VA	KD4GJH	Fido 1:275/40
Pat's Place	804-779-7373 VA	NØRDQ	
CVACC	805-499-5699 CA	KD6AEJ	Fido 1:102/1002
Simi Valley Am. Radio BBS	805-583-2282 CA	N6VDV	
The Barn	809-781-9805 PR	WP4KVF	
Indiana On-Line	812-332-7227 IN	WB9LWQ/KC9HI	
KA9LQM's Ham Shack BBS	812 428-3352 IN	KA9LQM	
DataCOM Systems BBS	813-791-1454 FL	N4WAK	
! The Cat House	813-796-2486 FL	KA3DBK	Fido 1:3603/110
Pac-Comm, Inc. BBS	813-874-3078 FL		
Solomon's Portico	813-954-5139 FL		FIDO 137/19
KCATVG Support BBS	816-459-9752 MO	NØOXV	Fido 1:280/12
THe Archive BBS	817-447-1969 TX	AA5MM	Fido 1:130/22
Revelstone HST	817-732-1767 TX	N5SUV	Fido 1:130/402.0
N6MBR BBS	818-597-0641 CA	N6MBR	Fido 1:102/1004
Anime Lane BBS	818-762-3695 CA	KC6VMP	Fido 1:102/833
R&D BBS	819-772-2952 Canada		Fido 1:163/506
Database of Tenn. BBS	901-749-5256 TN	KD4PYN	
Marty's Place	903-753-0485 TX	N5KBP	Fido 1:19/62
The Hot Muddy Duck BBS	904-651-8684 FL	N4HMD	
COPS-N-HAMS BBS	904-678-5998 FL	KK4KF	
Ancient City Wireless	904-823-3513 FL	KC4TDS	Fido 1:3620/9
Project X BBS	905-792-7890 Ontario	VE3SVE	
Ham Radio BBS	905-827-0704 Ontario		Fido 1:259/304
Discovery Software BBS	907-248-8130	NL7EL	
The Micro Room	908-245-6614 NJ	WA2BFW	Fido 1:107/919
Sunset - L	908-324-1211 NJ	WB6JBL	
The ByteWise BBS	908-363-2760 NJ	N2CKH	
The Attic BBS	908-396-0790 NJ	N2WXH	
Planet Shadowstar TBBS	908-494-3417 NJ	N2HGY	Fido 1:107/344
Microphone TBBS	908-494 3649 NJ	K2SHY	
Sunset - East	908-964-4357 NJ	WB2JIO	
ATTENTION to Details BBS	909-681-6221 CA	KE6LCS	71167.2176@compuserv.com
Albany Amateur Radio Club	912-435-9466 GA	W4MM	Fido 1:3617/6
Namu BBS	913-273-1550 KS	WVØS	
The Boarding House BBS	913-827-0744 KS	KØFPC	
Red Onion,Express	914-342-4585 NY	N2UBP	Fido 1:272/31.0
Coynet Amateur Radio BBS	914-485-3393 NY	WB2COY	
Joe Brown's BBS	914-667-9385 NY	KB2NBN	Fido 1:272/39
Hamnet BBS	915-653-9077 TX	N5JZZ	Fido 1:383/4
The CD-Rom BBS	915-673-8014 TX	AA5WF	Fido 1:392/6
Galatic Crossroads BBS	916-334-5641 CA	KC6IRK	Fido 1:203/57
WA6RDH	916-678-1535 CA	WA6RDH	
QST	916-920-1288 CA	WA6AXZ	Fido 1:203/730
The Ham Radio Emporium	918-272-4327 OK	WA4BFE	Fido 1:170/801
Dark Oak BBS	919-380-7194 NC	WD4DKT	
Raleigh ARS BBS	919-772-9176 NC	WD4RDT	Fido 1:151/102
CADIC Philippe BBS	+33-47-679-189 FRANCE		
FREE BBS	+39-11-482-751 Italy	SWL/I1-21171	
ComNet Luxembourg BBS	+352-222534 Luxembourg	LX1DQ	V.34
Ham-Box BBS	+358-28-23168 Aaland IsLands	OHØNC	20-N 20-E
Radio Amateur BBS	+41-61-996969 Switzerland	HB9MKQ	Fido 2:301/249
! SweDX	+46 40-973280 Sweden		Fido 2:200/110
Radio Active	+61 2 399 9268 Australia	VK2XGK	Fido: 3:712/412
COOL WORLD	+61 3 432 0716 Australia	VK3KBL	

** Listing criteria: HAM radio oriented, general / public access non-commercial. One number only listed with your FidoNet address or InterNet address NOT Both. Please provide suggestions and changes to The Roy Hobbs BBS 201-641-7307 cono@hobbs.com. Issued monthly on SaltAir BBS Murry,Utah 801.261.8976

PLEASE KEEP IN MIND that most remote access systems are privately owned and operated. Although many BBSes listed here are verified every month to have a modem on-line, this is no guarantee that a particular system will be operating at all times.

AMATEUR RADIO AND THE INTERNET

Guide to the Personal Radio Newsgroups
Revision: 2.1 1994/11/01 03:50:23

This message describes the rec.radio.amateur.*, rec.radio.cb, rec.radio.info, and rec.radio.swap newsgroups, as well as their Internet mailing list counterparts and complements. It is intended to serve as a guide for the new reader on what to find where. Questions and comments may be directed to the author, Jay Maynard, K5ZC, by Internet electronic mail at jmaynard@admin5.hsc.uth.tmc.edu. This message was last changed on 31 October 1994 to add the discussion of non-Usenet mailing lists and streamline most of the history discussion, and to change the moderator information for rec.radio.info.

History

Way back when, before there was a Usenet, the Internet hosted a mailing list for hams, called (appropriately enough) INFO-HAMS. Ham radio discussions were held on the mailing list, and sent to the mailboxes of those who had signed up for it. When the Usenet software was created, and net news as we now know it was developed, a newsgroup was created for hams: net.hamradio. The mailing list and the newsgroup were gatewayed together, eventually.

Over the years, as the net grew, the volume of discussion became progressively higher. First one by one, and then as part of two reorganizations, what was once one group became many. In the process, developments elsewhere on the net were reflected in the groups as they were created, most notably the change to place all of the ham radio groups in one hierarchy.

The collection of newsgroups continues to grow as more people join the net, and as more topics of discussion gain volume, I expect to see more groups be created as well. This follows what is happening on the rest of the net.

Nearly all of the radio newsgroups have corresponding mailing lists, the notable exception being rec.radio.swap. There are also a few mailing lists that don't have newsgroups.

The Current Groups

It's important to post messages to the group that's appropriate for them, and not to the groups that aren't. The whole idea of having different newsgroups is so that folks who aren't interested in, say, homebrewing, don't have to wade through messages about homebrewing on the way to read about Field Day. Posting appropriately is just good etiquette.

The rec.radio.amateur.misc group is the catchall. It is what rec.ham-radio was renamed to during the first major reorganization. Any message that's not more appropriate in one of the other groups belongs here, from contesting to DX to ragchewing on VHF to information on becoming a ham.

The group rec.radio.amateur.digital.misc is for discussions related to (surprise!) digital amateur radio. This doesn't have to be the common two-meter AX.25 variety of packet radio, either; some of the most knowledgeable folks in radio digital communications can be found here, and anything in the general area is welcome. The name was changed to emphasize this, and to encourage discus-

sion not only of other text-based digital modes, such as AMTOR, RTTY, and Clover, but things like digital voice and video as well. The former group, rec.radio.amateur.packet, was removed on September 21st, 1993. It is obsolete, and you should use .digital.misc instead (or the appropriate new mailing list, mentioned below). The group has .misc as part of the name to allow further specialization if the users wish it, such as .digital.tcp-ip.

The swap group is rec.radio.swap. This recognizes a fact that became evident shortly after the original group was formed: Hams don't just swap ham radio gear, and other folks besides hams swap ham equipment. If you have radio equipment, or test gear, or computer stuff that hams would be interested in, here's the place. Equipment wanted postings belong here too. Discussions about the equipment generally don't; if you wish to discuss a particular posting with the buyer, email is a much better way to do it, and the other groups, especially .equipment and .homebrew, are the place for public discussions. There is now a regular posting with information on how to go about buying and selling items in rec.radio.swap; please refer to it before you post there. To answer a frequently asked question: No, there is no mailing list that goes along with this group. If you can't read Usenet news directly, you're out of luck.

The group rec.radio.amateur.policy was created as a place for all the discussions that seem to drag on interminably about the many rules, regulations, legalities, and policies that surround amateur radio, both existing and proposed. Recent changes to the Amateur Radio Rules (FCC Part 97) have finally laid to rest the Great Usenet Pizza Autopatch Debate - it's now legal to order a pizza on the autopatch, if you're not in the pizza business - as well as complaints about now-preempted local scanner laws hostile to amateurs, but plenty of discussion about what a bunch of rotten no-goodniks the local frequency coordinating body is, as well as the neverending no-code debate, may still be found here.

The group rec.radio.cb is the place for all discussion about the Citizens' Band radio service. Such discussions have been very inflammatory in rec.ham-radio in the past; please do not cross-post to both rec.radio.cb and rec.radio.amateur.* unless the topic is genuinely of interest to both hams and CBers - and very few topics are.

The rec.radio.info group is just what its name implies: it's the place where informational messages from across rec.radio.* may be found, regardless of where else they're posted. As of this writing, information posted to the group includes Cary Oler's daily solar progagation bulletins, ARRL bulletins, the Frequently Asked Questions files for the various groups, and radio modification instructions. This group is moderated, so you cannot post to it directly; if you try, even if your message is crossposted to one of the other groups, your message will be mailed to the moderator, who is currently David Dodell, WB7TPY. The email address for submissions to the group is rec-radio-info@stat.com. Inquires and other administrivia should be directed to rec-radio-info-request@stat.com. For more information about rec.radio.info, consult the introduction and posting guidelines that are regularly posted to that newsgroup.

The groups rec.radio.amateur.antenna, .equipment, .homebrew, and .space are for more specialized areas of ham radio: discussions about antennas, commercially-made equipment, homebrewing, and amateur radio space operations. The .equipment group is not the place for buying or selling equipment; that's what rec.radio.swap is for. Similarly, the .space group is specifically about amateur radio in space, such as the OSCAR program and SAREX, the Shuttle Amateur Radio EXperiment; other groups cover other aspects of satellites and space. Homebrewing isn't about making your own alcoholic beverages at home (that's rec.crafts.brewing), but rather construction of radio and electronic equipment by the amateur experimenter.

Except for rec.radio.swap and rec.radio.cb, all of these newsgroups are available by Internet electronic mail in digest format; send a mail message containing "help" on a line by itself to listserv@ucsd.edu for instructions on how to use the mail server.

All of the groups can be posted to by electronic mail, though, by using a gateway at the University of Texas at Austin. To post a message this way, change the name of the group you wish to post to by replacing all of the '.'s with '-'s - for example, rec.radio.swap becomes rec-radio-swap - and send to that name@cs.utexas.edu (rec-radio-swap@cs.utexas.edu, for example). You may crosspost by including multiple addresses as Cc: entries (but see below). This gateway's continued availability is at the pleasure of the admins at UT-Austin, and is subject to going away at any time - and especially if forgeries and other net.abuses become a problem. You have been warned.

Mailing Lists

In addition to the mailing lists that mirror the Usenet newsgroups, there also are a few that stand alone. These cover specific areas of ham radio, and discussion is focused on just those areas.

The cq-contest mailing list is for discussions of contesting in ham radio. To join, send email with the word "subscribe" on a line by itself to cq-contest-request@tgv.com.

The DX mailing list covers the finer points of DXing. This one is also joined by mailing "subscribe" on a line by itself, this time to dx-request@ve7tcp.ampr.org.

There's also a VHF mailing list, for VHF operators of the weak signal persuasion. You can join this one be sending "subscribe vhf" on a line by itself to vhf-request@w6yx.stanford.edu.

All of the following mailing lists are sponsored to the Boston Amateur Radio Club, as well as some others of local interest. Thanks to N1IST for the information.

qrp-l: This is the qrp mailing list, previously maintained by Bruce Walker at Think.com. It is for discussions about the design, construction, and use of qrp (low power) radios and related equipment.

arrl-ve-list: This is a one-way list, run by Bart Jahnke of the ARRL VEC, for announcements to VEs and VE teams.

w1aw-list: ARRL bulletins, news, and information

newsline-list: Redistribution of Amateur Radio Newsline

letter-list: Redistribution of the ARRL Letter

fox-list: Fox hunting and Radio Direction Finding

fieldorg-l: ARRL field organization discussions

ham-tech: Technical discussions and questions about Amateur Radio

arrl-exam-list: Amateur radio license examinations scheduled in the US and in some foreign areas.

To sign up or inquire about these lists, send mail to listserv@netcom.com with the following in the body (subject is ignored) of the message. <listname> is the name of the list to subscribe to.

To subscribe: subscribe <listname>

To unsubscribe: unsubscribe <listname>

For more information: help

To post (to the two-way lists), send your message to <listname>@netcom.com

Please do NOT send subscription requests to the mailing lists themselves; that doesn't work very often, and is very annoying to those on the list. Also, please keep your electronic mail address current with any mailing lists you subscribe to, as dealing with returned email is a nuisance for the person maintaining the mailing list.

A Few Words on Crossposting

Please do not crosspost messages to two or more groups unless there is genuine interest in both groups in the topic being discussed, and when you do, please include a header line of the form "Followup-To: group.name" in your article's headers (before the first blank line). This will cause followups to your article to go to the group listed in the Followup-To: line. If you wish to have replies to go to you by email, rather than be posted, use the word "poster" instead of the name of a group. Such a line appears in the headers of this article.

One of the few examples of productive cross-posting is with the rec.radio.info newsgroup. To provide a filtered presentation of information articles, while still maintaining visibility in their home newsgroups, the moderator strongly encourages cross-posting. All information articles should be submitted to the rec.radio.info moderator so that he may simultaneously cross-post your information to the appropriate newsgroups. Most newsreaders will only present the article once, and network bandwidth is conserved since only one article is propagated. If you make regular informational postings, and have made arrangements with the moderator to post directly to the group, please cross-post as appropriate.

A Guide to Buying and Selling on Usenet

This message is a guide to buying and selling over Usenet. It is intended to serve as a guide for users unfamiliar with common conventions used in the Usenet marketplace. Questions and comments may be directed to the author, Jay Maynard, K5ZC, via Internet electronic mail at jmaynard@admin5.hsc.uth.tmc.edu. This message was last changed on 29 June 1994 to add the discussion about how UPS does COD remittals. Thanks go to readers of the personal radio newsgroups, who provided feedback to the net about proper use of this forum, and especially Paul W. Schleck, KD3FU, pschleck@unomaha.edu, who compiled most of the net wisdom and suggested the creation of this article.

Usenet has proven to be a valuable resource for many folks. Along with lots of discussion, argument, and good, solid information, it's also a good place to buy or sell equipment, and many people have done so successfully. As with any other medium, though, there are conventions that make everyone's life easier if they're followed as much as possible.

The following are some suggested guidelines for using the rec.radio.swap forum, based on general net-wisdom from users. Most of it is basic common sense, but it is unfortunate that some users have consistently abused this forum by not following such basic common sense. The general guidelines will serve as well for other groups on the net, such as misc.forsale and just.about.anything.marketplace.

What is appropriate to post in rec.radio.swap?

Any offer to buy or sell radio and electronics equipment, such as transmitters, receivers, antennas, electronics parts, and radio-related computer equipment is appropriate for this forum. Posts concerning non-hardware (but still radio-related) items such as documentation manuals, books, radio-related software, and publications, are also welcome.

Please do not post discussion articles to this group. If you really must post, please do so to the appropriate discussion group. Use email whenever possible, especially if you feel someone has committed a breach of etiquette.

Articles concerning illegal equipment (such as CB linear amplifiers and police radar jammers) are not welcome. Not only will you be severely "flamed", you are also opening yourself (and possibly the owners and administrators of your news site) up to civil and criminal liability. Individuals who are involved in the regular business of buying and selling for profit are requested not to abuse this forum by using it as a "free advertisement" service for their business, although they are welcome to participate as individuals. The distinction here is that there is a cultural bias on Usenet, and an actual prohibition on some networks that carry Usenet traffic, against using the net for commercial purposes. Let your conscience be your guide.

Doesn't this article violate its own guidelines?

Well, yes and no. In the strictest sense, this article violates the rule that only buying and selling advertisements belong in the rec.radio.swap newsgroup. However, since those using this newsgroup are most likely to see articles in the same newsgroup, and since this newsgroup serves readers of the rec.radio.amateur.*, rec.radio.cb, rec.radio.shortwave, and rec.radio.scanner newsgroups, posting it here provides the greatest visibility with the least intrusion. Other suggestions which achieve the same goals are welcome.

If you are looking for something specific...

Try to first find the item through other channels before resorting to the net. If the manufacturer is still in business, you may be pleasantly surprised that they still have the items on the shelf. Other companies specialize in discontinued and surplus parts and equipment and are your best source for tracking down items. Consult the mail-order electronics list, available from ftp.cs.buffalo.edu in file ~/pub/ham-radio/mail_order, or the advertising sections of most popular radio and electronics publications.

Once you have exhausted all other channels, then certainly do post. State clearly what you are looking for (e.g. "a part# 345X56 Bakelite Frobnicator for an American Hawk Fubar 2000, circa 1968-1970"), and how much you are willing to pay (or that you're willing to negotiate). Avoid sending out "equipment-wanted" posts unless you are willing to pay for shipping from wherever it may turn up (this newsgroup is read throughout the world), or

state clearly where you're willing to accept items from. Use the Distribution: header line to limit where your posting will go, but be aware that it's far from an absolute restriction; articles with ba (San Francisco Bay area) distribution, for example, are imported to places like Boston, London, and Singapore regularly.

If you are selling equipment...

Be specific in your first post about what you are selling and how much you want for it (or that you're willing to negotiate). State clearly whether or not the price includes shipping, and if it does, be sure to allow yourself a reasonable amount to cover the cost. Avoid sending out "for sale" posts unless you are willing to arrange for shipping to whomever in the message distribution wants to buy it (and remember the comment above about Distribution: headers...); if you cannot limit the posting's distribution for one reason or another, be clear in your message about where you will and will not ship. The US Postal Service has a 50-pound limit on the weight of packages sent through them, and United Parcel Service has a 150-pound limit; other carriers have similar limits. Check with your carrier before shipping. Anything heavier will have to go by motor-freight (read: EXPENSIVE). Don't advertise equipment that you cannot ship within a reasonable amount of time.

Once you have made a deal, state clearly your intentions and follow through on them. Nothing angers a buyer more than delays and excuses. Once you do ship, have it securely packaged (insurance is strongly recommended). Payment terms should be whatever you and the buyer are comfortable with, and commonly include options such as money up-front, COD (Collect on Delivery), or payment upon receipt and inspection. Don't be offended if the buyer wants to take steps to protect his position, since he probably doesn't know you. Most readers of this forum are basically honest and want to maintain their net-image, but the few bad apples should encourage you to only deal with honest, reputable people and to reasonably protect your position in any transaction.

Remember that COD stands for "Collect on Delivery" and not necessarily "Cash on Delivery." The carrier collects the funds from the buyer, and then hands him the package; they then send the payment on to you. They are not a party to the transaction, and so they don't care if the buyer gives you a bad check. Therefore, you may want to specify the collection of cash, money order, or other certified funds for your COD. Check with your carrier for exact COD options and policies. If you choose this option, make sure the buyer knows up front so that he can make the necessary arrangements. One thing to remember is that UPS, at least will send whatever is Collected on Delivery to the shipper's address as recorded in their files, and NOT to the return address on the package. If you use a commercial packing and shipping service, you'll have to go back there to pick up your payment; if you send from your office, make sure the shipping department knows what to do with the check they'll get from UPS in the mail.

If you are buying equipment...

Respond to an advertisement in a prompt manner. (The item may not be available if you don't!) Don't skip a message just because you think the price is too high; offer the seller a price you think is reasonable instead. You might be pleasantly surprised. State clearly your terms and intentions and follow through on them. Nothing angers a seller more than delays and excuses. As radio equipment is generally bulky and fragile, allow for a reasonable amount of money to package, insure, and ship your purchase properly. Payment terms should be similar to those suggested under seller's guidelines, and should reasonably protect your position (remember, you are probably buying equipment sight-unseen from a relative stranger), but remember that he needs to protect his position as well. If you are unsure of a given seller, ask a net-regular discreetly via E-mail. He or she will be more than happy to either ease your concerns or confirm your suspicions.

In general...

When you post to rec.radio.swap, be sure to use a meaningful Subject: line. "FOR SALE" or "WANTED", by themselves, give little information to the person skimming through the group by looking at the message subjects. "IC-32AT dual band 144/440 handheld for sale, $400" is much more useful; if the reader is looking for HF transceivers, he can skip right past your message. If you have lots of different things for sale, try to give as much information as you can, but remember that most systems get unhappy at Subject: lines longer than 80 characters, and a few older ones truncate them at 40.

It's generally a good idea to include your geographic location and a phone number where you can be reached somewhere in your posting as well. Besides reassuring your potential buyer or seller that you are a real person, it's often easier to bargain and make other arrangements on the telephone than through a protracted electronic mail exchange. Some buyers prefer dealing with folks in their local area, too, as that makes it easier for them to inspect the equipment before paying money.

The Usenet marketplace groups in general, and rec.radio.swap in particular, are a great place to buy that piece of gear you've had your eye on. Items go quickly for reasonable prices. I've sold a radio within three hours of posting the for sale message. The usefulness of these groups depends to a large extent on the people who inhabit them, though, and a few unscrupulous users can easily sink the whole thing. Whether you are a buyer, seller, or seeker of equipment, remember that your honesty and integrity reflects on the general reputation and usefulness of this forum and amateur radio in general.

Amateur Radio: Elmers List Info and Administrivia
Revision: 1.14 06/14/95 23:45:31

In order to standardize the Internet resource notation used in this Directory, I've decided to adopt Uniform Resource Locator (URL) format throughout. In addition to being a straightforward, human-readable format for specifying File Transfer Protocol (FTP) archives, Gopher and World-Wide Web (WWW) servers, and Usenet newsgroups, it is also amenable to formatting as hypertext links in Hypertext Markup Language (HTML). For example, users viewing this document at designated World-Wide Web Servers (see "How may I obtain the latest copy of the Elmers List?" below) will see all URL's converted to hypertext links on their WWW client. Rather than fumbling with the various conventions

of FTP, Gopher, WWW, and Usenet News software, the document or directory referenced by a hypertext link is but one mouse- click (or key-click) away.

Scott Ehrlich has graciously agreed to be a WWW Elmer and provide further information about how to obtain and use WWW client software such as Lynx and Mosaic (see his entry). While I'm happy to incorporate the latest and most popular information-formatting standards into the documents that I maintain, I really can't allow myself to be dragged into the role of a WWW help-desk for all of the Internet. Please understand this when I politely refer you to other Elmers, easily-obtainable on-line documentation, or even local expertise such as resident gurus, consultants, or help-desks at your school, company, or information service provider.

I expect there to be a bit of controversy regarding my adoption of this somewhat radical new standard, especially to many users who can't or won't use WWW. Those users should be assured that I wouldn't have adopted a standard unless it was easily human-readable by those accessing this document as straight ASCII (which is one of the main reasons why the entire Elmers list isn't HTML, MIME, MMDF, or one of many other competing, mostly non-compatible, information formats). The URL format is easily mapped into human FTP, Telnet, Gopher, and Usenet News reader commands. For example:

ftp://ftp.cs.buffalo.edu/pub/ham-radio/README

Anonymous FTP to ftp.cs.buffalo.edu and get the file README under the /pub/ham-radio directory.

gopher://oes1.oes.ca.gov:5555/

Access the Gopher root page at oes1.oes.ca.gov via non-standard port 5555 (if the standard Gopher port of 70 was used, the ":5555" part would be replaced by ":70" or most likely not appear at all).

telnet://callsign.cs.buffalo.edu:2000/

Initiate a Telnet (remote terminal) session with call-sign.cs.buffalo.edu via non-standard port 2000 (if the standard Telnet port of 23 was used, the ":2000" part would be replaced by ":23" or most likely not appear at all).

news:rec.radio.info

Access the rec.radio.info newsgroup on your Usenet newsreader from your local news server.

URL's that start with:

http:

Are only accessible via WWW client software (which is why almost all http: URL's in this directory also have corresponding ftp: or gopher: URL's).

For more information about URL formats, see:
http://www.cc.ukans.edu/lynx_help/URL_guide.html
rfc1630.txt">ftp://nis.nsf.net/documents/rfc/rfc1630.txt
http://info.cern.ch/hypertext/WWW/Addressing/URL/
URI_Overview.html

Disclaimer: While I have personally confirmed the accuracy of all URL's through the Lynx WWW client, any referenced documents external to this document are subject to future changes beyond my control. In addition, with networks and their administration being what it is, many services with up-to-date URL's may be temporarily unreachable. Please consult with your local gurus, consulting staff, or help-desk to confirm that it's a non-local problem, then ask the Elmer him or herself (me in the case of URL's which appear outside of individual entries). The currency and accuracy of URL's should be at least no worse than that of the Elmers entries themselves (which are each individually confirmed by me every 2 years, and updated within 1 month on request of the Elmer).

A Brief Historical Overview:

If there is any one constant in the changing state of the communications art, it is that "Hams" (Amateur Radio Operators) have always been on the forefront of it. Rumors abound where the term "Ham" came from. Some of the more amusing are described in the list of Frequently Asked Questions for this newsgroup.

Regardless of origin of the name, a "Ham" is universally recognizable as one who experiments in radio and communications.

Whether it be constructing a low-power CW radio with vacuum tubes, or designing TCP/IP packet networks, such experimentation has historically spilled over into the mainstream such as was the case with Howard Armstrong, who developed the regenerative oscillator and FM radio, or General Curtis LeMay (W6EZV) who was instrumental in making Single- Sideband the communications standard for the Strategic Air Command (1947-1992, now reorganized into a joint command called USSTRATCOM) and eventually the U.S. Air Force. Although packet-switching techniques originated from DARPA (Defense Advanced Research Projects Agency) and the ARPANet, no one can deny the tremendous influence that amateurs have had in demonstrating the viability of TCP/IP and AX.25 communications via radio links. The efforts of AMSAT (the Amateur Satellite Corporation), including the development of many ham satellites and the low-orbiting Microsats (communications satellites no bigger than a breadbox that use store-and forward packet techniques), have certainly advanced the state-of-the-art in communications, one of the defined purposes of the Amateur Radio Service, as recognized by international treaty.

Since in many cases hams are writing "the book", there is often no "book" or other established reference for a beginner to refer to. Traditionally, information has been passed on from ham to ham via word- of- mouth. Like many of the traditional crafts, a variation of the Master-Apprentice system has emerged, the Elmer-Novice relationship. Called "Elmers" because they are usually older and wiser, having the benefit of many years in the hobby, including several failed projects, and an electric shock or two, they have traditionally been the mainstay of amateur radio, and the source of many new hams, particularly those interested in working on emerging technologies.

Even more importantly, Elmers provided an outlet for the impatient newcomer who wanted "to know everything, and right away." Faced with such a request, a good Elmer will smile and proceed to lead the novice through some project or operating experience. Several hours, days, or weeks later, the novice would have his answers, but would have earned them. Even better, the sense of accomplishment would boost the novice's confidence and nudge him or her down the road to being a model, experienced ham operator.

Many present hams feel that such an experience is missing today. In today's hustle-bustle world, the response to such natural curiosity and desire to learn is, more often than not, "I'm too busy" or "RTFM." As a result, the quality of new hams declines and the knowledge and operating habits they develop in their first for-

mative months and years leave much to be desired. And the very same hams who claim that they "can't understand the new generation" also, in almost the same breath, lament about the "decline of amateur radio."

What is an Elmer today?

An Elmer today is of any age, male or female, who has some expertise and is willing to share it with beginners. Elmers don't even need to be licensed amateurs, just people with knowledge in some area of electronics or communications technology.

What is a Usenet Elmer?

With the ever-widening scope of the Internet, and the amateur radio newsgroups on Usenet, the potential for Elmers to share their knowledge to a wide audience has never been greater. To that end, I maintain a list of such Elmers. Volunteers need only send me their name, E-mail address, and area of expertise. I have set up an administrivia mailbox for this purpose (elmers-request@gonix.com, the default Reply-To: of this message).

Those desiring a more extensive list, or who need more specific assistance, are encouraged to contact Rosalie White, WA1STO, Educational Services Manager at the American Radio Relay League, 225 Main St., Newington, CT 06111 or via electronic mail addressed to rwhite@arrl.org.

How may I obtain the latest copy of the Elmers List?

There are currently 7 ways of obtaining the Elmers List. Any site at least reachable by Internet E-mail can use options 3 or 4:

1. Usenet News: The latest copy of the list can be found in the companion postings to this message, "Amateur Radio: Elmers Resource Directory [A-*]" and "Amateur Radio Elmers Resource Directory [*-Z]" (The * represents a wildcard character, as the need to split the list evenly to keep both parts under the 64K message limit for Usenet requires that I adjust the split occasionally). Since the list is cross-posted to the following newsgroups:

> news:rec.radio.amateur.misc
> news:rec.radio.info
> news:rec.answers
> news:news.answers

on the 1st of each month, with an expiration date 6 weeks into the future, there should always be a copy available at most news sites. Check your newsreader documentation for information about reading previously-read articles or articles that are "threaded" to this one. Also complain to your local news administrator (E-mail to "news" or "usenet" on your local host) if your local news server is configured to ignore Expires headers (and thus prematurely delete the articles) in worthwhile, mostly moderated, information newsgroups like those listed above.

2. Anonymous FTP: If your site is directly connected to the Internet, you may retrieve the latest copy via File Transfer Protocol (FTP) from the following sites:

> ftp://ftp.cs.buffalo.edu/pub/ham-radio/
> ftp://rtfm.mit.edu/pub/usenet/news.answers/radio/ham-radio/elmers/
> ftp://ftp.uu.net/usenet/news.answers/radio/ham-radio/elmers/

3. Mailing-List: Since the list is cross-posted to rec.radio.info, the latest copy may be obtained from the

mailing-list gateway for that newsgroup (along with many other informational articles about radio) when it is published each month. To subscribe, send E-mail to:

> listserv@ucsd.edu

and in the BODY (not the Subject) of the message, write:

> subscribe radio-info

The server may not be able to determine your return address. In that case write:

> subscribe radio-info (your E-mail address)

You should get an acknowledgement very shortly.

4. Mail-Server: If you don't want to read through the entire gateway of rec.radio.info, or want a copy of the list right away, send E-mail to:

> mail-server@rtfm.mit.edu

and in the BODY (not the Subject) of the message, write:

> send usenet/news.answers/radio/ham-radio/elmers/admin
> send usenet/news.answers/radio/ham-radio/elmers/index
> send usenet/news.answers/radio/ham-radio/elmers/list/part1
> send usenet/news.answers/radio/ham-radio/elmers/list/part2
> send usenet/news.answers/radio/ham-radio/elmers/diff

and the latest copy of the list should be sent to you E-mail within 24 hours (the mail-server uses batch priority to reduce system demand).

The last three services are experimental. I'm not terribly familiar with them, and cannot offer much technical support regarding their use. (I'd appreciate feedback on whether or not you find them useful, though.)

5. Internet Gopher: The latest copy of the list should be available from the following Gopher sites:

> gopher://cc1.kuleuven.ac.be/
> gopher://jupiter.sun.csd.unb.ca/
> gopher://gopher.univ-lyon1.fr/
> gopher://ftp.win.tue.nl/
> gopher://gopher.win.tue.nl/

see also news:comp.infosystems.gopher

6. World-Wide Web (WWW): The latest copy of the list should be available from the following WWW sites:

> http://www.cis.ohio-state.edu/hyper-text/faq/usenet/radio/ham-radio/elmers/top.html
> http://www.lib.ox.ac.uk/internet/news/faq/rec.radio.amateur.misc.html

The advantage of reading the Elmers list at these sites via WWW client software is that all URL's are converted to hypertext links..See also the newsgroups under comp.infosystems.www.* (I haven't listed URL's for all of the newsgroups under this hierarchy as there are now about a dozen of them (!!!).)

7. Wide-Area Information Service (WAIS): The latest copy of the list should be available from the WAIS server at:

> wais://rtfm.mit.edu/usenet

see also news:comp.infosystems.wais

How may I contribute to the Elmers List?

By using this resource, you are benefitting the net by obtaining assistance in the fastest and most efficient way possible. By volunteering to appear on this list, you are contributing to the good reputation of the radio-related newsgroups.

Thanks to all the volunteer Elmers, as well as courteous list users, for making this service a success.

73, Paul W. Schleck, KD3FU

pschleck@gonix.com (personal mail)
elmers-request@gonix.com (Elmers List administrivia)

USENET NEWS GROUPS WITH AMATEUR RADIO TOPICS (UPDATED AUGUST 5, 1995):

rec.radio.amateur.antenna	Antennas: theory and practical
rec.radio.amateur.digital.misc	Packet radio and other digital modes
rec.radio.amateur.equipment	Commercial amateur equipment
rec.radio.amateur.homebrew	Homebrew amateur equipment
rec.radio.amateur.misc	Contests, DX, general discussion
rec.radio.amateur.packet	Packet radio
rec.radio.amateur.policy	Regulation and FCC policy discussion
rec.radio.amateur.space	Satellite and space communications
rec.radio.amateur.swap	Flea market

Radio related newsgroups:

rec.radio.broadcasting	broadcast radio
rec.radio.info	FAQs and Informational postings.
rec.radio.scanner	Utility and above 30 MHz receiving
rec.radio.shortwave	Shortwave radio
rec.radio.noncomm	non-commercial radio discussion
rec.radio.cb	Citizen-band radio
rec.radio.swap	Flea market
alt.radio.pirate	Pirate radio discussion
alt.radio.scanner	Scanner discussion
rec.antiques.radio+phono	Antique Radios

Regional newsgroups:

aus.radio	Australia
aus.radio.amsat	Australia
aus.radio.amateur.digital	Australia
aus.radio.packet	Australia
aus.radio.amateur.misc	Australia
de.comm.ham	Germany
hannover.funk.amateurfunk	Hannover, Germany
z-netz.telecom.amateurfunk	Germany
fj.comm.ham	Japan
fj.rec.radio	Japan
francom.radio_amateur	France
uk.radio.amateur	United Kingdom
alt.radio.scanner.uk	United Kingdom
uwarwick.societies.amateur-radio	Warwick, UK
pt.rec.radio.amadorismo	Portugal
in.ham-radio	Indiana, USA
sbay.hams	South Bay Area/Silicon Valley, CA, USA
triangle.radio	Research Triangle area, NC, USA
su.org.ham-radio	Stanford, CA, USA
slac.rec.ham_radio	SLAC, CA, USA
alt.radio.amateur.club.clarc	

Amateurs who have access to the World Wide Web can search for usenet messages by key word using the database and search software at DejaNews:

> http://www.dejanews.com

FTP SITES CONTAINING AMATEUR RADIO DATA (UPDATED: AUGUST 3, 1995)

The latest copy of this document can be obtained from:

> FTP: ftp://ftp.netcom.com/VE/VE3SUN/
> WWW: http://mall.turnpike.net/~jc/

If you notice any errors or omissions, please let me know.

Amateurs with Internet ftp have access to an incredible wealth of technical information and shareware programs.

The following ftp sites contain amateur radio files. The topics listed with each site give an indication of the kind of information that is stored there, but it is far from complete, and it is recommended that you do some exploring on your own. Many gems are hidden in subdirectories on these and other sites.

If you do not have direct ftp access, but do have email, you can obtain files from these sites using email gateways such as:

    bitftp@pucc.princeton.edu
    ftpmail@decwrl.dec.com

Send a message with the subject: HELP to get the current instructions directly from the gateway site.

ftp://rtfm.mit.edu:/pub/usenet/news.answers/radio/
    Ham radio FAQs (frequently asked questions) gathered from the usenet news groups.
    FAQs from rec.radio.info
    FAQ on buying and selling via usenet.
ftp://dingus.n5lyt.datapoint.com/tapr
    TAPR archives Tucson Amateur Packet Radio
    ftp://oak.oakland.edu/pub/hamradio/arrl/
    ARRL Infoserver Mirror
    PCB artwork and other information from QST articles
    Logo artwork
ftp://ftp.qrz.com/qrz
    QRZ CD-ROM (AA7BQ)
ftp://SunSITE.unc.edu/pub/academic/agriculture/agronomy/ham/
    homebrew projects
ftp://archive.afit.af.mil/pub/space/amateur.tle
    OSCAR keplerian elements
ftp://ftp.cs.buffalo.edu:/pub/ham-radio (KA2NRC)
ftp://nic.funet.fi:/pub/ham/info (mirror)
    Radio mods
    Callsign databases
    Exam questions
    FCC regulations
    Technical information
    Clubs
    Introductory information
    Repeater information
    ARRL information
    and much more.
ftp://ftp.ucsd.edu:/hamradio
    ampr.org information and archives
    Technical Information Archives
    Slow Scan TV
    Satellite Information UoSats, Arsene,
    ARRL
    DSP
    NEC
    RACES
    RTTY
ftp://ftp.coast.net/SimTel/msdos/hamradio
    SimTel primary site.
    A huge collection of DOS and Windows shareware.
ftp://ftp.apple.com:/pub/ham-radio
    Macintosh ham software
ftp://ftp.fcc.gov:/pub
    FCC information for public distribution, such as:
    Daily Digest
    News Releases
    Public Notices
    Reports

Speeches
A complete index is in the file index.txt.
ftp://think.com:/pub/radio/ham
    FCC regs
    Repeater info
    QRP archives [major site for information, plans, FAQs]
ftp://nic.funet.fi:/pub/ham
    SimTel ham radio files mirror
    Technical information
    DSP
    Amiga shareware
    Macintosh shareware
    Radio modifications information
    VHF technical information
ftp://nic.funet.fi:/pub/dx
    SWL information, SW broadcast schedules
    Archives of many newsletters
    Mac, Amiga and PC software for DXers
ftp://qed.laser.ee.es.osaka-u.ac.jp:/pub/radio/ped/ped411i.zip
    PED CW contest trainer program
    ftp://col.hp.com:/hamradio
    Clip art
    Packet
    SSTV
ftp://bubba.business.uwo.ca:/SYS/HAMSTER/ham
    Information and software for digital modes.
    Radio mods.
ftp://vax.cs.pitt.edu:/pub/hamradio
    Software for packet BBS operation
ftp://helios.tn.cornell.edu:/pub/PMP
    Poor Man's Packet Baycom TNC software and source
ftp://wuarchive.wustl.edu
    Mirrors of SimTel and other sites.
ftp://ftp.demon.co.uk:/pub/ham
    Digital mode information and software.
    UK amateur FAQ.
ftp://sics.se:/archive/packet
    Digital mode and satellite information.
ftp://mgate.arrl.org
ftp://oak.oakland.edu/pub/hamradio/
    Boston Amateur Radio Club
ftp://plan9.njit.edu:/pub/hamradio
    Mininec and other antenna modeling software.
    Exam information.
ftp://grivel.une.edu.au:/pub/ham-radio
    Australian amateur radio information.
    Macintosh software.
    UCSD mirror.
    Buffalo mirror.
ftp://iraun1.ira.uka.de:/pub/ham-radio
    German amateur radio information.
ftp://nic.switch.ch:/software/hamradio
    Swiss amateur radio information.
    Alef/Null amateur radio DSP board software and information.
ftp://nic.switch.ch:/software/mac/
    Macintosh ham radio software
ftp://akutaktak.andrew.cmu.edu:/aw0g
    Macintosh Softkiss software.
ftp://gandalf.umcs.maine.edu:/pub/ham-radio
    ARES information.
    NTS information.

ftp://vixen.cso.uiuc.edu:/pub/ham-radio
    Question Pools
    Hypercard Hamstacks

Other sites with ham radio archives:
    ftp://suntan.tandem.com:/hamradio
    ftp://brolga.cc.uq.oz.au:/pub/ka9q
    ftp://gatekeeper.dec.com:/pub/net/ka9q
    ftp://ftp.waseda.ac.jp:/pub/ham-radio

WORLD WIDE WEB
    The World Wide Web is a goldmine of information on every subject, and amateur radio is no exception. However, finding the information you are looking for can be a major challenge. Because new sites are being added daily, and old ones change their addresses rapidly, the best strategy is to know how to use the various search engines available.

Yahoo
    Yahoo is one of the best indexes of WWW URLs. A new amateur radio site is added to this index almost every day.
    http://www.yahoo.com (General index)
    http://www.yahoo.com/Entertainment/Radio/Amateur_Radio/

Tradewave
    Another major index of WWW pages is Tradewave. These indexes are updated by editors who check the validity of the addresses before adding new sites to the list.
    http://www.einet.net (General index)
    http://www.einet.net/galaxy/Leisure-and-Recreation/Amateur-Radio.html

Lycos
    In addition to indexes like Yahoo and Tradewave, there are robot programs which scan the web and create indexes based on the text on individual pages. In order to use these indexes, you enter search terms that describe what you are looking for. Lycos and Webcrawler are two of the most popular indexes of this type.
    http://www.lycos.com
    http://www.webcrawler.com

Yahoo Entertainment:Radio:Amateur Radio Index
    * #HamRadio IRC Channel
    * Amateur Radio & Funny Stories - Amateur Radio & Funny Stories - Funny adventures in Ham radio plus travel info. and links to the Indianapolis Area.
    * Amateur Radio [hawaii.edu]
    * Amateur Radio [mcc.ac.uk] [*]
    * Amateur Radio [ncsu.edu] - A full featured Amateur Radio server, with an online Repeater Database (which you can add repeaters to), Newsline News in html format, Real time satellite pass predictions and more!
    * Amateur Radio [qrz.com]
    * Amateur Radio in Arizona
    * Amateur Radio in Utah County - Utah County has one of the most active ARES groups in the country. Drop in and see what is going on.
    * Amateur Radio Operator-Online Magazine - featuring news and information from the entire Amateur Radio World.On-Line Internet Hamfest On-Line Classifieds for

Amateur Radio
    * Amateur Radio Slow Scan TV
    * Amateur Radio station NL7J page - Information on SSTV and DX'ing links. Some hard to find SSTV software. Amateur Radio Clubs on internet in Alaska by NL7J
    * Amateur Radio World - Local, regional, state, country, and global amateur radio information.
    * Amateur Satellite
    * American Radio Relay League
    * AMSAT - Radio Amateur Satellite Corporation - a 501(c)(3) not-for-profit corporation.
    * AmSoft - World of Ham Radio CD-ROM
    * Bookworm Amateur Radio Page
    * Callsign Servers (9)
    * Clubs (64)
    * Companies@ (10)
    * CQWW Contest - CQWW Contest news and useful data
    * Deltatango
    * Ham Homepage
    * Ham Radio
    * Ham Radio and More
    * Ham Radio On The Internet
    * HF, VHF, and UHF Propagation
    * HR Showcase
    * KB5VQI (local DFW,Texas) Amateur Radio Information
    * Linux Amateur Radio Software List
    * Listing of Ham Radio, Computer, and Electronics Swap Meets
    * Low Profile Amateur Radio - Devoted to low profile radio operation
    * Mailing Lists and Newsgroups
    * N0ARY/BBS - a full service amateur radio packet bbs that also provides a bidirectional gateway to internet.
    * NF2G Scannist Pages - Contains regional frequency updates, links to other scanner sites, recorded traffic in .WAV format
    * No Spam...Ham! - a link devoted to resources for people interested in Ham Radio.
    * Ottawa-Area Amateur Radio
    * Packet Radio (7)
    * People (16) [[new]
    * Quarter Century Wireless Association, Inc. (QCWA) - non-profit association of amateur radio operators.
    * Shows (3)
    * Shuttle Amateur Radio Experiment - SAREX
    * Signal Radio - details and info on the company, programmes, presenters and much much more
    * The Society of Amateur Radio Astronomers
    * www.guy.com - The Grand Master - www.guy.com helping you learn to surf ham radio amateur radio games no sex
    * YS1ZKR Satellite DXpedition
    * FAQ - Radio
    * FAQ - rec.radio.amateur
    * Index - Amateur Radio - WWW Virtual Library
    * Index - Other Ham Radio Web Servers
    * Index - World Wide Web Sites
    * Usenet (14)

Tradewave Amateur Radio Index
    Galaxy | Add | Help | Search | What's New | About
TradeWave
    Amateur Radio - Leisure and Recreation
    New Items - less than 7 days old
    * KaWin Home Page
    * Woodbridge Wireless, Inc.
Articles
    * Sample U.S. Amateur Radio Exams Page
    * Valdosta Amateur Radio Club
Books
    * Amateur Radio Slow Scan TV Handbook
    * Thanks to Amateur Radio
Announcements
    * Marconi Centennial Celebrations Home Page
    * W8GYH - St. Xavier H.S. ARC
Software
    * KaWin Home Page
    * Super-Duper by EI5DI
Collections
    * Amateur Radio (UK)
    * Amateur radio - Ham Radio Callsign Database
    * Amateur Radio Server, North Carolina State
    * CHMA- FM 107 - Mount Allison University,
Sackville, New Brunswick, Canada
    * Josh Tuel's Amateur Radio Page
    * KB7UV's Amateur Radio Resources on the 'Net
Discussion Groups
    * #HamRadio IRC Channel Web page
    * Ham Opinion Poll
Directories
    * Amateur Radio Information - University of Maine
System
    * Ham Radio Callbook Server / SUNY at Buffalo
    * Ham Radio Pages
    * HR Showcase Amateur Radio WWW
    * The World-Wide Web Virtual Library: Amateur
Radio
    * Utah County (Utah) Amateur Radio
    * VE3SUN Ham ftp sites and mailing lists.
    * WB5AOX Home Page (Packet, Homebrew, etc)
    * WWW index of Noncommercial Radio sites
Organizations
    * Bradley University Amatuer Radio Club
    * cAVe California Amateur Vhf Enthusiasts
    * Embry-Riddle Aeronautical University Amateur
Radio Association's Home Page
    * KC HamNet
    * METU Radio Society
    * Northern Alberta Radio Club
    * The Radio Amateur Satellite Corporation - AMSAT
    * University of Central Florida Amateur Radio Club
    * W6BHZ - Cal Poly Amateur Radio Club
    * Woodbridge Wireless, Inc.
Academic Organizations
    * Harvard Wireless Club - W1AF
    * K9IU Indiana University ARC
    * Naval Postgraduate School Amateur Radio Club
    * W2SZ - Rensselaer Polytechnic Institute - Amateur
Radio Club
    * W3EAX - University of Maryland
    * W5AC - Texas A&M Amateur Radio Club
    * W6YX Stanford Amateur Radio Club
    * WB5FND - Univ of Houston Amateur Radio Club

Government Organizations
    * SAREX - Shuttle Amateur Radio Experiment
Non-profit Organizations
    * Clear Lake Amateur Radio Club
    * Lakehead Amateur Radio Club
    * North Shore Repeater Association
    * Radio Amateur Telecommunications Society
    * The New Hampshire Amateur Radio Page
    You can add information to this page!

    Galaxy | Add | Help | Search | What's New | About
TradeWave | Top of Page
    Reprinted with permission from Tradewave Galaxy.

THE ARRL E-MAIL SERVER

ARRL is the American Radio Relay League, representing and promoting Amateur Radio in the USA. They
have established an automated file server which responds
to information requests via electronic mail. To use the
server, send mail to info@arrl.org with any number of
one-line commands in your message. The server program
will respond to the commands, each in a separate message. The help command will retrieve the following file:

**HELP on using the ARRL's Automated Information
Server (info@arrl.org)**
    PLEASE Note: When you send messages to the
ARRL Email Information Server, you should use the
address:
        info@arrl.org   (NOT info-serv@arrl.org)
    Messages sent from the server are addressed from
info-serv@arrl.org in order to prevent bounced messages
from creating a mail loop.
    Send a new message to info@arrl.org for each batch
of requests.
    Each line of the message should contain a single com-
mand as shown below. You may place as many commands in a message as you want. Each file you request
will be sent to you in a separate message. Only ASCII
text files are supported at present.
    If you have FTP access, all of these files can be
obtained from oak.oakland.edu by anonymous FTP in the
files area pub/hamradio/arrl/infoserver. Retrieve the file
index.txt in this directory for a complete listing of available files listed by subdirectory with descriptions.
    W1AW bulletins featuring news of interest to all ama-
teurs (such as FCC actions, DX news, satellite orbit data,
propagation trends, SAREX mission information and
other items) are available by email from the Netcom List
Server.
    To subscribe to this mailing list, send an email mes-
sage to:
        listserv@netcom.com   (NOT info@arrl.org)
    Your message subject can be anything (it is ignored).
Put the following text in the body of your message:
        subscribe w1aw-list
        quit
    In addition, W1AW bulletins can also be FTPed from
oak.oakland.edu in the directory pub/hamradio/barc/
w1aw-list. There are several subdirectories under this for
each bulletin catagory. See the W1AW bulletin description files (.des extension) for more information on each
catagory.
    Binary files mentioned in QST that accompany arti-
cles are *NOT* available via the server. This is an email

service only and can not send binary files. HOWEVER, these files are also available from oak.oakland.edu, in the directory pub/hamradio/arrl/qst-binaries.

Valid INFO email commands:

help   Sends this help file

index   Sends an index of the files available from INFO

reply <address>   Sends the response to the specified address. Put this at the BEGINNING of your message if your From: address is not a valid Internet address.

send <FILENAME>   Sends "FILENAME" example: send PROSPECT.TXT

quit   Terminates the transaction (use this if you have a signature or other text at the end of the message.)

Note: your message will *not* be read by a human! Do not include any requests or questions except by way of the above commands. Retrieve the "HQ-EMAIL.TXT" file for a list of email addresses of ARRL HQ staff.

Your From: field or Reply-to: field in your header should contain a valid Internet address, including full domain name. If your From: field does not contain a valid Internet address, the answer will not reach you. However, we have recently added a reply function as a server command. If needed, the REPLY command should be the first command in your message.

syntax:

reply mailaddr

Where mailaddr is a valid Internet mail address (either user@domain or bang address accepted.) An invalid address generates an error. A wrong address results in non-delivery of your response.

The address given in the reply command is the address to which all subsequent requests in the message will be sent.

If an error message is generated, it will be sent to the last reply address given.

If anyone needs some help with the server, or has ascii information files that they would like to archive on our server, additional information or updates for any of our files, or suggestions for improvements, please contact Michael Tracy, KC1SX, mtracy@arrl.org at ARRL Headquarters via email or telephone (203) 594-0200.

Sample of files available from INFO: (There are lots more!)

FILENAME   SIZE DESCRIPTION

#Note - If you are not yet an Amateur Radio operator retrieve the #file prospect (send prospect) for information on how to easily get #started in this fun hobby.

PROSPECT.TXT	2k	How to get your Amateur Radio license
EXAMS.TXT	52k	Current exam schedule info - updated bi-weekly
EXAMINFO.TXT	9k	Examinations - what to bring - requirements
HQ-EMAIL.TXT	6k	List of HQ Email addresses
ARRLCAT.TXT	39k	Catalog of ARRL Publications - commercial content
JOIN.TXT	2k	How become an ARRL member
SERVICES.TXT	5k	A condensed list of ARRL membership services
TOUR.TXT	28k	An electronic tour of ARRL Headquarters
DIR.HQ	5k	Visiting ARRL HQ - directions and tour information
HFBANDS.TXT	7k	Breakdown of users of HF spectrum
W1AW.SKD	2k	W1AW schedule of transmissions and operation
PRODREV.TXT	25k	Listing of Product Reviews that appeared in QST
RFIGEN.TXT	37k	How to solve an EMI/RFI problem - QST Lab Notes
RFISOURC.TXT	13k	Where to buy filters - EMI-proof telephones etc.
ADDRESS.TXT	16k	Lots and lots of ham/electonic company addresses
KITS.TXT	6k	List of companies that sell kits
BBS.TXT	12k	List of ham-radio land-line bulletin boards

HAMFAQ1.TXT	25k	Introduction to the FAQ and Amateur Radio
HAMFAQ2.TXT	45k	Amateur Radio Orgs, Services and Info Sources
HAMFAQ3.TXT	32k	Amateur Radio Advanced and Technical Questions

Enjoy this ARRL service. Please direct comments or suggestions (or flames) to mtracy@arrl.org. 73 from ARRL HQ, Michael Tracy, KC1SX

AMATEUR RADIO MAILING LISTS

Revised: August 5, 1995

Mailing lists are used to disseminate information about a common topic of interest to a list of subscribers by email. There are two kinds of mailing list: reflectors, which simply reflect message from one subscriber to all of the other subscribers on the list; and distribution lists, which disseminate information from a single source to all of the subscribers on the list.

To participate in the traffic on these lists, you need to be able to receive email from the Internet, either directly, or through a gateway. In general, all of the lists operate in the same way. There is an address used for list maintenance commands, and another address used for the dissemination of messages, which is shown below as the reflector address.

It is important that list commands (the most common commands are SUBSCRIBE and UNSUBSCRIBE) not be sent to the reflector address, for two reasons. The first reason is that sending the message to the reflector will not accomplish anything. The command must be sent to the right address to take effect. The second reason is that the message will be broadcast to all of the subscribers on the list. Many subscribers pay for each message they receive, and seeing a message that says SUBSCRIBE or UNSUBSCRIBE joe@blow.com, for which they have just paid a quarter, is not likely to make their day.

If you are interested in joining one of the mailing lists below, the first thing you should do is send a message to the Subscribe address containing the word HELP on the first line. Within a few minutes, or perhaps a few hours in some cases, you will receive back a full description of the list's purpose and a set of commands that can be used to subscribe, unsubscribe, or get other information out of the list server.

There are many different flavors of list server software, and each has its own command structure. For reference, the subscribe command is shown in the list below. However, it is not unknown for the list manager to change software from time to time and the commands will change when this happens. Send the HELP message first.

In the list below, substitute YOUR email address for <user@site> and substitute your name for <name>.

For example, to subscribe to the Contest forum, send a message to CQ-Contest-Request@tgv.com with the word SUBSCRIBE in the body of the message. To send a message to all of the other subscribers, send the message to CQ-Contest@tgv.com.

Lists with Listserv@ or Majordomo@ have utility commands that you may wish to use. LISTS in the body of the message will give you a list of the mailing lists available on that server. INFO <listname> will tell you about the mailing list.

Contest Forum

Subscribe CQ-Contest-Request@tgv.com
Message SUBSCRIBE
Reflector CQ-Contest@tgv.com

DX Forum
    Subscribe DX-request@ve7tcp.ampr.org
    Message subscribe <name> (manual system)
    Reflector dx@ve7tcp.ampr.org
QRP Forum
    Subscribe listserv@netcom.com
    Message subscribe qrp-l
    Reflector qrp-l@netcom.com
VHF forum
    Subscribe Listserv@W6YX.Stanford.EDU
    Message Subscribe VHF
    Reflector VHF@W6YX.Stanford.EDU
N6TR Contest Logging program forum
    Subscribe N6TRLOG-REQUEST@CMICRO.COM
    Message SUBSCRIBE
    Reflector N6TRLOG@CMicro.COM
K1EA CT Contest Logging Program
    Subscribe ct-user-request@mlo.dec.com
    Message subscribe
    Reflector ct-user@mlo.dec.com
Ham Technical Forum
    Subscribe listserv@netcom.com
    Message subscribe ham-tech
    Reflector ham-tech@netcom.com
Tucson Amateur Packet Radio - Packet BBS Info
    Subscribe listserv@tapr.org
    Message subscribe bbssig FirstName LastName
    Reflector bbssig@tapr.org
Tucson Amateur Packet Radio - APRS
    Subscribe listserv@tapr.org
    Message subscribe aprssig FirstName LastName
    Reflector aprssig@tapr.org
Tucson Amateur Packet Radio - DSP
    Subscribe listserv@tapr.org
    Message subscribe dsp-93 FirstName LastName
    Reflector dsp-93@tapr.org
Tucson Amateur Packet Radio - HF Packet
    Subscribe listserv@tapr.org
    Message subscribe hfsig FirstName LastName
    Reflector hfsig@tapr.org
Tucson Amateur Packet Radio - Announcements and Bulletins
    Subscribe listserv@tapr.org
    Message subscribe tapr-bb FirstName LastName
    Reflector tapr-bbtapr.org
Email gateway to rec.radio.amateur.antenna
    Subscribe Listserv@ucsd.EDU
    Message SUBSCRIBE ham-ant
    Reflector ham-ant@UCSD.EDU
Email gateway to rec.radio.amateur.digital
    Subscribe Listserv@ucsd.EDU
    Message SUBSCRIBE ham-digital
    Reflector ham-digital@UCSD.EDU
Email gateway to newsgroup rec.radio.amateur .equipment
    Subscribe Listserv@ucsd.EDU
    Message SUBSCRIBE ham-equip
    Reflector ham-equip@UCSD.EDU
Email gateway to newsgroup rec.radio.amateur. homebrew
    Subscribe Listserv@ucsd.EDU
    Message SUBSCRIBE ham-homebrew
    Reflector ham-homebrew@UCSD.EDU
Email gateway to newsgroup rec.radio.amateur.policy
    Subscribe Listserv@ucsd.EDU

    Message SUBSCRIBE ham-policy
    Reflector ham-policy@UCSD.EDU
Email gateway to newsgroup rec.radio.amateur.space
    Subscribe Listserv@ucsd.EDU
    Message SUBSCRIBE ham-space
    Reflector ham-space@UCSD.EDU
Email gateway to newsgroup rec.radio.amateur.misc
    Subscribe Listserv@ucsd.EDU
    Message SUBSCRIBE info-hams
    Reflector info-hams@UCSD.EDU
Moderated email gateway to newsgroup rec.radio.info
    Subscribe listserv@ucsd.edu
    Message sub radio-info
    Reflector radio-info@ve6mgs.ampr.ab.ca
                Or
    radio-info@ucsd.edu

    The **HAM-RADIO mailing list** (an experimental digest using subject grouping and MIME encapsulation to provide a daily dose of ham radio related traffic from the Usenet)
    Subscribe Listserv@ucsd.EDU
    Message SUBSCRIBE ham-radio
    Reflector ham-radio@UCSD.EDU
Packet Radio List
    Subscribe Listserv@ucsd.EDU
    Message SUBSCRIBE packet-radio
    Reflector packet-radio@UCSD.EDU
KA9Q Unix Users Mailing List
    Subscribe listserv@knuth.mtsu.edu
    Message subscribe ka9q-unix <your name>
    Reflector ka9q-unix@knuth.mtsu.edu
Raider Amateur Radio Club Mailing List
    Subscribe listserv@knuth.mtsu.edu
    Message subscribe rarc-l@knuth.mtsu.edu
    Reflector rarc-l@knuth.mtsu.edu
KA9Q NOS BBS List
    Subscribe nos-bbs-request@hydra.carleton.ca
    Message add <user@site> nos-bbs
    Reflector nos-bbs@hydra.carleton.ca
Old Radio Equipment
    Subscribe boatanchors-request@theporch.com
    Message subscribe boatanchors <name>
    Reflector boatanchors@theporch.com
Shuttle Elements List
    Subscribe listserv@thomsoft.com
    Message subscribe <name>
ARRL Field Org List
    Subscribe listserv@netcom.com
    Message subscribe fieldorg-l
    Reflector fieldorg-l@netcom.com
ARRL Exam List
    Subscribe listserv@netcom.com
    Message subscribe arrl-exam-list
ARRL VE List
    Subscribe listserv@netcom.com
    Message subscribe arrl-ve-list
ARRL New England Division Mailing List
    Subscrive listserv@netcom.com
    Message subscribe arrl-nediv-list
Boston Amateur Radio Club List
    Subscribe listserv@netcom.com
    Message subscribe barc-list
Eastern Massachusetts ARRL List
    Subscribe listserv@netcom.com
    Message subscribe ema-arrl

ARRL Official News and Information List
Subscribe listserv@netcom.com
Message subscribe w1aw-list
WY1Z Amateur Radio Newsline List
Subscribe listserv@netcom.com
Message subscribe newsline-list
ARRL Letter Redistribution List
Subscribe listserv@netcom.com
Message subscribe letter-list
BARC RACES Mailing List
Subscribe listserv@netcom.com
Message subscribe barc-races
Fox Hunt Mailing List
Subscribe listserv@netcom.com
Message subscribe fox-list
KY1N New England VEC Information List
Subscribe listserv@netcom.com
Message subscribe ky1n-list
University Amateur Radio Clubs
Subscribe Listserv@W6YX.Stanford.EDU
Message Subscribe ham-univ
Reflector ham-univ@W6YX.Stanford.EDU
Stanford Amateur Radio Club (aka su.org.hamradio)
Subscribe Listserv@W6YX.Stanford.EDU
Message Subscribe w6yx-club
Reflector w6yx-clubW6YX.Stanford.EDU

(Compiled by AB6WM)

**CQ-CONTEST@TGV.COM Frequently Asked
Questions (FAQ) List**
Revised: May 12, 1995

What is CQ-CONTEST?

CQ-CONTEST@TGV.COM is an electronic mail reflector dedicated to hams interested in all types of amateur radio contesting. This is a good place for score reports, expedition rumors, and other contest-related discussion or announcements. This forum is more like the NCJ than QST; INFO-HAMS@UCSD.EDU and rec.radio.amateur.misc are good places to look for a more rounded discussion of the hobby.

Although there is overlap between contesters and DXers, CQ-CONTEST is not a DX-oriented group. DX@UNBC.EDU is an electronic mail mailing list dedicated to the discussion of DXing. For details on how to subscribe to this and other mailing lists, consult the List of Lists at the end of this message.

Each message you send to CQ-CONTEST@ TGV.COM will be sent out to all the other subscribers, kinda like a 2-meter repeater that has a coverage radius of 12,000 miles or so. Think of sending mail to the list as the equivalent of an ANNOUNCE/FULL message on PacketCluster. Use regular email to send a message to a specific individual.

Electronic mail is also different from packet radio, in that many subscribers receive their email through commercial services such as CompuServe and MCImail. In essence, many people are paying for each byte of every message sent to CQ-CONTEST. In order to minimize spurious messages, follow the operating hints detailed below.

How do I join CQ-CONTEST?

Subscription management is handled automatically by a program that answers mail send to CQ-CONTEST-

REQUEST@TGV.COM. Send a message to CQ-CON-TEST-REQUEST@TGV.COM that says SUBSCRIBE if you wish to join the group, or UNSUBSCRIBE if you want to drop out. The Subject: line is ignored. Messages sent to CQ-CONTEST@TGV.COM are broadcast to *all* readers, so don't send subscription requests there.

What are the suggested "operating practices" for CQ-CONTEST?

Put your name and call sign on every message you send. We don't all know everyone by just a call or a nickname.

Use a subject line that indicates the true subject of your message.

Wait a while before answering someone's question. Six other people have probably answered it already. Most answers should go directly to the person who posed the question, rather than to the list.

Unlike PacketCluster, many people pay $$$ when they receive messages. Some people pay per message, some per byte. Therefore, please take this into consideration when writing a response. Would you pay $0.50 to read the message that you just wrote?

Eschew flamage. If someone sends a flame to the list and you can't bite your tongue, send your flaming reply directly back to the flaming individual, not back to the list. No one wants to pay $1.00 to read these messages (the original flame + your reply). Treat flamers the way you would 2-meter repeater jammers - ignore them.

Make sure there is something of value in each message you send to the list. Avoid messages that are a complete reprint of someone else's message, with nothing but "I agree" or "Me too" added to the bottom — not much value there.

Some people pay by the byte, so when following up to someone else's message, be sure to include only the essential pieces or thread of the note. Don't include those 20 extra header lines that your mail gateway tacked onto the original message.

How can I get CQ-Contest in digest form?

Tack (je1cka@nal.go.jp) has graciously offered to redistribute CQ-Contest messages in digest form. This means that all messages posted to CQ-Contest on a given day will be bundled together and resent as a single message to the subscribers of Tack's list. This is useful for people with Internet providers that place a limit on the number of messages you can have in your mailbox at once. This is the case for many of the JA subscribers.

To subscribe to JE1CKA's CQ-Contest-Digest list, send a message to Contest-Request@DUMP-TY.NAL.GO.JP that says:

SUBSCRIBE cq-contest-digest your_callsign <your_email_address>

If you are subscribed to CQ-Contest, remember to send a message to CQ-Contest-Request@TGV.COM that says

SET NOMAIL

Since you will be getting the messages in digest form, you won't need to get them directly from CQ-Contest@TGV.COM, but you will need to remain subscribed if you still want to post messages.

How can I fetch messages from the CQ-Contest archive?

You can fetch messages from the CQ-Contest archive by sending a message to FileServ@TGV.COM that says

SENDME CQ-CONTEST-ARCHIVE.yyyy-mm where yyyy-mm is the year and month of the archive desired. For additional information, you can send a message to FileServ@TGV.COM that says HELP.

How can I find out the email address of a particular contester?

George Fremin, WB5VZL (geoiii@bga.com), maintains a fairly current list of contester email addresses. Send him a note to them asking for his lists. You can also get a list of registered CQ-Contest subscribers by sending a message to CQ-Contest-Request@TGV.COM that says REVIEW. If that doesn't work, trying calling the person you seek on the telephone and asking for their email address directly. Do not post a message to CQ-Contest that says "Does anyone have the email address for _____?"

How can I find out more about the Internet?

Pick up a copy of the book _The Internet Companion_ by Tracy LaQuey, Addison-Wesley, ISBN 0-201-62224-6. If your local technical book store doesn't carry it, you can order from Computer Literacy, 2590 North First Street, San Jose, CA 95131. Their phone number is 408-435-0744.

73, The Wouff Hong

The Internet DX Mailing List: Operations Manual
Comments or additions should be sent to:
lyndon@ve7tcp.ampr.org.
Last updated: June 18, 1995

Introduction

What is the Internet DX Mailing List? It is an electronic mailing list dedicated to the discussion of Amateur Radio DXing. This is the place to exchange tips and techniques, discuss hardware (rigs, antennas, DSP add-ons), awards, DX related software, announce upcoming DXpeditions, etc. Anything and everything pertaining to DX is fair game, with the following exceptions:

* Park your flames (and egos) at home. This list is for constructive commentary. If you don't like the way something is done, we don't want to hear about it unless you have a positive and constructive solution or alternative. (Perpetual whiners will be banished to 160 KHz with a 10 mW transmitter, and will not be admitted back to the list until they achieve DXCC honour roll on that band.)

* Don't send requests for QSL info to the DX mailing list. You can obtain QSL information via e-mail by sending a message to qsl-info@aug3.augsburg.edu. Include in the message body the callsigns you want QSL information for - one callsign per line. This service is operated by Ray Rocker (WQ5L). Please direct questions or comments to him at rrocker@rock.b11.ingr.com.

* Discussion about contests and contesting should be kept to a minimum. The CQ-Contest list already provides an excellent forum for contest related discussions. (To subscribe to CQ-Contest, send an e-mail request to cq-contest-request@tgv.com)

Subscribing, Unsubscribing, and Administrivia

How do I subscribe to or unsubscribe from the list? A robot maintains the mailing list subscriptions. To subscribe or unsubscribe you need to send a message to dx-request@ve7tcp.ampr.org. The body of the message contains commands for the robot. To subscribe to the list include the following command in the body of your mail message:
subscribe [address]
To unsubscribe use this command:
unsubscribe [address]
[address] refers to your e-mail address, and is optional. If you don't specify an explicit address the robot will extract it from your message. You don't need to specify an explicit address unless your mail software mashes up your return address, or you're changing your subscription from a different address than you usually use. If you do specify an address, leave out the '[' and ']' characters. If you get stuck you can send a message with only the word HELP in the body. This will tell the robot to send you help about the commands it understands.

Mail from the list might not cease immediately after you unsubscribe. Due to the way mail is handled, you will receive any messages in the delivery queue (but not yet delivered to you specifically) at the time you unsubscribe. This should only amount to two or three messages.

What about these duplicate messages?

Before you panic, check the mail headers. Many complaints about duplicates are really messages sent to more than one list. If you are subscribed to both lists (for example, the DX list and the CQ-Contest list) you will receive two copies of the message - one from each mailing list. This is not a bug.

Unfortunately, duplicate messages can also be caused by mailers at the receiving end of the mailing list. I take precautions to prevent error messages and the like from escaping back to the mailing list, but you would not believe the amount of totally broken mail software out there on the Internet these days.

If you receive one or two duplicate messages, ignore them.

If you receive more than one or two duplicate messages, and if you can extract the Received: headers from the message, you can forward copies of the message with the Recieved: headers intact to lyndon@ve7tcp.ampr.org. If you cannot include the Received: headers DON'T forward the messages. I can't debug the problem without them.

DO NOT send mail about the problem to the entire mailing list! The list subscribers cannot do anything to fix the problem, and have probably figured out for themselves that something is wrong.

Is the mailing list archived?

Yes. Archives (starting from Mid-June 1995) are available via anonymous FTP from ftp://ve7tcp.ampr.org/mailing-lists/archives/dx/.

If you don't have FTP capabilities you can retrieve the archives from the mail server. Send a message to listserv@ve7tcp.ampr.org with the command
INDEX DX
In the message body. In return you will receive a message listing all the files available for retrieval using the GET command.

How do I send a message to the list?

It's easy! Just send an e-mail message to:

dx@ve7tcp.ampr.org. Your message will be automatically resent to all the mailing list subscribers. Depending on the load on our mail system this could take anywhere from a few minutes to several hours. If you don't see your message within a day or so, send a message to lyndon@ve7tcp.ampr.org (NOT the general submission address) and I will look into the problem.

Here are a few things to keep in mind when sending a message to the list. If everyone follows these guidelines it will help the list operate smoothly.

* Make sure your subject line is appropriate. If a conversation wanders off the original topic, change the subject line to reflect the new line of discussion.

* Proofread your message before sending it! Spelling mistakes and poor formatting do nothing to help you get your message across.

* Keep it short! Many subscribers pay for their e-mail access. Don't include 100 lines of the message you're replying to and then add a three linecontribution. Only include as much text as is necessary to establish the context for your reply.

* Watch your language! This list is gatewayed onto packet radio. Don't say anything that you wouldn't (or shouldn't) say over the air. Anyone violating this rule will be dropped from the list immediately!

* If someone asks a question, send your answer to them directly. The person asking the question should collect the responses, then send a summary to the list. This helps cut down on unnecessary list traffic.

* Eschew flamage. If someone sends a flame to the list and you can't bite your tongue, send your flaming reply directly back to the flaming individual, not back to the list. No one wants to pay $1.00 to read these messages (the original flame + your reply). Treat flamers the way you would two-meter repeater jammers - ignore them.

* Make sure there is something of value in each message you send to the list.

When I reply to a message using my mailer's REPLY command I never see a copy of the message come back from the list. Why?

There are two possible causes. One is that your mailer sends the reply only to the originator of the message. In this case you will have to manually CC the mailing list. The second possibility is that your mail reader is so ancient and broken that it sends replies to the "envelope" from address instead of the From: or Reply-To: header values. If you are in the att.com or microsoft.com domains you want to double check for this second problem. Note that replies sent to the wrong address get filed in the bit bucket.

Copyright and Other Legal Muck

Can I reproduce material from the mailing list? No compilation copyright is asserted over the mailing list. The individual submissions to the mailing list may be (and probably are) covered by national copyright law, and international copyright conventions. When in doubt, consult a lawyer.

If you do reproduce information obtained from the list in a published newsletter please give credit to the message author and the mailing list.

Other Related Internet Services

Where can I obtain back issues of the various DX bulletins?

An incomplete archive of DX related bulletins is available via anonymous FTP from ve7tcp.ampr.org:/bulletins/. I add back issues as I come across them. Feel free to contact me if you can help fill in the holes. At some point I hope to set up a WAIS search index that will let you do keyword searches on the bulletins in the archive.

Where can I find out about current solar and propagation conditions?

One of the best sources is Cary Oler's Solar Terrestrial Dispatch. He provides both daily reviews of solar activity and a report of current solar conditions updated every three hours. Sites on the Internet can obtain these reports by fingering the following addresses:

    * solar@xi.uleth.ca (3-hourly report)
    * aurora@xi.uleth.ca (hourly auroral activity report)

Lyndon Nerenberg - VE7TCP - lyndon@ve7tcp.ampr.org

THE PACKETCLUSTER NETWORK

The Packetcluster network is the brainchild of Dick Newell, AK1A. Several years ago, he developed a networking system originally known as the Packet Conference Board System, which allowed several packet radio BBSs to link together. As time went on, the capability of linking stations to nodes and nodes to each other in clusters evolved into the present Packetcluster system. The primary application for this system was (and is) DX spotting and alerting. For example, if you hear a rare DX station on the air, you can alert all other hams connected to your local Packetcluster network by a simple announcement (called a "spot"). Packetcluster has grown extensively over the last several years, with numerous added features, including mail store–and–forward service, and interfaces to popular logging software packages. To get started using Packetcluster, you'll need a terminal or terminal emulation program, a suitable radio, and a TNC. Locate the node nearest to your location, set your radio to the frequency indicated, and type CONNECT <callsign of node>. If you are successful in connecting, the system will prompt you for various information. Typing HELP will access a list of Packetcluster commands.

DXNODES Database
31-Aug-95 Ver 7.3

Ron Rueter, NV6Z, 2113-108th Street S.E., Everett, WA 98208-5108
Please send updates, corrections or comments, to the above address or to NV6Z via:
    CompuServe: 72200,1347
    or DX-BBS (502) 898-8864
    or Internet nv6z@nwrain.com

Please Include date/version of database.
PacketCluster is a trademark of Pavillion Software.

FRC	Frankford Radio Club
GCDXC	Gulf Coast DX Cluster
GLPS	Great Lakes PacketCluster System
INWDXN	Inland NorthWest DX Network
LIDXA	Long Island DX Association
MDXPA	Midwest DX Packet Association
NCC	North Coast Contesters
NCN	Northeast Cluster Network
NFIDXA	Niagara Frontier International DX Assoc.
NJDXA	North Jersey DX Association
NWDXSN	Northwest DX Spotting Network
PVDXSN	Potomac Valley DX Spotting Network
PVRC	Potomac Valley Radio Club
TSCN	Tristate Cluster Network
YCCC	Yankee Clipper Contest Club

Alabama

Birmingham	K4FHQ-1	BHMDX	144.930 (Network node)
Birmingham	K4FHQ-2	HAMDX	145.790 (Network node)
Ft. Payne	W4BRE-1	FTPDX1	448.400 (Network node)
Ft. Payne	W4BRE-2	FTPDX2	145.750 (2400 baud network node)
Hoover	W4AXO		145.750
Huntsville	K4CEF		144.930
Huntsville	K4CEF-1	HSVDX	144.930 (Network node)
Huntsville	K4CEF-4	HSVDX4	448.400 (Network node)
Jacksonville	N4QLB-2	JAXDX	144.930 (Network node)
Loachapoka	KA4PKB		144.910
Mobile	WA4OSR		144.970 #
Tuscumbia	WB4VKW-1	TUSDX	144.930 (Network node)

= High Speed backbone port 9.6k or 19.2k

Alaska

Eagle River	NL7C	145.030 440.050 Backbone (Anchorage)

Arizona

Carefree	K5VT	K5VT	145.030
Chandler	K7NO	CADXA	145.090
Phoenix (NW)	N7CIX	CADXA	144.930
Tucson	N7BXX	SADXA	144.930

Access to N7US by public NET/ROM AZCTL on 450: Type C 3 N7CIX after connecting to AZCTL.

Arkansas

Batesville	K5EJ	145.690 ——> Link to KN4F	
		446.050 ——> Link to WD5B	
Fayetteville	WA5IED	144.950	
Little Rock	WD5B	144.950 ——> Link to WU3V	

California (Northern) Nevada DXPSN: 02/95
Network Node alias shown in brackets, i.e. [DX4]

Hanford	K6OZL		144.950
	K6OZL	[DX6]	144.950 (Bear Mtn, Fresno area)
	K6OZL	[DX7]	145.770 (Mt. Adelaide, Bakersfield area)
Los Gatos	N6ST		146.595 (Santa Cruz Mtns, Monterey Bay)
	N6ST	[DXF]	146.595 (Santa Cruz/Los Gatos)
Mountain View	K6LLK		144.950 (Mtn View, Ntwk Node and Hub)
Oakdale	K6AYA		146.580 (Modesto area)
Pittsburg	AH0U		146.580 (Walnut Creek area)
	AH0U	[DX4]	146.580 (Sugarloaf Mt.- Napa valley)
Palmdale	K6GXO		145.690 (Antelope Valley area)
Redwood City	W6OAT		145.770 (Redwood City North & East Bay)
	W6OAT	[DX5]	145.770 (Mt. Diablo)
Reno, Nevada	KI3V		144.950 also 146.58,441.500 (2400 baud), 51.7
	KI3V	[PCDX]	144.950 (Virginia City, NV)
	KI3V	[DX2400]	441.500 (2400 baud)
Rio Linda	W6GO		144.950 (Sacramento, Woodland, Davis)
	W6GO	[DXC]	144.950 (Cohasset Ridge - Chico area)
	W6GO	[DXR]	144.930 (South Fork Mtn - Redding area)
San Francisco	W6OTC		145.670 (East Bay and North)
Santa Rosa	WB2CHO		144.950
Milpitas	KJ6NN	[DXFMT]	1299.890 (San Jose - Fremont area)

California (Southern)

Bermuda Dunes	W6EEN		145.690 (Palm Springs)
Chino	N7QQ		145.660 , 144.490 (Inland Empire West)
El Segundo	K6EXO		144.490, 145.690 (South Bay, Central LA, San Gabriel Valley)
Fullerton	K6QEH		145.680
Hemet	KM6LM		144.950 (Hemet/San Jacinto Valley)
Oxnard/Ventura	N6VR		144.915
Palmdale	K6GXO		145.660 (Antelope Valley)
Ramona	KD6HRO		145.050 (Limited Cluster)
Redondo Beach	AA6SF		145.690 & 144.950
San Diego	K6JYO		144.950 (San Marcos)
San Diego	KM6K		144.950 (Kearney Mesa)
San Diego	KA5Q		145.690 (Fallbrook)
San Diego	N6ND		144.360 (Pt. Loma)
Santa Ana	N6HVZ		145.605 (Serving West Orange County)
Santa Maria	N6MB		144.950

Colorado

Aurora	KYØA		144.950
Boulder	KE6LT		144.930
Englewood	WØCP		144.970
Fort Smith	AB5K		144.910

Connecticut

Burlington	W1RM		144.950
Dayville	KB1H		144.930
——> Also see YCCC			

Delaware

Middletown	N3FDL-1	DEFRC	144.950 (KANODE to WB2YOF)
——> Also see FRC			

District of Columbia
——> See PVDXSN

Florida

Apopka	N4WW		144.950
Boca Raton	N4TL	SFDXA	145.590
Bradenton	WB4FNH		144.970
Clermont	WD4IXD		UNKNOWN
Ft Lauderdale	WA4YLD		145.590
Ft Walton Beach	WD4HDT	NWFDG	144.970
Jacksonville	NO4J	NFDXA	144.950, 145.690
Melbourne	K3ARV		144.950, 145.590
Orlando	N4WW		144.950
Palm Harbor	N4UYO		145.510
St. Leo	AB4PY		145.510
St. Petersburg	KO4J		144.950

Georgia

Atlanta	W8ZF	145.630 (2400 baud)
Atlanta	N4UCK	145.650
Atlanta	ATL	145.750 (2400 baud network node)
Atlanta	ATL1	145.750 (2400 baud network node)
Atlanta	ATL2	145.750 (2400 baud network node)
Douglasville	K4KG	145.910 (2400 baud)
Lawrenceville	KK4JF	145.710
	COMDX	145.710
	COMDX2	145.710 (2400 baud)
	COMDX3	440.750 (9600 baud)

Hawaii

Kauai	NH6AF	144.970
Oahu	AH6IO	144.930
Oahu (windward)	NH6UY	144.910
Maui	AH6LE	144.950
Big Island, Kona	NH6M	144.910

You can also connect to AH6LX 808326 on 145.050 from Kona, either thru KOA network node or direct for normal PBBS, or then to NH6M for access to the Cluster. Amazing system, great guys and they welcome visitors to check in.

Idaho

Boise	NK7U	145.150 BOIDX Digi
Boise	AA7FT	145.070 Linked to Spokane, WA

Illinois

Chicago (Southside)	N9QX	145.51, 223.42
Crete (Far S Chicago)	K9AJ	145.51, 223.42, 144.91*
Crystal Lake (Far NW Chicago)	KS9W	145.51, 223.42, 430.05*
Elgin (NW Chicago)	K9EC	145.51, 223.46, 144.91* , 430.05*
Godfrey	K9SD	144.91, 440.??
Lockport (SW Chicago)	N9AOL	144.91*, 223.42
Prairie View (N Chicago)	KS9U	223.44, 430.05*
Rockford	KA9LTR	144.91*, 147.435, 223.40*
Sterling	N4RR	145.51, 223.40*, 440.??
Springfield (planned)	WS9V	

* node-node link frequency only (no user access)
——> *Also see MDXPA*

Indiana

Fort Wayne	KR9U	144.910 (User)
Fort Wayne	KR9U	445.125 (User)
Fort Wayne	KR9U-1	144.910 (BPQ Switch)
Fort Wayne	KR9U-1	223.540 (Link)
Greenhill	WB9QPG-1	144.910 (Pure Network Node)
Hammond	KE9I	147.435, 223.420
Indianapolis	KJ9D	145.530 (User)
Indianapilis	KJ9D-1	224.440 (Link)
Indianapolis	KJ9D-1	145.530 (BPQ Switch)
Muncie	WB9DFD	145.530 (User)
Muncie	WB9DFD-1	145.690 (User)
Muncie	WB9DFD-1	223.540 (Link)
Valparaiso	WX9X	223.420
West Lafayette	K9FN	144.910 (User)
West Lafayette	K9FN-1	144.910 (BPQ Switch)
West Lafayette	K9FN-1	223.440 (Link)
Westfield	W9ZRX-2	223.440 (Pure Network Node)
Westfield	W9ZRX-2	433.500 (19.2 kB East)

——> *Also see MDXPA*

Iowa

Algona	WD8PKF	144.910
Bettendorf	G3WJN	144.910
Ely	WB8ZRL	144.910 223.400 (Cedar Rapids)
Mitchelville	NCØP	144.910 147.510 (Des Moines)
Sioux City	AA0DQ	144.910

Kansas

Leavenworth	NØFFH	144.930
Olathe	WBØSW	144.950 (Kansas City, KS, MO and suburbs)

Kentucky

Bowling Green	WB4FLB	144.930
Georgetown	N4TY	144.910
Louisville	N0KFO	145.590
Paducah	AE4AP	147.495 (Linked to KB4HU Nashville, TN)
Taulor Hill	N4XSQ	145.770

Louisiana

Abbeville	WU3V	144.970, 446.200 #*
Abbeville	WU3V	7.089, 14.101 #*
Baton Rouge	WB5TDD	144.950
Lafayette	N5JUW	145.680
Metairie	N5UXT-8	145.010 #*
New Orleans	K5NV	144.970 #
Shreveprot	KF5XV	145.050, 144.930

= High Speed backbone port 9.6k or 19.2k
* = Connects ot Internet Gulfnet or Texnet

Maine

Windham	K1OT	144.930 (KANODE to NK1K)

Maryland

Frederick	K3NA	145.630 (HBG220 or HBURG, then Elkton)
Glenwood	W3LPL	145.590, 441.250
Glenwood	W3LPL	144.950 (Digi to DEFRC KANODE/FRC)
Lothian	KE3Q	145.570, 440.950

——>Also see PVDXSN

Massachusetts

Bolton	AK1A	145.570
Groton	NO1A	144.990, 145.650
Harvard	K1EA	144.950
Peru	AA1AS	145.690
Roslindale	WA1G	145.690
Spencer	W1BIM	145.030, 145.090

——> Also see YCCC

Michigan

Ada	WJ8R	144.910
Adrian	K8AQM	144.970
Clarkston	W8WD	145.710
Grand Blanc	KB8S	145.710
Jackson	KD8B	147.540
Kalamazoo/Battle Creek	AB8Y	144.950
Port Huron	K8DD	144.970
St. Joseph	N8LXQ	147.435
St. Joseph	K9RHY	Freq Unknown
Toledo	W8HHF	144.970
Troy	K8NA	145.950
Ypsilanti	K8CC	144.910

Minnesota

Bemidji	WA0PUJ	(PCBJI)	144.910, 145.010
Duluth	N0BIL	(PCDLH)	144.910, 145.070
Minneapolis	NJ0M	(PCMSP)	144.950
Plymouth	N0AT	(PCPLY)	144.970
Rochester	W0MXW	(PCRST)	144.970
St. Cloud	K0IR	(PCSTC)	144.910
St. Paul	K0RC	(PCSTP)	144.930
Virginia	WB0QXM	(PCVIR)	144.970

Mississippi

Gulfport	WN5IJZ	144.950 #
Hattiesburg	W5VSZ	144.950 #*
Laurel	KB5IXI-1	144.950 Digipeater to W5VSZ

High Speed backbone port 9k6 or 19k2
* Connects to Internet Gulfnet or Texnet

Missouri

Godfrey, Il	K9SD	Freq unknown (St. Louis MO. area)

Montana

No known stations

Nebraska

Fremont	AJØI	144.910
Hoskins	WBØYWO	144.910 (Norfolk)
Omaha	KNØL	144.910

New Hampshire

Exeter	NK1K	144.930
Hollis	K1GQ	144.970
Rindge	K1XX	145.710

———> *Also see YCCC*

New Jersey

Moorestown	WB2YOF	144.950
Parsippany	W2JT	144.930

———> *Also see FRC*

New Mexico

Corrales	KN5D	Freq unknown
Santa Fe	W7LHO	Freq unknow

New York

Altamont	K2TR	145.690
Buffalo	KD2YP	144.930
Cottekill	K5NA	145.670
E. Fishkill	K2EK	145.750
Hamburg	KB2YJ-3	144.930
Huntington	W2HAP-2	144.930 (Digi to KD1F)
Pleasantville	KE2AY	144.950
Port Jefferson Sta.	KD1F	144.930
Rochester	NG2P	144.910
Utica	WA2TVE	145.050
Vista	KC2QF	144.990
Woodside	K2GX	145.690

———> *Also see LIDXA & YCCC*

Nevada

Reno	KI3V	144.950, 51.700

———> *Also see CA listings*

North Carolina

Asheville	N4AA		144.950
Charlotte	WD4R		144.910
Elizabethy City	WA4VTX		144.970
Hickory	N4AZ-5	DXHKY	144.930 (link to N4ZC)
Newport	K4DHZ		144.930
Raleigh	NOØT-2	SWAN	144.930 (Link to W4DW)
Raleigh	W4DW		147.570
Stanley	N4ZC		144.930
Wilmington	N4WI		144.950
Winston-Salem	W4NC-2	DXINT	145.630 (1200 Link Node to DXBVDS)
Winston-Salem	KC4LWI-3	DXNC	145.750 (2400 Link From DXRKE TO
DXSAL)			
Young Mt	KØSD-5	DXYNG	144.930 (Link to N4ZC)

North Dakota
No known stations

Ohio

Akron	WØCG	145.590*	222.060**
Chesterland	K8AZ	144.930*** 222.060**	
Cincinnati	K4ZLE	145.670* 221.110**	
Columbus	N8HTT	145.090* 223.420* 445.500**	
Columbus	KC8MK	144.950* 445.500** 447.000*	
Dayton	N8BJQ	145.030* 221.110**	
Findlay	N8ET	144.95	
Mentor	KQ8M	144.910*** 144.930*** 145.590* 222.030**	
Parma	K8MR	144.910*** 144.930*** 145.590* 222.060**	
Toledo	W8HHF	Freq Unknown	
Westerville	KC8MK	144.950* 145.590* 446.975* 445.500**	
Youngstown	N8IZR	144.910* 144.930**	

* = *Users Only;* ** = *Node Connects Only;* *** = *Node & User Connects*

Oklahoma

Claremore	N5OK	144.770
Oklahoma City	AD1S	144.950
Poteau	N5JKN	144.910

Oregon

Baker	NK7U	145.15 BKEDX Digi, See INWDXN
Veneta	KO7N	145.69 (Eugene area)
N. Portland	NØJO	145.71
E. Portland	WR7D	145.69
Salem	N7AVK	145.67 (Mid-Willamette Valley)
West Portland	WR7D-4	145.73 (Buckmaster database server)

——-> *Also see NWDXSN*

Pennsylvania

Camp Hill	NE3H	145.770 DX Packetcluster node
Erie	K3TUP	144.930
Mt Cobb	W3IQS	145.690
Perkasie	K3WW	144.950
Phoenixville	W3FRY	144.930
Pittsburgh	AD8J	144.930
Williamsport	KD3CR	145.010

——-> *Also see FRC & PVDXSN*

Puerto Rico

San Juan	WP4K	145.030

Rhode Island

Cranston	KC1CE	145.090
Johnston	NE1R	144.970

——-> *Also see YCCC*

South Carolina

Aiken	KN4UE	147.550
Columbia	K4AVU	144.970 PACKET CLUSTER PC-G8BPQ
Columbia	WX4M	145.610 PACKET CLUSTER PC-G8BPQ
Greenville	AC4TN	145.650 PACKET CLUSTER
Monnetta	W4MPY	145.790 PACKET CLUSTER
Charleston	KO4BR	145.610 PACKET CLUSTER
Orangeburg	KO4BR-15	145.610 (1200 BAUD USER PORT) DE-G8BPQ
Orangeburg	KO4BR-15	223.580 (9600 BAUD NETWORK NODE) DE-G8BPQ
Columbia	KO4BR-5	145.610 (1200 BAUD USER PORT) PC-DE-G8BPQ
Columbia	N4OBF-15	144.970 (2400 BAUD USER PORT) KISS NODE
Columbia	N4OBF-14	145.790 (1200 BAUD NETWORK NODE) DE-G8BPQ
Columbia	N4OBF-13	223.580 (9600 BAUD NETWORK NODE) DE-G8BPQ
Florence	N4JJ	147.585
Little Mtn.	K4RWN-5	145.790 (1200 BAUD NETWORK NODE) DE-G8BPQ
Little Mtn	K4RWN-5	223.580 (9600 BAUD NETWORK NODE) DE-G8BPQ
West Springs	KN4CW-5	223.580 (9600 BAUD NETWORK NODE) DE-G8BPQ
Spartenburg	KN4CW-1	223.580 (9600 BAUD NETWORK NODE) DE-G8BPQ
Sumter	KO4BR-10	223.580 (9600 BAUD NETWORK NODE) DE-G8BPQ
Greenville	N4JPN-5	223.580 (9600 BAUD NETWORK NODE) DE-G8BPQ
Greenville	N4JPN-5	145.790 (2400 BAUD NETWORK NODE) DE-G8BPQ
Six Mile Mtn.	WO4G-1	223.580 (9600 BAUD NETWORK NODE) DE-G8BPQ
Six Mile Mtn.	WO4G-1	145.750 (2400 BAUD NETWORK NODE) DE-G8BPQ
Sasafrass Mtn.	W4FX-2	223.580 (9600 BAUD NETWORK NODE) NETROM
Sasafrass Mtn.	W4FX-1	145.650 (1200 BAUD USER ACCESS) NETROM

NETWORK NODES ALL SHARE COMMON BACKBONE (9600 BAUD) PROVIDED BY THE PIONEER AMATEUR RADIO CLUB OF SOUTH CAROLINA.(P.A.R.C) SYSOP'S KO4BR@N4SZ

South Dakota

Sioux Falls	WØZWY	145.070, 145.010, 442.100

Tennessee

Athens	W4RJC-1	ATHDX	144.910
Chattanooga	WD4ALH		144.990
Cleveland	AG4M-2	WOM2	147.730 (Network node)
Cleveland	AG4M-4	WOM4	448.400 (Network node)
Concord	K4IUV		441.000
Elora	SMTNDX	145.730 (Network node)	
Jackson	W5HVV		144.850
Juliet	AA4DO		147.495
Kingsport	KA4IWG		145.750z

Knoxville	AG4M-3	ETDXA	147.585
Nashville	KB4HU		144.970
Nashville	KB4HU		145.585
Nashville	KB4HU		145.750
Memphis	KN4F		144.950, 446.050
Murfreesboro	KB4HU-2	NASDX	145.750 (Network node)
Murfreesboro	KB4HU-1	NASDX1	144.930 (Network node)
Oak Ridge	NS4W		144.910
Oak Ridge	NS4W		147.585
Ridgetop	K4UVH-4	NASDX4	145.750 (Network node)

Texas

Abilene	N5HRG	144.910
Amarillo	WA4NXI	145.010, 145.050
Austin	KG5ND	144.950 (AUSDXC)
Cibolo	WB5DDP	145.010, 223.580, 446.100 (CIBDXC)
Dickinson	AB5A	145.030 , 7089.0 (DKNDX)
Edinburg	K5TSQ	144.910
El Paso	WA5PIE	145.010 ELP ELP2
Farmersville	WA5QBX	145.790
Ft. Worth	WC5P	144.950
Haslet	K9MK	144.910 MK
Hempstead	KE5IV	144.910, 144.990 (HMPDXC)
Houston	K5DX	144.970 TDXS97
Irving	KS1G	145.730
Irving	W5AH	144.790
Kingsville	KA5SWC	145.070
La Marque	KC5SC	144.930, 144.970 (LMQDXC)
Lubbock	KA5EJX	145.010, 145.050
Midland	WF5E	144.910 (MAFDXC)
Rockwall	W5XJ	144.930
San Antonio	KA5IAU	144.970
Tyler	KD5GD	144.990

Utah

Salt Lake	K7CU	144.950

Vermont

Westford	KD1R	145.030

Virginia

Chesapeake	WY7C		145.010 DX PacketCluster node (SEVA)
Chesapeake	WY7C		145.070 DX PacketCluster node
Culpeper	AB4N-6	DXCULP	Planned 145.630 (Network node)
Dale City	N4WYH-2#	VADX1	221.010 (Network node link: DCA1 to DXWOOD)
Dale City	N4WYH-6	DXWOOD	145.630 (Network node to K3NA)
Elliston	KC4DY		Connects to N4SR & WB8CQV DX nodes
Elliston	KC4DY	POOR	145.090 DX PacketCluster Node
Elliston	KC4DY		145.650 DX PacketCluster Node
Elliston	KC4DY	DXPOOR	147.510 DX PacketCluster Node
Fair Port	WD4EBY-6	DXFAIR	Planned (Network node)
Great Falls	WØYVA		145.510 DX PacketCluster node
Hampton	WA4OHX		145.010 DX PacketCluster node (SEVA)
Hampton	WA4OHX		145.070 DX PacketCluster node
Lorton	N4SR		145.530 DX PacketCluster node
Richmond	WU4G		145.530 (Link port to N4SR)
Richmond	WU4G	CVCC	145.070 (DX PacketCluster node LAN)
Roanoke	N4MGQ	POOR	145.090 (Network Node to KC4DY)
Roanoke	KC4DY-1	DXSWVA	145.530 (Link Port to N4SR Back Bone Only)
Roanoke	KC4DY-3	DXPOOR	147.510 (Network Node to KC4DY)
Ringgold	KM4AU		144.910 , 145.610
Warrenton	WA6YOU-6	DXWARR	145.630 (Network node to K3NA)
Waynesboro	KC4DY-2	DXBRDN	145.530 (Link Node to N4SR Back Bone Only)

——> Also see PVDXSN

Virgin Islands
No known stations

Washington

N. Seattle	NV6Z	145.730 (Linked to N7FSW & WY7I)
S. Seattle	N7FSW	145.770
Marysville	WY7I	145.750

Tacoma	K7JF	145.710 (Linked to NV6Z)
Spokane	WS7I	144.930 , EWARG Digi, 144.930
——> Also see NWDXSN		
Beverly	NY7T	145.030 (BVEDX)
Boise, ID	AA7TF	145.070
Pasco	NY7T	145.030 (EWDXA)
Spokane	NQ7M	144.970
Spokane	WV7Y	144.930 DataBase Server
Spokane	WS7I	144.930 (EWARG), 145.01, 223.46, 10.018
——> Also see INWDXN		

West Virginia

Charleston	WB8CQV	145.090, 446.500

Wisconsin

Eau Claire	N9ISN	145.510 (1200b users)
Eau Claire	N9ISN	223.420 (backbone)
Fox Lake	FOX (KA-Node)	223.460
Freedom	AA9A	223.440 (Near Green Bay, WI)
Madison	MSNDX (KA-Node)	223.440
Madison	WI9L	223.440 (1200b/users)
Madison	WI9L	430.050 (backbone)
McFarland	WI9L	223.440
Milwaukee	NB9C	223.460 as backbone frequency)
North Freedom	WB9RNF	223.440 (1200b users)
North Freedom	WB9RNF	430.050 (Backbone)
Oshkosh	OSHDX (KA-Node)	144.970
Pepin	KB9S-5	223.420 (NetRom Node/backbone)
Plymouth	NF9R-2 (KA-Node)	223.460
Racine	KS9K	223.480
Sheboygan	KR9S	223.420
Stevens Point	WB9QFW	145.510 (1200b users)
Stevens Point	WB9QFW-1	223.440 (backbone)
Thorp	N9WKA-7	145.510 (KA-Node/users)
Wausau	KC9NW-2	145.510 (NetRom node/users)

Wyoming

Jackson	N7NG	145.010

Canada

Comox, BC	VE7GCE	7.101, 145.010, 145.710, 147.56
Fredricton, NB	VE1NH	145.010
Halifax, NS	VE1DXC	145.010
Middleport, ON	VE3CDX	145.710
Montreal, PQ	VE2PWI	145.010
Orillia, ON	VE3FJB	145.070
Ottawa, ON	VE3XDX	145.110
Prince George, BC	VE7AV	145.010
Toronto, ON	VE3TDX	144.970
Toronto, ON	VE3ZRB	145.770
Vancouver, BC	VE7CQD	145.710, 144.930 +600, 145.070 +600

Africa

Kinshasa	9Q5XO	145.775
Kinshasa	9Q5XO-7	28.103 (KAnode/Gateway to 9Q5XO)
Pretoria	ZS6AI-2	28.103 (BPQ Switch - Issue Command: C 2 9Q5XO-7, then X 9Q5XO to Connect to the 9Q5XO Node)
Pretoria	ZS6WGH	[DXPTA] 10.149 MHz LSB
Pretoria	ZS6WGH	[DXPTA] 21.109 MHz LSB
Pretoria	ZS6WGH	[DXPTA] 50.875 MHz
Pretoria	ZS6WGH	[DXPTA] 144.475 MHz (VHF BACKBONE)
Pretoria	ZS6WGH	[DXPTA] 144.625 MHz (JOHANNESBURG LAN)
Pretoria	ZS6WGH	[DXPTA] 144.675 MHz (PRETORIA LAN)
Pretoria	ZS6WGH	[DXPTA] 430.600 MHz (UHF BACKBONE)
Pretoria	ZS6WGH	[DXPTA] 438.025 MHz (ROSE BACKBONE)

Europe

Munchen	DBØBCC	Freq unknown
Gau Bischofsheim	DJØRX	Freq unknown
Cheltenham	GB7DXC	Freq unknown
Hanko	OH2RBG	Freq unknown
Skovde	SK6EI-2	144.625
	SK6EI-7	443.650
	SL6AL-1	145.350

Tielt	ON1CED-5	144.650, 433.650 (connect via ON1ABT only)
City unknown	SK3BG-6	Frequency Unknown
City unknown	SM3SJN-1	28.103 (KAnode/gateway to SK3BG-6)
Wokingham	GB7DXI	

Asia

Seoul, Korea	HL9OB	Freq unknown

Japan
Call Area 0, 1, & 2

QTH	Callsign	User Port Freq.
Abiko/Chiba	JR2BNF/1	431.16MHz
Amagi/Shizuoka	JG2ZCG	431.24/52.66MHz
Chichibu/Saitama	JK1ZMK	No User Port
Fujisawa/Kanagawa	JA1SYY	431.26MHz
Hamamatsu/Shizuoka	JF2ZOJ	1298.88MHz
Hanno/Saitama	JR1MAF	431.38/(HF)14.105MHz
Higashimatsuyama/Saitama	JA1NWD	438.38MHz
Kamakura/Kanagawa	JA1BRK	(HF)10.144MHz
Kamakura/Kanagawa	JA1UQP	431.40MHz
Kawaguchi/Saitama	JH1ROJ	431.44MHz
Kawasaki/Kanagawa	JA1SNF	No user port
Kita-ku/Tokyo	JE2ERH/1	431.36MHz
Koganei/Tokyo	JH1GTV	431.08/431.28/52.60MHz
Komoro/Nagano	JR0YEO	438.68MHz
Nagano/Nagano	JA0TBJ	431.36MHz
Nagaoka/Shizuoka	JH2TIP	431.38MHz
Odawara/Kanagawa	JK1GOK	431.14MHz
Omiya/Saitama	JK1FVP	438.44MHz
Oosato/Saitama	JH1XYR	431.22MHz
Oyama/Tochigi	JA1PEJ	431.26/431.44MHz
Setagaya/Tokyo	JA1SQD	438.42/21.123MHz
Setagaya/Tokyo	JF1MGI	Unknown
Shimada/Shizuoka	JE2UFF	144.62/431.04MHz
Shirone/Niigata	JA0YRH	438.68MHz
Takasaki/Gunma	JE1CAY	431.14MHz
Yokohama/Kanagawa	JA1AVV	Unknown
Yokohama/Kanagawa	JE1GMM	431.22MHz
Yokohama/Kanagawa	JR1MLU	431.30MHz

Japan-Call Area 3

Kusatsu	JA3YIH	43?.??? MHz
Kyoto	JA3ZDX	1293.240
Osaka	JJ3ZKD	1293.260
Shiga	JA3YIH	43?.???
Toyanaka	JJ3ZKD	1293.260

All Node Links are on 431.080 MHz

Japan-Call Area 7

Akita	JA7XRO	431.020 MHz
Iwate	JA7RHJ	431.260
Miyagi	JA7MYQ	438.980
Morioka	JA7RHJ	431.260
Sendai	JA7MYQ	438.980
Yokote	JA7XRO	431.020

No Node Links in JA7

FRC-(Frankford Radio Club)

Nodecall	QRG	QTH
W3FRY	144.930	Phoenixville, PA
WB2YOF	144.950	Moorestown, NJ
K3WW	145.530	Perkasie, PA
K2TW	145.570	Flemmington, NJ
KD3CN	?.??	Strasburg, PA

In addition, we have six feeder NETROMs/TheNets:

Nodecall	QRG	QTH	Type
DEFRC (N3FDL-1)	144.930	Middletown, DE	NETROM
NAR (KA3PIT-1)	144.930	Narvon, PA	TheNet
RDG (K3NW-1)	144.930	Reading, PA	TheNet
LANCO (N3BNA-1)	144.930	Lititz, PA	NETROM
FRCNET (K3RL-1)	145.540	Coopersburg, PA	TheNet
NNJ (K2SG-1)	145.540	White Plains, NJ	TheNet

GCDXC (Gulf Coast DX Cluster)

Nodecall	QRG	QTH
WU3V	144.970	Abbeville, LA
N5JUW	145.680	Lafayette, LA
WB5TDD	144.950	Baton Rouge, LA
N5UXT-8	145.010	Metairie, LA
K5NV	144.970	New Orleans, LA
KC5SC	144.930	LaMarque, TX
K5WA	144.950	Houston, TX
KE5IV	144.910	Hempstead, TX
W5VSZ	144.950	Hattiesburg, MS
WN5IJZ	144.950	Gulfport, MS
WA4OSR	144.970	Mobile, AL
WD4HDT	144.970	Ft. Walton Bch, FL

INWDXN (Inland NorthWest DX Network)

Nodecall	QRG	QTH
NK7U (BKEDX)	145.150	Baker, OR
NY7T (BEVDX) Digi to NY7T	145.150	Beverly, WA
NK7U (BOIDX) Digi to NK7U	145.150	Boise, ID
NY7T (EWADX)	145.030	Pasco, WA
WV7Y DataBase Server	144.930	Spokane, WA
WS7I (EWARG) Note 1	144.930	Spokane, WA

NOTE 1 - WS7I Linked to VE3CDX, N7NG, KE7X, K7CU on a frequent basis.

GLPS (Great Lakes PacketCluster System)

Nodecall	QRG	QTH
AB8Y	144.950	Kalamazoo/Battle Creek, MI
K8AQM	144.970	Adrian, MI
K8CC	144.910	Ypsilanti, MI
K8DD	144.970	Port Huron, MI
K8NLD	144.950	Fraser, MI
KB8S	145.710	Grand Blanc, MI
KD8B	147.540	Jackson, MI
N8LXQ	147.435	St. Joseph, MI
W8HHF	144.970	Toledo, OH
W8WD	145.710	Clarkston, MI
WI8L	145.690	E. Detroit, MI
WJ8R	144.910	Ada, MI

LIDXA (Long Island DX Association)

Nodecall	QRG	QTH
KD1F	144.930	Port Jefferson Station, NY
K2GX	145.690	Woodside, NY
KC2QF	144.990	Vista, NY

MDXPA (Midwest DX Packet Association)

The MDXPA network covers Northern Illinois, Eastern Wisconsin, Northwestern Indiana, and Eastern Iowa.

Nodecall	QRG *	QTH
K9AJ	145.510 MHz	Crete, IL (Far S Chicago)
N9AOL	223.420 MHz	Lockport, IL (SW Chicago)
K9EC	145.510, 223.460 MHz	Elgin, IL (NW Chicago)
KS9U	223.460 MHz	Prairie View, IL (N Chicago)
KS9W	145.510, 223.420 MHz	Crystal Lake, IL (Far NW Chicago)
N9QX	145.510, 223.420 MHz	Chicago, IL (Southside)
N4RR	145.510 MHz	Sterling, IL
KA9LTR	147.435 MHz	Rockford, IL
KS9K	223.480 MHz	Racine, WI
NB9C	223.460 MHz	Milwaukee, WI
WI9L	223.440 MHZ	McFarland, WI
AA9A	223.440 MHz	Freedom, WI
KR9S	223.420 MHz	Sheboygan, WI
KE9I	147.435 MHz	Hammond, IN
G3WJN	144.910 MHz	Bettendorf, IA
WB8ZRL	144.910, 223.400 MHz	Ely, IA

* primary user frequencies-see state listing for other frequencies

NCC (North Coast Contesters)

Nodecall	QRG	QTH
AD8J	144.930	Pittsburgh, PA
K3TUP	144.930	Erie, PA
K8AZ	144.930	Chesterland, OH
KQ8M	145.590	Mentor, OH

N8IZR	144.930	Youngstown, OH
W0CG	145.590	Akron/Canton, OH
VE3CDX	147.480	Middleport, ON
VE3CDX	145.670	Middleport, ON

NCN (Northeast Cluster Network)

Nodecall	QRG	QTH
KB1H	144.930	Dayville, CT
AK1A	145.570	Bolton, MA
K1GQ	144.970	Hollis, NH
AA1AS***	145.690	Peru, MA
W1RM	144.950	Burlington, CT
NO1A	144.990	Groton, MA
NO1A	145.650	Groton, MA
NG2P	144.910	Rochester, NY
K2TR**	144.970	Altamont, NY
KD2YP	144.930	Tonawanda, NY
KC8PE	145.710	Cheshire, CT
K2EK	145.750	E. Fishkill, NY
KD3CR	145.010	Williamsport, PA
K1EA	144.950	Harvard, MA
K5NA	145.670	Cottekill, NY
NK1K	144.930	Exeter, NH
W1BIM	145.030	Spencer, MA
W1BIM	145.090	Spencer, MA
KD1R	145.030	Westford, VT
K1XX	145.710	Rindge, NH
WA1G	145.690	Roslindale, MA
WA2TVE*	145.050	Utica, NY
W3IQS	145.690	Mt. Cobb, PA
KK4L	144.910	Johnson City, NY

*WA2TVE also on 147.42 for local connects. Connect to UTICA2 then DXCLUS, UTICA2 is LAN for Utica. On NEDA network just connect to 'DXCLUS' to get WA2TVE. 'UTICA' is local NEDA node in that area of WA2TVE.
** 'KNOX' is local neda node for K2TR. If you connect to KNOX then you can connect to DXKNOX, then to K2TR.
*** AA1AS reachable from node 'WMA220'.
+ KK4L also on 145.63, 147.435, 443.0125(+)

NCXDA (National Capital DX Association)
——> See PVDXSN

NFIDXA (Niagara Frontier International DX Assoc.)

Nodecall	QRG	QTH
K3TUP	144.930	Erie, Pa.
KD2YP	144.930	Buffalo, Ny.
KE6LT	144.930	Boulder, Co.
KN4F	144.950	Memphis Tn.
KQ8M	144.950	Mentor, Oh.
N0BIL	144.910	Duluth, Mn.
N4WW	144.950	Apopka, Fla.
NG2P	144.970	Rochester, Ny.
VE2PWI	145.010	Montreal, Pq.
VE3CDX	145.710	Middleport, On.
VE3FJB	145.070	Orillia, On.
VE3TDX	144.970	Toronto, On.
VE3XDX	144.930	Ottawa, On.
VE3ZRB	145.770	Toronto, On.
WA0PUJ	145.010	Bemidji, Mn.
K3ARV	144.950	Melbourne, Fl
KD2YP	144.930	Tonawanda, NY
VE1DXC	145.010	Halifax, NS

NJDXA (North Jersey DX Association)

Nodecall	QRG	QTH
W2JT	144.930	Parsippany

NWDXSN (Northwest DX Spotting Network)

Nodecall	QRG	QTH
N7FSW	145.770	Burien, WA
NV6Z	145.730	Kenmore, WA (N. Seattle)
WY7I	145.750	Marysville, WA
K7JF	145.710	Tacoma, WA
NK7U	145.150	Baker, OR
KO7N	145.690	Veneta, OR (Eugene area)

NØJO	145.710	N. Portland, OR	
WR7D	145.690	E. Portland, OR	
N7AVK	145.670	Salem, OR (Mid-Willamette Valley)	
WR7D-4	145.730	W. Portland, OR	
VE7CQD	145.710	Vancouver, BC	

PVDXSN (Potomac Valley DX Spotting Network)

Nodecall	QRG	QTH	NOTES
K3NA	145.630	Frederick, MD	HBURG or HBG220, then Elkton
W3LPL	144.950	Glenwood, MD	Digi to DEFRC KANODE/FRC
W3LPL	145.590	Glenwood, MD	
W3LPL	441.250	Glenwood, MD	
KE3Q	145.570	Lothian, MD	
KE3Q	440.950	Lothian, MD	
NE3H	145.770	Camp Hill, PA	
WY7C	145.010	Chesapeake, VA	SEVA
WY7C	145.070	Chesapeake, VA	
AB4N-6	145.630	Clupeper, VA	
N4WYH-2	221.010	Dale City, VA	Node link to DCA1 to DXWOOD
N4WYH-6	145.630	Dale City, VA	Node to K3NA
WØYVA	145.510	Great Falls, VA	
WA4OHX	145.010	Hampton, VA	SEVA
WA4OHX	145.070	Hampton, VA	DX PacketCluster Node
N4SR	145.530	Lorton, VA	
WU4G	145.530	Richmond, VA	Link port ot N4SR
WU4G	145.070	Richmond, VA	DX PacketCluster Node
N4LSP-3	145.530	Sumerduck, VA	KANODE linking N4SR & N4EHJ
WA6YOU-6	145.630	Warrenton, VA	Network Node to K3NA

PVRC (Potomac Valley Radio Club)
——> See PVDXSN

TSCN (Tristate Cluster Network)

Nodecall	QRG	QTH
K2GX	145.690	Queens, NY
K2SG	145.530	Morris plains, NJ
KD1R	144.930	Pt. Jefferson Station, NY
KE2AY	144.950	Pleasantville, NY
KE2CG	144.990	Piscataway, NJ
N1DVS	144.990	Vista, NY
W2JT	144.930	Parsippany, NJ
WB2KXA	145.590	Allentown, NJ
K2RW	File server	
WA2UEC	Tri-state routing	

YCCC (Yankee Clipper Contest Club) (Page 1 of 2)

Nodecall	QRG	QTH
K2TR	144.970	Altamont, NY
AK1A	144.950	Bolton, MA
W1RM	145.950	Burlington, Ct
K5NA	145.670	Cottekill, NY
KB1H	144.930	Dayville, CT
NK1K	144.930	Exeter, NH
NO1A	144.990	Groton, MA
NO1A	145.650	Groton, MA
K1EA	144.950	Harvard, MA
K1GQ	144.970	Hollis, NH
NE1R	144.970	North Scituate, RI
W2JT	144.930	Parsippany, NJ
AA1AS	145.690	Peru, MA
KE2AY	145.950	Pleasantville, NY
KD1F	144.930	Port Jefferson Station, NY
K2GX	145.690	Queens (NYC), NY
K1XX	145.710	Ringe, NH
WA1G	145.690	Roslindale, MA
W1BIM	145.030/090	Spencer, MA
N1DVS	144.990	Vista, NY
YCCCWF	147.430	West Falmouth, MA See Note 1.
KD1R	145.030	Westford, VT

Note 1: KA-NODE - After connecting, use command "XC NE1R"

SOURCES OF COMMERCIAL SOFTWARE OF SPECIAL INTEREST TO ACTIVE AMATEUR RADIO OPERATORS

This list is maintained by VE3SUN. The latest version can be obtained by ftp from ftp.netcom.com:/pub/VE3SUN/software.ham. Please send any updates or changes to peterj@netcom.com. Information contained in this list may be reproduced without limitation.

Chris Smolenski N3JLY
40 South Lake Way
Reisterstown MD 21136
MacMININEC Macintosh antenna modeling software.

Austin Antenna
10 Main St
Gonic NH 03839
Phone: 603 335–6339
dB Comp antenna gain calculation for verticals.

VALTEK Software
P O Box 1266
Smiths AL 36877
Phone: 205 480–9494
DX Lumberjack logging and award tracking software.

Paul O'Kane
36 Cookhill
Sandyford
Dublin 18
Ireland
EI5DI Super–Duper Contest Logger.

Mike Cook AF9Y
501 E Cedar Canyon Rd
Huntertown IN 46782
Phone: 219 637–3399
FFTDSP weak signal correlation analysis software.

Tucson Amateur Packet Radio
8987–309 E Tanque Verde Rd #337
Tucson, AZ 85749–9399
Phone: 817 393–0000
FAX: 817 566–2544
Packet radio related software, hardware and firmware.
Buckmaster
Route 4, Box 1630
Mineral, VA 23117
Phone: 800 282–5628
or: 703 894–5777
FAX: 703 894–9141
HamCall CD–Rom: 642,000 US listings, 350,000 International listings. Electronics Software Compendium CD–Rom.

Scientific Solutions, Inc
736 Cedar Creek Way
Woodstock, GA 30188
DXbase logging and award tracking software.

WJ2O Software
P O Box 16
McConnellsville, NY 13401
WJ2O Master QSO Contest logging and award tracking software.

Sensible Solutions
P O Box 474
Middletown NJ 07748
Phone: 800 538–0001
or: 908 495–5066
BBS: 908 787–2982
WB2OPA LogMaster logging and award tracking software.

Octavia Company
P O Box 40
352700 Maikop
Russia
Russian Callsign Database.

Brinson Microware Corporation
114 SE 4th St
Mooreland OK 73852
Phone: 800 874–0771
Ham Companion propagation prediction and database software.

EQF Software
396 Sautter Drive
Coraopolis PA 15108
Phone: 800 995–1605
or: 619 685–7291
FAX: 619 558–7850
N3EQF Logging and radio controls programs.
Debco Electronics Inc
4025 Edwards Rd
Cincinnati OH 45209
Phone: 800 423–4499
FAX: 513 531–4455
DX Desktop logging and rig control programs.

Personal Database Applications
2616 Meadow Ridge Dr
Duluth GA 30136
Phone: 404 242–0887
FAX: 404 449–6687
LOGic 4 logging and award tracking software for Windows and DOS. PDA QSL Route List. 30,000 QSL Managers.

PDK Inc
46 Oak St
Dunstable MA 01827
Phone: 508 649–4360
NX1P Ham View logging and rig control software for Windows.

Electronics Enterprises
6851 2nd St
Rio Linda CA 95673
Phone: 916 991–7263
FAX: 916 991–1000
BBS: 916 992–0923
SH/GO and GOQSL QSL manager lists for Pavilion Packetcluster systems.

Pavillion Software
5 Mount Royal Avenue
Marlborough MA 01752
Packetcluster multi–user, multi–node DX alerting software. Cluster Companion terminal program. G3WGV TurboLog logging and awards tracking software.

Kurt Andress NI6W
2538 S Center St
Santa Ana CA 92704
Phone: 714 957–3371
Yagi Stress antenna wind load modeling software.

AEA Advanced Electronic Applications
2006 196th St SW
P O Box 2160
Lynnwood WA 98036
Phone: 206 774–5554
FAX: 206 775–7373
Pakratt II packet radio software

AMSAT Radio Amateur Satellite Corp
P O Box 27
Washington DC 20044
Phone: 301 589–6062
InstantTrack satellite tracking software.
Telemetry decoding software.

AMSOFT
P O Box 666
New Cumberland PA 17070
Phone: 717 938–8249
AmSoft CD–ROM. Callsign database. 7,000 shareware programs. 1,000 radio mods.

Hardy Data Systems
P O Box 7304
Tifton GA 31793
Phone: 912 387–7373
10–10 award tracking software for county hunting and contesting.

Namlulu Communications
1120 Meadowview Rd
Willard OH 44890
Phone: 419 935–0270
Inexpensive contest logging and award tracking software. QTH location database.
LUCAS Radio/Kangaroo Tabor Software
2900 Valmont Rd, Suite H
Boulder CO 80301
IONCAP and CAPMAN propagation programs.

Larry Kebel
P O Box 2010F
Sparks NV 89432
KBØZP Contest Log real time contest logging software.

Jacques d'Avignon VE3VIA
965 Lincoln Drive
Kingston ON Canada K7M 4Z3
Phone: 613 634–1519
ASAPS propagation forecasting software.

Bob Brown NM7M
504 Channel View Dr
Anacortes WA 98221
Solar Max contesting/propagation game.

Softline
P O Box 29
SF 20101 Turku
Finland
OH1AA logging and awards tracking software.

Gote Lofstedt
Sim–Data Linko
Djurdsgatan 71, 6tr,
S 582 29 Linkoping
Sweden
CW QSO simulator.

Aerospace Consulting
P O Box 536
Buckingham PA 18912
Phone: 215 345–7184
FAX: 215 345–1309
Logwrite general purpose logging software.

Ashton ITC
P O Box 830
Dandridge TN 37725
Phone: 615 397–0742
FAX: 615 397–0466
Aries–2 logging and terminal program.

Base 2 Systems
2534 Nebraska St
Saginaw MI 48601
Phone: 517 777–5613
MufMap II propagation prediction program.

California Software
HamWindows logging and rig control software for Windows.

Data Communications International
7678 Venetian St
Miramar FL 33023
Phone: 305 987–9505
FAX: 305 987–4026
Datacom IV radio control software with automated logging.

Epsilon Co.
Box 715
Trumbull CT 06611
Phone: 203 261–7694
Long Wire Pro antenna modeling program.

Gemradio
17 Coborn St
Guelph ON Canada N1G 2MA
Phone: 519 821–2257
Real–time logging program.

J–Com
793 Canning Pkwy
Victor NY 14564
Phone: 800 446–2295
or: 716 924–0422
FAX: 716 924–4555
HamBase callsign database with grid squares, bearing and distance. DXQSL database of QSL managers.

K1EA Software
5 Mount Royal Ave.
Marlborough, MA 01752–1935
Orders: 508–779–5054
Support: 508–460–8873
BBS: 508–460–8877
FAX: 508–460–6211
CT Contest Logging software.

K6STI, Brian Beezley
507 1/2 Taylor
Vista CA 92084
Phone: 619 945–9824
MN, YO, NEC antenna modeling and optimization software.

Kantronics
1202 E 23rd St
Lawrence KS 66046
Phone: 913 842–7745
FAX: 913 842–2021
Host Master II+ multimode packet terminal program. SuperFax II weather fax reception software.

MFJ
P O Box 494
Mississippi State, MS 39792
Phone: 800 647–8324
or: 601 323–5869
FAX: 800 647–8324
or: 601 323–6551
Packet radio software. Grayline DX Advantage clock and terminator display. Easy DX logging and packet cluster software with awards tracking.

MacTrak Software
P O Box 1590
Port Orchard WA 98366
Phone: 206 871–1700
Satellite tracking program for Macintosh.

Northern Lights Software
Star Route, Box 60
Canton NY 13617
RealTrak satellite tracking and moonbounce software.

Milestone Technologies
3551 S Monaco Pkwy, Suite 323
Denver CO 80237
Phone: 303 752–3382
Logmaster dBASE III–compatible logging software.

PAYL
P O Box 926
Levittown PA 19058
Phone: 215 945–4404
DXLog logging and awards tracking program.

GRF Computing Services
6170 Downey Ave
Long Beach CA 90805
Phone: 310 531–4852
N6RJ Second Op logging and award tracking program.

Skymoon
9102 Kings Drive
Manvel TX 77578
Phone: 713 331–4200
EME moon tracking program.
Jack L Schultz, W2GGE
Contest logging software.

System One Control
FDLog Macintosh software for Field Day logging.

W6EL Software
11058 Queensland St
Los Angeles CA 90034
Miniprop propagation prediction software.

Unified Microsystems
P O Box 133
Slinger WI 53086
W9ST PSQSL QSL label generating software.

RT Systems
P O Box 8
Laceys Spring AL 35754
SAM US callsign database.

Viking Business Systems
10310 Main St, Ste 106
Fairfax VA 22030
W3YY LogPro integrated logging, packet and award tracking
software.

Austin Code Works
11100 Leafwood Lane
Austin TX 78750
Phone: 512 258–0785
FAX: 512 258–1342
WriteLog contest logging software for Windows with C source.

Bill Mullin AA4M/6
3042 Larkin Place
San Diego CA 92123
QQSL Quick QSL label program.

Swisslog
10 Robbins Ave
Amityville NY 11701
Phone: 516 598–0011
Swisslog logging software.

Xantek, Inc
P O Box 834
Madison Square Station
New York NY 10159
DX–Edge PC and Commodore propagation and grayline termina-
tor software.

ZCO
QLog Macintosh logging program. Sun Clock Macintosh gray line
program. DX Helper Macintosh propagation program. Satellite Pro
Macintosh satellite tracking software.

WB2DND Software
250 Standish St
Duxbury MA 02332
Phone: 617 934–7158
Logging software for PC compatibles.

WB5M
133 Light Falls Drive
Wake Forest NC 27587
FlexPac packetcluster terminal software.

Walnut Creek CD–Rom
1547 Palos Verdes Mall, #260
Walnut Creek CA 94596
Phone: 800 786–9907
QRZ! Ham Radio CD–Rom. US and Canadian callsign database.
Thousands of shareware programs. Usenet news files.
Ivanhoe Software Inc
944 Cedars Road
Lewisberry PA 17339
Phone: 717 766–6361
CLUSTER DX Packetcluster terminal software.

Roy Lewallen W7EL
P O Box 6658
Beaverton OR 97007
ELNEC antenna analysis software.

Wyvern Technology Inc
35 Colvintown Rd
Coventry RI 02816
Phone: 401 823–RTTY
WF1B RTTY contest software.

Skywave Technologies
17 Pine Knoll Rd
Lexington MA 02173
W1FM IONSOUND ionospheric propagation prediction program.
RAI Enterprises
4508 N 48th Dr
Phoenix AZ 85031
Quickyagi yagi design and optimization software. Autolog logging
and database software.

Geoclock
P O Box 5112
Arlington VA 22205
Grayline and great circle mapping software.

LTA
P O Box 77
New Bedford PA 16140
K8CC NA contest logging software.

Larry Tyree
15125 SE Bartel Rd
Boring OR 97009
N6TR contest logging program.

TOP 25 HAM–POPULATED CITIES IN THE UNITED STATES

This list was compiled by using a tally by ZIP code, and using some commercial software to map each ZIP code to a city and create a total for each city. Of the roughly 36,000 ham–populated ZIP codes in the United States, some 1,800 ZIP codes could not be translated into cities. There are several reasons for this. With the license term now being 10 years, some ZIP codes have simply been merged into other ZIP codes, since the Postal Service eliminates ZIP codes from time to time. Another source of errors is that licenses are added in ZIP codes that the commercial ZIP–to–city software database may not yet recognize. And, of course, a large number of errors arise from inaccurately–completed Form 610 license applications or data–entry errors at the FCC. Incidentally, these last two error sources make it virtually impossible to get an accurate count by merely sorting the FCC database by looking up city names. In any case, the numbers listed below should be quite close to the actual count, and the relative rankings are unlikely to shift as a result of these error sources.

Rank	City	State	Count								
1.	San Diego	CA	3910	9.	Tucson	AZ	2288	18.	Staten Island	NY	1627
2.	Houston	TX	3666	10.	Cincinnati	OH	2100	19.	Long Beach	CA	1593
3.	Los Angeles	CA	3628	11.	Portland	OR	2047	20.	Fort Worth	TX	1565
4.	Miami	FL	3493	12.	Chicago	IL	1970	21.	Albuquerque	NM	1530
5.	Phoenix	AZ	2818	13.	Dallas	TX	1969	22.	Jacksonville	FL	1488
6.	Seattle	WA	2790	14.	Indianapolis	IN	1773	23.	Colorado Springs	CO	1469
7.	San Jose	CA	2757	15.	Austin	TX	1710	24.	Las Vegas	NV	1420
8.	San Antonio	TX	2391	16.	Sacramento	CA	1687	25.	New York	NY	1406
				17.	Brooklyn	NY	1635				

TOP TWENTY-FIVE HAM–POPULATED U.S. CITIES AND TOWNS BY STATE

Rank	City	Count
Alabama		
1	Birmingham:	1156
2	Huntsville:	1085
3	Mobile:	539
4	Montgomery:	389
5	Madison:	212
6	Decatur:	197
7	Gadsden:	196
8	Tuscaloosa:	193
9	Dothan:	193
10	Florence:	190
11	Bessemer:	140
12	Anniston:	136
13	Athens:	134
14	Selma:	128
15	Auburn:	92
16	Jasper:	87
17	Hueytown:	86
18	Trussville:	80
19	Northport:	75
20	Enterprise:	71
21	Prattville:	70
22	Opelika:	70
23	Talladega:	67
24	Albertville:	66
25	Pinson:	61
Alaska		
1	Anchorage:	881
2	Fairbanks:	533
3	Juneau	200
4	Wasilla	114
5	North Pole	110
6	Eagle River:	101
7	Palmer:	78
8	Bethel:	73
9	Soldotna:	72
10	Kodiak:	71
11	Chugiak:	57
12	Ketchikan:	54
13	Sitka:	48
14	Kenai:	42
15	Petersburg:	31
16	Delta Junction:	29
17	Homer:	26
18	Auke Bay:	24
19	Elmendorf Afb:	22
20	Valdez:	21
21	Trapper Creek:	21
22	Wrangell:	20
23	Douglas:	20
24	Willow:	19
25	Seward:	18
Arizona		
1	Phoenix:	2818
2	Tucson:	2288
3	Mesa:	1253
4	Scottsdale:	782
5	Glendale:	708
6	Tempe:	559
7	Chandler:	340
8	Prescott:	328
9	Yuma:	327
10	Sierra Vista:	308
11	Sun City:	282
12	Flagstaff:	244
13	Lake Havasu City:	224
14	Kingman:	211
15	Peoria:	196
16	Apache Junction:	180
17	Sun City West:	178
18	Payson:	175
19	Green Valley:	145
20	Gilbert:	143
21	Prescott Valley:	120
22	Cottonwood:	102
23	Sedona:	82
24	Bullhead City:	71
25	Safford:	67
Arkansas		
1	Little Rock:	533
2	Fort Smith:	219
3	Hot Springs:	206
4	Paragould:	200
5	North Little Rock:	190
6	Jonesboro:	183
7	Fayetteville:	171
8	Russellville:	148
9	Benton:	145
10	Mountain Home:	137

11	Conway:	136	4	Bridgeport:	190
12	Rogers:	133	5	Milford:	182
13	Harrison:	130	6	Norwalk:	172
14	Springdale:	119	7	Newington:	168
15	Malvern:	106	8	Meriden:	168
16	Pine Bluff:	102	9	West Hartford:	165
17	Jacksonville:	100	10	Wallingford:	136
18	Searcy:	81	11	West Haven:	131
19	Sherwood:	74	12	Groton:	131
20	Batesville:	72	13	Danbury:	130
21	West Memphis:	69	14	Stratford:	129
22	El Dorado:	69	15	Hamden:	127
23	Van Buren:	63	16	Waterbury:	126
24	Arkadelphia:	59	17	Fairfield:	125
25	Cabot:	58	18	Trumbull:	122
			19	Enfield:	116
California			20	Southington:	114
1	San Diego:	3910	21	Manchester:	114
2	Los Angeles:	3628	22	New Britain:	110
3	San Jose:	2757	23	Torrington:	109
4	Sacramento:	1687	24	East Hartford:	106
5	Long Beach:	1593	25	Shelton:	100
6	Huntington Beach:	1319			
7	Riverside:	1174	**Delaware**		
8	San Francisco:	1127	1	Wilmington:	453
9	Anaheim:	1025	2	Newark:	227
10	Fresno:	871	3	Dover:	111
11	Sunnyvale:	858	4	New Castle:	71
12	Torrance:	795	5	Seaford:	62
13	Santa Ana:	766	6	Hockessin:	44
14	Fremont:	703	7	Milford:	36
15	Costa Mesa:	685	8	Bear:	36
16	Santa Barbara:	684	9	Claymont:	33
17	Bakersfield:	676	10	Lewes:	31
18	Garden Grove:	665	11	Millsboro:	23
19	Irvine:	653	12	Georgetown:	22
20	Santa Rosa:	605	13	Felton:	22
21	Pasadena:	601	14	Middletown:	20
22	Orange:	589	15	Laurel:	19
23	Glendale:	578	16	Rehoboth Beach:	17
24	Whittier:	576	17	Selbyville:	15
25	Oakland:	569	18	Hartly:	15
			19	Smyrna:	14
Colorado			20	Delmar:	14
1	Colorado Springs:	1469	21	Milton:	13
2	Denver:	949	22	Harrington:	12
3	Aurora:	752	23	Ocean View:	11
4	Littleton:	642	24	Dagsboro:	11
5	Boulder:	619	25	Camden:	11
6	Fort Collins:	480			
7	Lakewood:	381	**Florida**		
8	Longmont:	356	1	Miami:	3493
9	Pueblo:	327	2	Jacksonville:	1488
10	Arvada:	315	3	Orlando:	1316
11	Loveland:	305	4	Tampa:	1130
12	Grand Junction:	268	5	Pensacola:	888
13	Englewood:	253	6	Saint Petersburg:	875
14	Golden:	211	7	Fort Lauderdale:	857
15	Westminster:	161	8	Hialeah:	768
16	Greeley:	153	9	Sarasota:	684
17	Broomfield:	137	10	Boca Raton:	611
18	Thornton:	130	11	Clearwater:	573
19	Durango:	128	12	Melbourne:	548
20	Parker:	116	13	Bradenton:	502
21	Canon City:	114	14	West Palm Beach:	488
22	Woodland Park:	112	15	Tallahassee:	488
23	Sterling:	101	16	Ocala:	461
24	Montrose:	97	17	Naples:	419
25	Evergreen:	90	18	Largo:	415
			19	Lakeland:	412
Connecticut			20	Hollywood:	402
1	Bethel:	454	21	Gainesville:	398
2	Stamford:	208	22	Titusville:	380
3	Bristol:	190	23	Palm Bay:	374

24	Fort Myers:	374	17		Blackfoot:	41
25	Port Charlotte:	326	18		Orofino:	32
			19		Shelley:	29
Georgia			20		Rathdrum:	27
1	Atlanta:	930	21		Priest River:	27
2	Marietta:	789	22		Hayden Lake:	27
3	Savannah:	347	23		Chubbuck:	27
4	Lawrenceville:	335	24		Eagle:	26
5	Macon:	313	25		Burley:	26
6	Decatur:	305				
7	Stone Mountain:	287	**Illinois**			
8	Warner Robins:	241	1		Chicago:	1970
9	Roswell:	240	2		Rockford:	502
10	Augusta:	238	3		Naperville:	363
11	Lilburn:	228	4		Springfield:	332
12	Norcross:	220	5		Peoria:	327
13	Columbus:	203	6		Arlington Heights:	288
14	Albany:	192	7		Schaumburg:	251
15	Gainesville:	177	8		Aurora:	251
16	Douglasville:	169	9		Decatur:	245
17	Smyrna:	162	10		Belleville:	230
18	Athens:	161	11		Wheaton:	225
19	Acworth:	157	12		Skokie:	211
20	Conyers:	153	13		Des Plaines:	210
21	Kennesaw:	151	14		Elgin:	206
22	Duluth:	147	15		Champaign:	203
23	Alpharetta:	147	16		Downers Grove:	202
24	Cumming:	139	17		Bloomington:	202
25	Newnan:	128	18		Quincy:	197
			19		Palatine:	196
Hawaii			20		Saint Charles:	164
1	Honolulu:	1157	21		Evanston:	156
2	Hilo:	200	22		Lombard:	152
3	Kaneohe:	143	23		Joliet:	150
4	Kailua:	137	24		Elmhurst:	146
5	Pearl City:	123	25		Zion:	143
6	Aiea:	106				
7	Kailua Kona:	98	**Indiana**			
8	Wahiawa:	95	1		Indianapolis:	1773
9	Ewa Beach:	83	2		Fort Wayne:	958
10	Waianae:	67	3		Evansville:	406
11	Waipahu:	64	4		Kokomo:	405
12	Mililani:	63	5		South Bend:	371
13	Kahului:	50	6		Bloomington:	281
14	Keaau:	46	7		Terre Haute:	266
15	Wailuku:	41	8		Anderson:	255
16	Kaunakakai:	40	9		Muncie:	244
17	Lihue:	35	10		Elkhart:	238
18	Kapaa:	34	11		Lafayette:	211
19	Kalaheo:	32	12		Hammond:	165
20	Haleiwa:	31	13		Valparaiso:	158
21	Pahoa:	30	14		Carmel:	155
22	Kihei:	28	15		Richmond:	154
23	Laie:	25	16		Mishawaka:	146
24	Lahaina:	25	17		Columbus:	141
25	Kula:	25	18		West Lafayette:	140
			19		Marion:	140
Idaho			20		Huntington:	137
1	Boise:	762	21		Greenwood:	132
2	Idaho Falls:	324	22		La Porte:	130
3	Nampa:	202	23		Goshen:	121
4	Pocatello:	164	24		Greenfield:	116
5	Coeur D Alene:	156	25		Martinsville:	113
6	Meridian:	106				
7	Moscow:	104	**Iowa**			
8	Caldwell:	89	1		Cedar Rapids:	564
9	Twin Falls:	83	2		Des Moines:	452
10	Lewiston:	79	3		Davenport:	265
11	Post Falls:	77	4		Burlington:	240
12	Rexburg:	61	5		Sioux City:	237
13	Sandpoint:	52	6		Marion:	185
14	Rigby:	51	7		Ames:	154
15	Mountain Home:	49	8		Council Bluffs:	146
16	Emmett:	44	9		Dubuque:	139

10	Waterloo:	133
11	Iowa City:	124
12	Clinton:	118
13	Ottumwa:	106
14	Marshalltown:	100
15	Muscatine:	95
16	Cedar Falls:	95
17	Bettendorf:	87
18	Mason City:	73
19	Ankeny:	69
20	Fort Dodge:	65
21	West Des Moines:	63
22	Spencer:	60
23	Fort Madison:	57
24	Fairfield:	57
25	Boone:	56

Kansas

1	Wichita:	959
2	Topeka:	495
3	Overland Park:	367
4	Kansas City:	276
5	Olathe:	220
6	Salina:	200
7	Lawrence:	192
8	Leavenworth:	160
9	Manhattan:	157
10	Hutchinson:	126
11	Shawnee:	108
12	Hesston:	104
13	Winfield:	90
14	Prairie Village:	89
15	Lenexa:	88
16	Derby:	87
17	Newton:	84
18	Great Bend:	84
19	Shawnee Mission:	77
20	Emporia:	74
21	Leawood:	71
22	Coffeyville:	71
23	Pittsburg:	70
24	Junction City:	67
25	Independence:	66

Kentucky

1	Louisville:	1358
2	Lexington:	716
3	Bowling Gree:	168
4	Ashland:	166
5	Owensboro:	163
6	Paducah:	158
7	Bardstown:	142
8	Frankfort:	117
9	Pikeville:	114
10	Henderson:	113
11	Winchester:	108
12	Nicholasville:	90
13	Murray:	89
14	Glasgow:	89
15	London:	88
16	Benton:	85
17	Elizabethtown:	83
18	Hopkinsville:	81
19	Somerset:	80
20	Richmond:	76
21	Madisonville:	76
22	Florence:	76
23	Radcliff:	74
24	Covington:	69
25	Versailles:	63

Louisiana

1	Baton Rouge:	838
2	Shreveport:	520
3	New Orleans:	494
4	Metairie:	388
5	Slidell:	298
6	Lafayette:	229
7	Monroe:	153
8	Lake Charles:	149
9	Kenner:	142
10	Alexandria:	133
11	West Monroe:	114
12	Bossier City:	114
13	Houma:	110
14	Denham Springs:	110
15	New Iberia:	94
16	Gretna:	91
17	Pineville:	85
18	Ruston:	78
19	Chalmette:	74
20	Covington:	71
21	Springhill:	58
22	Baker:	56
23	Minden:	53
24	Hammond:	53
25	Marrero:	52

Maine

1	Bangor:	142
2	Portland:	112
3	Yarmouth:	97
4	Augusta:	92
5	Brunswick:	73
6	Lewiston:	68
7	Auburn:	68
8	South Portland:	64
9	Westbrook:	53
10	Skowhegan:	52
11	Presque Isle:	52
12	Bath:	52
13	Waterville:	47
14	Scarborough:	46
15	Kittery:	43
16	Gardiner:	43
17	Ellsworth:	43
18	Brewer:	42
19	Sanford:	41
20	Topsham:	40
21	Kennebunk:	39
22	Biddeford:	38
23	Saco:	37
24	Bridgton:	37
25	Rockland:	36

Maryland

1	Baltimore:	1221
2	Silver Spring:	627
3	Annapolis:	392
4	Rockville:	390
5	Gaithersburg:	328
6	Columbia:	320
7	Bethesda:	293
8	Laurel:	263
9	Glen Burnie:	235
10	Frederick:	221
11	Bowie:	213
12	Ellicott City:	208
13	Pasadena:	192
14	Hagerstown:	177
15	Potomac:	136
16	Westminster:	135
17	Severna Park:	127
18	Waldorf:	119
19	Salisbury:	118
20	Germantown:	115
21	Cumberland:	109
22	Mount Airy:	99

23	Arnold:	96
24	Bel Air:	95
25	Fort Washington:	94

Massachusetts

1	Worcester:	260
2	Cambridge:	239
3	Framingham:	238
4	Springfield:	190
5	Pittsfield:	179
6	New Bedford:	172
7	Lexington:	172
8	Chelmsford:	169
9	Boston:	168
10	Fall River:	143
11	Andover:	141
12	Newton:	139
13	Lynn:	138
14	Acton:	137
15	Eastham:	136
16	Plymouth:	135
17	Arlington:	134
18	Westford:	132
19	Peabody:	128
20	Lowell:	126
21	Waltham	125
22	Chicopee:	125
23	Brockton:	125
24	Sudbury:	122
25	North Eastham:	117

Michigan

1	Grand Rapids:	553
2	Detroit:	486
3	Kalamazoo:	387
4	Flint:	387
5	Ann Arbor:	348
6	Livonia:	329
7	Lansing:	308
8	Battle Creek:	284
9	Warren:	267
10	Jackson:	266
11	Saginaw:	256
12	Traverse City:	241
13	Sterling Heights:	241
14	Midland:	237
15	Muskegon:	234
16	Dearborn:	207
17	Troy:	189
18	Wyoming:	180
19	Holland:	176
20	Royal Oak:	174
21	Southfield:	167
22	Westland:	158
23	Ypsilanti:	157
24	Farmington Hills:	157
25	Mount Clemens::	153

Minnesota

1	Minneapolis:	803
2	Saint Paul:	689
3	Rochester:	409
4	Duluth:	366
5	Bloomington:	286
6	Minnetonka:	157
7	Burnsville:	145
8	Coon Rapids:	143
9	Plymouth:	142
10	Apple Valley:	129
11	Brooklyn Park:	122
12	Eagan:	119
13	Winona:	111
14	Eden Prairie:	110
15	White Bear Lake:	109
16	Saint Cloud:	105
17	Edina:	104
18	Richfield:	102
19	Mankato:	102
20	Bemidji:	96
21	Moorhead:	95
22	Roseville:	91
23	Saint Louis Park:	88
24	Maple Grove:	86
25	Blaine:	83

Mississippi

1	Jackson:	295
2	Gulfport:	199
3	Biloxi:	198
4	Hattiesburg:	171
5	Vicksburg:	165
6	Meridian:	124
7	Ocean Springs:	113
8	Starkville:	100
9	Laurel:	89
10	Pascagoula:	85
11	Brandon:	83
12	Corinth:	82
13	Tupelo:	80
15	Bay Saint Louis:	78
15	Clinton:	69
16	Lucedale:	67
17	Columbus:	64
18	Long Beach:	60
19	Booneville:	57
20	Natchez:	56
21	Picayune:	49
22	Southaven:	47
23	Amory:	47
24	Gautier:	45
25	Greenville:	44

Missouri

1	Saint Louis:	1182
2	Kansas City:	916
3	Springfield:	520
4	Independence:	418
5	Saint Charles:	321
6	Florissant:	263
7	Joplin:	217
8	Columbia:	210
9	Lees Summit:	205
10	Saint Joseph:	190
11	Jefferson City:	176
12	Ballwin:	151
13	Saint Peters:	135
14	Raytown:	134
15	Chesterfield:	124
16	Blue Springs:	115
17	Hannibal:	110
18	Rolla:	108
19	Fenton:	101
20	Liberty:	100
21	Poplar Bluff:	94
22	Gladstone:	90
23	Cape Girardeau:	88
24	Kirkwood:	78
25	Sedalia:	68

Montana

1	Billings:	266
2	Bozeman:	261
3	Great Falls:	233
4	Missoula:	212
5	Helena:	162
6	Butte:	132
7	Kalispell:	109
8	Havre:	58

9	Livingston:	54
10	Miles City:	53
11	Hamilton:	53
12	Glendive:	48
13	Libby:	42
14	Anaconda:	32
15	Colstrip:	31
16	Whitefish:	30
17	Belgrade:	30
18	Wolf Point:	29
19	Forsyth:	29
20	Stevensville:	27
21	Lewistown:	27
22	Sidney:	26
23	Columbia Falls:	26
24	Polson:	24
25	East Helena:	23

Nebraska

1	Omaha:	927
2	Lincoln:	701
3	North Platte:	129
4	Grand Island:	118
5	Fremont:	118
6	Bellevue:	104
7	Norfolk:	96
8	Kearney:	77
9	Hastings:	57
10	Columbus:	50
11	Papillion:	49
12	Plattsmouth:	41
13	Chadron:	41
14	Scottsbluff:	38
15	Mc Cook:	37
16	Sidney:	34
17	Blair:	31
18	Nebraska City:	26
19	South Sioux City:	25
20	Schuyler:	25
21	Ogallala:	25
22	Gering:	25
23	Beatrice:	24
24	Crete:	22
25	Albion:	22

Nevada

1	Las Vegas:	1420
2	Reno:	679
3	Sparks:	269
4	Carson City:	254
5	Henderson:	223
6	Pahrump:	137
7	Boulder City:	121
8	Gardnerville:	85
9	North Las Vegas:	79
10	Minden:	65
11	Elko:	59
12	Incline Village:	46
13	Fallon:	43
14	Dayton:	36
15	Fernley:	34
16	Winnemucca:	28
17	Yerington:	26
18	Tonopah:	26
19	Sun Valley:	25
20	Silver Springs:	22
21	Laughlin:	22
22	Stateline:	21
23	Zephyr Cove:	16
24	Ely:	13
25	Crystal Bay:	13

New Hampshire

1	Nashua:	427
2	Manchester:	323
3	Derry:	183
4	Merrimack:	129
5	Concord:	127
6	Londonderry:	118
7	Hudson:	112
8	Salem:	103
9	Rochester:	92
10	Bedford:	83
11	Amherst:	83
12	Dover:	77
13	Portsmouth:	76
14	Laconia:	65
15	Windham:	64
16	Keene:	64
17	Milford:	56
18	Claremont:	55
19	Hollis:	52
20	Newport:	44
21	Gilford:	44
22	Hampton:	43
23	Exeter:	40
24	Bow:	40
25	Pelham:	38

New Jersey

1	Toms River:	312
2	Cherry Hill:	285
3	Trenton:	275
4	Clifton:	184
5	Edison:	168
6	Jersey City:	161
7	Wayne:	154
8	Newark:	148
9	Lakewood:	145
10	Elizabeth:	120
11	Bridgewater:	114
12	Vineland:	113
13	Union:	113
14	Freehold:	113
15	Piscataway:	112
16	Brick:	106
17	East Brunswick:	105
18	Millville:	104
19	Westfield:	99
20	Wyckoff:	98
21	Perth Amboy:	95
22	Moorestown:	94
23	Mount Laurel:	93
24	Princeton:	92
25	Pennsauken:	89

New Mexico

1	Albuquerque:	1530
2	Las Cruces:	434
3	Los Alamos:	342
4	Santa Fe:	214
5	Alamogordo:	184
6	Rio Rancho:	170
7	Roswell:	146
8	Silver City:	128
9	Deming:	128
10	Farmington:	113
11	Clovis:	109
12	Hobbs:	99
13	Socorro:	94
14	Carlsbad:	63
15	Los Lunas:	57
16	Tijeras:	56
17	Sandia Park:	40
18	Belen:	36
19	Edgewood:	33
20	Grants:	32
21	Corrales:	31
22	Artesia:	28

#	City		#	City	
23	Gallup:	27	16	Bowman	15
24	Lovington:	25	17	Mayville:	14
25	Ruidoso:	24	18	Grafton:	14
			19	Harwood:	13
New York			20	Cavalier:	11
1	Brooklyn:	1635	21	Center:	10
2	Staten Island:	1627	22	Binford:	10
3	New York:	1406	23	Walhalla:	9
4	Rochester:	1257	24	New Salem:	9
5	Bronx:	746	25	Lisbon:	9
6	Syracuse:	421			
7	Schenectady:	411	**Ohio**		
8	Buffalo:	406	1	Cincinnati:	2100
9	Flushing:	369	2	Columbus:	1276
10	Apo N Y:	278	3	Dayton:	1034
11	Poughkeepsie:	271	4	Toledo:	769
12	Yonkers:	270	5	Akron:	659
13	Albany:	266	6	Cleveland:	571
14	Binghamton:	264	7	Hamilton:	445
15	Ithaca:	191	8	Canton:	375
16	Tonawanda:	184	9	Youngstown:	322
17	Liverpool:	179	10	Mansfield:	309
18	Fairport:	179	11	Springfield:	269
19	Utica:	178	12	Middletown:	256
20	Rome:	178	13	Kettering:	242
21	Huntington:	174	14	Parma:	237
22	Lockport:	169	15	Elyria:	229
23	Webster:	160	16	Mentor:	224
24	Jamaica:	151	17	Newark:	218
25	Pittsford:	145	18	Warren:	207
			19	Westerville:	200
North Carolina			20	Lima:	196
1	Raleigh:	1078	21	North Canton:	180
2	Charlotte:	953	22	Lorain:	177
3	Greensboro:	604	23	Findlay:	173
4	Winston Salem:	528	24	Centerville:	173
5	Fayetteville:	498	25	Cuyahoga Falls:	166
6	Durham:	369			
7	Asheville:	363	**Oklahoma**		
8	Cary:	320	1	Tulsa:	1191
9	Wilmington:	305	2	Oklahoma City:	1085
10	Greenville:	241	3	Norman:	352
11	Hendersonville:	216	4	Broken Arrow:	281
12	High Point:	215	5	Edmond:	278
13	Hickory:	210	6	Lawton:	276
14	Gastonia:	204	7	Bartlesville:	220
15	Chapel Hill:	201	8	Midwest City:	177
16	Shelby	187	9	Enid:	155
17	New Bern:	184	10	Moore:	145
18	Lexington:	163	11	Stillwater:	118
19	Fort Bragg	161	12	Ponca City:	114
20	Salisbury:	153	13	Bethany:	110
21	Matthews:	135	14	Altus:	108
22	Concord:	135	15	Yukon:	100
23	Jacksonville:	131	16	Owasso:	99
24	Goldsboro:	131	17	Shawnee	97
25	Waxhaw:	130	18	Duncan:	97
			19	Ardmore:	94
North Dakota			20	Muskogee:	89
1	Fargo:	253	21	Claremore:	85
2	Bismarck:	225	22	Sapulpa:	82
3	Minot:	125	23	Sand Springs:	80
4	Grand Forks:	125	24	Mc Alester:	76
5	Dickinson:	100	25	Durant:	75
6	Jamestown:	58			
7	Williston:	38	**Oregon**		
8	Mandan:	35	1	Portland:	2047
9	Devils Lake:	35	2	Eugene:	602
10	West Fargo:	26	3	Salem:	588
11	Wahpeton:	20	4	Klamath Falls:	392
12	Grand Forks Afb:	18	5	Beaverton:	384
13	Valley City:	17	6	Grants Pass:	349
14	Minot Afb:	17	7	Medford:	304
15	Maddock:	15	8	Corvallis:	283

9	Hillsboro:	265
10	Bend:	259
11	Gresham:	226
12	Aloha:	223
13	Roseburg:	219
14	Albany:	189
15	Milwaukie:	187
16	Springfield:	178
17	Tigard:	151
18	Oregon City:	150
19	Lake Oswego:	148
20	Florence:	127
21	Pendleton:	113
22	Astoria:	109
23	Coos Bay:	108
24	Lebanon:	107
25	Ashland:	101

Pennsylvania

1	Philadelphia:	1387
2	Pittsburgh:	1308
3	Erie:	445
4	York:	341
5	Allentown:	302
6	Lancaster:	295
7	Harrisburg:	267
8	Bethlehem:	251
9	West Chester:	224
10	Reading:	220
11	Butler:	179
12	State College:	177
13	Johnstown:	168
14	Levittown:	164
15	Norristown:	151
16	Lansdale:	151
17	Williamsport:	140
18	Warminster:	137
19	Easton:	135
20	New Castle:	134
21	Mechanicsburg:	132
22	Hanover:	130
23	Scranton:	128
24	Altoona:	128
25	Aliquippa:	120

Rhode Island

1	Warwick:	271
2	Providence:	173
3	Cranston:	167
4	Coventry:	124
5	Pawtucket:	118
6	Portsmouth:	113
7	East Providence:	86
8	Cumberland:	85
9	Woonsocket:	82
10	Newport:	82
11	Johnston:	81
12	West Warwick:	72
13	Tiverton:	68
14	North Kingstown:	63
15	Middletown:	62
16	Bristol:	61
17	Westerly:	60
18	East Greenwich:	56
19	Barrington:	54
20	North Providence:	52
21	Lincoln:	42
22	Wakefield:	41
23	Narragansett:	30
24	Foster:	29
25	Warren:	28

South Carolina

1	Columbia:	430
2	Greenville:	365
3	Charleston:	234
4	Spartanburg:	206
5	Anderson:	191
6	Sumter:	184
7	Rock Hill:	173
8	Summerville:	153
9	Aiken:	130
10	Lexington:	125
11	Easley:	116
12	West Columbia:	110
13	Florence:	107
14	Myrtle Beach:	106
15	Simpsonville:	105
16	Taylors:	103
17	Greenwood:	101
18	Greer:	93
19	Fort Mill:	89
20	Lancaster:	87
21	Seneca:	84
22	Goose Creek:	82
23	North Augusta:	77
24	Mount Pleasant:	73
25	North Charleston:	65

South Dakota

1	Rapid City:	198
2	Sioux Falls:	192
3	Aberdeen:	86
4	Watertown:	79
5	Hot Springs:	70
6	Brookings:	63
7	Pierre:	59
8	Huron:	44
9	Yankton:	41
10	Vermillion:	40
11	Mitchell:	27
12	Canton:	24
13	Spearfish:	22
14	Sturgis:	17
15	Redfield:	15
16	Custer:	14
17	Black Hawk:	14
18	Milbank:	13
19	Madison:	13
20	Brandon:	13
21	Mobridge:	12
22	Lead:	11
23	Harrisburg:	11
24	Freeman:	11
25	Wilmot:	10

Tennessee

1	Memphis:	1325
2	Knoxville:	1231
3	Nashville:	691
4	Chattanooga:	488
5	Cleveland:	327
6	Kingsport:	325
7	Oak Ridge:	307
8	Johnson City:	252
9	Maryville:	225
10	Jackson:	161
11	Bartlett:	159
12	Murfreesboro:	156
13	Clarksville:	155
14	Morristown:	150
15	Hixson:	150
16	Cookeville:	140
17	Germantown:	139
18	Hendersonville:	126
19	Bristol:	126
20	Greeneville:	124
21	Elizabethton:	124

22	Clinton:	104
23	Franklin:	102
24	Lenoir City:	99
25	La Follette:	98

Texas

1	Houston:	3666
2	San Antonio:	2391
3	Dallas:	1969
4	Austin:	1710
5	Fort Worth:	1565
6	El Paso:	945
7	Arlington:	901
8	Plano:	728
9	Garland:	693
10	Corpus Christi:	546
11	Lubbock:	514
12	Richardson:	500
13	Amarillo:	365
14	Abilene:	357
15	Irving:	348
16	Waco:	340
17	Midland:	332
18	Spring:	326
19	Wichita Falls:	321
20	Tyler:	308
21	Mesquite:	289
22	Carrollton:	281
23	San Angelo:	278
24	Denton:	247
25	Beaumont:	240

Utah

1	Salt Lake City:	1124
2	Orem:	517
3	Provo:	419
4	Sandy:	329
5	Ogden:	271
6	Logan:	193
7	West Jordan:	160
8	Bountiful:	154
9	Layton:	153
10	Saint George:	130
11	Cedar City:	116
12	West Valley City:	106
13	Spanish Fork:	105
14	Kaysville:	101
15	Payson:	96
16	Murray:	96
17	American Fork:	95
18	Roosevelt:	86
19	Price:	82
20	Roy:	81
21	Springville:	75
22	Pleasant Grove:	74
23	Kearns:	71
24	Salt Lake:	66
25	Brigham City:	63

Vermont

1	Essex Junction:	124
2	Burlington:	94
3	South Burlington:	73
4	Bennington:	67
5	Rutland:	65
6	Montpelier:	63
7	Colchester:	55
8	Barre:	48
9	Milton:	45
10	Shelburne:	41
11	Williston:	39
12	Springfield:	36
13	Underhill:	33
14	Jericho:	32
15	Brattleboro:	32
16	Saint Albans:	30
17	Bellows Falls:	27
18	Putney:	26
19	Middlebury:	26
20	Windsor:	25
21	Randolph:	25
22	White River Junct.:	24
23	Vergennes:	23
24	Swanton:	23
25	Northfield:	22

Virginia

1	Richmond:	947
2	Virginia Beach:	888
3	Alexandria:	658
4	Roanoke:	465
5	Arlington:	463
6	Norfolk:	393
7	Chesapeake:	389
8	Fairfax:	365
9	Newport News:	362
10	Lynchburg:	357
11	Woodbridge:	341
12	Springfield:	328
13	Vienna:	306
14	Manassas:	296
15	Hampton:	292
16	Falls Church:	269
17	Charlottesville:	215
18	Herndon:	201
19	Annandale:	193
20	Reston:	192
21	Sterling:	185
22	Mc Lean:	182
23	Portsmouth:	180
24	Fredericksburg:	180
25	Winchester:	174

Washington

1	Seattle:	2790
2	Spokane:	1120
3	Tacoma:	1091
4	Vancouver:	772
5	Everett:	627
6	Bellevue:	626
7	Olympia:	532
8	Renton:	479
9	Kent:	463
10	Bremerton:	453
11	Redmond:	391
12	Bothell:	379
13	Yakima:	351
14	Puyallup:	331
15	Bellingham:	330
16	Kirkland:	329
17	Auburn:	323
18	Port Orchard:	293
19	Gig Harbor:	288
20	Lynnwood:	261
21	Federal Way:	246
22	Richland:	236
23	Issaquah:	236
24	Sedro Woolley:	233
25	Edmonds:	233

West Virginia

1	Huntington:	277
2	Charleston:	247
3	Morgantown:	214
4	Parkersburg:	168
5	Princeton:	156
6	Wheeling:	144
7	Saint Albans:	126

8	Bluefield:	122
9	Clarksburg:	114
10	Fairmont:	113
11	Beckley:	107
12	South Charleston:	96
13	Buckhannon:	94
14	Martinsburg:	91
15	New Martinsville:	89
16	Hurricane:	78
17	Weston:	68
18	Moundsville:	67
19	Bridgeport:	58
20	Weirton:	55
21	Barboursville:	50
22	Oak Hill:	47
23	Hinton:	46
24	Elkins:	46
25	Vienna:	42

Wisconsin

1	Milwaukee:	824
2	Madison:	574
3	Green Bay:	308
4	Racine:	300
5	Waukesha:	213
6	Eau Claire:	205
7	Appleton:	197
8	Kenosha:	170
9	Wausau:	154
10	Chippewa Falls:	147
11	Oshkosh:	146
12	Brookfield:	137
13	West Allis:	123
14	Sheboygan:	114
15	Superior:	110
16	Janesville:	110

17	Stevens Point:	107
18	Fond Du Lac:	107
19	Wauwatosa:	105
20	Wisconsin Rapids:	104
21	New Berlin:	100
22	Rhinelander:	95
23	Manitowoc:	95
24	La Crosse:	88
25	Beloit:	86

Wyoming

1	Cheyenne:	307
2	Casper:	211
3	Laramie:	138
4	Sheridan:	88
5	Gillette:	72
6	Rock Springs:	61
7	Green River:	54
8	Torrington:	53
9	Riverton:	49
10	Rawlins:	44
11	Evanston:	43
12	Cody:	42
13	Lander:	30
14	Worland:	25
15	Jackson:	24
16	Wheatland:	19
17	Powell:	19
18	Kemmerer:	18
19	Douglas:	17
20	Mills:	12
21	Afton:	12
22	Thermopolis:	11
23	Buffalo:	10
24	Sundance:	8
25	Lusk:	8

U.S. HAM CENSUS BY STATE/CLASS
(FCC Records as of 9/1/95)

State	Extra	Advanced	General	Novice	Tech Plus	Technician	Grand Total
AK	327	544	638	438	542	682	3171
AL	1114	1729	1785	1010	2130	2561	10329
AR	726	1046	1053	613	1234	1771	6443
AZ	1488	2583	2685	1238	2815	3751	14560
CA	8679	16260	16642	16881	22408	26127	106997
CO	1219	2131	2158	1311	2241	2377	11437
CT	1111	1587	1941	1668	1768	1355	9430
DC	84	95	133	80	61	74	527
DE	200	236	289	209	302	245	1481
FL	4305	7997	9360	6788	7824	6766	43040
GA	1540	2579	2626	1495	3005	3015	14260
HI	312	519	566	689	679	609	3374
IA	732	1442	1525	1065	1192	1154	7110
ID	333	596	733	409	695	923	3689
IL	2589	4238	4845	3548	4860	4575	24655
IN	1512	2441	2782	2052	3357	3218	15362
KS	745	1172	1605	1032	1359	1683	7596
KY	852	1219	1462	1198	1684	2089	8504
LA	845	1371	1421	915	1410	1543	7505
MA	2064	2813	3355	2267	3251	2404	16154
MD	1509	2308	2271	1484	2149	2296	12017
ME	498	728	1043	558	771	806	4404
MI	2200	3686	4321	2617	4284	4319	21427
MN	1166	2015	2341	1298	2004	2011	10835
MO	1384	2295	2644	1535	2333	2497	12688
MS	490	831	852	538	821	1087	4619

MT	307	461	589	367	456	656	2836
NC	1809	2901	3075	1980	3302	3978	17045
ND	162	248	378	246	312	347	**1693**
NE	395	788	977	485	771	671	4087
NH	643	751	985	580	998	904	4861
NJ	2191	3242	3544	2613	3686	2696	17972
NM	601	942	903	379	834	1310	4969
NV	405	694	824	367	710	991	3991
NY	3831	5918	6854	6958	7166	6534	37261
OH	3125	5044	5637	4101	7442	6465	31814
OK	938	1533	1490	1123	1842	2160	9086
OR	1223	2194	2729	1607	2354	2542	12649
PA	3065	4569	5192	3613	5149	4289	25877
RI	331	377	536	416	587	367	2614
SC	681	1102	1351	688	1275	1262	6359
SD	177	321	379	173	252	274	1576
TN	1495	2373	2298	1462	3013	3040	13681
TX	4672	7583	7791	4371	8267	8823	41507
UT	475	820	758	752	1607	2648	7060
VA	2072	3109	3091	2016	3149	3312	16749
VT	259	332	434	230	399	494	2148
WA	2358	3811	4529	3090	4790	5364	23942
WI	1161	1866	2202	1278	1875	2175	10557
WV	575	748	953	854	1202	1719	6051
WY	178	243	282	220	291	390	1604
Total	71153	116431	128857	92905	136908	143349	689603

NUMBER OF U.S. AMATEUR LICENSES BY YEAR
(excluding RACES licenses, as of 1 September in each year)

Year	Number	Year	Number
1995	700,031	1988	438,688
1994	676,289	1987	431,776
1993	628,142	1986	423,401
1992	580,291	1985	414,012
1991	531,860	1984	412,695
1990	493,848	1983	412,870
1989	465,349		

SOURCE: FCC, GETTYSBURG

Note: While it is true that the number of amateurs has been growing in recent years, some of the statistics above reflect the change in January 1984 from 5-year license terms to 10-year license terms. Thus, licenses that would normally have expired and not been renewed after a five-year term have stayed in effect for an additional five years. The first 10-year licenses began expiring in January 1994.

MOST RECENT CALLSIGNS

(++:All call signs in this group have been issued in this area)

District	GroupA Extra	GroupB Advanced	GroupC Tech/Gen	GroupD Novice
July 1,1994				
Ø	AAØRE	KGØNM	++	KBØNHL
1	AA1JH	KD1VH	N1SCZ	KB1BIG
2	AA2SM	KF2VW	N2ZDJ	KB2QZR
3	AA3HY	KE3NH	N3SFO	KB3BCL
4	AD4SV	KR4TT	++	KE4MXM
5	AB5UN	KJ5YG	++	KC5HFC
6	AC6CT	KO6DF	++	KE6IFY
7	AB7CQ	KI7ZD	++	KC7DBL
8	AA8OZ	KG8JG	++	KB8TAT
9	AA9KY	KF9WC	N9XGH	KB9IYT
N.Mariana	KHØD	AHØAS	KHØCS	WHØAAY
Guam	WH2E	AH2CU	KH2JW	WH2ANK
Midway	++	AH4AB	KH4AG	WH4AAH
Hawaii	++	AH6NI	WH6UZ	WH6CRG
Amer.Samoa	AH8J	AH8AG	KH8BF	WH8ABB
Alaska	++	AL7PQ	WL7ST	WL7CHQ
Virgin	WP2N	KP2CC	NP2HM	WP2AHU
PuertoRico	++	KP4WW	++	WP4MPP

August 1,1994

Ø	AAØRW	KGØOO	++	KBØOKY
1	AA1JV	KD1VZ	N1SUN	KB1BJW
2	AA2TB	KF2WQ	N2ZZY	KB2RFN
3	AA3IF	KE3NT	N3TBF	KB3BEH
4	AD4UV	KR4XT	++	KE4PKY
5	AB5VT	KJ5ZS	++	KC5IVA
6	AC6DS	KO6FG	++	KE6KYY
7	AB7DK	KJ7BA	++	KC7EUY
8	AA8PM	KG8KK	++	KB8UIY
9	AA9LL	KF9WU	N9YCQ	KB9JAN
N.Mariana	KHØI	AHØAU	KHØDL	WHØABA
Guam	WH2F	AH2CU	KH2KA	WH2ANK
Midway	++	AH4AA	KH4AG	WH4AAH
Hawaii	++	AH6NN	WH6WD	WH6CRH
Amer.Samoa	AH8K	AH8AG	KH8BG	WH8ABB
Alaska	++	AL7PS	WL7WJ	WL7CHS
Virgin	WP2O	KP2CD	NP2HQ	WP2AHU
PuertoRico	++	KP4XE	++	WP4MRJ

September 1,1994

Ø	AAØSS	KGØPI	++	KBØOTI
1	AA1KL	KD1WR	N1SZJ	KB1BKN
2	AA2TP	KF2XI	++	KB2RPC
3	AA3IO	KE3ON	N3TFQ	KB3BEO
4	AD4WN	KS4BI	++	KE4QGF
5	AB5WP	KK5BL	++	KC5JHZ
6	AC6EP	KO6HD	++	KE6LWT
7	AB7EI	KJ7CL	++	KC7FHZ
8	AA8QD	KG8LA	++	KB8UST
9	AA9LY	KF9XH	N9YKN	KB9JAW
N.Mariana	KHØK	++	KHØDM	++
Guam	WH2G	AH2CW	KH2KP	++
Midway	++	AH4AA	KH4AG	WH4AAH
Hawaii	++	AH6NN	WH6XO	WH6CRI
Amer.Samoa	AH8K	AH8AG	KH8BH	WH8ABB
Alaska	++	AL7PT	WL7WV	++
VirginIs	WP2O	KP2CD	NP2HQ	WP2AHU
PuertoRico	++	KP4XJ	++	WP4MSJ

October 1,1994

Ø	AAØTH	KGØQC	++	KBØPDL
1	AA1KV	KD1XA	N1TDL	KB1BKV
2	AA2UF	KF2XT	++	KB2RWI
3	AA3IR	KE3OY	N3TLT	KB3BEZ
4	AD4XP	KS4EO	++	KE4RJB
5	AB5XS	KK5CJ	++	KC5JYC
6	AC6FU	KO6IW	++	KE6MUJ
7	AB7FA	KJ7DP	++	KC7FXN
8	AA8QP	KG8LU	++	KB8VBV
9	AA9MK	KF9XX	N9YRJ	KB9JAY
N.Mariana	KHØK	++	KHØDM	++
Guam	WH2G	AH2CW	KH2KP	++
Midway	++	AH4AA	KH4AG	WH4AAH
Hawaii	++	AH6NQ	WH6YB	WH6CRK
Amer.Samoa	AH8K	AH8AG	KH8BH	WH8ABB
Alaska	++	AL7PT	WL7XL	++
VirginIs	WP2O	KP2CD	NP2HR	WP2AHU
PuertoRico	++	KP4XP	++	WP4MTG

November 1,1994

Ø	AAØUA	KGØRD	++	KBØPLX
1	AA1LH	KD1XS	N1TKM	KB1BLE
2	AA2UO	KF2YO	++	KB2SFP
3	AA3IY	KE3PW	N3TSG	KB3BFG
4	AD4ZD	KS4HB	++	KE4SMC
5	AB5YQ	KK5EK	++	KC5KPV
6	AC6GZ	KO6KX	++	KE6NVP
7	AB7FV	KJ7FF	++	KC7GMT
8	AA8QX	KG8MV	++	KB8VMB
9	AA9MW	KF9YU	N9YZZ	KB9JBE

N.Mariana	KHØO	++	KHØDO	++
Guam	WH2H	AH2CY	KH2LO	++
Midway	++	AH4AA	KH4AG	WH4AAH
Hawaii	++	AH6NQ	WH6ZC	WH6CRL
Amer.Samoa	AH8L	AH8AG	KH8BJ	WH8ABB
Alaska	++	AL7PV	WL7YU	WL7CHV
Virgin	WP2P	KP2CD	NP2HR	WP2AHU
PuertoRico	++	KP4XS	++	WP4MUA

December 1,1994

Ø	AAØUN	KGØRV	++	KBØPTK
1	AA1LM	KD1YE	N1TQD	KB1BLO
2	AA2UX	KF2ZD	++	KB2SOZ
3	AA3JD	KE3QE	N3TWX	KB3BFO
4	AE4AD	KS4IM	++	KE4TEA
5	AB5ZA	KK5FX	++	KC5LCE
6	AC6IA	KO6MH	++	KE6ONG
7	AB7GH	KJ7FZ	++	KC7GZS
8	AA8RE	KG8NH	++	KB8VTG
9	AA9NE	KF9ZB	N9ZHC	KB9JBM
N.Mariana	KHØO	++	KHØDO	++
Guam	WH2I	AH2CZ	KH2LP	++
Midway	++	AH4AA	KH4AG	WH4AAH
Hawaii	++	AH6NQ	WH6ZP	WH6CRM
Amer.Samoa	AH8L	AH8AG	KH8BJ	WH8ABB
Alaska	++	AL7PV	WL7ZA	WL7CHW
Virgin	WP2P	KP2CD	NP2HS	WP2AHV
PuertoRico	++	KP4XY	++	WP4MUL

January 1,1995

Ø	AAØVB	KGØSL	++	KBØQDK
1	AA1LQ	KD1YQ	N1TYE	KB1BMP
2	AA2VD	KF2ZP	++	KB2SZC
3	AA3JO	KE3QT	N3UDE	KB3BGA
4	AE4BG	KS4KC	++	KE4UEJ
5	AB5ZL	KK5HL	++	KC5LSC
6	AC6IR	KO6NL	++	KE6PQB
7	AB7HD	KJ7HC	++	KC7HWL
8	AA8RS	KG8NW	++	KB8WEE
9	AA9NM	KF9ZO	N9ZQK	KB9JCB
N.Mariana	KHØO	++	KHØDP	++
Guam	WH2J	AH2CZ	KH2LP	++
Midway	++	AH4AA	KH4AG	WH4AAH
Hawaii	++	AH6NS	WH6ZY	WH6CRV
Amer.Samoa	AH8M	AH8AH	KH8BJ	WH8ABB
Alaska	++	AL7PW	WL7ZY	WL7CJD
Virginls.	WP2Q	KP2CD	NP2HV	WP2AHV
PuertoRico	++	KP4YH	++	WP4MVN

February 1,1995

Ø	AAØVS	KGØTJ	++	KBØQVG
1	AA1MB	KD1ZH	N1UFI	KB1BNN
2	AA2VU	KG2AP	++	KB2TLZ
3	AA3KC	KE3RA	N3UIZ	KB3BGK
4	AE4CX	KS4MX	++	KE4VDD
5	AC5AH	KK5JD	++	KC5MKA
6	AC6JX	KO6PF	++	KE6QOX
7	AB7HT	KJ7JB	++	KC7IQZ
8	AA8SA	KG8OO	++	KB8WQX
9	AA9NS	KG9AJ	N9ZZZ	KB9JCR
N.Mariana	KHØQ	++	KHØDQ	++
Guam	WH2K	AH2CZ	KH2NB	++
Midway	++	AH4AA	KH4AG	WH4AAH
Hawaii	++	AH6NU	++	WH6CSL
Amer.Samoa	AH8M	AH8AH	KH8CF	WH8ABB
Alaska	++	AL7PW	++	WL7CKG
Virgin	WP2Q	KP2CD	NP2HY	WP2AHV
PuertoRico	++	KP4YM	++	WP4MWC

March 1,1995

Ø	AAØWM	KGØUD	++	KBØRFW
1	AA1MK	KD1ZZ	N1UKU	KB1BOC
2	AA2WK	KG2BJ	++	KB2TTD

3	AA3KJ	KE3RS	N3UPB	KB3BGN
4	AE4EC	KS4QB	++	KE4WEV
5	AC5AZ	KK5LJ	++	KC5MXP
6	AC6KX	KO6QY	++	KE6RKO
7	AB7IK	KJ7KR	++	KC7JFS
8	AA8SN	KG8PM	++	KB8YBA
9	AA9OA	KG9AS	++	KB9JLU
N.Mariana	KHØQ	++	KHØDT	++
Guam	WH2L	AH2CZ	KH2NC	++
Midway	++	AH4AA	KH4AG	WH4AAH
Hawaii	++	AH6NY	++	WH6CSX
Amer.Samoa	AH8M	AH8AH	KH8CG	WH8ABB
Alaska	++	AL7PY	++	WL7CLA
Virgin	WP2Q	KP2CD	NP2IA	WP2AHV
PuertoRico	++	KP4YW	++	WP4MWU

April 1,1995

Ø	AAØWZ	KGØVF	++	KBØRVU
1	AA1MX	KE1AZ	N1USY	KB1BOI
2	AA2WW	KG2CB	++	KB2UCB
3	AA3LD	KE3SJ	N3UXM	KB3BHC
4	AE4GC	KS4TL	++	KE4YMF
5	AC5BV	KK5NB	++	KC5NRI
6	AC6LV	KO6TA	++	KE6SNC
7	AB7JE	KJ7MD	++	KC7JYT
8	AA8TA	KG8QJ	++	KB8YQE
9	AA9OG	KG9BH	++	KB9JVL
N.Mariana	KHØQ	AHØAV	KHØDW	WHØABC
Guam	WH2M	AH2CZ	KH2NM	++
Midway	++	AH4AA	KH4AG	WH4AAH
Hawaii	++	AH6OB	++	WH6CUD
Amer.Samoa	AH8N	AH8AH	KH8CG	WH8ABB
Alaska	++	AL7PZ	++	WL7CLX
Virgin	WP2R	KP2CD	NP2IA	WP2AHV
PuertoRico	++	KP4ZC	++	WP4MXF

May 1,1995

Ø	AAØXH	KGØWA	++	KBØSIT
1	AA1NB	KE1BK	N1UZB	KB1BQA
2	AA2XG	KG2CN	++	KB2ULC
3	AA3LK	KE3SY	N3VEA	KB3BHX
4	AE4HK	KS4VU	++	KE4ZNV
5	AC5CG	KK5NW	++	KC5OEV
6	AC6ML	KO6UK	++	KE6TKU
7	AB7JU	KJ7NE	++	KC7KRF
8	AA8TL	KG8QY	++	KB8ZEI
9	AA9ON	KG9CB	++	KB9KFO
N.Mariana	KHØR	AHØAW	KHØDW	WHØABC
Guam	WH2O	AH2CZ	KH2NM	++
Midway	++	AH4AA	KH4AG	WH4AAH
Hawaii	++	AH6OC	++	WH6CVA
Amer.Samoa	AH8O	AH8AH	KH8CG	WH8ABB
Alaska	++	AL7QB	++	WL7CMN
Virgin	WP2R	KP2CD	NP2IF	WP2AHV
PuertoRico	++	KP4ZK	++	WP4MYB

June 1,1995

Ø	AAØXW	KGØXC	++	KBØSWQ
1	AA1NO	KE1BU	N1VGK	KB1BSD
2	AA2XP	KG2CX	++	KB2UXP
3	AA3LW	KE3TS	N3VOD	KB3BJL
4	AE4IU	KS4YB	++	KF4AQY
5	AC5DE	KK5PK	++	KC5OYU
6	AC6NN	KO6WN	++	KE6URO
7	AB7KN	KJ7OI	++	KC7LLW
8	AA8TS	KG8RR	++	KB8ZWA
9	AA9PB	KG9CQ	++	KB9KSE
N.Mariana	KHØR	AHØAW	KHØDW	WHØABC
Guam	WH2P	AH2CZ	KH2NT	WH2ANM
Midway	++	AH4AA	KH4AG	WH4AAH
Hawaii	++	AH6OD	++	WH6CVT

Amer.Samoa	AH8O	AH8AH	KH8CI	WH8ABB
Alaska	++	AL7QC	++	WL7CNA
Virgin	WP2R	KP2CE	NP2IG	WP2AHX
PuertoRico	++	KP4ZO	++	WP4MYZ

July 1,1995

Ø	AAØYM	KGØYF	++	KBØTFM
1	AA1NW	KE1CJ	N1VMF	KB1BSX
2	AA2YB	KG2DH	++	KB2VEV
3	AA3MC	KE3UD	N3VSJ	KB3BJX
4	AE4KB	KT4AM	++	KF4BKY
5	AC5DP	KK5QZ	++	KC5POU
6	AC6OK	KO6XW	++	KE6VTW
7	AB7LI	KJ7PS	++	KC7MBY
8	AA8UB	KG8SR	++	KC8AIK
9	AA9PJ	KG9DI	++	KB9KZC
N.Mariana	KHØS	AHØAW	KHØED	WHØABC
Guam	WH2P	AH2DA	KH2OF	WH2ANM
Midway	++	AH4AA	KH4AG	WH4AAH
Hawaii	++	AH6OD	++	WH6CWI
Amer.Samoa	AH8O	AH8AH	KH8CJ	WH8ABB
Alaska	++	AL7QC	++	WL7CNX
Virgin	WP2R	KP2CF	NP2IG	WP2AHY
PuertoRico	++	KP4ZU	++	WP4NAB

WELL-KNOWN PERSONALITIES HOLDING AMATEUR RADIO LICENSES

This list includes widely-known personalities who are (or were) public figures and who would be recognized by an average person. Many prominent businessmen, local radio/TV personalities, and mid-ranking government officials are also licensed amateurs, but they are not listed here, in the interest of protecting their privacy.

Several lists have propagated through the amateur community from time to time, often with errors. For example, some lists claim N6KGB was actor Jimmy Stewart, when the call was really held by another well-known actor: Stewart Granger (whose real name was James Stewart and who became a Silent Key in 1993). Many lists claim that Russian cosmonaut Yuri Gagarin was a ham (some lists even claim his call was UA1LO); there is no hard evidence that he was ever licensed. Another rumor insists that musician Jon Bon Jovi is a ham. He is *not* licensed but his cousin, record producer Tony Bongiovi, is KX2Z. Other lists indicate that musician Jose Feliciano is a ham; WP4CO's name is definitely Jose Feliciano, but he's not the famous musician. Donny Osmond was once licensed as KA7EVD, but his license has expired—another Donald Osmond holds the call WD4SKT, but it's not the same person. KD6OY insists that his name is really Garry Shankling, not Shandling, even though he lives in Los Angeles. And WA7WYV's name is indeed Andy Griffith, but he's never played either Andy of Mayberry or Matlock on television.

Royalty

Juan Carlos I de Borbon y Borbon, EAØJC, (1938-), King, Spain
Bhumiphol Adulayadej, HS1A, (1927-), King, Thailand
Hussein I, JY1, (1935-), King, Hashemite Kingdom of Jordan

Political and Religious Leaders

Francesco Cossiga, IØFCG, (1929-), President, Italy
Barry Goldwater, K7UGA, (1909-), former U. S. Senator (R-AZ)
Carlos Saul Menem, LU1SM, (1930-), President, Argentina
Rajiv Gandhi, VU2RG, (1944-1991), former Prime Minister, Republic of India
General Curtis LeMay, W6EZV, (1906-1991), Vice-Presidential Candidate (with George Wallace), 1968
General Anastasio Somoza Debayle, YN1AS, (1925-1980), former President, Republic of Nicaragua
Roger Cardinal Mahony, W6QYI, (1936-), Cardinal of the Roman Catholic Archdiocese of Los Angeles

Entertainers/Musicians

Marlon Brando, FO5GJ, (1924-), Actor
Arthur Godfrey, K4LIB, (1903-1983), Entertainer
Burl Ives, KA6HVK, (1909-1995), Singer
Donny Osmond, ex-KA7EVD (expired), (1957-), Entertainer
Patty Loveless (born Patricia Ramey), KD4WUJ, (1957-), Musician
Stu Gilliam, KI6M, (1933-), Comedian
Larry Junstrom, KN4UB, (1957-), Rock Musician (".38 Special")
Stewart Granger (born James Stewart), N6KGB, (1918-1993), Actor
Priscilla Presley, N6YOS, (1945-), Actress
Alvino Rey, W6UK, (1908-), Musician/Bandleader
Chet Atkins, WA4CZD (1924-), Musician
Ronnie Milsap, WB4KCG, (1943-), Musician
Joe Walsh, WB6ACU, (1947-), Rock Musician ("Eagles," "James Gang")
Andy Devine, WB6RER, (1905-1977), Actor

Writers/Authors

Jean Shepherd, K2ORS, (1921-), Writer/Humorist
Roy Neal, K6DUE, (1921-), Network Science Reporter, NBC (retired)
Walter Cronkite, KB2GSD, (1916-), Network News Anchorman, CBS (retired)
David French, N4KET, (1938-), former Network News Anchorman, CNN
David Ruben, M.D., TI2DR, (1933-), Author (*Everything You Always Wanted to Know About Sex - But Were Afraid to Ask*)
Dr. Alexander Comfort, KA6UXR, (1920-), Author (*The Joy of Sex*)

Others

Joe Rudi, NK7U (1946-), Major League Baseball Player (retired)
Dick Rutan, KB6LQS, (1938-), pilot of Voyager airplane non-stop around the world
Dr. Joseph Taylor, Jr., K1JT (1941-), winner 1958 VHF SS Contest and 1993 Nobel Prize in Physics
Jeana Yeager, KB6LQR, (1952-), co-pilot of Voyager airplane non-stop around the world

Numerous astronauts and cosmonauts also hold amateur licenses. A list appears in the "Amateur Radio in Space" section.

U.S. HAM CENSUS BY YEAR OF BIRTH
(As of 09/01/95)

Year	Count	Year	Count	Year	Count	Year	Count	Year	Count
1890	4	1910	2687	1931	9557	1952	15698	1973	4826
1891	11	1911	2861	1932	9337	1953	15422	1974	4881
1892	11	1912	3363	1933	8889	1954	15211	1975	4461
1893	21	1913	3967	1934	9131	1955	14450	1976	4448
1894	28	1914	4670	1935	9540	1956	13771	1977	4315
1895	39	1915	5314	1936	9783	1957	13094	1978	3825
1896	65	1916	5705	1937	10498	1958	12150	1979	3584
1897	100	1917	6624	1938	11721	1959	11837	1980	2783
1898	119	1918	7537	1939	11833	1960	11516	1981	2035
1899	167	1919	7371	1940	13062	1961	11459	1982	1223
1900	246	1920	8670	1941	14120	1962	10777	1983	744
1901	349	1921	9535	1942	16646	1963	10091	1984	380
1902	445	1922	9356	1943	16507	1964	9021	1985	166
1903	664	1923	9403	1944	14949	1965	7678	1986	68
1904	894	1924	9921	1945	14432	1966	6964	1987	23
1905	1284	1925	9970	1946	17246	1967	6369	1988	8
1906	1702	1926	9759	1947	19193	1968	6201	1989	6
1907	2053	1927	9685	1948	16612	1969	5987		
1908	2341	1928	9513	1949	16312	1970	5892		
1909	2376	1929	9094	1950	15326	1971	5565		
		1930	9725	1951	15530	1972	5227		

AMATEUR RADIO CENSUS BY U.S. COUNTIES
(As of September 1, 1995)

Alaska

County	Count
Aleutian Islands	5
Anchorage	1082
Bethel	83
Bristol Bay	8
Denali	17
Dillingham	12
Fairbanks North Star	715
Haines	17
Juneau	245
Kenai Peninsula	222
Ketchikan Gateway	59
Kodiak Island	78
Lake And Peninsula	6
Matanuska-Susitna	255
Nome	23
North Slope	14
Prince Of Wales	19
Sitka	48
Skagway-Yakutat-Angoon	14
Southeast Fairbanks	62
Valdez-Cordova	52
Wade Hampton	8
Wrangell-Petersburg	54
Yukon-Koyukuk	43

Alabama

County	Count
Autauga	78
Baldwin	332
Barbour	27
Bibb	34
Blount	88
Bullock	1
Butler	37
Calhoun	306
Chambers	44
Cherokee	35
Chilton	67
Choctaw	5
Clarke	29
Clay	26
Cleburne	19
Coffee	98
Colbert	176
Conecuh	12
Coosa	21
Covington	66
Crenshaw	10
Cullman	113
Dale	103
Dallas	131
De Kalb	114
Elmore	101
Escambia	44
Etowah	284
Fayette	30
Franklin	64
Geneva	32
Greene	7
Hale	16
Henry	19
Houston	219
Jackson	120
Jefferson	1926
Lamar	14
Lauderdale	271
Lawrence	45
Lee	190
Limestone	155
Lowndes	12
Macon	21
Madison	1502
Marengo	40
Marion	64
Marshall	234
Mobile	739
Monroe	54
Montgomery	405
Morgan	351
Perry	7
Pickens	22
Pike	42
Randolph	41
Russell	52
Saint Clair	137
Shelby	297
Sumter	11
Talladega	156
Tallapoosa	60
Tuscaloosa	316
Walker	159
Washington	14
Wilcox	4
Winston	51

Arizona

County	Count
Apache	102
Cochise	610
Coconino	443
Gila	277
Graham	112
Greenlee	33
La Paz	74
Maricopa	7779
Mohave	661
Navajo	278
Pima	2536
Pinal	372
Santa Cruz	88
Yavapai	819
Yuma	343

Arkansas

County	Count
Arkansas	20
Ashley	31
Baxter	189
Benton	350
Boone	143
Bowie	35
Bradley	9
Calhoun	1
Carroll	72
Chicot	14
Clark	73
Clay	26
Cleburne	69
Cleveland	9
Columbia	69
Conway	27

Craighead	223	Inyo	153
Crawford	125	Kern	1781
Crittenden	119	Kings	463
Cross	36	Lake	290
Dallas	8	Lassen	96
Desha	11	Los Angeles	27920
Drew	13	Madera	324
Faulkner	199	Marin	972
Franklin	29	Mariposa	108
Fulton	16	Mendocino	417
Garland	320	Merced	286
Grant	41	Modoc	28
Greene	222	Mono	60
Hempstead	16	Monterey	888
Hot Spring	140	Napa	420
Howard	40	Nevada	484
Independence	98	Orange	12045
Izard	33	Placer	839
Jackson	11	Plumas	91
Jefferson	123	Riverside	3885
Johnson	51	Sacramento	3196
Lafayette	8	San Benito	115
Lawrence	72	San Bernardino	6139
Lee	8	San Diego	8422
Lincoln	12	San Francisco	1102
Little River	23	San Joaquin	1063
Logan	51	San Luis Obispo	953
Lonoke	98	San Mateo	2647
Madison	32	Santa Barbara	1715
Marion	72	Santa Clara	6947
Miller	8	Santa Cruz	1296
Mississippi	63	Shasta	597
Monroe	5	Sierra	11
Montgomery	18	Siskiyou	296
Nevada	13	Solano	1073
Newton	20	Sonoma	1515
Ouachita	33	Stanislaus	808
Perry	23	Sutter	231
Phillips	28	Tehama	148
Pike	19	Trinity	112
Poinsett	46	Tulare	613
Polk	74	Tuolumne	304
Pope	228	Ventura	2881
Prairie	13	Yolo	401
Pulaski	997	Yuba	158
Randolph	31	**Colorado**	
Saint Francis	39	Adams	749
Saline	174	Alamosa	25
Scott	16	Arapahoe	1118
Searcy	19	Archuleta	33
Sebastian	277	Baca	12
Sevier	34	Bent	13
Sharp	53	Boulder	1353
Stone	36	Chaffee	73
Union	93	Cheyenne	5
Van Buren	78	Clear Creek	24
Washington	377	Conejos	6
White	137	Costilla	4
Woodruff	12	Crowley	6
Yell	83	Custer	14
Western	83	Delta	49
California		Denver	879
Alameda	3862	Douglas	299
Alpine	6	Eagle	50
Amador	208	El Paso	1708
Butte	844	Elbert	50
Calaveras	221	Fremont	167
Colusa	14	Garfield	107
Contra Costa	2633	Gilpin	17
Del Norte	141	Grand	101
El Dorado	683	Gunnison	31
Fresno	1377	Hinsdale	2
Glenn	37	Huerfano	16
Humboldt	500	Jackson	1
Imperial	170	Jefferson	1496

Kiowa	2	Hillsborough	2095	Cook	9
Kit Carson	11	Holmes	29	Coweta	204
La Plata	167	Indian River	350	Crawford	7
Lake	16	Jackson	119	Crisp	21
Larimer	926	Jefferson	16	Dade	34
Las Animas	17	Lafayette	4	Dawson	24
Lincoln	17	Lake	604	De Kalb	1176
Logan	116	Lee	1151	Decatur	10
Mesa	328	Leon	493	Dodge	16
Mineral	3	Levy	78	Dooly	12
Moffat	36	Liberty	5	Dougherty	195
Montezuma	43	Madison	37	Douglas	224
Montrose	121	Manatee	592	Early	8
Morgan	36	Marion	430	Effingham	36
Otero	49	Martin	421	Elbert	32
Ouray	9	Monroe	440	Emanuel	17
Park	45	Nassau	143	Evans	9
Phillips	14	Okaloosa	686	Fannin	21
Pitkin	50	Okeechobee	100	Fayette	237
Prowers	31	Orange	1863	Floyd	162
Pueblo	375	Osceola	255	Forsyth	139
Rio Blanco	12	Palm Beach	2662	Franklin	41
Rio Grande	12	Pasco	1197	Fulton	1400
Routt	48	Pinellas	2321	Gilmer	40
Saguache	4	Polk	944	Glascock	5
San Juan	2	Putnam	149	Glynn	158
San Miguel	9	Saint Johns	349	Gordon	75
Sedgwick	12	Saint Lucie	408	Grady	15
Summit	28	Santa Rosa	328	Greene	14
Teller	159	Sarasota	796	Gwinnett	1295
Washington	8	Seminole	958	Habersham	81
Weld	282	Sumter	86	Hall	280
Yuma	19	Suwannee	59	Hancock	8
Connecticut		Taylor	47	Haralson	32
Fairfield	2466	Union	7	Harris	34
Hartford	2318	Volusia	1051	Hart	50
Litchfield	653	Wakulla	54	Heard	7
Middlesex	483	Walton	63	Henry	164
New Haven	1882	Washington	31	Houston	361
New London	922	**Georgia**		Irwin	7
Tolland	438	Appling	8	Jackson	66
Windham	244	Atkinson	6	Jasper	13
Delaware		Bacon	8	Jeff Davis	9
Kent	227	Baker	4	Jefferson	16
New Castle	928	Baldwin	79	Jenkins	6
Sussex	322	Banks	14	Johnson	9
Florida		Barrow	76	Jones	26
Alachua	535	Bartow	102	Lamar	32
Baker	37	Ben Hill	11	Lanier	7
Bay	477	Berrien	15	Laurens	71
Bradford	29	Bibb	339	Lee	37
Brevard	2678	Bleckley	19	Liberty	81
Broward	3620	Brantley	5	Lincoln	7
Calhoun	17	Brooks	16	Long	2
Charlotte	593	Bryan	33	Lowndes	121
Citrus	217	Bulloch	60	Lumpkin	44
Clay	476	Burke	25	Macon	5
Collier	465	Butts	18	Madison	45
Columbia	90	Calhoun	3	Marion	2
Dade	5507	Camden	85	Mc Duffie	29
De Soto	91	Candler	13	Mc Intosh	11
Dixie	43	Carroll	138	Meriwether	33
Duval	1639	Catoosa	88	Mitchell	19
Escambia	1056	Charlton	15	Monroe	19
Flagler	91	Chatham	386	Montgomery	7
Franklin	21	Chattooga	54	Morgan	64
Gadsden	39	Cherokee	212	Murray	49
Gilchrist	20	Clarke	179	Muscogee	235
Glades	21	Clay	2	Newton	101
Gulf	37	Clayton	318	Oconee	55
Hamilton	8	Clinch	3	Oglethorpe	9
Hardee	24	Cobb	1495	Paulding	103
Hendry	45	Coffee	23	Peach	36
Hernando	390	Colquitt	38	Pickens	32
Highlands	319	Columbia	82	Pierce	9

Pike	28	Idaho	47	Macon	299
Polk	85	Jefferson	66	Macoupin	160
Pulaski	5	Jerome	22	Madison	704
Putnam	43	Kootenai	365	Marion	174
Quitman	5	Latah	141	Marshall	31
Rabun	38	Lemhi	16	Mason	43
Randolph	3	Lewis	29	Massac	46
Richmond	391	Lincoln	7	Mc Donough	95
Rockdale	154	Madison	64	Mc Henry	637
Schley	4	Minidoka	28	Mc Lean	395
Screven	11	Nez Perce	87	Menard	22
Seminole	10	Oneida	5	Mercer	50
Spalding	117	Owyhee	15	Monroe	38
Stephens	140	Payette	56	Montgomery	106
Stewart	4	Power	7	Morgan	96
Sumter	34	Shoshone	56	Moultrie	33
Talbot	6	Teton	10	Ogle	106
Tattnall	7	Twin Falls	131	Peoria	482
Taylor	14	Valley	36	Perry	44
Telfair	10	Washington	26	Piatt	74
Terrell	18	**Illinois**		Pike	30
Thomas	83	Adams	243	Pope	10
Tift	45	Alexander	25	Pulaski	17
Toombs	26	Bond	80	Putnam	32
Towns	50	Boone	65	Randolph	50
Treutlen	2	Brown	17	Richland	41
Troup	95	Bureau	86	Rock Island	345
Turner	15	Calhoun	5	Saint Clair	551
Twiggs	9	Carroll	54	Saline	61
Union	41	Cass	31	Sangamon	478
Upson	54	Champaign	461	Schuyler	29
Walker	155	Christian	102	Scott	6
Walton	28	Clark	68	Shelby	55
Ware	58	Clay	18	Stark	22
Warren	5	Clinton	55	Stephenson	98
Washington	29	Coles	89	Tazewell	422
Wayne	49	Cook	7431	Union	47
Webster	8	Crawford	62	Vermilion	218
Wheeler	3	Cumberland	27	Wabash	56
White	84	De Kalb	232	Warren	41
Whitfield	184	De Witt	53	Washington	16
Wilcox	8	Douglas	46	Wayne	42
Wilkes	12	Du Page	2390	White	50
Wilkinson	13	Edgar	54	Whiteside	230
Worth	13	Edwards	19	Will	648
Hawaii		Effingham	103	Williamson	167
Hawaii	527	Fayette	48	Winnebago	685
Honolulu	2258	Ford	33	Woodford	89
Kauai	265	Franklin	116	**Indiana**	
Maui	304	Fulton	112	Adams	80
Idaho		Gallatin	8	Allen	1092
Ada	923	Greene	18	Bartholomew	164
Adams	13	Grundy	101	Benton	15
Bannock	204	Hamilton	40	Blackford	27
Bear Lake	7	Hancock	73	Boone	131
Benewah	25	Hardin	4	Brown	34
Bingham	87	Henderson	16	Carroll	37
Blaine	40	Henry	107	Cass	126
Boise	10	Iroquois	42	Clark	219
Bonner	142	Jackson	147	Clay	85
Bonneville	338	Jasper	20	Clinton	96
Boundary	41	Jefferson	102	Crawford	21
Butte	6	Jersey	38	Daviess	44
Camas	3	Jo Daviess	27	De Kalb	126
Canyon	331	Johnson	35	Dearborn	89
Caribou	7	Kane	998	Decatur	62
Cassia	33	Kankakee	153	Delaware	324
Clark	2	Kendall	79	Dubois	58
Clearwater	36	Knox	172	Elkhart	436
Custer	10	La Salle	240	Fayette	101
Elmore	77	Lake	1584	Floyd	136
Franklin	29	Lawrence	33	Fountain	52
Fremont	45	Lee	128	Franklin	33
Gem	44	Livingston	59	Fulton	41
Gooding	16	Logan	56	Gibson	94

County	#	County	#	County	#
Grant	210	Bremer	28	Sioux	28
Greene	106	Buchanan	45	Story	198
Hamilton	401	Buena Vista	46	Tama	50
Hancock	191	Butler	20	Taylor	22
Harrison	62	Calhoun	22	Union	32
Hendricks	240	Carroll	33	Van Buren	18
Henry	166	Cass	49	Wapello	119
Howard	469	Cedar	32	Warren	94
Huntington	184	Cerro Gordo	102	Washington	44
Jackson	66	Cherokee	22	Wayne	11
Jasper	55	Chickasaw	30	Webster	90
Jay	78	Clarke	7	Winnebago	27
Jefferson	74	Clay	74	Winneshiek	17
Jennings	23	Clayton	20	Woodbury	282
Johnson	265	Clinton	159	Worth	22
Knox	143	Crawford	24	Wright	37
Kosciusko	233	Dallas	57	**Kansas**	
La Porte	296	Davis	20	Allen	34
Lagrange	44	Decatur	35	Anderson	23
Lake	800	Delaware	22	Atchison	35
Lawrence	152	Des Moines	293	Barber	19
Madison	437	Dickinson	71	Barton	100
Marion	1848	Dubuque	161	Bourbon	32
Marshall	129	Emmet	29	Brown	35
Martin	54	Fayette	26	Butler	206
Miami	95	Floyd	29	Chase	2
Monroe	306	Franklin	6	Chautauqua	6
Montgomery	138	Fremont	10	Cherokee	44
Morgan	181	Greene	18	Cheyenne	2
Newton	10	Grundy	17	Clark	1
Noble	89	Guthrie	22	Clay	22
Ohio	4	Hamilton	50	Cloud	47
Orange	53	Hancock	17	Coffey	41
Owen	47	Hardin	21	Comanche	8
Parke	26	Harrison	41	Cowley	145
Perry	23	Henry	75	Crawford	106
Pike	53	Howard	43	Decatur	14
Porter	367	Humboldt	55	Dickinson	52
Posey	40	Ida	13	Doniphan	10
Pulaski	26	Iowa	25	Douglas	209
Putnam	54	Jackson	24	Edwards	3
Randolph	75	Jasper	49	Elk	10
Ripley	100	Jefferson	68	Ellis	71
Rush	65	Johnson	194	Ellsworth	19
Saint Joseph	691	Jones	61	Finney	55
Scott	30	Keokuk	19	Ford	37
Shelby	126	Kossuth	32	Franklin	65
Spencer	73	Lee	104	Geary	108
Starke	59	Linn	873	Gove	5
Steuben	107	Louisa	13	Graham	2
Sullivan	55	Lucas	51	Grant	17
Switzerland	9	Lyon	17	Gray	7
Tippecanoe	372	Madison	33	Greeley	6
Tipton	58	Mahaska	43	Greenwood	21
Union	18	Marion	78	Hamilton	11
Vanderburgh	406	Marshall	134	Harper	43
Vermillion	38	Mills	39	Harvey	219
Vigo	314	Mitchell	12	Haskell	14
Wabash	68	Monona	22	Hodgeman	2
Warren	9	Monroe	11	Jackson	28
Warrick	145	Montgomery	19	Jefferson	55
Washington	78	Muscatine	124	Jewell	8
Wayne	232	O'brien	27	Johnson	1223
Wells	62	Osceola	14	Kearny	7
White	51	Page	76	Kingman	10
Whitley	132	Palo Alto	33	Kiowa	10
Iowa		Plymouth	50	Labette	79
Adair	21	Pocahontas	9	Lane	5
Adams	13	Polk	695	Leavenworth	233
Allamakee	18	Pottawattamie	184	Lincoln	24
Appanoose	28	Poweshiek	38	Linn	35
Audubon	16	Ringgold	10	Logan	3
Benton	79	Sac	21	Lyon	84
Black Hawk	260	Scott	422	Marion	39
Boone	81	Shelby	15	Marshall	45

County	#	County	#	County	#
Mc Pherson	66	Crittenden	16	Rowan	30
Meade	14	Cumberland	5	Russell	13
Miami	76	Daviess	203	Scott	51
Mitchell	30	Edmonson	14	Shelby	75
Montgomery	168	Elliott	5	Simpson	9
Morris	14	Estill	49	Spencer	8
Morton	6	Fayette	714	Taylor	35
Nemaha	17	Fleming	15	Todd	7
Neosho	57	Floyd	172	Trigg	10
Ness	11	Franklin	118	Trimble	6
Norton	22	Fulton	14	Tuscola	1
Osage	58	Gallatin	3	Union	24
Osborne	37	Garrard	23	Warren	194
Ottawa	20	Grant	16	Washington	8
Pawnee	18	Graves	47	Wayne	17
Phillips	35	Grayson	47	Webster	43
Pottawatomie	88	Green	13	Whitley	99
Pratt	32	Greenup	120	Wolfe	14
Rawlins	1	Hancock	32	Woodford	68
Reno	156	Hardin	232	**Louisiana**	
Republic	31	Harlan	78	Acadia	50
Rice	25	Harrison	11	Allen	16
Riley	168	Hart	17	Ascension	113
Rooks	17	Henderson	126	Assumption	8
Rush	4	Henry	22	Avoyelles	23
Russell	22	Hickman	4	Beauregard	49
Saline	211	Hopkins	119	Bienville	19
Scott	29	Jackson	15	Bossier	190
Sedgwick	1162	Jefferson	1515	Caddo	602
Seward	30	Jessamine	111	Calcasieu	252
Shawnee	556	Johnson	49	Caldwell	26
Sheridan	8	Kenton	254	Cameron	3
Sherman	15	Knott	13	Catahoula	11
Smith	29	Knox	48	Claiborne	13
Stafford	13	Larue	19	Concordia	19
Stanton	8	Laurel	95	De Soto	32
Stevens	8	Lawrence	32	East Baton Rouge	986
Sumner	96	Lee	7	East Carroll	2
Thomas	40	Leslie	8	East Feliciana	36
Trego	11	Letcher	45	Evangeline	22
Wabaunsee	17	Lewis	23	Franklin	37
Wallace	4	Lincoln	19	Grant	32
Washington	14	Livingston	40	Iberia	103
Wichita	7	Logan	30	Iberville	72
Wilson	24	Lyon	14	Jackson	57
Woodson	6	Madison	122	Jefferson	891
Wyandotte	306	Magoffin	10	Jefferson Davis	29
Kentucky		Marion	14	La Salle	23
Adair	15	Marshall	128	Lafayette	298
Allen	13	Martin	12	Lafourche	87
Anderson	49	Mason	70	Lincoln	89
Ballard	15	Mc Cracken	165	Livingston	176
Barren	110	Mc Creary	15	Madison	7
Bath	30	Mc Lean	21	Morehouse	40
Bell	101	Meade	27	Natchitoches	55
Boone	151	Menifee	10	Orleans	491
Bourbon	22	Mercer	43	Ouachita	280
Boyd	226	Metcalfe	17	Plaquemines	36
Boyle	45	Monroe	10	Pointe Coupee	40
Bracken	21	Montgomery	56	Rapides	273
Breathitt	26	Morgan	11	Red River	11
Breckinridge	27	Muhlenberg	64	Richland	27
Bullitt	107	Nelson	163	Sabine	49
Butler	6	Nicholas	3	Saint Bernard	133
Caldwell	44	Ohio	35	Saint Charles	70
Calloway	111	Oldham	117	Saint Helena	12
Campbell	161	Owen	3	Saint James	17
Carlisle	9	Owsley	7	Saint John The Baptist	42
Carroll	8	Pendleton	12	Saint Landry	67
Carter	67	Perry	58	Saint Martin	46
Casey	16	Pike	210	Saint Mary	88
Christian	133	Powell	21	Saint Tammany	551
Clark	108	Pulaski	108	Tangipahoa	113
Clay	22	Robertson	2	Tensas	9
Clinton	7	Rockcastle	13	Terrebonne	127

County	Count	County	Count	County	Count
Union	27	Barry	148	Wexford	65
Vermilion	50	Bay	194	**Minnesota**	
Vernon	107	Benzie	29	Aitkin	29
Washington	80	Berrien	544	Anoka	611
Webster	159	Branch	106	Becker	49
West Baton Rouge	28	Calhoun	374	Beltrami	117
West Carroll	12	Cass	85	Benton	85
West Feliciana	29	Charlevoix	67	Big Stone	12
Winn	37	Cheboygan	63	Blue Earth	147
Maine		Chippewa	89	Brown	37
Androscoggin	251	Clare	81	Carlton	87
Aroostook	244	Clinton	88	Carver	95
Cumberland	852	Crawford	56	Cass	60
Franklin	107	Delta	156	Chippewa	9
Hancock	244	Dickinson	91	Chisago	88
Kennebec	407	Eaton	273	Clay	127
Knox	160	Emmet	102	Clearwater	12
Lincoln	165	Genesee	1172	Cook	18
Oxford	199	Gladwin	96	Cottonwood	16
Penobscot	509	Gogebic	46	Crow Wing	150
Piscataquis	57	Grand Traverse	309	Dakota	734
Sagadahoc	144	Gratiot	102	Dodge	21
Somerset	184	Hillsdale	75	Douglas	96
Waldo	139	Houghton	216	Faribault	23
Washington	185	Huron	77	Fillmore	32
York	532	Ingham	586	Freeborn	73
Maryland		Ionia	101	Goodhue	80
Allegany	206	Iosco	75	Grant	16
Anne Arundel	1784	Iron	65	Hennepin	2578
Baltimore	1437	Isabella	78	Houston	37
Baltimore (city)	557	Jackson	418	Hubbard	54
Calvert	177	Kalamazoo	607	Isanti	47
Caroline	61	Kalkaska	34	Itasca	122
Carroll	484	Kent	1254	Jackson	8
Cecil	183	Keweenaw	15	Kanabec	40
Charles	237	Lake	25	Kandiyohi	113
Dorchester	35	Lapeer	179	Kittson	20
Frederick	475	Leelanau	52	Koochiching	91
Garrett	80	Lenawee	288	Lac Qui Parle	12
Harford	433	Livingston	303	Lake	46
Howard	691	Luce	23	Lake Of The Woods	12
Kent	75	Mac Kinac	30	Le Sueur	52
Montgomery	2418	Macomb	1398	Lincoln	20
Prince George's	1405	Manistee	53	Lyon	71
Queen Anne's	86	Marquette	279	Mahnomen	1
Saint Mary's	320	Mason	96	Marshall	11
Somerset	43	Mecosta	103	Martin	33
Talbot	116	Menominee	45	Mc Leod	34
Washington	298	Midland	280	Meeker	22
Wicomico	184	Missaukee	36	Mille Lacs	49
Worcester	107	Monroe	342	Morrison	40
Massachusetts		Montcalm	171	Mower	90
Barnstable	1123	Montmorency	29	Murray	7
Berkshire	489	Muskegon	382	Nicollet	40
Bristol	1214	Newaygo	108	Nobles	49
Dukes	50	Oakland	2028	Norman	12
Essex	1852	Oceana	61	Olmsted	471
Franklin	209	Ogemaw	44	Otter Tail	152
Hampden	938	Ontonagon	23	Pennington	30
Hampshire	390	Osceola	40	Pine	37
Middlesex	4753	Oscoda	34	Pipestone	11
Nantucket	27	Otsego	107	Polk	70
Norfolk	1535	Ottawa	563	Pope	19
Plymouth	1058	Presque Isle	29	Ramsey	1218
Suffolk	568	Roscommon	97	Red Lake	5
Worcester	1893	Saginaw	340	Redwood	19
Michigan		Saint Clair	493	Renville	16
Alcona	24	Saint Joseph	137	Rice	89
Alger	32	Sanilac	92	Rock	21
Allegan	224	Schoolcraft	25	Roseau	57
Alpena	63	Shiawassee	169	Saint Louis	711
Antrim	70	Tuscola	91	Scott	124
Arenac	44	Van Buren	242	Sherburne	83
Baraga	26	Washtenaw	749	Sibley	10
		Wayne	2824	Stearns	191

Name	Count	Name	Count	Name	Count
Steele	62	Scott	42	Lafayette	77
Stevens	7	Sharkey	35	Lawrence	92
Swift	30	Simpson	39	Lewis	17
Todd	29	Smith	11	Lincoln	108
Traverse	3	Stone	23	Linn	34
Wabasha	38	Sunflower	23	Livingston	23
Wadena	31	Tallahatchie	9	Macon	80
Waseca	90	Tate	31	Madison	19
Washington	392	Tippah	39	Maries	12
Watonwan	10	Tishomingo	64	Marion	129
Wilkin	23	Tunica	5	Mc Donald	33
Winona	141	Union	44	Mercer	7
Wright	78	Walthall	12	Miller	33
Yellow Medicine	12	Warren	164	Mississippi	8
Mississippi		Washington	61	Moniteau	20
Adams	59	Wayne	18	Monroe	22
Alcorn	95	Webster	24	Montgomery	23
Amite	18	Wilkinson	6	Morgan	33
Attala	34	Winston	37	New Madrid	32
Benton	12	Yalobusha	16	Newton	100
Bolivar	33	Yazoo	14	Nodaway	30
Calhoun	10	**Missouri**		Oregon	41
Carroll	5	Adair	45	Osage	14
Chickasaw	19	Andrew	103	Ozark	26
Choctaw	24	Atchison	4	Pemiscot	13
Claiborne	10	Audrain	53	Perry	28
Clarke	15	Barry	156	Pettis	80
Clay	26	Barton	20	Phelps	143
Coahoma	17	Bates	30	Pike	32
Copiah	32	Benton	65	Platte	212
Covington	15	Bollinger	17	Polk	76
De Soto	174	Boone	278	Pulaski	73
Forrest	204	Buchanan	125	Putnam	5
Franklin	8	Butler	104	Ralls	20
George	68	Caldwell	20	Randolph	79
Greene	5	Callaway	75	Ray	42
Grenada	21	Camden	103	Reynolds	9
Hancock	115	Cape Girardeau	127	Ripley	15
Harrison	532	Carroll	32	Saint Charles	599
Hinds	401	Carter	14	Saint Clair	30
Holmes	12	Cass	213	Saint Francois	120
Humphreys	4	Cedar	68	Saint Louis	2426
Itawamba	37	Chariton	17	Saint Louis City (city)	319
Jackson	275	Christian	93	Sainte Genevieve	20
Jasper	21	Clark	7	Saline	49
Jefferson	4	Clay	606	Schuyler	13
Jefferson Davis	16	Clinton	45	Scotland	4
Jones	141	Cole	187	Scott	76
Kemper	10	Cooper	31	Shannon	7
Lafayette	25	Crawford	41	Shelby	11
Lamar	50	Dade	23	Stoddard	63
Lauderdale	147	Dallas	32	Stone	115
Lawrence	17	Daviess	23	Sullivan	7
Leake	12	De Kalb	24	Taney	135
Lee	143	Dent	56	Texas	60
Leflore	24	Douglas	18	Vernon	37
Lincoln	20	Dunklin	38	Warren	51
Lowndes	73	Franklin	197	Washington	24
Madison	65	Gasconade	24	Wayne	32
Marion	31	Gentry	10	Webster	69
Marshall	28	Greene	636	Worth	13
Monroe	81	Grundy	23	Wright	49
Montgomery	18	Harrison	7	**Montana**	
Neshoba	36	Henry	86	Beaverhead	19
Newton	62	Hickory	39	Big Horn	18
Noxubee	2	Holt	6	Blaine	4
Oktibbeha	116	Howard	21	Broadwater	6
Panola	8	Howell	87	Carbon	24
Pearl River	110	Iron	18	Cascade	259
Perry	9	Jackson	1565	Chouteau	21
Pike	48	Jasper	344	Custer	54
Pontotoc	37	Jefferson	354	Daniels	12
Prentiss	59	Johnson	91	Dawson	51
Quitman	7	Knox	2	Deer Lodge	33
Rankin	182	Laclede	79	Fallon	13

Fergus	33	Garfield	10	Cheshire	279
Flathead	224	Gosper	6	Coos	93
Gallatin	320	Grant	1	Grafton	248
Garfield	1	Greeley	2	Hillsborough	1690
Glacier	15	Hall	131	Merrimack	457
Golden Valley	5	Hamilton	23	Rockingham	1142
Granite	5	Harlan	13	Strafford	325
Hill	62	Hitchcock	9	Sullivan	202
Jefferson	37	Holt	38	**New Jersey**	
Judith Basin	7	Hooker	2	Atlantic	554
Lake	47	Howard	9	Bergen	2182
Lewis And Clark	198	Jefferson	23	Burlington	1067
Liberty	12	Johnson	4	Camden	1206
Lincoln	67	Kearney	15	Cape May	258
Madison	15	Keith	29	Cumberland	321
Mc Cone	2	Keya Paha	3	Essex	926
Mineral	10	Kimball	12	Gloucester	477
Missoula	260	Knox	18	Hudson	597
Musselshell	1	Lancaster	745	Hunterdon	374
Park	81	Lincoln	139	Mercer	733
Phillips	9	Logan	1	Middlesex	1387
Pondera	14	Loup	2	Monmouth	1660
Powder River	2	Madison	99	Morris	1339
Powell	14	Merrick	16	Ocean	1379
Prairie	4	Morrill	15	Passaic	857
Ravalli	141	Nance	8	Salem	166
Richland	44	Nemaha	17	Somerset	688
Roosevelt	35	Nuckolls	8	Sussex	509
Rosebud	63	Otoe	36	Union	983
Sanders	31	Pawnee	9	Warren	271
Sheridan	22	Perkins	3	**New Mexico**	
Silver Bow	138	Phelps	21	Bernalillo	1667
Stillwater	22	Pierce	5	Catron	9
Sweet Grass	4	Platte	53	Chaves	160
Teton	7	Polk	13	Cibola	34
Toole	16	Red Willow	43	Colfax	24
Treasure	2	Richardson	30	Curry	119
Valley	32	Rock	7	De Baca	4
Wheatland	3	Saline	40	Dona Ana	556
Wibaux	4	Sarpy	322	Eddy	98
Yellowstone	299	Saunders	43	Grant	168
Nebraska		Scotts Bluff	85	Guadalupe	1
Adams	68	Seward	35	Harding	2
Antelope	36	Sheridan	12	Hidalgo	12
Banner	2	Sherman	7	Lea	134
Boone	31	Stanton	24	Lincoln	35
Box Butte	23	Thayer	5	Los Alamos	354
Boyd	3	Thomas	2	Luna	137
Brown	8	Thurston	2	Mc Kinley	57
Buffalo	92	Valley	6	Otero	280
Burt	13	Washington	42	Quay	18
Butler	18	Wayne	22	Rio Arriba	36
Cass	69	Webster	5	Roosevelt	19
Cedar	5	York	31	San Juan	163
Chase	2	**Nevada**		San Miguel	38
Cherry	21	Carson City	225	Sandoval	235
Cheyenne	47	Churchill	43	Santa Fe	256
Clay	19	Clark	1924	Sierra	53
Colfax	37	Douglas	220	Socorro	104
Cuming	7	Elko	73	Taos	47
Custer	36	Esmeralda	13	Torrance	20
Dakota	35	Eureka	6	Union	4
Dawes	46	Humboldt	29	Valencia	112
Dawson	39	Lander	12	**New York**	
Deuel	1	Lincoln	9	Albany	658
Dixon	13	Lyon	141	Allegany	111
Dodge	150	Mineral	14	Bronx	841
Douglas	848	Nye	176	Broome	763
Dundy	4	Pershing	8	Cattaraugus	171
Fillmore	5	Storey	4	Cayuga	238
Franklin	6	Washoe	1064	Chautauqua	426
Frontier	7	White Pine	18	Chemung	297
Furnas	48	**New Hampshire**		Chenango	190
Gage	35	Belknap	263	Clinton	235
Garden	3	Carroll	142	Columbia	222

County		County		County	
Cortland	141	Columbus	45	Yadkin	78
Delaware	131	Craven	237	Yancey	119
Dutchess	956	Cumberland	786	**North Dakota**	
Erie	2030	Currituck	35	Adams	5
Essex	100	Dare	87	Barnes	37
Franklin	168	Davidson	277	Benson	22
Fulton	217	Davie	56	Billings	5
Genesee	204	Duplin	43	Bottineau	14
Greene	139	Durham	389	Bowman	23
Hamilton	22	Edgecombe	83	Burke	12
Herkimer	208	Forsyth	802	Burleigh	235
Jefferson	285	Franklin	62	Cass	319
Kings	1660	Gaston	385	Cavalier	3
Lewis	63	Gates	7	Dickey	10
Livingston	127	Graham	44	Divide	3
Madison	185	Granville	62	Dunn	2
Monroe	2267	Greene	14	Eddy	6
Montgomery	122	Guilford	1022	Emmons	4
Nassau	3020	Halifax	60	Foster	7
New York	1297	Harnett	93	Golden Valley	4
Niagara	574	Haywood	168	Grand Forks	177
Oneida	830	Henderson	342	Grant	2
Onondaga	1344	Hertford	23	Griggs	18
Ontario	307	Hoke	33	Hettinger	7
Orange	687	Hyde	6	Kidder	4
Orleans	104	Iredell	194	La Moure	8
Oswego	330	Jackson	97	Logan	6
Otsego	224	Johnston	142	Mc Henry	10
Putnam	239	Jones	11	Mc Intosh	5
Queens	2278	Lee	71	Mc Kenzie	10
Rensselaer	443	Lenoir	132	Mc Lean	10
Richmond	1653	Lincoln	123	Mercer	11
Rockland	630	Macon	136	Morton	49
Saint Lawrence	335	Madison	58	Mountrail	10
Saratoga	588	Martin	39	Nelson	12
Schenectady	608	Mc Dowell	156	Oliver	10
Schoharie	94	Mecklenburg	1200	Pembina	29
Schuyler	53	Mitchell	69	Pierce	9
Seneca	84	Montgomery	20	Ramsey	40
Steuben	268	Moore	112	Ransom	12
Suffolk	3849	Nash	94	Renville	4
Sullivan	186	New Hanover	350	Richland	33
Tioga	247	Northampton	17	Rolette	10
Tompkins	335	Onslow	260	Sargent	8
Ulster	536	Orange	304	Sheridan	2
Warren	173	Pamlico	41	Sioux	4
Washington	118	Pasquotank	77	Slope	3
Wayne	315	Pender	63	Stark	115
Westchester	1718	Perquimans	18	Steele	4
Wyoming	113	Person	31	Stutsman	68
Yates	87	Pitt	350	Towner	12
North Carolina		Polk	61	Traill	27
Alamance	270	Randolph	208	Walsh	27
Alexander	53	Richmond	82	Ward	169
Alleghany	35	Robeson	95	Wells	6
Anson	35	Rockingham	226	Williams	44
Ashe	51	Rowan	248	**Ohio**	
Avery	55	Rutherford	161	Adams	157
Beaufort	110	Sampson	48	Allen	284
Bertie	13	Scotland	41	Ashland	135
Bladen	37	Stanly	104	Ashtabula	287
Brunswick	140	Stokes	54	Athens	184
Buncombe	699	Surry	141	Auglaize	158
Burke	225	Swain	22	Belmont	250
Cabarrus	259	Transylvania	131	Brown	91
Caldwell	175	Tyrrell	6	Butler	1174
Camden	17	Union	235	Carroll	46
Carteret	180	Vance	25	Champaign	99
Caswell	21	Wake	1839	Clark	442
Catawba	383	Warren	12	Clermont	575
Chatham	79	Washington	22	Clinton	166
Cherokee	89	Watauga	68	Columbiana	431
Chowan	19	Wayne	160	Coshocton	119
Clay	35	Wilkes	104	Crawford	129
Cleveland	324	Wilson	86	Cuyahoga	2966

County	Count	County	Count	County	Count
Darke	176	Beckham	39	Benton	339
Defiance	108	Blaine	46	Clackamas	1118
Delaware	294	Bryan	121	Clatsop	198
Erie	242	Caddo	50	Columbia	180
Fairfield	306	Canadian	237	Coos	285
Fayette	82	Carter	112	Crook	40
Franklin	2310	Cherokee	58	Curry	159
Fulton	142	Choctaw	37	Deschutes	447
Gallia	70	Cimarron	9	Douglas	417
Geauga	250	Cleveland	614	Grant	14
Greene	534	Coal	16	Harney	29
Guernsey	118	Comanche	333	Hood River	88
Hamilton	2325	Cotton	13	Jackson	734
Hancock	222	Craig	65	Jefferson	51
Hardin	80	Creek	147	Josephine	453
Harrison	71	Custer	42	Klamath	498
Henry	71	Delaware	57	Lake	24
Highland	142	Dewey	20	Lane	1163
Hocking	64	Ellis	13	Lincoln	263
Holmes	76	Garfield	178	Linn	421
Huron	139	Garvin	52	Malheur	53
Jackson	77	Grady	98	Marion	773
Jefferson	233	Grant	12	Morrow	26
Knox	116	Greer	16	Multnomah	2047
Lake	713	Harmon	3	Polk	201
Lawrence	153	Harper	15	Sherman	7
Licking	418	Haskell	26	Tillamook	129
Logan	194	Hughes	22	Umatilla	323
Lorain	963	Jackson	127	Union	95
Lucas	1096	Jefferson	9	Wallowa	20
Madison	85	Johnston	9	Wacso	84
Mahoning	648	Kay	168	Washington	1588
Marion	207	Kingfisher	51	Wheeler	6
Medina	431	Kiowa	27	Yamhill	273
Meigs	54	Latimer	30	**Pennsylvania**	
Mercer	73	Le Flore	68	Adams	158
Miami	439	Lincoln	76	Allegheny	2546
Monroe	51	Logan	46	Armstrong	170
Montgomery	2270	Love	22	Beaver	633
Morgan	20	Major	33	Bedford	90
Morrow	72	Marshall	28	Berks	673
Muskingum	219	Mayes	88	Blair	271
Noble	20	Mc Clain	80	Bradford	201
Ottawa	136	Mc Curtain	29	Bucks	1600
Paulding	61	Mc Intosh	52	Butler	445
Perry	62	Murray	44	Cambria	276
Pickaway	105	Muskogee	134	Cameron	23
Pike	51	Noble	12	Carbon	120
Portage	450	Nowata	25	Centre	303
Preble	151	Okfuskee	22	Chester	1030
Putnam	81	Oklahoma	1869	Clarion	87
Richland	441	Okmulgee	112	Clearfield	160
Ross	196	Osage	101	Clinton	29
Sandusky	133	Ottawa	110	Columbia	200
Scioto	239	Pawnee	38	Crawford	236
Seneca	153	Payne	157	Cumberland	473
Shelby	106	Pittsburg	117	Dauphin	474
Stark	1095	Pontotoc	71	Delaware	1552
Summit	1649	Pottawatomie	150	Elk	73
Trumbull	604	Pushmataha	15	Erie	692
Tuscarawas	280	Roger Mills	13	Fayette	330
Union	75	Rogers	145	Forest	17
Van Wert	90	Seminole	37	Franklin	186
Vinton	29	Sequoyah	58	Fulton	14
Warren	470	Stephens	155	Greene	67
Washington	226	Texas	43	Huntingdon	126
Wayne	334	Tillman	24	Indiana	231
Williams	87	Tulsa	1665	Jefferson	141
Wood	348	Wagoner	116	Juniata	35
Wyandot	39	Washington	239	Lackawanna	421
Oklahoma		Washita	24	Lancaster	953
Adair	11	Woods	11	Lawrence	224
Alfalfa	5	Woodward	70	Lebanon	279
Atoka	54	**Oregon**		Lehigh	598
Beaver	19	Baker	76	Luzerne	654

Lycoming	279	Spartanburg	409	Bledsoe	15
Mc Kean	133	Sumter	222	Blount	316
Mercer	326	Union	37	Bradley	366
Mifflin	67	Williamsburg	10	Campbell	178
Monroe	302	York	370	Cannon	16
Montgomery	1896	**South Dakota**		Carroll	62
Montour	57	Aurora	1	Carter	159
Northampton	601	Beadle	52	Cheatham	32
Northumberland	177	Bennett	3	Chester	44
Perry	66	Bon Homme	12	Claiborne	56
Philadelphia	1398	Brookings	85	Clay	8
Pike	107	Brown	90	Cocke	79
Potter	40	Brule	11	Coffee	171
Schuylkill	293	Buffalo	2	Crockett	15
Snyder	58	Butte	8	Cumberland	97
Somerset	126	Charles Mix	5	Davidson	1020
Sullivan	10	Clark	15	De Kalb	16
Susquehanna	180	Clay	44	Decatur	17
Tioga	187	Codington	87	Dickson	81
Union	58	Corson	4	Dyer	51
Venango	157	Custer	16	Fayette	20
Warren	116	Davison	36	Fentress	35
Washington	444	Day	13	Franklin	74
Wayne	149	Deuel	17	Gibson	148
Westmoreland	786	Dewey	6	Giles	52
Wyoming	100	Douglas	5	Grainger	26
York	924	Edmunds	6	Greene	188
Rhode Island		Fall River	75	Grundy	18
Bristol	145	Faulk	5	Hamblen	184
Kent	552	Grant	23	Hamilton	1093
Newport	373	Gregory	1	Hancock	4
Providence	1192	Haakon	4	Hardeman	20
Washington	342	Hamlin	19	Hardin	38
South Carolina		Hand	5	Hawkins	86
Abbeville	31	Hanson	7	Haywood	20
Aiken	267	Harding	2	Henderson	19
Allendale	10	Hughes	60	Henry	100
Anderson	308	Hutchinson	16	Hickman	18
Bamberg	25	Hyde	4	Houston	9
Barnwell	34	Jackson	1	Humphreys	24
Beaufort	146	Jerauld	1	Jackson	15
Berkeley	203	Jones	1	Jefferson	101
Calhoun	19	Kingsbury	7	Johnson	24
Charleston	531	Lake	20	Knox	1404
Cherokee	77	Lawrence	43	Lake	5
Chester	46	Lincoln	48	Lauderdale	8
Chesterfield	70	Lyman	6	Lawrence	55
Clarendon	26	Marshall	5	Lewis	5
Colleton	46	Mc Cook	7	Lincoln	101
Darlington	46	Mc Pherson	1	Loudon	171
Dillon	19	Meade	52	Macon	13
Dorchester	182	Mellette	1	Madison	189
Edgefield	12	Miner	2	Marion	57
Fairfield	19	Minnehaha	223	Marshall	47
Florence	131	Moody	10	Maury	117
Georgetown	71	Pennington	218	Mc Minn	150
Greenville	786	Perkins	2	Mc Nairy	59
Greenwood	125	Potter	3	Meigs	37
Hampton	15	Roberts	29	Monroe	72
Horry	238	Sanborn	6	Montgomery	164
Jasper	7	Shannon	1	Moore	3
Kershaw	59	Spink	22	Morgan	140
Lancaster	114	Stanley	8	Obion	60
Laurens	76	Sully	5	Overton	20
Lee	20	Todd	2	Perry	15
Lexington	409	Tripp	7	Pickett	4
Marion	20	Turner	13	Polk	55
Marlboro	13	Union	17	Putnam	166
Mc Cormick	7	Walworth	17	Rhea	84
Newberry	50	Yankton	49	Roane	241
Oconee	176	Ziebach	1	Robertson	117
Orangeburg	92	**Tennessee**		Rutherford	244
Pickens	253	Anderson	461	Scott	33
Richland	493	Bedford	95	Sequatchie	15
Saluda	17	Benton	35	Sevier	176

Shelby	1936	De Witt	21	King	2
Smith	36	Deaf Smith	13	Kinney	14
Stewart	20	Delta	13	Kleberg	39
Sullivan	550	Denton	1000	Knox	7
Sumner	254	Dickens	1	La Salle	6
Tipton	84	Dimmit	3	Lamar	117
Trousdale	2	Donley	7	Lamb	15
Unicoi	49	Duval	10	Lampasas	37
Union	21	Eastland	47	Lavaca	29
Van Buren	11	Ector	221	Lee	31
Warren	56	Edwards	4	Leon	49
Washington	372	El Paso	988	Liberty	186
Wayne	16	Ellis	234	Limestone	25
Weakley	55	Erath	66	Lipscomb	14
White	42	Falls	36	Live Oak	6
Williamson	222	Fannin	121	Llano	61
Wilson	194	Fayette	55	Lubbock	554
Texas		Fisher	2	Lynn	20
Anderson	154	Floyd	6	Madison	27
Andrews	18	Foard	7	Marion	17
Angelina	145	Fort Bend	403	Martin	6
Aransas	72	Franklin	48	Mason	4
Archer	12	Freestone	46	Matagorda	89
Armstrong	5	Frio	15	Maverick	41
Atascosa	36	Gaines	31	Mc Culloch	21
Austin	48	Galveston	713	Mc Lennan	461
Bailey	10	Garza	9	Mc Mullen	1
Bandera	47	Gillespie	54	Medina	51
Bastrop	130	Glasscock	1	Menard	1
Baylor	10	Goliad	4	Midland	330
Bee	49	Gonzales	41	Milam	42
Bell	444	Gray	60	Mills	18
Bexar	2560	Grayson	341	Mitchell	7
Blanco	21	Gregg	297	Montague	57
Bosque	64	Grimes	25	Montgomery	525
Bowie	211	Guadalupe	138	Moore	17
Brazoria	642	Hale	73	Morris	60
Brazos	363	Hall	8	Motley	1
Brewster	67	Hamilton	25	Nacogdoches	94
Briscoe	4	Hansford	11	Navarro	111
Brooks	1	Hardeman	8	Newton	18
Brown	90	Hardin	123	Nolan	38
Burleson	21	Harris	5419	Nueces	582
Burnet	77	Harrison	88	Ochiltree	31
Caldwell	32	Hartley	1	Oldham	3
Calhoun	75	Haskell	20	Orange	186
Callahan	48	Hays	148	Palo Pinto	69
Cameron	337	Hemphill	5	Panola	72
Camp	24	Henderson	191	Parker	285
Carson	10	Hidalgo	357	Parmer	4
Cass	84	Hill	83	Pecos	64
Castro	8	Hockley	60	Polk	220
Chambers	32	Hood	154	Potter	203
Cherokee	95	Hopkins	61	Presidio	8
Childress	20	Houston	29	Rains	19
Clay	46	Howard	71	Randall	202
Cochran	1	Hudspeth	6	Real	13
Coke	4	Hunt	231	Red River	31
Coleman	22	Hutchinson	78	Reeves	12
Collin	1159	Irion	1	Refugio	38
Collingsworth	4	Jack	6	Roberts	4
Colorado	31	Jackson	19	Robertson	18
Comal	177	Jasper	65	Rockwall	78
Comanche	35	Jeff Davis	10	Runnels	10
Concho	4	Jefferson	561	Rusk	83
Cooke	79	Jim Hogg	9	Sabine	32
Coryell	105	Jim Wells	29	San Augustine	13
Cottle	1	Johnson	411	San Jacinto	31
Crane	5	Jones	28	San Patricio	110
Crockett	12	Karnes	12	San Saba	4
Crosby	7	Kaufman	201	Schleicher	5
Culberson	6	Kendall	113	Scurry	44
Dallam	9	Kent	1	Shackelford	10
Dallas	4867	Kerr	216	Shelby	24
Dawson	35	Kimble	9	Smith	462

Somervell	7	Essex	21	Isle Of Wight	53
Starr	11	Franklin	106	James City	43
Stephens	9	Grand Isle	25	King And Queen	5
Sterling	4	Lamoille	62	King George	56
Stonewall	3	Orange	110	King William	10
Sutton	4	Orleans	68	Lancaster	59
Swisher	16	Rutland	178	Lee	82
Tarrant	3609	Washington	231	Lexington (city)	49
Taylor	381	Windham	148	Loudoun	291
Terry	28	Windsor	204	Louisa	51
Titus	9	**Virginia**		Lunenburg	8
Tom Green	286	Accomack	136	Lynchburg (city)	357
Travis	1761	Albemarle	243	Madison	17
Trinity	26	Alexandria (city)	215	Manassas (city)	201
Tyler	30	Alleghany	3	Manassas Park (city)	103
Upshur	50	Amelia	16	Martinsville (city)	67
Upton	5	Amherst	72	Mathews	33
Uvalde	70	Appomattox	36	Mecklenburg	61
Val Verde	85	Arlington	465	Middlesex	36
Van Zandt	74	Augusta	114	Montgomery	215
Victoria	173	Bath	16	Nelson	20
Walker	83	Bedford	157	New Kent	28
Waller	31	Bedford (city)	70	Newport News (city)	385
Ward	26	Bland	25	Norfolk (city)	390
Washington	60	Botetourt	156	Northampton	19
Webb	128	Bristol (city)	88	Northumberland	41
Wharton	49	Brunswick	8	Norton (city)	17
Wheeler	14	Buchanan	33	Nottoway	18
Wichita	421	Buckingham	10	Orange	92
Wilbarger	27	Buena Vista (city)	25	Page	50
Willacy	21	Campbell	97	Patrick	31
Williamson	581	Caroline	32	Petersburg (city)	54
Wilson	34	Carroll	57	Pittsylvania	82
Winkler	18	Charles City	2	Poquoson (city)	47
Wise	88	Charlotte	23	Portsmouth (city)	180
Wood	76	Charlottesville (city)	64	Powhatan	36
Yoakum	6	Chesapeake City (city)	389	Prince Edward	43
Young	51	Chesterfield	599	Prince George	32
Zapata	22	Clarke	24	Prince William	575
Zavala	2	Clifton Forge (city)	15	Pulaski	73
Utah		Colonial Heights (city)	27	Radford (city)	52
Beaver	8	Covington (city)	66	Rappahannock	21
Box Elder	147	Craig	10	Richmond	15
Cache	418	Culpeper	94	Richmond (city)	175
Carbon	105	Cumberland	6	Roanoke	209
Daggett	9	Danville (city)	138	Roanoke (city)	233
Davis	714	Dickenson	48	Rockbridge	19
Duchesne	135	Dinwiddie	10	Rockingham	116
Emery	44	Emporia (city)	1	Russell	78
Garfield	4	Essex	20	Salem (city)	102
Grand	29	Fairfax	2787	Scott	79
Iron	136	Fairfax (city)	201	Shenandoah	57
Juab	20	Falls Church (city)	61	Smyth	64
Kane	51	Fauquier	203	South Boston (city)	52
Millard	65	Floyd	57	Southampton	7
Morgan	44	Fluvanna	35	Spotsylvania	166
Piute	2	Franklin	55	Stafford	233
Rich	4	Franklin (city)	6	Staunton (city)	86
Salt Lake	2284	Frederick	150	Suffolk (city)	67
San Juan	20	Fredericksburg (city)	4	Surry	5
Sanpete	75	Galax (city)	67	Sussex	11
Sevier	77	Giles	48	Tazewell	157
Summit	75	Gloucester	114	Virginia Beach (city)	890
Tooele	95	Goochland	18	Warren	74
Uintah	67	Grayson	24	Washington	77
Utah	1592	Greene	23	Waynesboro (city)	92
Wasatch	59	Greensville	1	Westmoreland	41
Washington	257	Halifax	66	Williamsburg (city)	5
Wayne	8	Hampton City (city)	304	Winchester (city)	109
Weber	489	Hanover	159	Wise	110
Vermont		Harrisonburg (city)	124	Wythe	115
Addison	103	Henrico	569	York	228
Bennington	155	Henry	72	**Washington**	
Caledonia	87	Highland	6	Adams	26
Chittenden	643	Hopewell (city)	54	Asotin	75

County	#	County	#	County	#
Benton	548	Mc Dowell	168	Langlade	70
Chelan	257	Mercer	317	Lincoln	123
Clallam	426	Mineral	71	Manitowoc	164
Clark	1163	Mingo	70	Marathon	255
Columbia	17	Monongalia	241	Marinette	114
Cowlitz	433	Monroe	17	Marquette	43
Douglas	119	Morgan	49	Menominee	3
Ferry	24	Nicholas	135	Milwaukee	1494
Franklin	146	Ohio	156	Monroe	72
Grant	214	Pendleton	95	Oconto	48
Grays Harbor	303	Pleasants	31	Oneida	164
Island	379	Pocahontas	31	Outagamie	297
Jefferson	280	Preston	79	Ozaukee	287
King	7021	Putnam	172	Pepin	14
Kitsap	1277	Raleigh	276	Pierce	66
Kittitas	167	Randolph	70	Polk	62
Klickitat	106	Ritchie	34	Portage	165
Lewis	248	Roane	34	Price	55
Lincoln	34	Summers	67	Racine	431
Mason	218	Taylor	49	Richland	20
Okanogan	145	Tucker	22	Rock	252
Pacific	120	Tyler	57	Rusk	22
Pend Oreille	205	Upshur	107	Saint Croix	127
Pierce	2478	Wayne	76	Sauk	84
San Juan	167	Webster	14	Sawyer	41
Skagit	654	Wetzel	136	Shawano	79
Skamania	19	Wirt	19	Sheboygan	204
Snohomish	2441	Wood	309	Taylor	39
Spokane	1500	Wyoming	91	Trempealeau	55
Stevens	125	**Wisconsin**		Vernon	35
Thurston	794	Adams	41	Vilas	66
Wahkiakum	12	Ashland	45	Walworth	152
Walla Walla	353	Barron	78	Washburn	50
Whatcom	600	Bayfield	33	Washington	187
Whitman	162	Brown	377	Waukesha	824
Yakima	592	Buffalo	18	Waupaca	81
West Virginia		Burnett	32	Waushara	45
Barbour	30	Calumet	38	Winnebago	283
Berkeley	166	Chippewa	187	Wood	231
Boone	60	Clark	54	**Wyoming**	
Braxton	48	Columbia	141	Albany	143
Brooke	50	Crawford	17	Big Horn	20
Cabell	382	Dane	941	Campbell	74
Calhoun	41	Dodge	125	Carbon	64
Clay	17	Door	84	Converse	23
Doddridge	17	Douglas	159	Crook	18
Fayette	162	Dunn	102	Fremont	97
Gilmer	30	Eau Claire	240	Goshen	65
Grant	23	Florence	15	Hot Springs	11
Greenbrier	101	Fond Du Lac	181	Johnson	10
Hampshire	33	Forest	11	Laramie	320
Hancock	94	Grant	53	Lincoln	51
Hardy	15	Green	55	Natrona	237
Harrison	272	Green Lake	38	Niobrara	8
Jackson	93	Iowa	23	Park	72
Jefferson	79	Iron	12	Platte	23
Kanawha	674	Jackson	18	Sheridan	105
Lewis	93	Jefferson	127	Sublette	14
Lincoln	39	Juneau	51	Sweetwater	122
Logan	172	Kenosha	222	Teton	37
Marion	158	Kewaunee	29	Uinta	52
Marshall	144	La Crosse	173	Washakie	28
Mason	40	Lafayette	12	Weston	9

A BRIEF HISTORY OF THE CQ WW CONTEST

Since the basis of much of the contest revolves around the concept of the WAZ Zones, it may be useful for the reader to review the history of the WAZ Award (see the Awards chapter in this *Almanac*).

The CQ World Wide DX Contest started in 1948, three years after *CQ* magazine was first published. But the very first "Worldwide DX Contest" grew out of an idea developed by Herb Becker, W6QD, the DX contributing editor for *Radio* magazine in 1939.

In the October 1939 issue, Herb announced the first *Radio* magazine World-Wide DX Contest. In Herb's words,

The World-Wide Contest which should be a break for the working man and those in schools, will be over two weekends, with 48 hours each. The starting time will be 0200 GMT November 25 and December 2. The competition will be divided into two divisions, CW. and phone. Each of these two divisions will be divided into two sections—the one-operator section and the more-than-one-operator section. CW. stations must work CW. stations, and phone stations must work phone stations. Competitors in the one-operator section may use one transmitter only, and competitors in the more-than-one-operator section may use any number of transmitters. Any number of receivers may be used by all competitors. Remember that the 7, 14, and 28 MHz bands are the only ones used in the contest.

This first WW DX contest was held in 1939 and reported in the June 1940 issue of *Radio*. It was in June of that same year that the US government suspended amateur radio activity due to World War II. Accordingly, W6QD reported in the July issue of *Radio* that the 1940 contest was canceled. WW II prevented the "World Wide Contest" from taking hold.

The thrust of *Radio* magazine shifted more and more towards the engineering side of radio, and in 1945, Radio Magazines, Inc. founded *CQ* to take up the needs of amateurs. *Radio* magazine itself continued until 1947 when it was renamed *Audio Engineering* and continues to this day as *Audio* magazine.

After the War, *CQ* showed up with a familiar layout—the old pre-war *Radio* layout. W6QD showed up as the DX and overseas editor for the new *CQ* magazine. Hams were back on the air in the summer of 1946 and, in January 1947, *CQ* announced the CQ WAZ award. The WAZ map was the same as first published in 1936. In the October 1948 issue of *CQ*, the announcement was made of "CQ's World-Wide DX Contest" to be held on October 29 to 31 (phone) and the CW contest on November 5 to 7. The rules read just like the earlier *Radio* magazine World-Wide DX Contest. The contest period was again from 0200 GMT Saturday to 0200 GMT Monday. Only the 7, 14, and 27/28 MHz bands were to be used. The competition was not separated into single or multi-operator and there was only an all-band category. The results were reported in the June, 1949 issue. It was reported in that first contest that all 40 zones were active! In the write-up, all the scores were listed by zone. Under each zone the call area or country

was listed. Both single operator and multi-operator were competing against one another! Only the point score was given with no further breakdown.

In the second year of the contest (1949), single and multi-operator stations now competed for separate awards. Another improvement was the single-band category. There were about 1,450 logs submitted in the second year of the CQWW. There was such an unexpected outpouring of interest in this contest that it caught *CQ* by surprise. The results actually had to be published over two months!

By the time the third CQ WW contest occurred in 1950, participation had grown to about 1,600 entrants. The scores were listed for the first time by country instead of by zone. Single-band scores were listed for the first time. The contest continued to be the last full weekend of October and the first weekend of November. The phone and CW sections were now reported in separate issues.

In the 1951 results, Herb had the assistance of his friend W6ENV in reporting the scores. He sums up the essence of the contest saying,

It seems that what you [DX/Contesters] want is a contest where there is DX...Every now and again a few of the boys think that too much publicity is given the high point men. There are a couple of ways of looking at it. In all hobbies, the man who scores the highest number of points or comes out ahead attracts the most attention. We would be sadly neglectful if we did not recognize it.

In 1952, W6QD was back as the sole author for the contest results. This was to be Herb's last year as the DX overseas editor for *CQ*. The DX column was handed over to Dick Spenceley, KV4AA. Dick was living in the Virgin Islands, and he could not take on the logistical problem of having contest logs mailed to him and then back to *CQ* in New York.

So in 1953, under the guidance of Larry LeKashman, W9IOP, *CQ* turned sponsorship over to a group in Indiana and Michigan. The contest was now sponsored by the International DX Club and they promptly renamed it the "International DX Contest," even saying that it was formerly known as the World-Wide CQ DX Contest. The International DX Club issued their own certificates for the contest. A picture of one of these certificates appears along with the 1953 results in August 1954 *CQ*. The International DX Club could not find the time to check the logs, however, so they turned the checking process over to the Potomac Valley Radio Club of suburban Washington, D.C. The reporting was back to the early days. In the write-up, only the entrant's category and score were listed. In addition, only the top scores were listed. Full results were available for a self-addressed stamped envelope.

In 1954, the rules for the International DX Contest had the phone weekend occurring on October 23 and the CW weekend on October 30. The 1954 rules make the first mention of a club competition. When the contest results were reported by W9IOP, it was apparent that something was wrong. The International DX Club was not capable of fully checking the results. Only calls and scores were listed under each category—nothing else. As *CQ* was searching for what to do with the contest, Bill Leonard, W2SKE, wrote an article in the December 1955 issue which summarized briefly the 1955 CQ WW Contest that had just occurred, encouraging hams to enter the 1956 contest. It was at this time that *CQ* took back the contest. W9IOP rewrote the rules and *CQ* Editor Wayne Green, W2NSD, recruited Frank Anzalone, W1WY, to check the 1955 results.

The 1955 CQ World Wide DX Contest, as it was now renamed, was reported by Frank, W1WY, in the May issue. Up until 1955, there were no power categories in the CQ WW. Starting in 1955 there were five power categories! A: up to 30 W; B: up to 125 W; C: up to 500 W and D: over 500 W. With W1WY at the helm, the format of the modern CQ WW reporting took on its familiar style. The *CQ* contest committee was now located in the New York area and the checking of logs took on a serious nature.

When the 1956 results were reported, there was a new department on the *CQ* masthead: The Contest Calendar, edited by W1WY. Contesters now had their own column and a concerned patron. The 1956 rules changed the power category of two of the power groups: 30 now became 35 W and 125 W became 150 W.

From 1955 to 1973, Frank worked to bring the CQ WW up to its present level. The 1959 rules indicated a change in the country multiplier list. As always, the DXCC list was used, but now the WAE country list was added. This year also saw the elimination of the novice category, which had been initiated in 1955 to give new hams a chance to compete among themselves. It was now realized that novice power limitation and frequency privileges were very different all over the world. There was no way to make the category a fair one. The rules of 1959 saw the multi-operator category divided into two categories. Due mainly to K2GL's station and his enthusiasm, there was now a multi-single and a multi-multi category for competitors.

The next significant rule changes occurred in the 1962 contest. Up to 1961, all the CQ WW contests began at 0200 UTC Saturday morning and ended at 0200 UTC Monday morning, convenient times for West Coast U.S. operators. In 1962, the contest now began and ended at 0000 UTC, a logical time to start a world wide event. This year also saw the elimination of the power categories. Lastly, 1962 was when the North American two-point rule came into existence. With this rule, it was hoped that more activity would occur in the Caribbean and Central American countries. All these changes were brought into the rules by Frank, W1WY.

From the middle 60s to the early 70s the new multi-multi category flourished. All over the U.S., big stations such as K2GL, W3MSK (W3AU), W6RW, W6VSS (K6UA), W7RM and W4BVV were assembled. The operators of these stations were enlisted to help in checking the logs. Between 1970 and 1975, K3EST, N2AA (K2KUR), WB2SQN (K2SS), N6AA, WA6EPQ (N6AR), K6NA and N6TJ all joined the *CQ* Contest Committee. Being in New York, K2GL contributed first to the contest committee with Fred, W2IWC (now K6SSS). Fred was an operator at K2GL and had volunteered to help Frank in checking the logs. After the results were reported for the 1973 contest, Frank asked Fred to take over as CQ WW contest director while he remained as overall contest chairman. Fred took over and reported the 1974 and 1975 results. It was under Fred's leadership that the contest committee took on its present profile—that is representative of both U.S. coasts. He invited several well known W6s who were contesters and DXers to join the committee. The contest was now being run from Southern California were it began some 35 years earlier.

After two years, Fred resigned as director and suggested WA6EPQ (now N6AR) as his replacement. Frank, W1WY, wanted more balance and they agreed to name two directors from the ranks of the Contest Committee. In 1976, K3EST and N6AR took over at the helm of the contest.

In 1978, the QRP category was initiated and the assisted category began in 1987. In the early 1980s, K3EST brought DX advisors onto the committee. These advisors helped resolve local problems and help the directors understand the ideas of their particular country.

In the early 1980s, the Contest Committee and top competitors realized that a more thorough checking of top scores was necessary in order to certify winners. Committee member K2SS developed a computer log checking procedure which, with a few variations and enhancements added by N6TR, remains the method used to check logs today.

So there you have the history of the CQ WW in brief. As you can see, the contest has a long tradition, beginning with the first *Radio* magazine World-Wide DX Contest in 1939. The *CQ* Contest Committee, with the help of amateurs all over the world, retains the original enthusiasm of W6QD.

This brief history is from The CQ World Wide Contest Handbook, *by K3EST, available from R2R International, 1816 Poplar Ln, Davis, CA 95616, USA.*

(In the following sections, parentheses indicate operator call sign, brackets indicate operator's current call sign.)

DX WINNERS OF CQ WORLDWIDE DX PHONE CONTEST BY YEAR/CATEGORY

YEAR	S/O	S/O Asst	QRP	M/S	MM	160	80	40	20	15	10
1994	P4ØE(CT1BOH)	P4ØW(W2GD)	NP2Q	HC8A	PJ1B	IR4T(I4JMY)	VP2EC(N5AU)	PJ9U(OH1VR)	PY0FM(PY5CC)	ZD8Z(N6TJ)	PQ0MM
1993	P4ØW(W2GD)	CH3EJ(VE3EJ)	7Z2AB(K2XR)	PJ1B	EA9UK	IV3PRK	DL3LAB	PJ9U(OH1VR)	ZXØF(PY5EG)	ZW5B(N5FA)	LU6ETB
1992	HC8A(N6KT)	WM5G(KRØY)	AA2U	VP2EC	PJ1B	9A1HCD	TI1C(TI2CF)	PJ9E(OH5BM)	PJ9P(OH6MW)	ZV5A(PY5EG)	ZW5B(N5FA)
1991	CR3A(OH2BH)	N4RJ(KM9P)	4M1G(YV1CLM)	IQ4A	VP9AD	IV3PRK	GW4OFQ	ZF2JR(N6RJ)	YW1A(YV1AVO)	ZX9A(PY5CO)	ZV5AJ(PY5EG)
1990	CT3BH(OH2BH)	K1ZM	KR2Q	EA8AGD	PJ1B	UG6GAW	HA8IE	ZF2JR(N6RJ)	YV3A(YV5IVB)	EI4ØR(KK3RV)	ZX9A(PY5CO)
1989	EA8RCT(N6KT)	YT3AA	YU2TY	KP2A	PJ1B	IH9/IV3PRK	IK5MAE	DJ4PT	YW1A(YV1AVO)	P4ØR(KU4EE)	Z76Y(PY5CY)
1988	CT9BZ(OH2BH)	*"A"*	ZY5EG(PY5CC)	VP2EC	P4ØV	YU3MM	IK5BAF	T11W(T2KD)	VP2ET(K5RX)	VP2ET(K5RX)	P4ØR(KU4EE)
1987	9Y4T(NQ4I)		PJ2FR(K7SS)	KP2A	PJ1B	UG7GWO	P4ØR(KU4EE)	EA8RCT(OH2MM)	P4ØSS(K2SS)	ZP5Y(PY5JCY)	VP2E(TK5RX)
1986	8R1X(NQ4I)		WP4G	4V2C	HC8X	LZ2CJ	VP4A	VP2ET(K5RX)	OH8PS	PY5EG	ZP5Y(PY5JCY)
1985	P2FR(N6KT)		WP4G	9Y4W	TI1C	LZ2CJ	YV3AZC	EA8AK	YZ9A(YT3AA)	ZZ5EG(N5FA)	LU1E(LU3AJW)
1984	P2FR(N6KT)		UP2BIM	9Y4W	TI1C	LZ2CJ	YV3AZC	YV2AMM	EA8AK	TI2CF	TI2KD
1983	P2FR(N6KT)		AA2Z/1	VP2MFW	OHØW	UP2BBT/U6V	4M3AZC	ZM1BIL	YV2AMM	HC1OT	CE6EZ
1982	P2FR(N6KT)		TG9GI	I4RYC	P41C	UP2BAW	YV3AZC	YS1X	ZM1BIL	KD7P/KH2	CE6EZ
1981	9Y4VT(N6AA)		I5NSR	HI8XWP	PJ2CC	YU3EF	4M3AZC	SP3DOI	YS1X	AHØAB(JA3ODC)	YV2AMM
1980	EA8AK		G4BUE	FY7BC	VP2KC	PA5ØHP	CT3BZ(OH2BH)	I5NPH	VP2KAA(N4PN)	VP2KAC(N4RJ)	ZZ5EG
1979	9Y4VT(N6AA)		W6POZ	FM0FC	PJ9JR	GM3ZSP	W1CF	I3MAU	I5NPH	NP4A	KH6XX
1978	9Y4VT(N6AA)		VP9AD	VP2M	EA8CR	DJ8WLA	KP4RF	KX6LA	UA6HZ	H31LR(HP1XOJ)	OH2MM/CT3
1977	9JØC(K1JX)		*#Q*	VP2M	P5M	KV4FZ	EA8CR	CX4CR	YV2AMM	YU3ZV	PY1MAG
1976	9JØG(K1JX)			PY2CAB	DLØPG	XJ3BMV	KV4FZ	OH5NW	FY7AK	CW3BR	CE6CZ
1975	FY7AK(F5OO)			VP2M	VP2M	PAØHIP	KV4FZ	CW3BR	CW3BR	G3WJN	CV4CR
1974	ZD3X(OH2MM)			PJ1AA	P9GIW	ZF1GS/VP7	KV4FZ	CR6WW	CR6WW	CR6OZ	CR6NO
1973	ZD3Z(OH2MM)			UK9ABA	ZD3X	KV4FZ	YV4AGP	KP4AST	CV4C	G3HCT	CR6CN
1972	4M4UA(N6TJ)			PA9AF	4M1A	VE3BS	YV3M/4X	HR1RF	CR8IK	G3HCT	KG6SL
1971	6D1AA(KØDQ)			ON4UN	GM3YCB	W6GDO	LAØAD	HR1RF	KV4FZ	CX1JM	XØYA
1970	KV4FZ			DL0WM	GM3YCB	DL9KRA	CT2AT	YV1BI	YV1LA	CW9AA	KP4AST
1969	9Y4AA(N6TJ)			I4GAD	OH5SM		W1FZJ/KP4	SM5BPJ	YV1LA	CW9AA	4X4JU
1968	ZD8Z(N6TJ)			ET3USA	PJØDX		OK1WGW	YV4UA	CW9AA	CX2CO	YV1LA
1967	VK2ADY/9(W9WNV)			YV9AA	OH2AM		OH5BTS	DL9XI	YV5ANF	DL6EN	LU1DAB
1966	VQ9AAD(W9WNV)			DL1KB	OF2AM		ON4JN	DL3GA	G5AAM	DL6EN	LU1DAB
1965	CX2CO			VQ4RF	YV9AA	GW3PMR	GI3GDF	OX3IV	YV5BIG	DL6EN	LU1DAB
1964	YV5BIG			4X4GB	YV5AKU		4X4DK	DJ2YA	FY7BL	WA2SFP[W2PV]	LU4ARR
1963	ZD7BW			HZ1AB	CX2AKU		4X4DK	W3PHL	CX2CO	W1IRIL	LU1DAB
1962	XT2Z(G1DP)			DJ3VM	DL3OU		ON4UN	W3PHL	HL9KH(W9WNV)	G3FXB	LU1DAB
1961	CX2CO			MI3US	ET2US		I1AM	4X4DK	MP4BBW	ZB1HC	LU1DAB
1960	VQ4DT			CN8ET	KA2RB			YO9CN	ZS7P	VQ4RF	LU1DAB
1959	4X4GB			HC2JR	*#MM*		G5MP		CX2CO	CE3DY	CX1AK
1958	F8PI			W6SA			G5MP	SP3PL	CO2ZS	CO2ZS	OH5NW
1957	4X4DK						W12BT	JA1EF	CE2CC	CE3DY	VQ4RF
1956	CX2CO						OK2KOD	JA1EF	OD5BZ	HC1ES	LU4AAP
1955	CX2CO						OZ8KR	JA1VP	OO2OZ	VQ4RF	VQ4RF
1954	CN8MM						I1CSP	JA1GV	W3JNN	OO5RU	LU1BK
1953	CT1FT						DL1LH	W8JIN	KA2CR	ZS6DW	LU4ARR
1952	CE3CZ						W9EDC		OD5AD	N/A	CX3BT
1951	CE3CZ						W8UKS	ZL1HY	PAØEEM	N/A	CE1AJ
1950	HC2JR								DL4WC	N/A	CE1AJ
1949	PY2AC							ZS5U	CN6EX		PY2GG
1948	PY2AC								ZS6JS		VQ4IMS
#1940											
1939	CE3AG										

Footnote # 1991. The Single-Operator Low Power class was added beginning with the 1991 event. Output power limited to 100 watts.

Footnote # A. The Single Operator Assisted Class was added beginning with the 1989 event. Rules permitted single operators to utilize any form "of DX spotting nets or DX alerting assistance."

Footnote # 1987. The 1987 RESULTS were the first time European TOP Scores were listed.

Footnote # Q. QRPp Single Operator was added to the 1978 event. Output power limited to 5 watts.

Footnote # MM. Multi-Operator Multi-Transmitter added to the 1959 event.

Footnote # 1948. CQ's original rules: 0200z start time for 48 hours. Fouth full weekend of October for Phone and first full weekend of November for CW, 80m, 40m, 20m and 27-29 MHz. One point per contact multiplied by the sum of zones per band and countries per band. (August 1948 CQ)

Footnote # MM. Multi-Operator Multi-Transmitter added to the 1959 event.
Single-Operator and Multi-Operator. No limit to the number of operators and receivers but you must observe the rules of your license. Exchange RS(T) and CQ W.A.Z. Zone number.

Footnote # 1940. The July, 1940 issue of *RADIO* magazine carried the rules and announcement for the Second Annual RADIO World-Wide DXContest. The changes included one weekend for CW, which was to begin 2300z October 4 to 0800z October 7, 1940 and one for Phone which was to begin 2300z October 11 to 0800z October 14, 1940. Log Deadline was November 4th! The available classes were changed: "Competitors will not be limited in the number of operators, or in equipment." **After the issue had been prepared, the FCC banned all amateur communications with foreign stations. Besides adding a Full Page Contest Cancellation, there were several boxes throughout the issue stating that the WWDX Contest had been cancelled.

U.S.A. WINNERS OF CQ WORLDWIDE DX PHONE CONTEST BY YEAR/CATEGORY

YEAR	S/O	S/O/Asst	QRP	M/S	M/M	160	80	40	20	15	10
1994	K1AR	N3AD	AA2U	KC1XX	W3LPL	K12M/2	W6RJ	K6NA	KM1H(KQ2M)	K4JPD	KE5FI
1993	K1AR	K1ZM/2	N1AFC	KC1XX	N2RM	WB9Z	WE3C	W7XR(W7WA)	KK9A	K1UO	W6AXX/3
1992	K1AR	WM5G(KR0Y)	AA2U	K1DG	N2RM	WB9Z	K12M	W7XR(W7WA)	W7XR(W7WA)	K5MR	N4RJ(KM9P)
1991	K1AR	N4R1(KM9P)	WA2UUK	K3LR	N2RM	AB4RU	K1UO	KC7EM	K4XS	K3RV/4	NR5M
1990	K1AR	K1ZM/2	KR2Q	K1AR	N2RM	K5UR	W6RJ	V0Q	KS1L	NG2X	NR5M
1989	WM5G(KR0Y)	*A*	KD2TT	KS9K	W3LPL	K5UR	W6RJ	WQ2M	W1RR	K3RV/4	K5GA
1988	N2NT		KR2Q	N2RM	W3LPL	K5UR	K4HJU	K5RR(WD5K)	W1RR	K3RV/4	K5GA
1987	KC1F		N3RS	K1AR	W3LPL	WB9HAD[WB9Z]	WB8JBM(NZ4K)	KV0Q	KV4P	K3RV/4	W0ZV
1986	KC1F		K3WS	KX4S	W3LPL	K5UR	W6RJ	K0RF	K2VV	K2EK	K6SVL
1985	K2TR		K7SS	K2BU	N5AU	AA1K/3	W6RJ	W7RM(W7WA)	K2EK	K2EK	K4JRB
1984	K1AR		AA2Z/1	WA0AW	N2AA	K5UR	W6RJ	KV0Q	K1OX(KC1F)	W6YA(N6NI)	KG1E
1983	AI6V(WA8VEF)		K8IA	K3LR	N2AA	WA2SPL	W1ZM(K1ZM)	KJ9D	K1UO	W0ZV	NU4Y
1982	W1ZM(K1ZM)		WB2ULI	K4VX	N2AA	W8LRL	N7DD	N7DD	K1KI	K1KI	NU4Y
1981	K1AR		AI6V	K5GA	W2PV	WB3GCG	W7IIX	K0RU	K2HFX	K6B	W6YA(N6NI)
1980	K7RI		W6PQZ	K5JA	N2AA	K5UR	N4XE	K0RU	K2HFX	K1RM	W0ZV
1979	N7DD		WA4IAR	W6ONV	N2AA	K1PBW	W0M	K4JRB	K0KX	WA6EKL	N7DD
1978	WA4DR(WA8ZDT)		*Q*	W3WJD(N3RS)	W7RM	K1PBW	W1CF	K0RF	K2HFX	WA6EKL	N7DD
1977	W3WJD(WA3LRO)			W7SFA	W2PV	W7RM	W2VP	W7KW	WB0LLR	K4YYL	WA2SPL
1976	W1ZM(K1ZM)			WB2SQN[K2SS]	W2PV	W8LRL	W7KW	AA6EPQ(N4AR)	W9PPY	K1JHX(K1RM]	K6OQ
1975	W7RM(K7JA)			WB2SQN[K2SS]	W2PV	WB8APH[W8LRL]	W1EBC(K1EB)	W3PHL	K2KUR[N2AA]	K6SOP[N6CW]	WB6PXP[AE6E]
1974	K6AHV			W4AXE	W2PV	WA4SGF	W1EBC(K1EB)	W3PHL	K2KUR[N2AA]	W4WSF[N4MM]	AB5HH
1973	W6RR			WA2CP/2	W3WJD[N3RS]	K1PBW	W1EBC(K1EB)	W4OCW(W4DR)	WB2OEU[K2TR]	K1VTM	WB5HH
1972	W2PV			WA6EPQ[N6AR]	W2PV	K1PBW	W5SZ	W3PHL	W2ONV	W4YHD[W4RX]	K4JRB
1971	K1KTH[N6JL]			K2HLB	W7RM	K1PBW	K6ERT	W3PHL	W4AXE	W1HQV	K4YYL
1970	W2PV			K2EG	WA2ZAA	W8GDQ	W4CRW	K2GXI	WA3GUL[K3OO]	W1HQV	W6QJW
1969	W4AXE			K2TMZ	K2GL	N4EA	K3UZE	W2DXL[W2IB]	W2ONV	WA8CZH	W2YT
1968	WA2SFP(W2PV)			WA8EPQ(N6AR)	W3MSK(W3AU)		W4AXE	K2GXI	K8YBU[K8RK]	WA4PXP	K3HPG
1967	WA2SFP(W2PV)			WA4XC	W3MSK(W3AU)		K2RBT	K2GXI	W3JNN	WA2SFP[W2PV]	W2SKE
1966	W3MSK(N4RV)			K6EVR	W3MSK(W3AU)		K8YWG	K2GXI	K2HWL	K6CT	W2BXA
1965	K3MSK(W2VCZ)			K6EVR	W7RM		W1AQH	K2GXI	K2HFX	W1RIL	W20KM
1964	K5MDX			K2GL	W3MSK(W3AU)		W1BU	W3PHL	K2HFX	K3ECE	WA4SUR
1963	K5MDX			K2GL	W3MSK(W3AU)			W3PHL	K2IEG	W2WZ	W2TVR
1962	W6GHM			K2GL	WBHGO		W1BU	W3PHL	W2HTI	W2WZ	KS0RH
1961	W1ONK			W2HJR[K2GL]	*#MM*		W1BU	K2DGT	W7HOC	W2ETD	W9PWU
1960	W3EWC			W6AM					W6VSS	W3AOH	W2VCG
1959	K2AAA			W7DL			W1ZBT	W6UED	W4KFC	W4NQM	K8AEK
1958	W6YY			W6AM					W3LOE	W2GFO	W1ONK
1957	W1ATE			W6RRG			W1ZBT	W3ECR	W6HNX	W4NQM	W3JTK
1956	W1ATE			W6AM				W8FXG	W6PWR	W2WZ	W3MDE
1955	W1ATE			W6SA			W7JLU	W7JLU	W6UYX	K6CZY	W4NQM
1954	W1ATE			W6SA				W8JIN	W1AFZ	W2JDE	W4NQM
1953	W1ATE						W9MEM		W6DI	N/A	W0GEK
1952	W4ESK(W7RM)						W9EDC			N/A	W4OBK
1951	W1ATE						W8UKS			N/A	W4VB
1950	W1ATE										W1BEQ
1949	W8KML										
1948											
#1940											
##1940	W6OCH										
1939											

Footnote #1991. The Single Operator Low Power class was added beginning with the 1991 event. Output power limited to 100 watts.
Footnote #A. The Single Operator Assisted Class was added beginning with the 1989 event. Rules permitted Single operators to utilize any form "of DX spotting nets or DX alerting assistance."
Footnote #1987. The 1987 RESULTS were the first time European TOP Scores were listed.
Footnote #Q. QRP, Single Operator was added to the 1978 event. Output power limited to 5 watts.
Footnote #MM. Multi-Operator Multi-Transmitter added to the 1959 event.
Footnote #1948. CQ's original rules: 0200z start time for 48 hours. Fouth full weekend of October for Phone and first full weekend of November for CW. 80m, 40m, 20m, and 27-29 MHz. Single-Operator and Multi-Operator. No limit to the number of transmitters and receivers but you must observe all the rules of your license. Exchange: RS(T) and CQ W.A.Z. Zone number. One point per contact multiplied by the sum of zones per band and countries per band. (August 1948 CQ)
Footnote #1940. The July, 1940 issue of *RADIO* magazine carried the rules and announcement for the Second Annual RADIO World-Wide DX Contest. The changes included one weekend for CW, which was to begin 2300z October 4 to 0800z October 7, 1940 and one for Phone which was to begin 2300z October 11 to 0800z October 14, 1940. Log Deadline was November 4th! The available classes were changed: "Competitors will not limited in the number of operators, or in equipment." **After preparing the issue, the FCC banned all amateur communications with foreign stations. Besides adding a Full Page Contest Cancellation, there were several boxes throughout the issue stating that the WW DX Contest had been cancelled.

DX WINNERS OF CQ WORLDWIDE DX CW CONTEST BY YEAR/CATEGORY

YEAR	S/O	S/OQRP	S/O/Asst	M/S	M/M	160	80	40	20	15	10
1994	CT3M(OH7JT)	T44KM(DK5WL)	P40W(W2GD)	IQ4A	9G5AA	4X4NJ	ZB2X(OH2KI)	EA9EO(EA7TL)	P40J(WX4G)	ZP0Y(LW9ELJ)	ZS6NW
1993	EA8EA(OH2MM)	AA2U	K8AZ	ZYC4Z	EA9EO	4X4NJ	K1ZM	PJ9U(OH1VR)	FY5EY(DH7XM)	CR3W(DF5UL)	D68GA(N6ZV)
1992	EA8EA(OH2MM)	722AB(K2XR)	VE3EJ	J6DX	EA9EO	4X4NJ	ON4UN	C41A(5B4ADA)	PZ5JR(OH6DO)	ZP0Y(K4UEE)	CV5A
1991	EA8EA(OH2MM)	HI8A(JA5DOH)	N4RJ(KM9P)	EA9EA	PJ9A	4X4NJ	ZB2X(OH2KI)	C42A(YU4OO)	P40V(N7NG)	ZW5B(LU8DQ)	ZS6BCR(ZS6EZ)
1990	CT3M(OH7JT)	HI8A(JA5DOH)	K3WW	TA5KA	PJ9A	UG6GAW	EA9EU	ED9ED(EA5BRA)	ED9ED(EA5BRA)	C56/OH7NH	CX0CW(CX8BBH)
1989	FY5YE(OH2MM)	W2TZ	K1DG	EA9EA	CT3M	LY2BTA	TA2BK	4Z8DX	4Z8DX	N7DF/NH2	ZP0Y(LU8DQ)
1988	FY5YE(OH2MM)	YU3BC	-#A-	EA9EA	PJ1B	TA2BK	NP4A(K1ZM)	ZS6BCR(ZS6EZ)	ZS6BCR(ZS6EZ)	CW8B(CX8BBH)	4M7A(YV7OP)
1987	FY5YE(OH2MM)	VE7DX(K7SS)		EA8AGD	KP2A	UP2NK/UF	ZC4D(4Z4DX)	EA8ID(DK3GI)	EA8ID(DK3GI)	LO8WW(LU8DQ)	4M7A(YV7OP)
1986	P4GD(W2GD)	YU3BC		KP4BZ	KP2N	HB9AMO	P40R(K4UEE)	P40N(N4PN)	P40N(N4PN)	PY5CA(N5FA)	4M7A(YV7OP)
1985	EA9IE(N6AA)	YU3BC		V3A	RF3V	YV3AGT	EA2IA	KP4FI	CX4CR	CX5AO	ZS6F
1984	9Y4VT(N6AA)	K8IA		FY0GA	EA9CE	LZ1KDP	UP2NK/UF	YX5A(YV5ANT)	CR6IK	9V4W	4M7QP(YV7QP)
1983	9Y4VT(N6AA)	UB5UCJ		HH2VP	P42E	EA8AK	VP2KAC(N4RJ)	YX5A(YV5ANT)	CV8B	YX5A	CS3T
1982	9Y4VT(N6AA)	UP2BIM		P41E	W2PV	EA8AK	EA2IA	YX5A(YV5ANT)	CR6IK	CX7QO(CX8BBH)	V3TV
1981	9Y4VT(N6AA)	AC2U		P41E	NP4A	OK3KFF(OK3CQW)	I4IND	UA1DZ	PY4AF	LU8DQ	Y8RW
1980	EA8AK(OH6DX)	YU3BC		RG6G	P3A	G3SJA	WB2FZO(KQ2M)	VP2KAA(N4PN)	KV4FZ	HD1E(K7CA)	KG6DX
1979	CT3BZ(OH2BH)	G4BUE		NP4A	NP4A	VR3AH	EA2OP	YU3ZV	PY40D	VP2MEE(N8BM)	LV8DQ
1978	9Y4VT(N6AA)	OA8V		4L6F	EA8CR	KT7BRW	CT3/OH1TV	KV4FZ	PY2SO	LU8DQ	LU1DZ
1977	EA4CST	-IQ-		PJ9MM	UK9AAN	KV4FZ	VR3AH	KV4FZ	1G5A(W9WNV)	KX6LA	YV4OB
1976	EA4CST			FY7AK	W3AU	YV1DB	KV4FZ	W5WIQ	CX2CO	5Z4NI	LU6EF
1975	KH6RS(K1GQ)			OD5IQ	W3AU	PA0HIP	YU3DBC	ZP5GS	VP8GQ	ZE8JN	WB4KSE/KW6
1974	ZD3X(OH2BH)			PJ1AA	CW3AA	KV4FZ	UB5CI	ZL1AMO	PY40D	CR6OZ	CX9BT
1973	ZD3Z(OH2MM)			PJ2VD	PJ9JT	ZF1GS	DL7AV	KP4AST(NP4A)	CX2CO	CV1B	ZE8JN
1972	KH6RS(K1GQ)			UK9ABA	PJ0FC	KG4CS	W3MFW	K6EBB	UA9DN	CW9BT	7O7AA
1971	ZS3AW			4M5ANT	PJ0CC	DL1CF	SM5BPJ	PY4AF	CX2CO	TJ1AW	CX1AAC
1970	9Y4AA(N6TJ)			DL0KF	PJ0CC	HB9NL	OZ5DX	W6ISA	VQ8CR	VQ8CR	CR6IK
1969	ZD8J			4L3A	PJ3CC	DL1CF	OM1BY	LZ1KPG	CR6GO	CR6GO	K1JGD
1968	ZD8J			4L7A	K2GL	ZC4RB	OK1ALW	4X4RD	G3HCT	G3HCT	PY2SO
1967	VR2EW(W9WNV)			ET3USA	K2GL	VO1FB	SM6MX	OK1ZQ	CX1AAC	CX1AAC	9J2BC
1966	9Y4OD			YV9AA	K2GL	W1BB/1	DJ3KR	K2STW	W8WZ	W8WZ	WA4WIP
1965	5A1TW(NZAA)			VK5NO	CX2CO	OK12C	4X4DH	5A1TW(N2AA)	PY2SO	ZS6IW	
1964	HL9KH(W9WNV)			CX2CO	4X9HQ	DJ2KS	UB5MZ	VK3AZZ	VP8GQ	ZS6IW	HK7ZI
1963	7G1A			VK5NQ	W3AOH	OK1ADX	OK1MG	W9WNV	PY40D	ZS6IW	HK7ZT
1962	UA9DN			UB5KAB	DJ3JZ	DL1WT	OK3DG	YF4GA	CX2CO	PY4GA	K2HWL
1961	CN8JX			W1BIH	DJ3JZ	DL1FF	SP2OX	K2DGT	VE3AW	DL6EN	4X4LC
1960	CN8JX			K2GL	-#MM-	OK2NRO	OK1AWJ	K2DGT	VE3AW	VP1HLF	OQ5IG
1959	V44FC			K2GL			WP8GX	W8FGX	G2AW	VE3AW	DL4CAP
1958	V44FC			KH6CBP			W3BVN	W3BVN	CX2CO	CX2CO	DL4AAP
1957	W4KFC			HA5KBA			W8FXG	W8FXG	WAVZQ	W8BKP	JA1CO
1956	W2XDJR(K2GL)			W4KVX			OK1JX	OK1JX	W4KFC	W2WZ	W3MDE
1955				ET2US			KH6ER	OK1JX	W3LOE	LU3EX	
1954	4X4RE			TA3AA			5A3TR	5A3TU	G2LB	DL3RM	
1953	4X4RE			ZC4XP			ZL3LL	ZL3LL	5A3TU	VK4FJ	G2BW
1952	4X4BX			CN8EG			KH6ZG	W3JTC	GW3ZV	-N/A-	VE7YR
1951	4X4RE			W6SZY			W4BRB	G2LB	W3JTC	-N/A-	PY1AJ
1950	4X4BX			ZL4GA					G2LB	-N/A-	G3DCU
1949	PA0UN										
1948	GI6TK										
#1940**	K6CGK(KH6IJ)										
1939											

Footnote #1991. The Single Operator Low Power class was added beginning with the 1991 event. Output power limited to 100 watts.
Footnote #A. The Single Operator Assisted Class was added beginning with the 1989 event. Rules permitted Single operators to utilize any form " of DX spotting nets or DX alerting assistance."
Footnote #1987. The 1987 RESULTS were the first time European TOP Scores were listed.
Footnote #Q. QRPp Single Operator was added in the 1978 event. Output power limited to 5 watts.
Footnote #MM. Multi-Operator Multi-Transmitter added to the 1959 event.
Footnote #1948. CQ's original rules: f/200z start time for 48 hours. Fouth full weekend of October for Phone and first full weekend of November for CW. 80m, 40m, 20m and 27-29 MHz. Single-Operator and Multi-Operator. No limit to the number of transmitters and receivers but you must observe all the rules of your license. Fouth full weekend of October for Phone and first full weekend of November for CW. 80m, 40m, 20m and 27-29 MHz. Single-Operator and Multi-Operator. No limit to the number of transmitters and receivers but you must observe all the rules of your license.
Footnote #1940. The July 1940 issue of RADIO magazine carried the rules and announcement for the Second Annual RADIO World-Wide DX Contest. The changes included one weekend for CW, which was to begin 2300z October 4 to 0800z October 7, 1940 and one for Phone which was to begin 2300z October 11 to 0800z October 14, 1940. Log Deadline was November 4th! The available classes were changed: "Competitors will not be limited in the number of operators, or in equipment." After the issue was prepared, the FCC banned all amateur communications with foreign stations. Besides adding a Full Page Contest Cancellation, there were several boxes throughout the issue stating that the WW DX Contest had been cancelled.

U.S.A. WINNERS OF CQ WORLDWIDE DX CW CONTEST BY YEAR/CATEGORY

YEAR	S/O	S/O/Ast	QRP	M/S	M/M	160	80	40	20	15	10
1994	K5ZD/1	K12M/2	AA2U	K1AR	W3LPL	W8BZ	W1MK	KC7EM	K3EST/6	KC2X/4	KE3Q
1993	K1KI	K3WW	AA2U	KC4XX	W3LPL	W1BYH	W1MK	W1RR	KM1H(KO2M)	K1ZZ	K4XS
1992	N4RJ(KM9P)	K3WW	AA2U	N3RS	K1AR	K2EK	K12M	K8PO/1	K2VV	K2SS/1	KA5W
1991	K1KI(KO2M)	N4RJ(KM9P)	AA2U	K5NA/2	N2RM	N6SS/7	W1FV	W0ZV	K4XS(WC4E)	K1RM	W0UN(N2IC)
1990	K1CC	K3WW	WZTZ	N3RS	N2RM	K5UR	W1FV	K12M	WY7I	KE3Q	W0ZV
1989	K3TUP(K5ZD)	K1DG	WA2HZR	K1AR	W3LPL	K5UR	W1FV	K2EK	K2EK	W7WA	K12M
		#A									
1988	W1KM		W8VSK	K1GQ	W3LPL	K5UR	W1FV	K1NA	K1NA	K5GO	N4EJW
1987	K1EA		K1CGP	K1GQ	N2AA	W1CF(WA2SPL)	W1FV	W6YA	W6YA	K3RV/4	N6BFM/4
1986	W1KM		K3WS	N4WW	N2AA	K5UR	W1FV	N6QR	K3RV/4	K3RV/4	WA3CGE
1985	K1AR		K8IA	K1KI	N2AA	K5UR	W1FV	K1OX(KC1F)	K2VV	K2EK	W8WPC(N9AG)
1984	N2LT		N3RS	K5RC	N2AA	K12M	K1PT	N5J	N5CR	NA5R(KN5H)	KZ5M
1983	W1KM		K8IA	K1GQ	N2AA	AE6U	W1ZMK(K12M)	W6AM(N6AW)	K1KI	W1RM	N4WW
1982	K1GQ		AC2U	K5RC	W2PV	W8LRL	WB2FZO(KO2M)	K1XM	N7UA	K6EWL	N4ZNC
1981	K1AR		W0KEA	N4AR	N2AA	W8LRL	WB3AVN	K0RF	W6VPH	W1RM	W0ZV
1980	K1AR		WA4LOF	K5RC	N2AA	K1PBW	K5NU	K0RF	W6VXX(KC1F)	N6CW	N4WW
1979	K1AR		W5ZY	N3RS	N2AA	K1PBW	W5UN	W5UN	K3M0RF	N4RJ	K8MFO
1978	W3RJ		"#Q"	N4AR	W2PV	K1PBW	W1MX(WA8WNU)	W5UN	K8IDE[W8TA]	WA4WSF[N4MM]	W5MYA
1977	W3LPL			AA5LES[K5RC]	W3WJD[N3RS]	K1PBW	W1MX(WA8WNU)	W5UZQ[W5UN]	K6SDR[N6CW]	W2H8TT[N2RM]	K5FVA[K5LM]
1976	W3LPL			W6GOUN	W3AU	K1PBW	W1MX	W5WZQ[W5UN]	W4AAV	W4KFC	W8WPC
1975	W3LPL			K6GSU[N4AR]	W3AU	K1PBW	W3MFW	WB5DTX[W8BJA]	WA1JUY	K1LWI	WA5RTG[K5GO]
1974	W3LPL			K6EBB	W3AU	WB8APH[W8LRL]	W3MFW	W5WZQ[W5UN]	W9BFJX	W4KFC	K1LWI
1973	W3LPL			K1DIR	W4BVV	WA4SGF	W8LT	K6EBB	K8MMM	K8WWU	W4KFC
1972	W1FV[N3RS]			K6EBB	W4BVV	W1HGT	W1SWX[W1NH]	K6ERT	W7VY	W5WWU/5	K5ABV
1971	W1FV[K1ZZ]			W3WJD[N3RS]	W4BVV	K1PBW	W1BB/1	W6ISA	W4AXE	WA8LYF[K8LX]	K5ABV
1970	K1KTH [N6JL]			W9EXE	W8LT	W1BB	W9NBK	K6NA	W2LXK	K1NCQL[K1NA]	K1LWI
1969	W4YHD[W4RX]			K1DIR	W3MSK[W3AU]	—	W9NBK	W2LXK	N6NA	W1MDQ[N1XX]	W8VSK
1968	W3GRF			W3WJD[N3RS]	W3MSK[W3AU]	W1BB/1	K6BPR	K2RBT	W7YGN	K9ECO	K1JGD
1967	K1DIR			W3MVB[W1ZM]	K2GL	W1BG/1	W2BU	W4BGO	W8WZ	W8WZ	W2MEL
1966	W3GRF			W2PCJ	K2GL	W2EQS	K6EIV	W9BFGX	W4KFC	K1NCQL[K1NA]	K1IMP
1965	W3GRF			WA6SBO	W3MSK[W3AU]	W2FYT	W1BU	W6AM(W9WNV)	W4KFC	W1WY	W4WIP
1964	W4KFC			W6RW	W3MSK[W3AU]	K7EKD	W9PNE	W9BFGX	W4KFC	W2HTI	—
1963	W3GRF			W1BIH	W3MSK[W3AU]	—	W1BU	K2DGT	W1BIH	W2WZ	K1LWI
1962	W3GRF			W9YT	W3AOH	W1BB	W3EIS	W9WNV	W4KFC	W2WZ	W6ID
1961	W4YHD[W4RX]			W6GHM/6	W3AOH	W7ZVY	W9PNE	K2DGT	W8QDFR	W8UPH[W4RX]	K2YFE
1960	W3GRF			W1BIH	*#MM*	—	W3EIS	W9BFGX	W8QDFR	W4YHD[W4RX]	K2HWL
1959	W8JIN			K2GL			W4HQN	W3BVN	W3BVN	W2WZ	W3LSG
1958	W4KFC			W6RW			W8KIA	W4VZQ	W4AIW	W4VZQ	W3LSG
1957	W4KVX			W6DFY			W6ZAT	W4FXG	W4AIW	W8BKP	K8AEK
1956	W2HJR[K2GL]			W6AM			W0NWX	W3MSK[W3AU]	W4AIW	W2WZ	W8WZ
1955	W4KFC			W6AM				K2EDL	W4VZQ	W8JIN	K4CTU
1954	W4KFC			W4VX				W6RW	W6WX	W8BHW	W3MDE
1953	W8JIN			W6AM				W1RWP	W6BAX	W3AYS	W6ITA
1952	W8JIN			W8WZ				W4BRB	W6BAX	N/A	
1951	W4KFC			W6AM				W4BRB	W6BAX	N/A	
1950	W4KFC			W6GAL					W3JTC	N/A	
1949	W4KFC			W6SZY					W9DUY		
1948	W4KFC			W2IQG							
#1940											
**#1939*											
1939	W6QD	W6GRL									

Footnote #991. The Single Operator Low Power class was added beginning with the 1991 event. Output power limited to 100 watts.

Footnote #A. The Single Operator Assisted Class was added beginning with the 1989 event. Rules permitted Single operators to utilize any form "of DX spotting nets or DX alerting assistance."

Footnote #1987. The 1987 RESULTS were the first time European TOP Scores were listed.

Footnote #Q. QRPp Single Operator class was added to the 1978 event. Output power limited to 5 watts.

Footnote #MM. Multi-Operator Multi-Transmitter added to the 1959 event.

CW, 80m, 40m, 20m and 27-29 MHz. Single-Operator and Multi-Operator. No limit to the number of transmitters and receivers but you must observe all the rules of your license. Exchange RS(T) and CQ W.A.Z. Zone number. One point per contact multiplied by the sum of zones per band and countries per band (August 1948 CQ).

Footnote #1948. CQ's original rules: 0200z start time for 48 hours. Fourth full weekend of October for Phone and first full weekend of November for CW, which was to begin 2300z October 7, 1940 and one for Phone which was to begin 2300z October 11 to 0800z October 14, 1940. Log Deadline was November 4th! The available classes were changed: "Competitors will not limited in the number of operators, or in equipment." After the issue was prepared, the FCC banned all amateur communications with foreign stations.

Footnote #1940. The July 1940 issue of RADIO magazine carried the rules and announcement for the Second Annual RADIO World-Wide DX Contest. The changes included one weekend for CW, which will not be limited in the number of operators, or in equipment." After the issue was prepared, the FCC banned all amateur communications with foreign stations. Besides adding a Full Page Contest Cancellation, there were several boxes throughout the issue stating that the WW DX Contest had been cancelled.

CQWW CQWW

DX Phone Records

Call	Year	Score
Single-Op All-Band (No Packet)		
AF CT3BH (OH2BH)	90	14,892,102
AS H20A(5B4ADA)	94	7,618,670
EU ZB2X (OH2KI)	91	7,128,646
NA KP2A(CT1BOH)	93	13,202,298
OC YJ1A (OH1RY)	90	9,516,731
SA HC8A (N6KT)	92	16,316,568
Single-Op Low Power (No Packet)		
AF TJ1GG (I2VXJ)	92	5,925,760
AS ZC4BS	91	5,244,877
EU EA7CEZ	94	2,121,693
NA VP2MBA	92	2,509,821
OC KH8/WB7RFA	92	4,301,640
SA HC1OT	93	4,194,840
Single-Band Low-Power		
160 HA8EK	94	36,780
80 CM3ZD	94	91,212
40 TG9AJR	92	395,488
20 5L2PP	94	1,989,144
15 5Z4BI	93	1,250,088
10 Z21HQ	91	1,118,611
Single-Op QRP <5 W (No Packet)		
AF No Entry		
AS JA6VZB	91	267,729
EU YU2TY	89	758,523
NA KR2Q	90	1,248,207
OC AH6EK	84	168,198
SA PJ2FR (K7SS)	87	3,171,166
Single-Band QRP		
160 UP2BKF	82	12,516
80 UB5IRN	88	15,704
40 UA4LC	90	76,131
20 YU1NR	85	114,912
15 UG/UB4WZZ	87	219,744
10 N6BFM/4	90	327,928
S/O 28 MHz		
AF ZD8Z (N6TJ)	91	2,341,866
AS JH1AJT	88	1,421,070
EU YU3ZV	88	1,541,603
NA VP2ET (K5RX)	88	2,432,880
OC KD7P/NH2	88	2,309,304
SA ZV5A (PY5EG)	91	2,984,166
S/O 21 MHz		
AF ZD8Z(N6TJ)	94	3,481,925
AS JHØJHA	92	1,430,856
EU CQ4A (CT1BOP)	90	1,757,780
NA V26N(KW8N)	93	2,159,460
OC AHØAB(JA3ODC)	82	1,923,840
SA ZW5B(N5FA)	93	2,834,228
S/O 14 MHz		
AF CT3DL	94	1,894,165
AS RFØFWW (UF6FFF)	87	1,447,128
EU OH2BH (OH2IW)	92	1,870,170
NA KP2A(KW8N)	94	2,255,250
OC ZM1BIL	83	1,334,232
SA PYØFM(PY5CC)	94	3,202,242
S/O 7 MHz		
AF EA8RCT (OH2MM)	87	859,362
AS JA8IXM	90	469,012
EU S59UN	92	875,875
NA TI1C(TI2CF)	94	1,108,140
OC 9M8R	94	1,077,440
SA PJ9U(OH1VR)	93	1,199,968
S/O 3.5 MHz		
AF IG9/IV3TAN	94	320,235
AS UW9AF	83	222,192
EU HA8IE	90	361,343
NA TI1C (TI2CF)	92	498,037
OC T32AF	85	222,768
SA P4ØR (K4UEE)	87	552,786
S/O 1.8 MHz		
AF IH9/IV3PRK	89	81,344
AS UG7GWO	87	255,852
EU LZ2CJ	84	107,818
NA VE3BMV	86	52,240
OC KH6CC	85	45,984
SA YV2IF	84	18,291
Single-Op WITH Packet		
AF ZS94F(ZS6YA)	94	1,890,350
AS JA8RWU	92	3,865,056
EU YT3AA	89	5,756,932
NA CH3EJ(VE3EJ)	93	8,167,096
OC NY6M/KH2	89	2,108,408
SA P4ØW(W2GD)	94	11,224,877
Multi-Op Single Transmitter		
AF EA8AGD	88	17,172,672
AS YM5KA	90	15,056,664
EU IQ4A	90	17,255,700
NA VP2EC	92	16,287,152
OC KH2S	91	11,095,392
SA PJ1B	93	22,596,570
Multi-Op Multi-Transmitter		
AF EA9UK	93	37,140,597
AS EW6V	82	18,746,136
EU LX7A	89	26,578,978
NA VP2KC	79	37,770,012
OC KHØAM	90	35,730,600
SA PJ1B	90	57,610,400

CQWW USA/VE Phone Records

Call Area	Call	Year	Score
Single-Op All-Band (No Packet)			
W1	K1AR	92	7,810,446
W2	N2LT	89	5,403,200
W3	K3TUP (K3LR)	88	6,004,683
W4	N4RJ (KM9P)	90	4,683,448
W5	WM5G (KRØY)	90	6,308,064
W6	K6NA	88	3,642,240
W7	N7DD	79	3,113,788
W8	NA8V	89	2,837,423
W9	W9RE	90	4,525,010
WØ	N2IC/Ø	88	3,656,100
VE	VB3XN	88	5,373,693
Single-Op Low Power (No Packet)			
W1	KG1D	92	1,160,764
W2	K2SG	93	1,576,995
W3	W3UJ	91	603,630
W4	KØEJ/4	92	1,207,288
W5	K5RX	91	993,806
W6	WB6JPY	92	754,974
W7	K7RI	91	1,286,946
W8	N8II	92	1,864,747
W9	WB9IQI	92	900,015
WØ	KX5Z/Ø	91	842,898
VE	VY2SS	92	2,035,230

Single-Op QRP<5 W (No Packet)

W1	KN1M	90	606,816
W2	KR2Q	90	1,248,207
W3	K3ZR	81	434,720
W4	WR4K	88	575,200
W5	K5RX	90	1,103,513
W6	AI6V	80	194,040
W7	K7RI	92	860,560
W8	K8IA	82	337,666
W9	N9AW	92	112,344
W0	W0UO	89	297,591
VE	VE1CBF	87	588,555

S/O 28 MHz

W1	KS1L	88	955,632
W2	WA2SPL	79	735,528
W3	WE3C	90	729,289
W4	K4XS	88	955,344
W5	NR5M	90	1,116,390
W6	W6NV (WB6SHD)	90	877,389
W7	K7RI	85	973,947
W8	K3ZJ/8	88	1,026,684
W9	W0AIH/9 (N0BSH)	89	793,507
W0	W0ZV	88	1,145,368
VE	VO1SA	91	1,381,913

S/O 21 MHz

W1	K1RM	79	870,237
W2	K2VV	88	1,172,760
W3	K3EST	83	380,540
W4	K3RV/4	88	1,270,478
W5	K5MR	92	1,168,044
W6	W6QHS	92	877,401
W7	N7DD	81	923,945
W8	K3ZJ/8	92	748,328
W9	K9DX	80	690,228
W0	W0UN(W0UA)	92	1,094,052
VE	VD7SV	92	1,662,660

S/O 14 MHz

W1	K1OX (KC1F)	85	1,131,328
W2	K2VV	91	939,624
W3	K3IPK	87	368,816
W4	K4XS	92	941,192
W5	W5WMU	82	305,660
W6	W6QHS	86	510,510
W7	K7RI(AA7FT)	93	697,151
W8	KW8N (WD8IXE)	87	563,313
W9	KK9A	93	733,698
W0	W0OG [WX3N]	90	734,290
VE	CH7SV	93	1,302,863

S/O 7 MHz

W1	K1UO	92	264,922
W2	K5NA/2	92	147,735
W3	W3GH	94	147,015
W4	N4ZC	92	135,314
W5	K5RR (WD5K)	88	177,280
W6	K6NA	94	233,105
W7	W7XR (W7WA)	92	363,900
W8	K8MJZ	90	102,368
W9	K9RN	94	145,676
W0	KV0Q	93	246,048
VE	XN3BMV	85	546,615

S/O 3.5 MHz

W1	W1ZM	83	177,862
W2	K1ZM/2	92	229,295
W3	WE3C	93	169,020
W4	WA4SVO	87	77,559
W5	W0MJ/5	84	91,952
W6	W6RJ	94	184,926
W7	K7SS	86	217,038
W8	WB8JBM (NZ4K)	87	105,138
W9	AI9J	82	67,068
W0	K0RF	76	54,096
VE	VE3BMV	85	383,040

S/O 1.8 MHz

W1	K1ZM	85	20,502
W2	WA2SPL	83	18,483
W3	AA1K/3	85	24,663
W4	W4DR	86	11,648
W5	K5UR	86	19,044
W6	AE6U	82	5,005
W7	K7IDX	87	7,080
W8	W8LRL	82	16,191
W9	WB9HAD [WB9Z]	87	27,181
W0	W0ZV	87	12,204
VE	VE3BMV	86	54,478

S/Op WITH Packet (Since 1989)

W1	W1PH	92	4,290,198
W2	K1ZM/2	93	4,436,796
W3	K3WW	92	5,228,249
W4	N4RJ (KM9P)	91	4,005,762
W5	WM5G (KR0Y)	92	6,631,513
W6	N3AHA/6	93	1,312,192
W7	KI3V/7	90	2,299,142
W8	K8MR	90	2,853,600
W9	KK9V	90	1,943,470
W0	K0RF	92	3,715,920
VE	CH3EJ(VA3EJ)	93	8,167,096

Multi-Op Single Transmitter

W1	K1AR	90	11,193,606
W2	N2RM	88	6,923,136
W3	N3RS	90	7,752,168
W4	K4ISV	92	6,098,814
W5	W5WMU	88	6,056,298
W6	W6GO	92	4,852,386
W7	KO7N	88	4,079,140
W8	K8AZ	92	6,039,200
W9	KS9K	90	7,190,746
W0	K4VX/0	88	6,227,259
VE	XM1DX	90	11,159,888

Multi-Op Multi-Transmitter

W1	NX1H	92	12,347,656
W2	N2RM	92	19,603,032
W3	W3LPL	90	16,517,214
W4	N4RJ	88	9,655,415
W5	N5AU	90	16,736,436
W6	N6ND	90	8,483,628
W7	W7XR	90	12,232,124
W8	K8CC	90	9,785,580
W9	W0AIH/9	92	6,700,464
W0	AA6TT/0	92	11,891,552
VE	VC1DX	91	15,458,874

CQWW DX
CW Records

	Call	Year	Score
Single-Op All-Band (No Packet)			
AF	EA8EA (OH2MM)	91	13,225,295
AS	JY8VJ (DL1VJ)	92	8,031,168
EU	ZB2X(OH2KI)	93	6,129,904
NA	TI1C(N6TR)	93	9,123,817
OC	AH3C	90	6,789,363
SA	P40F(KR0Y)	94	12,393,150

Single-Op Low Power (No Packet)			
AF	9X5EE(PA3DZN)	94	4,014,270
AS	9V1YC	92	2,679,948
EU	EA7CEZ	94	3,469,004
NA	NP4Z	93	3,948,966
OC	9M8DX (VK2DXI)	91	2,388,719
SA	OA4ZV	92	1,945,555

Single-Band Low-Meters

160	HA8EK	93	67,014
80	S59CAB	93	147,486
40	4N7N	93	609,738
20	PT7CB(YU1RL)	94	1,157,475
15	CX6VM	93	622,544
10	C56/G4ODV	91	902,967

Single-Op QRP <5W. (No Packet)

AF	ED8BIE	86	45,441
AS	7Z2AB(K2XR)	93	2,757,770
EU	DK3GI	91	1,359,280
NA	HI8A (JA5DQH)	91	3,316,768
OC	P29DK	91	272,630
SA	YV2BE	86	114,062

Single-Band QRP

160	UC2WAF	92	27,280
80	UA9CBM	94	72,051
40	RB5QNV	93	117,261
20	UW1BI	88	230,528
15	4Z4NUT	85	189,306
10	CX7CO	90	178,364

S/O 28 MHz

AF	D68GA (N6ZV)	92	1,281,656
AS	4Z5DX	90	826,759
EU	9H1EL	92	794,846
NA	J79DX (AA5DX)	89	859,360
OC	KD7P/NH2	88	1,037,608
SA	CX0CW (CX8BBH)	90	1,890,607

S/O 21 MHz

AF	CR3W (DF5UL)	92	1,652,170
AS	4Z4T (4Z4UT)	91	939,900
EU	OH6MCW	89	775,620
NA	V29W (KD6WW)	90	1,110,512
OC	N7DF/NH2	89	1,205,776
SA	ZP0Y(K4UEE)	93	1,869,978

S/O 14 MHz

AF	ED9ED	90	1,444,436
AS	7L1GVE	92	1,181,937
EU	OH0BH (OH2MAM)	94	1,003,353
NA	KP2A(KW8N)	94	1,332,460
OC	ZL3GQ	91	1,148,418
SA	P40V (N7NG)	91	1,883,700

S/O 7 MHz

AF	EA9EO(EA7TL)	94	1,122,506
AS	C41A(5B4ADA)	93	1,307,944
EU	S59UN	92	971,049
NA	ZF2TG (WQ5W)	92	1,087,862
OC	ZL3GQ	94	672,612
SA	PJ9U (OH1VR)	92	1,171,864

S/O 3.5 MHz

AF	EA8XS (OH5XT)	88	516,390
AS	ZC4DX (4Z4DX)	87	430,560
EU	ON4UN	93	630,568
NA	NP4A (K1ZM)	88	808,640
OC	VR3AH	76	178,560
SA	P40R (K4UEE)	86	576,725

S/O 1.8 MHz

AF	EA8AK	82	75,768
AS	4X4NJ	94	184,896
EU	GW4YDX	93	154,376
NA	VO1NA	93	148,050
OC	KH6CC	93	68,250
SA	YV3AGT	85	147,588

Single-Op WITH Packet (Since 1989)

AF	EA8NQ	94	113,580
AS	4X/S59PR	93	5,677,000
EU	OK1ALW	90	4,502,748
NA	VE3EJ	93	6,073,614
OC	VK2VM	93	450,448
SA	P40W(W2GD)	94	10,288,950

Multi-Op Single Transmitter

AF	EA9EA	91	13,096,080
AS	TA5KA	90	13,915,044
EU	LZ9A	89	9,962,386
NA	J6DX	93	11,691,029
OC	KH2S	92	7,249,952
SA	4M5I	93	11,222,746

Multi-Op Multi-Transmitter

AF	CN5N	90	33,659,256
AS	VS6WV	92	17,799,960
EU	LX7A	89	20,497,632
NA	KP2A	88	32,325,150
OC	KH0AM	92	23,951,385
SA	PJ1B	88	38,415,760

Club Score	Frankford Radio Club	92	389,564,535

CQWW U.S.A./VE CW Records

Call Area	Call	Year	Score
Single-Op All-Band (No Packet)			
W1	KM1H (KQ2M)	92	5,675,756
W2	N2NT	92	5,705,000
W3	K3TUP (K5ZD)	89	5,564,556
W4	N4RJ KM9P)	92	5,851,152
W5	N5AU (WN4KKN)	89	4,395,825
W6	N6UR (KR6X)	92	3,729,600
W7	K5MM/7	89	2,898,266
W8	KN8Z	92	5,436,795
W9	W9RE (WA8YVR)	90	4,755,206
W0	K4VX/0	93	3,832,281
VE	VE3EJ	92	5,011,815
S-Op Low Power (No Packet)			
W1	KC1SJ	93	1,326,332
W2	W2TZ	92	1,986,240
W3	W2UP/3	94	1,298,650
W4	K7SV/4	92	1,335,780
W5	N8RR/5	92	1,185,845
W6	W6JTI	91	912,216
W7	WA0RJY/7	91	423,073
W8	N8II	92	2,008,982
W9	K9QVB	91	814,484
W0	KZ6E/0	93	564,465
VE	VD2ZP	92	2,346,096
Single-Op QRP <5 W (No Packet)			
W1	K1CGJ	90	701,838
W2	AA2U	92	1,188,000
W3	N3RS	81	577,205
W4	N4KG	90	748,410
W5	K5RX	90	756,952
W6	WB6JMS	92	248,196
W7	NX7K	90	709,665
W8	K8IA	82	400,064
W9	W9KNI	82	345,417
W0	W0UO	90	675,840
VE	VE7DX (K7SS)	87	687,690
S/O 28 MHz			
W1	K1ZZ	88	537,568
W2	K1ZM/2	89	732,564
W3	K3UA	89	453,055
W4	K4XS	89	604,035
W5	KA5W	92	446,439
W6	W6YA	89	454,358
W7	KT7O	89	352,674

W8	N8CXX	89	653,868
W9	W9SU	89	550,565
WØ	WØZV	90	506,989
VE	CZ7SZ (WA6VEF)	90	530,208

S/O 21 MHz

W1	W1RM	90	698,257
W2	KR2Q	91	509,250
W3	KE3Q	90	753,660
W4	AA4NC	90	614,376
W5	W5VX	92	511,187
W6	N6RO (K3EST)	89	677,292
W7	W7WA	89	777,146
W8	K8MFO	92	519,750
W9	K9BG	89	534,240
WØ	WØUN (WØUA)	92	647,168
VE	VE7SZ (WA6VEF)	89	802,032

S/O 14 MHz

W1	KM1H(KQ2M)	93	1,001,035
W2	K2VV	92	943,920
W3	N3RS	84	314,505
W4	K3RV/4	85	634,293
W5	N5CR	89	700,135
W6	W6VPH	78	468,312
W7	WY7I	90	637,208
W8	KW8N (NZ4K)	92	815,948
W9	KC9T	90	573,501
WØ	NQØI	92	424,461
VE	CH7SZ(VE7NTT)	93	814,506

S/O 7 MHz

W1	K8PO/1	92	821,473
W2	K1ZM/2	90	839,520
W3	K3WX(KH2F)	93	542,300
W4	K4XS	92	683,459
W5	K5GO	90	542,108
W6	N6RO	93	505,938
W7	W7XR (K7SS)	92	794,992
W8	KV8H	90	319,501
W9	WA8YVR/9	85	256,080
WØ	WØZV	92	440,316
VE	XN3BMV	84	436,100

S/O 3.5 MHz

W1	W1MK	93	340,431
W2	K1ZM/2	92	416,160
W3	WE3C	93	218,163
W4	KT3Y/4	92	205,683
W5	K5RX	87	116,584
W6	W6RJ	86	106,600
W7	K9JF/7	85	45,056
W8	W9LT/8	93	204,472
W9	N4CC/9	93	137,372
WØ	KØRF	92	139,072
VE	CH7CC	93	213,853

S/O 1.8 MHz

W1	W1BYH	93	48,552
W2	K2EK	92	34,522
W3	AA1K/3	85	24,120
W4	K4TEA	93	27,115
W5	K5UR	85	47,005
W6	K6DDO	85	8,534
W7	N6SS/7	93	19,548
W8	W8LRL	90	20,000
W9	WB9Z	93	46,314
WØ	KVØQ	93	28,161
VE	VO1NA	93	148,050

Single Operator WITH Packet (Since 1989)

W1	K1DG	89	5,048,802
W2	K5NA/2	93	4,490,980
W3	K3WW	93	5,056,464
W4	N4RJ (KM9P)	91	4,481,038

W5	KA5W (KS1G)	90	2,642,640
W6	WZ6Z	90	1,738,110
W7	NK7U (KY7M)	90	2,556,400
W8	K8AZ	92	4,740,060
W9	KK9V	90	2,204,494
WØ	N2IC/Ø	90	3,728,264
VE	VE3EJ	93	6,073,614

Multi-Op Single Transmitter

W1	K1AR	89	9,383,458
W2	N2RM	89	8,512,840
W3	N3RS	92	8,585,380
W4	N4RJ	89	7,811,622
W5	W5WMU	92	5,361,675
W6	W6GO	92	5,252,702
W7	KO7N	88	3,405,324
W8	K8AZ	90	5,880,834
W9	KS9K	93	4,167,336
WØ	KØRF	89	5,544,000
VE	XN3EJ	91	6,955,345

Multi-Op Multi-Transmitter

W1	K1AR	92	19,473,615
W2	N2RM	92	18,408,663
W3	W3LPL	91	17,108,280
W4	K4JPD	88	7,853,620
W5	NR5M	89	12,302,131
W6	K6UA	90	9,265,088
W7	KO7N	89	7,752,240
W8	K8LX	80	7,601,825
W9	WØAIH/9	90	10,029,830
WØ	K4VX/Ø	89	11,988,957
VE	VE2CSI	92	8,184,910

CQ WW DX MISCELLANEOUS DATA
Phone Winners by Continent
Single-Op Continental Wins:
South America: 26
Africa: 14
Europe: 3
North America: 2
Asia: 2
Oceania: 1

Multi-Single Continental Wins:
South America: 14
Europe: 12
North America: 10
Africa: 6
Asia: 5
Oceania: 0

Multi-Multi Continental Wins
South America: 19
Europe: 7
North America: 5
Africa: 4
Asia: 1
Oceania: 0

CW Winners by Continent
Single-Op Continental Wins
Africa: 18
South America: 12
Asia: 8
North America: 4
Oceania: 4
Europe: 2

Multi-Single Continental Wins:
North America: 12
Asia: 11
South America: 9
Africa: 6
Europe: 5
Oceania: 4

Multi-Multi Continental Wins

South America: 13
North America: 11
Africa: 5
Asia: 4
Europe: 2
Oceania: 0

CQ WW Phone Contest (USA WINNERS)

Single-Ops

K1AR: 7
W1ATE: 5
WA2SFP[W2PV]: 4
W6RR: 3
KC1F: 2
K1ZM: 2
K5MDX: 2
W6YY: 2
Each of the following have one win:
K1HLB, K1KTH, W1ONK, K2AAA, N2NT, K2TR, W3LPL, W3MSK[W3AU], W3WJD/N3RS], W4AXE, W4DR, W4ESK[W7RM], WA6VEF, K6AHV, W6GHM, K7RI, W8KML, W9EWC, KRØY.

Multi-Single

W2HJR[K2GL]: 4
KX4S: 3
W6AM: 3
K1AR: 2
WB2SQN/K2SS: 2
K3LR: 2
K6EVR: 2
WA6EPQ[N6AR]: 2
W7SFA: 2
Each of the following have one win:
K1DG, KC1XX, K2BU, W2CP/2, K2HLB, K2IEG, N2RM, W3TMZ, W3WJD/N3RS, K4VX, W4ATE, W4HXC, W4QAW, K5GA, K5JA, W6OUN, W6ONV, W6RRG, WA6ZQU, W7DL, KS9K, KØUK.

Multi-Multi

K2GL/N2AA: 14
W3MSK/W3AU: 5
W2PV: 4
N2RM: 4
W3LPL: 4
W7RM: 2
W3WJD[N3RS]: 1
N5AU: 1
W8NGO: 1

CQWW DX CW Contest (USA Winners)

Single-Ops

W4KFC: 9
W3GRF: 6
K1AR: 4
W3LPL: 4
W1KM: 3
K5ZD: 2
W4YHD: 2
W8JIN: 2
Each of the following have one S/O win:
K1CC, K1DIR, K1EA, K1GQ, K1KI, K1KTH[N6JL], W1FBY[W1XT], W1BGD/2[W1RM], W2HJR[K2GL], N2LT, KQ2M, W3RJ, W3WJD/N3RS, W4KVX, KM9P.

Multi-Single

W3WJD[N3RS]: 5
K4GSU[N4AR]: 4
AA5LES[K5RC]: 4
W6AM: 3
K1AR: 3
K1DIR: 2
K1GQ: 2
W1BIH: 2
K6EBB: 2
W6RW: 2
Each of the following have one M/S win:
K1KI, K1GL, KC1XX, W2IQG, W2PCF, W3MVB, N4WW, W4KVX, K5NA, W6DFY, W6GAL, W6GHM/6, W6OUN, W6SZY, WA6SBO, W8WZ, W9EXE, W9YT.

Multi-Multi

K2GL[N2AA]: 11
W3MSK[W3AU]: 9
W3LPL: 5
W4BVV: 3
N2RM: 2
W2PV: 2
W3AOH: 2
K1AR: 1
W3WJD/N3RS: 1

Single-Band WAZ During CQ WW Contest

Callsign	Year	Category	Mode	Band
W6RW	1957	M/O	CW	20
K6EIV	1960	SO/SB	CW	20
W3MSK[W3AU]	1966	M/O	CW	20
W2ONV	1966	SO/SB	SSB	20
K2KUR	1968	SO/SB	SSB	20
K8MMM	1968	SO/SB	SSB	20
VE3LZ	1968	SO/SB	SSB	20
OH2AM	1968	M/O	SSB	20
W3MSK[W3AU]	1968	M/O	SSB	20
W4ETO	1968	M/O	SSB	20
OH2AM	1968	M/O	CW	20
W4BVV	1969	M/O	CW	20
WA2ZAA	1969	M/O	CW	20
W3MSK[W3AU]	1969	M/O	CW	20
K3AFO	1972	SO/SB	SSB	20
W4WSF	1972	SO/SB	SSB	20
W8JGU	1972	SO/SB	SSB	20
UW1AR	1972	SO/SB	SSB	20
G3RRS	1978	SO/SB	SSB	20
EX9A	1978	M/O	SSB	20
N2AA	1979	M/O	SSB	20
N2AA	1979	M/O	SSB	15
SK2KW	1979	M/O	SSB	10
YU3EY	1979	M/O	SSB	20
N2AA	1979	M/O	CW	20
WA2RLQ	1980	SO/SB	SSB	20
6D7LCH	1980	SO/SB	SSB	20
OH8SR	1980	SO/SB	SSB	20
N2AA	1980	M/O	SSB	20
K8LX	1980	M/O	SSB	20
K3WW	1980	M/O	SSB	20
SK2KW	1980	M/O	SSB	20
LZ7A	1980	M/O	SSB	10
YU3TU	1981	SO/SB	SSB	15

OH1AA	1981	M/O	SSB	20
K3KG/4	1982	SO/SB	SSB	20
YU3TWT	1982	SO/SB	SSB	20
YT3L	1982	SO/SB	SSB	20
OHØW	1982	M/O	SSB	20
OHØW	1982	M/O	SSB	15
WØZV	1983	SO/SB	SSB	20
YT3M	1984	SO/SB	SSB	20
K2HFX	1986	SO/SB	SSB	20
N2ME	1986	SO/SB	SSB	20
KX4S	1986	SO/SB	SSB	20
N4ZC	1986	SO/SB	SSB	20
RFØFWW	1987	SO/SB	SSB	20
IK8ETA	1987	SO/SB	SSB	20
IO1ZEU	1987	SO/SB	SSB	20
JAØQNJ	1987	SO/SB	SSB	20
JH8JWF	1987	SO/SB	SSB	20
4N7N	1987	SO/SB	SSB	20
JA7MHZ	1987	SO/SB	CW	40
YC2UKM	1988	SO/SB	SSB	15
IR4LCK	1988	M/O	SSB	10
VE2ZP	1988	S/O/S/B	CW	20
K1TO	1988	SO/AB	CW	20
N6GG	1988	SO/SB	CW	20
ZS6BCR	1988	SO/SB	CW	20
JA5DQH	1988	SO/SB	CW	15
JA7FTR	1988	SO/SB	CW	20
G3TXF	1988	SO/SB	CW	20
IT9GSF	1988	SO/SB	CW	20
UR2RDO	1988	SO/SB	CW	20
CW8B	1988	SO/SB	CW	15
K1AR	1988	M/O	CW	20
N3RS	1988	M/O	CW	20
K8AZ	1988	M/O	CW	20
K5NA/2	1988	M/O	CW	20
WM5G	1988	M/O	CW	20
W3LPL	1988	M/O	CW	20
NR5M	1988	M/O	CW	20
K1ST	1988	M/O	CW	20
K4JPD	1988	M/O	CW	20
K4VX/Ø	1988	M/O	CW	20
PJ1B	1988	M/O	CW	20
KP2A	1988	M/O	CW	15
EA9EA	1988	M/O	CW	20
OL8A	1988	M/O	CW	20
UL8LYA	1988	M/O	CW	20
HG5A	1988	M/O	CW	20
HG5A	1988	M/O	CW	15
LZ9A	1988	M/O	CW	20
LZ9A	1988	M/O	CW	15
JA7SGV	1989	SO/SB	SSB	20
DF8XC	1989	SO/SB	SSB	20
N3RS	1989	M/O	SSB	20
KS9K	1989	M/O	SSB	15
W3LPL	1989	M/O	SSB	20
NB1H	1989	M/O	SSB	20
K5NA/2	1989	M/O	SSB	20
PJ1B	1989	M/O	SSB	20
ZW5B	1989	M/O	SSB	20
ZW5B	1989	M/O	SSB	15
VP9AD	1989	M/O	SSB	20
LZ9A	1989	M/O	CW	15
NX7K	1990	SO/SB	SSB	20
N2RM	1990	M/O	SSB	20
N5AU	1990	M/O	SSB	20
KY1H	1990	M/O	SSB	20
KY1H	1990	M/O	SSB	15
LZ9A	1990	M/O	CW	10
ZAØRS	1991	M/M	SSB	15

ZW5B(LU8DQ)	1991	S/O/S/B	CW	15
ZD8Z(N6TJ)	1992	S/OAB	SSB	20
IR4T(I4JMY)	1992	SO/SB	SSB	20
LY5A(LY2BN)	1992	SO/SB	SSB	20
PJ1B	1992	M/M	SSB	20
K1RU	1992	SO/SB	SSB	20
N2RM	1992	M/M	SSB	20
N3RS	1992	M/S	SSB	20
W3LPL	1992	M/M	SSB	20
W5FO	1992	M/M	SSB	20
N2RM	1992	M/M	SSB	15
IT9A(IT9GSF)	1992	SO/SB	SSB	10
YL2KL	1992	SO/SB	CW	40
IQ4A	1992	M/S	CW	40
HG73DX	1992	M/M	CW	40
K2VV	1992	SO/SB	CW	20
7L1GVE	1992	SO/SB	CW	20
W8UVZ	1992	S/O/AS/B	CW	20
UB4HO	1992	SO/SB	CW	15
S59AA	1992	SO/SB	CW	10
KK9A	1993	SO/SB	SSB	20
CH7SV(VE7SV)	1993	SO/SB	SSB	20
JAØJHA	1993	SO/SB	SSB	20
LZ5W(LZ1YE)	1993	SO/SB	SSB	20
OK1RI	1993	SO/SB	SSB	20
OH1AA(OH1MYA)	1993	SO/SB	SSB	20
GM3WOJ(GM4YXI)	1993	SO/SB	SSB	20
IB9S(IT9BLB)	1993	SO/SB	SSB	20
SMØKV	1993	SO/SB	SSB	20
IN3QBR	1993	SO/SB	SSB	20
HG73DX	1993	M/M	SSB	20
UW2F	1993	M/M	SSB	20
EA9UK(N6KT)	1993	M/M	SSB	15
GØKPW	1993	M/M	SSB	15
AHØK	1993	M/M	CW	20
HG73DX	1993	M/M	CW	20
KN6M/5	1993	SO/SB	CW	20
YT7A(4N7DW)	1993	SO/SB	CW	20
DK8FD	1994	SO/SB	SSB	20
OH2PM	1994	SO/SB	CW	20
OH1NSJ	1994	SO/SB	CW	20
N6AW	1994	SO/SB	CW	40
N3RS	1994	M/O	CW	20

146 have achieved SB WAZ during CQ WW Contests!

71-N.A.	22-YRS	57-CW	6-10M
75-DX		89-PH	19-15M
			116-20M
			5-40M

*N2AA (Ex. K2KUR) has achieved 5 WAZ in CQ WW Contest: K2KUR (1968 CW), WA2ZAA (1969 CW), N2AA (1979 PH+CW, 1980 PH) all on 20M!

*LZ9A/LZ7A M/S entries have achieved five WAZ and LZ2CC and LZ2DF have been on each of those teams!

*K1TO is only S/O CW All-Bander to achieve CQ WW WAZ!

*ZD8Z (N6TJ) IS ONLY S/O SSB ALL BANDER TO ACHIEVE CQ WW WAZ!

*IT9GSF has achieved two WAZ under two different callsigns: IT9GSF, 1988 CW 20M and IT9A, 1992 SSB 10M!

Stations with three CQ WW WAZ (year third WAZ achieved indicated in parentheses):

W3MSK [W3AU] (1969), ZW5B (1991), PJ1B, N2RM, W3LPL (1992), HG73DX (1993), N3RS (1994)

Stations with two CQ WW WAZ:

OH2AM, OHØW, HG5A, K1ST/NB1H, K5NA, KY1H, IT9GSF, OH1AA

CQ WW THEORETICAL MAXIMUM SCORES

Although presented in the October 1989 CQ as part of K1AR's Contest column, surely many a contest operator has reviewed the final contest results to find interesting facts, trends and achievement. This is a compilation of the maximum achieved in each area: QSOs, zones and countries worked.

1994 CQ WW DX SSB Contest Maximums

Band	QSOs	Zones	Countries
160	841-PA3DFT	19-K3LR	69-IR4T
80	1766-VP2EC	33-W6RJ	105-W3LPL
40	2882-TI1C	38-9M8R, YT7A	140-HG1S, YT7A
20	5109-PYØFM	40-DK8FD, OH2PM	178-5L2PP
15	5535-ZD8Z	39-YZ1AU	179-ZD8Z
10	3413-LU6ETB	34-HAØUZ	145-PQØMM
Total	19,546	203	816

@2.93 Pts/QSO = 58,930,603

***PJ1B accomplished 5 Band DXCC!!!!!

	M/M World	M/M USA	M/S World	M/S USA
QSOs	PJ1B-15627	W3LPL-4242	HC8A-8479	K4ISV-2076
Zones	RU6L-175	W3LPL-160	IQ4A-174	K4ISV-150
Countries	GØKPW-769	W3LPL-664	OT4T-715	K1NG-575
Max. Score	= 43.22M	= 9.10M	= 22.08M	= 4.40M
Actual Score	PJ1B = 40.27	W3LPL = 9.10	HC8A =17.94	KC1XX =3.89

1994 CQ WW DX CW Contest Maximums

Band	QSOs	Zones	Countries
160	1046-OM7A	21-HG73DX	89-SP5GRM
80	2217-ZB2X	34-S58A, SN3A, UN2L	120-HG73DX
40	2994-VP5VW	40-N6AW	143-W3LPL
20	3610-9G5AA	40-OH1NSJ, N3RS	158-LZ1MC
15	3728-ZPØY	39-HG73DX	143-HG73DX
10	1323-9G5AA	32-HGØD	121-HGØD
Total	14,918	206	774

@2.93 Pts/QSO = 42,834,820!!!!!

***HG73DX and OT4T accomplish 5 Band DXCC!!!

	M/M World	M/M USA	M/S World	M/S USA
QSOs	VP5VW-11740	W3LPL-4196	NP4Z-5217	K1AR-3124
Zones	HG73DX-198	W3LPL-181	IQ4A-194	N2NU-168
Countries	HG73DX-743	W3LPL-666	OT4T-708	AR/NU-586
Max. Score	= 32.36M	= 10.41M	= 13.78M	= 6.90M
Actual Score	9G5AA= 22.96M	W3LPL = 9.69M	IQ4A= 8.84M	K1AR = 6.60M

1993 CQ WW DX SSB Contest Maximums

Band	QSOs	Zones	Countries
160	930-UW2F	16-VP2EC	69-UW2F
80	1663-HG73DX	32-W6RJ	106-UW2F,VP2EC
40	2584-EA9UK	39-GØKPW	149-GØKPW
20	3165-VP5L	40-MANY(12)	175-W3LPL
15	3916-VP2EC	40-EA9UK,G0KPW	173-GØKPW
10	3881-GØKPW	34-HG73DX	152-GØKPW
Total	16,139	201	824

@2.93 Pts/QSO = 48,469,175

***EA9UK, VP2EC, GØKPW, HG73DX, UW2F, N2RM, W3LPL accomplish 5 Band DXCC!

	M/M World	M/M USA	M/S World	M/S USA
QSOs	EA9UK-13547	N2RM-6533	P40L-9569	KC1XX-3368
Zones	UR8J-188	N2RM-174	LZ9A-179	KC1XX-162
Countries	GØKPW-797	N2RM-747	IQ4A-736	N3RS-646
Max. Score	= 39.1M	= 17.6M	= 25.6M	= 7.97M
Actual Score	EA9UK = 37.14	N2RM = 16.01	PJ1B = 22.59	KC1XX = 7.63

1993 CQ WW DX CW Contest Maximums

Band	QSOs	Zones	Countries
160	1155-OY9JD	24-K1AR	82-K1AR
80	2181-OJØ/OH1VR	36-SN3A	118-W3LPL
40	3013-EA9EO	39-ED6XXX	152-W3LPL
20	2759-AHØK	40-Several(4)	156-K1AR
15	3627-ZPØY	38-Many	141-K3LR
10	2187-CV5A	34-AHØK	116-CV5A
Total	14,922	211	765

@2.93 Pts/QSO = 42,671,696

	M/M World	M/M USA	M/S World	M/S USA
QSOs	EA9EO-11,049	W3LPL-6,157	J6DX-7,180	KC1XX-3,369
Zones	AH0K-198	K1AR-197	IQ4A-185	KC1XX-175
Countries	K1AR-720	K1AR-720	KC1XX-642	KC1XX-642
Max. Score	= 29.71M	= 16.54M	= 17.39M	= 8.06M
Actual Score	EA9AO = 27.55M	W3LPL = 15.33M	J6DX = 11.69M	KC1XX = 7.90M

1992 CQ WW DX SSB Contest Maximums

Band	QSOs	Zones	Countries
160	672-9A1HCD	16-VP2EC	65-9A1HCD
80	1695-TI1C(TI2CF)	31-TI1C(TI2CF)	108-TI1C(TI2CF)
40	2737-9A1A	38-G0KPW	137-G0KPW
20	4118-PJ1B	40-MANY(6)	178-N2RM, W3LPL
15	4381-PJ1B	40-N2RM	178-N2RM
10	4735-ZW5B (N5FA)	40-IT9A (IT9GSF)	170-N2RM
Total	18,338	205	836

@ 2.93 Pts/QSO = 55,932,930
***PJ1B Accomplishes 5 Band DXCC!

	M/M World	M/M USA	M/S World	M/S USA
QSOs	PJ1B-15,554	N2RM-7010	VP9AD-8138	K1DG-3396
Zones	HG73DX- 187	N2RM- 185	VP2EC- 183	K4ISV-175
Countries	IQ4A- 774	N2RM- 793	VP2EC- 685	K1DG- 701
Max. Score	= 43.79M	= 20.08M	= 20.69M	= 8.71M
Actual Score	PJ1B = 43.15M	N2RM = 19.60M	VP2EC = 16.28M	K1DG = 8.44M

1992 CQ WW DX CW Contest Maximums

Band	QSOs	Zones	Countries	
160	893-ON4UN	23-AA6TT	76-ON4UN	
80	1856-OE3GSA	31-LZ9A	107-K1AR	
40	2985-ZF2TG(WQ5W)	40-SEVERAL!	144-W3LPL	
20	3042-KH0AM	40-SEVERAL!	156-K1AR	
15	3092-CR3W(DF5UL)	40-UB4HO	154-K1AR	ZD8LII(G0LII)
10	2746-P40X(N6BT)	40-S59AA	139-N2RM	
Total	14,614	214	776	

@ 2.93 Pts/QSO = 42,390,810
***K1AR Accomplishes 5 Band DXCC!

	M/M World	M/M USA	M/S World	M/S USA
QSOs	EA9EA-12,305	K1AR- 7,127	ZC4Z-5,537	N3RS-3,478
Zones	UX1A- 211	N2RM,W3LPL-195	LZ9A- 194	N3RS- 186
Countries	UX1A- 718	K1AR- 753	IQ4A- 678	N3RS- 674
Max Score	= 33.49M	= 19.79M	= 14.14M	= 8.76M
Actual Score	EA9EA = 30.03M	K1AR = 19.47M	ZC4Z = 11.09M	N3RS = 8.58M

1991 CQ WW DX SSB Contest Maximums

Band	QSOs	Zones	Countries
160	559-VP9AD	18-K2TR	57-EA3ALD
80	1270-HG73DX	28-W7XR	96-HG73DX
40	2290-HG73DX	34-W7XR	128-IQ4A
20	4110-ZA0RS	39-MANY!	171-K2TR
15	4535-FM6A(F6HMQ)	40-ZA0RS	169-N2RM
10	5154-ZV5A(PY5EG)	38-FR5DX	162-N2RM
TOTAL	17,918	197	783

@ 2.93 Pts/QSO = 51,449,020

World Maximums

	M/M World	M/M USA	M/S World	M/S USA
QSOs	VP9AD-15,029	N2RM- 5465	PJ1B-9460	K3LR-2598
Zones	ZA0RS- 179	W3LPL- 181	LZ9A- 185	K3LR- 166
Countries	ZA0RS- 727	W3LPL- 703	IQ4A- 753	N3RS- 605
Max. Score	= 39.89M	= 14.00M	= 25.99M	= 5.80M
Actual High	VP9AD=28.08M	N2RM=12.88M	PJ1B=21.21M	K3LR=5.51M

1991 CQ WW DX CW Contest Maximums

Band	QSOs	Zones	Countries
160	620-J6DX	18-J6DX	63-HG73DX
80	1500-CT3M	33-RZ1A	102-RZ1A
40	2577-C42A(YU4OO)	39-K5NA	133-HG73DX
20	4036-PJ9A	39-SEVERAL!	145-K1AR
15	3951-J6DX	40-ZW5B(LU8DQ)	138-ZW5B(LU8DQ)
10	3534-PJ9A	38-IQ4A	133-W3LPL
Total	16,218	207	713

@ 2.93 Pts/QSO = 43,716,560
*** RZ1A Accomplishes 5 Band DXCC!

	M/M World	M/M USA	M/S World	M/S USA
QSOs	PJ9A-14,991	N2RM-5072	EA9EA-5854	K5NA-2960
Zones	HG73DX-191	N2RM- 187	RZ1A- 193	N3RS- 183
Countries	HG73DX- 664	W3LPL-663	RZ1A- 642	N3RS- 633
Max Score	= 37.55M	= 12.63	= 14.32M	= 7.07M
Actual Score	PJ9A = 35.32M	N2RM=12.37M	EA9EA=13.09M	K5NA=6.71M

1990 CQ WW DX SSB Contest Maximums

Band	QSOs	Zones	Countries
160	805-UG6GAW	20-VE1ZZ	56-IQ4A
80	1455-HA8IE	35-HA8IE	116-HA8IE
40	2275-PJ9W	37-IQ3A,IK5BAF	126-IQ3A
20	4925-PJ9W	40-SEVERAL(3)	179-PJ1B
15	5395-PJ1B	40-KY1H	176-PJ1B
10	5430-PJ1B	39-MANY!	182-PJ1B
Total	20,285	212	835

@ 2.93 Pts/QSO = 62,228,445!

	M/M World	M/M USA	M/S World	M/S USA
QSOs	PJ1B-19,655	N2RM-7321	6D2X-8169	K1AR-4437
Zones	PJ1B- 189	N5AU- 187	LZ9A- 202	KS9K- 177
Countries	PJ1B- 803	N2RM- 716	OM5W- 624	K1AR- 703
Max. Score	= 57.12M	= 19.36M	=19.77M	= 11.44M
Actual Score	PJ1B=57.61M	N2RM=18.14M	IQ4A=17.25M	K1AR=11.19M

1990 CQ WW DX CW Contest Maximums

Band	QSOs	Zones	Countries
160	936-ON4UN	21-V73AZ(NZ8B)	68-ON4UN
80	1873-PJ9A	28-W3LPL	99-PJ9A
40	2598-PJ9A	37-LZ9A	125-K1ZM
20	3936-CN5N	39-SEVERAL(4)	149-K1AR,N2RM,CN5N
15	3689-PJ9A	39-SEVERAL(3)	142-CN5N
10	3795-CXØCW(CX8BBH)	40-LZ9A	137-W3LPL
Total	16,827	204	720

@ 2.93 Pts/QSO = 45,555,972

	M/M World	M/M USA	M/S World	M/S USA
QSOs	PJ9A-15,242	N2RM-7018	TA5KA-7201	N3RS-3532
Zones	PA6DX- 185	N2RM- 189	LZ9A- 191	K5NA- 182
Countries	PJ9A- 659	K1AR- 683	LZ9A- 618	K5NA- 609
Max. Score	= 37.69M	= 17.93M	= 17.06M	= 8.18M
Actual Score	PJ9A=34.93M	N2RM=17.28M	TA5KA=13.91M	N3RS=8.024M

1989 CQ WW DX SSB Contest Maximums

Band	QSOs	Zones	Countries
160	1527-RB8M	14-VP9AD	57-VP9AD
80	1884-LX7A	25-HK3MAE	94-HK3MAE
40	2409-LX7A	35-LZ9A	121-PJ1B,LX7A
20	5420-VP9AD	40-MANY!	186-PJ1B
15	4445-EL2CX(K3RV)	40-ZW5B,KS9K	178-ZW5B
10	5714-PJ1B	39-ZW5B	178-PJ1B
Total	21,399	193	814

@ 2.93 Pts/QSO = 63,137,893

	M/M World	M/M USA	M/S World	M/S USA
QSOs	PJ1B-18,175	W3LPL-6061	PJ9W-9265	KS9K-2883
Zones	ZW5B- 189	W3LPL- 172	LZ9A- 180	KS9K- 167
Countries	PJ1B- 764	W3LPL- 706	LZ9A- 665	N3RS- 595
Max. Score	= 50.74M	= 15.59M	= 22.93M	= 6.43M
Actual Score	PJ1B=48.98M	W3LPL=15.02M	P40V=18.52M	KS9K=6.04M

1989 CQ WW DX CW Contest Maximums

Band	QSOs	Zones	Countries
160	882-LY2BTA	21-K5UR	68-LY2BTA
80	1859-LX7A	31-RB8M	100-LX7A
40	2720-4UØITU	37-LZ9A,K5NA	125-LX7A
20	4253-CT3M	39-SEVERAL(3)	148-LX7A
15	3652-NL7G	40-LZ9A	133-LX7A
10	3495-ZPØY(LU8DQ)	39-LZ9A	140-W3LPL
Total	16,861	207	714

@ 2.93 Pts/QSO = 45,499,242

	M/M World	M/M USA	M/S World	M/S USA
QSOs	CT3M-14,473	W3LPL-6866	EA9EA-5923	K1AR-4101
Zones	RB8M- 199	W3LPL- 190	LZ9A- 200	K1AR- 190
Countries	LX7A- 705	W3LPL- 619	LZ9A- 626	N2RM- 605
Max. Score	= 38.33M	= 16.27M	= 14.33M	= 9.55M
Actual Score	CT3M=32.03M	W3LPL=16.22M	EA9EA=11.75M	K1AR=9.38M

ALL-TIME MAXIMUM ACHIEVEMENTS DURING CQ WW DX PHONE AND CW

Maximum CQ WW DX Phone (results through 1994)

Band	QSOs/Stn./Yr.	Zones/Stn./Yr.	Countries/Stn./Yr.
160	1527-RB8M-'89	20-VE1ZZ-'90	69-UW2F-'93
		K5UR-'88	IR4T-'94
80	1884-LX7A-'89	35-HA8IE-'90	116-HA8IE-'90
40	2882-TI1C-'94	39-G0KPW-'93	149-G0KPW-'93
20	5420-VP9AD-'89	40-MANY!	186-PJ1B-'89
15	5417-VP2KC-'79	40-MANY!	178-N2RM-'92
			ZW5B-'89
10	5714-PJ1B-'89	40-SEVERAL!	182-PJ1B-'90
Total	22,971	214	880
	@ 2.93 Pts/QSO	= 73,631,702 !!!!	

All-time Category Theoretical Maximums

	M/M World	M/M USA	M/S World	M/S USA
QSOs	19,655-PJ1B-'90	7,321-N2RM-'90	9,569-P40L-'93	4,437-K1AR'90
Zones	189-Three	187-N5AU-'90	202-LZ9A-'90	177-KS9K-'90
Countries	803-PJ1B-'90	793-N2RM-'92	753-IQ4A-'91	703-K1AR-'90
Max Score	= 57.12 M	= 21.02 M	= 26.46 M	= 11.44 M

Maximum CQ WW DX CW (through 1994 results)

Band	QSOs	Zones	Countries
160	1046-OM7A-'94	26-K5UR-'87	89-SP5GRM-'94
80	2243-NP4A (K1ZM)-'88	34-SEVERAL	120-HG73DX-'94
40	3119-PJ1B-'88	40-SEVERAL!	144-W3LPL-'92
20	4253-CT3M-'89	40-MANY!	158-LZ1MC-'94
15	3951-J6DX-'91	40-SEVERAL!	154-K1AR-'92
10	3795-CX0CW-'90	40-S59AA-'92	140-W3LPL-'89
	(CX8BBH)	LZ9A-'90	
Total	18,407	220	805
	@ 2.93 Pts/QSO	= 55,280,822	

All-time Category Theoretical Maximums

	M/M World	M/M USA	M/S World	M/S USA
QSOs	15,242-PJ9A-'90	7,127-K1AR-'92	7,201-TA5KA-'90	4,101-K1AR-'89
Zones	211-UX1A-'92	195-N2RM, W3LPL-'92	200-LZ9A-'89	190-K1AR-'89
Countries	743-HG73DX-'94	753-K1AR-'92	708-OT4T-'94	674-N3RS-'92
Max. Score	= 42.6 M	= 19.79 M	= 19.16 M	= 10.38 M

A BRIEF HISTORY OF THE ARRL NOVEMBER SWEEPSTAKES CONTEST

As a direct result of the fervor by the participants of the Hiram Percy Maxim Birthday Relay of 1929 and the First International Relay Party, F.E. Handy, W1BDI, ARRL Communications Manager, noted that "...a number of amateurs have expressed increasing interest in taking part in more message-handling contests...." December 1929 QST carried the rules for the First All-Section Sweepstakes Contest, and the awards program was announced in the January 1930 issue. Sweep-brooms were offered to the top three scorers and were appropriately adorned with the ARRL emblem and "a symbolic vacuum tube...firmly affixed to the handle...."

But Why Was It Called the "Sweepstakes?"

In the January 1930 QST award announcement, ARRL Communications Manager F. E. Handy wrote of the Dutch commander Admiral Tromp who defeated the British fleet under Admiral Blake at Dover, England, in 1652. In keeping with the custom of the times, Tromp sailed up the English Channel with a broom hoisted at the masthead of his flagship, denoting that he had successfully "swept the seas." Handy felt that the broom was an appropriate prize (or "stakes") for a skilled operator having "swept the air." And so the ARRL's major domestic contest was named the "All-Section Sweepstakes Contest."

Planned for January 18-31, 1930, the first Sweepstakes was a 14-day, mixed mode event in which operators were permitted up to 90 hours of operation, exchanging actual messages. These messages were required to be in the proper form, containing city of origin, message number, date, address, a text message and signature. The score was determined by receiving one point for sending a message and one point for receiving a message, only then could you consider that ARRL section "worked." Total points multiplied by the number of sections worked (maximum of 68) was the final score. There were two classes of entry: above 100 watts and below 100 watts, and 116 logs were received at ARRL Headquarters! W1ADW was proclaimed the winner, submitting 153 QSOs and 43 sections worked. In second place was W9DEX with 142 QSOs and 43 sections. Third place was earned by W2BAI's 155 QSOs and 39 sections.

In 1931, with the addition of the San Joaquin Valley section, 276 entries arrived, some of which mentioned phone contacts. W8CHC came out on top, reporting 305 QSOs and 54 sections. The team at W5WF was second and W7AAT's 265 QSOs and 55 sections earned him third. Although phone operations occurred, they were but a small percentage of the total activity: 3 in 1932, 25 in 1933, 15 in 1934, 6 in 1935, 15 in 1936, 102 in 1937, 113 in 1938, 310 in 1939, and 277 in 1940. Finally a "Phone Only" Sweepstakes category was announced.

A power handicap was instituted in 1934. Three classes were offered. Under 25 watts input multiplied score by 3; above 25 watts input but below 75 multiplied score by 2; over 75 watts input received no handicap.

Separate CW and phone certificate awards were offered for the sixth ARRL Sweepstakes Contest. Additionally, the same 90-hour operating rule used in the International Relay Competitions was added to Sweepstakes. Those operating more than 90 hours had

their final score reported proportionally. The message was shortened to contain four or more words, compared to the previous rule of five or more. The three power classes were reduced to two: above and below 100 watts input. The below 100 watt class added a power multiplier of 1.5, which was later reduced to 1.25.

A club competition was instituted in 1936, and was dominated by the Frankford Radio Club for the first eight years. The Potomac Valley Radio Club began its reign in 1948, accumulating enough scores to win the gavel 21 years out of the next 27.

The big story in 1936 was W6ITH completing the very first "Clean Sweep"—working all sections during the contest. Everyone always expected that this would initially occur on CW, but it did not; it was achieved exclusively on AM! W6MVK completed the first CW Clean Sweep during the 1937 event. Note that the sections of 1937 included Northern and Southern Minnesota (two separate sections), "Utah-Wyoming" (one section), "Georgia-South Carolina-Cuba-Isle-of-Pines-Puerto Rico-Virgin Islands" (one section), and Philippines!

The 1937 Sweepstakes was held on two weekends, 33 hours each, with entrants limited to operating for a maximum of 40 hours. The 1941 event brought the first ARRL Sweepstakes exclusively for phone! ARRL November Sweepstakes was QRT for WWII from 1942 to 1945. The full score listings for the 13TH Sweepstakes (1946) were never published. Instead, there was a mailing to each entrant. Results were published in QST again yearly, beginning with the 1947 event.

Beginning with the 1963 event, the year in which the operator was first licensed became the message check number.

The 1964 Sweepstakes was the first 24-hour, Single-Mode event. The Low Power (150 watts or less) category power multiplier (1.25) was eliminated after the 1968 event. For the 1978 events, the Club Competition rules were changed to accommodate Large (Unlimited) clubs, Medium clubs and the smaller Local clubs to allow similar clubs to compete with their peers.

In 1963, the Maryland-Delaware Section "MDD" was split into the Maryland-D.C. ("MDC") section and the Delaware ("DEL") section. This brought the section total to 75, where it remained until 1978. When the U.S. relinquished control of the Canal Zone back to Panama in 1979, the section total was reduced to 74. The West Texas Section was added with the 1987 event, again raising the total sections to 75. The West Indies Section was split into the separate Puerto Rico and Virgin Islands Sections beginning with the 1988 event, raising the total number of sections to 76. The Washington Section was split into the East and West Washington Sections beginning with the 1989 event raising the total number of sections to 77.

To commemorate the 50th ARRL Sweepstake, H.Q. offered Clean Sweep whisk brooms to the fortunate operators who worked all sections. As a result, 219 earned one for CW and 198 did so on SSB. The QRP Single Operator (No Packet) class was added beginning with the 1987 event. The "P.I.N.S." (Participation IN Sweepstakes) Program began with the 1990 event and since then records have been falling!

ARRL SWEEPSTAKES CW WINNERS BY YEAR/CATEGORY

Year	SS#	S/O-HI	S/O-LOW-POWER	S/O-QRP	MULTI-OP	#Sweeps	Club: Unlimited/Med./Local
1994	#61	WM5G(KR0Y)	KP4WI(N0BSH)	KP4/KA9FOX	AA5B	123	NCCC-N.Tx.C.C.-River City C.
1993	#60	WM5G(KR0Y)	NP4A(K7JA)	N3SL/0	AA5B	84	NCCC/SCCC/River City C.
1992	#59	N4RJ(KM9P)		W9UP(KA9FOX)	AA5B	209	NCCC-N.Tx.C.C.-N.M. Big River C.C.
1991	#58	N5AU(WN4KKN/5)	N5RZ	NF7P	W6GO	127	FRC-N.Tx.C.C.-River City C.
1990	#57	N5AU(WN4KKN/5)	N5RZ	K7SS	WX0B	5	Mn W.A.-N.Tx.C.C.-River City C.
1989	#56	N5AU(WN4KKN/5)	K0EU	K0SCM	K5LZO	7	SMC-Mn.W.A.-Rubber Circle C.C.
1988	#55	K0RF(W0UA)	K0EU	N4RJ	N4RJ	60	—N.Tx.C.C.-River City C.
1987	#54	W5WMU(K5GA)	NF7P	N2IC	AA5B	73	Murphy's-N.Tx.C.C.-River City C.
1986	#53	NP4A(K5ZD)	K3RR	*#5*	K5CM	134	SMC-Mad River R.C.-Rubber Circle C.C.
1985	#52	K6LL	K4JPD(N4ZZ)		AA5B	108	SMC-N.Tx.C.C.-Rubber Circle C.C.
1984	#51	N5AU(K5ZD)	K7JA		N6BT	100	PVRC-N.Tx.C.C.-Colorado Contest Conspiracy
1983	#50	N5AU(K5ZD)	N0NO		K5CM	219	PVRC-Tx.DX.S.-Lincoln ARC
1982	#49	N5AU(K5ZD)	N6IG		WB8JBM	47	PVRC-NCCC-Alburquerque DX Assn.
1981	#48	N5AU(K5ZD)	W2TZ		N5DKG	55	NCCC-Tx.DX.S.-Willamette Valley (Ore) DXC.
1980	#47	K0RF(W0UA)	W1ZT		K1VTM	29	NCCC-YCCC-South Jersey Contest Coalition
1979	#46	KP2A(K2TR)	VE4VV		K6RLY	145	NCCC-Murphy's-Cascades ARC (MI)
1978	#45	W2GD	AA7A		WA2ECA	86	NCCC-*# CL*-*****
1977	#44	KP4RF(KP4EAJ)	K5LZO		W4BVV	27	NCCC
1976	#43	KV4FZ[K7VPF][K7JA]	AC8CQN[W8FJ]		W0ZLN	17	NCCC
1975	#42	WA5LES[K5RC]	WA5VDH[K5UR]		W6YRA	0	NCCC
1974	#41	W6RR[W6RTT][AH3C]	W8CQN[W8FJ]		KH6RS	62	PVRC
1973	#40	W7RM[K7VPF][K7JA]	WB2RJJ[N2OO]		WA1KID[W6XR]	25	MM
1972	#39	W7RM[K7VPF][K7JA]	WB0DLE		W7SFA[W7FU]	16	PVRC
1971	#38	W7RM[K7VPF][K7JA]	WA3DSZ		K4SXD[N4RG]	29	Mad River R.C.
1970	#37	W6HX[WB6OLD][N6NT]	K5RHZ		WB6WIT	14	PVRC
1969	#36	W6RW[W6DQX][N6ZZ]	WA3DSZ		K5LZO	22	PVRC
#LP							
1968	#35	KV4FZ			W6RW	18	PVRC
1967	#34	W9YT[K9ZMS][K6NA]			K4VDL	38	PVRC
1966	#33	W6CUF[W6CF]			K5LZO	17	FRC
1965	#32	W4KFC			K5LZO	3	PVRC
1964	#31	W4KFC			W3MWC	17	PVRC
#24							
1963	#30	W9IOP			WB2APG	24	FRC
#CK, CM							
1962	#29	W5WZQ[W5UN]			W3GQF	93	PVRC
1961	#28	W5WZQ[W5UN]			W3GQF	77	PVRC
1960	#27	W9IOP			W9WBY	69	FRC
1959	#26	W9IOP			W3FYS	93	PVRC
1958	#25	W4KFC			W3FYS	104	FRC
1957	#24	W2IOP			W3FYS	59	FRC
1956	#23	W4KVX			W9OCB	27	PVRC

Year	#	Callsign		Call	Score	Club
1955	#22	W7KVU		W9OCB	36	PVRC
1954	#21	W4KVX		W3FRY[K3WWVJ	37	PVRC
1953	#20	W9IOP		W3CTJ	17	PVRC
1952	#19	W4KFC		W3LTW	32	PVRC
1951	#18	W6BJU(W6CUF[W6CF])		W6EFW	9	PVRC
1950	#17	W4KFC		W3KT	8	PVRC
1949	#16	W2IOP			5	PVRC
1948	#15	W2IOP		W3CPV	3	PVRC
1947	#14	W4KFC		VE7ACS	2	FRC
1946	#13	W4FUH		*#YB*	1	FRC
#ZF						
1941	#12	W9FS		W4WE	0	FRC
1940	#11	W3BES		W8YX	0	FRC
1939	#10	W2IOP		W1AQ	6	FRC
1938	#9	W3BES		W8ADV	0	FRC
1937	#8	W6MVK		W8ADV	1	FRC*[W6MVK FIRST CW SWEEP!]
1936	#7	W1EZ		K5AC	0	FRC
1935	#6	W8JIN		W4IB	0	
1934	#5	W9HKC		W6BIP	0	
1933	#4	W9AUH		W9GDH	0	
1932	#3	W8ER		W5WF	0	
1931	#2	1ST-W8CHC-305Qx54; 2ND-W5WF-(4OPS)-?Qx58!;3RD-W7AAT-265x55; *276 Entrants some phone work reported.				
1930	#1	1ST-W1ADW-153Qx43; 2ND-W9DEX-142x43,3RD-W2BAI-155x390; **#ZZ**116 Entrants				

Footnotes to Sweepstakes

Footnote #1-The "P.I.N.S. (Participation IN Sweepstakes)" Program began with the 1990 event. CURRENT

Footnote #2-The WASHINGTON state section was split into the EAST Washington section and the WEST Washington section beginning with the 1989 event. (Total sections = 77). CURRENT

Footnote #3-The WEST INDIES section was split into the Puerto Rico section and the Virgin Is. section beginning with the 1988 event. (Total sections = 76).

Footnote #4-The WEST TEXAS section was created beginning with the 1987 event. (Total sections = 75).

Footnote #5-QRP Single Operator class was added beginning with the 1987 event. CURRENT

Footnote #CL-The Club Competition rules were changed to accommodate Large-Unlimited clubs, Medium clubs and Small,Local clubs. CURRENT

Footnote #LP-The Low Power (150 watts or less) category power multiplier was eliminated after the 1968 event. CURRENT

Footnote #24-The 1964 S.S. was the first 24Hour, Single Mode Event.

Footnote #CK-Beginning with the 1963 event, The year you were first licensed became your message check number. CURRENT

Footnote #CM-In 1963, the Maryland-Delaware Section, MDD was split into the Maryland, D.C., MDC section and the Delaware, DEL section.

Footnote #YB-The full scores listings for the 13TH Sweepstakes were never published. There was a mailing to each entrant.

Footnote #ZF-A.R.R.L. November Sweepstakes was QRT for WWII from 1942 to 1945.

Footnote #ZZ-There were two classes: Above and Below 100 watts. A 9 day, mixed mode event in which operators were permitted up to 90 hours of operation, exchanging actual messages.

ARRL SWEEPSTAKES PHONE WINNERS BY YEAR/CATEGORY

YEAR	SS#	S/OHI	S/OLOW POWER	S/OQRP	MULTI-OP	#Sweeps	Club: Unlimited/Medium/Local
1994	#61	W5KFT(WB5VZL)	NP4Z(WC4E)	W9UP(KA9FOX)	K9RS	305	NCCC/N.Tx.C.C./River City C.C.
1993	#60	KI3V/7	K7QQ	W9UP(KA9FOX)	N4ZZ	285	NCCC/SCCC/River City C.
1992	#59	WM5G(KR0Y)	VE4GV	WE9V	K7SS	458	NCCC/N.Tx.C.C./N.M. Big River C.C.
1991	#58	N5RZ	K4XS	K7SS	K9RS	214	FRC/N.Tx.C.C./River City C.
1990	#57	W6GO(N6IG)	K7QQ	K7SS	AA5B	215	Mn.W.A./N.Tx.C.C./River CityC.
1989	#56	W7XR(W7WA)	K0EU	K7SS	AI7B	141	SMC/Mn.W.A./Rubber CircleC.C.
1988	#55	KP2A(N2IC)	AA5B	W0LSD	W7EJ	177	—N.Tx.C.C./River CityC.
1987	#54	W7WA	K4XU	KE7X	K0UK	128	Murphy's/N.Tx.C.C./River City C.
1986	#53	N6BV	KE5CV	*#5*	K5LZO	143	SMC/Mad River CC/Rubber Circle C.C.
1985	#52	WA7NIN(W6OAT)	KE5CV		AA5B	146	SMC/N.Tx.C.C./Rubber Circle C.C.
1984	#51	KP4BZ	K4XS		KN6M	87	NCCC/T.DX.S./Rubber Circle C.C.
1983	#50	KP4BZ	K7RI(K7SS)		N4WW	198	PVRC/N.Tx.C.C./Colorado C.C.
1982	#49	NP4A(K1ZM)	K0UK		N6BT	160	PVRC/T.DX.S./Lincoln ARC
1981	#48	KV4FZ(WA6VEF)	W5MYA		K0WA	323	PVRC/Colorado.C.C./Albuquerque DXA.
1980	#47	N7DD(K7JA)	K7SS		W7ZR	175	NCCC/T.DX.S./Willamette Valley (Or) DX Club
1979	#46	KP4Q(K2TR)	W7XN		AF3P	333	NCCC/YCCC/South Jersey Contest Coalition
1978	#45	KP4RF[AH3C]	N5DX(K8CC)		K1GZL	269	NCCC/Murphy's/Cascades ARC (MI)
1977	#44	KP4RF	N4MM		W2SZ	168	NCCC
1976	#43	W0TR[[KR6X]	WB0MIV		W4BVV	109	NCCC
1975	#42	W7RM[[K7JA]	WA5VDH[K5UR]		W5GAD	24	NCCC
1974	#41	W6HX[[KR6X]	WB5FMJ		W5LUJ[W5XJ]	56	PVRC
1973	#40	W7RM[[K7JA]	K7JCA/6[K7SS]		W6ONV	75	Murphy's Marauders
1972	#39	W7RM[[K7JA]	WB0DSP		WB5DTX[N5DD]	102	PVRC
1971	#38	W7RM[[K7JA]	K1EUF		WA7ORM	125	MAD RIVER OHIO
1970	#37	K7VPF/7[K7JA]	K5RHZ		W1FBY[W1XT]	80	PVRC
1969	#36	K9LBQ/7	K4WAR		K5LZO	38	PVRC
#LP							
1968	#35	KV4FZ			K5LZO	26	PVRC
1967	#34	K4WJT			WA0CHH	21	PVRC
1966	#33	W7ESK[W7RM]			WA0CHH	9	PVRC
1965	#32	W4KFC			K5IIS	13	PVRC
1964	#31	K6EVR(W9WNV)			W3MWC	2	PVRC
#24							
1963	#30	K5MDX			K3KLQ	2	FRC
#CK,CM							
1962	#29	K6EVR			W0EEE	11	PVRC
1961	#28	W7ESK[W7RM]			WA2LPF	3	PVRC
1960	#27	K6EVR			W0YQ	6	FRC
1959	#26	K5MDX			W0YQ	15	PVRC
1958	#25	K6EVR			W1KBN	18	FRC
1957	#24	W0EDX			W0YQ	13	FRC
1956	#23	W6AM			K5EAT/	58	PVRC
1955	#22	K2AAA			W5ZED	3	PVRC

Year	Rank	Call	1ST / 2ND / 3RD		Call	#	Club	Notes
1954	#21	W6AM(W6FRW)			W9WHN	6	PVRC	
1953	#20	W9NDA			W4TEW/4	2	PVRC	
1952	#19	W0EDX			W9HHX	3	PVRC	
1951	#18	W6QEU	—			7	PVRC	
1950	#17	W6QEU			W6YX	3	PVRC	
1949	#16	W6QEU					PVRC	
1948	#15	W6QEU			VE7AKY	5	PVRC	
1947	#14	W6MLY			W4MOE	2	FRC	
1946	#13	W6ITH	*#YB*			1	FRC	
1941	#12	W9RBI				0	FRC	
1940	#11		1ST-W6ITH,2ND-W9YQN,3RD-W5BB		W1AW	1	FRC	274-OTHER PHONE SCORES REPORTED.
1939	#10		1ST-W6ITH,2ND-W6OCH,3RD-W9RBI		W4EQK	2	FRC	307-OTHER PHONE SCORES REPORTED.
1938	#9		1ST-W6ITH,2ND-W6OCH,3RD-VE5VO		W2JSE	0	FRC	110-OTHER PHONE SCORES REPORTED.
1937	#8		1ST-W6ITH,2ND-W9PWU,3RD-VE9AL(VE3BC)			0	FRC	99-OTHER PHONE SCORES REPORTED.
1936	#7		1ST-W6ITH,2ND-W9ATP,3RD-W9PWU			1	FRC	13-OTHER PHONE SCORES REPORTED. [1W6ITH FIRST S. S. PHONE SWEEP!]
1935	#6		1ST-VE3ER,2ND-W4BZA,3RD-W4DGS			0	---	3-OTHER PHONE SCORES NOTED.
1934	#5		1ST-W9GAF,2ND-W8DML,3RD-W9KBT			0	---	13-OTHER PHONE SCORES NOTED.
1933	#4		1ST-W6AHP,2ND-W9ACU,3RD-W8EMP			0	---	22-OTHER PHONE SCORES NOTED.W6AHP PLACED 8TH OVERALL!
1932	#3		1ST-W8ALC,2ND-W1APK,3RD-W9CUK			0	---	NO OTHER PHONE SCORES NOTED OR LISTED.
1931	#2		PHONE OPERATION NOTED BUT NOT REPORTED. ———			0	---	
1930	#1		ALL C.W.			0	---	

Footnotes to Sweepstakes

Footnote#1 -The 'P.I.N.S. (ParticipationIN Sweepstakes) Program' began with the 1990 event. CURRENT

Footnote#2-The WASHINGTON state section was split into the EAST Washington section and the WEST Washington section beginning with the 1989 event. (Total sections = 77). CURRENT

Footnote#3-The WEST INDIES section was split into the Peurto Rico section and the Virgin Is. section beginning with the 1988 event. (Total sections = 76).

Footnote#4-The WEST TEXAS section was created beginning with the 1987 event. (Total sections = 75).

Footnote#5-QRP Single Operator (NoPacket) class was added beginning with the 1987 event. CURRENT

Footnote#CL-The Club Competition rules were changed to accommodate Large-Unlimited clubs, Medium clubs and Small,Local clubs. CURRENT

Footnote#LP-The Low Power (150 watts or less) category power multiplier was eliminated after the 1968 event. CURRENT

Footnote#24-The 1964 S.S. was the first 24Hour, Single Mode Event. CURRENT

Footnote#CK-Beginning with the 1963 event, The year you were first licensed becams your message check number. CURRENT

Footnote#CM-In 1963, the Maryland-Delaware Section, MDD was split into the Maryland, D.C. (MDC) section and the Delaware (DEL) section.

Footnote#YB-The full Scores listings for the 13TH Sweepstakes were never published. There was a mailing to each entrant.

Footnote#ZF-A.R.R.L. November Sweepstakes was QRT for WWII from 1942 to 1945.

Footnote#ZZ-There were two classes: Above and Below 100 watts. A 9 day, mixed mode event in which operators were permitted up to 90 hours of operation, exchanging actual messages. Although 'Phone operations occurred, they were but a small percentage of the total activity; 3 ops in 1932, 25 ops in 1933.

ARRL SWEEPSTAKES CW SECTION RECORDS
Single-Operator, High Power

Section	Call	Year	Score	Section	Call	Year	Score
CT	K1TO	92	195,320	SJV	N6UR(KR6X)	94	203,434
EMA	K1AR	78	171,000	SV	N6IG	94	190,498
ME	N1SW	79	147,822	AK	AL7CQ	90	158,250
NH	K1DG(WZ1R)	93	180,424	AZ	K6LL	93	212,982
RI	K1IU	94	160,468	EWA	WA7EGA	91	132,088
VT	WB1GQR(WB2JSJ)	92	156,002	ID	K7NHV	83	172,568
WMA	K5ZD	94	187,800	MT	KØPP/7	93	168,476
ENY	K1ZM	90	180,950	NV	KI3V/7	94	201,432
NLI	NQ2D	90	167,092	OR	N6TR/7	92	208,054
NNJ	W2RQ	92	184,030	UT	W7CFL	91	159,544
SNJ	KZ2S	94	197,890	WWA	W7RM(N6TR)	93	218,834
WNY	K2ZJ	94	181,640	WYO	N7NG(AH3C)	92	193,648
DE	W3XU	81	157,176	MI	WA8ZDT	94	193,116
EPA	W3RJ	77	165,020	OH	KW8N(WD8IXE)	93	180,424
MDC	KE3Q	94	191,520	WV	N8II(KC8C)	86	154,906
WPA	K3LR(K3UA)	94	208,824	IL	K9FD	93	209,440
ALA	WZ4F	88	168,112	IN	W9RE	94	193,116
GA	N4RJ(KM9P)	94	232,232	WI	WØAIH(KØFVF)	93	162.624
KY	N4AR	92	174,020	CO	N2IC	93	216,524
NC	AA4NC	93	185,440	IA	NCØP(AG9A)	94	215,600
NFL	AC4NJ(WC4E)	94	197,582	KS	NØXA(KØVBU)	93	183,008
PR	NP4A(K5ZD)	86	195,656	MN	NØAT	91	187,110
SC	W4BTZ	93	126,900	MO	K4VX/Ø([WX3N])	94	219,296
SFL	K1ZX	94	203,832	ND	WBØO	87	166,204
TN	N4ZZ	92	173,588	NE	KØUP	78	133,298
VA	KT3Y	94	182,798	SD	KØZZ	78	171,828
VI	KP2/AB6FO	94	174,344	MAR	VE1QST(AK4L)	80	139,870
ARK	K5GO	92	182,182	PQ	VE2FU	91	115,808
LA	W5WMU(K5GA)	93	235,774	ONT	VE3EJ(K5ZD)	92	206,822
MS	W5XX	93	188,176	MB	VE4OY	78	135,648
NM	K7UP(KN5H)	94	203,126	SK	VE5DX	81	161,024
NTX	WM5G(KRØY)	94	240,702	AL	VE6UX	91	125,048
OK	WM4Z	91	160,776	BC	VE7SZ	92	202,048
STX	K5GN	93	221,452	NWT/YUK	NØTT/VE8	83	97,382
WTX	N5RZ	94	230,736				
EB	N6RO	93	212,212	Records for "Deleted Sections"			
LAX	W6AM(N6TJ)	77	182,646				
ORG	W6EEN(KA6SAR)	93	181,566	C.Z.	KZ5FR	78	138,262
PAC	AH3C	90	178,800	EFL	WA4NGO	63	235,875
SBR	N6TR	83	183,076	WFL	W4JJ(K4VFY)	63	140,799
SCV	N6TV	93	198,814	W.I.	KP4RF([AH3C])	77	193,732
SDG	K6XT(KY7M)	94	200,816	WA	K7RI([KC1F])	78	169,608
SF	W6NUT [K6BB]	74	153,750				

ARRL SWEEPSTAKES CW SECTION RECORDS
Single-Operator Low-Power

Section	Call	Year	Score	Section	Call	Year	Score
CT	K1CC	87	145,928	KY	N4TY	93	148,888
EMA	K1EA	82	129,204	NC	K7GM	94	159,448
ME	NX1T	91	97,532	NFL	WC4E	92	157,388
NH	W1PH	92	152,304	PR	NP4A(K7JA)	93	210,518
RI	K1IU	92	156,772	SC	KØEJ	91	132,088
VT	N4DW	91	137,522	SFL	N4BP	77	149,850
WMA	W1ZT	80	139,392	TN	N4ZZ	92	173,558
ENY	WA2STM	92	139,688	VA	N4DW	78	137,850
NLI	NQ2D	91	127,224	VI	N6OP/NP2	93	180,120
NNJ	W2GD	92	160,776	AR	KM5G	93	151,950
SNJ	N2MM	90	130,950	LA	K5SL	93	111,750
WNY	W2TZ	92	164,312	MS	W5XX	94	147,750
DE	K3WUW	84	114,108	NM	KT5X	92	141,208
EPA	K3NW	92	135,432	NTX	N5RZ	90	167,700
MDC	KN5H	92	127,224	OK	KM5H	87	122,840
WPA	WA3FYJ	83	115,884	STX	AD5Q	93	176,484
ALA	KC4ZV	89	114,912	WTX	N5RZ	93	196,840
GA	KM9P/4	90	155,040	EB	N6IG	82	136,802

ARRL SWEEPSTAKES CW SECTION RECORDS
Single-Operator, Low Power (Continued)

Section	Call	Year	Score	Section	Call	Year	Score
LAX	K6OY	93	173,712	WV	KD8G	82	115,778
ORG	K7JA	83	144,004	IL	K4XU	93	155,344
PAC	KH6HKM	72	91,542	IN	KO9Y	94	157,234
SBR	WA6FGV	91	119,550	WI	WB9YXY(N0BSH)	91	137,224
SCV	W6OAT	93	140,140	CO	K0EU	93	174,900
SDG	K6UA(AA5DX)	91	160,006	IA	K0LUZ	79	130,378
SF	W6JTI	89	121,360	KS	N0XA(K0VBU)	91	133,989
SJV	KD6WW	93	155,496	MN	N0AT	92	156,712
SV	W1FEA	93	166,936	MO	K4VX([WX3N])	88	144,900
AK	AA6DX	91	91,048	ND	WB0O	92	148,918
AZ	KY7M	93	179,700	NE	KV0I	90	135,864
EWA	K7MM	92	137,100	SD	WD0T	91	159,450
ID	W7ZRC	94	152,760	MAR	VY2SS	92	117,800
MT	KE7X	87	120,304	PQ	VE2UJ	92	77,404
NV	NF7P	87	152,400	ONT	VE3AKG	78	121,656
OR	KQ7I	92	152,400	MB	VE4VV	79	155,198
UT	W7CFL	93	150,920	SK	VE5MX	92	135,982
WWA	K7QQ	91	151,544	AL	VE6OU[VE3EJ]	83	98,420
WYO	K7MM	88	128,464	BC	VE7ARQ	88	98,988
MI	AA8AV	94	142,604	NWT/YUK	VY1JA	92	44,608
OH	K8CX	92	153,846				

ARRL SWEEPSTAKES CW SECTION RECORDS
Multi-Operator

Section	Call	Year	Score	Section	Call	Year	Score
CT	K1RU	94	165,242	SBR	N6VR	93	202,048
EMA	WA1JUY	72	111,672	SCV	N6BT	84	165,316
ME	K1RQ	92	150,300	SDG	K6XT	93	179,564
NH	KM1C	82	124,040	SF	W6BIP	93	161,084
RI	K1AD	77	111,148	SJV	K6AYA	78	125,208
VT	NW1S	94	146,916	SV	W6GO	92	193,424
WMA	KY1H	94	173,250	AK	AL7CQ	91	131,024
ENY	K5NA	94	166,782	AZ	N7CSC	81	86,664
NLI	AA2FB	94	144,144	EWA	K7WUW	90	32,240
NNJ	W2VJN	81	148,336	ID	W7UQ	79	109,668
SNJ	N2NU	91	148,456	MT	K7QA	80	130,640
WNY	W2HPF	94	172,368	NV	KI3V	91	170,324
DE	NW3Y	94	81,650	OR	KQ7I	91	143,700
EPA	K3WW	92	163,394	UT	K6XO	94	150,458
MDC	K1DQV	91	141,988	WWA	K7LXC	85	143,708
WPA	W3GH	94	188,480	WYO	K7MM	90	153,772
ALA	N4KG	85	140,896	MI	AA8U	94	164,934
GA	W4AQL	93	191,114	OH	WB8JBM	83	161,616
KY	N4TY	92	154,000	WV	KB8FJ	90	42,208
NC	AA4S	94	173,558	IL	KS9O	92	154,616
NFL	N4EEB	88	130,950	IN	KJ9D	85	161,468
PR	WP4BDS	81	46,944	WI	W0AIH/9	85	145,336
SC	WB4VJK	72	30,380	CO	K0RF	87	170,940
SFL	WA4LZR(K4XS)	77	98,420	IA	K9LUW/0	92	119,320
TN	N4ZZ	93	212,058	KS	K0WA	94	168,416
VA	W4BVV	77	159,840	MN	W0AA	93	178,024
VI	No Entry			MO	K0RWL	94	172,326
ARK	WC5N	88	105,228	ND	NJ9C	88	93,388
LA	K5MC	91	160,160	NE	K0GND	90	160,468
MS	W5TV	78	27,938	SD	WD0T	92	185,262
NM	AA5B	93	223,916	MAR	VE1NH	92	79,310
NTX	WX0B	93	192,280	PQ	VE2UN	82	68,340
OK	WM4Z	92	172,368	ONT	VE3ART	87	141,408
STX	K5LZO	89	167,808	MB	VE4RRC	78	43,560
WTX	WF5E	92	166,012	SK	VE5BBQ	79	16,932
EB	K6ZM	84	113,664	AL	VE6AO	90	60,582
LAX	W6UE	87	150,750	BC	VE7YU	91	90,244
ORG	N6PE	87	106,142	NWT/YUK	No Entry		
PAC	KH6RS	74	151,950				

ARRL SWEEPSTAKES CW SECTION RECORDS
Single-Operator QRP

Section	Call (Since 1987)	Year	Score	Section	Call (Since 1987)	Year	Score
CT	KH6CP/1	92	80,864	SBR	N6NMH	89	44,162
EMA	W1MJ	93	80,928	SCV	W6IO	89	88,200
ME	W1KX	91	26,280	SDG	K6ZH	92	72,072
NH	K1TR	92	106,552	SF	W6JTI	91	93,324
RI	WA1HYN	94	37,772	SJV	KI6PR	94	40,602
VT	WA1GUV	92	96,096	SV	N6WMF	93	50,616
WMA	KB1W	87	66,640	AK	NL7DU	87	9,840
ENY	KR2V	92	18,960	AZ	AA2U	92	110,850
NLI	KK2E	92	47,080	EWA	K7MM	93	84,096
NNJ	W2GD	94	120,274	ID	K2PO/7	94	83,804
SNJ	K2YY	91	84,150	MT	KE7X	88	90,150
WNY	NM2L	88	89,668	NV	NF7P	91	121,352
DE	NY3C	94	20,298	OR	W7YAQ	93	85,994
EPA	K3WW	93	92,872	UT	W7CFL	94	63,440
MDC	K3TM	92	96,000	WWA	K7LR	91	103,796
WPA	AG3H	94	58,400	WYO	K7SS	90	113,344
ALA	N4KG(NB9P)	91	89,850	MI	K8AQM	94	77,400
GA	KB4GID	94	90,450	OH	K8MFP	94	100,168
KY	KJ4VH	89	54,020	WV	W8DL	93	79,200
NC	K3PI	92	73,584	IL	K9ZO	92	95,000
NFL	KØLUZ	92	68,856	IN	W9RE	93	111,872
PR	KP4/KA9FOX	94	132,240	WI	W9UP(KA9FOX)	92	117,348
SC	KØEJ	92	100,320	CO	N3SL/Ø	92	110,264
SFL	K4MF	89	74,444	IA	WBØB	90	10,058
TN	KI4UZ	94	73,130	KS	NWØF	92	56,800
VA	WA4PGM	94	102,752	MN	N9CIQ	92	95,400
VI	NP2E	94	22,680	MO	NØEID	94	46,782
ARK	W9OBF	90	53,480	ND	WBØO	90	104,832
LA	NO5W	89	39,900	NE	KØSCM	92	112,480
MS	K4VFY/5(N4EVS)	88	89,644	SD	WAØZPT	91	66,816
NM	K5IID	91	81,400	MAR	No Entry		
NTX	AA5KK	91	76,320	PQ	VE2BLX	92	11,300
OK	WA5RES	94	56,560	ONT	VE3OOL	88	44,588
STX	AC5K	89	79,032	MB	No Entry		
WTX	W5VGX	87	44,460	SK	VE5VA	92	24,600
EB	KB6GK	93	374	AL	VE6SH	94	50,544
LAX	WB6JJE	90	60,088	BC	VE7YU	90	69,840
ORG	WB9AJZ/6	90	27,720	NWT/YUK	No Entry		
PAC	No Entry						

ARRL SWEEPSTAKES PHONE SECTION RECORDS
Single-Operator High Power

Section	Call	Year	Score	Section	Call	Year	Score
CT	K1KI	93	292,908	SC	KØEJ	92	235,774
EMA	N6BV/1	88	258,704	SFL	KC4VA	81	233,162
ME	K1RQ(NJ1F)	93	221,920	TN	N4ZZ	89	255,332
NH	N6BV	94	282,590	VA	W4MYA	93	273,966
RI	WA1TFF	78	196,200	VI	KP2A(N2IC)	88	398,088
VT	WB1GQR(WB2JSJ)	92	284,900	ARK	WA5RTG [K5GO]	75	222,300
WMA	WA1ABW [K1BW]	76	214,500	LA	W5WMU	90	339,416
ENY	K2TR	78	276,150	MS	W5VSZ	90	261,338
NLI	KD2RD	90	206,100	NM	K5TA	91	334,026
NNJ	WB2K	94	252,560	NTX	WM5G(KRØY)	92	374,682
SNJ	K2PS	91	228,228	OK	W5CCP(AH9B)	94	270,578
WNY	W2HPF	91	242,242	STX	NR5M	90	340,032
DE	K3WUW	90	216,062	WTX	N5RZ	91	374,374
EPA	K3ZA	92	239,316	EB	N6IG	88	320,720
MDC	KE3Q	94	288,496	LAX	W6HX(N6NT)	78	290,250
WPA	K3LR(K3UA)	93	328,328	ORG	K6RR(N6IN)	79	243,756
AL	WZ4F	94	282,898	PAC	K7SS/KH6	79	285,048
GA	K4JPD(WA4OZT)	77	229,200	SBR	KB6I(N6TR)	81	280,608
KY	ND4Y	93	261,646	SCV	W6QHS(KM9P)	93	338,030
NC	N4ZC(WB5M)	94	239,000	SDG	K6JYO	90	310,464
NFL	K4XS	90	341,880	SF	N6BV	78	246,300
PR	WJ2O/KP4	92	193,800	SJV	WC6H	90	304,304

ARRL SWEEPSTAKES PHONE SECTION RECORDS
Single-Operator High Power (Continued)

Section	Call	Year	Score	Section	Call	Year	Score
SV	W6GO(N6IG)	90	378,532	MN	NØAT	93	281,820
AK	AL7CQ	88	274,500	MO	K4VX/Ø([WX3N])	91	296,604
AZ	K6LL	89	363,286	ND	WBØO	90	337,722
EWA	K7FR	90	239,470	NE	KØSCM	79	213,416
ID	W7ZRC	91	262,262	SD	KØDD	88	319,808
MT	KØPP/7	93	289,982	MAR	VE1YX(AA2Z)	81	250,860
NV	WA7NIN(N6KT)	91	373,912	PQ	VA2UN([K3RV])	71	148,296
OR	W7NI(AI7B)	90	326,172	ONT	VE3RM	92	203,896
UT	W7CFL	91	301,224	MB	VE4RM	79	241,684
WWA	W7XR(W7WA)	90	363,440	SK	VE5DX	78	289,050
WYO	K7MM	90	311,696	AL	KS6H/VE6	91	175,098
MI	K8CC(WD8IJP)	93	294,294	BC	VE7SZ	92	300,608
OH	KW8N(NZ4K)	93	270,424	NWT/YUK	VY1JA	92	121,352
WV	N8II	81	217,686				
IL	AG9A	90	273,042	Records for "Deleted Sections"			
IN	W9RE	93	295,218	C.Z.	KZ5NO(K1MM)	78	163,950
WI	W9UP(NØBSH)	92	255,948	EFL	K4KXX	59	156,449
CO	KØRF(WØUA)	91	318,010	WFL	W4JLW	63	111,754
IA	WØEJ	91	245,168	WA	K7SS	77	286,950
KS	K5JZN/Ø	79	222,592	W.I.	KP4RF([AH3C])	78	333,600

ARRL SWEEPSTAKES PHONE SECTION RECORDS
Single-Operator Low Power

Section	Call	Year	Score	Section	Call	Year	Score
CT	KC1SJ	91	159,544	SBR	WA6FGV	91	188,188
EMA	WA1UZH	79	145,632	SCV	N6NF	92	201,740
ME	K1BZ(N1ATN)	89	119,776	SDG	W6UQF	91	182,028
NH	KM1C(WB8BTH)	82	184,836	SF	WA8LLY	89	132,450
RI	WA1TAQ	90	169,400	SJV	W9NQ/6	90	189,882
VT	WB2JSJ/1	77	116,772	SV	W1FEA/6	93	238,392
WMA	KZ1M	92	195,118	AK	KL7IVX	79	142,376
ENY	WA2STM	92	186,802	AZ	KY7M	92	219,296
NLI	KD2RD	91	161,100	EWA	K7IOO	91	131,838
NNJ	KY2P	85	159,248	ID	W7ZRC	93	210,518
SNJ	W2KI(K2PO)	78	151,900	MT	W7GKF	77	148,200
WNY	W2TZ	89	148,500	NV	ND7M	88	175,864
DE	K3WUW	84	133,776	OR	W7XN	79	180,264
EPA	K3MQH	91	161,424	UT	W7CFL	92	202,972
MDC	K3TM	80	125,504	WWA	K7QQ	93	278,740
WPA	K5ZD	91	190,190	WYO	K7MM	87	154,808
ALA	KC4ZV	88	135,280	MI	AA8AV	92	186,494
GA	KB4I(K2PO)	81	176,660	OH	K8BL	93	154,616
KY	AA4RX	94	235,448	WV	N8II	93	160,006
NC	AA4S(WQ4V)	92	165,088	IL	AG9A	91	230,538
NFL	K4XS	91	290,598	IN	AJ9C	92	233,618
PR	KØOO/KP4	80	98,124	WI	WØAIH(KA9FOX)	92	189,574
SC	KØEJ	87	133,200	CO	KØEU	93	273,658
SFL	K4VUD	93	251,328	IA	WDØEWD	81	175,972
TN	N4ZZ	90	237,468	KS	NOØY	92	182,336
VA	W4MYA	92	204,358	MN	KØHB	92	200,970
VI	NP2I	92	157,542	MO	KMØL	93	140,448
ARK	N5DX(K8CC)	78	197,100	ND	WDØECS	79	109,056
LA	N5PST	92	174,344	NE	KØSCM	92	215,840
MS	W5XX	90	197,120	SD	WDØT	93	213,752
NM	AA5B	88	254,904	MAR	VO1AA(K6ZH)	79	89,838
NTX	KN6M/5	91	267,036	PQ	VE2MS	89	66,880
OK	KM5H	93	231,616	ONT	VE3GP	81	89,644
STX	KE5FI	92	218,218	MB	VE4GV	92	254,870
WTX	N5DO	90	161,084	SK	VE5MX	92	226,380
EB	WA6DIL	81	191,660	AL	VE6ATT	89	157,916
LAX	N6HC	89	196,042	BC	VE7IN	92	119,504
ORG	N6RJ(N6PEQ)	92	202,818	NWT/YUK	VY1CC	82	57,084
PAC	K6GSS/KH6	89	177,870				

ARRL SWEEPSTAKES PHONE SECTION RECORDS
Multi-Operator

Section	Callsign	Year	Score	Section	Callsign	Year	Score
CT	N1MM	78	219,600	SBR	N6NB	77	188,352
EMA	W1AF	91	208,824	SCV	N6BT	82	245,134
ME	N1SD	79	146,584	SDG	N6ND	93	277,970
NH	K1GZL	78	231,150	SF	W6BIP	81	206,904
RI	K1NG	90	262,108	SJV	N6NB	93	241,318
VT	K1IK	80	208,236	SV	W6GO	92	336,490
WMA	W1YK	94	157,080	AK	AL7CQ	91	231,462
ENY	KY2J	94	238,392	AZ	N7EDV	83	179,524
NLI	N2MG	88	176,850	EWA	WA7EGA	92	240,397
NNJ	KZ2S	88	205,200	ID	W7LQT	80	178,044
SNJ	N2RM	94	181,720	MT	KB7SE	81	191,698
WNY	WA2ECA	78	209,850	NV	NF7P	91	289,366
DE	AC3T	83	132,568	OR	K7SS	92	367,906
EPA	K3WW	92	258,874	UT	W7MR	91	241,780
MDC	K1DQV	91	244,244	WWA	W7XR	88	282,872
WPA	W3GH	94	281,358	WYO	K7MM	88	228,760
ALA	K4BFT	92	173,250	MI	K8MJZ	93	246,092
GA	AB4RU	94	322,938	OH	KW8N	94	316,008
KY	N4TY	92	248,402	WV	W9LT	94	192,192
NC	AA4NC	88	213,408	IL	K9ZO	93	286,286
NFL	N4WW	83	268,472	IN	K9RS	89	242,242
PR	WA3FET/KP4	77	196,692	WI	W9XT	90	285,054
SC	N4QWL	90	168,014	CO	KØUK	87	281,550
SFL	AC4CT	91	251,790	IA	NØICI	90	196,812
TN	N4ZZ	93	325,094	KS	WØCEM	91	266,304
VA	WF7B	93	269,808	MN	NØAT	92	259,028
VI	NP2GM	94	253,022	MO	KMØL	94	269,554
ARK	WA5TCL	82	192,992	ND	AKØT	80	218,448
LA	N5WA	94	264,188	NE	KØDG	80	198,024
MS	K5TYP	91	250,712	SD	WDØT	90	285,054
NM	AA5B	90	376,684	MAR	VE9HF	94	73,458
NTX	WXØB	94	320,782	PQ	VE2CUA	94	172,420
OK	NJ1V	94	294,448	ONT	VE3XD	92	250,404
STX	K5LZO	86	267,288	MB	NJØC/VE4	87	176,700
WTX	KD5SP	92	246,862	SK	VE5FN	88	183,768
EB	K6OYE	79	230,732	AL	VE6SV	94	188,328
LAX	W6UE	92	281,358	BC	VE7DV	92	89,166
ORG	W6EEN	94	324,478	NWT/YUK	VY1QST	93	141,360
PAC	N7NR/KH6	91	273,042				

ARRL SWEEPSTAKES PHONE SECTION RECORDS
Single-Operator QRP (Category began 1987)

Section	Call	Year	Score	Section	Call	Year	Score
CT	KH6CP/1	88	58,380	NFL	K4VFY(N4EVS)	94	52,950
EMA	WB1EHL	90	18,762	PR	KP4DDB	92	24,090
ME	N1AFC	88	27,300	SC	KJ4KB	90	31,008
NH	K1DG(WZ1R)	93	119,504	SFL	W4PZV(WA4SVO)	94	116,424
RI	WZ1R	94	49,284	TN	N4TG	91	92,400
VT	WA1GUV	90	33,000	VA	WA4PGM	92	83,622
WMA	K1KNQ	88	59,472	VI	NP2E	94	45,540
ENY	WV2V	90	48,384	ARK	W9OBF	90	54,864
NLI	KD2TT	90	71,250	LA	K5KLA	88	36,984
NNJ	AA2U	93	94,556	MS	WB5KYK	88	9,400
SNJ	K2YY	92	87,764	NM	K5IID	91	145,992
WNY	WA2UUK	92	70,984	NTX	KF5QR	92	66,682
DE	K3WUW	87	61,200	OK	W5PAA(WC5D)	88	66,120
EPA	K3WW	93	78,540	STX	N5NMX	92	100,100
MDC	K3DI	90	64,380	WTX	W5VGX	88	79,804
WPA	W3SMX	91	31,892	EB	AA6GM	93	3,300
AL	KN4QS	93	20,538	LAX	WB2ODH/6	93	152,614
GA	KB4GID	90	67,144	ORG	WB9AJZ/6	90	7,740
KY	AA4RX	89	102,718	PAC	NH6WH	90	12,540
NC	AA4NC	91	51,300	SBR	K6VMN	92	91,250

ARRL SWEEPSTAKES PHONE SECTION RECORDS
SIngle-Operator QRP (Category began 1987) (Continued)

Section	Call	Year	Score	Section	Call	Year	Score
SCV	KD6PY	87	72,854	IN	N9DHX	89	53,808
SDG	K6ZH	89	43,632	WI	W9UP(KA9FOX)	93	161,546
SF	WA8LLY	91	16,740	CO	KØFRP	91	137,676
SJV	W9NQ	92	117,648	IA	WAØVBW	88	72,200
SV	N6WMF	93	99,484	KS	NOØY	91	39,520
AK	AL7MK	91	4,810	MN	KØTO(NØBSH)	93	95,850
AZ	N7IR	91	39,858	MO	NØEID	92	36,920
EWA	N7RWH	92	45,150	ND	NØAFW	93	70,224
ID	KK7A	92	45,158	NE	KØSCM	93	109,896
MT	KE7X	87	113,120	SD	KØWIU	94	20,160
NV	KB7GAP	92	11,024	MAR	No Entry		
OR	K7SS	87	103,200	PQ	No Entry		
UT	W7CFL	93	101,794	ONT	VE3POS	92	58,368
WY	K7MM	91	125,100	MB	VE4VV	92	115,654
WWA	K7SS	91	169,400	SK	No Entry		
MI	WB8G	93	100,408	AL	VE6SH	92	52,822
OH	WA8RJF	93	64,824	BC	VE7EKS	88	8,256
WV	K5IID	94	42,458	NWT/YUK	No Entry		
IL	K9ZO	91	117,040				

ARRL SWEEPSTAKES CW
All-Time Top 100 Scores

Call	Year	Score	Call	Year	Score
WM5G(KRØY)	94	240,702	KØRF(WØUA)	92	202,202
WM5G(KRØY)	93	236,512	VE7SZ(VE7NTT)	92	202,048
W5WMU(K5GA)	93	235,774	KI3V/7	94	201,432
N4RJ(KM9P)	94	232,232	K6XT(KY7M)	94	200,816
N4RJ(KM9P)	93	231,000	W5WMU(K5GA)	91	200,662
N5RZ	94	230,736	AD5Q	94	200,662
W5WMU(K5GA)	94	229,768	AA5BL	93	199,880
K5GN	93	221,452	K6LL	91	199,430
N4RJ(KM9P)	92	220,836	N6TV	93	198,814
K4VX/Ø(WX3N)	94	219,296	N5AU(WN4KKN)	90	198,750
W7RM(N6TR)	93	218,834	K3LR	92	198,660
K5GN	94	218,680	K4VX/Ø[WX3N]	91	197,890
N2IC	93	216,524	KZ2S	94	197,890
N5AU(WN4KKN)	92	215,600	K7UP(KN5H)	93	197,752
NCØP(AG9A)	94	215,600	K6LL	92	197,736
K5XI(NM5M)	94	214,368	KZ2S	93	197,600
WM5G(KRØY)	92	213,906	AC4NJ(WC4E)	94	197,582
NCØP(AG9A)	93	213,136	N5RZ L.P.	93	196,840
K6LL	93	212,982	NCØP(AG9A)	92	196,504
N2IC/Ø	94	212,520	N6UR(KR6X)	92	196,042
N6RO	93	212,212	K5GN	92	196,042
NM5M	93	212,212	W6QHS(W6CF)	94	195,888
N5AU(WN4KKN)	91	210,518	NP4A(K5ZD)	86	195,656
NP4A(K7JA) L.P.	93	210,518	K1TO	92	195,320
K5MR	94	209,748	WA8ZDT	94	195,000
K9FD(WX3N)	93	209,440	W1XE	93	194,194
W5WMU(K5GA)	92	209,440	KI3V/7	92	194,040
K5MR	93	209,304	KP4RF(KP4EAJ)	77	193,732
K3LR(K3UA)	94	208,824	N7NG(AH3C)	92	193,648
K3LR	93	208,544	KV4FZ(K7JA)	76	193,500
WM5G(KRØY)	91	208,544	VE7SZ	93	193,116
K6LL	94	208,516	W9RE	94	193,116
N6TR/7	92	208,054	AA6TT(K7JA)	92	192,500
K3LR	91	207,746	W6QHS	93	192,192
AH3C	93	207,438	WA7NIN(N6TV)	92	192,038
VE3EJ(K5ZD)	92	206,822	KE3Q	94	191,520
KA5W(K5ZD)	91	205,744	KZ2S	92	191,268
KØRF	94	205,282	N6RO	93	191,268
KØRF(WØUA)	91	204,358	K4VX/Ø (WX9E)	93	191,268
K1ZX	94	203,832	N8RR	93	191,268
AA5BL	94	203,588	KØRF(WØUA)	90	190,500
VE3EJ(K5ZD)	93	203,528	KI3L	93	190,498
N6UR(KR6X)	94	203,434	W7WA	92	190,456
K7UP(KN5H)	94	203,126	N6ND(NI6W)	92	190,190

ARRL SWEEPSTAKES PHONE
All-Time Top 100 Scores

Call	Year	Score	Call	Year	Score
KP2A(N2IC)	88	398,088	K6LL	94	317,240
W6GO(N6IG)	90	378,532	KI3V/7	91	316,624
WM5G(KRØY)	92	374,682	KA5W(KS1G)	92	316,470
N5RZ	91	374,374	K9RS(AA5B)	93	316,162
WA7NIN(N6KT)	91	373,912	KØKR	93	315,546
N5RZ	92	364,672	KI3V/7	88	315,096
K6LL	89	363,286	KØRF(WØUA)	89	314,776
W7XR(W7WA)	89	362,516	WC6H	90	314,468
K6LL	90	362,516	WC6H	93	314,468
NP4A(K1ZM)	82	361,416	WM5G(KRØY)	89	314,160
K6LL	89	359,364	KI3V/7	94	314,160
N5AU(WB5VZL)	90	353,430	W5KFT(WB5VZL)	93	313,852
N5AU(WB5VZL)	91	352,506	KØRF(WØUA)	81	313,612
KP4Q(K2TR)	79	351,796	WBØO	89	312,928
W6GO(K3EST)	89	348,194	K7MM	90	311,696
K6LL	91	345,192	K7UP(AA5B)	89	311,080
K4XS	90	341,880	K6JYO	90	310,464
NR5M	90	340,032	N7DD(K7JA)	80	309,320
W5WMU	90	339,416	K4XS	89	307,076
KI3V/7	93	338,492	W7EJ(NL7GP)	91	307,076
W6QHS(KM9P)	93	338,030	W7WA	92	307,076
WBØO	90	337,722	W7KW(W6TPH)	78	306,150
N5RZ	90	337,260	WA7NIN(N6OAT)	81	305,324
W5KFT(WB5VZL)	94	336,798	WC6H	90	304,304
K6LL	92	334,796	W7WA	87	304,050
W6GO(N6IG)	91	334,642	N6UR(KR6X)	92	303,842
K5TA	91	334,026	N5AU(K5ZD)	81	303,548
KP4RF([AH3C])	78	333,600	K5TA	89	302,764
KI3V/7	92	332,486	KØDD	87	302,550
W7WA	88	332,424	K6LL	88	301,720
KA5W(KS1G)	91	330,792	W7CFL	91	301,224
KA5W(KS1G)	94	328,428	VE7SZ	92	300,608
K3LR(K3UA)	93	328,328	KC7V	91	297,374
KV4FZ(WA6VEF)	81	327,820	K6LL	87	297,184
K6LL	93	327,096	WM5G(KRØY)	88	296,856
WA7NIN(N6KT)	92	326,942	W5WMU	92	296,758
W5WMU	91	326,480	WBØO	88	296,704
W7NI(AI7B)	90	326,172	K4VX/Ø([WX3N])	91	296,604
N5AU(WB5VZL)	92	325,864	KP4BZ	83	296,592
KA5W(KS1G)	90	324,632	W6GO(K3EST)	88	296,552
W6QHS(WN4KKN)	91	324,324	K5RX	94	296,296
K4XS	92	323,554	KØRF(WØUA)	88	295,944
AA5BL	94	323,092	W9RE	93	295,218
N6IG	88	320,720	W5WMU	93	295,064
W7RM(K7JA)	90	320,012	K8CC(WD8IJP)	93	294,294
KØDD	88	319,808	K3LR(K3UA)	94	293,678
WA7NIN	79	318,792	KØKR	94	293,524
KP4RF([AH3C])	77	318,450	K1KI	93	292,908
KØRF(WØUA)	91	318,010	K5RX	90	292,754
KY7M	93	317,240	N6TU	91	292,292

ARRL SWEEPSTAKES CW WINNERS

Single-Operator High Power
W2IOP/W9IOP: 8
W4KFC: 6
K5ZD: 5
K7JA/K7VPF: 4
WN4KKN: 3
W3BES: 2
W4KVX: 2
W5WZQ: 2
W6CF/W6CUF: 2
KRØY: 2
WØUA : 2
The following each have one win:
W1ADW, W1EZ, W2GD, K2TR, KP4EAJ, W4FU, KV4FZ, K5GA, K5RC, K6LL, WB6OLD, W6MVK, W6RTT, N6ZZ, W7KVU, W8CHC, W8ER, W8JIN, W9AUH, W9FS, W9HKC, KM9P, K9ZMS.

Single-Operator Low Power (150 W max.)
N5RZ: 3
WA3DSZ: 2
K7JA: 2
AC8CQN/W8CQN: 2
KØEU: 2
The following each have one win:
W1ZT, W2TZ, K3RR, N4ZZ, K5LZO, K5RHZ, WA5VDH, N6IG, AA7A, NF7P, WBØDLE, NØBSH, NØNO, VE4VV.

Single-Operator QRP (5 W or less)
KA9FOX: 2
The following each have one win:
N2IC/Ø, N3SL/Ø, K7SS, NF7P, KVØI, KØSCM.

Multi-Operator
AA5B: 5
K5LZO: 4
W3FYS: 3
W3GQF: 2
K5CM: 2
W5WF: 2
W8ADV: 2
W9OCB: 2
The following each have one win:
W1AQ, WA1KID, K1VTM, WB2APG, WA2ECA, W3CPV, W3CTJ, W3FRY, W3KT, W3LTW, W3MWC, W4BVV, W4IB, N4RJ, K4SXD, K4VDL, W4WE, K5AC, N5DKG, W6BIP, N6BT, W6EFW, W6GO, K6RLY, KH6RS, W6RW, WB6WIT, W6YRA, W7SFA, WB8JBM, W8YX, W9GDH, W9WBY, WX0B, W0ZLN, VE7ACS.

ARRL SWEEPSTAKES PHONE WINNERS

Single Operator
K7JA/K7VPF: 6 (4 in-a-row)
W6ITH: 6 (5 in-a-row)
W6QEU: (4 in-a-row) 4
K6EVR: 3
KP4BZ: 2
K5MDX: 2
W7RM/W7ESK: 2
W7WA: 2
4 OF K7JA/K7VPF wins were from W7RM's station
The following each have one win:
K1ZM, K2AAA, N2IC, K2TR, KI3V/7, KV4FZ, W4KFC, KP4RF, K4WJT, N5RZ, WB5VZL, W6AHP, W6AM, N6BV, N6CJ[AH3C], W6FRW, N6IG, W6MLY, W6OAT, WB6OLD, WA6VEF, W8ALC, W9GAF, K9LBQ/7, W9NDA, W9RBI, W9WNV, WBØDJY/WØUA, WØEDX, KRØY, VE3ER.

Single Operator Low Power (150 W or less)
K4XS: 2
KE5CV: 2
K7SS: 2
K7QQ: 2
The following each have one win:
K1EUF, N4MM, K4WAR, K4XU, WC4E, AA5B, WB5FMJ, W5MYA, K5RHZ, WA5VDH, K7JCA/6[K7JA], W7XN, K8CC, WBØDSP, KØEU, WBØMIV, KØUK, VE4GV.

Single Operator QRP
K7SS: 3
KA9FOX: 2
KE7X: 1
WE9V: 1
WØLSD: 1

Multi-Operator
K5LZO: 3
WØYQ: 3
AA5B: 2
K9RS: 2
WAØCHH: 2
The following each have one win:
W1AW, W1FBY, K1GZL, W1KBN, W2JSE, WA2LPF, W2SZ, K3KLQ, W3MWC, AF3P, W4BVV, W4EQK, W4MOE, W4TEW/4, N4WW, N4ZZ, WB5DTX, K5EAT/5, W5GAD, K5IIS, W5LUJ, W5ZED, N6BT, KN6M, W6ONV, W6YX, AI7B, W7EJ, WA7ORM, K7SS, W7ZR, W9HHX, W9WHN, WØEEE, KØUK, KØWA, VE7AKY.

Clubs

Unlimited	Medium	Local
PVRC: 24:	N.Tx.C.C.: 8	River City C.: 6
FRC: 13	Tx. DX.S.: 3	Rubber Circle C.: 4
NCCC: 10	Mad River: 1	NM Big River CC: 1
SMC: 3:	Mn W.A.: 1	Colorado C.C.: 1
Murphy's M.: 2:	YCCC: 1	Albuquerque DXA: 1
Mad River: 1	Murphy's M.: 1	Willamette VDXC: 1
Mn Wireless Assoc: 1	SCCC: 1	S. Jersey C.C.: 1
		Cascades, Mi ARC: 1

HISTORY OF THE ARRL INTERNATIONAL DX CONTEST

March 1927 QST carried an announcement for an "International Relay Party" (subtitled "A World-Wide Contact Contest") to be held in May of 1927. Previously, there had been competitive/experimental events such as the Transatlantic Tests of the early 1920s, where the goal was either one-way or two-way contact. The rules of this new "Relay Party" required international contacts and the exchange of short text messages during a two-week period from May 9 to May 23. Participants created their own 8-word (or longer) messages, and attempted to send them to foreign stations and receive a reply. Messages successfully transmitted from the U.S. to abroad counted 1 point, and messages received from abroad counted 3 points. For DX stations, messages recevied from the U.S. counted 1 point, and replies successfully transmitted to the U.S. counted 3 points. The stated purpose of the contest was to determine which stations were most reliable in their ability to relay messages from the U.S. to overseas locations. It was noted that "20 meters will undoubtedly play an interesting part in this contest".

The winners, as reported in October 1927 QST, were 2AHM for the U.S. with 90 points. Other top U.S scorers were 8GZ, 4IZ, 6AM (later W6AM), 6CUQ, 2APD, 8ADG, 2TP, 1ADM, and 1AUR. Top DX scorer was Chilean 3AG with 232 points. Other country winners were 1N (Mexico), 4UAH (Germany), 4SA (Puerto Rico), 2FG (Costa Rica), and 5BY (England). It was noted that the foreign country with the most reported activity was Australia (!) with 18 entries.

The Second International Relay Party, announced in December 1927 QST, was held for two weeks in February 1928. Entrants were required to pre-register for the event, and were assigned offical test messages to be used in the contest. The results were reported in August, 1928 QST. Prizes were offered to the highest scorers (mostly equipment donated by manufacturers, ranging from $2 coils and capacitors to $350 receivers), and these were listed in February 1928 QST. The entire prize list ran to almost three pages. Late additions were listed in August 1928 QST. The total value of all the prizes was over $4,000 (for comparison, a one-year subscription to QST in 1928 cost $2.50).

In the final results, the name of the event was changed to the "International Relay Contest of 1928". The USA winner was proclaimed to be 1ASR with 305 points. British station 5BY had 573 points, Belgian 4AU had 486 points, from Puerto Rico, 4SA had 405 points followed by Australian 7CW at 369 points and New Zealander 3AR at 328, as the top five DX scorers. The USA Top Ten: 1ASR-305, 2ALU-295, 2TP-259, 8GZ-253, 1CMP-241, 4FU-223, 9DNG-213, 1BHS-209, 5WZ-202, 8AHC-201. 6AM had 156 points and placed 15th. Disqualifications included: 1BW, 5KC, 1ALR, 2AVB, 2BFQ, 3AFW, 4EC, 9DPW, and 9AMN.

The Hiram Percy Maxim 60th Birthday Relay seems to have been the third major ARRL-sponsored contest. The planning for this event actually began during late 1927, when the ARRL Board of Directors decided to celebrate H.P.M.'s 60th birthday with a 'Birthday Relay' in 1928. Then it was discovered that H.P.M. would not be 60 years old until September 2, 1929! The Birthday Relay was put on hold and kept secret for over a year!

The event began August 31,1929, a Saturday, at 6 P.M. E.S.T. and ran until Monday, September 2nd at 6 P.M. E.S.T. During August of 1929, relay announcements were mailed to ARRL Official Relay Stations, Affiliated Clubs, Section Communications Managers, and Route Managers throughout the U.S. and Canada. During the ten days prior to the Birthday Relay, ARRL Official Broadcast Stations were broadcasting the details of the event. Luckily, H.P.M. never tuned in! ARRL Headquarters station W1MK was a multi-operator effort, manned by F. E Handy, W1BDI, chief operator R.P.(Robert Parmenter), and one other identified only as "E.V." for the full 48 hours! They received 356 messages on 3575 kcs and 7150 kcs. The first station worked had one message for H.P.M. The second station to contact W1MK had 10 messages for H.P.M. !! One station actually transmitted 40 messages to W1MK ! It was reported that '…..hardly two messages had the same wording.'

These three events were well-received by amateurs, who communicated their desire to have more of these "Parties" to the ARRL. In December 1929 QST, F.E. Handy, W1BDI, announced the new January contest and the new February contest. The January contest was to satisfy the W/VE crowd who enjoyed the HPM Relay and wanted a message-handling contest. This was combined with the ARRL's section structure. It's name was then changed to the All-Section contest and evolved over time to become the present ARRL November Sweepstakes. The February contest was to satisfy the amateurs who were displeased with Trans-Atlantic Tests and the like, where amateurs would transmit for 4 hours and listen for 2! These DXers wanted to work the DX, not merely listen to them. The February contest ultimately became the ARRL International DX Competition which is still held in February and March to this day.

The 3rd International Relay Competition was announced in December 1929 QST, to be held February 15-28, 1930. Requirements included requesting entry by mail before February 1, 1930.

Prospective entrants received entry forms by mail just days before the test began.

An interesting attempt was made in this 3rd International Relay Competition to level the playing field for American stations in different parts of the country: "A considerable amount of research has been attempted to answer the criticism expressed by several west coast stations that our last international test employed a method of scoring that automatically discriminated in favor of east coast amateurs, due to the preponderance of European amateurs and the larger number of European countries."

An attempt was made to weight the scores and equalize them with 'factors' in "a plan of balanced credits" and these were published for all to see. Contacts with stations on each continent were assigned different values for eastern and western stations. In the partial list that follows (from the original announcement), the first number is the point value for eastern stations (including Minnesota, Iowa, Missouri, Arkansas, Louisiana and points east), and the second number is for western stations:

Europe: 3, 11
Africa: 15, 10
South America: 3, 3
Asia: 20, 10

Section awards were also announced, in addition to the overall national awards.

The final results were published in August 1930 QST. It was a case of "he who lost the most sleep ran up the higher score" (still the same today!). W6BAX won with 3210 points. Next were W2CXL-83 QSOs, W1ASF, W8GZ, W6AAZ, W4FT, W2FP, W9UM, W9ECZ, and W1CMX. CM8UF was the DX winner with a score of 3564 from 98 QSOs, followed by HC1FG, K4KD, X9A, ZL2AC, G5BY, NJ2PA, PY1AW, VK7CH, VK5WR.

The 4th International Relay Contest was held March 8-22, 1931 and final results published in

August 1931 QST. W9UM was USA high scorer with 4,374 points and WAC! The rest of the top ten were W8BKP, W8GZ, W1AKV, W6EW, W1FM, W9MI, W6AQJ, W8ADQ, and W9ADN out of the 365 W/VE log entries. G5BY took the top spot for the DX participants followed by CM8UF, ZL2AC, CM2SH, VK3HL, ZL3AS, VK7CH, ZL4AO, G6RB, K4KB out of the 167 DX entries. Only total scores were reported.

January 1933 QST carried an announcement of the Fifth International Relay Competition. There would no longer be an advance-entry procedure or an assigned message or serial number. A participant could assign his own distinctive three digit serial number followed by the three digit number of his previous QSO. The 'factoring system' was also abandoned and the number of different countries worked became the score multiplier for U.S. stations.

The contest was held on March 11-19, 1933 and results reported in October 1933 QST. Nearly 1000 scores were submitted and it was determined that at least 300 foreign stations were on-the-air. Station EAR185 was proclaimed 'The World high-scorer (18,382) and winner of the gold charm for participants outside the U.S. and Canada'. In second was Miss Judy Leon at HC1FG, K4AAN, EAR96, ON4AU, VK3ML, G5BY, CT1AZ, F8PZ, and X9A. W3ZD won the stateside Gold award with a score of 14,976. Second was W4AJX, followed by W8CCW, W6CUH, W4ZH, W2BHZ, W6BYB, W8CRA, and W7BB.

The Sixth International Relay was held March 10-18, 1934 and final results were in September 1934 QST. W3ZJ worked 237 DX stations for a record score of 32,879. Next were W1SZ, W4AJX, W1ZI, W2BHZ, W5CBY, W6QD, W2DC, W1FH, W8CRA. The Top DX operator was NY1AB, followed by X1AA, HC1LC, EA5BE, ZL4AI, ZL2CI, G5BY, EA4AH, ZL4AO, and CM2JM. The final tabular score listings of this contest were missing numerous scores. The lapse begins after the score of EUROPEAN, AUSTRIA, OE1ER thru to NORTH AMERICAN, MEXICO, X1BC. Fortunately the identities of the top scorers were included in the text.

The Seventh International Relay Competition was announced with a few changes in February 1935 QST.

Entrants were to keep track of operating time and off time. Exceeding 90 hours of operation would not incur a penalty. A 100 hour effort would be scored as 90% of the total. Bonus points were added for using additional bands. An affiliated club competition was added.

The 8th event brought a name change - The International DX Competition - and a few more changes.

The exchange became RST plus a self-assigned three digit number. A quota was instituted for W/VE operators that allow them to work a maximum of three stations from any country. The band bonus points were rescinded and W/VE's scores were calculated as 3 points per QSO

multiplied by the sum of band-countries worked.

Although 'phone was in use long before 1937 and 'phone use in Sweepstakes was reported as early as 1932, the International DX Competition did not report any 'phone work until the Ninth event, when March 20-28, 1937 was for 'phone and March 6-14 for C.W. 90 hours of operation MAXIMUM.

The quota plan was not in effect for the 'phone contest.

The 1940 12th ARRL DX Competition had several changes: the contest period was 50 hours on two weekends, both modes simultaneously, W/VE's send RST and state and work other W/VE's, DX send serial number beginning with 001.

The DX Competition was, like most amateur radio activities, suspended for the duration of World War 2. The ARRL DX Competition was restarted in 1947. Back to exchanging RST and self-assigned number but there were now separate contest events for CW and 'phone, two weekends each. On CW, U.S. stations were allowed a quota of three QSOs per country per band maximum. There was no quota on phone, and no quota for DX stations. The quota system was modified over the years, eventually increasing to a 6-QSO maximum.

The Low Power Single Operator Class was added for the 1963 events.

Major changes went into effect for the 1967 DX Competition. The quota system was dropped completely. Multipliers for DX stations became the 48 continental states plus VO and VE1-8 (for a maximum of 57) instead of U.S. call areas. The first plaque awards were introduced for the top single-operator scorers on each continent (but none for W/VE!). KH6 and KL7 reverted to DX status. And new log forms were designed, including the first standardized dupe sheets (Form CD-175).

The Multi-Operator Multi-Transmitter Class beginning was added for the 1971 event, followed in 1983 by the Multi-Operator Two-Transmitter class. In 1975, new High-Band and Low-Band Single Operator Categories were added. These categories were only active from 1975 to 1979, when they were replaced by the present single-band categories. Beginning with the 1979 event, the contest period was reduced from two 96-hour weekends for each mode (total of 192 hours) to one 48-hour weekend for each mode (total of 96 hours).

The 1980 DX Competition rules included several major changes. First, a QRPp Single Operator Class was added. Second, the High-Band and Low-Band categories were dropped in favor of single-band categories. Third, a greatly-expanded awards program was instituted, with outside (non-ARRL) sponsorship of plaques. But the most controversial change was that DX-to-DX QSOs were permitted in the 1980 event as an experiment. This rule change was not communicated widely enough before the contest, and confusion ruled both contest weekends, with DX stations calling other DX stations who were unaware of the new rules and refusing to answer! This was rescinded for 1981, reverting back to the previous "world working W/VE" format. The other changes were all retained, and the awards program has continued to expand.

Recognizing the growing use of VHF packet spotting networks, the "Single-Operator-Assisted" category was added for 1990 event. This category allowed a single operator with packet spotting to transmit only one HF signal at a time, but permitted quick band changes to contact new multipliers located with the assistance of a spotting network.

DX WINNERS OF ARRL DX COMPETITION CW BY YEAR/CATEGORY

YEAR	Single-Op	QRP	Asst	M/S	M2	M/M	160	80	40	20	15	10
1995	P40R(K4UEE)	HP1AC	DK3GI	P49V	6E2T	9A1A	9A2TW	EA1AK/EA8	EA7CEZ	9Y4VU	ZW5B(YU1RL)	LU6BEG
1994	V26AS(YU1RL)	C6AHL(K3DI)	NP4Z	P49V	6D2X	9A1A	NO9M/KP4	ON4UN	9Y4VU	TU2MA	N6OP/NP2	V7A(V73C)
1993	VP5F(KR0Y)	V73C	ON4WW	P49V	ZL3GQ	9A1A	NO9M/KP4	YV1OB	T32AF(KH6UR)	OH8LO	VP5U(N4MO)	ZD8LII(G0LII)
1992	V27T(YU1RL)	G4BUE	JJ3YBB(JI3ERV)	UZ2FWA		HI8A	NO9M/KP4	G3LNS	ON4UN	OG1AA	LX/DL1VJ	FF0XX(FB1MUX)
1991	HK0/N3JT	YU2TY	HI8A(JA6DQH)	XE2FU		YT2R	VP2EXX	CU2AK	IO4IKW	YT1BB	9Y4VU	N6OP/NP2
1990	TG0AA(OH2BH)	K7SS/KH6	JA7YAB	ZF2KE		JA1YDU	YV1OB	OK3CBU	K0GVB/C6A	IA8A	G3FXB	N6OP/NP2
1989	P40GD(W2GD)	YU2TY	*#2*	ZF2KE		YT2R	YV1OB	K0GVB/C6A	I2VXJ	AI6V/VP9	KV4FZ(N6OP)	HC2G(HC2SL)
1988	P40GD(W2GD)	NP4Z		ZF2KE		6Y5L	K8WW/VP9	K8WW/VP9	KV4FZ(N6OP)	G3FXB	PY5ZBA	VP2ERA
1987	P40GD(W2GD)	YX3A	P40M	ZF2KE		JA1YXP	V3DA	EA8RL	KP4FI	LU8DQ	HK1KYR	HK3MAE
1986	9Y4A(N6TJ)	YV3AGT		KV4FZ		XE2FU	VP5GEX	NP4Z	I0JX	K7SS/KH6	ZP5XDW	PA0VDV/PJ2
1985	VP2MGD(W2GD)	C6AAA(KF1V)		N5RM/C6A		GB4DX	YV1OB	EA8RL	KP4FI	VP2EAG(KJ0D)	ZF2AY(K9LA)	LU5DVO
1984	HR1DAP(K8CC)	K7SS/KH6		AH6BK		IO3FIY	F8VJ	W9NXD/HR2	4M7OP	KP4EOF	YT3L(YU3EV)	ZP5XDW(N4PW)
1983	8P6J(N6TJ)	JA1MCU		VP2E		KH6XX	JA5DQH	CT2AK	9Y4VU	OH8AX/OH8PF)	ZY5XFR	LU5DVO
1982	V3MS(W0CP)	YV2BE		JA7YAA		XE2FU	OK3CPL	DL1KB	I0JX	YU4GD	LU8DQ	XE2BC(N6OP)
1981	HH2VP	G4BUE		DK0TU		YU1EXY	G3SZA	EA8RL	KP4EOF	OH5TQ	TF3YH	G4GIR
1980	VP2ML(K1ZZ)	G4BUE		XE2FU		V2AMK	LowBand:	UP2NK	I2XXG	YU3TU	YU7BCD	LU8DQ
1979	VP2MOC(K2YY)	*IQ*		DK0TU		NL7M	LowBand:	WA4UAZ/HC1	HighBand:	YU1KV		
1978	KP4EAJ			KP4EAS		JA7YAA	LowBand:	I0JX	HighBand:	KP4USN(K5PM)		
1977	KP4EAJ			ZF1AL		JA7YAA	Low Band	YV2OB	HighBand:	G3UJE		
1976	KP4EAJ			KH6GKD	VP5FUX	JA7YAA		AJ4EAS(KP4EAS)	AJ4EAS(KP4EAS)HighBand:	YU1BCD		
1975	PJ2VD			—	*#4**	YU1BCD		KP4EAS	High Band	KH6IGC		

World High S/O

YEAR	Single-Op	M/S	M/M
1975	PJ2VD	KH6GKD	VP2A
1974	LU5HF(K3ZO)	KZ5NG	ZF1AL
1973	KH6RS(K1GQ)	6Y5EE	JH1VVW
1972	XE1IL/(K0DQ)	YU3CV	OH1AD
1971	KH6RS(K1GQ)	LU2ES	K6AB
1970	PJ2VD	SK6AB	*#MM*
1969	ZD8Z(N6TJ)	LU8DQ	
1968	ZD8J(N6TJ)	OH2AM	
1967	HI8XAL(K3ZO)	PJ5ME	
1966	CO2BO	EL2AE	
1965	HP1IE	KZ5FC	
1964	HP1IE	PJ5ME	
1963	HC1DC	DL4QF	
1962	HC1AGI	KH6ECD	
1961	KV4AQ	KS4AZ	
1960	VP1-JH(W0INWX)	UR2KAA	
1959	VP7BT	KG6FAE	
1958	KH6IJ	KH6AYG	
1957	XF1A(XE1A)	KH6AYG	
1956	KV4AA	OE13USA	
1955	KH6MG	KG4AT	
1954	KV4AA	VP9BDA	
1953	KG4AT	CN8EX	
1952	KV4AA		

Year							
1951	KV4AA	KP4KB	EK1AO	4X4RE	GW3ZV	KV4AA	KH6DK	LU1DH
1950	XF1A(XE1A)	—	EK1AO	JA2CK	GW3ZV	XF1A(XE1A)	KH6IJ	CE3DZ
1949	XF1A(XE1A)	KZ5ER	EL3A	JA3AA	OK1FF	XF1A(XE1A)	KH6IJ	PY2AC
1948	XF1A(XE1A)	HC1JB	ZS2A	J3AAD	HB9CX	XF1A(XE1A)	KH6DD	CE3AG
1947	XE1A	"NONE*	ZS2AL	XU6GRL	HB9AW	XE1A	K6SCB	LU2FC
1940	XF1A(XE1A)	"NONE*	OQ5AB	XU8WS	"NONE*	XF1A(XE1A)	ZL1MR	LU7AZ
1939	XE2N(XE1A)	I6TK	ZS2AL	J2JJ	EI4J	XE2N(XE1A)	K6CGK[KH6IJ]	LU5AN
1938	XE1AG	EI8B	ZS1AH	J3FJ	OK1BC	XE1A	K6CGK[KH6IJ]	LU7AZ
1937	K5AY	—	ZS2A	J4CT	F8EO	K5AY	ZL2KK	OA4J
1936	XE2N(XE1A)	—	EA8AO	J2LO	EA4AO	XE2N(XE1A)	K6HLP	OA4J
1935	HC1FG(YL)	—	FM8BG	J2GX	ON4AU	X1AY	ZL4AI	HC1FG
1934	NY1AB	CM2OP	ZS2A	J2GX	EA5BE	NY1AB	VK3ML	HC1LC
1933	EAR185	G5BY	FM5IH	J1EC	EAR185	K4AAN	ZL2AC	HC1FG
1931	G5BY	—	—	—	G5BY	CM8UF	ZL2AC	K4KB
1930	CM8UF	—	—	—	G5BY	CM8UF		HC1FG
1929	"#ZY*	—	—					
1928	eg5BY[G]							
1927	3AG(CE)							

Footnote#1-There has not been a DX LowPower TopTen Box or Plaque/Trophy.

Footnote#2-Single Operator Assisted added for 1990 event. Single operator with Packet, with only ONE H.F. signal at a time. NO 10 minute rule.

Footnote#3-Changed format of event from 96 Hours during two weekends for each mode (Total of 192 hours) to 48 hours for one weekend for each mode (Total of 96 hours) beginning with the 1979 event.

Footnote#4-Added the Multi-Operator-TWO Transmitter class to the 1983 event. Each TX permitted only one signal and each must observe the 10 minute rule.

Footnote#Q-Added the QRPp Single Operator Class to the 1980 Event.

Footnote#1975-Added the Hi-Band and Lo-Band Single Operator Categories. These categories were only active from 1975-1979.

Footnote#MM-Added the Multi-Operator Single-Transmitter Class beginning with the 1971 event.

Footnote#1947- A stack of DX Phone logs was "lost, strayed or stolen". No awards will be issued until the affected amateurs have had a chance to resubmit copies of their 1947 DX Contest logs. Deadline was Feb 1,1948. Only 75 DX scores were included in the Nov 1947 Write-up. In the Nov 1948 issue of QST, after the final results of the 1948 DX contest, was an Addendum to the 1947 DX Contest. Resubmissions arrived from: OQ5AR, VQ4ERR, J3GNX, F8SI, G2PU, D4AEP, LA6J, LA7Y, GM8MN, TG9RC, ZL1MQ, LU5CK, PY2CK, PY2AJ, CE3AG, CE1AU.

Footnote#1937-Original ARRL DX Phone Competition Rules: March 20-28, 1937. 90 hours of operation MAXIMUM. Exchange RS(T) and a message number.

Footnote#1934-A gap was discovered in QST 1934 September Issue containing the final score listings of the Sixth International Relay Competition Results. The lapse begins after the score of EUROPEAN, AUS-TRIA, OE1ER thru to NORTH AMERICAN, MEXICO, X1BC. Missing results were taken from write-up.

Footnote#ZY-The Hiram Percy Maxim 60th Birthday Relay. Planning for this event actually began during late 1927, early 1928 when the ARRL Board of Directors decided to celebrate H.P.M.'s 60th birthday with a Birthday Relay. When it was discovered that H.P.M. would not be 60 years old until September 2, 1929 the Birthday Relay was put on hold and kept secret for over a year ! The event began August 31, a Saturday, at 6 P.M. E.S.T. until Monday, September 2nd at 6 P.M. E.S.T. During August of 1929, relay announcements were mailed to ARRL Official Relay Stations, Affiliated Clubs, Section Comm. Managers, and Route Managers throughout the U.S. and Canada. During the ten days prior to the Birthday Relay, ARRL Official Broadcast Stations were broadcasting the details of the event. (Luckily, H.P.M. never tuned in !) H.Q. station W1MK was manned by F.Edward Handy, R.P. and E.V. for the full 48 Hrs ! They received 356 messages on 3575kcs and 7150kcs. The first station worked had one message for H.P.M . The second station actually transmitted 40 messages to W1MK ! One station actually transmitted 40 messages for H.P.M. !! One station worked W1MK had 10 messages to contact W1MK ! It was reported that 'hardly two messages had the same wording.'

USA WINNERS OF ARRL DX COMPETITION CW BY YEAR/CATEGORY

YEAR	Single-Op	Low-Power	QRP	SOA	M/S	M2	M/M
1995	KM1H(KQ2M)	K7SV/4	AA2U	K1ZM/2	W3BGN	K1AR	W3LPL
1994	K5ZD	K7GM/4	AA2U	K3WW	AD1C	K1AR	M2RM
1993	K1ZM	W1PH	AA2U	K3WW	K3LR	K1AR	W3LPL
1992	WM5G(KRØY)	W1PH	AA2U	WB2Q	W3BGN	K1AR	N2RM
1991	KM1H(KQ2M)	KM9P	W2TZ	WB2Q	W3BGN	K1AR	W3LPL
1990	K1AR	W1PH	KR2Q	AA1K	W3BGN	N3RS	W3LPL
1989	KM1H(KQ2M)	VO1MP	KR2Q	*#2*	W3BGN	K1AR	W3LPL
1988	KM1H(KQ2M)	VO1MP	W8VSK		K3WW	N3RS	W3LPL
1987	KM1H(KQ2M)	VO1MP	NN4Q		K3KG	NR5M	W3LPL
1986	W3GRF(KØDQ)	VO1MP	NN4Q		N4WW	W2REH	W3LPL
1985	K1AR	VO1MP	K1UO		N4WW	K2TR	W3LPL
1984	K1AR	N8II	K3WS		W3BGN	W3GM	KN3O
1983	N2LT	N5AW	AA4AK		W3BGN	K5RC	N2AA
1982	K1GQ	K1JX	AC2U		K1ZZ	*#4*	K2UA
1981	W2PV(N2NT)	N5AW/1	N2AA		K5RC		K2UA
1980	K1PR	N4HI	N4BP		K5RC		W4BVV
1979	K1GQ	N5AW	*#Q*		W3AU		N2AA
1978	W3LPL	N4HI			N3RD		W4BVV
1977	W3LPL	WA1SSH[W1NG]			WA1NRF[K1PR]		N3RS
1976	W3LPL	WA1SSH[W1NG]			W2YD		W3WJD[N3RS]
1975	W7RM	WA1SSH[W1NG]			WA5LES[K5RC]		W3AU
1974	W3LPL	WA1SSH[W1NG]			K4VX		W3AU
1973	K1NOL[K1NA]	K2MFY			K1ZND[K1ZZ]		W3AU
1972	K1ZND[K1ZZ]	WA1ABW[K1BW]			WA8JUN[K8MD]		W3AU
1971	W1BGD/2[W1RM]	W4CRW			K6EBB		W3AU
1970	K1DIR	K5ABV			W4BVV		*#MM*
1969	W1BPW[W1UU]	W8QXQ[N8AA]			W3MSK[W3AU]		
1968	W9WNV/2	W4BRB			W3TMZ[W3OZ]		
1967	K1DIR	K1ZND[K1ZZ]			W4BVV		
1966	W4KFC	VE2NV			W3MSK[W3AU]		
1965	K2HLB	K1ZND[K1ZZ]			W3MSK[W3AU]		
1964	W3GRF	K1LPL[W3LPL]			W3MSK[W3AU]		
1963	W3GRF	K4TEA			W6RW		
1962	W4KFC				W3MSK[W3AU]		
1961	W3ECR(W3MFW)				W3MSK[W3AU]		
1960	W3ECR(W3MFW)				W3AOH		
1959	W3ECR(W3MFW)				W6RW		
1958	W3LOE				W3AOH		
1957	W4KFC				W3CTJ		
1956	W3DGM/3				W6RW		
1955	W2SAI([W3GM])				W3CTJ		
1954	W3CTJ(W3NOH)				W2SAI		
1953	W2SAI([W3GM])				W3MSK[W3AU]		
1952	W5ENE				W3BES[W3GM]		
1951	W3LOE				W3KT		
1950	W3LOE				W6EOG		
1949	W8BHW				W2IQG		
1948	W2GWE				W6LHN		
1947	W2GWE				———		
1940	W6GRL				———		
1939	W3CHE				W4YC		
1938	W2UK				W3HIL		
1937	W2UK				W1TW		
1936	W4DHZ				W6GRL		
1935	W3SI				W6GRL		
1934	W3ZJ				W4CBY		
1933	W3ZD				W1SZ		
1929	HPM Birthday Relay: W1MK-356,W1AOX-151,W8CHC-134						
1928	1ASF						
1927	2AHM						

Footnote#1-There has not been a DX Low Power Top Ten Box or Plaque/Trophy.

Footnote#2-Single Operator Assisted added for 1990 event. Single operator with Packet, with only ONE H.F. signal at a time. No 10 minute rule.

Footnote#3-Changed format of event from 96 Hours during two weekends for each mode (Total of 192 hours) to 48 hours for one weekend for each mode (Total of 96 hours) beginning with the 1979 event.

Footnote#4-Added the Multi-Operator-TWO Transmitter class to the 1983 event. Each TX permitted only one signal and each must observe the 10 minute rule.

Footnote#Q-Added the QRPp Single Operator Class to the 1980 Event.

Footnote#1980-DX-to DX QSOs were permitted in the 1980 event as an experiments. This was rescinded for 1981, reverting back to the previous form: "The world working the U.S." (Ed.—Things were chaotic that next year too!)

Footnote#1975-Added the Hi-Band and Lo-Band Single Operator Categories. These categories were only active from 1975-1979.

Footnote#MM-Added the Multi-Operator Single-Transmitter Class beginning with the 1971 event.

Footnote#1963-Added the Low Power Single Operator Class.

Footnote#1947- A stack of DX Phone logs was "lost, strayed or stolen". No awards will be issued until the affected amateurs have had a chance to resubmit copies of their 1947 DX Contest logs. Deadline was Feb 1,1948. Only 75 DX scores were included in the Nov 1947 Write-up. In the Nov 1948 issue of QST, after the final results of the 1948 DX contest, was an Addendum to the 1947 DX Contest. Resubmissions arrived from: OQ5AR, VQ4ERR, J3GNX, F8SI, G2PU, D4AEP, LA6J, LA7Y, GM8MN, TG9RC, ZL1MQ, LU5CK, PY2CK, PY2AJ, CE3AG, CE1AU.

Footnote#1937-Original ARRL DX Phone Competition Rules: March 20-28, 1937. 90 hours of operation MAXIMUM. Exchange RS(T) and a message number.

160	80	40	20	15	10	Club: Unlimited/Med./Loc.
W4MYA	W1MK	W3GH	K2SS/1	W5VX	K9OM	FRC-PVRC-Hudson Valley
K1ZM/2	W1MK	K8PO/1	W3USS(K8OQL)	N4CT	WS1M	FRC/PVRC/River City Contesters
W1FJH(AD8V)	W1FV	W6XX	KT3Y	KØLUZ/4	K5MR	FRC/Mad River/River City C.
K5UR	WE3C	K8PO	K1TO	K2VV	WØUN	FRC/NCC/RiverCityC.
W1NG	W1FV	K1ZM	W1RR	K4VX/Ø(WOØG)	WM5G(KR0Y)	FRC/NCC/C.Va.C.C.
W1NG	W1FV	WOØG	K4VX(W9WI)	K2VV	K1ZM	FRC/SMC/Overlook Mtn.
K1ZM	W1FV	KBØG	N2AA	K2VV	K1RM	FRC/NTCC/Overlook Mtn.
K1ZM	W6RJ	K2EK	K2VV	K1RM	N4BP	YCCC/PVRC/Overlook Mtn.
K1ZM	W1FV	K4XS	K1RM	W5VX	K9LA/5	YCCC/PVRC/Overlook Mtn.
K5UR	W1FV	N6QR	K2VV	K3RV	KR1R	FRC/PVRC/Overlook Mtn.
W1RR	W1FV	N4PN	N2AA	N2EK	WA7KLK	FRC/PVRC/CentralVaCC
W1RR	W1FV	W2YV(KQ2M)	K3UA	WB4TDH	W1WEF	FRC/PVRC/CentralVaCC
W8LRL	K1PT	NA5R(N5EA)	K8NA	WØZV	WB4TDH	FRC/N.Tx.C.C./CentralVaCC
N4WW	W1ZM(K1ZM)	W6XX	K1KI	K6LL/7	N4ZZ	FRC/Murphy's/ILLWINDC.
N4IN	W1ZM(K1ZM)	W5UN	K5IY(KA5CHW)	K1RM	WØZV	FRC/Murphy's/ILLWINDC.
W8LRL	N4AR	W5UN	K3TW	K6LL/7	N4WW	FRC/Murphy's/CentralVaCC
LowBand:	K7UR	HighBand:	AE2A			FRC/Murphy's/CentralVaCC
LowBand:	K3ZO	HighBand:	N2OO			FRC
LowBand:	N4PN	HighBand:	K1NOL[K1NA]			PVRC
LowBand:	K1NOL[K1NA]	HighBand:	W2DXL[W2IB]			FRC
						FRC
						MM
						FRC
						PVRC
						FRC
						PVRC
						PVRC
						PVRC
						PVRC
						FRC
						PVRC
						FRC
						FRC
						PVRC
						FRC
						FRC
						PVRC
						PVRC
						Southern California DX Club
						FRC
						FRC
						FRC
						FRC
						Southern California DX Club
						FRC
						FRC
						FRC
						S. Bay Am. Assoc. of L.A., Ca.
						S. Bay Am. Assoc. of L.A., Ca.
						S. Bay Am. Assoc. of L.A., Ca.
						Tri-Cnty. Am. Radio Assoc., N.J.
						Club Winner Not Announced
						Club Winner Not Announced

Footnote#1934-A gap was discovered in *QST* 1934 September Issue containing the final score listings of the Sixth International Relay Competition Results. The lapse begins after the score of EUROPEAN, AUSTRIA, OE1ER through to NORTH AMERICAN, MEXICO, X1BC. Missing results were taken from the write-up.

Footnote#ZY-The Hiram Percy Maxim 60th Birthday Relay. Planning for this event actually began during late 1927, early 1928 when the ARRL Board of Directors decided to celebrate H.P.M.'s 60th birthday with a Birthday Relay. This was put on hold when it was discovered that H.P.M. would not be 60 years old until September 2, 1929 ! The Birthday Relay was put on hold and kept secret for over a year ! The event began August 31, a Saturday, at 6 P.M. EST until Monday, September 2nd at 6 P.M.. EST. During August of 1929, relay announcements were mailed to ARRL Official Relay Stations, Affiliated Clubs, Section Comm. Managers, and Route Managers throughout the U.S. and Canada. During the ten days prior to the Birthday Relay, ARRL Official Broadcast Stations were broadcasting the details of the event. (Luckily, H.P.M. never tuned in!) H.Q. station W1MK was manned by F. Edward Handy, R.P. and E.V. for the full 48 Hrs! They received 356 messages on 3575kcs and 7150kcs. The first station worked had one message for H.P.M. The second station to contact W1MK had 10 messages for H.P.M.!! One station actually transmitted 40 messages to W1MK! It was reported that "hardly two messages had the same wording."

DX WINNERS OF ARRL DX PHONE COMPETITION BY YEAR/CATEGORY

YEAR	Single-Op	QRP	Asst	M/S	M2	M/M
1995	P4ØV(AI6V)	TG9GI	PY2EX	PJØB	VP2MFM	9A1A
1994	P4ØJ(WX4G)	PY3OC	EA3BT	P4ØV	WP4U	6D2X
1993	HC8A(N6KT)	V7A(V73C)	PYØFM	PJØB	6D2X	I3MAU
1992	TI1C(N6KT)	NH6T	EA3NY	PJØB	UZ2FWA	XE21
1991	EA8RCT(OH8XX)	TG9GI	HI8A	8P9X	XE2FU	I3MAU
1990	P4ØV(AI6V)	HB9ADD	SP2UUU	VP2E	K6GSS/KH6	4B2A
1989	V31C(KE5CV)	JA2JSF	*#2*	VP2MU	XE2FU	6D2DX
1988	PJ2FR(N6KT)	TG9GI		VP2MU	KP4BZ	I3MAU
1987	V31CV(KE5CV)	YX3A		NP4CC	K2SS/VP2V	VP9AD
1986	LU1BR	K7SS/KH6		ZF2HI	V47M	I3EVK
1985	ZF2FL(N6RJ)	AH6EK		HH2CQ	J87J	JA9YBA
1984	ZF2FL(N6RJ)	OK3CGP		ZF2HM	K9GL/VP2V	XE2FU
1983	ZF2FL(N6RJ)	TG9GI		KP4BZ	VP5KMX	VK2WU
1982	ZF2FL(N6RJ)	TG9GI		VP2E	*#4**	J3AVT
1981	VP2MP(K2YY)	I5NSR		VP2E		G4ANT
1980	VP2ML(WB2CHO)	TG9GI		VP1A		GB4ANT
1979	XE1OW(N5AU)	*#Q*		FY7AK		K1CO/PJ7
1978	KH6IJ			KH6DL		SK6JA
1977	KP4EAS			KH6GQW		PJ8CO
1976	KP4AST(NP4A)			WØNAR/6Y5		PJ8CO
1975	KZ5BC			OA4O		YU3DBC
1975	KZ5BC			OA4O		YU3DBC
1974	HR1RF			YV4AGP		KP4AXM
1973	6J9AA(KØDQ)			G4ANT		DL5AY
1972	XE1IIJ(KØDQ)			KZ5ZZ		KH6HCM
1971	XE1KS			HC1ARE		OH2BO
1970	KP4AST(NP4A)			KH6SP		*#MM*
1969	KV4FZ			6YØA		
1968	KH6IJ			XE2PTBC		
1967	HI8XAL(K3ZO)			KS6BV		
1966	KP4CKU			ZF1BP		
1965	HK3RQ			I1RB		
1964	YV5BIG			YV5AHG		
1963	XE1CCB			I1CWN		
1962	XE1CV			CO8RA		
1961	KP4AVQ			KH6ECD		
1960	VP2DX			GB2SM		
1959	KH6IJ			CN8JE		
1958	KH6IJ			KX6AF		
1957	KH6IJ			KH6CBP		
1956	TG9AD			KH6AYG		
1955	KH6IJ			ZB2A		
1954	KH6IJ			KH6AWM		
1953	KV4BB			VP9BDA		
1952	KH6IJ			XE1QB		
1951	VP6SD			MI3US		
1950	XF1A(XE1A)			OA1E		
1949	KP4ES			G3DFC		
1948	XF1A(XE1A)			VQ3HGE		
1947	XE1A	***#1947*		——		
1940	K4FKC			——		
1939	CO2WM			——		
1938	K4SA			——		
1937	XE2N(XE1A)			F8KW		

Although phone was in use long before 1937 and phone use in Sweepstakes was reported as early as 1932, the ARRL DX Competition did not report any phone work until the 1937 event, when March 6-14,1937 was for CW and March 20-28 was for phone.

Footnote#1-There has not been a DX Low Power Top Ten Box or Plaque/Trophy.

Footnote#2-Single Operator Assisted added for 1990 event. Single operator with Packet, with only ONE H.F. signal at a time. No 10 minute rule.

Footnote#3-Changed format of event from 96 Hours during two weekends for each mode (Total of 192 hours) to 48 hours for one weekend for each mode (Total of 96 hours) beginning with the 1979 event.

Footnote#4-Added the Multi-Operator-TWO Transmitter class to the 1983 event. Each TX permitted only one signal and each must observe the 10 minute rule.

Footnote#Q-Added the QRPp Single Operator Class to the 1980 Event.

Footnote#1980-DX-to DX QSOs were permitted in the 1980 event as an experiments. This was rescinded for 1981, reverting back to the previous form: "The world working the U.S." *(Ed.—Things were chaotic that next year too!)*

Footnote#1975-Added the Hi-Band and Lo-Band Single Operator Categories. These categories were only active from 1975-1979.

Footnote#MM-Added the Multi-Operator Single-Transmitter Class beginning with the 1971 event.

Footnote#1963-Added the Low Power Single Operator Class.

Footnote#1947- A stack of DX Phone logs was "lost, strayed or stolen". No awards will be issued until the affected amateurs have had a chance to resubmit copies of their 1947 DX Contest logs. Deadline was Feb 1,1948. Only 75 DX scores were included in the Nov 1947 Write-up. In the Nov 1948 issue of QST, after the final results of the 1948 DX contest, was an Addendum to the 1947 DX Contest. Resubmissions arrived from: OQ5AR, VQ4ERR, J3GNX, F8SI, G2PU, D4AEP, LA6J, LA7Y, GM8MN, TG9RC, ZL1MQ, LU5CK, PY2CK, PY2AJ, CE3AG, CE1AU.

160	80	40	20	15	10
FM5DN	VP2MFP(KK9A)	ZF2CF	HK3JJH	TGØAA(KA9FOX)	LU6ETB
PJ8H	ZF2CF(KE6CF)	FM5DN	ZF5ND	HC1HC	ZPØY(ZP5JCY)
CY2CE	4M3X	ZF2ND	IR2T(I4UFH)	HC1OT	LU6ETB
KV4FZ	ZF2ND	ON4UN(ON4WW)	IR4T(I4UFH)	9Y4VU	PJ9M(OH6RM)
CT1AOZ	YV1EQW	IO4VEQ	YT1BB	FM/F6HMQ	CT1BOP
YV1DRK	T32AF	4M5T(YV5JBI)	G3FXB	NP4CC	PJ9M(OH6RM)
HK3DFT	TE1L	ZF2MV	T32AF(KH6UR)	NP4CC	P4ØT(KB2HZ)
4U1UN(K2GM)	TI2LTA	XE2NQ(AA5B)	HC1HC	PT9ZZ(PY5CC)	LU1E(LU3AJW)
CT1AOZ	CU2AK	9Y4AA(N6TJ)	KK9A/VP2V	PY5IW	LU1E(LU3AJW)
IV3PRK	4U1UN(K2GM)	HC5EA	KP4FI	HK3MAE	LU6ETB
VP5SBX(AK8A)	KK9A/PJ7	T32AF(KH6UR)	HC1HC	ZV9ZZ(PY5IW)	HC1HC
AH6BK	HI8LC	JA2BAY	KP4EQF	KK9A/V2A	CP6EL
T32AF(KH6UR)	FM7WS	FM7CD	YT3A(YU3SO)	VP2MKD(K9MK)	G4GIR
——	YV1CD	HC1HC	F2SI	IØWDX	FGØFOO/FS(N6RA)
HI8JAG	HI8PGG	FMØFJE	OH2MM	YU3TU	H31LR(HP1XOJ)
G3SZA	CT2AK	PJ2FR	OH5LF	OH2FQ(OH2BH)	
LowBand:	HI8JAG	HighBand:	OA4ASX(K1MM)		
LowBand:	KP4WI	HighBand:	HC1BU		
LowBand:	G3TJW	HighBand:	HC1BU		
LowBand:	W4EV/VP9	HighBand:	YN1RWG		
LowBand:	EA4LH	HighBand:	HC1BU		
S/O AF	**S/O AS**	**S/O EU**	**S/O NA**	**S/O OC**	**S/O SA**
GW8DY	JA2JW	I3MAU	KZ5BC	KH6IJ	YV4YC
CR6NO	JA1BRK	CT2BG	HR1RF	KH6IJ	CE6EZ
ZS6DW	JA2IYJ	CT1BH	6J9AA	KH6IJ	4M4AGP
ZS6DW	JA2JW	CT1BH	XE1IIJ	KH6RS(K1GQ)	HC1RF
EL2CB	JH1CJQ	LAØAD	XE1KS	KH6RS(K1GQ)	YV5CVE
ZS6DW	JA1AEA	CT2AT	KP4AST	KH6IJ	HC1TH
ZD8Z([N6TJ)	JA1AEA	I1BAF	KV4FZ	KH6GPQ[K1GQ]	HC1TH
ZS6DW	KA7AB	EA3JE	YS1XEE	KH6IJ	HK3RQ
ZS6DW	KA7AB	DJ6QT	HI8XAL	KH6IJ	HK3RQ
ET3AC	JA1IBX	I1BAF	KP4CKU	VR1Z	HK4KL
ZD8HL	JA1CG	EA4GD	FG7XL	VK2APK	HK3RQ
6O6BW	JA1BRK	DL1KB	KP4AOO	KX6BU	YV5BIG
ET3USN	HL9KH(W9WNV)	EI4KA	XE2CCB	VR30	HC1DC
VQ2AT	KA2MA	OE1RZ	XE1CV	KW6DG	YV5AGD
EL8D	JA1BWA	OE1RZ	KP4AVQ	KW6DG	HC1KA
EL8D	JA3IS	ON4OC	VP2DX	ZL1MQ	VP3HAG
5A5TO	KA2UJ	DJ1BZ	VP9L	KH6IJ	VP3HAG
ZS6UR	KA2RB	F8PI	VP9L	KH6IJ	VP3HAG
ZS5JY	KA2FQ	F8PI	VP9L	KH6IJ	HC2BH
ZS6DW	JA3BB	EA4DL	TG9AD	KH6IJ	HC1ES
EL2X	KA2OJ	CT1SQ	VP7NX	KH6IJ	PJ2AF
ZS6DW	TA3AA(W1VQG)	VP7NX	HP3FL	KH6MG	PJ2AF
ZS6DW	TA3AA	G2PU	KV4BB	KH6IJ	PJ2AA
ZS6DW	JA8AB	CT1BS	VP6SD	ZL1MQ	PY2CK
ZS6DW	TA3GVU	I1BDV	VP6SD	KH6IJ	HC2OS
ZS6DW	XZ2SY	I1US	XF1A(XE1A)	KH6IJ	CE2CC
ZS6DW	JA3AA	G2PU	KP4ES	KG6AW/VK9	HC1KP
EL5A	J3GNX	G2PU	XF1A(XE1A)	KH6IJ	PY2AC
ZS6DW	J3GNX	G2PU	XE1A	K6PLZ	OA4M
OQ5AB	XU8AM	G2PU	K4FKC	——	LU5AN
ZS6DW	J3FZ	——	VP7NS	VK2IQ	LU9BV
ZS6DW	J2MI	GM6RG	K4SA	K6MVV	LU9BV
ZU6P	XU8HW	G6LK	XE2N(XE1A)		
		G6LK			

USA WINNERS OF ARRL DX PHONE COMPETITION BY YEAR/CATEGORY

YEAR	Single-Op	Low Power	QRP	Asst	M/S	M2	M/M
1995	KM1H(KQ2M)	W1PH	AA2U	K1ZM/2	K5NA/2	K1AR	W3LPL
1994	K1AR	K2SG	AA2U	K1DG(WZ1R)	K1RX	AD1C	N2RM
1993	K1DG	N8II	AA2U	N4RJ(KM9P)	K8AZ	K1AR	N2RM
1992	KM1H(KQ2M)	W1PH	AA2U	NZ7E(KI3V)	NQ4I	N2NT	W3LPL
1991	WM5G(KRØY)	K7RI	K5RX	K1AR	W3BGN	K2TR	W3LPL
1990	K1RU	W1PH	N4KG	WM5G(KRØY)	W3BGN	K2TR	W3LPL
1989	K3TUP(K3LR)	W2HPF	KR2Q	*#2*	AA1K	KC1F	W3LPL
1988	K3TUP(K3LR)	W2TZ	N6OJ		W3BGN	K2TR	W3LPL
1987	K1AR	W2TZ	K3WS		K3TUP	K5RX	KX4S
1986	K1KI	VE1NG	K3WS		K3TUP	KX4S	W3LPL
1985	K1RX	VE1NG	W8VSK		K4VX/Ø	K2TR	W3LPL
1984	W1ZM(K1ZM)	WA4PFN/2	K4JRB		K2BU	K2TR	KN3O
1983	AI6V(N6KT)	W2TZ	WB4BBH		W3BGN	W4QAW	N2AA
1982	W1ZM(K1ZM)	K1JX	KA1VQ		K1OX	*#4*	W3LPL
1981	K1VTM(K1JX)	W2TZ	N2AA		KØRF		W2PV
1980	K1VTM	VO2CW	VE3KKB		K4VX		N2AA
1979	K1PR	KB9ET	*#Q*		K5JA		W2PV
1978	W1ZM(K1ZM)	WB2SJG[KQ2M]			W1ZA		W2PV
1977	W2HMH[K2BU]	WA1SSH[W1NG]			W5TMN[K5JA]		W7RM
1976	W6HX	WA1SSH[W1NG]			WA1KID[W6XR]		AC2PV
1975	W3WJD[N3RS]	WA1SSH[W1NG]			W6ONV		W2PV
1975	***1975***						
1974	W6OAT	WA1SSH[W1NG]			WA1NRV		W3AU
1973	W7RM(K7VPF)	W4WRY			W1FBY[W1XT]		W3AU
1972	W7RM(K7VPF)	W8ECA			W4FDA		W3AU
1971	WA3HGV	WAØYAW			W5RER		W3AU
1970	W6RR	WA1DJG[K1CC]			W3ZKH/3[N4RV]		*#MM*
1969	W6RR	WA1DJG[K1CC]			W3MSK[W3AU]		
1968	K1DIR	WA1DJG[K1CC]			W3ZKH/3[N4RV]		
1967	K8DOC	W8ECA			W3MSK[W3AU]		
1966	WA4PXP	W8LXU			W3MSK[W3AU]		
1965	W4KFC	WA5ALB			W3MSK[W3AU]		
1964	K2HLB	WA2OOO			W3MSK[W3AU]		
1963	K5MDX	K9ECE			W3MSK[W3AU]		
1962	K2GXI	**#1963**			W3MSK[W3AU]		
1961	K2GXI				W1ETF		
1960	W1ONK				W8NWO		
1959	W3DHM				W3ECR		
1958	W8BKP				W3AOH		
1957	W2ATE				W3DHM		
1956	W2SKE/2				W6AM		
1955	W1ATE				W2SAI		
1954	W1ATE				W2SAI		
1953	W4ESK[W7RM]				W3LVF		
1952	W4ESK[W7RM]				W3BES		
1951	W4DCQ				W6AM		
1950	W1ATE				W1QDE		
1949	W2SAI				WØUQV		
1948	W2AFQ				W3BES		
1947	W2SAI				W5MLX		
1940	W6OCH						
1939	W3EMM				W4YC		
1938	W3EMM				W6GRL		
1937	W9ARA				W9PV		

Although phone was in use long before 1937 and phone use in Sweepstakes was reported as early as 1932, the ARRL DX Competition did not report any phone work until the 1937 event, when March 6-14,1937 was for CW and March 20-28 was for phone.

Footnote#1-There has not been a DX Low Power Top Ten Box or Plaque/Trophy.

Footnote#2-Single Operator Assisted added for 1990 event. Single operator with Packet, with only ONE H.F. signal at a time. No 10 minute rule.

Footnote#3-Changed format of event from 96 Hours during two weekends for each mode (Total of 192 hours) to 48 hours for one weekend for each mode (Total of 96 hours) beginning with the 1979 event.

Footnote#4-Added the Multi-Operator-TWO Transmitter class to the 1983 event. Each TX permitted only one signal and each must observe the 10 minute rule.

Footnote#Q-Added the QRPp Single Operator Class to the 1980 Event.

Footnote#1980-DX-to DX QSOs were permitted in the 1980 event as an experiments. This was rescinded for 1981, reverting back to the previous form: "The world working the U.S." (Ed.—Things were chaotic that next year too!)

Footnote#1975-Added the Hi-Band and Lo-Band Single Operator Categories. These categories were only active from 1975-1979.

Footnote#MM-Added the Multi-Operator Single-Transmitter Class beginning with the 1971 event.

Footnote#1963-Added the Low Power Single Operator Class.

Footnote#1947- A stack of DX Phone logs was "lost, strayed or stolen". No awards will be issued until the affected amateurs have had a chance to resubmit copies of their 1947 DX Contest logs. Deadline was Feb 1,1948. Only 75 DX scores were included in the Nov 1947 Write-up. In the Nov 1948 issue of QST, after the final results of the 1948 DX contest, was an Addendum to the 1947 DX Contest. Resubmissions arrived from: OQ5AR, VQ4ERR, J3GNX, F8SI, G2PU, D4AEP, LA6J, LA7Y, GM8MN, TG9RC, ZL1MQ, LU5CK, PY2CK, PY2AJ, CE3AG, CE1AU.

160	80	40	20	15	10
WØZV/4	KQ3V	K6NA	KS1L	K3ZJ/8	K6SVL
K5UR	K1ZM	KC7EM	KS1L	K1UO	KE5FI
WB9Z	WE3C	K1UO	K5MR	WØUN(WØUA)	K4XS
K1ZM	K1UO	KC7EM	KK9A	K2SS/1	WØUN(WØUA)
K1ZM	K8UR/1	WOØG	KS1L	W7WA	N1GLG
K1ZM	KA1XN	K4XS	WØZV	W7WA	K3ZJ
K1ZM	K4HJJ	KVØQ	AI7B	W7EJ	K4XS
WA4SVO	N2NT	K6NA	KS1L	K4XS(WC4E)	K5UR
K5UR	W5WMU	W6AQ(WA6OTU)	VO1SA	K6SVL	KE5FI
K1ZM	K2EK	NZ5I	K2VV	K3RV	K4JRB
K1ZM	WØMJ	KM6B	KS8S(KU8E)	W5XZ	WA3EEE
VE1YX	W1FC	K8NN/9	K1UO	WØZV	WA6DBC
VE1YX	KR2N	N6BV	N2PP	WØZV	WA6DBC
WA2SPL	K1PT	N5JJ	K1KI	W7RM(W7WA)	WØZV
W8LRL	W1CF(K8UR)	K7UR	K3KG	K7RI(W7WA)	K1UO
W4PZV	WA4SVO	WA7ZLC(N4EA)	K9DX	N7XX(WA1KKM)	VE6WQ
Low Band:	W1FC	High Band:	K9DX		
Low Band:	K3ZO	High Band:	N2LT		
Low Band:	K4YFQ[N4WW]	High Band:	N4MM		
Low Band:	WA8ZDF	High Band:	WB6PXP[AE6E]		
Low Band:	WA8ZDF	High Band:	K6SVL		

ARRL DX CONTEST

Continental CW Records

Continent	Call	Year	Score
Single-Op All-Band (NoPacket)			
AF	EA8EA(OH2MM)	95	4,976,016
AS	JA7FWR	89	1,748,682
EU	CR7M(CT1BOH)	93	3,468,867
NA	VP5F(KRØY)	93	6,489,216
OC	NH6T(WØZZ)	92	3,119,364
SA	P4ØGD(W2GD)	88	6,031,005
Single-Op QRP (NoPacket)			
AF	EA8BIE	87	72,891
AS	JA9RPU	92	231,840
EU	G4BUE	92	513,240
NA	C6AHL(K3DI)	94	1,715,538
OC	K7SS/KH6	90	1,394,271
SA	YV3AGT	86	769,986
Single-Op Low Power			
AF	6V6U(K3IPK)	95	4,542,060
AS	JR7OMD/2	93	577,404
EU	TM5GG(F6FGZ)	93	1,188,414
NA	ZF8BS(AA6KX)	94	3,992,028
OC	KH6RS(NØAX)	95	2,471,034
SA	PY2OU	95	1,176,480
S/O - 28 MHz			
AF	ZD8LII(GØLII)	93	378,000
AS	JHØKHR	89	215,655
EU	FFØXX(FB1MUX)	92	281,961
NA	N6OP/NP2	90	380,475
OC	VE7QO/AH6	89	247,800
SA	HC2G	89	306,900
S/O-21 MHz			
AF	ZD8LII(GØLII)	91	255,303
AS	7L1GVE(JAØJCJ)	91	217,674
EU	TF3YH	81	317,520
NA	N6OP/NP2	94	330,840
OC	KG6DX	89	187,587
SA	PY5ZBA	88	312,852
S/O-14 MHz			
AF	TU2MA	94	201,144
AS	RZ9UA	92	254,520
EU	OG1AA	92	315,945
NA	AI6V/VP9	89	342,684
OC	K7SS/KH6	85	258,042
SA	9Y4VU	95	302,400
S/O-7 MHz			
AF	EA9EU	82	113,679
AS	JAØJHA	89	108,528
EU	EA7CEZ	95	244,872
NA	KP4FI	86	353,115
OC	T32AF(KH6UR)	93	227,244
SA	4M7QP	83	223,776
S/O-3.5 MHz			
AF	EA1AK/EA8	95	161,616
AS	JH1OGC	94	41,952
EU	OT5T(ON4UN)	95	135,558
NA	K8WW/VP9	88	184,800
OC	W7AW/KH6	88	26,532
SA	YV1OB	93	114,348
S/O-1.8MHz			
AF	No Entry		
AS	JA1GTF	84	3,024
EU	CT1AOZ	88	34,440
NA	K8WW/VP9	86	87,363
OC	K9VV/KH6	93	8,064
SA	YV1OB	88	66,144
Single Op. WITH Packet (Since1990)			
AF	No Entry		
AS	JJ3YBB(JI3ERV)	92	1,229,256
EU	HG3DXC(HA4XT)	91	1,934,640
NA	NP4Z	94	4,013,862
OC	VK2VM	95	18,252
SA	PY2EX	95	149,040
Multi-Op Single Transmitter			
AF	EA4KR/EA8	93	3,270,939
AS	JA7YAA	81	2,729,124
EU	UX1A	92	4,422,888
NA	ZF2KE	90	5,980,590
OC	NH6T	93	4,633,263
SA	P49V	95	5,573,421
Multi-Op Two Transmitter			
AF	5H3XX	93	598,764
AS	JA8YBY	91	2,374,392
EU	UZ2FWA	92	4,097,184
NA	6D2X	94	8,011,644
OC	ZL3GQ	93	3,524,148
SA	HK3MAE	89	155,868
Multi-Op Multi-Transmitter			
AF	No Entry		
AS	JA8YBY	89	2,484,720
EU	9A1A	93	4,057,371
NA	XE2FU	83	8,064,153
OC	KH6XX	86	4,275,177
SA	No Entry		

U.S.A./Canada CW Records

Section	Call	Year	Score
Single-Op All-Band (No Packet)			
W1	KM1H(KQ2M)	91	4,001,790
W2	K1ZM/2	92	3,736,665
W3	K3TUP(K5ZD)	91	3,632,820
W4	N4RJ(KM9P)	95	2,919,300
W5	WM5G(KRØY)	92	3,954,411
W6	N6TV	91	1,694,628
W7	WA7NIN(N6TV)	92	2,415,969
W8	KW8N(WD8IXE)	92	2,919,168
W9	W9RE	91	2,737,788
WØ	KØRF(WØUA)	91	3,047,154
VE	VE3EJ	93	2,351,700
Single-Op Low Power (NoPacket) Since 1983			
W1	W1PH	91	1,631,817
W2	KZ2S	89	1,464,120
W3	W3UJ	91	732,450
W4	KM9P/4	91	1,826,181
W5	N5AW	89	1,080,108
W6	W6JTI	89	897,408
W7	W7YAQ	89	905,360
W8	N8II	84	981,360
W9	K9QVB	91	1,008,267
WØ	K3GWA/Ø	92	472,320
VE	VO1MP	89	1,601,538
Single-Op QRP<5 W (NoPacket)			
W1	KN1M	91	579,852
W2	AA2U	92	836,964
W3	W3EWL	91	496,701
W4	N4KG	91	685,482
W5	K5RX	91	659,178
W6	W6JTI	91	325,740
W7	NØAX/7	92	543,585
W8	W8VSK	89	389,340

W9	KS9U	89	227,835		W5	K5UR	85	19,902
WØ	WØUO	91	589,344		W6	K6DDO	86	4,392
VE	VE1NH	89	124,146		W7	N6SS/7	95	5,964
					W8	WD9INF/8	95	6,786
S/O-28 MHz					W9	WB9Z	95	10,332
W1	K1XA	92	487,740		WØ	WØZV	85	13,818
W2	N2KW	89	196,056		VE	VE3DO	91	7,437
W3	KA3SIO	89	305,025					
W4	K4XS(WC4E)	92	470,250		**Single Operator WITH Packet**			
W5	WM5G(KRØY)	91	423,000		W1	AA2Z	92	2,885,325
W6	W6YA	89	286,794		W2	WB2Q	92	4,604,574
W7	W7WA	89	294,570		W3	K3WW	92	4,158,168
W8	N8DCJ	89	422,718		W4	N4RJ(KM9P)	92	4,421,520
W9	K9QVB	92	312,708		W5	N6ZZ/5	95	943,740
WØ	WØUN(WØUA)	92	539,235		W6	N6ZZ	93	925,344
VE	VC4VV	92	216,384		W7	NZ7E(KI3V)	92	2,845,248
					W8	WR3G/8	92	2,842,800
S/O-21 MHz					W9	KK9V	91	1,720,950
W1	K1TO	89	504,495		WØ	NØAT	91	1,580,796
W2	K2VV	89	557,235		VE	VE3EJ	94	1,865,376
W3	KE3Q	89	510,600					
W4	K3RV/4	88	682,362		**Multi-Op Single Transmitter**			
W5	W5WMU	92	404,040		W1	K1VR	92	3,678,096
W6	WN4KKN/6	91	389,367		W2	N2NU	95	3,792,789
W7	K6LL/7	91	375,768		W3	K3LR	93	3,906,720
W8	KW8N	88	509,878		W4	N4RJ	93	3,389,463
W9	K9QVB	81	241,230		W5	K5RC	81	3,142,191
WØ	K4VX/Ø(WOØG)	91	478,332		W6	W6QHS	91	2,517,318
VE	VE7SZ(WA6VEF)	91	375,900		W7	N7NG	89	1,745,100
					W8	K8CC	92	2,903,040
S/O-14 MHz					W9	NA9J	89	1,810,215
W1	K1TO	92	609,504		WØ	KØRF	89	2,600,631
W2	N2AA	89	600,696		VE	VE6OU[VE3EJ]	81	1,462,500
W3	KE3Q	91	365,400					
W4	KT3Y/4	93	495,405		**Multi-Op Two Transmitter**			
W5	K5MR	92	450,780		W1	K1AR	92	8,295,039
W6	NI6W	89	365,904		W2	N2NT	93	6,968,592
W7	WA7RKJ	89	302,160		W3	N3RS	91	7,145,100
W8	W8TA	81	339,810		W4	N4AR	92	5,051,811
W9	K9QVB	89	281,340		W5	WM5G	89	4,831,083
WØ	K4VX/Ø(W9WI)	95	373,626		W6	N6RO	89	4,561,716
VE	VE9ST	95	471,240		W7	KO7N	89	3,249,900
					W8	K8AZ	91	6,114,051
S/O-7 MHz					W9	W9KDX	95	2,310,633
W1	K8PO/1	92	339,660		WØ	K4VX/Ø	89	5,496,942
W2	K1ZM/2	91	401,472		VE	VE1CR	83	43,281
W3	W3GH	95	235,521					
W4	K4JPD	92	320,985		**Multi-Op Multi-Transmitter**			
W5	K5GW	91	259,515		W1	K1ST	91	9,939,822
W6	W6XX	93	307,146		W2	N2RM	92	10,756,002
W7	K7QQ	93	205,884		W3	W3LPL	92	10,555,032
W8	K8LX	92	178,533		W4	W4BVV	79	5,778,663
W9	K9UWA	94	174,300		W5	NR5M	89	7,370,745
WØ	WØUN[WØUA]	94	291,924		W6	N6RO	90	4,516,740
VE	VE2FU	89	172,992		W7	W7FU	79	4,115,286
					W8	K8CC	93	5,037,210
S/O-3.5 MHz					W9	WØAIH/9	92	7,433,289
W1	W1MK	95	247,776		WØ	K4VX/Ø	93	5,710,944
W2	K2EK	90	123,360		VE	No Entry		
W3	WE3C	93	95,064					
W4	N4ZC	87	112,860		**Club**			
W5	AA5GY	95	45,600		Score	FRC	92	202,360,257
W6	W6RJ	88	70,584					
W7	KC7EM	90	33,276			**ARRL DX CONTEST**		
W8	K7EG/8	91	93,507					
W9	K9BGL	95	40,755			**Continental PHONE Records**		
WØ	AA6TT/Ø(AA5B)	95	57,828					
VE	VE3RM	95	23,550		**Continent**	**Call**	**Year**	**Score**
S/O-1.8 MHz					**Single-Op All-Band (No Packet)**			
W1	W1RR	85	25,020		AF	ZD8Z(N6TJ)	95	7,032,993
W2	K1ZM/2	88	21,960		AS	JH7DNO	92	2,125,416
W3	AA1K/3	85	22,134		EU	Y24UK	92	3,310,704
W4	W4MYA	95	26,796		NA	TI1C(N6KT)	92	9,619,584

OC	KH6/WR6R	93	5,329,242
SA	HC8A(N6KT)	88	10,114,020

Single-Op QRP (NoPacket)

AF	A22AA	89	39,600
AS	JA2JSF	91	131,670
EU	GØPAM(G4BUE)	92	335,943
NA	TG9GI	88	1,169,280
OC	V7A(V73C)	93	2,016,333
SA	4M1G	93	1,223,388

S/O-28MHz

AF	ZD8LII(GØLII)	93	516,309
AS	JAØJHA	90	376,362
EU	CT1BOP	91	583,965
NA	AL7CQ	92	586,401
OC	WR6R/KH6	89	469,104
SA	P4ØT(KB2HZ)	89	887,301

S/O-21MHz

AF	EA9UK	92	342,084
AS	7L1GVE	92	386,460
EU	IØWDX	82	495,432
NA	TGØAA(KA9FOX)	95	688,884
OC	AH6GQ	88	433,608
SA	PT9ZZ	88	637,188

S/O-14MHz

AF	TR8SA	88	208,620
AS	RZ9UA	93	276,834
EU	IR4T(I4UFH)	92	567,180
NA	ZF2ND	94	676,260
OC	T32AF(KB2HZ)	89	420,831
SA	HC1HC	88	486,330

S/O-7MHz

AF	CR3R(CT3BX)	94	179,550
AS	JA2BAY	92	99,372
EU	OT3T(ON4WW)	93	240,096
NA	ZF2CF	95	366,096
OC	T32AF(KH6UR)	84	201,135
SA	9Y4AA(N6TJ)	86	364,008

S/O-3.5MHz

AF	CQ3B(CT3EE)	93	67,680
AS	JH1HGC	92	19,200
EU	CT2CB	84	145,530
NA	KK9A/PJ7	84	403,389
OC	T32AF	90	141,075
SA	4M3X	93	170,316

S/O-1.8MHz

AF	NoEntry		
AS	NoEntry		
EU	CT1AOZ	88	34,515
NA	4U1UN(K2GM)	88	191,352
OC	AH6BK	83	13,392
SA	YV2IF	88	61,308

Single Operator WITH Packet

AF	NoEntry		
AS	JJ3YBB(JI3ERV)	92	1,229,256
EU	EA3NY	92	2,449,842
NA	HI8A(JA5DQH)	91	3,983,175
OC	NoEntry		
SA	PYØFM(PY5CC)	93	4,644,252

Multi-Op SingleTransmitter

AF	EA8ZS	95	4,547,457
AS	JA8YBY	92	2,122,926
EU	TM5C	92	5,469,690
NA	VP2E	90	9,764,490
OC	KH6XX	82	4,834,392
SA	P4ØV	94	9,859,860

Multi-Op TwoTransmitter

AF	6V6U	95	5,923,584
AS	JA8YBY	90	2,283,147
EU	UZ2FWA	92	5,664,375
NA	6D2X	93	14,466,708
OC	K6GSS/KH6	90	4,481,583
SA	PJØB	85	6,936,393

Multi-Op Multi-Transmitter

AF	EA9IE	81	86,838
AS	JE2YRD	91	3,173,904
EU	I3MAU	91	6,092,406
NA	6D2X	94	14,675,778
OC	KH6XX	86	6,062,463
SA	LU4FM	94	3,402,648

U.S.A./Canada Phone Records

Area	Call	Year	Score

Single-Op All-Band (No Packet)

W1	KM1H(KQ2M)	92	4,932,396
W2	N2LT	93	3,238,497
W3	K3OO	92	3,868,368
W4	K4XS	92	3,914,295
W5	WM5G(KRØY)	91	4,628,835
W6	AI6V	83	1,998,975
W7	N7TT	90	1,718,556
W8	NA8V	90	3,332,220
W9	W9RE	92	3,600,030
WØ	KØRF(WØUA)	91	2,721,510
VE	VE3EJ(K5ZD)	92	4,496,832

S-Op. Low Power (No Packet)

W1	W1PH	92	1,596,456
W2	W2TZ	93	1,055,610
W3	KQ3V	92	1,115,400
W4	KM9P/4	90	751,308
W5	K5RX	92	1,351,920
W6	K6SIK	89	729,195
W7	K7RI	91	1,716,075
W8	N8II	93	1,375,998
W9	KS9B(KB9BIB)	93	795,108
WØ	WD5COV/Ø	91	830,532
VE	VE3RM	93	1,156,680

Single-Op QRP<5 W (No Packet)

W1	KA1VQ	82	457,305
W2	AA2U	92	716,496
W3	W3EWL	91	529,842
W4	N4KG	91	628,056
W5	K5RX	91	653,760
W6	N6OJ	88	135,000
W7	KB7VD	93	168,480
W8	KN8P	82	428,028
W9	N9AW	92	304,704
WØ	WØUO	90	322,599
VE	VE3KKB	81	162,060

S/O-28 MHz

W1	NX1H	92	816,660
W2	KE2C	82	355,740
W3	K3ZJ	90	726,516
W4	K4XS	89	806,577
W5	W5WMU(KE5FI)	89	587,250
W6	N6TU	91	529,758
W7	AI7B	91	562,590
W8	N8CXX	90	631,158
W9	K9MDO	91	516,186
WØ	WØUN(WØUA)	92	871,125
VE	VE6WQ	80	650,268

S/O-21 MHz

W1	K2SS/1	92	907,200

W2	N2WT	82	584,052
W3	K3LR	82	689,040
W4	K3RV/4	92	866,502
W5	W5WMU(KZ5D)	93	645,795
W6	K3EST/6	92	832,866
W7	W7WA	91	808,860
W8	K3ZJ/8	94	368,625
W9	AI9J	82	500,502
WØ	WØUN(WØUA)	93	879,396
VE	VE7IN	81	491,892

S/O-14 MHz

W1	KS1L	94	718,428
W2	K2VV	86	625,485
W3	K3ZJ	92	526,350
W4	K3KG/4	81	338,742
W5	K5MR	93	359,910
W6	W6CCP	94	304,722
W7	W7WA	87	451,572
W8	WB8JBM(N8DCJ)	87	490,776
W9	KK9A	92	657,951
WØ	WØZV	90	544,236
VE	VO1SA	87	643,560

S/O-7 MHz

W1	K1UO	93	147,882
W2	KM2P	90	85,779
W3	W3GH	93	117,030
W4	K4XS	90	129,774
W5	W5KFT	93	114,750
W6	K6NA	95	164,016
W7	KC7EM	94	178,308
W8	KW8N	94	82,476
W9	WB9Z	91	106,677
WØ	WØUN[KØGU]	94	149,202
VE	VE7WJ	80	55,266

S/O-3.5 MHz

W1	KO1F	95	69,312
W2	K1ZM/2	94	130,410
W3	KQ3V	95	117,117
W4	W4PZV(WA4SVO)	95	106,800
W5	WØMJ/5	85	78,039
W6	W6RJ	93	62,016
W7	N7AVK	95	25,440
W8	W9LT/8	95	83,664
W9	K9BGL	95	30,396
WØ	KMØJ	93	27,450
VE	VE3POS	93	16,362

S/O-1.8 MHz

W1	W1RR	85	11,376
W2	K1ZM/2	85	11,592
W3	AA1K/3	84	5,832
W4	WA4SVO	88	17,658
W5	K5UR	87	14,310
W6	N6LL(WA6CDR)	88	1,176
W7	K7IDX	87	3,000

W8	N8ATR	87	3,600
W9	WB9HAD[WB9Z]	87	12,126
WØ	WØZV	87	4,896
VE	VE1YX	84	12,696

S/O WITH Packet [Since 1990]

W1	K1AR	90	3,831,828
W2	W2GD	91	2,315,640
W3	K3WW	93	3,415,464
W4	N4RJ(KM9P)	93	4,534,068
W5	WM5G(KRØY)	90	3,902,892
W6	N3AHA/6	93	2,052,648
W7	NZ7E(KI3V)	92	4,050,954
W8	WB3KKX/8[WR3G]	91	2,678,775
W9	KK9V	91	2,147,769
WØ	KØSCM	91	1,091,664
VE	VE5RA	95	483,132

Multi-Op Single Transmitter

W1	K1RU	91	4,237,038
W2	N2RM	91	4,505,814
W3	W3BGN	91	4,768,248
W4	NQ4I	92	4,412,880
W5	K5GA	81	3,758,637
W6	N6RO	90	2,590,199
W7	K7SP	81	2,537,115
W8	K8AZ	93	4,130,820
W9	KS9K	93	3,779,787
WØ	KØRF	81	3,878,175
VE	VE7ON	91	2,487,024

Multi-Op TwoTransmitter

W1	K1AR	93	9,851,358
W2	N2NT	92	8,668,605
W3	N3RS	92	8,496,798
W4	W4QAW	84	4,332,076
W5	N5AU	89	5,354,538
W6	N6RO	89	4,696,314
W7	NI7T	92	3,695,760
W8	K8CC	92	4,346,424
W9	KS9K	92	6,914,916
WØ	AA6TT/Ø	92	7,665,075
VE	VG3GCB	92	708,867

Multi-Op Multi-Transmitter

W1	KY1H	91	9,657,606
W2	N2RM	93	13,024,950
W3	W3LPL	92	12,412,320
W4	W4MYA	92	7,327,704
W5	N5AU	92	9,329,688
W6	N6RO	82	5,250,015
W7	NK7U	93	4,522,425
W8	K8CC	91	6,780,972
W9	WØAIH/9	92	7,133,049
WØ	KØRF	79	4,033,155
VE	No Entry		
Club	Frankford Radio Club	1992	202,360,257

ARRL DX CW, USA WINNERS

Single-Operator 1928–1993

KM1H(KQ2M op)	5
W3LPL	4
K1AR	3
W3ECR(W3MFW)	3
W3LOE	3
W4KFC	3
K1DIR	2
K1GQ	2
W2GWE	2
W2UK	2
W2SAI(W3BES op)[W3GM]	2

Each of the following have achieved one win:
1ASF, W1BGD/2[W1RM], W1BPW[W1UU], K1NOL[K1NA], K1PR, K1ZM, K1ZND[K1ZZ], K2HLB, N2LT, W2PV(N2NT op), W3CHE, W3CTJ, W3DGM/3, W3GRF, W3GRF(KØDQ op), W3SI, W3ZD, W3ZJ, W4DHZ, W5ENE, WM5G(KRØY op), K5ZD, W6BAX, W6GRL, W7RM, W8BHW, W9UM, W9WNV/2.

Single Operator Low Power 1964–1993

VO1MP	5	1985–1989
WA1SSH[W1NG]	4	1974-1977
W1PH	3	
N5AW	3	
K1ZND[K1ZZ]	2	
N4HI	2	

Each of the following have achieved one win:
WA1ABW[K1BW], K1JX, K1LPL[W3LPL], K2MFY, K7GM/4, W4BRB, W4CRW, K5ABV, K7SV, N8II, W8QXQ[N8AA], KM9P, VE2NV.

Multi-Operator Single Transmitter 1933–1993

W3MSK[W3AU]	8	
W3BGN	7	4-IN-A-ROW(1989–1992)
WA5LES/K5RC	3	
K1ZND/K1ZZ	2	
W3AOH	2	
W3CTJ	2	
W4BVV	2	
N4WW	2	
W6GRL	2	
W6RW	2	

Each of the following have achieved one win:
WA1NRF[K1PR], AD1C, W1SZ, W1TX, W2IQG, W2SAI, W2YD, W3BES[W3GM], W3HIL, K3KG, K3LR, W3KT, N3RD, W3TMZ[W3OZ], K3WW, W4CBY, K4VX, W4YC, K6EBB, W6EOG, W6HLN, WA8JUN[K8MD].

Multi-Operator TWO Transmitter 1983–1993

K1AR	6
N3RS	2

Each of the following have achieved one win:
W2REH, K2TR, W3GM, NR5M, K5RC.

Multi-Operator Multi-Transmitter 1971–1993

W3LPL	9	8 IN-A-ROW! (1985–1991)
W3AU	5	IN-A-ROW! (1971–1975)
W4BVV	2	
N2AA	2	
K2UA	2	
W3WJD/N3RS	2	
N2RM	2	

Each of the following have achieved one win:
KN3O.

USA PHONE WINNERS ARRL DX BY CATEGORY

Single Operator

W1ATE	4
K1AR	2
K1VTM	2
W1ZM(K1ZM)	2
K2GXI	2
KQ2M	2
W2SAI	2
W3EMM	2
K3TUP(K3LR)	2
W4ESK[W7RM]	2
W6RR	2
W7RM[K7JA]	2
K1AR	2

Each of the following have achieved one win:
K1DIR, K1DG, K1KI, W1ONK, K1PR, K1RU, K1RX, W2AFQ, K2HLB, W2HMH[K2BU], W2SKE/2 W3DHM, WA3HGV, W3WJD[N3RS], W4DCQ, W4KFC, WA4PXP, WM5G(KRØY op), W6HX, W6OAT, W6OCH, AI6V(N6KT op), W8BKP, K8DOC, W9ARE.

Single-Operator Low Power 1964-1993

WA1SSH[W1NG]	4	1974-1977!
W1PH	4	
W2TZ	4	
WA1DJG[K1CC]	3	1968-1970!
W8ECA	2	
VE1NG	2	

Each of the following have achieved one win:
K1JX, W2HPF, WA2OOO,K2SG, WB2SJG[KQ2M], W4WRY, WA4PFN/2, WA5ALB, K7RI, W8LXU, KB9ET, WAØYAW, VO2CW.

Multi-Operator Single Transmitter 1937-1993

W3MSK[W3AU]	6
W3BGN	4
W2SAI	2
W3BES[W3GM]	2
K3TUP	2
W3ZKH//3[N4RV]	2
K4VX	2
W5TMN/K5JA	2

Each of the following have acheived one win:
W1ETF, W1FBY[W1XT], AA1K, WA1KID[W6XR], WA1NRV, K1OX, W1QDE,K1RX, W1ZA, K2BU, W3AOH, W3DHM, W3ECR, W3LVF, NQ4I, W4FDA, W4YC, K5NA, W5MLX, W5RER, W6AM, W6GRL, W6ONV, K8AZ, W8NWO, W9PV, KØRF, WØUQV.

Multi-Operator TWO Transmitter 1983-1993

K2TR	5
K1AR	2

Each of the following have one win:
AD1C, KC1F, N2NT, W4QAW, KX4S, K5RX.

Multi-Operator Multi-Transmitter 1971-1993

W3LPL	9
W2PV	5
W3AU	4
N2AA	2

Each of the following have one win:
N2RM, KN3O, KX4S, W7RM.

A BRIEF HISTORY OF THE WAE DX CONTEST

In the 1950s the Deutscher Amateur Radio Club (D.A.R.C.) created the "Worked All Europe" awards program. To make it known and achievable for stations outside Europe, the "Worked All Europe DX Contest" (WAEDC) was started in 1954. It is believed that this contest was an invention of well-known DXer Rudi Hammer, DL7AA, who became a silent key in 1993, at the age of 82. Other call signs of those involved in the WAEDC committee work in the early years were DL7CW, DL7EN, DL6EN, DJ2BW and DL9OH, all of whom are still active DXers. To this day, the whole WAE DX Contest administration is done by volunteers.

The first contests were held in the spring of each year and used a 48-hour contest format for the CW and phone weekends. However, the change of the contest date to the second weekend in August for CW and second weekend in September for phone happened back in the early years and has been held at those times ever since.

Multipliers play an important role in any contest, as they also do in the WAEDC. While worldwide contests use the "everybody works everybody" format, the CW and SSB activity in this contest is "the world working Europe," which surely is challenging for the operator. To keep the contest interesting, this meant to create more multipliers both rewarding EU and Non-EU.

The solution was to divide large countries outside of Europe into separate multipliers by call areas. So the U.S. call areas, Canadian provinces, as well as the different JA, LU, PY, VK, ZL and ZS prefixes became multipliers. This multiplier system was kept for more than 20 years. Then, stations in the U.S. started to keep their call signs when they moved from one call area to another. Thus, a W6 could be a station in California or in Florida, and simply counting the call areas worked was no longer possible.

To give the U.S. stations still the attention they deserve as an important contribution to the WAEDC, it was decided to try using the U.S. states as multipliers and this was added to the exchange. Not many years later, an analysis of WAEDC-multipliers revealed, that many EU stations had begun to look for U.S. multipliers ONLY and there was hardly a counterpoise by other multipliers and the European DX Contest was about to change into a "European-USA" Contest. The contest committee felt that it was necessary to promote the participation of ALL DXCC countries and led over to the current multiplier system which seems to be widely accepted.

In the mid-80s an IARU-Region-1-recommendation suggested that contests should not exceed 24 hours. Although it was expected that the Deutscher Amateur Radio Club as the society with highest number of members in the Region 1 would set an example, it was unlikely that long time WAEDC contesters would have accepted a sudden change to a one-day-contest. So, in 1987, the 36-hour format and the High Band Class were announced. The High Band class was believed to be perfectly suited to those DXers who cannot work on 40m and/or 80m. But contesters have their own philosophy and soon it became evident, that many all-band contesters had restricted their activities to the high bands only. The low bands, always difficult in a summer time DX contest, became even more "contest free." It was surprising to see how high band scores began to overrun all-band scores. This led to the cancellation of the High Band Class after only three years, in 1989. Few contests depend on the sunspot cycle as much as the European DX Contest does. In years of low sunspot activity, 10m, and sometimes also 15m, are barely usable and activity is therefore limited. Additionally, multi-operator stations contribute a lot for a contest, but preparation and participation by these stations is not encouraged when conditions are expected to be bad during the 36-hour-contest. This is believed to be a cause for the reduced number of multi-operator station entries. So, for 1994 we are returning back again to the 48-hour-format, from which we hope to see more participation by multi-operator stations. As before, the single operators are allowed to work only 36 hours out of the 48 hour contest time.

By Herbert Ade-Thurow, DL2DN

CONTINENTAL WINNERS OF WAE CW BY YEAR/CATEGORY

YEAR	SINGLE OPERATOR						MULTI-OPERATOR SINGLE TX						MULTI MULTI				
	EU	AF	AS	NA	OC	SA	EU	AF	AS	NA	SA	EU	AF	AS	NA	OC	SA
1994	DK3GI	EA9LZ(YU1RL)	C47A	ZF2NE	VK2APK	L3CW (LU6BEG)	RU1A		C4C	KC1XX	PY2BW	R6L	UZØSXF				
1993	LY5R	EA9LZ	UH8EA	K2TW	ZL3GQ	YV1OB	RU1A		U29CWW	KC1XX	PY2OOM	E26L					
1992	LY5R	EA8AB	P30ADA	V21AS(YU1RL)	ZL3GQ	PP7JR	UX1A		U29CWW	N3RS		LY2ZO					
1991	UT4UZ	EA8AB	5B4ADA	K4XS	VK2DX/9M8	LU1ICX	LY2WW		U29CWA	K4XS							
1990	LY2BTA	5H3TW	UA9SA	N6AR	YB2FRR	PX5A	LY2WW	CN5A	UL8LYA	N5RZ							
1989	Y24UK	EA8AB	H27T(YU1RL)	K1EA	ZL3GQ	LU1EWL	LZ9A	CN2DX	U29FYR	N3BNA	ZW5B						
1988	Y24UK	EA5BS/EA8	P37T(YU1RL)	K1EA	ZL3GQ	OA42V	UP1BZZ	EA8RCT	U29WWH	W3GG	CE3AYZ						
1987	Y24UK	5L2GA	5B4OA	K1EA	VK2APK	LU8DQ	UP1BZZ	ZS1CT	U29WWH	K1IU	PY5ZBA			**MM**			
1986	Y24UK	EA8ABR	5B4TI	K1ZM	VK2APK	LU8DQ	DL6FBL	ZS1CT	U29WWH	K1AR	CX5AO						
1985	YZ9A		UH8EA	K1ZM	VK2APK	LU1EWL	DL6FBL	EA9CE	RF0FWW	K1AR							
1984	YU3EY	524MX	UA9SA	W2VV(KQ2M)	VK2APK	LU8DQ	HG5A	VQ9CI	U29CWW	KO3S/8	CX8BBH						
1983	YU3EY	ZS1CT	4X6FR	KC1F	VK2APK	LU9EIE	UK2PCR		UK9FER	K1AR	PY2BW						
1982	DK3GI	TU2IE	UF6CR	W1ZM(K1ZM)	VK2APK	LU8DQ	UK2BAS	TU2FZ	UK9AAN	VE2HQ	PY2BW						
1981	Y24UK	DL2VK/ST3	UM8MAO	K1GQ	VK2APK	LU8DQ	UK2BBB		UK9FER	W1IHN							
1980	Y24UK	EA8EY	UA9TS	K1VTM(K1JX)	VK3AEW	PY1BOA	UK2BBB		UK7LAH	K3EST							
1979	UB5JGR	C5AJ	4X4VE	K1PR	5W1BZ	PY1BOA	UK2BBB	EA9EO	UK9CAE	N1NA							
1978	DT2TUK	524ZW	UA9ACN	K1GQ	KH6IJ	LU8DQ	UK2PCR		UK6FAA	CG3BMV							
1977	UB5JIM	CT3BQ	UF6VAZ	N3RD	ZL3GQ	PY2BW	UK2BBB		UK6CAE	CY3IXE							
1976	YU3EY	D2AAI	UV9AX	W3LPL[K3RZ]	ZL3GQ	PJ2VD	UK2BAS		UK9CAE	WA1ABV/1							
1975	G3FXB	C9MIZ	4X4VE	WA1KID(K1ZZ)	KG6JAR	CX9BT	UK2BAS		UK9AAN	K3EST							
1974	DJ8SW	CR7IZ	UW9WL	K4VX(K4PQL)	9M2CX	LU1DZ	UK3AAO		UK9AAN	W3AU	LU1DZ						
1973	DJ8SW	ZS3AK	UA9ACN	WA1ABV[K1RX] K1JHX[K1RM]	KH6RS[K1GQ]	LU5HFI(K3ZO)	**MS**				CX3BH						
1972	YU3EY		4X4VE	W1BPW[W1UU]													
1971	DJ8SW	TJ1AW	UV9CU	W3AU													
1970	DL7AV	TJ1AW	UA9WS	W1BPW[W1UU]		PJ2VD											
1969	DJ8SW	CR6GO	UA9WS	W1BPW[W1UU]													
1968	DJ8SW	CR6GO	EP2BQ	W1BPW[W1UU]													
1967	SM2BJI	CR6CK	UA9WS	WB2CKS[K2LE]		CP5EZ											
1966	SM2BJI	CR6GO	4X4HF	W0GTA													
1965	DJ3KR	CR6AI	UA9DN	W3GRF													
1964	DJ3KR	CN8GB	EP2RC	W2JAE[W5VQ]		HC1DC											
1963	DJ3KR	6O1ND	UA9DN	W2JAE[W5VQ]		HC1DC											
1962	DJ3KR	ZS6IW	UA9DN	WA2WBH													
1961	DJ3KR	5A3TQ	EP2BK	W3GRF													
1960	DJ3KR		UA3DN	K2DGT													
1959	DL1KB	CR6AI	FA9UO	K2DGT		PY1ADA											
1958	DL1KB	CR6AI	4X4KK	W2EQS													
1957	DL1DX			W3GRF		CE3AG											
1956	DL1DX	CR6AI															
1955	OK1FF		4X4FS	W2WZ													

CONTINENTAL WINNERS OF WAE SSB BY YEAR/CATEGORY

| | SINGLE OPERATOR | | | | | | MULTI-OPERATOR SINGLE TX | | | | | | MULTI MULTI | | | | |
YEAR	EU	AF	AS	NA	OC	SA	EU	AF	AS	NA	OC	SA	EU	AS	NA	OC	SA
1994	IR2W (I2VXJ)	5U7Y	C47A	K1ZM	VK5GN	LU4AA (LU6BEG)	DFØHQ	ED9TQ	UN8PYL	N1AU		ZW5B	DLØDK	JRØXJO		DX1DBT	
1993	S52AA	EA1AK/EA8	RHØE	AK1A	YB6AVE	PRØR	DFØHQ	EA9LZ	P39C	KC1XX		ZW5B	DLØDK			DX1DBT	
1992	OE6MBG	5U7M	RØHE	K2TW	YB2FRR	ZW9A	Y34K		U29XWH	N1AU		YW1A	LY2ZO	JE2YRD		DX1DBF	LU1CF
1991	Y33UL	FR5DX	UA9CDT	K4XS	VK2APK	PX5A	LZ9A		5B4ES	N5KEA		ZW5B		JE2YRD	W3FV		
1990	YT9ØA	EA8/DL6FBL	RHØE	KM3T	YB2FRR	HC1OT	R6L		5B4ES	K4XS		AY6D	UP1BZZ	UL8LYA			LU1CF
1989	OK1RI	EA9IE	UA9SN	K1ZM		PY5EG	Y34K		5B4ES	N5KEA		YV1AVO *MM*					
1988	YT3T	EA9IE	UL7OB	K4YKZ	VK2FOC	LU3F	UP1BZZ		UL8LYA	NAØK		YV1AVO					
1987	YU3EO	EA9RM	UL7OB	KM1H	YB4FW	ZY5EG	Y34K		UL8LYA	8P6AW		YV1A					
1986	DK9IP	EA9IE	UW9CO	W1ZM(K1ZM)	YB4FW	LU6ETB	LZ2KTS		RL8PYL	WB3JRU		YV1TO					
1985	YZ9A	EA9LZ	UH8EA	W1ZM(K1ZM)	VK3EZ	ZZ5EG	LZ2KTS	ZS6TUK	RL8PYL	N1AU		YV1AVO					
1984	YZ3EK	EA9KF	UF6CR	W2YV	VK6AJW	PY5EG	LZ40KTS	ED9EA	RF0FWW	K6OKW							
1983	PA2TMS	CT3BM	UA9CKC	KC1F	VK6AJW	ZY5EG	Y34K	EA9IE	N2BZQ/4X	K9MWM/Ø	VK3FY						
1982	YU3MY	5N8ARY	UF6CR	K1KI	VK6AJW	ZY5EG	Y24UK		UK9AAN	N5AU		PY2BW					
1981	YU3EY	9G1JX	UF6DZ	W1ZM(K1ZM)	VK6AJW	ZY5EG	Y24UK		UK9AAN	K1AR		PY5EG					
1980	I4VEQ	EA8LS	4X4VE	W2YV(N2NT)	ZL2ACP	ZZ5EG	UK2BBB	3V8ONU	UK7PAL	KØUK		PY1BQK					
1979	DM2DUK	EA8LS	UL7MAR	VP2ML	VK6CT	DL2RL/YV6	UK2BBB		UK9CAE	W1RR		YX1DIG					
1978	G3FXB	9G1JW	UL7EAJ	W1ZM(K1ZM)	VK2APK	DL2RL/YV6	YU1BCD		UK6FAF	K2SS		YV1TO					
1977	DK2BI	9G1JX	4Z4MQ	W1ZM(K1ZM)	YBØACH	PJ2FR	YU1BCD		UK9ABA	WA4PYF		YV1AVO					
1976	DK1FW	9G1JX	4Z4MQ	W1ZM(K1ZM)	VK6CT	LU8AJG	YU1BCD	9GØARS	UK9AAN	K2IGW		PY1ZBJ					
1975	DK1FW	5L7F	4X4NJ	W1ZM(K1ZM)	VK4VU	YV4YC	G3WYX		UK9CAE	WA2MBM		PY1EMM					
1974	DK1FW	EA8CR	UW9AF	WB2OEU[K2TR]	VK4AK	PY4KL	YU1BCD	CR6OZ	4Z4HF	W9YT	9M2CJ	PY1EMM					
1973	DK4TP	EA8CR	UA9BE	WA2BVU	DU1FH	CE8AO	"MS*"										
1972	DJ4LK	IH9JT	UW9WR	WA2BVU													
1971	DJ4LK	9E3USA	JY8BI	WB2SQN[K2SS]	DU1FH	CX9CO											
1970	OH2BH	CR6GA	UW9AF	W3GM													
1969	ON8CT	CR6GO	EP2BQ			PY3BXW											
1968	DJ2YA	CR6FY	OD5BZ		VK2DKZ	PY1CAD											
1967	DJ6QT	7XØAH		W3GRF													
1966	DJ6QT	5A2TF	OD5BZ	WØGTA													
1965	DJ6QT	EA8CR	VS1LP	OX3JV													
1964	DJ6QT	EA8CR	VS9AAA	W3WJD[N3RS]		OA4KY											
1963	F9RY	5N2JKO	EP3RO	OX3JV		PZ1CE											
1962	DJ3KR	9G1YL	OD5BZ	OX3AI	VK2AOM	PY6MP											
1958	DL1KB	VQ4RF		TI2OE		CE3HL											
1957	DL1KB																
1956	DL1KB																
1955	OE5CK	VQ4RF		W2SKE	KC6CG	PY2CK											

WAE DX CONTEST

Continental CW Records

Continent	Call	Year	Score
Single-Op All-Band (With Packet)			
AF	EA9LZ(YU1RL)	94	2,342,550
AS	P37T (YU1RL)	88	2,142,476
EU	DK3GI	82	1,795,986
NA	V21AS	92	2,164,734
OC	ZL3GQ	89	595,907
SA	LU1SM (LU8DQ)	90	1,299,060
Multi-Op Single Transmitter			
AF	EA9CE	85	1,526,742
AS	C4C	94	2,108,204
EU	UK2BAS [LY2ZZ]	82	2,886,296
NA	KC1XX	93	1,808,352
OC	No Entry		
SA	ZW5B	90	1,102,469
Multi-Op Multi-Transmitter			
AF	No Entry		
AS	JE2YRD	89	1,045,384
EU	LY2ZO	91	1,353,604
NA	No Entry		
OC	No Entry		
SA	No Entry		

NOTE: As of 1991 ALL classes may use PacketClusters.

Continental Phone Records

Continent	Call	Year	Score
Single-Op All-Band (With Packet)			
AF	EA9IE	86	1,977,872
AS	C47A	94	2,725,800
EU	YU3MY	82	2,742,961
NA	KM3T	90	1,711,668
OC	YB2FRR	91	347,095
SA	PY5EG	89	1,371,111
Multi-Op Single Transmitter			
AF	ED9TQ	94	2,088,768
AS	RFØFWW	84	2,394,738
EU	Y24UK	82	5,094,720
NA	KC1XX	93	1,582,420
OC	VK3FY	83	59,112
SA	ZW5B	90	1,885,680
Multi-Op Multi-Transmitter			
AF	No Entry		
AS	JE2YRD	90	1,304,807
EU	LZ9A	90	2,524,158
NA	W3FV	91	167,668
OC	DX1DBF	91	87,016
SA	LU1CF	89	96,640

NOTE: As of 1991 all classes may use PacketClusters.

WAE DX CONTEST

U.S.A. Records

Call Area	Call	Year	Score
Single-Op All-Band (No Packet)			
W1	K1GQ	81	1,128,330
W2	W2YV (KQ2M)	84	936,240
W3	K3WW	90	732,096
W4	N6AR/4	90	958,848
W5	K5RC (K5ZD)	81	549,360
W6	N6AR	82	305,492
W7	W7IR	79	272,475
W8	WA8YVR	83	661,912
W9	W9RE	90	486,864
WØ	WØSF	79	229,840
VE	VE1FH	82	422,870
Multi-Op Single Transmitter			
W1	KC1XX	93	1,808,352
W2	NQ2D	90	1,103,925
W3	N3RS	93	1,471,058
W4	K4XS	90	1,151,871
W5	N5RZ	89	529,074
W6	N6AV	79	171,990
W7	WA7FAB	90	82,536
W8	W8LU	83	419,680
W9	No Entry		
WØ	NCØP	90	104,520
VE	VE2HQ	82	941,951
Multi-Op Multi-Transmitter			
W1	No Entry		
W2	No Entry		
W3	No Entry		
W4	No Entry		
W5	No Entry		
W6	No Entry		
W7	No Entry		
W8	No Entry		
W9	No Entry		
WØ	No Entry		
VE	No Entry		

NOTE: As of 1991 ALL classes may use PacketClusters.

U.S.A. Records

Section	Call	Year	Score
Single-Op All-Band (No Packet)			
W1	KM3T	90	1,711,668
W2	W2YV (N2NT)	80	1,404,480
W3	K3WW	93	474,249
W4	K4XS	91	1,336,929
W5	K5KLA	81	215,912
W6	K6SVL	90	305,500
W7	WA7JYW	81	43,884
W8	AB8K	80	280,388
W9	KBØC/9	90	381,260
WØ	WAØTKJ	79	51,246
VE	VE2DU	79	458,118
Multi-Op Single Transmitter			
W1	KC1XX	93	1,582,420
W2	K1ZM	90	1,025,398
W3	N3BNA	93	345,912
W4	N4ZC	90	1,348,925
W5	N5KEA	89	450,016
W6	K6OKW	81	140,592
W7	No Entry		
W8	W8CCI	91	125,020
W9	W9UP	91	326,598
WØ	KØUK	80	161,684
VE	No Entry		
Multi-Op Multi-Transmitter			
W1	No Entry		
W2	No Entry		
W3	W3FV	91	167,668
W4	No Entry		
W5	No Entry		
W6	No Entry		
W7	No Entry		
W8	No Entry		
W9	No Entry		
W0	No Entry		
VE	No Entry		

NOTE: As of 1991 ALL classes may use PacketClusters.

A BRIEF HISTORY OF THE CQ WPX CONTEST

The CQ WPX Contests are held the last full weekend of March (SSB) and May (CW) each year. Prefix competition is the name of the game with the goal being to score the highest based on the number of different prefixes worked times the number of QSO points. Prefix multipliers (N8, W8, ZA1, etc.) provide an almost unlimited number of available multipliers, with the number increasing each year.

The roots of the WPX contest go back to 1957 and the SSB DX Contest which was organized by then CQ SSB Column Editor Bob Adams, W3SW. In 1961, Irv and Dorothy Strauber, K2HEA and K2MGE, took over the SSB column and continued the contest. In 1964, the SSB column was phased out and the CQ Contest Committee picked up responsibility for the contest.

The year 1968 marked the official birth of the CQ WPX SSB Contest. Frank Anzalone, W1WY, and his committee ran the contest until 1970 when Bernie Welch, W8IMZ, became the first full time WPX Contest Director. Bernie held the job as Director from 1970 until his well-deserved retirement after the 1982 WPX Contests. During Bernie's tenure, the contest grew into the world's second largest DX contest and a CW WPX Contest was added to the schedule in 1979. Steve Bolia,

N8BJQ, has served as WPX Contest Director since 1983.

Several rule changes have been made over the years. In 1959, the call sign prefix was declared the multiplier. Several years later, double QSO points were added for contacts on 40, 80 and 160 to boost low band activity, especially on SSB. Several times in past years, contest winners have been those who took advantage of the low bands and the extra points. The WPX has mandatory off time (12 hours) which adds a bit of strategy to contest planning. Picking the wrong time to sleep can be devastating. There are categories for Multi Operator, QRP/p, Low Power, USA Novice/Technician and for single-band entries.

Each year the WPX contests have continued to grow. High QSO rates, coupled with lots of DX help to make it a fun contest for everyone. Licensing authorities in many countries have cooperated by issuing unique contest call signs for the contests. Contesters throughout the world strive to obtain a new contest call for the WPX contests each year. The fact that an N8 counts the same as a ZA1 in the multiplier column makes the contest fun and interesting for both big guns and little pistols.

(Contributed by Steve Bolia, N8BJQ - CQ WPX Contest Director.)

DX WINNERS OF CQ WW WPX PHONE CONTEST BY YEAR/CATEGORY

YEAR	S/O	QRP	M/S	M/M	160	80	40	20	15	10
1994	ZD8Z(N6TJ)	HC8A(N6KT)	ZX0F	VP2EC	IO3MAU	EA8/OH1MA	CT3BX	EA8AH(OH1RY)	ZP0Y(ZP5JCY)	L6ETB(LU6ETB)
1993	P40V(AI6V)	KR2Q	HC8A	LU4FM	OM3CQD	XK7CC	IO4VEQ	ED9LZ	KG6DX	ZP0Y(ZP5JCY)
1992	HC8A(N6KT)	HI50A(JA5DQH)	VP2EC	CT3M	EI7M	CT7N(CT4NH)	EA9LZ	ZV5A(PY5EG)	ZZ9A(PY5CC)	ZW5B(N5FA)
1991	ZW5B(N5FA)	VP2E	P40V	ED8ACH	UL7ACI	VA3EJ(VE3EJ)	YV5A(YV5ANF)	H2A(5B4SA)	ZX5C(PY5CC)	ZP50Y(ZP5JCY)
1990	P40V(AI6V)	VP2EXX	ZX5C	YT2A	UL7ACI	CF6OU/3	IO4VEQ	ID1V(I1ZEU)	ZP0Y(ZP5JCY)	ZX5C(PY5CC)
1989	KP2A(KW8N)	4X6IF	TX0A	LZ9A	VP9AD	TI2CC	TE1L(TI2LTA)	CQ1BOP	FG5R(W7EJ)	ZP0Y(ZP5JCY)
1988	P40V(AI6V)	4X6IF	PJ2FR	FK0AW	CT1AOZ	VE22P	HA9RE	ZZ5EG(PY5ZBA)	ZP5Y(ZP5JCY)	ZY5EG
1987	EA9AM(EA9IE)	TR8SA	NP4CC	KH6XX	OH1RY/CT3	CT3DL	H24LP(5B4LP)	ZP5JCY	CE6EZ	LU1E(LU3AJW)
1986	PJ2FR(N6KT)	K7SS/WH6	KD7P/NH4	ZZ5EG	LZ2CJ	VE3BMV	NP4A	VP2EC	ZZ5EG(N5FA)	XQ3D(CE3DPD)
1985	EA9IE	4X6IF	ZS5EG	KH6XX	CG3MFA	OH1RY/CT3	C23BMV	FO8JP	LU2FDR	LU1E(LU3AJW)
1984	VK2WU	H44R	VP2EC	KH6XX	EA9KF	4M3AZC	T32AF	ZY5EG	CX7BY	CE6EZ
1983	PJ2FR(N6KT)	G4GIR	VP2EC	VP2EC	LZ2RF	IO3MAU	FM7CD	KG6DX	OH0BH(OH2BH)	CE6EZ
1982	Y24UK	W8ILC	VP5FRS	YZ1EXY	VE3BBN	YV3BQS	DJ4PT	4N3ZV	AI7B	KB7U/KH2
1981	NP4A	N2AA	9A1ONU	NP4A	VE3MFT	DJ4PT	4M3AZC	YX2AMM	HC9A(HC1MD)	ZZ5EG
1980	PJ2CC(N4RV)	TG9GI	UK9AAN	KH6XX	VE3JAY	4M3AZC	I5NPH	VR3AH	HD0E(K7CA)	LU8DQ(LU6EF)
1979	OI1VR	W8ILC	KP4RF	VE7WJ	VE3BBN	HA9RU	I3MAU	ON4UN	9L1CA	KH6XX
1978	UA9ACN	4T8V	4J9B	CK7WJ	VE3BBN	YU3DBC	CG3IXE	OI2BA	CG3BMV	CW3BR
1977	UA9BE	-#3-	UK9AAN	UK9AAN	VE3BBN	I3MAU	CT4AT(K7ZZ)	YV4AGP	PY5EG	N6EE
1976	VP2G(W5MYA)		UK9ADT	CJ3DCB	XJ3FFA	YY4YC	W4BRB/C6A	PJ0JR	YV2AMM	CE6EZ
1975	UW9AF		PJ9JR	4J3A	DL8PC	I3MAU	YV4YC	YU2CDS	PT2ZBS	K4HWW
1974	CQ6LF		PY7BDX	VP5B	———	YU4EBL	OH2KI	CQ6LF	CR6OZ	CQ6OR
1973	9Y4VU		W4IZ/KP4	CV2T	———	YU3APR	DJ2YA	FL0QQ(F5QQ)	XX7IK	CR6OZ
1972	XE1IJ (K0DG)		G3WVX	WA3HRV	———	OH0AM(OH3XZ)	K6ERT	PJ9JR(W3ZKH)	PY2DSE	CR6LF
1971	PX3BXW		UK9AAN	4X4GV		4X4DK	WB6KBK	KV4FZ	W3AU(K3EST)	LU2DEK
1970	PJ9JR(W3ZKH)		PR2CQ	4X4NJ		YV5BTS	YV1BI	PU7APS	YV1SA	KG6AQY
1969	KV4FZ		KW6EJ	4Z4HF		LA5KG	G3NLY	JA1AEA	VP2MF	PR2AHJ
1968	ET2FMA		I9RB	CE6CA		YV5BTS	DJ4AX	JA1AEA	W4BVV	LU1DAB
1967	ZL1KG		4U1ITU	PJ5MM		YV5BTS	G3NLY	ET3AC	YV1LA	LU1DAB
1966	1M4A(W9WNV)		I1RB	-#MM-		ON4UN	DJ5BV	SM5BLA	YV1LA	LU1DAB
1965	DL3LL		9A1ZG			GI3CDF	G3NLY	HC2JT	4X4TP	HC1EW
1964	DL3LL		GB3RAF			GI3CDF	OH2TH	HL9KH(W9WNV)	W4RLS	WA4SUR
1963	DL3LL					4X4DK	K2GXI		KP4BMA	———
1962	4X4DK									
1961	ZS5JY									
1960	CN8JF									
1959	HB9IE									
1958	CN8MM									
1957	CN8MM									

Footnote#1-Changed the maximum allowable operating time for Single Operators from 30 Hours(18 hours off) to 36 Hours (12 hours off) beginning with the 1992 event.

Footnote#2-Added the Single Operator Low Power category beginning with the 1992 event.

Footnote#3-Added a Club Competition in 1978(Phone Only), this would encompass both the Phone and C.W. events in 1979. The 1978 Club Competition Winner was The Western Washington DX Club. Added the S/O QRPp Class in 1978.

Footnote#MM-Added the Multi-Multi Category beginning with the 1968 event.

USA WINNERS OF CQ WW WPX PHONE CONTEST BY YEAR/CATEGORY

YEAR	S/O	QRP	M/S	MM	160	80	40	20	15	10
1994	WM2H	WA1LNP	NX1H	WZ1R	K12M/2	WE3C	KC7EM	KK9A	KC1XX	KE5FI
1993	KM1H[KQ2M]	KR2Q	NZ5I	WZ1R	AA4MM	WE3C	WB2ULI/5	WD8LLD	KA2AEV	NU4Y
1992	KM1H[KQ2M]	KR2Q	WC4E	WZ6Z	WT3Q	K1ZM	KC7EM	KK9A	WN4KKN/6	KO4QW
1991	KM1H[KQ2M]	AA2U	WC4E	WZ6Z	AA4MM	KQ3V	KV0Q	K2VV	WN4KKN/6	NX1H
1990	KQ2M	KR2Q	WC4E	WZ6Z	W2GD	KE5FI	WB9Z	KK9A	WB5VZL	WN4KKN/5
1989	KM1H[KQ2M]	N1AFC	WC4E	WZ6Z	WA2IUO	K1ZM	WQ5W	K2VV	WB5VZL	WM5G
1988	KM1H[KQ2M]	WC7Q	WC4E	KC3EK	K5UR	NO2I	KV0Q	N8BJQ	K6SVL	KY5N
1987	KM1H[KQ2M]	N8CQA	KI6P	N5AU	K5UR	KN8R	KM6B	KM6B	W3BGN	K5MK
1986	N4WW	WB9HRO	KR0Y	N5AU	K5UR	KR2J	K5NW	K2VV	W3BGN	KR9G
1985	KI6P	WD4NBX	N5AU	NF5I	K5UR	KQ2M	NI6W	KQ2M	NA5R	N4EJV
1984	K2VV	KH6CP/3	N5AU	NG5X	WB3GCG	KQ2M	KI6P[WA6VEF]	KG1E	NA5R	KN5H
1983	K2VV	W8ILC	KJ9W	NA8V	W8LRL	N7DF/0	KR2Q	WA0TKJ	KC2X	A5R[K5GN]
1982	AB0I[WA6DGX]	W8ILC	K7RI	KN6M	W8LRL	KI6P[N6RO]	N6RO	AI7B	AI7B	N5AU[K5ZD]
1981	N6CW	N2AA	N4WW	AI6V	W8LRL	KO6G[N6NE]	KM5X	K8NA	AI7B	N42C[N5TR]
1980	K1AR	N2AA	K4VX	K3WW	W8LRL	WB2FZO[KR2N]	N4KE	K8NA	WA6EKL	N2RM[N2ME]
1979	K7RI	W8ILC	K4VX	K5JA	W8LRL	K5UR	N4KE	N4KE	N7XX	K6SVL
1978	K2SS	W8ILC	K3RV	N7XX	W8LRL	K8KX	W6AXX/3	W3GG	N6CW	K5JA
1977	W3LPL	#-3-	AC6YRA	A44IVL	W8LRL	N6VI	WA8LXJ	K8YZW	N6CW	N6EE
1976	AA9BWY[W9RE]		WA6GLD	K6BCE/6	WB8APH	W1CF[K1OME]	WSTMN[K5JA]	AC3CRE	AC4WSF[N4MM]	AB2CST
1975	W2PV		W1YK	W5NOP/5	WB8APH	AC1CF[W1CF]	AC3USS[WA1FEO]	AC3CRE	WA1NZT	K4HWW
1974	W2PV		WA3HRV	WA3HRV	WA4PXP	W4BVV	K6JAN	K4VX	K6SDR[N6CW]	K6SVL
1973	W2PV		W6HX	WB6GFJ/6		W7YTN[K7UR]	W6PAA[N6RO]	W3SS	W6BEJV	WA7PEZ
1972	W5OQQ/7		W6HX	W3SS		K9CUY	K6JAN	W1OKA	W6GFS	W3JSX[W3FV]
1971	W2SKE		W2SKE			K5PFL[K5NA]	K6ERT	K7HTZ[K7CW]	W3AU[K3EST]	K5MDX
1970	W5RER		W7SFA			K9CUY	W6BK[N6MG]	WB2SQN[K2SS]	W3AZD	K8HYC
1969	W3MSK[W3AU]		K6SEN/6			WA3KEG	K6AHV[W6RJ]	W4AXE	W3WJD[N3RS]	WA3JUS
1968	W3JNN		WB2FOV			WA2WMT[K0KE]	W8IHD	K8YBU[K8RK]	W4BVV	K4YYL
1967	WA2SFP[W2PV]					WA5ALB	W5ODJ	K2GXI	W3AZD	WA4WIP
1966	WA2SFP[W2PV]					K8DOC	K2GXI	WA4PXP	W2SKE/2	WA4SUR
1965	K2HLB					W2ZPO	WA4PXP	W3JNN	W5LGG	WA4SUR
1964	W6AM						K6AHV	W3JNN	W4RLS	
1963	W2VCZ						K6AHV[W6RJ]	K2HWL	W1AOL	
1962	W1ONK						K2GXI	W1ZFV		
1961	W2VCZ									
1960	W2SKE/2		W3MSK[W3AU]							
1959	W2SKE/2									
1958	W9EWC					W5AJY				
1957	UNKNOWN									

Footnote#1—Changed the maximum allowable operating time for Single Operators from 30 Hours(18 hours off) to 36 Hours (12 hours off) beginning with the 1992 event.

Footnote#2—Added the Single Operator Low Power category beginning with the 1992 event.

Footnote#3—Added a Club Competition in 1978(Phone Only), this would encompass both the Phone and C.W. events in 1979. The 1978 Club Competition Winner was The Western Washington DX Club. Added the S/O QRPp Class in 1978.

DX WINNERS CQWW WPX CW CONTEST BY YEAR/CATEGORY

YEAR-S/O	QRP	M/S	M/M	160	80	40	20	15	10	Club: World/USA
1994 P40W(W2GD)	UX8IX	ZX0F	9A1A	LY3BU	OK1DXS	AZ4F	UN2L	L50D	LP4F	NCCC/YCCC
1993 ZX0F(YU1RL)	W2GD	P44V	HG73DX	T99C	YT0T(YU1RA)	C47W(5B4WN)	YW1A(YV1DIG)	S58A(S59AM)	ZD8LII	NCCC/YCCC
1992 ZV5A(YU1RL)	W2GD	LU8DPM	HG73DX	4N2X(4N2LH)	GW8GT	AM9TY	LZ5W(LZ3ZZ)	TU45R(OH8SR)	ZS6BCR(ZS6EZ)	NCCC—
1991 4M2BYT(YU1RL)	VP2MU(KJ4VH)	4J0Q	HG73DX	YL2GVW	4N1A(YU1EA)	LZ5W(LZ3ZZ)	YW1A(YV1DIS)	4N3E(YT3AM)	YV3A(YV5ANT)	ARAUCARIA/NCCC
1990 ZW5B(OH2MM)	N3RS	YM3KA	HG73DX	OK5TOP	4N1A(YU1EA)	PY2RN	L25A(LZ1AX)	FS5T(AI7B)	CE3DNP	NCCC/NTCC
1989 V27T(YU1RL)	YU3BC	KP2A	Y34K	OK1DFP	YX3A(YV5ANT)	FS5R(W7EJ)	L25A(LZ1AX)	9Y3VU	ZY5ZBA(PY5CW)	NCCC/PVRC
1988 NH6J/NH0	4X1IF	RL1P	UP4A	UA2FF	I4AVQ	IO4HI	YY5A(YV5ANT)	ZY5ZBA(PY5CW)	4M7A(YV7QP)	NTCC/NCCC
1987 5L7T(YU1RL)	4X1IF	V31A	UP7A	UA2FF	UA9TS	YX5A(YV5ANT)	WC4E/KP4	YW7A(YV7QP)	YW7A(YV7QP)	NCCC/NCCC
1986 YU1AO	WP4F	L27A	IO3JSS	UB4FWC	CT5AT(K7ZZ)	VP2VCW(N6CW)	YX5A(YV5ANT)	ZY4OD	YO3KWJ	NCCC/NTCC
1985 YZ4GD	4X6IF	LZ2KTS	KH6XX	UP3BP/UF	UP2NK/UF	DF9ZP(DL6FBL)	KP4BZ(NP4Z)	YT3L(YU3OH)	OK3LZ	NTCC/TDXS
1984 UF6CR	NY4D(K4XS)	VP2EC	YZ1EXY	L22C(LZ2SC)	EA8RL	VE3BMV	CY3BMV	4N2CQ	JH7UJU	YUDXC/NTCC
1983 L8DG(LU8DO)	PA0PUR	UK6LAR	YZ1EXY	YU3EF	HA8KQX	I4INI	4N3DX	KP4EQF	PY1OBA	N.Oh.A.R.S./NTCC
1982 HK3A(K3ZO)	4X4UH	NP4A	YQ0A	YU3EF	UA9AJO	OA4AWD	YU4GD	5Z4CS(JE1JKL)	KG6DX	YUDXC/FRC
1981 YT2D	OA8V	UK2PCR	YT0R	YU3EF	4Z4DX	OA4AWD	4N1U	LU8DQ	LU8DQ	FraserVal./PVRC
1980 KP2A(N6CW)	N3RS	UZ9A	K3WW	SP5IXI	AN2OP	424DX	ZW4OD	HD0E		W.Wash.DXC/PVRC
1979 KG6SW(K7JA)	SM0GMG	4N4Y	HD1A	AE6U	UB5BAT	YT2D		VE3BMV	JA1PIG/PZ	

Footnote#1-Changed the maximum allowable operating time for Single Operators from 30 Hours (18 hours off) to 36 Hours (12 hours off) beginning with the 1992 event.

Footnote#2-Added the Single Operator Low Power category beginning with the 1992 event.

Footnote#3-Added a Club Competition in 1978(Phone Only), this would encompass both the Phone and C.W. events in 1979. The 1978 Club Competition Winner was The Western Washington DX Club. Added the S/O QRPp Class in 1978.

USA WINNERS OF CQWW WPX CW CONTEST BY YEAR/CATEGORY

YEAR	S/O	QRP	LP	M/S	M/M	160	80	40	20	15	10
1994	KF3P	AA2U	K7SV	WC4E	WW2Y	AA9AX	W3BGN	WU3V/5	K2VV	WR3G	K3UA
1993	KM1H(KQ2M)	W2GD	N5RZ	KQ8M	WZ1R	AD1G	K12M	KC1XX	K2VV	KT0F	WA2SYN
1992	KM1H(KQ2M)	W2GD	WR3G(K5ZD)	WC4E	N4WW	WT3Q	N6SS	W3BGN	K2VV	KA2AEV	WE3C
1991	KM1H(KQ2M)	N3RS	—	N4WW	—	W4VJN	K1ZM	NQ2D	NI8L	NE8T	WE3C
1990	KT3Y	N3RS	—	N4WW	AC6T	K1ZM	WA6VNR	KV0Q	K1TO	N3GB	K5RX
1989	KT3Y	KA2AEV	—	N4WW	WZ6Z	N6LL	NA3J	NM2Y/3	N2AA	K1XA	N5RZ
1988	KT3Y	W8VSK	—	N4WW	NS0Z	K5NA/2	NA3J	K1XA	K2VV	K6LL/7	WA6FGV
1987	KT3Y	N3RS	—	KR0Y	K3TUP	K5UR	N3AD	KM3T	K1XA	N4VZ	NU4Y
1986	K3ZO	W8VSK	—	NA5R	NM5M	W0ZV	K5NA/2	N5RZ	K2VV	WZ4Z	KE5CV
1985	KC1F	KA2AEV	—	NA5R	NM5M	K5UR	N2AU	N6CW	N2AA	KA5W	KE5CV
1984	N5AU(N5RZ)	NY4D(K4XS)	—	N4WW	NM5M	KA1SR	N6QR	N6QR	NA8V	KW8N(WD8LLD)	—
1983	KC1F	N3RS	—	NA5R	K0UK	W8LRL	NE6W	W6BIP	NBII	KZ5M	KT4W
1982	KC1F	WD6EWG	—	N4WW	K4CG	W8LRL	N6PE	K7RI	K5GA	WA6DBC	N4ZC
1981	KC1F	N3RS	—	N4WW	AI6V	W9CG	K5UR	N6ND	N2MM	K6LL/7	N4ZC
1980	K1GQ	N3RS	—	K3EST	K3WW	AE6U	K1MM	WA4DRU	W5FO	W3MM	K8WW
1979	AE4H	W9PNE	—	K3EST	N9MM	AE6U	K5UR		K5GA	K1RM	K5RC(K5ZD)

Footnote#1-Changed the maximum allowable operating time for Single Operators from 30 Hours(18 hours off) to 36 Hours(12 hours off) beginning with the 1992 event.

Footnote#2-Added the Single Operator Low Power category for the 1992 event.

Footnote#3-Added a Club Competition in 1978(Phone Only), this would encompass both the Phone and C.W. events in 1979. The 1978 Club Competition Winner was The Western Washington DX Club. Added the S/O QRPp Class in 1978.

CQ WPX CONTEST

Continental Phone Records

Continent	Call	Year	Score
Single-Op All-Band (No Packet)			
AF	ZD8Z(N6TJ)	94	18,118,880
AS	7Z2AB	92	9,177,296
EU	YZ9A	91	8,518,112
NA	KP2A(KW8N)	93	16,694,570
OC	WR6R/KH6	93	9,803,972
SA	HC8A	92	24,809,300
Single-Op Lo-Power (No Packet)			
100-Watt Output (Since 1992)			
AF	ZD88V(G4ZVJ)	93	14,501,145
AS	JF1SEK	92	2,779,975
EU	CU0WPX(KB3RG)	93	2,472,000
NA	VP5G(WB6CJE)	93	6,634,896
OC	NH6T	92	3,117,650
SA	LT1N(LU2NI)	94	4,112,703
Single-Op QRP < 5-W (No Packet)			
AF	TR8SA	87	1,041,112
AS	RV9C	94	1,362,975
EU	YU2TY	90	770,450
NA	VP2EXX	90	6,727,444
OC	K7SS/WH6	86	2,078,490
SA	HC8A(N6KT)	94	7,520,562
S/O 28 MHz			
AF	FR5DX	91	7,543,818
AS	JH1AJT	89	4,848,480
EU	9H1EL	89	5,882,825
NA	J68AX	92	4,709,985
OC	P20A	92	5,184,625
SA	ZW5B (N5FA)	92	13,006,917
QRP	NP4CC	90	2,165,940
L.P.	CX7BF	94	4,361,280
S/O 21 MHz			
AF	TR1G	90	6,788,925
AS	7L1GVE	92	6,848,136
EU	CT2A	92	6,029,559
NA	FG5R (W7EJ)	89	9,936,240
OC	AH0K	92	7,206,850
SA	ZP0Y	90	12,070,245
QRP	RW9AB	92	638,392
L.P.	ZF1CQ(W8BLA)	94	5,003,019
S/O 14 MHz			
AF	EA8AH(OH1RY)	94	8,194,536
AS	H2A (5B4SA)	91	6,297,464
EU	IU9S(IT9BLB)	94	5,677,177
NA	TI2CC	87	5,491,290
OC	KG6DX	90	4,558,527
SA	ZZ5EG (PY5ZBA)	88	8,219,627
QRP	RV3E/JT1BY	90	675,990
L.P.	7Q7XX	93	1,025,352
S/O 7 MHz			
AF	CT3BX	94	5,187,480
AS	H24LP (5B4LP)	87	5,348,975
EU	IO4VEQ	93	4,184,292
NA	NP4A	86	6,668,184
OC	T32AF(KH6UR)	93	3,995,928
SA	YV5A (YV5ANF)	91	3,460,900
QRP	LZ2QV	86	58,764
L.P.	9A2WV	94	1,126,634
S/O 3.5 MHz			
AF	OH1RY/CT3	85	2,816,754
AS	UA9CSS	94	1,074,780
EU	GW8GT(G41FB)	94	1,473,868
NA	VA3EJ	91	1,950,592
OC	N6VI/KH6	94	1,016,652
SA	YV3A (YV5IVB)	91	1,664,476
QRP	UX2MF	94	147,994
L.P.	UT7DX	94	515,200
S/O 1.8 MHz			
AF	OH1RY/CT3	87	290,140
AS	UL7ACI	91	331,008
EU	LZ2BE	84	261,504
NA	CG3MFA	85	319,140
OC	T32AF	83	16,872
SA	YV5JEA	84	40,320
QRP	UY5XE	84	25,110
L.P.	OZ3SK	93	104,876
Multi-Op Single Transmitter			
AF	EA8BR	94	15,311,851
AS	TA5/N0FYR	91	16,474,965
EU	LZ9A	89	14,399,625
NA	VP2EC	92	24,409,580
OC	P20K	93	13,440,570
SA	HC8A	93	32,502,677
Multi-Op Multi-Transmitter			
AF	ED8ACH	91	47,278,236
AS	VS6WO	94	16,555,040
EU	HG73DX	91	30,664,095
NA	VP2EC	94	39,530,455
OC	FK0AW	89	26,538,972
SA	ZZ5EG	87	38,096,250
Club	Northern California		
	Contest Club	92	97,527,906

U.S.A. Phone Records

Section	Call	Year	Score
Single-Op All-Band (No Packet)			
W1	KM1H (KQ2M)	92	7,854,840
W2	WM2H	94	6,360,736
W3	KF3P	94	6,204,605
W4	W4NT (KM9P)	89	4,867,629
W5	KM5X	93	4,380,993
W6	WN4KKN/6	94	5,912,144
W7	K7RI	93	5,102,520
W8	KW8N	92	4,652,883
W9	W9RE (WA8YVR)	89	5,057,994
W0	AA9AK/0	92	5,946,529
VE	XK3EJ	93	10,672,784
Single-Op Low Power (No Packet)			
W1	WS1A	94	1,867,131
W2	K2POF	92	1,079,884
W3	KQ3V	94	1,266,408
W4	KJ4TI	92	615,750
W5	WB5NXH (WB5VZL)	92	2,423,284
W6	W9NQ/6	93	1,712,340
W7	N7LOX	94	664,560
W8	ND8L(N6WLX)	94	1,461,680
W9	NG9L	92	520,080
W0	AC0W	93	863,154
VE	CY2C (VY2SS)	92	4,812,740
Single-Op QRP <5 W (No Packet)			
W1	WA1LNP	94	1,161,646
W2	KR2Q	92	1,269,960
W3	WT3W	92	382,800
W4	WA4PGM	94	398,748
W5	KY5N	92	845,598
W6	WB6JMS	92	438,426
W7	WC7Q	88	414,462
W8	W8ILC	82	1,044,012
W9	W9UP	91	139,755
W0	WA0VBW	87	47,250
VE	VE3KZ	79	507,210
S/O 28 MHz			
W1	NX1H	91	3,015,377
W2	KC2X	82	1,201,089
W3	KS3F	91	2,767,580
W4	KO4QW	92	2,828,120
W5	WM5G (KR0Y)	89	4,213,127
W6	K6OYE	80	2,029,131
W7	NA7P (K7SS)	89	3,085,779
W8	N8II	81	1,440,285
W9	W0AIH/9 (N0BSH)	89	1,929,660
W0	K0GU	81	1,260,441
VE	VG7NTT	92	4,105,570
S/O 21 MHz			
W1	KC1XX	94	2,951,454
W2	KA2AEV	92	4,278,888
W3	W3AU	79	1,739,410
W4	N4ZZ	89	2,230,452
W5	K5MR	92	4,443,048
W6	WN4KKN/6	92	4,538,050
W7	AI7B	82	4,151,232

W8	K3ZJ/8	92	3,063,720
W9	WE9V	94	2,649,340
WØ	KVØQ	83	1,675,242
VE	XL7SV	90	6,202,042

S/O 14 MHz

W1	KG1E	85	2,906,676
W2	K2VV	87	3,546,294
W3	KS3F	94	2,504,656
W4	KC2X	91	1,549,561
W5	WD5N	90	2,005,291
W6	WM2C/6	94	1,556,875
W7	N7TT	87	1,479,075
W8	K8NA	82	2,252,688
W9	KK9A	92	3,389,568
WØ	KØRWL	85	1,003,220
VE	VX7A(VE7SV)	94	4,534,944

S/O 7 MHz

W1	KA1DWX	91	164,352
W2	WQ2M	91	695,196
W3	W3GH	94	948,288
W4	WC4E (K4XS)	87	581,640
W5	WU3V/5(W5WMU)	94	1,154,560
W6	N6RO	94	1,307,332
W7	KC7EM	92	1,396,646
W8	K8NA	83	212,544
W9	KS9K(NØBSH)	94	798,160
WØ	KVØQ	91	1,068,144
VE	XL7SV	86	3,454,864

S/O 3.5MHz

W1	KE1Y	94	955,200
W2	K1ZM	92	1,266,514
W3	WE3C	94	1,008,268
W4	NI4P	94	515,328
W5	N5RZ	85	428,542
W6	KI6P	93	717,590
W7	K7GWK	85	133,952
W8	W9LT/8	94	522,704
W9	K9ZO	85	299,464
WØ	N7DF/Ø	84	148,930
VE	VA3EJ	91	1,950,592

S/O 1.8 MHz

W1	KA1SR	85	37,204
W2	K1ZM/2	94	88,560
W3	WB3GCG	84	43,368
W4	KG4W	85	74,112
W5	K5UR	85	122,664
W6	N6VI	86	18,300
W7	N6TR/7	86	24,926
W8	W8LRL	83	30,654
W9	WD9AHJ	82	3,564
WØ	N7DF/Ø	85	19,592
VE	CG3MFA	85	319,140

Multi-Op Single Transmitter

W1	NX1H	94	6,744,654
W2	WT2S	92	3,522,534
W3	K3EQ	93	3,916,890
W4	WC4E	92	11,611,929
W5	K5XI	94	6,681,816
W6	KI6P	87	5,825,328
W7	W7RM	84	5,800,090
W8	NE8T	92	6,333,020
W9	KJ9W [WX3N]	82	6,168,450
WØ	WOØG [WX3N]	89	7,900,395
VE	VG7SZ	92	10,313,226

Multi-Op Multi-Transmitter

W1	WZ1R	94	16,029,400
W2	KF2U	81	6,594,258
W3	K3WW	80	6,385,880
W4	WK4Y	92	11,226,946
W5	N5AU	86	9,733,248
W6	WZ6Z	89	18,737,170
W7	No Entry		
W8	KW8N	86	8,473,705
W9	W9ZRX	80	5,417,178
WØ	WOØG [WX3N]	90	10,682,362
VE	CZ7Z	92	20,228,367
Club	Northern California Contest Club	1992	97,527,906

Continental CW Records

Continent	Call	Year	Score
Single-Op All-Band (No Packet)			
AF	EA8EA(OH2MM)	93	10,693,146
AS	P31A (5B4ADA)	92	10,293,858
EU	CR7M(CT1BOH)	93	5,645,267
NA	V27T (YU1RL)	89	9,408,672
OC	NH6J/NHØ(JE1JKL)	88	4,484,760
SA	P4ØW(W2GD)	94	14,168,115
Single-Op Low Power (No Packet)			
AF	7Q7XX	93	4,665,770
AS	7Z2AB (K2XR)	92	3,231,025
EU	HA3UU	93	4,157,288
NA	C6AHY(KH6M)	94	4,712,408
OC	9V1YC	93	3,350,204
SA	FY5FY	92	3,317,918
Single-Op QRP <5W (No Packet)			
AF	5Y4FO	92	649,057
AS	4X4UH	82	1,028,904
EU	LZ2BE	91	1,137,488
NA	VP2MU (KJ4VH)	91	1,554,735
OC	FO8JP	86	572,131
SA	OA8V	81	441,768
S/O 28 MHz			
AF	ZS6BCR (ZS6EZ)	91	3,621,173
AS	4X4UH	81	1,081,262
EU	9H1EL	88	805,552
NA	HI8JKA	89	891,242
OC	KG6DX	81	1,238,806
SA	CE3DNP	89	2,857,038
QRP	ZLØAAH	89	256,665
L.P.	9M8DX (VK2DXI)	92	597,600
S/O 21 MHz			
AF	ZD8LII	91	5,118,527
AS	7L1GVE	91	2,811,478
EU	4N3E (YT3AM)	90	3,239,453
NA	FS5T (AI7B)	89	4,552,470
OC	N7DF/WH2	89	3,243,450
SA	LTØA (LU5UL)	91	4,290,988
QRP	4Z7U (4Z4UT)	91	1,031,400
L.P	3ZØKN(SP3RNZ)	93	1,719,852
S/O 14 MHz			
AF	ZD8LII(GØLII)	93	2,687,580
AS	4Z6DX	91	4,614,030
EU	LZ5W (LZ3ZZ)	92	4,222,665
NA	WC4E/KP4	86	3,613,248
OC	ZL3GQ	89	2,775,744
SA	YW1A (YV1DIS)	91	4,617,456
QRP	W8VSK	89	376,648
L.P.	H23W (5B4WN)	92	3,826,112
S/O 7 MHz			
AF	EA9LZ	93	2,829,276
AS	9K2ZZ	94	3,383,676
EU	S5ØA	94	3,293,004
NA	VP2VCW	86	4,641,120
OC	V7A(V73C)	93	2,205,922
SA	AZ4F(LW9EUJ)	94	4,496,980
QRP	OM3TUM	94	232,812
L.P	EA8CN	94	1,197,700
S/O 3.5 MHz			
AF	EA8RL	84	453,456
AS	UP2NK/UF	85	701,012
EU	OK1DXS	94	916,095
NA	XL7CC	94	609,730
OC	KX6DC (NZ8B)	89	258,258
SA	YX3A (YV5ANT)	89	1,004,060
QRP	4N7MOD	90	141,075
L.P.	S5ØC	94	379,050
S/O 1.8 MHz			
AF	ZS6BCR (ZS6EZ)	85	20
AS	UP3BP/UF	85	125,240
EU	LY3BU	94	132,112
NA	VE3BMV	86	43,428
OC	KX6DC (NZ8B)	88	12,240
SA	YV1OB	86	11,550
QRP	RA3AUU	87	39,050
L.P.	S57DX	94	104,920

Multi-Op Single Transmitter

AF	ZD8ØV	91	10,938,352
AS	YM5KA	90	13,098,790
EU	ZB2X	91	8,618,823
NA	KP2A	89	12,843,135
OC	AG9A/AH2	91	9,005,641
SA	ZXØF	94	12,280,162

Multi-Op Multi-Transmitter

AF	EA9CE	84	4,383,308
AS	JE2YRD	91	8,388,942
EU	HG73DX	93	16,543,420
NA	WL7E	88	12,397,316
OC	KH6XX	85	8,551,399
SA	LQ5A	89	8,290,016

U.S.A. CW Records

Call Area	Call	Year	Score
Single-Op All-Band (No Packet)			
W1	KM1H (KQ2M)	92	5,313,160
W2	KE2PF	91	3,537,675
W3	KF3P	94	4,197,134
W4	KT3Y	94	3,230,471
W5	K5GN	93	3,868,128
W6	N6TV	93	3,261,555
W7	NN7L	88	2,301,026
W8	KW8N(WD8IXE)	91	2,569,432
W9	WØAIH(NØBSH)	93	2,299,561
WØ	NS8Z(KM9P)	93	3,632,587
VE	VA8A (KQ8M)	92	5,962,176
S-Op Low Power (No Packet)			
W1	KG1D	93	722,240
W2	K2QMF	93	1,215,657
W3	WR3G (K5ZD)	92	1,397,682
W4	K7SV	94	1,970,916
W5	N5RZ	93	1,996,467
W6	AB6FO	93	1,110,795
W7	KX7L	93	392,450
W8	WS8O(N8LXS)	94	779,590
W9	K9LJN	93	1,155,018
WØ	KØGDS	93	460,464
VE	XL7NTT	94	1,570,995
S-Op QRP <5 W (No Packet)			
W1	KN1M	93	710,980
W2	W2GD	93	1,019,712
W3	N3RS	91	828,808
W4	NY4D (K4XS)	84	484,575
W5	K5RX	91	463,710
W6	NA6A (W6REC)	88	415,712
W7	WU7Q	94	515,338
W8	W8VSK	88	528,504
W9	WB9HRO	88	184,920
WØ	WØKEA	89	155,916
VE	VE3KP	93	520,514
S/O 28 MHz			
W1	K1TR	84	33,109
W2	K5NA/2	91	90,147
W3	WE3C	91	95,890
W4	N4ZC	89	153,081
W5	N5RZ	89	162,134
W6	WA6FGV	88	87,889
W7	WB7FDQ	86	4,148
W8	K8MFO	91	64,610
W9	KB9HG	81	16,200
WØ	WD9FTZ/Ø	89	15,566
VE	VE3BMV	80	113,412
S/O 21 MHz			
W1	K1XA	89	1,037,374
W2	KA2AEV	92	1,405,072
W3	N3RS	88	1,846,086
W4	K4BAI	88	1,255,626
W5	WM5K	88	890,568
W6	N6RO	81	1,184,612
W7	K6LL/7	88	2,163,388
W8	NE8T (K8JM)	91	1,848,300
W9	K9QVB	88	1,088,263
WØ	NØBSH	88	1,888,195
VE	VE3BMV	81	1,534,669
S/O 14 MHz			
W1	WS1M (YU3OH)	92	2,334,384
W2	K2VV	86	2,525,880
W3	Kt3Y	93	2,261,870
W4	K39A/4	93	1,735,356
W5	W5FO	88	1,640,745
W6	KI6P (WA6VEF)	83	1,417,565
W7	AI7B	86	1,278,011
W8	WD8IXE	86	1,713,162
W9	NØBSH/9	89	1,069,430
WØ	KØRWL	84	707,535
VE	VB7SZ	93	1,421,472
S/O 7 MHz			
W1	KC1XX	93	1,421,472
W2	KR2Q	85	1,555,048
W3	KM3T	86	1,640,820
W4	K1ZZI	90	719,280
W5	N5RZ	85	1,754,664
W6	NI6W	85	1,048,775
W7	KT7G	93	947,700
W8	K8MFO	92	502,740
W9	K9ZO	85	278,460
WØ	KVØQ	89	498,550
VE	VB7SV	93	2,626,704
S/O 3.5 MHz			
W1	K1ZM	93	406,080
W2	WW2Y	93	352,692
W3	W3BGN	94	355,348
W4	N4ZC	91	208,964
W5	K5UR	83	13,176
W6	NE6W (N6IG)	83	105,672
W7	N6SS	92	68,376
W8	KV8Q	94	163,200
W9	KJ9D (KK9V)	85	26,460
WØ	KMØL	86	52,800
VE	XL7CC	94	609,730
S/O 1.8 MHz			
W1	AG1D	93	1,100
W2	K5NA/2	88	8,052
W3	NA3J (WB2EKK)	88	13,392
W4	W4VJN	92	24
W5	K5UR	85	13,668
W6	N6LL (WA6CDR)	89	1,584
W7	KG7D	86	4,080
W8	W8LRL	83	1,482
W9	AA9AX	94	2,016
WØ	WØZV	86	7,140
VE	VE3BMV	86	43,428
Multi-Op Single Transmitter			
W1	WF1B	91	3,784,880
W2	KU2Q	89	4,091,568
W3	WD3I	91	2,059,442
W4	N4WW	92	7,168,734
W5	K5XI	93	4,735,077
W6	KI6P	88	5,117,778
W7	KE7V	86	2,671,482
W8	KQ8M	93	5,324,540
W9	KS9O	91	4,258,254
WØ	KRØY	86	4,516,974
VE	VE1DXA	83	3,728,256
Multi-Op Multi-Transmitter			
W1	WZ1R	93	8,165,315
W2	WW2Y	94	7,755,876
W3	K3TUP	87	3,780,081
W4	NQ4I	94	4,753,788
W5	WM5G	88	8,491,032
W6	AD6C	88	5,556,160
W7	WJ7R	94	3.511.200
W8	WD8LLD	92	3,071,900
W9	N9MM	79	2,124,150
WØ	NSØZ	88	10,870,380
VE	CY3PCA		83 4,977,817
Club Score	Northern California Contest Club	92	97,527,906

History of the CQWW RTTY Contest

In 1986 the publishers of the RTTY Journal and CQ magazine announced that they were joining forces and would be co-sponsoring the CQWW RTTY Contest. Dale Sinner, at that time the new publisher of the RTTY Journal, felt that an expansion of that magazine's old WW RTTY contest into the popular CQ format would be beneficial in both magazines.

Early in the Spring of 1987, both magazines announced the new rules. The fourth weekend in September was picked as the spot for the RTTY contest. This has proven to be a good time as conditions are usually very good.

In that first year a number of special contest Dxpeditions set forth on the quest to win the contest. HC8CQ went to the Galapogos Islands and featured some of the premier digital contesters. The site that they developed at El Junco on San Cristobal island became in later years the place from which all of the records of HC8A in SSB contests were established. The design of the CQWW rules put special emphasis on Dxpeditions. This was a first on RTTY. 1987's first event saw a very large turnout of RTTY contesters with wide spread support and enthusiasm around the globe.

RTTY had a long and rich history of contesting going back some number of years, but without a clear single big contest that was sustained year-after-year. The majority of contests traditionally had been put on by various national organizations such as BARTG, CARTG, SARTG, ANARTS, WAE, and VOLTA. These represented Britain, Canada, Scandinavia, Australia-New Zealand, Germany, and Italy. The U.S. had a number of different contests from time-to-time. There was always trouble with certificates and sponsorship in America. The joining of CQ magazine and the RTTY Journal ended the American problems and was the beginning of the "modern" era in RTTY contesting.

RTTY contests typically have two major differences from CW and SSB rules.

First, most RTTY contests are fairly short: 24 to 30 hours is the rule. The continuous duty cycle of RTTY equipment is very stressful on gear and shorter periods are better. Secondly, North America has always had a different multiplier basis than most non-digital contests. States/Provinces or call areas have always been multipliers. So the CQWW has States/Provinces, Zones, and Countries all as multipliers. The CQ 3-2-1 point multiplier was kept intact to keep foreign interest high and to make Dxpeditions a part of the contest.

There were about 300 logs submitted for the 1987 CQWW. The only problem that surfaced was that 160 meters was found to be essentially unusable by the digital stations. In later years this band was dropped. The first few years of the contest there was only a Multi-Single category and Multi-Multi was later added. In keeping with the other CQ contests, Single Operator Assisted was added. After the huge popularity of the low power class of 150 watts in the ARRL's RTTY Roundup, this class was also included in subsequent years in CQWW RTTY.

1994 brought the CQWW RTTY contest 48 hours of continuous operation by all operators. This will, over time, change the basis for the all-time records. However 1994 saw only a couple of new records.

Long time CQWW RTTY contest manager, Roy Gould, KT1N, added Ron Stailey, AB5KD, as his assistant manager in late 1994. The logs have grown over the years and the additional help will make sure things stay well organized and well done.

(Contributed by Jay Townsend, WS7I.)

CQWW RTTY Yearly Winners

Year	S/O	S.O./Asst	M/S	M/M	80	40	20	15	10
1987	VE1ASJ	-	HD8CQ	-	AB7Y	I2VXJ	8R1RPN	CT4UE	CE6EZ
1988	K5CKL	-	HD8EX	-	HB8DCQ	-	HC5EA	VE6ZX	CE6EZ
1989	TG9VT	-	HD8EX	-	SP3SUN	HB9DCQ	YU2W	KEØKB	JH1LBR
1990	HC5J (WS7I)	WA7EGA	RH7E	W3LPL	WA8TXT	UB4HQ	LZ5Z	CE6EZ	4U1ITU
1991	CT3M (DJ6QT)	AA4M/6	UZ9CWA	W3LPL	YT3HM	HJ4QIM	EA1QK	4M5RY	ZS6BCR
1992	GU3HFN (DJ6QT)	DK3GI	P4ØRY	W3LPL	YU3BQ	W2UP	4M5RY	ZP5JCY	ZD8LII
1993	CR3Y (CT3BX)	DLØWW	UZ9CWA	W3LPL	K1IU	W2UP	VY2SS	9A5Y	ZS6EZ
1994	HH2PK	DK3GI	RK9CWA	K1NG	9A1A	DJ2BW	9A2DQ	KP2N	-

Footnote#1 1987 160 was included and subsequently dropped
Footnote#2 1987 thru 1989 had only Multi-Op with rules like M/S.
Footnote#3 1990 Added Single Op. Assisted and Multi-Multi
Footnote#4 1993 Added High and Low Power Single All-Band and M/S.
Footnote#5 1994 Changed hours from 30 to 48 for Single Operators.

It is interesting to note that all contests are well represented in the single operator results. DJ6QT is the only repeat operator in the Single Operator unlimited class with two wins. Three M/S wins from the famous "El Junco" site in the Galapagos Islands in the early years and three wins for the crew from Asiatic Russia recently. The Multi/Multi class started with a string of victories by W3LPL from North America.

That ended in a close race in 1994 with K1NG unseating W3LPL. The new low power classes continue to grow in size.

CQ WORLD-WIDE DX CONTEST ALL-TIME RTTY RECORDS

Single Op/Single Band

Year	Band	Call	Score	QSO	QPts	Zone	Ctry	US/VE
Africa								
	3.5	No Entry						
1988	7	EA8AKQ	12	2	6	1	1	0
1990	14	EA8RA	104,451	315	941	25	46	40
1992	21	ZS6EZ	382,630	772	2,305	27	87	52
1992	28	ZD8LII	355,426	840	2,503	23	66	53
Asia								
1992	3.5	JR2CFD	153	10	17	4	3	2
1994	7	JA8BZL	5,916	48	116	18	23	10
1994	14	JR5EXW	157,615	377	1,087	30	74	41
1992	21	JE2UFF	84,588	259	742	25	55	34
1990	28	JR1IJV	123,066	328	954	28	59	42
Europe								
1994	3.5	9A1A(9A2RA)	47,894	291	622	9	52	16
1994	7	DJ2BW	135,168	414	1,024	22	69	41
1993	14	S51DX	293,433	700	1,869	30	77	50
1992	21	LZ1MC	247,950	623	1,653	27	70	53
1990	28	4U1ITU	236,842	547	1,499	32	79	47
North America								
1993	3.5	K1IU	39,710	273	418	10	37	48
1993	7	W2UP	125,656	489	904	22	68	49
1993	14	VY2SS	374,550	913	2,270	27	90	48
1994	21	KP2N	293,562	856	2,082	23	67	51
1990	28	AB8K	96,250	312	770	29	67	29
Oceania								
	3.5	No Entry						
	7	No Entry						
1990	14	VK3EBP	62,964	198	583	24	48	36
1990	21	YC1YMN	116,051	344	1,027	25	50	38
1989	28	KX6OI	49,572	206	612	18	37	26
South America								
	3.5	No Entry						
1993	7	PJ2MI	65,835	243	693	18	34	43
1992	14	4M5RY(YV5KAJ)	270,256	599	1,778	23	73	56
1992	21	ZP5JCY	433,532	871	2,596	30	85	52
1991	28	ZP5JCY	235,884	599	1,787	23	57	52

Single Operator/Single Band
World Record Holders

Year	Band	Call	Score	QSO	QPts	Zone	Ctry	US/VE
1994	3.5	9A1A(9A2RA)	47,894	291	622	9	52	16
1994	7	DJ2BW	135,168	414	1,024	22	69	41
1993	14	VY2SS	374,550	913	2,270	27	90	48
1992	21	ZP5JCY	433,532	871	2,596	30	85	52
1992	28	ZD8LII	355,426	840	2,503	23	66	53

Single Operator/All Band Assisted
World Record Holders

Year	Band	Call	Score	QSO	QPts	Zone	Ctry	US/VE
	AF	No	Entry					
1993	AS	JA3YBF	221,298	384	958	65	135	31
1994	EU	DK3GI	1,186,185	997	2,607	83	242	130
1992	NA	K1IU	971,412	911	2,028	96	222	161
	OC	No	Entry					
	SA	No	Entry					

World Record

Year	Band	Call	Score	QSO	QPts	Zone	Ctry	US/VE
1994	EU	DK3GI	1,186,185	997	2,607	83	242	130

Single Operator/All Band

Year	Band	Call	Score	QSO	QPts	Zone	Ctry	US/VE
1994	OC	ZL3GQ	737,741	802	2,357	72	133	108
1990	SA	HC5J(WS7I)	1,364,972	1,143	3,362	89	185	132
1991	AF	CT3M(DJ6QT)	1,075,584	941	2,801	82	213	89
1994	AS	9K2ZZ	962,104	1,068	3,016	61	179	79
1992	EU	GU3HFN	1,223,849	1,081	3,007	80	191	136
1994	NA	HH2PK	1,304,485	1,252	3,055	76	167	184

World Record

Year	Call		Score	QSO	QPts	Zone	Ctry	US/VE
1990	HC5J(WS7I)		1,364,972	1,143	3,362	89	185	132

Multi-Operator/Single Transmitter

1992	AF	EG8CMR	963,116	1,048	3,127	59	120	129
1993	AS	UZ9CWA	2,580,660	1,716	4,779	120	333	87
1992	EU	UW2F	2,847,220	1,767	4,909	106	271	203
1991	NA	V2/GØAZT	1,680,607	1,577	3,743	78	180	191
1993	OC	NH6T	1,138,070	1,042	3,118	83	130	152
1992	SA	P4ØRY	3,543,090	2,222	6,635	91	220	223

WORLD RECORD

1992		P4ØRY	3,543,090	2,222	6,635	91	220	223

Multi-Operator/Multi-Transmitter

	AF	No Entry						
1993	AS	JJ3YBB	586,249	684	1,873	79	187	47
1991	EU	LY2WW	927,710	916	2,285	87	236	83
1992	NA	W3LPL	3,111,748	2,233	4,556	116	326	241
1992	OC	T32RA	1,770,131	1,744	5,191	69	118	154
	SA	No	Entry					

World Record

1992	W3LPL	3,111,748	2,233	4,556	116	326	241

Please report any errors and omissions to WS7I.

(Contributed by Jay Townsend, WS7I.)

CLASSIC BOOKS ON AMATEUR RADIO

This is a short listing of books that are either classics, and should be in the well-stocked ham library, or books that are revised every year and should be replaced to keep your library current. You should check the advertisements in *CQ*, *QST*, or *73* or contact your local dealer for ordering information.

The Radio Amateur Callbook: The *Callbook* has been an institution in amateur radio since 1920! Published in two volumes (North American and International), it is the source for addresses of hams worldwide. With the growth in amateur radio, you really need to get a new set every year to keep current.

The Radio Amateur's Handbook: This mainstay of the ham's library includes sections on everything from semiconductor theory to Yagi construction, and everything in between. Written in a practical style without excessive math, it has served as an electronics reference book for hams and non-hams since the first edition was published in 1926. As the hobby has evolved, sections have been added to keep pace with the technology. If you haven't bought a *Handbook* in the last few years, you'll find a wealth of new material that you can use.

World Radio TV Handbook: Every SWL needs a current copy of this book. It includes schedules and frequencies for the world's short-wave broadcasters, and even listings of medium-wave, FM, and TV stations of the world. If you've got a general-coverage receiver, or a transceiver that tunes between the ham bands, you'll find this book an invaluable guide. Since broadcasters change their schedules, the WRTH is updated annually.

The ARRL Antenna Book: All hams need antennas, and most hams are interested in improving their antenna systems. *The ARRL Antenna Book* covers all types of antennas, from mobile whips to Yagis and loops. Recently updated to over 700 pages, it also includes information on transmission lines, antenna tuners, and more.

Radio Frequency Interference: Unfortunately, nearly every ham has to deal with RFI and TVI at some point in his life. This book is written by experts, and offers clear troubleshooting directions and suggestions on dealing with the diplomatic problems that come with TVI. It will definitely help you solve the problem.

The CQ Amateur Radio Almanac: This annual publication contains facts, figures, and data on amateur radio that you will be hard-pressed to find anywhere else. Much of the history of the hobby is captured, with a capsule history of the last year and complete listings of operating achievements. Reference tables, information on the number of hams by town in the U.S., and awards information is also found in this book. You'll want to get one every year.

AMATEUR RADIO BOOKS OF 1995

While not that many new books were released in 1995, those that were have created quite a bit of excitement. Here's a summary of what has happened in 1995 and what is coming up in 1996.

Mega-Highlights

On the very top of the list is the long awaited CD-ROM from the *Radio Callbook*. The Radio Callbook has for 76 years been the source of names and addresses most hams turn to. Their Foreign listings are by far the most comprehensive available to the Radio Amateur. When you throw in the US, you have over 1.3 million listings in over 250 countries, islands and dependencies. Callbook has designed the CD to work either in Windows or DOS (sorry not Mac compatible.) Includes tons more information including census data from each country, listings of clubs, new and old call cross-reference and much more. This CD is going to be a best seller in 1996!

CQ released this spring, a new and up-dated edition of the *Propagation Handbook*; the classic text by Cohen and Jacobs. Co-authored with Robert Rose, this revised edition covers in simple terms, the complex mechanisms of radio wave propagation. It makes extensive use of computer models and all the latest knowledge of our Ionosphere.

The final book to rank "top-billing" is *The Ultimate Scanner* by Bill Cheek. This long awaited book is full of scanner modifications covering all the latest models. It's a follow on to volumes 1 and 2 published by CRB Research. *The Ultimate Scanner* shows you how to modify your scanner so it will receive all those "forbidden" frequencies. Detailed drawings with simple easy-to-follow text make this a good book to have.

Now, here's what's going on in the world of books for 1996.

Antenna Books

While technology continues to get more complex, one area that hams are still quite involved in is antennas and antenna design. ARRL has released the fourth Volume of the *Antenna Compendium*. Included at no extra charge is a software disc (MS-DOS) that includes programs taken from the articles. Since we are at a Sunspot minimum, seven of the 38 articles deal with low band antennas for transmitting and receiving. These antennas will make your station perform better and help you work a greater share of DX. It also includes PC designed antennas and an inside look by K6STI at one of his designs that failed and why!

Troubleshooting Antennas and Feedlines. This book was written by Ralph Tyrrell, W1TF, one of the technical service reps from a major antenna manufacturer. It includes gobs of information on how to solve tricky antenna problems without a lot of extra work or hassle.

Study Guides and Licensing Aids

The new Advanced Class license exam went in to effect July 1, 1995. ARRL, Gordon West, Ameco and others have all released revised study manuals for the new exam. These books make the study process as painless as possible. The questions are fully explained with simple text and graphics where needed. The Extra class exam changes July 1, 1996 and there will be a whole new series of Extra class study guides out next Spring.

Jeff Otterson released a Windows based program, *Ham Exam*, that contains all of the questions and answers for Novice through Extra class. Jeff takes advantage of Windows graphic capabilities and gives the student drawings, charts and other graphics to help ensure that you learn, not memorize, the material. This is a great way to learn using your PC and the power Windows gives it.

The ARRL is issuing a new edition of the *FCC Rulebook*. This will include all of the new rules and regulations for the Part 97 Amateur Rules and Regulations.

Annual Releases (Amateur)

Of course, at the top of this list is the *CQ Amateur Radio Almanac*. This all purpose book is full of facts, records, superlatives and much more. Every ham should have a copy. Of course, there is the 1996 ARRL *Radio Amateur's Handbook*, which includes bundled software this year. The 1996 edition is crammed full of projects for all levels of Radio Amateur. The software includes programs to help design projects and perform many calculations plus much more. In addition to the new CD-ROM Callbook, there will also be the traditional *North American and International Editions of the Callbook* in printed form. Up-dated to include all the latest calls and addresses, these books are essential to ensuring you have the correct name and address of both domestic and international hams you work.

Both the ARRL and ARTSCI now have *Repeater Map Books*. These handy books list repeaters by frequency and map location. You'll find this great when you're traveling down the Highway and don't have time to look in the small pages and minute type found in the ARRL *Repeater Directory*.

Annual Release (SWL and Scanners)

Of course the 1996 *Passport to Worldband Radio* leads this list. Fully updated with all the latest schedules, program information and product reviews, *Passport* is a great tool to have while you are listening between the ham bands. The 1996 edition takes a critical look at radios, accessories and other gizmos and gadgets for the SWL beginner and expert alike.

If you listen to the AM broadcast bands, the 1996 edition of the *M-Street Directory* is a must. Cross referenced by frequency, callsign, and location, this book is an invaluable tool to identifying stations on the AM broadcast band. Fully revised with all the latest calls, formats and locations with full mailing addresses for QSLing.

World Radio-TV Handbooks have long been known as an excellent source of information for SWLs. In 1996 there is a new *Radio-TV Handbook* and a *WRTH Satellite Broadcasting Guide*. The *Radio-TV Handbook* has up-to-date schedules, formats and other helpful information not found in other books, such as stations below 200 kHz and some standard TV channels (data not found in other guides.) The *WRTH Satellite Broadcasting Guide* is chock-full of information about Satellites and the TV stations they retransmit from their geostationary orbits. It tells you about dishes, receivers and accessories, satellite "footprints", equipment tests, and key names and addresses from stations around the world. For those of you who wondered about the 1995 *WRTH's Buyer's Guide*, due to production difficulties, Billboard never published the edition. There are currently no plans to publish a 1996 edition either.

Operation

ARRL just released a new and revised edition of the *ARRL Operating Manual*. This comprehensive book covers almost every aspect of Amateur Radio operation from nets to traffic handling, DXing to 2 meter FM mobile operation. Great book for the beginner; perfect also for the expert operator. Ted Melinsoky has released his new *DX Awards Directory* for 1996. This book is the most comprehensive awards book available. It has over 240 pages of detailed rules and requirements for awards from around the world.

Your RTTY/Amtor Companion has been replaced by a brand new book, *Your HF Digital Companion* by Steve Ford of ARRL. Covers digital operation on the HF bands including RTTY, ASCII, Amtor, Pactor, G-Tor, Clover and more. Full of explanations and details on how to get on and use HF digital communications modes.

When the *Callbook* was bought by Watson Guptil, there were a number of popular Amateur products that were quietly put on the back shelf. One of the most popular was the *Callbook's World Atlas*. The German Amateur Radio Club, DARC, released an up-to-date version this spring at the Dayton Hamvention. This 20 page map book includes all the latest counties, calls and other important information. This is currently the most up-to-date Amateur Radio map available.

Until now, it appears there has never been a book dedicated to Six meter operation. Ken Neubeck's *SIX METERS Guide to the Magic Band* is the insider's look at this fascinating VHF band. Sometimes, the best you can do is talk across town. Other times, world-wide communications is possible on a frequency just below Channel 2 TV! That's awesome. Has equipment and antenna ideas as well as helpful hints and tips.

World Radio published a collection of Lenore Jensen's, W6NAZ, writing from back issues of World Radio Magazine. Lenore was a great writer who could capture exactly what it is that makes Amateur Radio so special. This is a great collection of stories about hams written by one of the best.

Xantek released an up-dated slide-rule and computerized version of their famous DX Edge. For those not familiar, the DX Edge shows you the path of the Sun over the Earth's surface. It is used extensively for 160/80 Grayline propagation forecasting. In fact, this is one of the handiest tools most low-band DXers have in their shacks.

SWL, Scanner and General Radio

Scanner Master has released a new edition of their all purpose scanner directory, *Monitor America Third Edition*. This comprehensive volume is the most detailed and accurate book ever written on VHF/UHF public service communications in America. This expanded 1,000 page volume covers over 250 cities, resorts, along with all state, inter-city, and Federal radio nets. Covers police, fire, hospital as well as hundreds of other services.

Don Schimmel's all new *Underground Frequency Guide* eliminates the mystery and intrigue of "numbers" stations and other clandestine spy stations. Schimmel gives the reader the "inside scoop" of what is going on, He tells you how to decipher "boops, beeps, piccolos and shavers" as well as number readers. This new edition includes known transmitter locations, languages, hundreds of active frequency listings and much more.

CRB, well known in the Scanner market has just released a new edition of their popular *Airscan Guide to Aeronautical Communications* by Tom Kneitel. Covers communications from HF to UHF and lists over 6,700 US and Canadian private, military, seaplane, commercial fields and frequencies. A complete, easy-to-understand introduction gives you a complete primer on how and where to tune aircraft stations.

Tuning In to RF Scanning by Bob Kay is written for the beginner who knows nothing about scanners, what they do and how to use them. Covers the basics of what radio to use, antennas, frequencies and much more.

Details cordless telephones, civilian and military aircraft, local and federal police, FBI and Secret Service, NASA and other interesting services.

Sneak Peek

There are a number of new books planned for early 1996. While complete details are not available at press time, what follows is a "sneak peak" at what's coming up.

As mentioned before, the *ARRL Handbook* will now include software. ARRL is also preparing a *Vertical Antenna Classic* book that will be perfect for the next couple of DXing seasons. In their "Companion Series" ARRL will have a *HF Antenna Companion* that will be written for the beginner setting up their first radio station. ARRL will also offer a Technician Class Q and A book that will list only questions and answers from the FCC exam syllabus.

CQ and Worldradio are also preparing several new titles of interest to hams. Keep an eye on the catalogs!

(Contributed by Craig Clark, NX1G, of Radio Bookstore - 1-800-457-7373)

PERIODICALS

The following list of publications contains a sampling of the most popular amateur radio bulletins and journals worldwide. Although every effort has been made to make it complete and useful, it's inevitable that some worthwhile ones will have been missed. Please send corrections to this list, as well as your suggestions for other publications to add to the list in the 1996 edition, to: Almanac Editor, c/o CQ Publishing, 76 North Broadway, Hicksville, NY 11801.

The most recent available pricing information is given for most, and unless stated otherwise, all prices are for full one-year subscriptions and are given in U.S. dollars. In some cases sample copies or trial subscriptions of less than one year are available, as are special rates for subscriptions of more than one year. Also, many publishers will now accept payment via credit cards, an especially convenient method of payment when subscriptions are quoted in foreign currency. A letter to the publisher asking for information on short- or long-term subscriptions or credit card payments will generally receive a swift response. In all cases, send along a self-addressed, stamped envelope or a self-addressed envelope with an IRC. Yes, publishers will usually respond without the help of SASEs and SAEs, but your response will generally appear more quickly if you make it easy for the publisher to answer. Good reading!

The 59(9) DX Report. Amateur radio DX information. Weekly. 1 year subscription: U.S. $36.00. Canada $41.00. Mexico $40.00. Elsewhere $58.00. The 59(9) Report, PO Box 73, Spring Brook, NY 14140.

73 Amateur Radio Today: Amateur radio general coverage. Monthly. 1-year subscription: U.S.: $24.97. Canada: $34.21. Elsewhere: $43.97 surface mail, $66.97 via air. *73 Amateur Radio Today*, PO Box 50330, Boulder, CO 80321-0330.

Amateur Radio. Journal of the Wireless Institute of Australia: Monthly. Contact WIA for membership information. The Membership Secretary, Wireless Institute of Australia, PO Box 300, Caulfield South, Vic 3162, Australia.

Amateur Radio Trader. Advertisements for amateur radio equipment. Twice monthly. Subscriptions: $14.95 3rd class, $34.95 first class. Amateur Radio Trader, 410 West 4th Street, P.O. Box 3729, Crossville, TN 38557.

The AMSAT Journal: Amateur radio satellite communications. Bi-monthly. 1-year subscription: U.S.: $30.00. Canada: $36.00. All others $45.00. AMSAT, 850 Sligo Avenue, Suite 600, Silver Spring, MD 20910-4703.

Antique Radio Classified: Antique radio collecting and history. Monthly. 1-year subscription: U.S.: $34.95 by 2nd class mail, $51.95 by 1st class mail. Canada: $45.00 by 2nd class mail, $56.00 by air. Mexico: $56.00. All others $50.00 surface mail, $98.00 air. *Antique Radio Classified*, P.O. Box 2, Carlisle, MA 01741.

The ARRL Letter: Newsletter covering amateur radio. Bi-weekly. 1-year subscription: U.S., Canada, Mexico: $19.50. All others $31.00 via air. Subscriptions are limited to ARRL members. American Radio Relay League, 225 Main St., Newington, CT 06111.

Break-In. Journal of the New Zealand Association of Radio Transmitters: Monthly. 1-year membership: New Zealand: $65.00 (New Zealand dollars). Overseas: $57.75 (New Zealand dollars.) NZART, PO Box 40-525, Upper Hutt, New Zealand.

The Canadian Amateur. Journal of the Radio Amateurs of Canada: Monthly. Canada: $36.00 (Canadian dollars). Elsewhere: $51.00 (U.S. dollars). Radio Amateurs of Canada, 614 Norris Court, Unit 6, Kingston, ON K7P 2R9, Canada.

Communications Quarterly: Amateur radio technical journal. Quarterly. 1-year subscription: U.S.: $29.95. Foreign: $39.95 surface, $60.95 via air. Canada/Mexico: $34.00. CQ Communications Inc., 76 North Broadway, Hicksville, NY 11801.

CQ-DL. Journal of the Deutscher Amateur-Radio-Club., e.V: In German. Monthly. Contact publisher with SASE or SAE with IRC for latest rates. Deutscher Amateur-Radio-Club, e.V., Postfach 1155, D-34216 Baunatal, Germany.

CQ Ham Radio. Amateur radio general coverage; in Japanese. Monthly. 1-year subscription: Contact publisher for latest information. CQ Publishing Co., Ltd., 14-2 Sugamo, 1-chome, Toshima, Tokyo 170, Japan.

CQ Contest: Amateur radio contesting. Ten issues. 1-year subscription: U.S.: $30.00, Canada/Mexico: $37.00, Elsewhere: $40.00 surface. CQ Communications, Inc., 76 North Broadway, HIcksville, NY 11801.

CQ VHF: Amateur radio general coverage. Monthly. 1-year subscription: U.S.: $21.95, Canada/Mexico: $31.95, Elsewhere: $33.95 surface, $81.95 via air. CQ Communications, Inc., 76 North Broadway, Hicksville, NY 11801.

CQ Amateur Radio: Amateur radio general coverage. Monthly. 1-year subscription: U.S.: $24.95. Canada/Mexico: $37.95. Elsewhere: $39.95 surface, $84.95 via air. CQ Communications, Inc., 76 North Broadway, Hicksville, NY 11801.

Digital Journal. Amateur radio digital communications. Monthly. 1-year subscription: North America: $25.00, Elsewhere: $32.00. American Digital Radio Society, 30 Rockefeller Plaza, New York, NY 10185.

Dots & Dashes: Historical Morse telegraphy. Quarterly. 1-year subscription: U.S.: $7.00 2nd class, $10 1st class and elsewhere. R.A. Iwasyk, 12350 W. Offner Rd., Manhattan, IL 60442.

The DX Bulletin: Amateur radio DX information. Weekly. 1-year subscription: U.S.: $34.00 2nd class mail, $44.00 1st class. Canada: $51.00. Mexico: $48.00. Elsewhere: $55.00 via air. *The DX Bulletin*, PO Box 50, Fulton, CA 95439-0050.

DX-Loggen. DX Bulletin, in Swedish. Weekly. Contact publisher with SASE or SAE with IRC for latest rates. Thomas Bevenheim, Gosta Ekmans Vag 5, S-12935 Hagersten, Sweden.

The DX Magazine: Amateur radio DX information. Bimonthly. 1-year subscription: U.S.: $15.00. Canada: $20.00. All others $25.00 by air. *The DX Magazine*, PO Box 50, Fulton, CA 95439.

DXMB: Amateur radio DX information; in German. Weekly. 1-year subscription: Contact publisher with SASE or SAE with IRC for latest rates. Bestellung DX-MB, DARC e. V., Postfach 11 55, W-34216 Baunatal, Germany.

The DX News Sheet: Amateur radio DX information. Weekly. 1-year subscription: Contact publisher with SASE or SAE with IRC for latest rates. Radio Society of Great Britain, Lambda House, Cranborne Road, Potters Bar EN6 3JE, England.

DXPress: Amateur radio DX information. Weekly. 1-year subscription: Contact publisher with SASE or SAE with IRC for latest rates. Central Bureau VERON, Box 1166, 6801 BD, Arnem, The Netherlands.

EA DX Bulletin: Amateur radio DX information; in Spanish. Weekly. 1-year subscription: Contact publisher with SASE or SAE with IRC for latest rates. Union de Radioaficionados Espanoles, Apartado Postal 220, 28080 Madrid, Spain.

Electric Radio: Collecting, restoring, and operating older tube-type equipment. Monthly. 1-year subscription: U.S.: $28.00 2nd class, $38.00 1st class. Canada: $39.00 by air. All other countries $70.00 by air. *Electric Radio*, PO Box 57, Hesperus, CO 81326.

Electronics Now: Hobby electronics. Monthly. 1-year subscription: U.S.: $19.97. Canada: $27.79. All other countries $28.97. *Electronics Now*, PO Box 51866, Boulder, CO 980321-1866.

Funk Amateur: Amateur radio, computers, hobby electronics; in German. 1-year subscription: Contact publisher with SASE or SAE with IRC for latest rates. Theuberger Verlag GmBH, Postfach 73, D-10122 Berlin, Germany.

Hollow State Newsletter. For lovers of vacuum tube radios. 3 issues per year. Subscriptions: $5.00 for 4 issues to U.S., Canada, and Mexico. $10.00 for 4 issues elsewhere. Ralph Sanserino, P.O. Box 1831, Perris, CA 92572-1831.

The GOLIST: QSL Manager List: QSLing information available from the publishers in both electronic and printed format. 3.5" disks, DOS format: Monthly, U.S. $62.00, $74.00 elsewhere; Quarterly, U.S. $32.00, $36.00 elsewhere. Printed format: Monthly, U.S. $30.00, $45.00 elsewhere. DX Enterprises, PO Box 2306, Paducah, KY 42002-2306.

ICOM Amateur Users' Exchange Group: Icom users' technical information. Bi-monthly. 1-year subscription: U.S. $12.00, 2nd class, $14.00 1st class. Mexico: $14.00 1st class. Canada: $15.00. Australia, Africa, New Zealand, Pacific Far East $19.50 via air. Elsewhere: $15.50 via air. International Radio and Computer, Inc., 3804 South U.S. 1, Fort Pierce, FL 34982. Inquire with publisher for special combined subscription rates with the

Kenwood *Amateur Users' Exchange Group* bulletin.

I.R.T.S. Newsletter: Journal of the Irish Radio Transmitters Society. Monthly. Contact publisher with SASE or SAE with IRC for latest rates. I.R.T.S., PO Box 462, Dublin 9, Ireland.

Japan DX News: Amateur radio DX information; in Japanese. Weekly. Contact publisher with SASE or SAE with IRC for latest rates. Japan DX News, PO Box 42, Urawa-Cyuo, Saitama 336, Japan.

Kenwood Amateur Users' Exchange Group: Kenwood users' technical information. Bi-monthly. 1-year subscription: U.S. $12.00, 2nd class, $14.00 1st class. Mexico: $14.00 1st class. Canada: $15.00. Australia, Africa, New Zealand, Pacific Far East $19.50 via air. Elsewhere: $15.50 via air. International Radio and Computer, Inc., 3804 South U.S. 1, Fort Pierce, FL 34982. Inquire with publisher for special combined subscription rates with the *ICOM Amateur Users' Exchange Group* bulletin.

KH6BZF Reports: Propagation newsletter. Weekly. 1-year subscription: Contact publisher with SASE or SAE with IRC for latest rates. Lee Wical, KH6BZF, 45-601 Luluki Road, CRT #44-25, Kane'Ohe, O'Ahu, HI 96744-1845.

Les Nouvelles DX: Amateur radio DX information; in French. Weekly. 1-year subscription: Contact publisher with SASE or SAE with IRC for latest rates. Jacky Billaud, F6BBJ, 14 avenue Massena, F-78600 Maisons Lafitte, France.

The Long Island DX Bulletin: Amateur Radio DX information. Bi-Weekly. 1-year subscription: U.S.: $25.00. Canada/Mexico: $30.00. Other foreign U.S. $35.00 via air mail. The Long Island DX Bulletin, PO Box 50, Fulton, CA 95939-0050.

Long Skip, Journal of the Canadian DX Association: Monthly. 1-year subscription: U.S. $25.00. Canada, $25.00 Canadian. Other countries: $25.00. CANAD-X, PO Box 717, Station "Q," Toronto, ON M4T 2N7, Canada.

The Low Band Monitor: Newsletter for 160, 80, and 40 meter enthusiasts. Monthly. 1-year subscriptions: U.S.: $24.00. Canada: $28.00. Elsewhere:$36.00. Low Band Monitor, P.O. Box 1047, Elizabeth, CO 80107.

The Lowdown: Long wave radio. Monthly. 1-year subscription: U.S.: $18.00. Canada: $19.00. Elsewhere: $26.00 via air. Longwave Club of America, 45 Wildflower Road, Levittown, PA 19057.

Lynx DX Bulletin: Amateur radio DX information; in Spanish. Weekly. 1-year subscription: Contact publisher with SASE or SAE with IRC for latest rates. Lynx DX Group, Apartado 20053, 48080 Bilbao, Spain.

Monitoring Times: Shortwave listening, scanning, hobby radio. Monthly. 1-year subscription: U.S.: $23.95. Canada/Mexico: $34.95. Elsewhere: $53.95. *Monitoring Times*, PO Box 98, 140 Dog Branch Rd., Brasstown, NC 28902-0098.

Morsum Magnificat: Morse code lore. Bi-monthly. 1-year subscription: UK £12.00. Europe: £12.75. Rest of the world: £15.50 via air. G. C. Arnold Partners, 9 Wetherby Close, Broadstone, Dorset BH18 8JB, England. (U.S. subscribers may remit in U.S. dollars: $26.50 via air to Owl Worldwide Publications, 4314 West 238th St., Torrance, CA 90505-4509.)

NCJ, The National Contest Journal: Amateur radio contesting. Bi-monthly. 1-year subscription: U.S.: $12.00 2nd class, $20.00 first class. Canada and Mexico, $20.00

first class. Elsewhere, $22.00 surface mail, $30.00 via air. The American Radio Relay League, 225 Main St., Newington, CT 06111.

OSCAR Satellite Report. Amateur radio satellite information. Twice monthly. 1-year subscription: U.S.: $32.00. Canada: $35.00. Elsewhere: $43.00. R Myers Communications, PO Box 17108, Fountain Hills, AZ 85269-7108.

The Old Timer's Bulletin: Antique radio collecting and history. Quarterly. 1-year subscription: U.S.: $12.00. All other countries $15.00 by air. Antique Wireless Association, c/o Joyce Peckham (Secretary), Box E, Breesport, NY 14816.

Popular Communications: Shortwave listening, scanning, hobby radio. Monthly. 1-year subscription: U.S.: $22.95. Canada/Mexico: $32.95. Elsewhere: $34.95 surface, $82.95 via air. *CQ Communications,* 76 North Broadway, Hicksville, NY 11801-2953.

Popular Electronics: Hobby electronics. Monthly. 1-year subscription: U.S.: $21.95. Canada: $28.84. Elsewhere: $29.45. Popular Electronics, Subscription Dept., PO Box 338, Mt. Morris, IL 61054-9935.

Practical Wireless: Amateur radio general coverage. Monthly. UK: £22.00. Europe: £25.00. Elsewhere: £27.00. U.S.: $45.00. Special rate for combined subscription with *Short Wave Magazine.* PW Publishing Ltd., Arrowsmith Ct., Station Approach, Bradstone, Dorset BH18 8PW, England.

Proceedings of the Radio Club of America. Journal of the Radio Club of America. Twice yearly, plus quarterly newsletters. Send SASE for membership info to Radio Club of America, PO Box 68, Glen Rock, NJ 07452.

QCWA Journal, Journal of the Quarter Century Wireless Association: Quarterly. Send SASE or SAE with IRC for latest membership information. Quarter Century Wireless Association, Inc., 159 E. 16th Ave., Eugene, OR 97401-4017.

QEX: Amateur radio experimenter's magazine. Monthly. U.S./Canada/Mexico: $25.00 for ARRL members, $37.00 for non-members, first class mail. Elsewhere: $25.00 for ARRL members, $32.00 for ARRL non-members, surface mail. American Radio Relay League, 225 Main St., Newington, CT 06111.

QTC. Journal of the Sveriges Saendareamatoerer (SSA); in Swedish, monthly. Contact publisher with SASE or SAE with IRC for latest rates. Hulander & Ask Information, Box 22002, 40072 Goeteborg, Sweden.

QRZ DX: Amateur radio DX information. Weekly. 1-year subscription: U.S.: $37.00 1st class. Canada: $40.00 1st class. Mexico: $38.00. Other countries: $55.00 via air. QRZ-DX, PO Box 16522, Asheville, NC 28816-6522.

QST, Journal of the American Radio Relay League: Monthly. 1-year membership: U.S.: $31.00. Elsewhere: $44.00. Age 65 and older, with proof of age, $25.00 U.S., $38.00 elsewhere.American Radio Relay League, 225 Main St., Newington, CT 06111.

Radio, Journal of the Reseau des Emetteurs Francais: in French. Monthly. 1-year membership: Contact publisher with SASE or SAE with IRC for latest rates. REF, Siege Social 2, Square Trudaine, F-75009 Paris, France.

Radio Byegones: Antique radio collecting and history. Bi-monthly. 1-year subscription: UK £17.00. Europe: £18.00. Middle East, Africa, Southeast Asia, Central & South America, U.S.A & Canada: £22.00 via air.

Australia, New Zealand, Far East & Pacific Regions: £23.00 via air. G. C. Arnold Partners, 9 Wetherby Close, Broadstone, Dorset BH18 8JB, England. (U.S. subscribers may remit in U.S. dollars: $37.50 via air to Owl Worldwide Publications, 4314 West 238th St., Torrance, CA 90505-4509.)

Radio Communication, Journal of the Radio Society of Great Britain: Monthly. 1-year membership: UK and overseas, £32.00, surface mail. Radio Society of Great Britain, Lambda House, Cranborne Road, Potters Bat, Herts EN63JE, England.

Radio Fun: Beginning amateur radio. Monthly. 1-year subscription: U.S.: $12.95. Canada: $20.97. Elsewhere: $24.97 surface, $44.97 via air. *Radio Fun,* PO Box 4926, Manchester, NH 03108-9839.

Radioaficionados. Journal of the Union de Radioaficionados Espanoles. Monthly. Contact publisher with SASE or SAE with IRC for latest rates. Union de Radioaficionados Espanoles, Box 220, E-28080 Madrid, Spain.

RadioRivista, Journal of the Associazione Radioamatori Italiani: in Italian. Monthly. 1-year membership: Contact publisher with SASE or SAE with IRC for latest rates. A.R.I., Ente Morale - via Domenico Scarlatti 31, I-20124 Milano MI, Italy.

Radio ZS, Journal of the South African Radio League: Monthly. Contact publisher with SASE or SAE with IRC for latest rates. South African Radio League, PO Box 807, Houghton, 2041, Republic of South Africa.

Revista Lynx: Amateur radio DX information. Monthly. 1-year subscription: Contact publisher with SASE or SAE with IRC for latest rates. Lynx DX Group, Apartado 20053, 48080 Bilbao, Spain.

Satellite Operator. Amateur radio satellite information. Monthly. 1-year subscription: U.S.: $37.00. Canada: $40.00. Elsewhere: $50.00. A special reduced rate is available to subscribers of OSCAR Satellite Report. R. Myers Communications, PO Box 17108, Fountain Hills, AZ 85269-7108.

Shortwave Magazine: Shortwave listening, computing, hobby radio. Monthly. UK: £22.00. Europe: £25.00. Elsewhere £27.00. U.S.: $45.00. Special rate for combined subscription with Practical Wireless. PW Publishing Ltd., Arrowsmith Ct., Station Approach, Bradstone, Dorset BH18 8PW, England.

The Spec-Com Journal. Specialized communications and journal of the U.S. ATV Society. Bi-monthly. 1-year subscription: U.S.: $20.00. Canada/Mexico: $25.00. Other foreign: $30.00. Spec-Com Journal, P.O. Box 1002, Dubuque, IA 52004-1002.

The Vail Correspondent: Telegraph key collecting and history. Quarterly. 1-year subscription: U.S.: $10.00. Canada or Mexico: $12.00 by air. All other countries: $16.00 by air. *The Vail Correspondent,* Box 88, Maynard, MA 01754.

The West Coast VHFer. VHF news. Monthly. 1-year subscription: U.S., $14.00. All other countries $18.00. West Coast VHFer, P.O. Box 685, Holbrook, AZ 86025.

The W5YI Report: Newsletter covering amateur radio, personal computing, and emerging electronics. Twice monthly. 1-year subscription: U.S., Canada and Mexico $24.50. All other countries $39.50 by air. The W5YI Group, 2000 E. Randol Mill Rd. #608A, Arlington, TX 76011.

Westlink Report: Newsletter covering amateur radio. Weekly. 1-year subscription: U.S., Canada and Mexico:

$24.50. Elsewhere $48.00. Westlink International, 28221 Stanley Ct., Canyon Country, CA 91351-3818.

The World Wireless Beacon. Newsletter of the Society of Wireless Pioneers, Inc. Quarterly. 1-year subscription: U.S.: $15.00. Send SASE for membership info and rates for other countries. Society of Wireless Pioneers, Inc., PO Box 86, Geyserville, CA 95441.

Worldradio: Amateur radio general coverage. Monthly. 1-year subscription: U.S. $14.00. All other countries $24.00 by surface mail. Worldradio, Inc., 520 Calvados Ave., Sacramento, CA 95815.

The Xtal Set Society Newsletter. A publication for crystal set enthusiasts. Bi-monthly. 1-year subscription: U.S.: $9.95. Canada: $11.00. Elsewhere: $16.00. The Xtal Set Society, c/o Rebecca Hewes, P.O. Box 3026, St. Louis, MO 63130.

Schematic Symbols Used in Circuit Diagrams

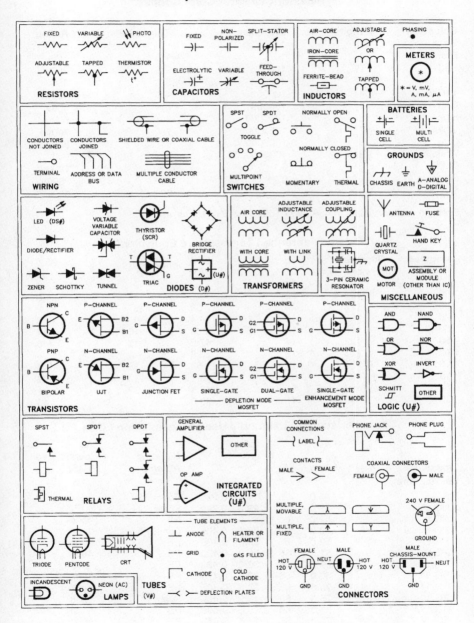

Resistor/Capacitor Color Codes

Color	Digit	Multiplier	Tolerance	Voltage Rating
Black	0	1	-	-
Brown	1	10	1%	100v
Red	2	100	2%	200v
Orange	3	1000	3%	300v
Yellow	4	10000	4%	400v
Green	5	100000	5%	500v
Blue	6	1000000	6%	600v
Violet	7	10000000	7%	700v
Gray	8	100000000	8%	800v
White	9	1000000000	9%	900v
Gold	-	0.1	5%	1000v
Silver	-	0.01	10%	2000v
No Color	-	-	20%	500v

Example: Yellow/Violet/Orange = 47x 1000 = 47,000 (Ohms, ef)

Celsius/Fahrenheit Temperature Equivalents

Degrees Celsius	Degrees Fahrenheit				
100	212	28	82.4	2	35.6
90	194	26	78.8	0	32
80	176	24	75.2	-2	28.4
70	158	22	71.6	-4	24.8
60	140	20	68	-6	21.2
50	122	18	64.4	-8	17.6
45	113	16	60.8	-10	14
40	104	14	57.2	-15	5
35	95	12	53.6	-20	-4
32	89.6	10	50	-25	-13
30	86	8	46.4	-30	-22
		6	42.8	-35	-31
		4	39.2	-40	-40

Characteristics of Commonly Used Transmission Lines

Type of line	Z_0 Ohms	VF %	pF per foot	OD	Dielectric Material	Max Operating Volts (RMS)
Coaxial Line						
RG-6	75.0	75	18.6	0.266	Foam PE	—
RG-8X	52.0	75	26.0	0.242	Foam PE	—
RG-8	52.0	66	29.5	0.405	PE	4000
RG-8 foam	50.0	80	25.4	0.405	Foam PE	1500
RG-8A	52.0	66	29.5	0.405	PE	5000
RG-9	51.0	66	30.0	0.420	PE	4000
RG-9A	51.0	66	30.0	0.420	PE	4000
RG-9B	50.0	66	30.8	0.420	PE	5000
RG-11	75.0	66	20.6	0.405	PE	4000
RG-11 foam	75.0	80	16.9	0.405	Foam PE	1600
RG-11A	75.0	66	20.6	0.405	PE	5000
RG-12	75.0	66	20.6	0.475	PE	4000
RG-12A	75.0	66	20.6	0.475	PE	5000
RG-17	52.0	66	29.5	0.870	PE	11000
RG-17A	52.0	66	29.5	0.870	PE	11000

Characteristics of Commonly Used Transmission Lines

Type					Dielectric	
RG-55	53.5	66	28.5	0.216	PE	1900
RG-55A	50.0	66	30.8	0.216	PE	1900
RG-55B	53.5	66	28.5	0.216	PE	1900
RG-58	53.5	66	28.5	0.195	PE	1900
RG-58 foam	53.5	79	28.5	0.195	Foam PE	600
RG-58A	53.5	66	28.5	0.195	PE	1900
RG-58B	53.5	66	28.5	0.195	PE	1900
RG-58C	50.0	66	30.8	0.195	PE	1900
RG-59	73.0	66	21.0	0.242	PE	2300
RG-59 foam	75.0	79	16.9	0.242	Foam PE	800
RG-59A	73.0	66	21.0	0.242	PE	2300
RG-62	93.0	86	13.5	0.242	Air space PE	750
RG-62 foam	95.0	79	13.4	0.242	Foam PE	700
RG-62A	93.0	86	13.5	0.242	Air space PE	750
RG-62B	93.0	86	13.5	0.242	Air space PE	750
RG-133A	95.0	66	16.2	0.405	PE	4000
RG-141	50.0	70	29.4	0.190	PTFE	1900
RG-141A	50.0	70	29.4	0.190	PTFE	1900
RG-142	50.0	70	29.4	0.206	PTFE	1900
RG-142A	50.0	70	29.4	0.206	PTFE	1900
RG-142B	50.0	70	29.4	0.195	PTFE	1900
RG-174	50.0	66	30.8	0.100	PE	1500
RG-213	50.0	66	30.8	0.405	PE	5000
RG-214*	50.0	66	30.8	0.425	PE	5000
RG-215	50.0	66	30.8	0.475	PE	5000
RG-216	75.0	66	20.6	0.425	PE	5000
RG-223*	50.0	66	30.8	0.212	PE	1900
9913 (Belden)*	50.0	84	24.0	0.405	Air space PE	—
9914 (Belden)*	50.0	78	26.0	0.405	Foam PE	—

Aluminum Jacket, Foam Dielectric

1/2 inch	50.0	81	25.0	0.500		2500
3/4 inch	50.0	81	25.0	0.750		4000
7/8 inch	50.0	81	25.0	0.875		4500
1/2 inch	75.0	81	16.7	0.500		2500
3/4 inch	75.0	81	16.7	0.750		3500
7/8 inch	75.0	81	16.7	0.875		4000

Parallel Line

Open wire	—	97	—	—		—
75-ohm transmitting twin lead	75.0	67	19.0	—		—
300-ohm twin lead	300.0	82	5.8	—		—
300-ohm tubular	300.0	80	4.6	—		—

Open Wire, TV Type

1/2 inch	300.0	95	—	—		—
1 inch	450.0	95	—	—		—

Dielectric Designation	Name	Temperature Limits
PE	Polyethylene	−65° to +80°C
Foam PE	Foamed polyethylene	−65° to +80°C
PTFE	Polytetrafluoroethylene (Teflon)	−250° to +250°C

*Double shield

OBITUARIES OF INTEREST TO
RADIO AMATEURS
September 1994–September 1995

Bill Leonard, W2SKE

Bill Leonard, W2SKE, a former president of CBS News, died October 23, 1994 in Laurel, Maryland. He was 78.

He received his first amateur license, W1JHV, while a student at Dartmouth College in 1934. After graduating in 1937, Leonard went to work as a reporter for the Bridgeport, Connecticut, *Post-Telegram*. After serving in the U.S. Navy from 1941 to 1945, he joined CBS News in 1945, where he stayed until his retirement in 1982

As CBS president, Leonard was credited with the selection of Dan Rather as Evening News anchor, as a member of the team that developed the "60 Minutes" newsmagazine, and of helping develop techniques to predict election outcomes.

Asked in 1981 about the future of Amateur Radio, W2SKE said, "I have a hunch that Amateur Radio is going to get more and more tangled up with amateur computer technology." He also said, "My bet is that ham radio, in one form or another, will be around 100 years from now."

Leonard, an avid DXer and contest operator in the 1960s and 1970s, was an advocate for Amateur Radio, writing, for example, an article for *Sports Illustrated* in 1958 entitled "The Battle of the Hams." It described the "sport of DXing" and Leonard's role in it from the contest super station of Buzz Reeves, K2GL. Leonard later provided the narration for the documentary on the K2GL, *Multi-Multi (To Win the World-N2AA)*. A profile of W2SKE appeared in March, 1981, *QST*.

A memorial service was held in December in New York City. Among those delivering eulogies were Dan Rather and Walter Cronkite, KB2GSD.

Tom Rutland, K3IPW

Tom Rutland, K3IPW, passed away on February 26, 1995. He was widely known in the VHF/UHF weak-signal community, and was the founder of Rutland Arrays, a highly-respected manufacturer of VHF/UHF antennas.

Stan Kaisel, K6UD

Stan Kaisel, K6UD, died June 22, 1995, in Portola Valley, California. He was 72 years old.

Kaisel, who became a licensed amateur at age 13, graduated from Washington University in St. Louis in 1943. He went to Cambridge, Massachusetts, and for two years was part of the Radio Research Lab there. Kaisel earned his doctorate from Stanford in 1949 and worked there as a researcher until forming a company, Microwave Electronics Corp, in 1959. Among his scientific accomplishments were helping to build the first linear electron accelerator and developing traveling-wave tubes.

Kaisel was active in the community, involved in fund-raising efforts and the founding of a Wellness Center at Stanford, just before his death. He also was an education activist, involved in supporting public school funding.

Stan Kaisel was an ARRL life member, a member of the board of directors of the Northern California DX Foundation (NCDXF), and a member of the Northern California DX Club (NCDXC). He was on the ARRL DXCC Honor Roll, both Mixed and Phone.

Leo Banigan, KB2XL

On Friday, April 7, 1995, Leo Banigan, KB2XL, passed away. Leo served many years as an Officer of the Radio Central Amateur Radio Club (Long Island, New York), helping to shape its growth and development as one of the most dynamic radio clubs in the Long Island area.

Dana Atchley, W1CF

Dana Atchley, W1CF (formerly W1HKK), of Lincoln, Massachusetts, died on Saturday, April 22, 1995, after a long illness. He was 77.

Dana was the son of the famed Columbia University/Presbyterian Hospital (New York City) physician of the same name, after whom the Atchley Pavilion was named. He was also a descendant of two American presidents, as he was related to the Adams family.

He was a founder of Microwave Associates, a Fortune 500 company for which he served as president and chairman of the board. Among the employees of Microwave Associates in years past was Sam Harris, W1FZJ, who engineered the Arecibo radiotelescope in Puerto Rico.

In the 1970s, Atchley created the "four square" vertical phased array, for which he was awarded a patent. He later described the antenna in *QST*. That antenna in recent years has seen a resurgence in interest among low-band HF DXers and has virtually become their antenna of choice on 80 meters.

Atchley, then W1HKK, held postwar DXCC number 3.

In addition to his fondness for low-band DXing, Atchley was an active promoter of amateur use of the microwave bands. He was instrumental in the creation of "The New Frontier" column in *QST*. He also arranged for a grant to the Smithsonian Institution from MA/COM in 1983 to help pay for renovations to NN3SI, the amateur exhibit station there.

A few years ago, he donated virtually his entire station to Harvard University, where much of it is in use today at the club station W1AF.

Bill Grenfell, W4GF

Bill Grenfell, W4GF, died on Friday, January 27, 1995, in Vienna, Virginia,. after an illness of several years. He was a former chief of the FCC's Amateur Branch and former chief of the Amateur Service Section. He was 82 years old.

Grenfell worked for the FCC from 1940 until retiring in 1971, except for several years in the U.S. Navy during World War II. He first joined the FCC's Seattle office as an assistant monitoring officer in the FCC Radio Intelligence Division in 1940. Following the war, Grenfell returned to the FCC as a radio engineer. In 1952 he became chief of the Amateur Branch and Amateur Service Section. From 1962 until 1966 he was chief of the FCC Rules and Standards Branch of the Amateur and Citizens Division and was chief of the Rules and Legal Branch of the Division from 1966 until retiring in 1971. He was one of the chief architects of the Incentive Licensing program for the Amateur Service. which took effect in the late 1960s.

An FCC news release announcing Bill Grenfell's retirement called him "a practicing (amateur) operator for almost 41 years" (he was first licensed in high school) and cited his contesting achievements. After retiring he wrote columns for the Washington, D.C.-area

Amateur Radio publication *Auto-Call* and for *WorldRadio*.

He was an ARRL life member, a Quarter Century Wireless Association life member, a fellow of the Radio Club of America, and a member and former president of the Potomac Valley Radio Club.

Mildred O'Brien, W6HTS

Mildred O'Brien, W6HTS, passed away in January 1995. She was 84. Hundreds learned the code from her and obtained their amateur licenses, and thousands benefited from her phone patch and MARS activities as MARS station supervisor at McClellan AFB during Vietnam and other military activities.

She was the mother of Jay O'Brien, W6GO, former publisher of the W6GO QSL List.

Marvin Camras, W9CSX

Marvin Camras, W9CSX, who is credited with the invention of magnetic tape recording, died June 23, 1995, in Evanston, Illinois. He was 79 years old and lived in Glencoe, Illinois.

According to *The New York Times*, Camras worked and taught at the Illinois Institute of Technology for more than 50 years. As a student in the late 1930s, he built a magnetic wire recorder and later discovered that making recordings on magnetic tape made splicing easier. In 1944 he was awarded a patent on "method and means of magnetic recording," *The Times* said.

Camras received the National Medal of Technology in 1990. He was awarded more than 500 patents for his work which were licensed to more than 100 manufacturers.

Marvin Camras was first licensed as W9CSX in the late 1930s and held that callsign until his death.

Harry A. Turner, W9YZE

Harry A. Turner, W9YZE, died December 21, 1994, in Alton, Illinois.

He was 88 years old.

While in the service in 1942, Turner was clocked at 35 wpm using a hand telegraph key, a record still standing and recorded in the current Guinness *Book of Records*. According to a *73* magazine article in 1976, Turner in 1964 applied to the Signal Corps for certification of his code speed record, which had been witnessed by General Ben Lear, and got it. Turner said he handled code under "business" conditions and learned to concentrate accordingly.

ADDITIONAL SILENT KEYS
September 1994–August 1995

WA1A	W1EMC	W1LLD	KA1TSF	WB2BFX	W2GGY
W1AJK	W1EQC	W1LQN	W1TUY	WA2BGM	W2GHK
W1AKG	WA1EQY	W1LWI	K1TZH	W2BKI	WB2GLB
W1ALT	KB1ER	KA1MKY	ND1U	W2BND	AA2GN
W1ASX	W1ESP	N1MMF	WA1UFT	N2BQE	WA2GSL
W1ATC	N1FEO	W1MMV	K1UFW	KA2BYP	N2GXO
K1ATU	N1FOL	W1MQA	KA1UHL	W2CH	N2HAR
KA1AVD	W1FRZ	W1MU	WA1UIV	W2CJN	W2HF
W1AVV	W1GAA	K1MZN	WA1UQC	KB2COA	W2HFI
W1AVY	WB1GBU	KO1N	W1UQN	WA2CSC	N2HPH
KB1AXP	W1GEN	NT1N	KA1VJS	N2CUI	N2HQX
W1AZY	KA1GLG	WJ1N	W1VN	N2CVK	W2HQY
N1BA	N1GLX	K1NBG	ND1W	N2CXE	KA2HRC
W1BBH	K1GUC	K1NEI	WA1WGR	W2DCO	N2HRW
W1BEO	W1GUC	W1NKS	W1WQU	W2DEO	W2HUO
N1BIO	KG1H	K1NSF	WA1YDD	W2DGV	K2HWP
K1BKB	W1HAD	W1NSN	K1YLH	WA2DIS	W2HYX
AA1BM	W1HCU	N1NWO	WA1YMW	W2DPV	W2IAK
W1BPO	K1HGY	K1NYT	W1YQ	WA2DTJ	WA2IDZ
W1BW	K1HMC	K1OCN	W1YRD	WB2DUQ	W2IJC
KA1BY	K1HNB	W1OPZ	W1YUB	KC2E	W2IMU
W1CAO	N1HOX	W1OTQ	W1YVM	NI2E	W2IQH
KA1CF	W1HRS	W1PBN	K1YXO	W2EA	W2IR
WB1CTD	KA1HWY	W1PHG	WG1Z	W2ECG	KA2ISZ
NN1D	K1HXB	KA1PJ	KA1ZES	N2ECU	WA2IUV
N1DGR	WA1HYC	W1PK	K1ZAJ	W2EDE	KJ2J
WA1DHM	W1IAO	W1PU	W2ADD	W2EDH	N2JDX
N1DIV	W1IIB	W1QEX	AA2AE	W2EJB	WB2JKU
KD1DN	K1IIM	W1QIS	WD2AFN	KB2EJW	KA2JTJ
K1DNC	KA1IKR	N1QJH	K2AJ	W2EM	W2JUJ
W1DOS	KA1IUA	W1QLF	WA2AJV	AA2EN	WA2JYR
K1DPG	KA1IYR	N1QMP	W2ANJ	W2EUR	KA2KBX
K1DPP	W1JRS	W1RY	K2ANL	N2EYV	W2KF
KA1DSK	W1JVZ	N1RZL	N2AQN	WA2FAU	K2KFE
K1DVA	W1KA	W1SFV	W2AXL	W2FC	K2KLJ
K1DVV	KA1KCC	W1SSF	W2AY	KB2FCG	WA2KNG
KA1DXS	WA1KEZ	W1TMO	W2AYJ	W2FFE	WA2KPF
KA1EDX	W1KGC	W1TOI	N2BAF	W2FFU	KB2KSL
KB1EF	W1LFB	W1TQG	KA2BBH	KA2FNQ	WC2L
WB1EKW	N1LIT	WA1TQP	W2BFH	KF2V	WW2L

KB2LCF	W2ZFE	W3OUX	KD4E	WA4JDH	KA4OHA
KB2LHF	WA2ZKP	K3QMR	W4EBO	N4JDY	W4OLJ
N2LLL	W2ZU	K3QYK	KC4EDW	KD4JGT	K4OOS
WA2LMW	WA3ACH	KA3PCV	K4EEJ	KC4JHN	KK4OQ
WA2LOC	W3ADV	W3PEV	N4EEQ	W4JI	K4OQM
W2LT	W3AEV	KA3PJZ	KC4EJI	K4JKG	N4OXJ
W2MBF	W3AQN	K3PKJ	WB4ELX	AA4JH	AE4P
W2MCC	WA3AXA	W3PKP	W4EPL	AA4JM	W4PCF
W2MCF	KA3BEI	W3PLK	W4EPZ	WB4JMG	W4PFP
WA2MHL	N3BOA	N3PNW	KD4ERP	WB4JMW	WB4PGD
N2MVD	W3BPX	W3QGK	KD4EST	AA4JO	WB4PHL
KD2NA	W3BSG	K3RHL	K4ETY	WB4JRL	W4PJF
W2NFP	WF3C	N3RPK	K4EUK	WA4JTA	K4PKM
N2NGJ	W3CCE	W3RWR	W4EXR	K4JV	W4PNQ
KA2NIW	WB3CCG	KA3SDM	WB4EYW	KA4JWD	W4PSS
W2NNH	N3CNB	W3ST	W4EZ	K4JZZ	KE4PW
KB2NXY	N3CNP	W3TQ	K4FIT	KT4K	K4PZV
K2OIU	W3CPF	W3UBO	N4FJL	W4KGF	KB4QDC
WA2OMI	K3DAK	W3UGI	K4FK	K4KHT	N4QIS
K2OSV	K3DD	KB3UK	N4FMF	N4KJI	KD4QMB
K2OX	W3DJF	WA3UTW	WA4FMS	WD4KKV	N4QGG
WA2OYV	W3DKE	W3UWT	W4FMV	WD4KLJ	K4QQL
W2PHF	N3DRY	W3VHV	KA4FOM	WA4KMG	NG4R
W2PIQ	AA3E	KA3VQG	W4FPT	W4KSL	K4RBN
KA2PPO	KA3EAO	KA3VSF	W4FRL	KC4KTG	W4RR
AA2PT	K3ECP	NY3W	KC4FXW	K4KTV	W4RWD
WA2PWV	K3EFG	W3WDD	WD4FXX	WB4KUD	WB4RXA
N2QBK	W3ESU	KA3WMB	KA4GDU	WB4KUO	W4SDM
WB2QIY	KA3EZJ	W3WRE	W4GEO	W4KVR	K4SM
W2QKN	W3FDY	W3YIY	W4GEW	KA4LAZ	W4SRK
W2QLP	W3FJI	NA3Z	W4GF	N4LBO	KJ4SS
KB2QWC	KB3FT	NY4A	W4GFZ	W4LBW	W4ST
WA2RAH	N3GTN	K4ABL	WA4GJW	W4LC	W4SUO
KA2RGG	WB3GUQ	W4ABP	KD4GKS	KK4LE	KD4T
W2RMJ	WB3GZU	W4AJN	K4GLM	WD4LGU	N4TAJ
WA2RMR	W3HGB	KB4AKX	WA4GOG	W4LKX	KC4TCI
AA2RU	KD3HI	KB4ALV	W4GPL	AA4LO	W4TJS
WA2RVM	W3HOT	AA4AN	K4GQB	K4LPG	KD4TJZ
W2RVW	KA3HVZ	WA4ANJ	K4GRO	W4LPO	W4TNX
W2RWN	W3HWD	WA4APE	WD4GSJ	KC4LTO	WB4TQI
K2RYI	WA3ICC	AD4AS	W4GT	W4LVD	WB4TVV
W2SAA	KA3III	K4AT	KE4GT	W4LVM	K4TU
WB2SHW	W3IMG	KC4AUO	K4GUR	KC4LWY	KD4TYB
W2SKE	K3IPW	KC4AWJ	KB4GVL	NB4M	KB4TYO
WA2SMN	KA3ITH	KD4AXD	K4GZZ	W4MAU	W4UBL
WA2SNV	WB3IVI	WD4BBX	KD4HAD	W4MDF	KC4UDJ
K2SQX	W3IXC	WD4BCX	W4HBL	W4MEB	K4UEC
K2SR	W3JBU	W4BNQ	W4HEI	AA4MF	KD4UIJ
KB2SR	N3JFK	N4BOA	KA4HEL	KD4MGT	KD4UIM
WA2SUL	WA3JHL	N4BOO	N4HEY	KR4MJ	KA4UIN
WB2SVH	KA3JVL	W4BUL	W4HIJ	WA4MMP	AB4UO
W2SVI	N3JVN	K4BWB	K4HJ	WA4MOB	KJ4UV
WA2SYI	W3KGX	WD4CCE	WD4HPJ	K4MPA	KB4UXF
WA2SYW	WB3KHD	K4CCW	K4HMI	WA4MQU	KR4UY
WA2TFF	K3KLU	WA4CGQ	K4HSC	KC4MTP	W4UYC
W2TXB	WA3KOO	K4CNK	K4HSJ	KA4MVJ	KB4UZV
W2UBS	KN3L	WA4CQT	WB4HXJ	WA4MVR	K4VID
W2UTT	WO3L	KI4CR	WB4HVQ	WA4MZD	N4VJ
W2UVS	K3LCQ	WA4CTY	WA4HZP	WD4NBD	WA4VKN
W2UYE	KB3LH	W4CVX	W4HZS	N4NBJ	WA4VLX
W2UYQ	K3LNL	WA4CVF	W4IFW	W4NEI	WA4VWP
NZ2V	N3LTU	KA4CWF	W4IND	W4NFJ	KI4WC
W2VHI	WB3LSW	WA4CZV	KA4IOW	KC4NH	KB4WGI
N2VJQ	K3LWP	KE4DIO	W4IOZ	KA4NNU	WA4WLA
W2VMT	W3MAK	KB4DIX	W4IPM	W4NPS	K4XA
W2VPL	W3MML	W4DLK	KK4IQ	K4NSB	KI4XI
WA2VZL	W3MNS	KE4DLT	KB4ISY	W4NWF	N4XWT
WA2WDM	W3NAP	K4DOW	WA4IVB	W4NWP	K4YAK
KE2WK	KA3NBQ	W4DPI	N4IXM	KA4NXO	WA4YJO
KE2YD	W3NGJ	W4DPY	WD4IZD	WU4O	WB4YKP
KA2YFL	K3OEN	W4DQU	N4JAH	KM4OA	W4YMI
W2YLV	KC3OF	KE4DVJ	KD4JAM	N4OCR	W4YOF
K2YMQ	K3ONF	KJ4DX	K4JBN	N4OCS	W4YPO
K2YQT	W3OQV	K4DXB	N4JCR	KB4ODA	KD4YUA

WA4YVW	W5HRN	WB5SVS	W6DDO	W6MLN	W6UBZ
W4YY	W5HU	W5SWH	KD6DFI	W6MOJ	KC6UNP
WT4Z	K5HXF	WB5SZP	WA6DFP	WB6MNM	KI6UQ
N4ZBH	W5IKU	KB5TCQ	W6DGG	KG6MQ	K6UV
W4ZBZ	AA5IQ	K5TKN	W6DGK	W6MSY	KE6UX
WB4ZDP	K5ITX	K5TL	WD6DKQ	N6MSZ	AA6UY
K4ZIR	AF5J	AA5TO	KF6DO	W6MUR	KA6V
WA4ZLX	WA5JCK	WA5TUY	W6DPJ	N6MYM	W6VCI
WB4ZPU	W5JFU	K5TYV	W6DTL	WA6NBW	KA6VHY
KE4ZS	AA5JG	WA5UIA	KE6DWL	KA6NDC	K6VIN
W4ZWZ	KG5JG	WA5ULP	KH6DYA	W6NI	W6VKF
KC4ZZD	N5JLO	W5URN	AF6E	W6NJG	KD6VOG
KB5A	WA5JQS	WA5UUQ	W6EBB	KC6NJW	WA6VQO
N5ACL	WD5KAM	WA5UWG	K6EEF	AA6NO	WA6VQV
WD5ADH	WD5KEH	W5VEU	WA6EJT	KD6NQT	KH6VS
W5AEJ	W5KJR	KF5VF	W6EKW	WB6NSZ	WA6VWL
KA5AGD	K5KOL	W5VMI	N6EON	W6NXK	K6VXO
W5ALS	WA5KSE	W5VSX	KE6EP	W6NYT	W6VZZ
K5ALY	WB5KSW	N5VUI	K6EWS	W6NZG	WA6VZZ
K5AO	W5KWP	WA5WHV	WA6EYX	KH6O	AA6W
N5APB	N5KWU	KB5WKS	W6EZM	KD6OBQ	KH6WC
K5ASZ	WB5LBC	KB5WN	W6FBB	KA6OGJ	W6WGF
W5AXA	W5LDR	WA5WQP	W6FEM	WA6OLA	W6WGG
W5BEY	WA5LOB	WB5WZW	W6FET	W6OMJ	WA6WHH
WB5BJG	W5LU	KB5XH	N6FFC	W6OQM	WA6WIH
K5BRL	W5LUT	W5XQ	N6FR	W6OYE	W6WOU
N5BT	W5LQZ	KE5YJ	N6FVJ	WB6PDP	KC6WQT
W5BWZ	W5MBB	W5YPN	KH6GDR	W6PFG	KC6WTR
KA5CHQ	K5MBS	KB5YYD	KE6GEW	WB6PHZ	KB6WU
W5CJV	W5MEK	NB5Z	K6GK	N6PL	KC6WUL
W5CKJ	AA5MF	KB5ZPL	KD6GLO	K6PN	KE6WZ
W5CNI	KA5MKZ	N5ZZT	WB6GMM	W6PN	NY6X
WB5CPA	N5MNB	NF6A	W6GQF	K6PNY	K6XP
W5CPC	N5MRU	W6AAC	KB6GVF	W6PPP	AG6Y
K5CPI	N5MTN	KA6ABL	NC6H	N6PQK	WG6Y
K5CRF	KC5NA	W6AID	WA6HAD	W6PWG	W6YFT
W5CUI	W5NCE	N6ANS	WB6HAF	K6PZL	KM6YI
AB5CW	WB5NFH	KE6APE	KD6HHZ	W6PZV	KC6YOR
KA5CYF	W5NHQ	W6APF	KB6HNG	KB6Q	WA6YRT
WN5D	KE5NI	K6ATC	W6HTS	NI6Q	K6YWF
W5DPO	W5NII	K6ATK	KB6HVA	KD6QBH	N6YXF
N5DR	N5NMT	W6AUC	KB6HYY	W6QBW	W6YY
KA5DUE	W5NRE	W6AUW	KB6I	W6QFN	KJ6YZ
KB5DUF	KA5NVM	W6AVJ	W6IDF	N6QPG	K6ZIU
K5EAO	KB5NW	W6AXA	K6IG	KK6QS	W6ZNZ
WD5EBW	W5NWH	N6AXQ	KB6INN	WA6QVS	WB6ZOR
KA5EFJ	NB5O	W6AZI	K6IU	KI6QX	KB6ZZF
WA5EIT	KI5OC	WB6BCE	KA6IXJ	WA6QXV	KW7A
W5EOH	N5ODS	AA6BD	W6JGH	WB6QZQ	N7ABP
KA5ESG	W5OFO	W6BEP	KD6JKL	WI6R	N7ADI
W5EVB	W5OMK	KH6BG	N6JLE	WA6RBI	W7AIX
N5EVI	W5OOF	K6BIY	WB6JLG	K6RMI	WL7ALH
W5EX	K5ORD	KH6BM	KA6JLR	W6RPA	KB7APM
W5FBJ	KA5OTE	KE6BNF	N6JUM	KE6RU	W7AWE
W5FFG	K5OZE	WA6BQG	WB6KER	W6SAU	W7AYJ
W5FJG	KJ5P	W6BSO	KA6KJK	WB6SJR	KZ7B
N5FKO	KB5PKE	K6BUU	W6KOK	N6SKS	WA7BAQ
W5FKU	W5PLE	N6BXF	KE6KYR	K6SLS	W7BBH
KF5FW	W5PSC	W6BYF	NQ6L	WA6SME	KG7BE
AB5G	KJ5Q	WA6BZF	N6LAD	W6SMP	W7BGH
W5GFM	W5QEY	W6CBR	W6LCB	K6SNP	W6BJG
KC5GIZ	W5QMS	W6CBN	W6LFD	N6SOR	WB7BJK
KC5GLD	W5QOU	KH6CHL	WA6LFN	W6STY	WA7BMJ
WB5HAV	KA5RBY	W6CIE	KB6LHN	W6SVP	KF7BT
W5HDK	KE5RJ	KD6CKM	W6LII	N6TFV	KA7BYO
WB5HGI	W5RJK	K6CNP	W6LIO	WB6TJZ	KA7C
KA5HGZ	N5ROS	N6COH	KA6LKN	KD6TKP	KB7CCQ
KG5HH	KB5RQE	W6CRG	KC6LMK	W6TNW	K7CEH
KC5HKH	W5RWQ	W6CTL	W6LXQ	W6TOL	W7CEV
WD5HLV	KA5SIO	KA6CUJ	W6MAZ	WA6TTC	W7CFH
WA5HMO	W5SIR	KD6CVV	WA6MBC	KM6TX	W7CGA
WB5HMZ	W5SKX	K6CXB	KM6MD	W6TZK	N7CMC
W5HQA	W5SM	W6CYL	W6MFH	W6TZQ	K7CXG
W5HR	W5SNY	K6CZI	N6MJR	W6UAW	KA7DNI

W7DPC	KF7SH	W8ELY	WD8RYX	WB9DIO	WA9PUH
N7DVZ	K7SLA	N8EVI	W8SAF	W9DKA	W9PVL
W7DZ	K7SMZ	W8EWM	N8SAR	KB9DKX	K9PVW
KD7ED	N7SPE	N8FFS	N8SBE	W9DLN	W9QBJ
W7EMP	KA7SQP	W8FGZ	W8SDE	W9DM	W9QEW
W7EQB	K7SRA	N8FNY	K8SDG	W9DNT	W9QOM
WA7EQC	AG7T	N8FOF	WA8SOT	W9DVW	W9RCV
WA7EQV	K7TAW	N8GEZ	WV8T	W9DW	WB9RCY
KL7ERC	N7THS	W8GNG	W8TFG	WB9EMC	K9RGF
N7EYX	WB7TIM	N8GRS	W8TIJ	W9ERC	WB9RFT
W7FFJ	W7TKI	WB8GTO	N8TOO	W9ERZ	WB9RGZ
W7FQ	WA7TMP	W8HAV	WA8TSV	W9ENT	W9SFB
WA7FVQ	N7TMU	WN8HFN	WA8TVE	WD9ESK	W9SMW
W7FY	N7TRH	WB8HMB	WA8TXD	N9EXF	W9SMY
KA7FYE	N7TRM	KA8HQC	W8UDB	W9FIX	WA9SOO
WA7FZP	WA7TUW	W8HQQ	KA8UFL	N9FTW	WB9SUU
W7GAA	WB7TVT	WB8HVR	W8UJL	W9FVC	W9SVT
N7GBA	WB7UDF	N8HWO	W8ULH	N9GLA	N9TN
N7GMN	W7UG	KB8HZF	W8UMI	WD9GNB	AA9U
KA7GOJ	WB7UKH	KK8I	WB8UNM	N9GOS	KC9U
K7GRC	W7ULC	WA8IBD	K8UTE	WD9GPN	NM9U
K7GSL	N7UXS	N8ICR	K8UZT	W9HB	KA9UBW
W7GSS	WA7VJL	W8INY	KA8VFM	W9HCJ	WA9UFO
N7GTZ	N7VWF	N8IRE	K8VIE	K9HGK	K9UJK
W7GUH	KF7WC	K8IRJ	K8VNK	W9HPQ	W9UL
W7HDC	N7WIV	WA8IYT	K8VXX	WB9HPR	K9UPD
KA7HFQ	N7WUD	WB8JDA	W8VYY	NX9I	KA9USA
W7HSE	N7WZB	N8JQN	W8WHX	W9IDG	W9VMG
AA7HV	W7YFT	W8JS	K8WLP	K9IFE	K9VRI
W7HXJ	W7YI	K8JTT	W8WNK	KB9IJC	K9VSY
KB7HYS	NL7ZB	W8JYG	WB8WOZ	K9IS	W9WEY
AL7I	KC7ZJ	WB8JYS	W8WUG	AA9JD	W9WSG
K7IDF	W7ZKL	W8KSX	W8XJ	WD9JFU	KC9YH
KE7IG	WA7ZUE	K8KVE	N8XKN	KF9JG	K9YRL
KA7IGY	K7ZWR	K8KWO	W8YGR	W9JJU	W9YYX
W7ISX	W8ACP	K8KYB	W8YHJ	W9JJV	W9YZE
WA7IVA	W8ACT	W8KZ	WA8YNU	WB9JOI	KU9Z
W7JE	W8ADY	WA8LAV	K8YWL	WB9JRC	W9ZAF
W7JMA	N8AGT	WD8LIZ	WI8Z	W9JWA	W9ZGC
KS7K	WB8AJC	W8LQN	N8ZA	KK9K	KE9ZI
W7KIC	W8AKM	WB8LZI	K8ZAF	WA9KAG	K9ZJV
W7KIN	N8ALO	W8MFN	W8ZAH	W9KBE	KD9ZU
W7KYV	W8AQ	W8MHM	W8ZKL	WA9KHG	KØABY
W7LDJ	W8AST	K8MMB	WA8ZPS	WA9KMD	WØACT
W7LHT	W8AXE	K8MNP	K8ZTC	N9KPV	KDØAE
W7LJA	W8AY	N8MOR	W8ZXC	K9KRK	WBØAHU
WA7LPQ	WB8BHT	WA8MXW	K9AA	K9KSA	KBØAKL
N7LQS	W8BLY	W8NBK	K9AGA	W9KSR	KØAQV
KB7LVW	WB8BMM	K8NCV	N9AHK	W9KSU	WDØAWO
W7NEJ	WB8BNT	WD8NTA	N9AJ	W9LKI	KØAZD
KB7NEP	W8BNW	KA8O	W9AJH	W9LKX	WVØB
W7NIB	KX8C	W8OAI	K9AOM	W9LSF	WDØBBD
WB7NWE	W8CDL	WB8OBF	WD9APJ	W9LSY	NØBMX
WN7NXC	WA8CEA	WA8OBU	N9AQJ	WA9MFP	KØBR
NB7O	WD8CET	K8OBX	N9AUA	W9MGQ	WØBVO
KA7OAO	WA8CFQ	KA8ODX	W9AZP	N9MMG	KAØBVO
K7OIR	K8CLC	W8OEA	WB9BNY	W9MRJ	KAØBWN
KB7OJE	W8CMM	WD8OLJ	WA9BPO	W9MT	AAØBX
WB7OKS	N8CON	W8OPC	W9BPT	KT9N	WBØBXN
K7ORA	WA8CQC	WB8OTO	W9BTQ	W9NNU	NJØC
W7OUG	WD8CRS	W8OUL	K9BUI	W9NOF	NNØC
W7OVE	K8CSO	N8OVX	NK9C	W9NQG	KAØCAG
WB7PRV	W8CUP	K8PE	W9CHK	W9NQN	WØCAW
W7PSN	K8CWI	KA8PEJ	W9CK	W9NUG	WBØCIY
KE7QE	W8CWX	K8PIP	W9CL	W9NUH	KBØCJS
KF7QN	KA8CYR	WD8PKT	K9CLF	AA9OB	NØCMS
W7QOB	W8DCJ	KA8PMT	W9CNI	N9OBD	AAØCN
KA7QQM	W8DJG	W8QDC	W9CNV	WB9OGD	WDØCOQ
W7RD	K8DPV	KE8QS	W9CWO	W9OVB	WØCR
KF7RM	K8DZA	W8QZK	WA9CYB	W9PAQ	NØCYJ
KD7RR	N8EEW	W8RDX	W9DCQ	WA9PKL	WDØDEA
KA7RUT	WD8EFD	WD8RFJ	K9DCX	W9PRO	WØDEE
W7SA	WD8EGV	K8RLT	KB9DEF	K9PTI	WØDGS
KL7SC	WA8ELC	K8RNT	KA9DFN	W9PTI	KØDIN

KØDIQ	KØHBP	NØLSX	WØRQJ	WBØWZY	TK5FF
WØDJD	KØHNX	KAØLXC	WØRWE	WAØYNP	VP9GQ
WDØDLW	WØHSV	WBØMGI	WØRWS	KØYRX	VE2FKI
WØDXC	KØHTO	WØMHR	NØRXG	WBØYSQ	VE3FBD
WØDYM	KØHZR	NØMIX	WAØRXS	WØYUY	VE3MC
KWØE	WØIDI	WØMLR	KEØSF	WØYYV	VE3PE
WDØEDA	WØIDW	WBØMUI	KAØSJU	WBØYZX	VE3RHR
WØEJG	NØIQP	NØNIJ	WØSOK	AGØZ	VE6DY
WDØEMS	WØJBM	WBØNPZ	WBØTEC		VE7ABF
WØENM	WBØJOF	NØNVJ	WBØTEX	DK1YZ	VE7AFP
WØEY	KAØJRY	WAØNWA	WAØTFO	DL1NG	VE7FHQ
NØEYY	ACØK	WØNZH	NØTGB	DL9PU	VE7LHR
KØFAE	WEØK	KFØO	WAØTMI	F9IR	VE7MZ
WØFBB	WØKET	WAØOFM	NØTXX	G2XM	VE7QR
WAØFIQ	KFØKK	WØOIP	WØUBI	G3VBM	VK3UK
KAØFTB	WAØKOX	NØOVN	WØUGN	GØJHO	VK3YL
WBØFWP	KØKOY	WØPJW	WØUFD	HB9AHA	VK4EF
NØFXX	WØKPK	KFØPT	WØVQ	HI8MEQ	VO1GW
KØGCS	AAØKZ	WØQDU	WØVQR	LA5N	VO1OM
KAØGEU	KBØLCG	WØQFY	WAØWBE	ON4LO	ZL3IS
AAØGH	KBOLDN	WØQJF	NØWCC	OY7ML	ZS1FD
KØGJR	WØLEF	WØRFP	WØWEN	OZ1D	
KBØGPR	WAØLIB	WØRJD	WBØWLG		
KØGVB	KEØLS	WØRKV	NØWTV	*(Source:* QST *listings)*	

Index

Index

Index